Study Guide and Solutions Manual

to Accompany

Organic Chemistry
Sixth Edition

Jim Parise
Department of Chemistry and Biochemistry
University of Notre Dame

Marc Loudon
Department of Medicinal Chemistry and Molecular Pharmacology
Purdue University

w.h.freeman
Macmillan Learning

New York

W. H. Freeman and Company

W. H. Freeman and Company

1 New York Plaza, Suite 4500, New York, NY 10004-1562

www.macmillanhighered.com

Cover art: Quade Paul

ISBN: 978-1-936221-86-8

10 9

Contents

28   PERICYCLIC REACTIONS   1021

Preface

The purpose of this preface is to give you an overview of the organization and features of this *Study Guide and Solutions Manual* so that you can use it more effectively.

Organization

Each chapter corresponds to the chapter of the text with the same number and title. The following sections are found within each chapter, each headed by its own icon.

STUDY GUIDE LINKS

Study Guide Links are short supplements to text material that are called out in the margins of the text by the icon depicted above. Study Guide Links provide hints for study, shortcuts, approaches to problem-solving, or additional explanations of topics that typically cause difficulty. For example, Study Guide Link 5.2, "How to Study Organic Reactions," on p. 108 of this manual, is called out at a point in the text (p. 200) at which some students begin to have difficulty learning reactions. It provides detailed hints on how to study and learn organic reactions efficiently.

FURTHER EXPLORATIONS

Further Explorations are short discussions that provide additional depth on certain topics; these are called out in the margins of the text by the icon above. For example, Further Exploration 13.2, called out with a margin icon on text p. 666, gives a more in-depth discussion of Fourier-Transform NMR, a method used for obtaining NMR spectra.

REACTION REVIEW

Reaction Reviews present the reactions in the chapter in detailed outline form. This summary consolidates the reactions of the chapter in one place and presents not only the reactions themselves but also the essential features of their mechanisms and stereochemistry for cases in which these issues are discussed in the text. Studying and learning reactions is a skill that should be cultivated, and Study Guide Link 5.2 on page 108 of this manual will help you to use these reaction summaries to best advantage.

The Reaction Review sections are found only in chapters in which specific reactions are covered in the text. For example, Chapters 12 and 13, which cover spectroscopy, do not have reaction reviews.

SOLUTIONS TO PROBLEMS

Undoubtedly, your instructor will stress the importance of problem solving for the mastery of organic chemistry. This text provides over 1600 problems of varying difficulty levels, many with multiple parts. The *Solutions* sections provide solutions to all problems in the text. The solutions to many of the problems are worked-out examples in which not only the answer is given, but also the detailed approach to the solution is discussed.

It is important not to rely too heavily on the solutions presented in this manual. You learn organic chemistry by *solving problems on your own*. Many students develop a false sense of understanding by trying a problem, giving up too soon, and then reading the detailed solution. The solution seems to make sense to them, they feel like they understand the concept, and move on. What they've actually done is short-circuited the learning process, and they're simply agreeing what the solution says; they haven't developed the capacity to solve problems on their own.

Here is a better approach if you get stuck on a problem. First, attempt to solve it by reviewing the text or your class notes, or by talking to a classmate, TA, or your instructor. Be sure to ask for *guidance* only, not the answer. If you're still stuck, glance quickly at the solution in this manual without reading for detail. Focus on the references within the solution—pages, equations, sections, etc.—where the answer might be found. Jot those down next to your work, and close the manual. Refer back to the text and see if you can solve the problem after referring to relevant sources. It may take longer to solve problems this way, but your time will be much better spent if you take this approach.

Solutions Icon Comments

Within the solutions, you will find two types of comments marked with icons.

 Comments marked with this icon provide additional information or depth about the solution.

 Comments marked with this icon provide cautionary warnings—things to be particularly careful about or common traps.

Acknowledgments

We would like to acknowledge Dr. Joseph Stowell for his contributions to previous editions of this manual. We are also indebted to our proofreader, Dr. Animesh Aditya of Purdue University.

We have tried very hard to eliminate as many errors as possible. Nevertheless, if errors are found, we would like to know about them so that we can correct them on reprint if possible. You can send these or any comments to us by email at *PariseOrganic6E@gmail.com*. A list of both textbook errors and study guide and solutions manual errors can be found at the link below.

http://people.pharmacy.purdue.edu/~loudonm/teaching/

We also enjoy hearing from students with comments and suggestions about the text or the solutions manual. We hope that you find this manual very useful in your study of organic chemistry.

Jim Parise and Marc Loudon
May 2015

About the Authors

Jim Parise

Jim Parise received his BS in chemistry from the State University of New York at Oswego in 2000 and his PhD in organic chemistry in 2007 from Duke University, where he worked with Professor Eric Toone. After a postdoctoral fellowship with Professor David Lawrence at the University of North Carolina at Chapel Hill, Dr. Parise joined the Department of Chemistry at Duke University, where he taught organic chemistry to both pre-professional students and science majors, and coordinated the organic chemistry laboratory courses. In 2011, he joined the Department of Chemistry and Biochemistry at the University of Notre Dame. He teaches primarily to pre-professional students and also oversees the pedagogy of the accompanying laboratory program. He has received a number of teaching awards. His research focuses on pedagogical techniques and the integration of classroom technology. He has also developed a teaching mentorship program for new instructors. Dr. Parise has authored and co-authored laboratory manuals and a peer-reviewed book chapter on writing in the laboratory, and he is co-author of the accompanying textbook, *Organic Chemistry, Sixth Edition*.

Marc Loudon

Marc Loudon received his BS (magna cum laude) in chemistry from Louisiana State University in Baton Rouge and his PhD in organic chemistry in 1968 from the University of California, Berkeley, where he worked with Professor Don Noyce. After a two-year postdoctoral fellowship with Professor Dan Koshland at Berkeley, Dr. Loudon joined the Department of Chemistry at Cornell University, where he taught organic chemistry to both pre-professional students and science majors. In 1977, he joined the Department of Medicinal Chemistry and Pharmacognosy (now the Department of Medicinal Chemistry and Molecular Pharmacology) at Purdue University in the School of Pharmacy and Pharmaceutical Sciences, where he teaches organic chemistry to pre-pharmacy students. Dr. Loudon served as Associate Dean for Research and Graduate Programs for the College of Pharmacy, Nursing, and Health Sciences from 1988 to 2007. He has received numerous teaching awards. Dr. Loudon was named the Gustav E. Cwalina Distinguished Professor in 1996, one of the first three faculty members to be recognized by Purdue as distinguished professors for teaching and teaching scholarship. Dr. Loudon participated in the HHMI-sponsored Nexus project from 2010–2014, one outcome of which was to completely redesign the first organic chemistry course to make is more relevant to the interests of students in pre-health profession programs. Dr. Loudon was named Cwalina Distinguished Professor emeritus following his retirement in 2015. In collaboration with Professor George Bodner, Dr. Loudon has developed and practices collaborative-learning techniques for teaching organic chemistry to large classes. Dr. Loudon is the author of numerous research articles, a co-author of an in-house laboratory manual, and he is the author of the accompanying textbook, *Organic Chemistry, Sixth Edition*. The first edition was published in 1984.

Chemical Bonding and Chemical Structure

STUDY GUIDE LINKS

1.1   Formal Charge

After you have a small amount of practice assigning formal charges, you'll begin to notice certain patterns that emerge. For example, an oxygen with three single bonds and one unshared electron pair always has a +1 formal charge (shown to the right).

Some other common patterns are shown below.

	carbon	nitrogen	oxygen

(Verify these patterns by calculating the formal charge in each one.) Recognition of these patterns can be a great time-saver in applying formal-charge rules. Conversely, when you calculate a formal charge, always double-check your calculation if you seem to be violating one of these common patterns.

1.2   Vector Addition Review

Vectors can be added together to produce a new vector, called a *resultant*. There are two ways to add two vectors. The first way is to line up the two vectors with their tails touching and construct a parallelogram. Form the resultant by connecting the vertex of the two vectors to be added to the opposite side of the parallelogram, as shown on the right.

resultant = R

Alternatively, let the head of one vector touch the tail of the second, and form the resultant by connecting the tail of the first to the head of the second to form a triangle, as shown on the right.

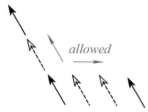

In either case, trigonometry can be used to calculate the magnitude of the resultant as well as its angle (consider the law of cosines).

To line up the vectors, you are allowed to move a vector in *any* direction parallel to, or along, its path. In other words, you can move a vector anywhere so long as its length (magnitude) doesn't change, and its angle (direction) relative to any arbitrary axis doesn't change.

allowed

 1.3   Structure Drawing Conventions

Don't be confused by typesetting conventions that seem to ignore what you've just learned about molecular geometry. For example, you might see methane (CH_4) written as follows:

You now know that methane is tetrahedral. But there is a good reason for using simpler but less accurate structures anyway. When geometrically accurate structures are not needed, there is no point in going to the additional trouble (and in the case of typesetting, the expense) of using them. In other words, we use the simplest structures that accomplish the purpose at hand, and so should you! When the shape of a molecule is an important issue, use line-and-wedge formulas or other types of structures that convey spatial information. Otherwise, use the simpler structures.

FURTHER EXPLORATIONS

 1.1   Dipole Moments

As noted in the text, the mathematical definition of the dipole moment $\boldsymbol{\mu}$ is as follows:

$$\boldsymbol{\mu} = q\mathbf{r} \tag{SG1.1}$$

where q is the magnitude of the separated charge and $\mathbf{r}$ is a vector from the site of positive charge to the site of negative charge. When $\boldsymbol{\mu}$ is a bond dipole, as in H—Cl, the magnitude of $\mathbf{r}$ is simply the bond length; in the case of H—Cl, it is the length of the H—Cl bond. In the HCl molecule the dipole moment vector $\boldsymbol{\mu}$ lies along the H—Cl bond, and it points from the H (the site of partial positive charge) to the Cl (the site of partial negative charge), as shown in the text. That is, $\boldsymbol{\mu}$ and $\mathbf{r}$ *have the same direction.*

The dimensions of the dipole moment, as suggested by Eq. SG1.1 above, correspond to charge × length. As the text indicates, the units of the dipole moment are called *debyes*. A dipole moment of one debye results when opposite charges, each with a magnitude of 1×10^{-10} electrostatic unit (esu), are separated by one Ångstrom. That is,

$$1 \text{ debye} = 1 \times 10^{-10} \text{ esu Å}$$

These units were established prior to the current trend toward the use of standard international units. To put these units on more familiar ground, the charge on an electron is 4.8×10^{-10} esu, and one Ångstrom $= 10^{-10}$ meter $= 100$ picometers. (In standard international (SI) units, in which the unit of length is the meter and the unit of charge is the coulomb (C), $1 \text{ D} = 3.34 \times 10^{-30}$ C m, or coulomb meter.)

An actual calculation of the charge separation in H—Cl can provide a more quantitative sense of the meaning of the dipole moment. A table of experimentally measured dipole moments gives $\mu = 1.08$ D for HCl. A table of bond lengths gives the length of the H—Cl bond as 1.274 Å; this is **r** in Eq. SG1.1. The only unknown remaining in Eq. SG1.1 is q. Because μ and **r** have the same direction, q can be calculated by dividing the magnitude of μ by that of **r**.

$$q = |\mu| \div |\mathbf{r}| = 1.08 \times 10^{-10} \text{ esu Å} \div 1.274 \text{ Å} = 0.848 \times 10^{-10} \text{ esu}$$

As noted above, the charge on an electron is 4.8×10^{-10} esu. Consequently, the charge separation in HCl in electronic charge units is

$$\begin{aligned} q \text{ (in electrons)} &= (0.848 \times 10^{-10} \text{ esu}) \div (4.8 \times 10^{-10} \text{ esu per electron}) \\ &= 0.18 \text{ electron} \end{aligned}$$

In other words, the dipole moment of HCl, 1.08 D, means that a partial negative charge of 0.18 units ("18% of an electron") is on the chlorine of HCl, and a partial positive charge of 0.18 units is on the hydrogen.

Notice that the definition of dipole moment in Eq. SG1.1 contains two elements: the *amount* of charge separated (q) and the *distance* between the separated charges (**r**). A smaller amount of charge separated by a long bond can result in as significant a dipole moment as a larger amount of charge separated by a shorter bond. (See Problem 1.36, text p. 43, in Chapter 1.)

 ## 1.2   Electron Density Distribution in Orbitals

The orbitals in the text (Figs. 1.7 and 1.8, p. 25, and Figs. 1.10–1.11, p. 26–27) are drawn as closed geometrical figures designed to encompass about 90% of the electron density within the orbital. These figures do not show the distribution of the density within the orbital. The purpose of this Further Exploration is to provide a better idea of this distribution.

In the text you learned that the wave motion of an electron in an orbital is described by a function ψ called a *wavefunction*. It turns out that the electron density at any point in space of an electron in an orbital ψ is proportional to the *square* of the wavefunction, ψ^2. Consequently, by plotting the square of the wavefunction as a function of spatial coordinates we can obtain a description of the electron density at any point in space. Because we are dealing with a function of three spatial dimensions, such a plot would require four dimensions, or at least a contour plot in three dimensions. A simpler plot that has all the information we need is obtained by plotting the electron density ψ^2 as a function of the radial distance r along any straight line originating at the nucleus. Consider, for example, the $1s$ orbital. Because the orbital is a sphere, we would get the same plot in any direction. The result for a $1s$ orbital is the plot in Fig. SG1.1a (dark line). From this plot you can see that the electron density is greatest at the nucleus and then falls off exponentially. There are no nodes—no regions of zero electron density—in this plot.

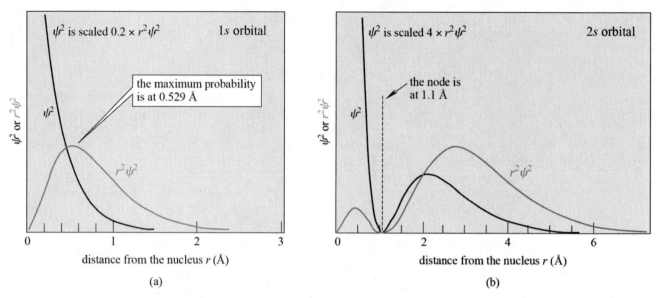

Figure SG1.1 Plots of wavefunctions (ψ^2) and their probability distribution functions ($r^2\psi^2$) for (a) a 1s orbital; (b) a 2s orbital. The square of the wavefunction (ψ^2) is scaled so that this quantity can be shown on the same plot. The "sizes" of the orbitals can be approximated by the distances at which $r^2\psi^2$ is a maximum. Thus, the 1s orbital has a maximum at 0.529 Å; the 2s orbital has a maximum around 3 Å.

The plot just described gives the electron density at a *point*. A different view of electron density is obtained if we plot the probability of finding the electron in a *spherical shell* at some distance r from the nucleus. The appropriate function for this probability is $r^2\psi^2$. The reason this makes sense is that the further away we are from the nucleus, the more surface area there is in a spherical shell. Another way of thinking about this is to imagine the dimples on two basketballs, one large and one small. Assume that the dimples on both are the same size. The large basketball, then, has more dimples. Then think of each dimple as one probability unit for finding an electron. The larger the sphere (that is, the greater is r), the larger is the probability of finding an electron. Since the surface area of a sphere increases as the square of the radius, that is, as r^2, then multiplying the probability at one point ψ^2 by r^2 gives the probability of finding the electron in a spherical shell of radius r. The gray curve in Fig. SG1.1a is a plot of this probability. Notice that this product has a maximum at 0.529 Å. This is called the *Bohr radius*. It is exactly the same as the radial dimension of the electron orbit calculated in an earlier, planetary model of the atom. This model said that the electron was in a fixed orbit with a radius of 0.529 Å; the quantum model says that the *maximum likelihood* of finding an electron is on a spherical shell 0.529 Å from the nucleus; but there is a finite probability at *any* distance from the nucleus.

A similar plot of electron density for the 2s orbital is given in Fig. SG1.1b. In the plot of ψ^2 you can clearly see the node—the region of zero electron density—at 1.1 Å. There is a small maximum in the spherical-shell electron probability at the Bohr radius, but the largest maximum is in the "outer ball" of electron density is at about 3 Å. If you contrast the value of r at which this maximum occurs with that for the 1s orbital you can get a very graphic sense of the different sizes of the two orbitals.

Plots of electron density for the 2p and 3p orbitals are given in Fig. SG1.2a and Fig. SG1.2b, respectively. These plots can be imagined to be taken on a line running through the nucleus and along the axis of the orbital. Both positive and negative values of r are shown to stress the lobe character of the orbitals. Both orbitals have planar nodes at the nucleus that separate the orbital into two lobes. The 3p orbital has, in addition, a spherical node separates the inner and outer regions of each lobe. This node is shown in Fig. SG1.2b as two dashed lines, which represent the $+r$ and $-r$ values, i.e., the radius, of the spherical node. The maximum in the spherical-shell electron probability $r^2\psi^2$ for the 2p orbital (grey line, Fig. SG1.2a) is at about 4 Bohr radii (2.1 Å), whereas the largest maximum in the 3p orbital occurs at about 12 Bohr radii (6.3 Å). Again, notice the larger size of the orbital with the larger quantum number.

These orbital pictures are derived from the *hydrogen atom*. Orbitals of atoms with higher atomic number have basically the same shape and nodal characteristics, but their sizes differ.

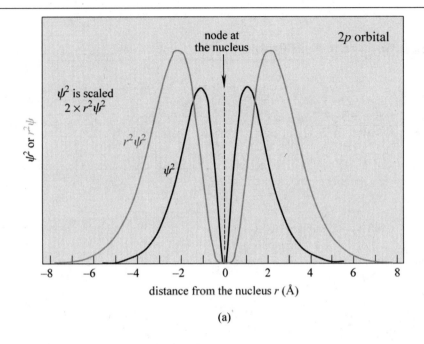

(a)

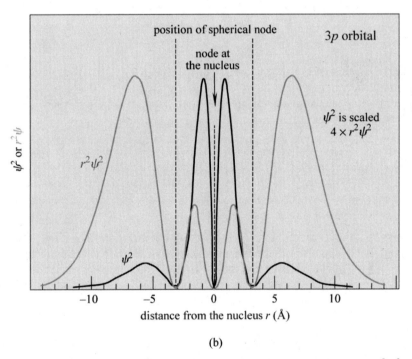

(b)

Figure SG1.2 Plots of wavefunctions (ψ^2) and their probability distribution functions ($r^2\psi^2$) for (a) a 2*p* orbital; (b) a 3*p* orbital.

SOLUTIONS TO PROBLEMS

Solutions to In-Text Problems

1.1   (a)   Because sodium (Na) is in Group 1A, it has one valence electron.
       (b)   The neutral calcium atom has a number of valence electrons equal to its group number, that is, 2.
       (c)   Oxygen has 6 valence electrons; hence, O^{2-} has 8.
       (d)   Neutral Br, being in Group 7A, has 7 valence electrons; therefore, Br^+ has 6.

1.2   (a)   Because neon has 10 electrons, the negative ion requested in the problem also has 10. The singly charged negative ion with ten electrons is F^-.
       (b)   The positive ion isoelectronic with neon must have 10 electrons and 11 protons, and therefore must have an atomic number = 11. This is the sodium ion, Na^+.
       (c)   The atomic number of Ar is 18; so, the dipositive ion with the same number of electrons must have 20 protons and atomic number = 20. This is the Ca^{2+} ion.
       (d)   Because Ne has atomic number = 10 and F has atomic number = 9, the neon species that has 9 electrons is Ne^+.

1.3   (a)   (b)   (c)   (d)

chloroform
(trichloromethane)   ammonia   ammonium ion   hydronium ion

1.4   One structure is that of ethanol, and the other is that of dimethyl ether.

ethanol   dimethyl ether

In all but the simplest cases there are many structures that have the same atomic composition. Compounds that have the same atomic composition, but different atomic connectivities, are called **constitutional isomers.** Thus, dimethyl ether and ethanol are constitutional isomers. For molecules of moderate size hundreds or even thousands of constitutional isomers are possible. You'll learn about isomers in Chapter 2.

1.5   (a)   The overall charge is 0.

       (b)   The overall charge is –2.

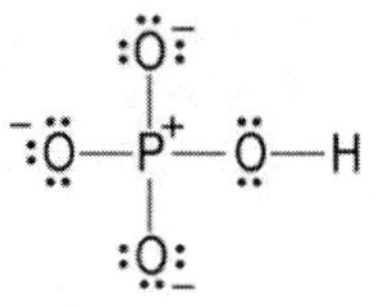

$$
\begin{array}{c}
:\!\overset{..}{\underset{}{O}}\!:^{-} \\
| \\
:\!\overset{..}{\underset{..}{O}}\!-\!\overset{+}{P}\!-\!\overset{..}{\underset{}{O}}\!-\!H \\
| \\
:\!\overset{..}{\underset{..}{O}}\!:^{-}
\end{array}
$$

1.6   The C—H bonds are the least polar, because carbon and hydrogen differ very little in electronegativity. The carbon that has the most partial positive character is the one indicated by the asterisk (*), because it is bound to two very electronegative atoms, O and Cl.

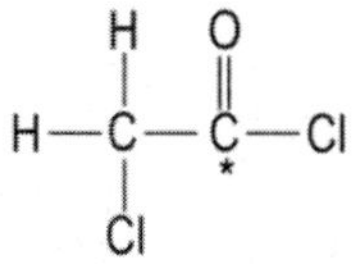

$$
\begin{array}{c}
\quad\; H \quad\;\; O \\
\quad\; | \quad\quad \| \\
H\!-\!C\!-\!\underset{*}{C}\!-\!Cl \\
\quad\; | \\
\quad\; Cl
\end{array}
$$

1.7   (a)   Formal charge does not give an accurate picture, because N is more electronegative than H; most of the positive charge is actually on the hydrogens.

(b)   Formal charge does give an accurate picture, because N is more electronegative than H; most of the negative charge is on the nitrogen.

(c)   An analysis of relative electronegativities would suggest that, because C is slightly more electronegative than H, a significant amount of the positive charge resides on the hydrogens. However, carbon does not have its full complement of valence electrons—that is, it is short of the octet by 2 electrons. In fact, both C and H share the positive charge about equally.

1.8   In addition to the structure shown on text p. 17, five other line-and-wedge depictions of dichloromethane exist. All are depictions of the same compound; think of them as the molecule tumbling in space, or as the view of the molecule from different perspectives.

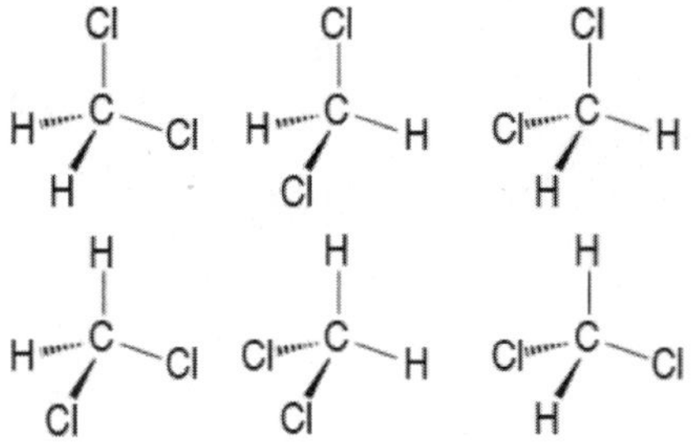

$$
\begin{array}{ccc}
Cl & Cl & Cl \\
| & | & | \\
H\!\cdots\!C\!-\!Cl & H\!\cdots\!C\!-\!H & Cl\!\cdots\!C\!-\!H \\
/ & / & / \\
H & Cl & H \\[2ex]
H & H & H \\
| & | & | \\
H\!\cdots\!C\!-\!Cl & Cl\!\cdots\!C\!-\!H & Cl\!\cdots\!C\!-\!Cl \\
/ & / & / \\
Cl & Cl & H
\end{array}
$$

1.9   (a)   The tetrafluoroborate anion is tetrahedral—that is, it has F—B—F bond angles of 109.5°.

(b)   Water has bent geometry; that is, the H—O—H bond angle is approximately tetrahedral. Repulsion between the lone pairs and the bonds reduces this bond angle somewhat. (The actual bond angle is 104.5°.)

(c)   The formaldehyde molecule has trigonal planar geometry. Thus, both the H—C—H bond angle and the H—C=O bond angle are about 120°.

(d)   In acetonitrile, each of the H—C—H bond angles, as well as the H—C—C bond angle, is tetrahedral (about 109.5°). The C—C≡N bond angle is linear (180°).

1.10   Bond angles: *aa, ab, bc, bd, cd, de, df,* and *ef* are all about 120°, because all are centered on atoms with trigonal planar geometry; *fg* is predicted to have the tetrahedral value of 109.5. The bond lengths increase in the order

$$
a \approx g < e < b < d \approx f < c
$$

(In Chapter 4, you'll learn that C—H bonds attached to carbons of double bonds are shorter than C—H bonds attached to carbons of single bonds. For this reason, *a < g*.)

1.11   A molecule contains dihedral angles only if it contains at least four atoms connected consecutively. Compound *A* fits this requirement (H, C, N, and H atoms bonded consecutively), and thus has dihedral angles. Imagine two planes, one containing a C—H bond and another containing a N—H bond. The dihedral angle is the angle between the two planes. The model with dihedral angle = 0° is as follows:

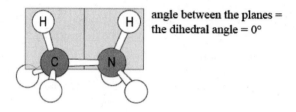

The model with dihedral angle = 60° is as follows.

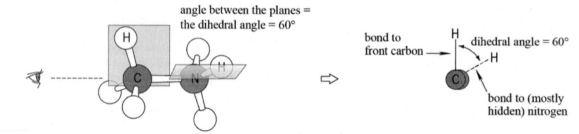

The model with dihedral angle = 180° is identical to the one with dihedral angle = 60°.

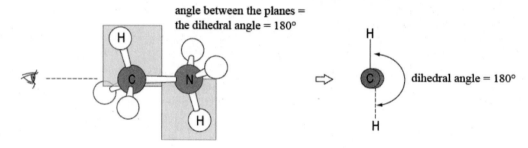

The most consecutive bonds present in ion *B* is three (F—P—F), so a dihedral angle is not possible. The structure of PF_6^- can be described completely by knowing bond lengths and bond angles. Similarly, ion *C* contains only three consecutive atoms (H—O—O) so a dihedral angle is not present.

1.12   The resonance structures of benzene:

$$
\left[
\begin{array}{c}
\text{resonance structure 1}
\end{array}
\longleftrightarrow
\begin{array}{c}
\text{resonance structure 2}
\end{array}
\right]
\qquad
\begin{array}{c}
\text{hybrid structure}
\end{array}
$$

resonance structures of benzene   hybrid structure of benzene

Each bond is a single bond in one structure and a double bond in the other. On average, each bond has a bond order of 1.5.

1.13   (a)   Structures of the allyl anion:

$$\left[H_2C\!\!=\!\!CH\!-\!\overset{..}{\overset{-}{C}}H_2 \quad \longleftrightarrow \quad H_2\overset{..}{\overset{-}{C}}\!-\!CH\!\!=\!\!CH_2 \right]$$

resonance structures of the allyl anion

Although the structure is shown in a linear representation, the central carbon is trigonal planar, and the carbon skeleton is bent, as application of the VSEPR rules will show.

(b)   Each CH_2 carbon contains 0.5 of a negative charge.

(c)   A hybrid structure:

$$H_2\overset{\delta-}{C}\!\cdots\!CH\!\cdots\!\overset{\delta-}{C}H_2$$

hybrid structure of the allyl anion

1.14   Fig. SG1.3 shows a plot of $\psi = \sin nx$ for the different values of n. Evidently, the number of nodes in this domain is one less than the "quantum number" n; that is, $nodes = n - 1$.

    Orbitals are described by mathematical functions of three dimensions, whereas $\psi = \sin nx$ is a function of one dimension. However, the quantum numbers of orbitals have the same effect as n in this simple equation: they control the number of nodes.

1.15   (a)   A 3s orbital is three concentric spheres of electron density, each separated by a node, as shown in the "cutaway" diagram in Fig. SG1.4a.

(b)   A 4s orbital is four concentric spheres of electron density, each separated by a node, as shown in "cutaway" diagram (b) in Fig. SG1.4b.

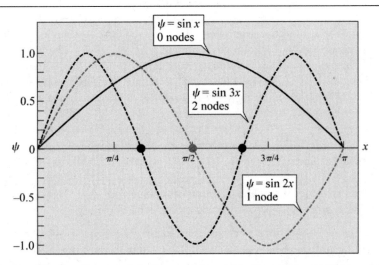

Figure SG1.3 A plot of the function $\psi = \sin nx$ for $n = 1, 2,$ and 3 for the solution to Problem 1.14. The nodes are given by a gray filled circle (for $\psi = \sin 2x$) and black filled circles (for $\psi = \sin 3x$).

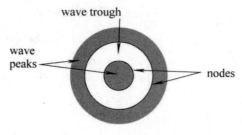

Figure SG1.4a Cross section of a 3s orbital in the solution to Problem 1.15a.

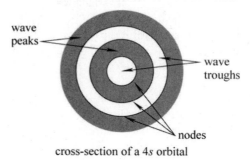

Figure SG1.4b Cross section of a 4s orbital in the solution to Problem 1.15b.

1.16 (a) The oxygen atom (atomic number = 8, therefore 8 electrons): $(1s)^2(2s)^2(2p_x)^2(2p_y)^1(2p_z)^1$. Two of the $2p$ electrons are unpaired; each of these is in a different $2p$ orbital. The valence orbitals are the $2s$ and $2p$ orbitals, and the valence electrons are the six electrons that occupy these orbitals.

(b) The chloride ion, Cl^- (atomic number = 17 and one negative charge, therefore 18 electrons): This ion has the same electronic configuration as argon: $(1s)^2(2s)^2(2p)^6(3s)^2(3p)^6$. The valence orbitals are the $3s$ and $3p$ orbitals, and the valence electrons are the eight electrons that occupy these orbitals.

(c) The potassium ion, K^+, has the same configuration as Ar, and therefore the same configuration as Cl^-: $(1s)^2(2s)^2(2p)^6(3s)^2(3p)^6$. The valence orbitals are the $3s$ and $3p$ orbitals, and the valence electrons are the eight electrons that occupy these orbitals.

(d) The sodium atom (atomic number = 11) has 11 electrons. Therefore its electronic configuration is $(1s)^2(2s)^2(2p)^6(3s)^1$. The valence orbital is the $3s$ orbital, and the valence electron is the one electron that occupies this orbital.

1.17 The pictures and energy levels of the molecular orbitals for parts (a), (b), (c), and (d) are essentially the same as they are for the dihydrogen molecular orbitals in Fig. 1.14 of the text.

(a) The He_2^+ ion contains three electrons. This can be conceived as the combination of a He^+ ion with a He atom. By the aufbau principle, two electrons occupy the bonding molecular orbital, and one occupies the antibonding molecular orbital. This is shown in the electron-occupancy diagram (a) of Fig. SG1.5. Because the bonding molecular orbital contains a greater number of electrons than the antibonding molecular orbital, this species is stable.

(b) The H_2^- ion contains three electrons. This can be conceived as the combination of a hydrogen atom with a hydride ion (H^-). By the aufbau principle, two occupy the bonding molecular orbital, and one occupies the antibonding molecular orbital. This is shown in the electron-occupancy diagram (b) of Fig. SG1.5. Because the bonding molecular orbital contains a greater number of electrons than the antibonding molecular orbital, this species is stable. Notice that in terms of electron occupancy, H_2^- and He_2^+ are identical.

(c) The H_2^{2-} ion can be conceived to result from the combination of two hydride ions (H^-). This species contains four electrons; two occupy the bonding molecular orbital, and two occupy the antibonding

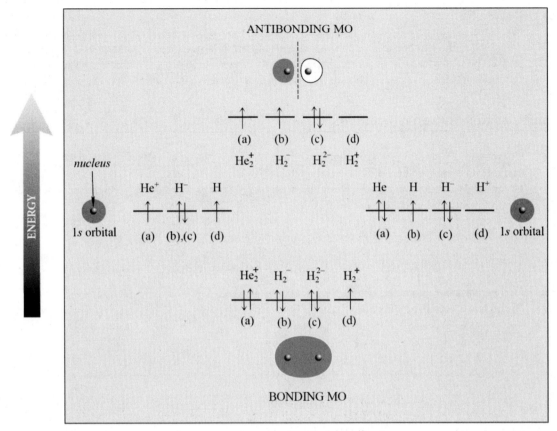

Figure SG1.5 Electron-occupancy diagrams for the solutions to Problems 1.17a-d. The electrons are represented by arrows.

molecular orbital. This is shown in the electron-occupancy diagram (c) of Fig. SG1.5. In this species, the energetic advantage of the electrons in the bonding molecular orbital is cancelled by the energetic disadvantage of the same number of electrons in the antibonding molecular orbital. This species consequently has no energetic advantage over two dissociated hydride (H^-) ions, and therefore it readily dissociates.

(d)   The H_2^+ ion is conceptually derived from the combination of a hydrogen atom with a proton. It contains one electron in a bonding molecular orbital. This is shown in the electron-occupancy diagram (d) of Figure SG1.5. Because the bonding molecular orbital contains a greater number of electrons than the antibonding molecular orbital, this species is stable.

1.18   Because the H_2^+ ion has one electron in a bonding molecular orbital, it should have about half the stability of H_2 itself relative to dissociated fragments. Thus, 217 kJ mol^{-1} (52 kcal mol^{-1}) is an estimate of the bond dissociation energy of this species.

1.19   Referring to Figure SG1.5d, the H_2^+ ion has 1 electron in the bonding MO and no electrons in the antibonding MO. Using Equation 1.8 in the text, the H_2^+ ion has a bond order of $(1 - 0)/2 = \frac{1}{2}$. This number corresponds with the answer to Problem 1.18, which states that the H_2^+ bond should be half as strong as the H_2 bond.

1.20   (a)   If the oxygen of water is sp^3-hybridized, then two of the sp^3 hybrid orbitals contain unshared electron pairs. The other two sp^3 hybrid orbitals contain one electron each; each of these overlaps with the $1s$ orbital of a hydrogen atom (which contains one electron) to give the two sp^3–$1s$ σ bonds (the O—H bonds) of water.

(b) Because unshared electron pairs prefer *s* orbitals, the orbitals containing the unshared pairs have more 2*s* character, and therefore the orbitals used to form the bonds have more 2*p* character. By the reasoning discussed in the text for ammonia, bonds involving orbitals with more 2*p* character are separated by smaller bond angles. Alternatively, VSEPR theory says that electron pairs have a larger spatial requirement. Either way, the H—O—H angle in water is less than 109.5°. (In fact, it is 104.5°.)

Solutions to Additional Problems

1.21 (a) Compound (3), sodium astatide, would be most likely to exist as an ionic species, because Na and At come from opposite sides of the periodic table.

(b) Compound (2), cesium fluoride, would be most likely to exist as an ionic species, because Cs and F come from opposite corners of the periodic table.

1.22 The formal charge on all the hydrogens is 0. For the other atoms:

(a) Carbon has a sextet and a formal charge of +1.
(b) Nitrogen has a complete octet and a formal charge of 0.
(c) Carbon has a complete octet and a formal charge of −1.
(d) Boron has a sextet and a formal charge of 0.
(e) Iodine has a sextet and a formal charge of +1.
(f) Boron has an octet and a formal charge of −1.

1.23 (a)   (b)   (c)

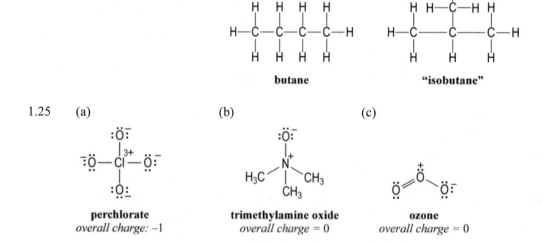

1.24 The lengths of the C—C bonds in isobutane have no significance; two are longer merely to accommodate the rest of the structure. Remember that these Lewis structures show only connectivity—not geometry.

1.25 (a)   (b)   (c)

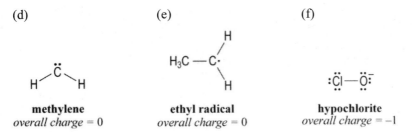

(d) **methylene**
overall charge = 0

(e) **ethyl radical**
overall charge = 0

(f) **hypochlorite**
overall charge = –1

1.26 (a) Chlorine atom (atomic number = 17; therefore 17 electrons): $(1s)^2(2s)^2(2p)^6(3s)^2(3p_x)^2(3p_y)^2(3p_z)^1$

 (b) Silicon atom (atomic number = 14; therefore 14 electrons): $(1s)^2(2s)^2(2p)^6(3s)^2(3p_x)^1(3p_y)^1$ or $(1s)^2(2s)^2(2p)^6(3s)^2(3p)^2$. The valence electrons in silicon are the two $3s$ and the two $3p$ electrons, and the valence orbitals are the $3s$ and the three $3p$ orbitals.

 (c) Argon atom (atomic number 18; therefore 18 electrons. This has the same configuration as chloride ion—that is, $(1s)^2(2s)^2(2p)^6(3s)^2(3p)^6$

 (d) Magnesium atom (atomic number = 12; therefore 12 electrons): $(1s)^2(2s)^2(2p)^6(3s)^2$

1.27 The $2d$ orbital is not permitted. The reason: d means that the l quantum number = 2; the value of the l quantum number cannot exceed *one less than* the value of the principal quantum number n, which in this case is 2. Thus, l cannot exceed 1 in principal quantum level 2.

1.28 (a) The carbon is bound to three "groups," the two Hs and an electron pair. Hence, this carbon is trigonal planar (if you count the electron pair) and the molecule is "bent" with an H—C—H angle of about 120° (or perhaps somewhat less because of the VSEPR rule that unshared pairs reduce bond angles).

 (b) The beryllium (Be) bears two groups, and therefore has linear geometry. The H—Be—H bond angle is 180°.

 (c) The carbon is bound to three groups; hence, this ion is trigonal planar, and the H—C—H bond angles are 120°.

 (d) The silicon is bound to four chlorines, and is therefore tetrahedral. The Cl—Si—Cl bond angle is 109.5°.

 (e) The central oxygen of ozone is bound to three "groups," two oxygens and an electron pair. Hence, this oxygen is trigonal planar (if you count the electron pair), and ozone is "bent" with an O—O—O bond angle somewhat less than 120°. (In fact, it is 117°.)

 (f) The terminal carbons are bound to three groups, and are therefore trigonal planar. The central carbon is bound to two groups, and is therefore linear. Hence, the H—C—H and H—C=C bond angles are 120°; the C=C=C bond angle is 180°.

 (g) The nitrogen is trigonal planar; the O=N—O bond angle is 120°, as are both of the C—N—O bond angles. The carbon is bound to four groups, and is tetrahedral; all bond angles centered on carbon are about 109.5°.

1.29 Bond angles: *bc, bd, cd, de, dh,* and *eh* are all about 120°, because all are centered on atoms with trigonal planar geometry; *ab, eg, ef,* and *fg* have the approximately tetrahedral value of 109.5. The bond lengths increase in the order

$$c < a < d < b \approx f < h < e < g$$

Notice that the rules on text pp. 14–15 are in order of importance. Thus, the row of the periodic table is paramount in determining bond length. Thus, bond *b* is between two period-2 elements (C and C), whereas bond *f* is between a period-1 and a period-3 element. The "average period" for both is 2. In fact, the two bonds have about the same length.

1.30 (a) The oxygen of the hydronium ion has approximately tetrahedral geometry and is therefore sp^3-hybridized. Two of the sp^3 hybrid orbitals contain an unshared electron pair. One of these becomes the lone pair in H_3O^+; the other overlaps with a proton (H^+) to give one of the O—H bonds. The other two sp^3 hybrid orbitals contain one electron each; each of these overlaps with the $1s$ orbital of a hydrogen atom (which contains one electron) to give the three sp^3–$1s$ σ bonds (the O—H bonds) of H_3O^+.

(b)   The H—O—H bond angles in H_3O^+ should be somewhat larger than those in water, because there is one less lone pair and a smaller associated repulsion.

1.31   (a)   The allyl cation is a hybrid of the two structures. Because the structures are equally important, each structure contributes equally to the structure of the hybrid. Because each carbon–carbon bond is a single bond (bond order = 1) in one resonance structure and a double bond (bond order = 2) in the other, the *average* bond order is $(1 + 2)/2 = 1.5$. In other words, each carbon–carbon bond is equivalent to a single bond plus "half" of a double bond.

(b)   By analogous reasoning, each of the terminal carbons bears +1 charge in one structure and 0 charge in the other. Hence, the charge on each carbon of the ion is the average of these two, or $(+1 + 0)/2 = +0.5$. The charge on the central carbon is zero in all resonance structures and therefore zero overall.

(c)   The resonance structure on the right violates the octet rule for nitrogen.

 Notice from this example that a positive charge does not necessarily mean that an atom lacks an octet. This point will be considered further in Chapter 3.

1.32   We use reasoning analogous to that used in the previous problem. This ion is a hybrid of three equally important structures. Hence, each property will be the sum of that property in each structure divided by 3.

(a)   Negative charge on each oxygen: $(-1 + 0 + -1)/3 = -2/3$.
(b)   Bond order of each carbon–oxygen bond: $(1 + 2 + 1)/3 = 4/3$ or 1.33.

1.33   (a)   Figure 1.8 on text p. 25 shows that a $2s$ orbital has a single spherical node; and Figure 1.10 on text p. 26 shows that the $2p$ orbital has one planar node. Figure 1.11 shows that the $3p$ orbital has two nodes—one planar node and one spherical node. In both cases, the total number of nodes is one less than the principal quantum number. In the $2s$ orbital, the value of l is 0, and the number of planar nodes is zero. In the $2p$ and $3p$ orbitals, the value of l is 1, and each orbital has one planar node.

(b)   The total number of nodes is $n - 1$, or 4. The value of l (0, which implies an s orbital) gives the number of planar nodes, which is 0. Hence, the $5s$ orbital has four spherical nodes. By analogous reasoning, a $3d$ orbital has two nodes, both of which are planar; hence, it has no spherical nodes. (Notice this property of the $3d$ orbital in the solution to the following problem.)

1.34   There are two planar nodes: one is the x–z plane, and the other is the y–z plane, where the z axis is perpendicular to the page.

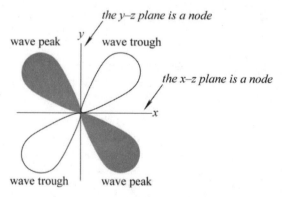

1.35   A $4p$ orbital is somewhat like a $3p$ orbital (Fig. 1.11 on text p. 27), except that it has one additional spherical node and one additional "mushroom cap" region of electron density. A cross section—that is, a planar slice along the axis of the orbital— is shown in Fig. SG1.6.

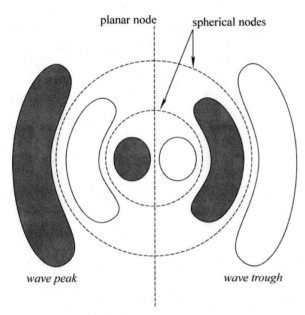

Figure SG1.6 A cross section of a *4p* orbital showing the nodes to accompany the solution to Problem 1.35. The actual shape of the orbital is the volume of revolution about the *x*-axis—that is, the volume "swept out" when the figure is rotated around the *x*-axis.

1.36   According to Eq. 1.4 (p. 11) in the text, the dipole moment is determined by *two* factors. The first is the separated charge, which should be greater for the compound containing the more electronegative atom, that is, for H_3C—F. However, the second factor is the bond length **r**, which is greater for the compound containing the *larger* atom, that is, for H_3C—Cl. Evidently, these two opposing factors just about compensate in their effects on the dipole moments of the two compounds.

1.37   (a)   The bond dipole for dimethylmagnesium should indicate that C is at the negative end of the C—Mg bond, because carbon is more electronegative than magnesium.

$$H_3C - Mg - CH_3$$

     (b)   The two carbon–magnesium bonds should be 180° from each other. VSEPR theory states that bonds should be oriented as far apart as possible.

     (c)   The two vectors indicating the bond dipoles of the C—Mg bonds are pointing in equal and opposite directions, cancelling each other, for an overall dipole moment of 0 for dimethylmagnesium.

1.38   No matter how any CH_2 group is turned, the resultant bond dipole is the same:

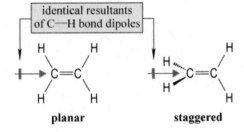

Hence, ethylenes with the two different dihedral angles should both have zero dipole moment because the resultants of the two CH_2 groups cancel each other in either arrangement. The observation of zero dipole moment does not permit a choice between these two dihedral angles. Note: Actually, ethylene is "locked" in the planar orientation under normal conditions, as will be discussed in Chapter 4.

1.39 (a) Because the carbon has trigonal-planar geometry, the H—C=O bond angle is 120°.

(b) The two structures are as follows:

O—C—O / C—O—C
dihedral angle = 0°

O—C—O / C—O—C
dihedral angle = 180°

1.40 A linear water molecule would have zero dipole moment, because the O—H bond dipoles would be oriented in opposite directions and would cancel. Hence, the nonzero dipole moment for water shows that it is a bent molecule. Out with Professor Szents!

1.41 To solve this problem, we first have to know the geometry of CCl_4. This should be easy: it's tetrahedral, because the carbon is bonded to four identical groups. Now, as suggested in the hint, let's do a vector addition for the bond dipoles of each Cl—C—Cl pair. (Use a dipole arrow of arbitrary length as long as it is the same for both. We'll use an arrow with the length of the C—Cl bonds.) Here's the result of the vector addition, which we've done graphically, for the first pair. Notice that, because the C—Cl bonds are identical, the resultant must *bisect* the angle between the two bonds.

resultant of
vector addition

Now let's do the resultant for the second pair. Because the C—Cl bonds are identical to those of the first pair, it follows that the resultant is the same length, and it again bisects the Cl—C—Cl bond angle. (We've rotated the molecule about an axis along the dipole so we are viewing the other two chlorines from the perspective we used to view the first two—in the plane of the page.)

resultant of
vector addition

The two resultants we have just obtained are oriented at an angle of 180°; that is, they are pointed in opposite directions. Since they are identical in magnitude, they cancel. Hence the dipole moment of CCl_4 is zero.

resultant dipole = 0

1.42 (a) When the O—H bonds lie at a dihedral angle of 180° their bond dipoles cancel; consequently, a hydrogen peroxide molecule in this conformation would have a dipole moment of zero. This conformation is *ruled out* by the observation of a significant dipole moment.

the O—H bond dipoles cancel

(b) The following construction for the O—O—H angle is useful for solving this problem:

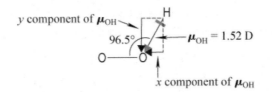

The x components of the two O—H bond dipoles point in opposite directions regardless of the bond angle or the dihedral angle. Hence, the x components cancel. Therefore, we can deal with the y component of the O—H bond dipole exclusively. First, we have to calculate the y component. This is just $1.52 \cos(6.5°) = 1.52\,(0.9935) = 1.51$ D.

Note that $96.5°$ is not the dihedral angle, but the bond angle. The idea is to calculate the dihedral angle. From the analysis in the previous paragraph, we know that the resultant dipole moment of the molecule (given as 2.13 D) is derived from vector addition of the y components of the two O—H bond dipoles, which we now know are 1.51 D. All we have to do is determine the angle θ that will give 2.13 D as the resultant from vector addition of two 1.51 D dipoles.

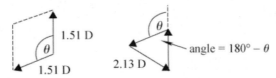

The proper construction is the arrangement on the left; because we can move a vector at will so long as its magnitude and direction remain unchanged, we derive the construction on the right. The resultant, 2.13D, is the vector that completes a head-to-tail connection of the two bond dipoles. From the law of cosines, this vector is

$$(2.13 \text{ D})^2 = (1.51 \text{ D})^2 + (1.51 \text{ D})^2 - 2(1.51 \text{ D})(1.51 \text{ D}) \cos(180° - \theta)$$

or

$$4.54 \text{ D}^2 = 2.28 \text{ D}^2 + 2.28 \text{ D}^2 - (4.56 \text{ D}^2) \cos(180° - \theta)$$

from which we have

$$\frac{-0.02 \text{ D}^2}{-4.56 \text{ D}^2} = \cos(180° - \theta) = 0.00439$$

Using the "arccos" function on a calculator, we find that $180° - \theta = 89.8°$. Thus, the desired dihedral angle is $180° - 89.8° = 90.2°$. Therefore, it appears that the dihedral angle in hydrogen peroxide is about $90°$. (The actual structure of hydrogen peroxide determined by electron diffraction shows a dihedral angle of $93.8°$!) Thus, the conformation of hydrogen peroxide is best represented as follows:

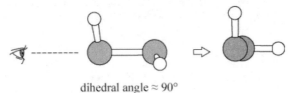

dihedral angle $\approx 90°$

1.43   This problem requires a construction similar to the one used in Problem 1.42 in which the resultant is known and the individual O—H bond dipoles are unknown.

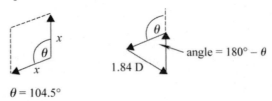

$\theta = 104.5°$

Applying the law of cosines, with x as the O—H bond dipoles,

$$1.84^2 = x^2 + x^2 - 2x^2 (\cos(180° - \theta)$$

$$3.39 = 2x^2(1 - \cos 75.5°) = 2x^2(1 - 0.25) = 1.50\,x^2$$

$$x = \sqrt{\frac{3.39}{1.50}} = 1.50\text{ D}$$

This is the calculated value of the O—H bond dipole, which is very similar to the value assumed in Problem 1.42.

1.44   Dilithium, Li_2, is derived from the combination of two lithium atoms. It contains two bonding electrons derived from two $2s$ valence electrons, one from each lithium atom. Each Li atom also has two $1s$ electrons, but these are not valence electrons and therefore are not involved in bond formation and are not shown in the interaction diagram (Fig. SG1.7). The interaction of the two $2s$ orbitals gives a bonding and an antibonding MO. The nodes present in the individual $2s$ orbitals are retained in the MOs. This is a stable species because the bonding molecular orbital contains a greater number of electrons than the antibonding molecular orbital.

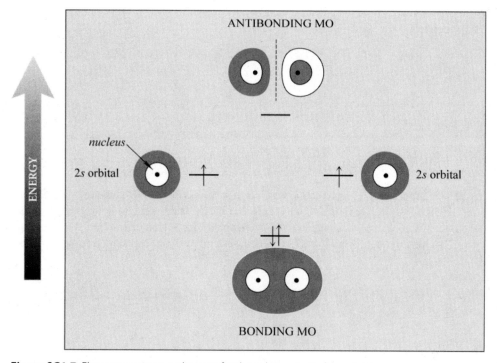

Figure SG1.7 Electron-occupancy diagram for the solution to Problem 1.44.

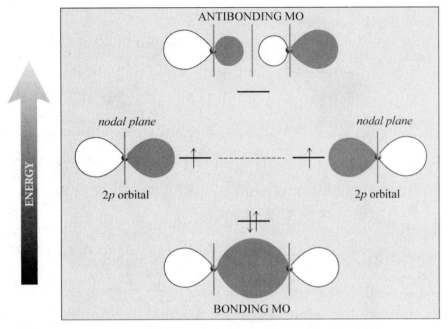

Figure SG1.8 An orbital interaction diagram for the head-to-head interaction of two $2p$ orbitals to accompany Problem 1.45. The small spheres are nuclei, and the vertical gray lines are nodal planes viewed end-on.

1.45   (a)   The bonding molecular orbital is derived by the constructive overlap of wave peaks. The antibonding molecular orbital is derived by changing the peak to a trough and the trough to a peak in one of the two $2p$ orbitals. The destructive overlap of a peak with a trough gives a node. The resulting molecular orbitals are shown in Fig. SG1.8.

(b)   The nodes are shown in Fig. SG1.8. Both of the MOs retain the nodes of the original $2p$ orbitals. The antibonding MO has an additional node that results from the destructive overlap of a peak and a trough.

(c)   Figure SG1.8 is the required interaction diagram.

(d)   The resulting bond is a σ bond because it is cylindrically symmetrical about the orbital axis.

1.46   (a)   The bonding molecular orbital is derived by the constructive "side-to-side" overlap of peaks with peaks and troughs with troughs. The antibonding molecular orbital is derived by changing the peak to a trough and the trough to a peak in one of the two $2p$ orbitals. Destructive overlap of troughs with peaks and peaks with troughs gives an additional node in the antibonding orbital. These molecular orbitals are shown in Fig. SG1.9.

(b)   The nodes are also shown in Fig. SG1.9. Both of the MOs have the nodes of the original $2p$ orbitals, which merge into a single node because the nodal plane is common to both orbitals. The antibonding MO has in addition a node between the original orbitals that results from the destructive overlap of a peak and a trough.

(c)   Fig. SG1.9 is the required interaction diagram.

(d)   The resulting bond is *not* a σ bond because it is not cylindrically symmetrical about the internuclear axis.

This type of bond, called a pi (π) bond, is important in the carbon–carbon double bond. Pi bonds are discussed in Chapter 4.

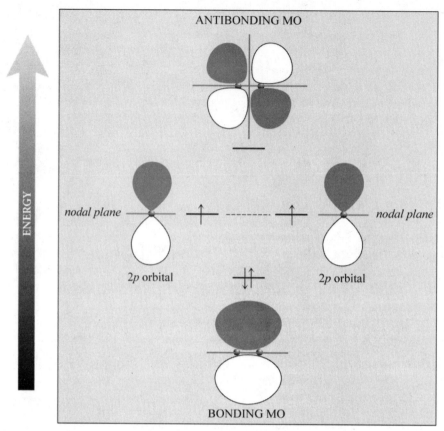

Figure SG1.9 An orbital interaction diagram for the side-to-side interaction of two 2*p* orbitals to accompany Problem 1.46. The small spheres are nuclei, and the gray lines are nodal planes viewed end-on.

1.47   (a)   Similar to Problem 1.44 (Fig. SG1.7), bringing two 2*s* orbitals together results in two new MOs, one bonding (2*sσ*) and one antibonding (2*sσ**), as shown in Figure SG1.10a.

(b)   Similar to Problem 1.45, (Fig. SG1.8), bringing the two $2p_x$ orbitals together results in two new MOs, one bonding (2*pσ*) and one antibonding (2*pσ**), as shown in Figure SG1.10b.

(c)   Similar to Problem 1.46, (Fig. SG1.9), bringing the two $2p_y$ orbitals together results in two new MOs, one bonding ($2p\pi_y$) and one antibonding ($2p\pi_y*$), as shown in Figure SG1.10c.

(d)   Bringing the two $2p_z$ orbitals together results in two new MOs, one bonding ($2p\pi_z$) and one antibonding ($2p\pi_z*$), as shown in Figure SG1.10d. These orbitals are identical to those obtained from overlapping the $2p_y$ orbitals as shown in part (c), except that they are oriented on the *z*-axis, rotated 90° from the *y*-axis.

(e)   The atomic orbitals of oxygen are combined according to parts (a–d) of this problem and are ordered in increasing energy as shown in Figure SG1.10e. Each oxygen atom has six valence electrons, two in the 2*s* orbital, and four in the three 2*p* orbitals. These electrons are placed in molecular orbitals according to Hund's rules.

(f)   The oxygen molecule has a net (or, nonzero) spin, and is magnetic. An unpaired electron generates a magnetic field due to its spin. The two unpaired electrons in the $2p\pi_y*$ and $2p\pi_z*$ orbitals do not have electrons of opposing spin to cancel them out, so the oxygen molecule is "paramagnetic" and liquid oxygen can be trapped between poles of a magnet.

(g)   We eliminate structure *A* because it does not have unpaired electrons. We also eliminate structure *D* because it has more than 12 total electrons. According to Eq. 1.8 (text p. 35), oxygen has a bond order of (8 – 4)/2 = 2. However, none of the remaining structures contains a double bond. Yet we can use an "average structure", or resonance hybrid, of the molecule containing the single bond and the one containing a triple bond. This average would have a bond order = 2.

$$\cdot\ddot{O}\!-\!\ddot{O}\cdot \quad\longleftrightarrow\quad \cdot\ddot{O}\!=\!\ddot{O}\cdot$$

You may notice that the second structure above violates the octet rule. But, as you perhaps learned when you worked Problems 1.17a and b, MO theory takes us beyond conventional Lewis structures and allows this violation provided that there is net bonding. However, there is an energetic cost: the bond energy of O_2 is about the same as that of a conventional *single* bond, but the bond length is about that of a conventional triple bond. This dichotomy requires the use of resonance structures if we insist on describing O_2 with Lewis structures!

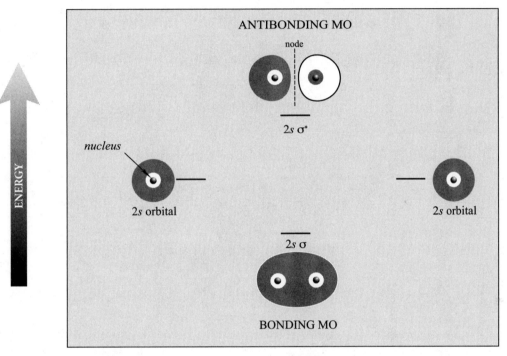

Figure SG1.10a An orbital interaction diagram for the interaction of two 2s orbitals for the solution to Problem 1.47a.

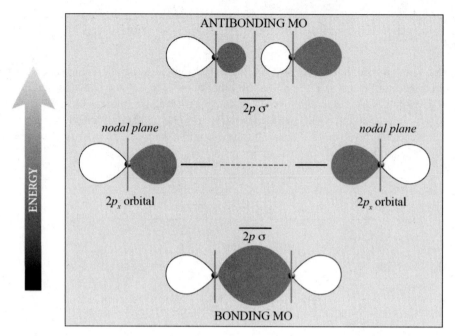

Figure SG1.10b An orbital interaction diagram for the head-to-head interaction of two $2p_x$ orbitals for the solution to Problem 1.47b. The small spheres are nuclei, and the vertical gray lines are nodal planes viewed end-on.

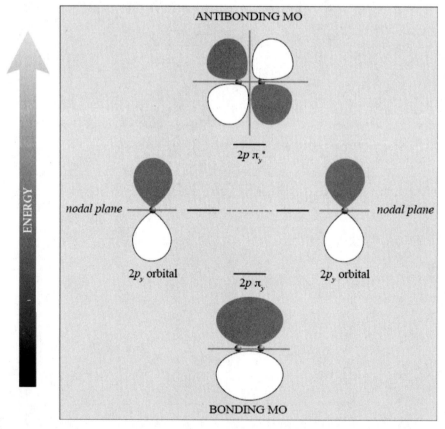

Figure SG1.10c An orbital interaction diagram for the side-to-side interaction of two $2p_y$ orbitals for the solution to Problem 1.47c. The small spheres are nuclei, and the gray lines are nodal planes viewed end-on.

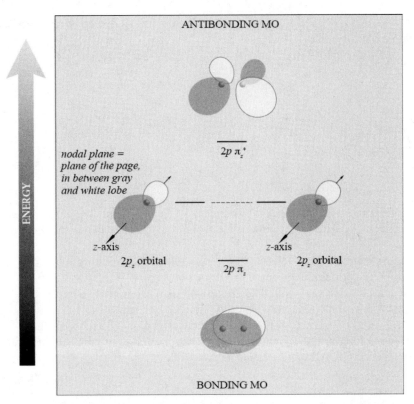

Figure SG1.10d An orbital interaction diagram for the side-to-side interaction of two $2p_z$ orbitals for the solution to Problem 1.47d. The small spheres are nuclei, and nodal planes are the plane of the page for all orbitals, with an additional plane perpendicular to the page (in the y/z plane) for the $2p\pi_z^*$ orbital.

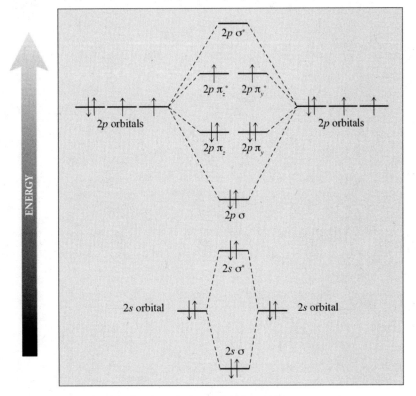

Figure SG1.10e Electron-occupancy diagram for the solution to Problem 1.47e.

1.48   When a hydrogen molecule absorbs light, an electron jumps from a bonding molecular orbital into an antibonding molecular orbital:

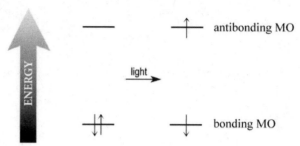

After light absorption, the hydrogen molecule has no excess of bonding electrons, and there is no energetic advantage to bonding. Hence, the molecule can readily dissociate into hydrogen atoms.

1.49   (a)   Because there are three values of spin, each orbital can hold three electrons.

(b)   The configuration of the atom with atomic number = 8: $(1s)^3(2s)^3(2p)^2$.

(c)   The second noble gas has a filled $2p$ orbital; but there are only two $2p$ orbitals ($m_l = 0, +1$); so the $2p$ level holds six electrons (three in each of the $2p$ orbitals). With filled $1s$, $2s$, and $2p$ levels, the second noble gas has $(3 + 3 + 6) = 12$ electrons; hence, its atomic number is 12.

(d)   Because a filled shell contains nine electrons (three electrons each in one s and two p orbitals), the corresponding rule might be called a "nonet rule."

Alkanes

STUDY GUIDE LINKS

 2.1 Nomenclature of Simple Branched Compounds

The IUPAC recognizes as valid the following older names for branched alkane isomers of four, five, and six carbon atoms because of their common historical usage.

H_3C
CH—CH_3 or $(CH_3)_3CH$
H_3C
isobutane

H_3C
CH—CH_2CH_3 or $(CH_3)_2CHCH_2CH_3$
H_3C
isopentane

H_3C
CH—$CH_2CH_2CH_3$ or $(CH_3)_2CHCH_2CH_2CH_3$
H_3C
isohexane

CH_3
H_3C—C—CH_3 or $(CH_3)_4C$
CH_3
neopentane

CH_3
H_3C—C—CH_2CH_3 or $(CH_3)_3CCH_2CH_3$
CH_3
neohexane

The distinguishing structural feature of an alkane named with the "iso" prefix is a pattern of two methyl branches at the end of a carbon chain. The distinguishing structural feature of an alkane with a "neo" prefix is a pattern of three methyl branches at the end of a carbon chain.

Table 2.2 on text p. 62 shows that the names of the common branched alkyl groups are derived from the names shown above for the corresponding alkanes.

FURTHER EXPLORATIONS

2.1   Atomic Radii and van der Waals Repulsion

The *atomic radius* is the effective size of an atom when it is *covalently bonded to another atom*. In this text, this term means the same thing as *covalent radius*. The van der Waals radius, in contrast, measures an atom's effective size when it is *not covalently bonded* to another atom.

The covalent radius of an atom is much smaller than its van der Waals radius. The reason is that because electrons are shared in a chemical bond, nuclei are drawn closer together than they would be in the absence of such sharing. In other words, electrons in a bond are attracted to both nuclei, and thus pull the nuclei together. *Nonbonded* atoms do not share their electrons, by definition.

Let's look at a simple example. The bond length of H_2 (H—H, or dihydrogen) is 0.746 Å. If we define the covalent radius as half of the separation of the two hydrogen nuclei, then the covalent radius of hydrogen is (0.746 Å)/2 = 0.378 Å. As the text states, the *van der Waals radius* of hydrogen is 1.2 Å. Clearly, a covalently bound hydrogen is separated much less from its bonded partner than it is from a nonbonded atom.

If nonbonded atoms are pushed together so that their electron clouds overlap, they resist. This resistance is called *van der Waals repulsion*. Pushing atoms together can result from applying pressure. We know if we apply pressure to a balloon, the instant we release the pressure, the balloon springs back to its original shape. This is because the molecules of gas inside the balloon resist being pushed together. In some structures, atoms that are not bonded to each other are forced to violate each other's van der Waals radii; *gauche*-butane is such a structure. The resulting van der Waals repulsions cause the molecule to have a greater internal energy as compared with *anti*-butane.

REACTION REVIEW

I. REACTION OF HYDROCARBONS

A. COMBUSTION

1. Combustion is the most important reaction of alkanes.
2. Complete combustion affords carbon dioxide and water as the only combustion products.

$$C_nH_{2n+2} \;+\; \frac{3n+1}{2}O_2 \;\longrightarrow\; n\,CO_2 \;+\; (n+1)\,H_2O$$

3. Combustion can be used in elemental analysis, which is the quantitative determination of elemental compositions.
   a. The mass percents of the elements obtained from a combustion analysis of a compound can be used, together with the molecular mass, to determine the molecular formula. (See Problems 2.44 and 2.45, text p. 85.)

SOLUTIONS TO PROBLEMS

Solutions to In-Text Problems

2.1   (a) The number of hydrogens in an alkane (branched or unbranched) is $2n + 2$, where n is the number of carbons. Consequently, there are $2(18) + 2 = 38$ hydrogens in the alkane with eighteen carbons.

     (b) No; all alkanes—indeed, all hydrocarbons—must have an even number of hydrogens. Thus, if $2n + 2 = 23$, then n would have to be 10.5—an impossible number of carbons.

2.2   Structural formulas of tridecane:

$$CH_3CH_2CH_2CH_2CH_2CH_2CH_2CH_2CH_2CH_2CH_2CH_2CH_3 \qquad CH_3(CH_2)_{11}CH_3$$

              structural formula of tridecane   condensed structural formula of tridecane

The difference in the boiling points of decane, undecane, and dodecane in Table 2.1, text p. 47, is about 20° per carbon. Consequently, the boiling point of tridecane is estimated by adding 20° to the boiling point of dodecane to obtain 236°. (The actual value is 235°.)

2.3   (a) The staggered conformations of isopentane are *A, C,* and *E;* the eclipsed conformations are *B, D,* and *F*.

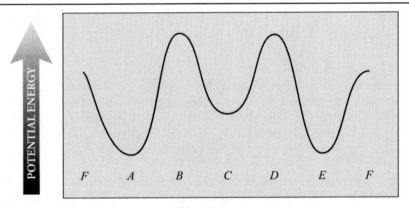

     (b) The curve of potential energy versus angle of rotation is shown in Fig. SG2.1. The staggered conformations *B* and *D* have the highest energy because they have eclipsed methyl groups.

Figure SG2.1 A diagram of potential energy versus angle of internal rotation in isopentane to accompany the solution to Problem 2.3b. The conformations are shown the solution to part (a). Each conformation differs from the adjacent ones by a rotational angle of ±60°.

(c)   Conformations *A* and *E* have the lowest energy because they have one less interaction in which methyl groups are "close together"—that is, the C—CH$_3$ bonds have a 60° dihedral angle—than conformation *C*; hence, either of these conformations is present in greater concentration than conformation *C*.

2.4   (a)   The projected bond is the carbon–carbon bond at the end of the butane molecule:

projected bond

H$_3$C —— CH$_2$CH$_2$CH$_3$

There are three identical staggered conformations and three identical eclipsed conformations.

A   *B*   *C*   *D*   *E*   *F*

(b)   The curve of potential energy versus angle of internal rotation is essentially identical to the curve for ethane (Fig. 2.3, text p. 51), except that the energy difference between staggered and eclipsed forms is slightly greater. That is, the energies of all staggered forms lie at equal minima, and the energies of all eclipsed forms lie at equal maxima.

(c)   All staggered conformations are present in equal amount because they are identical.

There is a subtlety here. Problem 2.4 has considered the conformation about the C1–C2 bond in isolation. Any staggered conformation about the C1–C2 bond contains a mixture of gauche and anti conformations about the C2–C3 bond.

2.5   (a)

(b)   Structures *B* and *C* are drawn correctly: two bonds in the plane, one solid wedge coming towards you and one dashed wedge going away. In structure *A*, the H–C–H bond angle appears to be 180°, or linear, which is inconsistent with the 109.5° bond angles in an *sp*^{3}-hybridized atom. In structure *D*, the wedge and dashed wedge are on opposite sites of the H–C–Br bonds, which is not possible for a tetrahedral molecule. You should build a model and prove to yourself that arrangements *A* and *D* cannot exist.

2.6   (a)   Observing the sawhorse structure from the perspective shown, the viewer would observe the compound as the Newman projection shown to the right. The propyl group is abbreviated as C$_3$H$_7$.

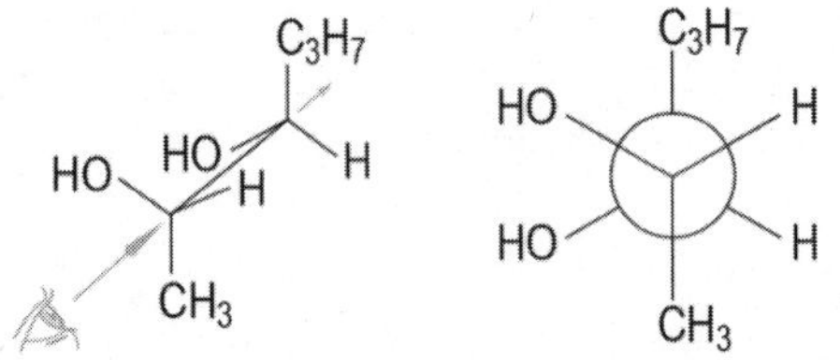

(b) Rotating C2, the front carbon, by 120° in the clockwise direction gives the structures shown below. View a model in each of the orientations shown so that you can relate the three-dimensional structures to your drawings.

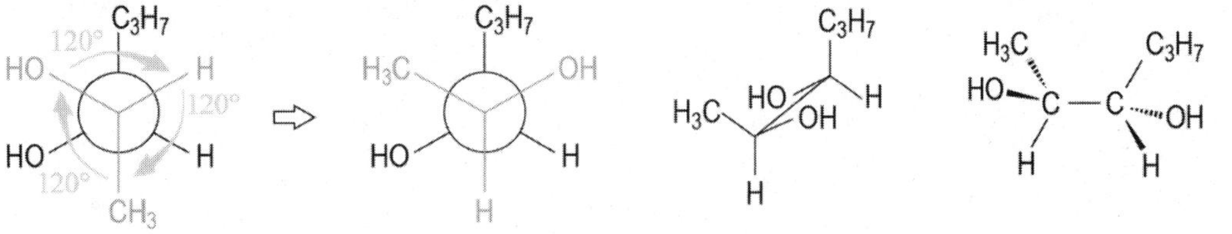

| Newman projection from part (a) | C2 rotated clockwise 120° | sawhorse structure | line-and-wedge structure |

(c) Rotating C2, the front carbon, by 180° gives the structures shown below. These conformations are eclipsed.

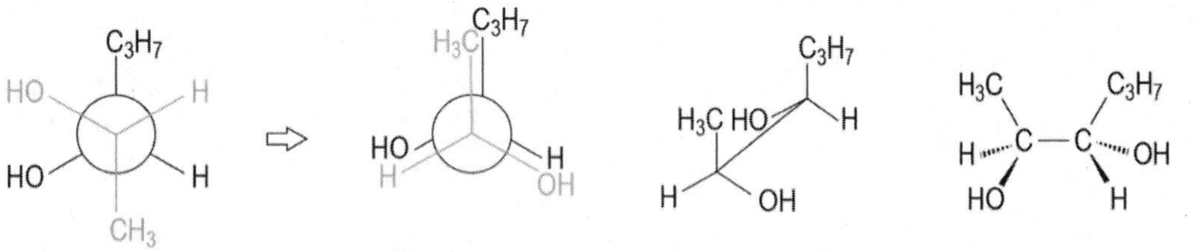

| Newman projection from part (a) | C2 rotated clockwise 180° | sawhorse structure | line-and-wedge structure |

2.7 **(a)** 2,3,5-Trimethylhexane. Numbering of the chain is determined by pairwise comparison of the possible numbering schemes: (2,3,5) versus (2,4,5). The first scheme has the lower number at first point of difference (rule 8, text p. 64).

(b) 2,4-Dimethylhexane

(c) 4-Ethyl-5-propyloctane. In this case, alphabetical citation governs the numbering of the principal chain, because numbering is ambiguous otherwise (Rule 10, text p. 64).

(d) 2,5,5-Trimethylheptane. Rule 8 rules out 3,3,6-trimethylheptane.

2.8 **(a)** This type of problem requires a systematic approach. First, draw the structure of hexane itself.

$$CH_3CH_2CH_2CH_2CH_2CH_3 \quad \textbf{hexane}$$

Next, draw the structures with a principal chain of five carbons and one methyl branch:

$$CH_3CHCH_2CH_2CH_3 \qquad CH_3CH_2CHCH_2CH_3$$
$$| \qquad\qquad\qquad\qquad |$$
$$CH_3 \qquad\qquad\qquad\qquad CH_3$$

2-methylpentane **3-methylpentane**

Finally, we draw the structures with two methyl branches.

$$
\begin{array}{cc}
\overset{\displaystyle CH_3}{\underset{\displaystyle |}{}} & \overset{\displaystyle CH_3}{\underset{\displaystyle |}{}} \\
CH_3CHCHCH_3 & CH_3CCH_2CH_3 \\
\underset{\displaystyle |}{} & \underset{\displaystyle |}{} \\
CH_3 & CH_3
\end{array}
$$

2,3-dimethylbutane   **2,2-dimethylbutane**

(b)   Following the systematic approach used in part (a), we first draw the structure of heptane itself.

$$CH_3CH_2CH_2CH_2CH_2CH_2CH_3 \quad \textbf{heptane}$$

Next, draw the structures with a principal chain of six carbons and one methyl branch:

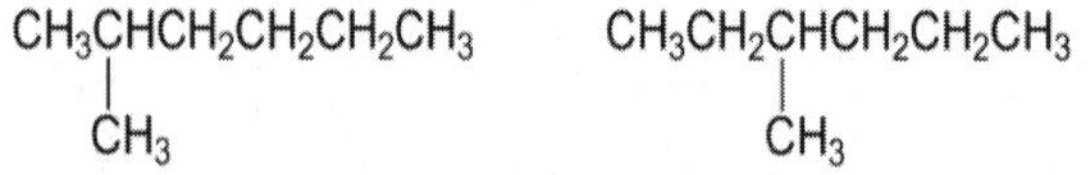

$$
\begin{array}{cc}
CH_3CHCH_2CH_2CH_2CH_3 & CH_3CH_2CHCH_2CH_2CH_3 \\
\underset{\displaystyle |}{} & \underset{\displaystyle |}{} \\
CH_3 & CH_3
\end{array}
$$

2-methylhexane   **3-methylhexane**

Notice that moving the methyl branch one or more carbons to the right gives identical structures. Next, draw the structures with five carbons in their principal chains and two methyl branches:

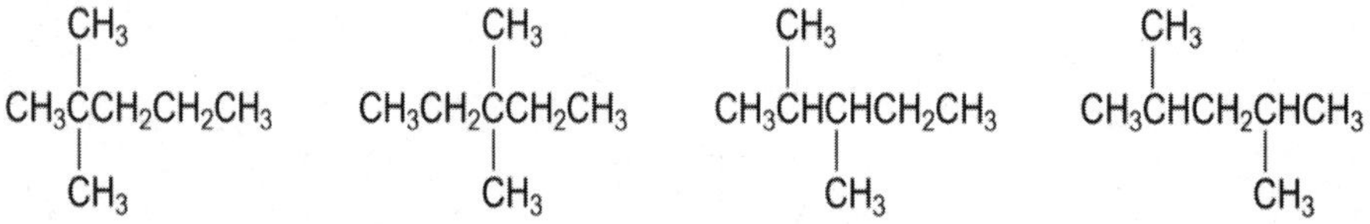

$$
\begin{array}{cccc}
\overset{\displaystyle CH_3}{\underset{\displaystyle |}{}} & \overset{\displaystyle CH_3}{\underset{\displaystyle |}{}} & \overset{\displaystyle CH_3}{\underset{\displaystyle |}{}} & \overset{\displaystyle CH_3}{\underset{\displaystyle |}{}} \\
CH_3CCH_2CH_2CH_3 & CH_3CH_2CCH_2CH_3 & CH_3CHCHCH_2CH_3 & CH_3CHCH_2CHCH_3 \\
\underset{\displaystyle |}{} & \underset{\displaystyle |}{} & \underset{\displaystyle |}{} & \underset{\displaystyle |}{} \\
CH_3 & CH_3 & CH_3 & CH_3
\end{array}
$$

2,2-dimethylpentane **3,3-dimethylpentane** **2,3-dimethylpentane** **2,4-dimethylpentane**

Then there is the structure with one ethyl branch, and the structure with a four-carbon principal chain:

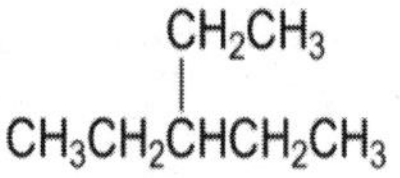

$$
\begin{array}{cc}
\overset{\displaystyle CH_2CH_3}{\underset{\displaystyle |}{}} & \overset{\displaystyle CH_3 \ CH_3}{\underset{\displaystyle | \ \ \ |}{}} \\
CH_3CH_2CHCH_2CH_3 & CH_3C\!-\!CHCH_3 \\
& \underset{\displaystyle |}{} \\
& CH_3
\end{array}
$$

3-ethylpentane

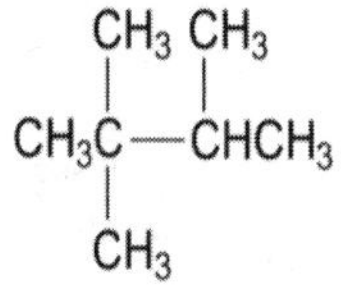

2,2,3-trimethylbutane

2.9   (a)   The designation and numbering of the principal chain are shown below.

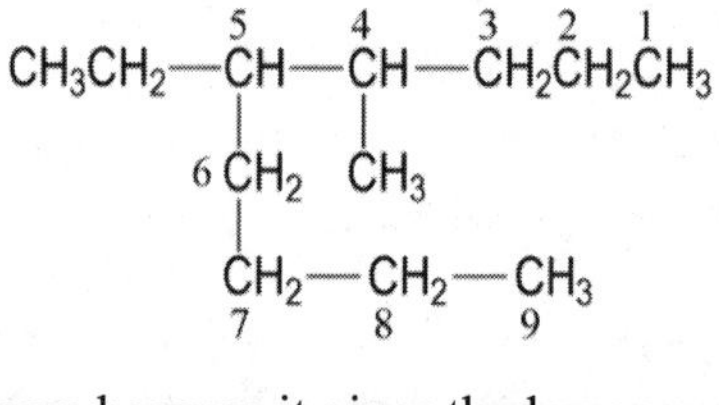

$$
\begin{array}{l}
CH_3CH_2\!-\!\overset{5}{C}H\!-\!\overset{4}{C}H\!-\!\overset{3}{C}H_2\overset{2}{C}H_2\overset{1}{C}H_3 \\
\qquad\quad \ \overset{6}{|} \quad\ \ | \\
\qquad\quad \ CH_2 \ \ CH_3 \\
\qquad\quad \ \ | \\
\qquad\quad \ \ CH_2\!-\!CH_2\!-\!CH_3 \\
\qquad\quad \ \ \ \ 7 \qquad 8 \qquad 9
\end{array}
$$

Numbering is in the direction shown because it gives the lower number at first point of difference. The compound is a substituted nonane; it has branches at carbon-4 and carbon-5. The name is 5-ethyl-4-methylnonane. (Notice the alphabetical citation of substituents.)

(b)   Numbering from either direction gives the branch numbers (4,5,5,6); for that reason, alphabetical citation (rule 10) provides the basis for numbering. The name is 4-ethyl-5,5,6-trimethylnonane.

2.10   The compound is 3-ethyl-4-propylheptane, and the expanded structure is as follows:

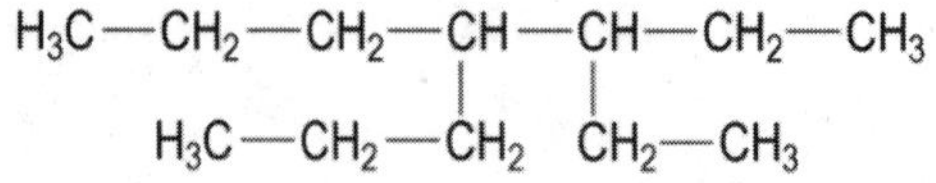

$$
\begin{array}{l}
H_3C\!-\!CH_2\!-\!CH_2\!-\!CH\!-\!CH\!-\!CH_2\!-\!CH_3 \\
\qquad\qquad\qquad\quad\ | \quad\ \ | \\
\qquad\ H_3C\!-\!CH_2\!-\!CH_2 \ \ CH_2\!-\!CH_3
\end{array}
$$

Your structure might look different from this one. Any structure with the same connectivity as this one is correct.

2.11   The structure of 4-isopropyl-2,4,5-trimethylheptane:

$$CH_3 \quad CH_3 \quad CH_3$$
$$CH_3CHCH_2C—CHCH_2CH_3$$
$$CH(CH_3)_2$$

4-isopropyl-2,4,5-trimethylheptane

2.12   (a)   In the following structure, p = primary, s = secondary, t = tertiary, and q = quaternary.

$$\overset{p}{CH_3} \quad \overset{p}{CH_3} \quad \overset{p}{CH_3}$$
$$\overset{p}{CH_3}\underset{t}{C}\underset{s}{H}CH_2\overset{q}{C}—\underset{t}{C}H\underset{s}{C}H_2\overset{p}{C}H_3$$
$$\underset{t \quad p}{CH(CH_3)_2}$$

4-isopropyl-2,4,5-trimethylheptane

(b)   The primary hydrogens are those in the structure in part (a) at the carbons labeled "p"; the secondary hydrogens are those at the carbons labeled "s"; and the tertiary hydrogens are those at the carbons labeled "t."

(c)   The methyl groups are the CH_3 groups in the structure of part (a) with carbons that are labeled with a "p". The other groups are outlined in the structures below.

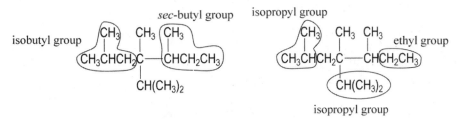

2.13   Ethyl groups are in the shaded circles; methyl groups are in the boxes. Each ethyl group by definition contains a methyl group.

4-sec-butyl-5-ethyl-3-methyloctane

2.14   (a)   (b)

2,2,3,4-tetramethylheptane   **ethylcyclopentane**

2.15

(a)

(b)

2.16 (a) The name is 1-ethyl-2,4-dimethylcyclopentane. Rule 10 (alphabetical citation) and rule 8 dictate the numbers of the substituents.

(b) The carbon with two branches receives the number 1 by rule 8, text p. 64. The name is 1,1,2,4-tetraethyl-3-methylcyclobutane. Tetraethyl begins with the letter *e* for citation purposes; that is, the prefix "tetra" is ignored in the alphabetical citation of prefixes.

2.17 (a) Every ring decreases the number of hydrogens by 2 from the fully saturated number of $2n + 2$ (n = the number of carbons). Therefore, the number of hydrogens is $2n + 2 - 4 = 2n - 2$.

(b) By the reasoning in part (a), the formula is $2n - 4$.

(c) By the reasoning in part (a), the formula is $2n + 2 - 2m = 2(n + 1 - m)$.

2.18 (a) By the reasoning used in the solution to problem 2.14(a), the number of hydrogens differs from the fully saturated number (18) by 8 (four pair). Therefore, the compound has four rings.

(b) The number of hydrogens differs from the fully saturated number (16) by 4, or two pair. Therefore, the compound has two rings.

2.19 (a) Because it is very symmetrical, the more branched and compact isomer 2,2,3,3-tetramethylbutane should have the higher melting point; its melting point is 100.7°.

 Note that this compound is a liquid over only about a 6° range; as a solid, it has a very high vapor pressure, especially at temperatures approaching its melting point. For this reason, this compound readily *sublimes* (is converted directly from a solid into a gas and vice-versa). If we gently heat the solid, it will evaporate and condense as a solid on a cooled condenser. This process, called *sublimation*, is used to purify solids that have a relatively high vapor pressure.

Branched alkanes pack together in the solid state less effectively than linear, straight chain alkanes, thus, the melting point of branched alkanes is lower (less energy is required to change the material from the solid to liquid state). So octane melts at –56°, and 2-methylheptane at –109°.

(b) Benzene is a very symmetrical compound. The carbon framework is essentially a planar hexagon. Because of its resonance structures, all carbon–carbon bonds have equal length. (See Problem 1.15, text p. 71.) The more symmetrical benzene has an usually high melting point (for a hydrocarbon), +5.5°. The less symmetrical toluene has a more typical melting point, –95°.

2.20 Water is not very effective in distinguishing a gasoline fire because the gasoline is not "smothered" by the water. Because it is less dense than water, gasoline always rises to the surface of water, where the air supports further combustion (see Sec. 8.6D, text p. 359, for further discussion).

2.21 (a) Balancing a general combustion equation starts with the fact that each carbon must generate a mole of CO_2. Water is then used to balance hydrogens and dioxygen to balance oxygens.

$$C_nH_{2n+2} + \left(\frac{3n + 1}{2}\right)O_2 \longrightarrow n\,CO_2 + (n+1)\,H_2O$$

(b) Follow the same procedure as in part (a), noting that there is one less mole of water formed per mole of CO_2 and one less atom of oxygen required.

$$C_nH_{2n} + \frac{3n}{2}O_2 \longrightarrow n\,CO_2 + n\,H_2O$$

2.22 Use the balanced equation in the solution to Problem 2.21a with $n = 8$. First, calculate the moles of octane in 15 gallons of octane. (The molecular mass of octane is 114.)

$$\text{moles } C_8H_{18} = (15\text{ gal})(3.785\text{ L gal}^{-1})(692\text{ g L}^{-1})([1/114]\text{ mol g}^{-1}) = 345$$

As shown in the solution to Problem 2.21a, eight moles of CO_2 are formed per mole of octane burned. Therefore, $(345)(8) = 2,760$ moles of CO_2 are formed.

Next, convert 2,760 moles of CO_2 into pounds of CO_2 to obtain the answer:

$$\text{lb } CO_2 = (2,760\text{ mol})(44\text{ g mol}^{-1})(0.001\text{ kg g}^{-1})(2.20\text{ lb kg}^{-1}) = 267$$

2.23 (a) A carboxylic acid—indeed, the *only* carboxylic acid—with this formula is acetic acid.

$$H_3C-\overset{\overset{\displaystyle O}{\|}}{C}-OH$$

acetic acid

(b) Any structure with five carbons, an —OH group, and *either* a ring or a double bond is correct. Two of many examples are the following:

⬠—OH

cyclopentanol

$$H_2C=CHCH_2CH_2CH_2-OH$$

4-penten-1-ol

2.24 With 12 hydrogens and 5 carbons, this compound cannot have any double bonds or rings. Because amides, carboxylic acids, phenols, and esters each contain one or more double bonds, these groups are ruled out. An amide is ruled out also because it contains nitrogen, which is not present in the compound. The remaining possibilities are a compound with two ether groups, a compound with two alcohol groups, or a compound with both an ether and an alcohol. Examples of each are the following:

$$HO-CH_2CH_2CH_2CH_2CH_2-OH$$

a compound with
two alcohol groups

$$HO-CH_2CH_2\overset{\overset{\displaystyle OCH_3}{|}}{C}HCH_3$$

a compound with
an alcohol and an ether group

$$CH_3O-CH_2\overset{\overset{\displaystyle OCH_3}{|}}{C}HCH_3$$

a compound with
two ether groups

Solutions to Additional Problems

2.25 (a) Use the rule of thumb that an additional carbon adds about 20–30 ° to the boiling point. Because the second compound has one more carbon than the first, its boiling point should be about 20–30 ° greater. The estimated boiling point is 175–185 °.

(b) The second compound has two more carbons than the first, and should have a boiling point that is 40–50 ° greater—approximately 168–178 °.

(c) Because the two compounds have the same functional groups and the same number of carbons, their boiling points should be very similar. An estimate is 152 °C.

2.26 (a) The isomers of octane with five carbons in their principal chains:

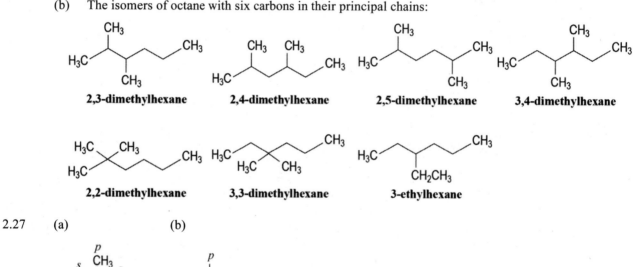

(b)   The isomers of octane with six carbons in their principal chains:

2,3-dimethylhexane   **2,4-dimethylhexane**   **2,5-dimethylhexane**   **3,4-dimethylhexane**

2,2-dimethylhexane   **3,3-dimethylhexane**   **3-ethylhexane**

2.27   (a)   (b)

2.28   (a)   Neopentane has five carbons, and all hydrogens are primary:

$$H_3C—\overset{\overset{\displaystyle CH_3}{|}}{\underset{\underset{\displaystyle CH_3}{|}}{C}}—CH_3$$

2,2-dimethylpropane

(b)   Cyclopentane has five carbons and only secondary hydrogens:

cyclopentane

(c)   The structures that fit this description have several rings fused together. Among these are the following, along with their trivial names.

prismane   cubane   tetrahedrane

(d)   The simplest composition has 6 carbons (72.06 atomic mass units) and 12 hydrogens (12.10 atomic mass units) for a molecular mass of 84.16. Cyclohexane or any other six-carbon hydrocarbon containing either one ring or one double bond would fit.

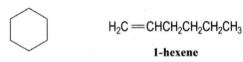

H_2C ═ $CHCH_2CH_2CH_2CH_3$

1-hexene

cyclohexane

2.29   (a)   5-Ethyl-4,4-dimethyloctane. Note that the numbering is governed by the "first point of difference" rule (rule 8, text p. 64).
         (b)   4-Ethyl-5-methyloctane
         (c)   2,3-Dimethylnonane

    In noncyclic skeletal structures, don't forget to count the carbons on the ends of the chains.

(d)   4-Isopropyloctane
(e)   In this example, alternative numbering schemes are compared: (1,2,2,3), (1,1,2,5), and (1,4,5,5). The first point of difference among these schemes occurs at the second number, and the *second* of these schemes has the lowest number at this point. The name, then, is 2,5-diethyl-1,1-dimethylcyclopentane.

2.30   (a)   (b)   (c)

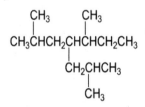

$CH_3CHCH_2CHCHCH_2CH_3$
         | | |
        CH₃   CH₃   CH₂CHCH₃
                              |
                             CH₃

4-isobutyl-2,5-dimethylheptane   **2,3,5-trimethyl-4-propylheptane**   **5-*sec*-butyl-6-*tert*-butyl-2,2-dimethylnonane**

2.31   (a)   This compound should be named 2,4,6,6-tetramethyloctane:

CH_3 CH_3
           | |
$CH_3CCH_2CHCH_2CHCH_3$
           | |
        CH_2CH_3 CH_3

2,4,6,6-tetramethyloctane

(b)   5-Neopentyldecane is a correct name.

$CH_3CH_2CH_2CH_2CHCH_2CH_2CH_2CH_2CH_3$
                            |
                    $CH_2C(CH_3)_3$

5-neopentyldecane

(c)   Ima's name ignored the "first-point-of-difference" numbering rule:

4-cyclopropyl-1,2-dimethylcyclohexane

(d)   The correct name is 4-*tert*-butyloctane.

4-*tert*-butyloctane

2.32   (a)   There are several possible candidates for the principal chain that have 7 carbons. Rule 3 on text p. 61 covers this situation: choose the chain with the greatest number of branches. The correct choice has 6 branches. (The other possibilities have 3 and 4 branches, respectively.) This is not named as a substituted cyclohexane because the principal chain has more than 6 carbons.

3,5-dicyclohexyl-5-ethyl-2,2-dimethyl-4-propylheptane

(b)   The systematic name for "pullane" is 1,1-di-*tert*-butyl-2,4-diethylcyclopentane.

2.33   (a)   The connectivity of A: the groups connected to the carbon of the projected bond nearest the observer are C_2H_5 and two CH_3 groups; the groups connected to the back carbon of the projected bond are H and two CH_3 groups. The connectivity of B and C are the same, as the following analysis shows.

B

C

Hence, structures B and C are Newman projections of the same compound, 3-ethyl-2-methylpentane. Compound A is 2,3,3-trimethylpentane.

A common mistake in interpreting Newman projections is to forget the carbon that is in the back, or "hidden." Be sure to remember that the Newman projection represents a bond to *two* carbons: the nearer one for which bonds are drawn to the center of the circle; and the farther one, for which bonds are drawn to the periphery of the circle. Naming the compounds is a lot easier if you draw out the structures rather than working directly from the Newman projections.

(b)   Structures *A* and *C* both have the connectivity [(CH₃)₂CH, CH₂, CH(CH₃), CH₂, CH₂, CH₃]. The connectivity of *B* is different. Consequently, structures *A* and *C* represent the same compound, 2,4-dimethylheptane. Compound *B* is 3-ethyl-2-methylhexane.

2.34   (a)   A skeletal structure of compound *B* in Problem 2.33(a), 2,3,3-trimethylpentane. (This structure can be drawn in many different ways.)

skeletal structure of 2,3,3-trimethylpentane

(b)   The ethyl group on carbon-4 has to be gauche to either the ethyl group or the isopropyl group on carbon-3. Because the isopropyl group is larger, the ethyl group is gauche to the other ethyl group in the most stable conformation.

The ambiguity in drawing the structure is that it has a mirror image, which is not the same, although it has the same energy. (The issue of mirror-image structures is considered in Chapter 6.) If you drew the mirror-image structure, it is equally correct. As noted in the solution to Problem 2.33(b), the name of this compound is 3-ethyl-2-methylhexane.

2.35   All staggered conformations (*A*) of chloroethane are identical and thus have identical energies; all eclipsed conformations (*B*) are identical, and also have identical energies. The energy barrier is the energy difference between conformations *B* and *A*. The diagram is shown in Fig. SG2.2.

conformations of chloroethane

2.36   Both curves would have the same general appearance, that is, three barriers of equal height. (See Fig. SG2.3.) However, the curve for 2,2,3,3-tetramethylbutane should have greater potential-energy barriers because each eclipsed conformation has three severe methyl–methyl eclipsing interactions, and because eclipsed methyl groups have greater repulsion than gauche methyl groups.

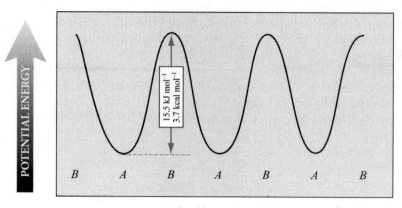

Figure SG2.2 A diagram of potential energy versus angle of internal rotation in chloroethane to accompany the solution to Problem 2.35. The conformations are shown in the solution. Each conformation differs from the adjacent ones by a rotational angle of ±60°.

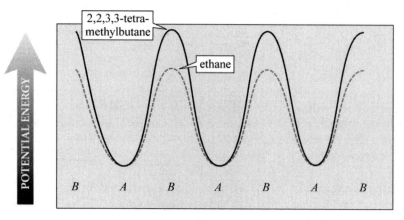

Figure SG2.3 Diagrams of potential energy versus angle of internal rotation in 2,2,3,3-tetramethylbutane and ethane to accompany the solution to Problem 2.36. The conformations are shown in the solution. The staggered conformations of ethane and 2,2,3,3-tetramethylbutane are placed at the same energy for comparison purposes. Each conformation differs from the adjacent ones by a rotational angle of ±60°.

2.37   (a)   The conformations of 1,2-dichloroethane are shown in Fig. SG2.4:

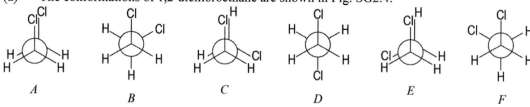

(b)   Conformation *D* is present in greatest amount, because it has the lowest energy.

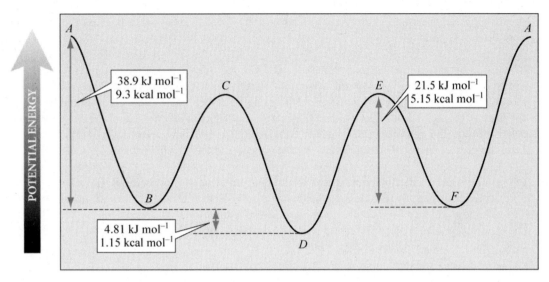

Figure SG2.4 A diagram of potential energy versus angle of internal rotation in 1,2-dichloroethane to accompany the solution to Problem 2.37. The conformations are shown in the solution. Each conformation differs from the adjacent ones by a rotational angle of ±60°. Notice that this curve is very similar to that of butane (Fig. 2.5, text p. 53).

2.38 (a) The principal chain of 2,2-dimethylpentane is a five carbon chain, so Newman projections are shown in the staggered conformations viewing down the C1—C2 bond, the C2—C3 bond, the C3—C4 bond, and the C4—C5 bond.

(b) The most stable conformation of 2,2-dimethylpentane will include all staggered conformations. The larger alkyl groups must be as far away from each other as possible, particularly with regards to the structure viewing down the C3—C4 bond. The *tert*-butyl group should be anti to the methyl group.

(c) Line-and-wedge structures of the most stable conformation of 2,2-dimethylpentane:

2.39 Because severe van der Waals repulsions occur between the *tert*-butyl groups in compound *A* at normal bond lengths and angles, the only way for the molecule to relieve these repulsions is to stretch the appropriate bonds and flatten itself by widening the C—C—C angles. The observed structure is a compromise between the Scylla of van der Waals repulsions and the Charybdis of nonoptimal bond lengths and angles.

2.40 A major contributor to the barrier to internal rotation is the van der Waals repulsions between the methyl groups in the eclipsed conformations. In compound *B* these repulsions are smaller than they are in compound *A* because

    If you find an error in this manual, please visit http://people.pharmacy.purdue.edu/~loudonm/teaching/

the distance between methyl groups is greater in *B*. This distance is greater because the Si—Si bond is longer than the C—C bond.

2.41   The C—O bond is somewhat shorter than the C—C bond (text p. 15). Consequently, when the methyl groups are gauche, the distance between them is smaller in the ether (compound *A*) than in butane (compound *B*). Consequently, van der Waals repulsions in the gauche conformation of the ether should be larger than those of the gauche conformation in butane, and the energy of the gauche conformation of the ether should be higher relative to the anti conformation, in which van der Waals repulsions are absent. In other words, the ether should have the greater energy difference between the gauche and anti conformations.

2.42   (a)   The expected dipole moment for the anti conformation is zero, because the polar C—Br bonds are oriented in opposite directions; hence their bond dipoles cancel. Likewise, corresponding to every C—H bond is another C—H bond pointing in the opposite direction.

(b)   It follows from part (a) that the observed dipole moment is due solely to the relatively small amount of gauche conformation present. Hence,

$$\mu = \mu_{anti}N_{anti} + \mu_{gauche}N_{gauche}$$

or, with $\mu_{anti} = 0$,

$$1.0 = (0)(0.82) + \mu_{gauche}(0.18) = \mu_{gauche}(0.18)$$

Solving for μ_{gauche} gives $\mu_{gauche} = 5.56$ D. This is a very large dipole moment. But this is what would be expected for a compound in which two C—Br bond dipoles (as well as the smaller C—H bond dipoles) are oriented in nearly the same direction.

2.43   The Oxhide family burns (12,000 miles/25 miles gal^{-1}) = 480 gallons of gasoline per year. The result from Problem 2.21 is based on 15 gallons; scale the result from that solution to 480 gallons by multiplying it by the factor (480/15) = 32. This gives a carbon footprint of 8,544 pounds per year of CO_2 for the Oxhide family car.

For about 250,000,000 vehicles in the United States, with an average mileage of 18 miles gal^{-1}, the carbon footprint is about 3×10^{12} (three *trillion*) pounds of CO_2!

2.44   (a)   The mass of carbon in the sample equals the mass of carbon in the CO_2 product.

mass of carbon in CO_2 = (mass of the CO_2 sample)(fraction of the CO_2 mass that is carbon)
= (24.60 mg)(12.01/44.01) = 6.71 mg

The mass of hydrogen in the sample equals the mass of hydrogen in the H_2O product:

mass of hydrogen in H_2O = (mass of the H_2O sample)(fraction of the H_2O mass that is hydrogen)

= (11.51 mg)(2.016/18.016) = 1.29 mg

The total mass of C and H is 8.00 mg, as it must be if C and H are the only elements present in the sample.

(b)   The millimoles (mmol) of C in the sample is 6.71/12.01 = 0.559, and the mmol of H is 1.29/1.008 = 1.28. Therefore, there are (1.28/0.559) = 2.29 moles of hydrogen per mole of carbon in the sample, or $C_1H_{2.29}$.

(c)   Multiplying 2.29 by 7 gives 16.0 hydrogens. Therefore, the formula of *X* is C_7H_{16}.

2.45   (a)   Using the procedure in Solution 2.44(b), the moles of carbon (in a 100 g sample) = (87.17/12.01) = 7.26, and the moles of hydrogen = (12.83/1.008) = 12.73. The ratio of hydrogen to carbon is 1.75. Multiplying by 4 gives a whole number of hydrogens (7), suggesting initially a formula of C_4H_7. However, all hydrocarbons must contain an even number of hydrogens. Therefore, multiply by 2 to obtain C_8H_{14} as the smallest possible formula. (Any multiple of this is also a possibility, but we'll stick with the simplest possibility.)

The formula that gives the smallest whole-number ratio of a compound is called the **empirical formula.** In this example, the empirical formula is C_4H_7. The empirical formula of ethane is CH_3. Combustion analysis gives the empirical formula. Sometimes, the empirical formula is the same as the molecular formula. More often, though, a molecular mass is needed to determine the molecular formula.

(b)   Because each ring subtracts two hydrogens from the number present in a noncylic hydrocarbon ($2n + 2$), the alkane must contain two rings. [See the solution to Problem 2.17(c)]. Two of several possibilities that meet the criteria are the following:

(c)   Because the compound contains **two rings, they must have a single** quaternary carbon in common. The following are examples are the only **possibilities.**

Compounds that contain rings joined at a single common atom are called *spirocyclic compounds*.

2.46   (a)   First list all the possibilities.

$$CH_3CH_2CH_2CH_2CH_2{-}Cl \qquad CH_3CH_2CH_2\overset{\overset{\displaystyle Cl}{|}}{C}HCH_3 \qquad CH_3CH_2\overset{\overset{\displaystyle Cl}{|}}{C}HCH_2CH_3$$

$$\qquad\qquad A \qquad\qquad\qquad\qquad B \qquad\qquad\qquad\qquad C$$

There is only one possible structure for each of the compounds *A* and *C*. However, building models will reveal that compound *B* can exist as two *nonsuperimposable* structures that are mirror images of each other. Note that every structure has a mirror image; the issue is the noncongruence (nonsuperimposability) of the two structures.

$$CH_3CH_2CH_2\overset{\overset{\displaystyle Cl}{|}}{C}HCH_3 \qquad\qquad CH_3CH_2CH_2{\cdots}\overset{\overset{\displaystyle CH_3}{|}}{\underset{\underset{\displaystyle H}{|}}{C}}{-}Cl \qquad Cl{-}\overset{\overset{\displaystyle CH_3}{|}}{\underset{\underset{\displaystyle H}{|}}{C}}{\cdots}CH_2CH_2CH_3$$

$$\qquad B$$

2-chloropentane   nonsuperimposable mirror-image structures of 2-chloropentane

Compounds that are nonsuperimposable, or noncongruent, mirror images are called *enantiomers*. We'll study enantiomers in Chapter 6.

(b)   The possibilities are as follows:

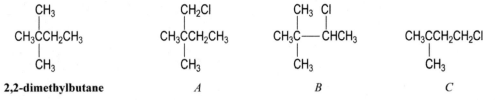

$$CH_3\overset{\overset{\displaystyle CH_3}{|}}{\underset{\underset{\displaystyle CH_3}{|}}{C}}CH_2CH_3 \qquad CH_3\overset{\overset{\displaystyle CH_2Cl}{|}}{\underset{\underset{\displaystyle CH_3}{|}}{C}}CH_2CH_3 \qquad CH_3\overset{\overset{\displaystyle CH_3}{|}}{\underset{\underset{\displaystyle CH_3}{|}}{C}}{-}\overset{\overset{\displaystyle Cl}{|}}{C}HCH_3 \qquad CH_3\overset{\overset{\displaystyle Cl}{|}}{\underset{\underset{\displaystyle CH_3}{|}}{C}}CH_2CH_2Cl$$

2,2-dimethylbutane   *A*   *B*   *C*

Compound *B*, like compound *B* in part (a), can exist as nonsuperimposable mirror-image structures.

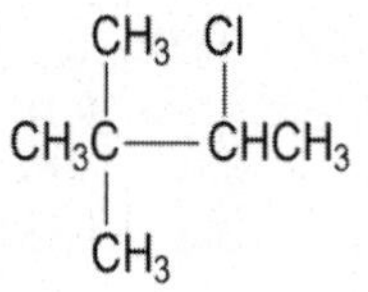

B

2,2-dimethyl-3-chlorobutane

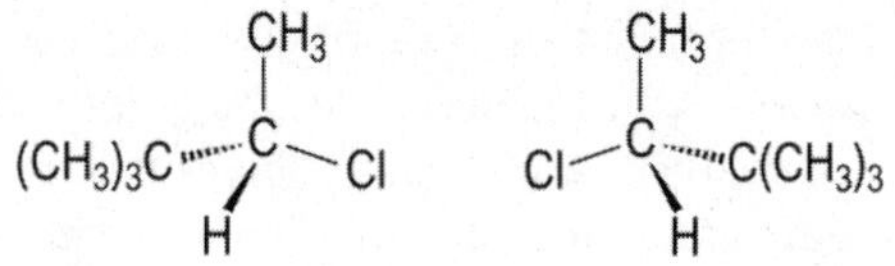

nonsuperimposable mirror-image structures of 2,2-dimethyl-3-chlorobutane

2.47   Consult the inside front cover of the text for a list of common functional groups.

(a) ketone   (b) nitrile   (c) alcohol   (d) ether

2.48   (a) amide   (b) phenol (an aryl alcohol)   (c) alcohol

2.49   The functional groups in acebutolol are identified and the compound classes are labeled in the following structure.

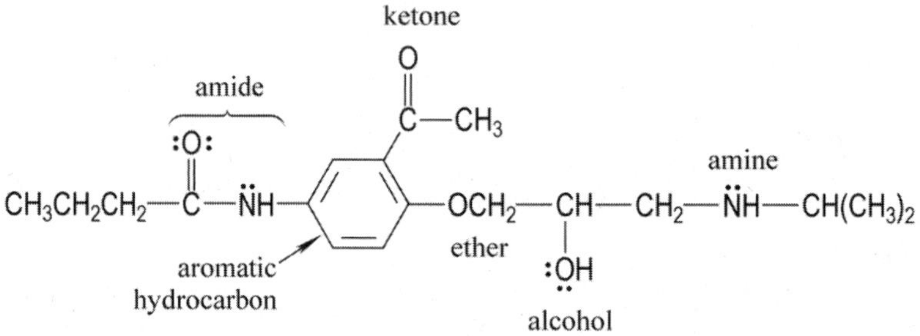

acebutolol

2.50   (a)   The isopropyl group can be connected to either the carbon or the nitrogen of the amide group.

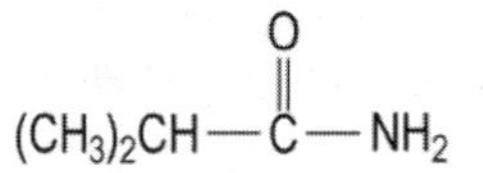

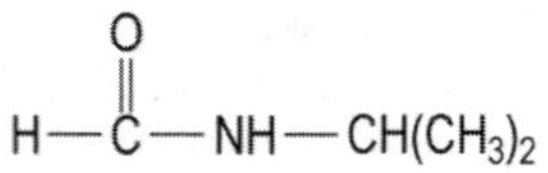

(b)   The following are three of several possibilities:

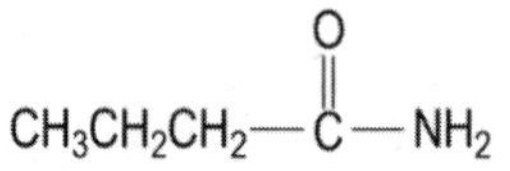

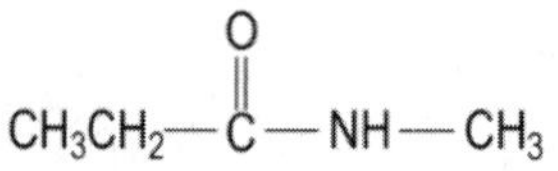

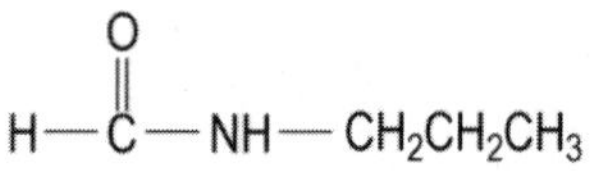

(c)   Such a compound must have a double bond to accommodate the formula. There are several possibilities.

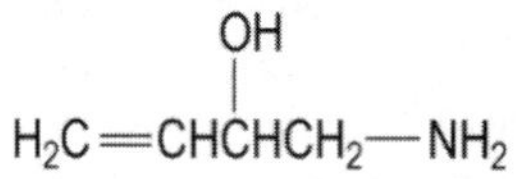

(d)   A compound with the formula C_4H_9NO could not be a nitrile. A nitrile contains a triple bond, which would reduce the number of hydrogens to 7.

Acids and Bases
The Curved-Arrow Notation

STUDY GUIDE LINKS

 ### 3.1   The Curved-Arrow Notation

In using the curved-arrow notation, some things matter and some things don't. In drawing curved arrows, the only things that matter are *the starting point and the destination of the arrows*. It doesn't matter whether an arrow curves up or down, or whether it goes to the top or bottom of the atom. Your curved arrows might appear different from the ones in the text or solutions manual; what matters is where they start and where they end. For example, in each of the following examples the curved arrows mean the same thing: breaking of a B—F bond with the electron pair moving onto the fluorine.

It also doesn't matter where the arrowhead points on an atom. For example, in the above example, the arrow can point directly to the F, or to an electron pair on the F. In each case, the meaning is that fluorine is taking on another electron pair.

The following is an example of the curved-arrow notation written in three different ways for an electron-pair displacement reaction. All of these are correct. They all refer to the same process: removal of a proton from the ammonium ion by hydroxide ion to give water and ammonia.

all describe the same process, which gives the following products:

$$HÖ—H \quad + \quad :N—H$$

with H above and below the N.

3.2   Rules for Use of the Curved-Arrow Notation

Now that you've seen several examples of the curved-arrow notation, let's summarize its use with six basic rules. If you follow these you will always use the curved-arrow notation correctly. The first two rules reiterate points that have been made repeatedly in the text. Examples of the last four rules are provided below.

Rules for use of the curved-arrow notation:

The first two rules are very important and general:

1. Curved arrows indicate the flow or "movement" of electron pairs, not atoms or nuclei.
2. A curved arrow is always drawn with its tail at the source of electrons (a bond or unshared pair) and its head at the destination.

The second two rules deal with the manipulation of unshared electron pairs:

3. When an atom donates an unshared electron pair, a curved arrow must indicate that the electron pair is used to form an additional bond to the *same atom.*
4. When an atom accepts an unshared electron pair, a curved arrow must show that the electron pair originates from a bond to the *same atom.*

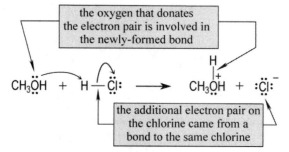

The last two rules deal with the manipulation of bonds:

5. When a bond is lost, a curved arrow must show its conversion either into an unshared electron pair on one of the two bonded atoms, or into a new bond involving one of the two original bonded atoms.
6. When a bond is formed, a curved arrow must show that it originates from either an electron pair on one of the two bonded atoms, or from a bond involving one of the two bonded atoms.

The following example illustrates these rules:

the bond lost from Br becomes an
electron pair on the same Br

$$HO: \quad H-O-CH_2-Br: \longrightarrow HO-H + O=CH_2 + :Br:$$

the bond between O and H becomes
a new bond between one of the atoms (O)

Don't try to memorize these rules. Instead, go back in the text and look at about three or four equations that show the curved-arrow notation and demonstrate to yourself that these rules are applied correctly. Then, the next few times you use the notation, use your intuition to answer the problem, but then check your result against these rules before you look at the answer. If you do this three or four times, you're not likely to misuse the notation.

3.3   Identification of Acids and Bases

This section stresses an important subtlety in dealing conceptually with acids and bases. An acid or a base—whether of the Lewis type or the Brønsted type—is defined by a *reaction,* and not by a *structure.* For example, some compounds react by donating a proton in certain reactions, in which case they are Brønsted acids; in other reactions, they react in other ways, in which case, they are *not* Brønsted acids.

You can sometimes tell from a structure whether a compound *might* react as an acid or a base. When someone looks at a structure and says, "That compound is an acid," what that person really means is "That compound is *capable of acting as an acid.*" It *is* an acid only in reactions in which it is acting like one. Thus, a compound with an unshared electron pair *might* react as a Lewis base; but if the same compound has an acidic hydrogen, it might also react as a Brønsted acid. The water molecule, with its amphoteric character, is an example of exactly this situation. Whether water is an acid or a base depends on the reaction. An electron-deficient compound certainly has the capability to react as a Lewis acid, and we might say, "That compound is a Lewis acid," but it is not a Lewis acid until it reacts like one. It might undergo other types of reactions in which it does not act as a Lewis acid. You have to see what happens to a compound in a reaction before you can classify the compound as an acid or base; and then the acid or base character is defined for that particular reaction.

Consider the following analogy. Suppose you have a friend who has just purchased a sleek, red convertible. He or she asks you, "Is that a fast car, or what?" Logically, you can't answer the question until you see the car perform, even though you suspect that it may have the capability of running at great speed. In the Indianapolis 500, your friend's car may *not* be fast at all. On the back roads of rural Indiana, it may be the fastest car anyone has ever seen. When someone shows you a structure and asks, "Is this a strong Brønsted acid or what?" your logical answer should be, "Show me the reaction and I'll tell you."

FURTHER EXPLORATIONS

  3.1   Inductive Effects

The name *inductive effect* originates from the idea that the polar effects of substituents are transmitted through bonds. To see how this works, consider the polar effect of a fluorine on the pK_a of fluoroacetic acid, the compound shown in the diagram in Eq. 3.43, text p. 116. This diagram shows the polar effect of the fluorine operating through *space*. In a through-bond model for polar effects, the positive end of the carbon–fluorine bond dipole pulls bonding electrons toward itself, thus inducing a positive charge on the neighboring carbon. In other words, one bond dipole induces dipoles in neighboring bonds. This induction continues throughout the molecule, resulting in an attractive interaction between the negative charge on the oxygens and the induced positive charge on the neighboring carbon:

Like the through-space model, this effect is stronger when there are more polar substituents, and it decreases in magnitude with increasing distance.

Opinions differ as to the importance of the "through-bond" and "through-space" models. However, the two models in most cases make the same qualitative predictions.

SOLUTIONS TO PROBLEMS

Solutions to In-Text Problems

3.1   (a)   The product results from donation of an unshared electron pair from the oxygen of H_2O to the electron-deficient carbon of the cation.

3.2   (a)   (b)   (c)

3.3

3.4

(b)   The nitrogen of ammonia donates an electron pair to electron-deficient boron of BF_3. (The electron pairs on the fluorine are not shown because they have no direct part in the reaction.)

3.5   (a, b, c)

electrophilic center   nucleophilic center

leaving group

$H_3C-\overset{+}{C}(CH_3)(CH_3)$ + $H\ddot{O}CH_3$ $\rightleftharpoons$ $H_3C-\overset{CH_3}{\underset{CH_3}{C}}-\overset{+}{\ddot{O}}CH_3 \atop H$

3.6   (a)

$\left[H_2\overset{+}{C}-CH{=}CH_2 \longleftrightarrow H_2C{=}CH-\overset{+}{C}H_2 \right]$

allyl cation

(b)

$\left[H_2\overset{\bar{\cdot\cdot}}{C}-CH{=}CH_2 \longleftrightarrow H_2C{=}CH-\overset{\bar{\cdot\cdot}}{C}H_2 \right]$

allyl anion

(c)

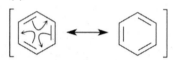

benzene

3.7   (a)

leaving group   electrophilic center
nucleophilic center

$H_3C-\overset{CH_3}{\underset{\cdot\cdot}{\overset{+}{O}}}-CH_3$ + $\bar{\ }:\!\ddot{S}-CH_3$ $\longrightarrow$ $H_3C-\ddot{\underset{\cdot\cdot}{O}}:$ + $CH_3-\ddot{S}-CH_3$

electrophile   **nucleophile**

$H_3C-\overset{CH_3}{\underset{\cdot\cdot}{\overset{+}{O}}}-H$ + $\bar{\ }:\!\ddot{S}-CH_3$ $\longrightarrow$ $H_3C-\ddot{\underset{\cdot\cdot}{O}}:$ + $H-\ddot{S}-CH_3$

Brønsted acid   **Brønsted base**

(b)

electrophilic center
nucleophilic center

$H_3C-CH_2-\ddot{Br}:$ + $\bar{\ }:\!C{\equiv}N:$ $\longrightarrow$ $H_3C-CH_2-C{\equiv}N:$ + $:\ddot{\underset{\cdot\cdot}{Br}}:^{\bar{\ }}$

electrophile   **nucleophile**

leaving group

$H-\ddot{Br}:$ + $\bar{\ }:\!C{\equiv}N:$ $\longrightarrow$ $H-C{\equiv}N:$ + $:\ddot{\underset{\cdot\cdot}{Br}}:^{\bar{\ }}$

Brønsted acid   Brønsted base

3.8   As noted on text p. 98, Brønsted acid–base reactions are generally *much* faster than the reactions of the same base with other electrophiles. Therefore, the reaction of hydroxide ion occurs with HBr almost exclusively. Therefore, the products are H_2O, Br^-, and unchanged H_3C—Br.

3.9   (a)

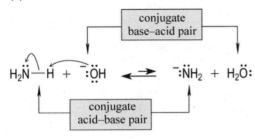

(b)

Let H_2O be the acid on the left. Then CH_3OH is the acid on the right. (If you chose CH_3OH as the acid on the left, then switch reactants and products in the reaction below.)

3.11   (a)   The reverse of reaction 3.16a is a Lewis acid–base association reaction; there is no leaving group.

(b)   The reverse of reaction 3.16b is a Lewis acid–base dissociation reaction. There is no nucleophile.

(c)   Like the forward reaction, the reverse of reaction 3.16c is a Brønsted acid–base reaction.

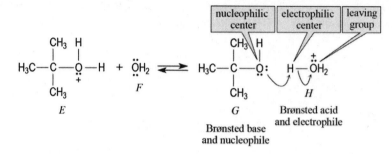

3.12   (a)   Two curved arrows with the proton of HBr as an electrophile makes this a Brønsted acid–base reaction.

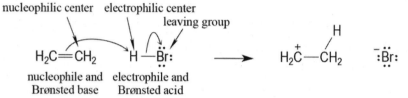

The reverse reaction is also a Brønsted acid–base reaction:

(b)   This is a Lewis acid–base association reaction.

The reverse reaction is a Lewis acid–base dissociation reaction:

3.13   Find the pK_a by taking the negative logarithm of the K_a.

(a)   pK_a = 3   (b)   pK_a = 5.24   (c)   pK_a = −1.7

3.14   Find the K_a by changing the sign of the pK_a and taking the antilog.

(a) $K_a = 10^{-4}$ (b) $K_a = 10^{-7.8} = 1.58 \times 10^{-8}$ (c) $K_a = 10^2 = 100$

3.15   (a)   The strongest acid in Problem 3.13 is the one with the lowest (most negative) pK_a—that is, (c).

(b)   The strongest acid in Problem 3.14 is the one with the largest K_a—that is, (c).

3.16   Basicities decrease with increasing pK_b.

3.17   (a)   According to Eq. 3.23, $pK_b = pK_w - pK_a = 14 - 4.67 = 9.33$.

(b)   When acetic acid is dissolved in water, a portion of the water molecules become protonated, according to the following equilibrium expression

$$CH_3CO_2H + H_2O \rightleftharpoons CH_3CO_2^- + H_3O^+$$

$$pK_a = 4.76 \qquad\qquad pK_a = -1.7$$
$$\text{weaker acid} \qquad\qquad \text{stronger acid}$$

Even though the equilibrium strongly favors the left side of the equation, some H_3O^+ ions have been created by adding the acetic acid to the water. So the pH of the solution is lowered at equilibrium.

The situation can be contrasted when sodium acetate is added to water as shown below.

$$CH_3CO_2^- + H_2O \rightleftharpoons CH_3CO_2H + HO^-$$

$$pK_a = 15.7 \qquad\qquad pK_a = 4.76$$
$$\text{weaker acid} \qquad\qquad \text{stronger acid}$$

Again, even though the left side is favored at equilibrium, there have been some OH^- ions generated, raising the pH of the solution.

3.18   (a)   The acid–base reaction is

$$H-CN + \ddot{N}H_3 \rightleftharpoons {}^-\!:CN + \overset{+}{N}H_4$$

When NH_3 acts as a base, it is converted into $^+NH_4$ (the ammonium ion), which becomes the acid on the right. The pK_a of the ammonium ion is 9.25. HCN is the acid on the left side of the equation, and its pK_a is 9.4. Therefore,

$$\log K_{eq} = 9.25 - 9.4 = -0.15$$

Therefore,

$$K_{eq} = 10^{-0.15} = 0.71$$

(b)   The acid–base reaction is

$$H-CN + :\ddot{\underset{..}{F}}: \rightleftharpoons {}^-\!:CN + H-\ddot{\underset{..}{F}}:$$

The acid on the right side of the equation is H—F ($pK_a = 3.2$), and the acid on the left side is H—CN ($pK_a = 9.4$). Using the procedure in part (a), we find that $\log K_{eq} = 3.2 - 9.4 = -6.2$, and $K_{eq} = 10^{-6.2} = 6.3 \times 10^{-7}$.

3.19   (a)   Ammonia acting as a base towards the acid water:

$$\ddot{N}H_3 + H_2\ddot{O}: \rightleftharpoons \overset{+}{N}H_4 + {}^-\!:\ddot{\underset{..}{O}}H$$
$$\text{base} \qquad \text{acid}$$

(b)   Ammonia acting as an acid towards the base water:

$$\ddot{N}H_3 + H_2\ddot{O}: \rightleftharpoons :\ddot{N}H_2^- + H_3\ddot{O}^+$$
$$\text{acid} \qquad \text{base}$$

    If you find an error in this manual, please visit http://people.pharmacy.purdue.edu/~loudonm/teaching/

Part (b) involves a stronger acid and stronger base on the right side *by far* than part (a). Therefore, the equilibrium constant for part (b) should be *much* smaller. To verify, apply Eqs. 3.26a and b. For part (a), the acid on the right is the ammonium ion ($pK_a = 9.25$). The acid on the left is water ($pK_a = 15.7$). The equilibrium constant is $10^{(9.25-15.7)} = 10^{-6.45} = 3.55 \times 10^{-7}$. For part (b), the acid on the right is H_3O^+ ($pK_a = -1.74$), and the acid on the left is ammonia ($pK_a = 35$). The equilibrium constant is $10^{(-1.74-35)} = 10^{-36.74} = 1.82 \times 10^{-37}$. The equation in part (a) is therefore more important in aqueous ammonia, as you might know from general chemistry. Even though this reaction is the more important of the two, its K_{eq} shows that there is nevertheless relatively little $^+NH_4\ ^-OH$ in aqueous ammonia.

3.20   Derivation of Eq. 3.31a from Eq. 3.30c:

$$f_A = \frac{10^{-pK_a}}{10^{-pK_a} + 10^{-pH}}$$

$$= 10^{-pK_a} \frac{1}{10^{-pK_a} + 10^{-pH}}$$

$$= \left(\frac{1}{10^{pK_a}}\right)\left(\frac{1}{10^{-pK_a} + 10^{-pH}}\right)$$

$$= \frac{1}{(10^{pK_a} \times 10^{-pK_a}) + (10^{pK_a} \times 10^{-pH})}$$

$$= \frac{1}{10^{pK_a - pK_a} + 10^{pK_a - pH}}$$

$$= \frac{1}{1 + 10^{pK_a - pH}}$$

$$= \frac{1}{1 + 10^{-pH + pK_a}}$$

$$= \frac{1}{1 + 10^{-(pH - pK_a)}}$$

$$= \frac{1}{1 + 10^{-\Delta}}$$

Derivation of Eq. 3.31b from Eq. 3.30d:

$$f_{AH} = \frac{10^{-pH}}{10^{-pK_a} + 10^{-pH}}$$

$$= \left(\frac{1}{10^{pH}}\right)\left(\frac{1}{10^{-pK_a} + 10^{-pH}}\right)$$

$$= \frac{1}{(10^{pH} \times 10^{-pK_a}) + (10^{pH} \times 10^{-pH})}$$

$$= \frac{1}{10^{pH - pK_a} + 10^{pH - pH}}$$

$$= \frac{1}{10^{pH - pK_a} + 1}$$

$$= \frac{1}{1 + 10^{\Delta}}$$

3.21   (a)   This part of the question is asking us to determine [AH] and [A]. The pK_a of ibuprofen is 4.43, which is less than the pH of the solution, pH = 5.0, therefore we expect more ibuprofen dissociated than undissociated, or [A] > [AH]. Use Eq. 3.31a to solve for the fraction dissociated: $f_A = 0.78$. In a 10^{-4} M solution, 78% dissociated makes [A] = 7.8×10^{-5} M, while the remaining is undissociated, and [AH] = 2.2×10^{-5} M.

(b)   Since the pK_a > pH, we predict that more ibuprofen is undissociated, or [A] < [AH]. Using Fig. 3.1 on text p. 108, we might predict that f_{AH} > 0.99 and that f_A < 0.01. When we use Eq. 3.31a we can precisely calculate f_A = 0.00370, or 0.370 %, which is consistent with our prediction.

(c)   Here, since pH < pK_a, there will be more ibuprofen dissociated than undissociated, nearly all dissociated if we consult Fig. 3.1. Solving for f_A gives 0.999, meaning that ibuprofen is over 99% dissociated in the bloodstream.

3.22   We want to again use Eq. 3.31a to solve for f_A. However, in order to solve for Δ, we need the pH, which is not given. So first, we must figure out the pH of a 0.1 M solution of acetic acid in water. We have the pK_a, which is the negative log of the K_a (Eq. 3.20a, text p. 101). We can use $K_a = 1.8 \times 10^{-5}$ to solve for the [H_3O^+] using Eq. 3.19 on text p. 101. You may recall from your general or high school chemistry courses the "ICE" method for solving problems such as these, using the approximation that x is very small in determining your answer.

$$K_a = \frac{[H_3O^+][A^-]}{[HA]}$$

$$1.8 \times 10^{-5} = \frac{(x)(x)}{0.1 - x}$$

Assuming x is very small,

$$1.8 \times 10^{-5} = \frac{x^2}{0.1}$$

$$1.8 \times 10^{-6} = x^2$$

$$1.34 \times 10^{-3} = x = [H_3O^+]$$

Now we can use Eq. 3.20b to show that the pH of a 0.1 M solution of acetic acid = $-\log[H_3O^+]$ = 2.87. Finally, we will employ Eq. 3.31a to calculate f_A = 0.0129, meaning that a 0.1 M solution of acetic acid is only about 1.3% dissociated in water, which is consistent with our classification of it as a weak acid.

3.23   (a)   Consider each species individually, and construct a plot similar to Fig. SG3.1. First, at low pH values, BH_2 (dotted line) is present in high concentrations. When the pK_a = pH (3.13), BH_2 is 50% dissociated. At a pH about 2 units away from this point (5.13), [BH] is nearly 0. Considering next BH (solid line), it begins to appear as BH_2 disappears, such that the sum of $f_{BH2} + f_{BH}$ always equals 1. At pH = 8.02, BH is 50% dissociated, and at pH = 10.2, it its concentration is nearly 0. Finally, considering B (grey line), it is the product of dissociation of BH, and is present at 50% concentration at pH = 8.02. At pH = 10, it is present at nearly 100%.

(b)   At a pH of about 5.5 f_{BH} will be at a maximum while $f_{BH2} + f_B$ are at minimums, as shown (b) in Fig. SG3.1.

(c)   In the bloodstream, pH = 7.4 (see problem 3.21), the major species present would be BH, as shown (c) in Fig. SG 3.1.

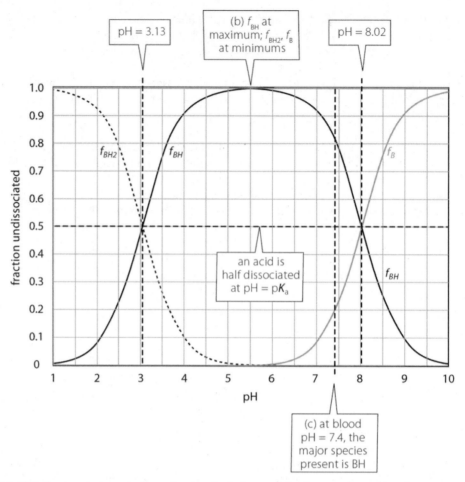

Figure SG3.1 Plot of percent concentrations of various ionic forms of nicotine at various pH vaules to accompany the solution to Problem 3.23. The dashed curve represents BH2, the solid black line BH, and the grey line B.

3.24   (a)   Use Eq. 3.35 on text p. 110 with $2.3RT = 5.71$ kJ mol^{-1} at 298 K:

$$K_{eq} = 10^{-(-14.6)/5.71} = 362$$

(b)   For this calculation use Eq. 3.35 on text p. 110.

$$\Delta G° = -2.30RT \log K_{eq} = -5.71 \log(305) \text{ kJ mol}^{-1} = -5.71(2.48) \text{ kJ mol}^{-1} = -14.2 \text{ kJ mol}^{-1}$$

 In this Solutions Manual, kJ mol^{-1} is used as the unit of energy. If you wish to convert kJ mol^{-1} to kcal mol^{-1}, simply divide by 4.184 kJ kcal^{-1}. Also, it is helpful to remember that $2.30RT$ at 298 K (25 °C) is 5.71 kJ mol^{-1}.

3.25   (a)   To complete this calculation, first calculate the equilibrium constant, and then use it to do a mass-action calculation.

$$K_{eq} = 10^{-\Delta G°/2.30RT} = 10^{-(-2.93)/5.71} = 10^{0.513} = 3.26$$

Make a table of concentrations for the three species in the equilibrium:

	A	B	C
Initially	0.1	0.2	0
Change:	−x	−x	+x
At equilibrium:	0.1 − x	0.2 − x	x

The equilibrium constant can then be written as

$$K_{eq} = \frac{[C]}{[A][B]} = \frac{x}{(0.1-x)(0.2-x)} = 3.26$$

from which is obtained the quadratic expression

$$x^2 - 0.607x + 0.02 = 0$$

The quadratic formula gives

$$x = \frac{0.607}{2} \pm \frac{\sqrt{0.368 - 0.080}}{2} = 0.303 \pm 0.268$$

Only $x = 0.035\ M$ is physically meaningful, because the alternate value, 0.571 M, is larger than the starting values of the reagents. Choose the physically meaningful value of x to obtain

$$[C] = 0.035\ M, [B] = 0.2 - x = 0.165\ M, \text{ and } [A] = 0.1 - x = 0.065\ M.$$

(b)   If the $\Delta G°$ is positive, the equilibrium constant is less than unity and the reaction is less favorable; consequently, there should be much less C and more A and B at equilibrium.

    An actual calculation shows that $[C] = 0.0061$. Then $[A] = 0.1 - x = 0.0939\ M$, and $[B] = 0.2 - x = 0.1939$. The concentration of C is 10% of that in part (a), a result consistent with the statement in part (b).

3.26   (a)   5.71

(b)   Let reaction (2) have the higher standard free energy. The answer then follows from the following equation:

$$\Delta G_2° - \Delta G_1° = 1 = -2.3RT \log \frac{K_2}{K_1} \quad \text{or} \quad -\frac{\Delta G_2° - \Delta G_1°}{2.3RT} = -\frac{1}{5.71} = -0.18 = \log \frac{K_2}{K_1}$$

or

$$\frac{K_2}{K_1} = 10^{-0.18} = 0.66 \quad \text{or} \quad \frac{K_1}{K_2} = 1.52$$

3.27   (a)   The ammonium ion, $^+NH_4$, is the more acidic species. You can verify that this is so in Table 3.1, text p. 102. This is a manifestation of the charge effect.

(b)   The more acidic species is $CH_3—^+SH_2$.

3.28   (a)   The element effect gives the acidity ranking $B > A > D$. The charge effect makes C the least acidic. The reason that B is more acidic than A is that HF is more acidic than CH_3SH. (Table 3.1, text p. 102.) The charge effect should make both more acidic, and we assume that it should not change their relative order.

Note that *increasing acidities* is the same as *decreasing* pK_a *values*.

(b)   Compound *A* is much more basic than compound *B* (charge effect). By the element effect, *C* is much more basic than *A*; this follows from the fact that if we rank their conjugate acids, the conjugate acid of *C*, ammonia, is less acidic than the conjugate acid of *A*, CH_3OH. Note that a comparison of *A* and *B* brings into play the amphoterism of methanol: like water, it can act as both an acid and a base in different reactions.

 A negatively charged atom is more basic than the same neutral atom. This can be understood on the basis of the charge effect for positively charged atoms by taking into account the relationships of conjugate acids and bases. By the charge effect, an acidic proton on a positively charged atom is more acidic than an acidic proton on the same neutral atom. It follows, then, that the neutral conjugate base of the positively charged acid is less basic than the negatively charged conjugate base of the neutral acid.

3.29   In each set, label the compounds from left to right as *A, B,* and *C*.

(a)   The order of acidities is $C < B < A$; hence, the order of pK_a values is $C > B > A$. The element effect differentiates *A* and *B*, and the polar effect differentiates *B* and *C*.

(b)   The order of pK_a values is *B* (least acidic) $> A > C$ (most acidic). The acidity of compound *A*, relative to compound *B*, is increased by the polar effect of one methoxy ($—OCH_3$) group. The acidity of compound *C* is increased by the polar effect of two methoxy groups.

(c)   *A* is more acidic than *C* because of the charge effect. (Notice that compound *C* is amphoteric—that is, it can act as both an acid and a base. *A* is the conjugate acid of *C*.) Compound *B* is the most acidic because of the polar effect of the —Cl. The order of pK_a values, then, is $C > A > B$; the order of increasing acidity is $C < A < B$.

3.30   (a)   From Eq. 3.42a, text p. 115, $\Delta G_a° = 2.3RT(pK_a) = (5.71)(2.66) = 15.2$ kJ mol^{-1}.

(b) $\Delta G_a° = 2.3RT(pK_a) = (5.71)(4.76) = 27.2$ kJ mol^{-1}. More energy is required to ionize the less acidic compound.

3.31   To say that acetic acid is weaker than fluoroacetic acid means that the conjugate base of acetic acid is less stable relative to the un-ionized acid than the conjugate base of fluoroacetic acid is to its un-ionized acid (Fig. 3.3). "Less stable" means that more free energy is required to convert the weaker acid into its conjugate base, and this reduced stability is reflected in the larger standard free energy of ionization for acetic acid. Just as it takes more work to climb a larger hill than it does to climb a smaller hill, it takes more "chemical work" (free energy) to form a more energetic species than to form a less energetic one.

Solutions to Additional Problems

3.32   (a)   The carbon of this ion is electron-deficient because it has a sextet of electrons.
(b)   Every atom has an octet; therefore, this is not an electron-deficient compound.
(c)   Every atom has an octet; therefore, this is not an electron-deficient compound.
(d)   The boron is electron-deficient because it has a sextet of electrons.
(e)   The nitrogen is electron-deficient because it has a sextet of electrons.

 Electron deficiency can occur without positive charge, as parts (d) and (e) demonstrate.

3.33   The general approach is to allow a nucleophile to react with an electron-deficient atom; then complete the resulting formal charges.

(a)

$$H_3C - \overset{..}{\underset{..}{O}} - CH_3 \quad BF_3 \longrightarrow H_3C - \overset{\overset{\displaystyle ^-BF_3}{|}}{\underset{..}{O}{}^+} - CH_3$$

(b)

$$\begin{array}{c} CH_3 \\ | \\ H_3C - \overset{+}{\underset{}{C}} - CH_3 \quad :\overset{..}{\underset{..}{Cl}}:{}^- \longrightarrow H_3C - \overset{\overset{\displaystyle CH_3}{|}}{\underset{\underset{\displaystyle :\overset{..}{\underset{..}{Cl}}:}{|}}{C}} - CH_3 \end{array}$$

(c)

$$H\overset{..}{\underset{..}{O}} - CH_2 - CH_2 - CH_2 - \overset{+}{CH} - CH_3 \quad \text{or} \quad$$

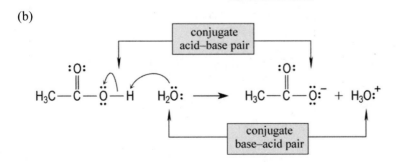

> Don't forget that carbon chains are flexible and can undergo internal rotations. Groups that appear far apart in structures that are drawn linearly can in many cases be brought into proximity by internal rotations. If this idea is not clear, construct a model for this case, and convince yourself that the —OH group can easily be brought within bonding distance of the positively charged carbon.

(d)

$$(CH_3)_3\overset{\frown}{B} \quad :\bar{C}{\equiv}\overset{+}{O}: \longrightarrow (CH_3)_3\bar{B} - C{\equiv}\overset{+}{O}:$$

(e)   We use the electron pair on the atom that has the complete octet. Even though the carbon of the :CH₂ contains an electron pair, this carbon is electron-deficient.

$$CH_3\overset{..}{N}H_2 \quad :\overset{\overset{\displaystyle H}{|}}{\underset{\underset{\displaystyle H}{|}}{C}} \longrightarrow H_2\overset{+}{N} - \overset{\overset{\displaystyle CH_3}{|}}{\underset{\underset{\displaystyle H}{|}}{C}}{:}^- \quad {}_{H}$$

3.34   (a)

(b)

(c)

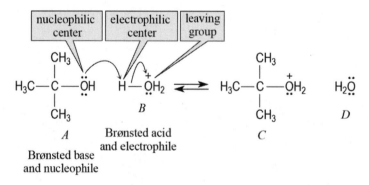

(d)

3.35   (a)   *Step 1* is (4)—a Brønsted acid–base reaction. Since every Brønsted acid–base reaction is a type of electron-pair displacement reaction, *Step 1* is also such a reaction (3). Step 2 is (2)—a Lewis acid–base dissociation reaction. *Step 3* is (4)—a Brønsted acid–base reaction—and (3)—an electron-pair displacement reaction.

       (b)   In *Step 1,* species B and D constitute a conjugate acid–base pair; and species A and C constitute a conjugate base–acid pair. In *Step 3,* species E and H constitute a conjugate acid–base pair, and species G and I constitute a conjugate base–acid pair.

       (c)   In the forward direction of *Step 1, A* (specifically the oxygen) is a Brønsted base; the transferred proton of B is an electrophile, and the OH$_2$ group of B is a leaving group (which becomes a water molecule D). In the forward direction of *Step 2,* the OH$_2$ group of C is a leaving group and, because it is accepting a pair of electrons, a Lewis acid. In *Step 3,* the water molecule G is a Brønsted base; the transferred proton of E is an electrophile; and the part of E other than the transferred proton is a leaving group.

       (d)   The curved–arrow notation, along with a summary of part (c), follows. This is for the forward direction only, as requested in the problem.

Step 1:

Step 2:

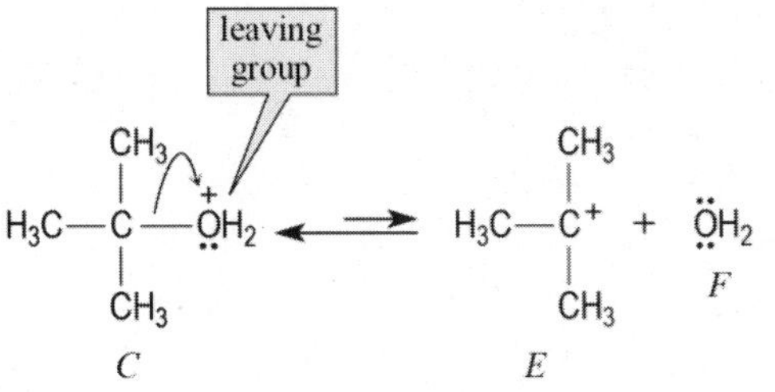

Step 3:

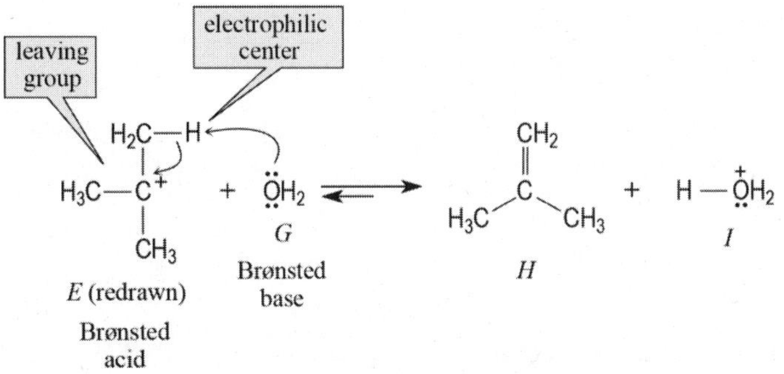

3.36   As in Problem 3.35, every reaction that involves the donation or acceptance of electron pairs, and therefore every reaction in this problem, is (1)—a Lewis acid–base reaction.

(a)   In general, if a reaction in one direction is a Brønsted acid–base reaction, it is also a Brønsted acid–base reaction in the reverse direction. Therefore, *Steps 1* and *3* are (4)—Brønsted acid–base reactions—and (4)—electron-pair displacement reactions.

(b)   The conjugate acid–base relationship is unchanged by the direction of the reaction. The answer is the same as for Problem 3.35(b).

(c)   See the equations in part (d).

(d)   The curved-arrow notation and the labeling of the various species:

Step 1:

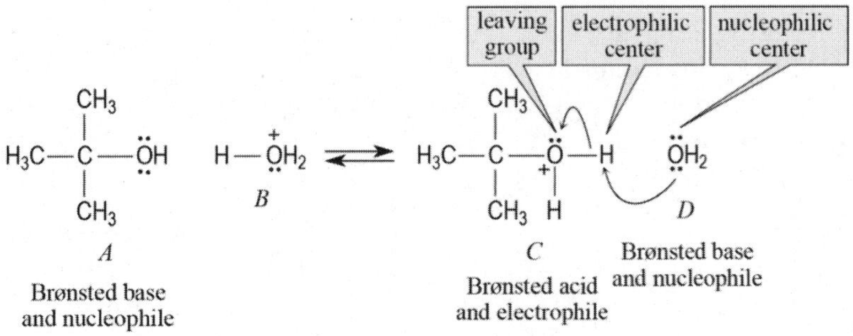

Step 2:

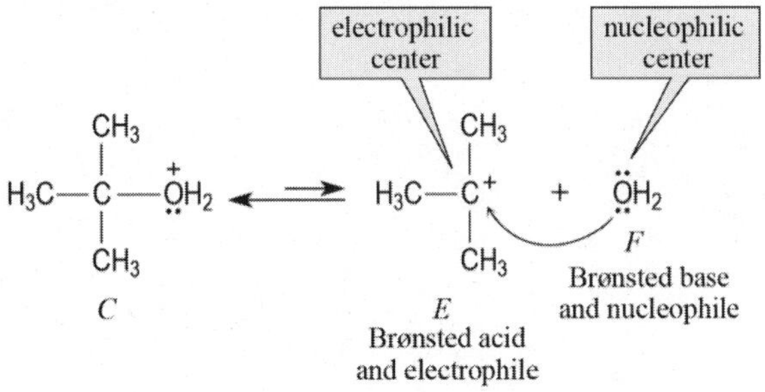

Step 3:

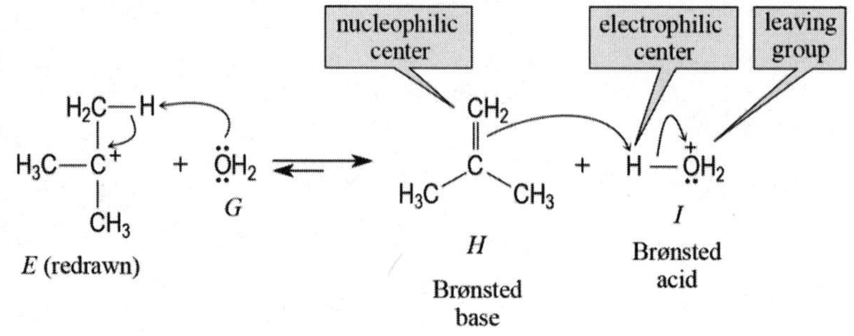

3.37   (a)   The resonance structures of protonated acetic acid:

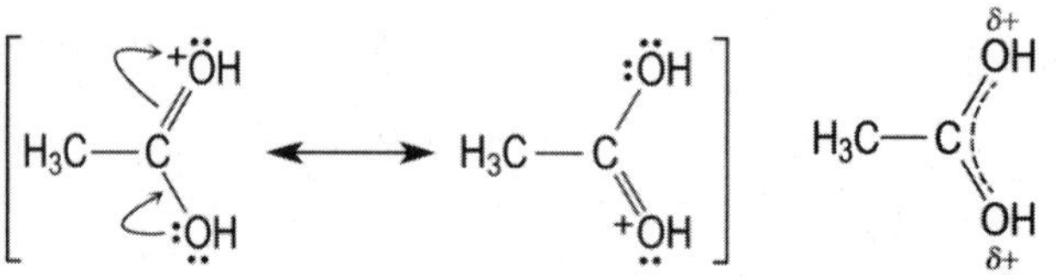

conjugate acid of acetic acid   hybrid structure

(b)   Interconversion of the resonance structures of carbon monoxide with the curved-arrow notation:

Because the carbon in the structure on the left has a sextet of electrons, it can accept another electron pair.

3.38   (a)   The structure of the N_5^+ cation:

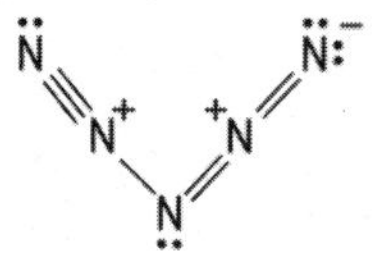

This structure is trigonal planar at the central nitrogen and therefore bent. The other two nonterminal nitrogens and one of the terminal nitrogens are bound to two groups, and thus have linear geometry. Although the other terminal nitrogen is surrounded by three "groups"—a nitrogen and two electron pairs—this has no effect on the shape.

(b)   Resonance structures of the N_5^+ cation:

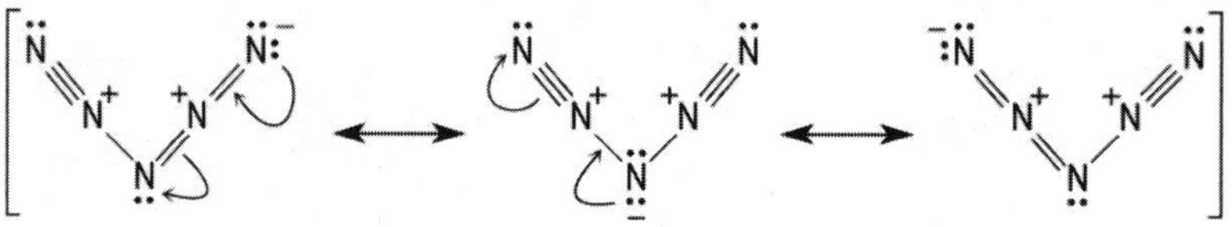

3.39   (a)   Resonance structures of ozone:

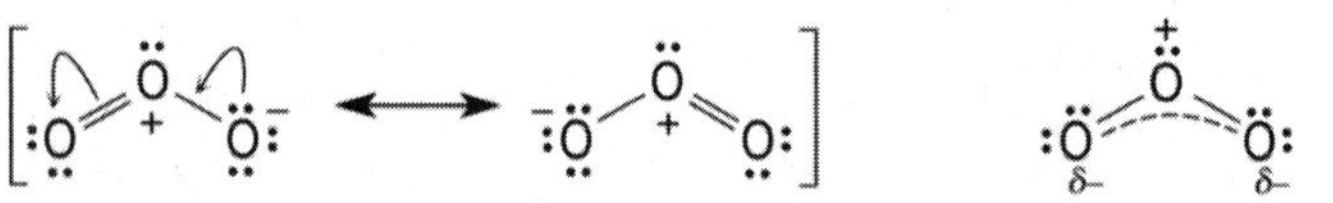

hybrid structure

(b)   The resonance structures of the carbonate ion are shown in Problem 1.32, text p. 42. The solution to that problem discusses the significance of these structures. The hybrid structure:

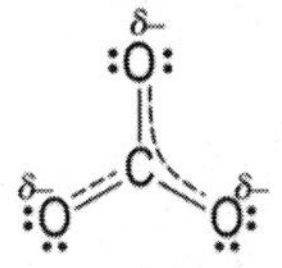

hybrid structure of the carbonate ion

(c)   Resonance structures of protonated formaldehyde:

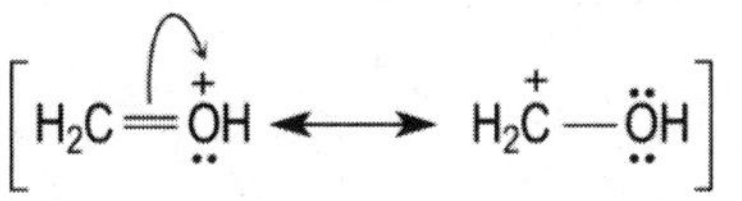

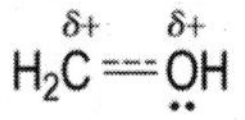

hybrid structure

3.40   (a)   (c)

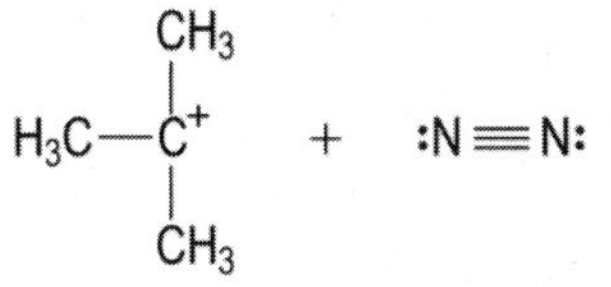

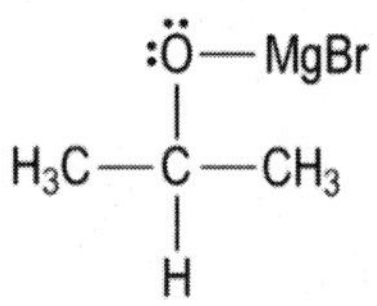

(b) 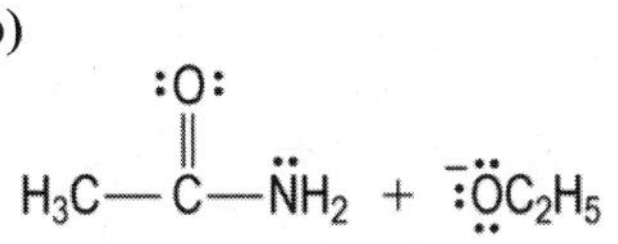   (d)

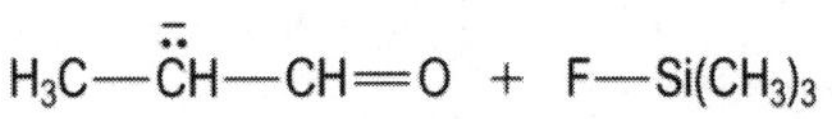

3.41   Remember that an arrow must originate at a *source* of electrons—a bond or an electron pair.

(a)

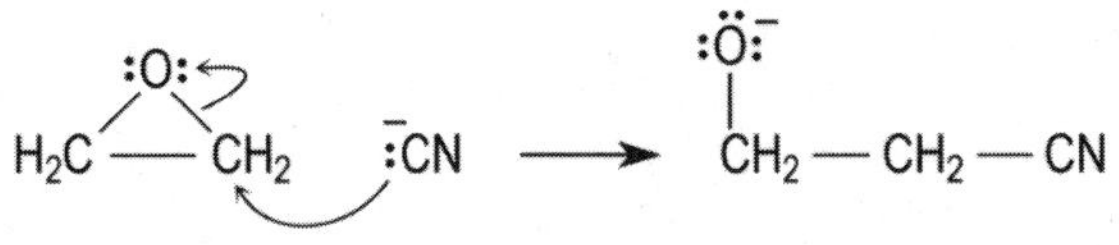

(b)

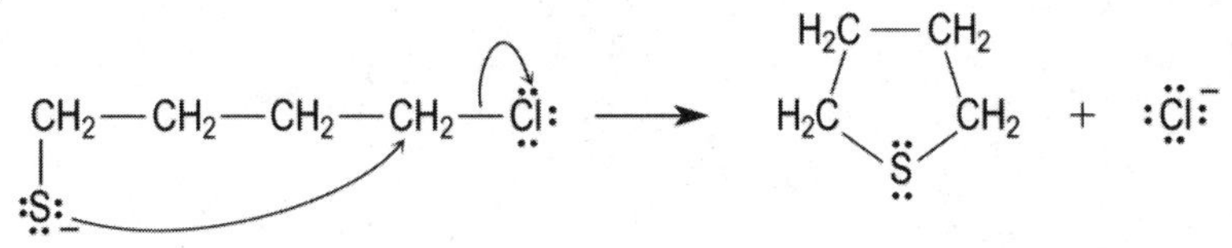

(c)

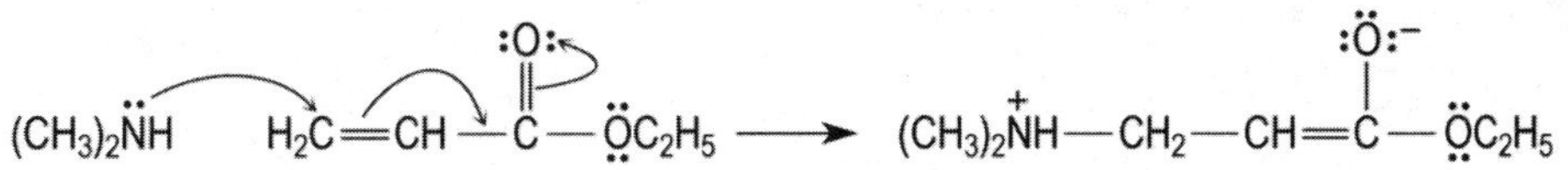

(d)

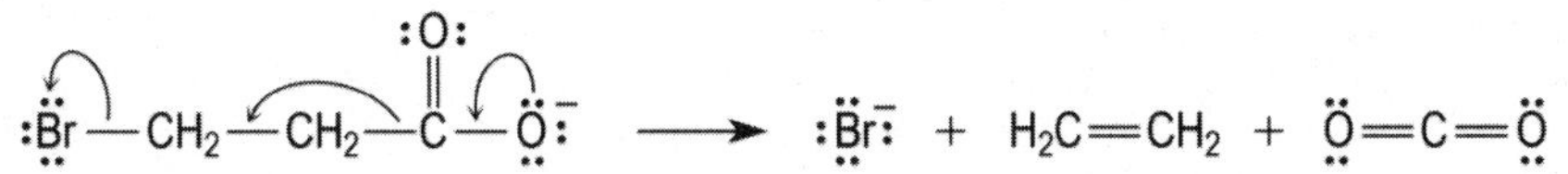

3.42   (a)   The $^-$OCH$_3$ base could react at the carbon to expel $^-$SH, or it could react at the proton to expel CH$_3$S$^-$. The latter reaction is the Brønsted acid–base reaction, which is faster.

$$CH_3\overset{..}{\underset{..}{S}}\!\!-\!\!H \qquad :\overset{..}{\underset{..}{O}}CH_3 \longrightarrow CH_3\overset{..}{\underset{..}{S}}{:}^- + H\!\!-\!\!\overset{..}{\underset{..}{O}}CH_3$$

(b)   The reaction is a Lewis acid–base association.

(c)   The reaction is a Lewis acid–base association.

(d)   The reaction is a Brønsted acid–base reaction with a proton:

3.43   (a)   Species *A* is a Brønsted base and nucleophile. The oxygen is the nucleophilic center. Proton *B* is an electrophilic center. Carbon *C* serves as a leaving group from the proton *B*. Carbon *D* is an electrophile. Species *E* is a leaving group.

(b)   Species *A* is a nucleophile and a Brønsted base. Carbon *B* is an electrophilic center. Carbon *C* is a leaving group toward *B* and also a nucleophile toward carbon *D*. Carbon *D* is an electrophilic center. The electrons between carbon *D* and oxygen *E* serve as a leaving group. When atoms in double or triple bonds serve as leaving groups, one bond remains intact. In these cases, the leaving group doesn't actually "leave," but remains tethered within the same molecule, which is the case with "leaving group" *E*—it remains attached because only one of the bonds of the double bond "left".

3.44   (a)   The curved arrow is going in the wrong direction; it should go from the electrons on the $^-$OH to the hydrogen of the carboxylic acid:

(b) The one curved arrow is shown correctly, but a second one is needed. Otherwise, carbon would have five bonds in the product, and this situation would violate the octet rule.

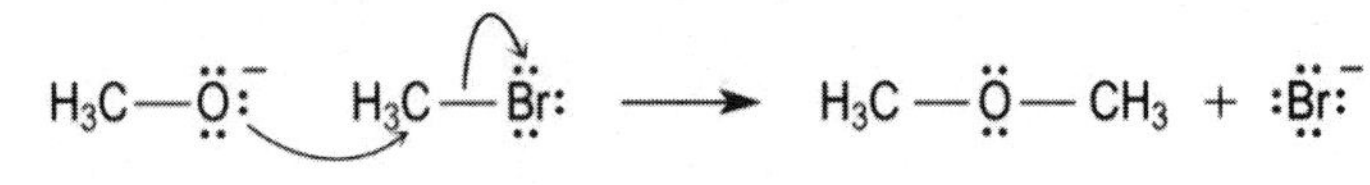

3.45 The resonance structures of naphthalene:

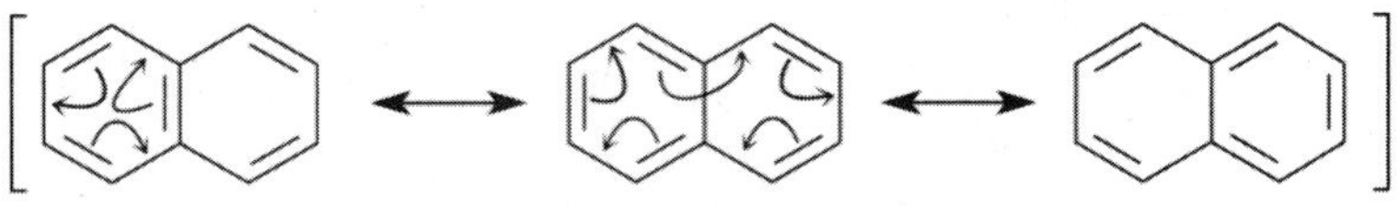

naphthalene

3.46 (a) For purposes of this solution let 2,2-dimethylpropane = D and pentane = P. Then, for the hypothetical reaction $P \rightleftharpoons D$, we define the equilibrium constant $K_{eq} = [D]/[P]$, and the standard free energy change $\Delta G°$ is therefore $G°(D) - G°(P) = -6.86$ kJ mol^{-1}. Using Eq. 3.36b on text p. 110,

$$K_{eq} = 10^{-(-6.86)/5.71} = 10^{1.20} = 15.9 = [D]/[P]$$

Taking the total hydrocarbon as 1.00, and noting from above that $[D] = 15.9[P]$,

$$1.00 \quad = \quad [D] + [P] = 15.9[P] + [P] = 16.9[P]$$

or $$[P] \quad = \quad 1.00/16.9 = 0.0592$$

and $$[D] \quad = \quad 1.00 - 0.0592 = 0.941$$

Consequently, at equilibrium the mixture contains 5.9% P and 94.1% D. Compound D is present in greater amount because it is more stable.

(b) Let K_{eq} be the ratio [anti]/[gauche]; then $\Delta G°$ for gauche $\rightleftharpoons$ anti is -2.8 kJ mol^{-1}. (Because gauche has higher energy than anti, $\Delta G°$ is negative.) Then use Eq. 3.36b in the text.

$$K_{eq} = 10^{-\Delta G°/2.3RT} = 10^{-(-2.8)/5.71} = 10^{0.49} = 3.1$$

This is the ratio of the anti conformation to any *one* gauche conformation (remember that in total, there are two gauche conformations and one anti conformation). Taking the total fraction of butane as 1.0, and noting from above that [anti] = 3.1[gauche],

$$1.0 = [anti] + 2[gauche] = 3.1[gauche] + 2[gauche] = 5.1[gauche]$$

or $$[gauche] = 1.0/5.1 = 0.20$$

Consequently, the total fraction of gauche conformations is $2 \times 0.20 = 0.40$ (that is, about 40% gauche conformations); and the total fraction of anti conformation is $1.0 - 0.40 = 0.60$ (that is, about 60% anti conformation) in a sample of butane.

3.47 To arrange in order of decreasing pK_a means to arrange in order of *increasing acidity*.

(a) CH_3CH_2OH is less acidic than $ClCH_2CH_2OH$, which is less acidic than Cl_2CHCH_2OH. Chlorine substituents increase acidity because of their electron-withdrawing polar effect; the more chlorines there are, the greater the effect.

(b) Because of the element effect, both thiols (the compounds with —SH groups) are more acidic than ethanol (CH_3CH_2OH). The compound $ClCH_2CH_2SH$ is the more acidic of the two thiols because of the polar effect of the chlorine substituent.

(c) The element effect operating from top to bottom of a column of the periodic table predicts that $H—N(CH_3)_2$ should be less acidic than $H—P(CH_3)_2$, which should be less acidic than $H—As(CH_3)_2$. Notice that these acidities refer to the reaction in which the *neutral* molecules lose a proton to give the *anionic* conjugate bases (M = N, P, or As):

$$H—\overset{..}{\underset{}{M}}(CH_3)_2 + base:^- \rightleftharpoons {}^-:\overset{..}{\underset{}{M}}(CH_3)_2 + base—H$$

The acidity of $H_2\overset{+}{As}(CH_3)_2$ refers to the reaction in which this positively charged species loses a proton to give the neutral conjugate base $H\overset{..}{As}(CH_3)_2$. The cationic acid $H_2\overset{+}{As}(CH_3)_2$ is *much* more acidic than the neutral acid $H\overset{..}{As}(CH_3)_2$ because of the charge effect.

(d) CH_3CH_2OH is less acidic than $(CH_3)_2NCH_2CH_2OH$ because of the polar effect of the electronegative nitrogen substituent. In the last compound, not only is the polar effect of the nitrogen present, but also a positive charge, which can stabilize the conjugate-base anion (charge effect). Furthermore, the positively-charged nitrogen is closer to the —OH group. Consequently, $(CH_3)_3N^+$—OH is the most acidic of the three compounds and it is thus the strongest acid of all.

3.48   In each part, we apply Eq. 3.26b on text p. 104 and the procedure used in Study Problem 3.6.

(a)   The information needed is the pK_a of H—CN, the acid on the left side of the equation. From Table 3.1 (text p. 102), the pK_a of H—CN is 9.40. The equilibrium favors the side with the weaker acid, namely the right side, but by very little, because the pK_a values of the two acids are similar. Applying Eq. 3.26b,

$$K_{eq} = 10^{(9.76-9.4)} = 10^{+0.36} = 2.3$$

(b)   Again we apply text Eq. 3.26b with $pK_a = 15.7$ for H_2O.

$$K_{eq} = 10^{15.7-10.5} = 10^{5.2} = 1.58 \times 10^5$$

 Don't confuse equilibrium constants (K_{eq}) for the reactions of acids and bases with dissociation constants (K_a). These equilibrium constants use dissociation constants in their calculation, but are not themselves dissociation constants. Similarly, log K_{eq} is related to the pK_a values of the two acid–base systems involves, but is not itself a pK_a.

3.49   (a)   Apply Eq. 3.35 (text p. 110) using the K_{eq} from the solution to Problem 3.48(a).

$$\Delta G° = -2.3RT \log K_{eq} = -5.71 \log (2.3) = -2.1 \text{ kJ mol}^{-1}$$

(b)   Use the same procedure as in part (a).

$$\Delta G° = -2.3RT \log K_{eq} = (-5.71)(5.2) = -29.7 \text{ kJ mol}^{-1}$$

(c)   If the equilibrium concentrations of $(CH_3)_3N$ and HCN are both x, then the equilibrium concentrations of $(CH_3)_3N^+H$ and ^-CN are both $(0.1 - x)$. Substituting these values into the equilibrium expression,

$$K_{eq} = 2.3 = \frac{[(CH_3)_3N^+H][^-CN]}{[(CH_3)_3N][HCN]} = \frac{(0.1-x)^2}{x^2}$$

This leads to the quadratic expression $x^2 + 0.154x - 0.00769 = 0$, which is solved to give $x = 0.0397$ *M*. Thus, at equilibrium, the concentrations of the species on the left side of the equation are both 0.0397 *M*, and the concentrations of the species on the right side are both $0.1000 - 0.0397 = 0.0603$ *M*.

3.50   (a)   By Eq. 3.42a, text p. 115, the compound with the smaller pK_a has the smaller standard free energy of dissociation $\Delta G_a°$. Consequently, phenylacetic acid has the smaller $\Delta G_a°$.

(b)   The problem is asking for the standard free energy of dissociation $\Delta G_a°$. Use Eq. 3.42a in the text. (Assume 25 °C or 298 K.)

$$\Delta G_a° = 2.3RT(pK_a) = 5.71(4.31) = 24.6 \text{ kJ mol}^{-1}$$

(c)   The fraction ionized (or, dissociated) is f_A and can be found by Eq. 3.31a (text p. 107).

$$f_A = 1/(1 + 10^{-\Delta}) = 1/(1 + 10^{-(4.5-4.31)}) = 0.910$$

Consequently, 91% of phenylacetic acid is dissociated (ionized) at pH = 4.5.

    If you find an error in this manual, please visit http://people.pharmacy.purdue.edu/~loudonm/teaching/

(d)   Phenylacetic acid should be more dissociated because it is the stronger acid (it has a lower pK_a). To be
sure, use the same procedure as in part (c) to figure out f_A for acetic acid.

$$f_A = 1/(1 + 10^{-\Delta}) = 1/(1 + 10^{+0.26}) = 0.355$$

So at pH = 4.5, phenylacetic acid is 91% ionized (part (c)) and acetic acid is only about 36% ionized.

(e)   Phenylacetic acid has a lower pK_a than acetic acid. Hence, the phenyl substituent is acid-strengthening; it
stabilizes the conjugate-base anion in the same sense that a chlorine does. Consequently, its polar effect,
like that of a chlorine, is evidently electron-withdrawing.

3.51   (a)   For the first ionization (pK_a = 2.86), either of the two equivalent carboxylic acid groups can ionize:

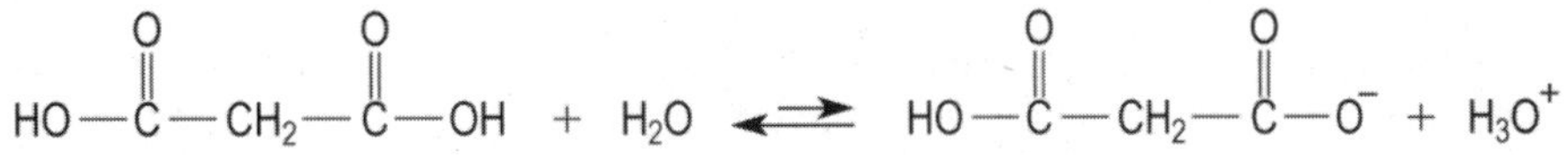

For the second ionization (pK_a = 5.70), the remaining carboxylic acid group ionizes:

(b)   Use a strategy similar to that used to solve Problem 3.23 and sketch a plot similar to Fig. SG3.1. At low
pH, when protons are abundant, both carboxylic acid groups will be protonated (undissociated), or AH_2
is present at nearly 100%. As the pH increases, AH_2 decreases and AH^- increases until the curves meet at
pH = 2.86 (50/50 AH_2/AH^-). At pH = 4.86, 2 pH units above the first pK_a, AH^- is present at
approximately 100%. Approaching the pH equal to the second pK_a, 5.70, AH^- decreases and A^{2-}
increases. At pH = 5.70, $[A^-] = [A^{2-}]$. Therefore, the ionization state at pH = 4.3, above the pK_a of AH_2
and below the pK_a of AH^-, is essentially all AH^- with only a few percent or less of AH_2 and the same few
percent of A^{2-}.

More precisely, you can use the instructions for creating an excel spreadsheet described in section 3.4F
(text p. 105) to construct a curve depicting the amounts of each species present at various pH values.

(c)   As reasoned in part (b), at pH = 4.3, malonic acid is essentially in the AH^- form, or in other words, one
of its carboxylic acid groups has been completely ionized. That means, that it takes 1 mole of sodium
hydroxide per mole of malonic acid to get to this pH.

(d)   The first pK_a is lower than that of acetic acid because, in the product of the first ionization, the electron-
withdrawing polar effect of the un-ionized carboxylic acid group stabilizes the conjugate-base anion on
the other. (The two electronegative oxygens are responsible for the polar effect of the carboxylic acid
group.) The second pK_a is greater than that of acetic acid because one negative charge in the dianionic
product of the second ionization interacts unfavorably (repulsively) with the second negative charge, and
the energy of the dianion is thereby raised. (See Fig. 3.3, text p. 116.)

(e)   As n becomes larger, the polar groups become more remote. Hence, *both* polar effects discussed in part
(d) become less significant, and *both* pK_a values approach that of acetic acid; that is, the pK_a values of the
two ionizations become more similar. (In terms of the hint, as r in the denominator of the electrostatic
law (Eq. 3.44, text p. 117) becomes larger, the energy of interaction of one charge with another becomes
smaller.)

Even if n is very large, the two pK_a values are not exactly equal. Theoretically, the first pK_a should be 0.6 smaller than the second pK_a (that is, the first K_a should be 4 times larger than the second K_a). The reason is that either of *two* equivalent acidic hydrogens can undergo the first ionization, but only one can undergo the second; this means that the first ionization is twice as probable statistically. Furthermore, either of *two* equivalent carboxylate ions can be protonated in the conjugate base of the monoacid, but only one is available in the conjugate base of the diacid. Thus, protonation of the dianion is twice as probable as protonation of the monoanion, which means that the second ionization is half as probable as the first—or again, that the first ionization is twice as probable as the second. Both of these factors multiply; thus, the first ionization is four times as likely as the second on purely statistical grounds; log (4) = 0.6.

3.52   (a)   Protons with pK_a values more than about 2 units below the pH are completely dissociated, and those with pK_a values more than about 2 units above the pH are completely associated. So at pH = 7.4, ascorbic acid has a net −1 charge.

(b)

3.53   The polar effect, governed by the electrostatic law (Eq. 3.44, text p. 117), dictates that two like charges in close proximity destabilize a molecule. The BH2 form of nicotine is a dication, so it is much less stable than the conjugate acid of 3-methylpyridine, which has only one positive charge. By destabilizing the molecule, it is more reactive, and that proton is more acidic, as shown in the free energy diagram in Fig. SG3.2.

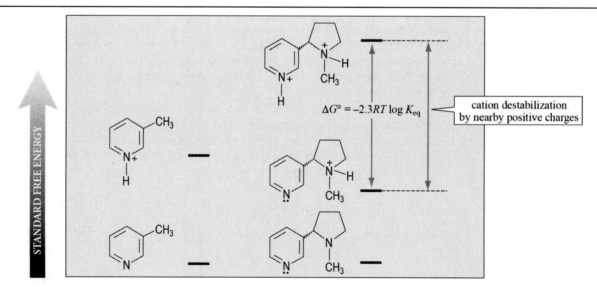

Figure SG3.2 Free-energy diagram to accompany the solution to Problem 3.53. The dication is destabilized by the two nearby positive charges. (The neutral compounds are arbitrarily placed at the same energy for comparison purposes.)

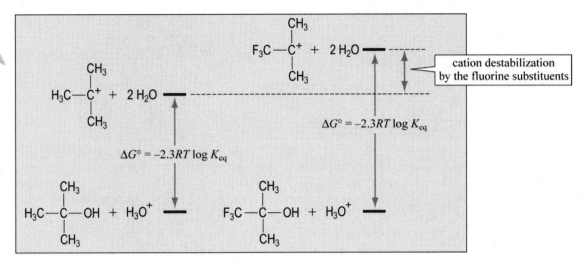

Figure SG3.3 Free-energy diagram to accompany the solution to Problem 3.54. The polar effect of the fluorine substituents raises the energy of the cation and therefore reduces the equilibrium constant for its formation. (The two alcohols, which are uncharged, are arbitrarily placed at the same energy for comparison purposes.)

3.54   The primary effect of a polar substituent is on the stability of a charged species. The charged species in these equilibria are the cations. The question then becomes, how does a fluorine substituent affect the stability of a cation? As Eq. 3.44 (text p. 117) shows, the polar effect of the fluorine *stabilizes negative charge;* however, it *destabilizes positive charge* because the positive ends of the C—F dipoles interact repulsively with the positive charge of the cation, as shown by the following diagram:

repulsive interaction

F—C—C⁺(CH₃)(CH₃)

Consequently, the cation product of the second equilibrium is destabilized relative to the neutral starting material. This situation is depicted in Fig. SG3.3. Because this cation has a higher energy, the energy required to form it is larger, and the equilibrium constant for its formation is smaller. In other words, Eq. (1) has the equilibrium constant most favorable to the right.

3.55   According to text Fig. 3.3 (text p. 116), selective destabilization of the conjugate acid form of an acid-base pair *raises* the energy of the acidic form relative to its conjugate base and *reduces* its pK_a.

   Now let's apply this reasoning to the specific example in the problem. This analysis is similar to the one used in the solution to Problem 3.54. First, assume that the chlorine affects the stability of the *charged species,* which in this case is the conjugate *acid*:

bond dipole of the C—Cl bond
interacts repulsively with the
positive charge

$Cl—CH_2CH_2—\overset{+}{N}H_3$

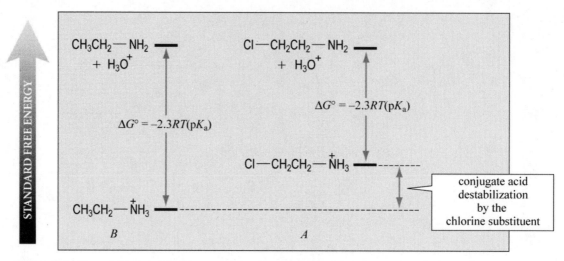

Figure SG3.4 The effect of chlorine substitution on the energy and pK_a of the conjugate acid of ethylamine to accompany the solution to Problem 3.55. Note that by raising the energy of the conjugate acid, chlorine substitution reduces the pK_a. (The two conjugate bases, which are uncharged, are arbitrarily placed at the same energy for comparison purposes.)

The repulsive interaction between the positive charge and the positive end of the C—Cl bond dipole selectively raises the energy of the conjugate acid. This should reduce the pK_a of the *conjugate acid* of A relative to the pK_a of the *conjugate acid* of B. This is shown diagrammatically in Fig SG3.4. This means that B is the more basic of the two conjugate bases. The actual data bear out this prediction: chlorine substitution reduces the pK_a of ethylammonium ion by almost two units:

Cl—CH$_2$CH$_2$—$\overset{+}{\text{N}}$H$_3$ CH$_3$CH$_2$—$\overset{+}{\text{N}}$H$_3$

2-chloroethylammonium ion   **ethylammonium ion**
conjugate acid of A conjugate acid of B
$pK_a = 8.8$ $pK_a = 10.6$

3.56   The equilibrium constant K_{eq} for the reaction of Eq. 3.25, text p. 104, is given by

$$K_{eq} = \frac{[BH][A^-]}{[AH][B^-]}$$

Following the hint, first show that this equals K_{AH}/K_{BH}. We do that by applying the definition of K_{AH} and K_{BH}.

$$K_{AH}\big/K_{BH} = \frac{[A^-][H_3O^+]}{[AH]} \times \frac{[BH]}{[H_3O^+][B^-]} = \frac{[BH][A^-]}{[AH][B^-]} = K_{eq}$$

or

$$K_{eq} = K_{AH}/K_{BH}$$

Then take logarithms of both sides of this equation and apply the definition of pK_a:

$$\log K_{eq} = \log K_{AH} - \log K_{BH} = pK_{BH} - pK_{AH}$$

Taking antilogs of both sides proves the assertion:

$$10^{(\log K_{eq})} = K_{eq} = 10^{(pK_{BH} - pK_{AH})}$$

3.57   The $\Delta G°$ for the reaction is found by taking the logarithm of the equilibrium constant and multiplying it by $-2.3RT$. From the solution to Problem 3.56,

$$\Delta G_{eq}^{\circ} = -2.3RT \log K_{eq} = -2.3RT(pK_{BH} - pK_{AH}) = \Delta G_{a,AH}^{\circ} - \Delta G_{a,BH}^{\circ}$$

3.58   (a)   Both HI and HCl are much stronger acids than H_3O^+. Consequently, the following equilibrium lies well to the right for both X = Cl and X = I:

$$HX + H_2O \rightleftharpoons H_3O^+ + X^-$$

Because the major acidic species in solution in each case is H_3O^+, and 10^{-3} M of this species is present in each case, both solutions have the same pH value of 3.

The most acidic species that can exist in a solvent is the conjugate acid of the solvent; in water, this is H_3O^+. Likewise, the most basic species that can exist in a solvent is the conjugate base of the solvent; in water this is ^-OH. Because acids or bases stronger than the solvent react to give the conjugate acid or base of the solvent, respectively, their greater acidity or basicity is not reflected in the pH of the resulting solution. This effect is sometimes termed the *leveling effect* of solvent.

(b)   The amide ion is strong enough to react completely with water, and this reaction results in a 10^{-3} M solution of hydroxide ion—that is, a solution pH = 11. This is identical to the solution made from 10^{-3} M hydroxide itself, except for the ammonia by-product, which reacts only slightly with water.

3.59   The first step in the reaction is shown below:

$$H-\overset{\overset{\textstyle H}{|}}{\underset{\underset{\textstyle H}{|}}{B}}-H \quad H-\ddot{O}H \longrightarrow H-B\overset{\textstyle H}{\underset{\textstyle H}{\big\langle}} + H-H + \ \ddot{:}\ddot{O}H^-$$

The second step involves donation of an electron pair from the nucleophile and Lewis base ^-OH to the electron-deficient electrophile and Lewis acid BH_3:

$$H-B\overset{\textstyle H}{\underset{\textstyle H}{\big\langle}} \quad ^-\ddot{:}\ddot{O}H \longrightarrow H-\overset{\overset{\textstyle H}{|}}{\underset{\underset{\textstyle H}{|}}{B}}-\ddot{O}H$$

3.60   To solve the various parts of this problem, recognize that astatine, At, is a halogen that is below iodine on the periodic table, and apply trends in the periodic table.

(a)   Because bond dissociation energy decreases down a column of the periodic table, the H—At bond should be weaker than the H—I bond. The bond dissociation energy of H—At is smaller.

(b)   Because electron affinities decrease down a column of the periodic table, the electron affinity of At should be smaller than that of I.

(c)   Because dissociation energies dominate the dissociation constant within a *column* of the periodic table, H—At should be a stronger acid than H—I.

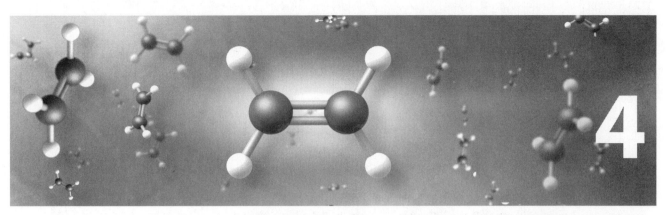

Introduction to Alkenes
Structure and Reactivity

STUDY GUIDE LINKS

 4.1 Different Ways to Draw the Same Structure

The discussion in the text requires you to realize that there are several different ways to draw the same alkene stereoisomer. For example, two equivalent ways to represent *cis*-2-butene are the following:

two ways of drawing *cis*-2- butene

If this is not clear, think of the structure as an object. Rotate the structure 180° about a horizontal axis through the C=C bond, as follows:

When we perform such a rotation, the atom labels, for example, CH_3, are maintained "right side up"; that is, they are not inverted.

Similarly, two equivalent ways of drawing *trans*-2-butene are as follows:

different ways of drawing *trans*-2-butene

Their equivalence can be demonstrated by a 180° rotation about a vertical axis.

The two structures in text Eq. 4.4 (Study Problem 4.1, text p. 133) are examples of identical structures drawn differently. Some students think that structures such as these are stereoisomers, but in fact they represent the same molecule. As these structures demonstrate, the ultimate test of identity of two structures is their congruence—the superimposability of each atom of one structure and an identical atom of the other.

4.2   Drawing Structures from Names

When asked to draw a complicated structure from a name, as in Problem 4.6(b), you should always strive to be *systematic*. The natural tendency is to try to write the finished structure immediately. Instead, you should take it one step at a time:

1. Write the carbon skeleton of the principal chain; do not be concerned with stereochemistry.
2. Add the substituents to the principal chain.
3. Add the hydrogens.
4. Decide on relative group priorities at the double bonds to which an *E* or *Z* configuration must be assigned. Write them on your structure.
5. Redraw your structure with proper stereochemistry. It helps to maintain proper 120° bond angles at the double bonds.

Let's illustrate with an example. Suppose you are asked to draw the structure of (2*E*,4*Z*)-3-isobutyl-2,4-hexadiene. Follow the above steps in order.

1. The principal chain contains six carbons with double bonds at carbons 2 and 4:

$$C—C{=}C—C{=}C—C$$

2. There is an isobutyl group at carbon 3. Add the hydrogens within this group, since you won't have to manipulate this group further.

$$C—C{=}C—C{=}C—C$$
$$|$$
$$CH_2CH(CH_3)_2$$
isobutyl group

3. Rewrite, adding the missing hydrogens.

$$H_3C—CH{=}C—CH{=}CH—CH_3$$
$$|$$
$$CH_2CH(CH_3)_2$$

4. Indicate the relative priorities of the groups at each double bond. For example, focusing on the C2–C3 double bond (carbons in boldface below), the relative priorities of the attached groups are as follows:

(You should complete the priorities for the C4–C5 double bond.)

5. Redraw the structure with proper stereochemistry.

(2E,4Z)-3-isobutyl-2,4-hexadiene

You should work Problem 4.6(b) using this step-by-step procedure.

In many cases, it is easier to work a complicated problem if it can be broken into smaller "chunks." When you learn to bypass the anxiety created by a complex problem and adopt this approach, you have made significant progress toward becoming a good problem solver.

4.3   Solving Structure Problems

A number of problems ask you to deduce structures that are initially unknown by piecing together chemical data. Problems 4.53 and 4.54 (text p. 178) are of this type. This study guide link illustrates a systematic approach to this type of problem by beginning the solution to Problem 4.53.

1. If the formulas of any of the unknowns are given, deduce all the information you can from the formulas. Begin with the unsaturation numbers (Sec. 4.3, text p. 144).

    In Problem 4.53, compound X has an unsaturation number $U = 2$, and compound Y has $U = 1$. This means that X has one ring and one double bond, or two double bonds. (It has to have at least one double bond, because it undergoes addition.) Compound Y results from addition of HBr, because the formula of Y is equal to that of X plus the elements of HBr.

2. Write all the information in the problem in equation form. This process gives you the entire problem at a glance.

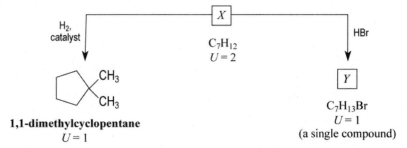

3. If there is a structure given explicitly anywhere in the problem, *even at the end of the problem*, deduce what you can from this structure.

    1,1-Dimethylcyclopentane is given explicitly in the problem as the catalytic hydrogenation product of X. Barring rearrangements, the structure provides the real key to solving the problem: it provides *the carbon skeleton of compound X.* It follows that one of the degrees of unsaturation of X must be a ring, and the other, as previously deduced, is a double bond. Thus, compound X must be a five-membered ring containing a double bond with two methyl groups on one carbon. The carbon bearing the methyl groups cannot be part of the double bond (why?).

4. Write out *all possibilities* and choose between (or among) them from the evidence presented, if possible.

This step is one that meets great resistance from many students for some reason. But it is essential. Suppose a detective arrests the first person he or she meets that fits the description of a burglar; chances are, the detective will make a lot of false arrests. Instead, he or she ideally should list *all* the suspects and decide which among them fit all the data. The only "suspects" for X are the following:

But what about the following structure?

It is the same as *X1*! Don't forget that most structures can be drawn in different ways.

To finish solving the problem, notice that compound Y is a *single* compound. Only one of the "suspects" above would give a *single compound* on addition of HBr; the other would give a mixture. What is the solution? Which compound is X?

FURTHER EXPLORATIONS

4.1   Relationship between Free Energy and Enthalpy

If you want to calculate an equilibrium constant, the *standard free energies* ($\Delta G°$ values) of the reactants and products are the relevant quantities. If you want to ask which of two molecules has a less energetic (or more stable) arrangement of bonds, then you need the *standard enthalpies of formation* ($\Delta H_f°$ values) of the two compounds. This section discusses why two different types of energy are needed for different purposes.

The discrepancy between the enthalpy change for any process (such as a reaction) and the free-energy change lies in something called the *entropy change* $\Delta S°$. The relationship between the free-energy change, the enthalpy change, and the entropy change is deceptively simple:

$$\Delta G ° = \Delta H° - T\Delta S° \qquad\qquad \text{(FE4.1)}$$

where T is the absolute temperature in kelvins.

This equation is essentially a statement of the division of energy into heat and work. $\Delta H°$ is the total energy change in a reaction (at constant temperature and pressure); $\Delta G°$ is the amount of work we could obtain from the reaction. A simple way to think about "chemical work" is charging a battery. If we could charge a battery and then harness it do some work—for example, by running an automobile (all with 100% efficiency)—then the $\Delta G°$ is the amount of work we could get from the reaction. The quantity $T\Delta S°$ is the amount of heat that would be produced when the reaction is harnessed with 100% efficiency to do work.

We can interpret the entropy change in more useful and descriptive terms than simply a discrepancy between total energy and work. Entropy measures molecular randomness, or *freedom of motion*. For example, consider the following reaction:

 $\longrightarrow$ $H_2C{=}CHCH_2CH_3$

cyclobutane   **1-butene**

The experimentally determined $\Delta H°$ for the reaction at 25 °C (or 298 K) is –26.53 kJ mol^{-1} (–6.34 kcal mol^{-1}), and the $\Delta G°$ for the reaction at 25 °C is –40.13 kJ mol^{-1} (–9.59 kcal mol^{-1}). Solving Eq. FE4.1, $T\Delta S°$ at T = 298 K is 13.60 kJ mol^{-1} (3.25 kcal mol^{-1}); that is, $\Delta S°$ is positive. A positive $\Delta S°$ means that there is more *randomness* or *freedom of motion* in the products than the reactants.

What is it about 1-butene that is more "random" than cyclobutane? 1-Butene has two carbon–carbon single bonds. Internal rotation can occur readily about these bonds, just as in butane. However, in cyclobutane, *internal rotations cannot occur* because the various carbon atoms are constrained into a small ring. (If this is not clear to you, make a model of cyclobutane and convince yourself that internal rotation about its carbon–carbon bonds cannot occur.) Thus, there is *greater freedom of motion* in 1-butene than in cyclobutane. Because of the internal rotations in 1-butene, the hydrogens on the carbons connected by single bonds can move through a greater volume of space than the same hydrogens in cyclobutane. Thus, *their positions are more random*. Hence, the randomness—or *entropy*—in 1-butene is greater than that in cyclobutane.

Because of entropy—the "randomness factor"—the energy that controls the position of a chemical equilibrium ($\Delta G°$) is different from the energy stored in chemical bonds ($\Delta H°$). Eq. SG4.1 shows that an increase of randomness in a reaction—a positive $\Delta S°$—gives the product an additional advantage in a chemical equilibrium *over and above* that which results from the formation of more stable bonds. In terms of the heat–work interpretation, we can get more work from this reaction than is available from bond changes, because heat is absorbed from the surroundings that goes toward the increased randomness of the product.

4.2   Sources of Heats of Formation

Heats of formation are not obtained by direct measurement. Rather, they are calculated from more readily available data by applying Hess's law. Two types of data that are commonly used are heats of combustion and heats of hydrogenation. To illustrate, suppose that we want to calculate the heat of formation of *trans*-2-butene from the following known heats of combustion: hydrogen, –241.8 kJ mol^{-1}; carbon, –393.5 kJ mol^{-1}; and *trans*-2-butene, –2530.0 kJ mol^{-1}. The combustion reaction (Sec. 2.7) is the reaction of each species with oxygen to give water (in the combustion of hydrogen), carbon dioxide (in the combustion of carbon), and both water and carbon dioxide (in the combustion of hydrocarbons).

Hess's law allows us to express the formation of *trans*-2-butene from its elements as the sum of three combustion reactions. Identical species on opposite sides of the equations cancel, and the resulting equations and their enthalpies are added algebraically:

Equations:		$\Delta H°$ (kJ mol^{-1})
$4\,H_2 \ + \ 2\!\!\!/\,O_2 \ \longrightarrow \ 4\!\!\!/\,H_2O$		$4(-241.8) = -967.2$
$4\,C \ + \ 4\,O_2 \ \longrightarrow \ 4\!\!\!/\,CO_2$		$4(-393.5) = -1574.0$
$4\!\!\!/\,CO_2 \ + \ 4\!\!\!/\,H_2O \ \longrightarrow \ 6\!\!\!/\,O_2 \ + \ trans\text{-2-butene}$		$+2529.6$
Sum: $\quad$ $4\,C \ + \ 4\,H_2 \ \longrightarrow \ trans\text{-2-butene}$		$\Delta H_f° = -11.6 \ \text{kJ mol}^{-1}$

Because four moles of both H_2 and C are required, their respective enthalpies of combustion must be multiplied by 4. Note also that the combustion of *trans*-2-butene must be written in reverse so that the formation equation comes out with *trans*-2-butene on the right. Consequently, the sign of the enthalpy of this combustion is also reversed.

As this example illustrates, heats of formation derived from combustion are in many cases small differences between large numbers and are therefore subject to considerable uncertainty. Another reaction that has been used for obtaining heats of formation is *catalytic hydrogenation.* In this reaction, hydrogen is added to an alkene double bond in the presence of a catalyst (Sec. 4.9A). For *trans*-2-butene, this reaction is as follows:

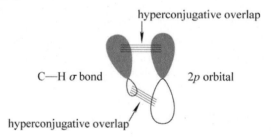

trans-**2-butene**

The overall $\Delta H°$ of this reaction can be measured with excellent precision, and is typically much smaller than a heat of combustion. (For example, the $\Delta H°$ of hydrogenation of *trans*-2-butene is -115.5 kJ mol^{-1}.) The heats of formation of simple alkanes are known with good precision. For example, the heat of formation of butane is -127.1 kJ mol^{-1}. The heat of formation of H_2 is zero by definition. Hence, it is possible to calculate the heat of formation of the alkene (try it!). See Problems 4.15 (text p. 149) and 4.17 (text p. 151) for calculations of heats of formation from combustion and hydrogenation data, respectively.

Fortunately, calculations like these are not necessary every time a heat of formation is needed, because such calculations have already been done to provide the data available in standard tables of heats of formation.

 ## 4.3   Molecular Orbital Description of Hyperconjugation

A molecular orbital treatment shows why hyperconjugation is a stabilizing effect. If you understand the molecular orbital treatment of π bonds (Sec. 4.1B, text pp. 128–130), you should have no difficulty with the treatment that follows.

We'll use as our starting orbitals the empty $2p$ orbital of the carbocation and the sp^2–$1s$ σ-bond orbital of the adjacent C—H bond. Both orbitals contain a node. When the axes of the two orbitals are coplanar, they can overlap, as shown by the overlap lines in the following diagram. This overlap *is* hyperconjugation.

To construct molecular orbitals from these two orbitals, imagine both bonding overlap and antibonding overlap, and construct an orbital interaction diagram as shown in Fig. FE4.3.1.

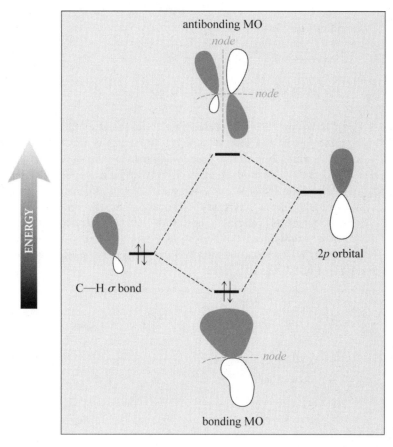

Figure FE4.1   An orbital interaction diagram showing a molecular-orbital interpretation of hyperconjugation. (The nodes are shown in the molecular orbitals only.)

Two electrons must be distributed between the two molecular orbitals. With opposite spin, both of these electrons can be accommodated in the bonding molecular orbital, as shown in the figure. Because this orbital has a lower energy than the original σ-bond orbital, molecular orbital formation (that is, hyperconjugation) lowers the energy of the system.

It is not strictly correct to use a bond orbital as a starting orbital for this treatment, because a bond orbital is not a proper molecular orbital of the carbocation. In fact, a full-fledged molecular-orbital treatment of the *tert*-butyl cation would start with all the carbons and hydrogens in the proper geometry and allow all their valence orbitals to interact. Adopting such an approach gives a more rigorous and complete molecular orbital picture, but, as noted in Sec 1.9A on text pp. 35–36, discrete bonds disappear and the descriptive value of the treatment is diminished. However, in such a full-fledged molecular-orbital approach, the overlap described here can be discerned as a component of the complete molecular orbitals.

4.4   A Stepwise View of Rearrangements

As noted in Sec 4.7, the methide or hydride shift is a one step rearrangement of a carbocation. However, it may be easier for you to understand this process by thinking of it as if it occurs in a stepwise fashion (it is important to note here that *this process is most certainly one step*, however, we are *imagining* it as if it were two steps).

If we reimagine the one-step reaction shown in Eq. 4.27a on p. 160 as a two step reaction, it would proceed as follows:

The CH_3 group "leaves", or detaches from, the main chain with the electron pair that was holding it to the rest of the ion. What's left behind is a dication (a "di-cation" which is exceptionally unstable, which is why this process is most certainly imaginary). The methide, or $^-:CH_3$, now has a choice: rebond with the carbon from which it came, or bond with the adjacent carbon. Since the original carbon looks like a tertiary carbocation, this carbon is more stable than the adjacent one, which looks like a secondary carbocation. So the methide ion donates its electron pair to the secondary carbon, yielding a more stable tertiary carbocation.

This stepwise method of thinking is especially useful when considering ring expansion reactions. When a carbocation can rearrange to one that is more hyperconjugated, has less ring strain, or both, it usually does.

secondary   tertiary
carbocation   carbocation

Again, it's best to write the mechanism in one step with one arrow showing the shift.

Remember that you should only consider a rearranged product as major if the rearranged carbocation intermediate is significantly more stable (through hyperconjugation, resonance, or ring strain relief) than its precursor.

4.5   Activation Energy

You may have learned in general chemistry about the *Arrhenius equation,* which treats rates in terms of an energy barrier called the *activation energy, E_a.*

$$k_{rate} = Ae^{-E_a/RT} = A(10^{-E_a/2.30RT}) \qquad \text{(FE4.2)}$$

In this equation, k_{rate} is the rate constant, which is the rate measured under standard conditions of 1 *M* concentration of all reactants and catalysts. (We'll discuss the rate constant in Chapter 9; but, for now, think of it as the rate.) The factor A is called the *preexponential factor.* The purpose of this Further Exploration is to relate the activation energy to the standard free energy of activation.

The Arrhenius equation originated as an empirical way to describe the variation of rate with temperature. The view was that the activation energy is an energy barrier, but its description in thermodynamic terms was not provided. Eyring theory, or transition-state theory, actually envisioned a barrier characterized by a structure—the transition state. The transition state can in principle be described in terms of its thermodynamic properties, such as free energy, enthalpy, and entropy.

As transition-state theory emerged as an important theoretical framework for discussing reaction rates, it became of interest to relate the variables in the Arrhenius equation to those in transition-state theory. It can be shown that the activation energy of the Arrhenius equation can be related to the *enthalpy* of activation:

$$E_a = \Delta H^{\circ\ddagger} + RT \tag{FE4.3}$$

The enthalpy of activation is the enthalpy of the transition state minus that of the starting materials. Because many typical organic reactions have $\Delta H^{\circ\ddagger}$ values that are substantially greater than RT (which equals about 2.5 kJ mol^{-1}), then, for these reactions,

$$E_a \approx \Delta H^{\circ\ddagger} \tag{FE4.4}$$

It can also be shown that the preexponential factor in the Arrhenius equation is related to the *entropy* of activation:

$$A = \frac{k_B T}{h} e^{\Delta S^{\circ\ddagger}/R} \tag{FE4.5}$$

where k_B is the Boltzmann constant (= 1.38×10^{-23} J K^{-1}=R/N, where N = Avogadro's number), T is the absolute temperature in kelvins, and h = Planck's constant = 6.63×10^{-34} J sec.

Because $\Delta G^{\circ\ddagger} = \Delta H^{\circ\ddagger} - T\Delta S^{\circ\ddagger}$, the standard free energy of activation can be calculated from the $\Delta H^{\circ\ddagger}$ and $T\Delta S^{\circ\ddagger}$ derived empirically from the Arrhenius equation. The entropy of activation $\Delta S^{\circ\ddagger}$ is a particularly interesting aspect of transition-state theory. (See the interpretation of entropy in Further Exploration 4.1.) The $\Delta S^{\circ\ddagger}$, like ΔS° in an equilibrium, can be interpreted in terms of molecular freedom, or randomness. This interpretation provides a "window" on the structure of the transition state.

REACTION REVIEW

I.   ADDITION REACTIONS OF ALKENES

A. ADDITION OF HYDROGEN HALIDES TO ALKENES

1. Hydrogen halides add to alkenes in a regioselective manner so that the hydrogen adds to the carbon of the double bond with fewer alkyl substituents and the halide group to the carbon of the double bond with more alkyl substituents (Markovnikov's rule).

2. The products of these reactions are called alkyl halides.

HBr + (cyclopentene)—CH₃ ⟶ (cyclopentane with H, Br, CH₃)

an alkyl halide

3. The addition of a hydrogen halide to an alkene is a regioselective reaction that occurs in two successive steps.

   a. In the first step, protonation of the alkene double bond occurs at the carbon with the fewer alkyl substituents so that the more stable carbocation is formed—the one with the greater number of alkyl substituents at the electron-deficient carbon.

   b. In the second step, the halide ion reacts at the electron-deficient carbon.

4. Hydrogen halide addition to an alkene is a regioselective reaction because addition involves the transition state that resembles the more stable of the two possible carbocation intermediates.

5. The formation of a carbocation from an alkene is a Brønsted acid–base reaction—that is, an electron-pair displacement reaction in which the π bond acts as a Brønsted base towards the Brønsted acid H—X.

6. In hydrogen halide addition to alkenes, rearranged products are formed if a rearrangement-prone carbocation intermediate is involved—for example, a secondary carbocation that could rearrange to a tertiary carbocation.

B. CATALYTIC HYDROGENATION OF ALKENES

1. Catalytic hydrogenation is an addition of hydrogen in the presence of a catalyst; it is one of the best ways to convert alkenes into alkanes.

   a. Many of the common hydrogenation catalysts are insoluble in the reaction solution; these are examples of heterogeneous catalysts.

   b. Benzene rings are inert to conditions under which normal double bonds react readily, but, under conditions of high temperature and pressure, they can be hydrogenated.

C. HYDRATION OF ALKENES

1. The addition of water to the alkene double bond, called alkene hydration, is an acid-catalyzed reaction.

   a. In many cases, the catalyzing acid is soluble in the reaction solution. Soluble catalysts are called homogeneous catalysts.

   b. Hydration is a regioselective reaction. A proton is added to the carbon of the double bond with fewer alkyl substituents, and the hydroxy (OH) group is added to the carbon of the double bond with more alkyl substituents.

2. Alkene hydration is a multistep reaction.

   a. In the first step of the reaction (the rate-limiting step), a Brønsted acid–base reaction, the double bond is protonated to give a carbocation.

   b. In the next step of the hydration reaction, the carbocation reacts with the Lewis base water in a Lewis acid–base association reaction.

   c. Finally, a proton is lost to solvent in another Brønsted acid–base reaction to give an alcohol product and regenerate the catalyzing acid.

3. In alkene hydration, rearranged hydration products are formed if a rearrangement-prone carbocation intermediate is involved—for example, a secondary carbocation that could rearrange to a tertiary carbocation.

4. Alkene hydration is a reversible reaction.

   a. Although hydration is used industrially, dehydration of alcohols to alkenes is the more common laboratory use of this reaction.

   b. The mechanism of alcohol dehydration is the reverse of the mechanism of alkene hydration (principle of microscopic reversibility).

SOLUTIONS TO PROBLEMS

Solutions to In-Text Problems

4.1   The order of increasing bond length is $d < a < c < e < b < f$. The two carbon-hydrogen bonds are shorter than any of the carbon-carbon bonds, so d and a are shorter than any of the others. The d bond is shortest because sp^2–$1s$ bond is slightly shorter than a sp^3–$1s$ bond, as in a. Bond c is longer, but the shortest carbon-carbon bond, because it is a double bond. Bond e is next longest because it is an sp^2–sp^2 sigma bond, followed by bond b because it is an sp^2–sp^3 sigma bond. Finally, bond f is the longest because it is an sp^3–sp^3 sigma bond.

4.2   (a)   1-Pentene cannot exist as stereoisomers.

   (b)   3-Hexene can exist as both cis and trans isomers; both carbons of the double bond (carbons 3 and 4) are stereocenters.

   (c)   1,3-Pentadiene can exist as cis and trans isomers at the bond between carbons 3 and 4; both of these carbons are stereocenters. (Cis-trans isomerism is *not* possible at the double bond between carbon-1 and carbon-2.)

stereoisomers of 1,3-pentadiene

   (d)   2-Methyl-2-pentene cannot exist as cis, trans isomers because two of the groups at one carbon of the double bond are the same (the methyl groups). This molecule contains no stereocenters.

   (e)   In principle, cyclobutene, which is a cis alkene, might also exist as a trans isomer. However, attempting to build a model of *trans*-cyclobutene should convince you that such an isomer is far too strained to exist. The carbons of the double bond are stereocenters.

4.3   (a)   (b)

2-methylpropene   **4-methyl-1,3-hexadiene**

   (c)   (d)

1-isopropenylcyclopentene   **5-(3-pentenyl)-1,3,6,8-decatetraene**

Some chemists were educated in nomenclature to understand that if the position of a substituent is unambiguous it need not be cited by number. Thus, compound (b) could be called methylpropene, because the methyl substituent *must* be on carbon-2. That is, if it is on carbon-1, the compound would not be named as a propene; it would be named as a butene. None of the formal "rule books" on nomenclature support this idea. Hence, we number *all* substituents. The 1993 nomenclature recommendations of the IUPAC went a step further and recommended numbering not only substituents, but also all double bonds, whether ambiguous or not; furthermore, the number is *always* cited before the suffix "ene." Thus, compound (b) in the 1993 system is called 2-methylprop-1-ene. (The 1993 recommendations are discussed in the sidebar on text p. 138.) A positive aspect of this recommendation is that it creates consistency in the way that all alkene names are constructed. However, this recommendation has not yet been generally accepted, and for that reason it has not been used in this text.

4.4   (a)   1,3-dimethylcyclopentene
        (b)   3-heptene
        (c)   5-allyl-2,6-nonadiene

 Remember, the 2-propenyl group is called the *allyl group;* see text page 137.

4.5   (a)   First, name the compound without considering stereochemistry: 3-isopropyl-2-hexene. Then, determine the stereochemistry at the double bond between carbons 2 and 3. The isopropyl group is higher priority than the *n*-propyl group. Since the isopropyl group is on the same side of the double bond as the higher priority methyl group, then compound's name is (*Z*)-3-isopropyl-2-hexene.

        (b)   First, draw out the compound completely, translating abbreviations like Me and *i*-Bu into appropriate skeletal forms. Next, name the compound without considering stereochemistry: 4-isobutyl-1,4-hexadiene. Then, determine the stereochemistry at the double bond between carbons 4 and 5. (*E,Z* isomerism is not possible at carbon-1). Start by writing the structure with the double bond between carbons 1 and 2 replicated, and decide on the path of highest priority within each group (shown in boldface in the following structure):

$$\text{higher priority at carbon-4} \left\{ \begin{array}{c} (C,H,H) \\ \mathbf{H_2C} - \mathbf{CH} - \mathbf{CH_2} \\ | \quad\quad | \quad\quad \backslash \\ C \quad\quad C \quad\quad \mathbf{C} = \mathbf{C} \\ / \quad\quad / \quad\quad \backslash \\ H_3C - CH - CH_2 \quad H \\ (H,H,H) \quad | \\ CH_3 \end{array} \right\} \text{ higher priority at carbon-5}$$

carbon-4 / higher priority at carbon-4

As the diagram above shows, the first point of difference occurs at the third carbon out; thus, the allyl group has higher priority than the isobutyl group. Because the groups of higher priority are on the same side of the double bond, the stereochemistry at this double bond is *Z*, and the full name is therefore (*Z*)-4-isobutyl-1,4-hexadiene. No number before the *Z* is necessary when only one double bond has a stereochemical designator.

4.6   (a)

(E)-4-allyl-1,5-octadiene

(b)

(2E,7Z)-5-[(E)-1-propenyl]-2,7-nonadiene

4.7   (a)

C(O,C,H)

$(CH_3)_3C$— versus H_3C—CH—

C(C,C,C) OCH_3

higher priority

(b)

CH_3CH_2 Cl_2HC

CH— versus   CH—

CH_3O HO

O(C)   O(H)

higher priority

(c)

C(C,H,H)   C(0,0,0)

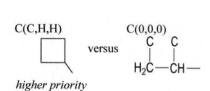

versus   C   C

H_2C—CH—

higher priority

(d)

C(H,H,H)   C(C,C,H)

↓

$(CH_3)_3C$— versus   C   C

HC—C—

C   C

higher priority

4.8   (a)   Use Eq. 4.6 on text p. 144 to obtain $U = 0$.
      (b)   Use Eq. 4.7 on text p. 144 to obtain $U = 3$.
      (c)   The prefix "cyclo" as well as the suffix "ane" indicate that the compound has one ring; consequently, $U = 1$.
      (d)   The suffix "triene" in the name indicates that the carbon has three double bonds; consequently, the unsaturation number is 3.

4.9   The unsaturation number is 4. Because there are no double bonds, the compound contains four rings. The stipulation of two methyl groups requires that there are 18 carbons in the four rings. The only way to have 18 carbons in four six-membered rings is for the rings to share carbon atoms. The following are three of many examples that meet the criteria:

4.10   (a)   A way to tell whether a formula is reasonable is to calculate its unsaturation number using Eq. 4.7 on text p. 144. A formula that gives a negative or fractional unsaturation number is not possible. Formula *A* cannot be a correct formula, because a compound containing only C, H, and an odd number of nitrogens must have an odd number of hydrogens. Formula *C* also could not be the formula of an organic compound, because the maximum number of hydrogens possible in a compound containing 10 carbons, 3 nitrogens, and any number of oxygens is 25. Formulas *B* ($C_{10}H_{20}N_2O_2$) and *D* ($C_{10}H_{16}O_2$) are both reasonable, and could be organic compounds.

 An early printing of the text had a formula *D* that was the same as formula *A* ($C_{10}H_{20}N_3$), which would make it unreasonable.

(b)   There are nine isomeric alcohols and one ketone with the formula C_4H_8O. The first three alcohols are called enols and, as you will learn in Chapter 14, they are in equilibrium with their corresponding ketone or aldehyde forms.

a ketone

alcohols

4.11   (a)   For simplicity, ignore the H—C bond dipoles. *cis*-2-Butene has the greater dipole moment because the bond dipoles of the H_3C—C bonds tend to add. In contrast, the dipole moment of *trans*-2-butene is zero because corresponding bonds are oriented in opposite directions, and their bond dipoles consequently cancel.

(b)   In propene, the H_3C—C bond dipole adds vectorially to the smaller H—C bond dipole; the resultant is smaller than the H_3C—C bond dipole itself, as the following diagram shows. In 2-methylpropene, the resultant of the two H_3C—C bond dipoles is about equal in magnitude to one H_3C—C bond dipole. (This can be shown analytically if we assume 120° bond angles.) 2-Methylpropene has a somewhat greater dipole moment. (In the following vector analysis, the H—C bond dipoles of the CH_2 groups are ignored because they cancel in the comparison of the two molecules.)

H$_3$C—C bond dipole

resultant (sum)

H—C bond dipole

propene

2-methylpropene

H$_3$C—C bond dipole

resultant (sum) of H$_3$C—C bond dipoles

4.12   As stated in the text above the problem, the polar effect of a double bond is electron-withdrawing. In Sec. 3.6C on text pp. 114–118, we learned that an electron-withdrawing polar effect increases the acidity of a nearby carboxylic acid group, and so it is in this case. (The pK_a of 3-butenoic acid is 4.37; that of butanoic acid is 4.78.)

4.13   (a)   The enthalpy of reaction is obtained by subtracting the ΔH_f° of the reactant, 1-butene, from that of the product, 2-methylpropene. Therefore, $\Delta H^\circ = -17.30 - (-0.30) = -17.0$ kJ mol^{-1} (-4.06 kcal mol^{-1}).

   (b)   The butene with the lower ΔH_f° is more stable. Since the conversion of 1-butene to 2-methylpropene is "energetically downhill"—that is, more negative—2-methylpropene is more stable.

4.14   (a)   The overall ΔH° of the process, $+15.3$ kJ mol^{-1}, must equal the difference in the ΔH_f° values of product and reactant. Hence,

$$+15.3 = \Delta H_f^\circ(\text{1-hexene}) - \Delta H_f^\circ(\text{2-ethyl-1-butene})$$
$$+15.3 = -40.5 - \Delta H_f^\circ(\text{2-ethyl-1-butene})$$

Solving, we have $\Delta H_f^\circ(\text{2-ethyl-1-butene}) = -55.8$ kJ mol^{-1}.

   (b)   Because the ΔH_f° of 2-ethyl-1-butene is more negative than that of 1-hexene, 2-ethyl-1-butene is the more stable isomer.

4.15   The combustion reaction for 1-heptene is

$$\text{H}_2\text{C}=\text{CHCH}_2\text{CH}_2\text{CH}_2\text{CH}_2\text{CH}_3 \ + \ 10.5\,\text{O}_2 \ \longrightarrow \ 7\,\text{CO}_2 \ + \ 7\,\text{H}_2\text{O}$$

The given ΔH° for this reaction is -4693.1 kJ mol^{-1}. This is equal to the heats of formation of products minus those of the reactants. The heats of formation of CO$_2$ and H$_2$O are the same as the heats of combustion of C and H$_2$, respectively, because these combustion reactions form CO$_2$ and H$_2$O from their elements.

$$-4693.1 \text{ kJ mol}^{-1} = 7\Delta H_f^\circ(\text{CO}_2) + 7\Delta H_f^\circ(\text{H}_2\text{O}) - \Delta H_f^\circ(\text{1-heptene}) - 10.5\Delta H_f^\circ(\text{O}_2)$$

Letting $\Delta H_f^\circ(\text{1-heptene}) = x$,

$$-4693.1 \text{ kJ mol}^{-1} = 7(-393.51) + 7(-285.83) - x - 10.5(0)$$

Solving,

$$\text{or } x = \Delta H_f^\circ(\text{1-heptene}) = -62.28 \text{ kJ mol}^{-1} \ (-14.89 \text{ kcal mol}^{-1})$$

    This calculation illustrates the point made in Further Exploration 4.1, that heats of formation derived from heats of combustion are small difference between large numbers and are therefore subject to some uncertainty.

4.16   (a)   Compound A is more stable than compound B because the double bond in A has one more alkyl substituent.

   (b)   Compound B is more stable than compound A, because compound A, which is a cis alkene, is destabilized by significant van der Waals repulsions between the *tert*-butyl group and the methyl group. Because compound B is a trans alkene, these destabilizing repulsions are absent.

4.17   Hydrogenation of either (E)- or (Z)-3-hexene gives the same product, hexane. This allows us to use the heats of hydrogenation in a Hess's law calculation:

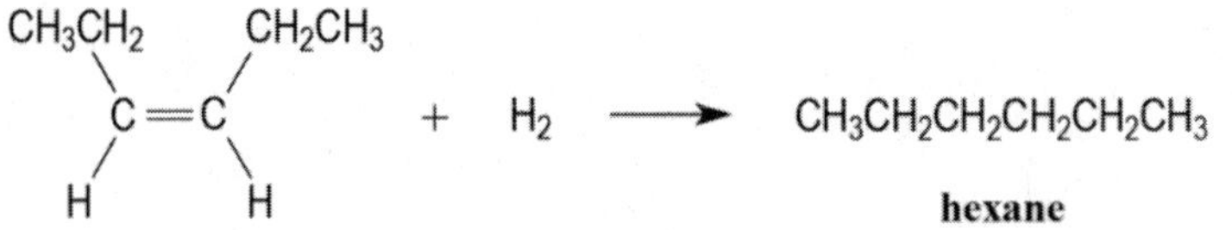

$$\Delta H^\circ = -121.6 = -167.2 - x$$

hexane

$$\Delta H_f^\circ = -167.2$$

(Z)-3-hexene

$$\Delta H_f^\circ = x$$

From this we obtain the ΔH_f° of (Z)-3-hexene as –45.6 kJ mol^{-1} (–10.9 kcal mol^{-1}). A similar calculation for (E)-3-hexene gives the ΔH_f° as –49.3 kJ mol^{-1} (–11.8 kcal mol^{-1}). Notice that the E isomer is more stable, as expected. (Why?)

4.18   (a) CH_3CH_2—I   (b)   Br—CH_2CH_2—Br   (c) $(CH_3CH_2)_3B$

4.19   In (a), the chlorine goes to the carbon of the alkene with the methyl branches. In (b), the bromine goes to the ring carbon with the methyl branch.

(a) $(CH_3)_3C$—Cl

(b)

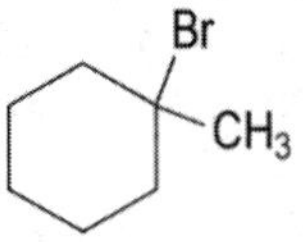

4.20   (a)

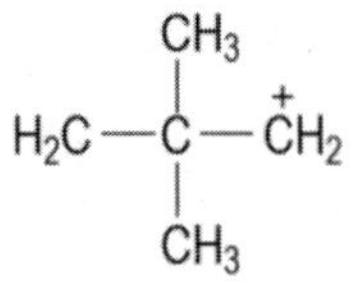

a primary carbocation
(least stable)

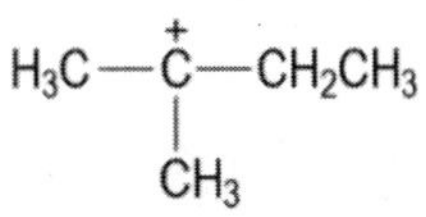

a tertiary carbocation
(most stable)

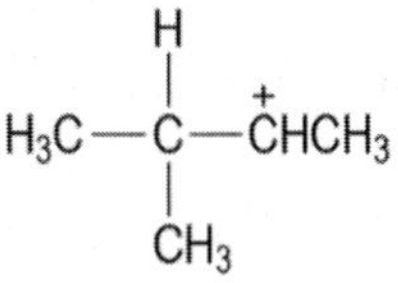

a secondary carbocation
(intermediate stability)

(b)

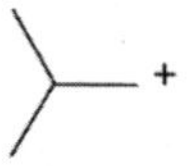

a primary carbocation
(least stable)

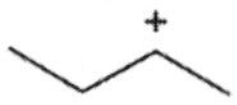

a secondary carbocation
(intermediate stability)

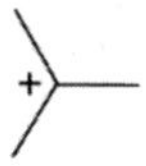

a tertiary carbocation
(most stable)

4.21   We follow the patterns in text Eqs. 4.18a and 4.18b. First the double bond is protonated on the carbon with *fewer alkyl substituents* by HBr to give a carbocation on the carbon with *more alkyl substituents* and Br$^-$. Note that even through a proton is transferred, we use the curved-arrow notation to show *the flow of electrons*. In the second step, the Br$^-$ reacts with the carbocation in a Lewis acid–base association reaction to give the product.

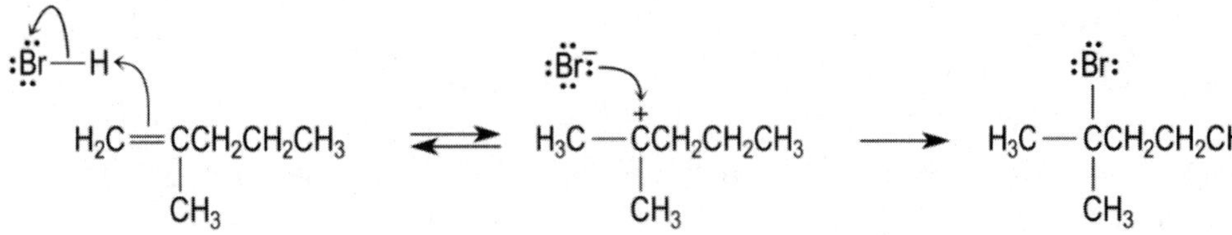

2-methyl-1-pentene   *a tertiary carbocation*

4.22   (a)

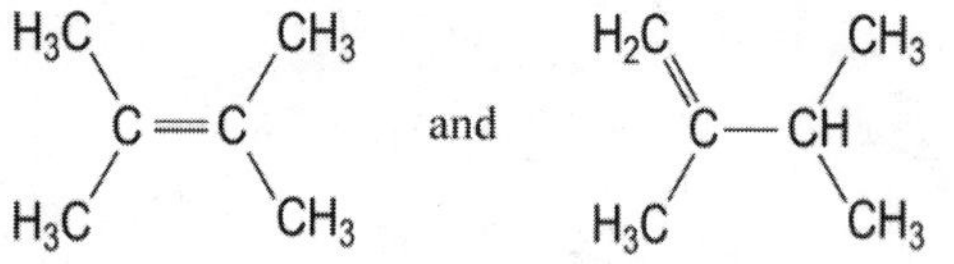

and

(b)

and

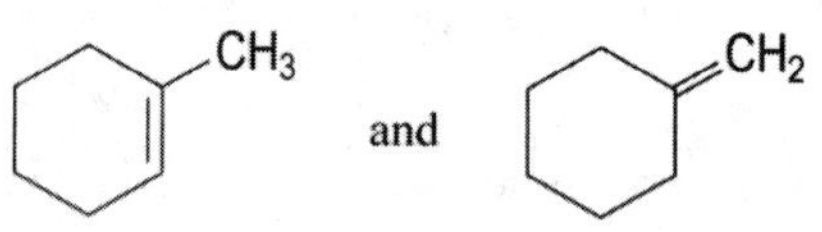

4.23   Rearrangement will occur when a carbocation can rearrange by the shift of an *adjacent* atom or group to give a more stable carbocation.

(a)   This secondary carbocation can rearrange to a tertiary carbocation by the shift of a hydride.

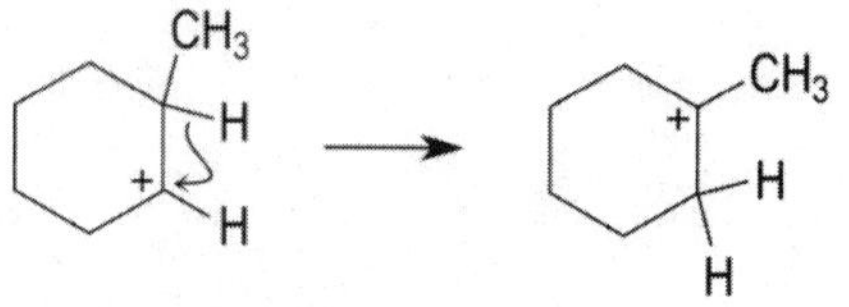

(b)   A rearrangement by the shift of a hydride from the adjacent tertiary carbon gives the same carbocation; hence, this carbocation does not rearrange.

(c)   Rearrangement of this secondary carbocation by a methide shift gives a tertiary carbocation. If you're still getting used to skeletal structures, redraw the structure using condensed or line bond structures.

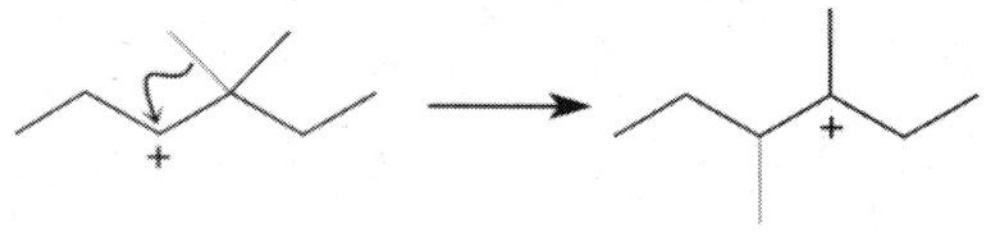

4.24   Protonation of the double bond gives a secondary carbocation:

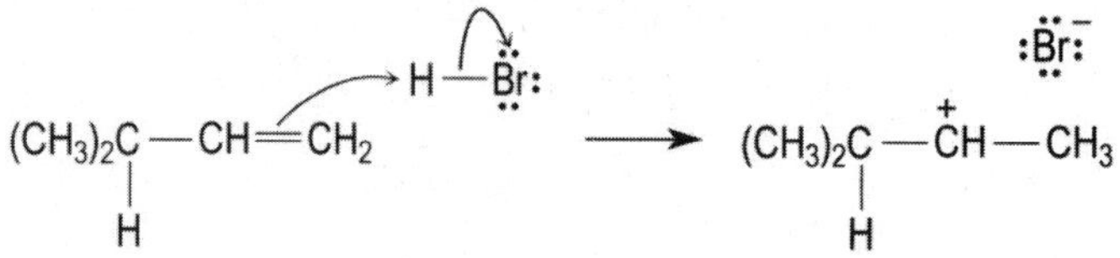

This secondary carbocation can react by two pathways: (a) rearrangement to a more stable tertiary carbocation followed by reaction of the rearranged carbocation with Br^-; (b) reaction with Br^-.

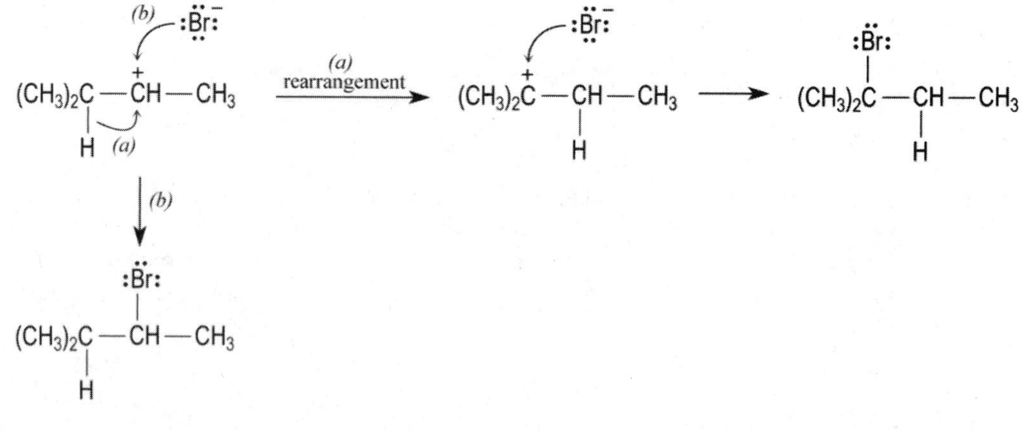

4.25   If these alkyl halides could be prepared from alkenes, the alkene starting materials would have to be *A1, B1,* and *C1* for *A, B,* and *C,* respectively:

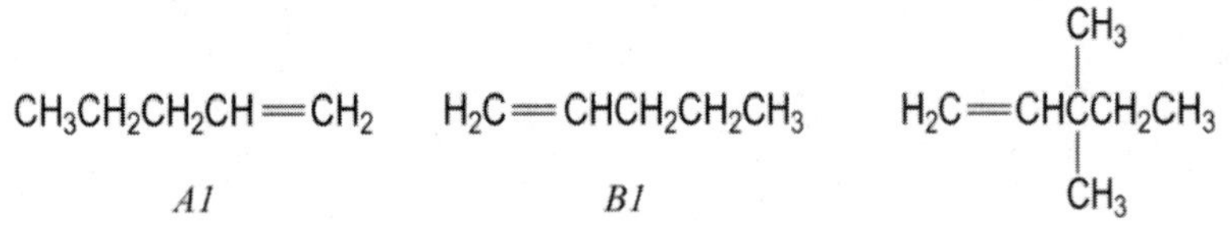

$$CH_3CH_2CH_2CH{=}CH_2 \qquad H_2C{=}CHCH_2CH_2CH_3 \qquad H_2C{=}CHCCH_2CH_3$$

with CH₃ substituents shown:

$$A1 \qquad\qquad B1 \qquad\qquad\qquad C1$$

(*B1* is the same as *A1;* it is drawn differently to correspond to the way the products are drawn.) Reaction of *A1* (or *B1*) with HBr would give *B* but not *A* because the bromine goes to the carbon of the double bond with the alkyl substituent. Hence, compound *B*, but not compound *A*, could be prepared by HBr addition to alkene *B1*. Addition of HBr to alkene *C1* would not give solely compound *C*, but would also give one or more rearrangement products. (Be sure to draw out the rearrangement and derive these products.)

4.26   (a)   This is an electron-pair displacement reaction.

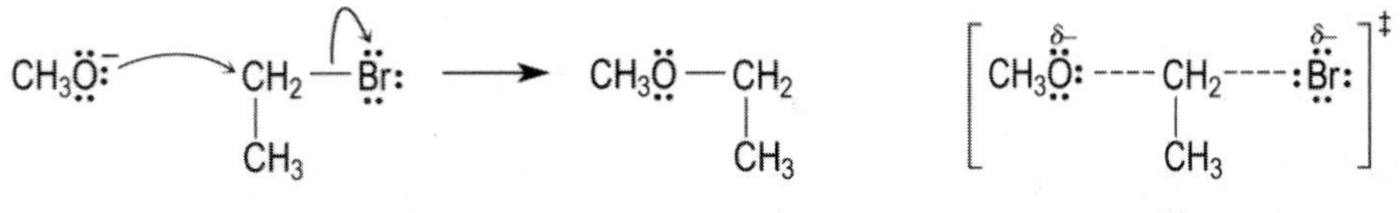

transition state

(b)   This is a Lewis acid–base dissociation reaction.

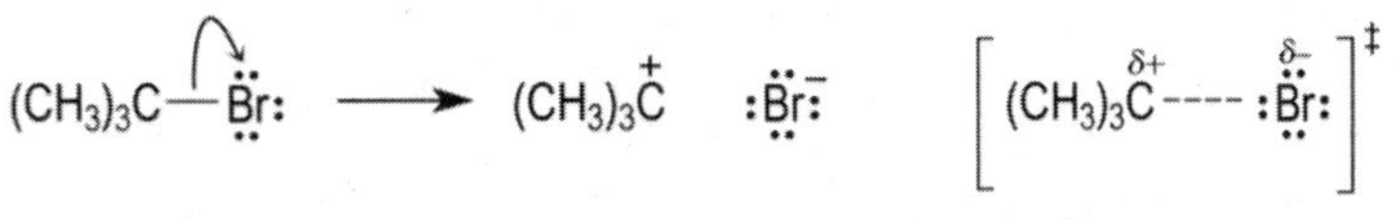

transition state

4.27   (a)   The curved-arrow notation and the transition state for the reverse of Eq. 4.30:

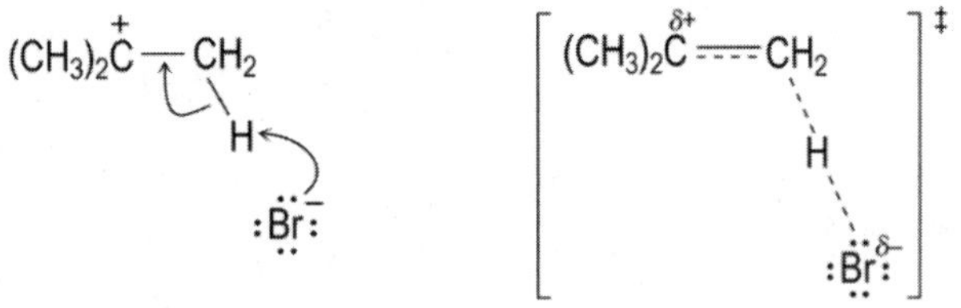

(b)   The transition state for any reaction and its reverse are the same.

4.28   (a)   Apply Eq. 4.33a on text p. 164 with $\Delta G_B^{\circ\ddagger} - \Delta G_A^{\circ\ddagger} = -15$ kJ mol^{-1}.

$$\log\!\left(\frac{\text{rate}_A}{\text{rate}_B}\right) = \frac{G_B^{\circ\ddagger} - G_A^{\circ\ddagger}}{2.30RT} = \frac{-15}{5.71} = -2.63$$

Therefore,

$$\frac{\text{rate}_A}{\text{rate}_B} = 10^{-2.63} = 2.36 \times 10^{-3}$$

Thus, the rate of reaction *A* is 0.0024 times that of reaction *B*, or reaction *B* is (1/0.0024) or 417 times as fast as reaction *A*.

(b)   The text states that a reaction rate approximately doubles for every 10° rise in temperature. The question is, then how many "doublings" in rate correspond to the rate factor 417 above?

$$417 \;=\; 2^n$$
$$\log 417 \;=\; n \log 2$$
$$2.63 \;=\; 0.301n$$
$$n \;=\; 8.74$$

Evidently, a rate factor of 417 corresponds to almost 9 doublings of the rate—8.74, to be exact. The temperature at which reaction A is run would have to be raised an estimated 87° to make it as fast as reaction B.

4.29   (a)   Apply Eq. 4.33b, with :

$$\log\!\left(\frac{\text{rate}_A}{\text{rate}_B}\right) = \log(10^{-6}) = -6 = \frac{G_B^{\circ\ddagger} - G_A^{\circ\ddagger}}{5.71} = \frac{G_B^{\circ\ddagger} - 90}{5.71}$$

Solving for ΔG_B°,

$$\Delta G_B^{\circ\ddagger} = 55.7 \text{ kJ mol}^{-1}$$

(b)   The reaction free-energy diagrams are shown in Fig. SG4.1.

(c)   The standard free energy of activation of a reverse reaction is the height of the energy barrier measured measured from the energy of the products. In each case, this is 10 kJ mol^{-1} greater than the standard free energy of activation of the forward reaction: 100 kJ mol^{-1} for reaction A, and 66 kJ mol^{-1} for reaction B. (See Fig. SG4.1.)

4.30   The reaction-free energy diagram shown in Fig. SG4.2 meets the criteria in the problem.

4.31   The diagram in Figure SG4.3 meets the criteria in the problem.

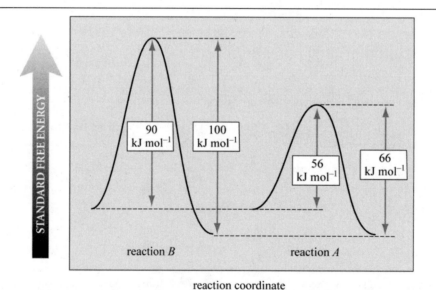

reaction coordinate

Figure SG4.1 A reaction free-energy diagram to accompany the solution to Problem 4.29b. The two curves are drawn to scale. The standard free energies of activation for the forward reactions are shown under their respective curves; the standard free energies of activation for the reverse reactions are shown to the right of their respective curves.

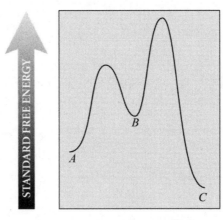

reaction coordinate

Figure SG4.2 A reaction free-energy diagram to accompany the solution to Problem 4.30.

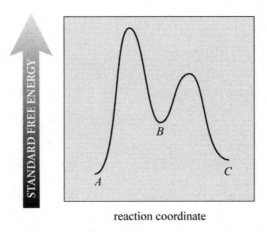

reaction coordinate

Figure SG4.3 The reaction-coordinate diagram to accompany the solution to Problem 4.31.

4.32   Because the transition states of the two reactions should resemble the respective carbocation intermediates (Hammond's postulate), examine the structures of the two carbocations and determine which should be more stable.

$$H_3C\overset{+}{\underset{\underset{\displaystyle CH_3}{|}}{C}}CH_3 \quad Br^-$$

carbocation intermediate
from HBr addition to
2-methylpropene

$$H_3C\overset{+}{-}CH-CH_2-CH_3 \quad Br^-$$

carbocation intermediate
from HBr addition to *trans*-2-butene

The carbocation intermediate in the addition of HBr to 2-methylpropene is tertiary, and is thus more stable than the carbocation intermediate in the addition of HBr to *trans*-2-butene, which is secondary. If the transition states for the two addition reactions resemble the carbocation intermediates, the transition state for the reaction of 2-methylpropene is also more stable. The relative rates of the two reactions are governed by the *differences* between the standard free energies of the transition states and their respective starting materials. Therefore, we have to consider the free energies of *both* the alkenes and the transition states. *If the two alkenes do not differ appreciably in energy, then the relative rates of the two reactions are governed only by the relative free energies of the two transition states.* (This is a fairly good assumption, because both alkenes have two alkyl branches.) Under this assumption, addition of HBr to 2-methylpropene is faster, because the transition state for this reaction has lower free energy.

4.33   (a)   The hydrogenation product of 1-pentene product is pentane, $CH_3CH_2CH_2CH_2CH_3$.
        (b)   The hydrogenation product of (E)-1,3-hexadiene is hexane, $CH_3CH_2CH_2CH_2CH_2CH_3$.

4.34   (a)   The formula restricts us to considering only alkenes with one double bond. Any alkene with a six-carbon unbranched carbon skeleton will give hexane as a product. Five alkenes meet this criterion.

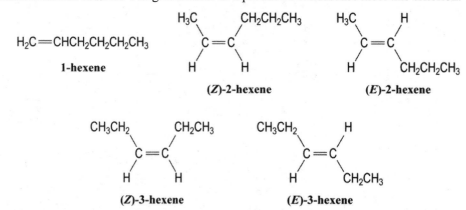

                        1-hexene   **(Z)-2-hexene**   **(E)-2-hexene**

                      (Z)-3-hexene   **(E)-3-hexene**

        (b)   Any alkene with the carbon skeleton of methylcyclopentane will undergo hydrogenation to give this cycloalkane. There are four such alkenes.

      1-methylcyclopentene   **3-methylcyclopentene**   **4-methylcyclopentene**   **methylenecyclopentane**

4.35   In the first step of the reaction, the double bond is protonated by the acid catalyst in a Brønsted acid–base reaction:

The resulting secondary carbocation *A* rearranges to a tertiary carbocation *B* by a hydride shift:

Carbocation *B* reacts with water in a Lewis acid–base association reaction, and the resulting adduct loses a proton to another water molecule in a Brønsted acid–base reaction to form the product alcohol and regenerate the catalyzing acid:

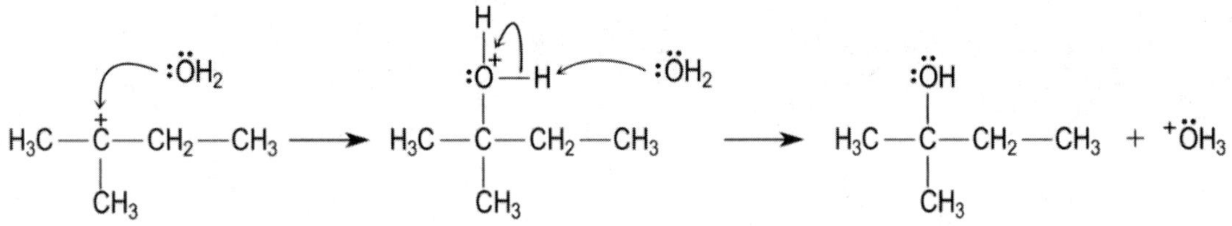

B

4.36 The reaction mechanism is exactly like the one in the solution to Problem 4.35, except that a methyl group migrates in the carbocation intermediate. The various intermediates are shown below; we leave it to you to fill in the other molecules or ions involved as well as the curved-arrow mechanism.

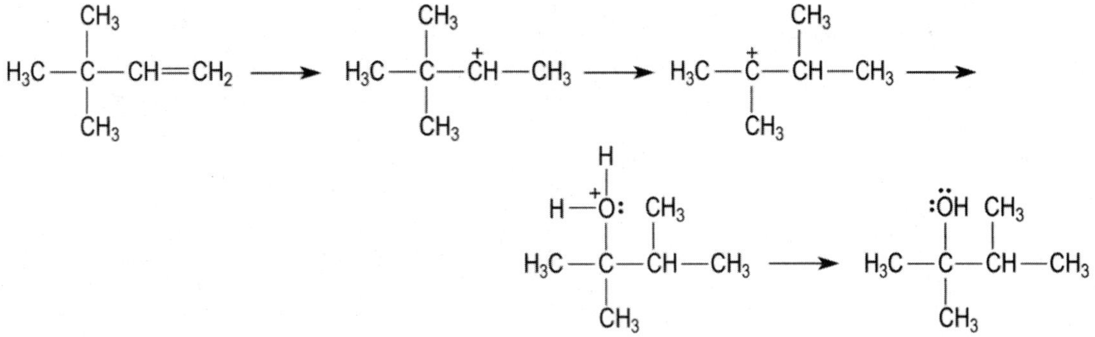

4.37 (a) The two alkenes result from loss of the two chemically nonequivalent protons on the carbons adjacent to the positively charged carbon in carbocation *B*:

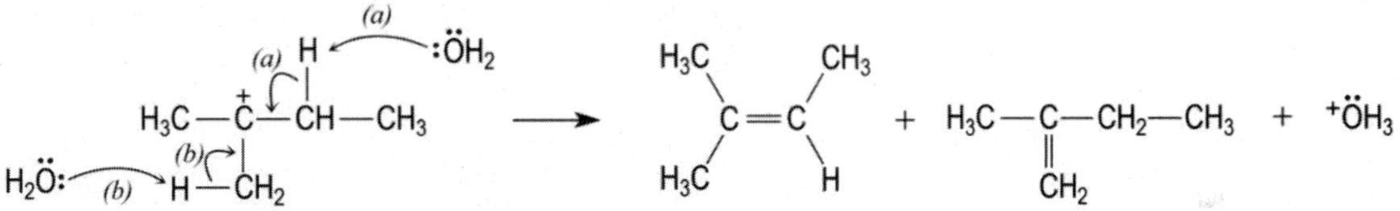

B   from *(a)*   from *(b)*

(b) In the solution to Problem 4.35, the rearrangement of carbocation *A* to carbocation *B* is completely irreversible because the much greater stability of *B*. Once this carbocation is formed, the only alkenes that can form are those that are derived from it—the ones shown in part (a).

4.38 The intermediate in the hydration of ethylene is a primary carbocation. Hence, the transition state resembles such a cation. As you have learned, primary carbocations are *much* less stable than secondary and tertiary carbocations; hence, transition states that resemble primary carbocations are also unstable. An unstable transition state means that the transition state has a very high energy. When a reaction has a transition state of very high energy, the reaction is slow.

4.39 The hydration of propene involves the usual mechanistic steps of alkene protonation to form a carbocation followed by the nucleophilic reaction of water with the carbocation. The structure of isopropyl alcohol follows from the structure of the carbocation. Notice that a secondary carbocation rather than a primary carbocation is involved. (Why?).

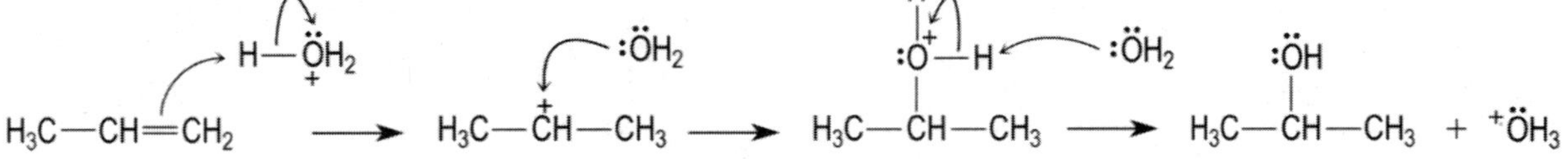

propene   **isopropyl alcohol**

Solutions to Additional Problems

4.40   The structures and substitutive names of the methylpentene isomers:

$$H_2C{=}CCH_2CH_2CH_3 \quad\quad H_2C{=}CHCHCH_2CH_3 \quad\quad H_2C{=}CHCH_2CHCH_3$$
$$\underset{CH_3}{|} \quad\quad\quad\quad \underset{CH_3}{|} \quad\quad\quad\quad \underset{CH_3}{|}$$

2-methyl-1-pentene   **3-methyl-1-pentene**   **4-methyl-1-pentene**

$$CH_3C{=}CHCH_2CH_3 \quad\quad CH_3CH{=}CCH_2CH_3 \quad\quad CH_3CH{=}CHCHCH_3$$
$$\underset{CH_3}{|} \quad\quad\quad\quad \underset{CH_3}{|} \quad\quad\quad\quad \underset{CH_3}{|}$$

2-methyl-2-pentene   **3-methyl-2-pentene**   **4-methyl-2-pentene**
                              E and *Z*   *E* and *Z*

4.41   The structures of the isomeric substituted butenes:

$$\underset{CH_3}{\overset{CH_3}{H_2C{=}CHCCH_3}} \quad\quad \underset{CH_3}{\overset{CH_3}{H_2C{=}CCHCH_3}} \quad\quad \underset{CH_2CH_3}{H_2C{=}CCHCH_3} \quad\quad \overset{H_3C}{\underset{H_3C}{}}C{=}C\overset{CH_3}{\underset{CH_3}{}}$$

3,3-dimethyl-1-butene   **2,3-dimethyl-1-butene**   **2-ethyl-1-butene**   **2,3-dimethyl-2-butene**

4.42   Each of the alkenes from the solution to Problem 4.41 *except* 3,3-dimethyl-1-butene should give one product when it reacts with HBr. 3,3-Dimethyl-1-butene should give one normal addition product and one rearranged product. (The corresponding addition of HCl to the same alkene is shown on text pp. 159–160.)

4.43   Arranging the alkenes in order of increasing heats of formation is the same as arranging them in order of decreasing stability. The major factors that govern alkene stability are (1) the number of alkyl substituents on the double bond; and whether the alkene is cis or trans. The order of increasing heats of formation:

$$\overset{H_3C}{\underset{H}{}}C{=}C\overset{CH_3}{\underset{CH_2CH_3}{}} \;<\; CH_3C{=}CHCH_2CH_3 \;\approx\; \overset{H_3C}{\underset{H}{}}C{=}C\overset{CH_2CH_3}{\underset{CH_3}{}} \;<\; H_2C{=}CCH_2CH_2CH_3 \;\approx\;$$
$$\underset{CH_3}{|} \qquad\qquad\qquad\qquad\qquad\qquad\qquad\qquad\qquad \underset{CH_3}{|}$$

(*E*)-3-methyl-2-pentene   2-methyl-2-pentene   (*Z*)-3-methyl-2-pentene   2-methyl-1-pentene

$$\overset{H_3C}{\underset{H}{}}C{=}C\overset{H}{\underset{CH(CH_3)_2}{}} \;<\; \overset{H_3C}{\underset{H}{}}C{=}C\overset{CH(CH_3)_2}{\underset{H}{}} \;<\; H_2C{=}CHCHCH_2CH_3 \;\approx\; H_2C{=}CHCH_2CHCH_3$$
$$\qquad\qquad\qquad\qquad\qquad\qquad\qquad\qquad\qquad \underset{CH_3}{|} \qquad\qquad\qquad \underset{CH_3}{|}$$

(*E*)-4-methyl-2-pentene   (*Z*)-4-methyl-2-pentene   3-methyl-1-pentene   4-methyl-1-pentene

2-Methyl-2-pentene should be somewhat less stable than (*E*)-3-methyl-2-pentene because the former has a cis methyl–ethyl interaction, whereas the latter has a cis methyl–methyl interaction. 2-Methyl-2-pentene has the same cis interactions as (*Z*)-3-methyl-2-pentene and the same number of substituents on the double bond.

4.44   (a)

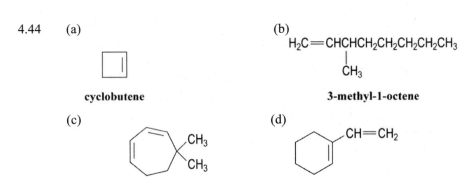

cyclobutene

(b)

$H_2C\!=\!CHCHCH_2CH_2CH_2CH_2CH_3$
|
CH_3

3-methyl-1-octene

(c)

5,5-dimethyl-1,3-cycloheptadiene

(d)

1-vinylcyclohexene

4.45   (a)   3-Ethylcyclopentene
(b)   6-Methyl-1-heptene
(c)   2-Isopropyl-1-methyl-1,3-cyclohexadiene
(d)   (*E*)-2-Butyl-1,3-pentadiene
(e)   1-(1-Cyclopentenyl)cyclopentene
(f)   (*E*)-5-[(*Z*)-1-propenyl]-1,6-nonadiene

4.46   (a)

$CH_3CH_2\!-\!CH\!=\!CH_2$

"3-butene"
1-butene

(b)

"*trans*-1-*tert*-butylpropene"
(*E*)-4,4-dimethyl-2-pentene

(c)

(Z)-2-hexene
(the correct name)

(d)

"6-methylcycloheptene"
4-methylcycloheptene

4.47   (a)   The stereochemistry is *E*. The Br has higher priority on one carbon stereocenter, but it has lower priority on the other.
(b)   *Z*; the decision is based on the preference for a heavy isotope over a lighter one in both cases.
(c)   The stereochemistry is *E*. The two branches of the ring are treated as separate substituents; the fact that the two branches are tied into a ring has no effect on their relative priorities.
(d)   The structure is drawn below with the double bond of the cyclobutenyl group in replicated form. Because groups of higher priority are on opposite sides of the double bond, this compound has the *E* configuration.

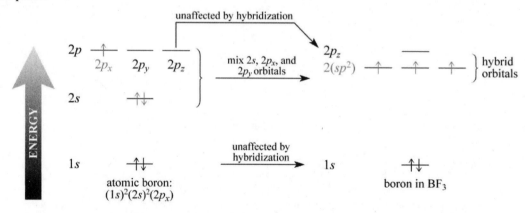

4.48   (a)   These compounds are constitutional isomers (C) because they have different connectivities.

         (b)   These are not isomers because they have different molecular formulas (N).

         (c)   These structures represent the same compound (I).

         (d)   These compounds are constitutional isomers (C) because they have different connectivities.

         (e)   These compounds are stereoisomers (S)—specifically, E and Z isomers of the same compound.

4.49   Boron trifluoride (trifluoroborane, BF_3) is trigonal planar with F—B—F bond angles of 120°. Because hybridization and geometry are related, and we know that other trigonal-planar atoms (e.g., carbon of alkenes or carbocations) are sp^2-hybridized, it is reasonable to suppose that boron of BF_3 is also sp^2-hybridized. An orbital diagram for boron is much like the one for carbon in Fig. 4.3 on p. 127 of the text, except that boron has three rather than four valence electrons. In this diagram, the valence electrons are shown in gray. To form BF_3, each of the sp^2 orbitals in boron, which contain a single electron, overlap with an orbital of a fluorine atom that contains a single electron. This makes three B—F electron-pair bonds. Because the boron orbitals used to form the bonds are oriented at 120° in a common plane, then the three B—F bonds are also oriented at 120° in a common plane.

The boron in BF_3 has an empty $2p$ orbital, shown as $2p_z$ in the diagram.

4.50   (a)   An $sp^3(C)$–$1s(H)$ σ bond

         (b)   An $sp^3(C)$–$sp^2(C)$ σ bond

         (c)   An $sp^2(C)$–$sp^2(C)$ σ bond and a carbon–carbon π bond

         (d)   An $sp^2(C)$–$sp^2(C)$ σ bond

         (e)   An $sp^2(C)$–$1s(H)$ σ bond

4.51 (a) Either of the following two alkenes would react with HBr to give the alkyl halide product shown in the problem:

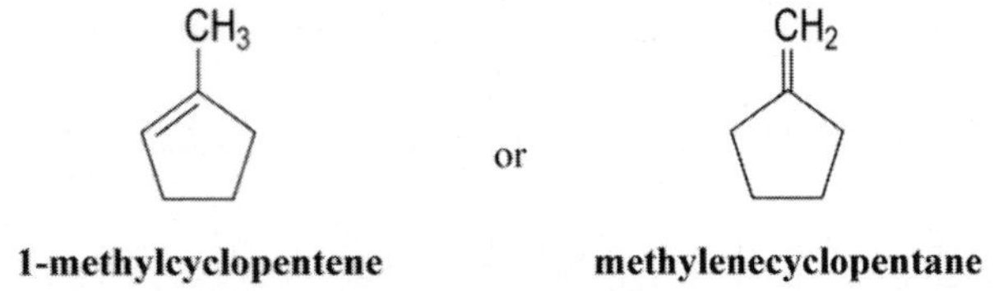
CH3
or
CH2
1-methylcyclopentene
methylenecyclopentane

(b) Although the bromine is on the same carbon in the products from the respective alkenes, the hydrogen from HBr (and hence the deuterium from DBr) is not. Consequently, different products are obtained from DBr addition to the two alkenes. In other words, DBr differentiates the proton that is added from the hydrogens that were in the alkene to begin with.

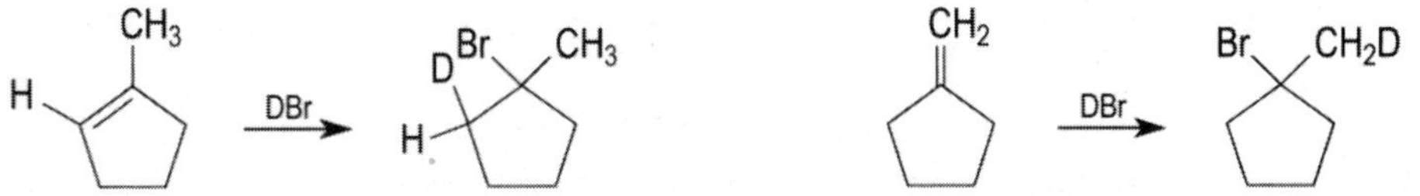
CH3
H
DBr
D
Br
CH3
H
CH2
DBr
Br
CH2D

4.52 Any one of the alkenes below, when reacted with hydrogen in the presence of a catalyst, will give propylcyclohexane.

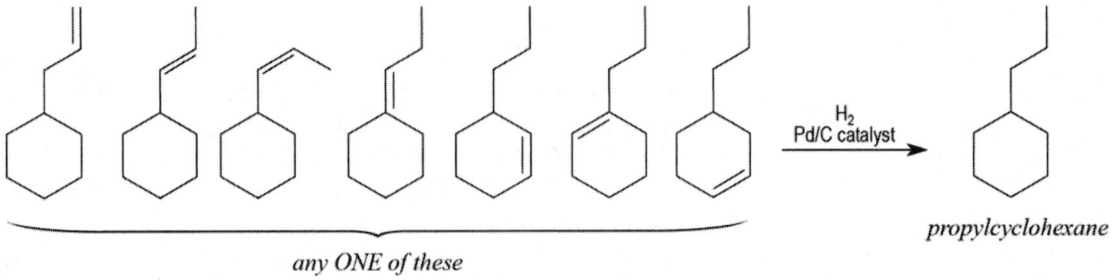
H2
Pd/C catalyst
any ONE of these
propylcyclohexane

4.53 The process to be used in working this problem is thoroughly explained in Study Guide Link 4.3 on p. 73 of this manual. The structures of *X* and *Y* are as follows:

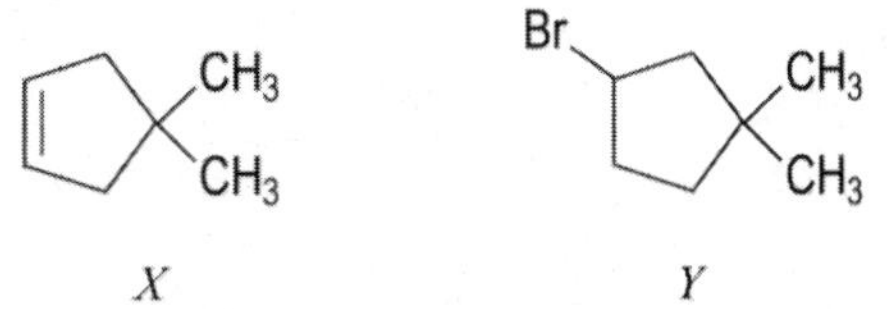
CH3
CH3
X
Br
CH3
CH3
Y

4.54 Both *cis*- and *trans*-3-hexene should give the same *single* alkyl halide product when they react with HI, and both would undergo catalytic hydrogenation to give hexane:

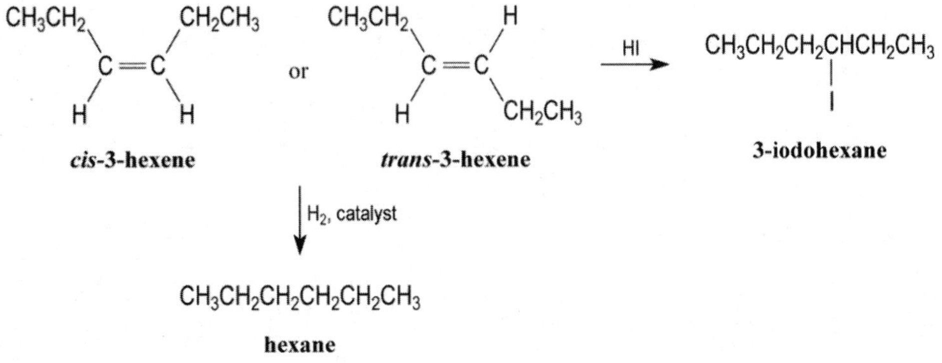
CH3CH2
CH2CH3
C=C
H
H
cis-3-hexene
or
CH3CH2
H
C=C
H
CH2CH3
trans-3-hexene
HI
CH3CH2CH2CHCH2CH3
I
3-iodohexane
H2, catalyst
CH3CH2CH2CH2CH2CH3
hexane

4.55   Hydration, because of its mechanism, yields an alcohol in which the —OH group is positioned at the carbon of the double bond with *more alkyl substituents*. Al Keyhall's proposal would require a different regioselectivity, and therefore would not work. (As you'll learn in Sec. 5.4B, there is a different way to prepare this alcohol from 3-methyl-1-butene.)

Al Keyhall's proposal

$$H_2O \ + \ (CH_3)_2CHCH{=}CH_2 \ \xrightarrow[]{\overset{H_3O^+}{\times}} \ (CH_3)_2CHCH_2CH_2OH$$

$$\downarrow H_3O^+$$

$$\underset{(CH_3)_2CHCHCH_3}{\overset{OH}{|}} \ + \ \underset{(CH_3)_2CCH_2CH_3}{\overset{OH}{|}}$$

observed products

(For the origin of the rearrangement product, see the solution to Problem 4.35.)

4.56   (a)   Before drawing the reaction-free energy diagram, convert K_{eq} into a free energy. (Assume 25 °C or 298 K.)

$$\Delta G° = -2.3RT \log K_{eq} = -5.71 \log(150) = -12.4 \text{ kJ mol}^{-1} \ (-2.97 \text{ kcal mol}^{-1})$$

This number is the energy difference between A and B. Compound B has the lower energy, because it is favored in the equilibrium. The reaction free-energy diagram is shown in Fig. SG4.4.

(b)   The standard free energy of activation for the reaction in the B to A direction is 12.4 kJ/mol + 96 kJ/mol = 108.4 kJ mol^{-1} (25.9 kcal mol^{-1}). This follows geometrically from the diagram in part (a); it is the difference between the standard free energy of B and the standard free energy of the transition state.

4.57   (a)   Compound D is present in greatest amount when the reaction comes to equilibrium, because it has the lowest standard free energy. Compound C is present in least amount because it has the highest standard free energy.

(b)   The rate-limiting step in this reaction is the $B \rightleftharpoons C$ step, because it has the transition state of highest standard free energy.

(c)   The standard free energy of activation is the energy difference between the transition state of highest free energy and the reactant. See Fig. SG4.5 for the $A \rightarrow D$ case. (What is $\Delta G°^{\ddagger}$ when the reactant is D, that is, in the case of the reverse reaction?)

(d)   The reaction of $C \rightarrow D$ is faster than the reaction $C \rightarrow B$ because the energy barrier for the $C \rightarrow D$ reaction is smaller. The two barriers are labeled in Fig. SG4.5.

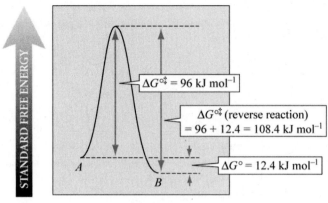

Figure SG4.4 A reaction free-energy diagram to accompany the solution to Problem 4.56.

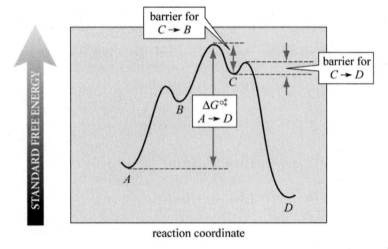

Figure SG4.5 Reaction free-energy diagrams to accompany the solutions to Problem 4.57c and 4.57d.

4.58   The first step of hydration, protonation of the double bond, is rate-limiting (text p. 173), and the reactive intermediate that results from this process is a carbocation. According to Hammond's postulate, the transition state closely resembles this intermediate.

carbocation intermediate
from protonation of 1-methylcyclohexene

4.59   (a)   The product from hydration of methylenecyclobutane:

methylenecyclobutane   **1-methylcyclobutanol**
                                            (compound X)

(b)   The reactive intermediate formed in the rate-limiting step is the carbocation:

(c)   The transition state of the rate-limiting step is the transition state for protonation of the double bond:

(d)   If the rate-limiting step is protonation of the double bond in the hydration (forward) direction, then the rate-limiting step in the reverse direction is the reverse of the same step: deprotonation of the carbocation intermediate to give the alkene. The transition state is identical to the one in part (c), because a reaction and its reverse have the same transition state.

4.60   (a)   The bonding arrangement in (*E*)-1,3-pentadiene is more stable because this diene has the smaller heat of formation.

 When two double bonds are separated by one single bond, the double bonds are said to be *conjugated*. The conjugated arrangement of double bonds is a particularly stable arrangement. The reason for this enhanced stability is discussed in Chapter 15.

(*E*)-1,3-pentadiene
(a conjugated diene)

$H_2C=CHCH_2CH=CH_2$
1,4-pentadiene

(b)   The relevant combustion equation is

$$(E)\text{-1,3-pentadiene} + 7\,O_2 \longrightarrow 5\,CO_2 + 4\,H_2O$$
$$(C_5H_8)$$

| heat of formation (kJ mol^{-1}) | 75.8 | 0 | 5(−393.5) | 4(−285.8) |

(Note that the heats of formation of CO_2 and H_2O are by definition identical to the heats of combustion of carbon and hydrogen, respectively.) The $\Delta H°$ for this reaction is

$$4(-285.8) + 5(-393.5) - 75.8 = -3186.5 \text{ kJ mol}^{-1}$$

The heat of combustion (always expressed as a positive number for heat liberated) is therefore 3186.5 kJ mol^{-1} (761.6 kcal mol^{-1}).

4.61   (a)   The enthalpies of the three alkenes relative to 2-methylbutane is diagrammed in Fig. SG4.6.

(b)   Since all three alkenes produce the same alkane (2-methylbutane) upon catalytic hydrogenation, we can use the heat liberated in each reaction to determine their relative stabilities. Since 2-methyl-2-butene gives off the least amount of heat (−112.6), it is the most stable. 3-Methyl-1-butene gives off the most heat, so it is the least stable.

(c)   Because 3-methyl-1-butene, 2-methyl-1-butene, and 2-methyl-2-butene are isomers, they are formed from the same elements (C_5H_{10}). The difference in heats of hydrogenation must equal the difference in their heats of formation. We don't know what the individual heats of formation are, but we know that they differ by between 14.2 and 6.6 kJ mol^{-1}.

(d)   As discussed in section 4.5B, the more alkyl groups that are attached to the carbons of the double bond, the more stable the alkene. This phenomenon arises from the greater number of sp^2–sp^3 bonds present in the more stable alkenes.

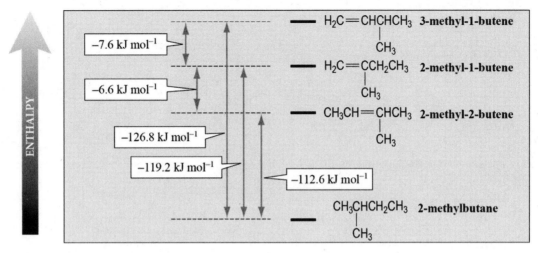

Figure SG4.6 A diagram of relative enthalpies (not to scale) to accompany the solution to Problem 4.61.

4.62   When we make the models as requested we find that it is very difficult to close the ring containing the double bond with the *E* (or trans) configuration, whereas the ring containing the *Z* (or cis) double bond can be closed without difficulty. (If this conclusion is not clear, be sure you built a model of the *cyclic* alkenes and not a pair of the acyclic heptenes.) Thus, (*E*)-cycloheptene is less stable than (*Z*)-cycloheptene. Because the more stable isomer of a pair has the lower heat of formation, it follows that (*Z*)-cycloheptene has the lower heat of formation.

4.63   (a)   The data show that methyl groups are more electron-donating than hydrogens. That is, the H_3C—CA bond dipole is larger than the H—CA bond dipole. This follows because each H_3C—C bond dipole in the first structure, and each H—C bond dipole in the second structure, is parallel to a C—Cl bond dipole within the same structure. As shown in the following vector diagram, in which R = CH_3 or R = H, the parallel dipoles reinforce to give a larger resultant **A**; and two such resultants add to give a resultant **B** for the entire molecule.

It follows from this diagram that the larger is the R—C bond dipole, the larger is the resultant dipole moment of the molecule. Because the compound containing the methyl groups has the larger dipole moment, it follows that the H_3C—C bond dipole is larger than the H—C bond dipole. In other words, a methyl group is more electron donating than a hydrogen towards a double bond.

(b)   Assume that the major contributors to the molecular dipole moment are the H_3C—C and the C—Cl bond dipoles, as indicated in part (a). These bond dipoles reinforce in the *E* isomer, and are nearly opposed in the *Z* isomer. Consequently, the *E* isomer has the greater dipole moment.

4.64   The mechanism is a sequence of two Brønsted acid–base reactions. In the first (step *a*), the double bond is protonated to give a tertiary carbocation; in the second (step *b*), the carbocation loses a different proton to give the alkene. The equilibrium is driven to the right by the greater stability of the alkene product. (Why is it more stable?)

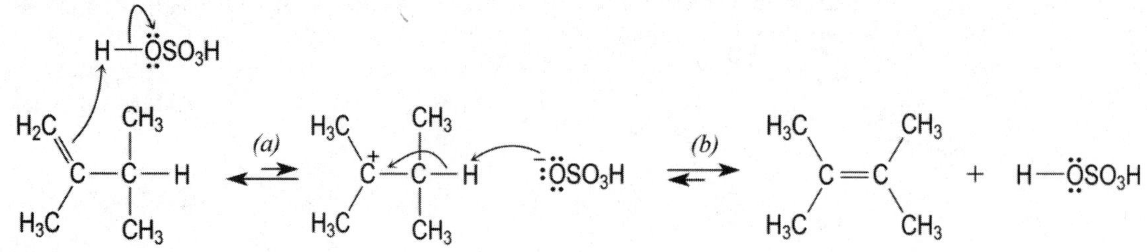

4.65 Follow the suggestions in the problem. First, we label corresponding atoms.

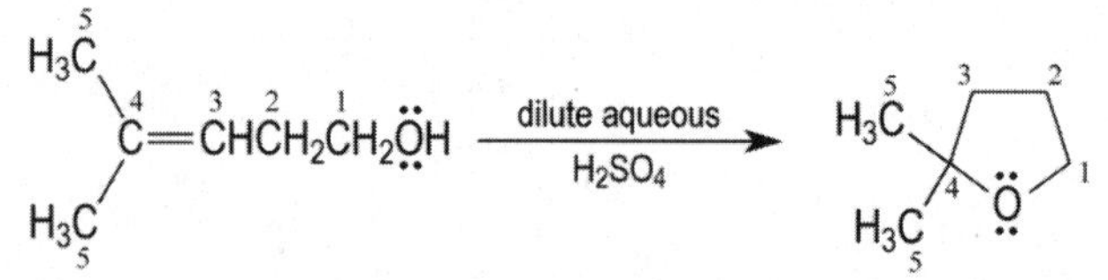

Next, consider the fate of the various functional groups. The double bond has disappeared. Carbon-3 has gained a proton, and carbon-4 has been joined to the oxygen. In other words, an *addition* of the O—H group to the double bond has occurred. We have seen two other additions involving acids: addition of HBr, and acid-catalyzed hydration. Because a catalyzing acid is present in this reaction, and an oxygen is involved in the addition, it is reasonable to think of acid-catalyzed hydration as a possible model for this reaction. Carbon-3 can gain a proton if the double bond reacts as a Brønsted base with the Brønsted acid H_3O^+ (step *a* in the mechanism below). This forms a tertiary carbocation, which then reacts in a Lewis acid–base association (step *b*). Loss of a proton to solvent H_2O in another Brønsted acid–base reaction (step *c*) regenerates the catalyzing acid and forms the product.

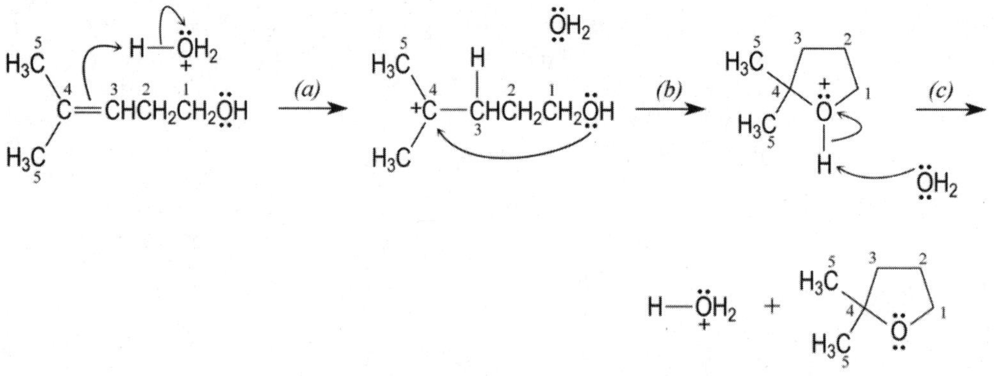

4.66 This is an acid-catalyzed addition to the double bond much like hydration, except that an alcohol rather than water is the nucleophile that reacts with the carbocation intermediate. Notice that the ionized form of H_2SO_4 in methanol is protonated methanol ($CH_3O^+H_2$).

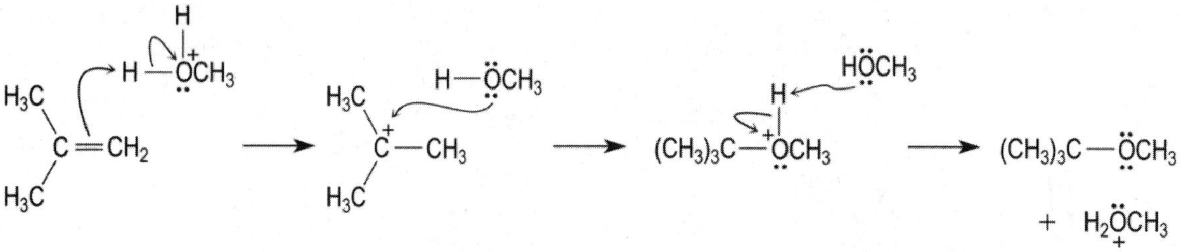

4.67 Protonation occurs at carbon-5 (see numbered structure below) to form a tertiary carbocation. (Protonation at carbon-1 would give a secondary carbocation.) Because the tertiary carbocation is more stable, it is formed more rapidly (Hammond's postulate). The tertiary carbocation resulting from protonation then adds to the other double bond (the only choice) to give a secondary carbocation (step *b*). This step can be viewed as a Lewis acid–base association reaction in which the π electrons of the double bond act as a nucleophile. The result is a secondary carbocation. A Lewis acid–base reaction of this carbocation with water (step *c*) followed by loss of a proton to another solvent water molecule (step *d*, another Brønsted acid–base reaction) gives the product.

Why. in step (*b*), would a tertiary carbocation be converted into a secondary one? Because in this particular process a π bond is also replaced by a σ bond. A σ bond is *much* stronger than a π bond—strong enough to compensate for the formation of a secondary carbocation. In other words, the energy reduction from σ-bond formation is enough to offset the energy increase from formation of a secondary carbocation.

4.68　(a)　We refer to the alkene on the left, 2,3-dimethyl-1-butene, as *A*, and the alkene on the right, 2,3-dimethyl-2-butene, as *B*. Compound *B* is favored at equilibrium because it has the lower ΔG_f°. (This alkene is more stable because it has more alkyl substituents on its double bond.) The equilibrium constant is obtained from Eq. 3.36b on text p. 110. Define K_{eq} as $[B]/[A]$:

$$K_{eq} = 10^{-\Delta G^{\circ}/2.30RT} = 10^{-(-3.10/5.71)} = 10^{0.543} = 3.49$$

(b)　The equilibrium constant tells us *nothing* about how rapidly this interconversion takes place. For example, in the absence of a catalyst, the rate of this interconversion is essentially zero. However, this reaction does occur at a convenient rate in the presence of an acid catalyst, as suggested by Problem 4.64.

4.69　(a)　First draw the structures of the two compounds:

$$H_2C{=}CHCH_2CH_3 \qquad H_2C{=}C{\overset{\displaystyle CH_3}{\underset{\displaystyle CH_3}{\big<}}}$$

1-butene
A

2-methylpropene
B

2-Methylpropene has two alkyl substituents on its double bond, whereas 1-butene has only one. Because alkyl substituents on the double bond stabilize an alkene, 2-methylpropene is the more stable of the two alkenes.

(b)　The reaction with the smaller standard free energy of activation ($\Delta G^{\circ\ddagger}$) is faster. Consequently, hydration of 2-methylpropene is faster.

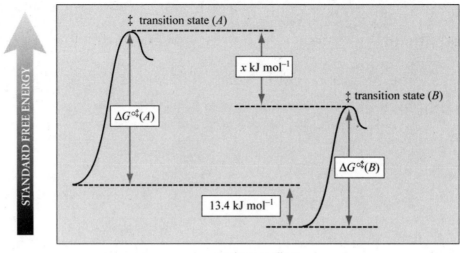

Figure SG4.7   Reaction free-energy diagrams to accompany the solution to Problem 4.69.

(c)   Let 1-butene = A, and let 2-methylpropene = B. The reaction-free energy diagrams for hydration are shown in Fig. SG4.7. (Only the first, rate-limiting, step of the mechanism is shown in each diagram.)

(d)   Let x represent the difference in transition-state free energies, as shown in Fig. SG4.7. The diagrams in this figure require the following equality:

$$13.4 + \Delta G^{\circ\ddagger}(A) = x + \Delta G^{\circ\ddagger}(B)$$

Rearranging, and recognizing that the difference between the standard free energies of activation (given in the problem) is 22.8 kJ mol^{-1}, we have

$$\Delta G^{\circ\ddagger}(A) - \Delta G^{\circ\ddagger}(B) = 22.8 = x - 13.4$$

Solving for x gives

$$x = 22.8 + 13.4 = 36.2 \text{ kJ mol}^{-1}$$

Thus, the transition state for hydration of 1-butene is 36.2 kJ mol^{-1} *less stable* than the transition state for hydration of 2-methylpropene. From Sec. 4.9B, the mechanism of hydration involves a carbocation intermediate, and the first step, protonation of the double bond to form this intermediate, is rate-limiting. In the case of 2-methylpropene, the carbocation intermediate is tertiary; in the hydration of 1-butene, it is secondary. Tertiary carbocations are more stable than secondary carbocations. Because the transition states resemble the respective carbocations in structure and energy (Hammond's postulate), the transition state for formation of the more stable carbocation is the more stable transition state, as observed.

4.70   This problem can be framed in terms of two reaction coordinate diagrams. To do this for methylenecyclobutane requires a knowledge of the standard free energy of activation for its hydration. Letting the hydration of methylenecyclobutane be reaction A, and the hydration of 2-methylpropene be reaction B, we use Eq. 4.33b on text p. 164 to obtain

$$\log\left(\frac{\text{rate}_A}{\text{rate}_B}\right) = \log(0.6) = -0.222 = \frac{\Delta G_B^{\circ\ddagger} - \Delta G_A^{\circ\ddagger}}{2.30RT} = -\frac{\Delta G_A^{\circ\ddagger} - \Delta G_B^{\circ\ddagger}}{5.71}$$

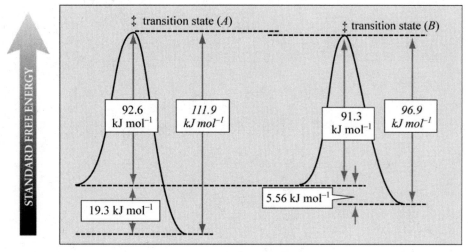

Figure SG4.8 Reaction coordinate diagrams to accompany the solution to Problem 4.70. The starting alkenes are placed at the same energy for comparison. The numbers in italics are calculated graphically from either the numbers given or calculated in the problem.

---

or $\Delta G_A^{\circ\ddagger} - \Delta G_B^{\circ\ddagger} = 1.27$ kJ mol^{-1}. With $\Delta G_B^{\circ\ddagger} = 91.3$ kJ mol^{-1}, then $\Delta G_A^{\circ\ddagger} = 92.6$ kJ mol^{-1}.

Next, we need to convert the relative hydration equilibrium constants for reactions A and B into relative standard free energies. A similar equation applies. (This can be derived by combining two equations like Eq. 3.36a on text p. 110.)

$$\log\left(\frac{K_A}{K_B}\right) = \log(250) = 2.40 = \frac{\Delta G_B^{\circ\ddagger} - \Delta G_A^{\circ\ddagger}}{2.30RT} = \frac{-5.56 - \Delta G_A^{\circ\ddagger}}{5.71}$$

or $\Delta G_A^{\circ} = -19.3$ kJ mol^{-1}. We can now construct reaction coordinate-free energy diagrams for the two reactions, which are given in Fig. SG4.8. We have enough information to calculate the standard free energies of activation for the reverse reactions, which are shown in italic type in the figure. Because the standard free energy of activation of the reverse of reaction A is larger, it is the slower reaction. The relative rates are calculated from Eq. 4.33b on text p. 164:

$$\log\left(\frac{\text{rate}_B}{\text{rate}_A}\right) = \frac{\Delta G_A^{\circ\ddagger} - \Delta G_B^{\circ\ddagger}}{2.30RT} = \frac{15}{5.71} = 2.63$$

The reverse of reaction B is therefore $10^{2.63}$, or 427, times faster than the reverse of reaction A.

Addition Reactions of Alkenes

STUDY GUIDE LINKS

 ### 5.1 Transition Elements and the Electron-Counting Rules

Equations 5.21a–e in the text involve the element mercury, which is a transition element. If you have been paying careful attention to formal charges and electron counts, you may be puzzled about how to deal with an element such as Hg. First, don't try to apply the usual rules of electron counting to transition elements, because these elements contain more than eight valence electrons, and the octet rule does not apply. (We'll discuss electron counting in transition elements in Sec. 18.5.) It's also hard to calculate formal charge by the rules given in Chapter 1 because transition elements have more than eight valence electrons, and we can't show them all without making a hopelessly complex-looking structure. In Eq. 5.21a, we've shown as an unshared pair only two of the *ten* valence electrons on mercury in Hg(OAc)$_2$. However, we can calculate the *change* in the formal charge on an element in a reaction by applying the following rules of formal-charge calculation:

1. Any electrons in unshared pairs on an atom are both assigned to the atom.
2. One electron from each electron-pair bond is assigned to an element.
3. The change in formal charge on an element in any reaction is equal to the number of electrons assigned to the element in the starting material minus the number in the product.

(These rules are completely consistent with the formal-charge rules you've already learned, and work on *any* atom.)

Let's apply these rules to calculate the formal charge on Hg in Eq. 5.21b (text p. 190). We are *given* that Hg in the structure on the left (and in structure *A*) has a formal charge of zero.

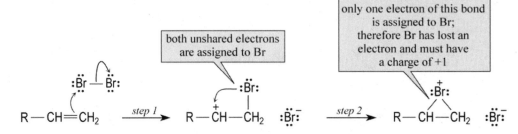

Now consider how the formal charge on Hg is affected by the reaction in step 2. In species *A*, both unshared electrons and one electron in each covalent bond are assigned to Hg, for a total of 4 electrons. In species *B*, 3 electrons (one from each covalent bond) are assigned to Hg. The change in charge is $4 - 3 = +1$. Given that Hg has zero charge in *A*, it must have a +1 charge in *B*.

Contrast this equation with the similar process for bromonium-ion formation:

Mechanistically, mercurinium-ion formation and bromonium-ion formation are identical.

 ### 5.2   How to Study Organic Reactions

You have now studied several organic reactions, and perhaps it is apparent that you are going to study many more. How can you keep all these reactions straight? How can you recall just the right reaction to use in solving a particular problem?

There are four keys to learning reactions; each of these points is discussed further below.

1.   Be active when you read the text.
2.   Be organized.
3.   Review frequently.
4.   Study in small chunks.

Be active when you read the text. To see what this means, let's go through one section of the text right now. Open your text to p. 190, Section 5.4A, "Oxymercuration–Reduction of Alkenes." This section is fairly typical of those in the text that describe important reactions. First comes a statement of the reaction. When you read this, *be active!* Try to make a hypothesis about where the different atoms come from. (The use of color in the equations is designed to help you with this.) For example, where does the —OH group in the product of text Eq. 5.20 originate? Where do the two —OAc groups end up? If you think you have the right idea, write the same reaction using a different alkene starting material.

Next, the text provides a few practical facts about the reaction, for example, some information about the solvent or the reaction conditions. These details can be very important, even crucial (in which case the text will say so), or they might be added for completeness or for your later use as reference material. In most cases, you should read them but not bother to memorize them, because the object of our first pass through the chapter is to understand the reaction itself.

Next follows a discussion of the *mechanism* of the reaction. What should you do with the mechanism? What you *should not* do is memorize it. What you *should do* is to follow each step of the mechanism to see whether your hypothesis about where the various atoms come from in the product is correct, and how they get there.

Here's an important point about curved-arrow mechanisms—a point we have already introduced in Chapter 3. Each step in a curved-arrow mechanism *must* be one of three types.

1.   a Lewis acid–base association (text p. 88)
2.   a Lewis acid–base dissociation (text p. 89)
3.   an electron-pair displacement reaction (text pp. 90–91)
     a.   a Brønsted acid–base reaction (text p. 96)
     b.   another electron-pair displacement (text pp. 90–91)
     c.   a combination of (a) and (b)

Now is a good time to review Sec. 3.4B on pp. 98–100. You will rely on this section *a lot* in studying reaction mechanisms.

Reactions of type (1) and (2) require *single* curved arrows. Reactions of type (3) require *two or more* curved arrows. Reactions of type (3a) involve by definition electron-pair displacements on a *proton.* As you examine a mechanism, *classify each mechanistic step* using the list above. Then *look for points in common with other mechanisms.* For example, *all* of the alkene reactions you have studied so far start with the π electrons of an alkene reacting as a Lewis base (a nucleophile or a Brønsted base) with an electrophile. Then consider the reactive intermediates, if any (for example, carbocations). Ask where you have seen such intermediates before.

One purpose of mechanisms is to help you see certain unifying ideas that seemingly different reactions have in common. (You may have noticed, for example, that carbocation theory keeps creeping into the discussion of many of the addition reactions you have studied.) As you go further along in the text, you might even try to write a mechanism on your own before looking at the one in the text. (Use the instructions in Problem 4.65, text p. 179, to help you.) Comparing your mechanism to the one in the text will help you to refine your mechanism-writing skills. Finally, after studying the mechanism, *be active!* Start with a different alkene starting material and write the mechanism on your own.

Once again: you should *not* have to memorize the mechanism. After analyzing each step as suggested, and after writing a few mechanisms on your own, you will begin to gain an intuition about mechanisms that will help you to avoid memorizing—and will help you eventually to predict reactions!

Sometimes there are further applications of the reactions or additional facts—sort of a "wrap-up." For example, Eq. 5.24 illustrates the point that no rearrangements are observed in oxymercuration.

Finally come the problems. *Working problems is the most important thing you can do to master organic chemistry.* The answers to the problems in the text are provided in this manual. You might be tempted to read the answers before you work the problems. Instead, work problems immediately when you come to them, and work them *with this manual closed!* These provide a valuable test over your understanding of the chemistry involved. Put a mark by the ones you find difficult; re-work these problems before an exam.

As you read the section, you may come across certain unfamiliar terms. For example, in the very first sentence of Section 5.4A is the term "addition." Do you know what this means? *The great temptation is to skip over unfamiliar terms, hoping that they will become clear in context.* This strategy often works when reading a novel, but you *can't* do this in scientific reading! You *must* understand each term before proceeding to the next. Ask yourself, "What is an addition reaction?" If you don't know, don't feel that you're intellectually deficient—just find out the answer. *Everyone* requires a number of repetitions in order to learn new ideas. (The author can't keep repeating the definition, because it would make the book even larger than it is!) How do you find out? *Use the index!* Or, in some cases, a cross-reference to an earlier section will be given. Look it up! If you get in the habit of taking this approach, you will find yourself retaining more, and you ultimately will have to do less work. If an unfamiliar term is one you have had to look up before, make yourself a glossary: Write the term and its definition *in your own words—not necessarily the words of the text*—and review it periodically. One prominent theory of learning says that students construct their own knowledge. What this means is that until you have processed something in *your own mind* you don't know it. Hence, when you see a new term, *read* the definition; *process it* (that is, frame it terms that *you* can understand, no matter how unsophisticated you might think your description is), and then *write it* (that is, be active). In summary, here's what you should be doing:

1. *Read*
2. *Process*
3. *Write*

If you do what has been suggested, you'll notice that you have been very active. You have a piece of scratch paper on your desk, and it should be filled with notes and structures.

Two other suggestions: First, *do not* underline (or highlight) large sections of the text! This represents an effort to be active, when in fact it represents in many cases *physical* activity coupled with *mental* passivity. If there is something you really want to emphasize, then by all means, highlight away! Otherwise, don't bother. It's a waste of time and effort. Second, if you outline the chapter, don't go into an "autopilot" mode; think about what you write, and don't passively copy sections of the text. Pretend you are a teacher and that you're writing the outline for someone who can't understand the text at all. This Study Guide has also provided outlines of the reactions for you. Use these for a quick review, but don't let them substitute for your own efforts to organize the material.

Be organized. After you have studied a reaction, write it down in a general form for later review. One good way is to make flash cards. Write the reactants on one side of the card and the products on the other side. Write two arrows, one on each side. Use "R" groups or "dangling bonds" to make the reaction as general as possible. Let's illustrate. Flash cards for oxymercuration–reduction might look as shown in Fig. SG5.1 on the following page.

Cards 1 and 2 summarize the individual steps of oxymercuration and reduction. Card 3 is a summary card for both steps. Card 4 is used to test whether you know the conditions for bringing about the reaction; or, in reverse, whether you know the transformation for which a given set of reagents is used. Card 5 is a "term" card, which you may want to separate out with other cards containing definitions of terms. Each time you learn a new reaction, make a new group of flash cards. Notice above that not only the individual reactions of oxymercuration and reduction have been entered, but also a summary reaction. The choice of one "R" group is arbitrary; you might instead use an alkene starting material $R_2C{=}CH{-}R$. The point here is to use enough "R" groups that the regioselectivity of the reaction is apparent. Notice also that the reactions are not balanced; the goal is to focus on what happens to the organic compound.

Each time you finish a study session, look at one side of each card, and complete the other side on a piece of scratch paper. Then, with the other side in view, complete the first side. (Notice that the arrow bearing the reaction conditions is repeated on both sides of the card for this purpose.) You should find that this process helps you learn reactions in both directions. This knowledge will become particularly valuable when you start combining reactions in multistep syntheses. Next, ask yourself why the reaction is reasonable. For the oxymercuration step, you should recognize that the reaction is an addition. Go through the mechanism. What is the electrophile? Why does it react with the alkene π electrons? Why is the reaction regioselective? (Note that the text does not cover the mechanism of the reduction step.) Then write a specific example of the reaction using R groups of your choice. Make examples that use cyclic compounds.

If you do this *every time you read,* and *if you understand what you read,* then what you are doing here is *not* memorizing the reactions, at least not in a rote fashion, because you will have already learned them by careful study. Rather, you are cataloging them for use as a future study tool. If you can't complete one side or the other of a reaction, look at the missing side. If it looks unfamiliar, return to the text and focus exclusively on the part that is unfamiliar.

Figure SG5.1 Five flash cards used to study oxymercuration–reduction.

Review frequently. Review your reaction sheet *every time you study!* Go back about three assignments for each review, dropping older material from your review. Then review the older material again once every week or two, or right before an exam or quiz. (This gives you something *active* to do in preparing for an examination.) Do the same thing with the list of terms. You will find that relatively few reactions and terms are covered in each assignment, and that learning each day's assignment takes relatively little time. If you get behind, however, the number of reactions and terms that must be learned will grow rapidly! A nice feature of flash cards is that they can be shuffled, so that you can review them out of context. (You might want to number your cards so you can return them to their original order.)

Many students and professors suffer from adherence to a theory that one of the author's colleagues calls the "immunization theory of learning." The student's version of this theory is: "I've studied this material once, and therefore I should never have to see it again … *ever.*" The professor's version of the theory is, "Students have studied this material once, and therefore they should automatically know it. There is no need for me to review, and a student should have total recall of everything he/she has ever been taught." Those who adhere to the "immunization theory" deny a fundamental caveat about learning: *Continued reinforcement is one of the best ways to learn.* The author finds that students are sometimes frustrated because they didn't learn something on the first pass. You should *expect* to forget material, and you should expect to re-learn it, probably more than once. Each relearning, however, takes less time and brings with it deeper understanding. What would happen to the college basketball team if the coach taught them a play *once* and they never reviewed it before a game? What would happen to a musician who practiced a piece once before a concert? Why should studying chemistry (or anything else, for that matter) be any different?

Study in small chunks. Set realistic goals for each study session and pursue them regularly. If you can only afford six hours per week, it is far better to study six times per week for one hour than one time for six hours. If you can allow twelve hours per week, then break it into six two-hour sessions. Notice how this strategy is built into the discussion of the flash cards above: you should study your flash cards over and over again, but confine your concentration to relatively small numbers of them except for general reviews.

If you have ever played an organized sport, or studied a musical instrument seriously, you should understand the "small chunk" strategy well. Does the coach of a winning team hold one ten-hour practice per week, or five two-hour practices? Why? Because the mental and physical learning capacities of most individuals erode significantly as fatigue sets in. Does a skilled performer on a musical instrument practice once per week for twelve hours, or six times per week for two hours?

What is remarkable is that when you *stay organized* and *study in small chunks*, your capacity for further study increases! Just as an athlete's endurance increases with regular workouts, your ability to concentrate for longer periods of time will increase. Just as an athlete's speed and skill increase with each workout, you will also find that you will become more efficient: you will accomplish much more in less time!

The suggestions given here for learning reactions are not the only ones that will work. But if you are at a loss about how to study organic chemistry on your own, try these suggestions, and then refine them to suit your own needs. The author has actually tested these suggestions on his class, and has determined that student grades are significantly improved when students follow these suggestions.

5.3   Solving Structure Problems

Before working Problem 5.47 in the text, be sure you have read Study Guide Link 4.3 (also called "Solving Structure Problems").

Problem 5.47 contains a lot of information. The temptation is to solve this problem from the top down. It bears repeating that this is not usually the best way to deal with this type of problem. What is known about *A*? Its formula indicates an unsaturation number of 3. Its reactions with Br_2, H_2/catalyst, and O_3 all suggest that *A* is an alkene. Immediately you should look for a *structure*. Two structures are given in the problem, one in words, the other explicitly. The hydrogenation product of *A* is 1-isopropyl-4-methylcyclohexane:

1-isopropyl-4-methylcyclohexane

This tells us *immediately* the carbon skeleton of *A*. Only the double bonds are missing. This structure further implies that *A* contains a ring. Remember that the ring accounts for one degree of unsaturation. How many double bonds does compound *A* have?

The second structure given in the problem is the ozonolysis product. Remember that ozonolysis produces *two* C=O groups for every double bond in the molecule. However, in this ozonolysis product, there are *three* C=O groups. But there are also *nine carbons*. Evidently, one carbon and one C=O group are missing. This means that the missing carbon is part of the missing C=O group, and that a one-carbon fragment was liberated by ozonolysis as a second product which was evidently not recovered. This conclusion is consistent with the unsaturation number, which indicates that there are two double bonds. Two double bonds, after ozonolysis, gives four $C=O$ groups.

This information establishes a partial structure for *A:* compound *A* must contain a $=CH_2$ group. Such a group can come from the known carbon skeleton in only two ways:

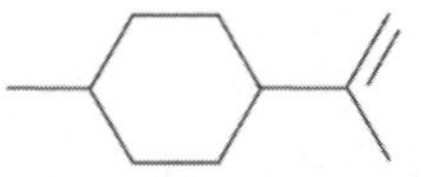

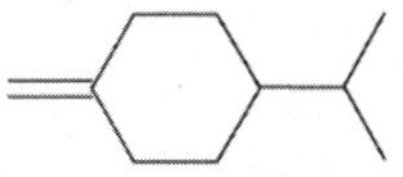

Because *A* contains two double bonds, one additional double bond must be provided to convert either *(1)* or *(2)* into possible structures for *A*. From the ozonolysis product, this double bond must be *within* the ring, because ozonolysis opens the ring. Now it is time to *write all the possibilities*. (Recall that this point was stressed in Study Guide Link 4.3.) From *(1)* the only possibilities for *A* are the following:

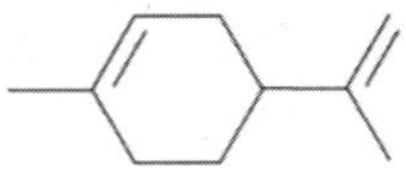

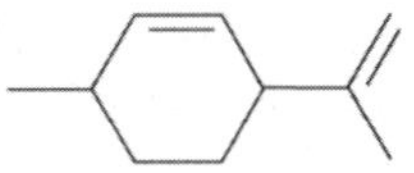

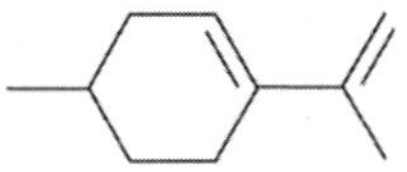

From *(2)* the only possibilities are the following:

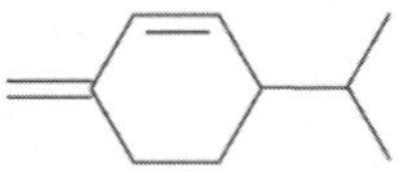

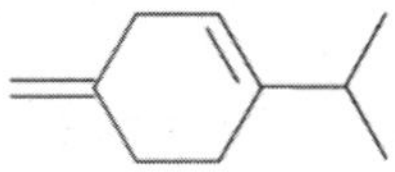

There are two ways to finish the problem. One is to apply ozonolysis to each of the five structures and see which one gives the observed product. Notice that the observed product contains two acetyl groups:

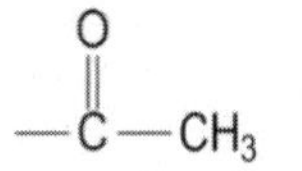

an acetyl group

Which structures would give *two* of these groups after ozonolysis? You should finish the problem by deciding between these.

The second way to finish the problem is to reverse ozonolysis mentally in all possible ways. The carbons involved in double bonds have to be the ones "labeled" with oxygens. These are carbons 1, 2, and 3 in the following structure of the ozonolysis product:

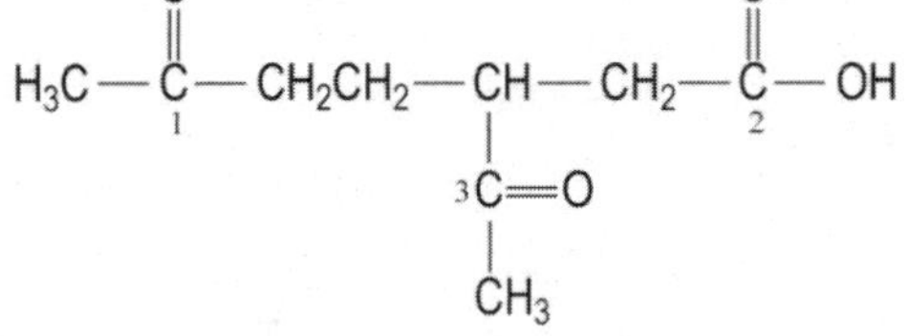

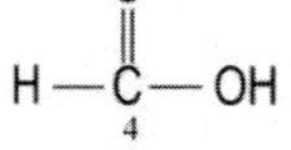

the one-carbon fragment

(The missing carbon, which evidently gives formic acid, is referred to as carbon-4.) Once again, *write out all possibilities.* All possible connections are:

1.   C1 with C2, C3 with C4
2.   C1 with C3, C2 with C4
3.   C1 with C4, C2 with C3

All but one of these possibilities can be ruled out by recalling that compound *A* has to contain a six-membered ring.   The only structure for *A* that has a six-membered ring is obtained by connecting C1 to C2. Hence, C3 and C4 are also connected. You should finish reconstructing compound *A*. (The final solution is in the "Solutions to Additional Problems" part of this chapter.

FURTHER EXPLORATIONS

 ### 5.1   Mechanism of Organoborane Oxidation

Although the oxidation step of hydroboration–oxidation may seem strange, it is nothing more than a series of Brønsted and Lewis acid–base reactions.

First, the hydroperoxide anion is produced by ionization of H_2O_2. As you can see from the pK_a values below, this ionization is quite favorable when ^-OH is used as the base:

$$HO:^- \;\; H—O—OH \longrightarrow HO—H \;+\; ^-:O—OH$$

$$pK_a = 11.6 \qquad\qquad pK_a = 15.7 \qquad \textbf{hydroperoxide anion}$$

Next, this anion reacts with the boron of the organoborane in a Lewis acid–base association reaction. (The alkyl groups on boron are abbreviated as R groups.)

$$R_3B \longleftarrow\; :O—OH \longrightarrow R_3\bar{B}—O—OH$$

hydroperoxide anion

An alkyl group on boron moves from boron to oxygen with breaking of the (very weak) bond between the two oxygens and loss of the ^-OH group:

$$R_2\bar{B}—O—OH \longrightarrow R_2B—O—R \;+\; ^-:OH$$
$$\qquad\;\; |$$
$$\qquad\;\; R$$

This is the step that replaces a carbon–boron bond with a carbon-oxygen bond. Why would such a reaction occur?

Let's use the device of imagining that this process occurs in a stepwise manner. The O—O bond is very weak, because two electronegative atoms are "pulling" at the bonding electrons. Imagine that ^-OH is expelled, leaving an *electron-deficient oxygen:*

$$R_2\bar{B}—O—OH \longrightarrow R_2\bar{B}—O^+ \;+\; ^-:OH$$
$$\qquad\;\; |\qquad\qquad\qquad |$$
$$\qquad\;\; R\qquad\qquad\qquad R$$

an electron-deficient oxygen

The electron-deficient oxygen is like a carbocation, except with *much* higher energy. (Notice: this is not the same situation that occurs in $H_3O:^+$, which is *not* electron deficient!) This very unstable cation can rearrange by taking electrons from the adjacent C—B bond:

$$R_2\bar{B}—O^+ \longrightarrow R_2B—O—R$$
$$\qquad\;\; |$$
$$\qquad\;\; R$$

Thus, the driving force for this rearrangement is the flow of electrons from the relatively electropositive boron to the very electronegative oxygen of ⁻OH, as well as the neutralization of charge.

In subsequent steps of the reaction, the bonds between the other R groups are converted in a similar manner to R—O bonds:

$$R_2B\!-\!\ddot{\underset{..}{O}}\!-\!R \xrightarrow{\text{H}_2\text{O}_2,\ ^-\text{OH}} RO\!-\!\underset{\underset{OR}{|}}{B}\!-\!OR$$

the same mechanism

a trialkyl borate

Finally, the resulting trialkyl borate undergoes hydrolysis to the alcohol product. The final hydrolysis breaks the boron–oxygen bonds but leaves the oxygen-carbon bonds unaffected.

$$4\ ^-\!\!:\!\ddot{\underset{..}{O}}\text{H}\ +\ \underset{RO}{\overset{OR}{\underset{}{\diagdown}}}B\overset{}{\diagup}OR \xrightarrow{\text{H}_2\text{O}}\ 3\ \text{H}\!-\!\ddot{\underset{..}{O}}\text{R}\ +\ \bar{B}(\text{OH})_4$$

Try to write a stepwise mechanism for this hydrolysis. The first step is the Lewis acid–base association of ⁻OH with the boron of the trialkyl borate.

5.2   Mechanism of Ozonide Conversion into Carbonyl Compounds

The conversion of ozonides into carbonyl compounds can also be rationalized by a series of electron-pair displacement reactions. Dimethyl sulfide acts as a nucleophile toward an oxygen electrophile in the O—O bond, displacing the other oxygen. A series of electron-pair displacements eventually liberates dimethyl sulfoxide, an oxidation product of dimethyl sulfide.

A similar series of reactions can occur with water as the nucleophile.

Hydrogen peroxide is produced as a by-product of this reaction. (Water is literally oxidized by its reaction with the ozonide.) Aldehydes are easily oxidized by hydrogen peroxide (or other oxidants, such as the ozonide) to carboxylic acids. This involves the nucleophilic reaction of hydrogen peroxide at the aldehyde carbonyl carbon.

The key step in the oxidation is the migration of the aldehyde hydrogen to an oxygen. This is much like the migration step in organoborane oxidation, which, as shown in Further Exploration 5.1, can be viewed as a migration to an electron-deficient oxygen. The hydrogen rather than the R group migrates because the resulting "carbocation" (which is simply a resonance structure of the carboxylic acid) is stabilized by the alkyl substituent.

5.3   Bond Dissociation Energies and Heats of Reaction

Study Problem 5.5 on text p. 214 shows how bond dissociation energies can be used to calculate heats of reactions: subtract the bond dissociation energies of the bonds formed from those of the bonds broken. This study guide link shows why this process works.

Essentially, this is a Hess's law calculation. This can be illustrated by using bond dissociation energies to calculate the $\Delta H°$ of the following gas-phase reaction:

$$\cdot CH_3 + Cl_2 \longrightarrow Cl-CH_3 + \cdot Cl \qquad (FE5.1)$$

The term "bond dissociation energy" will be abbreviated with the acronym "BDE" for convenience. The bond formed is a H_3C-Cl bond, with BDE = 350 kJ mol^{-1}. The bond broken is a $Cl-Cl$ bond, with BDE = 239 kJ mol^{-1}. The $\Delta H°$ for the reaction, then, is $239 - 350 = -111$ kJ mol^{-1}.

The reaction in Eq. FE5.1 can be viewed as the sum of the following two reactions and their associated $\Delta H°$ values.

			$\Delta H°$
$\cdot CH_3 + \cdot Cl$	$\longrightarrow$	$Cl-CH_3$	$-BDE$ of $Cl-CH_3$
Cl_2	$\longrightarrow$	$2\,Cl\cdot$	BDE of Cl_2
Sum: $\cdot CH_3 + Cl_2$	$\longrightarrow$	$Cl-CH_3 + \cdot Cl$	BDE of $Cl_2 - BDE$ of $Cl-CH_3$

As this equation shows, the $\Delta H°$ values are simply the BDEs of the bonds involved. Because the first reaction is simply the bond dissociation of H_3C—Cl written in reverse, its $\Delta H°$ is the negative BDE of H_3C—Cl. Thus, the total $\Delta H°$ is then the BDE of the bond broken less the BDE of the bond formed.

 # REACTION REVIEW

I. CONVERSION OF ALKENES INTO HALOGEN-CONTAINING COMPOUNDS

A. ADDITION OF CHLORINE AND BROMINE

1. Bromine and chlorine undergo addition to alkenes to give compounds with halogens on adjacent carbons called vicinal dihalides. (X = Br or Cl)

$$X_2 \; + \; \text{C=C} \; \longrightarrow \; X-C-C-X$$

2. The rapid disappearance of the red bromine color during bromine addition is a useful qualitative test for alkenes.

3. The equilibrium for addition of I_2 favors the alkene. Addition of F_2 occurs, but the reaction is violent and is accompanied by other reactions.

4. Bromine addition to alkenes in many cases involves a reactive intermediate called a bromonium ion.

    a. In a bromonium ion, a positively charged bromine with an octet of electrons is bonded to two other atoms in a three-membered ring.

    b. Reaction of the bromide ion as a nucleophile at either of the carbons bound to bromine in the bromonium ion completes the addition of bromine.

B. FORMATION OF HALOHYDRINS

1. When an alkene reacts with Cl_2, Br_2, or I_2 in a solvent containing a large amount of water, water reacts as a nucleophile with the cyclic ion. (X = Cl, Br, or I)

2. Loss of a proton from the oxygen in a subsequent Brønsted acid–base reaction with the solvent gives a compound containing both an —OH and a halogen (—X group) called a halohydrin.

a halohydrin

a. Halohydrins are compounds that contain both a halogen and an —OH group.
b. The reaction is completely regioselective when one carbon of the alkene is bonded to two alkyl substituents and the other is bonded to two hydrogens; in such cases, the OH group is goes to the carbon with two alkyl substituents, and the halogen goes to the carbon bearing the hydrogens.

C. FREE-RADICAL ADDITION OF HYDROGEN BROMIDE TO ALKENES

1. Alkenes react with hydrogen bromide in the presence of peroxides so that the bromine adds to the less branched carbon of the double bond. (The "less branched" carbon is the carbon with fewer alkyl substituents.)

a. This reversal of regioselectivity in HBr addition in the presence of peroxides is termed the peroxide effect.
b. This mechanism involves reactive intermediates called free radicals.

2. The bromine radical is formed by initiation reactions described on p. 205–6.
3. The regioselectivity of HI or HCl addition to alkenes is not affected by the presence of peroxides.

II. CONVERSION OF ALKENES INTO ALCOHOLS

A. OXYMERCURATION–REDUCTION OF ALKENES

1. Alkenes react with mercuric acetate in aqueous solution to give addition products in which an —HgOAc (acetoxymercuri) group and an —OH (hydroxy) group derived from water have added to the double bond; this reaction is called oxymercuration.
a. The first step of the reaction mechanism involves the formation of a cyclic ion called a mercurinium ion.

b. The solvent water reacts as a nucleophile with the mercurinium ion exclusively at the more branched carbon. (The "more branched" carbon is the carbon with the greater number of alkyl substituents.)

c. The addition is completed by a Brønsted acid–base reaction in which a proton is transferred to the acetate ion that is formed in the previous step.

2. Oxymercuration adducts are converted into alcohols by treatment with the reducing agent sodium borohydride ($NaBH_4$) in base. In this reaction, the carbon–mercury bond is replaced by a carbon–hydrogen bond.

3. The overall reaction sequence, called oxymercuration–reduction, results in the net addition of the elements of water (H and OH) to an alkene double bond in a regioselective manner.
   a. The —OH group is added to the more branched carbon of the double bond, and the hydrogen is added to the less branched carbon.
   b. The product is free of rearrangements and other side reactions that are encountered in hydration.

B. HYDROBORATION–OXIDATION OF ALKENES

1. Borane (BH_3) adds regioselectively to alkenes so that the boron goes to the less branched carbon of the double bond, and the hydrogen goes to the more branched carbon.
   a. The addition of BH_3 to alkenes, called hydroboration, occurs in a single step without intermediates— that is, by a concerted mechanism.
   b. In many cases, one borane molecule can add to three alkene molecules; the product is a trialkylborane.

2. Trialkylboranes are oxidized to alcohols with basic hydrogen peroxide, H_2O_2. The net result of this transformation is replacement of the boron by an —OH in each alkyl group.

3. The overall reaction sequence, called hydroboration–oxidation, results in the net addition of the elements of water (H and OH) to an alkene in a regioselective manner.

   a. The —OH group is added to the less branched carbon atom of the double bond; a hydrogen is added to the more branched carbon of the double bond.

   b. The product is free of rearrangements and other side reactions that are encountered in hydration.

C. COMPARISON OF METHODS FOR THE SYNTHESIS OF ALCOHOLS FROM ALKENES

   1. Hydration of alkenes is a useful industrial method for preparing a few alcohols, but it is not a good laboratory method.

   2. Hydroboration–oxidation is a good laboratory method that gives an alcohol in which the —OH group has been added to the less branched carbon of the double bond and a —H has been added to the more branched carbon.

   3. Oxymercuration–reduction is a good laboratory method that gives an alcohol in which the —OH group has been added to the more branched carbon of the double bond.

III. OTHER REACTIONS

A. OZONOLYSIS OF ALKENES

   1. The reaction of an alkene with ozone to yield products of double-bond cleavage is called ozonolysis.

   2. Ozone, O_3, adds to alkenes at low temperature in a concerted cycloaddition reaction to yield unstable cyclic compounds (you may see these compounds referred to as molozonides).

   3. This initial cycloaddition product is unstable and spontaneously forms the ozonide.

an ozonide

   4. If the ozonide is treated with dimethyl sulfide, $(CH_3)_2S$, the ozonide is split into products that contain a C=O group.

   a. If a carbon of the double bond in the starting material bears a hydrogen, an aldehyde is formed.

   b. If a carbon of the double bond bears no hydrogens, a ketone is formed.

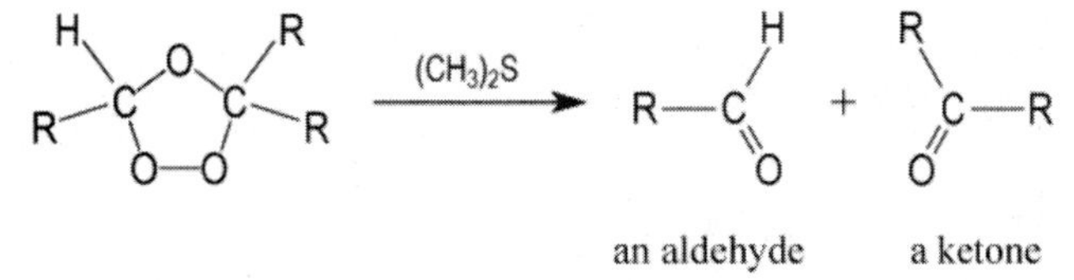

an aldehyde   a ketone

5. If the ozonide is treated with water and hydrogen peroxide (H_2O_2), carboxylic acids are formed instead of aldehydes.

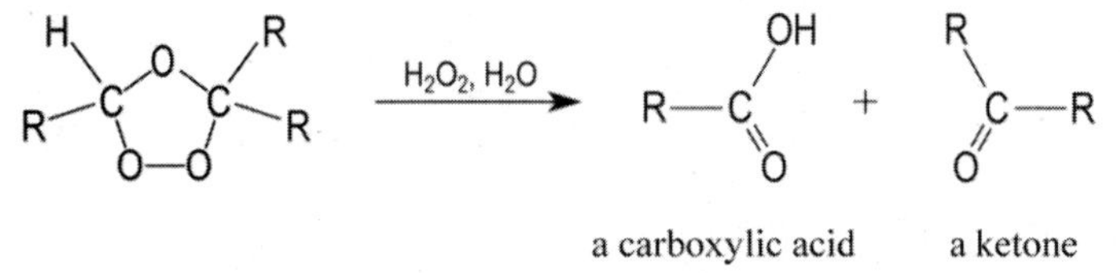

a carboxylic acid   a ketone

B. FREE-RADICAL POLYMERIZATION OF ALKENES

1. In the presence of free-radical initiators such as peroxides, many alkenes (monomers) react to form polymers in a reaction called polymerization.

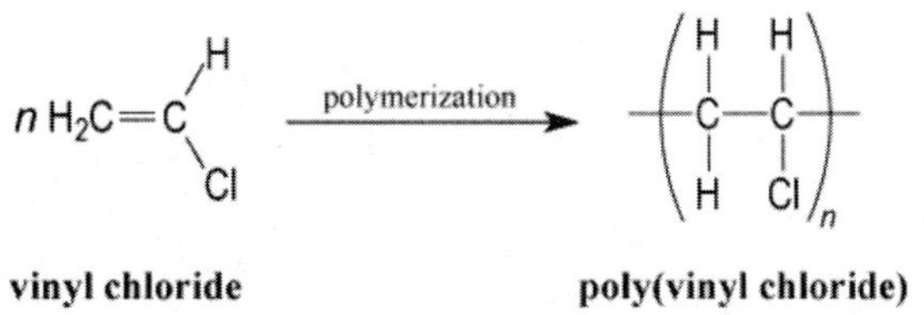

vinyl chloride   **poly(vinyl chloride)**

a. Polymers are very large molecules composed of repeating units.
b. Polyethylene is an example of an addition polymer—a polymer in which no atoms of the monomer unit have been lost as a result of the polymerization reaction.
c. Polyethylene formation occurs by a free-radical mechanism and is thus a free-radical polymerization.

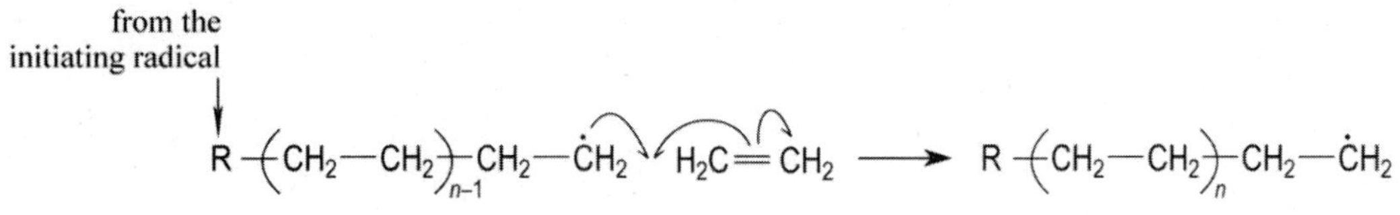

THERMAL CRACKING OF ALKENES

2. Simple alkenes are produced industrially from alkanes in a process called thermal cracking, which breaks larger alkanes into a mixture of smaller hydrocarbons, some of which are alkenes.
3. Ethylene, the alkene of greatest commercial importance, is produced by thermal cracking of ethane.

SOLUTIONS TO PROBLEMS

Solutions to In-Text Problems

5.1 (a) The iodine is the electrophile, because it adds to the carbon of the double bond with fewer alkyl substituents. This result does seem to fit the electronegativity pattern for electrophilic addition, as nitrogen is considerably more electronegative than iodine.

(b) The less electronegative group (I) adds to the carbon of the double bond with fewer substituents, and the more electronegative group (Br) adds to the carbon of the double bond with more substituents.

$$(CH_3)_2C \!-\! CH_2$$
$$\qquad\;\; | \qquad\; |$$
$$\qquad\;\; Br \quad\; I$$

5.2 (a) The reaction of Br_2 with 2-methyl-1-hexene follows the pattern in Eqs. 5.11 and 5.12 of the text.

2-methyl-1-hexene

(b) The mechanism involves a bromonium ion intermediate, formed as shown in Eqs. 5.11 on text p. 185. Water reacts as a nucleophile at the carbon of the bromonium ion that has the greater number of substituents. (See Eq. 5.16, text p. 187.)

2-methyl-1-hexene

In the last step, you might have used Br^- as the Brønsted base to give HBr as the product. This would constitute a satisfactory mechanism. However, $H_3O^+\,Br^-$ is the actual form of HBr in aqueous solution, and water, as the solvent, is present in much higher concentration than bromide ion. In other words, HBr is the product, but it is fully ionized to $H_3O^+\,Br^-$ in aqueous solution.

(c) Because iodine is the electrophile in iodine azide (as discussed in the solution to Problem 5.1(a)), an iodonium ion is formed. An azide ion reacts with this cyclic ion to give the product.

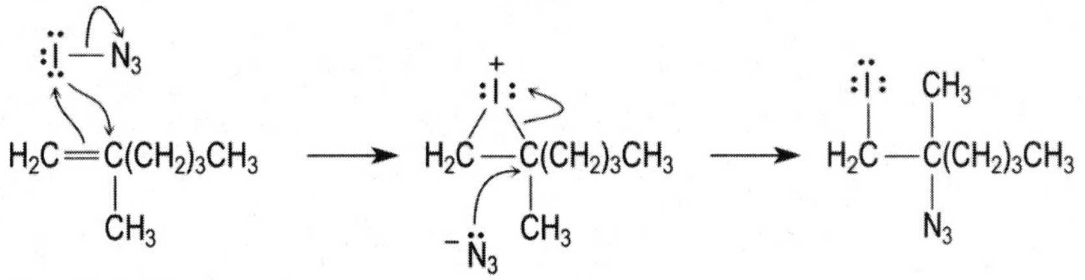

2-methyl-1-hexene

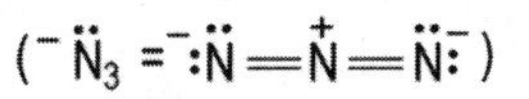

$$(^-\ddot{N}_3 = {}^-\!:\!\ddot{N}\!=\!\overset{+}{N}\!=\!\ddot{N}\!:{}^-)$$

structure of the azide ion

5.3   The alkene that could be used to form chlorohydrin *B* is methylenecyclohexane:

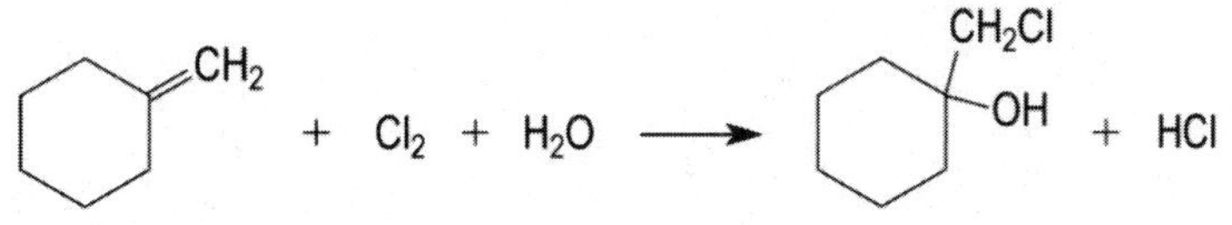

B

5.4   (a)   In this part, the double bond is symmetrically placed in the molecule; so, regioselectivity is not an issue.

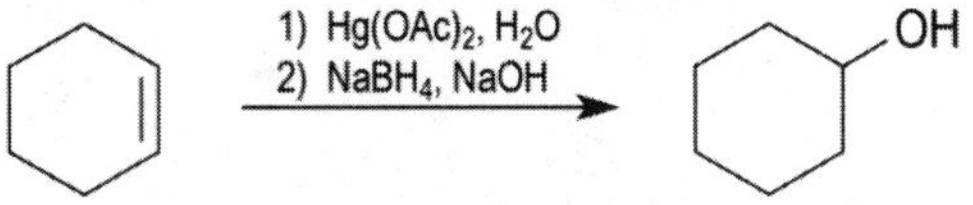

cyclohexene

(b)

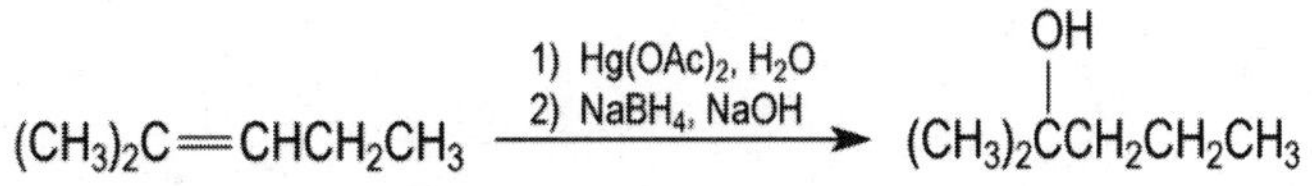

2-methyl-2-pentene

(c)   In this case, two constitutional isomers are formed because each carbon of the double bond has the same number of substituents—one.

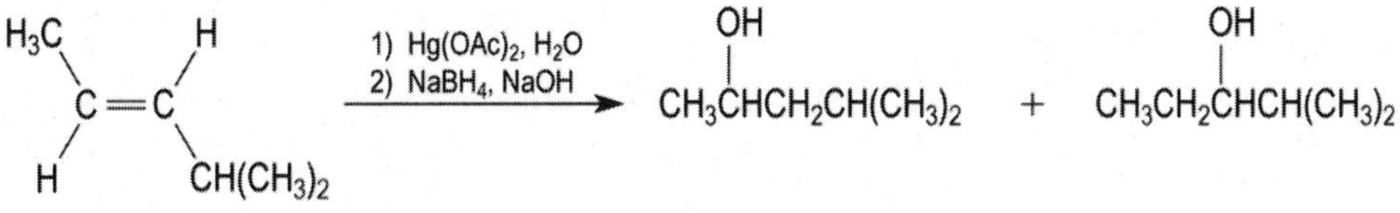

(*E*)-4-methyl-2-pentene

(d)

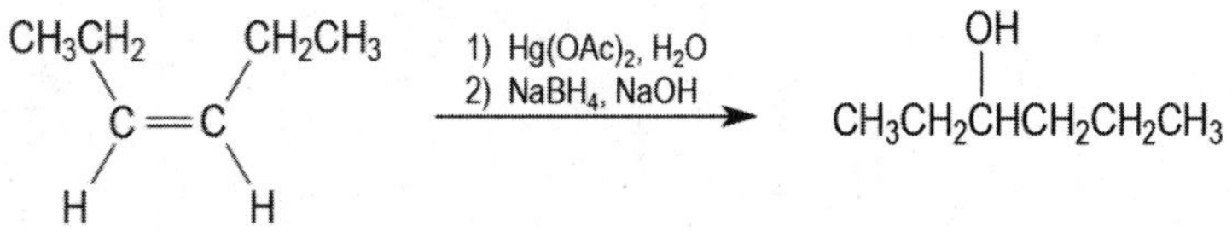

***cis*-3-hexene**

5.5   When 3-methyl-1-butene undergoes acid-catalyzed hydration, rearrangement occurs because carbocation intermediates are involved. (See the solution to Problem 4.36.) Oxymercuration–reduction gives addition without rearrangement because the intermediate is a cyclic mercurinium ion, which does not rearrange.

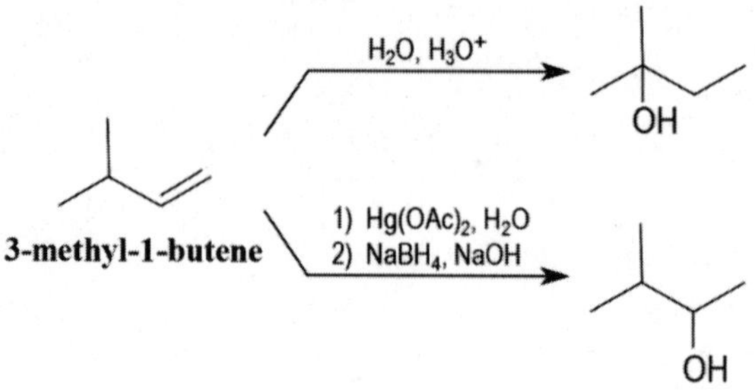

3-methyl-1-butene

5.6 (a)

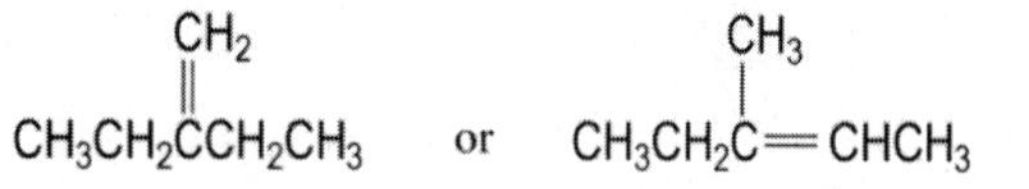

2-ethyl-1-butene or **3-methyl-2-pentene** (either *E* or *Z* stereoisomer)

(b)

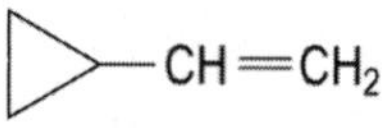

cyclopropylethylene

5.7 (a) In this part and in parts (c) and (d), the double bond is symmetrically placed in the molecule; so, regioselectivity is not an issue.

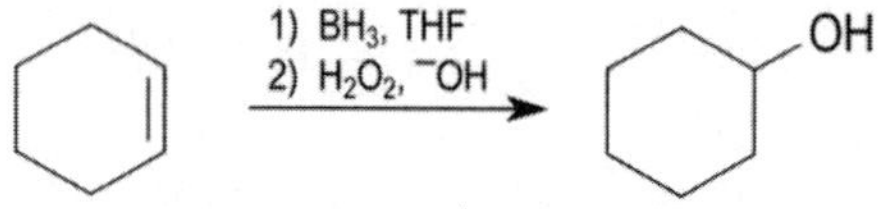

cyclohexene

(b)

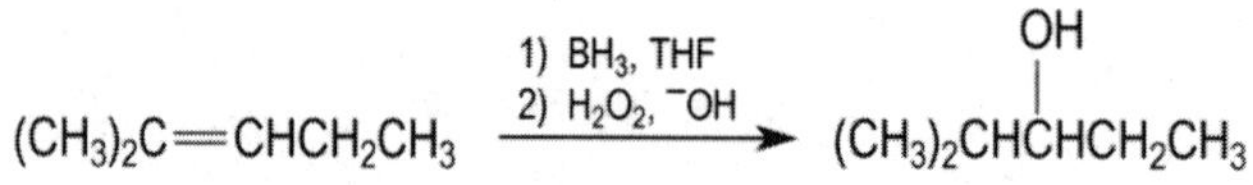

2-methyl-2-pentene

(c) In this case, two constitutional isomers are formed because each carbon of the double bond has the same number of substituents—one.

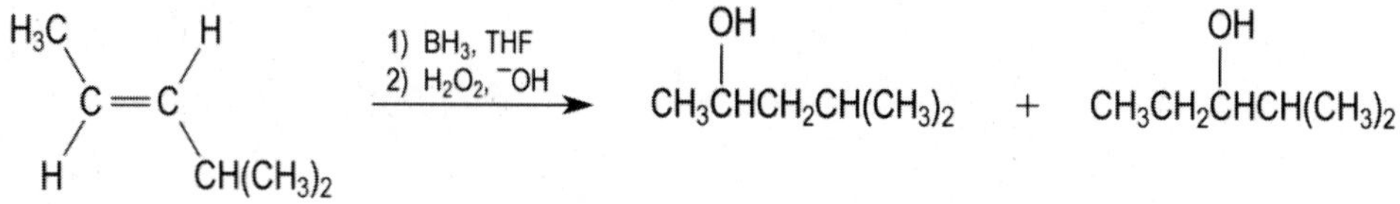

(*E*)-4-methyl-2-pentene

(d)

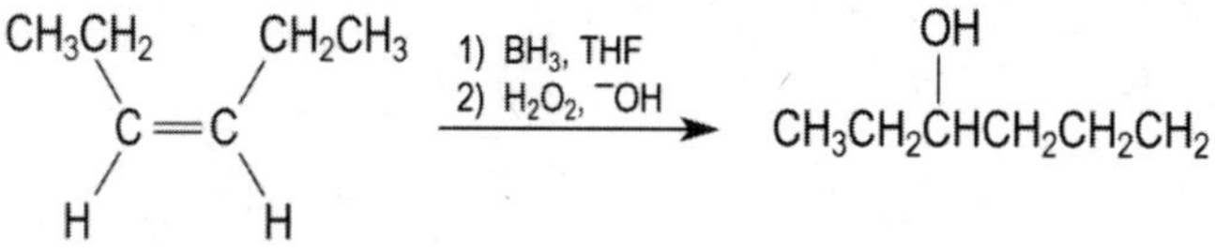

***cis*-3-hexene**

5.8 The products of hydroboration–oxidation are the same as the ones in parts (a) and (d) of Problem 5.4. In both, the alkene is symmetrical; consequently, regioselectivity has no meaning. In part (b), they are different because the carbons of the double bond have different numbers of alkyl substituents. In part (c), the carbons of the double bond have the same number of alkyl substituents; consequently, there is no reason to expect significant regioselectivity. However, the relative amounts of the products in each case could be different.

5.9   (a)

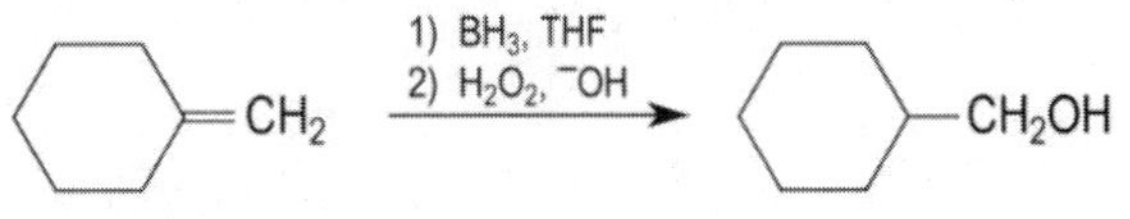

methylenecyclohexane

(b)

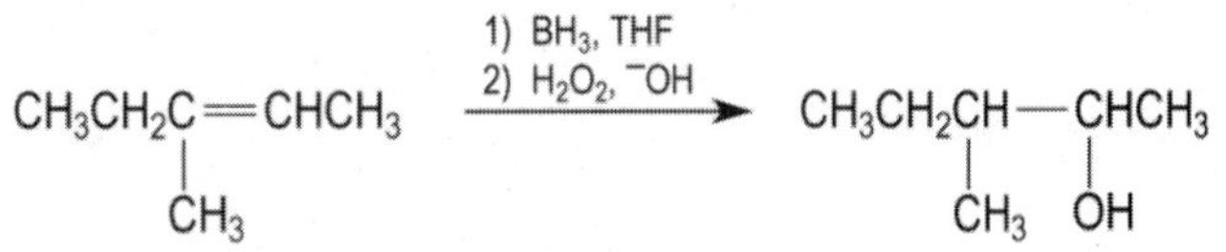

3-methyl-2-pentene

5.10   (a)   Because the OH is placed symmetrically in the alcohol, either reaction would work; cyclopentene is the starting alkene. (Acid-catalyzed hydration also works.)

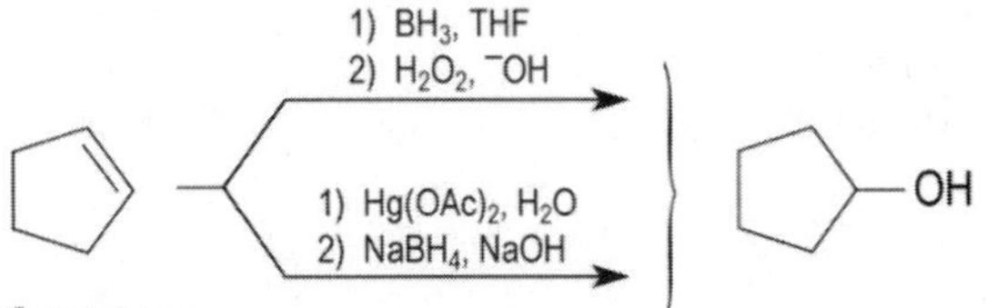

cyclopentene

(b)   The only possible alkene starting material is 5-methyl-1-hexene; only hydroboration–oxidation will put the OH on the carbon of the double bond with fewer substituents.

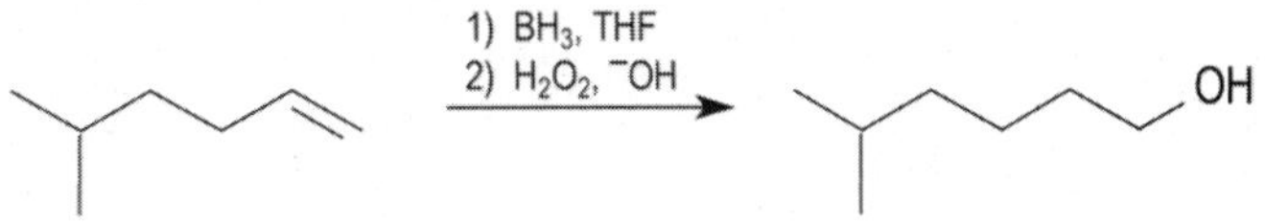

5-methyl-1-hexene

(c)   This alcohol has to come from 3-ethyl-2-pentene; only oxymercuration–reduction will bring about the desired conversion, because the OH group must end up on the carbon of the double bond with the greater number of substituents. (Acid-catalyzed hydration also works.)

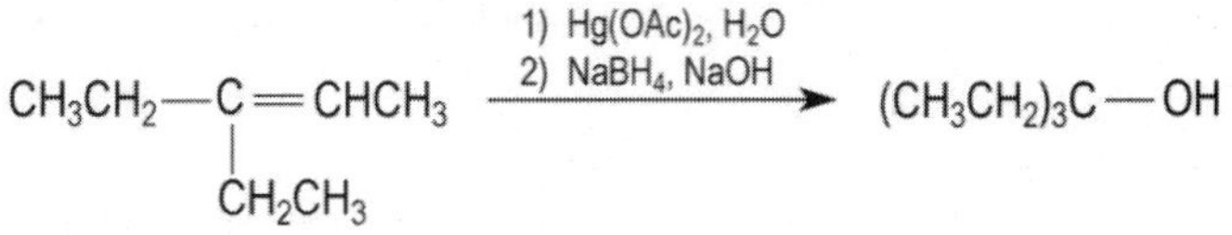

3-ethyl-2-pentene

5.11   (a)   Because *cis*-2-butene is a symmetrical alkene, it reacts to give the same product in either oxymercuration-reduction or hydroboration-oxidation.

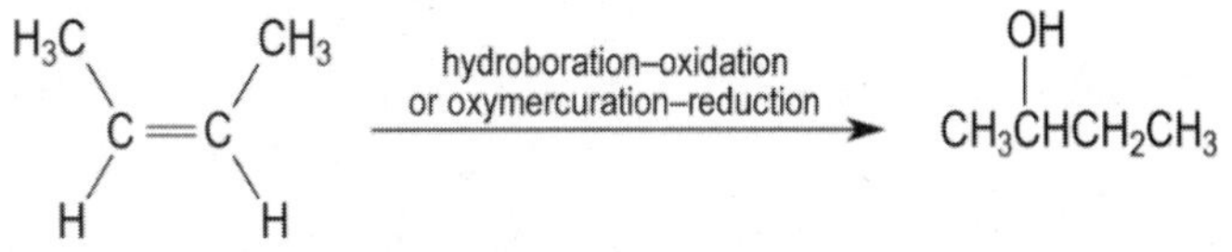

cis-2-butene

(b) Because 1-methylcyclohexene has different numbers of alkyl branches at the carbons of its double bond, it gives different products in the two reactions.

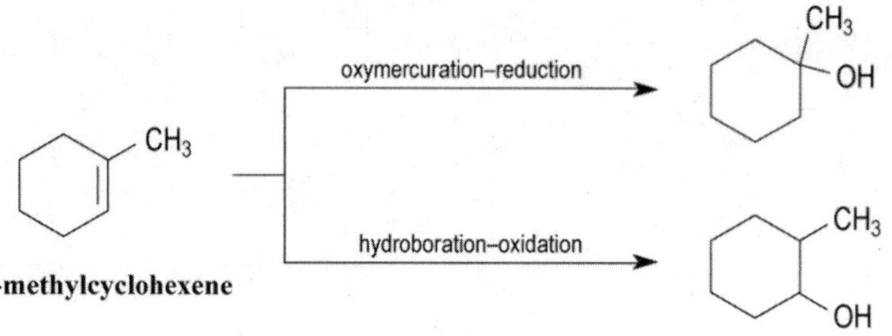
CH3
oxymercuration–reduction
OH
CH3
hydroboration–oxidation
CH3
OH

1-methylcyclohexene

5.12 (a)

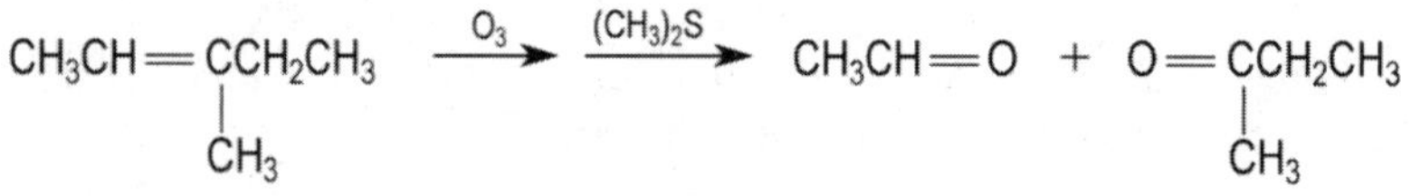
CH3CH=CCH2CH3
CH3
O3
(CH3)2S
CH3CH=O + O=CCH2CH3
CH3

3-methyl-2-pentene

(b)

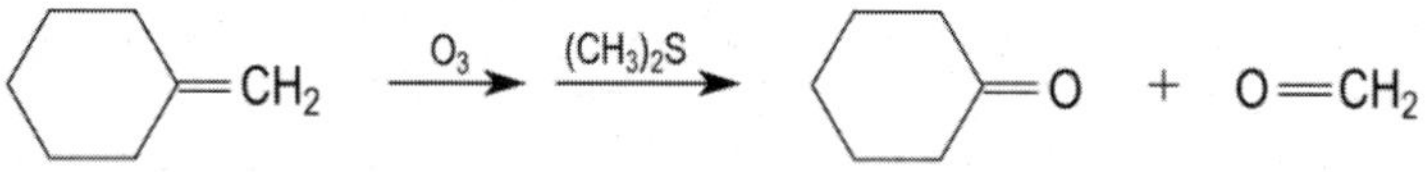
=CH2
O3
(CH3)2S
=O + O=CH2

methylenecyclohexane

(c)

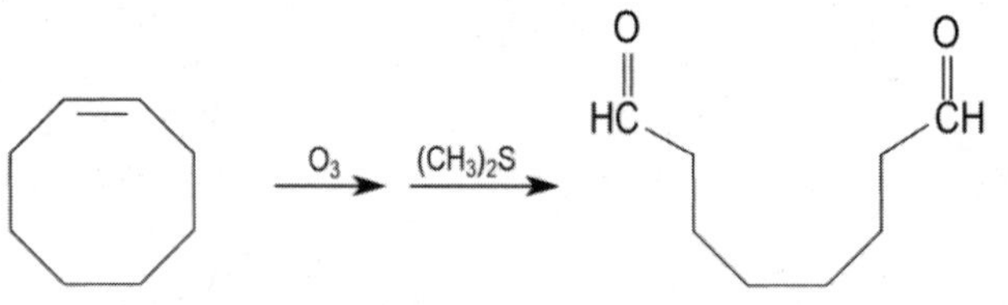
O3
(CH3)2S
O
HC
O
CH

cyclooctene

(d) 2-Methylpentane is an alkane; because it has no double bond, it does not undergo ozonolysis.

5.13 (a)

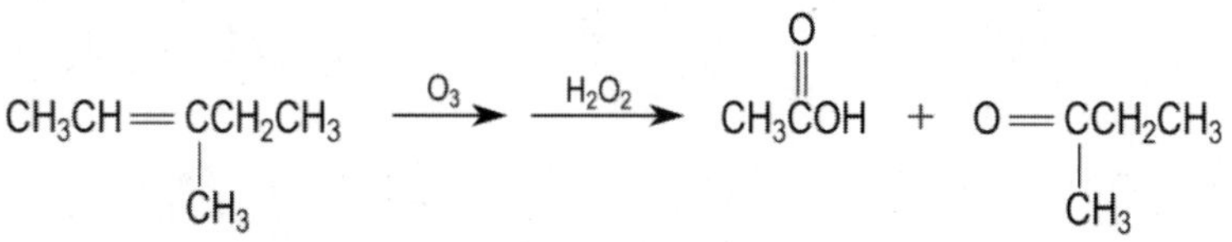
CH3CH=CCH2CH3
CH3
O3
H2O2
O
CH3COH + O=CCH2CH3
CH3

3-methyl-2-pentene

(b)

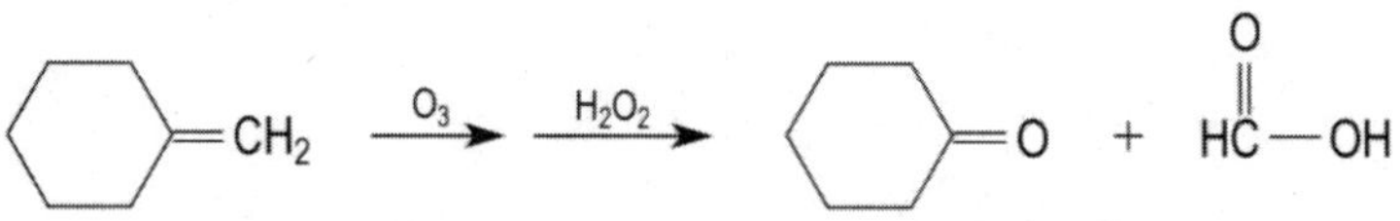
=CH2
O3
H2O2
=O + HC—OH
O

methylenecyclohexane

(c)

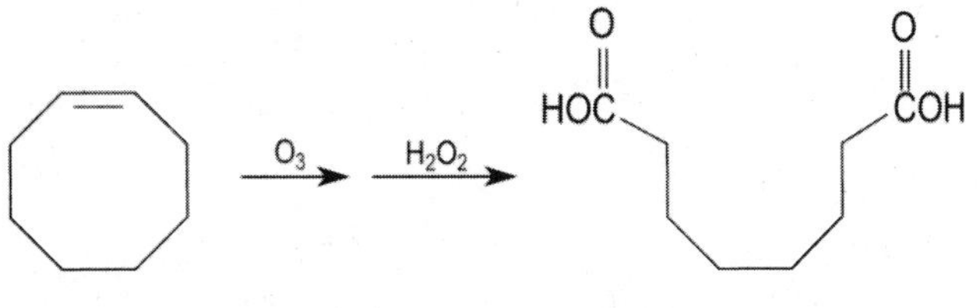

cyclooctene

(d)   2-Methylpentane is an alkane; because it has no double bond, it does not undergo ozonolysis.

5.14   **(a)**   **(b)**   **(c)**

$CH_3CH_2CH_2CH = CHCH_2CH_2CH_3$

cis- or *trans*-4-octene   **1-methylcycloheptene**

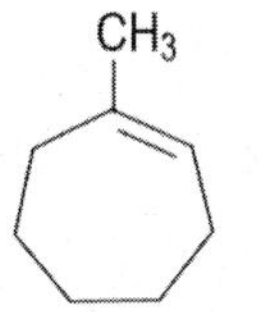

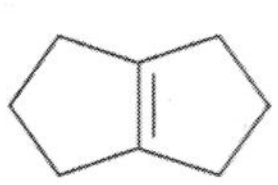

5.15   As the solution to Problem 5.14(b) illustrates, E and Z isomers of an alkene give the same ozonolysis products. Therefore, ozonolysis cannot be used to determine the stereochemistry at the alkene double bond.

5.16   The addition of ozone to 2-pentene gives an initial cycloaddition product, similar to Eq. 5.34.

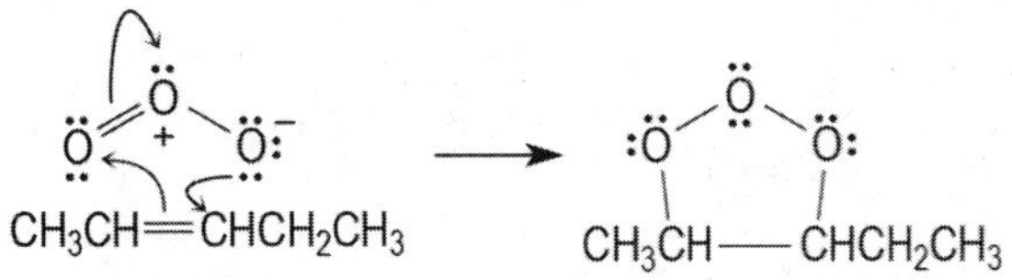

2-pentene   initial cycloaddition product

Eq. 5.36a shows how an initial cycloaddition product to break down into an aldehyde and an aldehyde oxide. This particular compound can break down in two ways, leading to two sets of products.

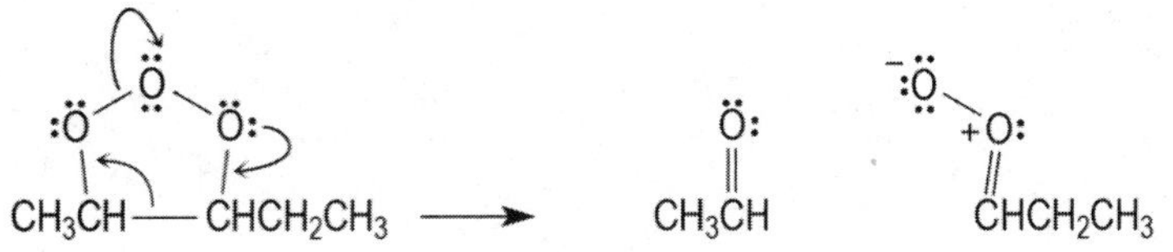

acetaldehyde   **propanal oxide**

or

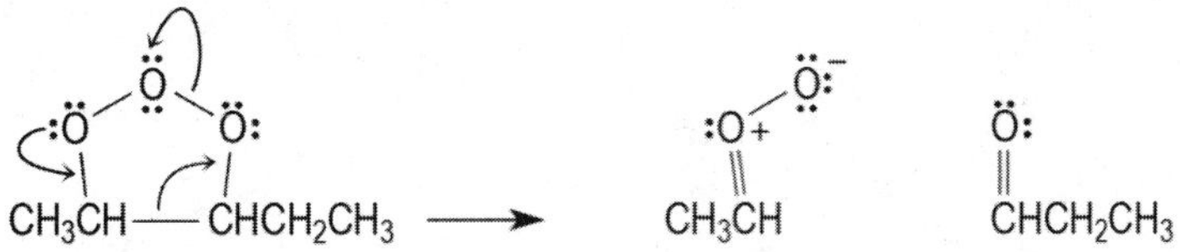

acetaldehyde oxide   **propanal**

According to Eq. 5.36b, turning an *aldehyde* over and recombining it with an *aldehyde oxide* gives an ozonide as a product. Acetaldehyde (an *aldehyde*) can be flipped over and combined with propanal oxide (an *aldehyde oxide*) to give ozonide *B*.

propanal oxide   acetaldehyde   **ozonide *B***

Propanal can combine with acetaldehyde oxide to produce the same product, *B*. Make sure you can use curved arrow formalism to illustrate this process.

Additionally, acetaldehyde can combine with acetaldehyde oxide to produce ozonide *A*. Finally, propanal and propanal oxide can combine to form ozonide *C*.

acetaldehyde oxide   acetaldehyde   **ozonide *A***

propanal oxide   propanal   **ozonide *C***

5.17   (a)

$$(CH_3)_3C \! - \! C(CH_3)_3 \longrightarrow 2\ (CH_3)_3C\cdot$$

(b)

$$(CH_3)_2C \! = \! CH_2 \quad \cdot \ddot{B}r \!: \quad \longrightarrow \quad (CH_3)_2\dot{C} \! - \! CH_2 \! - \! \ddot{B}r\!:$$

(c)

$$\left[H_2C \! = \! CH \! - \! CH \! = \! CH \! - \! CH_2\cdot \quad \longleftrightarrow \quad H_2\dot{C} \! - \! CH \! = \! CH \! - \! CH \! = \! CH_2 \right]$$

 As part (c) demonstrates, the fishhook notation, like the curved-arrow notation, can be used to derive resonance structures.

(d)

$$R \! - \! CH_2 \! - \! CH_2\cdot \longrightarrow R\cdot \ + \ H_2C \! = \! CH_2$$

5.18   (a)   This reaction is heterolytic, because it requires movement of electron *pairs*.

$$:N \! \equiv \! C\!: \quad CH_2CH_3 \longrightarrow :N \! \equiv \! C \! - \! CH_2CH_3 \ + \ :\ddot{B}r\!:^-$$
$$\overset{|}{:}\!\ddot{B}r\!:$$

(b)   This reaction is homolytic, because it involves movement of *single* electrons.

$$CH_3CHOH \longrightarrow CH_3\overset{\bullet}{C}HOH + H-\overset{\bullet\bullet}{\underset{\bullet\bullet}{O}}CH_3$$

5.19   The initiation step, shown in the problem, produces a bromine atom. This adds to the π bond in the first propagation step to give the free radical *A*.

$$H_2C=CH_2 \longrightarrow H_2\overset{\bullet}{C}-CH_2Br$$
$$\qquad\qquad\qquad\qquad A$$

In the second propagation step, the radical *A* reacts with Br_2 to give the addition product, 1,2-dibromoethane, and a bromine atom, which can begin the propagation cycle again.

$$Br-Br$$

$$H_2\overset{\bullet}{C}-CH_2Br \longrightarrow BrCH_2-CH_2Br + \cdot Br$$
$$\qquad A \qquad\qquad\qquad \textbf{1,2-dibromoethane}$$

 Radical *A* could form the product by reacting with another initiating radical Br·. But this does not happen! The reason is that the initiating radical is present in trace amounts, whereas the reactant Br_2 is present in large amounts. Be sure to read Study Problem 5.4 if you missed this point.

5.20   (a)   The first initiation step is shown in Eq. 5.50, text p. 206. In the second initiation step, the radical produced in the first initiation step abstracts a hydrogen atom from HBr:

$$H-Br$$

$$NC-\overset{\bullet}{C}(CH_3)_3 \longrightarrow NC-\overset{H}{\underset{|}{C}}(CH_3)_3 + \cdot Br$$

The bromine atom then adds to the π bond of cyclohexene in the first propagation step (*1*); the resulting radical then reacts with HBr in the second propagation step (*2*) to give the addition product.

(b) Three possible recombination products are *A*, *B*, and *C*, formed by the following recombination reactions. All of the radical reactants are chain-propagating radicals in the reaction mechanism of part (a). The chances of these steps occurring is very small, so these products are typically observed only in very small amounts at the end of the reaction.

$$2\ Br\cdot \longrightarrow Br\text{—}Br$$

A

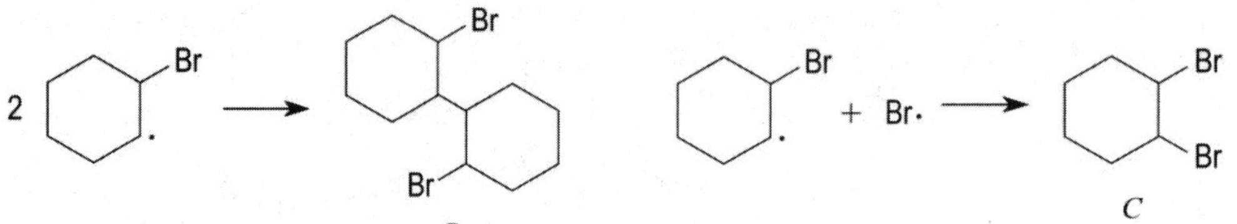

B *C*

5.21 (a)

$$H_2C{=}CHCH_2CH_2CH_3 + HBr \xrightarrow{\text{peroxides}} BrCH_2\text{—}CH_2CH_2CH_2CH_3$$

1-pentene **1-bromopentane**

(b) In this case, both of the two possible reaction pathways involve secondary free radicals. Recall, however, from the text that the regioselectivity of free-radical HBr addition is controlled in part by steric factors. Hence, the Br goes to the less congested side of the double bond.

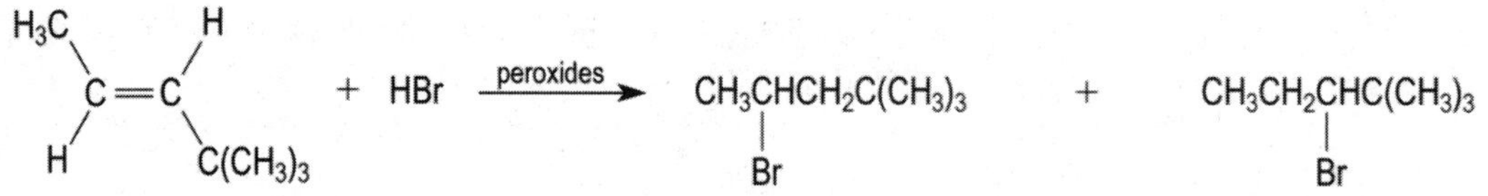

(*E*)-4,4-dimethyl-2-pentene **4-bromo-2,2-dimethylpentane** (the major product) **3-bromo-2,2-dimethylpentane**

5.22 (a) The free-radical intermediates in the propagation steps of the peroxide-promoted HBr addition to methylenecyclopentane:

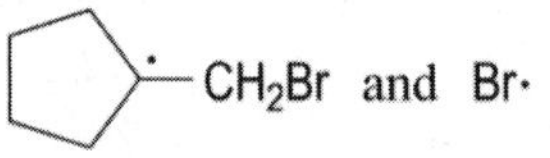

(b) The free-radical intermediates in the propagation steps of the peroxide-promoted HBr addition to 1-methylcyclopentene:

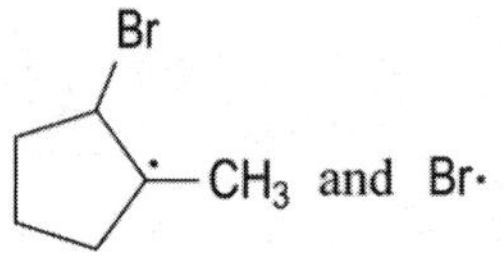

5.23 (a) One H—Cl bond (431 kJ mol^{-1}) and one H_3C—Cl bond (350 kJ mol^{-1}) are formed; and one H_3C—H bond (439 kJ mol^{-1}) and one Cl—Cl bond (239 kJ mol^{-1}) are broken. As shown by Eq. 5.69 on text p. 214, subtract the bond dissociation energies of the bonds formed from those of the bonds broken to obtain the desired estimate for $\Delta H°$.

$$\Delta H° = 239 + 439 - (431 + 350) = -103 \text{ kJ mol}^{-1}$$

(b) Two Cl—CH_2 bonds are formed (355 kJ mol^{-1} each); a Cl—Cl bond (239 kJ mol^{-1}) and a carbon–carbon π bond ($\approx$ 243 kJ mol^{-1}) are broken.

$$\Delta H° = 239 + 243 - 2(355) = -228 \text{ kJ mol}^{-1}$$

5.24   (a)   Calculation of the $\Delta H°$ values involves subtracting the bond dissociation energies of the bonds formed from those of the bonds broken, as in the solution to Problem 5.23. In the second propagation step, a secondary C—H bond is formed, and an H—Br bond is broken.

$$\Delta H° = 368 - 412 = -44 \text{ kJ mol}^{-1}$$

(b)   The corresponding calculation of $\Delta H°$ for HCl is

$$\Delta H° = 431 - 412 = +19 \text{ kJ mol}^{-1}$$

(c)   As noted in the text, endothermic propagation steps are much slower than exothermic ones. Because the second propagation step of the free-radical addition of HCl is endothermic, it is too slow to compete with the very exothermic recombination reactions that effectively shut down the chain mechanism.

5.25   The second propagation step in which Br is abstracted rather than H involves formation of a secondary C—Br bond rather than a secondary C—H bond; the bond broken (H—Br) is the same. Hence, we use the calculation in the solution of Problem 5.24(a), except that the bond dissociation energy of a C—Br bond is substituted for that of a C—H bond. The closest value in Table 5.3 is the bond dissociation energy of $(CH_3)_2CH$—Br, 309 kJ mol^{-1}. The required $\Delta H°$ is therefore

$$\Delta H° = 368 - 309 = 59 \text{ kJ mol}^{-1}$$

Hence, abstraction of Br is a very *unfavorable* process energetically ($\Delta H° \gg 0$), whereas the abstraction of H is a very favorable process ($\Delta H° < 0$). As is usually the case in free-radical reactions, the relative enthalpies of the processes governs their relative rates. Hence, abstraction of H is *much* more favorable, and therefore is much faster, than abstraction of Br.

5.26   (a)   (b)

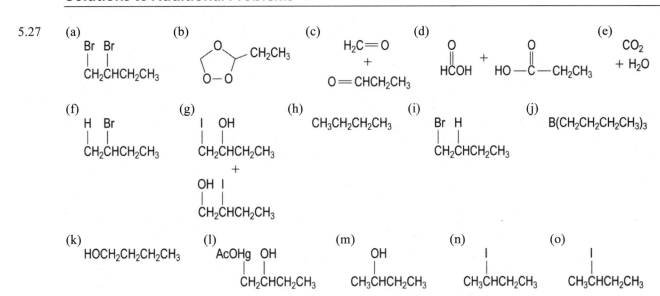

poly(methyl methacrylate)   poly(vinyl chloride)

Solutions to Additional Problems

5.27

In part (o), there is no "peroxide effect" on HI addition; that is, free radical initiators such as peroxides or AIBN have no effect on the regioselectivity of HI addition. Only HBr addition is subject to the peroxide effect.

5.28   (a)   (b)   (c)   (d)   (e) CO_2 + H_2O (f)

(g)   (h)   (i)   (j)   (k)

(l)   (m)   (n)   (o)

In part (o), there is no "peroxide effect" on HI addition; that is, free radical initiators such as peroxides or AIBN have no effect on the regioselectivity of HI addition. Only HBr addition is subject to the peroxide effect.

5.29   (a)   Any alkene with six carbons and a symmetrically-placed double bond is an acceptable answer. Three examples are the following:

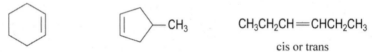

$CH_3CH_2CH{=}CHCH_2CH_3$

cis or trans

(b)   To answer this, you have to imagine double bonds at all possible places within the same carbon skeleton.

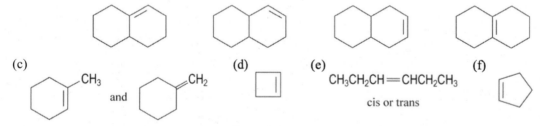

(c) CH_3 and CH_2 (d)   (e) $CH_3CH_2CH{=}CHCH_2CH_3$ (f)

cis or trans

5.30   (a)   (b)   (c) CH_2 CH_3 CH_3 CH_3

(d) $(CH_3)_2C{=}CHCH_2CH_2CH_3$

and $H_2C{=}C(CH_3)CH_2CH_2CH_2CH_3$

(e)
$$\begin{array}{c} H_3C \qquad\qquad CH_2CH_2CH_3 \\ C{=}C \\ CH_3CH_2CH_2 \qquad CH_3 \end{array}$$
E or *Z*

(f)   (g) $CH_3CH_2CH{=}CHCH_2CH_3$

cis or trans

5.31   (a)   (b)   (c)

(d)   CH₃CH₂   H
              C=C
        H₃C   CH₂CH₃

        (either stereoisomer)

(e)

5.32   (a)   This alcohol can be prepared by oxymercuration–reduction of 3-ethyl-2-pentene:

$$CH_3CH_2-C=CHCH_3 \quad \xrightarrow[\text{2) NaBH}_4\text{, NaOH}]{\text{1) Hg(OAc)}_2\text{, H}_2\text{O}} \quad CH_3CH_2-\underset{\underset{CH_2CH_3}{|}}{\overset{\overset{OH}{|}}{C}}-CH_2CH_3$$

CH₂CH₃

3-ethyl-2-pentene

Remember that reaction schemes in which the reagents are numbered over the arrow mean two *separate* reactions. Thus, in the reaction above, the starting alkene reacts with Hg(OAc)₂ and water; then, in a *separate reaction*, NaBH₄ is added. Because there are two reactions, the intermediate compound that reacts with NaBH₄ is not shown; what is it? When you write synthetic schemes this way, be sure to number the reagents. If you were to write the following:

$$CH_3CH_2-C=CHCH_3 \quad \xrightarrow[\text{NaBH}_4\text{, NaOH}]{\text{Hg(OAc)}_2\text{, H}_2\text{O}} \quad CH_3CH_2-\underset{\underset{CH_2CH_3}{|}}{\overset{\overset{OH}{|}}{C}}-CH_2CH_3$$

CH₂CH₃

it would mean that the Hg(OAc)₂, H₂O, NaBH₄, and NaOH are added to the reaction mixture at the same time. This would *not* give the desired product.

(b)   This alcohol can be prepared by hydroboration–oxidation of 1-hexene:

$$H_2C=CHCH_2CH_2CH_2CH_2 \quad \xrightarrow[\text{2) H}_2\text{O}_2\text{, }^-\text{OH}]{\text{1) BH}_3\text{, THF}} \quad HO-CH_2CH_2CH_2CH_2CH_2CH_3$$

1-hexene

(c)

$$H_2C=CHCH_2CH_2CH_3 \quad \xrightarrow{\text{HBr, peroxides}} \quad Br-CH_2CH_2CH_2CH_2CH_3$$

1-pentene

(d)   This bromohydrin can be prepared by the reaction of 2-ethyl-1-butene with Br₂ in aqueous solution.

$$CH_3CH_2-\underset{\underset{CH_2}{||}}{C}-CH_2CH_3 \quad \xrightarrow{\text{Br}_2\text{, H}_2\text{O}} \quad CH_3CH_2-\underset{\underset{CH_2Br}{|}}{\overset{\overset{OH}{|}}{C}}-CH_2CH_3$$

2-ethyl-1-butene

(e)

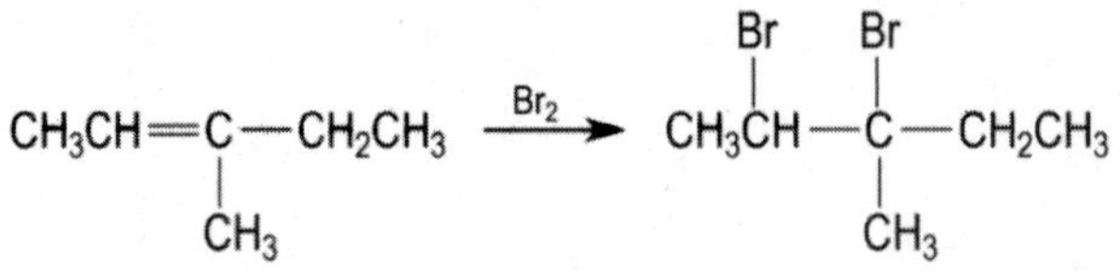

$$CH_3CH=\underset{\underset{CH_3}{|}}{C}-CH_2CH_3 \xrightarrow{Br_2} CH_3\underset{\underset{}{|}}{\overset{\overset{Br}{|}}{CH}}-\underset{\underset{CH_3}{|}}{\overset{\overset{Br}{|}}{C}}-CH_2CH_3$$

3-methyl-2-pentene

(f) Ozonolysis of 1-methylcyclohexene followed by treatment with $(CH_3)_2S$ would give the desired compound:

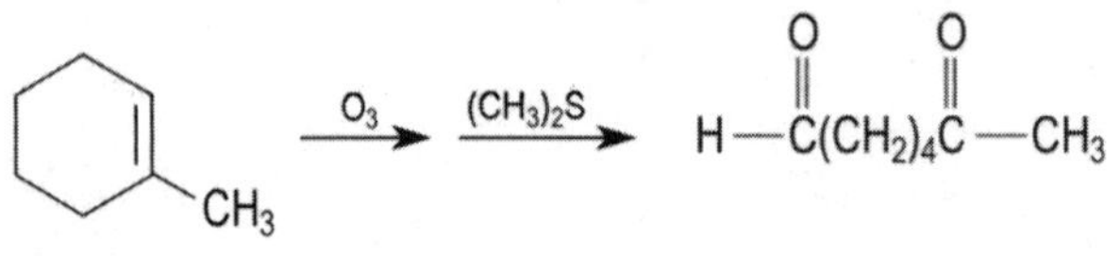

1-methylcyclohexene → O_3 → $(CH_3)_2S$ → $H-\overset{\overset{O}{||}}{C}(CH_2)_4\overset{\overset{O}{||}}{C}-CH_3$

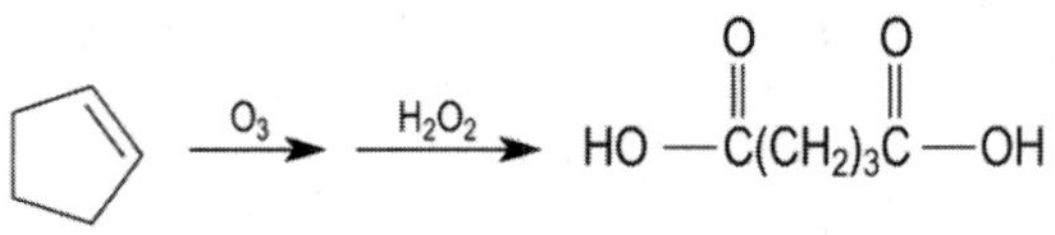

(g)

cyclopentene → O_3 → H_2O_2 → $HO-\overset{\overset{O}{||}}{C}(CH_2)_3\overset{\overset{O}{||}}{C}-OH$

(h) The addition of HBr to 1-pentene will give the desired alkyl bromide:

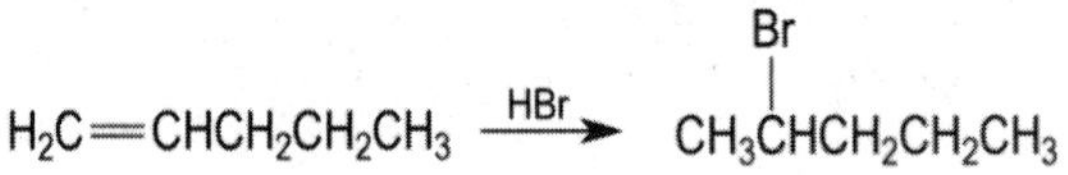

$$H_2C=CHCH_2CH_2CH_3 \xrightarrow{HBr} CH_3\underset{\underset{}{|}}{\overset{\overset{Br}{|}}{CH}}CH_2CH_2CH_3$$

1-pentene

(i) Hydrogenation of 3-ethyl-2-pentene (or 3-ethyl-1-pentene, not shown) would give the desired alkane.

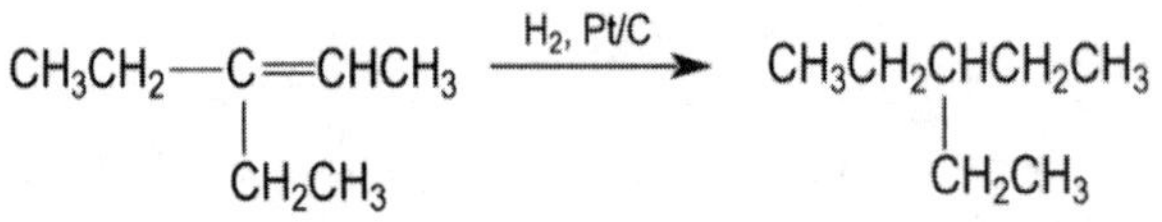

$$CH_3CH_2-\underset{\underset{CH_2CH_3}{|}}{C}=CHCH_3 \xrightarrow{H_2,\ Pt/C} CH_3CH_2\underset{\underset{CH_2CH_3}{|}}{CH}CH_2CH_3$$

3-ethyl-2-pentene

5.33 (a) This transformation can be carried out by hydroboration–oxidation in which the source of deuterium is BD_3.

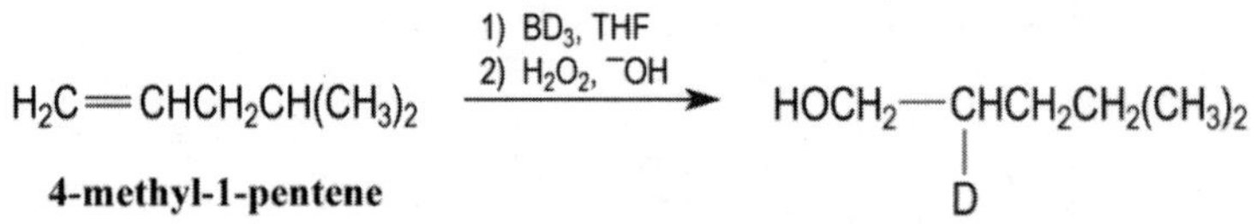

$$H_2C=CHCH_2CH(CH_3)_2 \xrightarrow[2)\ H_2O_2,\ ^-OH]{1)\ BD_3,\ THF} HOCH_2-\underset{\underset{D}{|}}{CH}CH_2CH_2(CH_3)_2$$

4-methyl-1-pentene

(b) An oxymercuration–reduction on the alkene used in part (a) gives the desired compound.

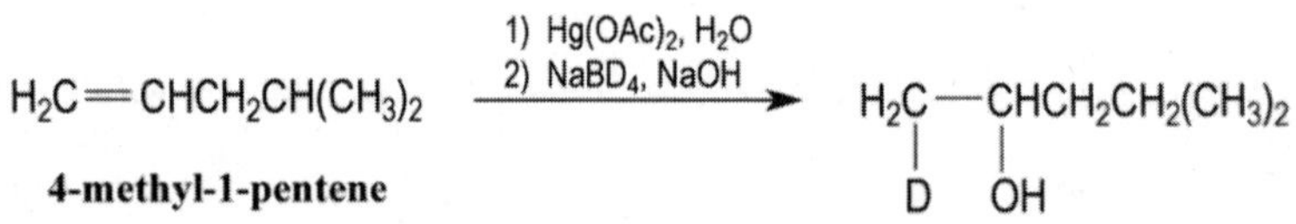

$$H_2C=CHCH_2CH(CH_3)_2 \xrightarrow[2)\ NaBD_4,\ NaOH]{1)\ Hg(OAc)_2,\ H_2O} H_2C-\underset{\underset{D}{|}}{\underset{\underset{}{}}{C}}\underset{\underset{OH}{|}}{CH}CH_2CH_2(CH_3)_2$$

4-methyl-1-pentene

5.34 The product depends on the nucleophile (Nuc:) that reacts with the bromonium ion, which is formed by the mechanism shown in Eq. 5.11, text p. 185:

(a)   The nucleophile is the bromide counter-ion; the product is the bromine addition product.

(b)   The nucleophile is a water molecule from the solvent, and the product is a bromohydrin.

(c)   The nucleophile is a molecule of the solvent methanol; the mechanism is analogous to the one shown in part (b). We leave it to you to complete the mechanism.

(d)   When the solution contains both methanol and bromide ion, both nucleophiles compete for the bromide ion. Which one "wins" the competition depends on their intrinsic effectiveness as nucleophiles and on their relative concentrations. If the bromide concentration is high enough, the product will be largely the dibromoalkane shown in the solution to part (a). At lower bromide concentrations the product will be a mixture of this dibromoalkane and the ether shown in the solution to part (c).

5.35   The reasoning is similar to that used in the solution to Problem 5.34. The product depends on the nucleophile that reacts with the mercurinium ion.

(a)   This is the usual oxymercuration–reduction product, in which water reacts with the mercurinium ion. (The mechanism is shown in Eqs. 5.21c–d, text p. 191.)

(b) Isopropyl alcohol is the nucleophile that reacts with the mercurinium ion, and the ultimate product is an ether.

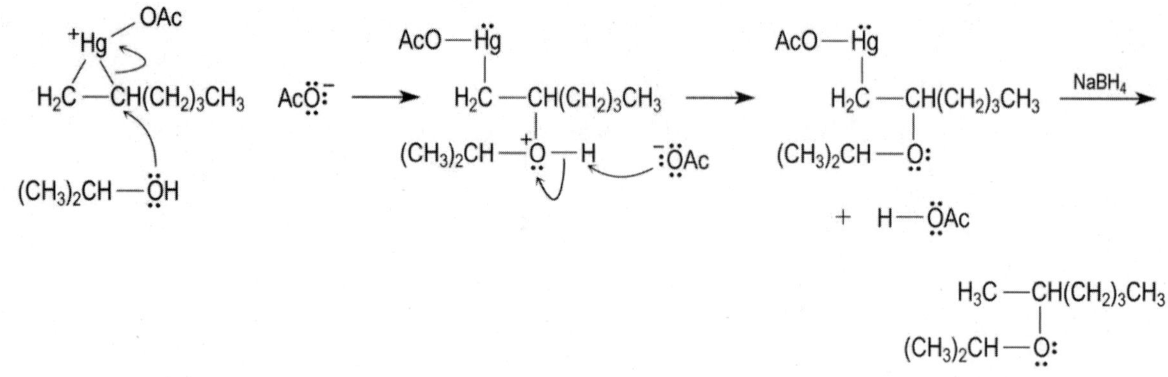

(c) In this case, the nucleophile is the either the acetate ion ($^-$OAc), the counter-ion of the mercurinium ion, which is the conjugate base of acetic acid, or acetic acid itself. The product is an ester. (We leave it to you to complete the mechanism.)

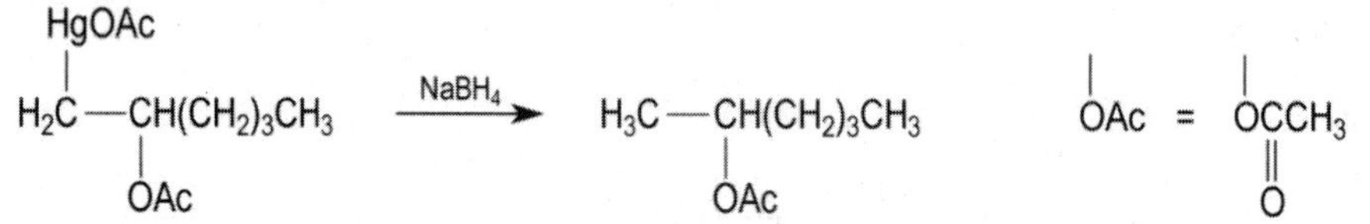

5.36 (a) Only the first two products are formed in the absence of peroxides, and only the third is formed in the presence of peroxides. Different products are formed because different mechanisms and reactive intermediates are involved under the different conditions.

(b) The mechanism for the formation of the first two products is identical to that shown for reaction of the same alkene with HCl in Eqs. 4.26 and 4.27a–b, text p. 159, except that HBr is used instead of HCl. The first product results from a carbocation rearrangement, and the second from normal regioselective ("Markovnikov") addition. The third product is the consequence of a free-radical addition mechanism, the propagation steps of which are as follows:

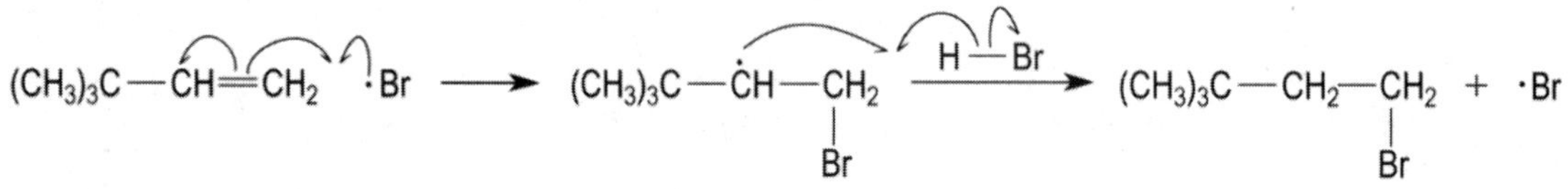

(The initiation steps are shown in Eqs. 5.49 and 5.51 on p. 206 of the text.)

(c) Peroxide-promoted addition is in competition with normal addition and rearrangement. The normal processes occur at the same rate at which they occur in the absence of peroxides. The fact that *only* the product of peroxide-promoted addition is observed, then, means that this process is much faster than the other, competing, processes.

5.37 (a)
reactive intermediate:

product:

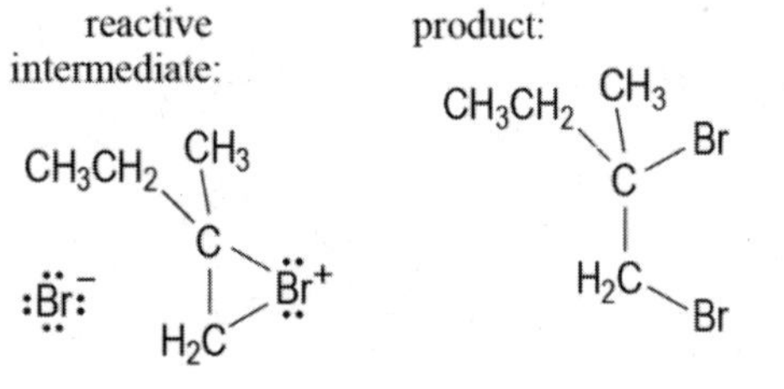

a bromonium ion

(b)
reactive intermediate:

product:

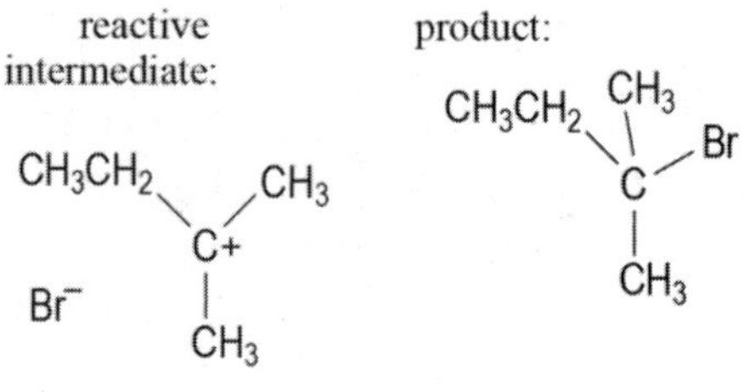

a carbocation

(c)

reactive intermediate:

a mercurinium ion

product:

(d)

reactive intermediate:

a free radical

product:

5.38   (Be sure to read Study Problem 5.4.) Because the CF_3 group has added to the carbon of the double bond with no alkyl branches, the radical $\cdot CF_3$ must be the first one that adds to the alkene. The radical that results from this addition abstracts an iodine atom from CF_3I to give the product.

5.39   (a)   Because carbon–carbon bonds are weaker than carbon–hydrogen bonds (see Table 5.3, text p. 216), a carbon–carbon bond should break more easily. Breaking of a bond to any of the methyl groups (pathway (1) below) gives a methyl radical and a tertiary free radical. Breaking of the central carbon–carbon bond (pathway (2) below) gives two tertiary free radicals. Because process (2) gives two tertiary radicals, it is the one that should occur most readily.

(b)   The cracking of 2,2,3,3-tetramethylbutane [as shown in part (a), pathway (2)] is faster because it gives two tertiary radicals; the cracking of ethane, in contrast, gives two methyl radicals. The bond that breaks to give the radical with the greater number of alkyl substituents is the weaker bond, and is therefore the bond that is more easily broken.

(c)   The $\Delta H°$ for the reaction in part (a) is the difference between the heats of formation of products and reactants. Thus,

$$\Delta H° = 2\,\Delta H_f°(\cdot C(CH_3)_3) - \Delta H_f°(2,2,3,3\text{-tetramethylbutane})$$
$$= 2(51.5) - (-225.9) = 328.9 \text{ kJ mol}^{-1}$$

The $\Delta H°$ for the cracking of ethane is

$$\Delta H° = 2\,\Delta H_f°(\cdot CH_3) - \Delta H_f°(\text{ethane})$$
$$= 2(146.6) - (-84.7) = 377.9 \text{ kJ mol}^{-1}$$

This calculation shows that the cracking of ethane is 49 kJ mol^{-1} less favorable than the cracking of 2,2,3,3-tetramethylbutane. This demonstrates quantitatively the assertion in part (b), namely, that 2,2,3,3-tetramethylbutane undergoes cracking more readily. The underlying assumption is that the free energies of activation and enthalpies of the cracking reaction are similar.

5.40   (a)   Ozonolysis breaks the polymer at its double bonds:

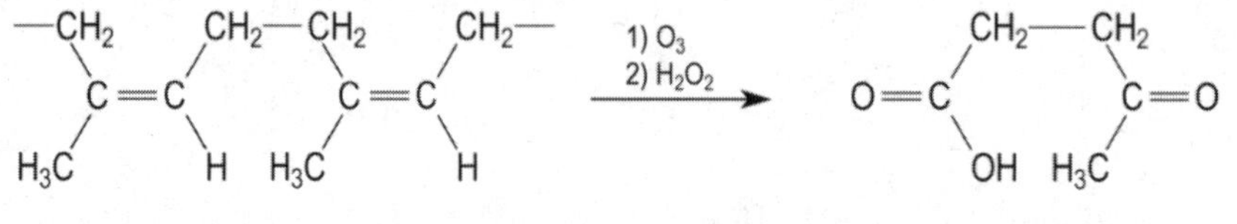

(b)   Because the E and Z isomers of alkenes give the same ozonolysis products, gutta-percha is evidently either the all-E stereoisomer of natural rubber or a polymer containing both E and Z alkene units. (In fact, it is the all-E stereoisomer.)

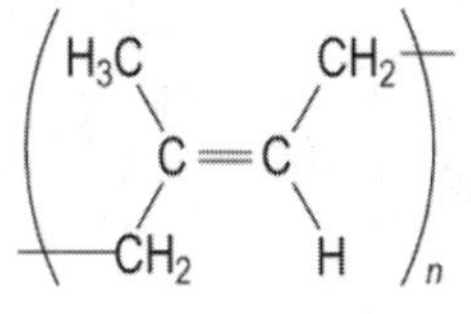

gutta-percha

5.41   (a)   The dissociation of ethanethiol at the C—S bond gives the •SH radical and the ethyl radical. From Table 5.2, text p. 212, we find that the $\Delta H_f°$ of the ethyl radical is 121.3 kJ mol^{-1}. Therefore, the bond dissociation energy is

$$\Delta H° = \Delta H_f°(\cdot SH) + \Delta H_f°(\cdot CH_2CH_3) - \Delta H_f°(\text{ethanethiol}) = (143.1 + 121.3) - (-46.2) = 310.6 \text{ kJ mol}^{-1}$$

(b)   The reaction of ethanethiol with the *tert*-butyloxy radical is

$$CH_3CH_2S\text{—}H \quad \cdot O\text{—}C(CH_3)_3 \longrightarrow CH_3CH_2S\cdot \quad + \quad H\text{—}O\text{—}C(CH_3)_3$$

ethanethiol   ***tert*-butyl alcohol**

By Eq. 5.69, text p. 214, the $\Delta H°$ of this reaction is the bond dissociation energy (BDE) of ethanethiol minus the BDE of *tert*-butyl alcohol. In Table 5.3, text p. 216, the closest BDE to that of *tert*-butyl alcohol is the BDE of H—OCH$_3$ (methanol), 438 kJ mol^{-1}. The BDE of H—SCH$_3$ (methanethiol), 366 kJ mol^{-1}, is the closest approximation to that of ethanethiol. The $\Delta H°$ of the reaction is then $366 - 438 = -72$ kJ mol^{-1}. Because the reaction is exothermic, it should be a good source of free radicals.

(c)   The first propagation step is (by analogy with Eq. 5.56b, text p. 209) addition of the •SR radical to a double bond, which breaks a carbon–carbon π bond (243 kJ mol^{-1}) and forms a carbon–sulfur bond (from part (a), -307.5 kJ mol^{-1}). The $\Delta H°$ of this step is $243 - 308 = -65$ kJ mol^{-1}. The second propagation step is (by analogy with Eq. 5.56c) formation of a carbon–hydrogen bond (-423 kJ mol^{-1}) and breaking of a sulfur–hydrogen bond (366 kJ mol^{-1}). The $\Delta H°$ of this step is $366 - 423 = -57$ kJ mol^{-1}. This calculation shows that both propagation steps are exothermic; therefore, both are reasonable propagation steps.

5.42   (a)   The H—CN bond is stronger than the O—H bond; the abstraction of a hydrogen atom from HCN by the *tert*-butoxy radical is endothermic by $528 - 438 = 90$ kJ mol^{-1}. (These numbers are from Table 5.3 with the bond energy of CH$_3$O—H as an approximation for that of (CH$_3$)$_3$CO—H.) The first initiation step, formation of the *tert*-butyloxy radical from a peroxide, is also endothermic. Because *both* initiation steps are highly endothermic, the reaction is not likely to generate a high enough concentration of radicals to initiate a chain reaction.

(b)   This propagation step involves breaking an H—CN bond (528 kJ mol^{-1}) and formation of a secondary carbon–hydrogen bond (-412 kJ mol^{-1}). The $\Delta H°$ of this step is $528 - 412 = +116$ kJ mol^{-1}. This step is highly endothermic and therefore is not reasonable as a propagation step.

5.43   (a)   In the general reaction, the C—H bond of methane and the X—X bond of the halogen are broken; the C—X bond of the alkyl halide and the H—X bond of the hydrogen halide are formed. The overall energetics of the reactions are as follows, in kJ mol^{-1}:

	Breaking of C—H bond	Breaking of X—X bond	Formation of H_3C—X bond	Formation of H—X bond	Overall energy change ($\Delta H°$)
X = Cl	439	239	−350	−431	−103
X = Br	439	190	−302	−368	−41
X = I	439	149	−241	−297	+50

These calculations show that chlorination and bromination are exothermic, whereas iodination is endothermic. Consequently, iodination does not occur because it is energetically unfavorable.

(b) Because the iodination reaction is energetically unfavorable, its reverse is energetically favorable. Consequently, H—I (from the trace of acid) reacts with H_3C—I as follows:

$$\text{H—I} \ + \ \text{H}_3\text{C—I} \ \longrightarrow \ \text{I—I} \ + \ \text{CH}_4$$

As this equation shows, iodine is formed in this reaction.

5.44

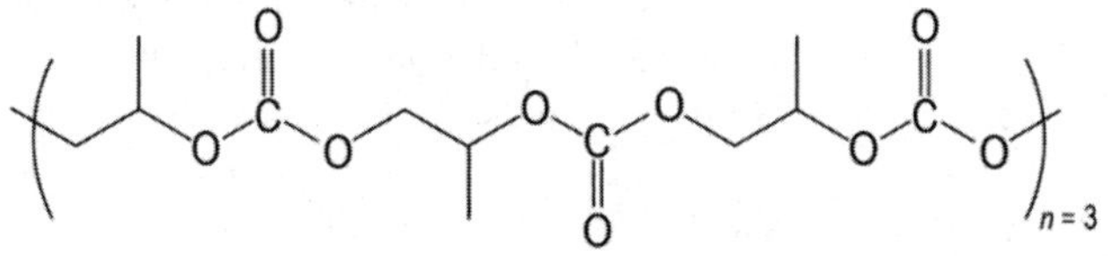

5.45 (a) As the bond dissociation energies in Table 5.3 of the text show, secondary C—H bonds are weaker than methyl C—H bonds (412 versus 423 kJ mol^{-1}). Therefore, this process is favorable by about 11 kJ mol^{-1}.

(b) The secondary radical becomes one of the chain propagating radicals, thus forming a branch.

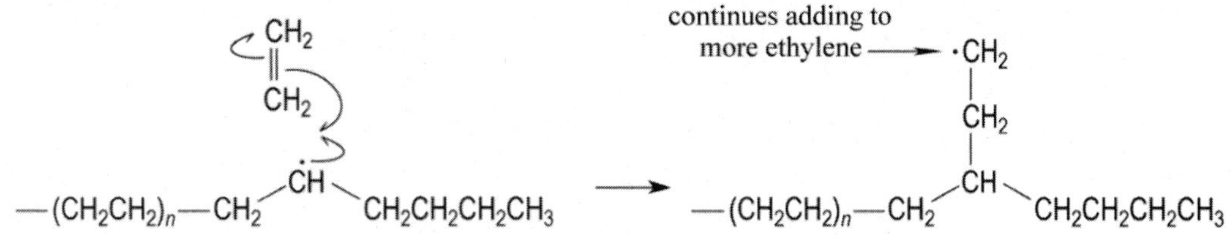

5.46 (a) The structure of polystyrene:

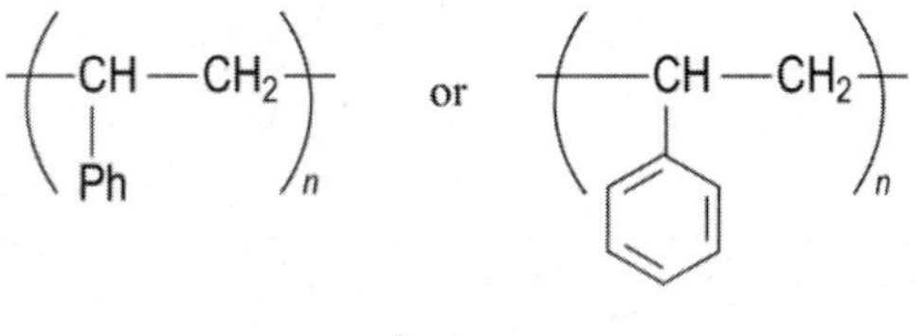

polystyrene

(b) Because both "ends" of 1,4-divinylbenzene can be involved in polymer formation, addition of 1,4-divinylbenzene serves to connect, or *crosslink,* polymer chains. Such a crosslink is shown with bolded bonds in the following structure:

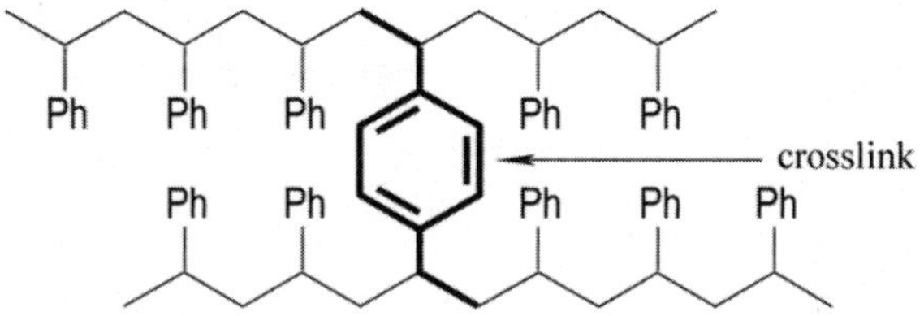

Notice that because only a small amount of 1,4-divinylbenzene is used, divinylbenzene does not polymerize with itself.

Crosslinks are introduced into polymers to increase their strength and rigidity.

5.47   A detailed discussion of the approach to solving this problem is found in Study Guide Link 5.3. Compound *A* is limonene, a natural product obtained from the oils of lemons and oranges.

limonene
(compound *A*)

5.48   Compound *A* has the connectivity of octane; because it has an unsaturation number $U = 1$ and reacts with bromine, it is an alkene. The ozonolysis results show that compound *A* is 4-octene, $CH_3CH_2CH_2CH{=}CHCH_2CH_2CH_3$. Because the double bond is located symmetrically, only one ozonolysis product is formed. Ozonolysis cannot determine whether the compound is cis or trans.

5.49   (a)   Protonation of the double bond gives a tertiary carbocation, which undergoes a Lewis acid–base association reaction with one of the oxygens within the same molecule.

(b)   The nitrogen within the same molecule reacts as a nucleophile with the bromonium ion intermediate. The bromonium ion is formed by the mechanism shown in Eq. 5.11 on text p. 185.

(c)   In the first part of the mechanism, a sulfonium ion is formed by a nucleophilic reaction of the π electrons of one double bond with the sulfur in sulfur dichloride. (This process is analogous to the formation of a bromonium ion.) The chloride ion liberated in this step then reacts at a ring carbon to open the sulfonium ion. The process is then repeated, except that the process is intramolecular.

An isomer of the product shown could also be formed. What is it? Consider the last step. Once you've tried to work it out for yourself, see part (d) below for an analogous explanation.

(d)   The oxygen of a hydroxy group introduced in the oxymercuration reaction of one double bond serves as the nucleophile in the opening of the mercurinium ion formed at the second double bond within the same molecule. Two products are formed because the oxygen can react with the mercurinium ion at either of two carbons.

[The acetate ion is a by-product of the first step of oxymercuration; its reaction with the protonated ether is shown only for *(b)*.] Treatment with $NaBH_4$ replaces the mercury with a hydrogen.

(e)   A ·CBr₃ radical formed in the initiation steps adds to the double bond; the resulting radical then abstracts a bromine atom from Br₃C—Br to propagate the chain.

(f)   The initiation step, reaction of the thiol with an alkoxy radical formed by homolysis of a peroxide, is shown in Eq. 5.56a, text p. 209. The radical produced in that step adds to the π bond of the alkene so as to produce the tertiary free radical, and this radical reacts with the thiol to propagate the chain.

Be sure that you did not form the product by a recombination of two radicals; see the discussion of Eq. 5.57 in Study Problem 5.4, text p. 209.

(g)   Homolysis of the O—Cl bond gives a chlorine atom plus an alkoxy radical, which then undergoes a reaction that opens the ring to give a new radical *A*.

Radical *A* then reacts with the starting material to regenerate the alkoxy radical, which propagates the chain.

5.50   (a)   The methyl radical reacts at the carbon of the double bond with fewer substituents because this mode of reaction gives the more substituted radical, and because the less-substituted carbon is the least sterically hindered site of reaction. (For ethylene, in reaction *A*, only one product is possible.)

(reaction *A*)   H₂C=CH₂ + ·CH₃ ⟶ ·CH₂CH₂CH₃

(reaction *B*)   (CH₃)₂C=CH₂ + ·CH₃ ⟶ (CH₃)₂Ċ—CH₂CH₃

(reaction *C*)   (CH₃)₂C=CHCH₃ + ·CH₃ ⟶ (CH₃)₂Ċ—CH(CH₃)₂

(b)   When we compare reactions *B* and *A*, we see that two methyl groups at the site of the radical accelerate the reaction modestly. This is probably due to the stability of the free radical (tertiary versus primary) and the reflection of this stability in the transition state for its formation (Hammond's postulate). It might be surprising that the effect on rate is so modest. However, if you consider the effect of substituents on alkene stability (Table 4.1, text p. 150) and the effect of substituents on free-radical stability (Table 5.2, text p. 212), you can see that these effects are not very different. Remember that relative rates reflect the *difference* in standard free energy of transition states and reactants. Because branching affects both starting material and transition state in the same way, and because the effect is of a similar magnitude in both alkenes and free radicals, the residual effect on the transition state is rather modest.

When we compare reactions *B* and *C*, we start with a *more* substituted alkene in *B* and go to an identically substituted radical in both cases. This should lower the energy of the alkene starting material more than that of the transition state. If so, reaction *C* should be slower—again, not because the transition state is *higher* in energy, but because the starting material is *lower* in energy. (When you climb two energy mountains, the top of the mountains may be at the same elevation above sea level, but if the base camp is lower for one mountain, then that mountain requires a greater climb.) In addition, a steric effect operates on reaction *C* that does not operate on reaction *B*, because the carbon at which reaction *C* occurs is more substituted. This would tend to raise the energy of the transition state for reaction *C* and lower the rate, as observed.

5.51   The structure of disiamylborane is shown below.

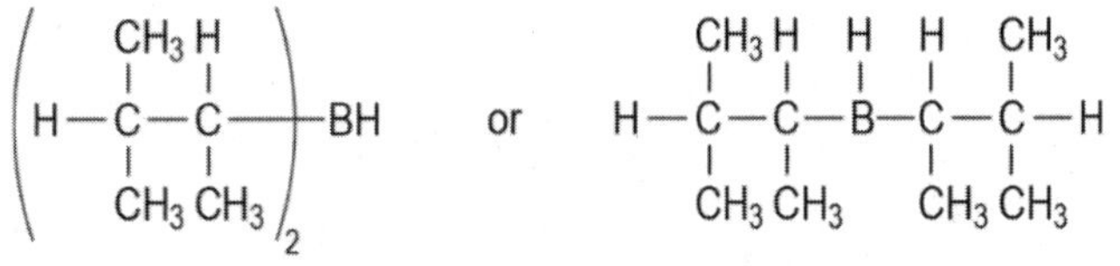

This is an example of steric effects. Placing three highly branched groups around a central boron results in van der Waals repulsions. These repulsions are severe enough that only two groups can be bound to boron in the case of disiamylborane. Now, alkyl branches stabilize sp^2-hybridized boron (see the solution to Problem 4.49 on p. 96 of this manual for a discussion of boron's hybridization) just as they stabilize sp^2-hybridized carbon; but if the alkyl branches are *themselves* branched, they form a thicket of methyl groups that interact repulsively with each other as more of these branches are accumulated. These repulsive interactions reduce the relative stability of the trialkylboranes—evidently, so much so that they cannot form.

5.52   The addition of ozone to *trans*-3-hexene gives an initial cycloaddition product, as illustrated by Eq. 5.34 on p. 199. This product can fragment into propanal and propanal oxide, similar to Eq. 5.36a on p. 200, and shown below.

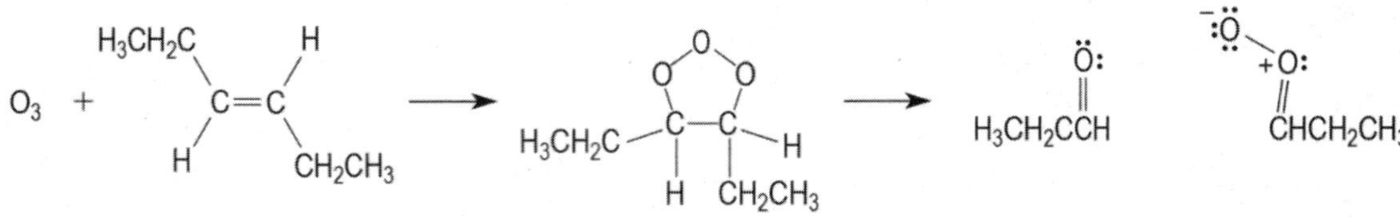

These two products can then recombine to form an ozonide, similar to Eq. 5.36b on p. 200. However, the problem states that there is an excess of ^{18}O-labeled acetaldehyde. (An *excess* of a compound means that there are many more equivalents of it present relative to anything else.) So it is more likely that the propanal oxide will react with the ^{18}O-labeled acetaldehyde, because there are many more equivalents of it, rather than the propanal. Note the final position of the isotope label in the ozonide product.

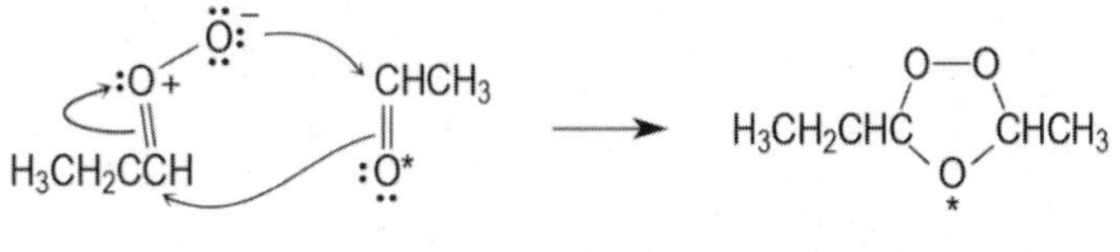

5.53   The structure of the product indicates that the *more substituted* carbon of one alkene molecule is connected to the *unsubstituted* carbon of another (or vice-versa). Because Lewis acids can react with double bonds, it would seem that, if one alkene molecule could be converted into a Lewis acid such as a carbocation, this could get the process started. The acidic conditions are just right for carbocation formation:

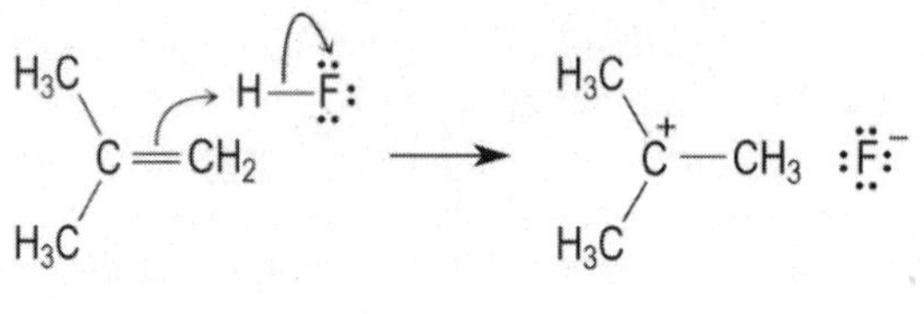

This carbocation reacts as a Lewis acid with a Lewis base—the π electrons of another alkene molecule:

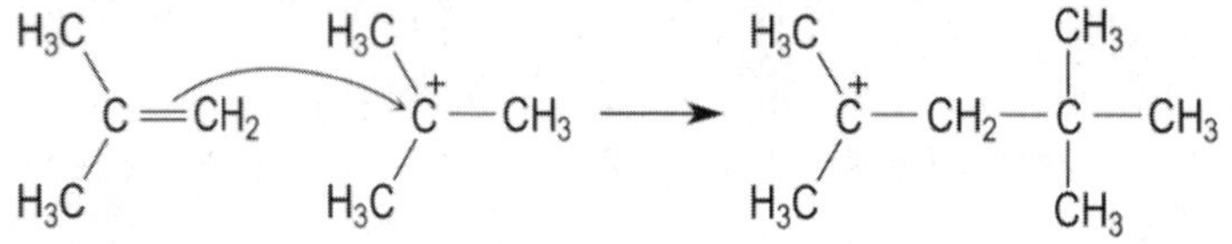

This type of process continues indefinitely:

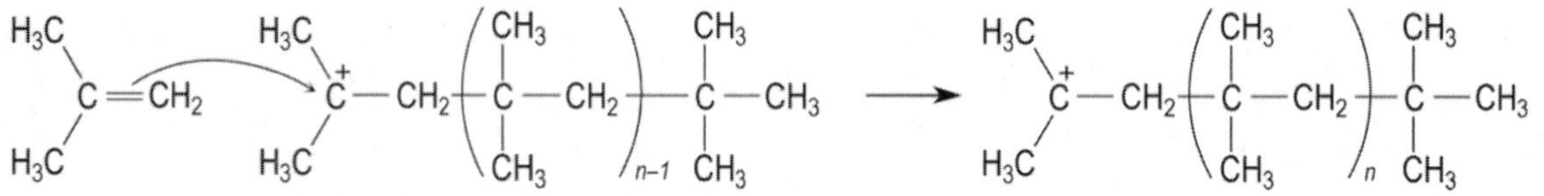

Eventually, the alkene concentration becomes so low that the fluoride ion competes as a Lewis base for the carbocations. The *tert*-butyl fluoride is formed by a Lewis acid–base association reaction of the *tert*-butyl carbocation with fluoride.

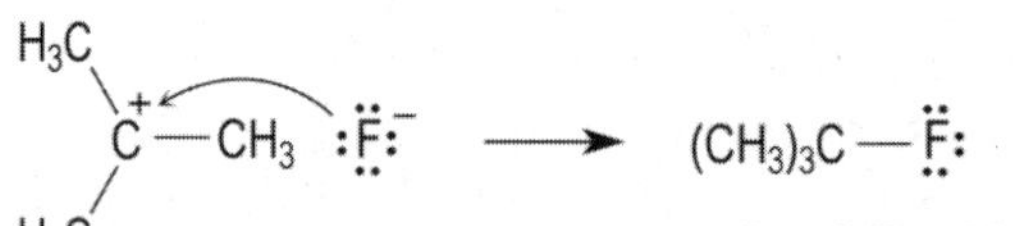

tert-butyl fluoride

The polymer chain can terminate either by a Lewis acid–base association reaction of the carbocation with fluoride ion or by loss of a proton to give an alkene. In either case, n is so large that the terminal groups are ignored in the final structure.

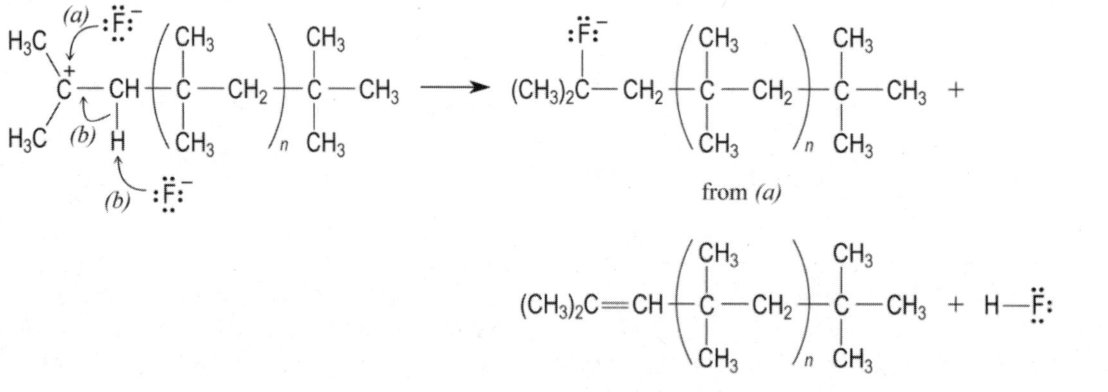

from *(a)*

from *(b)*

5.54   The starting alkene contains six carbons; acetone, the ozonolysis product of *B*, contains three carbons. Because acetone is the only product formed, it is reasonable to suppose that *two equivalents* of it are produced. The formula of *B* indicates one degree of unsaturation; because it undergoes ozonolysis and other reactions of alkenes, *B* is evidently an alkene that contains one double bond. From the ozonolysis data, compound *B* is 2,3-dimethyl-2-butene, $(CH_3)_2C{=}C(CH_3)_2$. A comparison of the carbon connectivities of compound *B* and the alkene starting material shows that a rearrangement has occurred during the transformation of the alkene starting material to *B*. A source of the rearrangement is a reaction that involves carbocations; the HBr addition is such a reaction. Indeed, the reactions shown in Eqs. 4.26–4.27, text p. 159–160, with HBr instead of HCl, shows that the structure of compound *A* is likely to be

$$Br$$
$$(CH_3)_2CH - C(CH_3)_2$$

2-bromo-2,3-dimethylbutane
(compound *A*)

This structure has the same carbon connectivity as compound *B*. In the reaction that converts *A* to *B*, the strong base in the problem removes a proton and a bromide ion is expelled as NaBr. Given the structure of compound *B* shown above, the following analysis is reasonable:

Br   lost as Br⁻
$$(CH_3)_2C - C(CH_3)_2$$
removed by the base H

A curved-arrow mechanism consistent with this analysis is the following:

:B̈r:
$$(CH_3)_2C - C(CH_3)_2 \longrightarrow (CH_3)_2C = C(CH_3)_2 + CH_3\ddot{O} - H + :\ddot{B}r:^-$$
$$CH_3\ddot{O}:^- \quad H$$

A similar process can and does occur by removal of a hydrogen from any of the methyl groups to give 2,3-dimethyl-1-butene. However, compound *B* is the major product, and the problem focused on that product for simplicity. (This type of reaction is called an E2 reaction, and will be discussed in detail in Sec. 9.5.)

Principles of Stereochemistry

STUDY GUIDE LINKS

6.1   Finding Asymmetric Carbons in Rings

Finding asymmetric carbons in rings requires that each arm of a ring be treated as a separate group. Consider the carbon below that is indicated with an asterisk.

CH_3

CH_3

CH_3

This carbon bears a CH_3 group and a hydrogen. (The hydrogen is not shown explicitly because this is a skeletal structure.) The other two groups are the two "arms" of the ring. *Even though the two groups are tied together into a ring, they can be considered as separate groups.* Proceeding counterclockwise from the asterisked carbon, the connectivity sequence is —CH_2—CH_2—; proceeding clockwise, the connectivity sequence is —CH_2—$C(CH_3)_2$—; a difference occurs at the second carbon out. Hence, all the groups on the asterisked carbon are different, and thus this carbon is an asymmetric carbon. Notice that we continue as far as necessary along each branch to find a difference.

The following structure, in contrast, has no asymmetric carbon:

CH_3

1

6   2

5   3

4

For example, carbon-1 bears a methyl, a hydrogen, and the two "arms" of the ring. Proceeding around the ring from carbon-1, at carbon-2 and carbon-6 the arms are the same (—CH_2— groups);

at carbon-3 and carbon-5 the arms are the same (again, —CH_2— groups); and the two arms join at carbon-4. If no difference is found when proceeding from the carbon in question to the point at which the arms of the ring join, the two arms of the ring are identical.

 ## 6.2   Stereocenters and Asymmetric Atoms

Stereochemical terminology changes as chemists acquire a more rigorous understanding of the geometrical aspects of stereochemistry. The purpose of this section is to alert you to a problem that still exists with the term *stereocenter*. The definition used in this text—"an atom at which interchange of two groups gives a stereoisomer"—is the one accepted by all experts in the field of stereochemistry. A number of texts, however, have defined *stereocenter* to be an atom with four different groups attached. This is the same definition that is used in this text for *asymmetric carbon*. Thus, some texts do not distinguish between the terms *asymmetric carbon* and *stereocenter*. As shown in the text, however, all asymmetric carbons are stereocenters, but the converse is not true: not all stereocenters are asymmetric carbons. The cases discussed in Sec. 4.1C on p. 131 of the text demonstrate this point: the carbons of the double bond in *E* and *Z* isomers are stereocenters but are not asymmetric carbons. Because so many texts misdefine this term, you should be alert to the fact that you may see it misused.

A number of chemists have objected to the use of the term *asymmetric carbon* because a carbon atom is not "asymmetric." In fact, many chiral *molecules* themselves are not necessarily "asymmetric;" many have certain types of symmetry. Perhaps a better term for asymmetric carbon would be *asymmetrically substituted carbon*. Other terms you might hear are "chiral center" and "chiral carbon," but objections have been raised to these also. The term *asymmetric carbon* has been so widely used that everyone knows what it means, and this text has continued to use it. The important point is that, whatever we call it, an asymmetric is *not* the same as a stereocenter; rather, asymmetric atoms constitute only one type of stereocenter.

One last point about the definition of asymmetric carbon: When we say that an asymmetric carbon has "four different groups" attached, we might have occasion to ask, "How different do the groups have to be for a carbon to be labeled *asymmetric*?" The differences among attached groups used to define "asymmetric carbon" generally refer to *constitutional* (that is, compositional) or *connectivity differences. Any* constitutional or connectivity difference between two attached groups makes them different. However, it is also possible for two groups to have the same connectivity but differ only in their *stereochemical configurations*. Problem 6.45 on text p. 270 explores how we deal with such a situation. However, this is a relatively rare situation for beginning students.

 ## 6.3   Using Perspective Structures

The most foolproof way to assign absolute configurations from a perspective structure (that is, a line-and-wedge structure) is to build a model and use it to assign configurations. However, once you gain experience with models, you will probably be able to use perspective structures themselves to assign configurations.

Before trying to use line-and-wedge structures, be sure you understand how to interpret them, as was discussed in Chapter 1, p. 16, of the text. To review:

this group (and the
dashed bond to it) are
behind the page

this group (and the
wedged bond to it)
are in front of the page

these three groups (and the
bonds to them) are
in the plane of the page

If you use journals, monographs, or other texts, you will find that there are variations on the conventions used for representing perspective. In some cases, solid bonds rather than solid wedges are used; dashed bonds rather than dashed wedges are used. Thus, all of the following structures mean the same thing:

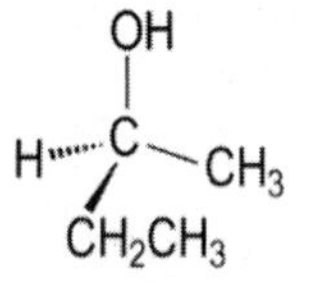

convention used in this text	bar/hatched bar convention	wedge/inverted dashed wedge convention	wedge/dashed line convention

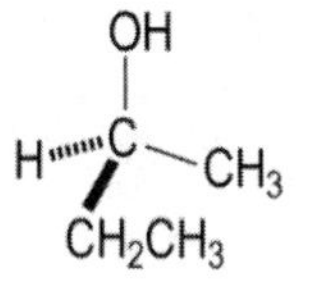

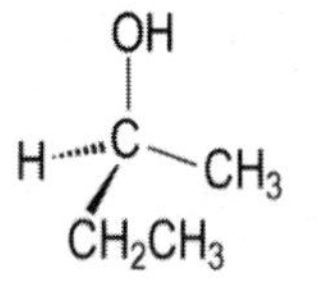

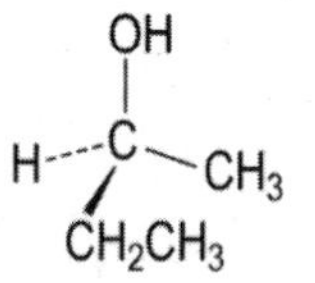

If you have a choice in the matter, it is easiest to assign the configuration of an asymmetric carbon using perspective formulas if you place the group of lowest priority behind the page—on a dashed wedge. When you do this, you are automatically viewing the molecule in the proper manner for assigning configuration:

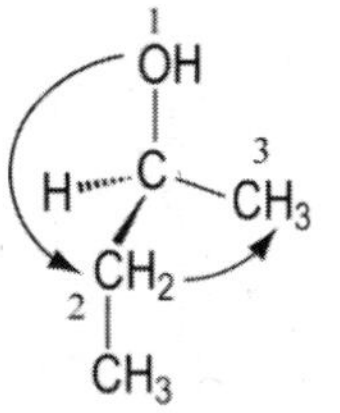

That is, you are looking from the asymmetric carbon down the bond to H. It doesn't matter whether the dashed bond is slanted to the right (as in Study Problem 6.2, text p. 234) or to the left, as above.

If you are confronted with a structure that is *not* drawn in this standard manner and you need to assign a configuration you have two options: build a model or learn to manipulate the perspective structure mentally. Let's consider these two situations in turn.

If you build a model, keep your model as simple as possible. *Don't build more of a model than necessary.* (For example, if a molecule contains a methyl group and you don't have to deal explicitly with the hydrogens, then simply use a carbon.) When you are asked to assign configuration, a handy device is to dedicate four atoms of your model set for this purpose. Take an atom and label it "1" with a marker, paint, or "white-out." Label three other atoms "2", "3", and "4". Then, when you see a perspective formula you can't interpret directly, first assign relative priorities to the groups in the structure. Then build a model using a carbon and only your four labeled atoms. Just make sure they go in the same positions of relative priority as the groups of the structure you are trying to interpret. Then manipulate your model so that you are looking *from* the asymmetric carbon to group "4," and then assign the configuration.

Alternatively, the ability to manipulate structures mentally gradually comes by working problems in stereochemistry with models. However, the following short exercise might help you develop this ability. Start with a tetrahedral model, and imagine turning it 120° in the plane of the page as shown by the arrows:

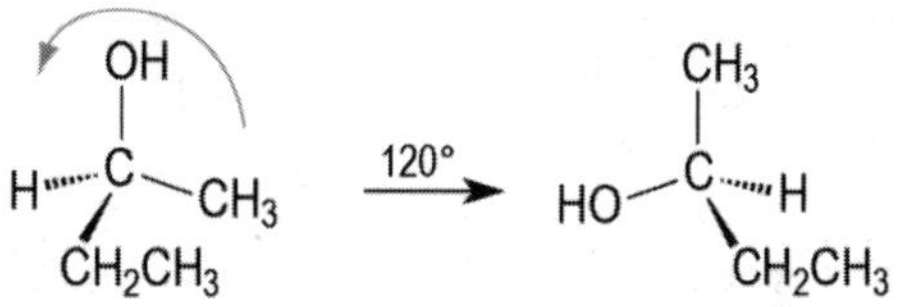

Now imagine rotating it 120° about each bond in turn so that each group moves into the adjacent position, as follows:

Then practice turning the structure about the other bonds and draw the result. Use a model to check yourself. If you spend a few minutes carrying out these or similar exercises, you should quickly become more proficient in visualizing perspective structures in three dimensions.

One last, very useful, trick in assigning configurations uses the fact that there are only two possible configurations, *R* and *S*. If you assign a configuration *and know you're wrong*, then you can reverse the assigned configuration to the correct one. For example, if you find yourself looking *from* the atom of lowest priority *to* the asymmetric carbon, you are looking down the bond in the "wrong" way for assigning configuration. Knowing you're wrong, assign the configuration anyway, but then reverse it.

Clockwise descending priorities implies *R*; but, because the view is wrong, the assignment is wrong. Therefore, the configuration is *S*.

assigning configuration
from the "backwards" view

In this example, the atom of lowest priority is the hydrogen; we are viewing the molecule incorrectly for assigning configuration; so, we assign the "wrong" configuration and then reverse it. In stereochemistry, two wrongs make a right!

FURTHER EXPLORATIONS

6.1   Terminology of Racemates

Nowadays the terms *racemic mixture* and *racemate* are often used interchangeably to indicate *any* equimolar mixture of enantiomers. However, the term *racemic mixture* also has a more precise use. This term is sometimes used to indicate a *solid state* of a racemate that is a mixture of two crystal types, one for each enantiomer. In other words, a racemic mixture is a conglomerate of two enantiomeric crystal forms.

The term *racemic compound* is sometimes used, and it, too, has a more precise definition. This term indicates a *single* crystal form of a racemate that contains equal numbers of molecules of each enantiomer. It is not a conglomerate, but rather a single crystal that contains equal amounts of each enantiomer within its crystal structure.

Racemic mixtures and racemic compounds can be distinguished by an analysis of their melting behavior as a function of composition.

6.2   Isolation of Conformational Enantiomers

Although butane undergoes internal rotation too rapidly for its enantiomeric gauche forms to be isolated, chemists have succeeded in preparing other compounds in which internal rotation about a single bond is so slow that conformational enantiomers can be isolated and purified. The following compound is an example:

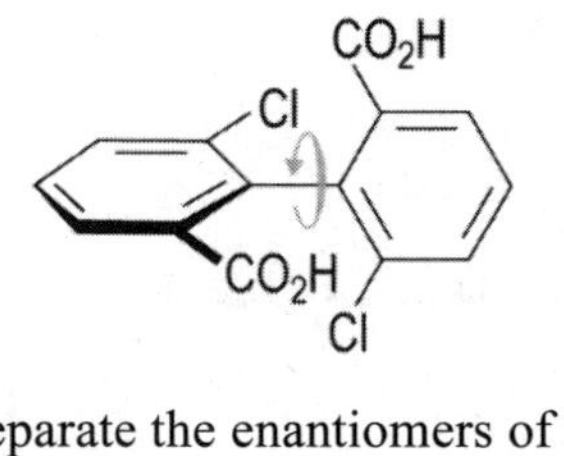

conformational enantiomers
can be isolated at room temperature

(How might you separate the enantiomers of this compound?) The two benzene rings are forced to occupy perpendicular planes; if they were to occupy the same plane, substituents on one ring would "bump into" those on the other, and very severe van der Waals repulsions would result. The conformation of the compound shown above is chiral. (You can verify its chirality by drawing or constructing its mirror image and showing that it is noncongruent.) Interconversion of the two enantiomeric conformations requires a rotation about the central bond, indicated by the circular arrow, which takes the rings into the same plane. Because of the van der Waals repulsions just noted, the energy barrier to this process is so high that rotation cannot occur at room temperature. In contrast, the compound in which the —CO$_2$H groups are replaced by hydrogens and the chlorine substituents are replaced by fluorines *cannot* be separated into its enantiomers at room temperature. In this case, racemization requires a planar conformation in which two small atoms (F and H) are close. Because van der Waals repulsions are much smaller here, this process requires less energy, and racemization occurs readily at room temperature.

    If you find an error in this manual, please visit http://people.pharmacy.purdue.edu/~loudonm/teaching/

SOLUTIONS TO PROBLEMS

Solutions to In-Text Problems

6.1 (a) This compound is chiral.
   (b) This compound is achiral.
   (c) This compound is chiral. Notice that the groups directly attached to the asymmetric carbon (CH_2) are the same; however, the groups are different because of more remote differences.
   (d) This compound is chiral; the asymmetric atom is the nitrogen.

6.2 (a) A shoe is chiral, because it has a noncongruent mirror image (the shoe of the opposite foot).
   (b) A book is achiral, if we neglect the writing.
   (c) A human body is achiral, if we neglect the internal organs and any superficial characteristics. However, if the hair is parted to one side or the other, or if there are other features that differentiate left from right, the body is chiral. If we "burrow down" to the molecular level, the amino acid and DNA building blocks of the body are chiral. Thus, the body itself is chiral.
   (d) The pair of shoes is achiral. Again, however, if we "burrow down" to the molecular level, we would find that the biomolecules that make up leather are chiral, and this would make the pair itself chiral.
   (e) Chiral. (If you're right handed, try using scissors to cut something with your left hand.)

6.3 (a) Planes of symmetry in methane bisect one set of H—C—H bonds and contain the other H—C—H bonds. (There are four such planes.)
   (b) The planes of symmetry are all the planes that bisect the base and include the tip.
   (c) One plane of symmetry in ethylene is the plane of the page; the two others are the planes perpendicular to the page. The center of symmetry is the point in the center of the C==C bond.
   (d) The plane of symmetry is the plane of the page, and the center of symmetry is the point in the center of the C==C bond.
   (e) The plane of symmetry in *cis*-2-butene is the plane of the page and the plane perpendicular to the page that bisects the C==C bond.
   (f) The plane of symmetry contains the three C—C bonds, and the center of symmetry is at the center of the C2–C3 bond.

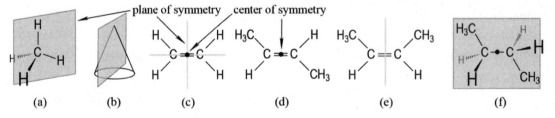

6.4 The asymmetric carbons are indicated with asterisks. (In part (c), there are no asymmetric carbons.)

   (a)   (b)

6.5   Remember that there are many different ways to draw a correct line-and-wedge structure. If your structures don't look like these, and if you're not sure whether yours is correct, make a model of both and check them for congruency. (When possible, we often adopt a "standard" representation in which the bond to the atom of lowest priority—hydrogen in (a) and (b)—is the dashed wedge bond (going away), and it is placed to the left of the asymmetric carbon. This makes it very easy to determine configuration. However, this "standard" representation is not necessary or always possible.)

(a)   (b) 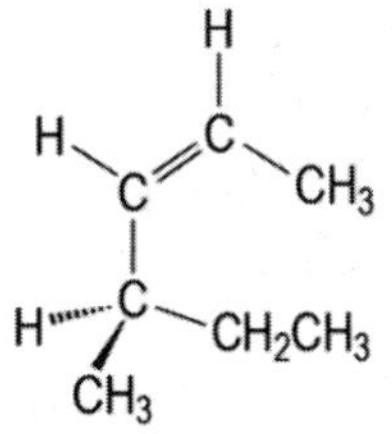   (c)

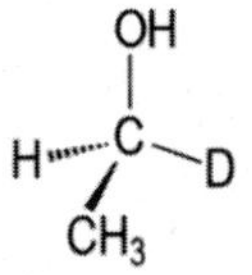

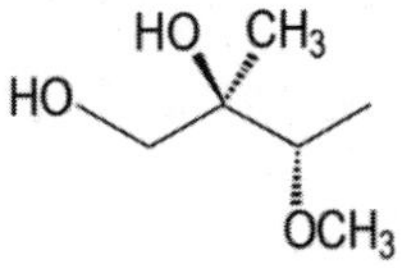

6.6   (a)   The asymmetric carbon in the given stereoisomer of alanine has the *S* configuration.
(b)   The asymmetric carbon in the given stereoisomer of malic acid has the *S* configuration.
(c)   The asymmetric atom is a nitrogen, and it has the *S* configuration. [See the solution to Problem 6.1(c)].

6.7   Halve the concentration. If the rotation at the first concentration is +10°, the rotation at the lower concentration should read +5°; if the rotation at the first concentration is −350°, the rotation at the lower concentration should be −175°, which is the same as +185°; if the rotation at the higher concentration is +370°, the rotation at the lower concentration should be +185°. To decide between the last two possibilities, halve the concentration again. Essentially, we should plot observed rotation against concentration; the slope of the line is the specific rotation.

6.8   (a)   Use Eq. 6.1 on text p. 239:

$$\alpha = [\alpha]cl = \frac{(66.1 \text{ deg mL g}^{-1} \text{ dm}^{-1})(1 \text{ dm})(5 \text{ g})}{100 \text{ mL}} = 3.31 \text{ deg}$$

(b)   The specific rotations of enantiomers must have the same magnitude but the opposite sign. Therefore, the enantiomer of sucrose must have specific rotation = −66.1 deg mL^{-1} g^{-1} dm^{-1}.

6.9   (a), (b)   The asymmetric carbon of thalidomide (marked with an asterisk, *) and the configuration of the *S* enantiomer:

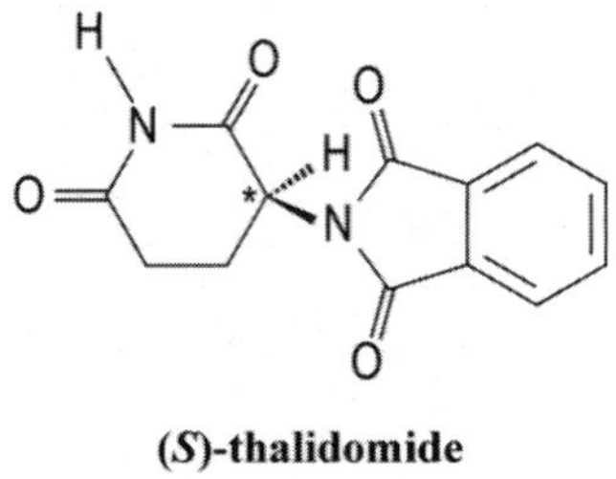

(*S*)-thalidomide

6.10   (a)   A concentration of 0.1 *M* is the same as a concentration of $(0.1 \text{ mol L}^{-1})(150 \text{ g mol}^{-1})(0.001 \text{ L mL}^{-1}) = 0.015 \text{ g mL}^{-1}$. Using this number for c in Eq. 6.1 on text p. 237.

$$[\alpha] = \frac{\alpha}{cl} = \frac{+0.20 \text{ deg}}{(0.015 \text{ g mL}^{-1})(1 \text{ dm})} = +13.3 \text{ deg g}^{-1} \text{ mL dm}^{-1}$$

(b)   Because the two enantiomers are present in equal amounts, their rotations cancel, and the observed rotation is zero.
(c)   Because the concentration is halved, the observed rotation is also halved. Thus, it is 0.10 deg.

(d)   The specific rotation is the same, because by definition this parameter is independent of concentration.

(e)   The specific rotation of L is -13.3 deg mL^{-1} g^{-1} dm^{-1}; enantiomers have specific rotations of equal magnitudes and opposite signs. Again, the specific rotation is the same regardless of concentration.

(f)   In this solution, the 0.005 mole of L cancels the rotation of 0.005 mole of D. The resulting rotation is as if only 0.005 mole, or a concentration of 0.05 mol L^{-1}, of D were present. This amount of D corresponds to a concentration of $(0.05 \text{ mol L}^{-1})(150 \text{ g mol}^{-1})(0.001 \text{ L mL}^{-1}) = 0.0075$ g mL^{-1}. Since this concentration is half of the original concentration, the observed rotation is also half of the original—that is, $+0.10$ deg.

(g)   Using Eq. 6.5b, we can calculate the EE by figuring out the % of the major enantiomer. Since there is 0.01 moles of D and 0.005 moles of L, then D is the major enantiomer and the solution is (0.01 D/0.015 D & L) 66.7% D. Substituting into Eq. 6.5b we find that the EE of D is 33%.

6.11   (a)

(*S*)-ibuprofen

(b)   Since the specific rotation of the mixture, $[\alpha]_{\text{mixture}}$, and the EE is given, use Eq. 6.4 and solve for $[\alpha]_{\text{pure}}$, which works out to $+57.4$ degrees mL g^{-1} dm^{-1}. The pure R enantiomer is then -57.4 degrees mL g^{-1} dm^{-1}.

(c)   Using Eq. 6.5b, solving for the percentage of the major, S-enantiomer, gives 95%. Thus the minor, R-enantiomer, must be present in the remaining 5%.

6.12   The racemate has no effect on the observed rotation other than to dilute the sample. Hence, after addition of the racemate, the concentration of the excess (*R*)-2-butanol is 0.75 M. This corresponds to $(0.75 \text{ mol L}^{-1})(74.12 \text{ g mol}^{-1})(0.001 \text{ L mL}^{-1}) = 0.055$ g mL^{-1}. Use this as the value of c in Eq. 6.1 with $[\alpha] = -14.7$ deg mL g^{-1} dm^{-1}:

$$\alpha = (-14.7 \text{ deg mL g}^{-1} \text{ dm}^{-1})(0.0556 \text{ g mL}^{-1})(1 \text{ dm}) = -0.817 \text{ deg}$$

6.13   Begin with the perspective structure of the reactant; in the product, corresponding groups are in corresponding positions, because the reaction breaks none of the bonds to the asymmetric carbon.

(*R*)-(–)-enantiomer   (–)-enantiomer;
                                        has the R configuration

6.14   Proceed in the manner suggested by the solution to Problem 6.13. The absolute configuration of the alkene in Eq. 6.6 is known. Carry out the following catalytic hydrogenation:

R enantiomer R enantiomer

If we assume that hydrogenation proceeds in the normal manner, then the product must have the R configuration. Determine the sign of its specific rotation. If positive, then the product shown is the (*R*)-(+)-

enantiomer, and it is the dextrorotatory enantiomer; if negative, then the product shown is the (R)-$(-)$-enantiomer, which means that the (S)-$(+)$-enantiomer is the dextrorotatory enantiomer.

6.15   (a)   For a molecule to have a meso stereoisomer, it must have more than one asymmetric atom, and it must be divisible into constitutionally identical halves (that is, halves that have the same connectivities relative to the dividing line). By these criteria, compound (a) does possess a meso stereoisomer.

Notice that the dividing line may pass through one or more atoms.

(b)   This compound does not have a meso stereoisomer. A meso compound must have at least two asymmetric carbons.

(c)   Although achiral, *trans*-2-hexene cannot exist as a meso stereoisomer because it has no asymmetric atoms.

6.16   As suggested by the hint, the internal mirror plane passes through the central carbon atom, the hydrogen, and the OH group. This carbon can have either of two configurations. The only requirement for a meso compound (in addition to the identical connectivities of the two half-structures relative to a dividing line) is that the asymmetric carbons at the end of the structure must have opposite configurations.

the internal mirror plane
passes through these three atoms

the internal mirror plane
passes through these three atoms

6.17   (a)   If the enantiomer of the CSP were used, the elution order of R- and S-Nirvanol would be reversed—that is, the S-enantiomer would elute first.

(b)   If the racemate of the CSP were used, the two enantiomers would not separate by this method—they would elute at the same time.

6.18   A resolving agent must be *chiral* and it must be *enantiomerically pure*. Only compound A meets both of these criteria.

6.19   In *anti*-butane, the two internal carbons are stereocenters, because interchange of a CH_3 group and an H at either carbon gives a stereoisomer (*gauche*-butane).

anti-butane

gauche-butane

Remember that we are just imagining these structures as if they were objects that cannot interconvert. Of course, butane has free rotation about the C–C bonds so these carbons are *not* stereocenters in butane under normal circumstances.

6.20 (a) The three conformations of *meso*-2,3-butanediol:

A	B	C
achiral; has a center of symmetry	chiral	chiral

enantiomers

Conformation *A* is achiral; it has a center of symmetry and is congruent to its mirror image. (See Fig. 6.4 on text p. 233.) Conformations *B* and *C* are enantiomers. This relationship can be seen from the following manipulation of conformation *C*:

mirror

C	C	C	B

enantiomers

(b) Because its conformations interconvert rapidly, *meso*-2,3-butanediol cannot be optically active. As the text indicates, molecules that consist of rapidly interconverting enantiomers are said to be achiral. However, at very low temperatures near absolute zero, conformations *B* and *C* could in principle be isolated; each would be optically active, and the two conformations would have rotations of equal magnitudes and opposite signs.

6.21 (a) All staggered conformations of propane are achiral (and identical); therefore, even at low temperature, propane could not be resolved into enantiomers.

(b) Like butane, 2,3-dimethylbutane, $(CH_3)_2CH-CH(CH_3)_2$, contains two enantomeric conformations and one achiral conformation. Therefore, it could in principle be resolved into enantiomers at very low temperatures.

(c) Like ethane, 2,2,3,3-tetramethylbutane, $(CH_3)_3C-C(CH_3)_3$ consists entirely of achiral (and identical) staggered conformations and therefore cannot be resolved into enantiomers even at very low temperature.

6.22 (a) The carbon stereocenter is fixed in the *S* configuration, but the nitrogen stereocenter undergoes inversion between *R* and *S* configurations. Consequently, the molecule is a mixture of C(*S*),N(*S*) and C(*S*),N(*R*) conformations; that is, it is a rapidly equilibrating mixture of diastereomers.

(b) Each of the rapidly interconverting species in part (a) has an enantiomer: C(*S*),N(*S*) has an enantiomer C(*R*),N(*R*), and C(*S*),N(*R*) has an enantiomer C(*R*),N(*S*). Because inversion of the nitrogen stereocenter does not affect the configuration of the carbon stereocenter, it would be possible to resolve the racemate of this compound into enantiomeric sets of rapidly interconverting *diastereomers*.

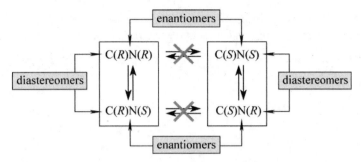

In other words, the set of compounds in one box could be resolved from the set in the other box.

6.23   Arsines should invert at an even slower rate than either amines or phosphines, for the same reason that phosphines invert at a slower rate than amines. The lone electrons on arsenic are in an orbital with an even larger degree of *s* character, meaning that it takes a very large amount of energy to convert them into a *p*-orbital, which is required for inversion.

6.24   Pasteur isolated the chiral stereoisomers, that is, the (2*R*,3*R*) and the (2*S*,3*S*) stereoisomers. (Because they are optically active, they must be chiral.) Yet to be isolated was the meso stereoisomer. Like all meso compounds, *meso*-tartaric acid is achiral and therefore is optically inactive. (The structures below are drawn in eclipsed conformations for ease in seeing the symmetry relationships.)

(2S,3S)-tartaric acid   **(2R,3R)-tartaric acid**   ***meso*-tartaric acid**

6.25   The "resolving agent" was the first crystal that Pasteur separated. The handedness of each subsequent crystal was either "like" that of the first one or "opposite" to it.

Solutions to Additional Problems

6.26   Asymmetric carbons are indicated with asterisks (*), and stereocenters with diamonds (◊).

(a)   4-Methyl-1-pentene, $H_2C{=}CHCH_2CH(CH_3)_2$, has no stereocenters or asymmetric carbons.

(b)

(E)-4-methyl-2-hexene

(c)

3-methylcyclohexene

The instability of the *E* stereoisomer of 3-methylcyclohexene doesn't alter the fact that the alkene carbons—carbons 1 and 2—of this compound are stereocenters.

(d)   2,4-Dimethyl-2-pentene, $(CH_3)_2C\!\!=\!\!CHCH(CH_3)_2$, has no stereocenters or asymmetric carbons.

6.27   The structure:

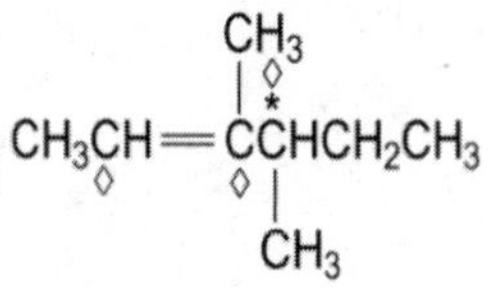

$$CH_3CH\!\!=\!\!\overset{\overset{\displaystyle CH_3}{\underset{\displaystyle *}{|}}}{C}\overset{}{\underset{\underset{\displaystyle CH_3}{|}}{C}}HCH_2CH_3$$

3,4-methyl-2-hexene

This compound exists as four stereoisomers: (2*Z*,4*S*), (2*E*,5*S*), (2*Z*,5*R*), and (2*E*,5*R*).

(a)   The carbon stereocenters are carbons 2, 3, and 4, indicated with diamonds (◊) in the structure in part (a).
(b)   Carbon 4 is an asymmetric carbon, indicated with an asterisk (*) in the structure in part (a).

6.28   The asymmetric carbons are indicated with asterisks (*).

(a)

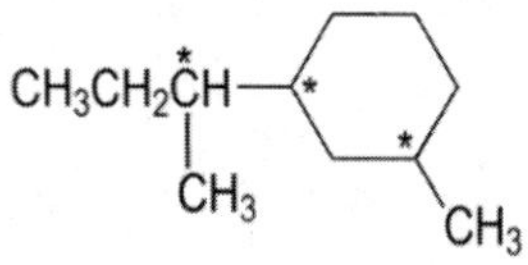

(b)

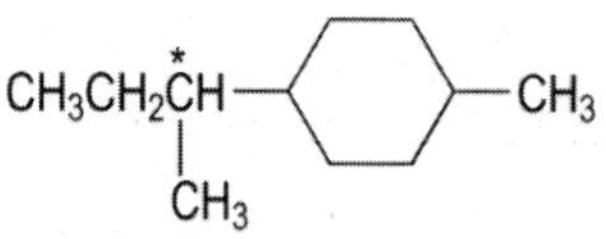

(c)

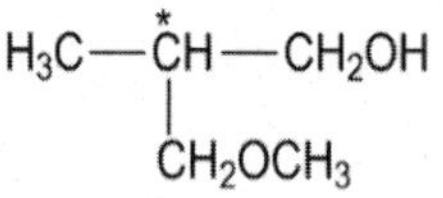

(d)

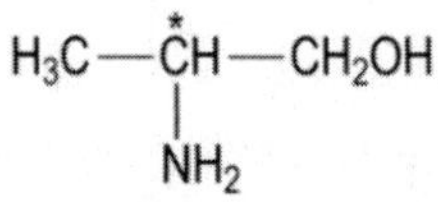

(e)

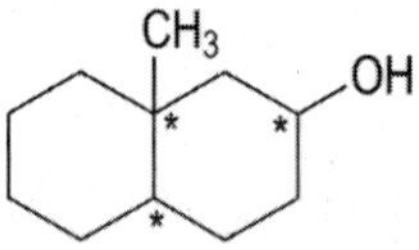

(f)   There are no asymmetric carbons in this structure.

(g)

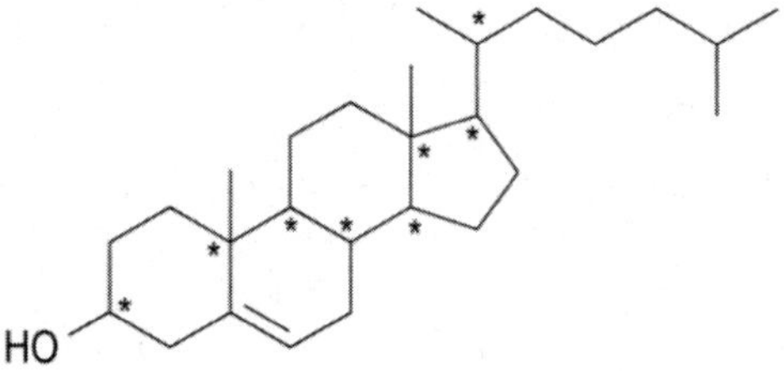

6.29   (a)

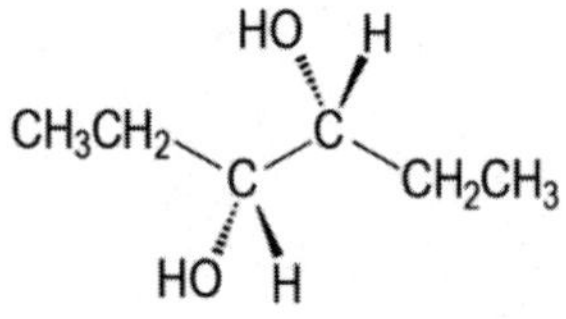

(b)

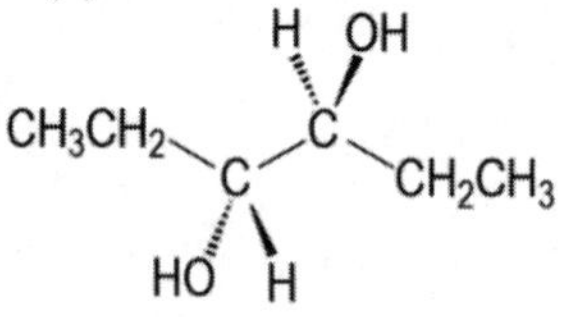

(c)

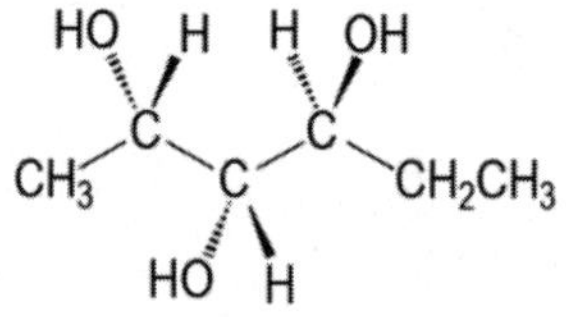

6.30   (a)

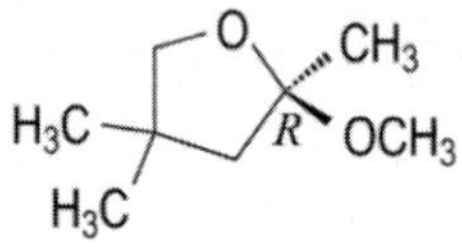

(b)

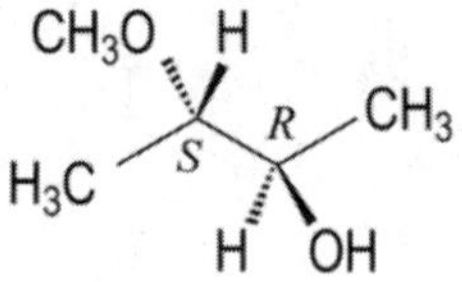

(c)

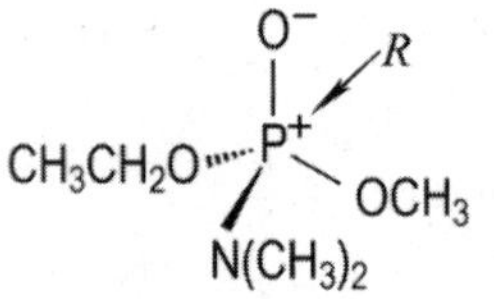

(d)   Any meso compound containing two asymmetric carbons must have opposite configurations at the two carbons. Therefore, one of the asymmetric carbons is *S* and the other is *R*.

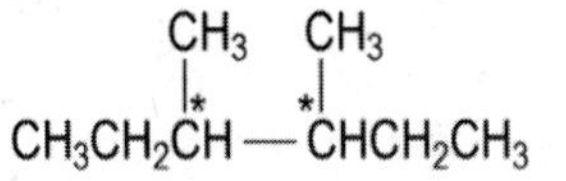

one of the asymmetric carbons (*) is *S* and the other is *R*

3,4-dimethylhexane

6.31   In answering this question, notice that, in deciding priorities at carbon-1, carbon-2 has a higher priority than the phenyl (Ph) group, because carbon-2 has an attached nitrogen.

(a)   If you put the methyl group on the left and the phenyl on the right, the positions of the other groups will be reversed from the following ones.

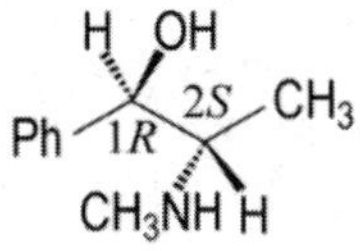

ephedrine 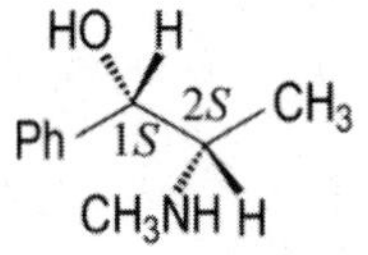   **pseudoephedrine**

(b)   Sawhorse and Newman projections:

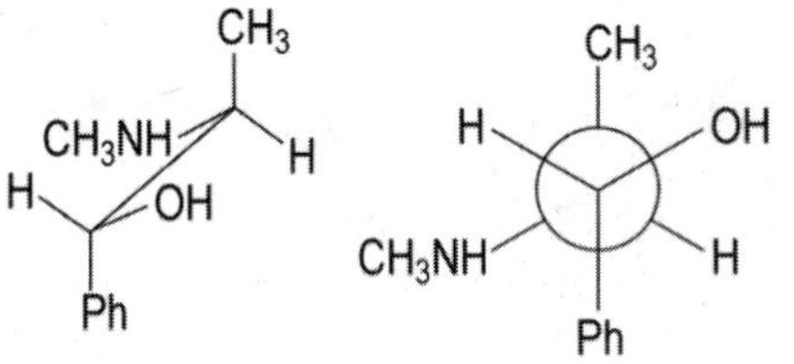

ephedrine 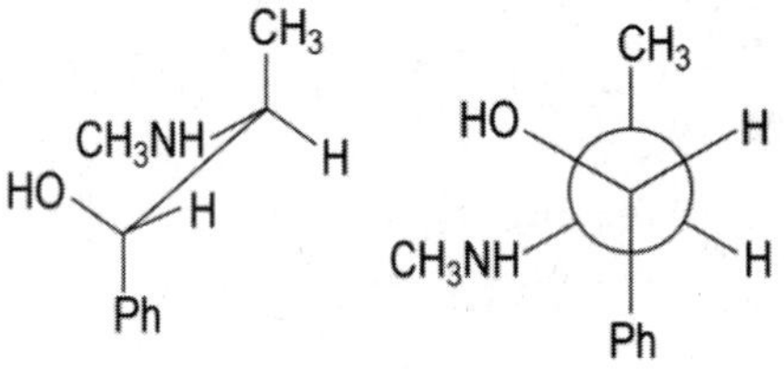   **pseudoephedrine**

(c)   These two compounds are not constitutional isomers because they have the same atomic connectivity. Since they are not identical, they are stereoisomers. If these two compounds were enantiomers, they would have opposite configurations at *both* asymmetric carbons. Because they are stereoisomers but not enantiomers, they are diastereomers.

(d)   In principle, their melting points should be different. In fact, they are. Ephedrine has a melting point of 38–39 °C; pseudoephedrine has a melting point of 116–119 °C.

(e)   One can say two things about the optical activities of these two compounds. First, they are both optically active, because they are chiral. Second, because they are diastereomers, their optical activities have no simple relationship—that is, their magnitudes are expected to be different, and one cannot say anything about their signs.

6.32   Either of the following isomeric heptanes is a correct answer:

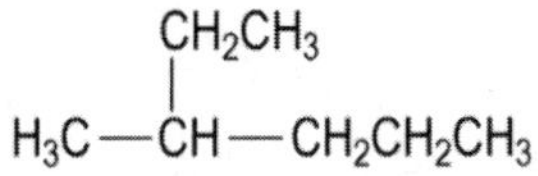

3-methylhexane 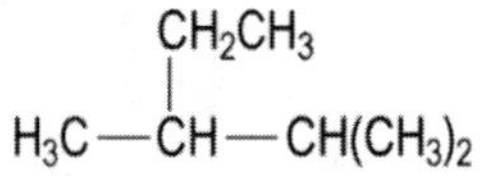   **2,3-dimethylpentane**

6.33   The following stereoisomer of 1,2-dimethylcyclopropane is chiral. If you're not convinced, build a model of it and another model of its mirror image and test them for congruence.

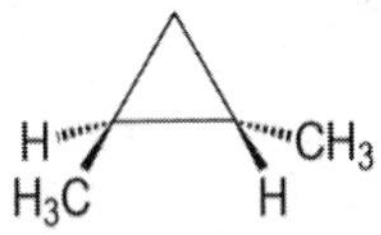

6.34   (a)   True. 2-Chlorohexane and 3-chlorohexane are examples of chiral constitutional isomers. (Note that while they are chiral, they are not enantiomers.)

      (b)   True by definition.

      (c)   False. Every object—chiral or achiral—has a mirror image. Only an achiral molecule is congruent to its mirror image. (See Fig. 6.1, text p. 230, for an example.)

      (d)   True by definition.

      (e)   False. For example, 2-chlorohexane has no diastereomer.

      (f)   False. Some E,Z isomers are not chiral (e.g., (E)- and (Z)-2-butene). Likewise, any meso compound is an achiral diastereomer of a compound containing asymmetric carbons.

      (g)   False. Meso compounds have asymmetric carbons and are not chiral. (A compound containing one *and only one* asymmetric carbon is chiral, however.)

      (h)   False, because some stereoisomerism is not associated with chirality—for example, E,Z stereoisomerism (double-bond stereoisomerism).

      (i)   True. By definition, interchanging two groups at a stereocenter generates a stereoisomer.

      (j)   False. Molecules with a mirror-image relationship must be either enantiomeric or identical.

      (k)   False. Every chiral compound is optically active, although it is possible in some cases for the optical activity to be too small to be detected experimentally.

      (l)   False. Optical rotation has no general relationship to R and S configuration.

      (m)   False. A chiral structure must have neither a plane of symmetry nor a center of symmetry. Therefore, a structure might not a plane of symmetry and yet be achiral because it has a center of symmetry. (For example, see the solution to Problem 6.20(a), conformation A.)

      (n)   True, because the presence of a plane of symmetry is sufficient (although not necessary) to eliminate chirality.

      (o)   True. An asymmetric carbon is one type of stereocenter. (Note that the converse is not true—not all stereocenters are asymmetric carbons.)

6.35   The isomeric alkyl chlorides $C_6H_{13}Cl$ with the carbon skeleton of 3-methylpentane:

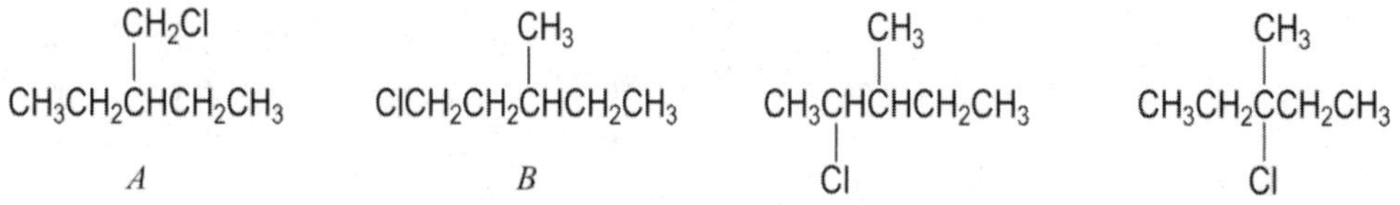

Compounds B and C are chiral. The four compounds $A–D$ are constitutional isomers. Compound B can exist as a pair of enantiomers; compound C can exist as two diastereomeric pairs of enantiomers (four stereoisomers total).

6.36   Meso compounds must be achiral compounds with at least two symmetrically-placed asymmetric carbons and symmetrical branching patterns.

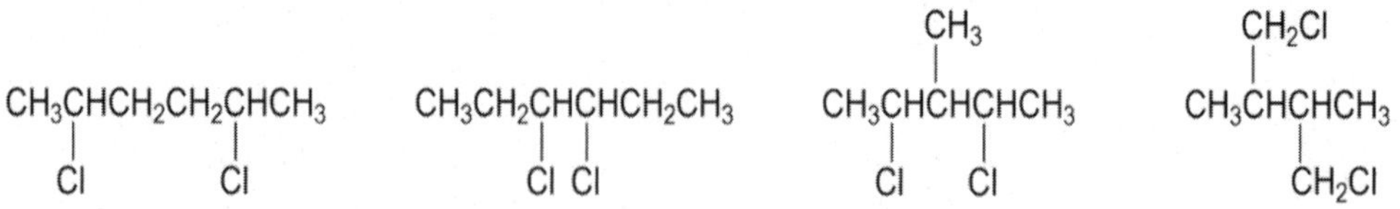

All exist as a single meso compound, except for compound C, which can exist as two different meso compounds.

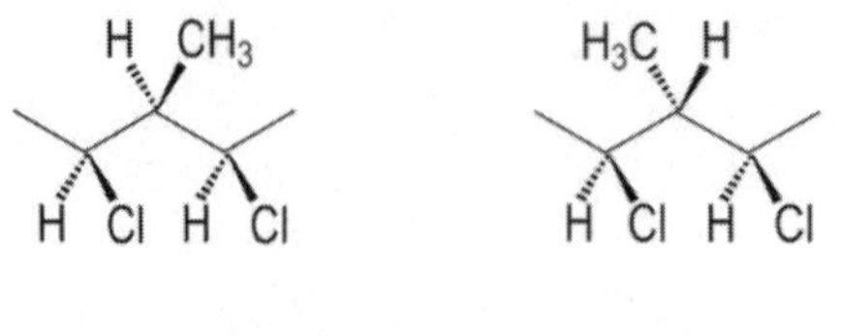

6.37   The Newman projections of 2-methylbutane about the C2–C3 bond (viewed from the C3 direction):

A   *B*   *C*

The sawhorse projections of 2-methylbutane:

A   *B*   *C*

(a)   Conformations *A* and *B* are chiral; they are conformational enantiomers.

(b)   These two conformations interconvert so rapidly that they cannot be isolated. As discussed in Sec. 6.9A, a molecule that consists of rapidly interconverting enantiomers is considered to be achiral.

(c)   The heats of formation of *A* and *B* are identical, and are lower than the heat of formation of *C*, because *C* contains two *gauche*-methyl interactions, whereas *A* and *B* each have only one. The heats of formation of *A* and *B* are equal because enantiomers have identical physical properties.

6.38   (a)   The sawhorse projections of ephedrine: (The projections labeled with double letters are eclipsed; the projections labeled with single letters are staggered.)

A   *B*   *C*

(b)   Each of these conformations is chiral. Had any one of them been achiral, ephedrine would not be chiral.

6.39   Compound *B* can undergo amine inversion, which rapidly interconverts its enantiomeric forms, as shown in Fig. 6.18, text p. 261. Compound *A* cannot undergo inversion because it does not have an unshared electron pair; therefore, its two enantiomeric forms cannot interconvert and can be isolated.

6.40   (a)   Any compound that is chiral can in principle be resolved into enantiomers. Compound *A* possesses a phosphorous atom containing 4 "groups" (the electron pair counts as a different group in this case), so it is chiral and could be resolved. Elements in the third period do not undergo rapid inversion (Sec. 6.10B). Compound *B* also contains a phosphorous atom with 4 different groups, so it can be resolved. The sulfur atom in compound *C* is attached to 4 groups and, like phosphorous, is too slow to interconvert, but two of its attached groups are identical. Thus, it is achiral and doesn't have an enantiomer. Compounds *D* and *E* can be resolved because they contain asymmetric carbons (the ones bearing the phenyl group in each). Compound *F* can be resolved into enantiomers for the same reasons as compounds *A* and *B*.

(b)   Sulfur is the asymmetric center in omeprazole. It has four different "groups" attached, if we include the lone pair of electrons. Since sulfur does not undergo rapid inversion (Sec. 6.10B) the two isomers can be isolated.

(c)   The simplified structure of the *S* enantiomer of omeprazole is below. The cyclic groups are abbreviated "Ar".

6.41   (a)   A 0.5 *M* solution of this alkene contains (0.5 mol L^{-1})(146.2 g mol^{-1})(0.001 L mL^{-1}) = 0.0731 g mL^{-1}. Using Eq. 6.1 on text p. 239, with $l = 0.5$ dm, the observed rotation is

$$(+79 \text{ deg mL g}^{-1} \text{ dm}^{-1})(0.0731 \text{ g mL}^{-1})(0.5 \text{ dm}) = +2.89 \text{ deg}$$

(b)   The rotation of the (−)-enantiomer cancels half of the rotation of the (+)-enantiomer, and the concentration of the (+)-enantiomer is, in addition, halved. Therefore the rotation is 25% of that in part (a)—that is, +0.72 deg.

(c)   Since there is twice as much of the major enantiomer, the % major enantiomer is about 66.6% to 33.3% of the minor enantiomer. Using Eq. 6.5a on text p. 242, we get:
$$EE = 2(\% \text{ major enantiomer}) - 100\% = 33.4\%$$

You can check your work by solving for the specific rotation, $[\alpha]$, of the mixture in part (b), then using Eq. 6.4 on text p. 242 to solve for the EE.

6.42   (a)   Since the specific rotation of the mixture is negative, we know that the major enantiomer is the (−)-enantiomer. The specific rotation of the (+)-enantiomer is given, so the specific rotation of the (−)-enantiomer is known as well. Using Eq. 6.4 on text p. 242, we find that the EE of the (−)-enantiomer is 68%.

(b)   Using Eq. 6.5b, the % major enantiomer is 84%, so the % minor enantiomer is 16%.

6.43   (a)   Using the equation given and a ratio $r = 0.064$, solving for t gives an age of 205 years.

(b)   Using Eq. 6.5a on text p. 242, the EE is 88%.

(c)   First, rearrange Eq. 6.4 to find the specific rotation of the mixture, $[\alpha]_{\text{mixture}}$. Then use this value, +21.56 degrees, in Eq. 6.1 on text p. 239 with a concentration $c = 0.01$ g mL^{-1}. Solving for the observed rotation, α, gives $\alpha = +0.216$ degrees.

6.44   Each compound contributes to the optical activity in proportion to the fraction present. If the fraction of the α form is P_α, and that of the β form is P_β,

$$\text{specific rotation} = 52.7° = P_\alpha(+112°) + P_\beta(+18.7°)$$

Because $P_\alpha + P_\beta = 1.0$, then

$$52.7° = P_\alpha(+112°) + (1.00 - P_\alpha)(+18.7°)$$

Solving for P_α gives $P_\alpha = 0.36$, or 36%, and the fraction of $P_\beta = 64\%$.

6.45   (a)   The two meso stereoisomers of 2,3,4-trichloropentane:

two meso stereoisomers of 2,3,4-trichloropentane

(b)   Each compound has a plane of symmetry that passes through carbon-3 as well as the C—Cl and C—H bonds to that carbon:

planes of symmetry

(c)   These two stereoisomers must be diastereomers, because they are not enantiomers.

(d)   Carbon-3 is a stereocenter, because interchanging the H and the Cl gives a stereoisomer—specifically, the other meso compound.

(e)   To specify the configuration of carbon-3, we would need a rule to specify the relative priorities of groups that differ only in their stereochemical configuration. If we arbitrarily let an *R* carbon take precedence over *S* (that is, precedence is alphabetical), then carbon-2 has precedence over carbon-4. This means that the stereoisomer on the left in parts (a) and (b) has the 3*S* configuration, and the one on the right has the 3*R* configuration.

Asymmetric carbons are generally diagnosed using connectivity relationships. Therefore, this type of asymmetric carbon is unusual, because the connectivity of two of the attached groups is the same. In recognition of this situation, a carbon that is "asymmetric" only because of the differing configurations of the attached groups is called a *pseudoasymmetric carbon*.

(f)   The two stereoisomers in parts (a) and (b) are the meso stereoisomers. There are also two enantiomers that have, respectively, the 2*R*,4*R* and the 2*S*,4*S* configurations. (Convince yourself that in these enantiomers, carbon-3 is *not* a stereocenter.) These are the only four stereoisomers.

6.46   (a)   Compounds *A* and *B* are diastereomers; both are meso. (See the solution to Problem 6.45 for a similar case.) Compound *C* is a diastereomer of both *A* and *B* because it differs in configuration at some, but not all, of the asymmetric carbons. The two asymmetric carbons in compounds *C* and *D* have the *R* configuration, so it appears that these are the same compound. To show this convincingly, first carry out an internal rotation in *D*, then rotate *D* 180° about an axis in the page:

D   *D*   *C*

Compound *E* is the enantiomer of *D* (and *C*), because both asymmetric carbons have the opposite (*S*) configuration. To summarize:

A and *B:*   diastereomers   *A* and *C:*   diastereomers
A and *D:*   diastereomers   *A* and *E:*   diastereomers
B and *C:*   diastereomers   *B* and *D:*   diastereomers
B and *E:*   diastereomers   *C* and *D:*   identical
C and *E:*   enantiomers   *D* and *E:*   enantiomers

(b)   Compounds *A* and *B* are both meso; they are diastereomers.

(c)   Compounds *C* (and its identity *D*) and *E* are chiral and therefore optically active.

6.47   (a)   Hydrogenation makes two of the branches to the asymmetric carbon of the starting alkene identical. Therefore, the product is achiral and is incapable of optical activity.

$$H_2C=\!\!=\!\!CH-\overset{\underset{\displaystyle |}{CH_3}}{CH}-CH_2CH_3 \quad \xrightarrow{\text{H}_2,\ \text{catalyst}} \quad CH_3CH_2-\overset{\underset{\displaystyle |}{CH_3}}{CH}-CH_2CH_3$$

3-methyl-1-pentene
(either enantiomer)

3-methylpentane
(achiral and therefore
optically inactive)

(b)   Hydrogenation of (*S*)-3-methyl-1-hexene does not break any of the bonds to the asymmetric carbon, and we are given the experimental result that the product, 3-methylhexane, has (–) rotation. Therefore, we can deduce the absolute configuration of (–)-3-methylhexane. Notice that *the relative priorities of the groups at the asymmetric carbon change as a result of the reaction.*

CH₂CH₂CH₃ ... H₂, catalyst ... CH₂CH₂CH₃ ... therefore: ... CH₂CH₂CH₃

(S)-(+)-3-methyl-1-hexene   **(R)-(–)-3-methylhexane**   **(S)-(+)-3-methylhexane**

Because (–)-3-methylhexane must have the *R* configuration, it follows that its enantiomer, (+)-3-methylhexane, must have the *S* configuration.

6.48   First, draw the (*S*)-enantiomer of the starting material using line and wedge bonds. Then, reduce the vinyl group to an ethyl group, but make sure the product substituents are drawn in the same relative positions as they were in the starting material. When the vinyl group is reduced to an ethyl group, it goes from being the second priority group to being the third priority group. This serves to reverse the absolute configuration of the product; it is the (*R*)-enantiomer. This is an example of a rare case where, even though there was no chemical reaction at the asymmetric carbon, the reaction caused the attached groups to swap priorities.

6.49   Salts that are either identical or enantiomers should have identical solubilities. The solution to the problem then hinges on determining the relationship between each pair of salts. If they have the same configurations at corresponding asymmetric carbons, they are identical and have identical solubilities. If they have different configurations at *both* of the corresponding asymmetric carbons, they are enantiomers and have identical solubilities, because enantiomers have identical properties. If they have different configurations at one asymmetric carbon and the same configuration at the other, they are diastereomers and have different solubilities, because diastereomers have different properties.

   Salt *A* has the *S* configuration at both asymmetric carbons, and salt *D* has the *R* configuration at both asymmetric carbons. Therefore, salts *A* and *D* are enantiomers and have identical solubilities. Salt *B* has the *S* configuration at the asymmetric carbon of the ammonium salt and the *R* configuration at the asymmetric carbon of the carboxylate. Salt *C* has the opposite configuration at both of these carbons, that is, (*R*,*S*). Therefore, salts *B* and *C* are enantiomers and have identical solubilities. However, the solubilities of *A* and *D* are different in principle from the solubilities of *B* and *C*.

   Once you have deduced the configuration of one salt, the configuration of the corresponding carbon in the others can be determined by interchanging groups to reproduce the original. Each interchange changes the configuration; even numbers of interchanges leave the configuration unchanged. Alternatively, you can rotate about a single bond freely; this does not change the configuration. Thus, interchanging the H and the Ph gives the ammonium ion of *C* from that of *A*; hence, the configuration is different, and it must therefore be *R*. An internal rotation of the asymmetric carbon 120° counterclockwise about the methyl—C bond converts the ammonium ion of *A* into that of *B*; hence, these two ions must be identical.

6.50   (a)   We should know by now that if the geometry is tetrahedral, there are two enantiomers:

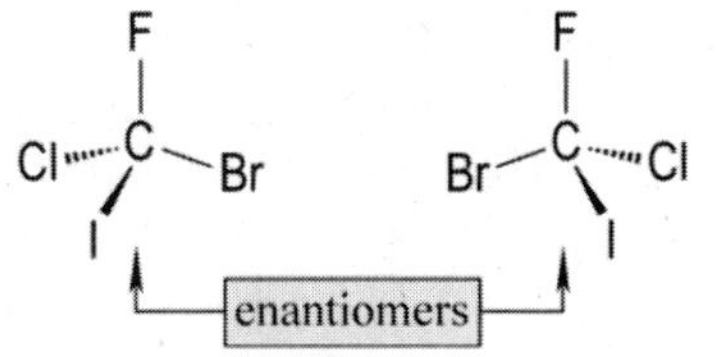

(b)   For square-planar geometry, there are three diastereomers. These are found by making all possible pairwise switches between adjacent groups and ruling out identities. Note that, because of the square planar geometry, mirror images must be congruent, and therefore identical. There can be no enantiomers.

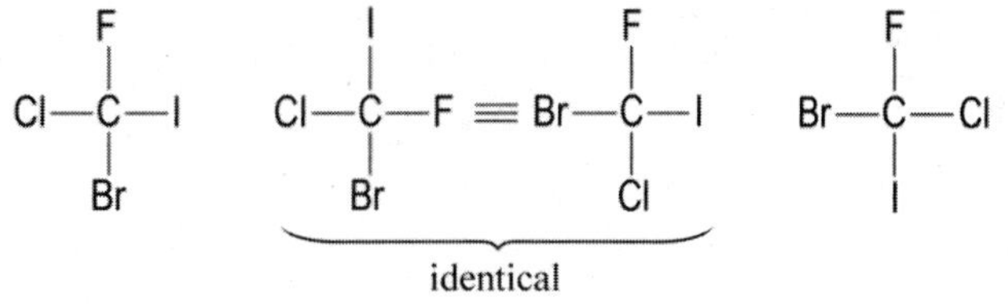

(c)   For pyramidal geometry, there are three diastereomeric sets of enantiomers (six stereoisomers total):

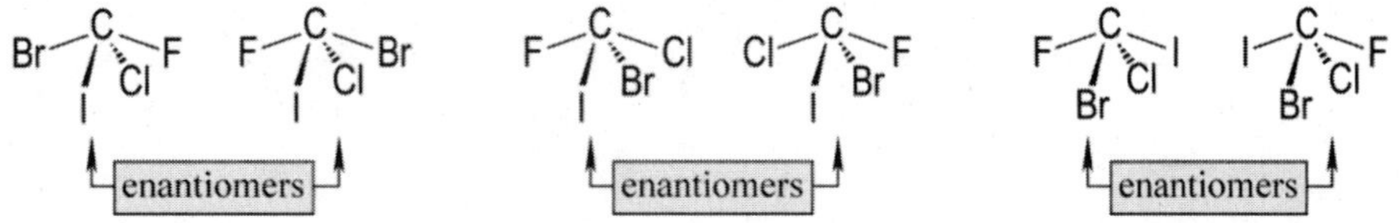

6.51   In a compound of the form X_2ZY_2 with square-planar geometry, there are two ways to arrange groups X and Y about atom Z: with like groups in adjacent corners, or with like groups in opposite corners:

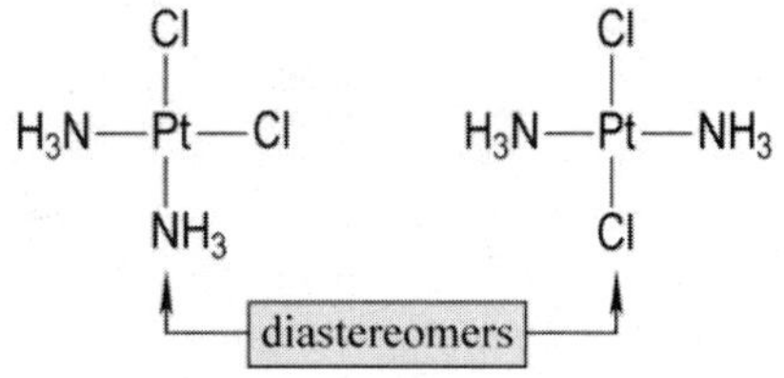

Because these are stereoisomers, and they are not enantiomers, they must be diastereomers. Tetrahedral compounds of the form X_2ZY_2 (for example, H_2CCl_2) *cannot* exist as stereoisomers. Hence, the fact that $Cl_2Pt(NH_3)_2$ exists as stereoisomers with different properties shows that these stereoisomers are diastereomers, and hence that their geometry about platinum is square-planar.

6.52   The ultimate test for chirality is to make the mirror image and test it for congruence:

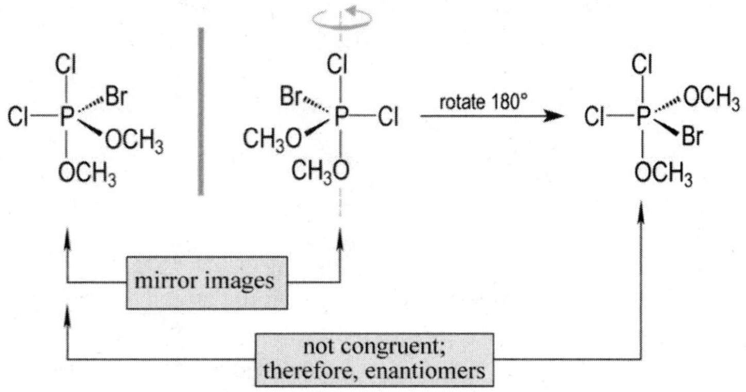

Because the structure is not congruent to its mirror image, the compound is chiral.

6.53   The optical rotations and melting points show that Fischer obtained enantiomers as a result of this transformation. Consider what each of the two possible geometries predicts for an interchange of the two groups. For tetrahedral carbon, the interchange of any two groups at an asymmetric carbon gives enantiomers. Consequently, the experimental result is consistent with tetrahedral geometry.

　　　　If the molecule were pyramidal, and if Fischer *happened* to be dealing with a stereoisomer in which the —CO_2H and —$CONH_2$ groups were at opposite (rather than adjacent) corners of the pyramid, he would also have obtained enantiomers from his experiment:

enantiomers

But if Fischer *happened* to be dealing with a stereoisomer in which the —CO_2H and —$CONH_2$ groups were at adjacent corners, the exchange should have given diastereomers. Thus, the conversion of one stereoisomer into its enantiomer could have been the result of two things occurring simultaneously: (1) pyramidal geometry, and (2) the fortuitous choice of a particular stereoisomer. Consequently, Fischer's result was consistent with either tetrahedral or pyramidal geometry.

diastereomers

6.54   Neither result taken alone rules out pyramidal geometry. However, for an atom with two identical groups and *pyramidal* geometry to be achiral, the two carboxy groups in the product, and hence, the carboxy and carboxamido group in the starting material, must be at the opposite corners of the pyramid.

carboxy group　　carboxamido group　　　　　　achiral

If these two groups had been at opposite corners of the pyramid in the *first* experiment, then in the *second experiment with the same starting material,* the carboxy group and the methyl group must have been at *adjacent* corners, and this transformation would have yielded a chiral compound:

carboxy group　　carboxamido group　　　　　　a chiral compound

(Be sure to convince yourself that the molecule on the right is chiral. Do this by drawing (or making a model of) its mirror image and show that the two are noncongruent.) On the other hand, if the starting compound had tetrahedral geometry, then making *any* two groups identical would give an achiral compound. Since this was the experimental result, pyramidal geometry was ruled out, and tetrahedral geometry thus remained the only (and, as it turned out, correct) possibility.

Cyclic Compounds
Stereochemistry of Reactions

STUDY GUIDE LINKS

  7.1   Relating Cyclohexane Conformations

Although you can often resort to models to visualize the stereochemical relationships between chair conformations of substituted cyclohexanes, it is useful to be able to carry out on paper manipulations like the ones used in the text. To gain greater facility with such manipulations, try the following exercises, using *cis*-1,2-dimethylcyclohexane as an example. Draw the result of each manipulation, and then check your drawing against the results shown. *Be sure to follow each manipulation with a model if there is any question about it!*

First, turn the structure 180° about a horizontal (*"x"*) axis:

Now turn the original structure 180° about a "*y*" axis:

Practice turning the structure in 60° increments about the same axis:

Finally, turn the original structure 180° about an axis perpendicular to the page (a *"z"* axis):

If you can readily perform manipulations like these without recourse to models, you can solve virtually any problem involving cyclohexane stereochemistry.

Let's use these skills to solve the following problem. What is the stereochemical relationship between the two chair conformations of *trans*-1,3-dimethylcyclohexane?

(SG7.1)

There are three possible relationships between these conformations: they could be conformational diastereomers; they could be conformational enantiomers; or they could be identical. The strategy here is to orient one of the methyl groups in one structure in the same way as a methyl group in the other structure. Then compare the relative positions of the remaining methyl group. One way to perform this manipulation is to rotate the second structure about an axis perpendicular to the page, as follows:

Finally, rotate the resulting structure 60° about an axis in the page, as follows:

The resulting structure is identical to the first conformation in Eq. SG7.1. Consequently, the chair interconversion in this case results in *identical molecules*.

It is worth noting that some people, even after practice, have difficulty with manipulations like these because they have inherent difficulty with spatial perception. (Don't assume you are such a person until you have practiced!) This has nothing to do with your intelligence! If you are such a person, don't hesitate to use models. The answer you get may take a little more time, but it is just as correct.

 ## 7.2   Reactions of Chiral Molecules

Sec. 7.7A of the text deals only with reactions of *achiral molecules* that give enantiomeric pairs of products. What happens when *chiral* molecules react to give stereoisomeric products? No general principles cover this situation. In some cases, stereoisomers of the product may be formed in equal amounts; in other cases, one stereoisomer may predominate. The result can be determined by the reaction mechanism—that is, the stereochemical nature of chemical changes that occur at stereocenters. (A situation of this sort is discussed in Sec. 7.8B.) In other cases reactants or intermediates are racemized in the reaction, in which case the principles of Sec. 7.7A apply to each enantiomer of the racemate.

It is even possible for chiral molecules to react in some cases to give achiral products even when a stereocenter does not undergo a chemical change. Thus, the following hydrogenation reaction involving a chiral alkene gives an achiral alkane because the asymmetric carbon in the alkene is no longer an asymmetric carbon in the alkane product:

$$H_2C=CH\text{---}\underset{\underset{H}{|}}{\overset{\overset{CH_2CH_3}{|}}{C}}\text{---}CH_3 \quad + \quad H_2 \quad \xrightarrow{\text{catalyst}} \quad CH_3CH_2\text{---}\underset{\underset{H}{|}}{\overset{\overset{CH_2CH_3}{|}}{C}}\text{---}CH_3$$

       chiral and optically active   achiral and optically inactive

Optical activity (and chirality) is lost in this reaction because the product is not a chiral molecule. In *this particular case,* the reaction causes a plane of symmetry to occur within the structure of the product that does not exist in the reactant. Optical activity would *not* be lost in the same reaction of a different alkene in which the product is chiral:

$$H_2C=CHCH_2\text{---}\underset{\underset{H}{|}}{\overset{\overset{CH_2CH_3}{|}}{C}}\text{---}CH_3 \quad + \quad H_2 \quad \xrightarrow{\text{catalyst}} \quad CH_3CH_2CH_2\text{---}\underset{\underset{H}{|}}{\overset{\overset{CH_2CH_3}{|}}{C}}\text{---}CH_3$$

       chiral and optically active   chiral and optically active

(See also Problem 6.47(a), text p. 270.)

7.3   Analysis of Reaction Stereochemistry

A given type of reaction might yield diastereomers with one starting material, enantiomers with another, and no stereoisomers at all with a third. For example, the bromine addition to cyclohexene in Study Problem 7.5 on text p. 306 gives a pair of diastereomers, and one of these is formed as a pair of enantiomers. In bromine addition to propene, only a pair of enantiomers is formed. In bromine addition to ethylene, no stereoisomers at all are formed. For a given reaction, no single stereochemical result applies to all compounds. Each reaction must be analyzed individually.

7.4   Stereoselective and Stereospecific Reactions

Some texts define the term *stereospecific reaction* to mean "100% stereoselective within limits of detection." This idea can be conveyed by saying simply that a reaction is "highly stereoselective." This definition of *stereospecific* is not generally accepted; be careful not to confuse the terms *stereospecific* and *stereoselective* in this way.

7.5   When Stereoselectivity Matters

The stereoselectivity of a reaction (or lack of it) is of no concern if stereochemically different modes of a reaction give the same products. Let's see how this statement would apply in the case of bromine addition to alkenes.

    The addition of bromine to ethylene does not give stereoisomers; hence, it makes no difference in the outcome of the reaction whether the reaction is a syn- or anti-addition or a mixture of the two.

$$H_2C=CH_2 \xrightarrow{\ Br_2\ } Br\text{---}CH_2\text{---}CH_2\text{---}Br$$

Furthermore, we cannot even tell whether the addition is syn or anti. It is often assumed (reasonably) that additions which occur in one particular way on certain compounds occur in the same way on similar compounds for which the stereoselectivity cannot be determined. So, because addition of bromine is an anti-addition with other alkenes, we assume that it is an anti-addition for ethylene as well. In fact, this assumption has been tested: the stereochemistry of bromine addition to isotopically substituted ethylenes was investigated in 1998:

$$\underset{D}{\overset{H}{\diagdown}} C = C \underset{D}{\overset{H}{\diagup}} \qquad \text{and} \qquad \underset{D}{\overset{H}{\diagdown}} C = C \underset{H}{\overset{D}{\diagup}}$$

This experiment established that anti-addition does in fact occur. Most chemists would accept the anti-addition of bromine to the deuterium-substituted ethylenes as proof that ethylene itself undergoes anti-addition of bromine.

In the addition of bromine to propene, stereoisomers are formed:

$$H_3C-CH\!=\!CH_2 \xrightarrow{\;Br_2\;} H_3C-\underset{\underset{Br}{|}}{CH}-CH_2-Br$$

Two *enantiomers* are possible, but they *must* be formed in the same amounts, because the reagents are achiral. Because carbon-1 of the product is not a stereocenter, the question of a syn- or anti-addition is irrelevant to the composition of the product mixture in the addition of bromine to propene.

Always remember a point made in Sec. 7.8A of the text: the stereochemistry of an addition is relevant only when *both* carbons of the double bond become stereocenters as a result of the addition. Thus, one can study the mode of bromine addition to *cis*- and *trans*-2-butene, as in Sec. 7.8C, because the two carbons of the double bond in the addition product are stereocenters. In a practical sense, bromine addition would not be as useful when applied to the 2-butenes or to cyclohexene if it were not stereoselective, because a mixture of diastereomers would be formed.

In contrast, the text shows that oxymercuration–reduction is not a stereoselective reaction sequence. The lack of stereoselectivity matters only when it is applied to an alkene that would give a mixture of diastereomeric products. Many reactions that are not stereoselective are useful as long as their use is confined to situations in which stereoselectivity is not an issue.

FURTHER EXPLORATIONS

  ## 7.1   Other Ways of Designating Relative Configuration

The use of *cis* and *trans* for designating relative stereochemistry in disubstituted rings has been adopted by *Chemical Abstracts,* the index to the world's chemical literature.

   Chemical Abstracts adopted a somewhat different system for designating absolute configuration of chiral cyclic compounds than the one used in the text, which is based on the original IUPAC rules. The *Chemical Abstracts* system involves the use of relative configurations. Consider the following compound:

The IUPAC name for this compound is (1*S*,2*R*)-1-chloro-2-methylcyclohexane. The *Chemical Abstracts* name is (1*S*-*cis*)-1-chloro-2-methylcyclohexane. In this system, the absolute configuration is given for the stereocenter of lower number (in this case the carbon bearing the chlorine), and then the configuration of the other stereocenter is given with the relative descriptor *cis* or *trans*.

   The 1,4-disubstituted cyclohexanes are particularly interesting in this context.

A   *B*

These compounds are *achiral*, but each one contains two carbon stereocenters. There are *no* asymmetric carbons. (This is another illustration of the point that not all carbon stereocenters are asymmetric carbons.) Hence, the *R* and *S* system cannot be applied. The stereochemistry of these compounds is fully conveyed by the designation *trans* or *cis*. Thus, compound *A* is *trans*-1-chloro-4-methylcyclohexane; compound *B* is *cis*-1-chloro-4-methylcyclohexane.

   As the text states, the cis–trans system is cumbersome when a ring contains more than two substituents. Although various extensions of this system have been proposed, *Chemical Abstracts* uses the terms α and β to describe relative stereochemistry. The term α means "on the same side of the ring in a planar projection," and β means "on the opposite side of the ring in a planar projection." The positions of all substituents are referred to the position of the substituent on a reference carbon, which, in the simplest cases, is the lowest-numbered carbon. The substituent at the reference carbon by definition is in the α position. Thus we have the following example.

[1*S*-(1α,2β,4α)]-1,4-dichloro-2-methylcyclopentane

The chlorine at carbon-1 is α by definition. Since it is "up", the other "up" substituent (the other chlorine) is also α, and the "down" substituent (the methyl group) is β. If we did not know the absolute configuration, but only the relative configuration, we would leave off the initial "1*S*" and name the compound as (1α,2β,4α)-1,4-dichloro-2-methylcyclopentane.

The following two cases are examples of achiral compounds with three stereocenters:

(1α,3α,5α)-1,3,5-trimethylcyclohexane   **(1α,3α,5β)-1,3,5-trimethylcyclohexane**

In the compound on the left, the three methyl groups are all designated as α because they are on the same "side" of the ring—that is, all "down."

The *Chemical Abstracts* system has additional layers of complexity for more complicated cases, but the above introduction covers most of the cases that one would ordinarily encounter.

7.2   Alkenelike Behavior of Cyclopropanes

The large amount of strain in cyclopropanes causes them to be more reactive than ordinary cycloalkanes. For example, cyclopropanes can be hydrogenated, although not so readily as alkenes.

Although reactions like this suggest that cyclopropanes react like alkenes, an alternative view is that alkenes are cycloalkanes with "two-membered rings."

ethylene viewed as a "cyclic alkane"

In fact, it can be argued from hybrid-orbital considerations that a view of the alkene double bond as two "bent bonds" (similar to those in cyclopropane) is just as valid as the picture adopted by this text in which the two bonds are different, one a σ bond and one a π bond. (Some model sets represent double bonds as "bent bonds.")

If ethylene indeed is visualized as a "two-membered ring"—that is, *cycloethane*—it would be expected to be very strained, and that this strain would be reflected in a very high heat of formation. Indeed, the heat of formation of ethylene per CH_2 is even greater than that of cyclopropane.

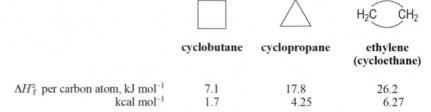

	cyclobutane	cyclopropane	ethylene (cycloethane)
ΔH_f° per carbon atom, kJ mol^{-1}	7.1	17.8	26.2
kcal mol^{-1}	1.7	4.25	6.27

In the conventional π model of the alkene double bond, the high energy of the bonds in ethylene resides completely in the relatively weak π bond. In the bent-bond model, the two bonds are equivalent. They are stronger than a π bond, but weaker than a σ bond.

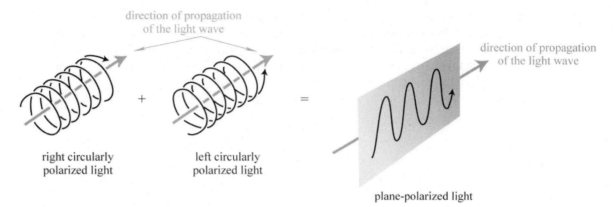

Figure SG7.1 The electric fields of right and left circularly polarized light propagate through space as helical vectors, and the two vectors add to give plane-polarized light. The two helices are noncongruent mirror images; that is, they are enantiomers. Consequently, plane-polarized light, their vector-addition product, is a "racemate."

 ## 7.3   Optical Activity

Optical activity is an important example of the principles discussed in the text. Plane-polarized light is actually the *vector sum of right and left circularly polarized light,* two forms of light in which the electric field vector propagates as a right- and left-handed helix, respectively. (See Fig. SG7.1.) These two forms of light are noncongruent mirror images—that is, they are enantiomers. In other words, plane-polarized light is "enantiomeric light." Let's call the right-handed form of light *R,* and the left-handed form *S.* Suppose that plane polarized light—consisting of equal amounts of *R* and *S* forms—is passed through an enantiomerically pure sample that has the *S* configuration. When the light interacts with the sample, two types of interaction occur: *S* sample with *S* light, and *S* sample with *R* light. *These two interactions are diastereomeric,* because an *S,S* combination is the diastereomer of an *S,R* combination. Because diastereomers have different energies, these two interactions have different energies—one is stronger than the other. The form of light with the stronger interaction is retarded in its passage through the sample. This in turn causes the plane of the polarized light to rotate—that is, it causes optical activity. *The enantiomeric forms of light differ when they interact with a chiral object.*

 ## 7.4   Stereochemistry of Organoborane Oxidation

The mechanism of organoborane oxidation was discussed in Further Exploration 5.1 (p. 114 of this manual). Retention of stereochemistry in this reaction can be understood in terms of this mechanism. An alkyl group on boron moves from boron to oxygen in such a way that the configuration of the migrating carbon is unaffected:

As the second part of this equation shows, this process occurs at each of the carbon–boron bonds to yield ultimately a trialkyl borate. Thus, a carbon–boron bond is replaced by a carbon–oxygen bond without changing the stereochemical configuration of the carbon.

Finally, the resulting trialkyl borate undergoes hydrolysis to the alcohol product. The final hydrolysis breaks the boron–oxygen bonds but leaves the oxygen–carbon bonds unaffected. Because these bonds are not affected, the configuration at carbon does not change.

SOLUTIONS TO PROBLEMS

Solutions to In-Text Problems

7.1   First, the half-chair conformation is strained because the C—C—C bond angles in the flattened part of the molecule are widened to nearly 120°, whereas the optimal angle is 109°. (Your model may pop back from a half-chair to a boat or chair if you try to place it on a table without holding it.) Second, there is significant eclipsing between some of the hydrogens. As we learned when we studied the conformations of ethane, eclipsing interactions are destabilizing.

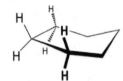

half-chair cyclohexane
(eclipsed hydrogens are shown)

7.2   We apply text Eq. 3.36b, text p. 110, with $\Delta G° = 7.4$ kJ mol^{-1}.

$$K_{eq} = 10^{-\Delta G°/2.30RT} = 10^{-7.4/5.71} = 10^{-1.30} = 5.1 \times 10^{-2}$$

Letting the axial form be A and the equatorial form be E,

$$K_{eq} = \frac{[A]}{[E]} = 0.051 \text{ or } [A] = 0.051[E]$$

Recognizing that, at 1 mol L^{-1} total concentration, $[A] + [E] = 1.0$, or (incorporating the above equation),

$$1.051[E] = 1.0$$

from which we obtain $[E] = 0.95$ and $[A] = 0.05$. A sample of methylcyclohexane contains 95% equatorial and 5% axial conformation.

7.3   Following the procedure in the solution to Problem 7.2, we first calculate the K_{eq} for *tert*-butylcyclohexane:

$$K_{eq} = 10^{-\Delta G°/2.30RT} = 10^{-20/5.71} = 10^{-3.5} = 3.1 \times 10^{-4}$$

From this we calculate, in a total concentration of mol L^{-1}, $[A] = (3.1 \times 10^{-4})[E] \approx 0.00031$. There is about $0.051/0.00031 = 168$ times more axial conformation of methylcyclohexane than there is axial conformation of *tert*-butylcyclohexane per mole.

7.4   (a)   Because the axial conformation of fluorocyclohexane has *two* fluorine–hydrogen 1,3-diaxial interactions (just as the axial conformation of methylcyclohexane has two methyl–hydrogen 1,3-diaxial interactions), the energetic cost of a single 1,3-diaxial interaction in fluorocyclohexane is 0.50 kJ mol^{-1} (0.13 kcal mol^{-1}).

     (b)   In the case of methylcyclohexane, the energy cost of an axial methyl group is slightly greater than twice that of the methyl–methyl interaction in *gauche*-butane. If we take the same approach with fluorocyclohexane and 1-fluoropropane, the energy cost of an axial fluorine in fluorocyclohexane [1.0 kJ mol^{-1}, from part (a)] should be about twice the energy cost of a methyl–fluorine interaction in the gauche conformation of 1-fluoropropane. Therefore, such a gauche interaction should be about 0.50 kJ mol^{-1}.

7.5   (a)   The two chair conformations of *cis*-1,3-dimethylcyclohexane:

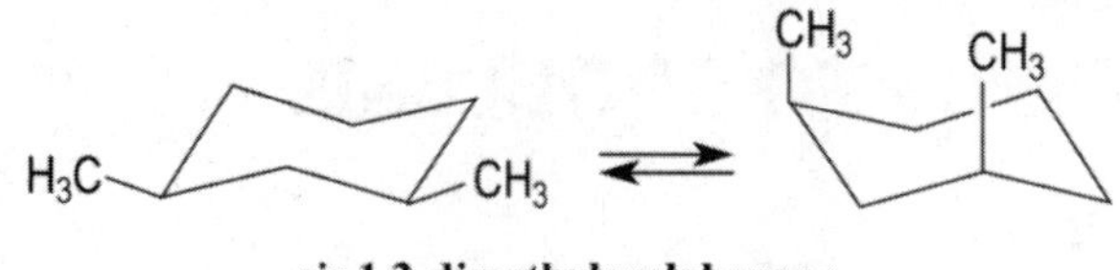

cis-1,3-dimethylcyclohexane

(b)   The conformations of *trans*-1-ethyl-4-isopropylcyclohexane:

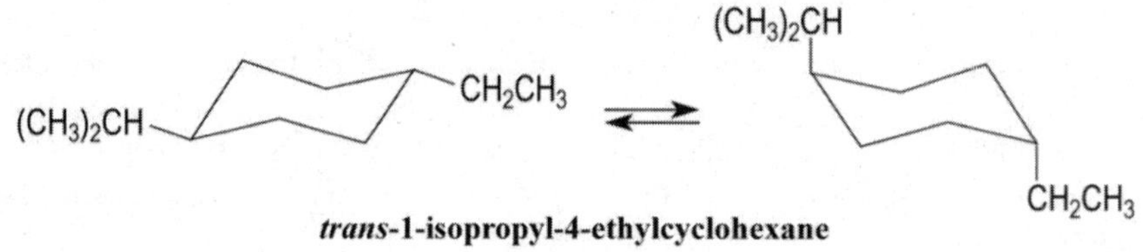

***trans*-1-isopropyl-4-ethylcyclohexane**

> Don't forget that a substituted cyclohexane can be drawn in many valid ways. If the structures you drew look different from those above, either verify that they are correct with models, or try to manipulate your structures mentally to see whether they can be made to look identical to the ones above. (Study Guide Link 7.1 provides some exercises that should be helpful.)

7.6   (a)   A boat conformation of *cis*-1,3-dimethylcyclohexane:

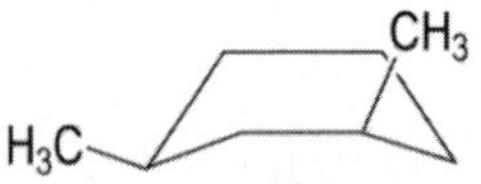

cis-1,3-dimethylcyclohexane
(a boat conformation)

(b)   Remember that there are several valid ways to draw this structure.

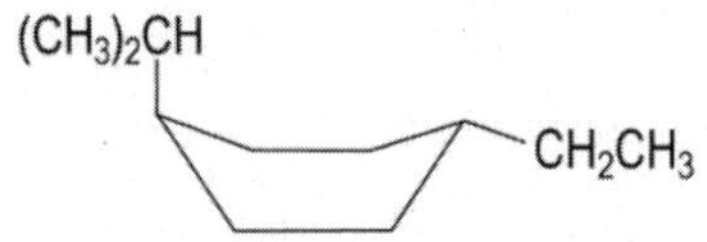

***trans*-1-isopropyl-4-ethylcyclohexane**
(a boat conformation)

7.7   (a)   Planar ring structures of *cis*-1-bromo-3-methylcyclohexane (both enantiomers):

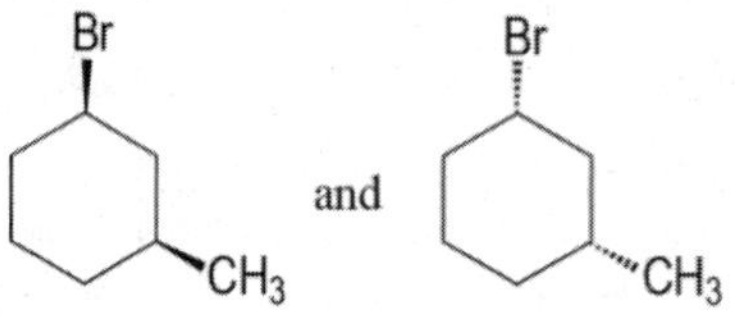

and

(b)   Planar ring structure of (1*R*,2*S*,3*R*)-2-chloro-1-ethyl-3-methylcyclohexane:

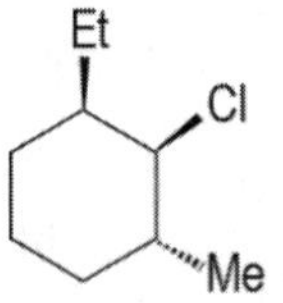

7.8   (a)

and

(b)

7.9   (a)   (b)

7.10   (a)   (b)

7.11   (a)   An achiral dichlorodimethylcyclohexane that undergoes the chair interconversion to give conformational enantiomers:

120°

conformational enantiomers

(b)   A chiral dichlorodimethylcyclohexane that undergoes the chair interconversion to give identical chiral structures:

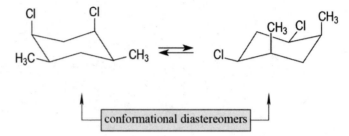

(c)   An achiral dichlorodimethylcyclohexane that undergoes the chair interconversion to give conformational diastereomers:

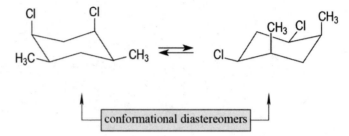

7.12   As the discussion in this section of the text has shown, a molecule is chiral (that is, can be isolated in optically active form) when its planar structure is not congruent to its mirror image.

(a)   *Trans*-1,2-dimethylcyclohexane can be isolated in optically active form, because it can exist as a pair of enantiomers:

(b)   1,1-Dimethylcyclohexane is achiral, and therefore cannot be optically active.

(c)   *Cis*-1-ethyl-4-methylcyclohexane is achiral, and therefore cannot be optically active.

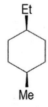

(d)   *Cis*-1-ethyl-3-methylcyclohexane can be isolated in optically active form, because it can exist as a pair of enantiomers:

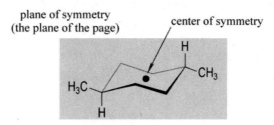

7.13 **(a)** *Trans*-1,4-dimethylcyclohexane does *not* have asymmetric carbons because none exist that are tetrahedral and attached to four different groups.

(b) The starred carbons are *stereo*centers because if you change the configuration at either carbon, you get its *stereo*isomer, *cis*-1,4-dimethylcyclohexane.

(c) *Trans*-1,4-dimethylcyclohexane is not chiral because it has a plane of symmetry (that is, an internal mirror plane) and a center of symmetry. These symmetry elements are shown here for one chair conformation; the other has the same symmetry. (The ultimate test of chirality is that it is not congruent to its mirror image.)

(d) The two chair conformations are diastereomers.

7.14 **(a)** First, draw the two chair conformations of *trans*-1,4-dimethylcyclohexane:

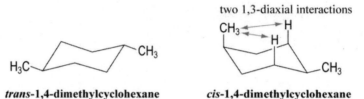

trans-1,4-dimethylcyclohexane

Conformation *A* has no 1,3-diaxial methyl–hydrogen interactions. Conformation *B* has four—two for each methyl group. At 3.7 kJ mol⁻¹ each, these interactions make conformation *B* 14.8 kJ mol⁻¹ less stable than conformation *A*.

(b)   The more stable conformations of the two 1,4-dimethylcyclohexanes:

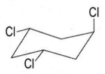

trans-1,4-dimethylcyclohexane   **cis-1,4-dimethylcyclohexane**

The cis isomer has the same number of 1,4 methyl–hydrogen interactions—two—as the axial conformation of methylcyclohexane itself, and thus has a destabilizing contribution of 7.4 kJ mol⁻¹. There are no destabilizing interactions in the trans isomer.

7.15   (a)   The larger *tert*-butyl group assumes the equatorial position:

(b)   The most stable conformation is the one that has the greater number of groups in the equatorial position:

(c)   The larger isopropyl group assumes the equatorial position. The presence of oxygen does not alter the fact that the ring exists in a chair conformation.

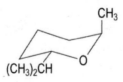

7.16   (a)   If *trans*-1,3-dibromocyclobutane were planar, the carbon–bromine bonds would be directed in opposite directions; hence, their bond dipoles would cancel, and the net dipole moment of the molecule would be zero. In puckered *trans*-1,3-dibromocyclobutane, the bond dipoles do not cancel.

planar *trans*-1,3-dibromocyclobutane:
C—Br bond dipoles cancel

puckered *trans*-1,3-dibromocyclobutane:
C—Br bond dipoles cancel

(b)   Cyclobutane undergoes an interconversion of puckered forms analogous to the chair interconversion of cyclohexane. This interchanges axial and equatorial groups. Therefore, one conformation of *trans*-1,2-dimethylcyclobutane—the more stable conformation—has diequatorial substituents, and the other—the less stable conformation—has diaxial substituents.

more stable conformation

***trans*-1,2-dimethylcyclobutane**

7.17   (a)   *Cis*-1,2-dimethylcyclopropane is achiral; it is a meso compound.

internal mirror plane

***cis*-1,2-dimethylcyclopropane**

(b)   *Trans*-1,2-dimethylcyclopropane is chiral.

***trans*-1,2-dimethylcyclopropane**
noncongruent mirror images

7.18   (a)   Bicyclo[2.2.2]octane, a bridged bicyclic compound
         (b)   Bicyclo[3.2.0]heptane, a fused bicyclic compound

7.19   A fused bicyclic compound, by definition, has a "zero-carbon bridge," and will have a zero in the number associated with the name. Consequently, compound *B* is a fused bicyclic compound; compound *A* is a bridged bicyclic compound.

7.20   The two indicated carbons in the following diagram of *cis*-decalin are axial substituents in the other ring; consequently, the hydrogens on these carbons are involved in 1,3-diaxial interactions (double-headed arrows in the following structure.). There are, as in other axially-substituted alkylcyclohexanes, two 1,3-diaxial interactions per axial carbon, except that one is common to both carbons; hence, there are a total of three 1,3-diaxial interactions. Because there are no 1,3-diaxial interactions in *trans*-decalin, it is more stable than *cis*-decalin by 3 × 3.7 = 11.1 kJ mol^{-1} (2.65 kcal mol^{-1}).

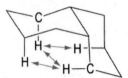

7.21 (a) The model of *cis*-bicyclo[3.1.0]hexane is much more easily built. The smaller is a ring, the closer is the dihedral angle of the trans bonds to 180°. Forcing this dihedral angle to much smaller values introduces strain within the ring. Such a large dihedral angle causes the ends of trans bonds to be too far apart to be readily bridged by only one carbon.

cis-**bicyclo[3.1.0]hexane**

 Be sure to build the models called for in this and the following problem. Models are much more revealing, and provide you with much greater insight, than simply reading verbal descriptions.

(b) The model of *trans*-bicyclo[5.3.0]decane is easier to build. The larger is a ring, the easier it is to compress the dihedral angle of the trans bonds at the ring junction without introducing significant strain in the ring. This angle must be very close to 0° in order to accommodate a fused cyclopropane, that is, to bridge the ends of trans bonds with only one carbon.

a dihedral angle close to 0°
is required for the cyclopropane ring

strain in the larger ring
decreases with
increasing ring size

7.22 (a) Compound *B* is less stable because it has a bridgehead double bond within a small ring. Consequently, it would have the greater (more positive or less negative) heat of formation.

(b) Although both molecules have bridgehead double bonds, the double bond in compound *B* is more twisted, and a model of this molecule is more difficult to build. Consequently, compound *B* is less stable and therefore would have the greater (more positive or less negative) heat of formation.

7.23 (a) Because a nail and a hammer are achiral objects, all interactions of these objects with any chiral object (such as a hand) and its enantiomer are identical. Hence, assuming equal strength, a right and a left hand should be equally adept at driving a nail. (This assumes, of course, that the person driving the nail doesn't accidentally bend it into some chiral shape.) In contrast, when it drives a screw, a chiral hand interacts with the helical (and chiral) threads of the screw. For this reason, the right and left hands of the same person should differ in their ability to drive a screw.

(b) Premeds could take out their stethoscopes and determine which side of each person the heart is on. Or, you could offer to shake hands. Mr. L would extend what he calls his right hand, but to us it would be his left. Or, you could ask them to smell a spearmint leaf and describe the odor. (*R*)-Carvone, the active principle of spearmint, has a spearmint odor, but its enantiomer (*S*)-carvone smells like caraway (the odor of rye bread). (*R*)-Carvone would smell like caraway to Mr. L.

(R)-(−)-carvone

In any case, the point is that the two mirror images are distinguished by comparing a chiral reference element (our bodies, our right hands, or the odors or tastes of enantiomers) with the corresponding elements of the two people.

7.24   (a)   Because enantiomers react with achiral compounds at the same rate, enantiomers are formed in equal amounts.

enantiomers; formed at the same rate

    (b)   Enantiomers react at different rates with chiral reagents, so the products are diastereomers formed in unequal amounts.

diastereomers; formed at the different rates

7.25   Three stereoisomers of 2,3-dibromobutane—(2S,3S), (2R,3R), and meso—could form in bromine addition to *cis*-2-butene. The enantiomers, (2S,3S) and (2R,3R), must be formed in identical amounts (that is, as a racemic mixture) because the starting materials are achiral. The meso diastereomer is formed in different amounts from the racemate.

7.26   Hydroboration-oxidation of *trans*-2-butene gives racemic 2-butanol—that is, two enantiomers formed in equal amounts.

racemic 2-butanol

7.27   Write all possible diastereomers resulting from addition to one enantiomer of the starting material. These are in principle formed in different amounts. (Problem 7.29 considers this case further.)

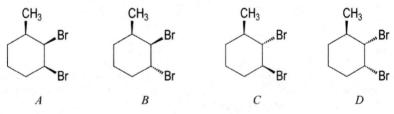

A   *B*   *C*   *D*

Each of these structures has an enantiomer; drawing these gives four additional diastereomers, which are also formed in different amounts. (Diastereomers are *always* formed in different amounts in principle.)

A' B' C' D'

Each point in the process that gives A (that is, each reactive intermediate, each transition state, etc.) is enantiomeric to the corresponding point in the process that gives A'; therefore the two processes must occur with identical rates. Because the starting alkene is racemic, the same amounts of the enantiomeric alkenes are involved in each process; therefore, the enantiomers of each pair are formed in identical amounts. Thus, A and A' are formed in identical amounts; B and B' are formed in identical amounts; and so on.

7.28   Because the bromonium-ion mechanism involves anti-addition, bromine addition to cyclohexene must give *trans*-1,2-dibromocyclohexane as a product.

(1R,2R)-1,2-dibromocyclohexane

Reaction of the bromide ion at the other carbon of the bromonium ion gives the enantiomer of *trans*-1,2-dibromocyclohexane, and it is formed in equal amounts.

(1S,2S)-1,2-dibromocyclohexane

7.29   (Refer to the structures in the solution to Problem 7.27.) Because the bromines of the product are trans, products B and C, and their respective enantiomers B' and C', are the major products. Compounds B and B' are formed in identical amounts, because they are enantiomers, and compounds C and C' are formed in identical amounts for the same reason. Compounds B and C, as well as compounds B' and C', are formed in different amounts, because they are diastereomers.

7.30   (a)   Because hydroboration-oxidation is a net syn-addition, the H and OH are added to the same face of the alkene. This addition can occur in two different (and enantiomeric) ways to give enantiomeric products, which are formed in identical amounts. For ease in seeing the mode of addition, the products are shown in an eclipsed conformation.

(2S,3S)   (2R,3R)

(b)   It is always true that, for a stereospecific reaction, use of a diastereomeric starting material gives the diastereomeric set of enantiomeric products, which are formed in identical amounts. As in part (a), the products are shown in eclipsed conformations.

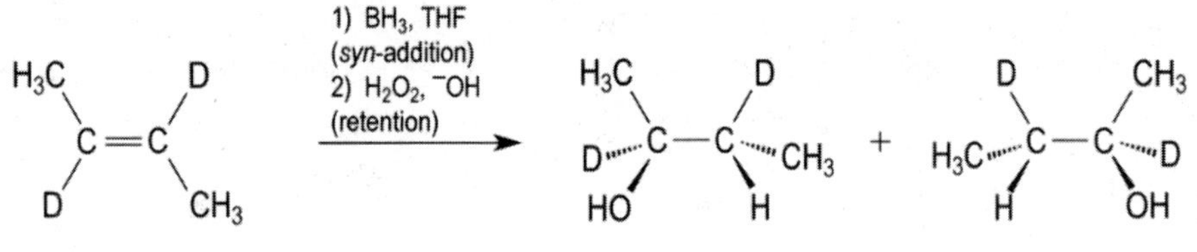

(2*S*,3*R*) (2*R*,3*S*)

7.31 Two products are formed, resulting from whether the BH_3 adds from the top or bottom face of the alkene. The two products are enantiomers, formed in equal amounts.

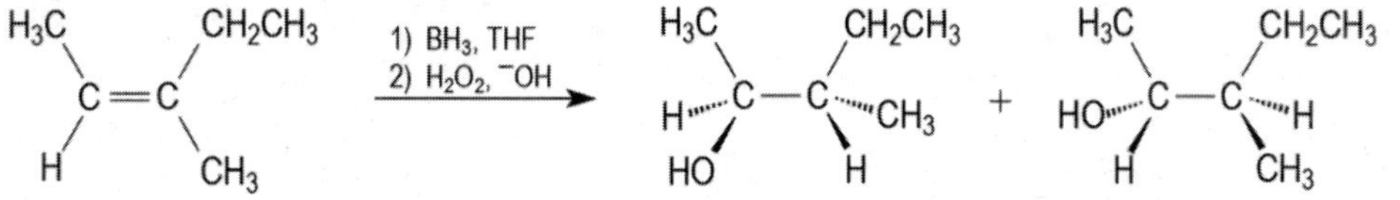

2*S*, 3*S* 2*R*, 3*R*

enantiomers; formed in identical amounts

7.32 (a) Oxymercuration is an anti-addition. Two enantiomeric products are formed in equal amounts.

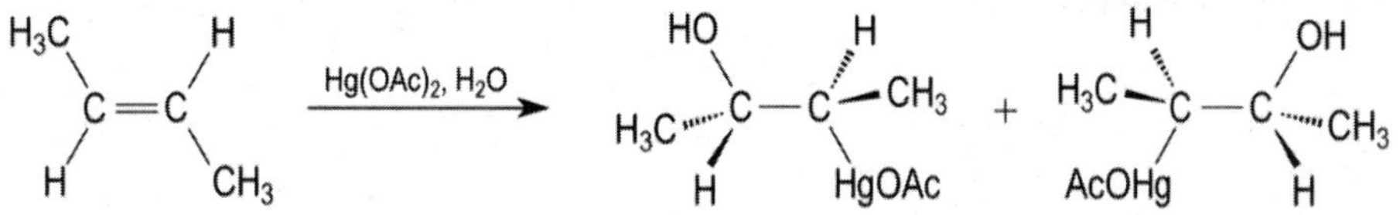

enantiomers; equal amounts are formed

(b) The enantiomeric products of the reaction in part (a) are shown below as *P* and *P*′. Because the $NaBD_4$/NaOH reaction occurs with *loss of stereochemistry,* two diastereomers are formed from each of the enantiomers *P* and *P*′. As a result, all four possible stereoisomers are formed.

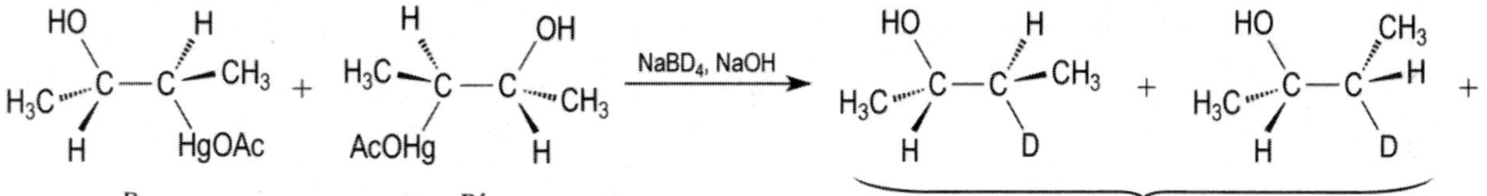

P *P*′

products from *P*

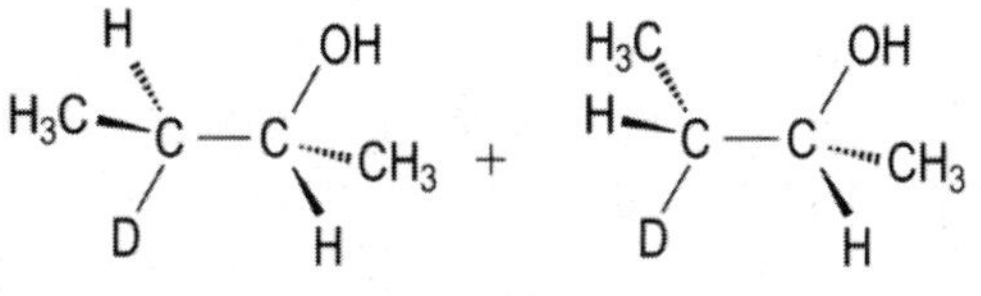

products from *P*′

7.33 (a) Alkenes *A* and *B* give a single compound; these products are achiral and have no stereocenters.

product from *A* product from *B*

(b) Alkene *C* gives two diastereomers. (Each a mixture of two enantiomers formed in equal amounts, because the overall reaction is not stereospecific.)

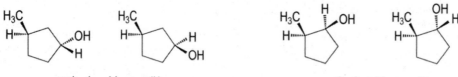

products from *C*
(each one is racemic)

(c)   Alkene *D* gives two constitutional isomers, each of which is a mixture of two diastereomers.

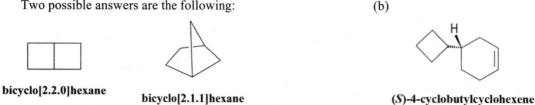

constitutional isomer #1   constitutional isomer #2
(a mixture of diastereomers)   (a mixture of diastereomers)

Solutions to Additional Problems

7.34   (a)   Two possible answers are the following: (b)

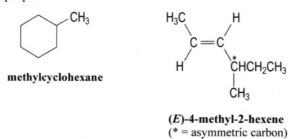

bicyclo[2.2.0]hexane

bicyclo[2.1.1]hexane

(*S*)-4-cyclobutylcyclohexene

7.35   Methylcyclohexane and (*E*)-4-methyl-2-hexene are *constitutional isomers*. Hence, they have the same molecular mass but different properties.

methylcyclohexane

(*E*)-4-methyl-2-hexene
(* = asymmetric carbon)

   (a)   A molecular mass determination would not distinguish between the two compounds because their molecular masses are identical.
   (b)   Uptake of H_2 over a catalyst would distinguish between the two compounds because the alkene would take up hydrogen in the presence of a catalyst, whereas the cycloalkane would not.
   (c)   The reaction with Br_2 would distinguish between the two compounds because the alkene would decolorize Br_2 solution (because alkenes react with Br_2), whereas the alkane would not.
   (d)   Determination of the molecular formula would not distinguish between the two compounds because both compounds are isomers; they have the same formula.
   (e)   Because constitutional isomers have different physical properties, determination of the heat of formation would distinguish between the two compounds. (Which would have the greater heat of formation?)
   (f)   Enantiomeric resolution would distinguish between the two because (*E*)-4-methyl-2-hexene is chiral, and can be resolved into enantiomers, whereas methylcyclohexane, an achiral compound, cannot be.

7.36   As the problem indicates, 2-pentanol is a chiral molecule.

$$\overset{\displaystyle OH}{\underset{\displaystyle CH_3CHCH_2CH_2CH_3}{|}}$$

2-pentanol

Distinguishing between enantiomers requires a technique that has a chiral attribute.

(a)   Boiling point does not distinguish between two enantiomers.

(b)   Optical rotation does have a chiral aspect, and therefore can distinguish between two enantiomers. (See Further Exploration 7.3.)

(c)   Because hexane is an achiral solvent, enantiomers have identical solubilities in hexane.

(d)   Density is not a chiral property, and therefore the two enantiomers have identical densities.

(e)   The solubilities of two enantiomers in principle differ in an enantiomerically pure chiral solvent.

(f)   Dipole moment is not a chiral property, and therefore the two enantiomers have identical dipole moments.

(g)   Because taste buds are chiral, they in principle differentiate between two enantiomers.

7.37   (a)   (b)   (c)

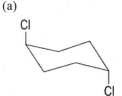

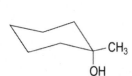

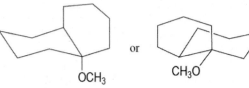

or

7.38   (a)   In this compound the large *tert*-butyl group must assume the equatorial position. The positions of the other groups follow from their positions relative to the *tert*-butyl group.

(b)   In choosing which conformation to draw, the goal is to put the maximum number of methyl groups possible in equatorial positions.

7.39   (a)   The equilibrium constant is given as 2.07 for the ratio [equatorial]/[axial]. The standard free-energy difference is calculated by applying Eq. 3.35, text p. 110:

$$\Delta G° = -2.30RT \log (2.07) = -(5.71 \text{ kJ mol}^{-1})(0.316) = -1.80 \text{ kJ mol}^{-1}$$

In other words, for the equilibrium axial $\rightleftharpoons$ equatorial, the equatorial form has the lower standard free energy, and is thus more stable.

(b)   The standard free energy change for the reaction equatorial $\rightleftharpoons$ axial is +9.2 kJ mol^{-1}; the axial conformation has the higher standard free energy. The equilibrium constant is calculated by applying Eq. 3.36b, text p. 110.

$$K_{eq} = 10^{-\Delta G°/2.30RT} = 10^{-9.2/5.71} = 10^{-1.61} = 2.4 \times 10^{-2}$$

That is, the ratio [axial]/[equatorial] is 0.024. This means that there is 1/0.024 = 42 times as much equatorial conformation as there is axial conformation at equilibrium.

7.40   A rapid rotation can occur about the bond between the cyclohexane ring and the ethyl group so that the van der Waals radii of the methyl group and the axial hydrogens do not overlap.

like a CH_3–H interaction

A   *B*   *B'*

In the conformations *B* and *B'*, the interactions of the axial hydrogens with the hydrogens of the ethyl group are about the same as the corresponding interactions in a methyl–hydrogen 1,3-diaxial interaction. As a result, the ethyl–hydrogen interaction has about the same energy cost as a methyl–hydrogen interaction.

There is a little more to this situation than meets the eye. Because conformation *A* has relatively high energy, it is effectively "locked out" of the conformational equilibrium. This means that axial ethylcyclohexane has *two* rather than three conformations about the ring–ethyl bond. In other words, axial ethylcyclohexane has *reduced freedom of motion*. This conformational constraint would be expected to show up in its *entropy,* and, in fact, it does. Axial ethylcyclohexane has a lower $\Delta S°$ than axial methylcyclohexane; this lower entropy increases its standard free energy, because $\Delta G° = \Delta H° - T\Delta S°$. However, in equatorial ethylcyclohexane, there are two gauche methyl–CH_2 interactions with the ring that are not present in equatorial methylcyclohexane.

B   like a *gauche*-butane CH_3–H interaction

These interactions raise the enthalpy of the equatorial form of ethylcyclohexane relative to methylcyclohexane and raise its free energy. These two changes—decreased entropy in the axial conformation, and increased enthalpy in the equatorial conformation—almost exactly compensate.

7.41   (a)   Compound *A* can be prepared by hydroboration–oxidation from either *cis*- or *trans*-3-hexene because the alkene is symmetrical and, for that reason, regioselectivity and stereoselectivity are not relevant.

$$CH_3CH_2CH{=}CHCH_2CH_3 \xrightarrow[\text{2) } H_2O_2,\ ^-OH]{\text{1) } BH_3,\ THF} CH_3CH_2\overset{\overset{\displaystyle OH}{|}}{C}HCH_2CH_2CH_3$$

3-hexene   **3-hexanol**
                              (compound *A*)

Hydroboration–oxidation of 1-methylcyclopentene would give compound *D*, not its diastereomer *B*, because this reaction sequence results in a syn-addition.

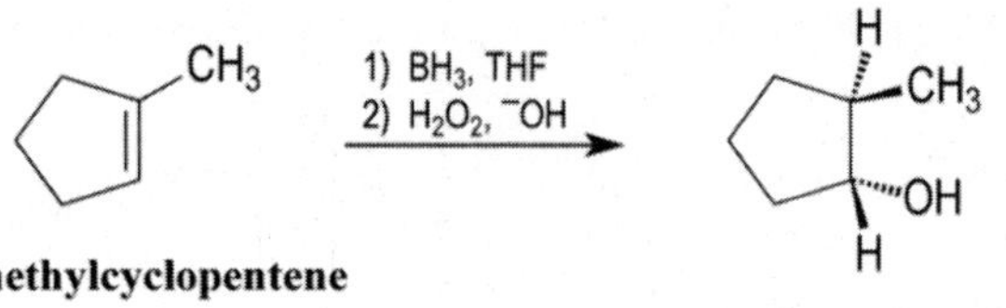

1-methylcyclopentene

1) BH₃, THF
2) H₂O₂, ⁻OH

(±)-*trans*-2-methyl-1-cyclohexanol
(compound *D*)

Compound *D* is racemic, and is therefore a 50:50 mixture of the stereoisomer shown above and its enantiomer.

Compound *C* cannot be produced by hydroboration–oxidation of an alkene because the OH group would have to end up at the more branched carbon of any possible alkene starting material.

(b)   Compound *A* could be formed by oxymercuration–reduction from either *cis*- or *trans*-3-hexene because the alkene is symmetrical and, for that reason, regioselectivity and stereoselectivity are not relevant.

Oxymercuration–reduction of 1-methylcyclopentene gives compound *C*; therefore compounds *B* and *D* are not formed.

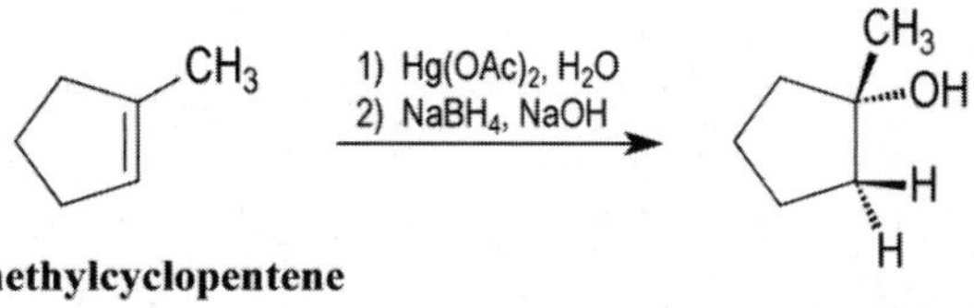

1-methylcyclopentene

1) Hg(OAc)₂, H₂O
2) NaBH₄, NaOH

(±)-1-methyl-1-cyclohexanol
(compound *C*)

7.42   *For reaction (1):*

(a)   The products:

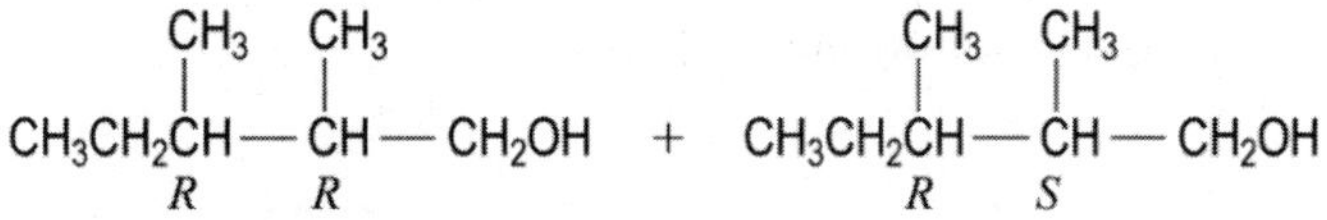

$$CH_3CH_2\underset{R}{\overset{\overset{\displaystyle CH_3}{|}}{CH}}-\underset{R}{\overset{\overset{\displaystyle CH_3}{|}}{CH}}-CH_2OH \quad + \quad CH_3CH_2\underset{R}{\overset{\overset{\displaystyle CH_3}{|}}{CH}}-\underset{S}{\overset{\overset{\displaystyle CH_3}{|}}{CH}}-CH_2OH$$

(b)   The two products are diastereomers.
(c)   The two products are formed in different amounts.
(d)   The two products have different boiling points and melting points, and could therefore be separated (in principle) by fractional distillation or fractional crystallization.

For reaction (2):

(a)   The products:

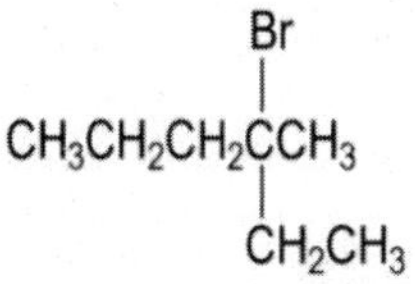

$$CH_3CH_2CH_2\underset{\underset{\displaystyle CH_2CH_3}{|}}{\overset{\overset{\displaystyle Br}{|}}{C}}CH_3$$

R and *S*

(b)   The two products are enantiomers.
(c)   The two products are formed in identical amounts.
(d)   The two products have identical boiling points and identical melting points. An enantiomeric resolution would be required to separate the two compounds.

For reaction (3):

(a)   The products:

$$\overset{\displaystyle Br}{\underset{|}{CH_3CH_2CHCH_2Br}}$$

R and *S*

(b)   The two products are enantiomers.

(c)   The two products are formed in identical amounts.

(d)   The two products have identical boiling points and identical melting points. An enantiomeric resolution would be required to separate the two compounds.

For reaction (4):

(a)   The starting alkene is racemic, and is therefore an equimolar mixture of two enantiomers. To predict the products, make the prediction for each enantiomer separately and then combine the results.

$$\underset{A}{\overset{Ph}{\underset{\displaystyle Br}{\underset{S}{CH_3\overset{S}{C}HCHCH_2Br}}}} \;+\; \underset{B}{\overset{Ph}{\underset{\displaystyle Br}{\underset{S}{CH_3\overset{R}{C}HCHCH_2Br}}}} \;+\; \underset{C}{\overset{Ph}{\underset{\displaystyle Br}{\underset{R}{CH_3\overset{S}{C}HCHCH_2Br}}}} \;+\; \underset{D}{\overset{Ph}{\underset{\displaystyle Br}{\underset{R}{CH_3\overset{R}{C}HCHCH_2Br}}}}$$

from the 3*S* alkene   from the 3*R* alkene

(b)   Compounds *A* and *B*, which come from the *S* enantiomer of the starting material, are diastereomers, as are compounds *C* and *D*, which come from the *R* enantiomer of the starting material. Compounds *A* and *C* and compounds *B* and *D* are also diastereomers. Compounds *A* and *D*, as well as compounds *B* and *C*, are enantiomers.

(c)   The diastereomers are formed in different amounts; the enantiomers are formed in identical amounts.

(d)   Any pair of diastereomers have different melting points and boiling points; any pair of enantiomers have identical melting points and boiling points.

For reaction (5):

(a)   The product:

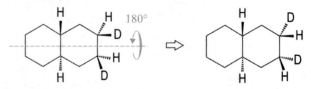

(b)–(d)   Only one product is formed.

For reaction (6):

(a)   The starting alkene is a single enantiomer of a chiral compound. Two syn-additions of D_2 are possible: one from the upper face of the double bond, and one from the lower face.

addition from the upper face   addition from the lower face

(b)–(d)   As the foregoing structures show, the two modes of addition give the same product; so, only one structure is possible.

7.43   (a)   An achiral tetramethylcyclohexane that undergoes the chair interconversion to give identical molecules:

(b)   An achiral trimethylcyclohexane that undergoes the chair interconversion to give conformational diastereomers:

(c)   A chiral trimethylcyclohexane that undergoes the chair interconversion to give conformational diastereomers:

(d)   An achiral tetramethylcyclohexane that undergoes the chair interconversion to give conformational enantiomers:

7.44   The conformational representation of the steroid:

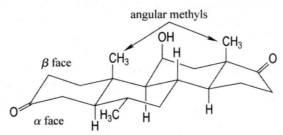

7.45   The chair conformations of α-(+)-glucopyranose:

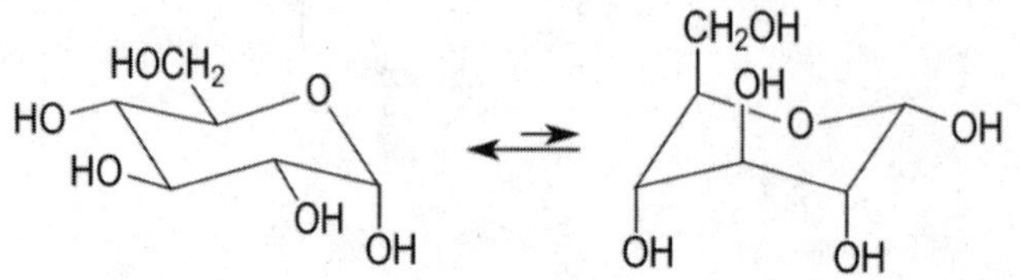

Conformation A is more stable because all the substituents (except one) are equatorial.

7.46   (a)   Anti-addition can occur to give two chiral diastereomers, which will be formed in different amounts. Each is a single enantiomer because the starting material is a single enantiomer.

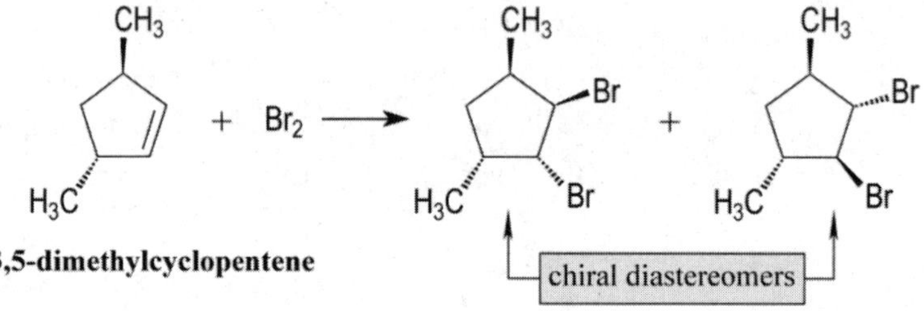

(3R,5R)-3,5-dimethylcyclopentene   chiral diastereomers

(b)   The bromonium ion derived from the reaction of cyclopentene with bromine undergoes backside substitution by water to give the *trans*-bromohydrin, which is chiral. Because the starting materials are achiral, the chiral product is obtained as a racemate.

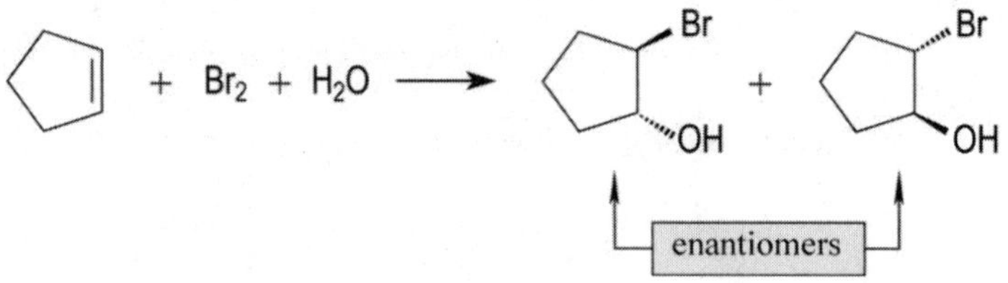

enantiomers

7.47   The two dibromides are the two diastereomeric *trans*-disubstituted *trans*-decalin derivatives:

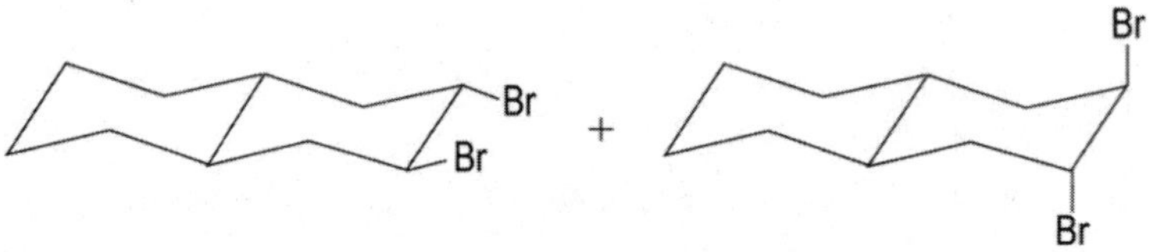

The different compounds result from the reaction of bromide ion with the bromonium ion at the two possible carbons. (Be sure to work through this mechanism.) Because the starting alkene is chiral, each diastereomer is obtained as a single enantiomer. Also, remember that the rings of a *trans*-decalin do not undergo the chair interconversion.

7.48   Given that addition of bromine at each double bond is anti, two diastereomers of 1,2,4,5-tetrabromocyclohexane can be formed. One is the achiral meso compound, and the other is the racemate.

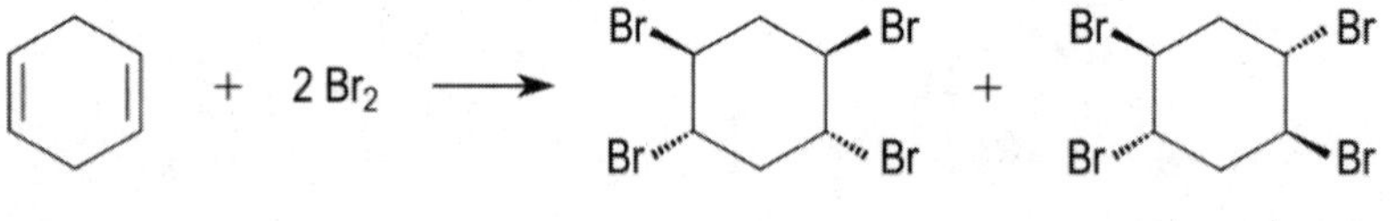

meso   racemate

Because these are diastereomers, they have different physical properties, including melting points.

Assuming you had samples of the two compounds but didn't know which was which, what experiment could you do that might identify the two compounds? (*Answer:* Carry out an enantiomeric resolution. The compound that can be separated into optically active components—its two enantiomers—is identified as the racemate.) It turns out that the meso compound is the higher-melting stereoisomer.

7.49   Compound *A* is *not* consistent with the data because it is meso and cannot be optically active. Compound *B* is consistent with the data because it is chiral and would hydrogenate to give *trans*-1,4-dimethylcyclohexane, an achiral product. Compound *C* is consistent with the data because it is chiral and would hydrogenate to give a mixture of *cis*- and *trans*-1,4-dimethylcyclohexane, both of which are achiral products. Compound *D* is *not* consistent with the data, because, although it is chiral and thus optically active, it would hydrogenate to give a chiral and optically active product, (1*S*,3*S*)-1,3-dimethylcyclohexane. Compound *E* is *not* consistent with the data because it is not chiral and therefore could not be optically active.

7.50   Because this is an amine, both the chair interconversion and amine inversion can occur. The amine inversion interchanges the axial and equatorial positions of the electron pair and the hydrogen on the nitrogen *without* interconverting the chair conformations of the ring. Thus, four chair conformations are in rapid equilibrium:

7.51   (a)   The only stereocenter in the molecule is the nitrogen, which rapidly undergoes inversion, a process that rapidly interconverts enantiomers. Consequently, this compound cannot be resolved into enantiomers at room temperature.

(b)   Because nitrogen inversion interconverts enantiomers, the compound cannot be resolved into enantiomers.

(c)   The bicyclic structure prevents the nitrogens from inverting (see the solution to Problem 7.75), and both nitrogens are stereocenters. Consequently, the molecule is chiral and can be resolved into enantiomers.

(d)   The asymmetric ring carbon causes the molecule to be chiral; inversion of the nitrogen interconverts diastereomers. Hence, the compound can be resolved into two enantiomeric sets of rapidly interconverting diastereomers. (This case is much like the one in Problem 6.22.)

7.52   In the transition state for inversion of both amines, the nitrogen is sp^2-hybridized. The C—N—C angle required by this hybridization is 120°. This angle is much more difficult to achieve within a three-membered ring, which has an optimal C—N—C angle of 60°, than it is in within a five-membered ring, which has an optimal C—N—C angle of 109°. Hence, there is much more angle strain in the transition state for inversion of 1-methylaziridine than there is in the transition state for inversion of 1-methylpyrrolidine. For this reason, the transition state for inversion of the aziridine derivative has a very high energy, and nitrogen inversion in this derivative is very slow. (This is a very unusual situation, since inversion of most amines is very rapid; see Sec. 6.10B.)

ring angle ≈ 60°; sp^2 hybridization
requires 120°; therefore this transition state
contains a large amount of angle strain

ring angle ≈ 109°; sp^2 hybridization
requires 120°; therefore this transition state
contains less angle strain

sp^2-hybridized
transition states
for amine inversion

7.53   The cis isomer of the starting alkene was used. This is shown by starting with the cis isomer and imagining the syn-addition of the two OH groups to the same face of the double bond.

KMnO₄, ⁻OH   R— = CH₃(CH₂)₇—

addition to a symmetrical
cis alkene

a meso stereoisomer

By a similar analysis, syn-addition to the trans isomer of the alkene gives the racemate. (You should demonstrate this to yourself.)

As this example and the examples in Secs. 7.8C and 7.8D of the text show, there is a relationship between the *stereochemistry of the alkene*, the *stereochemistry of the reaction*, and the *stereochemistry of the product*. If we specify any two of these, the other is determined.

7.54   **(a)**   Imagine both syn- and anti-additions to fumarate with the OD entering from the face of the alkene that leads to the product malate-3-*d* in which carbon-2 has the *S* configuration. A syn-addition will lead to (2*S*,3*S*)-malate-3-*d*, whereas an anti-addition will lead to the observed 2*S*,3*R* product. Therefore, the reaction is an anti-addition.

syn-addition

(2S,3S)-malate-3-*d*

anti-addition

(2S,3R)-malate-3-*d*
(observed)

There are experimental techniques—for example, NMR spectroscopy, which is the subject of Chapter 13—that can make the subtle stereochemical distinctions required to solve stereochemical problems like this.

(b)   The use of D₂O allows us to differentiate the protons of the solvent from those of the starting material. The stereochemistry of the addition cannot be determined without this distinction.

7.55   (a)   There are two enantiomeric (and therefore chemically equivalent) modes of anti-addition that give enantiomeric products as a racemate, and therefore equal amounts.

(2R,3S)-2,3-dibromopentane

(2S,3R)-2,3-dibromopentane

The difference in this addition and the addition to *trans*-2-butene is that there is no meso stereoisomer possible here. Meso products are only possible if the alkene is symmetrical.

(b)   This is essentially like the addition in part (a), except that the nucleophile that reacts with the bromonium ion is water; a bromohydrin is formed rather than a dibromide. (See Sec. 5.2B in the text to review this reaction.) The product is a racemic mixture.

(3S,4R)-4-bromo-3-hexanol

(3R,4S)-4-bromo-3-hexanol

(c)   Catalytic hydrogenation is a syn-addition. Analyzing this case in the same manner as part (a) gives the result that the product is the meso stereoisomer of hexane-3,4-d_2.

meso-hexane-3,4-d_2

(Convince yourself that addition from the lower face of the alkene gives the same product.)

As with other syn-additions, showing the product in an eclipsed conformation is the easiest way to derive its stereochemistry and to see the symmetry relationships involved. The molecule doesn't exist in this conformation. The rapid internal rotations that occur in molecules like this allow us to use whatever conformation we like in deducing stereochemistry. This point was discussed in Sec. 6.9A in the text.

(d)   Analyze the problem as in part (c). The reaction is a net syn-addition that gives a racemic product.

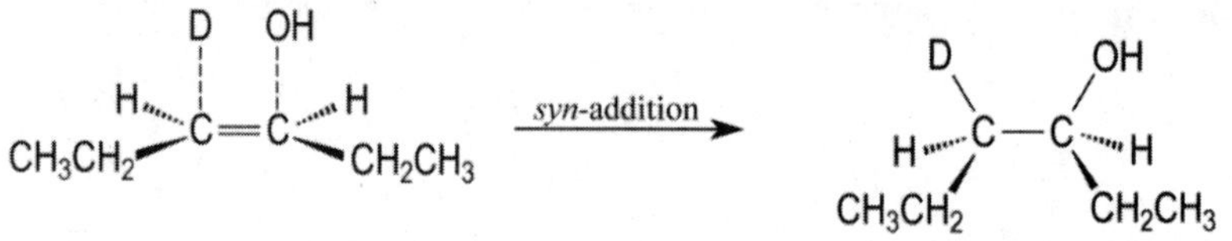

(3S,4R)-3-hexanol-4-d

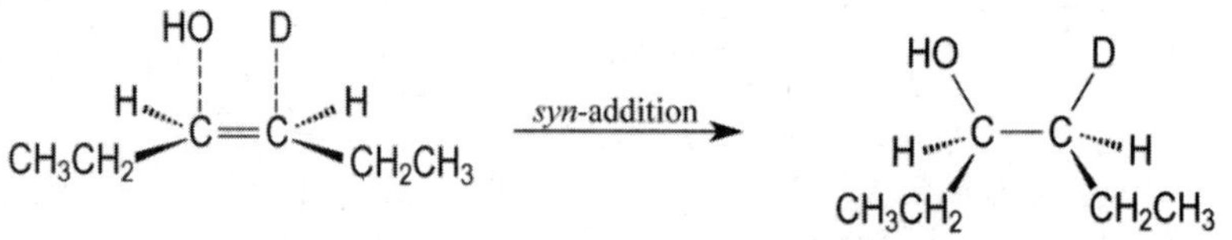

(3R,4S)-3-hexanol-4-d

(Convince yourself that the two modes of addition from the bottom face of the alkene give the same products.)

7.56   To follow the hint in the problem, interpret each chair structure as a planar "line-and-wedge" structure. If these two structures are congruent, then the two chair structures are *either* identical *or* are conformational stereoisomers. Decide between these by using the chair structures themselves. If the two chair structures are not congruent, then use their planar representations to decide on their relationship.

(a)   The planar representations are identical:

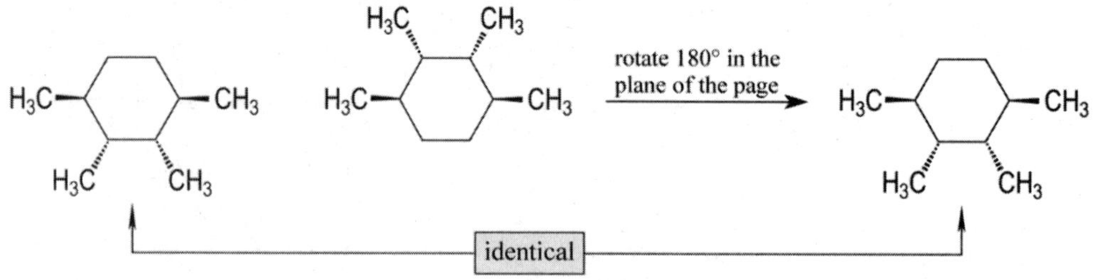

Therefore, two chair structures are either identical or they are conformational stereoisomers. Examination of the chair structures is necessary to reveal which alternative is correct. Let's rotate the second structure in the problem 180° about the axis shown to place all substituents in the same relative positions.

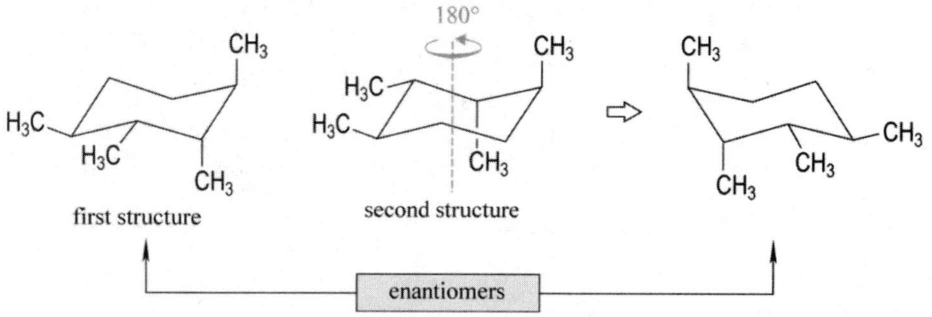

These two chair structures are mirror images, and you should convince yourself that they are noncongruent mirror images—that is, enantiomers. The chair interconversion converts the first structure into the second; hence, the two chair structures are conformational enantiomers.

first structure   second structure (as redrawn)

Remember that the use of models to solve problems in which two structures are compared requires *two models,* one of the original structure and another of the structure with which it is compared.

(b)   Following the procedure in part (a) reveals that these two molecules are enantiomers. (Convince yourself that these are noncongruent mirror images.)

Their enantiomeric relationship might not be obvious from the two chair structures without subjecting one of them to the chair interconversion.

Use of the planar structures avoids this additional manipulation.

(c)   The types of manipulations are similar to the ones in parts (a) and (b). The two molecules yield planar structures that, after a little manipulation, can be shown to be identical. The chair interconversion of one yields the other; hence, the two chair conformations are in fact identical; demonstration of the identity requires some manipulation of the two chair structures. (We leave it to you to demonstrate these points.)

Problem 7.56 is a particularly good test of your ability to manipulate cyclohexane rings. If you were able to solve it without a great deal of difficulty, you have probably mastered the art of handling chair conformations.

7.57   (a)   Hydration involves a carbocation intermediate. Because a carbocation can react with a nucleophile at either face of its vacant $2p$ orbital, a mixture of stereoisomeric hydration products A and B is expected.

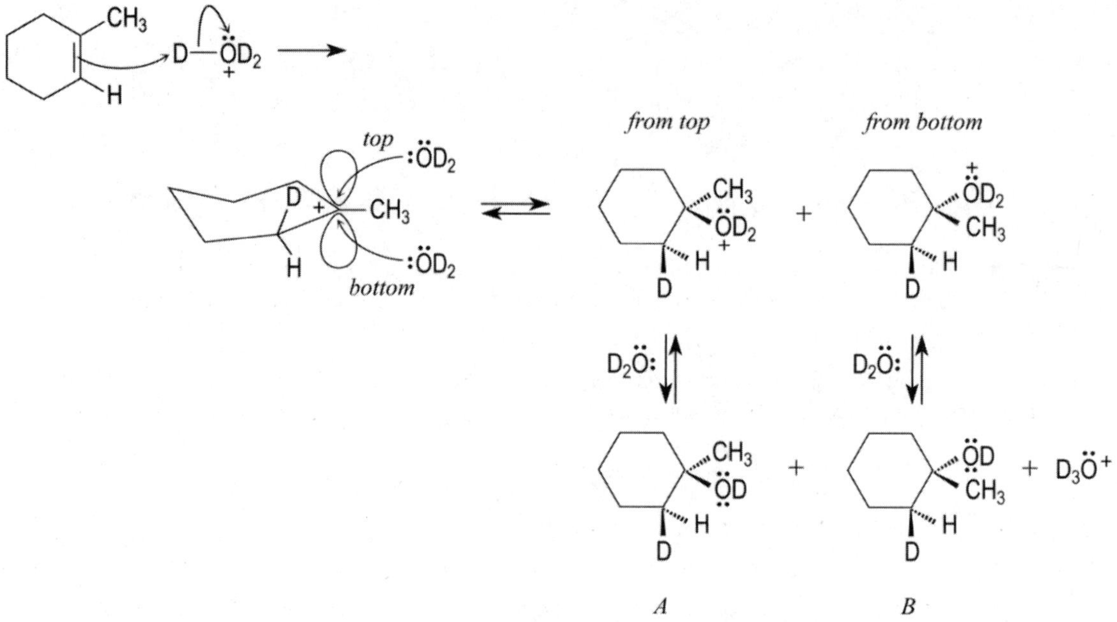

(b) As in the solution to Problem 7.54(b), the use of D_2O allows us to differentiate the protons of the solvent from those of the starting material.

(c) The two hydrogens referred to in part (b) could also be distinguished if the one in the starting alkene is replaced by a deuterium, and H_3O^+ is used as the acid—that is, if an alkene of the following structure is used:

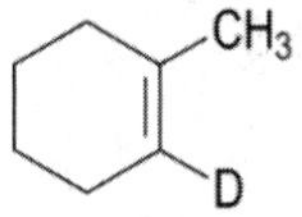

You should work out the detailed mechanism, as in part (a), to demonstrate this point.

7.58 (a) The two chiral stereoisomers are the following enantiomeric pair:

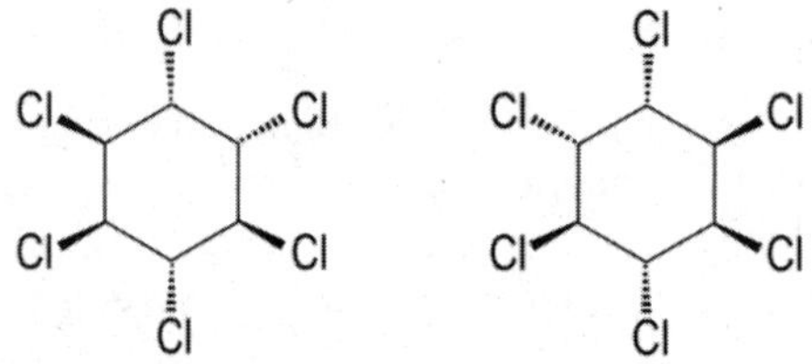

(b) The only candidate structures are those in which the number of axial chlorines equals the number of equatorial chlorines. In addition, the two chlorines within each of the three pairs that have a 1,4-relationship must differ in whether their positions are axial or equatorial. Thus, each of the following stereoisomers has two identical chair conformations. (The numbers refer to corresponding carbons.)

7.59 A systematic way to work this problem is to start with all methyl groups in a cis arrangement; then change the stereochemistry one group at a time.

We could go on to start with stereoisomer *A* and change two groups at a time in all possible ways, and then change all three groups. But when we do this, we find that all "new" possibilities are identical to one of the foregoing compounds *A–D*. Therefore, there are four stereoisomers—two meso compounds and two enantiomers.

7.60 (a) False. They are diastereomers, and they are not interconvertible without breaking bonds.
(b) False. As noted in (a), they are stereoisomers; they have the same connectivity.
(c) True, as noted in part (a).
(d) True, as noted in part (a).
(e) False, because they are not mirror images (and also because they are diastereomers).
(f) False. They do not interconvert at all.

7.61 Olean can exist as a pair of non-congruent mirror images, therefore it is a chiral compound. Additionally, it contains no symmetry elements. Although it may appear that there is a plane of symmetry based on the way that it is presented in the problem, you must remember to think about the chair conformations three dimensionally (you may wish to build a model). The compound may also be presented as follows; a clear lack of symmetry is visible this way, and the two enantiomers can be seen as mirror images.

(a) False. It is chiral, there are no symmetry elements.
(b) False, as noted in part (a).

(c)   True, the compound is chiral and it contains an asymmetric atom: the carbon at the bridgehead where the two rings connect.

(d)   True, the carbon in noted in part (c) is also a stereocenter, because exchanging the attachment of any two groups would produce a stereoisomer (specifically, the enantiomer).

(e)   False, the compound has a stereocenter as noted in part (d).

 Evidence suggests that the female fruit fly secretes olean as a racemate (mixture of both enantiomers), but laboratory tests show that male fruit flies are only attracted to the (R)-(−)-enantiomer, and ignore the (S)-(+)-enantiomer. Another example of the chirality of many natural systems!

7.62   (a)   True. This compound is achiral, as a plane of symmetry exists. Its mirror image is congruent. This compound also contains a stereocenter, two actually. Each of the bridgehead carbons are stereocenters, because switching two groups at either one would produce a stereoisomer, specifically a diastereomer of the one shown.

(b)   False, as noted in part (a), the compound contains stereocenters.

(c)   False, the compound is achiral and it contains no asymmetric atoms.

(d)   False, the compound is achiral.

(e)   False, the compound is achiral.

7.63   If the chirality of structures (a)–(c) is not clear, draw the mirror image in each case and test it for congruence. This process is essential for cases of this nature. (Stereocenters are indicated with a ◊.)

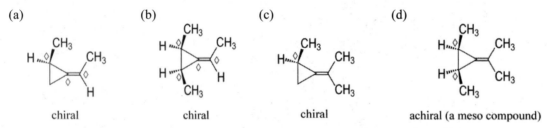

(a)   (b)   (c)   (d)

chiral   chiral   chiral   achiral (a meso compound)

7.64   In order for a compound to exist in a relatively high-energy twist-boat conformation, there must be a good reason for it to avoid the chair. Indeed, compounds B and D have very severe 1,3-diaxial interactions between a *tert*-butyl group and a methyl group in the chair conformations shown. Compound D, a *cis*-decalin derivative, can avoid its 1,3-diaxial interaction by undergoing a chair interconversion. However, compound B, a *trans*-decalin derivative, cannot undergo the chair interconversion (see Eq. 7.18, text p. 296). Hence, the only way it can avoid the 1,3-diaxial interaction is for the ring containing the *tert*-butyl group to assume a twist-boat conformation. The correct answer, then, is B.

7.65   *Cis*-1,3-di-*tert*-butylcyclohexane can exist in a chair conformation (A) in which both *tert*-butyl groups are equatorial. However, in either chair conformation of *trans*-1,3-di-*tert*-butylcyclohexane (B), there is an axial *tert*-butyl group. The axial *tert*-butyl group can be avoided if compound B exists in a twist-boat conformation C. in which both *tert*-butyl groups are equatorial. (This conformation is shown as a boat for simplicity.) Evidently the twist-boat conformation is more stable than either chair conformation containing an axial *tert*-butyl group. The interaction of the methyl groups with the ring hydrogens should be about the same in both conformations A and C, so that these cancel in the comparison.

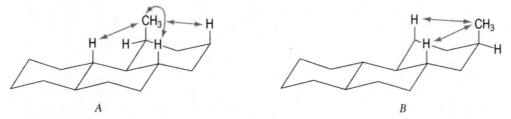

7.66   Within each structure, count the 1,3-diaxial interactions between the methyl group and axial ring hydrogens four carbons away (counting the methyl as one carbon). These are shown with double-headed arrows in the structures of *A* and *B* below. (There are no methyl-hydrogen 1,3-diaxial interactions in compound *C*.)

A   *B*

Each interaction shown adds 3.7 kJ mol^{-1} to the heat of formation relative to that of *C*. Consequently, *C* has the lowest heat of formation; the heat of formation of *B* is $(2 \times 3.7) = 7.4$ kJ mol^{-1} greater; and the heat of formation of *A* is $(3 \times 3.7) = 11.1$ kJ mol^{-1} greater than that of *C*.

7.67   (a)   *Trans*-decalin, *C*, is more stable than *cis*-decalin, *A* (see Problem 7.20, text p. 296). However, neither compound has angle strain. Compound *B* is least stable because of the strain in its four-membered ring. You can see from Table 7.1, text p. 273, that the strain in a cyclobutane ring $(4 \times 7.1 = 28.4$ kJ mol$^{-1})$ is far more destabilizing that the three 1,3-diaxial interactions $(3 \times 3.7 = 11.1$ kJ mol$^{-1})$ in *cis*-decalin. Therefore, the order of increasing ΔH_f° is $C < A < B$.

   (b)   The ring strain in *B* makes it less stable than *C*; and the twisted double bond in *A* (violation of Bredt's rule) makes it so unstable that it cannot be isolated. Therefore, the order of increasing ΔH_f° is $C < B \ll A$.

7.68   (a)

cis-2-butene   **borolane**

   (b)   The products in part (a) are enantiomers. Since they were formed from achiral reagents, they are formed in equal amounts.

(c)

cis-2-butene   (2R,5R)-2,5-dimethyl-
borolane

(d) The products in part (c) are diastereomers. Diastereomeric transition states with different energies lead to the products, so they were formed at different rates, and thus, in different amounts.

(e) The organoborane can approach the π bond of cis-2-butene molecule in two distinguishable ways. Approach from one face gives a transition state in which there is severe van der Waals repulsions between a methyl group of the alkene and a methyl group on the organoborane (structure A). Approach from the opposite face gives a transition state in which the van der Waals repulsion is between a *hydrogen* of the alkene and a methyl group of the borane (structure B).

Given that transition state B has less severe van der Waals repulsions, the product borane derived from that transition state is formed more rapidly than the product from transition state A.

(f) When allowed to react with basic H_2O_2, it gives the corresponding alcohol, which has the S configuration.

product from transition state B   (S)-2-butanol

7.69 (a) The question states that the predominant diastereomer is formed as a result of anti-addition of H and Br to the alkene. That means that the two components are added on opposite sides, ultimately leading to the Br and the CH_3 being on the same side of the ring. Thus, the predominant product is named as *cis*, because the two larger groups are on the same side of the ring. (It's easy to get confused on this type of a problem, as the name of the compound seems to contradict the chemistry that's happening. The best way to approach the problem is to solve the chemistry first: draw the product that results from anti-addition, then name the products that you've formed.)

anti-addition

The H and Br added to opposite sides (*anti*-addition), but the product is named *cis*- because the methyl and Br ended up on the same side.

cis-1-bromo-2-
methylcyclohexane

(b)   After the bromine radical adds to the double bond, the first reactive intermediate in the reaction is a tertiary radical. You've learned that radical carbons are approximately planar, or sp^2-hybridized, so that would suggest that the hydrogen radical should have about equal access to the upper and lower faces of the ring, and the product would be a mixture of *cis-* and *trans-*. However, the problem states that the reaction occurs with anti stereochemistry. A reasonable explanation might be that the bromine blocks one face of the ring, forcing the HBr to add a hydrogen radical from the opposite side. It has been postulated that an unusual cyclic bromonium radical (similar to a cyclic bromonium ion) is involved.

bottom face blocked; HBr
must approach from above

7.70   (a)   The two diastereomeric products result from addition at the upper (U) and (L) lower faces of the double bond.

(b)   The methyl group shields the upper face (in steroid terminology, the α face) of the molecule like an umbrella. Therefore, product L is the predominant product.

the steric effect of the methyl group forces the reaction to the opposite (lower) face of the double bond

7.71   Because the product contains one Br, it appears that a reaction with the double bond—most reasonably, an addition reaction—has taken place. The product X has the same degree of unsaturation as the starting material. If the double bond has disappeared, then there must be an additional ring. This deduction, in turn, suggests that a nucleophile other than bromide ion has reacted with the bromonium ion to form a ring, and the only available nucleophile is the —OH group. The following two possibilities for X are consistent with this deduction, in which the bromonium ion is formed at the less crowded upper face of the double bond, and the nucleophilic oxygen is then set up to react at the opposite (lower) side.

If you got to either of these possibilities, you've made a solid analysis. Product *A* is the correct one. The mechanism starts with the bromonium ion intermediate, formed as shown in Eq. 5.11, text p. 185. The OH group reacts as a nucleophile with this ion:

Product *B* would be formed by a similar mechanism, with the nucleophilic reaction of the OH group occurring at the other carbon of the bromonium ion.

Product *A* is formed for two reasons. First, as shown in Eq. 5.16 and Fig. 5.1, text p. 187, nucleophiles react at the tertiary carbon of a bromonium ion, because it has a high degree of carbocation character. Second, the nucleophilic reaction has to take place so that the nucleophilic oxygen approaches the carbon "backside" to the leaving bromine. This creates, in the *transition state*, a 90° O—C—C angle within the six-membered ring, which results in considerable strain. (This point can be verified with a model.)

 We might ask why bromide ion does not react with the bromonium ion to complete the addition of bromine. It is often found that intramolecular reactions that form 5- and 6-membered rings are thousands of times faster than competing intermolecular reactions. This topic will be considered in Chapter 11.

7.72   (a)   Conformation *A* contains four 1,3-methyl–hydrogen diaxial interactions; the energy cost of these is (4 × 3.7) = 14.8 kJ mol^{-1}. In addition, conformation *A* contains a *gauche*-butane interaction for an additional 2.8 kJ mol^{-1}, for a total energy cost of 2.8 + 14.8 = 17.6 kJ mol^{-1}. Conformation *B* contains two methyl–hydrogen 1,3-diaxial interactions, at an energy cost of (2 × 3.7) = 7.4 kJ mol^{-1}; a *gauche*-butane interaction at 2.8 kJ mol^{-1}; and the methyl–methyl 1,3-diaxial interaction, at an energy cost to be determined. Let the energy cost of a methyl–hydrogen 1,3-diaxial interaction be $\Delta G°$(Me–H) = 3.7 kJ mol^{-1}, that of a methyl–hydrogen *gauche*-butane interaction be $\Delta G°$(gauche) = 3.7 kJ mol^{-1}, and let the energy cost of a methyl–methyl 1,3-diaxial interaction be $\Delta G°$(Me–Me). It is given that the $\Delta G°$ of the overall equilibrium, $\Delta G°$(eq), equals 8.4 kJ mol^{-1} and that conformation *A* has lower energy. Then

$$\Delta G°(\text{eq}) = \Delta G°(B) - \Delta G°(A) = 8.4 = 7.4 + 3.7 + \Delta G° (\text{Me–Me}) - 17.6$$

with all quantities in kJ mol^{-1}. Solving for $\Delta G°$ (Me–Me),

$$\Delta G° (\text{Me–Me}) = 14.9 \text{ kJ mol}^{-1} (3.6 \text{ kcal mol}^{-1})$$

This calculation shows that a methyl–methyl 1,3-diaxial interaction is more than four times as costly in energy terms as a methyl–hydrogen 1,3-diaxial interaction. This is consistent with the greater size of a methyl group relative to a hydrogen.

(b)   Conformation *C* has no 1,3-diaxial interactions, whereas conformation *D* has one methyl–methyl 1,3-diaxial interaction, which, from part (a), costs 15.8 kJ mol^{-1}, and two methyl–hydrogen 1,3-diaxial interactions, which cost (2 × 3.7) = 7.4 kJ mol^{-1}. Thus, conformation *D* is (15.8 + 7.4) = 23.2 kJ mol^{-1}

less stable than conformation C. The standard free energy change for the equilibrium is therefore +23.2 kJ mol^{-1} (5.54 kcal mol^{-1}).

> In order to measure an equilibrium constant, there must be measurable amounts of both components of the equilibrium. For the interconversion of C and D, the equilibrium constant can't be measured directly because too little of compound D is present. This problem illustrates the "counterpoise" method for making estimates of very unfavorable equilibrium constants. In part (a), the methyl–methyl diaxial interaction on the right is counterbalanced by the two axial methyl groups on the left. This strategy raises the energy of the left side of the reaction and thus leads to measurable amounts of both components. The "cost" of the counterbalance can then be subtracted because its effect is known from the conformational analysis of methylcyclohexane.

7.73   (a)   The approach to this problem is similar to that used in Problem 7.72. The $\Delta G°$ for this equilibrium, $\Delta G°(AB)$, balances two phenyl–hydrogen 1,3-diaxial interactions, on the right, against two methyl–hydrogen 1,3-diaxial interactions, on the left. Since $\Delta G° > 0$, the equilibrium is unfavorable; evidently a Ph–H diaxial interaction has a greater energy cost than a CH$_3$–H diaxial interaction. This deduction is consistent with a larger size for the Ph group.

(b)   From the analysis in part (a),

$$\Delta G°(AB) = 4.73 = 2\Delta G°(\text{Ph–H}) - 2\Delta G°(\text{CH}_3\text{–H}) = 2\Delta G°(\text{Ph–H}) - 7.4$$

Solving,

$$\Delta G°(\text{Ph–H}) = (4.73 + 7.4)/2 = 6.1 \text{ kJ mol}^{-1}$$

In Equation (1) of the problem, two Ph–H diaxial interactions and balanced against an equatorial Ph. As we have just calculated, the energy cost, and therefore $\Delta G°$ for equation (1), is 2×6.1 kJ mol^{-1} = 12.2 kJ mol^{-1}.

In Equation (2), two Ph–H diaxial interactions (13.2 kJ mol^{-1}) and a *gauche*-butane interaction (3.7 kJ mol^{-1}) on the left side are balanced against four methyl–hydrogen diaxial interactions on the right (14.8 kJ mol^{-1}). The overall $\Delta G°$ for equation (2) is therefore $14.8 - 16.9 = -2.1$ kJ mol^{-1}.

7.74   (a)   These compounds are noncongruent mirror images and are therefore enantiomers.

(b)   We know that in most common cases (for example, *cis*- and *trans*-2-butene), E,Z isomers are diastereomers. However, all that is needed to apply the E,Z nomenclature is a difference in priorities between the two groups attached to *each* carbon of a double bond. Using compound A as an example, the issue, then, is whether the two groups beginning with the carbons *(a)* and *(b)* have different priorities.

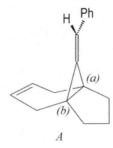

A

There is no differences in *connectivity* if we begin at carbons (a) and (b) and move outwards along the path of highest priority within each group. However, there is a difference at the carbons themselves, and that difference is in their *stereochemical configurations*. Because we did not cover this possibility in Chapter 4, you would have been justified in making up your own priority rule; but Cahn, Ingold and Prelog decided arbitrarily that an R configuration receives the higher priority. (See the solution to Problem 6.45, which involves the same issue.) Because carbon (b) has the R configuration, and it is on the side of the double bond opposite to the higher-priority phenyl (Ph) group, then alkene A is the E isomer. It follows that alkene B is the Z isomer. This, then, is an amusing and unusual situation in which E,Z isomers are actually enantiomers.

7.75   (a)   Two rings in a bridged bicyclic compound can be joined in three stereochemically different ways which might be termed (*out,out*), (*in,out*), and (*in,in*). Assuming hydrogens at both bridgeheads, these ways can be represented schematically as follows. (The large dots are the bridgehead carbons.)

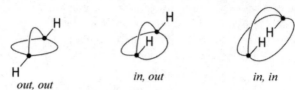

out, out   in, out   in, in

(b)   As these simple diagrams suggest, the *in, out* and *in, in* patterns require that one and two hydrogens, respectively, occupy the region of space *within* a ring. This is sterically impossible for small rings. For small rings, this would actually require such a distortion of the tetrahedral structure that all four bonds to the bridgehead carbon with the *in* configuration would lie on one side of a plane through this carbon. (If the impossibility of achieving this sort of structure is not clear, try to build a model of (*in, out*)-bicyclo[2.2.2]octane.) The larger the rings, the larger the space within the ring, and for very large rings, the "inner" hydrogens can be accommodated easily within the ring and the tetrahedral bonding geometry of the bridgehead carbons can be achieved without significant strain. Hence, the (*in, out*) and (*in, in*) stereoisomers of the 77-carbon bicyclic alkane are stable.

Noncovalent Intermolecular Interactions

STUDY GUIDE LINKS

8.1 Common Nomenclature and the *n*-Prefix

Sometimes you may see an *n* prefix used in the common nomenclature of some organic compounds, as in the following example:

$$CH_3CH_2CH_2CH_2CH_3 \qquad\qquad CH_3CH_2CH_2CH_2 \text{—Br}$$

pentane   **1-bromobutane** (substitutive nomenclature)
(sometimes called *n*-pentane)   **butyl bromide** (common nomenclature)
                         (sometimes called *n*-butyl bromide)

The *n* prefix stands for "normal." At one time, the prefix *n* and the word normal were used to indicate an isomer containing a functional group at the end of an unbranched carbon chain. However, this prefix is superfluous and unnecessary. For example, the name butyl bromide itself means the structure shown on the right above; an additional prefix is unnecessary. Branched-chain isomers have other names, such as isobutyl bromide, *sec*-butyl bromide, or *tert*-butyl bromide. Because common names are unambiguous without the prefix *n-,* the IUPAC recommended abandoning it. Despite this recommendation, the prefix continues to be used.

FURTHER EXPLORATIONS

8.1 Trouton's Rule

The purpose of this exploration is to understand why the boiling point is a good measure of intermolecular interactions in the liquid state. The basis of this discussion is *Trouton's rule,* which came about from the observations by Frederick Trouton of Trinity College, Dublin, reported in

1884, that the boiling points of different liquids are related to their heats of vaporization in a simple way.

First, let's consider the equilibrium between a pure liquid and its vapor. As in any equilibrium, the free energy difference between the final state (in this case, a vapor) and the initial state (in this case, a liquid) is described by the familiar equation

$$\Delta G = \Delta H - T\Delta S \tag{SG8.1}$$

Let the boiling temperature be T_b (in kelvins). At this temperature, liquid and vapor are in equilibrium, and therefore $\Delta G = 0$. We can then re-write Eq. SG8.1 as

$$T_b = \frac{\Delta H^{\bullet}_{vap}}{\Delta S^{\bullet}_{vap}} \tag{SG8.2}$$

where the black-dot superscripts ($\bullet$) mean that these are molar quantities.

At the boiling point, $\Delta H^{\bullet}_{vap}$ is the "latent heat of vaporization," the energy required to convert a liquid to its vapor at its boiling point. If we assume that there are *no* interactions between molecules in the gas phase (that is, the gas is ideal), then $\Delta H^{\bullet}_{vap}$ is a *direct measure* of the energies of the noncovalent interactions in the liquid.

The $\Delta S^{\bullet}_{vap}$ is the entropy difference between gaseous and liquid states. We'll explore the details of entropy later in the chapter, but a simple interpretation in this context is that it measures *freedom of motion.* Our image of a liquid is that the molecules are moving randomly but are packed together rather closely and have a limited degree of motional freedom. (In other words, a molecule can't move very far without colliding with another molecule—or several other molecules.) In a gas, molecules are much farther apart and have a much greater freedom of motion. Therefore, we expect (and find) that $\Delta S^{\bullet}_{vap} > 0$. What is remarkable is that $\Delta S^{\bullet}_{vap}$ is actually *nearly the same* —about $+85$ J mol^{-1} K^{-1} —for a large number of organic liquids. If this is so, then Eq. SG8.2 becomes

$$T_b \text{ (in kelvins)} \approx \frac{\Delta H^{\bullet}_{vap}}{85} \tag{SG8.3}$$

This is the "modern version" of Trouton's rule. Because $\Delta H^{\bullet}_{vap}$ is a direct measure of the energies of the noncovalent interactions in the liquid, this proves the assertion that the boiling point (in K) is a quantitative measure of these interactions. Notice that Trouton's rule justifies not only the qualitative notion that we can think about boiling points in terms of intermolecular interactions in the liquid, but also gives us a way to estimate the magnitude of these interactions.

As we might expect, there are some liquids that don't follow Trouton's rule. Water is one. The $\Delta S^{\bullet}_{vap}$ for water is $+109.1$ J mol^{-1} K^{-1}. This is because water has a "more ordered" (lower-entropy) liquid state than (for example) hydrocarbons. This more ordered liquid state is caused by hydrogen bonding, which constrains the motion of water molecules in the liquid. Thus, when water is converted into a gas, in which such constraints are gone, more entropy is gained. However, accompanying this larger entropy gain is a larger $\Delta H^{\bullet}_{vap}$. Because of the hydrogen bonds in the liquid, $\Delta H^{\bullet}_{vap}$ is also unusually large because the hydrogen bonds have to be broken to form the gas. In other words, $\Delta H^{\bullet}_{vap}$ and $\Delta S^{\bullet}_{vap}$ tend to compensate in cases like this. Thus, using the boiling point as a crude measure of intermolecular interactions is justified even for liquids that do not follow Trouton's rule exactly.

SOLUTIONS TO PROBLEMS

Solutions to In-Text Problems

8.1 (a) Isobutyl fluoride is a primary alkyl halide.
 (b) Hexyl iodide is a primary alkyl halide.
 (c) Cyclopentyl bromide is a secondary alkyl halide.
 (d) Neopentyl chloride is a primary alkyl halide.

8.2 (a) (b)

2,2-dichloro-5-methylhexane   chlorocyclopropane

 (c) (d)

6-bromo-1-chloro-3-methylcyclohexene   methylene iodide

8.3 (a) 3-bromo-3-chloro-1-methylcyclopropene
 (b) (*Z*)-3-chloro-2-pentene
 (c) 3-bromo-1,1,1-trichloro-2-fluorobutane
 (d) chloroform (HCCl$_3$) is the traditional name for trichloromethane.
 (e) neopentyl bromide [(CH$_3$)$_3$C—CH$_2$—Br] is the common name for 1-bromo-2,2-dimethylpropane.
 (f) *trans*-1,3-dibromocyclobutane
 (g) 1-chloro-3-isopropyl-1-methylcyclohexane

8.4 (a)   (b)   (c)   (d)   (e)

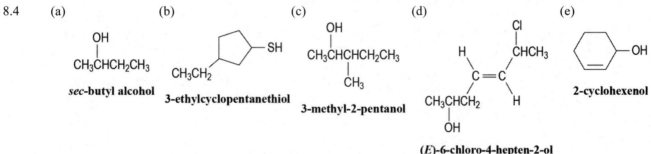

sec-butyl alcohol   3-ethylcyclopentanethiol   3-methyl-2-pentanol   (*E*)-6-chloro-4-hepten-2-ol   2-cyclohexenol

8.5 (a) 1-butanol
 (b) 3-bromo-1-butanol
 (c) (*E*)-1-chloro-3-methyl-3-penten-2-ol
 (d) (1*S*,5*R*)-2-chloro-5-methyl-2-cyclopenten-1-thiol. The 2-refers to the position of the double bond; the position of the —SH group is at the 1-position because it is the principal group. The stereochemical configurations (*R*, *S*) come first in the name (See Sec. 6.2, text p, 236).

    (e)   (2*R*,3*R*,4*R*)-3-butyl-5-hexene-2,4-diol (See rules 2a and 3a on text p. 329.)

    (f)   2,5-cyclohexadienol

    (g)   1-mercapto-2-pentanol

    (h)   2-methyl-2-propanethiol

8.6   (a)   (b)   (c)

$CH_3CH_2OCH_2CH_2CH_3$

ethyl propyl ether   **dicyclohexyl ether** $(CH_3)_3C—S—CH(CH_3)_2$

                                                          tert-butyl isopropyl sulfide

    (d)   (e)

$H_2C{=}CHCH_2—O—CH_2$⟨⟩   ⟨⟩$—O—CH{=}CH_2$

allyl benzyl ether   **phenyl vinyl ether**

    (f)   (g)

$CH_3CHCH_2CH_2CHCH_2CH_3$

                                                   CH_3 SCH_2CH_3

(2R,3R)-2,3-dimethyloxirane   **5-(ethylthio)-2-methylheptane**

8.7   (a)   2-methoxy-2-methylpropane (commonly called MTBE, for methyl *tert*-butyl ether)

    (b)   2-ethoxyethanol (or 2-ethoxy-1-ethanol)

    (c)   1-(isobutylthio)-2-methylpropane

    (d)   (*E*)-5-methoxy-3-penten-1-ol

8.8   (a)   The structure of the compound and the correct name is as follows. The ring is numbered to give the substituent the lower number.

O $CH_2CH_2CH_2CH_3$

"3-butyl-1,4-dioxane" (incorrect name)
2-butyl-1,4-dioxane (correct name)

    (b)   The structure of 2-butoxyethanol is $HO—CH_2CH_2—OCH_2CH_2CH_2CH_3$.

8.9   In Table 8.1, the bond lengths in a given period are about 0.04 Å greater in the column containing oxygen and sulfur than in the column containing the halogens. Assuming that this trend holds true for CH_3SeH, then the predicted carbon–selenium bond length in this compound is about 1.98 Å. The prediction for $(CH_3)_2Se$ can be refined by noting in Fig. 8.1, text p. 335, that the carbon–sulfur bond length in dimethyl sulfide is about 0.02 Å smaller than that in methanethiol. Assuming a similar trend in the corresponding selenium derivatives, the predicted bond length for $(CH_3)_2Se$ is 1.96 Å.

    The point of this problem is that more often than not, molecular properties vary in a regular way throughout the periodic table, and that reasonably good predictions can be made within one group from trends in a nearby group.

8.10   If bonds to an atom involved *only* its *p* orbitals, then the bond angle would be 90°, because the inter-orbital angle of *p* orbitals is 90°. It follows, then, that the closer the bond angle is to 90°, the more *p* character in the bond. Figure 8.1 on text p. 335 shows that the C—S—H bond angle is much closer to 90° than is the C—O—H bond angle of methanol. Hence, bonds to sulfur (that is, the sulfur orbitals used to make these bonds) have more *p* character than bonds to oxygen. A corollary is that the unshared pairs on sulfur have more *s* character than those on oxygen.

8.11 Because of the molecular geometry of *trans*-1,2-dichloroethylene, the C—Cl bonds are oriented in opposite directions, as are the C—H bonds. Consequently, their bond dipoles cancel, and the molecular dipole moment of *trans*-1,2-dichloroethylene is zero. (The cancellation of the C—Cl bond dipoles is shown in the diagram below; the much weaker C—H bond dipoles also cancel for the same reason.) The corresponding bond dipoles of *cis*-1,2-dichloroethylene do not cancel; consequently it has a significant dipole moment. The interaction of the molecular dipole moments of *cis*-1,2-dichloroethylene molecules provides a cohesive force in the liquid state that is not present in the trans stereoisomer. (See text p. 341.) Because such cohesive forces enhance boiling point, *cis*-1,2-dichloroethylene has the higher boiling point of 60.3°.

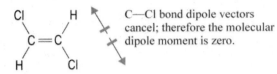

trans-1,2-dichloroethylene   **cis-1,2-dichloroethylene**

8.12 Similar to the relationship between pentane and neopentane, shown on text p. 338, octane and 2,2,3,3-tetramethylbutane have the same molecular weights but very different boiling points. Like neopentane, 2,2,3,3-tetramethylbutane is branched, more spherical, and thus has a smaller surface area and less van der Waals interactions than octane, which is linear. The energy required to force octane molecules into the gas phase is larger, so octane boils at the higher temperature, 126 °C.

8.13 (a) Acetaldehyde has a higher boiling point (21 °C) than propene (–47 °C) because there are stronger dipole-dipole interactions between acetaldehyde molecules.

(b) The curve for boiling points of aldehydes should show up somewhere between the curves for alkanes/ethers/alkyl chlorides and nitriles on Fig. 8.2b. Aldehydes have dipole moments greater than alkanes (and alkenes, which as stated in the problem are about the same as alkanes) so the curve would be higher.

8.14 (a) Recall that a bromine contributes about the same molecular mass (80 units) as a pentyl group (73 units), and that alkyl bromides have boiling points *lower* than alkanes of about the same molecular mass; then 2-bromopropane should have a boiling point that is lower than that of an eight-carbon alkane. Because the two alkanes listed have more than eight carbons, their boiling points are correspondingly higher. Because the two alkanes have the same type of branching, their boiling points should be in the same order as their molecular masses. These considerations suggest that the order of increasing boiling points is 2-bromo-propane < 4-ethylheptane < 4-ethyloctane. (The actual boiling points are 59°, 141°, and 168°, respectively.)

(b) A chlorine contributes about the same molecular mass (35 units) as an ethyl group (29 units), and chloroalkanes have about the same boiling points as alkanes of the same molecular mass. Hence, chloromethane has about the same boiling point as propane, which has a lower boiling point than the five-carbon alkene 1-pentene. The alcohol has the highest boiling point because it has about the same molecular mass as 1-pentene, but can donate and accept hydrogen bonds. Consequently, it has the highest boiling point of all. The order of increasing boiling points is, therefore, chloromethane (–42°) < 1-pentene (30°) < 1-butanol (118°).

8.15 (a) HBr is primarily a hydrogen-bond donor because it is a strong acid; a neutral —Br is a poor acceptor.

(b) The hydrogen in hydrogen fluoride is an excellent hydrogen-bond donor, and the fluorine is an excellent hydrogen-bond acceptor.

(c) Acetone is a hydrogen-bond acceptor; the oxygen can accept hydrogen bonds at its unshared electron pairs.

accepts hydrogen bonds

$$\begin{array}{c} :\!O:\\ \parallel\\ H_3C-C-CH_3 \end{array}$$

acetone

(d)   *N*-methylacetamide can serve as both a hydrogen-bond donor and a hydrogen-bond acceptor.

the oxygen can accept
hydrogen bonds

$$\begin{array}{c} :\!O:\\ \parallel\\ H_3C\quad C\quad N\!-\!CH_3\\ \qquad\qquad | \\ \qquad\qquad H \end{array}$$

the nitrogen accept
a hydrogen bond and
can donate its hydrogen to
a hydrogen bond

***N*-methylacetamide**

(e)   Phenol is both a hydrogen-bond donor and a hydrogen-bond acceptor. The O—H can participate in hydrogen-bond donation, and the oxygen can accept hydrogen bonds at its unshared electron pairs.

the oxygen can accept
hydrogen bonds

Ö—H ---- the oxygen can donate
its hydrogen to a
hydrogen bond

phenol

(f)   The ethylammonium ion can donate its N—H hydrogens to hydrogen bonds, but it cannot accept hydrogen bonds because it has no unshared electron pairs.

$$\begin{array}{c} H\\ |\\ CH_3CH_2-\overset{+}{N}-H \cdots\cdots\\ |\\ H \end{array}$$

the nitrogen can donate its
hydrogens to hydrogen bonds

ethylammonium ion

8.16   The two carboxylic acids are compounds *B* and *C*, and since the problem states that they are solids at room temperature, the two highest melting points belong to them. Compound *C* possesses more symmetry than compound *B*, which allows it to pack the most tightly together in the solid phase and thus possess the highest melting point (168 – 172 °C). Compound *B*'s melting point is then 143 – 147 °C. The same analysis can be applied to the two alkenes: compound *A* is more symmetrical so it has the higher melting point of the two (–74.5 °C); compound *D* melts at –175 °C.

8.17   A similar "sawtooth" pattern should also be expected for the densities of the solid unbranched alkanes, as their packing is more efficient with even numbered carbons. More efficient packing results in higher density.

8.18   (a)   An outcome where two are heads and two are tails has greater entropy (more possible arrangements) than when one is heads and three are tails. There are six 'two heads two tails' arrangements: HHTT, TTHH, HTHT, THTH, HTTH, THHT. There are only four 'one heads three tails' arrangements: HTTT, THTT, TTHT, TTTH.

        (b)   Both of these arrangements have the same number of possible outcomes, so they have equal entropy.

(c)   Use Eq. 8.6 on text p. 351 to solve for the entropy in each situation. The 1 mole situation is solved in the text; $+41.9$ J K^{-1} mol^{-1}. When there are 5 moles of solute, the ΔG_{mixing} increases to $+143$ J K^{-1} mol^{-1}.

8.19   In order to calculate the free energy of mixing (ΔG_{mixing}) we first must calculate the entropy of mixing (ΔS_{mixing}) using Eq. 8.6. The isotopically labeled hexane, hexane*, will be the solute, and ordinary hexane will be the solvent. We know that there is 1 mole of hexane* to be dissolved in 1 L of hexane. 1 L of hexane is equal to 5.79 moles of hexane. So, for hexane*, $n_1 = 1$ mol, and the mole fraction $x_1 = (1/1 + 5.79) = 0.147$. For hexane, $n_2 = 5.79$ mol, and the mole fraction $x_2 = (5.79/1 + 5.79) = 0.853$. Plugging these values into Eq. 8.6 gives:

$$\Delta S_{mixing} = -2.3R(n_1 \log x_1 + n_2 \log x_2)$$
$$= -2.3(8.31 \text{ J K}^{-1}\text{mol}^{-1})[(1 \times \log 1.47) + (5.79 \times \log 8.53)]$$
$$= 23.6 \text{ J K}^{-1}\text{mol}^{-1}$$

Now use Eq. 8.7 to solve for ΔG_{mixing} at room temperature (298 K):

$$\Delta G_{mixing} = -T\Delta S_{mixing}$$
$$= -(298)(23.6)$$
$$= -7033 \text{ J} = -7.033 \text{ kJ}$$

This is an energetically favorable process.

8.20   (a)   2-Methoxyethanol is a polar, protic, donor solvent.
        (b)   2,2,4-Trimethylpentane (a major component of gasoline) is an nonpolar, aprotic, nondonor solvent.
        (c)   This compound, 2-butanone, is a polar, aprotic, donor solvent.

8.21   The lower layer is dichloromethane because it has a higher density than water. The long hydrocarbon chain of 1-octanol would both reduce its solubility in water and promote its solubility in dichloromethane. The data on text p. 357 show that 1-hexanol has a very low solubility in water. The solubility of 1-octanol would be much lower still. The 1-octanol would be found primarily in the dichloromethane layer.

8.22   Water can accept a hydrogen bond from the hydrogen of the N—H bond, and can donate a hydrogen bond to the oxygen of the C=O bond as well as to the nitrogen. This hydrogen bonding tends to solubilize acetanilide in water. In contrast, the CH$_3$ group and the phenyl ring cannot form hydrogen bonds with water; consequently, these nonpolar ("greasy") groups tend to make acetanilide insoluble.

    Groups such as alkyl and phenyl groups that reduce water solubility are sometimes termed *hydrophobic groups* (see Sec. 8.6D).

8.23   The lower molecular weight alcohols are the most "water-like"—they can hydrogen bond with water molecules and have shorter hydrophobic chains—as evidenced by the miscibility data on text p. 357. Once the chain starts getting longer (greater than four carbons), the miscibility decreases with increasing chain length. This decrease in miscibility is associated with a decrease in standard entropy of the solution. As the carbon chains get longer, the water molecules need to form a larger cavity around the individual molecules, increasing "order" and decreasing entropy, which is unfavorable energetically.

8.24   (See structures on the next page.)
        (a)   Dimethyl sulfoxide has an ion–dipole interaction and a donor interaction with a sodium ion.
        (b)   Water can donate hydrogen bonds to the oxygen of DMSO and can have a favorable dipole–dipole interaction with the sulfur.
        (c)   Dimethyl sulfoxide also has an ion–dipole interaction with a chloride ion. (In both parts (a) and (c), multiple DMSO molecules cluster around the ions; only one is shown here.)

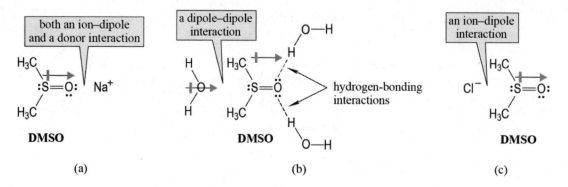

8.25   (a)   Addition of HCl protonates the triethylamine, forming triethylammonium chloride, an ionic compound that is soluble in water.

$$CH_3CH_2—N: \xrightarrow{\text{HCl}} CH_3CH_2—\overset{+}{N}—H \quad Cl^-$$

triethylamine
(insoluble in water)

triethylammonium chloride
(an ionic salt, soluble in water)

(b)   Adding a base to the solution, such as sodium hydroxide (NaOH), would deprontonate the triethylammonium ion, converting it back into insoluble triethylamine. Sodium chloride, another water soluble salt, would be formed in the process.

$$CH_3CH_2—\overset{+}{N}—H \quad Cl^- \xrightarrow{\text{NaOH}} CH_3CH_2—N: \quad + \quad Na^+Cl^- \quad + \quad H_2O$$

triethylammonium chloride
(an ionic salt, soluble in water)

triethylamine
(insoluble in water)

8.26   (a)   One or both of the $C_{17}H_{35}$ groups could be substituted by any long, unbranched alkyl group with an odd number of carbons; this group can also contain one or more cis double bonds.

phosphatidylserine

(b)   One or both of the $C_{17}H_{35}$ groups could be substituted by any long, unbranched alkyl group with an odd number of carbons; this group can also contain one or more cis double bonds.

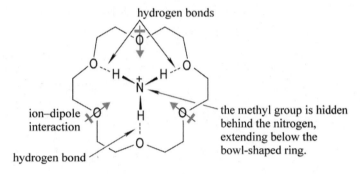

phosphatidylcholine
(a lecithin)

8.27   The three hydrogens of the methylammonium ion are perfectly positioned to interact with three of the oxygens of the crown ether by hydrogen bonding. The other three oxygens of the crown ether can stabilize the complex by ion–dipole interactions. In the following structure, we are viewing the nitrogen as in a Newman projection down the N—C bond, with the methyl group hidden from view.)

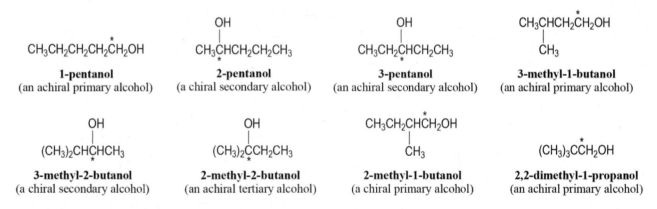

Solutions to Additional Problems

8.28   The answers to all five parts are as follows (α-carbons are noted with an *):

$CH_3CH_2CH_2\overset{*}{C}H_2OH$

1-pentanol
(an achiral primary alcohol)

$CH_3\overset{|}{\underset{*}{C}}HCH_2CH_2CH_3$ (OH above)

2-pentanol
(a chiral secondary alcohol)

$CH_3CH_2\overset{|}{\underset{*}{C}}HCH_2CH_3$ (OH above)

3-pentanol
(an achiral secondary alcohol)

$CH_3CHCH_2\overset{*}{C}H_2OH$ (CH_3 below)

3-methyl-1-butanol
(an achiral primary alcohol)

$(CH_3)_2CH\overset{*}{C}HCH_3$ (OH above)

3-methyl-2-butanol
(a chiral secondary alcohol)

$(CH_3)_2\overset{|}{\underset{*}{C}}CH_2CH_3$ (OH above)

2-methyl-2-butanol
(an achiral tertiary alcohol)

$CH_3CH_2\overset{*}{C}HCH_2OH$ (CH_3 below)

2-methyl-1-butanol
(a chiral primary alcohol)

$(CH_3)_3\overset{*}{C}CH_2OH$

2,2-dimethyl-1-propanol
(an achiral primary alcohol)

8.29   The answers to all four parts are as follows (α-carbons are noted with an *); common nomenclature is listed first, followed by IUPAC substitutive nomenclature:

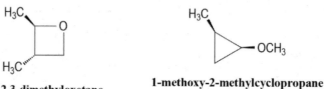

$CH_3CH_2\overset{*}{C}H_2O\overset{*}{C}H_2CH_3$

ethyl propyl ether
or 1-ethoxypropane

$CH_3CH_2\overset{*}{C}H_2\overset{*}{C}H_2O\overset{*}{C}H_3$

butyl methyl ether
or 1-methoxybutane

$CH_3\overset{\underset{|}{CH_3}}{\underset{*}{C}H}O\overset{*}{C}H_2CH_3$

ethyl isopropyl ether
or 2-ethoxypropane

$(CH_3)_3\overset{*}{C}O\overset{*}{C}H_3$

methyl *tert*-butyl ether
or 2-methoxy-2-methylpropane

$(CH_3)_2CH\overset{*}{C}H_2O\overset{*}{C}H_3$

isobutyl methyl ether or
1-methoxy-2-methylpropane

$CH_3CH_2\overset{\underset{|}{CH_3}}{\underset{*}{C}H}O\overset{*}{C}H_3$

***sec*-butyl methyl ether**
or 2-methoxybutane
(a chiral ether)

8.30   (a)   The unsaturation number of the ether is 1; because it has no double bonds, it must contain a ring. Two of the many possibilities are

***trans*-2,3-dimethyloxetane**   **1-methoxy-2-methylcyclopropane**

(b)   The unsaturation number is 2, and both rings and/or multiple bonds are allowed in this case. Two of several possibilities are

(c)   The following glycol, although it has asymmetric carbons, is achiral because it is meso.

***cis*-4-cyclohexen-1,2-diol**

(d)   2,3-Butanediol exists as a meso stereoisomer and two enantiomers.

$$CH_3\overset{\underset{|}{OH}}{CH} - \overset{\underset{|}{OH}}{CH}CH_3$$

2,3-butanediol

(e)   1,2-Butanediol exists as a pair of enantiomers:

$$HOCH_2\overset{\underset{|}{OH}}{CH}CH_2CH_3$$

1,2-butanediol

(f)   The six epoxides with the formula C_4H_8O:

cis-2,3-dimethyloxirane
(a meso compound)

trans-2,3-dimethyloxirane
(exists as two enantiomers)

2,2-dimethyloxirane

2-ethyloxirane
(exists as two enantiomers)

8.31   (a)   The systematic name of halothane is 2-bromo-2-chloro-1,1,1-trifluoroethane.

      (b)   The systematic name of methoxyflurane is 2,2-dichloro-1,1-difluoro-1-methoxyethane.

8.32   (a)   2-butene-1-thiol

      (b)   3-methyl-1-butanethiol

8.33   (a)   The IUPAC name of the compound is 4-butyl-1-cyclohexyl-3-cyclopentyl-3-(methoxymethyl)octane.

      (b)   The IUPAC name of the compound is (3*R*,*Z*)-4-butyl-3-cyclopentyl-3-(2-propenyl)oct-5-en-1-ol. Note that the propenyl group is also called allyl (text p. 137, top).

8.34   (a)   The order of boiling points is *tert*-butyl alcohol < 2-pentanol < 1-hexanol. (The actual boiling points are 82°, 119°, 158°.) This follows the order of molecular masses. The increased branching of 2-pentanol relative to 1-hexanol, and *tert*-butyl alcohol relative to 2-pentanol, makes the differences between these boiling points even greater than they would be for unbranched alcohols of the same molecular masses.

      (b)   The order of boiling points is 1-hexene < 1-chloropentane < 1-hexanol. (The actual boiling points are 63°, 108°, 158°.) Use the fact that the mass of a chlorine (35 units) is about the same as that of an ethyl group (29 units), and that alkyl chlorides have about the same boiling points as alkanes (and alkenes) of the same molecular mass. Consequently, 1-chloropentane has about the same boiling point as 1-heptene, which, in turn, has a higher boiling point than 1-hexene. 1-Hexanol has about the same molecular mass as heptane, and its boiling point is considerably higher because of hydrogen bonding.

      (c)   The order of boiling points is propane < diethyl ether < 1,2-propanediol. (The actual boiling points are –42°, 37°, 189°.) Diethyl ether has a higher boiling point than propane because diethyl ether is more polar and because it has a greater molecular mass. 1,2-Propanediol has the highest boiling point because of hydrogen bonding.

      (d)   The order of boiling points is cyclobutane < chlorocyclobutane < cyclooctane. Use the facts that the chlorine has about the same mass as an ethyl group, and that alkyl chlorides and alkanes of the same molecular mass have about the same boiling points. Then order the boiling points by molecular mass. (The actual boiling points are 12°, 83°, and 151°, respectively.)

8.35   (a)   The first compound, acetic acid, has a higher boiling point than the second, ethyl acetate, because hydrogen bonding between acetic acid molecules can take place; the O—H hydrogen on one molecule can be donated to form a hydrogen bond to either of the oxygens on another molecule. Because ethyl acetate has no O—H, molecules of ethyl acetate cannot associate by hydrogen bonding. The intermolecular hydrogen bonds must be broken in order to vaporize acetic acid, and the energy required to break these hydrogen bonds is reflected in a higher boiling point.

      (b)   Use the same reasoning as in part (a). The N—H hydrogens of the first compound, acetamide, can be donated in hydrogen bonds between molecules that involve the oxygen or the nitrogen as an acceptor. The second compound, *N,N*-dimethylacetamide, has no hydrogens that can be involved in hydrogen bonding. The hydrogen bonding in the liquid state of acetamide is reflected in its higher boiling point, despite the higher molecular mass of the second compound.

8.36   Despite the fact that nitromethane cannot form hydrogen bonds, it has a higher boiling point than 2-propanol most likely because nitromethane is a very polar molecule. Apparently, the dipole-dipole interactions of nitromethane are strong enough that its boiling point is higher than a compound that can hydrogen bond. Looking up the dipole moments of both molecules supports this conclusion: the dipole moment of nitromethane is 3.5 D and the dipole moment of propanol is 1.6 D.

8.37   (a)   Solvents become less miscible as the difference in their dipole moments increases. From Table 8.2, DMSO has a dipole moment of 4.0, and of the solvents listed, we can expect that the least miscible solvent will be the one with the smallest dipole moment. In this case, hexane (dipole moment = 0.08) would be immiscible in DMSO.

        (b)   Hexane has a very small dipole moment (0.08). Methanol has the highest dipole moment (2.9) of this group of solvents, so we would expect it to be immiscible in hexane. However, the dipole moment of acetone is also relatively high (2.7), and close to methanol's. Why might acetone be miscible in hexane? We've learned that dipoles can be induced in hydrocarbons, and which can justify acetone's miscibility in hexane. But trying to dissolve methanol in hexane would require breaking of the high energy hydrogen bonds in methanol, which is energetically unfavorable. So in this case, methanol would be immiscible.

8.38   (a)   Compound A is soluble in hydrocarbon solvents by the "like-dissolves-like" criterion. The alkyl groups, particularly the large cetyl ($C_{16}H_{33}$) group, resemble the hydrocarbon chains of the solvent. In contrast, ammonium chloride would require a hydrogen bond acceptor solvent to dissolve; hydrogen bonds cannot be provided by hydrocarbons.

        (b)   Although compound A is soluble in hydrocarbon solvents, it exists in such solvents as ion pairs and higher aggregates rather than as free ions. The reason is that hydrocarbon solvents have very low dielectric constants ($\epsilon \approx 2$), whereas a high dielectric constant is required to separate ionic aggregates into free ions. Furthermore, a hydrocarbon offers no solvation by hydrogen bonding to the bromide counter-ion. Hence, this anion remains in proximity to its positive partner.

8.39   Water forms a Lewis acid–Lewis base complex with Cu^{2+} ion:

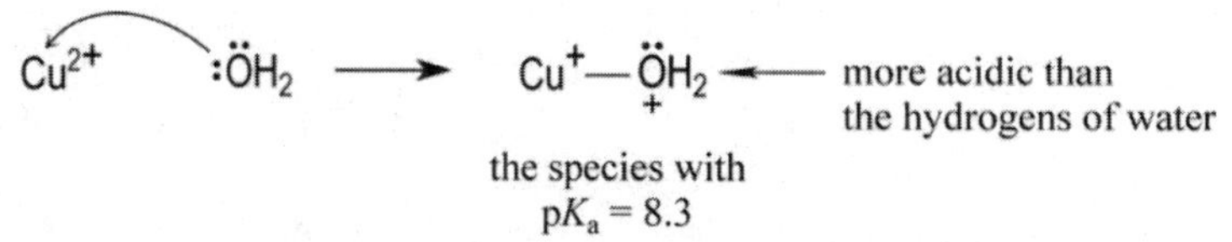

the species with
$pK_a = 8.3$

        Ionization of a hydrogen of the complexed water removes a positive charge from the oxygen and eliminates the charge–charge repulsion in the complex. Ionization of water itself does not have this driving force. Hence, water that is complexed to copper is more acidic.

8.40   In the presence of concentrated acid, dibutyl ether is protonated. The protonated ether is an ionic compound, and ionic compounds are soluble in water:

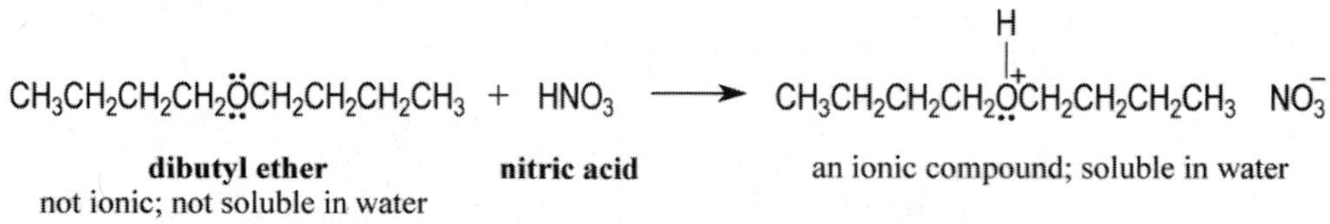

        dibutyl ether   **nitric acid**   an ionic compound; soluble in water
        not ionic; not soluble in water

8.41   (a)   The ammonium ion is solvated by hydrogen bonding with the solvent ethanol; the ion is the donor and the ethanol is the acceptor. Ethanol also stabilizes the dissolved ion by ion–dipole interactions. The chloride ion is also solvated by hydrogen bonding in which ethanol is the donor. (Only one interaction of each type is shown.)

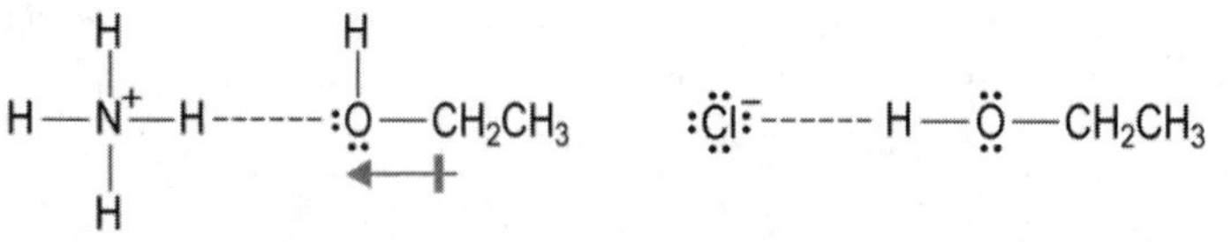

        (b)   The sodium ion is stabilized by the dipole of the carbonyl of N-methylpyrrolidone. By considering a resonance structure of N-methylpyrrolidone, there is an ion-ion interaction between the chloride and the partial positive charge on nitrogen.

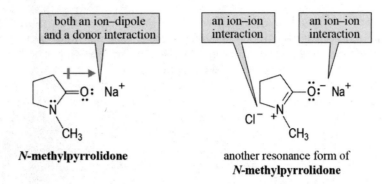

8.42   Think of the dissociation reaction of the crown ether–K^+ complex as you would any other equilibrium:

crown ether–K^+ +   solvent   ⇌   crown ether   + K^+ (solvated)

Anything that stabilizes the species on the right side of the equation will increase the dissociation constant of the complex. Because potassium ion, a typical positive metal ion, is solvated more effectively by water than by ether, the dissociation constant for the complex is greater in water.

8.43   The ammonium ion will interact by donating hydrogen bonds to the oxygens of nonactin. (See the solution to Problem 8.27 on p. 215 of this manual, which describes the similar interaction of ammonium ion with a crown ether.) It is interesting that both the crown ether and nonactin are selective for binding of the potassium ion, and both also bond the ammonium ion. Evidently, the spatial requirements for both ions is similar. The cavity that fits the potassium ion equally well accommodates a nitrogen and its bound hydrogens.

8.44   The following compound is a constitutional isomer of the one given in the problem. It is more soluble in water because it contains ionic groups, and ionic compounds are generally more soluble in water than neutral compounds. It is part of a general class of compounds called amino acids, which contain both an amino group and a carboxylic acid group. Look up the pK_a values of a general carboxylic acid and a general amine and see if you can justify why it exists in ionic form, as opposed to neutral form. Amino acids are covered in more detail in Chapter 27.

$$\underset{\text{HOCCH}_2\text{CH}_2\text{CH}_2\text{CH}_2\text{CH}_2\text{NH}_2}{\overset{\overset{\text{O}}{\|}}{}} \rightleftharpoons \underset{{}^-\text{OCCH}_2\text{CH}_2\text{CH}_2\text{CH}_2\text{CH}_2\overset{+}{\text{N}}\text{H}_3}{\overset{\overset{\text{O}}{\|}}{}}$$

There are many other examples; a few are listed below.

leucine   **isoleucine**   **N-methyl pentanohydroxamic acid**

8.45   Hexethal should be (and is) the more potent sedative, because it is more soluble in membranes, and therefore can readily pass through them. It is more soluble because the longer alkyl chain is more like the interior of a membrane than is the shorter ethyl chain. Furthermore, the long alkyl chain makes hexethal *less* soluble in water, because large alkyl groups are not effectively solvated by water.

8.46   Lecithin, a phospholipid, will form phospholipid vesicles in sufficient concentrations, as explained on text p. 370. These vesicles contain phospholipid bilayers that incorporate molecules of the oil within the "hydrocarbon" part of the bilayer. As with cell membranes, the aqueous sides of the vesicles—the polar head groups—interact well with water. Thus, the vesicles containing the oil intermix with the water and do not form a separate layer.

8.47   The structure of propofol is largely hydrocarbon-like, even though it contains a polar OH group, which is why it is insoluble in water. The lecithin/soybean oil mixture forms phospholipid vesicles. The propofol is incorporated into the vesicles near the aqueous interface so that the OH group can H-bond to solvent water and the apolar portion can interact with the apolar region of the vesicles.

8.48   (a)   Vitamin C and the vitamin Bs shown are likely water soluble due to the large number of hydroxy groups they possess (for hydrogen-bonding with water) and other polar attributes (causing dipoles, again making them soluble in water). Vitamin A and D_3 have much more hydrophobic character, making them more likely soluble in fat. Even though they each posses one hydroxyl group, the molecules are sufficiently large that their properties are dominated more by their hydrocarbon portions.

        (b)   Overdosing on a fat-soluble vitamin would be more dangerous, because it is more difficult to clear the body of non-water soluble compounds through natural excretion processes.

8.49   (a)   By Le Châtelier's principle, when the concentration of ethanol increases, dimer concentration increases.
        (b)   Use the usual relationship between $\Delta G°$ and K_{eq}; recall that $2.3RT$ at 298 K is 5.71 kJ mol^{-1}.

$$\Delta G° = -2.30RT(\log K_{eq}) = -5.71(\log 11) = -5.95 \text{ kJ mol}^{-1}$$

        (c)   Neglect the volume of ethanol and assume that 1 mole of ethanol is dissolved in 1 L of solution so that the initial concentration of ethanol is 1 M. (This introduces an error of only 2%.) Assume that x M of ethanol react to form dimer. The concentration of ethanol remaining is then $(1.0 - x)$ M, and the concentration of dimer is $x/2$ M. (Remember, it takes two ethanol molecules to form one molecule of dimer.) The equilibrium-constant expression for dimerization becomes

$$K_{eq} = 11 = \frac{[\text{dimer}]}{[\text{ethanol}]} = \frac{x/2}{(1-x)^2}$$

                From the quadratic formula, $x = 0.808$ M. Thus, the concentration of dimer is $x/2 = 0.404$ M; that of free ethanol is $(1 - x)$ or 0.192 M. Thus, about 80% of the ethanol is dimerized in a 1 M solution.
        (d)   Because dimerization occurs through hydrogen-bond formation, then the greater equilibrium constant for dimerization of ethanol shows that ethanol has the stronger hydrogen bond.
        (e)   The alcohol $CH_3SCH_2CH_2OH$ would have the greater solubility, because it can both donate and accept hydrogen bonds to water. An —SH group is a much poorer hydrogen bond acceptor and hydrogen bond donor than —OH.

8.50   (a)   Compound *B* has the higher melting point because of its symmetry (see text Sec. 8.5D). [Their boiling points are similar because this property is largely determined by molecular weight (see text Sec. 8.5A).]
        (b)   Compound *A* has the higher melting point, because of its higher degree of symmetry. Draw both compounds in their chair forms. While both contain symmetry elements, compound *A* contains more.

8.51   (a)   Even though they have the same molecular weight, 1-butanol has a higher boiling point because of its larger surface area. *tert*-Butyl alcohol is more spherical and has less van der Waals forces to hold it together (see Study Problem 8.4 text p. 346). Additionally, hydrogen bonding is probably more effective in 1-butanol due to less steric crowding around the hydroxyl group.
        (b)   The entropy of solution would be lower (less favorable) for 1-butanol. Again, the longer, tube like structure of 1-butanol means that it has more surface area per molecule than the more spherical *tert*-butanol. Thus, there would be more low entropy (or more, highly organized) water molecules needed to surround 1-butanol.
        (c)   As the hint suggests, the boiling point is a measure of the tendency of a molecule to escape from its liquid state. The solubility in water is a combination of the escaping tendency from the pure alcohol phase and solvation in the aqueous phase. Both of these factors work in the same direction. The tertiary alcohol has a greater escaping tendency *and* requires a less negative entropy of solvation in water.

8.52   (a)   Ethanol and 1-propanol both contain a "hydrophilic end" (the OH group) and a "hydrophobic end" (the alkyl chain). The bifunctionality of these compounds allows them to intercalate into cell membranes which are similarly composed of molecules with hydrophilic and hydrophobic ends. However, since the alcohols' alkyl chains are much shorter, they disrupt the packing of the lipids and change the properties of the bilayer, leading to disruption.

(b)   Sugars, despite having six carbon atoms, are surrounded by hydrophilic hydroxyl groups. They cannot penetrate the highly "greasy" (hydrophobic) inner layer of a lipid bilayer.

(c)   The tetrabutylammonium cation is interesting in that it has a net +1 charge but it is largely composed of hydrophobic alkyl chains. In water, solvation of this ion is difficult because of the large cavity that must be formed between water molecules, organizing the water around the ion (negative, unfavorable ΔS) and disrupting the hydrogen bonding network. DMSO, on the other hand, is both a non-hydrogen-bonding solvent and it has a greater dipole moment (4.0 to 1.9 for water, from Table 8.2). Both of these properties make DMSO better suited to stabilize the tetrabutylammonium cation, resulting in a strongly positive, or favorable, ΔS.

(d)   As described in the solution to 8.51 (c), solubility in water is composed of two factors: escape of a molecule from its own liquid phase *and* solvation by water. Although 1-pentanol can be both a hydrogen bond donor and acceptor, this favorable interaction must be outweighed by the fact that the molecules are attracted to each other through hydrophobic interactions of the five-carbon alkyl chain. The relative boiling points tell us that the intrinsic escaping tendency of the ether from its own liquid phase is much greater than that of the alcohol. Even though the alcohol can donate hydrogen bonds, it seems that the escaping tendency is more important. Thus, diethyl ether, with its hydrogen bond-accepting ability, smaller alkyl chains, and lower boiling point, is more soluble in water.

8.53   As the temperature increases, the $T\Delta S_{mixing}$ term of Eq. 8.9 on text p. 352 becomes more negative. A large negative $T\Delta S_{mixing}$ term contributes to a negative ΔG_s. The more negative ΔG_s, the larger the equilibrium constant K (Eq. 3.33, text p. 100), and the more favorable the reaction.

8.54   (a)   A greater partition coefficient in an octanol/water mixture (P_{ow}) corresponds to more of the drug in the octanol phase. Based on our understanding of solubility, the more hydrophobic of the two drugs would prefer the octanol phase. We can see that atenolol contains many more polar bonds and hydrogen bonding groups, so it would be more soluble in water. Propranolol contains fewer polar bonds and hydrogen bonding groups, and contains an additional aromatic (hydrocarbon) ring that atenolol doesn't. These features make propranolol more hydrophobic and more soluble in octanol.

(b)   If $\log P_{ow} = 0.22$, then $P_{ow} = 1.66$. This number is the ratio of drug in octanol to water, or we can say that there is a 1.66:1 ratio of the drug in octanol versus in the water. We need the denominator and the numerator to add up to 1 g, and so we need to find the factor that we can multiply ratio by. If we set the factor to x/x, we get $1.66x/x$. Since $1.66x + x = 1$ (numerator + denominator), we can solve for $x = 0.375$. This is the factor that we can use to solve for the two concentrations:

$$P_{ow} = 1.66 = 1.66x/x = 0.624 \text{ g}/0.375 \text{ g}$$

So, there is 0.624 g of the drug in the octanol phase and 0.375 g in the water phase.

(c)   We may conclude on the basis of "like dissolves like" that atenolol is more soluble in water because it is more polar than propranolol. However, atenolol's melting point is very high relative to propranolol. According to Eq. 8.14, the solubility of a solid is equal to the solubility of the liquid plus the free energy of fusion, which is proportional to the melting point. So, it would not be surprising if the solubilities are close, or if atenolol is even slightly less soluble. (The discussion of the solubilities of nifedipine and 9,10-dihydroanthracene on text p. 362 illustrates a similar situation.)

(d)

amide nitrogen, much less basic than amine nitrogen

(e) At a pH of 7.4, an acid with a pK_a of 9.5 is more than 99% protonated (Eq. 3.31b, text p. 107). The protonated forms are ionic, less soluble in octanol, and more soluble in water. Therefore, the partition coefficient would decrease, as the denominator increases relative to the numerator.

(f) Physiological pH is close to 7.4, so administering the drugs in the protonated form [see part (e)] increases their water solubility.

8.55 (a) In compound *A*, the equatorial position of the —OH group is preferred for the usual reason: it avoids 1,3-diaxial interactions with hydrogens on the ring. However, compound *B* prefers to have an axial —OH group because it affords the opportunity for intramolecular hydrogen bonding; this additional bonding is a stabilizing effect, and lowers the energy of the axial conformation below that of the equatorial conformation.

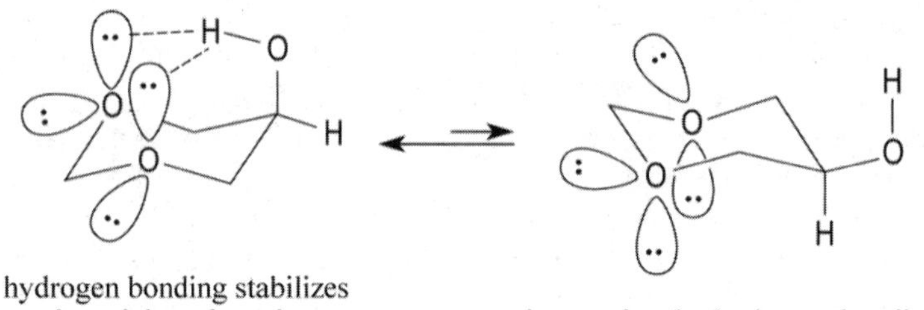

hydrogen bonding stabilizes
the axial confomation

intramolecular hydrogen bonding
is not possible in the
equatorial conformation

(b) The solution to part (a) shows that intramolecular hydrogen bonding can stabilize conformations that otherwise might be less stable. In this case, intramolecular hydrogen bonding can stabilize the *gauche* conformations of both stereoisomers. Indeed, in either enantiomer of the racemate, such hydrogen bonding can occur in one conformation in which the large *tert*-butyl groups are anti to each other. However, in the meso stereoisomer, the necessity that the hydroxy groups be gauche in order for hydrogen bonding to occur also means that the large *tert*-butyl groups must also be gauche. The stabilizing effect of intramolecular hydrogen bonding cannot compensate for the magnitude of the resulting van der Waals repulsions between the *tert*-butyl groups.

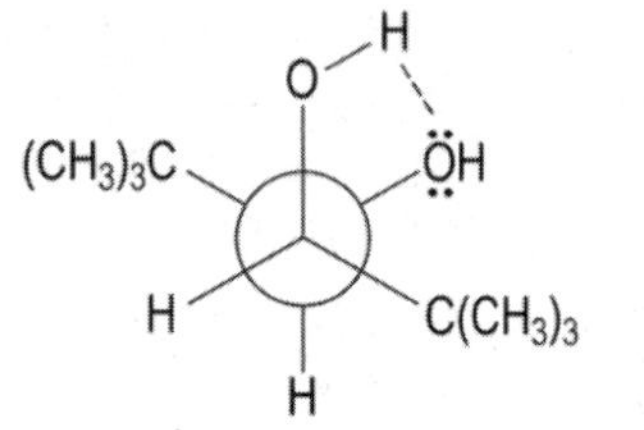

one of the conformations of
(3*R*,4*R*)-2,2,5,5-tetramethyl-3,4-hexanediol
is stabilized by internal hydrogen bonding, and
the *tert*-butyl groups are anti

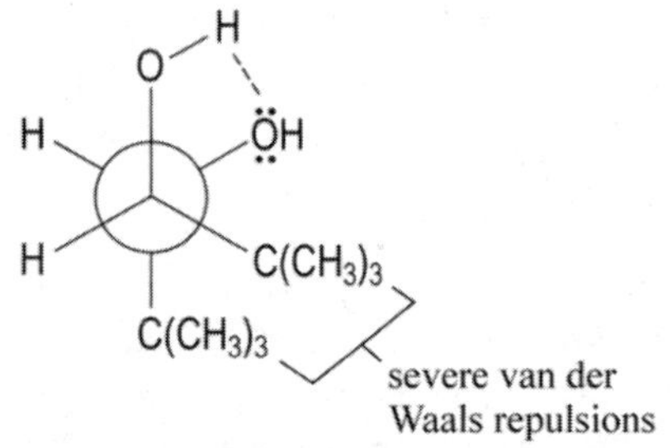

any conformation of
meso-2,2,5,5-tetramethyl-3,4-hexanediol
in which internal hydrogen bonding is possible
is destabilized by van der Waals repulsions between
the *tert*-butyl groups

Convert the Newman projection of the meso stereoisomer to an eclipsed conformation if the meso stereochemistry is not clear.

8.56 (a) In the chair conformation, for every C—O bond dipole in a given direction, there is another C—O bond dipole of the same magnitude pointing in the opposite direction. Thus, in the following diagram, the gray dipoles cancel each other, and the black dipoles cancel each other.

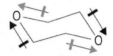

chair conformation of 1,4-dioxane

Because pairs of dipoles cancel, the overall dipole moment is zero.

(b)   To the extent that the twist-boat conformation is present, it will contribute a nonzero dipole moment, because the C—O bond dipoles do not cancel in this conformation. (We use the simpler boat conformation to illustrate this idea.)

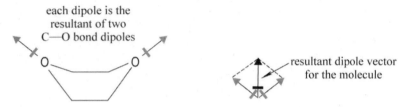

boat conformation of 1,4-dioxane

The dipole moment of the twist-boat conformation is actually rather large, but it is present in very small concentration. The dipole moment of any molecule is the weighted average of the dipole moments of individual conformations. In other words, the nonzero dipole moment of 1,4-dioxane results from the presence of a very small amount of a conformation that has a large dipole moment.

8.57   (a)   Equilibrium (2) contains more of the conformation with the methyl group in an axial position than equilibrium (1) does; that is, equilibrium (2) lies farther to the right. The reason is that, because C—O bonds are shorter than C—C bonds, the axial methyl group in equilibrium (1) is closer to the axial hydrogens than the axial methyl group in equilibrium (2). Therefore, the 1,3-diaxial interactions in the axial-methyl conformation of equilibrium (1) are somewhat more severe than those in the axial-methyl conformation of equilibrium (2).

smaller distance;
greater van der Waals repulsions

larger distance;
smaller van der Waals repulsions

axial conformation in equilibrium (1)

axial conformation in equilibrium (2)

(b)   The reasoning is much the same as that in part (a). Because the C—O bonds are shorter in the ether *A* than the corresponding C—C bonds in butane *B*, the methyl groups are brought closer together in the gauche conformation of the ether than they are in the gauche conformation of butane. Consequently, van der Waals repulsions in the gauche conformation of the ether are somewhat greater than they are in the gauche conformation of butane. In contrast, the shorter bonds of the ether should have little effect on the energy of the anti conformation, in which the methyl groups are far apart. Hence, the greater energy of the gauche conformation of the ether (relative to the anti conformation) causes less of the gauche conformation to be present at equilibrium. Thus, butane contains more gauche conformation at equilibrium.

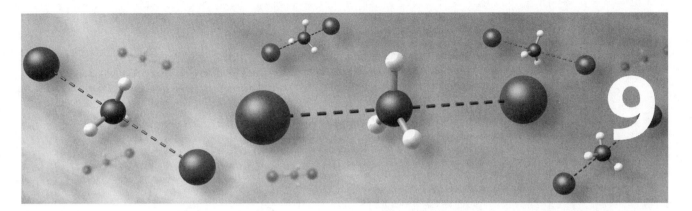

The Chemistry of Alkyl Halides

STUDY GUIDE LINKS

 9.1   Deducing Mechanisms from Rate Laws

Suppose you are an expert in traffic safety, and you are asked to determine as much as you can about traffic accidents on a bridge. In fact, your job is to come up with a description for a "typical" accident and formulate a strategy to eliminate the accidents. Of course, you would interview eyewitnesses and read police reports to gather your data. But suppose eyewitness accounts are not available to you for some reason. This is the situation in deducing reaction mechanisms. We cannot directly observe transition states (that is, molecular collisions).

Imagine that the traffic data you have available are the number of accidents as a function of the traffic density on the bridge in the two directions. Suppose you find that the number of accidents is proportional to the number of cars traveling east *times* the number of cars traveling west. From this information you can construct a mental picture—a mechanism—for a typical collision. Because the number of accidents in a given time is proportional to the traffic flow in *both* directions, you could reasonably conclude that a typical accident is a head-on collision between a car in one lane and a car in the other. An *unreasonable* picture would be two cars in the same lane having a rear-end collision, because, in this case, the number of accidents in a given time would be proportional to some function of the traffic density in only that lane. The head-on collision is as good a picture of an accident as any other until other data force you to refine it. In fact, it suggests an experiment for reducing the number of accidents: build a wall between the two lanes of traffic. If this "experiment" reduces the number of accidents, it would support your picture of a typical accident. Notice that it would *not prove your picture to be correct;* it would only *show that your picture is not incorrect.* Nothing short of eyewitnesses can prove you correct.

Similarly, a starting point for deducing a chemical mechanism is the rate law. The rate law tells us about "molecular collisions" that lead to a reaction, but does not tell us whether these collisions occur by a head-on mechanism, a rear-end mechanism, or some less direct pathway. The simplest mechanism possible is adopted, and a modification is considered only when required by new data.

9.2   Ring Carbons as Alkyl Substituents

Study Problem 9.2 on text p. 417 refers to alkyl substituents in alkene *C*. Some students have difficulty in seeing ring carbons as "substituents," even though they have no problem with the idea of substituents in acyclic compounds. The reason for this difficulty is that the "substituents" in a cyclic compound are "tied together" into a ring. For example, "the substituents at the double bond" in compound *C* should be viewed as follows:

(The methyl group is a third substituent on the double bond that is not tied back.)

A similar approach is taken to the analysis of *β* substituents in alkyl halide *B*:

The fact that the substituent carbons are linked in a ring doesn't change the fact that they are substituents, just as tying your hands together wouldn't change the fact that you still have two hands.

 9.3   Diagnosing Reactivity Patterns in Substitution and Elimination Reactions

The last case study in Study Problem 9.3 (text p. 428) points out a danger in using a guide such as Table 9.7 on text p. 427: *you can sometimes expect special cases that are not covered by the table or borderline cases that are hard to predict.* You can't diagnose every possible reaction with a summary table any more than a physician can diagnose a patient with a table of symptoms and diseases! When you encounter borderline cases that are difficult to predict, you should be able to narrow the possibilities of what can happen. The important thing is for you to be able to predict trends and to understand the reasoning behind them. Table 9.7 will prove most useful if you *make the effort to understand each entry in terms of the principles that underlie it,* and then *think* about each problem you are asked to solve in terms of these principles rather than attack it by rote.

A number of professors of organic chemistry, the author included, find that the section of our organic chemistry courses that deals with substitution and elimination reactions—the section covered by this chapter—is a point at which a number of students have particular difficulty. We have discerned at least two reasons for this problem. First, this is a point at which the number of reactions and ideas have accumulated to the stage that the student who is not carefully organized is overwhelmed. If you are in this situation, you should read (or re-read) Study Guide Link 5.2, "How to Study Organic Reactions," on p. 108 of this manual, and follow the suggestions there. The second reason for difficulty is the fact that all four mechanisms of substitution and elimination—S_N1, S_N2, E1, and E2—can potentially happen simultaneously, and it can be difficult to keep straight exactly when each process can take place. Again, to use an analogy from the medical world, when a physician diagnoses a problem with a patient, he/she can't focus simply on one system, such as the heart; the physician must understand that the many systems in the body interact, and that there can be many root causes contributing to a given symptom. In some cases, the problems are so complex that they can't be readily solved! Table 9.7 is designed to help you keep all the possibilities in substitution and elimination reactions organized.

A broader philosophical issue is relevant to this discussion. The temptation is for students to believe that there are "recipes" or set formulas for solving every problem. This section on substitutions and eliminations shows that there are *principles* that can guide you into the right "ballpark," but inevitably, uncertainties will arise in some situations. Again: Your goal should be not to memorize every conceivable case, but rather to bring to bear your knowledge of the principles so as to narrow the possibilities.

If you were a laboratory scientist trying to predict the outcome of a reaction, you would make your best prediction using the principles that you know, and then you would go into the laboratory and run an experiment to see whether you are correct. A physician facing such uncertainty runs tests to gather as many facts as possible, adopts a hypothesis, and checks it by administering appropriate medication or carrying out other tests. Two people with difficulties in a relationship adopt a hypothesis as to the nature of the problem and, if they are behaving rationally, change their behavior and see whether the problem is solved. For most problems there aren't magic recipes and tables that allow us to solve problems by rote, nor are there simple formulas into which we "plug" data for an answer. Uncertainty is in the very nature of science, and indeed, of life itself. If you think about it, life would be awfully dull without the challenge of such uncertainty. *One reason that organic chemistry is valued as a prerequisite course of study by various disciplines is that a student must, in order to succeed, develop a body of principles that can be applied to situations in which there is inherent uncertainty.* The good news is that if you master this skill in organic chemistry, you'll find it easier to do so in your other endeavors as well.

FURTHER EXPLORATIONS

9.1   Reaction Rates

Because rates are usually functions of concentration, and because concentrations change as a reaction proceeds, you can perhaps appreciate that the rate of a reaction changes over its course. If the rate of a reaction is proportional to the concentrations of one or more reactants, the rate diminishes during a reaction because the concentration of reactants diminishes. (See Problem 9.8 on p. 391 of the text and its solution.)

Reaction rates, like other rates of change, are more precisely expressed in the language of calculus. Suppose, for example, we have a reaction $S \rightarrow P$ that follows a rate law that is first order in S. Then the rate can be written

$$\text{rate} = \frac{d[P]}{dt} = -\frac{d[S]}{dt} = k[S] \tag{FE9.1}$$

This is a first-order differential equation that can be solved for $[S]$ as a function of time. If S_0 is the concentration of S at the start of the reaction ($t = 0$), then the solution to this equation is

$$[S] = S_0 e^{-kt} \tag{FE9.2}$$

The concentration of product as a function of time is

$$[P] = S_0(1 - e^{-kt}) \tag{FE9.3}$$

(If you are familiar with differential equations, you can easily verify this solution by differentiating Eq. FE9.2, as follows.

$$d[S]/dt = -kS_0 e^{-kt} \tag{FE9.4}$$

Substituting Eq. FE9.2 into Eq. FE9.1 yields Eq. FE9.4 and thus verifies the solution. You can, in a similar manner, verify by differentiating Eq. FE9.3 that $d[P]/dt = -d[S]/dt$.)

Equation FE9.2 gives the expression for the actual concentration of $[S]$ as a function of time, and shows that it is a decaying exponential. Each rate law has its own characteristic solution for the dependence of concentration on time. Computer-fitting actual concentration versus time data for a reaction of known kinetic order to the appropriate rate law allows the experimentalist to determine the value of the rate constant k. This is how rate constants are usually derived in practice.

9.2   Absolute Rate Theory

The dependence of the first-order rate constant k on the standard free energy of activation can be calculated from Eyring theory, or absolute rate theory. (Henry Eyring [1901–1981] was for many years a professor at the University of Utah, and he also served as President of the American Chemical Society in 1963. Eyring developed absolute rate theory, also called "transition-state theory.") A result of absolute rate theory is that a first-order rate constant is given by

$$k = \frac{\kappa T}{h} e^{-\Delta G^{\circ\ddagger}/RT} = (6.21 \times 10^{12} \text{ sec}^{-1})(e^{-\Delta G^{\circ\ddagger}/RT}) \tag{FE9.5}$$

where κ = the Boltzmann constant (1.381×10^{-23} J K^{-1}), h = the Planck constant (6.626×10^{-34} J s^{-1}), and T is the absolute temperature in kelvins. The evaluation of the $\kappa T/h$ pre-exponential factor in Eq. FE9.5 is at $T = 298$ K, or approximately room temperature. This is the equation used to give the relationship between the standard free energy of activation and rate constant in Table 9.2 on text p. 390.

The time required to complete a reaction is related to its half-life—the time required for 50% completion. The half-life of a reaction, $t_{1/2}$, can be calculated by setting $[S] = S_0/2$ in Eq. FE9.2 and calculating t. When we make this calculation, we find

$$t_{1/2} = \ln 2/k = 0.693/k \qquad \text{(FE9.6)}$$

In each successive half-life of a reaction, the percent completion of a reaction decreases by half. Thus, a reaction is 50% complete after one half-life; 75% complete after two half-lives; and $100 \times (1.00 - 0.5^n)\%$ complete after n half-lives. Because the concentration of products is a decaying exponential, a reaction is technically never fully completed; but it is generally accepted that a reaction is considered practically complete after 10 half-lives, after which a reaction is about 99% complete. From Eq. FE9.6, ten half lives = $(10)(0.693)/k$, or about $7/k$. This is the source of the "time to completion" numbers in Table 9.2.

9.3   Mechanism of Formation of Grignard Reagents

Despite being discovered over 100 years ago and studied extensively ever since, the details of the formation of organomagnesium compounds (Grignard reagents) are still debated. It is largely agreed that an alkyl radical is formed on the surface of the metal but it is unclear how "free" this radical is: does it stay adsorbed to the magnesium surface, or diffuse away before it combines with magnesium ions and the halide? A detailed investigation of these theories is beyond the scope of our consideration here. In either scenario, the mechanism of formation of Grignard reagents can be summarized as follows. First, an electron is transferred from the magnesium to the alkyl halide (X = I, Br, or Cl). The alkyl halide fragments into the alkyl radical and a halide anion. Finally, the radical combines with the magnesium ion and halide, which then bonds with the halide, to form the Grignard reagent.

$$R\!-\!X \;+\; \ddot{M}g \;\longrightarrow\; R\!-\!X^{\cdot -} \;+\; Mg^{\cdot +}$$
radical anion

$$R\!-\!X^{\cdot -} \;\longrightarrow\; R\!\cdot \;+\; X^-$$

$$R\!\cdot \;+\; \cdot Mg^+ \;\longrightarrow\; R\!-\!Mg^+$$

$$R\!-\!Mg^+ + X^- \;\longrightarrow\; R\!-\!Mg\!-\!X$$
Grignard reagent

As noted in the text, these reagents are likely part of a complex with solvent molecules (Sec. 9.8B), typically diethyl ether, which can stabilize the electron-deficient magnesium.

REACTION REVIEW

I. REACTIONS OF ALKYL HALIDES

A. NUCLEOPHILIC SUBSTITUTION REACTIONS

1. A nucleophilic substitution reaction, or nucleophilic displacement reaction, is a very general type of reaction:

   a. It is a *substitution* because one group is substituted for (or displaces) another group, which departs.

      i. The group that is displaced in a nucleophilic substitution reaction is called the leaving group.

      ii. The leaving group in a nucleophilic displacement reaction is the nucleophile in the reverse reaction, and vice-versa.

      iii. The best leaving groups give the weakest bases as products.

   b. It is a *nucleophilic substitution* because the substituting group acts as a nucleophile, or Lewis base.

      i. Many nucleophiles are anions and others are uncharged.

      ii. The roles of the nucleophile and the leaving group are reversed if the reaction can be run in the reverse direction.

   c. Nucleophilic substitution reactions occur by two common mechanisms:

      i. The S_N2 mechanism occurs in a single step with inversion of stereochemical configuration; it is characterized by a second-order rate law.

$$Nuc:^- \quad C \quad X \longrightarrow Nuc-C + X:^- \qquad rate = k\left[Nuc:^-\right]\left[C-X \right]$$

   (a) An S_N2 reaction is an electron-pair displacement reaction that is conceptually similar to a Brønsted acid–base reaction; the nucleophile is analogous to the Brønsted base and the electrophile is something other than a proton. The leaving group is displaced from the electrophile in an S_N2 reaction, just as a leaving group is displaced from a proton in a Brønsted acid–base reaction.

$$Nuc:^- \quad C \quad X \longrightarrow Nuc-C + X:^- \qquad \text{an } S_N2 \text{ reaction}$$

$$Base:^- \quad H \quad X \longrightarrow Base-H + X:^- \qquad \begin{array}{l}\text{a Brønsted}\\\text{acid–base reaction}\end{array}$$

   (b) S_N2 reactions are typically carried out in polar protic solvents, polar aprotic solvents, or mixtures of these. Polar aprotic solvents in many cases accelerate S_N2 reactions because they do not form hydrogen bonds to the nucleophile.

      ii. The S_N1 mechanism is characterized by a first-order rate law that contains only a term in alkyl halide concentration. The S_N1 reaction can involve significant loss of stereochemical configuration.

$$-\overset{|}{\underset{|}{C}}-X \xrightarrow{\text{(rate-limiting)}} -\overset{/}{C}+ + X:^- \xrightarrow[\substack{\text{(very}\\\text{fast)}}]{Nuc:^-} -\overset{|}{\underset{|}{C}}-Nuc \qquad rate = k\left[-\overset{|}{\underset{|}{C}}-X \right]$$

2. The reaction of an alkyl halide with a protic solvent in which no base or nucleophile has been added is called a solvolysis.

a. The solvolysis reactions of tertiary alkyl halides are fastest in polar, protic donor solvents.

b. Tertiary and secondary alkyl halides undergo solvolysis reactions by the S_N1 and E1 mechanisms; tertiary alkyl halides are much more reactive.

c. Rearrangements are observed in S_N1 solvolysis reactions that involve rearrangement-prone carbocations as intermediates.

3. Substitution reactions can be intramolecular—that is, the nucleophile, the electrophile, and the leaving group are part of the same molecule.

4. Nucleophilic substitution reactions can be used to transform alkyl halides into a wide variety of other functional groups. (See Table 9.1 on text p. 385.)

B. ELIMINATION REACTIONS

1. An elimination reaction is a reaction in which two or more groups are lost from within the same molecule.

a. In an alkyl halide, the carbon bearing the halogen is often referred to as the α-carbon, and the adjacent carbons are referred to as the β-carbons.

b. An elimination that involves loss of two groups from adjacent carbons is called a β-elimination.

c. A β-elimination reaction is conceptually the reverse of an addition to an alkene.

2. An α-elimination, as in the formation of dichloromethylene from chloroform, is an elimination of two groups (in this case the elements of HCl) from the same atom.

3. β-Elimination reactions occur by two mechanisms:

a. The E2 mechanism, which competes with the S_N2 mechanism, has a second-order rate law and typically occurs with anti stereochemistry.

anti-elimination

b. The E1 mechanism is an alternative product-determining step of the S_N1 mechanism in which a carbocation intermediate loses a β-hydrogen to form an alkene.

4. Base-promoted β-elimination reactions in many cases occur by E2 mechanisms; they typically follow a rate law that is second order overall and first order in each reactant.

a. A Brønsted base reacts at a β-hydrogen of the alkyl halide, not at a carbon atom as in a nucleophilic substitution reaction.

b. If the reacting alkyl halide has more than one type of β-hydrogen atom, then more than one β-elimination reaction is possible.

c. Strong bases, such as sodium ethoxide [Na⁺ ⁻OCH₂CH₃] and potassium *tert*-butoxide [K⁺ ⁻OC(CH₃)₃], promote the β-elimination reactions of alkyl halides.

d. Often, the conjugate-acid alcohols of these bases are used as solvents, for example, sodium ethoxide in ethanol or potassium *tert*-butoxide in *tert*-butyl alcohol.

C. FORMATION OF GRIGNARD AND ORGANOLITHIUM REAGENTS

1. Compounds that contain carbon–metal bonds are called organometallic compounds.

a. A Grignard reagent is an organometallic compound of the form R—MgX, where X = Br, Cl, or I.

b. Organolithium reagents are organometallic compounds of the form R—Li.

2. Both Grignard and organolithium reagents are formed by adding the corresponding alkyl or aryl halides to rapidly stirred suspensions of the appropriate metal.

$$R-X + Mg \xrightarrow{\text{ether}} R-Mg-X \quad (X = Cl, Br, or I)$$

Grignard reagent

$$R-X + 2Li \xrightarrow{\text{hydrocarbon}} R-Li + LiX \quad (X = Cl, Br, or I)$$

organolithium reagent

a. Reaction of alkyl halides with magnesium metal yields Grignard reagents.

i. Ether solvents must be used for the formation of Grignard reagents; the ether solvent plays a crucial role in their formation because the ether associates with the metal in a Lewis acid–base interaction.

ii. Grignard reagents are formed on the surface of the magnesium metal.

b. Reaction of alkyl halides with lithium in hydrocarbon solvents yields organolithium reagents.

c. Because they react vigorously with water and oxygen, Grignard and organolithium reagents must be prepared under rigorously moisture-free and oxygen-free conditions.

3. Grignard and organolithium reagents behave as strong Brønsted bases and react violently with acids, including water and alcohols, to give alkanes. (See Part III.A of this section.)

4. All reactions of Grignard and organolithium reagents can be understood in terms of the polarity of the carbon–metal bond.

a. Because carbon is more electronegative than either magnesium or lithium, the negative end of the carbon-metal bond is the carbon atom.

i. A carbon bearing three bonds, an unshared electron pair, and a negative formal charge is termed a carbon anion, or carbanion.

ii. Carbanions are powerful Brønsted bases because their conjugate acids, the corresponding alkanes, are extremely weak acids.

b. Grignard and organolithium reagents react as if they were carbanions; however, they are not true carbanions because they have covalent carbon–metal bonds.

$$-\overset{|}{\underset{|}{C}}\overset{\delta-}{\underset{}{}}\overset{\delta+}{MgBr} \qquad \text{reacts as if it were} \qquad -\overset{}{\underset{}{C}}{:}^{-} + \overset{+}{MgBr}$$

polarized bond   carbanion

II. REACTION OF ALKANES

A. FREE-RADICAL HALOGENATION OF ALKANES

1. Alkanes react with bromine and chlorine in the presence of heat or light in free-radical substitution reactions to give alkyl halides.

2. The mechanism of this reaction in fact follows the typical pattern of other free-radical chain reactions; it has initiation, propagation, and termination steps.

   a. The reaction is initiated when a halogen molecule absorbs energy from heat or light and dissociates homolytically into halogen atoms.

   $$X\frown X \xrightarrow[\text{or heat}]{\text{light}} X\cdot + \cdot X$$

   b. The ensuing chain reaction has the following propagation steps.

   $$R\frown H \frown X \longrightarrow R\cdot + H-X$$
   $$R\cdot \frown X \frown X \longrightarrow R-X + \cdot X$$

   propagation steps

   c. Termination steps result from the recombination of radical species.

   $$X\cdot \frown X \longrightarrow X-X \qquad \text{one of several possible termination steps}$$

3. Free-radical halogenations with chlorine and bromine proceed smoothly, halogenation with fluorine is violent, and halogenation with iodine does not occur.

III. ACID–BASE REACTIONS OF GRIGNARD REAGENTS AND ORGANOLITHIUM REAGENTS

A. PROTONOLYSIS OF ORGANOMETALLIC REAGENTS

1. All Grignard or organolithium reagents react vigorously with even relatively weak acids, such as water and alcohols, to give the conjugate-base hydroxide or alkoxides and the conjugate-acid hydrocarbon of the carbanion; this reaction is an example of protonolysis.

$$R-MgX + R'-OH \longrightarrow R-H + Mg^{2+} + X^- + R'O^-$$

2. A protonolysis is a reaction with the proton of an acid that breaks chemical bonds.

   a. Protonolysis of Grignard reagents is also useful, because it provides a method for the preparation of hydrocarbons from alkyl halides.

   b. A particularly useful variation of this reaction is the preparation of hydrocarbons labeled with the hydrogen isotopes deuterium (D, or ^2H) or tritium (T, or ^3H) by reaction of a Grignard reagent with the corresponding isotopically labeled water.

$$+ Mg^{2+} + Br^- + DO^-$$

IV. FORMATION OF CYCLOPROPANES

A. DIHALOCARBENES

1. Reaction of a haloform with base in the presence of an alkene yields a 1,1-dihalocyclopropane.
2. Dihalomethylene is a reactive intermediate in this reaction. The addition of a dihalomethylene to an alkene is a concerted syn-addition reaction.

B. CARBENOIDS: THE SIMMONS–SMITH REACTION

1. Cyclopropane without halogen atoms can be prepared by allowing alkenes to react with methylene iodide (CH_2I_2) in the presence of a zinc–copper couple (a copper-activated zinc preparation) in a reaction called the Simmons–Smith reaction.

    a. The active reagent in the Simmons–Smith reaction is believed to be an α-halo organometallic compound, a compound with halogen and a metal on the same carbon.

    b. The formation of this species is analogous to formation of a Grignard reagent.

    c. The Simmons–Smith reagent can be conceptualized as methylene ($:CH_2$, the simplest carbene) coordinated to the zinc atom.

    d. Because they show carbenelike reactivity, α-halo organometallic compounds are sometimes called carbenoids. A carbenoid is a reagent that is not a free carbene but has carbenelike reactivity.

2. Addition reactions of methylene from the Simmons–Smith reagent to alkenes, like the reactions of dichloromethylene, are syn-additions. (See the example under IV.B.1.)

SOLUTIONS TO PROBLEMS

Solutions to In-Text Problems

9.1   (a)   The expected nucleophilic substitution product is butyl methyl sulfide, $CH_3CH_2CH_2CH_2—S—CH_3$.

(b)   The product is ethylammonium iodide.

$$CH_3CH_2\overset{+}{N}H_3 \quad I^-$$

ethylammonium iodide

9.2

$$^-SCH_2CH_2CH_2CH_3 \ + \ I—CH_3 \longrightarrow CH_3SCH_2CH_2CH_2CH_3 + I^-$$

9.3   (a)   Two elimination products are expected because there are two types of β-hydrogens in the starting alkyl halide.

2-bromo-2,3-dimethylbutane $\xrightarrow{\beta\text{-elimination}}$

2,3-dimethyl-1-butene
(from elimination of
a hydrogen of type 1)

\+

2,3-dimethyl-2-butene
(from elimination of
the hydrogen of type 2)

(b)   As in part (a), because there are two types of β-hydrogens, two alkenes can be formed:

1-chloro-1-methylcyclohexane $\xrightarrow{\beta\text{-elimination}}$

1-methylcyclohexene
(from elimination of a
hydrogen of type 1)

\+

methylenecyclohexane
(from elimination of a
hydrogen of type 2)

9.4   (a)   The substitution product can in principle be obtained as two stereoisomers. (Sec. 9.4C of the text shows which of these is formed and why.)

$$\xrightarrow[CH_3OH]{Na^+ CH_3O^-}$$

substitution products   elimination products

(b)   Methyl iodide, $H_3C—I$, can form only a substitution product, dimethyl ether, $H_3C—O—CH_3$. There can be no elimination products because there are no β-protons.

(c)   (Bromomethyl)cyclopentane can form one substitution product and one elimination product.

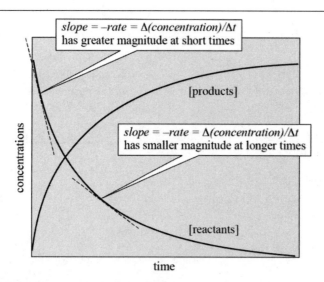

(bromomethyl)cyclopentane

9.5 (a) Because fluoride ion is a stronger base than chloride ion, the equilibrium lies to the right.
(b) Because azide ion is a stronger base than chloride ion, the equilibrium lies to the right.
(c) Because methoxide ion is a much stronger base than chloride ion, the equilibrium lies to the right.

9.6 (a) The reaction is third order overall; the reaction is first order in alkene and second order in Br_2. The dimensions of the rate constant are such that the overall rate has the dimensions mol L^{-1} sec^{-1}; thus, the rate constant has the dimensions L^2 mol^{-2} sec^{-1}, or M^{-2} sec^{-1}.
(b) The reaction is first order overall, and first order in alkyl halide. The rate constant has the dimensions of sec^{-1}.

9.7 (a) Apply Eq. 9.20b:

$$\frac{k_A}{k_B} = 10^{(\Delta G^{\circ\ddagger}_B - \Delta G^{\circ\ddagger}_A)/2.30RT} = 10^{14.0/5.71} = 10^{2.45} = 283$$

(b) We transform Eq. 9.20c to get the difference between the standard free energies of activation.

$$\Delta G^{\circ\ddagger}_A - \Delta G^{\circ\ddagger}_B = 2.30RT \log\left(\frac{k_B}{k_A}\right) = (5.71)\log(450) = 15.2 \text{ kJ mol}^{-1}$$

Therefore, reaction A has the higher $\Delta G^{\circ\ddagger}$ by 15.2 kJ mol^{-1}.

9.8 Because D and E are converted into F, their concentrations decrease with time; hence, the rate also decreases with time. Since the rate is the *slope* of the curves of [reactants] or [products] with time, then the slopes of these curves continually decrease with time, and approach zero at long times. A plot of these concentrations with time is shown in Fig. SG9.1.

Figure SG9.1. A diagram for the change of concentration with time in a first-order reaction. The rate is the instantaneous slope of the concentration versus time curve. The magnitude of the slope decreases with time because the concentration decreases.

9.9   A mechanism consistent with the rate law is a bimolecular process in which ammonia displaces acetate ion from the proton:

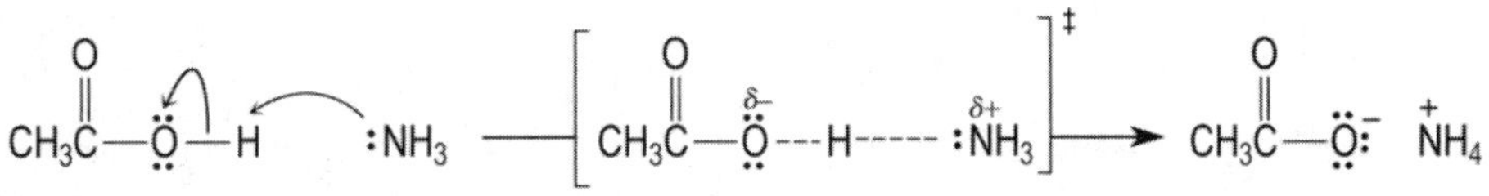

acetic acid   **acetate ion**

9.10   Because the S_N2 mechanism involves a molecule of alkyl halide and a molecule of nucleophile in a bimolecular reaction, the expected rate law is second order, first order in alkyl halide and first order in cyanide:

$$\text{rate} = k[C_2H_5Br][^-CN]$$

9.11   Sodium ethoxide reacts orders of magnitude more rapidly with the Brønsted acid than with the alkyl halide. Because this reaction consumes all of the ethoxide, only the HI reacts. The products are ethanol, CH_3CH_2OH, and sodium iodide, $Na^+\ I^-$. The methyl iodide is unchanged.

9.12   There is more than enough sodium cyanide to react with both the acid HBr and the alkyl halide. Therefore, the products are hydrogen cyanide (HCN, 0.1 *M*), "ethyl cyanide" (propionitrile, CH_3CH_2CN, 0.1 *M*), sodium bromide (NaBr, 0.2 *M,* half from the reaction with HBr and half from the reaction with ethyl bromide), and unreacted sodium cyanide (0.8 *M*). However, the HCN and 0.1 *M* NaBr is formed instantaneously. The rest of the NaBr, as well as the nitrile, are formed much more slowly, because Brønsted bases react much more rapidly with Brønsted acids than with alkyl halides. In other words, Brønsted acid-base reactions are much faster than S_N2 reactions.

9.13   The reaction is an S_N2 reaction with inversion of configuration. Because the relative priorities of the groups attached to the asymmetric carbon are not changed, the product has the *S* configuration.

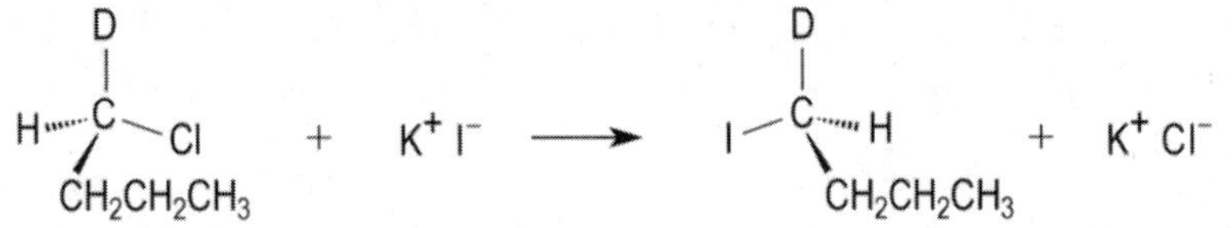

(*R*)-chlorobutane-1-*d*   **(*S*)-iodobutane-1-*d***

9.14   In an S_N2 reaction, the stronger base is the better nucleophile when the same attacking atom is involved. Because ethoxide is a much stronger base than ethanol, it is the better nucleophile.

9.15   (a)   The products of the S_N2 reaction between potassium acetate and ethyl iodide:

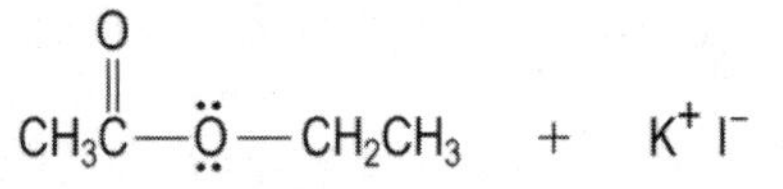

ethyl acetate   **potassium iodide**

(b)   Potassium acetate is a better nucleophile in acetone because ethanol is a protic solvent and reduces the nucleophilicity of potassium acetate by hydrogen-bond donation. Consequently, potassium acetate in acetone reacts more rapidly with ethyl iodide than a solution of the same nucleophile in ethanol.

9.16   The discussion of this section shows that elements in higher periods (lower on the periodic table) within a group are better nucleophiles in protic solvents than elements in lower periods of the same group. Thus, triethylphosphine, $:P(C_2H_5)_3$, is a better nucleophile than triethylamine, $:N(C_2H_5)_3$. The products are as follows:

X = P   X = N

9.17 The formation of the carbon–anion (carbanion) intermediate, its resonance structure, and the mechanism of its breakdown to product, and are as follows:

resonance structures of the anion

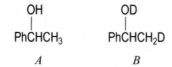

elimination reaction of the anion

The carbanion is more stable than a typical carbanion because it is stabilized by the polar effect of the nearby oxygen, and because it is resonance-stabilized. This stabilization lowers the pK_a for its formation.

 Ionization of the O—H group also occurs, but this does not lead to product.

9.18 (a) The stepwise process involves formation of a methyl cation, which is very unstable. The instability of this cation, by Hammond's postulate, raises the energy of the transition state and retards the reaction. The concerted mechanism avoids formation of this high-energy intermediate.

(b) A tertiary alkyl halide such as *tert*-butyl bromide, $(CH_3)_3C$—Br, can undergo the stepwise mechanism, because ionization gives a relatively stable tertiary carbocation—in this case, $(CH_3)_3C^+$. This lowers the energy of the transition state (by Hammond's postulate). Section 9.6 describes this mechanism.

9.19 (a) A primary deuterium isotope effect retards the reaction of B relative to C; and the better leaving group accelerates the reaction of A relative to B and C. The reactivity order is therefore $B < C < A$.

(b) Compound B has a better leaving group. The rates are in the order $A < B$.

9.20 (a) The hydration of styrene should be (and is) slower in D_2O, because the transferred proton is a deuterium, and a rate-retarding primary deuterium isotope effect operates when the proton transferred in the rate-limiting step is isotopically substituted. Be sure to write this mechanism (see Sec. 4.9B, text p. 172).

The products differ by isotopic substitution. In H_2O/H_3O^+, the hydration product is compound A below; in D_2O/D_3O^+, the hydration product is compound B.

$$\underset{A}{\overset{\overset{\displaystyle OH}{|}}{PhCHCH_3}} \qquad \underset{B}{\overset{\overset{\displaystyle OD}{|}}{PhCHCH_2D}}$$

(b) The hydration rate of the deuterium-substituted styrene should differ very little, if at all, from that of styrene itself, because the deuteriums are not transferred in the rate-limiting step.

A small effect of isotopic substitution occurs in this case because of the differential effect of deuterium and hydrogen on the rehybridization of carbon in the transition state. (The carbon bearing the deuteriums rehybridizes from sp^2 to sp^3 in the rate-limiting step.) However, this effect on rate amounts to only a few percent. Effects of this sort are called *secondary deuterium isotope effects.*

9.21 (a) Analyze one enantiomer of (±)-stilbene dibromide; the result is the same for the other. First, draw an eclipsed conformational projection. How do we know that it is one of the enantiomers? The two asymmetric carbons have the same configuration—in this case, R. Also, in the eclipsed conformation shown, with the two hydrogens in identical positions, there is no internal mirror plane. (The meso diastereomer would have an internal mirror plane.) Next, internally rotate about the central carbon–carbon bond until the H and Br that are eliminated are anti (*step 2* below). Finally, draw the alkene that is formed and note its stereochemistry (*step 3* below).

the alkene product has
the Z configuration

(b)   Reasoning identical to that used in part (a) shows that the alkene formed has the *E* configuration. (We leave it to you to draw the appropriate structures.)

> In a stereospecific reaction with a given stereochemistry—anti-elimination, in this case—diastereomeric starting materials must give diastereomeric products.

9.22   In a stereospecific reaction with a given stereochemistry—anti-elimination, in this case—a diastereomeric product requires a diastereomeric starting material (either enantiomer). The easiest path to the answer is to convert the starting material in Eq. 9.40a into its diastereomer by the interchange of any two groups at one of the carbons. Either the following compound or its enantiomer would give the product of *E* configuration.

$$Ph \cdots C(H)(CH_3) - C(H)(Br) Ph$$

9.23   (a)   Let isobutyl bromide react with dimethyl sulfide, preferably in a polar aprotic solvent (although the reaction of a third-period nucleophile would occur at a useful rate in a polar solvent such as ethanol).

$$(CH_3)_2CHCH_2Br \ + \ :S(CH_3)_2 \ \xrightarrow{\text{acetone}} \ (CH_3)_2CHCH_2\overset{+}{S}(CH_3)_2 \ Br^-$$

isobutyl bromide   **dimethyl sulfide**   **isobutyldimethylsulfonium bromide**

(b)   In this case, the nucleophile is the conjugate base of ethanethiol, sodium ethanethiolate. Although a polar aprotic solvent could be used, ethanol would probably be the most convenient solvent (for solubility reasons). Because the nucleophile is from the third period, the reaction would occur at a convenient rate in a protic solvent; hence, experimental convenience determines the choice of solvent. The nucleophile would be easily formed from ethanethiol with one equivalent of sodium ethoxide, as shown in Eq. 10.4, text p. 454.

$$(CH_3)_2CHCH_2Br \ + \ Na^+ \ C_2H_5S^- \ \xrightarrow{C_2H_5OH} \ (CH_3)_2CHCH_2SC_2H_5 \ + \ Na^+ \ Br^-$$

isobutyl bromide   **sodium**   **isobutyl ethyl sulfide**
                         ethanethiolate

(c)   In contrast, this reaction is an elimination that could be brought about by the reaction of isobutyl bromide with any strong base, such as $Na^+ \ CH_3CH_2O^-$ (sodium ethoxide) or $K^+ \ (CH_3)_3C{-}O^-$ (potassium *tert*-butoxide).

$$(CH_3)_2CHCH_2Br \ + \ K^+ \ (CH_3)_3CO^- \ \xrightarrow{(CH_3)_3COH} \ (CH_3)_2C{=}CH_2 \ + \ (CH_3)_3COH \ + \ K^+ \ Br^-$$

isobutyl bromide   **potassium**   **2-methylpropene**
                         ***tert*-butoxide**   **(isobutylene)**

9.24   Analyze the alkyl substitution pattern in each alkyl halide: alkyl substitution at both the α- and β-positions increases the ratio of elimination to substitution. Compound D is secondary and it has two β substituents; compound B has two β substituents; compound C has one β substituent; and compound A, of course, has no β hydrogens and therefore cannot undergo β-elimination. Therefore, the ratio of E2 elimination to S_N2 substitution decreases in the order $D > B > C > A$.

9.25   Chloride ion is a weaker base than all of the alkoxide bases. Because α-substitution in the alkoxide promotes a greater proportion of elimination, the order is $C > A > B > D$.

9.26   Products A and B result from the Lewis acid–base association reactions of water and ethanol, respectively, with the carbocation intermediate that is formed by ionization of the alkyl halide starting material. Products C and D result from the Lewis acid–base association reactions of water and ethanol, respectively, with the rearranged carbocation intermediate. (Compound D is the product shown in Eq. 9.60.)

Alkenes can form by loss of a β-proton from either of the two carbocation intermediates. (The solvent, indicated below by ROH, acts as the base that removes the proton.)

Alkene F would be formed in greatest amount because it has the greatest amount of alkyl substitution at the double bond. Alkene G would be formed in smaller amount; alkene E would be formed in least amount for two reasons: it comes from a carbocation that can rearrange, and it has only one alkyl substituent on the double bond.

9.27   The mechanism for the formation of the rearranged product in Eq. 9.60 (product D in the solution to Problem 9.26):

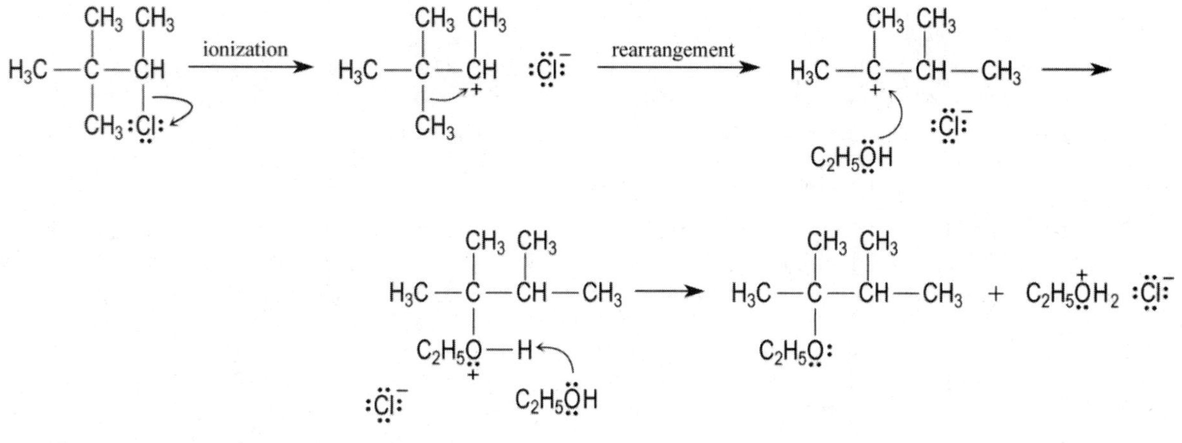

9.28 The amount of racemization, 66%, corresponds to 33% of the *S* product and 33% of the *R* product. The amount of inversion corresponds to 34% of the *S* product. Hence, the product mixture consists of 67% of the *S* methyl ether and 33% of the *R* methyl ether.

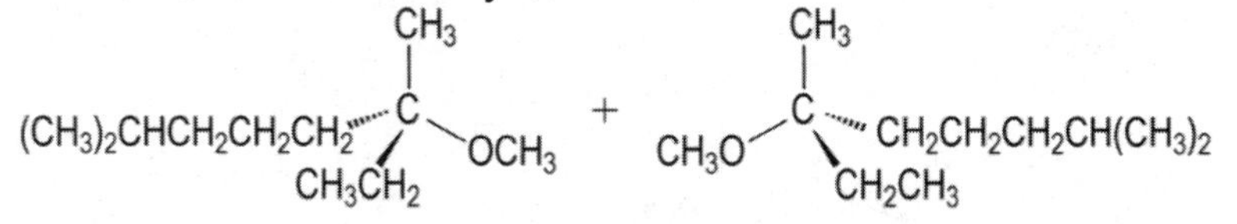

3-methoxy-3,7-dimethyloctane
(product *A*)

9.29 (a) There should be an increasing percentage of racemization for more stable carbocations, because their stability increases their lifetimes; a greater fraction survives past the ion-pair stage to form fully solvated, achiral carbocations, as shown in Fig. 9.13.

(b) A less stable carbocation, by the same reasoning as in part (a), should have a shorter lifetime; a greater fraction will not last past the ion-pair stage, which reacts with the solvent by inversion.

9.30 (a) Entry 3 of Table 9.7 on text p. 427 covers this case; the major product is the substitution product 1-methoxybutane , which is formed by an S_N2 mechanism:

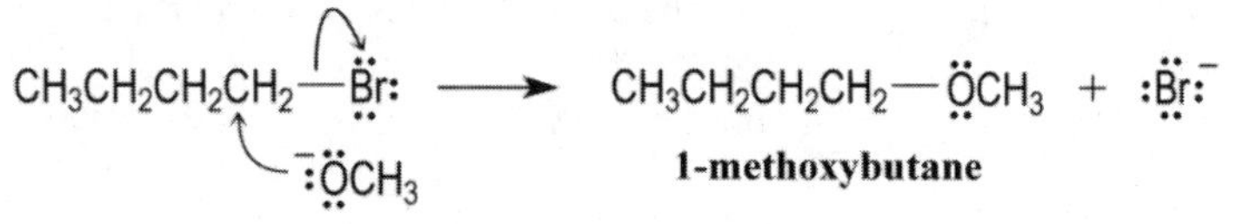

1-methoxybutane

A small amount of the alkene 1-butene will also be formed in an E2 reaction:

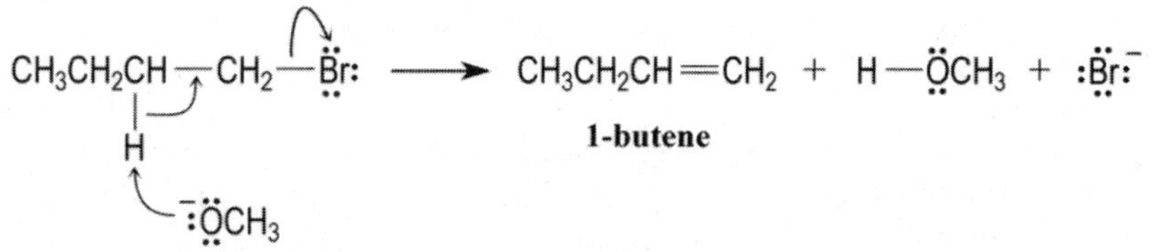

1-butene

(b) 2-Bromobutane is a secondary alkyl halide, and potassium *tert*-butoxide is a strong, highly branched base. Entry 7 of Table 9.7 covers this case. Thus, the E2 reaction is the mechanism that occurs. Two possible alkenes, 1-butene and 2-butene, can form. Either or both can be considered as correct answers. A significant amount of 1-butene is formed because the large base molecule reacts at the least sterically hindered hydrogen.

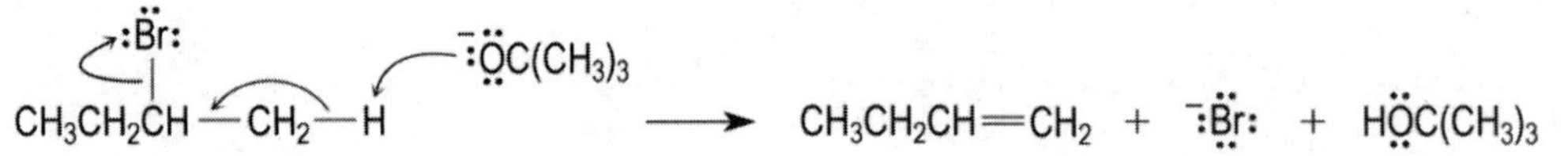

2-bromobutane **1-butene**

We leave it to you to show the formation of the 2-butenes.

(c) 2-Bromo-1,1-dimethylcyclopentane is a secondary alkyl halide with significant β-alkyl substitution; a protic solvent is used without a strong base. Entry 9 in Table 9.7 covers this case. Rearrangement from the initially formed secondary carbocation *a* to the tertiary carbocation *b* is likely to occur.

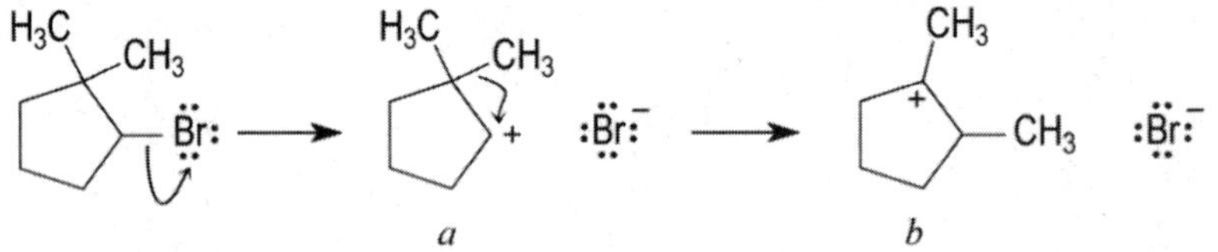

a *b*

Each carbocation can react to give both S_N1 substitution and E1 elimination products. Carbocation *a* gives the substitution product *A* and the elimination product *B*.

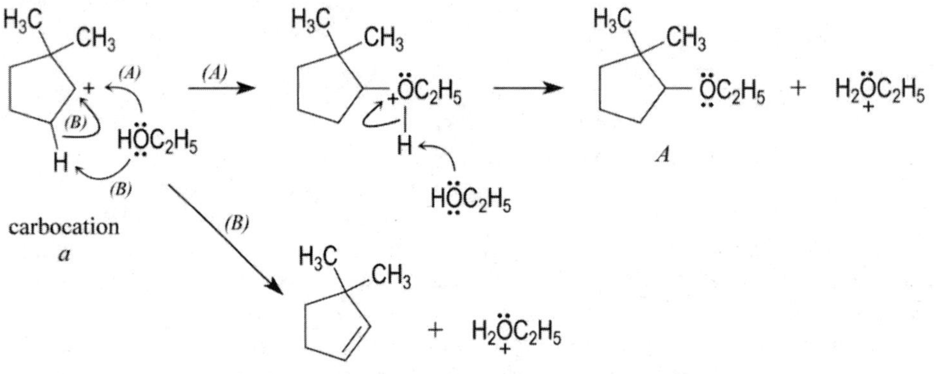

carbocation *a*

A

B

Carbocation *b* gives the substitution product *C* and the elimination products *D*–*F*. (The curved-arrow mechanism for formation of *C* is analogous to the mechanism shown above for the formation of *A*; and the curved-arrow mechanisms for the formation of alkenes *D*–*F* (by the E1 pathway) are analogous to the mechanism shown above for the formation of *B*.) Products *C*–*E* are likely the major products; alkene *F* would be formed in the smallest amount because it is the least substituted alkene.

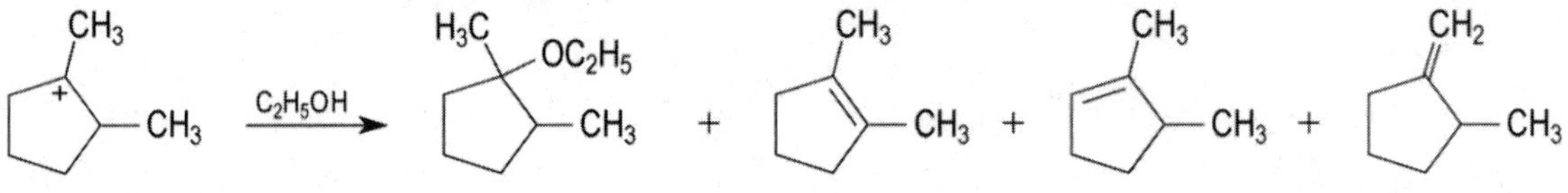

carbocation *b* *C* *D* *E* *F*

(d) Bromocyclohexane is a secondary alkyl halide; methanol is a polar, protic solvent; and there is no strong base present. Entry 9 of Table 9.7 covers this situation; both S_N1 product *A* and E1 product *B* are formed.

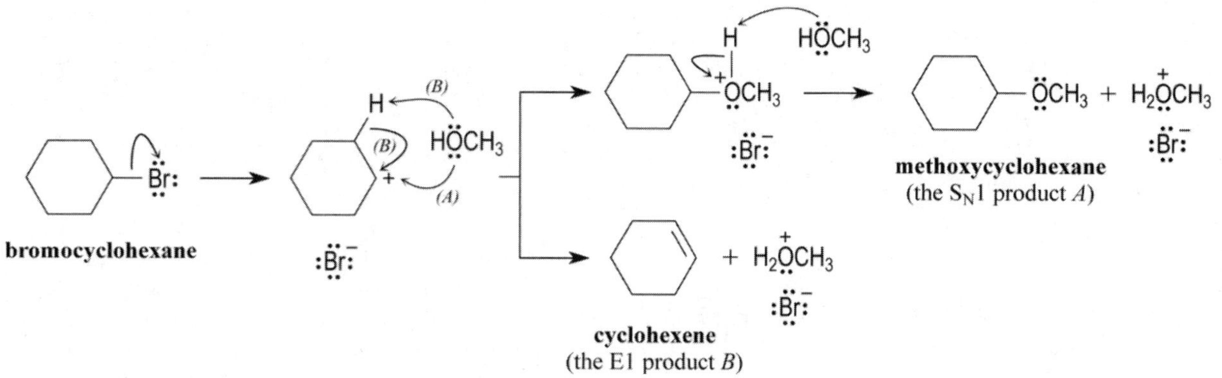

bromocyclohexane   :B̈r:⁻   HÖCH₃   **methoxycyclohexane**
(the S$_N$1 product *A*)

cyclohexene
(the E1 product *B*)

9.31   (a)

$(CH_3)_2CH—Br$ + Mg $\xrightarrow{\text{ether}}$ $(CH_3)_2CH—MgBr$

isopropyl bromide   **isopropylmagnesium bromide**
(a Grignard reagent)

(b)

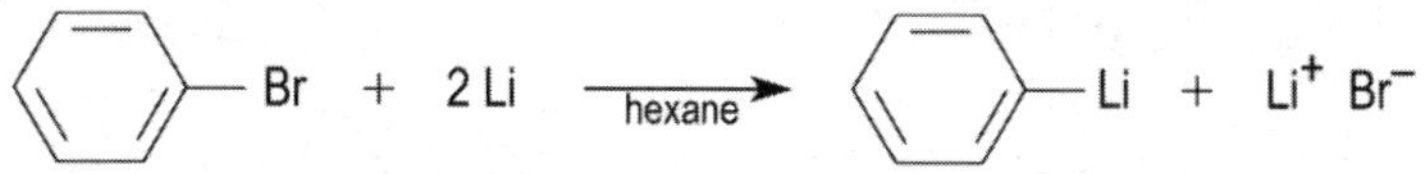

Br + 2 Li $\xrightarrow{\text{hexane}}$ Li + $Li^+\ Br^-$

(c)   Not all organometallic reagents are prepared from alkyl halides. Here's one from Chapter 5:

BH_3 + $3\,(CH_3)_2C{=}CH_2$ $\xrightarrow{\text{THF}}$ $[(CH_3)_2CCH_2]_3B$

9.32   (a)

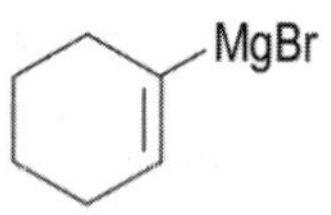

MgBr

1-cyclohexenylmagnesium bromide

(b)   The product of the reaction is *tert*-butyllithium, $(CH_3)_3C—Li$, and lithium chloride, $Li^+\ Cl^-$.

9.33   (a)   The product of the reaction results from protonolysis of the lithium reagent (text p. 431) at the carbon of the C—Li bond to give methane, CH_4, and lithium methoxide, $CH_3O^-\ Li^+$. The curved-arrow notation:

$CH_3\ddot{O}—H$ $CH_3—Li$ $\longrightarrow$ $CH_3\ddot{O}$:⁻ Li^+ + CH_4

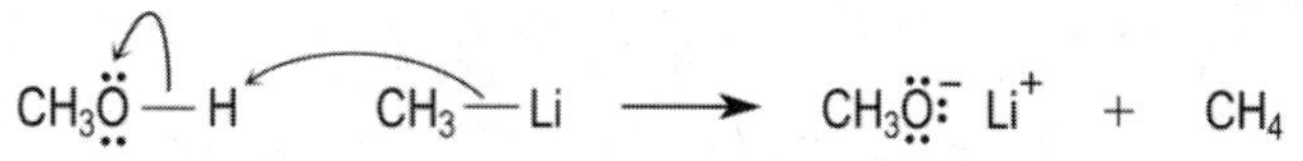

methane

(b)   The products result from protonolysis of the C—Mg bond: isobutane, $(CH_3)_2CHCH_3$, and HOMgCl, which, under the aqueous reaction conditions, is ionized to Mg^{2+}, HO^-, and Cl^-. The curved-arrow notation for the protonolysis:

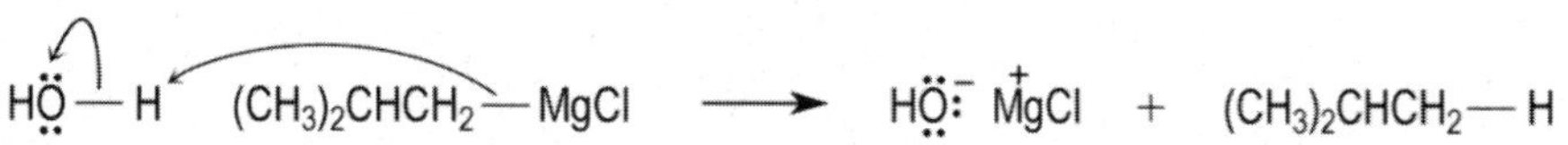

$H\ddot{O}—H$ $(CH_3)_2CHCH_2—MgCl$ $\longrightarrow$ $H\ddot{O}$:⁻ $\overset{+}{M}gCl$ + $(CH_3)_2CHCH_2—H$

isobutane

9.34   (a) $(CH_3)_2CH—MgBr$ and $CH_3CH_2CH_2—MgBr$ both react with water to give propane.
        (b)   The compounds formed in the reactions of the Grignard reagents in part (a) with D_2O are $(CH_3)_2CH—D$ and $CH_3CH_2CH_2—D$, respectively. You should convince yourself of these results by writing a curved-arrow mechanism as shown in problem 9.33(b).

9.35   (a)

cyclohexene   +   HCBr₃

bromoform

(b)

2,3-dimethyl-2-butene   +   PhCH₂Br

benzyl bromide

9.36   (a)

cyclopentene   +   HCCl₃   +   K⁺ ⁻OC(CH₃)₃   ⟶ $\triangleright$CCl₂   +   HOC(CH₃)₃   +   K⁺ Cl⁻

(b)   Products *A* and *B*, since they are diastereomers, are formed in different amounts.

CH₃   +   HCCl₃   +   K⁺ ⁻OC(CH₃)₃   ⟶   CH₃ ...CCl₂   +   CH₃ ...CCl₂   +   HOC(CH₃)₃

+   K⁺ Cl⁻

A   *B*

9.37   (a)

(*Z*)-3-methyl-2-pentene $\xrightarrow[\text{CH}_2\text{I}_2]{\text{Zn/Cu}}$ (racemic)

(b)

$=$CHCH₃ $\xrightarrow[\text{CH}_2\text{I}_2]{\text{Zn/Cu}}$ —CH₃ H

9.38   (a)   Cyclopropane formation occurs at the face of the ring opposite to the methyl group for steric reasons.

(b)

CH₂

CH₂

9.39   The free-radical chain mechanism (initiation and propagation steps) for bromination of ethane:

Initiation: Br —Br $\xrightarrow{\text{light}}$ 2 Br·

Propagation: Br· H — CH$_2$CH$_3$ $\longrightarrow$ Br — H + ·CH$_2$CH$_3$

ethyl radical

Br — Br ·CH$_2$CH$_3$ $\longrightarrow$ Br — CH$_2$CH$_3$ + Br·

**ethyl bromide
(bromoethane)**

9.40   In the bromination of ethane, the ethyl radical, CH$_3$CH$_2$·, is one of the chain-propagating radicals (Problem 9.39). The recombination of two ethyl radicals in a termination step of ethane bromination forms a small amount of butane.

CH$_3$CH$_2$·   ·CH$_2$CH$_3$ $\longrightarrow$ CH$_3$CH$_2$CH$_2$CH$_3$

butane

9.41   In this reaction, there is only 36% of the tertiary alkyl chloride product, and 64% of the primary chloride. While it would appear that abstraction of a primary hydrogen is favored, remember that there is a 9/1 statistical advantage to abstracting one of these hydrogens relative to the tertiary one. Therefore, we can calculate the relative rate of the two reactions. There is a little more than half, or 0.56 times, as much *tert*-butyl chloride as there is isobutyl chloride (36% ÷ 64%). But on a per hydrogen basis, the tertiary hydrogen is (0.56 ÷ 1/9) = 5 times as reactive as a primary hydrogen. This discrepancy is not as dramatic as in the reaction with bromine (Eq. 9.86, p. 439), but still favors abstraction of a tertiary hydrogen over a primary one.

9.42   The two major products formed when pentane reacts with bromine in light are 2-bromopentane and 3-bromopentane. These products are formed in roughly equal amounts, because they both come from similarly stable secondary radicals. A third possible product, 1-bromopentane (not shown) is only a minor product because it would have to be produced from a much less stable primary radical.

pentane   **2-bromopentane   3-bromopentane**

9.43   (a)   The major bromination product is the one that forms from the most stable radical intermediate, in this case, a tertiary radical. Other products would form from less stable secondary or primary radicals, at a slower rate, and in minor amounts.

methylcyclohexane

(b)   The first step in solving a problem like this is to draw the structures of the starting material and the products. There are three different monochlorinated products that can form in this reaction. In 2,2,3-trimethylbutane, there are fifteen primary hydrogens, and only one tertiary, which means the statistical chances of abstracting a primary hydrogen and forming the corresponding monochlorinated product is 15/1. However, as you know from the discussion on selectivity in Sec. 9.10A, tertiary radicals are much more stable than primary ones, and are formed at a faster rate. We're told that the relative reactivity of the tertiary hydrogen under these conditions is 4.5 times greater than any of the primary hydrogens. So, if we multiply 4.5 by 1/15, we get 0.3.

The ratio then of the tertiary alkyl chloride product to the primary alkyl chloride products is 0.3:1. To convert that ratio into percentages, we add 0.3 + 1 = 1.3, and then divide 0.3 by 1.3 (and multiply by 100%) to get 23.1%, which is the percentage of the tertiary product formed.

The remaining 76.9% must be the two primary products. Since we can assume that there's no chemical preference for any one of the primary hydrogens over any of the others, we assume that there is a 9:6, or 3:2 ratio of these two products to one another, based on the number of hydrogens available that would lead to each product. Thus, there is 30.8% of one of the primary alkyl chlorides, and 46.1% of the other.

$$
\begin{array}{c}
\text{H}_3\text{C} \quad \text{CH}_3 \\
| \quad\quad | \\
\text{H}_3\text{C}-\text{C}-\text{C}-\text{H} \\
| \quad\quad | \\
\text{H}_3\text{C} \quad \text{CH}_3
\end{array}
+ \text{Cl}_2 \xrightarrow[-15\,°\text{C}]{\text{light,}}
\begin{array}{c}
\text{H}_3\text{C} \quad \text{CH}_3 \\
| \quad\quad | \\
\text{H}_3\text{C}-\text{C}-\text{C}-\text{Cl} \\
| \quad\quad | \\
\text{H}_3\text{C} \quad \text{CH}_3
\end{array}
+
\begin{array}{c}
\text{H}_3\text{C} \quad \text{CH}_2\text{Cl} \\
| \quad\quad | \\
\text{H}_3\text{C}-\text{C}-\text{C}-\text{H} \\
| \quad\quad | \\
\text{H}_3\text{C} \quad \text{CH}_3
\end{array}
+
\begin{array}{c}
\text{ClH}_2\text{C} \quad \text{CH}_3 \\
| \quad\quad | \\
\text{H}_3\text{C}-\text{C}-\text{C}-\text{H} \\
| \quad\quad | \\
\text{H}_3\text{C} \quad \text{CH}_3
\end{array}
$$

2,2,3-trimethylbutane (23.1%) (30.8%) (46.1%)

Solutions to Additional Problems

9.44   The first step in any problem that requires structures is to draw the structures:

$$
\text{CH}_3(\text{CH}_2)_4\text{CH}_2\text{Br} \qquad
\begin{array}{c}
\text{CH}_3 \\
| \\
\text{CH}_3\text{CH}_2\text{CCH}_2\text{CH}_3 \\
| \\
\text{Br}
\end{array}
\qquad
\begin{array}{c}
\text{CH}_3 \\
| \\
\text{BrCH}_2\text{CCH}_2\text{CH}_3 \\
| \\
\text{CH}_3
\end{array}
\qquad
\begin{array}{c}
\text{CH}_3 \\
| \\
\text{CH}_3\text{CHCHCH}_2\text{CH}_3 \\
| \\
\text{Br}
\end{array}
\qquad
\begin{array}{c}
\text{Br} \\
| \\
\text{CH}_3\text{CHCHCH}_2\text{CH}_3 \\
| \\
\text{CH}_3
\end{array}
$$

1-bromohexane (1)   **3-bromo-3-methyl-pentane** (2)   **1-bromo-2,2-dimethyl-butane** (3)   **3-bromo-2-methyl-pentane** (4)   **2-bromo-3-methyl-pentane** (5)

(a)   Compounds (4) and (5) can exist as enantiomers.

(b)   Compound (5) can exist as diastereomers because it has two asymmetric carbons.

(c)   Compound (1) gives the fastest S_N2 reaction with sodium methoxide because it is the only primary alkyl halide with no alkyl branching—that is, only one substituent—at the β-carbon.

(d)   Compound (3) is least reactive to sodium methoxide in methanol because it cannot undergo a β-elimination—it has no β-hydrogens—and the three β-substituents make it virtually unreactive in the S_N2 reaction, much like neopentyl bromide.

(e)   Only compound (1) can give only one alkene in the E2 reaction with potassium *tert*-butoxide.

(f)   Compound (2) will give an E2 but no S_N2 reaction with sodium methoxide in methanol.

(g)   Compounds (4) and (5) undergo the S_N1 reaction with rearrangement. If compound (3) is forced to react, it will also undergo the S_N1 reaction with rearrangement, but under ordinary conditions it will not react.

(h)   Compound (2) will give the fastest S_N1 reaction because it is the only tertiary alkyl halide.

9.45   The products of the reactions of isopentyl bromide, $(\text{CH}_3)_2\text{CHCH}_2\text{CH}_2\text{Br}$: (When the solvent is a by-product it is not shown.)

(a)
$(\text{CH}_3)_2\text{CHCH}_2\text{CH}_2\text{I}$

$+\ \text{K}^+\ \text{Br}^-$

(b)
$(\text{CH}_3)_2\text{CHCH}_2\text{CH}_2\text{OH}\ +\ (\text{CH}_3)_2\text{CHCH}_2\text{CH}_2\text{OC}_2\text{H}_5$

$+\ (\text{CH}_3)_2\text{CHCH}=\text{CH}_2\ +\ \text{K}^+\ \text{Br}^-$
(small amount)

(c)
$(\text{CH}_3)_2\text{CHCH}=\text{CH}_2$

$+\ \text{K}^+\ \text{Br}^-$

(d)

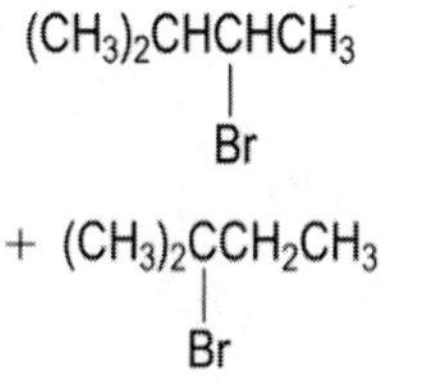

$(CH_3)_2CHCHCH_3$
          |
          Br

+ $(CH_3)_2CCH_2CH_3$
            |
            Br

(from rearrangement)

(e)

$(CH_3)_2CHCH_2CH_2F$

+ Cs^+ Br^-

(f)

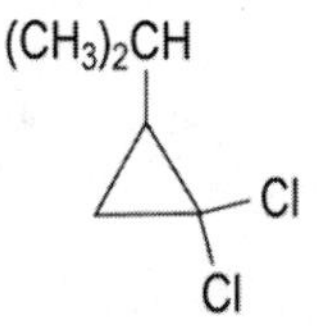

$(CH_3)_2CH$

 Cl
 Cl

+ $(CH_3)_3COH$
+ K^+ Cl^-

(g)

$(CH_3)_2CH$

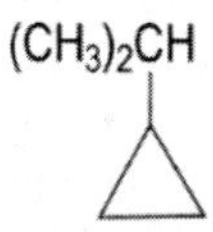

+ ZnI_2

(h)

$(CH_3)_2CHCH_2CH_3$

+ Li^+ Br^- + Li^+ $^-OC_2H_5$

(i)

$(CH_3)_2CHCH_2CH_2OCH_3$

+ $(CH_3)_2CHCH$=CH_2 + Na^+ Br^-
(small amount)

(j)

$(CH_3)_2CHCH_2CH_2D$

+ Mg^{2+} + ^-OD + Br^-

9.46   The products of the reactions of 2-bromo-2-methylhexane:

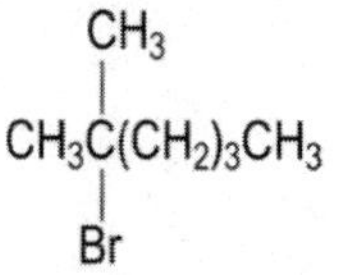

CH_3
  |
$CH_3C(CH_2)_3CH_3$
  |
  Br

2-bromo-2-methylhexane

(a)

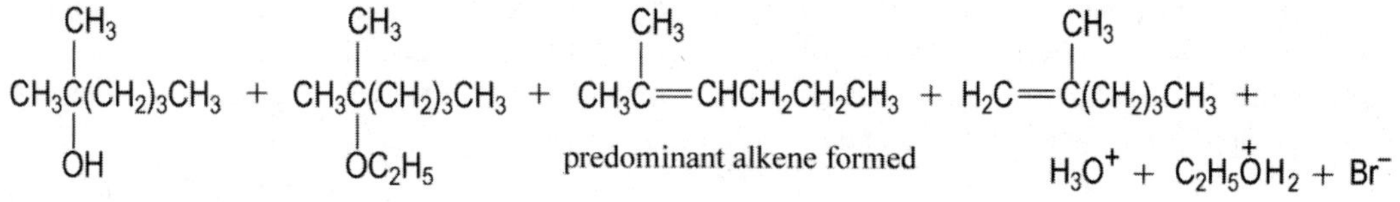

CH_3 CH_3 CH_3

$CH_3C(CH_2)_3CH_3$ + $CH_3C(CH_2)_3CH_3$ + CH_3C=$CHCH_2CH_2CH_3$ + H_2C=$C(CH_2)_3CH_3$ +
      | | predominant alkene formed
      OH   OC_2H_5

H_3O^+ + $C_2H_5\overset{+}{O}H_2$ + Br^-

(b)

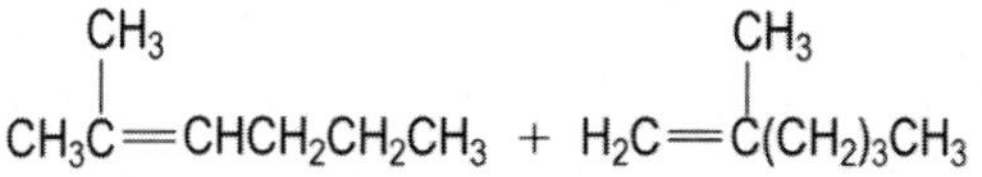

CH_3 CH_3
  | |
CH_3C=$CHCH_2CH_2CH_3$ + H_2C=$C(CH_2)_3CH_3$

+ Na^+ Br^-

(c)

CH_3 CH_3
  | |
$CH_3C(CH_2)_3CH_3$ + $CH_3C(CH_2)_3CH_3$ + K^+ Br^-
  | |
  OH   I

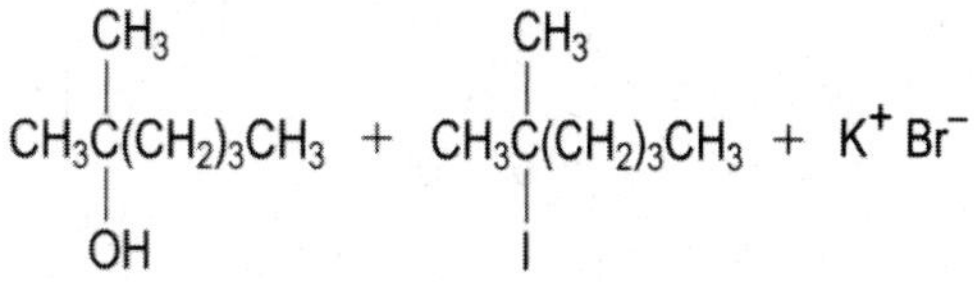

+ the alkenes in part (b) + H_3O^+ Br^-

(d)

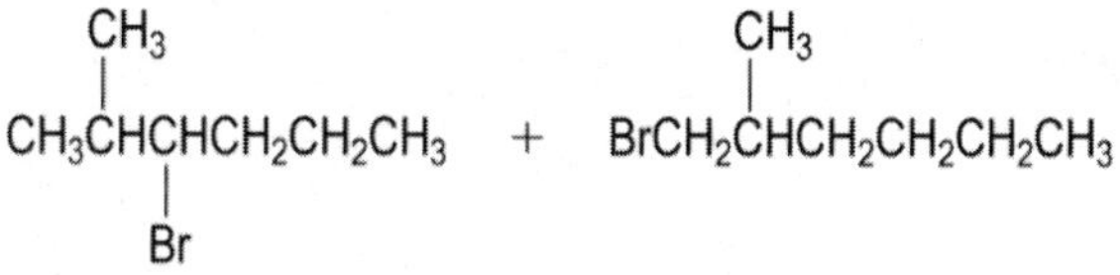

CH_3 CH_3
  | |
$CH_3CHCHCH_2CH_2CH_3$ + $BrCH_2CHCH_2CH_2CH_2CH_3$
      |
      Br

(e)

CH_3
  |
$CH_3C(CH_2)_3CH_3$
  |
  OH

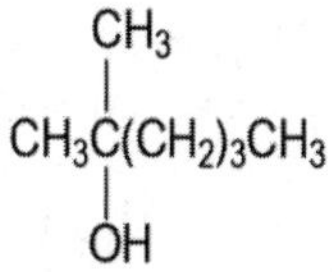

+ the by-products
of oxymercuration–reduction
shown in Eq. 5.22, text p. 191

(f)

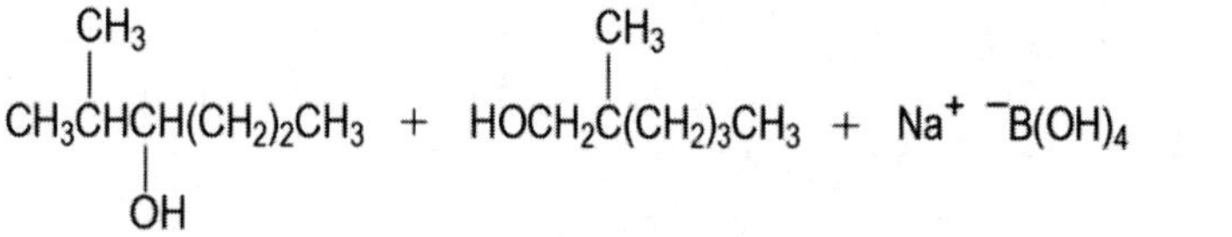

CH_3 CH_3
  | |
$CH_3CHCH(CH_2)_2CH_3$ + $HOCH_2C(CH_2)_3CH_3$ + Na^+ $^-B(OH)_4$
      |
      OH

9.47   (a)   The gas formed is ethane, CH_3CH_3, along with the by-product $HO^-\ ^+MgBr$.

(b)   The gas formed is propane, $CH_3CH_2CH_3$, along with the by-product $(CH_3)_2CH—O^-\ ^+MgBr$

9.48   (a)   The compound must be a saturated alkane, because its formula is C_nH_{2n+2}. So, trial and error works here, start drawing different compounds that meet the criteria. The only one that gives only two achiral monochlorinated products is 2,3-dimethylbutane.

2,3-dimethylbutane

(b)   The four compounds must be stereoisomers, and all must be chiral. Additionally, they must contain a ring because they possess one degree of unsaturation but the problem states that they do not contain a double bond. Finally, they must contain an acidic proton, one that can be removed by the Grignard reagent to yield ethane, a gas (see problem 9.47 (a)). Here, use trial and error to construct structures that meet all of the above criteria.

9.49   The order of increasing S_N2 reaction with KI in acetone is $A < B < C < E < D$. Tertiary alkyl halides such as A react slowly, if at all, in S_N2 reactions; secondary alkyl halides such as B react more rapidly; primary alkyl halides with two β-substituents (C) react even more rapidly; and unbranched primary alkyl halides (D and E) react most rapidly. Of the two primary alkyl halides shown, the one with the better leaving group (D, with the bromide leaving group) reacts more rapidly.

9.50   In all cases, the increase in polarizability is due to an increase in how easily distorted the electron cloud of an atom is going down a column on the periodic table (see Sec. 9.4E, text p. 403).

(a)   O, S, Se

(b)   fluoroform, chloroform, iodoform

(c) F^-, Cl^-, Br^-, I^-

9.51   The first thing to do is to draw out the structures.

The order of increasing S_N2 reaction rates is $C < B < E < D < A$. Alkyl halides with three β-substituents (C) are virtually unreactive in S_N2 reactions. Secondary alkyl halides with no β- substituents (B) react more slowly than primary alkyl halides with two β- substituents (E), and the latter react more slowly than unbranched primary alkyl halides (D). Methyl halides (A) react most rapidly.

9.52   (a)   (b)   (c)

$Na^+\ \overset{..}{\underset{..}{O}}CH_2CH_2CH_3$ $Na^+\ ^-:CN$ $Na^+\ \overset{..}{\underset{..}{O}}CH_2CH_2CH_2OCH_3$

(d)

(e)

Na$^+$ $\ddot{\underset{\cdot\cdot}{S}}$—[cyclopentane ring with CH$_3$]

(CH$_3$)$_3$N:

9.53   (a)

CH$_2$CH$_2$CH$_3$

CH$_3$CH$_2$O—C$^{\cdots}$CH$_3$

H

(S)-2-ethoxypentane

(b)

CH$_2$CH$_2$CH$_3$

CH$_3$CH$_2$O—C$^{\cdots}$H

D

9.54   (a)

CH$_3$CH=C(CH$_3$)$_2$ + CH$_3$CH$_2$C=CH$_2$

CH$_3$

(b)

[cyclohexene structure]

(c)

[cyclohexyl]—CH=C(CH$_3$)$_2$ + [cyclohexyl]—CH$_2$C=CH$_2$

CH$_3$

(d)

CH$_3$CH$_2$C=CH$_2$

CH$_3$

9.55   Dimma has attempted the S$_N$2 reaction of a *tertiary* alkyl halide with a strong base. As Dimma would have realized, had he consulted a superior organic chemistry text such as yours, that these are precisely the conditions that promote the E2 reaction. Consequently, Dimma formed the alkene isobutylene (2-methylpropene, (CH$_3$)$_2$C=CH$_2$), which bubbled merrily out of the reaction flask. Ima, on the other hand, allowed a strong base to react with methyl iodide, which cannot undergo elimination; the only alternative is the desired S$_N$2 reaction.

9.56   Allow the appropriate alkoxide, (CH$_3$)$_3$CCH$_2$—O$^-$ Na$^+$, to react with ethyl iodide, CH$_3$CH$_2$—I. Using sodium ethoxide with neopentyl bromide, (CH$_3$)$_3$CCH$_2$—Br, won't work; why?

9.57   The different boiling points indicate that the three alkyl halides are either constitutional isomers or diastereomers. The outcome of the Grignard protonolysis shows that all of the alkyl halides have the same carbon skeleton, that of 2,4-dimethylpentane. The protonolysis in D$_2$O confirms the fact that the bromines are bound at different places on the carbon skeleton. The only three possibilities for the alkyl halides are

CH$_3$ CH$_3$

CH$_3$CCH$_2$CHCH$_3$

Br

A

CH$_3$ CH$_3$

BrCH$_2$CHCH$_2$CHCH$_3$

B

CH$_3$ CH$_3$

CH$_3$CHCHCHCH$_3$

Br

C

Compound *B* is the chiral alkyl halide, and compounds *A* and *C* are the other two. (The absolute configuration of *B* (that is, whether it is *R* or *S*) is not determined by the data.)

The protonolysis products in H$_2$O and D$_2$O are

In H$_2$O:

CH$_3$ CH$_3$

CH$_3$CHCH$_2$CHCH$_3$

from *A, B,* and *C*

In D$_2$O:

CH$_3$ CH$_3$

CH$_3$CCH$_2$CHCH$_3$

D

from *A*

CH$_3$ CH$_3$

DCH$_2$CHCH$_2$CHCH$_3$

from *B*

CH$_3$ CH$_3$

CH$_3$CHCHCHCH$_3$

D

from *C*

9.58   2,3-Dimethylbutane contains two different types of hydrogens, six primary and two tertiary. The bromine radical will selectively remove tertiary protons over primary protons (Sec. 9.10). So compound *A* will have one bromine incorporated at a tertiary carbon. Compound *B* still has one more tertiary proton, which is more reactive than any of the primary protons, so the second bromine gets incorporated at the other tertiary carbon.

2,3-dimethylbutane   *A*   *B*

9.59   In base, some of the chlorines are eliminated as chloride ion in E2 reactions. One such reaction is as follows; can you find another? (S_N2 reactions probably do not occur; why?)

9.60   (a)   The presence of a strong base favors an E2 reaction. Two different protons could be lost, along with a bromide ion. Proton *(a)* is lost by an anti-elimination. To determine the stereochemistry, draw the structure with this proton and the bromine in an anti relationship.

(2S,3R)-2-bromo-3-methylpentane

As the equation above shows, the loss of hydrogen *(b)* gives a different alkene. Because this alkene has no stereoisomers, the stereochemistry of the elimination is not relevant. Methanol and Na^+ Br^- are formed as by-products in both eliminations.

(b)   The reaction conditions favor a solvolysis reaction, which occurs by an S_N1–E1 mechanism accompanied by rearrangement.

A	*B*	*C*	*D*
substitution product from the first-formed carbocation	substitution product from the rearranged carbocation	E1 product from the first-formed carbocation (small amount)	E1 product from both carbocations (*E* and *Z* isomers)

$+ CH_3\overset{+}{O}H_2$ Br^-

The S_N1 product *A* is derived from the Lewis acid–base association reaction of solvent with the first-formed carbocation. It should be partly racemic, although the exact stereochemical outcome is difficult to predict exactly. It should have largely inverted configuration at carbon-2, because the carbocation will be a backside-solvated ion pair; this carbocation will react faster with the solvent molecule than the bromide ion is replaced by a second solvent molecule. (See Fig. 9.13, text p. 426.). The stereochemistry of carbon-3 is unaffected. The substitution product *B* is derived from the rearranged tertiary carbocation. Compounds *C* and *D* (*D* is a mixture of *E* and *Z* isomers) are E1 products. The stereochemistry at carbon-3 of alkene *C* is the same as in the starting alkyl halide. Alkene *D* should be the major alkene product because it has the larger number of substituents on the double bond.

9.61   (a)   Redraw the structures in a sawhorse or line-and-wedge projection in which the hydrogen or deuterium that is eliminated is anti to the bromine. The stereochemistry of the alkene follows by looking at the relationship of the substituents—gauche or anti. Anti substituents are trans in the product alkene.

   (b)   Proceed the same way as in part (a). Because the starting materials are diastereomers, the products must also be diastereomers if the elimination is anti.

9.62   In the transition state of the S$_N$2 reaction, the carbon at which substitution occurs is sp^2-hybridized; consequently, the ideal bond angles at this carbon are 120°. (See Fig. 9.2 on text p. 395.) This requirement for a 120° bond angle means that as cyclopropyl iodide is converted into the S$_N$2 transition state, the bond angle within the cyclopropane ring should increase. The increasing angle causes additional angle strain in the cyclopropane ring that raises the energy of the transition state. Consequently, S$_N$2 reactions of cyclopropyl halides are unusually slow.

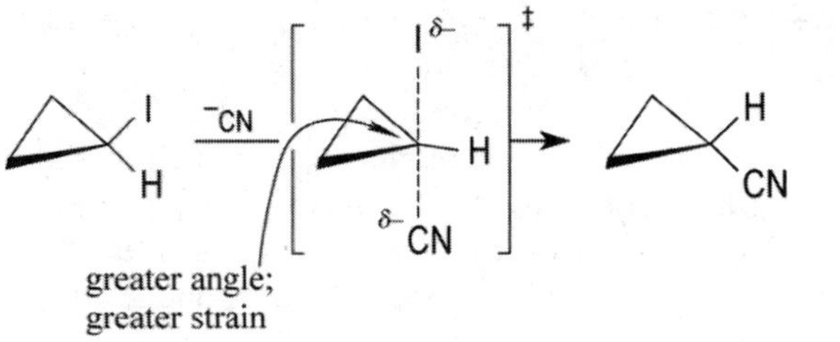

greater angle;
greater strain

Such rate-retarding angle strain is present in neither compound (b), isopropyl iodide, nor its S_N2 transition state. Consequently, the S_N2 reactions of isopropyl iodide are faster than those of cyclopropyl iodide.

9.63   If this compound were to undergo solvolysis, it would have to form carbocation A:

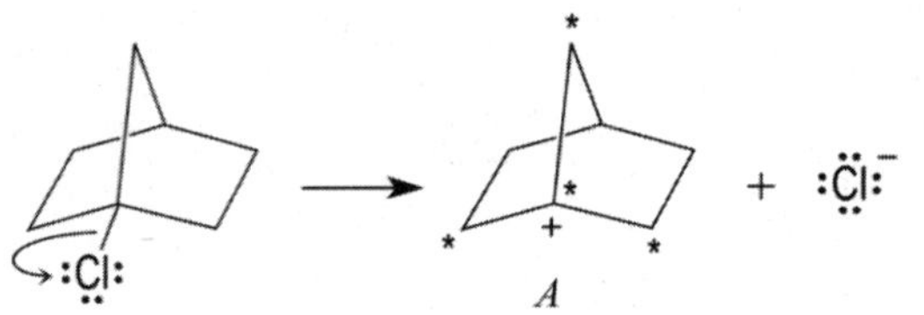

As a model will verify, the four asterisked carbons cannot become coplanar because of the constraints of the bicyclic ring system. Yet sp^2 hybridization requires trigonal-planar geometry. Remember: *hybridization and geometry are connected.* If a molecule can't achieve the geometry for a given hybridization, then that hybridization will not occur. Because sp^2 hybridization is the lowest-energy hybridization for a carbocation, the inability to achieve this hybridization raises the energy of the carbocation.

But that's not all. Solvation of the carbocation can only occur from one side, because the opposite side of the electron-deficient carbon is blocked by part of the ring system. The absence of effective solvation, then, also raises the energy of this carbocation. Any solvolysis reaction involving such an unstable carbocation is slow.

9.64   (a)   Potassium fluoride, an ionic compound, is not soluble in hydrocarbons such as benzene. Consequently, it will not react in such solvents because its dissolved concentration is essentially zero. When the crown ether is added, the crown ether forms a benzene-soluble complex with the potassium ion, which, by electrical neutrality, includes an accompanying "naked" (that is, essentially unsolvated) fluoride ion. This soluble but unsolvated fluoride ion is a very good nucleophile, and reacts with benzyl bromide to yield benzyl fluoride, $PhCH_2F$, plus "naked" bromide ion, which then exchanges with insoluble potassium fluoride to give more dissolved fluoride and solid KBr. The newly dissolved fluoride can react with more benzyl bromide. The cycle is repeated until one of the reactants is exhausted. To summarize:

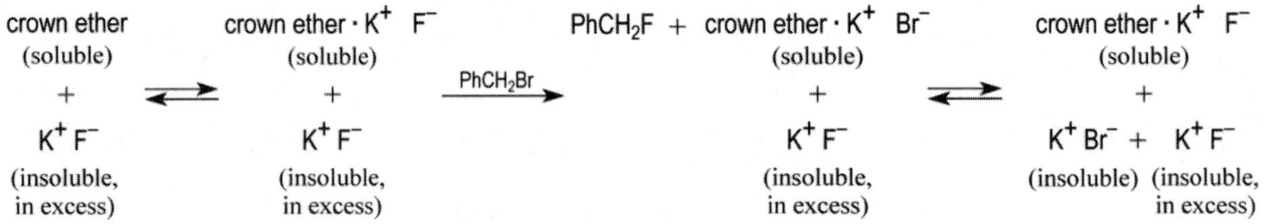

(b)   The scheme in part (a) shows that complexation of the cation by the crown ether is essential for the reaction to occur, because this is the only way to obtain the dissolved anion. Because [18]-crown-6 does not bind the smaller lithium cation—it is selective for the larger potassium cation—lithium fluoride is not solubilized by the crown ether, and is therefore unreactive whether the crown ether is present or not.

9.65   (a)   The S_N1 solvolysis products result from the Lewis acid–base association reaction of the respective solvent molecules on the carbocation intermediate, followed by loss of a proton.

(b) Both solvents are protic and donor solvents, and therefore differ very little in their ability to solvate ions by hydrogen bonding or donor interactions. The major difference between the two solvents is their dielectric constants. The reaction is faster in formic acid because its higher dielectric constant promotes the separation of the chloride-ion leaving group from the carbocation. (This effect is discussed in Sec. 9.6C on text p. 422.) Because this is the rate-limiting step in the S_N1 reaction, lowering the standard free energy of activation of this step increases the rate.

9.66 (a) In a hydrogen-bonding solvent such as ethanol, reactions of alkyl halides with strong-base nucleophiles are retarded by hydrogen bonding of the solvent to the nucleophile. Hence, within a column of the periodic table, the element effect on nucleophilicity is that weaker bases are better nucleophiles. Hence, the thiolate ion (which is a weaker base than the alkoxide ion) will out-compete the alkoxide ion for a limited amount of alkyl halide. As a result, the product of the reaction is ethyl methyl sulfide, $CH_3CH_2S—CH_3$.

(b) In polar aprotic solvents, which cannot form hydrogen bonds to strong bases, stronger bases within a column of the periodic table are the better nucleophiles. In this case, the alkoxide ion out-competes the thiolate for the alkyl halide, and the product is thus ethyl methyl ether, $CH_3CH_2O—CH_3$.

9.67 (a) The thiosulfate dianion contains two types of nucleophilic atoms: the anionic oxygen and the anionic sulfur. Hence, alkylation could occur at the oxygen or the sulfur:

A
product of *O*-alkylation

B
product of *S*-alkylation

(b) In a hydrogen-bonding solvent such as methanol, the more weakly basic atom is the better nucleophile. (The principle is the same as in the solution to Problem 9.66.) Hence, product *B* is the major one observed. (What solvent change could you make to obtain more of product *A*?)

9.68 (a) The principle to apply is that the right side of the equilibrium is favored by the solvent that better separates and solvates ions. Because ethanol is the more polar solvent—it has the higher dielectric constant—it better separates ions. Because it is a protic, donor solvent, it better solvates ions; consequently, the equilibrium lies farther to the right in ethanol. Furthermore, ethanol's ability to donate hydrogen bonds allows it to provide better solvation for the bromide ion than does diethyl ether.

(b) In this case, the dielectric constants of the two solvents are the same. Consequently, the ability of the solvent to solvate the ions determines the relative position of the equilibrium. Because dimethylacetamide is *aprotic,* it does not solvate anions as well as the aqueous methanol solvent, which has a hydrogen-bonding solvation mechanism available. Consequently, the equilibrium lies farther to the right in aqueous methanol.

9.69   As Eq. 9.64 (text p. 431) shows, ethers solubilize Grignard reagents by forming Lewis acid–base complexes with them. Because a tertiary amine is also a Lewis base, it can solubilize Grignard reagents by the same type of interaction:

$$R_3N\!:\overset{\delta+}{\longrightarrow} \underset{\underset{Br}{|}}{\overset{\overset{R}{|}}{\underset{\delta-}{Mg}}} \overset{\delta+}{\longleftarrow} :NR_3$$

9.70   The reaction between methyl iodide and sodium ethoxide is an S_N2 process that has the following rate law:

$$\text{rate} = k[CH_3I][CH_3CH_2O^-]$$

This means that the rate of the reaction depends on the concentration of the nucleophile. In the first case, the nucleophile concentration changes during the reaction from 0.1 M to 0 M; in the second case, the nucleophile concentration changes from 0.5 M to 0.4 M. In the second case, the reaction is faster because the concentration of the nucleophile is higher at all times during the reaction. However, in either case, there is sufficient nucleophile present to react completely with the alkyl halide. If we wait long enough, the yield of the reaction will be the same in either case.

9.71   In methanol, the iodide ion is an excellent nucleophile but, because of its weak basicity, also a good leaving group. Hence, the iodide ion reacts rapidly with methyl bromide to give methyl iodide. The methyl iodide then reacts more slowly with the methanol solvent in a slow S_N2 solvolysis reaction.

9.72   Because it is a tertiary alkyl halide without β-hydrogens, trityl chloride can react only by the S_N1 mechanism. A characteristic feature of this mechanism is that the first step is rate-limiting, whereas the Lewis acid–base association reaction of the nucleophile(s) with the carbocation intermediate occurs in the second, product-determining, step. Hence, the rate of reaction of trityl chloride is *independent of the nucleophile concentration,* and therefore independent of the nucleophile that reacts with the carbocation.

The product-determining steps for the various reactions are different. The principle to apply is that the better nucleophile reacts more rapidly with the carbocation intermediate; hence, the major product is derived from the better nucleophile.

    Thus, in reaction (1), the only nucleophile available is water. The reaction of water with the carbocation and loss of a proton to solvent gives trityl alcohol, Ph_3C—OH, as the solvolysis product. (The solvent acetone can also in principle react with the carbocation, but the reaction product is unstable under the reaction conditions for reasons that we won't discuss here.) In reaction (2), both water and azide ion are in competition for the trityl cation. The basicities of water and azide are determined by the pK_a values of their respective conjugate acids. The conjugate acid of water, H_3O^+, has a pK_a value of −1.7. The conjugate acid of azide ion,

HN_3, has a pK_a value of 4.7. Consequently, azide is the more basic nucleophile, and is evidently the better nucleophile. When, in reaction (2), sodium azide is added to the reaction mixture, trityl chloride ionizes at the same rate, but the carbocation intermediate reacts preferentially with azide ion.

When hydroxide ion is added in reaction (3), the rate of ionization of trityl chloride again remains unchanged; however, because both hydroxide ion and azide are present, both nucleophiles react with the carbocation to give a mixture of trityl alcohol and trityl azide.

9.73　(a)　If retention is the stereochemical scenario, each act of substitution has no affect on the optical activity, and $k°/k_S = 0$. If inversion is the stereochemical scenario, each act of substitution inverts the stereochemical configuration of one alkyl halide molecule, which gives it the opposite sign but the same magnitude of optical rotation. In this case, the optical activities of *two* molecules cancel; so, two molecules' worth of the original optical activity are lost. Therefore, $k°/k_S = 2$. This situation can be envisioned with the following "cartoon:"

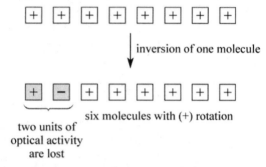

If equal amounts of retention and inversion occur, then for every event that results in the loss of two units of optical activity, we have another event that results in the loss of no optical activity; on average, $k°/k_S = 1$.

(b)　The results show that $k°/k_S = 2$, thus establishing inversion as the stereochemical outcome.

9.74　The fact that protonolysis reactions of the corresponding Grignard reagents give the same hydrocarbon indicates that the two compounds have the same carbon skeleton. The conditions of ethanol and no added base are S_N1 conditions. Since compound *A* reacts rapidly to give a solution containing bromide ion, it must be an alkyl halide that readily undergoes an S_N1 reaction, and therefore it is probably a tertiary alkyl bromide. Because the two alkyl halides give the same ether, the product from compound *B* must be formed in a rearrangement. The only tertiary alkyl halide with the formula $C_5H_{11}Br$ is 2-bromo-2-methylbutane, and this is therefore compound *A:*

$$
\begin{array}{ccc}
\overset{\overset{\textstyle CH_3}{|}}{\underset{\underset{\textstyle Br}{|}}{CH_3CCH_2CH_3}} + CH_3CH_2OH & \longrightarrow & \overset{\overset{\textstyle CH_3}{|}}{\underset{\underset{\textstyle OCH_2CH_3}{|}}{CH_3CCH_2CH_3}} + HBr
\end{array}
$$

2-bromo-2-methylbutane　　　　　　　**2-ethoxy-2-methylbutane**
(A)　　　　　　　　　　　　　　　　　　　*(C)*

Two possible alkyl halides with the same carbon skeleton as *A* could rearrange in respective S_N1 reactions to give the same carbocation, and hence the same ether product, as *A*; these are labeled *B1* and *B2* below. (An S_N1 reaction of *B1* would be very slow, if it occurred at all.)

$$
\overset{\overset{\textstyle CH_3}{|}}{BrCH_2CHCH_2CH_3} \qquad\qquad \overset{\overset{\textstyle CH_3}{|}}{\underset{\underset{\textstyle Br}{|}}{CH_3CHCHCH_3}}
$$

B1　　　　　　　　　　　　　　　　*B2*

However, only *B2* can react in an E2 reaction with sodium ethoxide to give an alkene that furnishes acetone as one of its ozonolysis products:

Consequently, compound *B2* is compound *B*, 2-bromo-3-methylbutane.

9.75 Remember that the key to working problems of this type is to find a known structure and work from it. (See point 3 in Study Guide Link 4.3, p. 73 of this manual.) Compound *B* is an alkene, because it decolorizes Br_2 in an inert solvent. The structure of the ozonolysis product *C* leads to the identity of *B*.

The molecular formula of *A*, $C_8H_{13}Br$, is equivalent to that of the alkene *B* plus the elements of HBr. Because the structure of *C* shows that the carbon skeleton must be bicyclic, there can be no additional unsaturation in *A*, a point confirmed by the fact that it does not add bromine. Evidently, compound *A* is an alkyl halide that undergoes a reaction with a strong base to give the alkene *B;* this reaction is then an E2 reaction. The alkyl halide *A*, and its E2 reaction, are as follows:

9.76 Compound *A* is an alkyl halide with two degrees of unsaturation; because one unsaturation is accounted for by the cyclohexane ring found in subsequent products, the other must be a double bond. Subsequent elimination reactions of *A* yield two dienes *B* and *C*. The catalytic hydrogenation of these two compounds establishes that compounds *A–C* and 1-isopropyl-4-methylcyclohexane have the same carbon skeleton. Because ozonolysis leaves the cyclohexane ring intact, the double bond must be external to the ring; and the identity of product *F* shows that both double bonds are in fact attached to the ring. There is only one possibility for compound *B :*

compound *B*

The following are possibilities for *A*:

 If you find an error in this manual, please visit http://people.pharmacy.purdue.edu/~loudonm/teaching/

Any other possibility for *A* is ruled out both by the structure of *B* and by the fact that compound *A* is achiral. Compound *A1* is ruled out by the fact that it can only give one diene in the E2 reaction; yet compound *A* gives two dienes, compounds *B* and *C*. That compound *A* undergoes S_N1 solvolysis rapidly also rules out compound *A1*, because it is a primary alkyl halide that would not react under S_N1 conditions. Because compound *A* undergoes ozonolysis to give acetone—and because structure *A1* was ruled out—*only structure A2 fits all the data.* The structure of alkene *B* is given above; the structure of alkene *C*, the other E2 reaction product of compound *A* = *A2*, is as follows:

compound *C*

Now that we know compound *A*, we are in a position to derive the structures of compounds *D* and *E*. These are the bromine-addition products of compound *A*. The two products are the two possible diastereomers:

D and *E*

9.77   Compound *A* has no conformation in which the *β*-hydrogens and the bromine are anti, whereas in compound *B*, the *β*-hydrogens are anti to the bromine in the more stable conformation (shown). For this reason, compound *B* should more readily undergo the E2 reaction.

Compound *A* has no Hs anti to the Br in either conformation

Compound *B* has two Hs anti to the Br in the more stable conformation

9.78   The important keys to solving this problem are (1) that the E2 reaction is fastest when the hydrogen and chlorine are anti; and (2) the substituted cyclohexanes can undergo the chair interconversion rapidly. In order for menthyl chloride to undergo anti-elimination, it must undergo the chair interconversion so that the chlorine is axial:

menthyl chloride
(this conformation cannot undergo *anti*-elimination)

anti
to the Cl

2-menthene

The only alkene that can be formed by anti-elimination is 2-menthene. In contrast, neomenthyl chloride has two *β*-hydrogens, H^a and H^b, that can be lost in anti-eliminations along with the chlorine to give 2-menthene and 3-menthene, respectively:

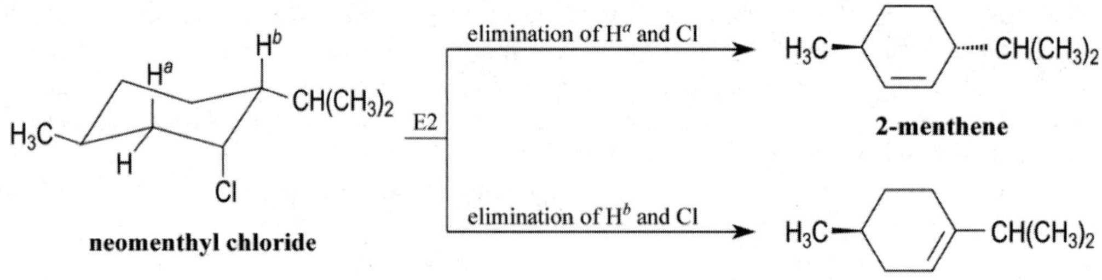

2-menthene

3-menthene

neomenthyl chloride

More 3-menthene is formed because it is the alkene with the greater number of alkyl substituents at its double bond.

9.79   (a)   Because the alkene product is cis, the transition state for the elimination in reaction (1) must have the two methyl groups on the same side of the molecule—that is, the two carbon–methyl bonds must be gauche. Drawing the molecule in a conformation with gauche methyl groups shows that the bromines are eliminated from opposite sides of the molecule; that is, the elimination is anti. In this reaction, I⁻ is a nucleophile, one Br is an electrophile, and the other is a leaving group.

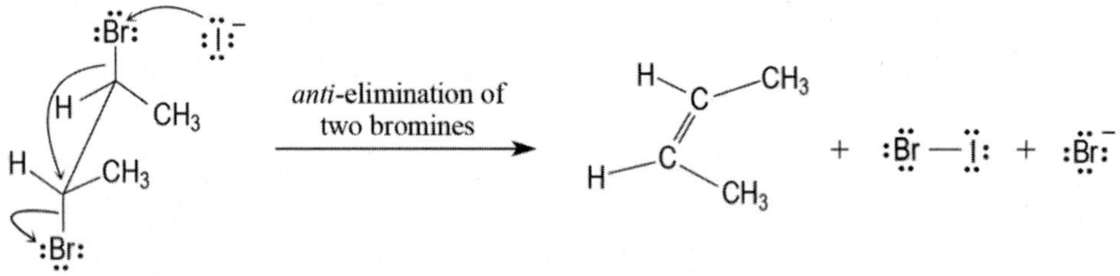

anti-elimination of two bromines

In reaction (2), because the phenyl groups are on opposite sides of the product alkene, the transition state for elimination must also have the phenyl groups on opposite sides. Drawing a conformation that meets this condition reveals that the hydrogen and the acetoxy group are eliminated from the same side of the molecule; that is, the elimination is syn.

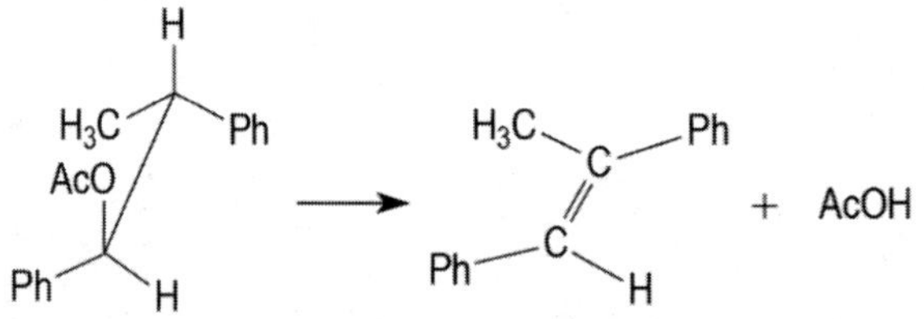

(b)   The first-order kinetics suggests a unimolecular process—that is, the absence of any other molecule in the transition state. The following mechanism would accommodate this observation as well as the stereochemistry in part (a).

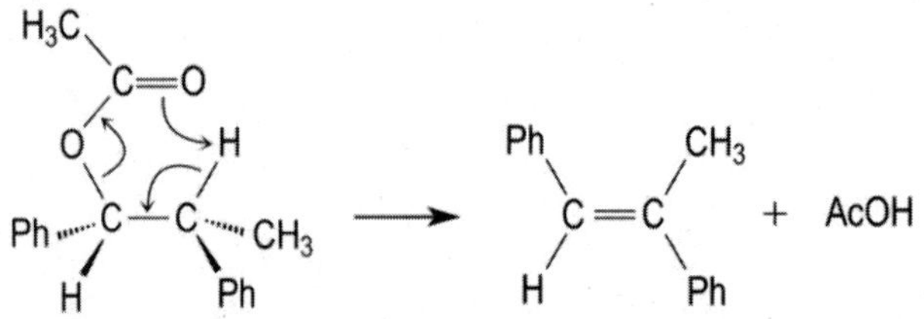

An intramolecular anti-elimination involving a similar proton transfer is stereochemically impossible.

9.80   (a)   The key to the solution is to assume that E2 reactions are faster when they can occur with anti stereochemistry. In the first compound, two β-hydrogens can be eliminated with the bromine: the hydrogen at the ring junction, and a hydrogen of the methyl group. The hydrogen at the ring junction is eliminated because it gives the alkene with the greater number of alkyl substituents on the double bond.

anti to the Br

(b)   In the second compound, the hydrogen at the ring junction cannot be anti to the bromine; hence, the only anti hydrogen available for elimination is a hydrogen of the methyl group:

syn (gauche) to the Br

anti to the Br

9.81 (a)   If radicals rearranged similarly to carbocations, then we would expect primary radicals to rearrange to more stable tertiary radicals. We know that primary radicals exist as intermediates in this experiment because we see chlorine incorporated into the methyl groups (product *B*). However, since we don't see the deuterium incorporated into the methyl groups in the products in this experiment, we can deduce that the primary radical is not rearranging.

isobutyl radical   *tert*-butyl radical

(b)   In Eq. 9.88, *tert*-butyl chloride is formed in 36%, and isobutyl chloride is formed in 64%. The first step in any of these reactions is the abstraction of a proton (or deuterium). To form *tert*-butyl chloride, a hydrogen must be removed from the tertiary carbon. To form product *A* in Fig. P9.81, a deuterium must be removed from the tertiary carbon. Since the C–H bond is weaker than the C–D bond, we expect the former reaction to be faster, and the latter to be slower. Thus, product *A* would be formed in some amount less than the 36% we see for *tert*-butyl chloride, and product *B* would make up the difference, forming in greater than the 64% we see for isobutyl chloride.

(c)   Starting with isobutane, carry out a radical bromination to create *tert*-butyl bromide (Eq. 9.86, text p. 439). Then treat the product with magnesium metal in ether to create a Grignard reagent, which can then be reacted with D_2O, DCl, or other reagents containing an acidic deuterium to produce isobutane-2*d*.

isobutane   *tert*-butyl bromide   Grignard reagent   isobutane-2*d*

9.82   (a)   Initiation of the reaction by AIBN (see Eq. 5.50, text p. 206) suggests a free-radical chain mechanism. The first initiation step is shown in that equation. Let the free-radical product of that reaction be abbreviated as R•, and let the butyl group be abbreviated as Bu. The second initiation step is

$$R• \frown H \frown Sn(Bu)_3 \longrightarrow R—H + •Sn(Bu)_3$$

The propagation steps are as follows:

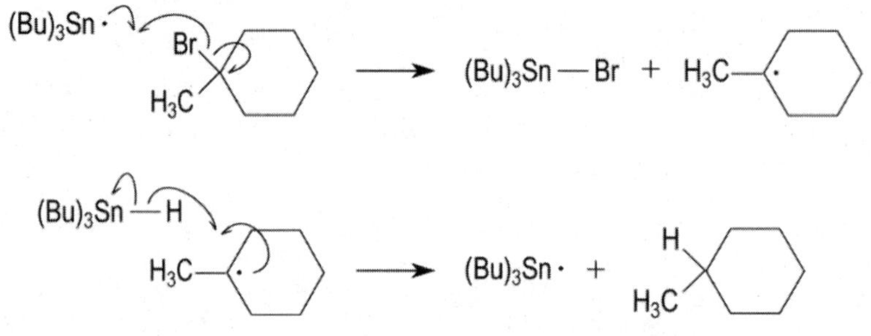

(b)   One sequence of reactions is to convert the alkyl halide into a Grignard reagent, which is then treated with water:

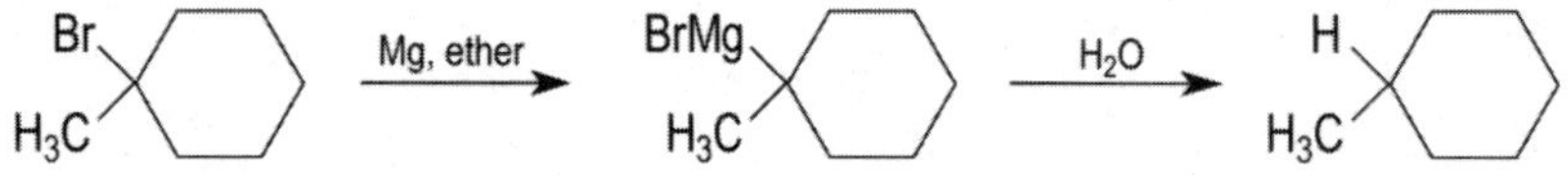

Another is to carry out an E2 reaction to obtain an alkene (or a mixture of alkenes), which is then subjected to catalytic hydrogenation to give methylcyclohexane.

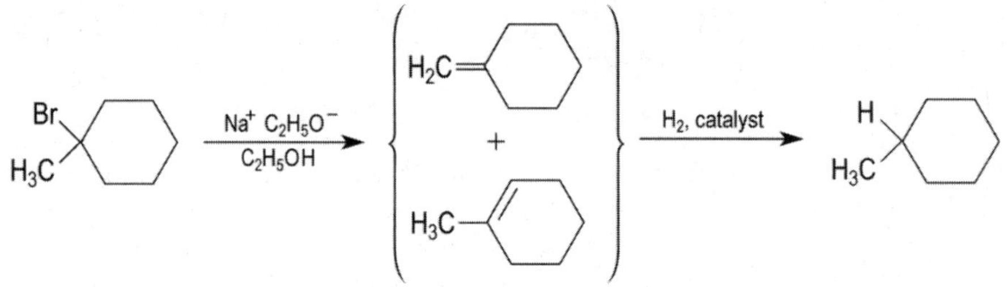

9.83   (a)   Product A results from anti-elimination by a conventional E2 reaction:

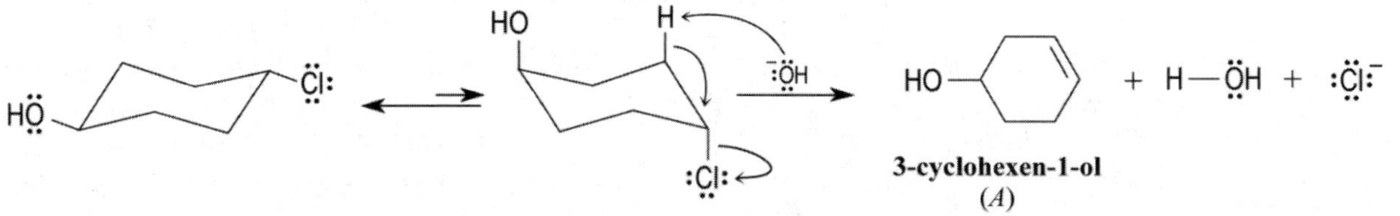

3-cyclohexen-1-ol
(A)

(The same sort of elimination reaction gives compound A from *cis*-4-chlorocyclohexanol.) Product B is formed by an intramolecular nucleophilic substitution reaction of the alkoxide that occurs in the twist-boat conformation (shown here for simplicity as a boat):

Product *C* results from an ordinary S_N2 reaction:

(b)   The intramolecular substitution that yields the bicyclic compound *B* requires, like all nucleophilic substitutions, approach of the nucleophile to the backside of the electrophilic carbon—the side opposite to the departing chloride. This can happen only in the boat form of the trans stereoisomer. Because a conformation that permits backside substitution of the intramolecular alkoxide is not available to the cis stereoisomer, it reacts instead by an ordinary S_N2 reaction with the external nucleophile ⁻OH.

Notice that the formation of product *B* instead of *cis*-1,4-cyclohexanediol from the trans stereoisomer of the starting material means that the *intramolecular* nucleophilic substitution reaction is much faster than the S_N2 reaction of the same compound with hydroxide. Intramolecular reactions are in many cases faster than analogous intermolecular reactions. This point is explored in more detail in Sec. 11.8 of the text.

9.84   The reaction of butylamine with 1-bromobutane is a typical S_N2 reaction. (Bu— = the butyl group = $CH_3CH_2CH_2CH_2$—):

$$Bu—\ddot{N}H_2 \quad :\ddot{Br}—Bu \longrightarrow Bu—\overset{+}{N}H_2—Bu \quad :\ddot{Br}:^-$$

This mechanism is consistent with the second-order rate law, because the rate law requires one molecule of amine and one molecule of alkyl halide in the transition state.

The second reaction is also a nucleophilic substitution reaction, but, because it is intramolecular (that is, the nucleophile and carbon at which it reacts are part of the same molecule), the reaction is first-order.

9.85   **(a)**   The reaction occurs largely with retention of configuration and only a small amount of inversion. This is inconsistent with stereochemical inversion observed with other S_N2 reactions, and this result caused a stir for that reason. (By the time this study was done, stereochemical inversion in the S_N2 reaction had become well established.)

   (b)   The initial formation of *trans-P* occurs by the usual S_N2 mechanism with inversion. The subsequent formation of *cis-P* occurs when iodide ion in solution displaces iodide, again with inversion, from *trans-*

P by a second S_N2 reaction with inversion. Starting with either *cis-* or *trans-P* gives the same equilibrium mixture in which *cis-P* is the major component of the equilibrium.

(c)   *Cis-P* is favored because both substituents can exist in equatorial arrangements. (See Sec. 7.5B and Problem 7.16 for a discussion of the puckering of cyclobutane rings.) As in substituted cyclohexanes, substituted cyclobutanes are destabilized by 1,3-diaxial interactions.

cis-P
both substituents are equatorial
(more stable)

trans-P
one substituent is axial
(less stable)

9.86   (a)   Bromine addition to alkenes is anti; consequently, the stereochemistry of compound *B* is as follows:

(racemate)

(b)   Draw the structure of compound *B* in a conformation in which the butyl (Bu) group and the Br that remains after the elimination are on opposite sides of the molecule, because this is the way they are in the alkene product. This shows that the trimethylsilyl group and the bromine are anti; consequently, the elimination shown is an anti-elimination.

(c)   Diastereomeric starting materials must give diastereomeric products if the stereochemistry of the two reactions remains the same. The *E* stereoisomer of compound *A* therefore would give the diastereomer of compound *B* (as the racemate), and the subsequent elimination would give the *Z* stereoisomer of compound *C*. This can be shown by the analysis used in parts (a) and (b) with the positions of the trimethylsilyl group and the hydrogen interchanged.

9.87   (a)   Sodium hydroxide acts as a base to form trichloromethyl anion, which then forms dichlorocarbene. (See Eq. 9.72a–b on text p. 433.) Dichlorocarbene reacts with iodide ion to form a new anion, which is protonated by water. The reaction is not observed in the absence of NaOH because a base is required to form trichloromethyl anion.

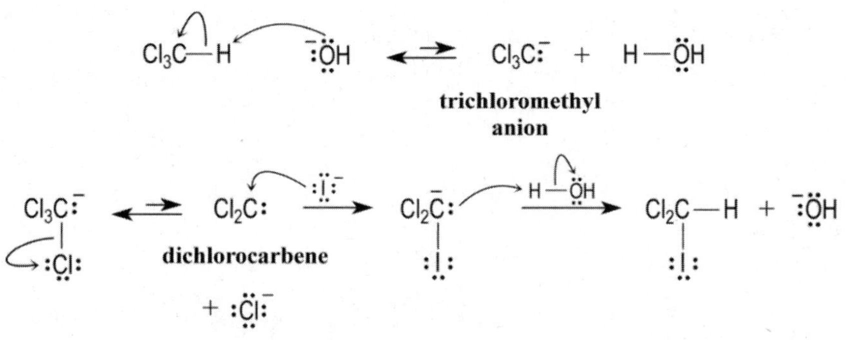

$$Cl_3C\!-\!H \quad \ddot{:}\!\ddot{O}H \quad \rightleftarrows \quad Cl_3C\!:^- \quad + \quad H\!-\!\ddot{O}H$$

trichloromethyl
anion

$$Cl_3C\!:^- \quad \rightleftarrows \quad Cl_2C\!: \quad \xrightarrow{:\ddot{I}\!:^-} \quad Cl_2\bar{C}\!: \quad \xrightarrow{H-\ddot{O}H} \quad Cl_2C\!-\!H \quad + \quad \ddot{:}\!\ddot{O}H$$

dichlorocarbene

$$+ \quad :\ddot{C}l\!:^-$$

(b)   Butyllithium acts as a base to form an anion that decomposes to phenylcarbene, which then reacts with the alkene to form the cyclopropane.

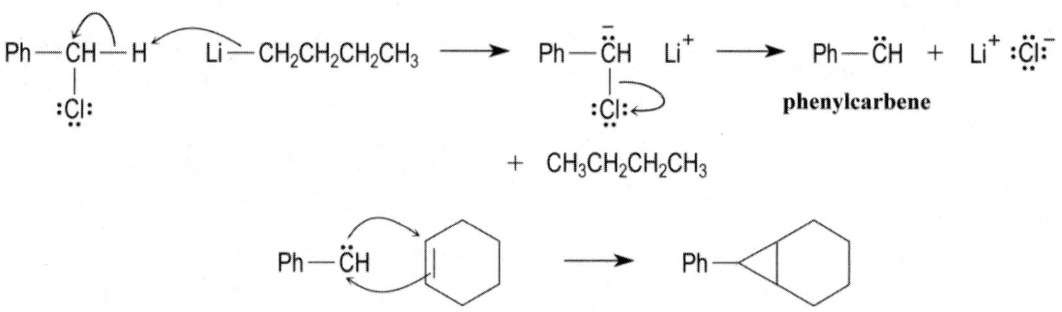

$$Ph\!-\!\overset{|}{C}H\!-\!H \quad Li\!-\!CH_2CH_2CH_2CH_3 \quad \longrightarrow \quad Ph\!-\!\overset{|}{\ddot{C}}H \quad Li^+ \quad \longrightarrow \quad Ph\!-\!\ddot{C}H \quad + \quad Li^+ \; :\ddot{C}l\!:^-$$

phenylcarbene

$$+ \quad CH_3CH_2CH_2CH_3$$

$$Ph\!-\!\ddot{C}H \quad \bigcirc \quad \longrightarrow \quad Ph\!-\!\triangleleft\!\bigcirc$$

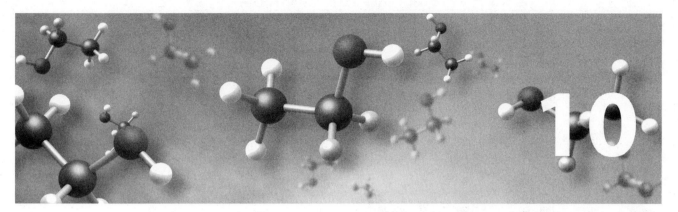

10

The Chemistry of Alcohols and Thiols

STUDY GUIDE LINKS

 ### 10.1 More on Half-Reactions

The technique used for balancing the half-reaction in Eq. 10.40a–c of the text, shown in Study Problem 10.3, text p. 476, is relatively straightforward when oxygens are involved. How can half-reactions be written when oxygens are not involved? For example, what is the half-reaction in the following oxidation?

$$H_2C=CH_2 \xrightarrow{Br_2} Br-CH_2-CH_2-Br$$

Since oxygens are balanced with H_2O, the bromines can be balanced with HBr. Taking this approach results in the following half-reaction:

$$2\,HBr + H_2C=CH_2 \longrightarrow Br-CH_2-CH_2-Br + 2e^- + 2H^+$$

The corresponding inorganic half-reaction is

$$2e^- + 2H^+ + Br_2 \longrightarrow 2\,HBr$$

If the two half-reactions are added, and if the "free electrons" and the HBr molecules on each side of the equation are canceled, the sum is the overall bromine addition reaction:

$$2\,\cancel{HBr} + H_2C=CH_2 \longrightarrow Br-CH_2-CH_2-Br + \cancel{2e^-} + \cancel{2H^+}$$

$$\cancel{2e^-} + \cancel{2H^+} + Br_2 \longrightarrow 2\,\cancel{HBr}$$

$$\rule{6cm}{0.4pt}$$

$$H_2C=CH_2 + Br_2 \longrightarrow Br-CH_2-CH_2-Br$$

Although the notion of a half-reaction can thus be applied to bromine addition, it isn't very useful because not only "free electrons" but also fictitious HBr molecules have to be used to make it work; HBr is not involved in the bromine addition reaction of alkenes. On the other hand, water and protons *really are* involved in the oxidation of ethanol to acetic acid, so that the idea of a half-reaction is somewhat more realistic.

When you want to determine the number of "electrons lost" and "electrons gained" in an oxidation or reduction reaction, you'll find it much simpler as a rule to calculate the change in oxidation number than to balance a half-reaction. The only time half-reactions are worth the effort in organic chemistry is when you have to balance a complicated oxidation–reduction equation, as demonstrated in Study Problem 10.6 (text p. 480). (One certainly doesn't need half-reactions to balance the bromine-addition reaction above.) Nevertheless, it is important for you to understand the concept of a half-reaction because biochemists frequently discuss redox processes in terms of half-reactions. If you take biochemistry, you'll probably be glad you learned (or re-learned) these concepts here.

FURTHER EXPLORATIONS

 ## 10.1   Solvation of Tertiary Alkoxides

The discussion in the text shows that the relative basicities of alkoxides in solution is clearly governed by solvation, because the basicity order in solution (tertiary > secondary > primary) is different from that in the gas phase (tertiary < secondary < primary). How can we visualize this solvation effect?

At first, it was thought that this might be a steric effect; that is, the alkyl branches in a tertiary alkoxide "get in the way" of the solvent that would normally be solvating the alkoxide oxygen. However, this was shown not to be the case in the following way. There is some indication that the $\Delta H°$ of solvation for tertiary alkoxides is actually negative. Steric effects normally are reflected by *increases* in $\Delta H°$. (For example, this is why *cis*-2-butene has a greater $\Delta H_f°$ than *trans*-2-butene.) The unfavorable $\Delta G°$ for solvation, then, resides in a negative $\Delta S°$. (Read about the solubility of hydrocarbons in water in the text, Sec. 8.6D.) The negative $\Delta S°$ means that the solvent is becoming more ordered, less chaotic, when the tertiary alkoxide dissolves in it.

When the tertiary alkoxide dissolves, the solvent surrounding the hydrocarbon groups cannot form hydrogen bonds to the hydrocarbon groups, so they form stronger hydrogen bonds to each other. This releases energy; therefore, the $\Delta H°$ decreases. However, the tighter hydrogen bonds reduce the random motions of the solvent molecules; because randomness decreases, $\Delta S°$ decreases. In effect, the solvent around the alkoxide becomes more like a solid (ice, if the solvent were water). Because of the tightened hydrogen-bond structure of the solvent, the solvent resists disruption by interaction with the alkoxide oxygen, and the alkoxide oxygen is left "naked," that is, unsolvated. This is what we mean when we say that the hydrocarbon groups adversely affect the solvation of the alkoxide ion.

Think of the following analogy. Imagine a large number of little children running around wildly on a playground, more or less at random. A new child (let's call him Josh) enters the playground carrying candy. A few of the children see this and run up to Josh and interact "favorably" with him by accepting some of his candy.

In a second scenario, let's imagine that a teacher is directing a game in which the students have to hold hands tightly and skip around her in a very ordered circle. The teacher has a bag of candy, but the students can't interact with her "candy" because they are holding hands. The "candy" is analogous to the alkoxide; the teacher is analogous to the hydrocarbon groups; and the children are analogous to solvent molecules. Because the students are forced to interact (hold hands) with each other, they can't reach out for the candy.

10.2   Mechanism of Sulfonate Ester Formation

As the text suggests, the formation of sulfonate esters is another nucleophilic substitution reaction. The role of the pyridine is to form a small amount of the conjugate-base alkoxide ion of the alcohol. Removal of a proton also avoids the formation of free HCl and thus avoids strongly acidic conditions, under which some alcohols and tosylates are reactive.

This alkoxide, acting as a nucleophile, displaces chloride ion from the sulfonyl chloride in a nucleophilic substitution reaction to form the sulfonate ester.

There is really nothing new in this mechanism; the only difference is that the sulfur of a sulfonyl chloride instead of the α-carbon of an alkyl halide is the electrophile.

 10.3   Symmetry Relationships Among Constitutionally Equivalent Groups

In Sec. 6.1A, you learned that enantiomers are molecules that are noncongruent mirror images. You also learned in Sec. 6.1C that enantiomers lack certain *symmetry elements,* such as internal mirror planes and centers of symmetry. The purpose of this Further Exploration is to show that the different types of constitutionally equivalent groups—homotopic, enantiotopic, and diastereotopic groups—can also be classified by their symmetry relationships.

Homotopic groups are interchanged by a rotation of the molecule that gives an indistinguishable structure. For example, rotation of the dichloromethane (methylene chloride) molecule 180° about an axis (dashed line) bisecting the two C—Cl (or the two C—H) bonds exchanges the hydrogens and the chlorines, yet leaves the molecule looking exactly as it did before the rotations:

Consequently, the hydrogens are homotopic, and the chlorines are also homotopic. (The hydrogens of CH_3Cl are also homotopic, as shown in the text; demonstrate to yourself the rotation of this molecule that interchanges the three hydrogens while leaving the molecule invariant; see Eq. 10.60 on text p. 490.)

Enantiotopic groups cannot be interchanged by such a rotation. For example, a rotation that interchanges the two enantiotopic α-protons of ethanol gives a molecule that looks different:

That is, the CH_3 group is "up" before the rotation and the OH group is "down." However, after the rotation, these groups have changed positions. The second structure is not congruent to the first without turning it over again.

Enantiotopic groups are related as object and mirror image by an internal mirror plane of the molecule. The internal mirror plane of the ethanol molecule contains the CH_3, the OH, and the central carbon, and it bisects the angle between the two C—H bonds at the α-carbon.

The two hydrogens of ethanol are related as object and mirror image by this plane; hence, these two hydrogens are enantiotopic. Notice that two homotopic groups might also be related by an internal mirror plane, but in order to be homotopic, they must *in addition* be related by a rotation of the molecule as described above. Since a chiral molecule by definition cannot contain an internal mirror plane, it follows that a chiral structure *cannot* contain enantiotopic groups (but it can contain homotopic and diastereotopic groups!).

Finally, diastereotopic protons have no symmetry relationship whatsoever.

 # REACTION REVIEW

I. ACID–BASE REACTIONS OF ALCOHOLS AND THIOLS

A. FORMATION OF ALKOXIDES AND MERCAPTIDES

1. Alkoxides can be formed from alcohols with stronger bases such as sodium hydride, NaH, which is a source of the hydride ion, $H{:}^-$.

$$R{-}OH \; + \; NaH \; \longrightarrow \; R{-}\ddot{O}{:}^- \; + \; Na^+ \; + \; H_2$$

2. Sodium metal reacts with an alcohol to afford a solution of the corresponding sodium alkoxide; the rate of this reaction depends strongly on the alcohol.

$$2\,R{-}OH \; + \; 2\,Na \; \longrightarrow \; 2\,R{-}\ddot{O}{:}^-\,Na^+ \; + \; H_2$$

3. Because thiols are much more acidic than water or alcohols, they, unlike alcohols, can be converted completely into their conjugate-base mercaptide anions by reaction with one equivalent of hydroxide or alkoxide.

$$R-SH + NaOH \xrightarrow[H_2O]{} R-\ddot{\underset{..}{S}}:^- + Na^+ + H_2O$$

4. Thiols form insoluble mercaptides (sulfides) with many heavy-metal ions, such as Hg^{2+}, Cu^{2+}, and Pb^{2+}.

$$2R-\ddot{\underset{..}{S}}H + Hg^{2+} \xrightarrow[H_2O]{} R\ddot{\underset{..}{S}}-Hg-\ddot{\underset{..}{S}}R + H_3O^+$$

II. SUBSTITUTION AND ELIMINATION REACTIONS OF ALCOHOLS AND THEIR DERIVATIVES

A. SULFONATE ESTERS AND INORGANIC ESTERS OF ALCOHOLS

1. Sulfonate esters are derivatives of sulfonic acids, which are compounds of the form $R-SO_3H$.

   a. A sulfonate ester is a compound in which the acidic hydrogen of a sulfonic acid is replaced by an alkyl or aryl group.

p-toluenesulfonic acid   **ethyl p-toluenesulfonate**
                                               (ethyl tosylate)
                                               a sulfonate ester

   b. Sulfonate esters are prepared from alcohols and other sulfonic acid derivatives called sulfonyl chlorides.

p-toluenesulfonyl chloride
   a sulfonyl chloride

      i. This is a nucleophilic substitution reaction in which the oxygen of the alcohol displaces chloride ion from the tosyl chloride. (See Further Exploration 10.2 on p. 267 of this manual.)
      ii. The pyridine used as the solvent is a base; besides catalyzing the reaction, it also prevents HCl from forming in the reaction.

2. Sulfonate esters have approximately the same reactivities as the corresponding alkyl bromides in substitution and elimination reactions.

   a. Sulfonate anions, like bromide ions, are good leaving groups.

   b. Sulfonate anions are weak bases; they are the conjugate bases of sulfonic acids, which are strong acids.

3. Sulfonate esters prepared from primary and secondary alcohols, like the corresponding alkyl halides, undergo S_N2 reactions in which a sulfonate ion serves as the leaving group.

a nucleophile

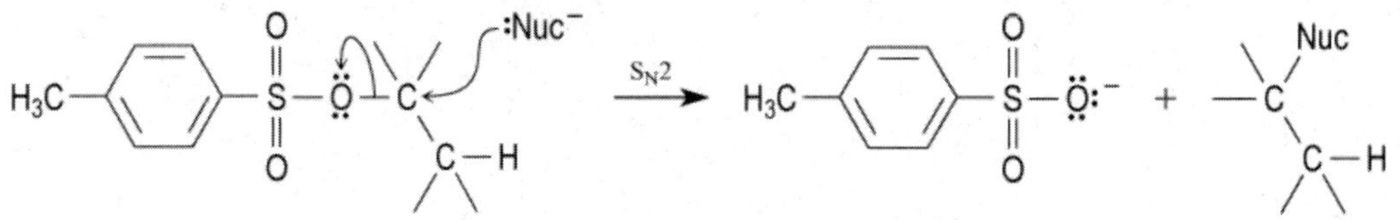

a tosylate ester

p-toluenesulfonate
(tosylate) anion
(a weak base)

4. Secondary and tertiary sulfonate esters, like the corresponding alkyl halides, also undergo E2 reactions with strong bases.

a base

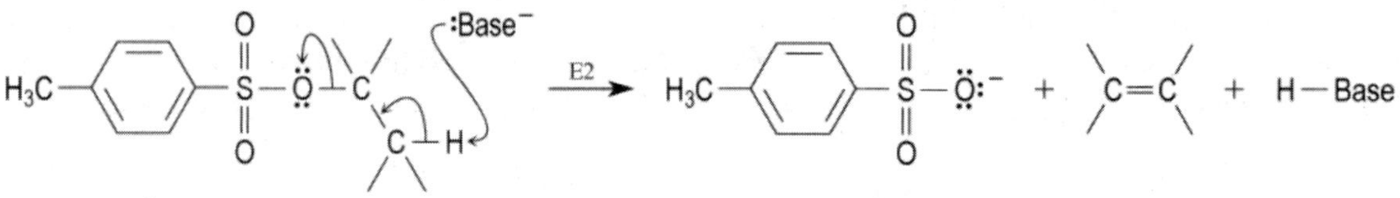

a tosylate ester

p-toluenesulfonate
(tosylate) anion
(a weak base)

5. Secondary and tertiary sulfonate esters, like the corresponding alkyl halides, also undergo S_N1-E1 solvolysis reactions in polar protic solvents.

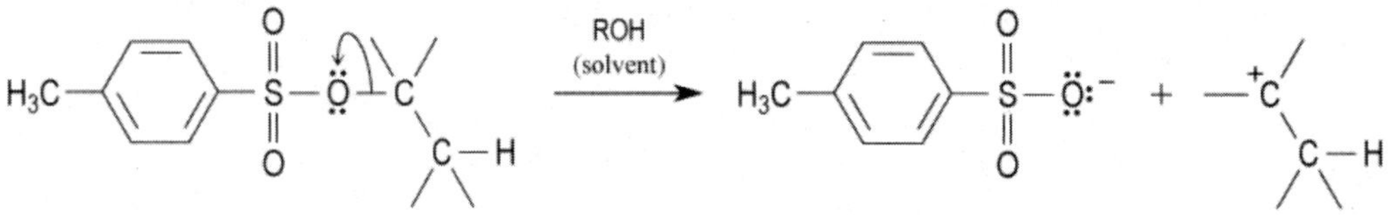

a tosylate ester

p-toluenesulfonate
(tosylate) anion
(a weak base)

carbocation
intermediate

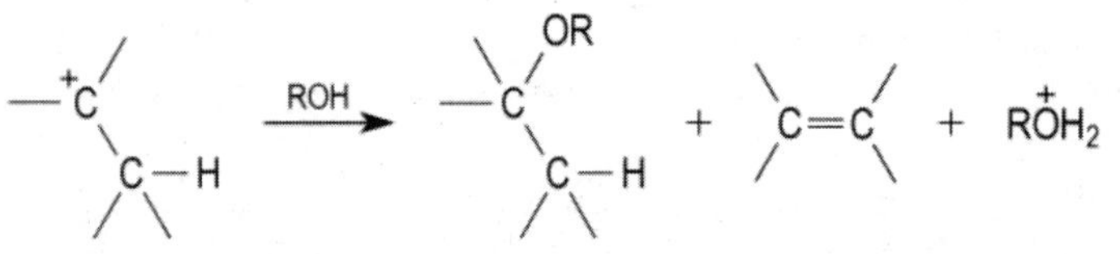

carbocation
intermediate

solvolysis
(S_N1) product

elimination
(E1) product

B. PREPARATION OF PRIMARY ALKYL HALIDES

1. Primary alkyl chlorides are best prepared from primary alcohols by the thionyl chloride method.

$$CH_3(CH_2)_2CH_2OH \xrightarrow[CH_2Cl_2]{SOCl_2} CH_3(CH_2)_2CH_2Cl$$

2. Primary alkyl bromides are prepared from primary alcohols by the reaction of the alcohol with concentrated HBr or with Ph_3PBr_2.

$$CH_3(CH_2)_2CH_2OH \xrightarrow{\text{HBr or} \atop \text{Ph}_3\text{PBr}_2} CH_3(CH_2)_2CH_2Br$$

3. Primary alkyl iodides are prepared from primary alcohols by the reaction of the alcohol with concentrated HI, which is often provided by mixing KI and H_3PO_4.

$$CH_3(CH_2)_2CH_2OH \xrightarrow{\text{KI, H}_3\text{PO}_4} CH_3(CH_2)_2CH_2I$$

4. The sulfonate ester method works well for primary alkyl halides but requires two separate reactions:
   a. formation of the sulfonate ester
   b. reaction of the ester with halide ion.

$$CH_3(CH_2)_2CH_2OH \xrightarrow[\text{pyridine}]{\text{RSO}_2\text{Cl}} CH_3(CH_2)_2CH_2OSO_2R \xrightarrow[\text{acetone}]{X^-} CH_3(CH_2)_2CH_2X$$

a sulfonate ester

5. Because all these methods have an S_N2 mechanism as their basis, alcohols with a significant amount of β-branching (that is, alkyl substituents at the β-position) will react more slowly or not at all, with one exception. The exception is Ph_3PBr_2, as described on p. 471, will react even with neopentyl alcohols.

C. PREPARATION OF SECONDARY ALKYL HALIDES

1. Secondary alkyl chlorides can be prepared from secondary alcohols by the thionyl chloride method if the secondary alcohol has relatively little β-branching, or by the Ph_3PCl_2 method for more hindered alcohols.

$$(CH_3)_2CH - OH \xrightarrow{\text{SOCl}_2 \text{ or} \atop \text{PPh}_3\text{Cl}_2} (CH_3)_2CH - Cl$$

2. Secondary alkyl bromides can be prepared from alcohols by the Ph_3PBr_2 method even if the desired alkyl halide product has significant β-branching.

$$H_3C - CH_2 - \overset{\displaystyle OH}{\underset{\displaystyle |}{CH}} - CH_3 \xrightarrow{\text{Ph}_3\text{PBr}_2} H_3C - CH_2 - \overset{\displaystyle Br}{\underset{\displaystyle |}{CH}} - CH_3$$

3. Secondary alkyl halides can be prepared from secondary alcohols via the sulfonate ester, which should be treated with the appropriate halide ion in a polar aprotic solvent to avoid rearrangements.

D. PREPARATION OF TERTIARY ALKYL HALIDES

1. Tertiary alkyl chlorides can be prepared from the corresponding tertiary alcohols by reaction with HCl under mild conditions.

$$HO - \overset{\displaystyle CH_3}{\underset{\displaystyle CH_3}{\overset{\displaystyle |}{\underset{\displaystyle |}{C}}}} - CH_3 \xrightarrow[\text{H}_2\text{O}]{\text{HCl}} Cl - \overset{\displaystyle CH_3}{\underset{\displaystyle CH_3}{\overset{\displaystyle |}{\underset{\displaystyle |}{C}}}} - CH_3$$

2. Tertiary alkyl bromides can be prepared from the corresponding tertiary alcohols by reaction with HBr under mild conditions.

3. Tertiary sulfonates do not undergo S_N2 reactions.

E. PREPARATION OF ALKENES

1. Alkenes can be prepared by the acid-catalyzed dehydration of alcohols.

2. The equilibrium in this reaction is driven towards the alkene by removal of the volatile alkene, removal of water, or both.
3. The relative rates of alcohol dehydration are in the order tertiary > secondary >> primary.
4. Sulfonate esters readily undergo the E2 reaction with alkoxide bases instead of undergoing S_N2 reactions.

5. This reaction is especially useful
   a. when the acidic conditions of alcohol dehydration lead to rearrangements or other side reactions.
   b. for primary alcohols in which dehydration is not an option.

III. OXIDATION OF ALCOHOLS AND THIOLS

A. OXIDATION OF ALCOHOLS TO ALDEHYDES AND KETONES

1. Primary and secondary alcohols are oxidized by reagents containing Cr(VI), that is, chromium in the +6 oxidation state, to give carbonyl compounds (compounds containing the carbonyl group, C=O).
   a. Many forms of chromium are commonly used for these oxidations, including chromate ion (CrO_4^{2-}), dichromate ion ($Cr_2O_7^{2-}$), and chromic anhydride or chromium trioxide (CrO_3). Specific reagents include chromic acid (H_2CrO_4) and sodium or potassium dichromate ($Na_2Cr_2O_7$ or $K_2Cr_2O_7$).
   b. The α-carbon atom of an alcohol must bear a hydrogen atom for oxidation to an aldehyde or ketone to occur.
   c. Primary alcohols react with Cr(VI) reagents to give aldehydes, but if water is present, aldehydes are further oxidized to carboxylic acids.

d. A complex prepared from CrO_3, pyridine, and HCl, called pyridine chlorochromate (PCC), is often used as an anhydrous Cr(VI) preparation for the oxidation of primary alcohols to aldehydes.

e. The general mechanism for these oxidations is E2-like (see Eqs. 10.52–10.54, text p. 484).

B. OXIDATION OF ALCOHOLS TO CARBOXYLIC ACIDS

1. Primary alcohols can be oxidized to carboxylic acids using aqueous solutions of Cr(VI) such as aqueous potassium dichromate ($K_2Cr_2O_7$) in acid.

2. Primary alcohols can be oxidized to carboxylic acids with potassium permanganate ($KMnO_4$) in basic solution, followed by acidification.

1-nonanol
a primary alcohol

nonanoic acid
a carboxylic acid

3. Because $KMnO_4$ reacts with alkene double bonds, Cr(VI) is preferred for the oxidation of alcohols that contain double or triple bonds.

4. Potassium permanganate is not used for the oxidation of secondary alcohols to ketones because many ketones react further with the alkaline permanganate reagent.

C. OXIDATION OF THIOLS

1. Oxidation of a thiol takes place not at the carbon, but at the sulfur.

a. The most commonly occurring oxidation products of thiols are disulfides and sulfonic acids.

general structure of
a disulfide

general structure of
a sulfonic acid

b. The Lewis structures of these derivatives require either violation of the octet rule or separation of formal charge.

c. Sulfur can accommodate more than eight valence electrons because, in addition to its $3s$ and $3p$ orbitals, it has unfilled $3d$ orbitals of relatively low energy.

d. The same oxidation-number formalism used for carbon can be applied to oxidation at sulfur.

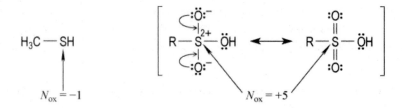

2. Sulfonic acids are formed by vigorous oxidation of thiols or disulfides with $KMnO_4$ or nitric acid (HNO_3).

$$CH_3SH \quad \xrightarrow{\text{HNO}_3} \quad CH_3SO_3H$$

methanethiol   **methanesulfonic acid**

3. Many thiols spontaneously oxidize to disulfides merely on standing in air (O_2).

4. Thiols can also be converted into disulfides by mild oxidants such as I_2 in base or Br_2 in CCl_4.

   a. These reactions can be viewed as a series of S_N2 reactions in which halogen and sulfur are the electrophiles that react with nucleophilic thiolates anions.

$$H_3C-S-H \quad \xrightarrow[-H_2O]{^{-}OH} \quad H_3C-S^{-} \quad \xrightarrow[-I^{-}]{I_2} \quad H_3C-S-I \quad \xrightarrow[-I^{-}]{H_3C-S^{-}} \quad H_3C-S-S-CH_3$$

methanethiol   **methyl disulfide**

   b. When thiols and disulfides are present together in the same solution, an equilibrium among them is rapidly established.

$$CH_3SH \quad + \quad CH_3CH_2S-SCH_2CH_3 \quad \rightleftharpoons \quad CH_3CH_2SH \quad + \quad CH_3S-SCH_2CH_3$$

methanethiol   **diethyl disulfide**   **ethanethiol**   **ethyl methyl disulfide**

SOLUTIONS TO PROBLEMS

Solutions to In-Text Problems

10.1   (a)   (b)   (c)

Na^+ $^-OCH(CH_3)_2$ K^+ $^-OC(CH_3)_3$

sodium isopropoxide   **potassium *tert*-butoxide**

Mg^{2+} $\left(^-OCH_2\underset{\underset{CH_3}{|}}{\overset{\overset{CH_3}{|}}{C}}CH_2CH_3\right)_2$

magnesium 2,2-dimethyl-1-butanolate

10.2   (a)   Calcium methoxide or calcium methanolate
          (b)   Cuprous ethanethiolate [or copper(I) ethanethiolate]

10.3   (a)   The principle is that the polar effect of a chloro substituent enhances acidity; the effect is greater when more chloro substituents are present; and the effect diminishes as the chloro group is separated from the hydroxy group by more carbons. Thus, 3-chloro-1-propanol, $ClCH_2CH_2CH_2$—OH, is least acidic and has the greatest pK_a; 2-chloroethanol, $ClCH_2CH_2$—OH, is more acidic; and 2,2-dichloroethanol, Cl_2CHCH_2—OH, is most acidic and has the lowest pK_a.
          (b)   Thiols are more acidic than alcohols, other things being equal (element effect); and a chloro substituent enhances acidity by a polar effect. Thus, ethanol (CH_3CH_2OH) is least acidic and has the greatest pK_a; 2-chloroethanol is more acidic; and 2-chloroethanethiol, $ClCH_2CH_2$—SH, is most acidic and has the lowest pK_a.
          (c)   An alkoxy group enhances acidity because the polar effect of the electronegative oxygen. Thus, 1-butanol, $CH_3CH_2CH_2CH_2OH$, is least acidic and has the greatest pK_a; and 2-methoxyethanol, $CH_3OCH_2CH_2OH$, is most acidic and has the lowest pK_a.

10.4   (a)

$\underset{\substack{A \\ (E\text{ and }Z)}}{CH_3CH{=}\underset{\overset{|}{CH_3}}{\overset{\overset{CH_3}{|}}{C}}CH_2CH_2CH_2CH_3}$ + $\underset{\substack{B \\ (E\text{ and }Z)}}{CH_3CH_2\underset{}{\overset{\overset{CH_3}{|}}{C}}{=}CHCH_2CH_2CH_3}$ + $\underset{C}{CH_3CH_2\overset{\overset{CH_2}{||}}{C}CH_2CH_2CH_2CH_3}$

(b)   The major, *E*-alkene is shown below. The *Z*-alkene is a minor product (why?).

$$\underset{H}{\overset{Ph}{\diagdown}}C{=}C\underset{Ph}{\overset{H}{\diagup}}$$

10.5   The OH group of the alcohol is protonated in a Brønsted acid–base reaction to form the conjugate acid of the alcohol. This loses water to form a carbocation in a Lewis acid–base dissociation reaction. Finally, in a Brønsted acid–base reaction, water acts as a Brønsted base to remove a β-proton from the carbocation, which acts as a Brønsted acid, to give the alkene. The formation of product *A* by removal of proton *(a)* is shown here; the formation of products *B* and *C* occurs in an analogous manner by removal of β-protons *(b)* and *(c)*, respectively.

$$CH_3CH_2\overset{\overset{\displaystyle CH_3}{|}}{\underset{\underset{\displaystyle :OH}{|}}{C}}CH_2CH_2CH_2CH_3 \rightleftharpoons CH_3CH_2\overset{\overset{\displaystyle CH_3}{|}}{\underset{\underset{\displaystyle \overset{+}{:}OH_2}{|}}{C}}CH_2CH_2CH_2CH_3 + :\ddot{O}H_2 \rightleftharpoons CH_3CH_2\overset{\overset{\displaystyle CH_3}{|}}{\underset{\underset{\displaystyle +}{}}{C}}CH_2CH_2CH_2CH_3 + :\ddot{O}H_2$$

$$CH_3CH\overset{\overset{\displaystyle CH_3\,(b)}{|}}{\underset{\underset{\underset{\displaystyle :\ddot{O}H_2}{}}{H\,(a)}}{C}}CH_2CH_2CH_2CH_3 \longrightarrow CH_3CH=\overset{\overset{\displaystyle CH_3}{|}}{C}CH_2CH_2CH_2CH_3 + H\overset{+}{-}\ddot{O}H_2$$
$$A$$

10.6   The carbocation intermediate in the dehydration of 3-ethyl-3-pentanol is derived by protonation of the OH group and loss of water:

$$\underset{\textbf{3-ethyl-3-pentanol}}{CH_3CH_2\overset{\overset{\displaystyle CH_2CH_3}{|}}{\underset{\underset{\displaystyle :OH}{|}}{C}}CH_2CH_3} \rightleftharpoons CH_3CH_2\overset{\overset{\displaystyle CH_2CH_3}{|}}{\underset{\underset{\displaystyle \overset{+}{:}OH_2}{|}}{C}}CH_2CH_3 + :\ddot{O}H_2 \rightleftharpoons CH_3CH_2\overset{\overset{\displaystyle CH_2CH_3}{|}}{\underset{\underset{\displaystyle +}{}}{C}}CH_2CH_3 + :\ddot{O}H_2$$

10.7   The *E* isomers of *A* and *B* will predominate over the *Z* isomers because the *E* isomers are considerably more stable. Because alkenes *A* and *B* have more alkyl substituents at the double bond than *C*, the *E* isomers of *A* and *B* will be formed in greater amount than alkene *C*. (Recall that the E1 reaction, of which dehydration is an example, gives the most stable alkene isomers.)

10.8   (a)   1-Methylcyclohexanol and either *cis*- or *trans*-2-methylcyclohexanol should give 1-methylcyclohexene as the major product of dehydration, because this is the most stable alkene that could be formed in each case. The tertiary alcohol 1-methylcyclohexanol should react most rapidly because it involves a tertiary carbocation intermediate; dehydrations of the 2-methylcyclohexanols involve a less stable secondary carbocation intermediate (which would probably rearrange to a tertiary carbocation).

1-methylcyclohexanol

2-methylcyclohexanol

H_3O^+

1-methylcyclohexene

(b)   Both 3-methyl-3-pentanol and 3-methyl-2-pentanol should give 3-methyl-2-pentene as the major product. The tertiary alcohol 3-methyl-3-pentanol should dehydrate more rapidly.

$$CH_3CH_2\overset{\underset{\displaystyle OH}{|}}{\underset{\displaystyle |}{\overset{\displaystyle CH_3}{\overset{\displaystyle |}{C}}}}CH_2CH_3$$

3-methyl-3-pentanol

$$CH_3\overset{\underset{\displaystyle OH}{|}}{\underset{\displaystyle |}{CH}}CHCH_2CH_3$$

3-methyl-2-pentanol

H_3O^+ →

$$CH_3CH=\overset{\underset{\displaystyle |}{CH_3}}{\underset{\displaystyle |}{\overset{\displaystyle |}{C}}}CH_2CH_3$$

3-methyl-2-pentene

10.9   Protonation of the OH group and loss of water as shown in several of the previous solutions, as well as in Eqs. 10.13a–b on p. 458–9 of the text, gives a secondary carbocation. As the text discussion of Eq. 10.16 suggests, the mechanism involves a rearrangement of the initially formed secondary carbocation to a more stable tertiary carbocation.

initially formed
secondary carbocation

rearranged
tertiary carbocation

Loss of the two possible β-protons gives the two alkene products.

2,3-dimethyl-1-butene
A

2,3-dimethyl-2-butene
B

10.10   **(a)**   The Brønsted acid present in highest concentration in methanol is the conjugate acid of methanol:

$$CH_3\ddot{O}H + H_2SO_4 \rightleftarrows CH_3\overset{+}{\ddot{O}}H_2 \quad HSO_4^-$$

$$pK_a \approx -3 \qquad\qquad pK_a \approx -2$$

The relative pK_a values show that the protonation of methanol is favorable and has $K_{eq} \approx 10$.

(b)   If a base is involved, it must be methanol itself, or, to a very small extent, bisulfate ion. The base is *not* sodium methoxide, Na^+ CH_3O^-. Sodium methoxide would be protonated instantaneously under the acidic reaction conditions.

10.11   **(a)**

$$CH_3\overset{\underset{\displaystyle OH}{|}}{\underset{\displaystyle |}{CH}}CH_2CH_3 + HBr \longrightarrow CH_3\overset{\underset{\displaystyle Br}{|}}{\underset{\displaystyle |}{CH}}CH_2CH_3$$

(b)

(c)

$$HO—CH_2CH_2CH_2CH_2CH_2—OH + HI \text{ (concd.)} \longrightarrow I—CH_2CH_2CH_2CH_2CH_2—I$$

Because concentrated HI solutions are not very stable, "concentrated HI" is usually formed by combining potassium iodide (KI) or sodium iodide (NaI) and phosphoric acid.

10.12   This reaction involves a carbocation rearrangement. We use HBr as the acid, although, because water is generated as a product, H_3O^+ could also be used.

10.13   (a)

$$CH_3CH_2CH_2—Br$$

1-bromopropane

(b)   The product is $I—CH_2CH_2CH_2—I$

(c)   The product in part (c) results from a carbocation rearrangement.

$$(CH_3)_2\overset{\overset{\displaystyle Br}{|}}{C}—CH(CH_3)_2$$

2-bromo-2,3-dimethylbutane

(d)   The compound, neopentyl alcohol, is a primary alkyl halide and cannot react by the S_N1 mechanism; and it has too many β-substituents to react by the S_N2 mechanism. Consequently, there is no reaction.

10.14   (a)   (b)

isopropyl methanesulfonate
(isopropyl mesylate

methyl *p*-toluenesulfonate
(methyl tosylate)

(c)   (d)

phenyl tosylate
(phenyl *p*-toluenesulfonate)

cyclohexyl mesylate
(cyclohexyl methanesulfonate)

10.15

cyclohexanol + methanesulfonyl chloride → cyclohexyl mesylate

(reaction: cyclohexanol-OH + Cl—SO$_2$CH$_3$ — pyridine → cyclohexyl—O—SO$_2$CH$_3$)

> ⚠ Remember that when we outline a synthesis we provide key starting materials and products, but do not necessarily show by-products and do not necessarily balance the reaction. For example, by-products are missing in the previous solution. This is acceptable as long as the request is for a synthesis rather than for a balanced reaction.

10.16   (a)

$(CH_3)_2CHCH_2CH_2CH_2OH$ —TsCl, pyridine→ $(CH_3)_2CHCH_2CH_2CH_2OTs$ —Na$^+$ I$^-$ / polar aprotic solvent→ $(CH_3)_2CHCH_2CH_2CH_2I$

(b)

cyclopentyl—$CH_2CH_2CH_2OH$ —NaH→ cyclopentyl—$CH_2CH_2CH_2O^-$ Na$^+$ + H$_2$ —tosyl chloride (Cl—SO$_2$—C$_6$H$_4$—CH$_3$)→

cyclopentyl—$CH_2CH_2CH_2OTs$ —Na$^+$ CH$_3$S$^-$→ cyclopentyl—$CH_2CH_2CH_2SCH_3$

Although a polar aprotic solvent would accelerate the last step, it would probably work in an alcohol solvent. The nucleophile, CH$_3$S$^-$, can be generated by allowing the thiol CH$_3$—SH to react with one equivalent of sodium ethoxide in ethanol.

10.17   (a)   Cyanide ion displaces the tosylate ester formed in the first step.

$$CH_3\overset{\underset{|}{CN}}{CH}CH_2CH_3$$

(b)   The triflate ester formed in the first step is displaced by the fluoride ion. The nucleophilicity of fluoride ion is increased by complexation of the counter-ion K+ with [18]-crown-6, and by the use of acetonitrile, a polar aprotic solvent.

3-fluorohexane

10.18   (a)   (b)

$$CH_3O-\underset{\underset{OCH_3}{|}}{\overset{\overset{O}{||}}{P}}-OCH_3$$

trimethyl phosphate

$$CH_3CH_2O-\underset{\underset{OH}{|}}{\overset{\overset{O}{||}}{P}}-OH$$

monoethyl phosphate

10.19   (a)   (b)

$$H_3C-\overset{+}{N}H_2-CH_3 \quad {}^-\underset{\underset{O}{||}}{\overset{\overset{O}{||}}{OS}}OCH_3$$

dimethylammonium methylsulfate

$$CH_3OH \quad + \quad Na^+ \; {}^-\underset{\underset{O}{||}}{\overset{\overset{O}{||}}{OS}}OCH_3$$

methanol

(c)   (d)

$$CH_3CH_2OCH_3 \quad + \quad Na^+ \; {}^-\underset{\underset{O}{||}}{\overset{\overset{O}{||}}{OS}}OCH_3$$

ethyl methyl ether

$$CH_3CH_2CH_2SCH_3 \quad + \quad Na^+ \; {}^-\underset{\underset{O}{||}}{\overset{\overset{O}{||}}{OS}}OCH_3$$

methyl propyl sulfide

10.20   1-Butanol can be converted into 1-bromobutane by treating it with triphenylphosphine dibromide (Ph_3PBr_2). (See Eq. 10.35–10.36b on text p. 471.) 1-Butanol can also be converted into 1-bromobutane by first converting it into a sulfonate ester such as butyl tosylate, then treating that ester with sodium bromide in a polar aprotic solvent such as DMSO. Finally, and perhaps most simply, 1-butanol can be converted into 1-bromobutane by treatment with concentrated HBr and H_2SO_4.

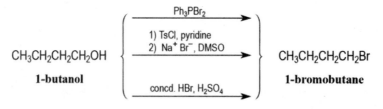

10.21   (a)   Eq. 10.34 of the text shows that the nucleophilic reaction of the chloride ion on the chlorosulfite ester is an S_N2 reaction; consequently, it should take place with inversion. It follows that the product should be the R enantiomer of $CH_3CH_2CH_2CHD$—Cl.

(b)   Eqs. 10.36a and 10.36b of the text shows that the nucleophilic reaction of the bromide ion on the reactive intermediate occurs by a concerted (S_N2) substitution reaction. Therefore, the reaction should occur with inversion of stereochemistry, and the product would then be (S)-2-bromopentane. The S_N2 reaction occurs at an acceptable rate on a secondary carbon in the absence of β substituents. In addition, the reaction is fast because the leaving group is a *very* weak base, and because the reaction is essentially intramolecular.

10.22   We can predict the structure of an *acyl phospate* by inspecting the structures shown in the text. An acyl chloride, shown in the problem, contains a chloride, so the rest of the compound must be an acyl group. The phosphate group is shown on p. 472. Combining the two gives us an acyl phosphate.

an acyl phosphate

Magnesium binding as shown in Eq. 10.38 (or alternatively, hydrogen bonding provided by the enzyme) can stabilize the negative charges on the oxygen. Presumably the bond between the carbon and the oxygen is also stretched, weakening it, making the phosphate group a very good leaving group.

10.23   (a)   Because the alcohol groups are primary, concentrated HBr/H_2SO_4 will bring about the desired reaction. Treatment with PPh_3Br_2 will also work.

(b)   The simplest method for effecting the conversion shown is to treat the alcohol with thionyl chloride and pyridine. Conversion of the alcohol to a sulfonate ester and treatment of the ester with sodium chloride in a polar aprotic solvent would also work, but involves more steps.

(c)   The alcohol reacts with HBr to give a carbocation intermediate that rearranges. However, this is precisely the reaction desired; reaction of the rearranged carbocation with bromide ion gives the desired product. (See Study Problem 10.1 on text p. 461.)

(d)   Because a carbocation intermediate and hence rearrangements are a distinct possibility if this alcohol is exposed to acidic reagents such as HBr, the sulfonate ester method should be used. Thus, treat the alcohol with tosyl chloride and pyridine, and treat the resulting tosylate with sodium bromide in a polar aprotic solvent. This type of solvent suppresses carbocation formation. Alternatively, PPh_3Br_2 gives similar results.

10.24   2-Propanol, cyclopentanol, and cyclohexanol are three secondary alcohols (among others) that could be converted into the corresponding alkyl bromides without rearrangements. If hydride shifts occur within a carbocation intermediate derived from these alcohols, the same carbocation would be formed.

10.25   (a)   Treatment with HBr/H_2SO_4 would protonate the OH group, converting it into a good leaving group. A secondary carbocation is produced, which would then rearrange via a hydride shift to a more stable tertiary carbocation. The S_N1 reaction would conclude by attack of the bromide at the positive center, leaving the rearranged product 2-bromo-2-methylpentane. (See answer 10.12 in this manual.)

(b)   Treating secondary alcohols with triphenylphoshine dibromide also converts the OH group into a good leaving group, but not one that is as prone to leave on its own though an S_N1 mechanism. Instead, the bromide displaces the leaving group through an S_N2 mechanism, producing the unrearranged 3-bromo-2-methylpentane.

3-bromo-2-methylpentane

10.26   (a)   The bromination of methane to give methyl bromide is a two-electron oxidation.

      (b)   The conversion of toluene into benzoic acid is a six-electron oxidation.

      (c)   The conversion of an alkyl iodide into an alkane is a two-electron reduction.

      (d)   The dihydroxylation of an alkene by $KMnO_4$ is a two-electron oxidation.

      (e)   The ozonolysis of 2-methyl-2-butene is a six-electron oxidation. Notice that both carbons of the double bond must be considered, even though the molecule is "split in two" as a result of the reaction.

      (f)   The addition of HBr to an alkene is neither an oxidation nor a reduction. (One carbon of the alkene is formally oxidized and the other is reduced by the same amount.)

      (g)   This is a two-electron reduction; the negative charge contributes –1 to the oxidation number of the product.

10.27   The half-reaction of Problem 10.26, part (b):

$$2\,H_2O \;+\; Ph-CH_3 \;\longrightarrow\; Ph-\overset{\overset{\textstyle O}{\|}}{C}-OH \;+\; 6e^- \;+\; 6H^+$$

Electrons are "lost" in this half-reaction. This observation is consistent with the classification of this reaction in the solution to Problem 10.26(b) as an oxidation. To complete the sentence, "This reaction is a six-electron oxidation."

10.28   (a)   The alkene is reduced, and the H_2 is oxidized.

      (b)   The organic compound is reduced, and the $^-AlH_4$ is oxidized.

      (c)   The alkene is oxidized, and the Br_2 is reduced.

      (d)   The organic compound is oxidized, and oxygen is reduced.

10.29   The "glib" way to work this problem is to balance the changes in oxidation states of the oxidizing agent and the organic compound. The manganese is reduced from Mn(VII) to Mn(IV); this is a three-electron reduction. Toluene undergoes a six-electron oxidation. Therefore, it will take two equivalents of MnO_4^- to oxidize one equivalent of toluene, or two moles of MnO_4^- per mole of toluene, to reconcile the electrons lost and the electrons gained.

    Let's now prove that this works by balancing the equations, multiplying the lower one by 2 so that the "free" electrons cancel, and adding the two, canceling H_2O and H^+:

$$2\,H_2O \;+\; PhCH_3 \;\longrightarrow\; PhCO_2H \;+\; 6e^- \;+\; 6H^+$$

multiply by 2 $\longrightarrow$ $\overset{8}{\cancel{4}H^+} \;+\; \overset{2}{MnO_4^-} \;+\; \overset{6}{\cancel{3}e^-} \;\longrightarrow\; \overset{2}{MnO_2} \;+\; \overset{4}{\cancel{2}H_2O}$

$$\overline{PhCH_3 \;+\; 2\,MnO_4^- \;+\; 2H^+ \;\longrightarrow\; PhCO_2H \;+\; 2\,MnO_2 \;+\; 2\,H_2O}$$

As this shows, and as we deduced above, two permanganates are required per mole of toluene for this oxidation.

10.30   (a)   On the assumption that sufficient PCC has been added, both primary alcohols are oxidized:

$$HO-CH_2CH_2CH_2-OH \xrightarrow{PCC} O{=}CHCH_2CH{=}O$$

(b)   Only the primary alcohol is oxidized:

10.31   (a)   The carboxylic acid could be prepared from $(CH_3)_2CHCH_2CH_2CH_2CH_2OH$ (5-methyl-1-hexanol) by oxidation either with aqueous chromic acid [or any other form of aqueous Cr(VI)], or with aqueous potassium permanganate followed by acid.

(b)   This compound (3-pentanone) can be prepared by a PCC oxidation of the corresponding alcohol, 3-pentanol. (Aqueous dichromate could also be used.)

3-pentanol   **3-pentanone**

(c)   The aldehyde can be prepared from 5-methyl-1,5-hexanediol by the PCC oxidation or other variation of nonaqueous oxidation with Cr(VI).

5-methyl-1,5-hexanediol   **5-methyl-5-hydroxyhexanal**

(d)   This aldehyde can be prepared by a PCC oxidation of the corresponding alcohol.

10.32   This oxidation occurs in much the same way that NAD^+ oxidations occur: by transfer of a hydride from the alcohol to the carbocation. (Notice that it is a *hydride* (a hydrogen with two electrons), not a proton, that is transferred.)

10.33   (a)   Hydrogens *a* and *b* are constitutionally equivalent and enantiotopic. (The analysis of this case is essentially identical to the analysis of the α-hydrogens of ethanol; see Eq. 10.61, text p. 490.)

(b)   The two methyl groups *a* and *b* are constitutionally equivalent and diastereotopic. The pair of methyl groups *a* and *c* are constitutionally nonequivalent, as are the pair *b* and *c*.

When investigating the stereochemical relationships among several groups, the appropriate technique is to consider *each pair*. Thus, three relationships must be categorized within three groups (although it is possible that some relationships might be the same.) Here's an amusing little puzzle that uses the same sort of reasoning. Man *A* looks at a picture and comments about the person *B* in the picture, saying, "Brothers and sisters have I none; this man's father is my father's son." What is the relationship between *A* and *B*?

(c) The two fluorines are constitutionally equivalent and diastereotopic.

(d) Hydrogens *a* and *b* within a particular chair conformation are constitutionally equivalent and diastereotopic. However, the chair interconversion interchanges the positions of these two hydrogens and makes them completely equivalent, that is, homotopic, *over time.*

(e) Replacing H^a and H^b in turn with a "circled H" shows that these hydrogens are constitutionally equivalent and enantiotopic, as are H^c and H^d. H^a and H^c are also constitutionally equivalent and enantiotopic, as are H^b and H^d. Finally, H^a and H^d are constitutionally equivalent and homotopic, as are H^b and H^c.

(f) Carbon-2 and carbon-4 are constitutionally equivalent and enantiotopic. Hydrogens *a* and *d* are constitutionally equivalent and diastereotopic, as are hydrogens *b* and *c*. Hydrogens *a* and *c* are constitutionally equivalent and enantiotopic, as are hydrogens *b* and *d*. Finally, hydrogens *a* and *b* are constitutionally equivalent and diastereotopic, as are hydrogens *c* and *d*.

These relationships are best seen with substitution tests—one for each pair. For example, the substitution test for carbons 2 and 4 shows that the two molecules formed by circling each carbon in turn are enantiomers; therefore, these carbons are enantiotopic.

10.34 (a) In both, the hydrogen (or deuterium) is delivered from the pro-*(R)* position on NADH (the "up" position when NADH is drawn as it is in the problem) so that it ends up in the pro-*(R)* position of deuterated ethanol. In part (a), the resulting isotopically substituted ethanol is chiral.

(S)-ethanol-1-*d*

(b) Because deuterium is delivered, the α-carbon of the resulting ethanol bears two deuteriums—that is, the product is CH$_3$CD$_2$OH—and it therefore has no asymmetric carbon; hence, the molecule is achiral.

10.35 (a) In a thiol, the oxidation number of sulfur is –1; in a sulfonic acid, it is +5; hence, the change in oxidation number is +6. The oxidation of a thiol to a sulfonic acid is a 6-electron oxidation. (This can be verified by writing a balanced half-reaction.)

(b) This is a 2-electron oxidation, because a hydrogen (which contributes –1 to the oxidation number of sulfur) is replaced by an OH (which contributes +1).

10.36 (a) Following the methodology in Sec. 10.6A on p. 495, first write the half reaction. Then, balance the oxygens with water, the hydrogens with protons, and finally the positive charges with electrons. The reaction is a two electron oxidation.

$$Ph_3P: + H_2O \longrightarrow Ph_3\overset{+}{P}{=}O + 2H^+ + 2e^-$$

(b)

$$\begin{array}{c} :\ddot{\text{O}}:^- \\ | \\ Ph{-}\overset{+}{\underset{|}{P}}{-}Ph \\ | \\ Ph \end{array}$$

(c)   A resonance structure containing a N–O double bond is not possible because nitrogen cannot violate the octet rule. In trimethylphosphene oxide, however, a resonance structure containing a double bond is possible because phosphorous possesses *d* orbitals that can accommodate additional electrons.

$$\begin{array}{ccc} :\ddot{\text{O}}:^- & :\text{O}: & :\ddot{\text{O}}:^- \\ | & || & | \\ Me{-}\overset{+}{\underset{|}{P}}{-}Me \longleftrightarrow Me{-}\underset{|}{P}{-}Me & & Me{-}\overset{+}{\underset{|}{N}}{-}Me \\ Me & Me & Me \end{array}$$

P–O bond has partial double   N–O bond has no
      bond character   double bond character

10.37   The reaction is a nucleophilic substitution in which one sulfur acts as a nucleophile toward another sulfur, which acts as an electrophile. Because a thiolate is a much better nucleophile than a thiol—thiolates are *much* more basic than neutral thiols—a thiolate serves as the nucleophile in this reaction. A base is required to form the thiolate anion. The reaction is faster in the presence of base because the reaction rate depends on the nucleophile concentration, and base increases the concentration of the nucleophile.

$$CH_3CH_2\ddot{S}{-}H \quad \bar{:}\ddot{O}CH_2CH_3 \rightleftharpoons CH_3CH_2\ddot{S}:^- + H{-}\ddot{O}CH_2CH_3$$

$$CH_3CH_2\ddot{S}:^- \quad S{-}\ddot{S}CH_2CH_2CH_3 \longrightarrow CH_3CH_2\ddot{S}{-}SCH_2CH_2CH_3 + {}^-:\ddot{S}CH_2CH_2CH_3$$
$$\underset{|}{\overset{|}{CH_2CH_2CH_3}}$$

10.38   (a)   2-Methyl-3-pentanol can be prepared by hydroboration–oxidation of 2-methyl-2-pentene, which can be prepared by dehydration of 2-methyl-2-pentanol:

$$\begin{array}{ccc} \underset{\underset{CH_3}{|}}{\overset{\overset{OH}{|}}{CH_3\underset{|}{C}CH_2CH_2CH_3}} & \xrightarrow{H_2SO_4} & \underset{\underset{CH_3}{|}}{CH_3C{=}CHCH_2CH_3} \xrightarrow[\text{2) } H_2O_2,\ {}^-OH]{\text{1) } BH_3,\ THF} & \underset{\underset{CH_3}{|}}{\overset{\overset{OH}{|}}{CH_3CHCHCH_2CH_3}} \end{array}$$

   2-methyl-2-pentanol   **2-methyl-2-pentene**   **2-methyl-3-pentanol**

(b)   The deuterium-containing alkane can be prepared by protonolysis of a Grignard reagent in D_2O; the Grignard reagent can be prepared from an alkyl halide; and the alkyl halide can be prepared from an alcohol.

$$CH_3(CH_2)_4CH_2OH \xrightarrow{HBr,\ H_2SO_4} CH_3(CH_2)_4CH_2Br \xrightarrow[\text{ether}]{Mg} CH_3(CH_2)_4CH_2MgBr \xrightarrow{D_2O} CH_3(CH_2)_4CH_2{-}D$$

(c)   The carboxylic acid can be prepared by oxidation of a primary alcohol, which, in turn, can be prepared from the alkene starting material by hydroboration–oxidation.

$$\text{methylenecyclohexane} \xrightarrow[\text{2) H}_2\text{O}_2,\ ^-\text{OH}]{\text{1) BH}_3,\ \text{THF}} \text{cyclohexylmethanol} \xrightarrow[\text{2) H}_3\text{O}^+]{\text{1) KMnO}_4/\text{NaOH}} \text{cyclohexanecarboxylic acid}$$

(d)   The aldehyde can be prepared by oxidation of a primary alcohol; the required primary alcohol can be prepared by hydroboration–oxidation of an alkene; and the required alkene can be prepared by an E2 reaction of a primary alkyl halide using a branched base.

$$\underset{\overset{|}{CH_3}}{CH_3CH_2CH_2CHCH_2Br} \xrightarrow[\text{(CH}_3)_3\text{COH}]{\text{K}^+\ \text{(CH}_3)_3\text{CO}^-} \underset{\overset{|}{CH_3}}{CH_3CH_2CH_2C}{=}CH_2 \xrightarrow[\text{2) H}_2\text{O}_2,\ ^-\text{OH}]{\text{1) BH}_3,\ \text{THF}}$$

$$\underset{\overset{|}{CH_3}}{CH_3CH_2CH_2CHCH_2OH} \xrightarrow{\text{PCC}} \underset{\overset{|}{CH_3}}{CH_3CH_2CH_2CHCH}{=}O$$

Solutions to Additional Problems

10.39   (a) $CH_3CH_2CH_2CH_2Br$

(b) $CH_3CH_2CH_2CH_2\overset{+}{O}H_2$
   HSO_4^-

(c) $CH_3CH_2CH_2CH{=}O$

(d) $CH_3CH_2CH_2CH_2O^-\ Na^+$
   $+\ H_2$

(e) $CH_3CH_2CH_2CH_2OCH_3$
   $Na^+\ I^-$

(f) $CH_3CH_2CH_2CH_2OTs$
   $+$ [pyridinium] $\overset{+}{N}H\ Cl^-$

(g) $CH_3CH_2CH_2CH_2O^-\ \overset{+}{M}gBr$
   $+\ CH_3CH_2CH_3$

(h) $CH_3CH_2CH_2CH_2Cl$
   $+\ SO_2\ +$ [pyridinium] $\overset{+}{N}H\ Cl^-$

(i) $CH_3CH_2CH_2CH_2Br$

(j) $CH_3CH_2CH_2CH_2MgBr$

(k) $CH_3CH_2CH{=}CH_2\ +\ CH_3CH_2CH_2CH_2OC(CH_3)_3$
   $+\ (CH_3)_3COH\ +\ K^+\ Br^-$ (small amount)

(l) $CH_3CH_2CH_2CH_2{-}\overset{\overset{O}{\|}}{\underset{\underset{O}{\|}}{O}SCF_3}\ +$ [pyridinium] $\overset{+}{N}H\ ^-\overset{\overset{O}{\|}}{\underset{\underset{O}{\|}}{O}SCF_3}$

(m) $CH_3CH_2CH_2CH_2F\ +\ K^+\ ^-\overset{\overset{O}{\|}}{\underset{\underset{O}{\|}}{O}SCF_3}$

10.40   (a) $(CH_3)_3CCl$

(b) no reaction

(c) $(CH_3)_2C{=}CH_2$

(d) no reaction

(e) $(CH_3)_3CO^-\ K^+$
   $+\ H_2$

(f) $(CH_3)_3COMs\ +$ [pyridinium] $\overset{+}{N}H\ Cl^-$

(g) $(CH_3)_2C{=}CH_2\ +\ Na^+\ ^-OMs$

(h) $(CH_3)_2C{=}CH_2\ +\ K^+\ Cl^-$
   $+\ (CH_3)_3COH$

10.41   (a)

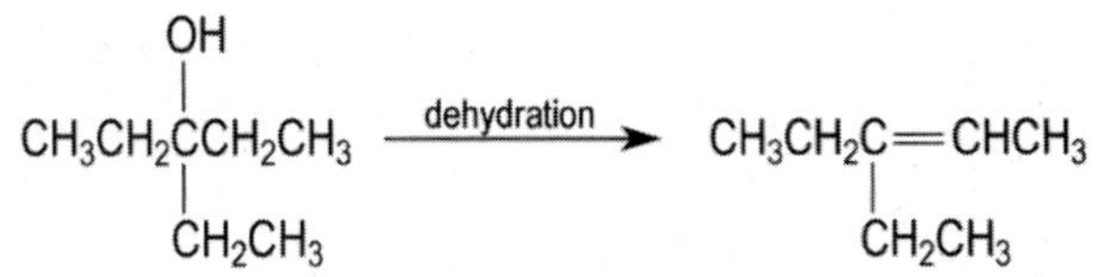

3-ethyl-3-pentanol   →(dehydration)   **3-ethyl-2-pentene**

(b)

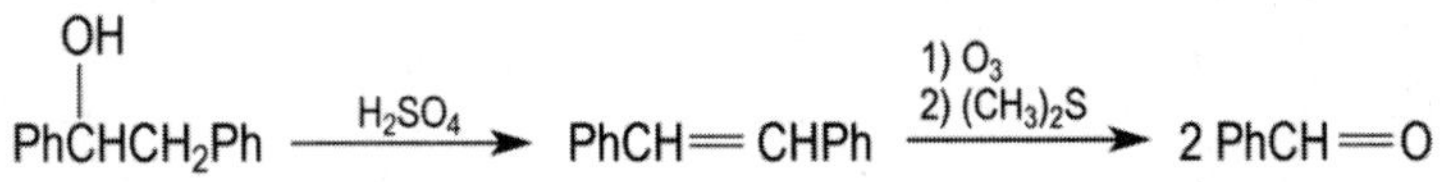

1,2-diphenylethanol   **stilbene**
(mostly trans)   **benzaldehyde**

(c)

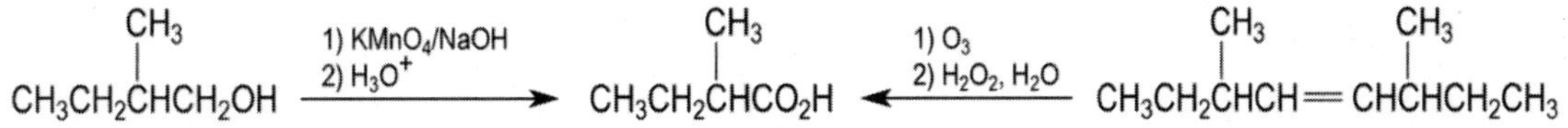

2-methyl-1-butanol   **3,6-dimethyl-4-octene**

10.42   This is a triester of 1,2,3-propanetriol (glycerol) and nitric acid, known by its traditional name of nitroglycerin. It is derived from glycerol (a triol) and nitric acid:

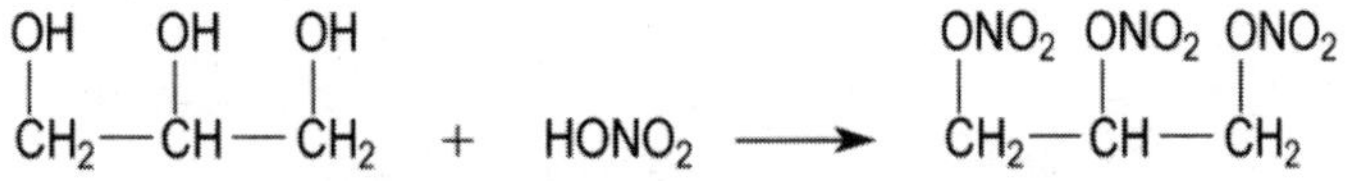

1,2,3-propanetriol   **nitric acid**   **glyceryl trinitrate**
(nitrogylcerin)

10.43   In the following answers, chemically equivalent items have the same letter.

(a)   In this compound, 1,1,1,2,2-pentafluoroethane, fluorines *a* are constitutionally equivalent and homotopic; fluorines *b* are constitutionally equivalent and enantiotopic; fluorines *a* are constitutionally nonequivalent to fluorines *b*.

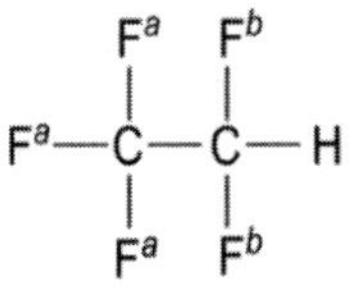

(b)   Fluorines *a* are constitutionally equivalent and diastereotopic; fluorines *a* are constitutionally nonequivalent to fluorine *b*.

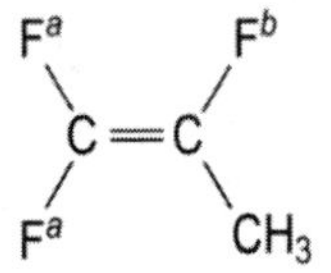

(c)   Fluorines *a* are constitutionally equivalent and diastereotopic; fluorines *a* are constitutionally nonequivalent to fluorine *b*.

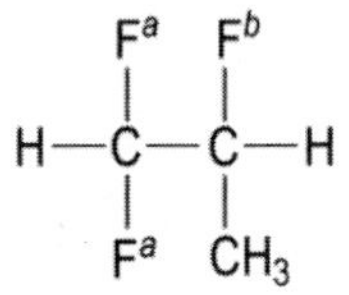

10.44   (a)   (b)

three chemically
nonequivalent
sets of Hs

four chemically
nonequivalent
sets of Hs

(c)   (d)

five chemically
nonequivalent
sets of Hs

five chemically
nonequivalent
sets of Hs

10.45   Protonation of the —OH group, a Lewis acid–base dissociation reaction to form a carbocation, and a Lewis acid–base association reaction of isotopically labeled water with this carbocation are the key steps in this mechanism. (*O represents ^{18}O.)

10.46   (a)   *Tert*-butyl alcohol < isopropyl alcohol < propyl alcohol < 1-propanethiol. Alcohols with a greater degree of alkyl substitution near the OH group are less acidic because their conjugate bases are more poorly solvated; and thiols are substantially more acidic than alcohols (element effect).

(b)   2-Chloroethanol < 3-chloro-1-propanethiol < 2-chloro-1-propanethiol. Thiols are more acidic than alcohols (element effect). 2-Chloro-1-propanethiol is the more acidic thiol because the electronegative chlorine is closer to the site of negative charge in the conjugate-base thiolate anion (polar effect).

(c) $CH_3NHCH_2CH_2CH_2OH$ < $CH_3NHCH_2CH_2OH$ < $(CH_3)_3N^+CH_2CH_2OH$. The last compound is most acidic because it has a full-fledged positive charge that stabilizes a negative charge in the conjugate-base anion. In the other two compounds, the nitrogen stabilizes the conjugate-base anion by its electron-withdrawing polar effect; in the more acidic of these compounds, the nitrogen is closer to the oxygen, which is the site of negative charge in the conjugate-base alkoxide anion (polar effect).

(d) ^-O—CH_2CH_2—OH < $CH_3CH_2CH_2$—OH < CH_3O—CH_2CH_2—OH < CH_3O—CH_2CH_2—OH_2^+. The protonated alcohol is the most acidic because it has an oxygen with a full positive charge. 2-Methoxyethanol is most acidic because of the electron-withdrawing polar effect of the oxygen. The anion is least acidic because the negative charge on the oxygen interacts repulsively with a second negative charge formed on ionization of the O—H group:

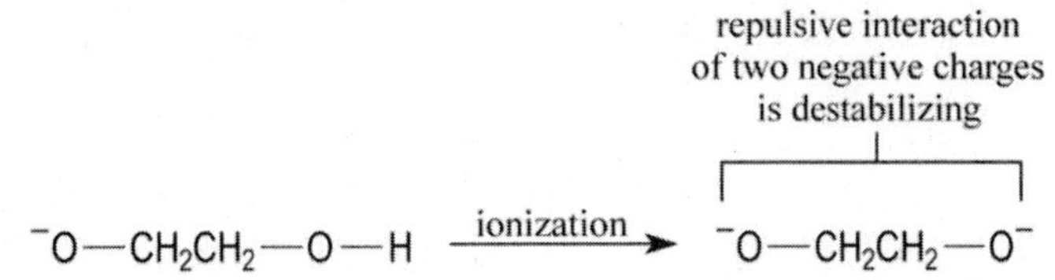

repulsive interaction
of two negative charges
is destabilizing

$$^-O-CH_2CH_2-O-H \xrightarrow{\text{ionization}} {}^-O-CH_2CH_2-O^-$$

(e) Relative to the acidity of ethanol, the acidity of the protonated ether is enhanced by the charge effect; and the acidity of the protonated sulfide is enhanced by both the charge effect and the element effect. The order of increasing acidity is

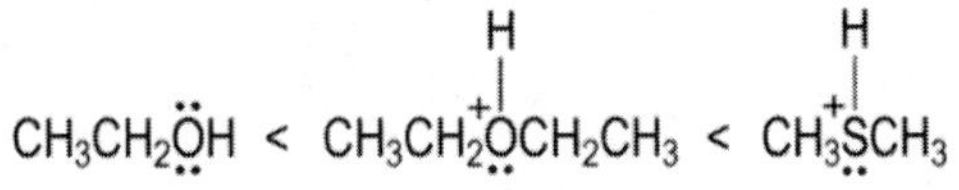

$$CH_3CH_2\overset{..}{\underset{..}{O}}H \; < \; CH_3CH_2\overset{H}{\underset{..}{\overset{+}{O}}}CH_2CH_3 \; < \; CH_3\overset{H}{\overset{+}{S}}CH_3$$

10.47 (a) This exchange occurs essentially through a series of Brønsted acid–base reactions. As shown below, once the deuterium is incorporated into the solvent, it is significantly diluted, so that its probability of reaction with the alkoxide is very small. In addition, such a reaction is retarded by a significant primary deuterium isotope effect and competes less effectively with the corresponding reaction of water.

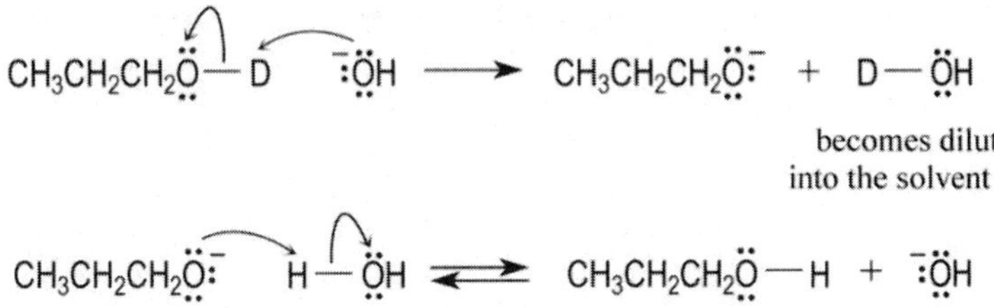

$$CH_3CH_2CH_2\overset{..}{\underset{..}{O}}-D \quad \overset{..}{:}\overset{..}{O}H \longrightarrow CH_3CH_2CH_2\overset{..}{\underset{..}{O}}{}^- \; + \; D-\overset{..}{\underset{..}{O}}H$$

becomes diluted
into the solvent H_2O

$$CH_3CH_2CH_2\overset{..}{\underset{..}{O}}{}^- \quad H-\overset{..}{\underset{..}{O}}H \rightleftharpoons CH_3CH_2CH_2\overset{..}{\underset{..}{O}}-H \; + \; \overset{..}{:}\overset{..}{O}H$$

(b) To prepare $CH_3CH_2CH_2-OD$ from $CH_3CH_2CH_2-OH$, use the same reaction with $D_2O/NaOD$ as the solvent.

10.48 (a) Neither an oxidation nor a reduction. An internal oxidation–reduction has occurred, but there is no *net* oxidation or reduction.

(b) This reaction is a two-electron reduction.

(c) This reaction is a two-electron oxidation.

(d) This reaction is a two-electron oxidation.

(e) This reaction is neither an oxidation nor a reduction; it is a simple ionization.

10.49 (a) We have to use a method that involves one inversion of configuration. Thus, we convert the alcohol to a tosylate; this reaction does not affect the C—O bond. Then we displace the tosylate with inversion using isotopic $^-$OH (that is, $^-O^{18}$—H; the isotopic oxygen is indicated by an asterisk: *O.)

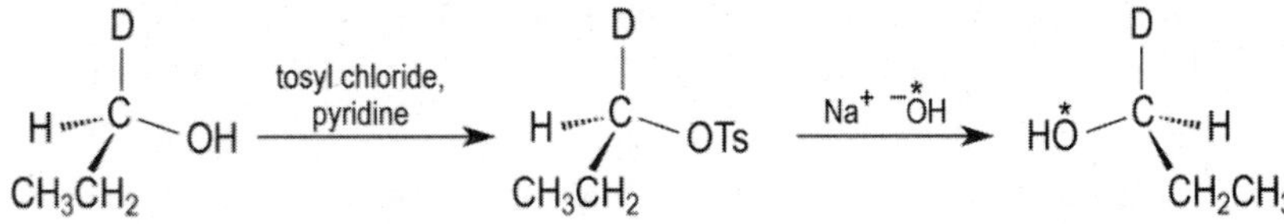

(b) In this case, convert the alcohol into a bromide using a method that involves an inversion of configuration. (Either Ph_3PBr_2 or the two-step alcohol → tosylate → alkyl bromide sequence shown below will work.) Then, in a second inversion step, displace the bromide with $^-O^{18}$—H to provide the alcohol with the desired configuration.

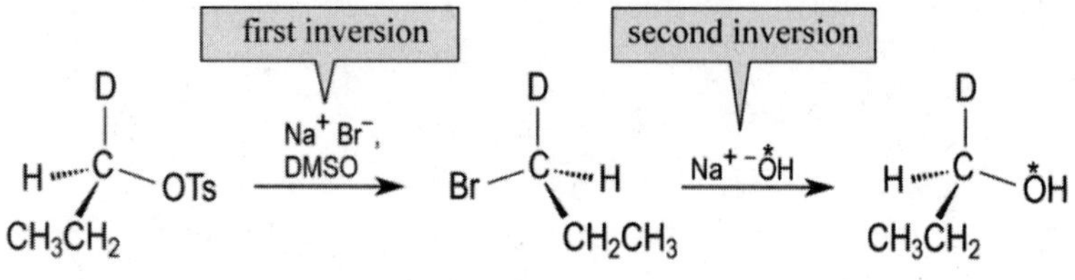

from part (a)

10.50 (a) To introduce a deuterium we need to start with a Grignard reagent, which, in turn, requires an alkyl bromide. Alkyl bromides can be prepared from tertiary alcohols with concentrated HBr, but only after we move the —OH group from carbon-1 to carbon-2 via the alkene.

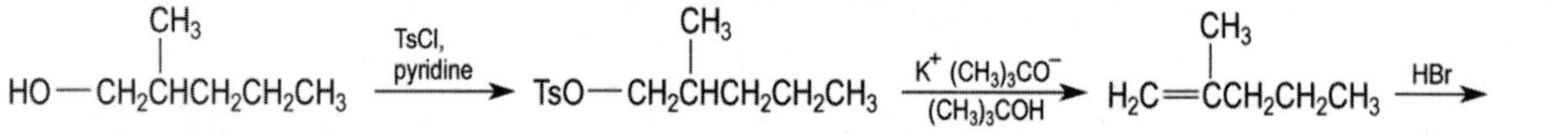

2-methyl-1-pentanol

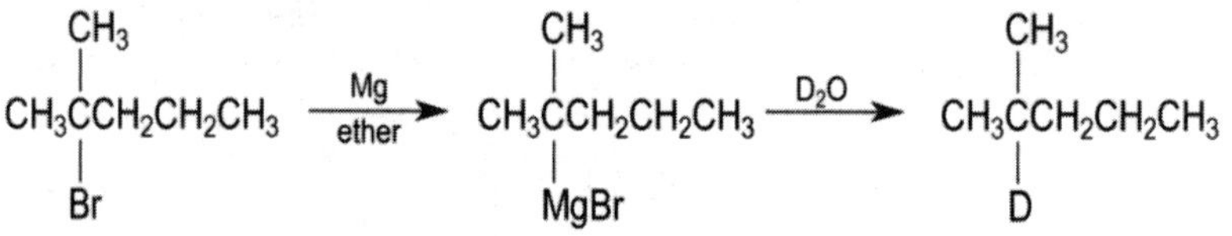

(b)

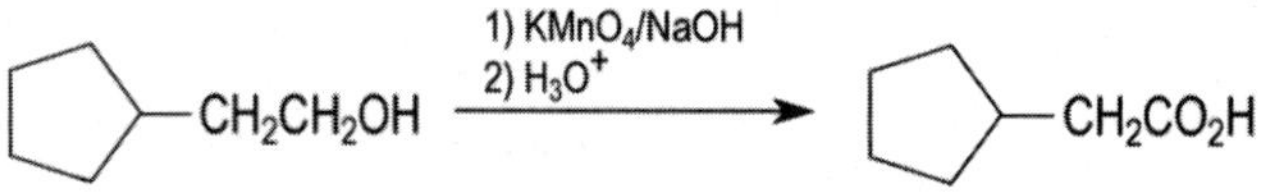

from part (c)

(c) Aldehydes are prepared by oxidation of primary alcohols; hydroboration–oxidation provides the required alcohol from the alkene.

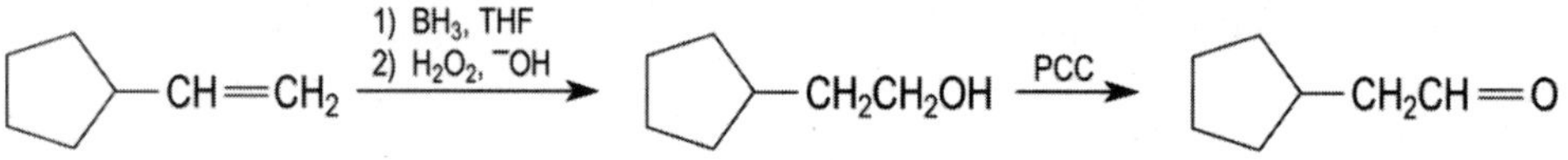

(d) In this part, we have to "throw away" a carbon; ozonolysis comes to mind:

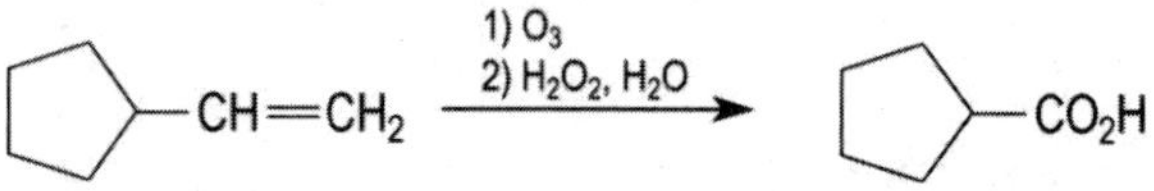

(e) In this question, stereochemistry is important. The product is a racemic mixture of the cis isomers. We can make bromides from the corresponding alcohols, in this case the non-Markovnikov alcohol. Treatment with triphenylphosphine dibromide inverts the configuration at that carbon, yielding the cis products.

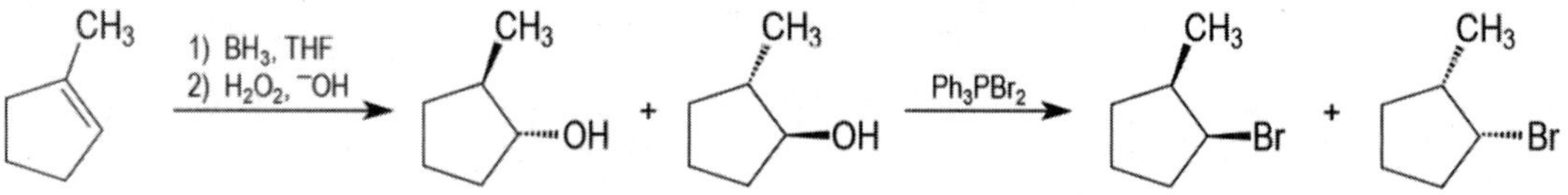

(f)

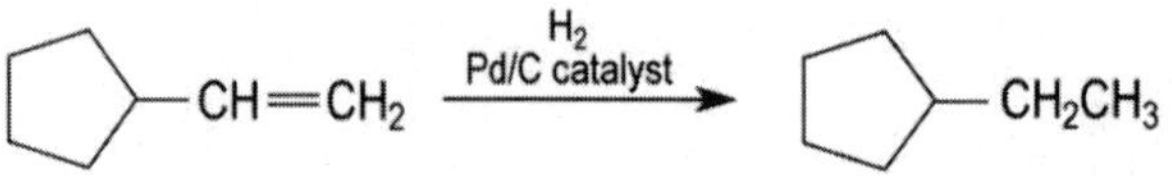

ethylcyclopentane

10.51 (a) The sulfonate ester serves as a leaving group in either case:

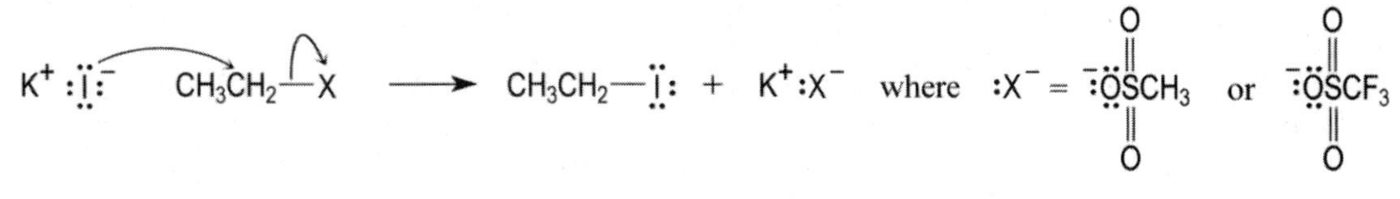

$$K^+ \; :\ddot{I}:^- \quad CH_3CH_2 {-} X \longrightarrow CH_3CH_2{-}\ddot{I}: \; + \; K^+{:}X^- \quad \text{where} \quad :X^- = \; ^-{:}\ddot{O}SCH_3 \quad \text{or} \quad ^-{:}\ddot{O}SCF_3$$

mesylate   triflate

(b) The triflate anion is a weaker base than the mesylate anion because the *polar effect* of the fluorines stabilizes the negative charge in the triflate anion and thereby lowers the pK_a of the conjugate sulfonic acid. (See Sec. 3.6C of the text.)

(c) The principle to apply is that the better leaving group is the weaker base. This is true because the leaving group is accepting a negative charge and breaking a covalent bond in both the Brønsted acid–base reaction with a base and an electron-pair displacement (S_N2) reaction with a nucleophile. The polar effect of the fluorines should operate in the same way on both processes, because the processes are so similar.

10.52 The oxidation of a secondary alcohol to a ketone is a 2-electron oxidation. In the process, CrO_3, a form of Cr(VI), is converted into Cr^{3+}, a form of Cr(III); hence the chromium half-reaction is a 3-electron reduction. Therefore, 2/3 mole of CrO_3 is required to oxidize 1 mole of the alcohol.

The molecular mass of the alcohol is 116; therefore, 10.0 g = 0.0862 mole. Consequently, (0.667)(0.0862) = 0.0575 mole of CrO_3 is required for the oxidation. The molecular mass of CrO_3 = 100; therefore, 5.75 g of CrO_3 is required for the oxidation.

10.53 First, write a balanced equation using the method of half-reactions:

Multiply by 3 →
$$3\,H_2O + 3\,CH_3OCH_2CH_2OH \longrightarrow 3\,CH_3OCH_2CO_2H + 12\,e^- + 12\,H^+$$

Multiply by 4 →
$$12\,e^- + 12\,H^+ + 4\,HNO_3 \longrightarrow 4\,NO + 8\,H_2O$$

Sum: $3\,CH_3OCH_2CH_2OH + 4\,HNO_3 \longrightarrow 3\,CH_3OCH_2CO_2H + 4\,NO + 5\,H_2O$

Because the alcohol oxidation is a four-electron oxidation, and the reduction of nitric acid to NO is a three-electron reduction, it takes four moles of nitric acid to oxidize three moles of the alcohol, that is, 4/3 mole of nitric acid per mole of alcohol. Therefore, it takes (4/3)(0.100) = 0.133 moles of nitric acid to oxidize 0.1 mole of the alcohol.

10.54 The essence of this solution is a balanced equation for the oxidation of ethanol by dichromate and conversion of the stoichiometry into a value for percent blood alcohol. Because the dichromate is aqueous, assume that ethanol is oxidized to acetic acid (a four-electron oxidation). A balanced equation for this process is as follows (verify this):

$$16\,H^+ + 3\,CH_3CH_2OH + 2\,Cr_2O_7^{2-} \longrightarrow 3\,CH_3CO_2H + 11\,H_2O + 4\,Cr^{3+}$$

This equation shows that 3/2 mole of ethanol is oxidized for every mole of $Cr_2O_7^{2-}$ ion consumed. Required to solve the problem are the grams of ethanol oxidized and the mL of blood in which that ethanol is contained. The process can be diagrammed as follows:

moles of dichromate ⇨ moles of ethanol ⇨ grams of ethanol ⎫ divide and
                                                            ⎬ multiply the
                                     mL of air ⇨ mL of blood ⎭ result by 100

The amount of ethanol oxidized is (3/2 mol ethanol per mol of dichromate)(0.507×10^{-6} mol of dichromate) = 0.761×10^{-6} mol of ethanol. The molecular mass of ethanol is 46.07 g mol^{-1}; hence, the grams of ethanol in the sample of blood is (46.07 g mol^{-1})(0.761×10^{-6} mol) = 3.50×10^{-5}.

The 52.5 mL of air collected is equivalent to 0.0250 mL of blood:

$$\text{mL blood} = \frac{52.5 \text{ mL of air}}{2100 \text{ mL of air per mL of blood}} = 0.0250$$

Finally, the grams of ethanol per mL of blood is $(3.50 \times 10^{-5}$ g of ethanol$) \div (0.0250$ mL of blood$) = 1.40 \times 10^{-3}$, and, using the formula in the problem, the percent blood alcohol content is this number times 100, or 0.14%. Consequently, Bobbin is legally intoxicated and Officer Order should make the arrest.

10.55 (a) The glycol is oxidized. This follows from the fact that a bond to carbon is replaced by a bond to oxygen at each carbon of the glycol. The other participant in the reaction, periodate (IO_4^-) must therefore be reduced. Indeed, the ionic product iodate (IO_3^-) contains one less oxygen bound to the iodine.

(b) The number of electrons involved in the oxidation half-reaction is determined from the oxidation numbers of the carbons that change:

The number of electrons lost is $[(+1) + (+1)] - [0 + 0] = +2$. (This result could also be determined from a balanced half-reaction.)

The iodine can be assigned an oxidation number of +7 in periodate and +5 in iodate. How do we know this? Assign +2 to every oxygen because oxygen is divalent and presumably has two bonds to the iodine. (See the top of text p. 480 for a similar case.) Assign a –1 for every negative charge. Hence, the reduction of periodate is a two-electron reduction. (You can verify this with a balanced half-reaction.) Another way to reach the same conclusion is to note that one mole of periodate is required per mole of diol. Because the diol undergoes a two-electron oxidation, periodate must undergo a two-electron reduction.

(c) From the balanced equation, shown in the problem, 0.1 mole of periodate is required to oxidize 0.1 mole of the diol.

10.56 Because thiols and disulfides equilibrate to give a mixture of all possible thiols and disulfides (see Eq. 10.71a–b, text p. 498), Stench has evidently obtained a complex mixture of di(2-octyl) disulfide, dibutyl disulfide, and butyl 2-octyl disulfide. Although the desired product is *statistically* favored (it is twice as likely to be formed as the other disulfides, other things being equal), it is not likely to be more than 50% of the reaction mixture.

10.57 The reactivity data and the molecular formula of *A* indicate that compound *A* is an alkene with one double bond. The identity of compound *D* follows from the oxidation of 3-hexanol; it can only be 3-hexanone (see following equation). It is given that 3-hexanone is an ozonolysis product of alkene *A* (along with H—CO_2H (formic acid), not shown in the following equation). Since alkene *A* has seven carbons and one double bond, and 3-hexanone has six carbons, the carbon of alkene *A* not accounted for by 3-hexanone must be part of a =CH_2 group. Therefore, the identity of alkene *A* is established as 2-ethyl-1-pentene. The identities of compounds *B* and *C* follow from the reactions of *A*.

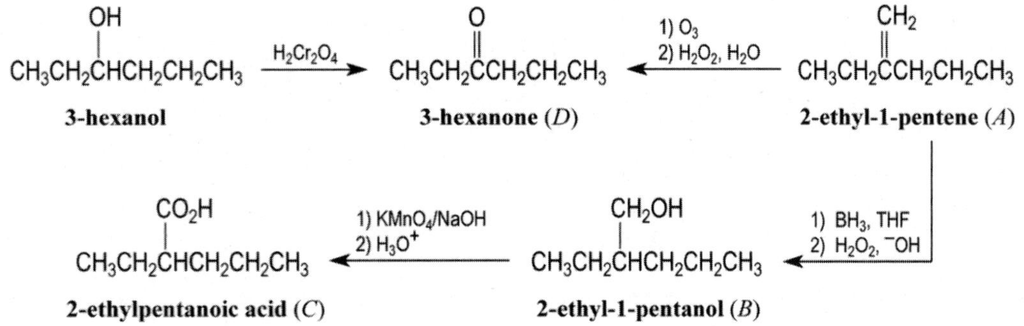

3-hexanol → **3-hexanone (D)** ← **2-ethyl-1-pentene (A)**

2-ethylpentanoic acid (C) ← **2-ethyl-1-pentanol (B)**

10.58 Their reactions with NaH, their formulas, and the fact that they can be resolved into enantiomers show that compounds *A* and *B* are chiral isomeric alcohols. Their eventual conversion into methylcyclohexane shows that they have the same carbon skeleton as methylcyclohexane. When *optically active A* and *B* are converted into tosylates and subjected to the E2 reaction with potassium *tert*-butoxide they give *optically active,* and therefore *chiral,* alkenes. The only possible chiral alkenes with the carbon skeleton of methylcyclohexane are 4-methyl-cyclohexene and 3-methylcyclohexene, which are compounds *C* and *D;* the data do not determine which is which. To summarize:

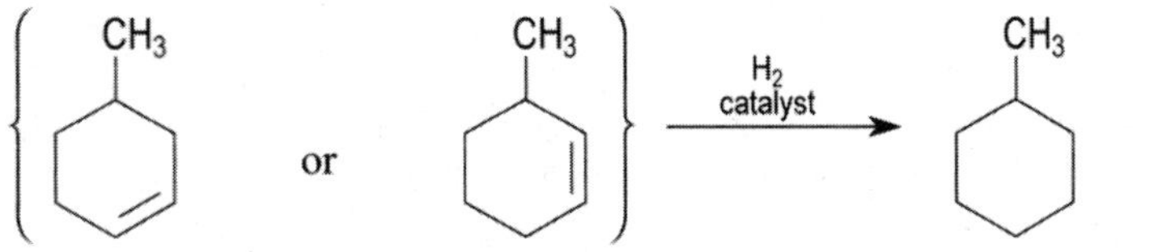

4-methylcyclohexene **3-methylcyclohexene**   **methylcyclohexane**

compounds *C* and *D*
(the data do not determine which is which)

Because compounds *A* and *B* are chiral, their tosylates must also be chiral. The two chiral tosylates that would both give a mixture of compounds *C* and *D* are the tosylates of the stereoisomeric 3-methylcyclohexanols. Because alcohols *A* and *B* have different melting points, they must be the *diastereomeric* 3-methylcyclohex-anols. (Both stereoisomers of 4-methylcyclohexanol are ruled out because both are achiral.) The individual identities of these alcohols follow from an analysis of the E2 reactions of their tosylates. An anti-elimination can take place only when the tosylate group is in an axial position. In the tosylate of *cis*-3-methylcyclohexanol, the tosylate group can only assume the axial position required for E2 elimination when the molecule is in a very unstable conformation in which the methyl group is also axial:

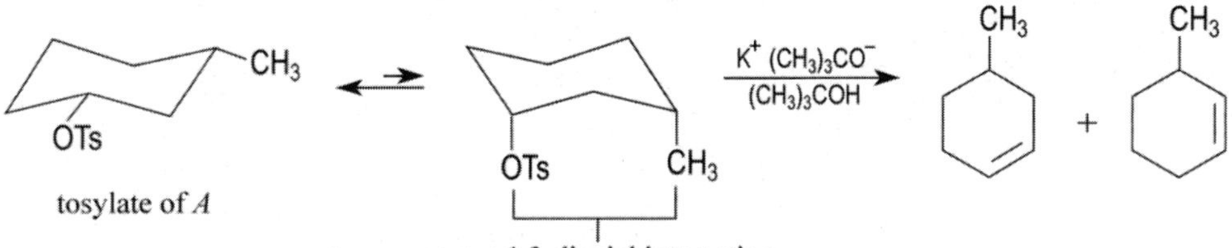

tosylate of *A*

severe 1,3-diaxial interaction

In the tosylate of *trans*-3-methylcyclohexanol, the tosylate group can assume the axial position in a confor-mation in which the methyl group is equatorial. Furthermore, the presence of one axial group in both chair conformations tends to make their energies more nearly equal than in the tosylate of compound *A*, in which the diaxial conformation is *much* less stable.

The compound containing the greater amount of the conformation in which the tosylate group is in the axial position—the tosylate of *B*—undergoes elimination more rapidly. Consequently, compound *A* is *cis*-3-methylcyclohexanol, and compound *B* is *trans*-3-methylcyclohexanol.

cis-3-methylcyclohexanol
(compound *A*)

trans-3-methylcyclohexanol
(compound *B*)

10.59 (a)   First, convert the poor leaving group (the hydroxy group) into a good leaving group (the tosylate group). Then displace the leaving group with bromine to give the product.

(b)   Treatment with HI protonates the OH, turning it into a good leaving group ($-OH_2^+$). Iodide then displaces the H_2O by an S_N2 mechanism, which gives the product below.

(c)   For the structure and mode of action of dimethyl sulfate, see Eq. 10.31 on text p. 469.

$$(CH_3)_2CH-SH \ + \ CH_3O^- \longrightarrow (CH_3)_2CH-S^- \xrightarrow{\text{dimethyl sulfate}} (CH_3)_2CH-SCH_3$$
$$+ \ CH_3OH$$

(d)   The hydroxy group is converted into a good leaving group, $-OS(O)Cl$, which is displaced by chloride as shown by the mechanism on Eq. 10.34, text p. 470, to give the product below.

$$\overset{\text{Cl}}{\underset{}{\underset{\displaystyle CH_3CHCH_2CH_2Ph}{|}}}$$

(e)   Similar to the reaction of triphenylphosphine dibromide (Eqs. 10.36a–b, text p. 471) the reaction of alcohols with triphenylphosphine dichloride proceeds by an S_N2 mechanism.

(f)   Treatment of the alcohol with strong acid protonates the hydroxy group making it a good leaving group. Once the water leaves, there are two different but equivalent protons adjacent to the positively charged carbon. Since there is an equal opportunity to remove either one leading to one of two enantiomers, a racemic mixture of E1 elimination products is generated.

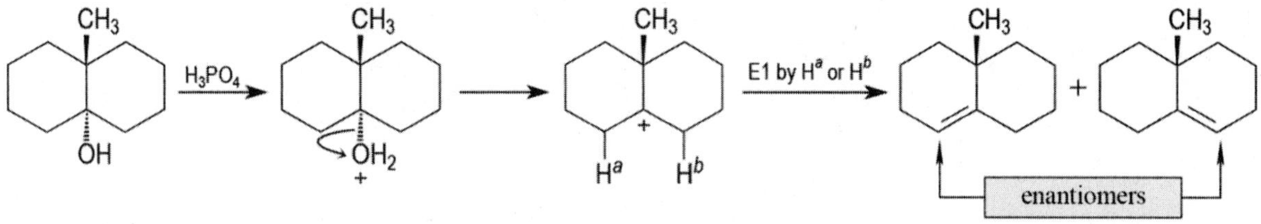

(racemic mixture)

(g)   The first step occurs by an S_N1 or S_N2 mechanism to give the alkyl bromide, which then undergoes an E2 elimination to give the alkene.

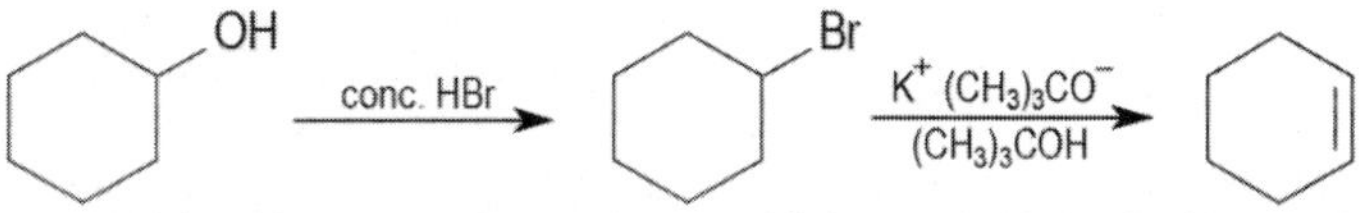

(h)   An equilibrium is established between the various sulfides and thiol, similar to the solution to problem 10.56.

$(CH_3)_2CHCH_2CH_2$—S—S—CH_2CH_3 + $(CH_3)_2CHCH_2CH_2$—S—S—$CH_2CH_2CH(CH_3)_2$ +

CH_3CH_2—S—S—CH_2CH_3 + $(CH_3)_2CHCH_2CH_2$—SH + CH_3CH_2—SH

10.60   (a)   The primary deuterium kinetic isotope effect of 6.6 suggests that the step of the mechanism shown in Eq. 10.52c, text p. 484, is rate-limiting, because the isotopically substituted hydrogen is transferred in this step. (Review the discussion of the *deuterium kinetic isotope effect*, Sec. 9.5D.)

(b)   Because of the primary kinetic isotope effect in part (a), the deuterium is removed more slowly than the hydrogen (about seven times more slowly). Thus, more of the aldehyde containing deuterium is formed. Because PCC is an achiral reagent, it cannot differentiate between the two enantiotopic α-hydrogens of ethanol; each is removed with equal frequency. Consequently, the isotope effect causes isotopic discrimination *regardless* of the stereochemical positions of the hydrogen and deuterium.

(c)   The enzyme alcohol dehydrogenase, which is a chiral catalyst, distinguishes between the α-hydrogens of ethanol so completely that it removes *only* the pro-(R) hydrogen whether this hydrogen is isotopically substituted or not. Thus, when this hydrogen is substituted by deuterium, the deuterium is removed, and, as a result, the aldehyde contains no deuterium. Although there may be an isotope effect associated with removal of this hydrogen, it is not reflected in the relative amounts of H and D removed from a given enantiomer because the stereochemical discrimination (which may be a factor of 1000 or more) is a much larger effect than the primary deuterium isotope effect (which could be a factor only as large as 7 or so).

10.61   (a)   The oxidation of the alcohol RCH_2OH to the aldehyde $RCH{=}O$ is a two-electron oxidation. The reduction of DMSO to the sulfide is a two-electron reduction. Although the carbon of carbon monoxide has undergone a net one-electron reduction, the carbon of CO_2 has undergone a compensating net one-electron oxidation.

(b)   The oxidizing agent is the sulfur atom in DMSO; it is reduced (because it oxidized the alcohol carbon).

(c)   To solve this problem, first write out the reaction that occurs with deuterium-labeled alcohol:

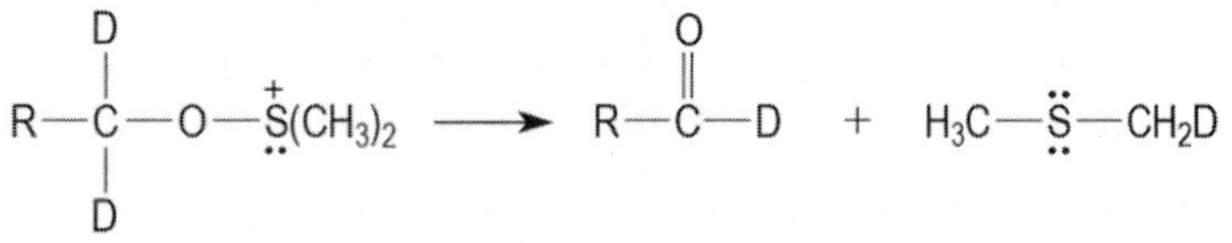

Since deuterium ends up incorporated into one of the methyl groups on sulfur, one of those hydrogens must have been removed at some point. Use triethylamine to remove a proton, generating an intermediate called a sulfur ylide. The sulfur ylide then can react *intramolecularly* to remove a deuterium from the nearby carbon, oxidizing that carbon, and eliminating dimethyl sulfide containing deuterium. The sulfur ultimately accepts electrons so it is oxidized, as noted in the solution to part (b).

a sulfur ylide

10.62   The two branches of citrate at the central carbon are *enantiotopic.* Hence, they are chemically distinguishable to a chiral catalyst such as an enzyme. Evidently, the difference is such that the dehydration occurs into the unlabeled branch, as shown in Fig. P10.62.

One difference in H_2SO_4 solution is that the carboxylate groups are not ionized, but this is not the key difference. The point of the problem is that an achiral laboratory reagent will not make the distinction between enantiotopic groups. Hence, equal amounts of dehydration should occur into each branch, and there is no reason to expect exclusively the *Z* stereochemistry observed in the product of the enzyme-catalyzed reaction.

+ the (*E*)-stereoisomers

10.63   First, draw the 2*S*,3*R* stereoisomer of the product so that the stereochemistry of the addition can be deduced.

(2S,3R)-3-deuteriomalate

Only a malate stereoisomer with the 2*S* configuration will dehydrate; if the enzyme is stereospecific in one direction, it must be stereospecific in the other. (See Eq. 7.21, text p. 302.) The pro-*R* hydrogen at carbon-3 comes from the solvent; if the reaction had been run in H_2O, this would be a hydrogen. This is an anti-addition; we leave it to you to confirm this point. (See Problem 7.54(a), text p. 319.)

(a)   Because this stereoisomer has the 2*S* configuration, it will dehydrate. Although the pro-*R* hydrogen at carbon-3 will be lost to solvent, the stereochemistry at this position is irrelevant because there is no isotopic label. The deuterium in the starting material is retained, because this deuterium is not part of the reaction. The product is

(E)-2-deuteriofumarate

What if the reaction, when it reverses back to malate, gives the product with the OH group on the carbon *without* the deuterium? This is possible, because deuterium and hydrogen differ in size by only a tiny amount, and the enzyme will not distinguish them by size alone. This reaction, because of the stereospecificity of the enzyme, must also form the 2*S* enantiomer, and it must be formed in an anti-addition. This malate product, then, must be the diastereomer of the 2*S*,3*R* stereoisomer we started with:

$$^-O_2C-C-C-CO_2^-$$

(2S,3S)-3-deuteriomalate

To the extent that any malate remains at equilibrium, it will ultimately be a mixture of the 2S,3S and 2S,3R diastereomers. Now, what if this stereoisomer then reverses back to fumarate? Will the deuterium be washed out? It will not be, and we leave it to you to apply the principle of microscopic reversibility to understand why.

(b)   The fumarate stereoisomer obtained in this reaction is the same as in part (a). The same comments apply to the dynamic reversal of the reaction; the starting material, to the extent that it remains at equilibrium, will be a mixture of the 2S,3S and 2S,3R diastereomers, and the deuterium will not wash out.

(c)   Because this malate stereoisomer has the 2R configuration, and because the enzyme is specific for 2S-malate, this isomer will not react.

10.64   (a)   Following the steps in Sec. 10.5A, we can balance the equation. First, we don't need to add any water because no oxygens are involved. Next, we add one H^+ to the right side to replace the one that was lost from serotonin. Finally, we add two electrons to the right side to cancel the plus charges. Thus, two electrons were lost, or, we can say this reaction is a two-electron oxidation.

(b)   The way that the serotonin structure is shown, the hydrogen pointing away is the pro-R hydrogen (see Sec. 10.9B).

pro-R hydrogen

Since the enzyme is specific and can specifically remove the pro-R hydrogen instead of the pro-S hydrogen, different products will result depending on whether compound A or B is fed to the enzyme.

A

B

(c)   Compound A reacts faster, due to the deuterium isotope effect, discussed in Sec. 9.5D. When the pro-R hydrogen is H (as in A) it is abstracted faster than when it is D (as in B).

10.65   Unlike HBr, HCN is a rather weak acid ($pK_a = 9.4$), and it provides very little cyanide ion because it is essentially un-ionized. In addition, it is not acidic enough to protonate the alcohol to any reasonable extent, because the pK_a of a protonated alcohol is –2. Because the OH group of the alcohol is not protonated, it is not a good leaving group. With a nucleophile concentration of nearly zero and a poor leaving group, a substitution reaction cannot take place.

Addition of H_2SO_4 to the reaction mixture would indeed solve the leaving-group problem, because the H_2SO_4 would protonate the OH group. However, added protons suppress the ionization of HCN even more than it is suppressed normally. Without a reasonable concentration of nucleophile, the reaction is doomed. And, unless he is working in a very efficient fume hood, Bluelip is doomed, too, because addition of H_2SO_4 to a solution of cyanide ion produces lethal HCN gas.

10.66   Ethanol is partially protonated under the acidic conditions. Water is displaced from a protonated ethanol molecule by another ethanol molecule:

CH₃CH₂—ÖH₂ ⟶ CH₃CH₂—Ö⁺—CH₂CH₃ + ÖH₂ ⇌ CH₃CH₂—Ö—CH₂CH₃ + H₂ÖCH₂CH₃

protonated ethanol
(the conjugate acid
of ethanol)

Ethanol is a weak base, and it is therefore not a very potent nucleophile. This is why the reaction requires high temperature.

The outcome of the reaction is dependent on the reaction conditions. The reaction is run in a sealed pressure vessel, because the temperature is well above the boiling points of the reactants, products, and solvent. If the vessel were not sealed, ethylene would be formed by alcohol dehydration and, as the lowest-boiling component, would escape from the vessel. In a sealed vessel, and ethylene that is formed remains in solution and can undergo acid-catalyzed addition of ethanol.

10.67   (a)

CH₃CH₂CHCH₂CH₃ + Br—P—Br ⟶ CH₃CH₂CHCH₂CH₃ ⟶ CH₃CH₂CHCH₂CH₃ + HO—PBr₂

(b)

(PhO)₃P: + H₃C—I ⟶ PhO—P⁺—OPh ⁻I ⟶ ⟶
                                    X

⟶ I + (PhO)₂P—CH₃ + PhOH

(c)

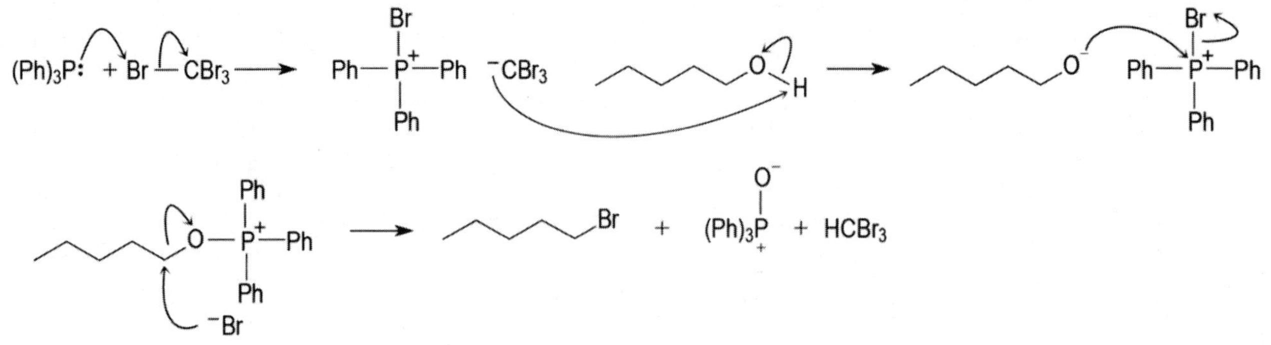

10.68   (a)   The carbocation rearrangement in this solution, in which a tertiary carbocation is converted into a secondary carbocation, is driven by relief of ring strain.

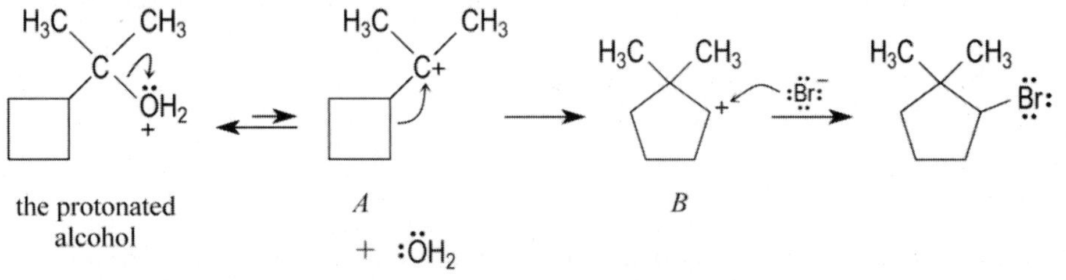

the protonated
alcohol

A

+ :ÖH$_2$

B

The product shown above is the major product (74%). But this mechanism begs a question: Why isn't the major product one that forms from carbocation *B* rearranging again by a methyl shift to a tertiary carbocation? [What would the product(s) be?] The answer may be that the empty 2*p* orbital of this carbocation is not aligned optimally with the ring–methyl bonds for rearrangement to occur. (For rearrangement to occur, overlap must occur between the bonding orbitals and the empty orbital.)

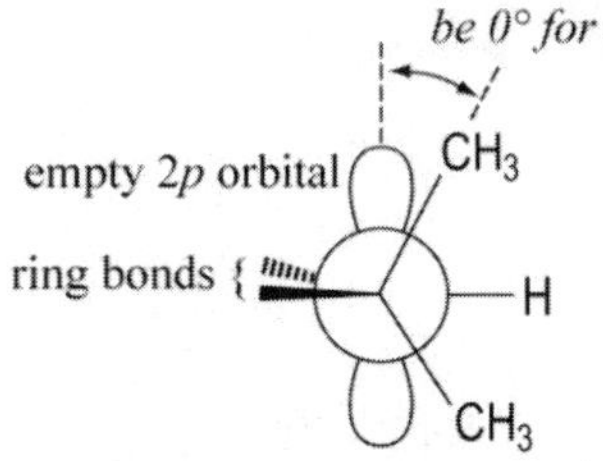

*this angle should
be 0° for optimum overlap*

empty 2*p* orbital

ring bonds {

If the rate of this rearrangement of carbocation *B* is decreased, then the large excess of bromide ion (from ionization of HBr) completely consumes carbocation *B* before it can rearrange. The alkyl bromide, which is very insoluble in the partially aqueous solvent, forms a separate layer and is removed from the reaction. Undoubtedly, in a solvolytic medium such as ethanol, in which the alkyl halide would be soluble, rearrangement of carbocation *B* would be observed.

Carbocation *A* has no problem with orbital alignment because there is more or less free (or less constrained) rotation about the bonds to the cationic carbon. The rearrangement is evidently much faster than the capture of this ion by bromide ion.

(b)   The carbocation rearrangement in this mechanism also involves a ring expansion. In this case, a tertiary carbocation is converted into another tertiary carbocation. The reaction occurs because a more stable six-membered ring forms from the slightly less stable five-membered ring. Other products are likely observed, but the question only asks us to provide a mechanism for the one shown.

the protonated alcohol   + :ÖH₂

(c)   The product contains eight carbons; hence, it involves the reaction of two 2-methylpropene carbon skeletons, joined as follows:

this carbon has
an additional H

this carbon has
an additional H

The first step in this process is the reaction of one alkene molecule with HF to give a *tert*-butyl cation *A*:

(We won't show the fluoride counterion in subsequent steps.) Carbocation *A* then reacts in a Lewis acid–base association reaction with a second molecule of alkene to give carbocation *B*:

$(CH_3)_3\overset{+}{C}$ $H_2C=C(CH_3)_2$ ⟶ $(CH_3)_3C-CH_2-\overset{+}{C}(CH_3)_2$

B

Finally, carbocation *B* reacts with isobutane by removing a *hydride* (a hydrogen with two electrons). This is exactly like a carbocation rearrangement by hydride shift, except that the reaction occurs across two different molecules:

H—C(CH₃)₃

$(CH_3)_3C-CH_2-\overset{+}{C}(CH_3)_2$ ⟶ $(CH_3)_3C-CH_2-\overset{H}{\underset{|}{C}}(CH_3)_2$ + $(CH_3)_3C+$

B   **2,2,4-trimethylpentane**   *A*

The product is accompanied by a *tert*-butyl cation *A*, which is the cation produced in the initial protonation step. This cation can thus react with another molecule of alkene to give another carbocation *B*, and so on. Notice that if a *tert*-butyl cation *A* reacts with a molecule of isobutane by a hydride shift, another *tert*-butyl cation *A* and another isobutane are formed, that is, such a "side reaction" (which undoubtedly occurs) has no net effect. Finally, the hydride-shift mechanism accounts for the results with deuterium.

This reaction is actually a *chain reaction*, in which the *tert*-butyl cation propagates the chain. Although most chain reactions are *free-radical* chain reactions, this is an interesting example of a *cationic* chain reaction.

You might compare this problem with Problem 5.53 (text p. 228) and see if you can tell what it is about the conditions of the two reactions, both of which involve 2-methylpropene and HF, that causes the outcome to be different.

(d)   This is an addition to the alkene that is conceptually similar to hydration or HBr addition. Trifluoromethanesulfonic acid (triflic acid), a strong acid, protonates the alkene double bond to give a carbocation, which then undergoes a Lewis acid–base association reaction to give the product.

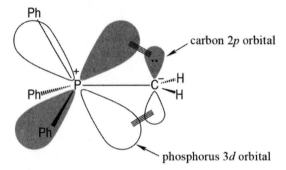

10.69   (a)   The electrons on carbon are in a 2*p*-orbital, since carbanions are *sp*²-hybridized. Phosphorus has no empty p-orbitals, but it does have empty 3*d*-orbitals. One of its 3*d*-orbitals shares enough symmetry with the 2*p*-orbital to sufficiently overlap. The π bond between phosphorus and carbon must come from overlap between these two orbitals (see Sec. 10.10B).

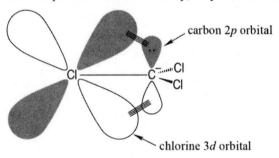

   (b)   Both carbon and phosphorus possess eight electrons, either as lone pairs or shared.

   (c)   Carbanions are very strong bases because they are the conjugate bases of very weak acids. This compound possesses a partial negative charge on carbon, (partial because the structures in parts (a) and (b) are averaged) which makes this carbon very basic.

10.70   Similar to the answer to 10.69a, the chlorines provide empty 3*d*-orbitals to overlap with the 2*p*-orbital of the carbon containing the lone electron pair, helping to delocalize the negative charge across the whole molecule. Note that while each chlorine atom can provide the same overlap, only one such interaction is shown.

11

The Chemistry of Ethers, Epoxides, Glycols, and Sulfides

STUDY GUIDE LINKS

 11.1   Learning New Reactions from Earlier Reactions

Quite often the text points out the close relationship between two reactions—a new one, and one you have already studied. Such connections will help you to understand new reactions in terms of what you already know and will substantially reduce the amount of *really new* material that you must master. Alkoxymercuration–reduction is such a case; this reaction is closely related to oxymercuration–reduction. In order to make the connection between the two reactions, it almost goes without saying that you have to know what oxymercuration–reduction is! If you did not know the meaning of this term when you read this section, did you go back and find out?

The question at the end of Study Problem 11.1 on text p. 514—why *tert*-butyl bromide and sodium methoxide will not work in a Williamson ether synthesis—is another effort to get you to think about earlier material, in this case the reactivities of alkyl halides in the presence of a strong base. Did you answer this question? (If you can't answer it, review Sec 9.5G as suggested in the problem.)

Remember that *continued review* is one of the keys to successful study in organic chemistry. You must try constantly to remain active rather than passive when you study. Answering questions such as the ones posed in the text will assist you. If you can't answer one of these questions, write it down and get help at the next available opportunity!

 11.2   Common Intermediates from Different Starting Materials

An important thing to notice about this section is that the same reactive intermediate—in this case, a carbocation—in some cases can be generated in several different ways. For example, consider the many different ways that a *tert*-butyl cation can be formed as a reactive intermediate:

Whether a carbocation is formed by protonation of an alkene, by dehydration of an alcohol, or by sovolysis of an alkyl halide, the same product is obtained if a large excess of the same nucleophile is present in each case. Thus, if methanol is used as the solvent, *tert*-butyl methyl ether is the product regardless of the origin of the carbocation. On the other hand, if concentrated HBr is used, *tert*-butyl bromide is the product.

It is natural at first to think of hydrogen halide addition to alkenes, alcohol dehydration, alkene hydration, and solvolysis of tertiary alkyl halides as different reactions. What you now should be able to see is that these reactions are all linked through a common intermediate, the carbocation. Which product is obtained depends on the conditions used.

FURTHER EXPLORATIONS

 11.1   Mechanism of OsO₄ Addition

Even though the OsO₄ addition is a concerted reaction—a reaction that occurs in a single step—let's break the reaction into fictitious individual steps so that we can see the relationship of this reaction to other electrophilic additions we have studied.

First, let's think about the electronic character of OsO₄. Remember that osmium in a +8 oxidation state is very electronegative. When a nucleophile reacts at any of the oxygens, electrons are transferred onto the osmium.

In this sense, osmium serves as a "leaving group" in the nucleophilic substitution reaction on oxygen—except that it remains tethered to that oxygen by its one remaining bond. The mechanism of OsO₄ addition can now be viewed as a close analogy to the mechanism of ozone addition to alkenes (Further Exploration 5.2). The electrophilic oxygen accepts the pair of π electrons from the alkene, and a carbocation is generated from the alkene. We can write a resonance structure for the osmium species in which the electron pair (and negative charge) on osmium is delocalized onto an oxygen. This oxygen undergoes a Lewis acid–base association to complete the addition.

The reaction mechanism diagram showing osmate ester formation:

$$\text{an osmate ester}$$

Compare this to the "fictitious" ozonolysis mechanism, in which the central oxygen serves as an "electron sink" and the terminal, negatively-charged, oxygen reacts with the carbocation:

Even though these mechanisms are concerted, thinking of them as two-step processes, as shown above, helps to understand why these reactions take place, and helps to see the connection between these reactions and other electrophilic addition reactions that involve carbocations explicitly.

REACTION REVIEW

I. SYNTHESIS OF ETHERS, EPOXIDES, GLYCOLS, AND SULFIDES

A. WILLIAMSON ETHER SYNTHESIS

1. The Williamson ether synthesis is an S_N2 reaction in which the conjugate base of an alcohol (or thiol) reacts as a nucleophile with an unhindered alkyl halide or sulfonate ester to form an ether (or sulfide).

$$(CH_3)_3C-O^-\,Na^+ \ + \ H_3C-OTs \ \longrightarrow \ (CH_3)_3C-O-CH_3 \ + \ TsO^-\,Na^+$$

| sodium 2-methyl-2-propanolate | methyl tosylate | 2-methoxy-2-methylpropane (or *tert*-butyl methyl ether) | sodium tosylate |

$$Ph-S^-\,Na^+ \ + \ CH_3CH_2Br \ \longrightarrow \ Ph-S-CH_2CH_3 \ + \ Na^+\,Br^-$$

| sodium thiophenolate | bromoethane | (ethylthio)benzene (or ethyl phenyl sulfide) | sodium bromide |

   a. Methyl halides, primary alkyl halides, or the corresponding sulfonate esters can be used in a Williamson synthesis.
   b. Tertiary and many secondary alkyl halides cannot be used in a Williamson synthesis because the E2 (elimination) reaction occurs instead.

$$H_3C-O^-\,Na^+ \ + \ (CH_3)_3C-I \ \longrightarrow \ H_3C-O-H \ + \ Na^+\,I^- \ + \ (CH_3)_2C=CH_2$$

| sodium methanolate | *tert*-butyl iodide (or 2-iodo-2-methylpropane) | methanol | sodium iodide | 2-methylpropene |

2. In principle, two different Williamson syntheses are possible for any ether with two different alkyl groups.
   a. The preferred synthesis is usually the one that involves the alkyl halide with the greater S_N2 reactivity.
   b. Tertiary and secondary alkyl groups should be derived from the alkoxide.

B. ALKOXYMERCURATION–REDUCTION OF ALKENES

1. An alkoxymercuration–reduction reaction occurs when oxymercuration–reduction is carried out in an alcohol solvent instead of water.

2. The mechanism of the alkoxymercuration reaction is completely analogous to the mechanism of oxymercuration, except that an alcohol instead of water is the nucleophile that reacts with the mercurinium ion intermediate.

3. The alkoxymercuration–reduction reaction can be used to prepare ethers that cannot be prepared by a Williamson ether synthesis.

C. ETHERS FROM ALCOHOL DEHYDRATION AND ALKENE ADDITION

1. Two molecules of an alcohol can undergo dehydration to give an ether (a reaction used primarily in industry).

a. This is generally restricted to the preparation of symmetrical ethers derived from primary alcohols.

b. Secondary and tertiary alcohols cannot be used because they undergo dehydration to alkenes.

c. The formation of ethers from primary alcohols is an S_N2 reaction in which one alcohol displaces water from another molecule of protonated alcohol.

2. Tertiary alcohols can be converted into unsymmetrical ethers by treating them with dilute solutions of strong acids in an alcohol solvent.

a. The formation of an ether from a tertiary alcohol and a primary alcohol solvent is an S_N1 solvolysis reaction in which the protonated alcohol loses water to give a relatively stable carbocation that reacts with solvent to give the product.

b. One of the alcohol starting materials must readily lose water after protonation to form a relatively stable carbocation; typically this is a tertiary alcohol.

c. The alcohol that is used in excess must be one that either cannot lose water after protonation to give a carbocation or should form a carbocation much less readily than the tertiary alcohol. Typically this is a primary alcohol.

d. Any alkene that does form (by an E1 pathway) is not removed but is reprotonated to give back a carbocation, which eventually reacts with the alcohol solvent.

e. Ether formation from tertiary alcohols and the dehydration of tertiary alcohols are alternative branches of a common mechanism. (See Study Guide Link 11.2 and Sec 10.2 in the text.)

3. Treatment of an alkene with a large excess of alcohol in the presence of an acid catalyst gives an ether provided that a relatively stable carbocation intermediate is involved.

1-methylcyclohexene   **1-ethoxy-1-methylcycloxexane**

D. OXIDATION OF ALKENES WITH PEROXYCARBOXYLIC ACIDS

1. One of the best laboratory preparations of epoxides involves the direct oxidation of alkenes with peroxycarboxylic acids.

cyclopentene   a peroxycarboxylic acid   a carboxylic acid

cyclopentene oxide
(1,2-epoxycyclohexane)

a. The oxidizing agent, a peroxycarboxylic acid, is a carboxylic acid that contains an —O—O—H (hydroperoxy) group rather than an —OH (hydroxy) group.

b. The general terms *peroxyacid* or *peracid*, denoting any acid containing a hydroperoxy group, are sometimes used instead of peroxycarboxylic acid.

c. Many peroxycarboxylic acids are unstable and are formed just prior to use by mixing a carboxylic acid with hydrogen peroxide.

2. The formation of an epoxide from an alkene and a peroxycarboxylic acid is a concerted addition reaction. (See Sec. 11.3A in the text for a further discussion of this mechanism.)

alkene   peroxycarboxylic   epoxide   carboxylic
          acid   acid

a. This mechanism is very similar to that for the formation of a bromonium ion in bromine addition to alkenes.

b. The formation of epoxides with peroxycarboxylic acids is a stereoselective reaction; it takes place with complete retention of the alkene stereochemistry. (See example under point 1.)

E. CYCLIZATION OF HALOHYDRINS

1. Epoxides can be synthesized by the treatment of halohydrins with base.

a. This reaction is an intramolecular variation of the Williamson ether synthesis in which the alcohol and the alkyl halide are part of the same molecule.

b. The alkoxide anion, formed reversibly by reaction of the alcohol with NaOH, displaces halide ion from the neighboring carbon.

2. Like other S_N2 reactions, this reaction takes place by backside substitution of the nucleophilic oxygen anion at the halide-bearing carbon.

a. The nucleophilic oxygen anion and the leaving halide assume an anti relationship in the transition state of the reaction.

b. Halohydrins derived from cyclic compounds must be able to assume the required anti relationship through a conformational change if epoxide formation is to succeed.

forms an epoxide in basic solution

does not form an epoxide in basic solution

F. ASYMMETRIC EPOXIDATION

1. Allylic alcohols react with *t*-butyl hydroperoxide in the presence of titanium(IV) isopropoxide catalyst to form an epoxide at the double bond of the allylic alcohol functionality.

an allylic alcohol   **t-butyl hydroperoxide**   an epoxide   **t-butyl alcohol**

a. Only double bonds with an allylic —OH group form an epoxide; other double bonds usually do not react.

b. In the presence of (2R,3R)-(+)-diethyl tartrate, (+)-DET, the epoxide form is a preponderance of only one enantiomer is formed; in the presence of (2S,3S)-(−)-diethyl tartrate, (−)-DET, the epoxide form is a preponderance of the other enantiomer is formed.

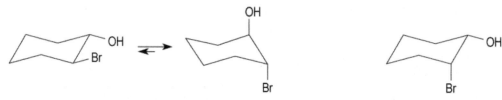

"O" adds from below

"O" adds from above

G. PREPARATION OF GLYCOLS

1. Some glycols can be prepared by the acid-catalyzed reaction of water with epoxides. (See also Section II.B.)

an epoxide   a glycol

2. Alkenes can be oxidized to glycols with OsO_4.

   a. The osmium in OsO_4 is in a high oxidation state, Os(VIII), and thus attracts electrons.

   i. This electron-attracting ability results in a concerted cycloaddition reaction between OsO_4 and an alkene that reduces the Os(VIII) to Os(VI) and gives an intermediate called an osmate ester. (See Further Exploration 11.1 on p. 304 of this manual for more on the mechanism.)

an osmate ester

   ii. A glycol is formed when the cyclic osmate ester is treated with water.

an osmate ester   a glycol

   b. A mild reducing agent such as sodium bisulfite ($NaHSO_3$) is added in many cases to convert the osmium-containing by-products into reduced forms of osmium that are easily removed by filtration.

   c. Osmium and its compounds are very toxic and are quite expensive. By including an amine oxide (another oxidizing agent) in the reaction mixture, only catalytic amounts of OsO_4 are required, because the amine oxide converts the osmium by-products back into OsO_4.

an alkene   (catalytic   water/tert-butyl   an osmate ester   a glycol   Os(VI)
               amount)   alcohol (solvent)
                                 pyridine

Os(VI)   an amine oxide   catalyst   an
                                           regenerated   amine

d. The formation of glycols from alkenes is a stereospecific syn-addition. The five-membered osmate ester ring is easily formed when two oxygens of OsO_4 are added to the same face of the double bond by a concerted mechanism.

osmate ester   cis-1,2-cyclohexanediol

e. Hydrolysis of epoxides and the OsO_4 oxidation of alkenes are complementary reactions because they provide glycols with different stereochemistry.

3. Glycol formation from alkenes can also be carried out with potassium permanganate ($KMnO_4$), usually under aqueous alkaline conditions.

cis-1,2-cyclohexanediol

a. This reaction is also a stereospecific syn-addition, and its mechanism is probably similar to that of OsO_4 addition.

   i. Visually, when oxidation occurs, the brilliant purple color of the permanganate ion is replaced by a murky brown precipitate of magnesium dioxide (MnO_2).

   ii. This color change can be used as a test (called the Baeyer test) for functional groups that can be oxidized by $KMnO_4$.

b. A problem with the use of $KMnO_4$ is that yields are low in many cases because over-oxidation occurs.

II. REACTIONS OF ETHERS, EPOXIDES, GLYCOLS, AND SULFIDES

A. BASICITY OF ETHERS AND SULFIDES

1. Ethers are relatively unreactive compounds; however, the oxygen of an ether can be protonated under acidic conditions.

$$R—O—R' \; + \; H—A \; \longrightarrow \; R—\overset{+}{\underset{H}{O}}—R'$$

$$pK_a = -2 \text{ to } -3$$

2. Thiols are less basic than ethers, but react similarly in acid.

$$R—S—R' \; + \; H—A \; \longrightarrow \; R—\overset{+}{\underset{H}{S}}—R'$$

$$pK_a = -5 \text{ to } -7$$

3. Ethers can also act as Lewis bases with electron deficient Lewis acids.

$$R-O-R' \quad + \quad H-B-H \quad \longrightarrow \quad \begin{array}{c} \overset{+}{R}-O-R' \\ | \\ H-\overset{-}{B}-H \\ | \\ H \end{array}$$

an ether	borane
a Lewis base	a Lewis acid

B. CLEAVAGE OF ETHERS

1. The C—O bonds of ethers react under acidic conditions (HI or HBr) to give alcohols and alkyl halides in a reaction called ether cleavage.

$$R-O-R' \; + \; H-I \; \longrightarrow \; R-OH \; + \; R'-I$$
$$\xrightarrow[\text{(in some cases)}]{H-I} \; R-I \; + \; H_2O$$

   a. Ethers containing only methyl and/or primary alkyl groups require strong acid and relatively harsh conditions to bring about ether cleavage.

   b. Secondary and tertiary ethers cleave more readily than primary or methyl ethers because the S_N1 mechanism intervenes and carbocation intermediates can be formed.

   c. The alcohol formed in the cleavage of an ether can go on to react with HI or HBr to give a second molecule of alkyl halide.

2. The cleavage mechanism of ethers containing only methyl and/or primary alkyl groups involves

   a. protonation of the ether oxygen, followed by

   b. a substitution reaction of the protonated ether by the iodide or bromide ion (a good nucleophile) in an S_N2 reaction to form an alkyl halide, which liberates an alcohol as a leaving group.

$$CH_3(CH_2)_4CH_2-\overset{..}{\underset{..}{O}}-CH_3 \; \underset{\text{heat}}{\overset{\text{HBr}}{\rightleftharpoons}} \; CH_3(CH_2)_4CH_2-\overset{\overset{\displaystyle H}{|}}{\underset{+}{O}}-CH_3 \; \longrightarrow \; CH_3(CH_2)_4CH_2-\overset{..}{\underset{..}{O}}H \; + \; H_3C-\overset{..}{\underset{..}{Br}}:$$

1-methoxyhexane ⟶ **1-hexanol**   **bromomethane**

$$CH_3(CH_2)_4CH_2-OH \; \xrightarrow[\text{heat}]{\text{HBr}} \; CH_3(CH_2)_4CH_2-Br$$

1-hexanol   **1-bromohexane**

3. The mechanism of cleavage in ethers containing tertiary alkyl groups (which occurs under milder conditions) involves

   a. protonation of the ether oxygen, followed by

   b. formation of a carbocation by loss of the alcohol leaving group, and finally

   c. a Lewis acid–base association reaction of the carbocation with the halide ion.

4. Because the S_N1 reaction is faster than competing S_N2 processes, none of the primary alkyl halide is formed.

$$H_3C-\overset{..}{\underset{..}{O}}-C(CH_3)_3 \; \underset{}{\overset{\text{HI}}{\rightleftharpoons}} \; H_3C-\overset{\overset{\displaystyle H}{|}}{\underset{+}{O}}-C(CH_3)_3 \; \longrightarrow \; H_3C-\overset{..}{\underset{..}{O}}H \; + \; (CH_3)_3C^+ \; :\overset{..}{\underset{..}{I}}:^- \; \longrightarrow \; (CH_3)_3C-\overset{..}{\underset{..}{I}}:$$

***tert*-butyl methyl ether**   **methyl alcohol**   a tertiary carbocation   ***tert*-butyl iodide**

5. Secondary alcohols can react by both mechanisms.

C. REACTIONS OF EPOXIDES WITH ACID OR BASE

1. Because of their ring strain, epoxides undergo ring-opening reactions with ease; the opening of an epoxide relieves the strain of the three-membered ring.

2. The ring opening of epoxides by bases involves backside substitution by the nucleophile at the epoxide carbon.

a. A reaction of this type is essentially an S_N2 reaction in which the epoxide oxygen serves as the leaving group.

b. Under basic or neutral conditions, nucleophiles typically react with unsymmetrical epoxides at the carbon with fewer alkyl substituents.

   i. This regioselectivity is expected from the effect of branching on the rates of S_N2 reactions.

   ii. Alkyl substitution at the epoxide carbon retards the rate of nucleophilic substitution; hence, a reaction at an unbranched carbon is faster and leads to the observed product.

c. Inversion of configuration is observed if the nucleophilic reaction occurs at a stereocenter.

3. The regioselectivity of the ring-opening reaction is different under acidic conditions.

a. The nucleophile reacts at the more branched carbon of the epoxide.

b. If neither carbon is tertiary, a mixture of products is formed in most cases.

   i. Bonds to tertiary carbon atoms are weaker than bonds to primary carbon atoms and the protonated oxygen is a good leaving group.

   ii. The regioselectivity of acid-catalyzed epoxide ring opening is very similar to the nucleophilic reactions of water and alcohols with bromonium ions.

4. Acid-catalyzed epoxide hydrolysis is generally a useful way to prepare glycols.

an epoxide   a glycol

a. This reaction occurs by opposite-side substitution of water on the protonated epoxide.

b. This reaction, when combined with epoxide formation from an alkene, gives glycols with different stereochemistry from that obtained from the OsO_4 reaction with alkenes. (Compare with the reaction in Sec. I.G.2.d.)

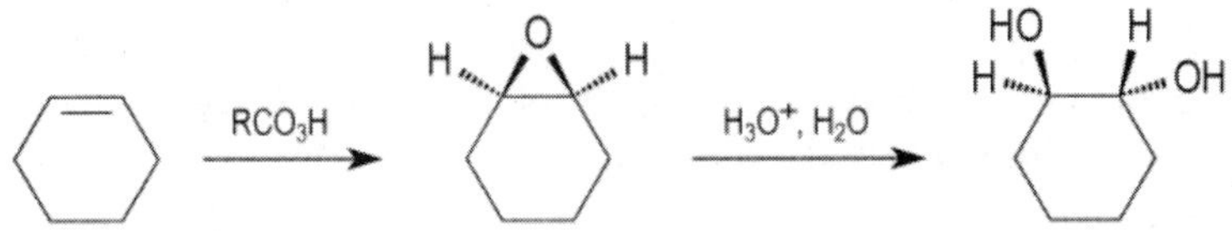

(±)-*trans*-1,2-cyclohexanediol

5. Base-catalyzed hydrolysis of epoxides also gives glycols, although, in some cases, polymerization occurs as a side reaction under the basic conditions.

D. REACTION OF ETHYLENE OXIDE WITH GRIGNARD REAGENTS

1. Grignard reagents react with ethylene oxide to give, after a protonation step, primary alcohols.

a. The carbon of the C—Mg bond of the Grignard reagent has carbanion character and can react with the epoxide as a nucleophile.

b. The magnesium of the Grignard reagent is a Lewis acid and coordinates with the epoxide oxygen.

c. Coordination of an oxygen to a Lewis acid makes the oxygen a better leaving group.

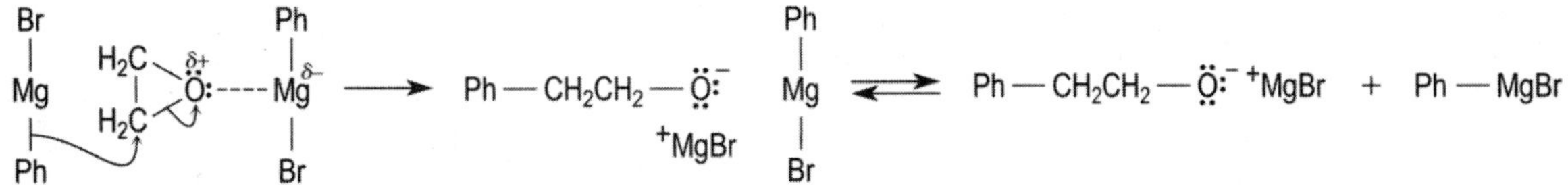

d. After the Grignard reagent has reacted, the alkoxide is converted into the alcohol product in a separate step by the addition of water or dilute acid.

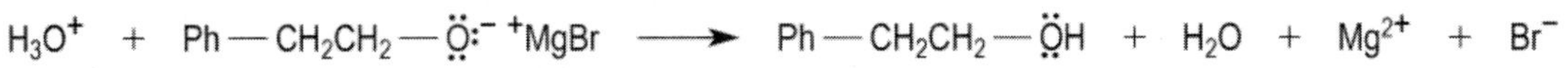

2. The reaction of a Grignard reagent with ethylene oxide provides a method for the formation of carbon–carbon bonds.

3. Grignard reagents react with unsymmetrical epoxides to give mixtures of products caused by rearrangements and other side reactions.

E. OXIDATIVE CLEAVAGE OF GLYCOLS

1. The carbon–carbon bond between the —OH groups of a vicinal glycol can be cleaved with periodic acid to give two carbonyl compounds.

a. The cleavage of glycols with periodic acid takes place through a cyclic periodate ester intermediate that forms when the glycol displaces two —OH groups from H_5IO_6.

b. The cyclic ester spontaneously breaks down by a cyclic flow of electrons in which the iodine accepts an electron pair.

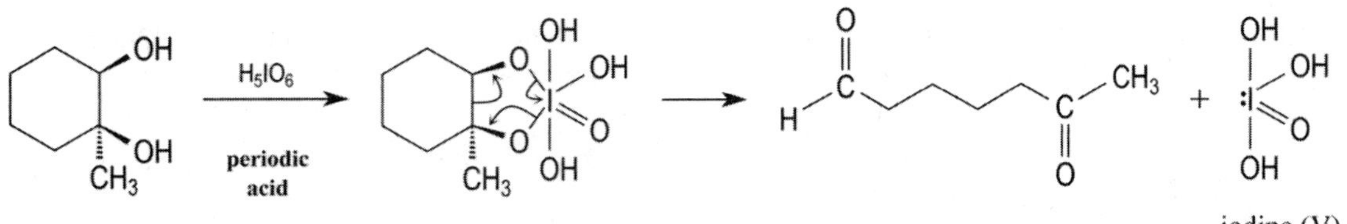

a cyclic periodate ester;
iodine (VII)

+ 2 H₂O

iodine (V)

2. A glycol that cannot form a cyclic ester intermediate is not cleaved by periodic acid.

3. The periodate cleavage reaction has been used as a test for glycols as well as for synthesis.

F. OXIDATION OF ETHERS AND SULFIDES

1. Except for peroxide formation (which occurs when ethers are allowed to stand in air) and combustion, ethers are relatively inert towards oxidizing conditions.

2. Sulfides oxidize at sulfur rather than at carbon when they react with many common oxidizing agents; sulfides are readily oxidized to sulfoxides and sulfones.

   a. Sulfoxides and sulfones can be prepared by the direct oxidation of sulfides with one and two equivalents, respectively, of hydrogen peroxide, H_2O_2.

   b. Other common oxidizing agents, such as $KMnO_4$, HNO_3, and peroxyacids, readily oxidize sulfides.

3. Nonionic Lewis structures for sulfoxides and sulfones cannot be written without violating the octet rule.

G. REACTIONS OF OXONIUM AND SULFONIUM SALTS

1. If the acidic hydrogen of a protonated ether is replaced with an alkyl group, the resulting compound is called an oxonium salt; the sulfur analog of an oxonium salt is a sulfonium salt.

$$(CH_3)_3O^+ \ ^-BF_4 \qquad\qquad (C_2H_5)_3S^+ \ ^-BF_4$$

trimethyloxonium tetrafluoroborate   **triethylsulfonium tetrafluoroborate**

2. Oxonium and sulfonium salts react with nucleophiles in S_N2 reactions.

$$R\!-\!OH \ + \ (CH_3)_3O^+ \ ^-BF_4 \ \longrightarrow \ R\!-\!O\!-\!CH_3 \ + \ H_3C\!-\!O\!-\!CH_3 \ + \ HBF_4$$

an alcohol   **trimethyloxonium**   a methyl ether   **dimethyl ether**   **tetrafluoro-**
         tetrafluoroborate   **boric acid**
         (an oxonium salt)

   a. Oxonium salts are among the most reactive alkylating agents known (they react very rapidly with most nucleophiles). They are usually isolated as $^-BF_4$ salts because the $^-BF_4$ counterion is not nucleophilic. These salts must be stored in the absence of moisture.

   b. Sulfonium salts are considerably less reactive than the corresponding alkyl chlorides in S_N2 reactions and therefore are handled more easily than oxonium salts.

   c. Sulfonium salts are somewhat less reactive than the corresponding alkyl chlorides in S_N2 reactions.

3. The important biological sulfonium salt *S*-adenosylmethionine (SAM) is a methylating agent for biological nucleophiles (see Fig. 11.1, text p. 537).

SOLUTIONS TO PROBLEMS

Solutions to In-Text Problems

11.1   Compound B is the weakest acid; there are only protons attached to carbon atoms and none of them are very acidic. The strength of the sulfur-hydrogen bond is much less than the oxygen-hydrogen bond, so compound A is a stronger acid than compound D. The greater electronegativity of oxygen relative to nitrogen makes compound D a stronger acid than compound C. (See Sec. 3.6A for a detailed description of the *element effect* on acidity.) Thus, the compounds arranged in order of increasing acidity are:

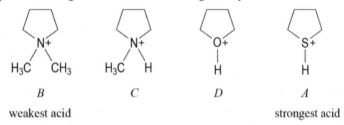

$$\begin{array}{cccc} B & C & D & A \\ \text{weakest acid} & & & \text{strongest acid} \end{array}$$

11.2   **(a)**   In this reaction, an alkoxide, formed by reaction of the alcohol with sodium, is alkylated by methyl iodide to give isopropyl methyl ether:

$$(CH_3)_2CHOH \xrightarrow{\text{Na}} (CH_3)_2CHO^- \text{ Na}^+ \xrightarrow{CH_3I} (CH_3)_2CHOCH_3 + \text{Na}^+ \text{ I}^-$$
$$+ \text{ H}_2 \qquad \textbf{isopropyl methyl ether}$$
$$\textbf{(2-methoxymropane)}$$

(b)   A thiolate ion, formed by reaction of NaOH with the thiol, is alkylated by allyl chloride to give $H_2CACH—CH_2—S—CH_3$ (allyl methyl sulfide) $+ \text{Na}^+ \text{Cl}^-$. NaOH is a strong enough base to form the thiolate anion, but would not be strong enough to form the conjugate base of an alcohol.

(c)   Alkoxide bases react with tertiary alkyl halides to give elimination products rather than substitution products. (See Sec. 9.5G.) Consequently, the products are $(CH_3)_2CACH_2$ (2-methylpropene), $\text{Na}^+ \text{Br}^-$ (sodium bromide), and CH_3OH (methanol).

(d)   Neopentyl halides and sulfonate esters do not undergo S_N2 reactions at room temperature; furthermore, they do not undergo β-elimination because there are no β-hydrogens. Thus, no reaction occurs.

11.3   **(a)**

$$\langle \rangle—CH_2CH_2OH \xrightarrow{\text{NaH}} \langle \rangle—CH_2CH_2O^- \text{ Na}^+ \xrightarrow{CH_3CH_2I} \langle \rangle—CH_2CH_2OCH_2CH_3$$

A second Williamson synthesis, the reaction of sodium ethoxide with (2-bromoethyl)cyclohexane, is in principle possible; however, because the alkyl halide has more alkyl branching this reaction would give, in addition to the desired product, a significant amount of alkene resulting from elimination.

(b)

$$(CH_3)_2CHSH \xrightarrow[C_2H_5OH]{\text{Na}^+ \text{ } C_2H_5O^-} (CH_3)_2CHS^- \text{ Na}^+ \xrightarrow{CH_3I} (CH_3)_2CHSCH_3$$

The alternative synthesis, reaction of the methanethiolate ion with isopropyl bromide, is less desirable because secondary alkyl halides react more slowly than methyl halides and give some elimination products.

(c)   Di-*tert*-butyl ether cannot be prepared by a Williamson synthesis, because its preparation would require an S_N2 reaction of *tert*-butoxide with a *tert*-butyl halide. With this combination of reagents, an E2 reaction would occur instead, and 2-methylpropene and *tert*-butyl alcohol would be the only organic products formed. (See the solution to part (c) of Problem 11.2.)

11.4   (a)   Let —R = —$CH_2CH_2CH_2CH_3$. The mechanism is essentially identical to that shown in Eq. 5.21b–d on text pp. 190-91, except that the nucleophile is isopropyl alcohol rather than water.

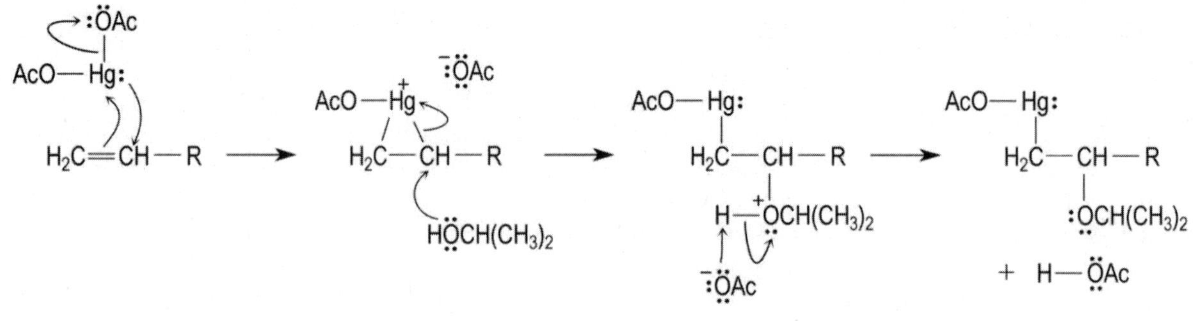

The nucleophilic reaction of the alcohol with the mercurinium ion intermediate occurs at the carbon with the greater number of alkyl substituents because this carbon–mercury bond is weaker (that is, easier to break) than the bond to primary carbon; alternatively, this carbon has the greater degree of carbocation character.

(b)   Synthesis of the ether in Eq. 11.8b would require a secondary alkyl halide and an alkoxide with α-substituents. A significant, if not predominant, amount of alkene by-product would be obtained by the E2 mechanism.

11.5   The product results from alkoxymercuration of the double bond followed by reduction (by $NaBH_4$) to remove the mercury. The net effect is overall addition of the elements of ethanol to the double bond with the ethoxy group on the carbon with the greater number of alkyl substituents.

$$(CH_3)_2CHCHCH_3$$
$$|$$
$$OCH_2CH_3$$

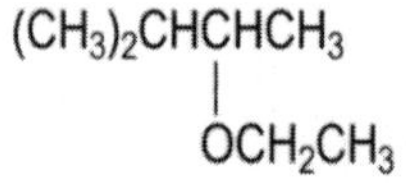

2-ethoxy-3-methylbutane

11.6   Because each carbon of the alkene double bond has one alkene substituent, there is no strong preference for the reaction of methanol at either carbon of the resulting mercurinium ion. Consequently, two constitutional isomers of the product ether are formed.

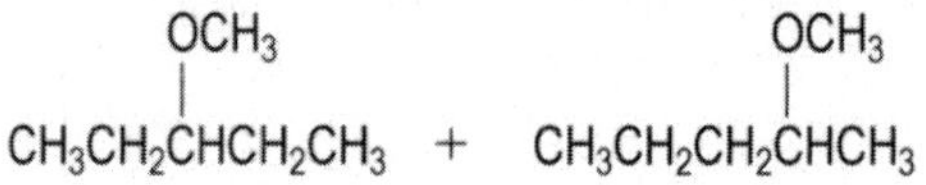

$$OCH_3 \qquad\qquad OCH_3$$
$$| \qquad\qquad\qquad |$$
$$CH_3CH_2CHCH_2CH_3 \ + \ CH_3CH_2CH_2CHCH_3$$

3-methoxypentane   2-methoxypentane

(The same result would be obtained regardless of the stereochemistry of the alkene starting material.)

11.7   (a)   Cyclohexene is subjected to oxymercuration in cyclohexanol, and the product is then reduced with $NaBH_4$:

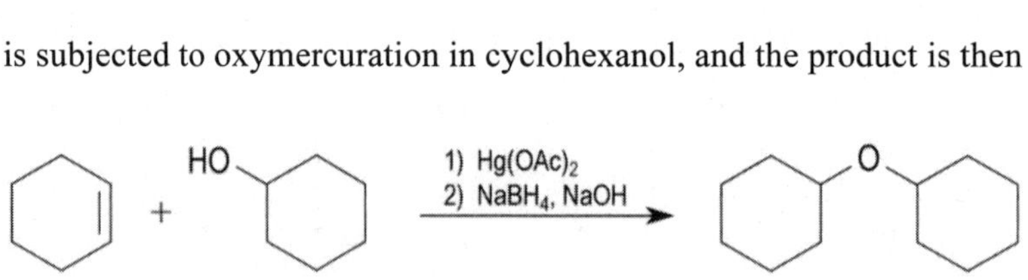

dicyclohexyl ether

(b)   Isobutylene (2-methylpropene) is subjected to alkoxymercuration in isobutyl alcohol, and the resulting organomercury compound is reduced with $NaBH_4$.

$$(CH_3)_2C=CH_2 \ + \ HOCH_2CH(CH_3)_2 \ \xrightarrow[\text{2) } NaBH_4, \, NaOH]{\text{1) } Hg(OAc)_2} \ (CH_3)_3C-OCH_2CH(CH_3)_2$$

isobutylene   **isobutyl alcohol**   ***tert*-butyl isobutyl ether**

11.8   If we start with two primary alcohols, ROH and R′OH, we would expect them to have similar basicities and similar nucleophilicities. Each alcohol could react with each protonated alcohol. Consequently, three possible products would be formed: R—O—R, R′—O—R, and R′—O—R′. None of the alcohols would be formed in very high yield, and separation of the products could be quite laborious.

The reason we can let a *tertiary* alcohol react with a primary alcohol to give an unsymmetrical ether is that the tertiary alcohol forms a carbocation in acidic solution *much* faster than either it or the primary ether react by the S_N2 mechanism; and, once the carbocation is formed, it is rapidly consumed by its Lewis acid–base association reaction with the large excess of primary alcohol that is present.

11.9   The carbocation formed from the tertiary alcohol reacts with ethanol to give the following ether:

11.10   (a)   Use the discussion in Study Guide Link 11.2: What alkene would give the same carbocation as the starting alcohol in Problem 11.9?

carbocation intermediate

11.11   (a)   Because this ether is symmetrical, and because both alkyl groups are primary, alcohol dehydration is the appropriate method. This ether can be prepared by treating $ClCH_2CH_2OH$ (2-chloroethanol) with H_2SO_4 at high temperature.

(b)   Because this is an ether with one tertiary alkyl group and one methyl group, it can be prepared by dehydration of the tertiary alcohol in the presence of methanol, or by acid-catalyzed addition of methanol to either of two possible alkenes.

2-methyl-2-butanol   2-methyl-1-butene   2-methyl-2-butene   2-methoxy-2-methylbutane

(c)   Alkoxymercuration–reduction of isobutylene using isopropyl alcohol as the solvent in the alkoxymercuration step is preferred. (Acid-catalyzed addition of isopropyl alcohol to 2-methylpropeme might also work.)

$$(CH_3)_2CHOH \ + \ (CH_3)_2C{=}CH_2 \ \xrightarrow[\text{2) NaBH}_4\text{, NaOH}]{\text{1) Hg(OAc)}_2} \ (CH_3)_3C{-}OCH(CH_3)_2$$

isopropyl alcohol
(solvent)

***tert*-butyl isopropyl ether**

(d)   Dibutyl ether, $CH_3CH_2CH_2CH_2OCH_2CH_2CH_2CH_3$, is a symmetrical ether that can be prepared by the acid-catalyzed dehydration of 1-butanol, $CH_3CH_2CH_2CH_2OH$.

11.12   (a)   (b) $(CH_3)_2C{=}CH_2$ (c)   (d)

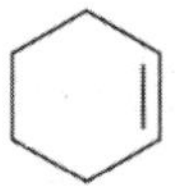

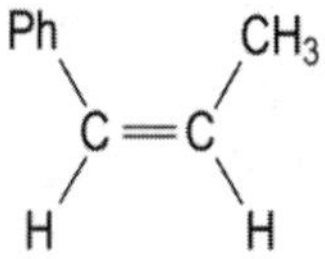

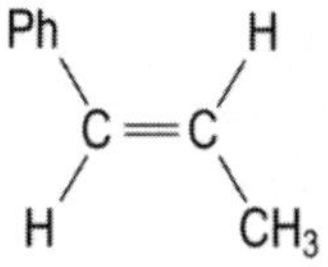

In part (b), the stereochemistry of the product and the net syn-addition dictate the stereochemistry of the alkene starting material.

In part (c), the stereochemistry of the product and the net syn-addition dictate the stereochemistry of the alkene starting material.

11.13   (a)   (b)

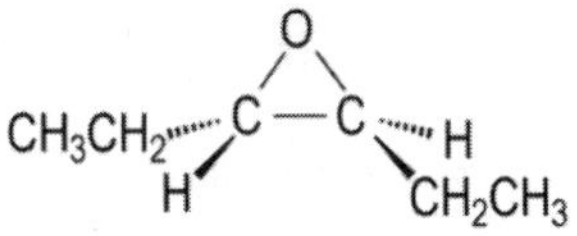

(the racemate)

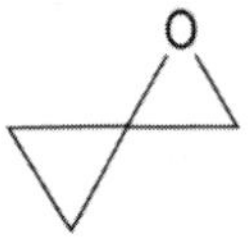

11.14   The key to solving this problem is to realize that, in order for epoxide formation to occur, the oxygen of the conjugate-base alkoxide must be anti to the bromine in the transition state so that it can execute a backside substitution. Draw each of the stereoisomeric alkoxides in such a conformation and evaluate the relative energies of the resulting transition states.

In the transition state for the reaction of the alkoxide conjugate base of *A* there are two significant gauche interactions (brackets).

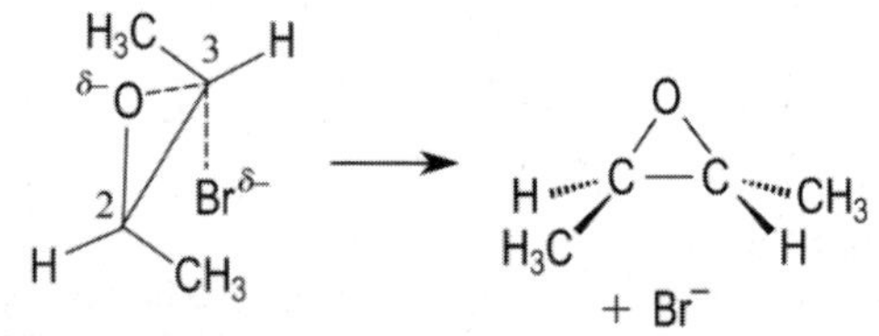

(2R,3S) diastereomer (*A*)
(or its enantiomer)

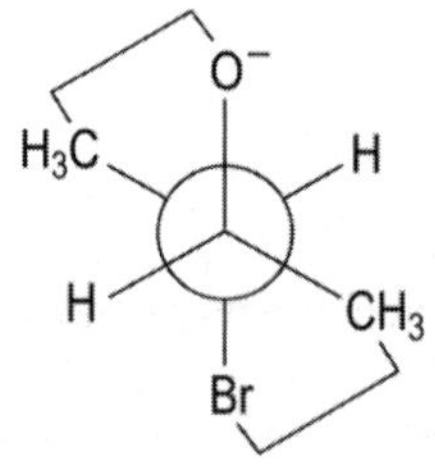

+ Br⁻

the transition-state conformation
is the most stable conformation

The transition-state conformation is very close to the most favorable conformation of *A*. The transition-state conformation of *B*, however, has three significant gauche interactions, but its more stable conformation has two (brackets).

(2R,3R) diastereomer (B)
(or its enantiomer)

transition-state
conformation

most stable
conformation

Consequently, isomer B must gain additional energy associated with a methyl–methyl gauche interaction in order for it to react. This represents an additional component to the activation free energy of B that is not present in the reaction of A. Consequently, the reaction of A is faster.

11.15   As in the solution to Problem 11.14, the alkoxide oxygen and the halogen leaving group must be anti in the transition state. This anti relationship is possible in a chair conformation only in the reaction of the trans stereoisomer (see Eq. 11.23 on text p. 521). In the cis stereoisomer, such an anti relationship is not possible. Because the cis stereoisomer cannot achieve the appropriate transition-state conformation for epoxide formation, it is unreactive.

trans-2-chlorocyclohexanol
(conjugate-base alkoxide)
an anti conformation is accessible

cis-2-chlorocyclohexanol
(conjugate-base alkoxide)
an anti conformation is not accessible

11.16   (a)   In the protonated ether (see Eq. 11.28, text p. 523, for a typical protonation reaction) the iodide ion can react as a nucleophile at either the carbon of the methyl group or the α-carbon of the butyl group. Because this is an S_N2 reaction, and because S_N2 reactions of methyl compounds are faster than those of butyl compounds (why?), the nucleophilic reaction occurs at the methyl carbon to give methyl iodide and 1-butanol.

(b)   Because alcohols react with HI to give alkyl iodides, the initially formed 1-butanol would be expected to give 1-iodobutane.

(c)   The discussion in Sec. 11.4 of the text indicates that the cleavage reactions of tertiary ethers, which are S_N1 reactions, occur under milder conditions than the S_N2 cleavage reactions of methyl and primary ethers. Hence, cleavage of the protonated ether occurs so as to give the *tert*-butyl cation and methanol. (See Eq. 11.29 on text p. 523 for the mechanism of a similar reaction.) The Lewis acid–base association reaction of iodide with the *tert*-butyl cation gives *tert*-butyl iodide.

(d)   Tertiary ethers and sulfides cleave by an S_N1 mechanism. The rate-limiting step in S_N1 reactions is formation of the carbocation by Lewis acid–base dissociation. The rate of this dissociation reaction is proportional to the concentration of the reacting species—in this case, the conjugate acid of the ether or sulfide. Because sulfides are much less basic than ethers (Sec. 11.1, text p. 512), the concentration of the reacting species is much smaller in the case of sulfides. Consequently, sulfides are less reactive than ethers toward acid-promoted cleavage.

(e)   This case speaks to the balance between S_N1 or S_N2 cleavage of an ether at a secondary alkyl group vs. S_N2 cleavage of the same ether at a methyl group. Notice that the asymmetric carbon (*) is the α-carbon of the ether in the secondary alkyl group:

conjugate acid

Thus, optical activity serves as a barometer of what is going on mechanistically at this carbon. If cleavage of the conjugate acid occurs by a nucleophilic reaction of Br⁻ at the *sec*-butyl group by an S_N2 mechanism, or even by an S_N1 mechanism, the product would be *sec*-butyl bromide and methanol, and the product would be optically active *with inverted configuration*. (An S_N1 mechanism would be expected to show significant inversion; see Fig. 9.13 on text p. 426.) Neither the configuration of the products nor their identity fits the data. Hence, this cannot be what is happening.

Finally, if bromide ion reacts at the methyl carbon of the conjugate acid, then the products would be methyl bromide and optically pure 2-butanol with *retained configuration*. The observations are consistent with this prediction.

For completeness, we should consider S_N1 cleavage of the conjugate acid at the methyl group; but this reaction would require the formation of $^+CH_3$, a very unstable carbocation. This can be rejected on the basis of what we know about carbocations.

11.17   (a)   Diisopropyl ether reacts initially to give isopropyl alcohol and isopropyl iodide by either an S_N1 or S_N2 mechanism, or by both mechanisms operating simultaneously. Once the alcohol is formed, it reacts with concentrated HI to give isopropyl iodide as well. Therefore, the product is isopropyl iodide.

diisopropyl ether

(b)   This tertiary ether reacts rapidly by a carbocation (S_N1) mechanism to give, initially, the tertiary iodide and ethanol. The ethanol subsequently reacts more slowly to give ethyl iodide.

2-ethoxy-2,3-dimethylbutane

11.18   Under basic or neutral conditions under which the epoxide is not protonated, the nucleophiles react at the carbon of the CH_2 group. The configuration of the asymmetric carbon is not affected because no bonds to this carbon are broken.

(a)

(b)

11.19   The strategy in this problem is to let the —OH group originate from the epoxide oxygen. In the starting material, this oxygen must be attached to the same carbon as the —OH group in the product as well as to an adjacent carbon; the nucleophile becomes attached to the adjacent carbon. To summarize:

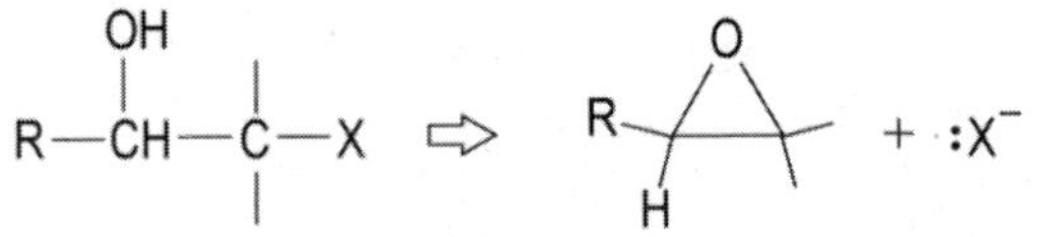

(a)

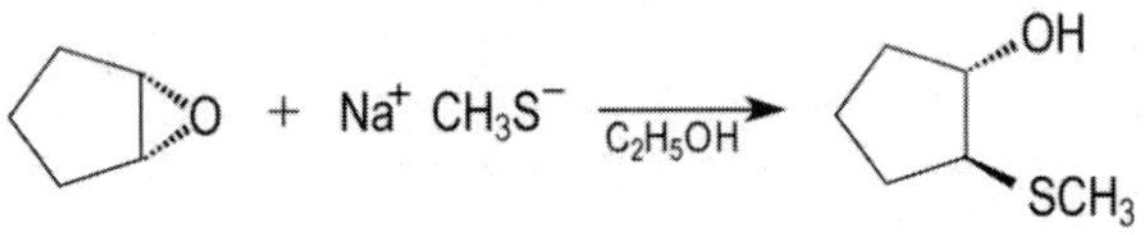

(b)

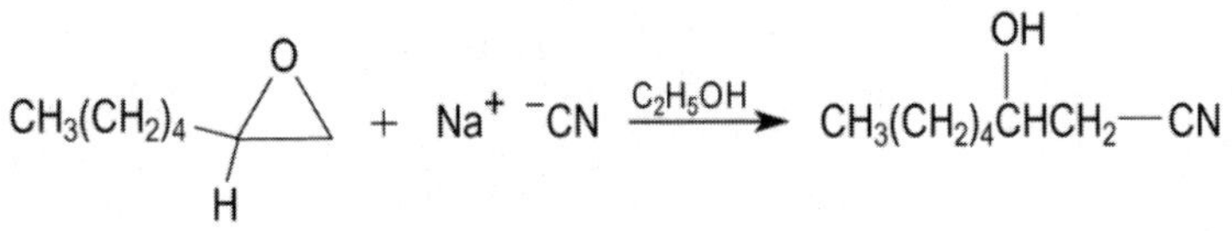

The strategy outlined above suggests another possibility:

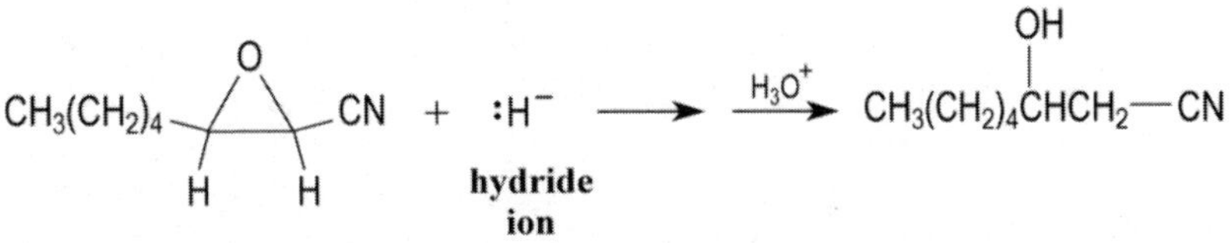

If you came up with this idea, you are reasoning correctly. However, every good idea has to be tempered by practical reality. In this case, the strategy will not work because the most common source of nucleophilic hydride, LiAlH$_4$ (lithium aluminum hydride), also reacts with the cyano (CN) group.

11.20 (a) The nucleophilic reaction of methanol on the protonated epoxide occurs with inversion to give (3S,4R)-4-methoxy-3-hexanol:

(3S,4R)-4-methoxy-3-hexanol

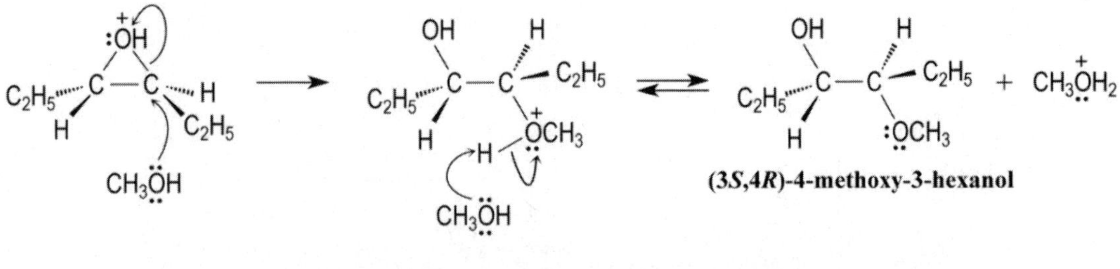

(You should verify that the nucleophilic reaction of methanol at the other asymmetric carbon of the protonated epoxide gives the same result.)

(b) The enantiomer of the epoxide in part (a) gives the enantiomer of the product formed in part (a), (3R,4S)-4-methoxy-3-hexanol. (Verify this by writing the mechanism.)

11.21 (a) Under acidic conditions, the nucleophilic reaction of water occurs at the carbon of the epoxide ring with the methyl substituents (on the protonated epoxide); consequently, the OH group at that carbon is enriched in ^{18}O. In basic solution, the reaction of hydroxide occurs at the CH$_2$ carbon of the epoxide, and the OH group at that carbon is enriched in ^{18}O. (*O = ^{18}O.)

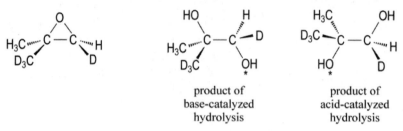

(b)   Inversion of configuration would occur at the carbon of the epoxide (or protonated epoxide) at which the nucleophilic reaction occurs. (The nucleophilic oxygen is labeled with an asterisk.)

product of base-catalyzed hydrolysis   product of acid-catalyzed hydrolysis

11.22   (a)

$$CH_3CH_2CH\text{—}MgBr + \overset{O}{\triangle} \xrightarrow{H_3O^+} CH_3CH_2CH\text{—}CH_2CH_2OH$$

with CH₃ substituents on both the MgBr carbon and the product carbon.

3-methyl-1-pentanol

(b)   When an epoxide is opened by an organometallic reagent, the OH group of the product alcohol is on the carbon *adjacent to* the carbon that is the electrophilic target of the "carbon anion" from the organometallic reagent. This strategy reveals two choices for possible reagent combinations, with the alkyl groups that could be provided by the complex organocuprate shown in boxes

$$\boxed{CH_3}\text{—}CH_2\text{—}\underset{\underset{CH_3CH_2}{|}}{\overset{\overset{OH}{|}}{C}}\text{—}CH_2\text{—}\boxed{CH_2CH_2CH_3}$$

3-ethyl-3-heptanol

$$(CH_3)_2Cu(CN)Li_2 + H_2C\text{—}\underset{\underset{CH_2CH_2CH_2CH_3}{|}}{\overset{O}{C}}\text{—}CH_2CH_3 \xrightarrow[H_3\overset{+}{O}]{H_3\overset{+}{O}} (CH_3CH_2)_2\overset{O}{C}\text{—}CH_2 + (CH_3CH_2CH_2)_2Cu(CN)Li_2$$

11.23   (a)   (b)

(racemate)

The racemate is formed in (b) because the phenyl (Ph) group can react at either asymmetric carbon with equal probability. Because the epoxide is a meso compound, and is therefore achiral, chiral products must be formed as racemates.

11.24   (a)

1-methylcyclopentene $\xrightarrow{\text{NMMO, H}_2\text{O} \atop \text{OsO}_4 \text{ (cat.)}}$ (racemate)

1-methylcyclopentene   (racemate)

(b)

trans-2-butene $\xrightarrow{\text{NMMO, H}_2\text{O} \atop \text{OsO}_4 \text{ (cat.)}}$

trans-2-butene   **(±)-2,3-butanediol**
(the racemate)

11.25   (a)   The alkene required is 4-ethoxy-1-butene, $CH_3CH_2OCH_2CH_2CH{=}CH_2$.

(b)

methylenecyclobutane

(c)

$CH_3CH_2CH_2$ $CH_2CH_2CH_3$
C=C
H   H

***cis*-4-octene**

(d)

$CH_3CH_2CH_2$ H
C=C
H $CH_2CH_2CH_3$

***trans*-4-octene**

In part (c), the syn-addition of two identical groups—in this case, OH groups—to a symmetrical *cis*-diene gives a meso diol. Alternatively, in part (d), the syn-addition of two identical OH groups to a symmetrical *trans*-diene gives a racemic diol. Be sure to analyze these additions in detail if these points are not clear. (See Sec. 7.9C for a similar type of analysis.)

11.26   This mechanism is much like the one shown for OsO_4 in Eq. 11.48a, text p. 532. A cyclic electron flow that gives either of the resonance structures of the cyclic intermediate is correct.

permanganate ion

11.27   (a)

+ $\overset{O}{\overset{\|}{C}}HCH_3$

(c)

CH=O
CH=O

(b)   The products are phenylacetaldehyde, $PhCH_2{=}O$, and formaldehyde, $O{=}CH_2$.

11.28   (a)

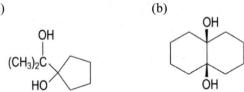

$(CH_3)_2C$
OH
HO

(b)

OH
OH

11.29   Because iodide ion is a good nucleophile and the trimethyloxonium ion is an excellent alkylating agent, alkylation of the iodide ion occurs to give dimethyl ether and methyl iodide.

$$(CH_3)_2\overset{+}{O}\!\!-\!\!CH_3 \quad :\overset{..}{\underset{..}{I}}:^- \quad \longrightarrow \quad (CH_3)_2\overset{..}{\underset{..}{O}} \quad + \quad H_3C\!\!-\!\!\overset{..}{\underset{..}{I}}:$$

trimethyloxonium iodide   **dimethyl ether   methyl iodide**

11.30   In each case, a nucleophile reacts with a methyl group of the oxonium salt, expelling dimethyl ether as a leaving group.

(a)   (b)

$$(CH_3)_3\overset{+}{\underset{..}{S}} \quad \bar{B}F_4 \quad + \quad (CH_3)_2\overset{..}{\underset{..}{O}}$$

11.31   An acid-base reaction must precede every methylation to reveal a lone pair of electrons on the nitrogen to make them available for an S_N2 reaction with the methyl group on SAM.

SAM   ***N*-methyllysine**

***N*-methyllysine   SAM**   ***N,N*-dimethyllysine**

11.32   The intramolecular product tetrahydrofuran results from an internal nucleophilic substitution reaction of the alkoxide on the alkyl halide. The intermolecular product 1,4-butanediol results from the S_N2 reaction of hydroxide ion with the alkyl halide.

$$HO\!\!-\!\!CH_2CH_2CH_2CH_2\!\!-\!\!Br \xrightarrow{\text{NaOH}} \quad \text{(THF ring)} \quad \text{or} \quad HO\!\!-\!\!CH_2CH_2CH_2CH_2\!\!-\!\!OH$$

4-bromo-1-butanol   **tetrahydrofuran**   **1,4-butanediol**
                                        (the intramolecular   (the intermolecular
                                        substitution product)   substitution product)

To form the cyclic product tetrahydrofuran, ⁻OH must react with the alcohol to ionize it; the conjugate base anion of the alcohol then cyclizes in an *internal* substitution reaction. In contrast, to form 1,4-butanediol, ⁻OH must react with the alkyl halide in an S_N2 reaction. Ionization reactions are much faster than S_N2 reactions; and once the ionization has occurred, the cyclization, because it is intermolecular, is *much* faster than the S_N2 reaction between hydroxide and the alkyl halide. Hence, tetrahydrofuran would be the major product.

Another reasonable *intermolecular* possibility you may have considered is following:

$$HO\!\!-\!\!CH_2CH_2CH_2CH_2\!\!-\!\!Br \xrightarrow{\text{NaOH}} HO\!\!-\!\!CH_2CH_2CH_2CH_2\!\!-\!\!O\!\!-\!\!CH_2CH_2CH_2CH_2\!\!-\!\!Br$$

4-(4-bromobutoxy)-1-butanol

To form this product, ionized 4-bromobutanol must react with the alkyl halide "end" of another molecule of 4-bromobutanol in an S_N2 reaction. Because such a reaction is bimolecular, the intramolecular reaction of the same ion to give tetrahydrofuran is *much* faster.

11.33   Because the $\Delta H^{o\ddagger}$ of the two reactions is the same, the difference in $\Delta G^{o\ddagger}$ is the negative of the difference in the $T\Delta S^{o\ddagger}$:

$$\Delta G^{o\ddagger}{}_B - \Delta G^{o\ddagger}{}_A = T(0.180 - 0.030) = 44.7 \text{ kJ mol}^{-1}$$

Reaction B has the higher $\Delta G^{o\ddagger}$. To convert this different into a relative rate, apply Eq. 9.20b:

$$\frac{\text{rate}_A}{\text{rate}_B} = 10^{(\Delta G^{o\ddagger}_B - \Delta G^{o\ddagger}_A)/2.30RT} = 6.7 \times 10^7$$

In other words, reaction A is about 70 million times faster than reaction B.

11.34   (a)   The formation of product A "freezes" fewer bond rotations, and therefore has the less negative entropy of activation. This would make the formation of product A faster, except for the fact that it has more strain—that is, a greater enthalpy—than product B. Which product is formed more rapidly depends on the balance between the entropy and enthalpy of activation.

(b)   The first reaction is a bimolecular reaction, whereas the second is an intramolecular reaction. Forming the transition state of the first reaction requires "freezing" the translation of one molecule relative to the other, whereas forming the transition state of the second reaction requires no freezing of translations. Because the bond changes in the two reactions are about the same, we suspect that the second reaction will be much faster because its $\Delta S^{o\ddagger}$ is less negative (or more positive).

11.35   (a)   The S_N2 mechanism with $^-$OH as the nucleophile:

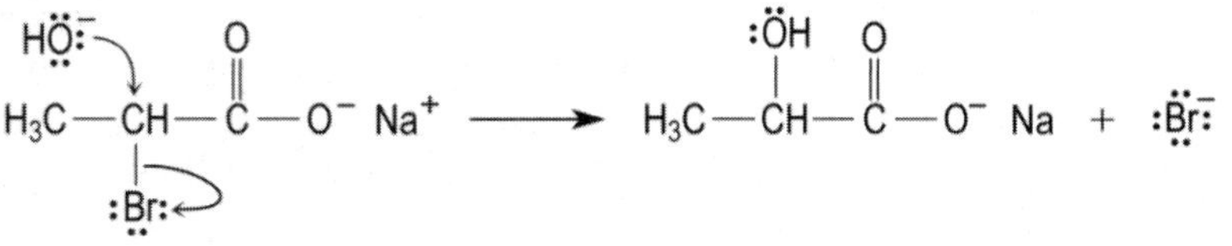

(b)   The intramolecular reaction mechanism, in the rate-limiting step, results in an unstable intermediate called an α-lactone.

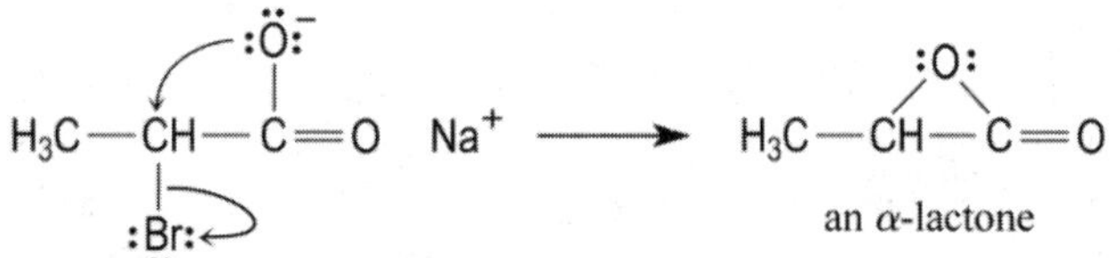

an α-lactone

The α-lactone reacts rapidly with $^-$OH in another S_N2 reaction:

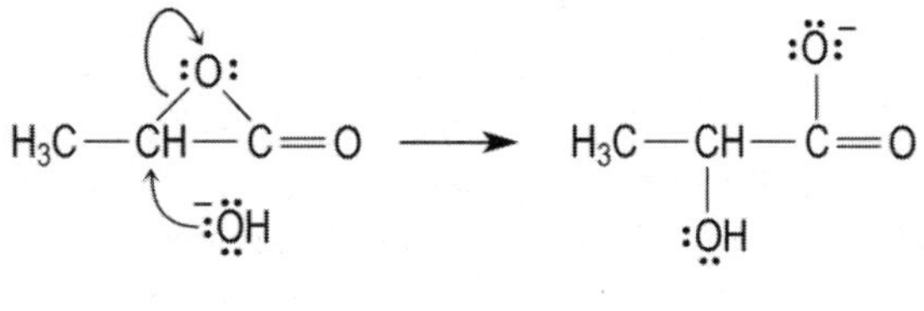

(c)   The proximity effect is

$$\frac{k_1}{k_2} = \frac{1.2 \times 10^{-4} \text{ s}^{-1}}{6.4 \times 10^{-4} \, M^{-1}\text{s}^{-1}} = 0.19 \, M$$

where k_1 is the first-order rate constant, and k_2 is the second-order rate constant. This is a rather low effective molarity.

(d)   The intramolecular rate law is

$$\text{rate}_1 = k_1[S]$$

where S is the starting material; and the second-order rate law for the reaction of hydroxide with the alkyl halide is

$$\text{rate}_2 = k_2[S][^-OH]$$

From the ratio of the two rate constants calculated in part (c), the two rates are the same when $[^-OH] = 0.19\ M$.

(e)   The ratio of the two rates, $\text{rate}_2/\text{rate}_1$, in 1 M NaOH is $1.0/0.19 = 5.3$. Therefore, the second-order reaction accounts for about 84% of the product in 1 M NaOH.

(f)   The carbonyl carbon ideally has bond angles of 120°. When it is forced into a 3-membered ring, its orbitals have to be "bent" to accommodate an interatomic angle closer to 60°. In an epoxide, a saturated carbon atom has a preferred bond angle of 109.5°. Consequently, there is more angle strain in the α-lactone; this accounts for its greater instability. The instability of the α-lactone intermediate is the reason that the intramolecular reaction has a rather small effective molarity despite its entropic advantage.

11.36   An analysis identical to the one shown in Study Problem 11.6, text pp. 547–548, results in the following episulfonium ion, which is the diastereomer of the one shown in Eq. 11.68b.

In this case, however, the episulfonium ion is meso and therefore achiral. The nucleophilic reactions of water at C-2 and C-3 give enantiomers, which must be formed in identical amounts.

11.37   The intramolecular mechanism shown in the solution to Problem 11.35(b) indicates that *two* inversions of configuration, which is equivalent to net retention of configuration, occur at the same carbon in reaching the final product. Because this mechanism predominates at low NaOH concentration, its stereochemical consequences—retention of configuration—are manifest under the same conditions. As the NaOH concentration is increased, the intermolecular conventional S_N2 mechanism takes over and becomes the predominant one at high NaOH concentration. In this mechanism, inversion of configuration is expected and is observed.

11.38   If intramolecular participation occurs, an intermediate episulfonium ion is formed. In this ion, the two carbons in the three-membered ring are equivalent and indistinguishable except for the label. The reaction with water is equally probable at each of these carbons. Consequently, 50% of the label ends up at each carbon.

If, however, water reacts as a nucleophile directly at the labeled carbon to displace the chloride ion in a conventional S_N2 reaction, an episulfonium ion would not be involved, and the label would not scramble. That is, product *A* would form exclusively.

11.39   The two reactions proceed through a common episulfonium ion intermediate that results from an intramolecular substitution of the protonated OH group by the sulfur.

Evidently, the nucleophilic reaction of the chloride ion with this intermediate takes place at the carbon with the methyl substituent. Because the same intermediate is formed in both reactions, the product is the same.

A separate question is why the chloride ion should react only at the more branched carbon. The cyclic intermediate is much like a protonated epoxide (or a bromonium ion or mercurinium ion); hence, the nucleophilic reaction of chloride ion occurs at the more branched carbon for the same reason it does in a protonated epoxide. (See discussion of Eq. 11.34b on text p. 526.).

11.40   (a)   Lengthening a carbon chain by two carbons is required; what better way than the reaction of a Grignard reagent with ethylene oxide?

$(CH_3)_2C{=}CH_2$ $\xrightarrow[\text{2) H}_2\text{O}_2,\ ^-\text{OH}]{\text{1) BH}_3,\ \text{THF}}$ $(CH_3)_2CHCH_2OH$ $\xrightarrow{\text{HBr, H}_2\text{SO}_4}$ $(CH_3)_2CHCH_2Br$ $\xrightarrow[\text{ether}]{\text{Mg}}$

2-methylpropene   **isobutyl alcohol**   **isobutyl bromide**

$(CH_3)_2CHCH_2MgBr$ $\xrightarrow[\text{2) H}_3\text{O}^+]{\text{1)}\ \triangle}$ $(CH_3)_2CHCH_2CH_2CH_2OH$ $\xrightarrow[\text{2) H}_3\text{O}^+]{\text{1) KMnO}_4/\text{NaOH}}$ $(CH_3)_2CHCH_2CH_2CO_2H$

An alternate synthesis of isobutyl bromide is to treat 2-methylpropene with HBr in the presence of peroxides.

(b)   Oxidize the isobutyl alcohol, which is obtained in the hydroboration–oxidation reaction in part (a).

$(CH_3)_2CHCH_2OH$ $\xrightarrow[\text{2) H}_3\text{O}^+]{\text{1) KMnO}_4/\text{NaOH}}$ $(CH_3)_2CHCO_2H$

isobutyl alcohol
from part (a)

(c)   The sulfone must be obtained by oxidation of dibutyl sulfide; and the sulfide is obtained from the S_N2 reaction of the thiolate conjugate base of 1-butanethiol with 1-bromobutane.

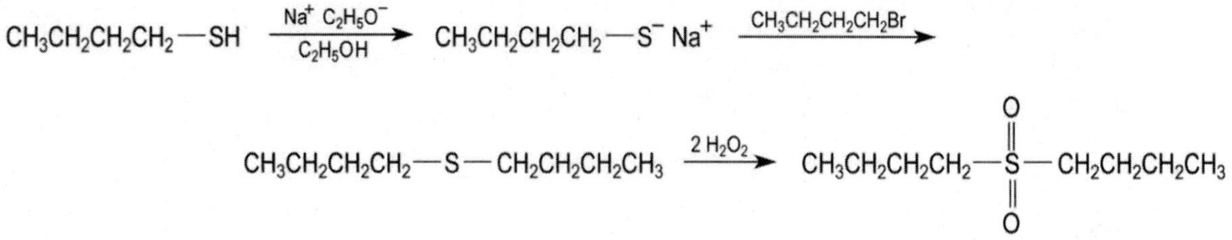

$$CH_3CH_2CH_2CH_2-SH \xrightarrow[C_2H_5OH]{Na^+ C_2H_5O^-} CH_3CH_2CH_2CH_2-S^- Na^+ \xrightarrow{CH_3CH_2CH_2CH_2Br}$$

$$CH_3CH_2CH_2CH_2-S-CH_2CH_2CH_2CH_3 \xrightarrow{2 H_2O_2} CH_3CH_2CH_2CH_2-\overset{\overset{O}{\|}}{\underset{\underset{O}{\|}}{S}}-CH_2CH_2CH_2CH_3$$

dibutyl sulfone

(d)   The approach is similar to that used in Study Problem 11.8 on text p. 554, except that a carbon–carbon bond is formed in this case; a complex organocuprate is used for this step.

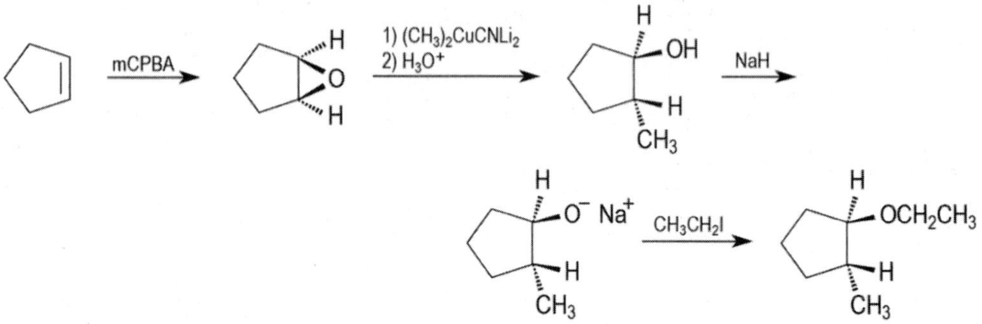

(±)-*trans*-1-ethoxy-
2-methylcyclopentane

11.41   (a)   (b)   (c)

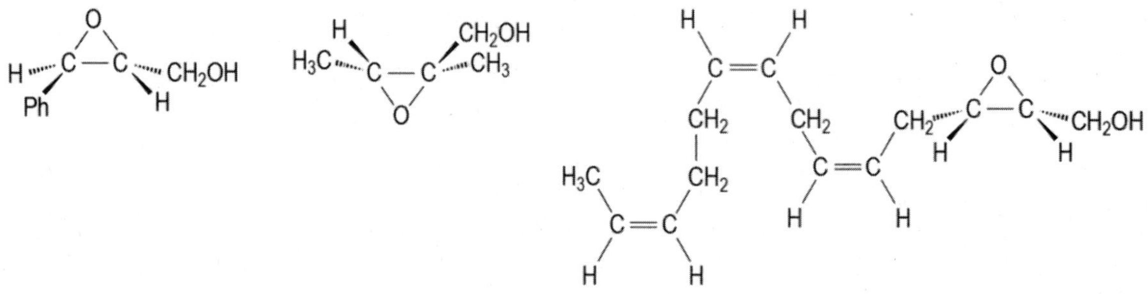

11.42   (a)

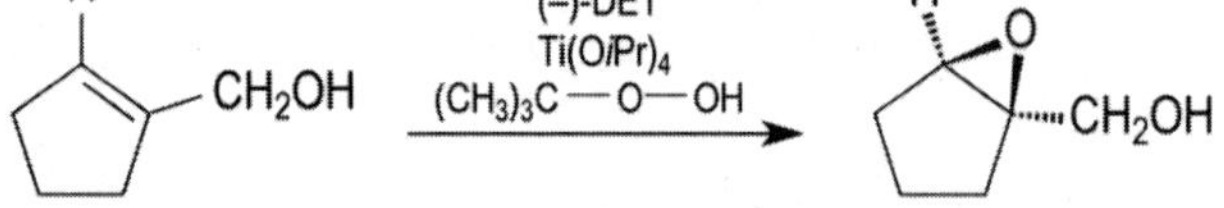

(b)

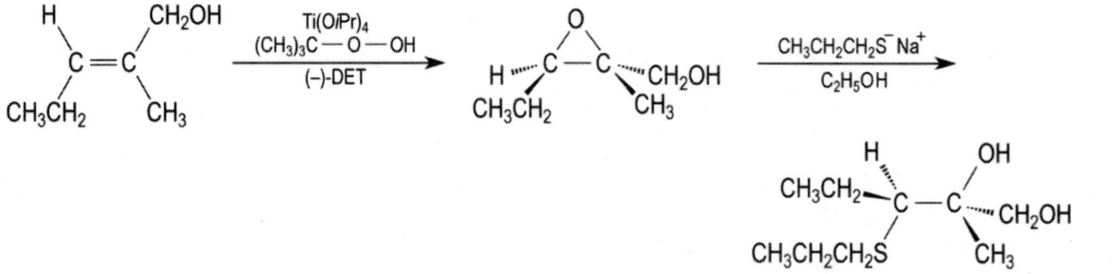

11.43   (a)   A substituent that is cis to the —CH₂OH group points back into the catalyst, and the resulting steric repulsions can either reduce the binding affinity of the compound for the catalyst, or, more likely, push the alkene group out so that it is not optimally aligned with the peroxy —OH group for epoxidation.

(b)   The same phenomenon would be observed with (–)-DET, because the catalyst is enantiomeric to the one formed from (+)-DET and the alkene is not chiral; hence, the same energetics should apply.

Solutions to Additional Problems

11.44   (a)   Two of the many nine-carbon ethers that cannot be prepared by the Williamson ether synthesis:

2-methyl-2-*tert*-butoxybutane   ***tert*-butyl neopentyl ether**

The Williamson ether synthesis involves an S_N2 reaction between an alkoxide ion and an alkyl halide or sulfonate ester. Thus, one of the alkyl groups of an ether to be prepared by the Williamson synthesis must be derived from an alkyl halide that will undergo the S_N2 reaction. The foregoing ethers contain only tertiary and neopentyl alkyl groups, and neither tertiary nor neopentyl alkyl halides undergo the S_N2 reaction.

(b)   Any ether containing only methyl or primary alkyl groups is a correct answer. Two of several possible examples are CH₃O(CH₂)₇CH₃ (methyl octyl ether, or 1-methoxyoctane) and CH₃(CH₂)₃O(CH₂)₄CH₃ (butyl pentyl ether, or 1-butoxypentane).

(c)   Because the alkyl halides formed are in the same molecule, the ether is cyclic:

tetrahydrofuran

(d)   Dipropyl ether (CH₃CH₂CH₂)₂O, would give propyl bromide as the only alkyl halide.

(e)   Each of the following alkenes would give diastereomeric glycols as a result of the different syntheses. The KMnO₄ treatment would give *meso*-2,3-dimethyl-2,3-butanediol from *cis*-2-butene and would give (±)-2,3-dimethyl-2,3-butanediol from *trans*-2-butene; the epoxidation–hydration sequence would give the opposite results with the two alkenes. The KMnO₄ treatment would give *cis*-1,2-cyclobutanediol from cyclobutene, and the epoxidation–hydration sequence would give *trans*-1,2-butanediol.

cis-**2-butene**   *trans*-**2-butene**   **cyclobutene**

(f)   1-Butene, H₂C=CHCH₂CH₃, gives racemic 1,2-butanediol with either reagent.

(g)   Since the alkene has an unsaturation number = 3 and forms di-epoxides, it must also have a ring.

1,3-Cyclohexadiene is not a correct answer because it gives two stereoisomeric mono-epoxides and three stereoisomeric di-epoxides.

(h)   A correct answer must be an alkene that gives a glycol that cannot exist as diastereomers. The alkene 1-hexene (H₂C=CHCH₂CH₂CH₂CH₃) is one such compound.

11.45   (a)

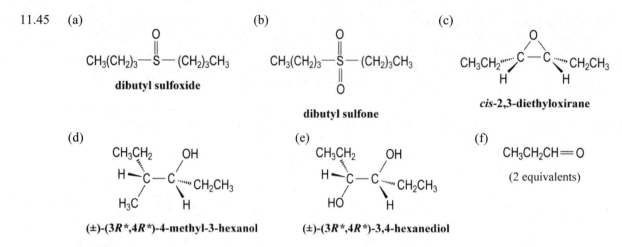

(b)

dibutyl sulfone

dibutyl sulfoxide

(c)

***cis*-2,3-diethyloxirane**

(d)

(±)-(3*R**,4*R**)-4-methyl-3-hexanol

(e)

(±)-(3*R**,4*R**)-3,4-hexanediol

(f)

CH₃CH₂CH＝O

(2 equivalents)

    The asterisks in the name indicate *relative* rather than absolute configurations. In the case of parts (d) and (e), the (3*R**,4*R**) notation means that the two carbons have the same configuration; if one carbon is *R*, the other is also *R*; but if one is *S*, the other is also *S*. This notation tells us which diastereomeric pair we are dealing with. (See also Further Exploration 6.2 on p. 148 of this manual for a discussion of the same notation in another context.)

(g)

(±)-(3*R,4*R**)-3-methoxy-
4-methylhexane**

(h)

**(±)-3-ethoxy-
3-methylhexane**

(i)

**(±)-3-methoxy-
3-methylhexane**

(j)   All of the compounds in this sequence have the *R* configuration at the asymmetric carbon. (The configuration of the asymmetric carbon is not shown explicitly because it is not affected by the sequence.) In the last step, the alkoxide produced from the Grignard reaction is not protonated with H₃O⁺, as is usual, but rather is alkylated directly with CH₃I.

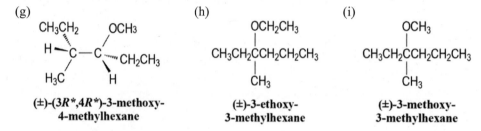

(*R*)-1-bromo-3-methylpentane

(*R*)-1-methoxy-5-methylheptane

11.46   It is assumed that the starting material is racemic. As an additional exercise, you should consider the stereochemical course of the reactions if the starting material is a single enantiomer.

2-ethyl-2-methyloxirane

(a) and (b)

$$\underset{\underset{CH_3}{|}}{CH_3CH_2\overset{\underset{|}{OH}}{C}CH_2OH}$$

(c)

$$\underset{\underset{CH_3}{|}}{CH_3CH_2\overset{\underset{|}{OH}}{C}CH_2OCH_3}$$

(d)

$$\underset{\underset{CH_3}{|}}{CH_3CH_2\overset{\underset{|}{OCH_3}}{C}CH_2OH}$$

(e)

$$\underset{\underset{CH_3}{|}}{CH_3CH_2\overset{\underset{|}{OH}}{C}CH_2CH_3}$$

(f)

$$\underset{\underset{CH_3}{|}}{CH_3CH_2\overset{\underset{|}{Br}}{C}CH_2OCH_3}$$

(g)

$$\underset{\underset{CH_3}{|}}{CH_3CH_2\overset{\underset{|}{Br}}{C}CH_2OH}$$

(h)

$$\underset{\underset{CH_3}{|}}{CH_3CH_2\overset{\underset{|}{OCH_3}}{C}CH_2OCH_3}$$

(i)

$$\underset{\underset{CH_3}{|}}{CH_3CH_2\overset{\underset{|}{OCH_3}}{C}CH_2OCH_2CH_3}$$

(j)

$$\underset{\underset{}{}}{CH_3CH_2\overset{\overset{O}{||}}{C}CH_3} \; + \; \overset{\overset{O}{||}}{CH_2}$$

(k)

$$\underset{\underset{CH_3}{|}}{CH_3CH_2\overset{\underset{|}{MgBr}}{C}CH_2OCH_3}$$

(l)

$$\underset{\underset{CH_3}{|}}{CH_3CH_2\overset{\underset{|}{CH_2CH_2OH}}{C}CH_2OCH_3}$$

11.47   Because the C—S bond is weaker than the C—O and C—C bonds, and because a thiolate ion is less basic than an alkoxide ion or carbanion, the thiirane (the sulfur compound) reacts most rapidly.

11.48   The first reaction, opening of the epoxide, should occur most rapidly, because the epoxide is considerably more strained than tetrahydrofuran. It is the ring strain that causes epoxide opening to be so fast.

11.49   (a)   3-Ethoxypropene undergoes bromine addition, and therefore decolorizes a solution of Br_2 in an inert solvent; 1-ethoxypropane does not.

(b)   1-Pentanol, the alcohol, is fairly soluble in water, whereas the ether is not. Alternatively, the alcohol evolves H_2 when treated with NaH or Na, or evolves methane (CH_4) when treated with the Grignard reagent CH_3MgBr; the ether, which has no O—H group, does not.

(c)   1-Methoxy-2-chloro-2-methylpropane, a tertiary alkyl halide, undergoes the S_N1 reaction on gentle heating in a protic solvent such as aqueous acetone. This results in an acidic solution of aqueous HCl, the formation of which can be detected with litmus paper. Alternatively, the chloride ion produced can be precipitated with silver ion.

$$\underset{\substack{\textbf{1-methoxy-2-chloro-}\\\textbf{2-methylpropane}}}{\underset{\underset{CH_3}{|}}{CH_3OCH_2\overset{\underset{|}{Cl}}{C}CH_3}} \xrightarrow{H_2O} \underset{\underset{CH_3}{|}}{CH_3OCH_2\overset{\underset{|}{OH}}{C}CH_3} + HCl \xrightarrow{AgNO_3} AgCl$$

The first compound, 1-methoxy-2-methylpropane, is not an alkyl halide and therefore does not undergo any of these tests.

11.50   Epoxides add HCl to give neutral chloro alcohols. This reaction is so fast and so favorable that HCl can be quantitatively (that is, completely) removed by adding 2-methyloxirane to a reaction mixture.

$$\underset{H_3C}{\overset{O}{\triangle}} + HCl \longrightarrow H_3C-\overset{\overset{\displaystyle Cl}{|}}{CH}-CH_2OH$$

11.51 (a) Sodium ethoxide reacts with water to give ethanol and sodium hydroxide. Although the pK_a values of water and ethanol are similar, water is present in excess because it is the solvent, and the equilibrium therefore favors sodium hydroxide. Consequently, the alcohol $(CH_3)_2CHCH_2CH_2OH$ rather than the ether will be formed as the major substitution product.

$$Na^+ \ C_2H_5O^- \ + \ H_2O \ \rightleftharpoons \ Na^+ \ ^-OH \ + \ C_2H_5OH$$

solvent;   reacts with
in large excess   the alkyl halide

Another problem is that the alkyl halide is insoluble in water. This insolubility makes any reaction that does occur very slow because the concentration of alkyl halide is limited to the very small amount that will dissolve. Changing the solvent to ethanol would solve all of these problems.

(b) The alkyl halide from which the Grignard reagent is formed is also an alcohol. Because alcohols react rapidly with Grignard reagents in Brønsted acid–base reactions, the Grignard reagent will be destroyed. (See Eq. 9.69, text p. 432.)

11.52 (a) 2-Pentanol cannot be prepared by the reaction of a Grignard reagent with ethylene oxide, because this method can only be used for the preparation of certain *primary* alcohols.

(b) 1-Pentanol, a primary alcohol, can be prepared by the reaction of propylmagnesium bromide with ethylene oxide followed by protonolysis.

$$CH_3CH_2CH_2-MgBr \xrightarrow[\text{2) } H_3O^+]{\text{1) } \triangle} CH_3CH_2CH_2-CH_2CH_2-OH$$

1-pentanol

11.53 (a) Reaction of ethylene ($H_2C{=}CH_2$) with OsO_4 in the presence of water gives ethylene glycol ($HO-CH_2CH_2-OH$). This compound is achiral and thus cannot be resolved into enantiomers.

(b) The reaction of OsO_4 with any alkene is a syn-addition. The reaction with *cis*-2-butene gives *meso*-2,3-butanediol. Because meso compounds are achiral, this compound cannot be resolved into enantiomers.

meso-2,3-butanediol

(c) In contrast, the syn-addition of two OH groups to *trans*-2-butene gives (±)-2,3-butanediol, that is, the racemate. Any racemate can in principle be resolved into enantiomers.

(±)-2,3-butanediol

 If you find an error in this manual, please visit http://people.pharmacy.purdue.edu/~loudonm/teaching/

(d)   The syn-addition of two OH groups to *cis*-2-pentene gives (±)-(2*R**,2*S**)-2,3-pentanediol—that is, the racemate of 2,3-pentanediol in which the two asymmetric carbons have opposite configurations. (See the solutions to Problems 11.45(d)–(e) for an explanation of this notation.) Because this compound is a racemate, it can be resolved into enantiomers.

(±)-(2*R**,3*S**)-2,3-pentanediol
(the 2*S*,3*R* enantiomer is shown)

This outcome is different from that of the reaction in (b) because *cis*-2-butene and 2,3-butanediol have symmetry not present in *cis*-2-pentene and 2,3-pentanediol—that is, no meso stereoisomer is possible in 2,3-pentanediol.

11.54   Because the hydroxide ion reacts with an epoxide at the CH_2 group, the configuration at the asymmetric carbon is unaffected. Consequently, (+)-2-methyloxirane has the *R* configuration.

(*R*)-(+)-2-methyloxirane   **(*R*)-(–)-1,2-propanediol**

11.55   In this reaction the nucleophile (water) reacts at the protonated epoxide at the more substituted carbon with inversion of stereochemical configuration.

(*S*)-2-ethyl-2-methyloxirane   **(*R*)-2-methyl-1,2-butanediol**

11.56   The formula of the product indicates one degree of unsaturation. Because the double bond undergoes addition with mercuric acetate, this unsaturation cannot be due to a double bond and therefore is accounted for by a ring. As the hint and the molecular formula of the product suggest, this is an alkoxymercuration reaction in which the —OH group from within the same molecule serves as the nucleophile:

$C_8H_{16}O$

11.57   As a result of this reaction, carbon-2 becomes asymmetric, and diastereomers are formed corresponding to the two possible configurations of this carbon. The product with the *S* configuration at carbon-2 (compound *A*) is optically active; however, the other product (compound *B*) is a meso compound, and hence, is optically inactive.

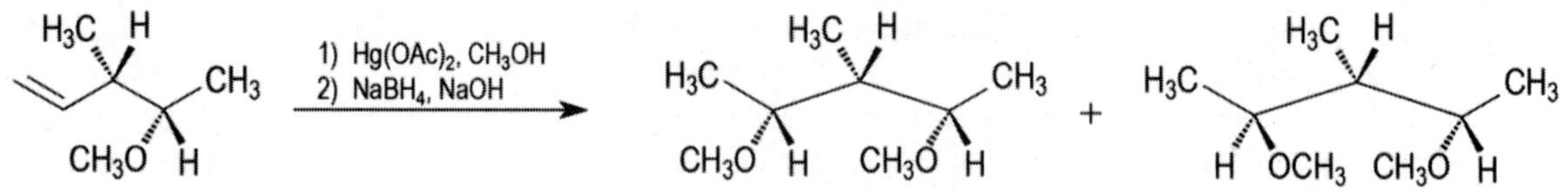

(3S,4S)-4-methoxy-3-methyl-1-pentene

compound A
(meso; therefore achiral and optically inactive)

+

compound B
(optically active)

11.58   Synthesis (1) will not accomplish the desired objective, because the —OH group of the alcohol would be lost as water. (See the mechanism in Eqs. 11.13–11.14, text p. 516.) In contrast, synthesis (2) would result in complete incorporation of the isotopic oxygen into the ether. Because this is a conventional Williamson synthesis, the carbon–oxygen bond to the ring is not broken. (See Eq. 11.5, text p. 513.) Consequently, you should advise Flaskclamper to use the second synthesis.

11.59   The structures of the compounds without stereochemical detail are as follows. Compounds (2) and (3) are diastereomers; compound (2) is the meso diastereomer. Although we do not know which enantiomer of compound (3) is the (+)-enantiomer, all we need to know to solve the problem is that it is one or the other.

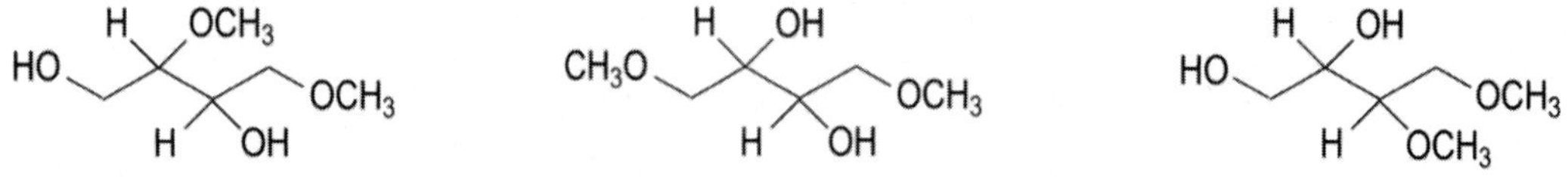

2,4-dimethoxy-1,3-butanediol
compound (1)

1,4-dimethoxy-2,3-butanediol
compounds (2) = the meso diastereomer;
compound (3) = a chiral disatereomer

3,4-dimethoxy-1,3-butanediol
compound (4)

Because compound (2) is meso, and no meso compound can be optically active, compound (2) = D. All of the other compounds are enantiomerically pure chiral compounds and are therefore optically active. Diastereomers (2) and (3) would give the same products of periodic acid oxidation; hence, compound (3) = C. Because compound (1) is not a vicinal glycol, it would not react with periodic acid; hence, compound (1) = A. Therefore, compound (4) = B; and, indeed, it would react to give periodate cleavage products that are different from those obtained from C (= compound 3) or D (= compound 2).

11.60   (a)   This is a Williamson ether synthesis by the S_N2 mechanism that gives $CH_3CH_2CH_2$—O—CH_2CH_3 (ethyl propyl ether, or 1-ethoxypropane) + $Na^+ Br^-$.

(b)   A strong base with a tertiary halide gives elimination by the E2 mechanism.

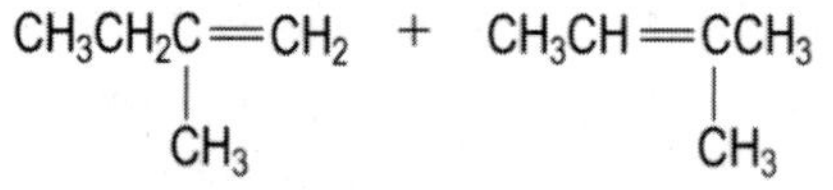

2-methyl-1-butene   **2-methyl-2-butene**

(c)   This is an alkoxymercuration–reduction that gives the following ether:

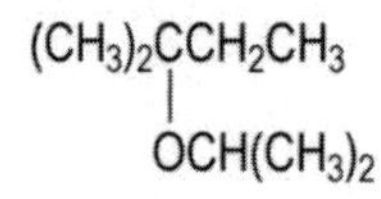

2-methyl-2-isopropoxybutane

(d)   This is an epoxidation with retention of alkene configuration:

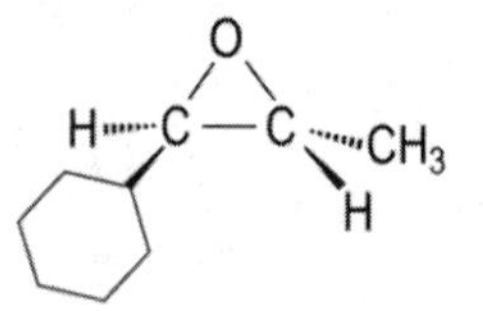

(e)   A bromohydrin is formed in the first step, and this is oxidized to an α-bromo ketone by the CrO_3.

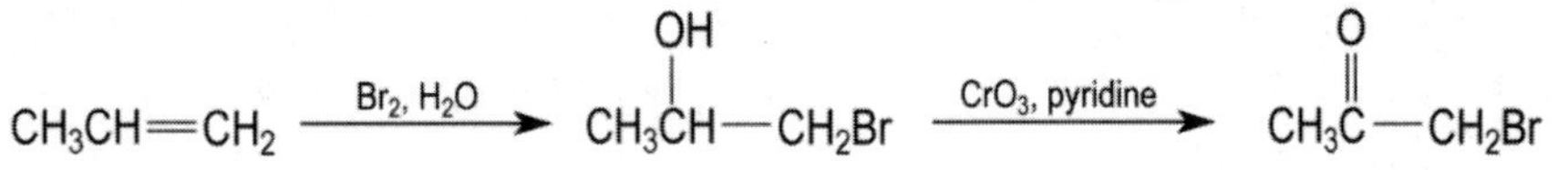

$$CH_3CH{=}CH_2 \xrightarrow{Br_2,\ H_2O} CH_3\underset{OH}{CH}{-}CH_2Br \xrightarrow{CrO_3,\ pyridine} CH_3\overset{O}{C}{-}CH_2Br$$

an α-bromo ketone

(f)   The epoxide is opened to a glycol by the acidic aqueous reaction conditions, and the glycol is cleaved as it is formed by the periodic acid.

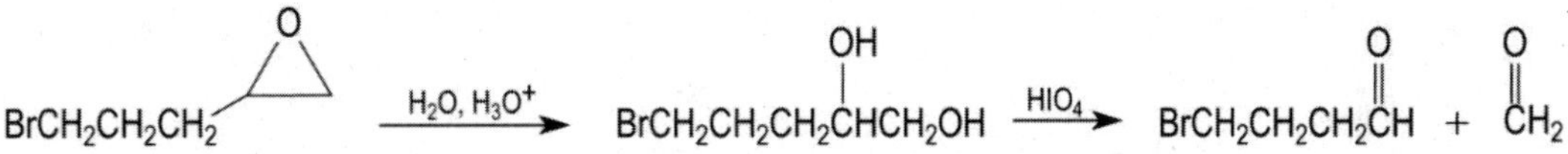

$$BrCH_2CH_2CH_2\text{—epoxide} \xrightarrow{H_2O,\ H_3O^+} BrCH_2CH_2CH_2\underset{OH}{CH}CH_2OH \xrightarrow{HIO_4} BrCH_2CH_2CH_2\overset{O}{CH} + \overset{O}{CH_2}$$

(g)   This is glycol formation with $KMnO_4$. The syn stereochemistry of the addition and the cis stereochemistry of the starting material determine the stereochemistry of the product, which is the racemate:

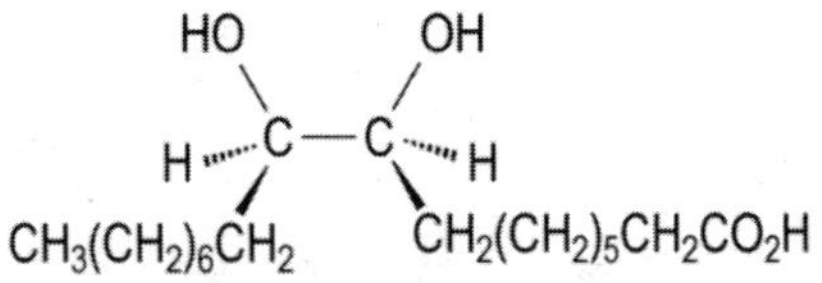

$$H{\cdots}\underset{CH_3(CH_2)_6CH_2}{\overset{HO}{C}}{-}\underset{CH_2(CH_2)_5CH_2CO_2H}{\overset{OH}{C}}{\cdots}H$$

(h)   The azide nucleophile opens the epoxide with inversion of configuration to give the racemate of the trans azido alcohol:

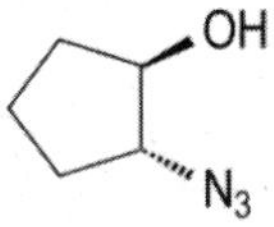

(i)   The nucleophilic "methyl anion" of the organocuprate reagent opens the epoxide to give the racemate of the trans alcohol:

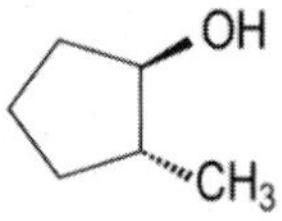

trans-2-methylcyclopentanol

(j)   Sulfide ion displaces one chloride to give a thiolate ion, which then reacts in an internal nucleophilic substitution to give the cyclic sulfide tetrahydrothiophene.

tetrahydrothiophene

(k)   A tosylate of one OH group is formed. (Introduction of a second tosylate is much slower because of the resulting van der Waals repulsions caused by the introduction of a second large tosylate group.) The resulting hydroxy tosylate reacts like a bromohydrin. (Remember, you can think of the tosylate group as a "fat halide.") Thus, in base, the OH group ionizes to an alkoxide, which undergoes an internal nucleophilic substitution reaction with the tosylate expelled as a leaving group to give an epoxide. This is very much like epoxide formation from a bromohydrin.

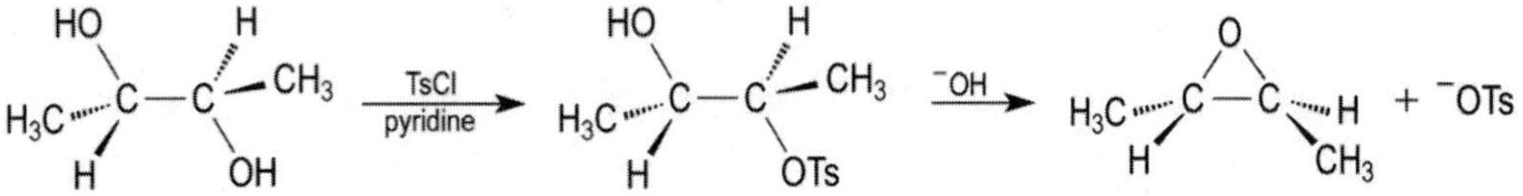

meso-2,3-butanediol

Because the starting diol is achiral (it is meso) and the reagents are all achiral, all chiral compounds in this reaction scheme are racemates.

11.61 (a)

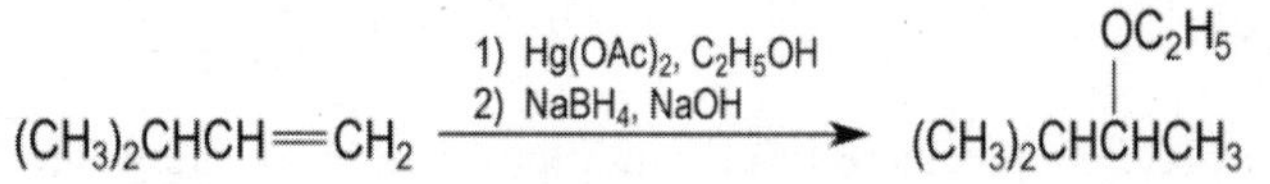

$(CH_3)_2CHCH=CH_2$ $\xrightarrow[\text{2) NaBH}_4, \text{NaOH}]{\text{1) Hg(OAc)}_2, \text{C}_2\text{H}_5\text{OH}}$ $(CH_3)_2CHCHCH_3$ with OC_2H_5

3-methyl-1-butene   **2-ethoxy-3-methylbutane**

(b)

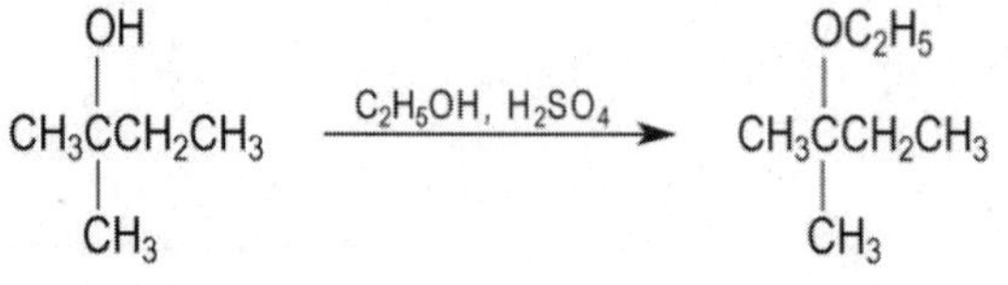

$CH_3CCH_2CH_3$ (with OH and CH₃) $\xrightarrow{\text{C}_2\text{H}_5\text{OH, H}_2\text{SO}_4}$ $CH_3CCH_2CH_3$ (with OC₂H₅ and CH₃)

2-methyl-2-butanol   **2-ethoxy-2-methylbutane**

(c)

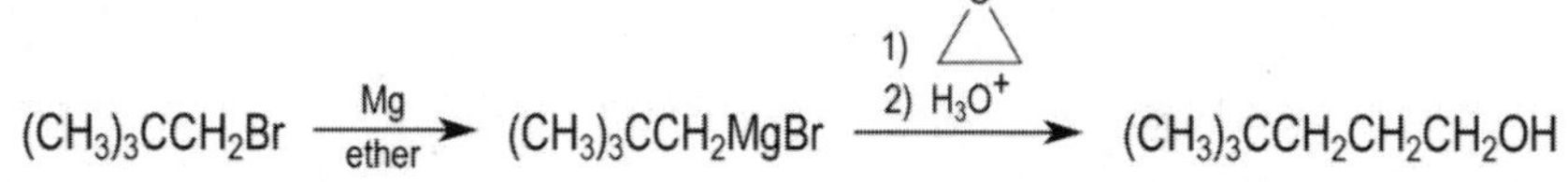

$(CH_3)_3CCH_2Br$ $\xrightarrow[\text{ether}]{\text{Mg}}$ $(CH_3)_3CCH_2MgBr$ $\xrightarrow[\text{2) H}_3\text{O}^+]{\text{1) epoxide}}$ $(CH_3)_3CCH_2CH_2CH_2OH$

4,4-dimethyl-1-pentanol

(d)

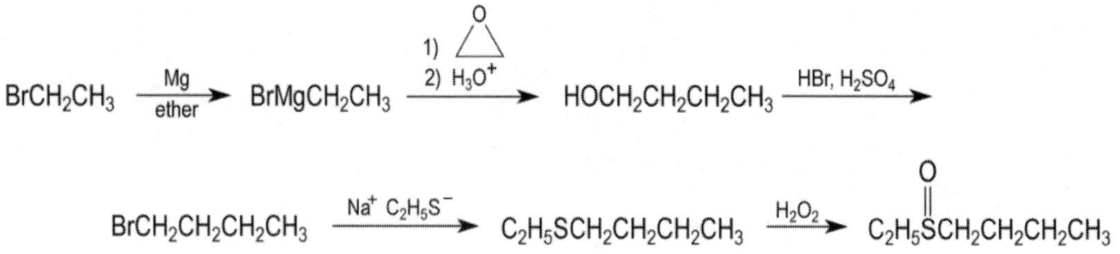

$BrCH_2CH_3$ $\xrightarrow[\text{ether}]{\text{Mg}}$ $BrMgCH_2CH_3$ $\xrightarrow[\text{2) H}_3\text{O}^+]{\text{1) epoxide}}$ $HOCH_2CH_2CH_2CH_3$ $\xrightarrow{\text{HBr, H}_2\text{SO}_4}$

$BrCH_2CH_2CH_2CH_3$ $\xrightarrow{\text{Na}^+ \text{C}_2\text{H}_6\text{S}^-}$ $C_2H_5SCH_2CH_2CH_2CH_3$ $\xrightarrow{\text{H}_2\text{O}_2}$ $C_2H_5\overset{O}{\underset{}{S}}CH_2CH_2CH_2CH_3$

(e)

$\xrightarrow{\text{mCPBA}}$ $\xrightarrow{\text{H}_2\text{O, H}_3\text{O}^+}$

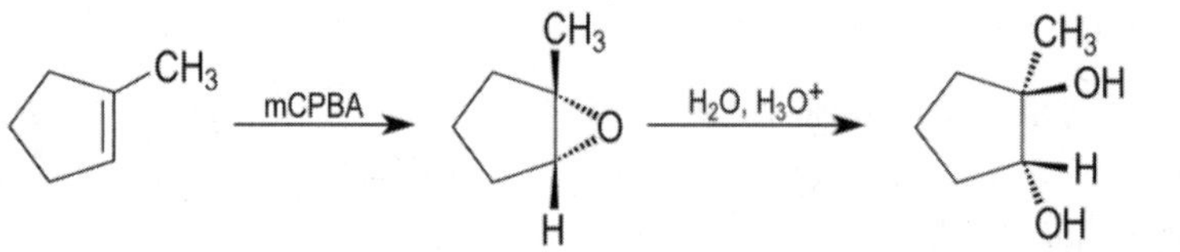

Remember that all chiral compounds, when prepared from achiral starting materials, are racemates.

(f)

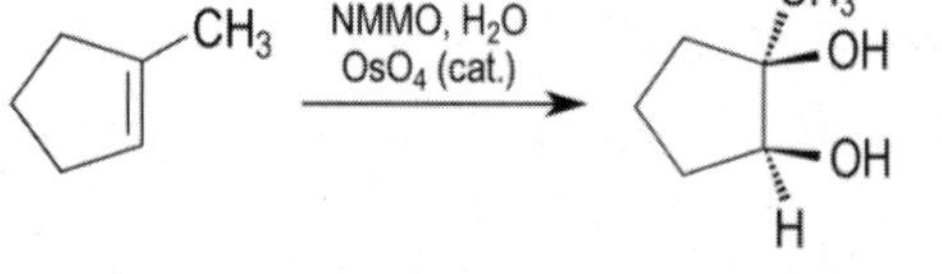

$\xrightarrow[\text{OsO}_4 \text{ (cat.)}]{\text{NMMO, H}_2\text{O}}$

(g)

$\xrightarrow[\text{2) NaBH}_4, \text{NaOH}]{\text{1) Hg(OAc)}_2, \text{(CH}_3)_2\text{CHOH}}$ with $OCH(CH_3)_2$

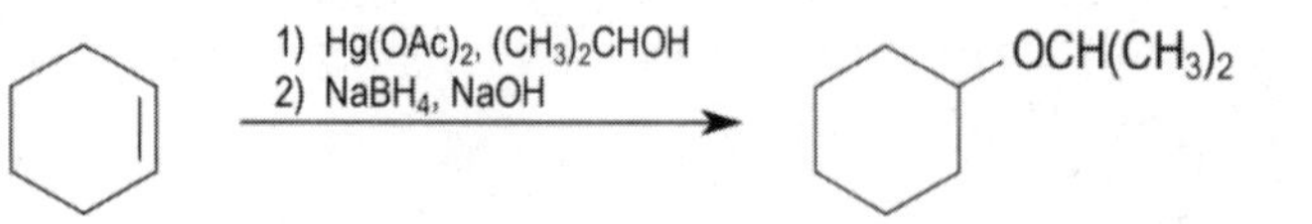

cyclohexyl isopropyl ether

(h)

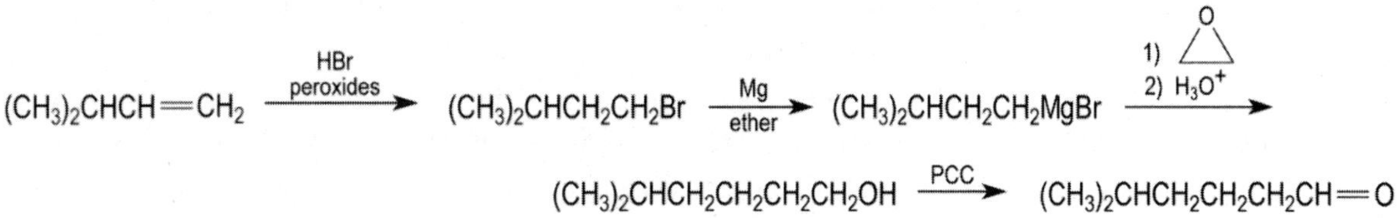

$(CH_3)_2CHCH{=}CH_2 \xrightarrow[\text{peroxides}]{HBr} (CH_3)_2CHCH_2CH_2Br \xrightarrow[\text{ether}]{Mg} (CH_3)_2CHCH_2CH_2MgBr \xrightarrow[\text{2) } H_3O^+]{\text{1) } \triangle\!O}$

$(CH_3)_2CHCH_2CH_2CH_2CH_2OH \xrightarrow{PCC} (CH_3)_2CHCH_2CH_2CH_2CH{=}O$

(i)

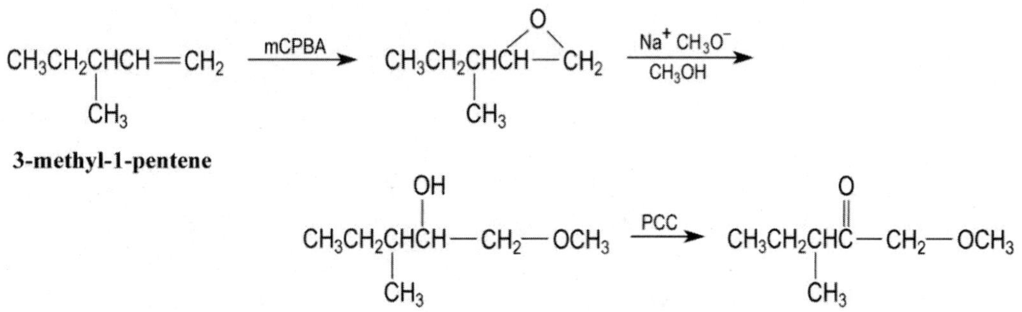

$\underset{\underset{CH_3}{|}}{CH_3CH_2CHCH}{=}CH_2 \xrightarrow{mCPBA} \underset{\underset{CH_3}{|}}{CH_3CH_2CHCH}\overset{O}{\frown}CH_2 \xrightarrow[CH_3OH]{Na^+ CH_3O^-}$

3-methyl-1-pentene

$\underset{\underset{CH_3}{|}}{CH_3CH_2\overset{\overset{OH}{|}}{C}HCH}{-}CH_2{-}OCH_3 \xrightarrow{PCC} \underset{\underset{CH_3}{|}}{CH_3CH_2\overset{\overset{O}{||}}{C}HC}{-}CH_2{-}OCH_3$

(j)

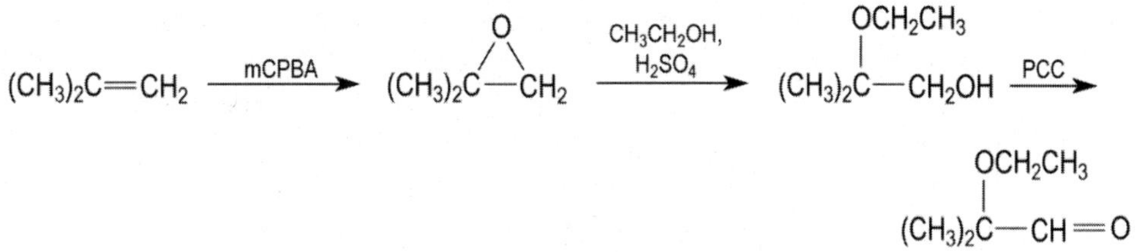

$(CH_3)_2C{=}CH_2 \xrightarrow{mCPBA} (CH_3)_2C\overset{O}{\frown}CH_2 \xrightarrow[H_2SO_4]{CH_3CH_2OH,} (CH_3)_2\overset{\overset{OCH_2CH_3}{|}}{C}{-}CH_2OH \xrightarrow{PCC}$

$(CH_3)_2\overset{\overset{OCH_2CH_3}{|}}{C}{-}CH{=}O$

(k)

$H_2C{=}CHCH_2Cl \xrightarrow[EtOH]{NaOEt} H_2C{=}CHCH_2OEt \xrightarrow[\text{2) } (CH_3)_2S]{\text{1) } O_3} O{=}CHCH_2OEt$

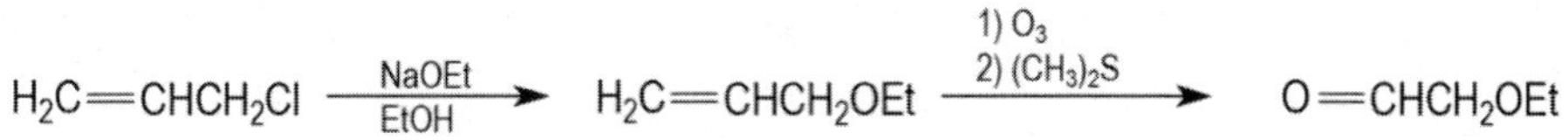

(l)

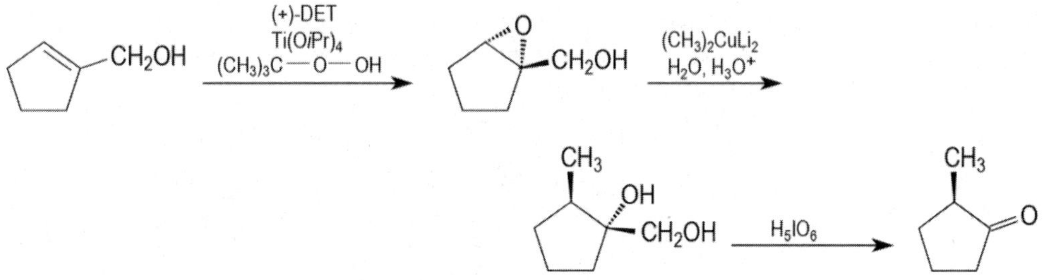

An extra equivalent of the organometallic reagent must be added because the first equivalent is consumed by its reaction with the OH group.

(m)

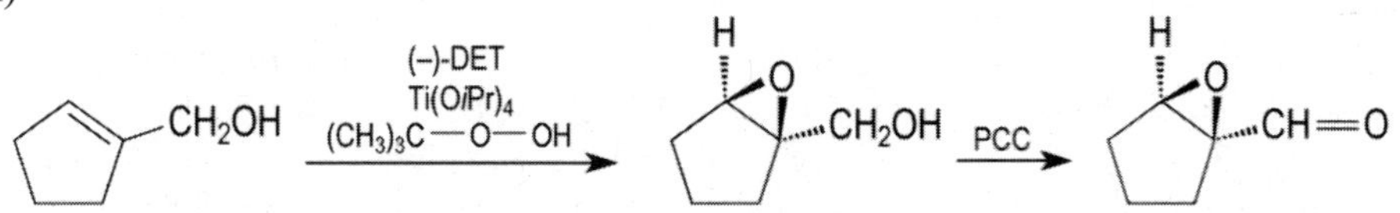

11.62   (a)   The desired compound can be prepared by a ring-opening reaction of (2R,3R)-2,3-dimethyloxirane with acidic methanol, because epoxide ring-opening occurs with inversion of configuration:

(2R,3S)-3-methoxy-2-butanol

(Be sure to convince yourself that the nucleophilic reaction of methanol at either carbon of the epoxide ring gives the same stereoisomer of the product.)

(b)   Carbon-3, the asymmetric carbon, has the *S* configuration, which is the same configuration that carbon-3 has in the product of part (a). Consequently, oxidation of the alcohol product of part (a) gives the desired compound:

(2R,3S)-3-methoxy-2-butanol
from part (a)

(c)   If the OH group can be converted into an ethyl ether without changing the configuration of either carbon, the problem will be solved. The Williamson synthesis is a good way to accomplish this objective, because it does not break a carbon–oxygen bond.

(2R,3S)-3-methoxy-2-butanol
from part (a)

(d)   Because the carbon bearing the ethoxy group is the one that must be inverted, use the ring-opening reaction in part (a), except substitute ethanol for methanol. Then carry out the Williamson synthesis used in part (c) with methyl iodide instead of ethyl iodide.

(2R,3S)-3-ethoxy-2-butanol

Don't be confused by the fact that the carbon numbering changes because of numbering conventions. Thus, carbon-2 in the alcohol becomes carbon-3 in the product, and vice versa.

11.63   Compound *A* is an octene because it undergoes typical alkene reactions and it gives octane on catalytic hydrogenation. Compounds *C* and *D* are stereoisomeric glycols, and glycol *D* is achiral. The only octene isomer that would give an achiral glycol is one of the stereoisomers of 4-octene. *Cis*-4-octene reacts with mCPBA to give an *achiral* (meso) epoxide *B*, which, upon ring opening in aqueous acid, gives *racemic* 4,5-octanediol. The same alkene reacts in a syn-addition with OsO_4 to give *meso*-4,5-octanediol, an achiral compound. Therefore, compound *A* is *cis*-4-octene. To summarize:

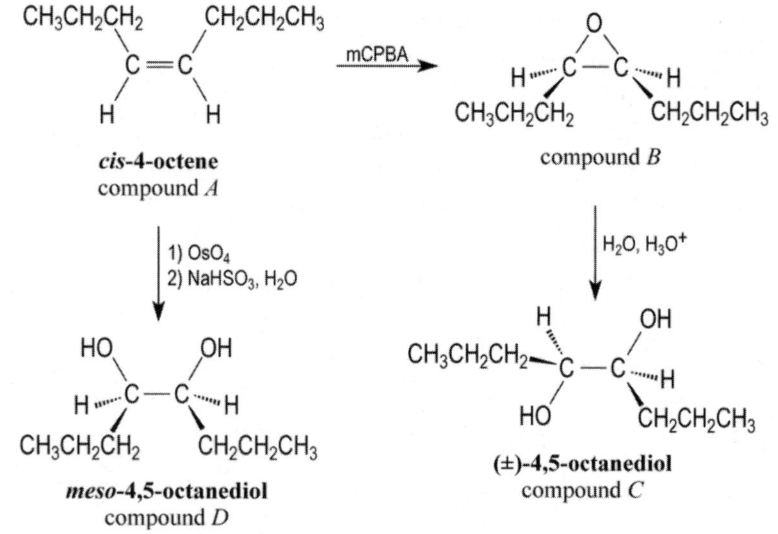

CH_3CH_2CH_2 ... CH_2CH_2CH_3 (cis-4-octene, compound A) →(mCPBA)→ epoxide compound B

cis-4-octene
compound A

compound B

1) OsO_4
2) NaHSO_3, H_2O

H_2O, H_3O^+

meso-4,5-octanediol
compound D

(±)-4,5-octanediol
compound C

11.64   (a)   The OsO_4 (in the presence of water) effects glycol formation from the alkene starting material. This produces Os(VI), which is re-oxidized to Os(VIII) by periodic acid. One equivalent of the periodic acid effects this oxidation, continually recycling the osmium until the alkene is fully utilized; this role of periodic acid is the same as that of NMMO in that method. The second equivalent of the periodic acid then brings about cleavage of the glycol. To summarize:

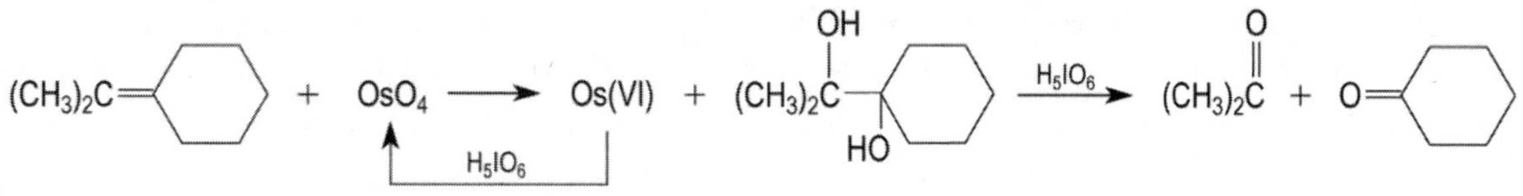

$(CH_3)_2C=$⬡ + OsO_4 → Os(VI) + $(CH_3)_2C$(OH)(HO)⬡ →(H_5IO_6)→ $(CH_3)_2C=O$ + O=⬡

(H_5IO_6)

(b)

⬡ + OsO_4 → Os(VI) + ⬡(OH)(OH) →(H_5IO_6)→ CH=O, CH=O = ⬡ with two CH=O groups

(H_5IO_6)

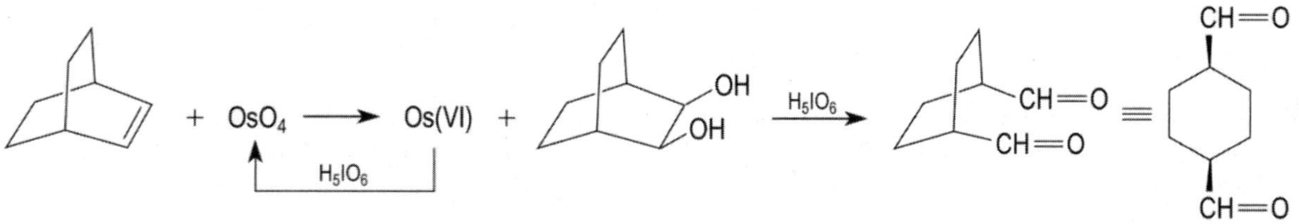

11.65   In both parts the mCPBA reacts at the side of the alkene π-bond that involves the less severe van der Waals repulsions.

(a)   One of the methyl groups must be axial, and therefore it creates a 1,3-diaxial interaction with the incoming reagent on one face of the ring. Therefore, the reaction occurs at the other face.

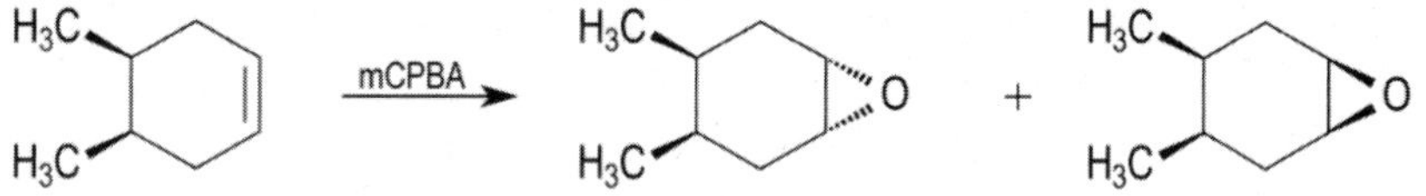

H_3C... ⬡ ...H_3C →(mCPBA)→ H_3C... ⬡ ...O (A, major product) + H_3C... ⬡ ...O (B)

A
(major product)

B

(b) The face of the ring on the *same side* of the methyl group is blocked; hence, reaction occurs at the opposite face.

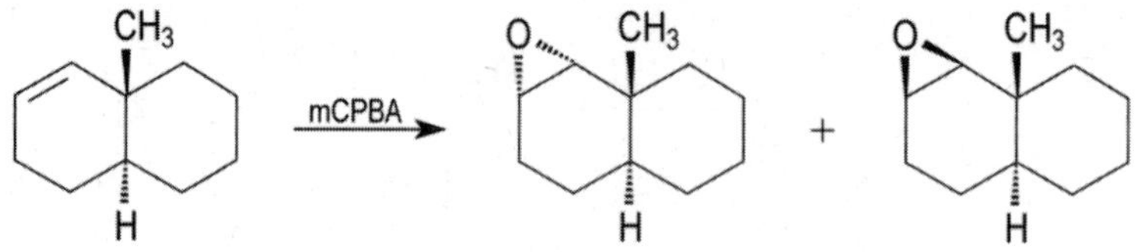

A
(major product)

B

11.66 (a) The curved-arrow mechanism for salt formation: (The two reactions occur separately.)

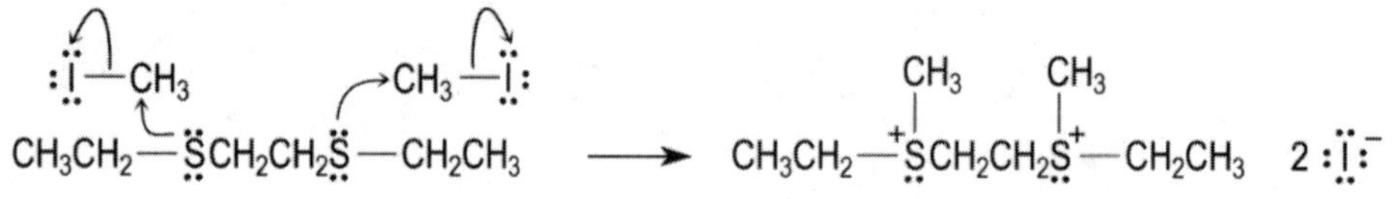

(b) Counting the electron pair, each sulfur in the product bears four different groups. Consequently, the sulfurs are stereocenters. Because the sulfurs do not undergo rapid inversion, stereoisomeric sulfonium salts that differ in configuration at *sulfur* can be isolated. Because each salt contains two sulfur stereocenters, the two salts are diastereomers . One diastereomer is the meso, or (*R,S*), stereoisomer; the other is the racemate, a 1:1 mixture of the (*R,R*) and (*S,S*) stereoisomers.

11.67 If sulfur does invert over a long period of time, its configuration goes from *S* to *R*. Since there are other asymmetric carbons in SAM which retain their configuration, this new compound is a diastereomer of naturally occurring SAM.

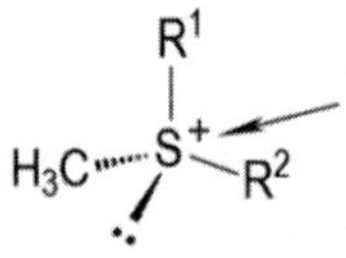

sulfur has the *R* configuration
($R^2 > R^1 > CH_3$. Electrons are lowest priority.
See Fig. 11.1 in the text for the structure of SAM.)

11.68 Because sulfur in a sulfoxide does not undergo inversion, it is an asymmetric atom. Because there are two asymmetric carbons in methionine (asterisks in the structure below), methionine sulfoxide can exist as diastereomers. (See the solution to Problem 11.67.)

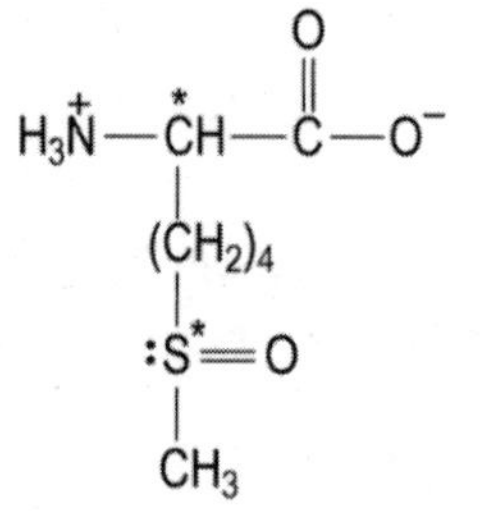

methionine sulfoxide

11.69 The anion that results from the ring-opening reaction with hydroxide reacts as a nucleophile in another epoxide ring-opening reaction; and so on.

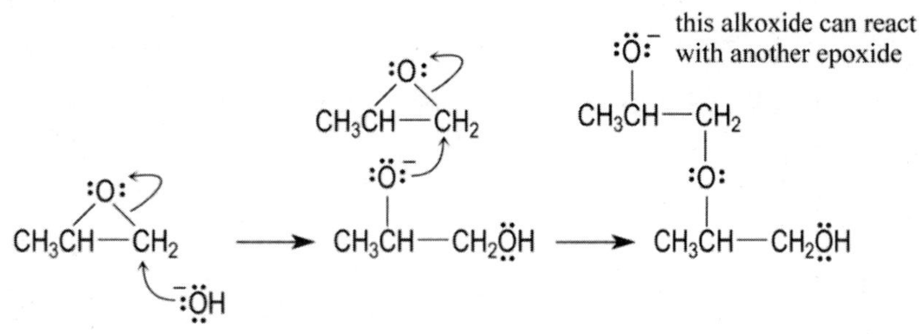

this alkoxide can react
with another epoxide

CH₃CH—CH₂
       |
      :Ö:⁻

CH₃CH—CH₂
       |
      :Ö:

CH₃CH—CH₂
       |
  :Ö:⁻

CH₃CH—CH₂ÖH ⟶ CH₃CH—CH₂ÖH

CH₃CH—CH₂
       |
     :ÖH

⟶ CH₃CH—CH₂ÖH

11.70 **(a)** The role of the ⁺NH₄ Cl⁻ is to protonate the nitrogen of the aziridine. Without this protonation, the aziridine nitrogen is too basic to serve as a good leaving group.

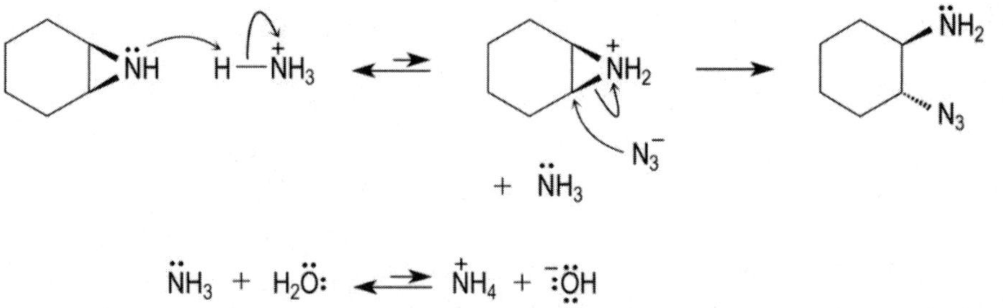

+ ṄH₃

ṄH₃ + H₂Ö: ⇌ ⁺ṄH₄ + ⁻:ÖH

Like the ring opening of epoxides, the ring opening of aziridines involves backside substitution and therefore gives the trans stereoisomer of the product.

(b) Cleavage of an epoxide generates an alkoxide leaving group (conjugate-acid $pK_a \approx 15$). Although this is quite basic, it is much less basic (by about 17 pK_a units) than the amide (—⁻NH) leaving group in an aziridine (conjugate-acid $pK_a \approx 32$). Normally, alkoxides do not make very good leaving groups, because they are too basic; but, as explained in Sec. 11.5A in the text, epoxides are opened because of the great ring strain of the three-membered ring. The leaving group in an aziridine is too poor a base to be opened without protonation despite its ring strain.

(c) An HCl solution at 0.01 M has a pH = 2. This is acidic enough to protonate the azide ion, thus eliminating its nucleophilicity. The ammonium ion is not acidic enough to protonate the azide ion significantly.

11.71 **(a)** The rate acceleration suggests a mechanism involving neighboring-group participation. The product of this reaction results from the net substitution of the chlorines by —OH groups to give a diol. The mechanism for the first substitution is shown below in detail; you should write the mechanism for the second.

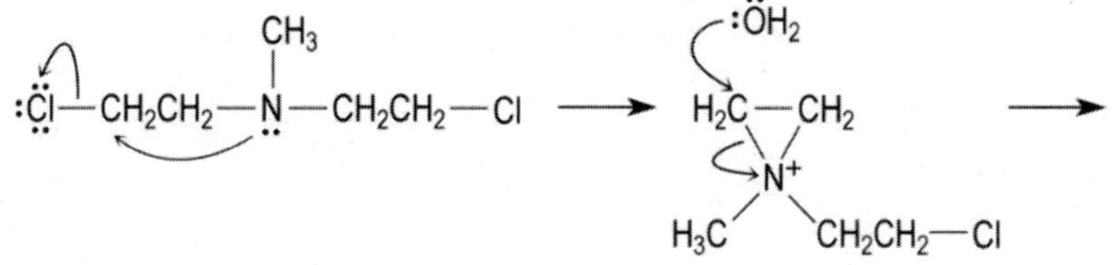

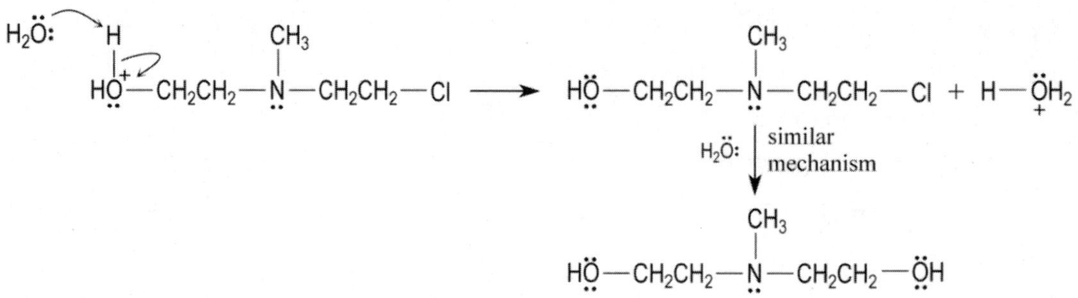

(b) The reaction of mechlorethamine with an amine R_3N: follows a similar pattern in which the amine rather than water serves as the ultimate nucleophile.

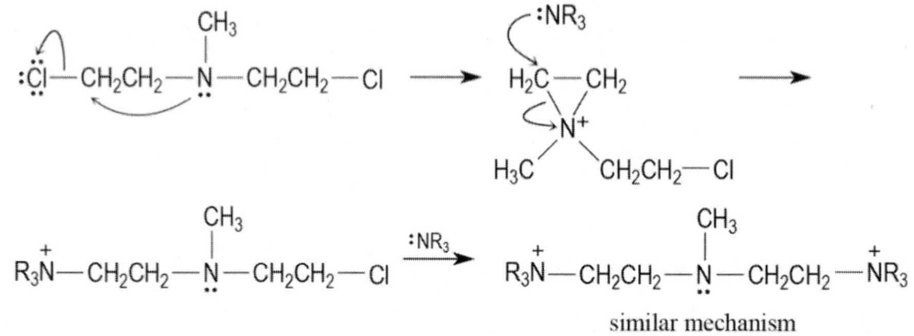

similar mechanism

11.72 (a) Because the large *tert*-butyl group virtually locks the molecule into the chair conformation in which this group is equatorial, the two —OH groups in the compound *A* are locked into a trans-diaxial arrangement. The cyclic periodate ester intermediate cannot be formed from two hydroxy groups in this arrangement without introducing significant ring strain. Because the required intermediate cannot form, the oxidation of compound *A* cannot occur. In contrast, compound *B* can be oxidized with periodic acid because the two —OH groups are trans-diequatorial.

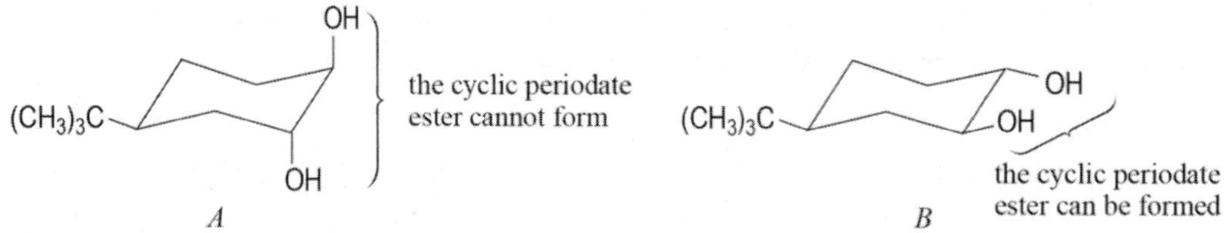

the cyclic periodate ester cannot form

A

the cyclic periodate ester can be formed

B

(b) Compound *B* is locked into a conformation in which the two hydroxy groups are trans-diaxial; consequently, this compound cannot be oxidized for the same reason that compound *A* cannot be oxidized in part (a).

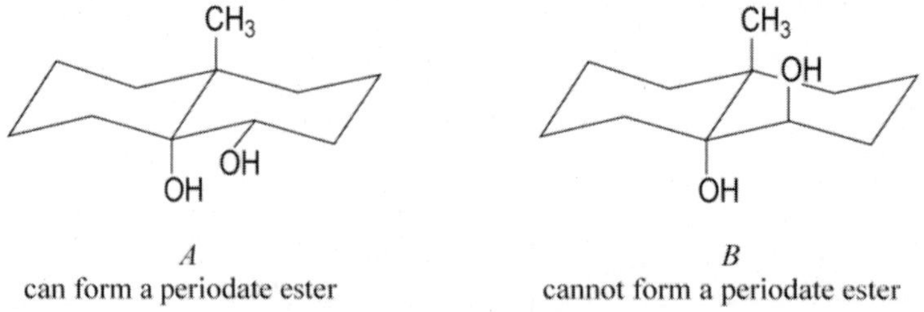

A
can form a periodate ester

B
cannot form a periodate ester

11.73 The transformation of compound *A* to compound *B* occurs with *retention of configuration*. A retention of configuration can be the result of *two successive inversions of configuration*. This fact and fact (3) suggest that neighboring-group participation is occurring. Indeed, a neighboring-group mechanism involves a meso and therefore *achiral* cyclic sulfonium-ion intermediate *D*, formed by an intramolecular substitution with inversion of configuration:

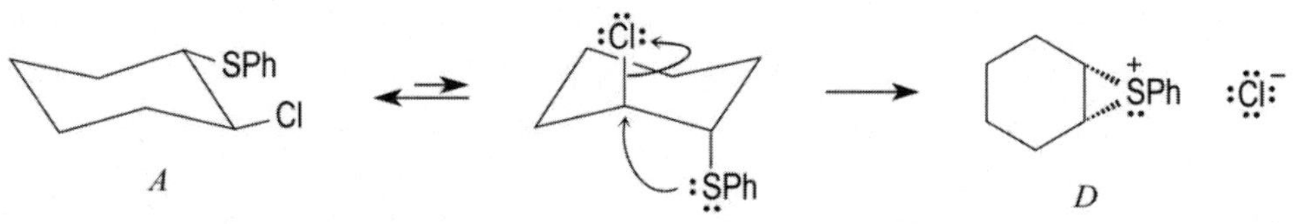

Because intermediate *D* is achiral, any chiral products that result from it must be formed as racemates (Sec. 7.7A); this accounts for fact (2). Reactions of intermediate *D* with ethanol at the two carbons of the three-membered ring give respectively the two enantiomers of product *B*. (The formation of one enantiomer is shown below; you should show formation of the other.) This substitution reaction occurs with inversion of configuration; coupled with the first inversion that takes place in the formation of *D*, this accounts for the overall retention noted in fact (1).

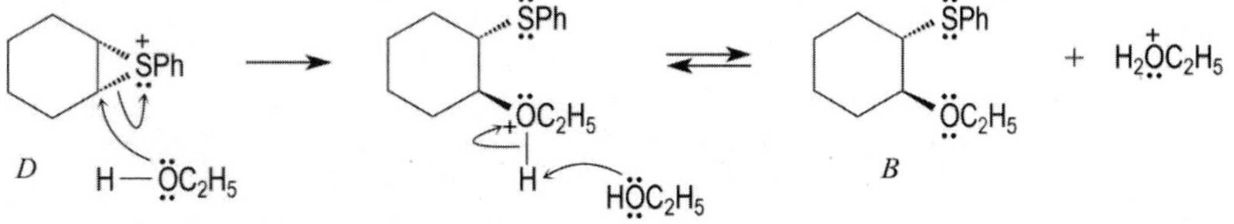

11.74   (a)   The substitution with retention of configuration suggests a neighboring-group mechanism involving a boat-like intermediate *X*.

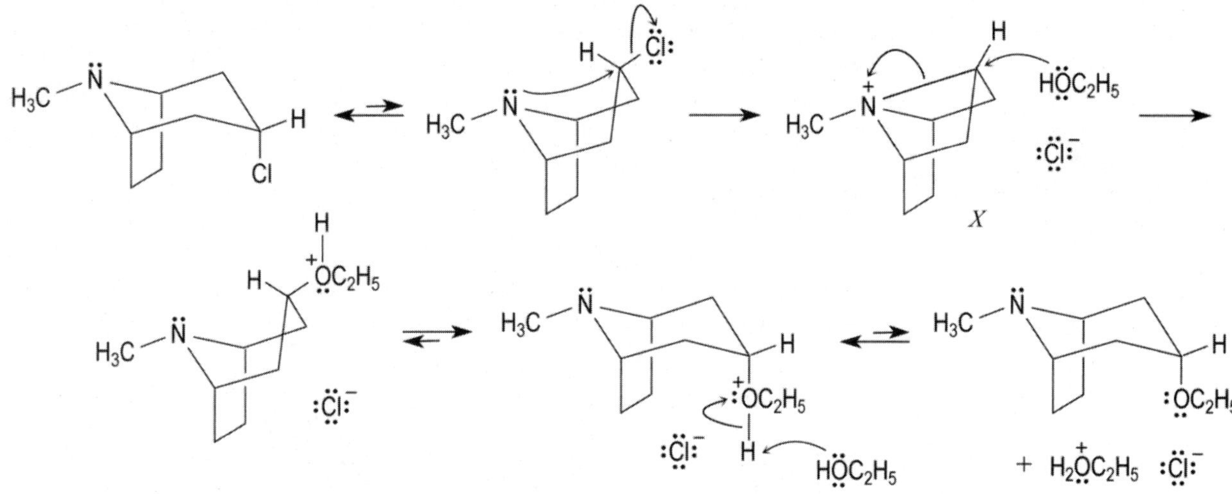

This intermediate is unstable because it incorporates two fused four-membered rings and a boat cyclohexane; it is highly strained, and thus opens readily.

(b)   The substitution with retention of configuration suggests a neighboring-group mechanism involving a bicyclic episulfonium ion *Y*.

11.75 (a) The sulfur can serve as a nucleophile in an intramolecular nucleophilic substitution reaction. The product is a cyclic sulfonium salt.

(b) The assumption is that the reaction of water with the starting sulfide occurs at about the same rate as it does with 1-chlorohexane. If so, then the relative rate of the intramolecular reaction above and the competing intermolecular S_N2 reaction of water with the same compound is 21, as given in the problem. A calculation very similar to the one in Study Problem 11.5, text p. 544, shows that $k_1/k_2 = (21)(20\ M) = 420\ M$. This is the proximity effect.

11.76 Reaction (1) is an S_N2 reaction in which oppositely charged species react. This reaction should be promoted by conditions that increase the interaction between charges, that is, by a smaller dielectric constant. Therefore, reaction (1) is faster in ethanol ($\epsilon = 32$) than it is in water ($\epsilon = 80$). (Solvent dielectric constants are found in Table 8.2 on text p. 355.) Reaction (2) is an S_N1 reaction in which the rate-limiting step involves dissociation of an alkyl halide to a carbocation and a chloride ion; this reaction is promoted by conditions that foster separation of charged groups, that is, by a greater dielectric constant. Consequently, reaction (2) is faster in water. Reaction (3) is an S_N1 reaction, but, because the starting sulfonium ion is positively charged, formation of the carbocation intermediate does not involve any change in the number of charges. Consequently, the rate of this reaction is not affected very much by a change of solvent.

11.77 (a) In this reaction, the initially formed bromonium ion reacts intramolecularly with the oxygen of the OH group. (The mechanism for the formation of a bromonium ion is shown in Eq. 5.11, text p. 185.)

(b)   The alkoxide, formed by ionization of the alcohol OH group, opens the epoxide intramolecularly. The hydroxide catalyst is regenerated by protonation of the resulting alkoxide ion.

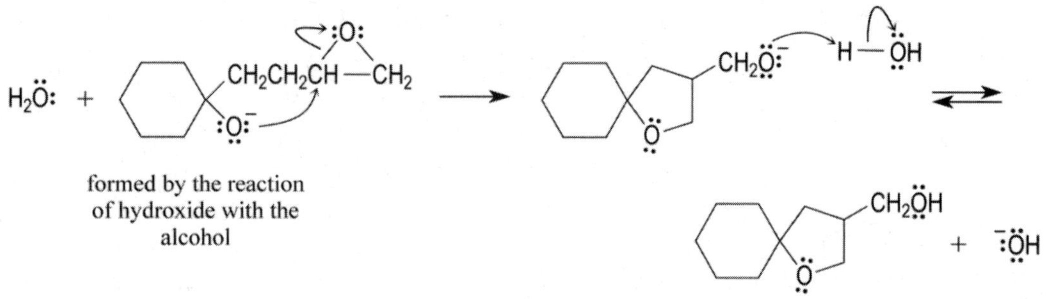

formed by the reaction
of hydroxide with the
alcohol

(c)   This mechanism is very similar to the mechanism for the intramolecular substitution involved in the formation of product *B* in Problem 9.83 on text p. 450. The thiolate anion, formed by ionization of the thiol (not shown), displaces the tosylate group within a twist-boat conformation (shown here for simplicity as a boat).

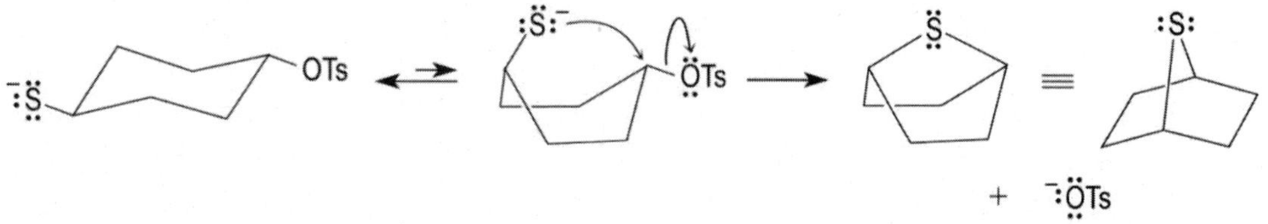

(d)   The epoxide chemistry in this chapter has focused on ring-opening substitution reactions; however, you have learned that elimination competes with substitution. Thus, this reaction is an E2-like elimination reaction in which either of the two distinguishable β-hydrogens is removed and the epoxide oxygen serves as a leaving group. The mechanism for the formation of the major alkene product is shown below; the minor product is formed by an essentially identical mechanism involving the other β-hydrogen. Substitution (that is, nucleophilic ring-opening) evidently does not occur because of the van der Waals repulsions that would result in the transition state of such a reaction between the alkyl branches of the base and those of the epoxide.

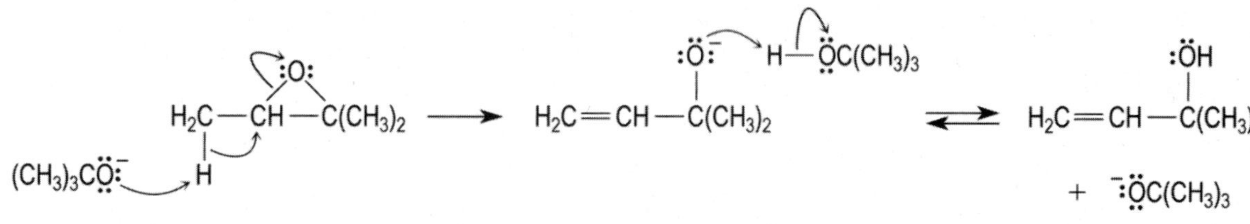

(e)   The carbocation intermediate, which is formed by protonation of the double bond, undergoes an intramolecular Lewis acid–base association reaction with the sulfur nucleophile:

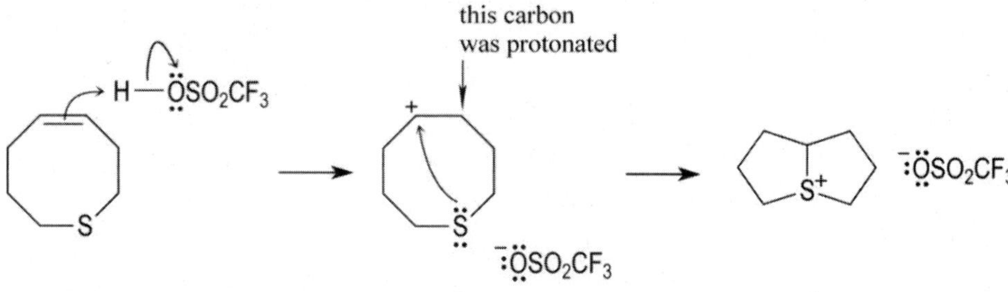

this carbon
was protonated

(f)

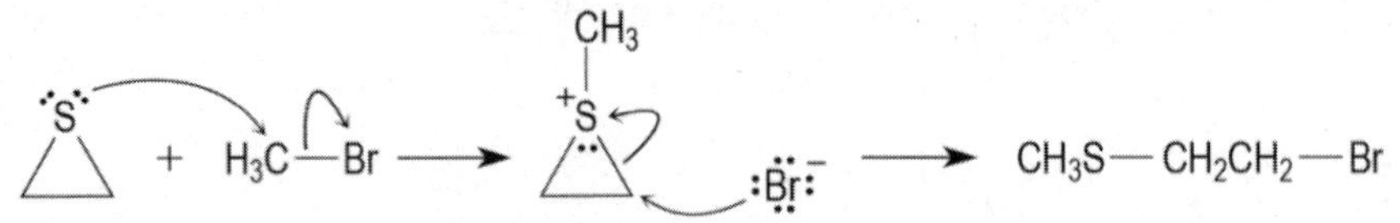

$$CH_3S—CH_2CH_2—Br$$

(g)   The key to this mechanism is to notice that inversion of configuration has occurred at carbon-2, and the cyanide ion has reacted at carbon-1. The formation of another epoxide by neighboring-group participation with inversion of configuration at C-2 is followed by reaction of the cyanide ion as a nucleophile at C-1, which is the less substituted carbon of the epoxide. (The rearrangement of one epoxide to another by the intramolecular reaction is called the *Payne rearrangement.*)

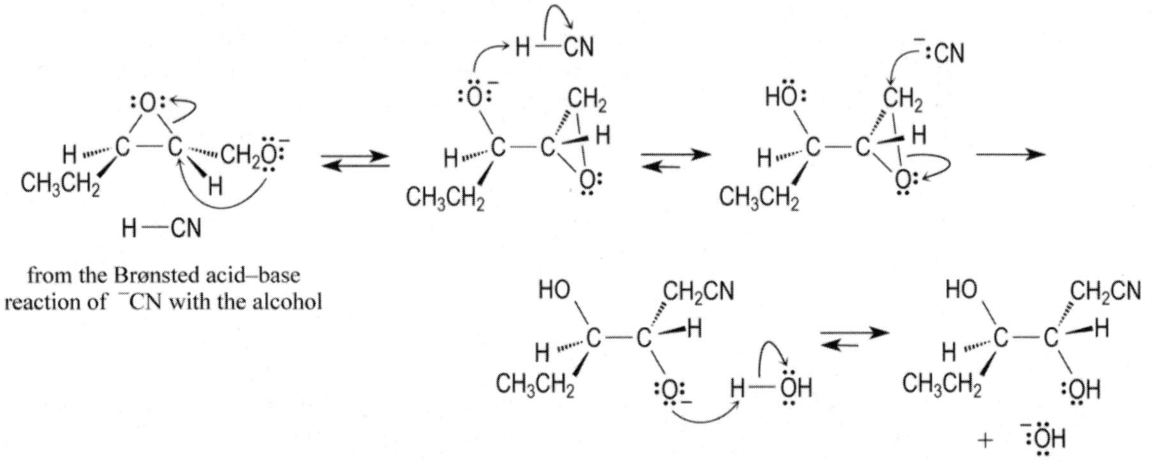

from the Brønsted acid–base reaction of ⁻CN with the alcohol

(h)   This appears to be a relatively rare instance of neighboring-group participation involving a four-membered ring. The ring is opened by the nucleophilic reaction of methanol at the more substituted carbon.

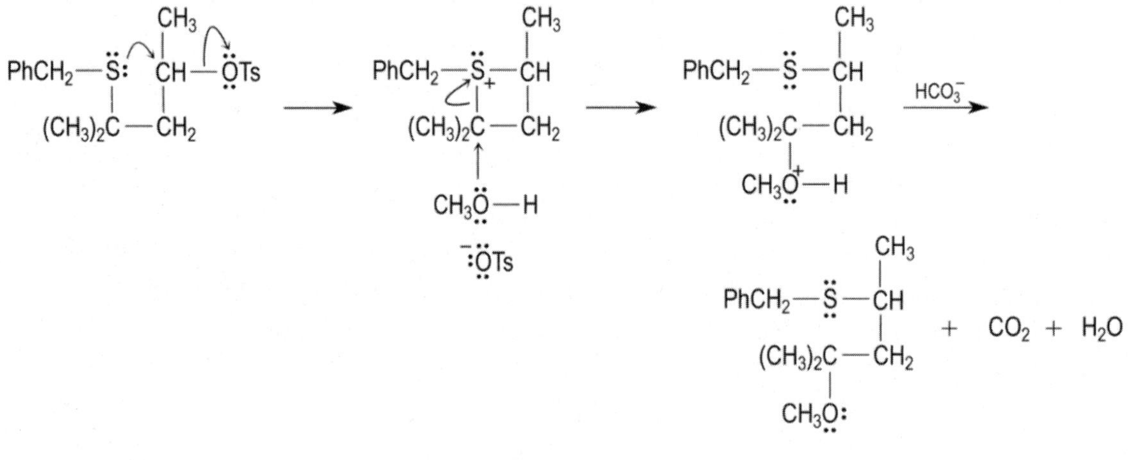

(i)

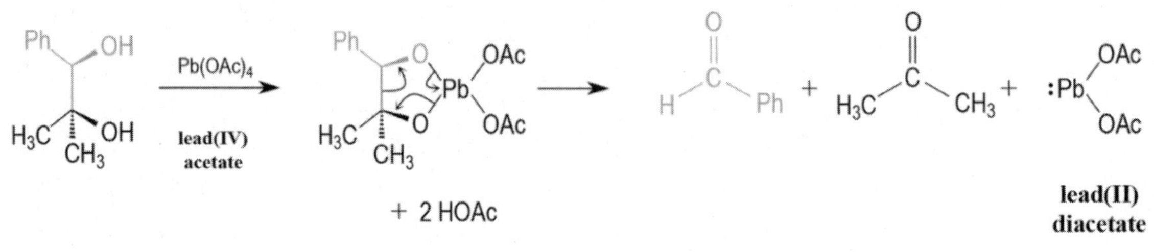

11.78   (a)   An episulfonium ion (analogous to a bromonium ion) X is formed at one double bond; reaction with the liberated chloride ion results in a net anti addition.

The sulfur in this product then serves as the electrophile for the reaction at the other double bond, which involves another episulfonium ion Y. (Because of the pucker of the cyclooctene ring, as a model will show, the distance from the double bond to the sulfur is not as far as it might seem.)

(b)   The substitution with retention of stereochemical configuration suggests that neighboring-group participation has taken place. The first substitution involves the formation of episulfonium ion B from part (a) with loss of chloride ion under the rather extreme conditions. This ion is opened by the azide ion.

The second substitution occurs by an essentially identical mechanism.

11.79   The principles to use in solving this problem are that (*a*) the *tert*-butyl group, because it is so large, must assume the equatorial position; and (*b*) the —OH and —Cl groups must be able to achieve an anti relationship in order for backside substitution (and thus epoxide formation) to occur. The required anti relationship is impossible for compound (3) because the —OH and —Cl groups are cis and are thus fixed in either chair conformation at a dihedral angle of about 60°. Therefore, compound (3) is A. In the chair conformation in which the *tert*-butyl group is equatorial, the —OH and —Cl groups in compound (1) are trans-diaxial, and are ideally set up for the formation of an epoxide. Consequently, compound (1) is B, and the epoxide formed from it is D.

Finally, consider compound (2): the —OH and —Cl groups are trans, but they are diequatorial. The only way that they can achieve an anti relationship is either for the ring to undergo a chair interconversion, or for it to assume a twist-boat conformation. In either case, the conformation required for epoxide formation has a very high energy, and very little of it will be present. Because the transition-state conformation has very high energy, epoxide formation from compound (2), although possible, is very slow. Therefore, compound (2) = C and the epoxide formed from it is E. Epoxides D and E are stereoisomers, as required by the problem.

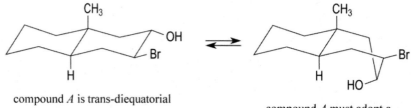

11.80   As illustrated by Eq. 11.23 on text p. 521, and as discussed in the solution to the previous solution, the —O⁻ and —Br groups must be able to assume a trans-diaxial arrangement in the transition state for backside substitution to occur.

Immediately we rule out compound D, as the —OH and —Br groups are cis.

The —OH and —Br groups in compounds A–C are trans. Because all of the compounds shown are trans-decalin derivatives, they cannot undergo the chair interconversion. Hence, the —OH and —Br groups could be trans-diequatorial or trans-diaxial. Examine each one in turn.

Compound A must assume a twist-boat conformation in order to undergo epoxide formation by backside substitution. Because this conformation, and hence the transition state of the reaction, has a very high energy, epoxide formation is likely to be very slow.

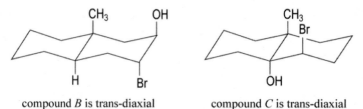

The remaining two bromohydrins B and C have trans-diaxial arrangements and thus react readily:

11.81   (a)   Because the product is meso, it must have opposite configurations at the two asymmetric carbons—in other words, it has the 2S,3R configuration. (In a meso compound, this is the same as the 2R,3S configuration.) Substitution of the OH by a Br must occur, therefore, with retention of configuration, and such a stereochemical result suggests that neighboring-group participation takes place. The protonated OH group departs with concomitant backside substitution by the bromine within the same molecule to form a bromonium ion. The bromonium ion then reacts with bromide ion (from the ionized HBr in solution) with inversion to give the *meso*-2,3-dibromobutane.

The center of symmetry shows that the product is meso. Alternatively, you can form an eclipsed conformation to see an internal mirror plane.

     Reaction of the bromide ion at the other carbon of the bromonium ion is equally probable. We leave it to you to show that this reaction gives the same result.

(b)   The similar reaction of the 2*S*,3*S* stereoisomer gives a meso bromonium ion, which is achiral. Reactions of bromide ion at the two carbons give respectively (2*S*,3*S*)- and (2*R*,3*R*)-2,3-dibromobutane in equal amounts—that is, the racemate.

12

Introduction to Spectroscopy
Infrared Spectroscopy and
Mass Spectrometry

FURTHER EXPLORATIONS

12.1   The Vibrating Bond in Quantum Theory

In the text we've used Hooke's law to describe bond vibrations. If the vibrating bond were like a classical vibrating spring, its energy (and therefore its frequency) could be increased by any arbitrary amount. However, vibrating bonds are subject to quantum theory, which requires that bond vibrations must have discrete energy levels. The energy levels of a vibrating bond are given by

$$E_n = (n + \tfrac{1}{2})h\nu_0 \qquad \text{(SG12.1)}$$

where n is a *vibrational quantum number,* and ν_0 is the fundamental vibrational frequency. This quantum number can assume the integral values 0, 1, 2, ... The lowest energy level, E_0, is obtained by substituting $n = 0$ into this equation:

$$E_0 = \tfrac{1}{2}h\nu_0 \qquad \text{(SG12.2)}$$

This is called the *zero-point energy*. The wavenumber of this lowest-energy vibration is given by the Hooke's law expression shown in Eq. 12.11 on p. 579 of the text, and the frequency ν_0 is obtained by multiplying the wavenumber by the velocity of light c.

There is a certain length of a spring (or bond), r_m, at which the potential energy of a spring is a minimum. In terms of simple springs, this is the length of the "relaxed" spring—the length at which the spring is neither stretched nor compressed. Hooke's law assumes that the potential energy stored in a spring (bond) is proportional to the square of the amount by which the spring is stretched or compressed from this distance:

$$E_{\text{potential}} = \tfrac{1}{2}k(r - r_m)^2 = \tfrac{1}{2}k\Delta r^2 \qquad \text{(SG12.3)}$$

An oscillating spring gathers potential energy when it is stretched or compressed, and this potential energy is released as kinetic energy as the spring returns to r_m. The conservation of energy requires that the sum of the potential and kinetic energies of the spring remains constant (barring frictional losses). The potential energy function in Eq. SG12.3 describes a parabola, and this is not a bad approximation for the potential energy of a vibrating bond for the first few energy levels. Such a parabola is shown in Fig. SG12.1. Superimposed on this parabola are the discrete energy levels required by quantum theory given by Eq. SG12.1. The total energy of the vibrating bond—potential + kinetic—in a particular energy level is given by the position of the energy-level line.

This brief discussion of quantum theory can be used to justify three points made in the text. The first point is that the frequency of light that brings about energy absorption must equal the fundamental frequency of the bond vibration ν_0. For a bond to absorb energy from light, it must absorb an amount of energy exactly equal to $E_1 - E_0$; that is, the energy of the vibration must "jump" to the next energy level. From Eq. SG12.1, the energy difference between E_1 and E_0 is

$$E_1 - E_0 = \Delta E_{1,0} = \frac{3}{2}h\nu_0 - \frac{1}{2}h\nu_0 = h\nu_0 \tag{SG12.4}$$

This equation shows that promotion of the bond to energy level E_1 from E_0 requires light with energy $= \Delta E_{1,0} = h\nu_0$ and thus a frequency ν_0. But this is just the fundamental frequency of the bond vibration, as asserted.

The second point made in the text is that the frequency of the bond vibration doubles as a result of light absorption. The transition to a higher energy level is associated with a change in vibrational frequency from ν_0 to ν_1.

$$E_1 - E_0 = h\nu_1 - h\nu_0 \tag{SG12.5}$$

But Eq. 12.4 states that $E_1 - E_0 = h\nu_0$. Equations SG12.4 and SG12.5 can be true simultaneously only if $\nu_1 = 2\nu_0$; that is, the frequency after absorption is double the original frequency.

Finally, the assertion was made that the amplitude of the bond vibration increases as a result of light absorption. This point can be understood from Fig. SG12.1. The intersections of the energy levels E_1 and E_0 with the potential-energy curve define the vibrational amplitudes Δr_1 and Δr_0. The vibrational amplitudes Δr_n increase as the molecule transitions to higher energy levels.

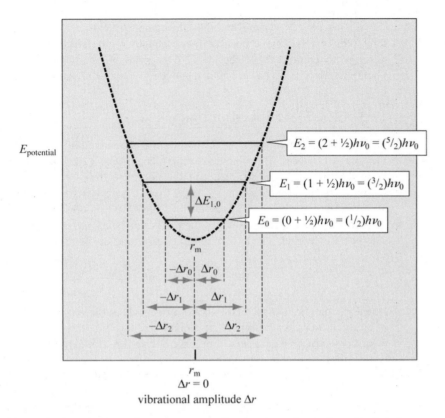

Figure SG12.1 A plot of potential energy versus vibrational amplitude for a Hooke's law potential. The potential energy curve is the heavy dashed line. The vibrational quantum levels are superimposed on this curve. (Only the first three levels are shown.) $\Delta E_{1,0}$ is the energy required for excitation of a vibration in the lowest energy level, E_0, to the next level, E_1.

 ## 12.2   FT-IR Spectroscopy

This section will give you a better idea how an FT-IR spectrometer works. In an FT-IR spectrometer, radiation to be passed through the sample is split and sent over two paths simultaneously by a system consisting of two mirrors, one moving and one fixed (a Michelson interferometer). The difference between the lengths of the two light paths at various times is precisely calibrated with a laser. The superimposed beams of light from the two paths is passed through the sample. The interference patterns, or interferograms, that result can be thought of as superimposed plots of absorption *vs.* time from both light beams in which all wavelengths in the spectrum contribute simultaneously. The interferograms are stored in a small computer and analyzed by *Fourier transformation.* Fourier transformation is a mathematical technique for decomposing time-based wave motions into their contributing frequency components.   (For example, the unique sound of a bell as it dies out over time might be analyzed in terms of the different contributing pitches by the same technique.) Fourier analysis converts the interferograms into a plot of absorption versus wavenumber—the IR spectrum. The entire process takes just a few seconds.

Besides saving time, FT-IR has much better wavenumber resolution than conventional IR spectroscopy. (This means that two closely-spaced peaks would appear as separate peaks rather than as a single broad peak.) In addition, multiple spectra from the same sample can be digitally added; because "noise" is random, it decreases when multiple spectra are summed, and the signal due to the IR spectrum grows. Consequently, many spectra can be accumulated from the same sample in a very short time, and thus strong spectra can be obtained from very dilute samples. The principles of FT-IR have been known for many years, but only in relatively recent times has the technique become practical because of the availability of inexpensive, high-powered computers. The digital revolution has come to spectroscopy just as it has to sound and video technology!

 ## 12.3   The Mass Spectrometer

This section will give you a more quantitative understanding of how a mass spectrometer operates. The magnetic-sector instrument diagrammed in Fig. 12.21, text p. 603, is the simplest type of mass spectrometer; it is still used for the mass spectrometry of small molecules. When an ion in such a mass spectrometer is accelerated to a voltage V, its kinetic energy is $mv^2/2$, where m is its mass, and v is its velocity. Since all ions have, to a good approximation, the same velocity as a result of the accelerating voltage, then it is the *ionic mass* that gives one ion a different kinetic energy from another. The basic equation describing the bending of an ion by a magnetic field $\mathbf{B}$ is

$$m/z = k\frac{\mathbf{B}^2 r^2}{V} \qquad\qquad (SG12.6)$$

where r is the radius of the circular path over which the ion is bent, z is the charge on the ion, and k is a proportionality constant. This equation shows that the paths of ions of a larger mass are bent over a circle of greater radius. Similarly, for an ion of a given mass, increasing the magnetic field $\mathbf{B}$ decreases the radius of its path.

In the magnetic-sector mass spectrometer, a collector is positioned to sense only the ions traversing a path of a particular radius r. By increasing the magnetic field $\mathbf{B}$ in a regular way, ions of progressively increasing mass are focused onto the collector. The relative intensity of the ion beam of a given mass is recorded as an ion current. Thus is obtained a plot of ion current versus m/z—the mass spectrum.

Other types of ion sorting are used in modern spectrometers. In FT-ICR (Fourier Transform-Ion Cyclotron Resonance) mass spectrometry, ions are trapped in a chamber by a magnetic field, and the trapped ions move in closed, circular loops perpendicular to the magnetic field. The angular frequency of this motion is inversely proportional to the m/z of the ion. An analysis of the angular frequencies is tantamount to an analysis of the masses present. FT-ICR mass spectrometry gives very precise values of m/z.

In TOF (time-of-flight) mass spectrometers, ions are accelerated down a tube by an electric field. The time that it takes an ion to traverse the tube is proportional to the square root of its m/z. The measurement of the times at which ions reach the detector at the end of a flight tube provides essentially a measurement of the ion masses.

SOLUTIONS TO PROBLEMS

Solutions to In-Text Problems

12.1 (a) Apply Eq. 12.1.

$$\nu = \frac{3\times10^8 \text{ m sec}^{-1}}{9\times10^{-6}\text{m}} = 3.33\times10^{13}\text{sec}^{-1}$$

(b) Apply Eq. 12.1 and include the conversion factor 10^{-10} m Å^{-1}.

$$\nu = \frac{3.00\times10^8 \text{ m sec}^{-1}}{(4800\text{ Å})(10^{-10}\text{ m Å}^{-1})} = 6.25\times10^{14}\text{ sec}^{-1}$$

12.2 (a) Multiply the frequency obtained in the solution to Problem 12.1(a) times Planck's constant:

$$E = h\nu = (3.99\times10^{-13}\text{ kJ sec mol}^{-1})(3.33\times10^{13}\text{ sec}^{-1}) = 13.3\text{ kJ mol}^{-1}$$

(b) Use the same procedure as in part (a) of this problem:

$$E = h\nu = (3.99\times10^{-13}\text{ kJ sec mol}^{-1})(6.25\times10^{14}\text{ sec}^{-1}) = 249\text{ kJ mol}^{-1}$$

12.3 (a) The energy of X-rays is greater than that of any visible light, including blue light. (In fact, the energy is so much greater that prolonged exposure to X-rays is harmful.)

 If the energy of microwaves is much less than that of red light, why is it that microwave energy rapidly cooks food, whereas red light does not? The reason is that microwaves are absorbed by molecules in food, and the absorbed energy is rapidly converted into heat. In contrast, red light is reflected or transmitted, but it is mostly not absorbed, and is therefore not converted into heat.

(b) The energy of radar (that is, microwaves) is less than that of any visible light, including red light.

12.4 (a) Apply Eq. 12.7a on text p. 574:

$$\tilde{\nu} = \text{wavenumber} = (1\times10^4\ \mu\text{m cm}^{-1}) \div 6.0\ \mu\text{m} = 1667\text{ cm}^{-1}$$

(b) Apply Eq. 12.7b on text p. 574:

$$\lambda = \text{wavelength} = (1\times10^4\ \mu\text{m cm}^{-1}) \div 1720\text{ cm}^{-1} = 5.81\ \mu\text{m}$$

12.5 The peak at 1380 cm^{-1} has a percent transmittance (%T) of about 41. By Eq. 12.9B, its T is 0.41. Using Eq. 12.9c, solve for absorbance, A, which is the negative log of T, or 0.39.

12.6 From Eq. 12.8a on text p. 575 we know that

$$\nu = c\tilde{\nu}$$

Convert the frequency into wavenumber and then consult Fig. 12.4 on text p. 573.

$$\tilde{\nu} = \frac{\nu}{c} = (9 \times 10^{13} \ sec^{-1}) / (3 \times 10^{10} \ cm \ sec^{-1}) = 3000 \ cm^{-1}$$

The group of peaks around 3000 cm^{-1} are due to the C—H stretching vibrations in nonane.

12.7   Using Eq. 12.8 on text p. 582, convert the wavenumber to a frequency, which is the "times per second" equivalent of wavelength or wavenumber.

$$\nu = c\tilde{\nu} = (3 \times 10^{10} cm \ sec^{-1})(2143 \ cm^{-1}) = 6.43 \times 10^{13} sec^{-1}$$

12.8   According to the text, bond strength is the major determining factor of IR absorption frequency as we go across a row of the periodic table. Table 5.3 on text p. 216 is essentially a table of bond strengths. From this table, we find that bond strengths are in the order

$$H—F > H—OH > H—NH_2 > H—CH_3$$

Because the strongest bonds vibrate with the highest frequencies, the order of stretching frequencies is the same; that is, H—F has the highest frequency, and H—CH$_3$ the lowest.

12.9   Take the ratio of two equations like Eq. 12.13 on text p. 581, one for the C—H bond, and the other for the C—D bond. Everything cancels except the wavenumbers and the square roots of the masses.

$$\frac{\tilde{\nu}_D}{\tilde{\nu}_H} = \sqrt{\frac{m_H}{m_D}}$$

or

$$\tilde{\nu}_D = \tilde{\nu}_H \sqrt{\frac{m_H}{m_D}} = (3090 \ cm^{-1}) \sqrt{\frac{1}{2}} = 2185 \ cm^{-1}$$

In fact, C—D vibrations appear in the IR at lower energy than the corresponding C—H vibrations (to the right in conventional IR spectra).

12.10   (a)   Active. The triple bond is located asymmetrically; hence, even a small dipole moment will be affected by the stretching vibration.

(b)   Active. The C=O dipole is increased by the stretch because its length changes. Recall (Eq. 1.4, text p. 11) that dipole moment is proportional to length.

(c)   Inactive. The zero dipole moment of cyclohexane is unaffected by this symmetrical vibration.

(d)   Inactive. The zero dipole moment of this alkyne is not changed by stretching the triple bond.

(e)   Active. Because the nitro group is bent, stretching the two N—O bonds symmetrically will increase these bond dipoles and will also increase the resultant dipole of the molecule.

(f)   Active. The dipole moment of the molecule is due mostly to the bond dipole of the C—Cl bond, which is the only bond in the molecule with significant polarity. Increasing the length of this bond dipole will increase the dipole moment of the molecule.

(g)   Inactive. The dipole moment of this molecule is zero, and it remains zero when the C≡C bond is stretched.

12.11   The carbon skeleton of the hydrogenation product defines the following alkenes as possibilities.

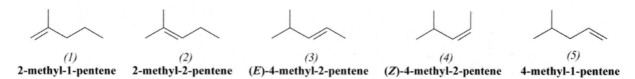

| (1) | (2) | (3) | (4) | (5) |
| 2-methyl-1-pentene | 2-methyl-2-pentene | (E)-4-methyl-2-pentene | (Z)-4-methyl-2-pentene | 4-methyl-1-pentene |

The strong 912 and 994 cm^{-1} C—H bending absorptions are very close to the standard values of 910 and 990 cm^{-1}, which are typical of —CH=CH$_2$ groups. Therefore compound A is (5). The weak or absent C=C stretching absorption in the 1660–1670 cm^{-1} region suggest that the candidates for B, C, and E are compounds (2), (3), and (4). The strong 967 cm^{-1} C—H bending absorption is the definitive absorption for trans alkenes. Therefore, compound E is (3). Compound C must be the cis alkene (4), from the C—H bending absorption at 714 cm^{-1}, because the C—H bending absorption for cis alkenes occurs at lower wavenumber than the C—H bending absorption of other alkene types. The 1650 cm^{-1} C=C stretching absorption as well as the 885 cm^{-1} C—H bending absorption point to (1) as the structure of compound D. This leaves only compound B unassigned, and it must therefore have structure (2). Its C—H bending absorption is consistent with this assignment.

12.12   The moderately strong C=C stretching absorption at about 1650 cm^{-1} and the C—H bending vibration at 890 cm^{-1} suggest that (a) is the spectrum of 2-methyl-1-hexene. The C—H bending absorption at about 970 cm^{-1} and the virtual absence of a C=C stretching absorption in spectrum (b) confirms that this is the spectrum of *trans*-2-heptene.

12.13   The absence of a C=C stretching absorption near 1650 cm^{-1} and an O—H stretching absorption near 3300 cm^{-1}, and the presence of a strong absorption near 1100 cm^{-1} (C—O stretching absorption), show that this is the spectrum of the ether.

12.14   The two C—O bonds of an ether can undergo both symmetrical and unsymmetrical stretching vibrations; each of these normal vibrational modes has an associated infrared absorption. (These vibrations are described on the first two lines of Fig. 12.8 on text p. 582.)

12.15   Think of molecular association by hydrogen bonding as a reaction like any other and apply Le Châtelier's principle.

$$2\,RO—H \rightleftharpoons RO—H----:\ddot{O}R$$

unassociated
alcohols

hydrogen-bonded
alcohols

Increasing the concentration of alcohol drives this equilibrium to the right, and decreasing the concentration of the alcohol drives this equilibrium to the left. At sufficiently low concentrations of alcohol, the spectrum is that of the unassociated alcohol. The O—H bond in a hydrogen-bonded O—H absorbs at a different frequency than the unassociated O—H.

 The broadness of the absorption for a hydrogen-bonded O—H is due to the presence of hydrogen bonds of many different strengths. Not only does the absorption change position when the sample is diluted, but it also becomes much sharper.

12.16   Table 12.3 on text p. 595 shows that Si has the following isotopic abundances: ^{28}Si, 92.21; ^{29}Si, 4.67; and ^{30}Si, 3.10. These abundances result in the following relative intensities for the *m/z* = 74 and 75 peaks:

m/z = 74: relative intensity (4.67)/(92.21) = 0.0506, or 5.06%
m/z = 75: relative intensity (3.10)/(92.21) = 0.0336, or 3.36%

To this must be added the contribution due to ^{13}C. The molecular mass of $(CH_3)_4Si$ is 88; therefore, the base peak at $m/z = 73$ results from loss of 15 mass units (CH_3). Hence, the $m/z = 74$ peak will have a contribution from three isotopic carbons equal to $3(1.11)/(92.21) = 0.0361$ (3.61%). Add this contribution to that calculated above to get the final value for the $m/z = 74$ peak:

$m/z = 74$: relative intensity = 5.06% + 3.61% = 8.67%

The contribution of ^{13}C to the $m/z = 75$ peak is negligible. Hence, the $m/z = 74$ peak has a relative intensity of 8.67%, and the $m/z = 75$ peak has a relative intensity of 3.36%.

12.17   Assume the molecular ion (base) peak at $m/z = 50$ has an abundance of 100%. This peak is due to molecules that contain ^{12}C, ^{1}H, and ^{35}Cl. The M + 1 peak is due to molecules that contain either ^{13}C, ^{1}H, and ^{35}Cl or ^{12}C, ^{2}H, and ^{35}Cl. The intensity of the M + 1 peak due to ^{13}C *relative to the base peak* is $(0.0111/0.989) = 1.12\%$, and the relative intensity due to ^{2}H is $3(0.00015)/(0.99985) = 0.0004$, or 0.04%. Hence, the M + 1 peak at $m/z = 51$ is due almost entirely to molecules that contain ^{13}C, ^{1}H, and ^{35}Cl, but we'll include the contribution of ^{2}H for completeness. Because there is one carbon, the intensity of the $m/z = 51$ peak is 1.16% (1.12% + 0.04%). The M + 2 peak at $m/z = 52$ is due almost entirely to ^{12}C, ^{1}H, and ^{37}Cl; according to Table 12.3, the ratio of this peak to the base peak should be $0.2423/0.7577 = 0.320$, or 32%. (This peak has a contribution from molecules that contain ^{13}C, ^{2}H, and ^{35}C equal to $3 \times (0.0111/0.989)(0.00015/0.99985) = 0.00017\%$ that can be ignored). Finally, there is a peak at M + 3, or $m/z = 53$, which is due to molecules that contain either ^{13}C, ^{1}H, and ^{37}Cl or to ^{12}C, ^{2}H, and ^{37}Cl. The contribution of ^{2}H is only 0.0004. Thus, the relative abundance from the *simultaneous* presence of ^{13}C and ^{37}Cl is $(0.0112)(0.320) = 0.00358$, or 0.36%. This peak is almost negligible.

12.18   (a)   An $m/z = 71$ corresponds to the formula C_5H_{11}. The only rational fragmentation that could give this formula is between the two CH_2 carbons:

(b)   The mechanism of this fragmentation:

$m/z = 71$

(c)   A rearrangement to a more stable tertiary carbocation occurs.

$m/z = 71$

Actually, this rearrangement is probably concerted with the fragmentation shown in part (b) so that formation of a primary carbocation is avoided.

12.19   When the molecule contains only C, H, and O, odd-electron ions have even mass, and even-electron ions have odd mass. Therefore, (a) and (d) are even-electron ions, and (b) and (c) are odd-electron ions.

12.20   (a)   Because even-electron ions of compounds containing only C, H, and Cl have odd mass, the ion with $m/z = 57$ is the even-electron ion and the ion with $m/z = 56$ is the odd-electron ion.

(b)   HCl can be lost to give the odd-electron ion.

(c)   The molecular mass of 2-chlorobutane is 92 (for the ^{35}Cl isotope). The $m/z = 57$ peak is due to loss of a chlorine atom from the molecular ion by an give the *sec*-butyl cation. (You'll learn in Sec. 12.6D that this is called *inductive cleavage*. It's the mass-spectrometric analog of an S_N1 reaction!)

$$CH_3\overset{+}{C}HCH_2CH_3 \longrightarrow CH_3\overset{+}{C}HCH_2CH_3 + :\overset{..}{\underset{..}{C}}l:$$

sec-butyl cation

The $m/z = 56$ peak is due to hydrogen-atom transfer within the molecular ion followed by loss of HCl in a process analogous to that shown for loss of water in Eq. 12.26a on text p. 597:

$$H_3C-\overset{:\overset{..+}{C}l}{\underset{|}{C}H}-\overset{H}{\underset{|}{C}H}-CH_3 \longrightarrow H_3C-\overset{:\overset{..+}{C}l-H}{\underset{|}{C}H}-\overset{.}{C}H-CH_3 \longrightarrow :\overset{..}{\underset{..}{C}}l-H + H_3C-\overset{+}{C}H-\overset{.}{C}H-CH_3$$

$m/z = 56$

12.21   (a)   An m/z of 45 indicates a loss of 43 mass units, which corresponds to an isopropyl group. An α-cleavage mechanism could give such a loss:

$$H_3C-\overset{+}{\underset{..}{O}}-CH_2-CH(CH_3)_2 \longrightarrow H_3C-\overset{+}{\underset{..}{O}}=CH_2 + \cdot CH(CH_3)_2$$

molecular ion $m/z = 45$

(b)   Inductive cleavage at either side of the oxygen would give a primary carbocation. With di-*sec*-butyl ether, in contrast, a secondary carbocation is formed by inductive cleavage. Because of the greater stability of secondary carbocations, more inductive cleavage occurs in the fragmentation of di-*sec*- butyl ether.

(c)   The major difference would be a large abundance of molecular ion and fewer low-mass fragments.

12.22   (a)   A fragment ion with $m/z = 101$ corresponds to a loss of 30 mass units, which corresponds to two methyl groups—that is, a molecule of ethane:

$$\begin{array}{c} CH_3CH_2 \quad H \quad CH_2CH_3 \\ \underset{|}{C}H-\overset{+}{\underset{..}{O}}-\underset{|}{C}H \\ CH_3 \quad\quad CH_3 \end{array} \longrightarrow CH_3CH_3 + \begin{array}{c} CH_2CH_3 \\ CH=\overset{+}{\underset{..}{O}}-\underset{|}{C}H \\ CH_3 \end{array}$$

CI molecular ion $m/z = 101$
$m/z = 131$

(b)   The same process occurs, except that 16 mass units—that is, methane—is eliminated.

$$\begin{array}{c} CH_3CH_2 \quad\quad CH_2CH_3 \\ \underset{|}{C}H-\overset{..+}{\underset{|}{O}}-\underset{|}{C}H \\ CH_3 \quad H \quad CH_3 \end{array} \longrightarrow CH_4 + \begin{array}{c} CH_3CH_2 \quad CH_2CH_3 \\ CH=\overset{..+}{\underset{|}{O}}-\underset{|}{C}H \\ CH_3 \end{array}$$

CI molecular ion $m/z = 115$
$m/z = 131$

(c)   As long as we're eliminating groups on the α-carbon, we might as well also eliminate the hydrogen:

$$CH_2CH_3$$
$$H \curvearrowright \overset{|}{CH}-CH_3$$
$$H_3C-\overset{|}{\underset{|}{C}}-\overset{+}{\underset{\cdot\cdot}{O}}-H \longrightarrow H-\overset{|}{CH}-CH_3 + H_3C-\overset{|}{\underset{|}{C}}=\overset{+}{\underset{\cdot\cdot}{O}}-H$$
$$CH_2CH_3 \qquad\qquad\qquad\qquad CH_2CH_3$$

CI molecular ion
m/z = 131

m/z = 73

Solutions to Additional Problems

12.23 The three factors that determine the wavenumber, or position, of an infrared absorption are: strength of the bonds involved, masses of the atoms involved, and the type of vibration—that is, stretching or bending.

12.24 The intensity of an infrared absorption is affected by (1) the number of absorbing groups and (2) the size of the dipole moment change when the molecule undergoes the bond vibration. The number of absorbing groups, in turn, depends on (1) the number of groups of interest within a given molecule and (2) the concentration of molecules in the sample.

12.25 (a) Catalytically hydrogenate the alkene, and confirm the reaction by observing loss of the C=C stretching absorption at 1660–1675 cm^{-1}.

1-methylcyclohexene → **methylcyclohexane**

(b) Carry out a Williamson ether synthesis. Confirm the reaction by observing loss of the O—H stretch of the alcohol in the 3200–3400 cm^{-1} region of the IR spectrum, and by intensification of the C—O stretching absorption.

$$CH_3(CH_2)_4CH_2OH \xrightarrow{NaH} CH_3(CH_2)_4CH_2O^- \ Na^+ \xrightarrow{CH_3I} CH_3(CH_2)_4CH_2OCH_3$$

1-hexanol → **1-methoxyhexane (hexyl methyl ether)**

12.26 The principle is that compounds that have different physical properties have different IR spectra.

(a) Because 3-pentanol and racemic 2-pentanol are constitutional isomers, they have different IR spectra.

(b) Because enantiomers have identical physical properties, the IR spectra of (*R*)- and (*S*)-2-pentanol are identical.

(c) This one is a little tricky! The two chair conformations of cyclohexanol are diastereomers and, because diastereomers have different physical properties, the two conformations also have different IR spectra. The problem is *observing* the two conformations independently. Any sample of cyclohexanol at room temperature is a mixture of these diastereomeric conformations, and the IR spectrum of cyclohexanol is therefore a spectrum of the mixture, with the spectrum of each conformation present in proportion to its concentration.

Even though the two chair conformations of cyclohexanol are rapidly interconverting, the IR absorption process is "fast enough" to record the spectra of the two diastereomeric conformations independently. Therefore, the spectrum observed is a superposition of the spectra of the two compounds.

12.27   Spectrum (1) has no peaks characteristic of alkenes or alcohols, but it does have a strong peak near 1100 cm^{-1} that is a C—O stretching absorption. This is the spectrum of compound (d), dipropyl ether.

Spectrum (2) shows the typical absorptions of a —CH=CH$_2$ group: the double C—H bending absorption at 910 and 990 cm^{-1}, and the strong C=C stretching absorption at 1640 cm^{-1}. This is the spectrum of compound (a), 1,5-hexadiene.

Spectrum (3) is virtually featureless; the only strong absorptions are the C—H stretching absorptions in the 2800–3000 cm^{-1} region. This is the spectrum of compound (f), cyclohexane.

Spectrum (4) shows both O—H stretching and C=C stretching absorptions, and is the spectrum of compound (c), 1-hexen-3-ol. The double peaks at 910 and 990 cm^{-1} confirm the presence of a —CH=CH$_2$ group. Compound (g), 3-hexanol, is ruled out because it should not have alkene absorptions in its IR spectrum.

Spectrum (5) has C=C stretching absorption at 1675 cm^{-1} and shows the typical high-wavenumber shoulder at about 3050 cm^{-1} for the alkene C—H stretching absorption. The position of the C=C stretching absorption along with the C—H bending absorption near 800 cm^{-1} indicates an alkene with three alkyl substituents on its double bond. This is the spectrum of compound (b), 1-methylcyclopentene. The definite absence of a C—H bending absorption near 970 cm^{-1} rules out compound (e), *trans*-4-octene.

12.28   Only spectrum (2) has the C=C stretching absorption near 1640 cm^{-1} and the high-wavenumber shoulder at about 3050 cm^{-1} for the alkene C—H stretching absorption; this is therefore the spectrum of compound *A,* and spectrum (1) is the spectrum of compound *B.*

12.29   (a)   Because stronger bonds absorb at higher wavenumber (text pp. 579–580), the order of increasing C—H bond strength is alkane < alkene < alkyne. This deduction can be verified by consulting the relevant C—H bond dissociation energies in Table 5.3, text p. 216; these bond energies are alkane (423 kJ mol^{-1}), alkene (463 kJ mol^{-1}), and alkyne (558 kJ mol^{-1}).

(b)   From Eq. 12.11 (text p. 579) and the associated arguments, the stretching frequencies are proportional to the square roots of the force constants κ. The force constants, in turn, are roughly proportional to the bond strengths. We then have

$$\frac{\tilde{\nu}_{\text{alkyne C—H}}}{\tilde{\nu}_{\text{alkane C—H}}} = \frac{3300}{2850} = 1.157 = \sqrt{\frac{558 \text{ kJ mol}^{-1}}{x \text{ kJ mol}^{-1}}}$$

Solving,

$$x = (558 \text{ kJ mol}^{-1})/(1.157)^2 = 417 \text{ kJ mol}^{-1}$$

The value in Table 5.3 is 423 kJ mol^{-1}. This calculation thus gives the correct answer within 5% of the difference between the two bond energies.

12.30   The solution to the previous problem noted the correlation between bond dissociation energies and IR absorption frequencies. Triple bonds are clearly stronger than double bonds, which are stronger than single bonds. The question is which of the two double bonds is stronger. Table 5.3 on text p. 216 shows that the bond dissociation energy of a C=C double bond is 728 kJ mol^{-1}, and that of a C=O double bond is 749 kJ mol^{-1}. On this basis, the C=O bond is predicted to absorb at higher frequency. (This is realized in practice; typical C=O absorptions occur in the 1710 cm^{-1} region, whereas C=C absorptions occur in the 1650 cm^{-1} region.) In summary, then, the order of increasing bond strengths and increasing IR absorption frequencies is:

$$\text{C—C} < \text{C=C} < \text{C=O} < \text{C≡C}$$

12.31   (a)   The vibrational modes of water must transform the molecule in a way that makes the molecule distinguishable from the original. The three modes that meet this criterion are

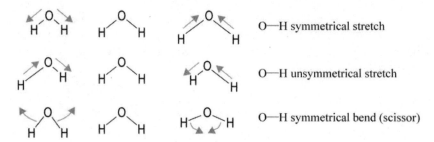

O—H symmetrical stretch

O—H unsymmetrical stretch

O—H symmetrical bend (scissor)

(b) The classifications are shown in the diagram in part (a).

(c) Stretching vibrations typically occur at higher frequency than bending vibrations. Therefore, the 1595 cm^{-1} vibration is the bending vibration; the other two vibrations are the stretching vibrations.

12.32 The two stretching vibrations of the nitro group correspond to the symmetrical and unsymmetrical stretching modes.

12.33 (a) The bond dipole of the S—H bond is much less than that of the O—H bond because the electronegativities of sulfur and hydrogen differ less than do the electronegativities of oxygen and hydrogen (Table 1.1, text p. 9). As a result, the S—H vibration is less active in the infrared, and the corresponding absorption is less intense.

(b) The S—H absorption occurs at lower frequency mostly because the S—H bond is much weaker than the O—H bond; compare the bond dissociation energies for H—SH and H—OH in Table 5.3, text p. 216. The difference between the masses of sulfur and oxygen has little effect on the vibration frequency because the mass of the smaller atom, hydrogen, is the major determinant of the vibration frequency, as shown by Eq. 12.13 on text p. 581.

(c) Compound B has a typical strong, broad O—H stretching absorption in the 3200–3400 cm^{-1} region, and is therefore the alcohol. Compound A, then, is the thiol. Note the considerably weaker S—H stretching absorption in the spectrum of compound A at about 2530 cm^{-1}, as suggested by part (a) of the problem.

12.34 (a) Both compounds have the same absorptions, except that the absorptions of compound C in Fig. P12.33 are displaced to lower frequency, an observation that implies a higher mass for the absorbing group; see the discussion of the mass effect associated with Eq. 12.13 on text p. 581. In particular, a peak at 3000 cm^{-1} in compound D, undoubtedly a C—H stretching absorption, is displaced to 2240 cm^{-1} in compound C. Hence, compound C is CDCl$_3$.

 Eq. 12.13 on text p. 581 gives the quantitative basis of the mass effect. If the force constants of C—H and C—D bonds are nearly the same (and they are), then the ratio of stretching frequencies of these two bonds should be equal to the square root of the ratios of the mass terms. Eq. 12.13 predicts a ratio of $\sqrt{2} = 1.414$; the more exact equation, Eq. 12.11 (text p. 579), predicts a ratio of 1.36. The actual ratio, (3000/2240) = 1.34, is very close to this prediction.

(b) The two compounds could be distinguished by mass spectrometry by the masses of their parent ions. Both CHCl$_3$ and CDCl$_3$ have four molecular ions. (Why four? See Table 12.3, text p. 595.) Each molecular ion peak of CDCl$_3$ lies at one unit higher mass than the corresponding molecular ion peak of CHCl$_3$.

12.35 (a) The parent compound has molecular mass = 115. The fragment with m/z = 72 corresponds to a loss of 43 mass units, which is accounted for by loss of a propyl radical by an α-cleavage mechanism.

molecular ion $m/z = 72$

(b)   Loss of a propyl radical from the molecular ion of 3-methyl-3-hexanol (molecular mass = 116) by α-cleavage mechanism similar to that in part (a) gives a fragment with $m/z = 73$.

molecular ion $m/z = 73$

(c)   1-Pentanol has molecular mass = 88; therefore, a fragment with $m/z = 70$, which must be an odd-electron ion, could result from the loss of H_2O (18 mass units.) A hydrogen-transfer mechanism followed by loss of H_2O, exactly as shown for 1-heptanol in Eq. 12.26a on text p. 598, accounts for this ion.

molecular ion of $m/z = 70$
1-pentanol

(d)   Fragmentation of neopentane at any one of its four carbon–carbon bonds gives a methyl radical and the *tert*-butyl cation, which has the correct mass:

neopentane   ***tert*-butyl cation**
                                   $m/z = 57$

12.36   (a)   A mass of 28 units could be accounted for by a molecule of ethylene, $H_2C{=}CH_2$.
         (b)   Loss of 18 mass units could indicate loss of H_2O. (See the solution to Problem 12.35(c) for an example of such a loss.)
         (c)   The data indicate the presence of chlorine, which has a mass of 35; loss of 36 mass units, then, could correspond to loss of HCl. (See the solution to Problem 12.20(c) for an example of such a process.)

12.37   The reaction with NaH and CH_3I is a Williamson ether synthesis to form a methyl ether.

$$R{-}OH \xrightarrow{\text{NaH}} \xrightarrow{\text{CH}_3\text{I}} R{-}OCH_3$$

mass = 102   mass = 116
   A   B

Whatever the mass of the starting alcohol A, the methylation adds 14 mass units. Because the CI mass spectrum of B has $m/z = 117$, ether B itself has $m/z = 116$, and alcohol A then has a mass of $116 - 14 = 102$. The $m/z = 87$ peak in the mass spectrum of A represents a loss of 15 mass units, which is a methyl group. The $m/z = 73$ peak represents a loss of 29 mass units, which is an ethyl group. The tertiary alcohol C—OH group has a mass of 29. The mass of 102 can be accounted for if the carbon of this alcohol group bears two ethyl groups (58 mass units) and one methyl group (15 mass units). Compound A is therefore 3-methyl-3-pentanol, and its fragmentation with loss of an ethyl group occurs by α-cleavage:

$$\text{CH}_3\text{CH}_2\overset{\displaystyle :\overset{..}{\text{O}}\text{H}}{\underset{\displaystyle \text{CH}_3\text{CH}_2}{\overset{\displaystyle |}{\underset{\displaystyle |}{\text{C}}}}}\text{CH}_3 \xrightarrow{-e^-} \text{CH}_3\text{CH}_2\overset{\displaystyle \overset{+..}{\text{O}}\text{H}}{\underset{\displaystyle \text{CH}_3\text{CH}_2}{\overset{\displaystyle |}{\underset{\displaystyle |}{\text{C}}}}}\text{CH}_3 \longrightarrow \text{CH}_3\text{CH}_2\overset{\displaystyle \overset{+..}{\text{O}}\text{H}}{\overset{\displaystyle ||}{\text{C}}}\text{CH}_3 \ + \ \text{CH}_3\overset{.}{\text{C}}\text{H}_2$$

mass = 102   molecular ion   *m/z* = 73

A

3-methyl-3-pentanol

(We leave it to you to show the formation of the *m/z* = 87 fragment.)

12.38   First ask what products are expected from the reaction. Two constitutional isomers, both with molecular mass = 88, are anticipated:

$$\underset{\textit{trans}\text{-2-pentene}}{\overset{\displaystyle \text{H}_3\text{C}\qquad\quad\text{H}}{\underset{\displaystyle \text{H}\qquad\quad\text{CH}_2\text{CH}_3}{\text{C}=\text{C}}}} \xrightarrow[\text{2) H}_2\text{O}_2/\ ^-\text{OH}]{\text{1) B}_2\text{H}_6/\text{THF}} \underset{\textbf{2-pentanol}}{\overset{\displaystyle \text{OH}}{\underset{\displaystyle |}{\text{CH}_3\text{CHCH}_2\text{CH}_2\text{CH}_3}}} \ + \ \underset{\textbf{3-pentanol}}{\overset{\displaystyle \text{OH}}{\underset{\displaystyle |}{\text{CH}_3\text{CH}_2\text{CHCH}_2\text{CH}_3}}}$$

Spectrum (b) is consistent with 3-pentanol; the base peak at *m/z* = 59 corresponds to loss of CH_3CH_2—, and there are two ethyl groups in this compound that could be lost by α-cleavage. Spectrum (a) has a base peak at *m/z* = 45 that corresponds to loss of 43 units (a propyl group). 2-Pentanol has a propyl branch that could be lost as a radical by α-cleavage.

When unknown compounds come from a chemical reaction, use what you know about the reaction as a starting point for postulating structures.

12.39   (a)   The molecular mass of 1-methoxybutane is 88. It can lose a propyl group (43 mass units) by α-cleavage to give a peak at *m/z* = 45:

$$\text{CH}_3\text{CH}_2\text{CH}_2\text{—CH}_2\text{—}\overset{+}{\underset{..}{\text{O}}}\text{CH}_3 \longrightarrow \text{CH}_3\text{CH}_2\overset{.}{\text{C}}\text{H}_2 \ + \ \text{CH}_2=\overset{+}{\text{O}}\text{CH}_3$$

molecular ion of 1-methoxybutane   *m/z* = 45

It can also lose methanol following a hydrogen transfer. (This mechanism is analogous to the one by which water is lost from primary alcohols, shown in Eq. 12.26a, text p. 597.)

$$\underset{\substack{\text{molecular ion of}\\\text{1-methoxybutane}}}{\overset{\displaystyle \text{H}\quad\overset{+}{\overset{..}{\text{O}}}\text{CH}_3}{\text{CH}_3\text{CH}_2\text{CH—CH}_2}} \longrightarrow \underset{}{\overset{\displaystyle \text{H}\text{—}\overset{+}{\overset{..}{\text{O}}}\text{CH}_3}{\text{CH}_3\text{CH}_2\overset{.}{\text{C}}\text{H—CH}_2}} \longrightarrow \underset{\textit{m/z} = 56}{\text{CH}_3\text{CH}_2\overset{.}{\text{C}}\text{H—}\overset{+}{\text{C}}\text{H}_2} \ + \ \text{H—}\overset{..}{\underset{..}{\text{O}}}\text{CH}_3$$

(b)   2-Methoxybutane, molecular mass = 88, can lose either a methyl group (15 mass units) or an ethyl group (29 mass units) by α-cleavage. Such losses would give rise to peaks at *m/z* = 73 and *m/z* = 59, respectively. The ethyl group is lost preferentially because of the relative stability of the ethyl versus the methyl radical.

$$\underset{\substack{\text{molecular ion of}\\\text{2-methoxybutane}}}{\overset{\displaystyle \overset{+..}{\text{O}}\text{CH}_3}{\text{CH}_3\text{CH}_2\text{—CH—CH}_3}} \longrightarrow \text{CH}_3\overset{.}{\text{C}}\text{H}_2 \ + \ \underset{\textit{m/z} = 59}{\overset{\displaystyle \overset{+..}{\text{O}}\text{CH}_3}{\overset{\displaystyle ||}{\text{CH—CH}_3}}}$$

12.40   The three peaks correspond to dibromomethane molecules that have, respectively, two atoms of ^{79}Br, one atom each of ^{79}Br and ^{81}Br, and two atoms of ^{81}Br. The probability of any combination of two isotopes is the product of their separate probabilities. Make a table of the possibilities using the relative abundances from Table 12.3, text p. 595:

	Br #1	Br #2	
$m/z = 172$	79	79	relative probability $= (0.5069)^2 = 0.257$
$m/z = 174 \}$	79	81	relative probability $= (0.5069)(0.4931) = 0.250$
	81	79	relative probability $= (0.5069)(0.4931) = 0.250$
			total relative probability 0.500
$m/z = 176$	81	81	relative probability $= (0.4931)^2 = 0.243$

The probabilities, and therefore the relative abundances, of the molecular ions in dibromomethane are in the ratio 0.257:0.500:0.243, or 1.00:1.95:0.95, or about 1:2:1, as asserted in the problem.

 The middle entry of the table above shows that two *different* isotopes can occur within the same compound in two ways. An analogy is the combinations that can be rolled with a pair of six-sided dice. There is only one way to roll a "2" (a "1" on each of the dice), but there are two ways to roll a "3" (a "1" on one of the dice and a "2" on the other, and vice-versa).

12.41   Follow the procedure used in the solution to Problem 12.40, except that the relative abundances are those of the chlorine isotopes.

	Cl #1	Cl #2	
$m/z = 84$	35	35	relative probability $= (0.7577)^2 = 0.574$
$m/z = 86 \}$	35	37	relative probability $= (0.7577)(0.2423) = 0.184$
	37	35	relative probability $= (0.2423)(0.7577) = 0.184$
			total relative probability 0.368
$m/z = 88$	37	37	relative probability $= (0.2423)^2 = 0.059$

Taking the peak at $m/z = 84$ as 100%, the ratios of the peaks are 100%, 64.1%, and 10.3%, respectively.

12.42   (a)   The molecular mass of ethyl bromide (CH_3CH_2Br) is 108 (for ^{79}Br) and 110 (for ^{81}Br). Consequently, the ion with $m/z = 110$ is the molecular ion of the molecule containing the heavier isotope. It is formed by ejection of an electron from one of the bromine unshared pairs:

$$CH_3CH_2-\ddot{B}r: \xrightarrow{-e^-} CH_3CH_2-\overset{+}{\dot{B}r}:$$

$$m/z = 110, 108$$

(b)   The ion with $m/z = 108$ is the molecular ion containing the lighter bromine isotope, and it is formed by the process shown in part (a).

(c)   The fragment at $m/z = 81$ corresponds to the formation of a bromine cation and an ethyl radical from the compound containing the heavier bromine isotope:

$$CH_3\dot{C}H_2-\overset{+}{B}r: \longrightarrow CH_3\dot{C}H_2 + {}^+\ddot{B}r:$$

$$m/z = 81, 79$$

(d)   The ion with $m/z = 79$ is the bromine cation formed from the molecular ion containing the lighter bromine isotope, and it is formed by the mechanism shown in part (c).

(e)   The fragment at $m/z = 29$ is the ethyl cation formed by inductive cleavage at the carbon-bromine bond:

$$CH_3CH_2 - \overset{+}{Br}: \longrightarrow CH_3\overset{+}{C}H_2 + :\overset{.}{Br}:$$

$$m/z = 29$$

(f)   The ion with $m/z = 28$ is an odd-electron ion formed from the molecular ion by hydrogen transfer followed by loss of HBr:

$$H_2C - CH_2 \longrightarrow H_2\overset{.}{C} - CH_2 \longrightarrow H_2\overset{.}{C} - \overset{+}{C}H_2 + H - \overset{.}{Br}:$$

$$m/z = 28$$

(g)   The ion at $m/z = 27$ could arise by elimination of a hydrogen atom by α-cleavage from the ion with $m/z = 28$ formed in part (f).

$$H_2\overset{.}{C} - \overset{+}{C}H \longrightarrow H_2C = \overset{+}{C}H + H\cdot$$

$$m/z = 28 \qquad\qquad m/z = 27$$

12.43   The presence of an odd number of nitrogens in a molecule containing, as the other atoms, any combination of C, H, O, and halogen reverses the odd-electron/even-electron mass correlations in Sec. 12.6C on text p. 596 because such a molecule must have an odd molecular mass. (Molecules containing no nitrogen or an even number of nitrogens have even molecular masses.)

(a)   The molecular ion of any molecule, because it is formed by ejection of a single electron, must be an odd-electron ion.

(b)   A fragment of even mass containing one nitrogen is an even-electron ion. (The ion in the solution to Problem 12.35(a) is an example.)

(c)   A fragment ion of odd mass containing a single nitrogen must be an odd-electron ion.

12.44   (a)   Sigma (σ) electrons are held closer to the nucleus than π electrons; by the electrostatic law, attraction between particles of opposite charge (such as an electron and a nucleus) is greater when the particles are closer. Hence, there is a greater energy of attraction of the nucleus for σ electrons. This means that more energy must be used to remove a σ electron; hence, a π electron is removed more easily. An equivalent explanation is that the bond dissociation energy of a typical π bond is considerably less than that of a σ bond (see Table 5.3, text p. 216). It should take less energy to eject an electron from a weaker bond—a π bond—than it does from a stronger bond—a σ bond.

(b)   The structure of the ion formed by loss of a π-electron from 1-heptene:

$$CH_3(CH_2)_4CH = CH_2 \xrightarrow{-e^-} \left[CH_3(CH_2)_4\overset{.}{C}H - \overset{+}{C}H_2 \longleftrightarrow CH_3(CH_2)_4\overset{+}{C}H - \overset{.}{C}H_2 \right]$$

1-heptene

  Because the unpaired electron is in a π molecular orbital and is therefore shared between the two carbons, we can put the unshared electron (and therefore the positive charge) on either carbon.

(c)

$$CH_3CH_2CH_2\overset{.}{C}H_2 - CH_2 - \overset{.}{C}H - \overset{+}{C}H_2 \longrightarrow CH_3CH_2CH_2\overset{.}{C}H_2 + CH_2 = CH - \overset{+}{C}H_2$$

1-butyl radical   **allyl cation**

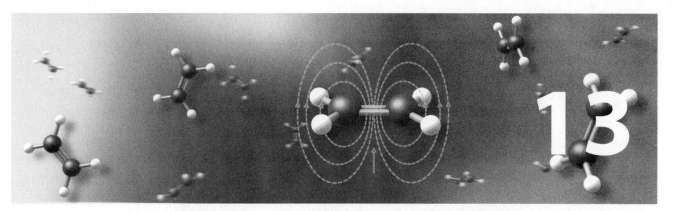

Nuclear Magnetic Resonance Spectroscopy

STUDY GUIDE LINKS

 13.1   Approaches to Problem Solving

The approach given on pp. 621–622 in the text for solving NMR structure problems is not necessarily the only approach. For example, you might take step 5 first. The point is that *you should be systematic and extract all the information you can from each piece of evidence in the spectrum before starting to write complete structures.*

A good analogy (suggested to the author by a reviewer) is the various approaches to working a jigsaw puzzle. Some people start by working around the borders, and others by looking for unique patterns or colors in the picture on the box in which the puzzle is packaged. The important point is that fitting together pieces of the puzzle at random is not likely to yield a solution in a reasonable amount of time; *your approach must be systematic.*

The dangers of using "recipes" for problem solving were discussed in Study Guide Link 9.3 on p. 226 of this manual. Remember that exceptions to any recipe will occur, and you will be best equipped to deal with such situations by developing your own approach and understanding the basis of each step. Do not let any "recipe" inhibit you from trying your own style of problem solving. Ultimately the style that works for you is the right one. The way to become proficient in developing your own style of problem solving is to practice. As you develop your problem-solving techniques, you are dong more than learning about NMR (or other aspects of organic chemistry); you are developing a life skill.

 13.2   More NMR Problem-Solving Hints

When you solve structure problems with NMR (or any sort of spectroscopy), it is very important that you be *intellectually honest* with yourself. Do *not* force a structure to fit the data. That is, write down each observation from the spectrum and then write down what it tells you. This point can be illustrated with the jigsaw-puzzle analogy in Study Guide Link 13.1. Anyone who has worked a jigsaw puzzle has encountered pieces that look as if they should go together, but do not quite fit; the temptation to force-fit the pieces is often great—but force-fitting never solves the puzzle. When you propose a structure, be able to defend it on the basis of *evidence that you can cite*. If you must guess a structure, then you haven't really solved the problem. If you are stymied by a structural problem, write down *all possible structures* that seem to fit the data. Ask yourself carefully whether each of these structures fits all the information available: chemical shift, integration, and splitting. Sometimes the act of writing down and analyzing the information in this way will lead to a solution. If you must consult the solution without having completely solved the problem, then at least you have defined the points of uncertainty. If, after looking at the solution, you still don't see why your structures are ruled out, seek assistance! You can sometimes learn more from a situation like this, as frustrating as it is, as you can from a correctly worked problem!

Now to some specific tips. The author has observed a tendency among some students to assume that protons that are physically adjacent in a structure must also have absorptions that are adjacent in their NMR spectra. Don't be caught in this trap. Although this may be true in specific cases, in general it is not something that you can rely on.

The last hint is to remind you about common splitting patterns in NMR spectra that appear over and over again. These patterns are so easy to recognize that they *immediately* suggest the presence of certain groups. Let's consider a few of these.

A three-proton triplet at high field along with a two-proton quartet at lower field [Fig. SG13.1(a)] is the earmark of an ethyl group in a structure CH_3CH_2—X, in which the group X has no coupled protons. Ethyl esters and ethyl ethers show this type of pattern. The chemical shift of the quartet provides information about the nature of the group X.

A one-proton septet or multiplet at low field along with a six-proton doublet at high field [Fig. SG13.1(b)] is characteristic of an isopropyl group. Sometimes the one-proton pattern, because of its complexity, is buried among other absorptions, or is otherwise difficult to distinguish. In such cases, the six-proton doublet alone can be used to diagnose an isopropyl group. Likewise, a three-proton doublet at relatively high field is characteristic of the methyl group in the following partial structure:

$$\overset{\diagdown}{\underset{\diagup}{CH}}-CH_3 \longleftarrow \text{a 3-proton doublet}$$

A three-proton singlet almost always suggests a methyl group. For example, a three-proton singlet in the δ 3–3.5 region in many cases corresponds to a CH_3O— group. A nine-proton singlet a *tert*-butyl group. Be on the lookout for these common patterns when you solve unknowns with NMR.

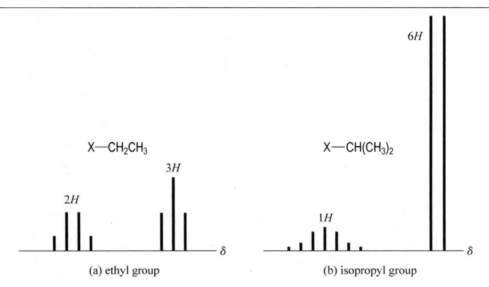

Figure SG13.1 Two splitting patterns commonly observed in proton NMR spectra. In this diagram, absorption peaks are represented as lines.

Several collections of known spectra are available in many libraries. One excellent collection is the *Aldrich Library of Carbon-13 and 300 MHz Proton NMR Spectra.* Although these volumes contain a limited number of spectra, they do encompass a large number of simpler compounds. A more extensive collection is the *Sadtler Standard NMR Spectra,* which includes literally thousands of spectra, most of which are 60 MHz spectra. Because these spectra are taken at lower field, they are in some cases more complex than spectra of the same compounds taken at higher field. (This point is discussed in Sec. 13.5B.) However, this collection nevertheless contains a wealth of relatively simple spectra. (Sadtler has also published a more limited catalog of 300 MHz spectra.) A corresponding collection of carbon-13 spectra also exists. (Carbon NMR is discussed in Sec. 13.9 of the text.) All of these collections contain good indexes, including formula indexes, that allow you to find spectra quickly. A good way to improve your ability to interpret NMR spectra is to look through the formula indexes of these collections for compounds with relatively few carbons—say, eight carbons or fewer for aliphatic compounds—and try to write the structure from the name. Then predict the spectrum, and, finally, check your prediction against the actual spectrum. If your prediction doesn't match the actual spectrum, try to explain why.

FURTHER EXPLORATIONS

 ## 13.1   Quantitative Estimation of Chemical Shifts

This section provides some tools for making more quantitative estimates of chemical shifts than Fig. 13.4 in the text. The basis of all these methods is that the influence of surrounding groups on the chemical shift of a set of protons is largely additive.

The chemical shifts of compounds H_3C—G, where G is a substituent group, can be read directly from the second column of Table SG13.1. For example, the chemical shift of CH_3I is $\delta\ 2.2$. *(Tables are found at the end of this section.)*

Table SG13.1 also provides the data that can be used for calculating the chemical shifts of methylene groups in a compound G^1—CH_2—G^2, where G^1 and G^2 are attached groups. To estimate the chemical shift, we simply add the *shift contributions* of the two groups to 0.2. That is,

$$\delta\,(G^1\text{—}CH_2\text{—}G^2) = 0.2 + \sigma(G^1) + \sigma(G^2) \tag{SG13.3}$$

where $\sigma(G^1)$ and $\sigma(G^2)$ are the shift contributions of the two groups. These shift contributions are given in the last column of Table SG13.1.

For example, suppose you want to calculate the chemical shift of the methylene protons of methylene chloride (CH_2Cl_2). In this example, $G^1 = G^2 = Cl$. Applying Eq. SG13.1, with $G^1 = G^2 = 2.5$, we have δ (Cl—CH_2—Cl) = 0.2 + 2.5 + 2.5 = 5.2. The actual chemical shift of methylene chloride is $\delta\ 5.3$. (This method gives results that are generally accurate to within a few tenths of a ppm.)

Here are some other examples:

$H_3COCH_2C{\equiv}N$: 0.2 + 2.4 + 1.7 = δ 4.3; the observed shift is δ 4.25.
$H_3COCH_2OCH_3$: 0.2 + 2.4 + 2.4 = δ 5.0. (Figure 13.1 shows that the actual shift is δ 4.6.)
H_3COCH_2Cl: 0.2 + 2.5 + 2.4 = δ 5.1; the observed shift is δ 5.5.

When using this method, the effect of a β-halogen, β-ether, or β-OH group can be taken into account by adding another 0.5 ppm. For example, let's estimate the chemical shift of the boldfaced protons in the following compound:

$$Br\text{—}\mathbf{CH_2}\text{—}\mathbf{CH_2}\text{—}Cl$$

In this example, the chlorine is β to the protons of interest. Using Eq. SG13.1, we have

$$\delta = 0.2 + 0.6 + 2.3 + 0.5 = 3.6$$

where the last 0.5 is the contribution for the β-halogen. (The observed shift is δ 3.3.)

A review of this approach can be found in two articles in the *Journal of Chemical Education*, which you should be able to find online or in your college library:

E. C. Friedrich and K. G. Runkle, *J. Chem. Educ.*, **1984**, *61*, 830–832; ibid., **1986**, *63*, 127–129.

The second of these articles extends the method discussed in the text to the chemical shifts of methine protons.

Another method for estimating chemical shifts makes use of a very useful tabulation of chemical-shift data that was assembled by Tom Curphey (who became a professor in the Dartmouth School of Medicine) while a graduate student at Harvard in 1961, and subsequently modified by Harry Morrison (now Professor Emeritus of Chemistry and Dean Emeritus of Science at Purdue). The Curphey-Morrison data are given in Table SG13.2 on pp. 373. *(Tables follow at the end of this section.)*

The use of this table can be illustrated by a calculation of the chemical shifts of the various protons of the following compound.

$$\underset{a}{H_3C}-\underset{b}{\overset{\overset{\displaystyle Cl}{|}}{CH}}-\underset{c}{CH_2}-\underset{d}{CH_2}-Cl$$

Protons *a* are methyl protons; therefore begin the calculation with a base shift of δ 0.9. To this add δ 0.63 for a chlorine β to a methyl group (first line, last column of the chlorine entry). The predicted chemical shift of protons *a* is δ 1.53. (The observed shift is δ 1.60.)

Proton *b* is a methine proton; begin with a base shift of δ 1.55. To this add δ 2.55 for an α-chlorine contribution (third line for chlorine under "Alpha shift") for a predicted chemical shift of δ 4.10. (The observed shift is δ 4.27.)

Protons *c* are methylene protons with *two* β-chlorines. The predicted shift is δ 1.20 + δ 0.53 + δ 0.53 = δ 2.26. (The observed shift is δ 2.15.)

Try to predict the chemical shift of the protons *d*. (The observed shift is δ 3.72.)

The comparisons given above between calculated and observed shifts illustrate the point that the chemical shifts calculated by this (or any other) method are estimates; do not rely on them for accuracy to more than a few tenths of a part per million.

The Curphey-Morrison table gives explicitly the contributions for β shifts as well as the shifts for methine protons. Its advantage over the Table SG13.1 is that it can be used to calculate the shifts of methine protons. However, it does not predict very well the shifts of CH_2 or CH protons bound to two or more electronegative groups. (For example, the predicted shift of CH_2Cl_2 is δ 5.8; the method above using Eq. SG13.1 predicts δ 5.2; the observed shift is δ 5.3.)

 ## 13.2 Fourier-Transform NMR

In FT-NMR, the sample is pulsed with a very intense, short pulse of rf radiation that causes all the nuclear spins to absorb energy simultaneously. The decay of absorption is then monitored for a second or two. During this time, all the nuclear spins in the sample return to their normal equilibrium distribution of $+1/2$ and $-1/2$ spins and, in doing so, emit radiofrequency energy, each at its characteristic frequency. An analogy is the ringing of a bell. When we "pulse" (strike) the bell, all frequencies (pitches) in the bell are activated simultaneously. As the bell rings, the pitches add together to give the characteristic sound of the bell, which dies out over time. An actual graph of this rf energy decay for dimethoxymethane (the compound whose spectrum is shown in Fig. 13.1 on p. 612 of the text) is shown in Fig. SG13.2. Remember, this is the total rf emitted from all the different proton frequencies simultaneously. You can see the die-out over time, as well as the periodic pulsing of the rf, in the same sense that you can hear pulsations in the sound of a bell when it rings. (In fact, this trace is sometimes called a "ringdown.") A mathematical technique called *Fourier transformation* can be used to translate the ringing of the bell over time into a plot of sound intensity versus pitch (frequency). Such a plot would be a spectrum showing the different pitches of which the sound is composed and their relative intensities. The same technique can be applied to the decaying NMR signal over time to convert it into a plot of absorption intensity versus frequency (or field), which is the conventional NMR spectrum. Thus, Fourier transformation of the data in Fig. SG13.2 provides the actual NMR spectrum of dimethoxymethane shown in Fig. 13.1 on p. 612 of the text. This mathematical process, once very laborious, is performed in a fraction of a second on a desktop computer.

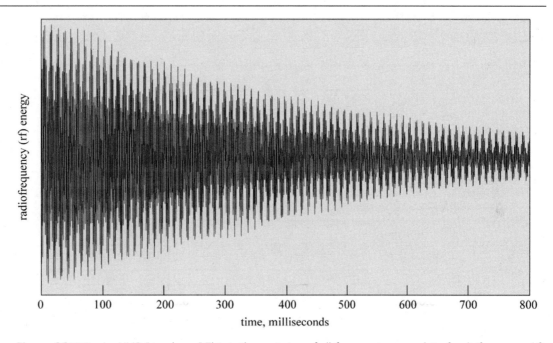

Figure SG13.2. An NMR "ringdown." This is the emission of all frequencies as a plot of radiofrequency (rf) energy versus time following the pulse of dimethoxymethane with a short, powerful burst of rf energy. Fourier transformation of these data give the spectrum in Fig. 13.1 on p. 612 of the text.

Table SG13.1	Group Contributions for Estimating Chemical Shifts of CH_2 Protons Using SG13.1	
Functional group G	**δ for CH_3—G**	**$\sigma(G)$ for —CH_2—G**
—H	0.2	0.0
—CR_3 or —CR_2— (R = H, alkyl)	0.9	0.6
—F	4.3	3.6
—Cl	3.0	2.5
—Br	2.7	2.3
—I	2.2	1.8
—CR=CR_2 (R = H, alkyl)	1.8	1.3
—C≡CR (R = H, alkyl)	2.0	1.4
—OH	3.5	2.6
—OR (R = alkyl)	3.3	2.4
—OR (R = aryl)	3.7	2.9
—SH, —SR	2.4	1.6
(Ph, phenyl)	2.3	1.8
O‖ —C—R	2.1 (R = alkyl) 2.6 (R = H)	1.5
O‖ —C—OR (R = alkyl, H)	2.1	1.5
O‖ —O—C—OR	3.6 (R = alkyl) 3.8 (R = aryl)	3.0
O‖ —C—NR_2 (R = alkyl, H)	2.0	1.5
R O| ‖ —N—C—R (R = alkyl, H)	2.8	—
—NR_2 (R = alkyl, H)	2.2	1.6
—N—R (R = aryl)	2.9	—
—C≡N	2.0	1.7

Table SG13.2	The Curphey–Morrison Table for Estimating Chemical Shifts*

Effect of a Substituent G on the Chemical Shifts of H^α and H^β

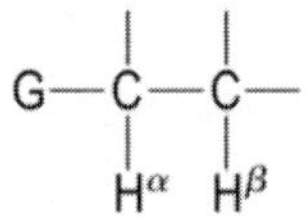

$$G—C—C—$$
$$H^\alpha \quad H^\beta$$

Base Values of Shift Positions: —CH_3 δ 0.90   —CH_2— δ 1.20

—CH— δ 1.55

Functional group G	Type of hydrogen	Alpha shift	Beta shift
—Cl	CH_3	2.43	0.63
	CH_2	2.30	0.53
	CH	2.55	0.03
—Br	CH_3	1.80	0.83
	CH_2	2.18	0.60
	CH	2.68	0.25
—I	CH_3	1.28	1.23
	CH_2	1.95	0.58
	CH	2.75	0.00
—Aryl	CH_3	1.40	0.35
	CH_2	1.45	0.53
	CH	1.33	0.22
—CH=O and —CR=O	CH_3	1.23	0.25
	CH_2	1.10	0.30
	CH	0.95	—
—CO_2H and —CO_2R	CH_3	1.23	0.25
	CH_2	1.05	0.30
	CH	1.05	—
—C≡C—	CH_3	0.78	0.05
	CH_2	0.75	0.10
	CH	1.25	0.00
—OH and —OR	CH_3	2.45	0.35
	CH_2	2.30	0.15
	CH	2.20	—
—O—Aryl	CH_3	2.95	0.40
	CH_2	3.00	0.45
	CH	3.30	—
—O—CO_2R	CH_3	2.88	0.38
	CH_2	2.98	0.43
	CH	3.45	0.36

(Table continues)

Functional group G	Type of hydrogen	Alpha shift	Beta shift
—NR$_3$	CH$_3$ CH$_2$ CH	1.25 1.40 1.35	0.13 0.13 0.00
—C≡N	CH$_3$ CH$_2$ CH	1.10 1.10 1.05	0.45 0.40 0.47
—NO$_2$	CH$_3$ CH$_2$ CH	3.50 3.15 3.05	0.65 0.85 —
—C≡C—	CH$_3$ CH$_2$ CH	0.90 0.80 0.35	0.15 0.05 —

*Reprinted by permission from the Ph. D. dissertation of T. J. Curphey, Harvard University, 1961, as modified by T. J. Curphey and H. A. Morrison

SOLUTIONS TO PROBLEMS

Solutions to In-Text Problems

13.1   (a)   Use Eq. 13.4 on text p. 615:

$$\nu = \frac{\gamma_H}{2\pi}\mathbf{B} = \frac{26{,}753 \text{ rad gauss}^{-1} \text{ sec}^{-1}}{2\pi \text{ rad}}(117{,}430 \text{ gauss}) = 500 \times 10^6 \text{ sec}^{-1} = 500 \text{ MHz}$$

A 500 megahertz NMR spectrometer would be required.

 Whenever π enters into a formula, its units are "radians." Remember that the hertz (Hz) is another way of representing the unit sec^{-1}.

(b)   Use the same equation and solve for **B**:

$$\nu = 900 \times 10^6 \text{ Hz} = \frac{\gamma_H}{2\pi}\mathbf{B} = \frac{26{,}753 \text{ rad gauss}^{-1} \text{ sec}^{-1}}{2\pi \text{ rad}}\mathbf{B}$$

Solving,

$$\mathbf{B} = 211{,}373 \text{ gauss}$$

 Because ν and **B** are proportional, we could also have used the result of the last part and multiplied the value of **B** in the last part by the ratio of frequencies (= 900 MHz/500 MHz).

13.2   (a)   The reduction in shielding at protons a is the field equivalent of, the chemical shift *difference*, 1005 Hz. Using Eq. 13.4, text p. 615, for both protons a and for the TMS protons, and then subtracting the two, we obtain

$$\nu_a - \nu_{TMS} = 1005 \text{ Hz} = \frac{\gamma_H}{2\pi}(\mathbf{B}_a - \mathbf{B}_{TMS}) = \frac{26{,}753 \text{ rad gauss}^{-1} \text{ sec}^{-1}}{2\pi \text{ rad}}(\mathbf{B}_a - \mathbf{B}_{TMS})$$

Solving for $\mathbf{B}_a - \mathbf{B}_{TMS}$,

$$\mathbf{B}_a - \mathbf{B}_{TMS} = 0.24 \text{ gauss}$$

This is 3.35×10^{-6} of the applied field $\mathbf{B}_0$, which is exactly the chemical shift in parts per million. From this example, we see that we could have calculated the chemical shift in gauss by multiplying the applied field (70,500 gauss) by the chemical shift in ppm and including the factor 10^{-6}. The point of this problem is to show how small chemical shifts are as a percentage of the applied field. This is why the units are "parts per million."

(b)   Following the procedure used in part (a),

$$\mathbf{B}_b - \mathbf{B}_{TMS} = (10^{-6})(4.56 \text{ ppm})(\mathbf{B}_0) = 0.32 \text{ gauss}$$

13.3   The greater the chemical shift, the *less shielded* are the protons. Therefore, the protons at δ 5.5 are least shielded), and those at δ 1.3 are the most shielded.

13.4   (a)   Using Eq. 13.7, we can calculate that the operating frequency ν_0 at 21,100 gauss is 89.8 MHz. Applying Eq. 13.1 for the chemical shift of the methyl protons—protons a—of dimethoxymethane (3.35 ppm), we obtain $\Delta\nu$ = the chemical shift in Hz = (3.35 ppm)(89.8 MHz) = 301 Hz.

       (b)   Following the procedure in part (a), we calculate that the operating frequency ν_0 at 141,000 gauss is 600 MHz. Applying Eq. 13.1 for the chemical shift of the methyl protons of dimethoxymethane (3.35 ppm), we obtain $\Delta\nu$ = the chemical shift in Hz = (3.35 ppm)(600 MHz) = 2011 Hz.

13.5   (a)   Subtract the two chemical shifts in ppm and apply Eq. 13.1. That is,

$$\text{chemical-shift difference} = \delta_2 - \delta_1 = \frac{\Delta\nu}{\nu_0} = \frac{45}{60} = 0.75 \text{ ppm}$$

       (b)   Using the procedure from part (a) gives

$$\text{chemical-shift difference} = 45/300 = 0.15 \text{ ppm.}$$

13.6   Parts (a)–(c) are answered in the text discussion that follows the problem (p. 620).

       (d)   Because Si is more electropositive (less electronegative) than any of the other atoms in the table, hydrogens near the Si are *more shielded*. Because chemical shift decreases with increased shielding, this means that $(CH_3)_4Si$ has a smaller chemical shift. A derivative $(CH_3)_xM$, in which M is an element more electropositive than Si, should have a negative chemical shift. $(CH_3)_2Mg$ and $(CH_3)_4Sn$ are two of several possible correct answers.

13.7   (a)   The order is $C < B < A$. The protons of methylene chloride (dichloromethane, CH_2Cl_2) have the greatest chemical shift, because chlorine is more electronegative than iodine. The chemical shift of methylene iodide is greater than that of methyl iodide because two iodines have a greater chemical shift contribution than one.

       (b)   The order is $B < A < D < C$. The protons in compound C (1,1,2,2-tetrachloroethane) have the greatest chemical shift of compounds A–C because they are adjacent to two chlorines. In comparing C and D, compound C has the greater chemical shift because of the β chlorines in C, as well as the fact that the proton in C is a methine proton, whereas the protons in D are methylene protons. In comparing A to B, the protons in A are methylene protons that are α to a chlorine, whereas in B the protons are methyl protons that are β to chlorine. Finally, in comparing A with D, the protons in D are α to two chlorines, whereas the protons in A are α to only one chlorine. The effect of an α-chlorine is much greater than that of a β-chlorine, because chemical shift falls off with the distance of the electronegative group from the proton.

       (c)   The order is $B < C < A$. This follows the electronegativity order of the central atom; the protons in the compound containing the least electronegative element have the smallest chemical shift.

13.8   (a)   Because protons H^a and H^b are diastereotopic, their chemical shifts are different.
       (b)   Because protons H^a and H^b are constitutionally nonequivalent, their chemical shifts are different.
       (c)   Because protons H^a and H^b are diastereotopic, their chemical shifts are different. Because the chair interconversion switches the positions of these two protons, their chemical shifts are the same when averaged out over time. (Sec. 13.8 addresses this point.)
       (d)   Because protons H^a and H^b are enantiotopic, their chemical shifts are identical.

13.9   The question asks essentially how many chemically nonequivalent sets of protons there are in each case.

       (a)   This compound has four chemically nonequivalent sets of protons; hence, its NMR spectrum consists of four resonances. Only the methyl groups of the *tert*-butyl group are equivalent.
       (b)   This compound has two chemically nonequivalent sets of protons; hence, its NMR spectrum consists of two resonances. δ
       (c)   Three resonances are observed. Notice that the two methyl groups are diastereotopic and hence chemically nonequivalent.

13.10   (a)   This is the structure given in Problem 13.9(b).

$$\underset{\substack{\displaystyle (CH_3)_3C-\overset{\displaystyle \overset{\displaystyle Cl}{|}}{C}(CH_3)_2 \\ \delta\,1.59}}{\overset{\delta\,1.08}{}}$$

     Another structure consistent with the data is the following:

$$CH_3CH_2-\overset{\overset{\displaystyle Cl}{|}}{\underset{\underset{\displaystyle CH_2CH_3}{|}}{C}}-CH_2CH_3$$

This is a satisfactory answer at this point in the chapter. However, as you will learn in Sec. 13.4 of the text, the spectrum of this compound will consist of a number of lines that result from a phenomenon called splitting, which occurs when protons are on *adjacent carbons*.

(b)

$$\underset{\substack{\displaystyle \underset{\delta\,4.31}{CH_2Cl}}}{\overset{\delta\,6.55}{Cl_2CH}}-\overset{\overset{\displaystyle \overset{\delta\,1.99}{CH_3}}{|}}{\underset{|}{C}}-CH_3$$

13.11   The spectrum will contain three resonances, which arise from the protons labeled *a, b,* and *c.*

$$\underset{\substack{\displaystyle \underset{c}{CH_2}-OC\overset{a}{H_3}}}{\overset{a}{CH_3}O-\overset{c}{CH_2}-C-\overset{c}{CH_2}-OC\overset{a}{H_3}}$$

with $OC\overset{b}{H_3}$ on the central carbon.

Protons *a*, integrating for 9*H*, occur near δ 3.2. Protons *b*, integration for 3*H*, will have nearly the same chemical shift. Because these protons are chemically nonequivalent to protons *a*, their chemical shift should be different, but only slightly. Protons *c*, integrating for 6*H,* should have a somewhat greater chemical shift than protons *a* or *b*—probably about δ 3.5—because methylene protons have a slightly greater chemical shift than methyl protons in similar chemical environments.

13.12   (a)   This is a quantitative analysis problem that can be solved by assigning the resonances and using their integrals. The resonance at δ 1.8 is that of *tert*-butyl bromide, and the resonance at δ 2.2 is that of methyl iodide. (These assignments follow from Table 13.1 and Fig. 13.4 on text pp. 619 and 621, respectively.) The ratio of methyl iodide to *tert*-butyl bromide is 15:1. This follows from the fact that there are three times as many protons in the *tert*-butyl bromide molecule as in the methyl iodide molecule; thus, for a given concentration, the *tert*-butyl bromide gives a resonance that is three times as intense as that of methyl iodide.

$$\frac{5\text{-proton resonance for } CH_3I}{3 \text{ protons per molecule}} \times \frac{9 \text{ protons per molecule}}{1\text{-proton resonance for } (CH_3)_3Br} = \frac{15 \text{ molecules of } CH_3I}{1 \text{ molecule of } (CH_3)_3Br}$$

The mole percent of CH_3I is the number of moles of CH_3I (15) divided by the total moles of both compounds (16) expressed as a percent. Therefore, the mole percent of CH_3I is (15/16) × 100, or 94 mole percent; the mole percent of $(CH_3)_3CBr$ is (1/16) × 100, or 6 mole percent.

(b)   The $(CH_3)_3CBr$ impurity in CH_3I is more easily detected, because a given mole fraction of $(CH_3)_3CBr$ gives a resonance that is three times as strong as the resonance for the same amount of CH_3I, as the solution to part (a) of this problem demonstrated.

13.13   In the following discussions, the integral is not mentioned. It corresponds to the number of protons under observation in each case.

(a)   The resonance of the CH_3 protons should be a doublet (one neighboring proton) in the alkyl region, which is δ 0.7–δ 1.7. Because of the β-chlorines, this doublet will probably be at the higher end of this range. The —$CHCl_2$ proton is a quartet at considerably greater chemical shift. Fig. 13.4 (text p. 621) puts δ 4.2 as the limit on the chemical shift of a CHCl proton; because there are two chlorines, the resonance would be at higher shift. (The shift of CH_2Cl_2 in Table 13.1 is δ 5.3; a methine proton would be shifted even more.)

(b)   The protons of the $ClCH_2$— group would be a doublet at δ 4.2 or higher. The protons of the —$CHCl_2$ group would be a triplet at considerably greater chemical shift. This chemical shift would be similar to, and probably somewhat greater than, the chemical shift of the —$CHCl_2$ protons in part (a). (The greater shift would be caused by the β chlorine.)

(c)   The methyl groups would be a large doublet in the δ 0.7–1.7 range, probably near the middle. (The β halogen would increase the shift, but methyl protons are usually near the low end of the ranges given in Fig. 13.4, p. 621.) The methine proton would be a septet near δ 3.0.

(d)   This compound, 1,3-dimethoxypropane, should have three resonances. The methylene resonances for the boldfaced protons $CH_3O\mathbf{CH_2}CH_2\mathbf{CH_2}OCH_3$ should be a triplet in the δ 3.5 region, because there are two neighboring protons. The central —CH_2— should be a quintet (a five-line pattern), because there are four neighboring protons. This would normally be in the δ 1.0 region, but should appear at somewhat greater chemical shift, perhaps around δ 1.8, because of the β-oxygens. The methyl groups should be one large singlet near δ 3.2.

(e)   This compound (oxetane) is really a "tied-down" version of the compound in part (d) without the methyl groups. The methylenes closest to the oxygen are a triplet near δ 3.5, and the central —CH_2— should be a quintet (a five-line pattern) near δ 1.8.

(f)   The spectrum of this compound consists of two singlets. The —CH_2— resonance occurs at a chemical shift greater than δ 4.2 (because of the β-chlorine), and the methyl groups should appear in the δ 0.7–1.7 region, probably towards the higher end because of the chlorines.

 More precise estimates of chemical shift can be made using the methods discussed in Further Exploration 13.1 on p. 369 of this manual.

13.14   There are four possibilities for the spin of three neighboring equivalent protons b; these are shown in the following table. There are three ways in which two spins can be the same. This table shows that the resonance for protons a would be split by the three protons b into a quartet whose lines are in the intensity ratio 1:3:3:1.

	b protons			
	1 2 3 all (+)	1 2 3 two (+), one (−)	1 2 3 one (+), two (−)	1 2 3 all (−)
spin combinations	+½ +½ +½	+½ +½ −½ +½ −½ +½ −½ +½ +½	+½ −½ −½ −½ +½ −½ −½ −½ +½	−½ −½ −½

13.15   (a)   This structure is ruled out by several facts. First, there are *eight* nonequivalent sets of protons in this candidate; notice that the two CH_3O— groups on the left, as well as the CH_2 protons, are diastereotopic, and hence, nonequivalent. The actual NMR spectrum shows five sets of protons. Also, this candidate should have three methoxy singlets unless two of them accidentally overlap. Furthermore, the methyl group on the right should be a doublet, whereas it is actually a singlet.

(b)   This candidate would also be expected to have *eight* nonequivalent sets of protons. (Can you find them all?) This would also not show two triplets, and there should be four methoxy singlets (three if the two diastereotopic ones accidentally overlapped), but certainly not two.

13.16 (a)

$\delta\,1.03$ $\delta\,3.40$
$$H_3C-CH_2-CH_2-Br$$
$\delta\,1.88$

(b)

$\delta\,3.98$ $\delta\,5.87$
$$ClCH_2-CHCl_2$$

(c)

$$\begin{array}{c} CH_2Br \\ | \\ BrCH_2CCH_2Br \\ | \\ CH_2Br \end{array}$$

13.17 The molecular formula indicates no degrees of unsaturation; this rules out rings or double bonds. The IR spectrum rules out an alcohol. All of the oxygens, then, are associated with ether groups. The $9H$ singlet at $\delta\,3.2$ requires three equivalent CH_3O groups. The unsplit $3H$ resonance at $\delta\,1.2$ indicates a methyl group on a quaternary carbon. The $6H$ triplet at $\delta\,3.4$ suggests three equivalent CH_2 groups attached to an oxygen. These CH_2 groups, from their splitting, are also adjacent to three other CH_2 groups, which must be the six protons at $\delta\,1.4$, and these are also attached to a quaternary carbon. because they are split only by the CH_2 groups. A structure with this number of carbons and only four resonances must involve a high degree of symmetry. A structure that fits all the data is the following:

$$\underset{\delta\,1.4}{\underset{\nearrow}{}}\ CH_3OCH_2CH_2-\overset{\overset{\delta\,1.2}{\overset{CH_3}{|}}}{\underset{\underset{\delta\,3.4}{\underset{\nwarrow}{}}}{\underset{CH_2CH_2OCH_3}{|}}}{C}-\overset{\delta\,3.2}{CH_2CH_2OCH_3}$$

13.18 (a) The reaction is the peroxide-promoted HBr addition to the alkene (Sec. 5.6 in the text):

$$BrCH_2CH{=}CH_2 + H-Br \xrightarrow{\text{peroxides}} BrCH_2CH_2CH_2Br$$

3-bromo-1-propene   **1,3-dibromopropane**
                                                  compound A

(b) The chemical shifts and splittings are indeed consistent with this structure. The resonance of the central methylene group is predicted to be a quintet in the $\delta\,0.7–1.7$ range (Fig. 13.4); the β-bromines will probably bring this to the high end of this range or beyond; and the resonance for the two methylenes α to the bromines is predicted to be a triplet near $\delta\,3$. The splittings agree exactly, and the observed chemical shifts are acceptably close.

13.19 (a) The resonance for protons H^d will be a singlet. The resonances for protons H^a, H^b, and H^c will all be triplets at their respective chemical shifts since the coupling constants for protons $a-c$ are all equal, the splitting will follow the $n+1$ rule.

(b) The resonance for protons H^d will be a singlet as in part (a). The resonance for proton H^a will be split into a triplet by the two H^b protons, by whatever the coupling constant between those two protons is. The resonance for protons H^c will also be split into a triplet by the two H^b protons, but the triplet will have different spacing between its peaks due to the different coupling constant between protons b and c. The resonance for the H^b protons will be split into a doublet of doublets instead of a triplet as in part (a), due to the unequal coupling constants.

          If you find an error in this manual, please visit http://people.pharmacy.purdue.edu/~loudonm/teaching/

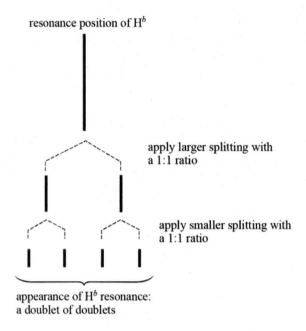

Note that in the above solution, the larger of the two coupling constants was applied first, followed by the smaller. It is arbitrary which gets applied first; doing the reverse (smaller followed by larger) will result in the same splitting pattern. Prove it to yourself!

13.20   The resonances for protons H^c and H^a will be doublets at their respective chemical shifts, and the resonance of proton H^b will be split into a quartet ($J = 6.6$ Hz) by the three protons H^c, and each line of this quartet will be split into a doublet ($J = 3.6$ Hz) by the single proton H^a. The resonance for H^b can be analyzed with a splitting diagram as shown in Fig. SG13.3(a). The complete spectrum is sketched in Fig. SG13.3(b).

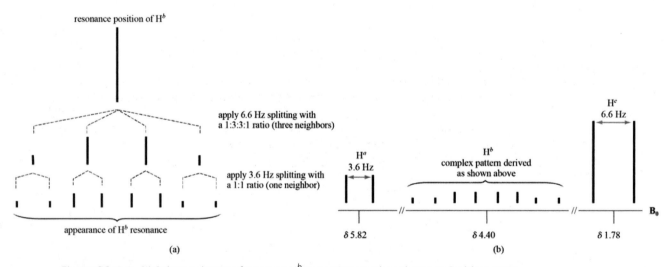

Figure SG13.3. (a) Splitting diagram for proton H^b to accompany the solution to Problem 13.20. (b) The complete NMR spectrum for 1,1,2-trichloropropane. Compared with the diagram in part (a), horizontal and vertical scales have been compressed.

13.21 (a) Because protons H^b and H^c are diastereotopic, they are chemically nonequivalent; therefore, counting the methyl group, there are four chemically nonequivalent sets of protons. The methyl group is split into a doublet by H^a. The resonance for proton H^a is split into a quartet by the methyl protons, and each line of this quartet is split into a triplet by H^b and H^c, to give a "quartet of triplets," or twelve lines, for proton H^a. (The triplet is the result of two overlapping doublets with $J_{ab} = J_{ac}$; see Fig. 13.11 on text p. 639 for a related case.) Because H^b and H^c are diastereotopic, they absorb at different chemical shifts. Proton H^b is split by proton H^a into a doublet, and each line of this doublet is split into another doublet by H^c. Therefore, proton H^b is a doublet of doublets, or four equally intense lines. The same is true of H^c —its resonance will be four equally intense lines at a slightly different chemical shift from the resonance of H^b. Several of these patterns may overlap to create a very complex-looking spectrum.

     (b) The difference from part (a) is that, in the quartet resonance for proton H^a, each line of this quartet is split into a *doublet* by H^b and into *another doublet* by H^c, to give a "quartet of doublet of doublets," or sixteen lines. Accidental overlaps could reduce this number.

13.22 (a) The integration of the δ 3.42 resonance and its splitting show that compound A has the partial structure —CH_2CH_2Br. The δ 0.91 resonance indicates the two methyls of an isopropyl group. This means that compound A can only be

$$(CH_3)_2CHCH_2CH_2—Br$$

1-bromo-3-methylbutane
(isoamyl bromide; compound A**)**

Compound B, from the chemical shifts, has no hydrogens α to the bromine. Hence, this is a tertiary alkyl halide. We see two methyl singlets and an ethyl quartet, which determine the structure.

$$\begin{array}{c} CH_3 \\ | \\ CH_3CH_2—C—Br \\ | \\ CH_3 \end{array}$$

2-bromo-2-methylbutane

     (b) Running the spectrum of compound A at higher field could simplify the complexity of the δ 1.7-1.8 resonance. This region contains the combined resonances for the following protons:

$$\delta\ 1.7–1.8$$
$$(CH_3)_2CH—CH_2—CH_2—Br$$

The difference in chemical shift between these protons, in Hz, would increase (double), but the coupling constants would remain unchanged. Thus, the likelihood that Eq. 13.12 (text p. 642) is valid would be increased. In the event that all resonances are first-order, and all coupling constants are nearly equal, the —CH_2— group would become a quartet and the methine CH of the isopropyl group would remain complex (21 lines even if first-order).

13.23 The δ 1.2–1.5 resonance corresponds to the following protons in 1-chlorohexane:

$$\delta\ 1.2–1.5$$
$$H_3C—CH_2—CH_2—CH_2—CH_2—CH_2—Cl$$

1-chlorohexane

The complexity and hence, the inability to assign an exact chemical shift, arises because of complex splitting of the protons of interest. To solve this problem, we would prepare the deuterium-substituted analogs, in which this splitting is not present.

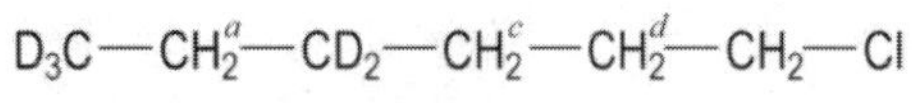

$$D_3C\!-\!CH_2^a\!-\!CD_2\!-\!CH_2^c\!-\!CH_2^d\!-\!CH_2\!-\!Cl \qquad H_3C\!-\!CD_2\!-\!CH_2^b\!-\!CD_2\!-\!CH_2\!-\!CH_2\!-\!Cl$$

compound A 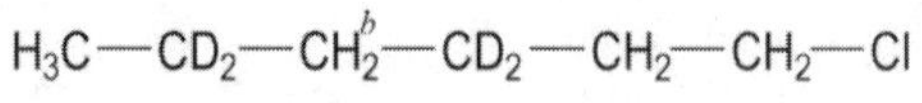   compound B

In the NMR spectrum of compound *A*, protons *a* would be a singlet, and protons *c* would be a simple triplet. (These are well separated from protons *d*; see Fig. 13.13, text p. 642.) In the NMR spectrum of compound *B*, protons *b* would be a singlet.

13.24   (a)   Assuming that in the absence of D_2O, the sample is very dry, the spectrum of 3-methyl-2-buten-1-ol would change as follows as the result of a D_2O shake:

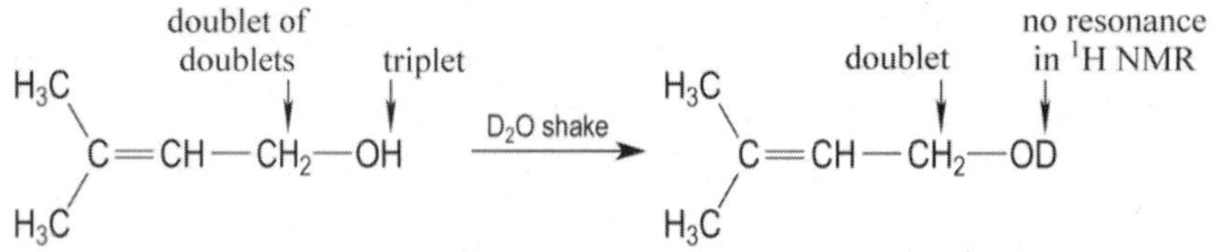

3-methyl-2-buten-1-ol

      (b)   Assuming that in the absence of D_2O, the sample is very dry, the spectrum of 1,2,2-trimethyl-1-propanol would change as follows as the result of a D_2O shake:

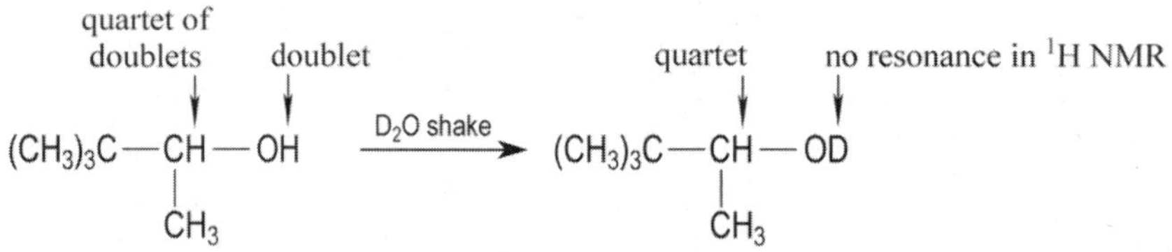

(A "quartet of doublets" can also be referred to as a "doublet of quartets.")

13.25   The two protons in the δ 4.6–4.8 region are vinylic protons, and they are not chemically equivalent. Their very small splitting is consistent only with geminal (that is, H_2CA) protons. Thus, we have the partial structure

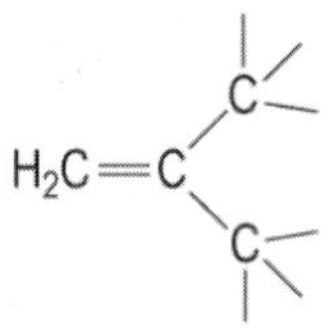

The resonance at δ 1.8 integrates for 3 protons, and it is in the allylic region. This means that there is an allylic methyl group. The resonance at δ 1.03 integrates for nine protons. Because it is a singlet, this can only be a *tert*-butyl group. This completes the structure:

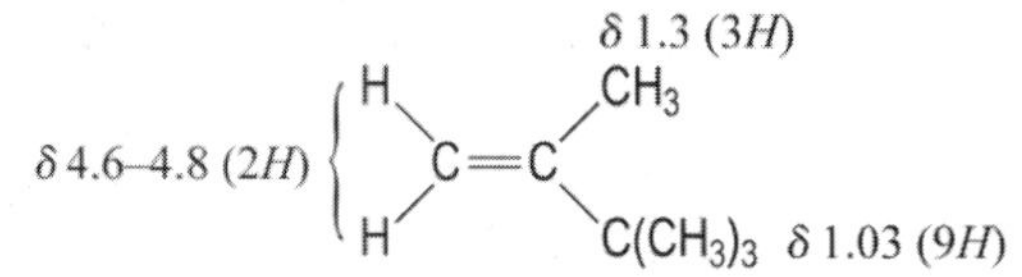

2,2,3-trimethyl-1-butene

13.26   (a)   Nothing but ethyls and one oxygen! This can only be diethyl ether, $CH_3CH_2\!-\!O\!-\!CH_2CH_3$.

      (b)   The large coupling constants for both sets of protons indicate splitting by fluorine. The fact that they are triplets suggest that they are adjacent to the carbon bearing both fluorines, but not on carbons adjacent to each other; otherwise they would split each other as well. The δ 3.63 protons are α to the chlorine. These considerations demand the following structure:

$$\delta 1.75 \quad \overset{\overset{\displaystyle F}{|}}{\underset{\underset{\displaystyle F}{|}}{H_3C-C-CH_2-Cl}} \quad \delta 3.63$$

1-chloro-2,2-difluoropropane

Remember that the fluorines are not observed in proton NMR. What would the ^{19}F NMR spectrum look like? *Answer:* a quartet of triplets (or triplet of quartets) with two splittings, coupling constants $J = 13$ Hz and 17.5 Hz, respectively.

13.27   The spectrum of ethyl chloride would resemble that of ethyl bromide (Fig. 13.6, text p. 628). It would consist of a typical ethyl pattern—a triplet (for the CH_3) and a quartet (for the CH_2), except that the quartet for the CH_2 in the ethyl chloride spectrum would be at a somewhat greater chemical shift than it is in ethyl bromide. In ethyl fluoride, the resonance of the methyl group would be a *triplet of doublets*. (Splitting by the CH_2 group gives three lines; each of these are split into two by the fluorine for a total of six.) The resonance of the CH_2 group would be a *doublet of quartets*. (Splitting by the fluorine gives a widely spaced doublet; each line of the doublet is split into a quartet by the CH_3 group.) The chemical shift of the CH_2 resonance in ethyl fluoride would be expected to be somewhat greater than that of the CH_2 group in ethyl chloride.

13.28   (a)   The unknown is *tert*-butyl alcohol, $(CH_3)_3C$—OH. Notice the absence of an α-proton.

(b)   The exchangeable proton suggests the presence of an alcohol. The resonance at $\delta 5.40$ is a vinylic proton adjacent to two other protons, and the resonance at $\delta 4.10$ arises from two protons adjacent to both the oxygen and the vinyl group. (Their chemical shift is too great to be caused by either the oxygen or the vinyl group alone.) Evidently the sample is wet enough that the splitting between the —OH proton and the α-proton is not observed; this situation is very common. (Contrast this with the solution to Problem 13.24a, which involves the same compound.) 3-Methyl-2-buten-1-ol, $(CH_3)_2C$=$CHCH_2OH$, is the structure consistent with the data.

13.29   At room temperature, the three conformations of 1-bromo-1,1,2-trichloroethane are in rapid equilibrium. Hence, the NMR spectrum of this compound is a singlet whose chemical shift reflects the environment of the proton averaged over time across all conformations. Any changes observed at low temperature would be related to the fact that the interconversion between the different conformations is retarded, and thus certain conformations can be individually observed at the lower temperature. There are three such conformations, two of which are enantiomers:

enantiomeric conformations
have identical NMR spectra

In conformation A, protons H^a and H^b are equivalent and have the same chemical shift. In either one of the enantiomeric conformations, H^a and H^b are chemically nonequivalent because they are diastereotopic. In either of these conformations, the resonance of H^a is split by H^b and is therefore a doublet; likewise, the resonance of H^a is split by H^b and is also a doublet. Hence, the spectrum of conformation B should consist of two doublets, that is, four lines of equal intensity, assuming no overlap. Because conformations B and C are enantiomeric, their spectra will be identical. Hence, the total spectrum should consist of one line for conformation A, and two doublets for conformations B and C, for a total of five lines at low temperature, assuming no overlaps. The relative integrals depend on the amount of form *A* relative to the amounts of forms *B* and *C*; the latter two forms will be present in equal amounts.

13.30   The resonance for the methyl group at room temperature is a singlet. When the temperature is lowered, the resonance of this methyl group should consist of two singlets, one for the conformation of 1-chloro-1-methylcyclohexane in which the methyl group is axial, and one for the conformation in which the methyl group is equatorial. The relative integrals of the two singlets will be proportional to the relative amounts of the two conformations. Because chlorine and methyl are about the same size, there should be about equal amounts of the two conformations.

1-chloro-1-methylcyclohexane

13.31   Because of its symmetry, 4-heptanol has only four chemically nonequivalent sets of carbons, and hence its spectrum consists of four lines, and is therefore spectrum 1. All of the carbons of 3-heptanol are chemically nonequivalent; hence, the ^{13}C NMR spectrum of this compound consists of seven lines, and is therefore spectrum 2. The following structures show equivalent carbons with the same numbers.

OH
|
CH₃CH₂CH₂CHCH₂CH₂CH₃
 1   2   3   4   3   2   1

4-heptanol

OH
|
CH₃CH₂CHCH₂CH₂CH₂CH₃
 1   2   3   4   5   6   7

3-heptanol

13.32   The ^{13}C NMR spectrum of 1,1-dichlorocyclohexane should consist of four lines, whereas that of *cis*-1,2-dimethylcyclohexane should consist of three lines. (Equivalent carbons have the same numbers in the structures below.)

1,1-dichlorocyclohexane   ***cis*-1,2-dichlorocyclohexane**

     Because the chair interconversion is rapid, we can use the planar structures of cyclohexanes in resolving questions of symmetry. Can you deduce how the ^{13}C NMR spectra of these compounds would differ—if at all—at very low temperature?

13.33   (a)   This compound has five nonequivalent sets of carbons and therefore should give five lines, not three as observed.

(b)   This compound should have a ^{13}C NMR spectrum with three lines, as observed. Furthermore, the attached hydrogen ratio is also consistent with this structure. What is *not* consistent, however, is the chemical shift information. The δ 112.9 resonance suggests a carbon bound to *more than one* oxygen. In this structure, no carbon is bound to more than one oxygen, whereas in the correct structure, the methine (CH) carbon is bound to three oxygens.

13.34   (a)   First, draw the structures!

trans-1-chloro-2-butene   2-chloro-1-butene

Two reasons that *trans*-1-chloro-2-butene (*A*) is not consistent with the data are that this alkene would have a spectrum in which the integral indicates a total of *two,* not three, vinylic protons, and *two,* not one, protons on the carbon α to the chlorine.

   Among the reasons that 2-chloro-1-butene (*B*) does not fit the data are that it offers no explanation for the δ 4.5 absorption, and it has only two vinylic protons, whereas the observed NMR spectrum indicates three.

When considering possible structures for consistency with NMR data, some beginning students try to wrestle with splitting data before they have considered other aspects of the data that are more straightforward and at least as powerful. Notice in the foregoing solution that the alternative structures can be clearly ruled out on the basis of relative integral or number of absorptions.

(b)   The assignments for the vinylic protons are determined by both relative integrals and chemical shifts. Terminal vinylic protons always occur, other things being equal, at a smaller chemical shift than internal vinylic protons, because alkyl substitution results in an increased chemical shift. Secondly, the integral for the δ 5.0–5.3 resonances is twice that of the δ 5.9–6.0 resonances, and there are twice as many terminal vinylic protons as internal ones. Finally, the δ 5.0–5.3 resonances show splittings consistent with *both* a cis and a trans relationship to other vinylic protons, and the only way this can occur is if there are two such protons—one cis to the remaining vinylic proton and the other trans.

   And this brings us to the issue of detailed assignment. Trans splittings are larger than cis splittings. Because the vinylic proton at lower chemical shift (about δ 5.1) in the δ 5.0–5.3 set has the smaller splitting, it is the one that has the cis relationship to the internal vinylic proton. The vinylic proton at δ 5.3 has the trans relationship. This assignment also makes sense because the vinylic proton with the larger chemical shift is closer to the chlorine. In summary,

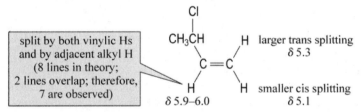

13.35   It would be easy to believe that the methyl groups of *A* could account for the twelve-proton singlet at δ 1.22, and the two OH groups for the two-proton exchangeable resonance at δ 1.96; but this leaves no hydrogens to account for the δ 1.57 resonance. In other words, structure *A* has two nonequivalent sets of hydrogens, whereas the NMR spectrum indicates at least three. Structure *A* also has two nonequivalent sets of carbons, whereas the ^{13}C NMR spectrum indicates three.

   Structure *B* is ruled out by its molecular formula, which is $C_8H_{16}O_2$. But the NMR spectrum is also not consistent with structure *B*, which should have two methyl singlets, each integrating for 6*H*, and one methyl singlet integrating for 3*H*, along with one exchangeable OH proton. Although it would have the required three resonances, the integration ratio (6:3:1) is different from that observed (12:4:2 or 6:2:1). This difference is well within the ability of the spectrometer to differentiate. But the ^{13}C NMR is even more definitive. Structure *B* has five nonequivalent sets of carbons, whereas the ^{13}C NMR spectrum indicates three sets. Even if the methyl resonances overlapped accidentally, which is unlikely, the attached-hydrogen analysis for *B* would predict no carbon with two attached protons.

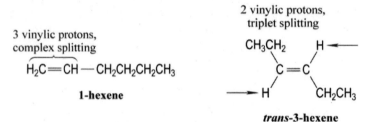

structure *B*   equivalent

Solutions to Additional Problems

13.36   The four pieces of information are

1.   the number of individual resonances, or peaks (not counting splitting), that are determined by how many different sets of inquivalent protons are present;

2.   the chemical shift, which is used to determine what functional groups are near the proton(s) under observation;

3.   the integral, which is used to determine how many protons are being observed relative to the total number in the molecule;

4.   splitting, which is used to determine connectivity—that is, how many protons are on atoms adjacent to the carbon that bears the proton being observed.

13.37   (a)   The spectrum of cyclohexane is a singlet. The spectrum of its constitutional isomer *trans*-2-hexene has many resonances, including vinylic resonances in the δ 5–δ 6 region that are not present in cyclohexane.

(b)   Only 1-hexene will have a complex vinylic proton absorption that integrates for 25% of the total absorption (that is, 3 protons); the vinylic proton absorption of *trans*-3-hexene will consist of a triplet integrating for 17% of the total absorption (that is, 2 protons).

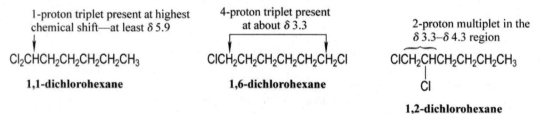

(c)   The resonance at greatest chemical shift in the NMR spectrum of 1,1-dichlorohexane would be a one-proton triplet; this resonance would be at a chemical shift of at least δ 5.9. The resonance at greatest chemical shift in the NMR spectrum of 1,6-dichlorohexane would be a four-proton triplet at about δ 3.3. The resonances at highest chemical shift in the spectrum of 1,2-dichlorohexane would be a complex pattern in the δ 3.3–4.3 region. How complex depends on whether the diastereotopic protons at carbon-1 have different chemical shifts and whether they have different coupling constants with the methine proton at carbon-2.

1-proton triplet present at highest   4-proton triplet present   2-proton multiplet in the
chemical shift—at least δ 5.9   at about δ 3.3 δ 3.3–δ 4.3 region

$Cl_2CHCH_2CH_2CH_2CH_2CH_3$ $ClCH_2CH_2CH_2CH_2CH_2CH_2Cl$ $ClCH_2CHCH_2CH_2CH_2CH_3$

1,1-dichlorohexane   **1,6-dichlorohexane**   Cl

1,2-dichlorohexane

(d)   The spectrum of *tert*-butyl methyl ether consists of two singlets; that of isopropyl methyl ether contains a singlet for the methoxy group, but a more complex doublet-septet pattern for the isopropyl group.

1*H* septet

6*H* doublet   at δ 3.7

9*H* singlet   3*H* singlet   at δ 1.2   3*H* singlet
at δ 1.2   at δ 3.2   at δ 3.2

(CH₃)₃C—OCH₃   (CH₃)₂CH—OCH₃

tert-butyl methyl ether   **isopropyl methyl ether**

(e)   The first compound, Cl₃CCH₂CH₂CHF₂, has the absorption at greatest chemical shift and the more complex spectrum; furthermore, this high-shift absorption will show not only H–H splitting, but also the typically large H–F splitting, and will thus appear as a triplet of triplets—that is, a triplet in which each individual line is a more closely spaced triplet. Only the second compound, CH₃CH₂CCl₂CClF₂, will show a three-proton triplet at about δ 1 as part of a typical ethyl triplet-quartet pattern.

13.38   (a)   NMR spectroscopy requires that the sample be situated in a magnetic field before absorption of electromagnetic radiation can occur. Other forms of absorption spectroscopy do not require the presence of a magnetic field. The field establishes the energy difference between nuclear spins. NMR spectroscopy also differs conceptually in the phenomenon responsible for absorption: the "flipping" of nuclear spins.

(b)   Some of the protons change from a lower-energy spin to a higher-energy spin.

(c)   The chemical shift in Hz is proportional to the magnitude of the applied field.

(d)   The coupling constant J is independent of the applied field (and also the operating frequency).

(e)   The coupling constant of vicinal protons varies as the angle between them changes, according to the Karplus relationship (Eq. 13.10, p. 635).

(f)   The chemical shift in ppm (δ) does not change with operating frequency ν_0 because it is *defined* as the proportionality constant that relates the chemical shift in Hz, $\Delta\nu$, to the operating frequency ν_0. (See Eq. 13.1 on text p. 612.) Because $\Delta\nu$ and ν_0 are proportional, δ for a given resonance is the same at any operating frequency.

(g)   The chemical shift difference *in frequency units* (Hz) between two protons must be considerably greater than the coupling constant between these protons (in the same units). This relationship is embodied in Eq. 13.12 on text p. 642. Or, the splitting patterns of the two coupled protons should not overlap.

13.39   (a)   (c)   (e)

(CH₃)₃C—O—C(CH₃)₃   H₃C   CH₃

di-tert-butyl ether   H₃C   CH₃

cyclohexane   **1,1,2,2,tetramethylcyclopropane**

(b)   (d)

H₃C   CH₃   (CH₃)₃C—CH₂—C(CH₃)₃

C=C   **2,2,4,4-tetramethylpentane**

H₃C   CH₃

2,3-dimethyl-2-butene

13.40   (a)   The compound is neopentane (2,2-dimethylpropane), (CH₃)₄C.

(b)   The compound is cyclopentane. (2,3-Dimethyl-2-butene would not be a bad answer; see the solution to Problem 13.38b. However, δ 1.5 is somewhat low for a chemical shift of the allylic methyl groups.)

cyclopentane

(c)   The compound is 1,1-dimethoxyethane, $(CH_3O)_2CH$—CH_3.

(d)   The D_2O exchange result indicates the presence of an alcohol. Given that the splitting indicates an ethyl group, the integral indicates three equivalent ethyl groups. Because no resonance occurs in the δ 3–4 region, there is no α proton, that is, no CH—O—. Hence, the alcohol is a tertiary alcohol. The compound is

CH₃CH₂
|
CH₃CH₂—C—OH
|
CH₃CH₂

3-ethyl-3-pentanol

(e)   The spectrum shows indicates vinylic protons with no cis or trans splittings. Hence, a partial structure is

H_2C=C—

The resonances at δ 2.0 and δ 1.8 are nearly singlets and integrate for 2 and 3 protons, respectively. These protons, from their chemical shifts, must be allylic. We now have the partial structure

H_2C=C—CH_2—
|
CH_3

The very small splittings of these groups are evidently allylic splittings (Table 13.3, text p. 647). The remaining resonance is a nine-proton singlet, which can only be a *tert*-butyl group. This completes the structure, which is consistent with the result of catalytic hydrogenation.

H_2C=C—CH_2—$C(CH_3)_3$ $\xrightarrow{H_2,\ catalyst}$ H_3C—CH—CH_2—$C(CH_3)_3$
| |
CH_3 CH_3

2,4,4-trimethyl-1-pentene   **2,2,4-trimethyl-1-pentane**

(f)   The compound has one degree of unsaturation and a *tert*-butyl group. This, plus the requirement for two chlorines and a partial structure CH_2—CH required by the splitting leaves only the following possibility:

$(CH_3)_3C$—CH_2—CH=CCl_2

1,1-dichloro-4,4-dimethyl-1-pentene

(g)   The two protons are evidently equivalent (a single chemical shift), and, from the coupling constants, they must be close to the fluorines, but not on the same carbon. The compound is $BrCH_2CF_2Br$, 1,2-dibromo-1,1-difluoroethane.

   Why not the following compound?

Br—CH—CH—Br
| |
F   F

1,2-dibromo-1,2-difluoroethane

The proton NMR spectrum of this compound would be a doublet of doublets. (Why?) In addition, with *two* halogens on the same carbon, the chemical shift would be much greater. Furthermore, one of the H—F coupling constants would be considerably larger because of the presence of a fluorine on the same carbon.

(h)   The compound is F_3C—CH_2—I (1,1,1-trifluoro-2-iodoethane). The splitting of the protons is caused by the fluorines.

(i)   Since there are sixteen protons, the integrals account for 2, 12, and 2 protons, respectively. The compound is $(CH_3O)_2CHCH_2CH(OCH_3)_2$ (1,1,3,3-tetramethoxypropane), better known to chemists as malonaldehyde dimethyl acetal. What makes this problem tricky—and the whole point of the problem—

is that not all of the equivalent protons of a given type are on the same carbon. For example, the two CH protons are equivalent and therefore have exactly the same chemical shift; they are split into a triplet by the adjacent CH_2 protons. Likewise, the CH_2 protons are split by the two adjacent protons into a triplet.

(j)   The D_2O exchange indicates an alcohol, which must be tertiary, because of the absence of a resonance in the δ 3–4 region. The ^{13}C NMR spectrum indicates that there are four nonequivalent sets of carbons. The doublet-septet pattern indicates an isopropyl group, and the 6-proton singlet indicates two methyl groups. The compound, with assigned ^{13}C NMR chemical shifts, is

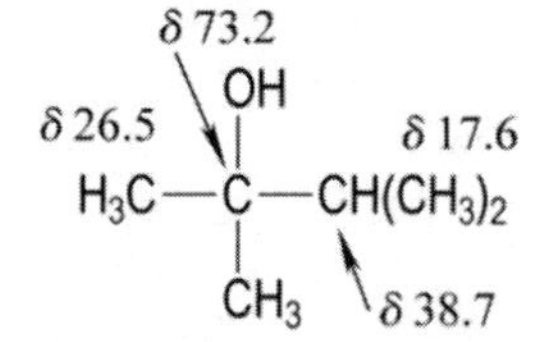

$$\delta 73.2$$

$$\delta 26.5 \quad\quad OH \quad\quad \delta 17.6$$
$$H_3C - C - CH(CH_3)_2$$
$$CH_3 \quad \delta 38.7$$

2,3-dimethyl-2-butanol

13.41   (a)   The NMR spectrum of the starting material has two sets of resonances, one of which is in the vinylic-proton region (δ 4.5–5.0). The other would be a singlet in the allylic methyl range, perhaps around δ 1.8. The NMR of the product is a singlet near δ 1.7. To follow the reaction, look for the disappearance of the vinylic proton resonance and one singlet and the appearance of the singlet arising from the product.

(b)   The NMR spectrum of the starting material is a singlet at δ 1.7; see the solution to Problem 13.39(b). The NMR spectrum of the product should contain an isopropyl doublet-septet pattern at about δ 0.9 and δ 2, respectively, as well as a six-proton methyl singlet near δ 1.4. To follow the reaction, look for the disappearance of the singlet at δ 1.7 and the appearance of a doublet around δ 0.9.

13.42   The most obvious difference is that the spectrum of 1-methylcyclohexene should show *one* vinylic proton and a three-proton singlet (neglecting any allylic splitting); whereas the spectrum for 3-methylcyclohexene should show *two* vinylic protons and a three-proton doublet.

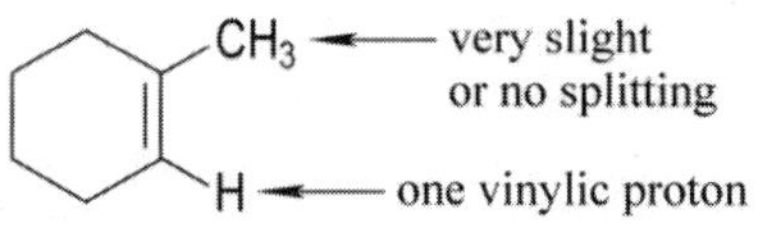

1-methylcyclohexene 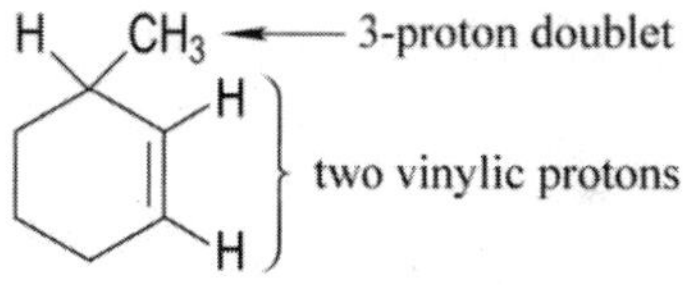   **3-methylcyclohexene**

13.43   (a)   All methyl groups are homotopic; all methine carbons are homotopic; and all methylene carbons are homotopic. (Prove this by a substitution test if this isn't clear.) Each carbon of one type is constitutionally nonequivalent to all carbons of other types. Consequently, the ^{13}C NMR spectrum of this compound should consist of three resonances.

(b)   Because the chair interconversion is rapid, the relationships of the carbons in these compounds can be decided from the planar projections shown. The compound shown in this part should have six ^{13}C NMR absorptions because there are six chemically nonequivalent sets of carbons (unless there are accidental overlaps between some absorptions). In the structure of this compound shown below, carbons with the same number are enantiotopic, and therefore chemically equivalent. (Notice the plane of symmetry.) The following pairs of carbons are diastereotopic, and therefore chemically nonequivalent: 3 and 4; 5 and 6; 1 and 2. All other pairs are constitutionally nonequivalent.

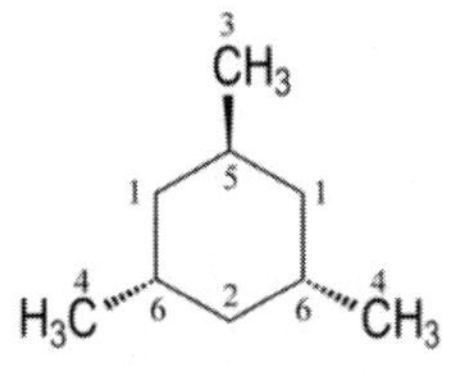

13.44   (a)   The NMR spectrum of the first compound, $(CH_3)_2CH$—Cl (isopropyl chloride), should consist of the doublet-septet pattern characteristic of isopropyl groups, with the doublet at about $\delta\ 1.2$ and the septet at about $\delta\ 3.7$. The NMR spectrum of the deuterium-substituted analog, $(CH_3)_2CD$—Cl, should consist of a singlet at essentially the same chemical shift as the doublet in the first compound. (The splitting between H and D nuclei on adjacent carbons is nearly zero.)

(b)   The NMR spectrum of the compound containing deuterium should consist of two triplets of equal intensity. (The coupling constant between H and D nuclei on adjacent carbons is nearly zero.) The NMR spectrum of the other compound, as shown in Fig. 13.7 on text p. 631, consists of a triplet at low field and a quintet at higher field; the triplet should have twice the integrated intensity of the quintet. Each set of resonances in one spectrum should have the same chemical shift as the corresponding set of resonances in the other.

(c)   To decide on the spectra for the two diastereomers, first analyze the third compound. The analysis is shown on the bottom of Fig. SG13.4. The resonance of the methyl group will be a three-proton singlet. The two methylene hydrogens are diastereotopic and are therefore chemically nonequivalent. Hence, they should have different (if only slightly different) chemical shifts, and each will split the other into a doublet. Thus, the resonances for these two protons should consist of a closely spaced doublet of doublets.

   The diastereomers should have different NMR spectra. Their analysis is shown as the top two compounds in Fig. SG13.4. To begin with, neglect the splitting by deuterium. (Deuterium splitting, which is considered in Problem 13.50(c), can be eliminated by an instrumental technique.) Each spectrum should consist of a three-proton singlet for the methyl group and a one-proton singlet for the methylene hydrogen. Although the methyl singlets will likely have identical chemical shifts in the two compounds, the chemical shifts of the other hydrogens will differ by essentially the same amount as the chemical shifts of the two methylene hydrogens in the third compound. The resonance of the methylene hydrogen in the 1S,2R diastereomer will have the same chemical shift as the resonance of the pro-S hydrogen in the third compound; and the resonance of the methylene hydrogen in the 1R,2R diastereomer will have the same chemical shift as the resonance of the pro-R hydrogen in the third

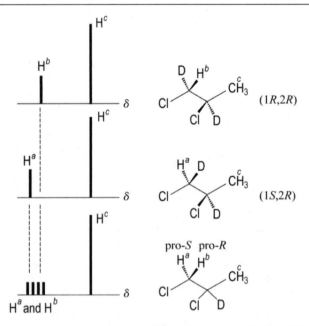

Figure SG13.4 Schematic spectra to illustrate the differences expected in the NMR spectra of the three compounds in Problem 13.44(c). In these spectra, the pro-S hydrogen is arbitrarily assigned the greater chemical shift, but the chemical shifts of the two diastereotopic hydrogens might be reversed. Splitting by deuterium is neglected.

compound. In each compound, both hydrogens will be singlets, neglecting any splitting by deuterium. The order of the chemical shifts of the diastereotopic hydrogens is arbitrary; the order could be reversed.

13.45   Compound A is 2,3-dimethyl-2-butene, $(CH_3)_2CAC(CH_3)_2$. [See Problem 13.39(b).]

   The IR spectrum of compound B indicates a H_2CAC group. The NMR spectrum shows a typical isopropyl pattern: a septet/doublet pattern (δ 2.2, δ 1.0, respectively); from its chemical shift, the isopropyl CH group is allylic. The isopropyl group must be attached to a carbon that bears no protons, since the CH of the isopropyl group is split only by the adjacent methyls. The presence of two vinylic hydrogens (δ 4.6–4.7) with very weak splitting is in agreement with the geminal vinylic hydrogens suggested by the IR spectrum, and the three equivalent hydrogens remaining probably correspond to a methyl group. Compound B is

$$(CH_3)_2CH - C = CH_2$$
$$\overset{|}{\underset{CH_3}{}}$$

2,3-dimethyl-1-butene
(compound B)

   Compound C shows a nine-proton singlet characteristic of a *tert*-butyl group. The remaining three protons are vinylic. The IR spectrum shows CAC stretching and C—H bending absorptions that confirm the presence of a —CHACH$_2$ group. Although it is not necessary to interpret the splitting to define the structure, it is very similar to that in Figs. 13.16–13.17 on text pp. 648–649. Compound C is therefore $(CH_3)_3C$—CHACH$_2$, 3,3-dimethyl-1-butene.

   That compound D is not an alkene follows from both the chemical data and the NMR absorption. A six-carbon hydrocarbon that is a singlet in the NMR, that has one degree of unsaturation, and that is not an alkene can be only cyclohexane. [You may have seen this one before: Problem 13.39(a).]

cyclohexane
(compound D)

 The structures of compound A and D would be difficult to distinguish on the basis of their proton NMR spectra alone. However, ^{13}C NMR would readily distinguish them. Can you see why?

13.46   The spectrum shows at least four nonequivalent sets of protons in the ratio 1:1:4:6. Let's see how the possible structures stack up against this analysis.

CH$_3$CH$_2$ CH$_2$CH$_3$ CH$_3$CH$_2$O   CH$_2$CH$_3$ H   CH$_2$CH$_3$
        C=C   C=C   C=C
    H   H   H   H   H   CH$_2$CH$_3$

cis-**3-hexene** (A)   (**Z**)-**1-ethoxy-1-butene** (B)   **2-ethyl-1-butene** (C)

Cl$_2$CH —CH(OCH$_2$CH$_3$)$_2$ Cl—CH—CH—Cl
                                                    | |
1,1-dichloro-2,2-diethoxyethane (D)   CH$_3$CH$_2$O   OCH$_2$CH$_3$

1,2-dichloro-1,2-diethoxyethane (E)

Compounds A–C are ruled out immediately. The vinylic hydrogens in A would be a single triplet. In compound B, there would be two methyl triplets and a more complex splitting for the two vinylic hydrogens. In compound C, the vinylic hydrogens would be a singlet (neglecting allylic splitting). The two methine hydrogens of D would definitely give a pair of doublets at fairly high chemical shift, as observed. The two methine hydrogens of E are completely equivalent and would give a singlet as the absorption at greatest chemical shift. Therefore, compound E is ruled out, and compound D is the remaining possibility.

As to the complexity of the δ 3.7 resonance, the two CH_2 hydrogens in either ethyl group on both D and E are *diastereotopic,* and hence, chemically nonequivalent. For example, in the case of D,

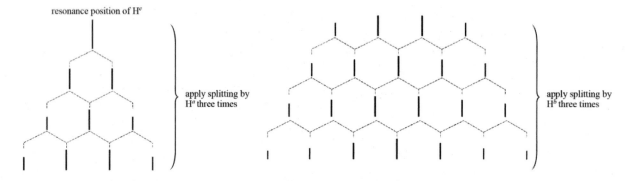

H^a and H^b are diastereotopic; H^c and H^d are also diastereotopic. On the other hand, H^a is enantiotopic to H^c, and H^b is enantiotopic to H^d. (You should verify these statements by a substitution test.) Therefore, H^a and H^c are chemically equivalent, as are H^b and H^d. But H^a and H^b are chemically *nonequivalent*, as are H^c and H^d. Hence, there are *five* nonequivalent sets of hydrogens. It is perhaps not surprising that the chemical shifts of the diastereotopic protons are not very different, because their chemical environments are very similar.

Let's now consider the splitting patterns of these diastereotopic protons. We've already said that H^a and H^b occur at slightly different chemical shifts. This would account for two absorptions. Each of these is split into a doublet by the other to give four lines. Each of these is split into a quartet by the neighboring methyl group. This gives $4 \times 4 = 16$ lines. Because H^c is chemically equivalent to H^a, and H^d is chemically equivalent to H^b, the analysis is complete. You can actually see all sixteen lines for these diastereotopic hydrogens in the spectrum.

13.47   (a)   Protons a and b are diastereotopic protons and thus give different chemical shifts as discussed in Sec. 13.3D. You can verify for yourself that they are diastereotopic by performing the substitution test (Sec. 10.9A) and comparing the virtual products.

         (b)   Proton a is split both by proton b and the proton on the adjacent carbon. Because the two coupling constants are different, the $n + 1$ rule fails here. Instead of a triplet, a doublet of doublets is observed instead (see solution to problem 13.19(b) in this manual for a similar situation).

13.48   (a)   Protons a and b are diastereotopic and thus give different chemical shifts as discussed in Sec. 13.3D. You can verify for yourself that they are diastereotopic by performing the substitution test (Sec. 10.9A) and comparing the virtual products. They are each split into a doublet by proton c.

         (b)   The signal for proton c is split three times by 7.0 Hz into a quartet by protons a, and that quartet is split three more times, again by 7.0 Hz, into a septet by protons b.

resonance position of H^c

apply splitting by H^a three times

apply splitting by H^b three times

Finally, each peak in that septet is split into a doublet by 4.3 Hz by proton d.

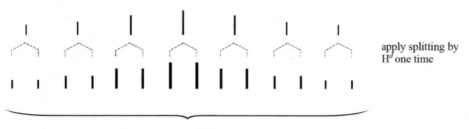

apply splitting by
H^d one time

appearance of H^c resonance:
a septet of doublets

The signal for proton c appears on a spectrum as a "septet of doublets", though one would most likely just refer to it as a "multiplet".

13.49   Compounds A and B have four sets of nonequivalent carbons and compound C has three. Because the given spectrum indicates four nonequivalent sets of carbons we can rule out compound C. We now have to decide between compounds A and B. An important difference in these structures is that only A contains a carbon connected to *two* oxygens. From the chemical-shift chart (Fig. 13.22 on text p. 658), we see that a typical carbon in the environment —C—O— has a chemical shift between δ 40 and δ 80. The maximum chemical shift in the spectrum, however, is nearly δ 100, and this would be consistent with a compound in which a carbon is bound to two oxygens. Hence, the compound is A. To further rule out B, note that the spectrum contains no methyl (3-proton) carbon in the ether region.

13.50   (a)   The nitrogen can take on three values of spin with equal probability: $+1$, 0, and -1. Hence, the *protons* near a nitrogen of each spin state are shifted accordingly (or not shifted, in the case of spin $= 0$.) The *proton* NMR spectrum of $^+NH_4$ should (and does) consist of three lines of equal intensity.

(b)   The reason the ^{13}C NMR spectrum of $CDCl_3$ is a 1:1:1 triplet is the same reason that the proton NMR spectrum of $^+NH_4$ is a 1:1:1 triplet [see part (a)]: deuterium, like nitrogen, can have spins of $+1$, 0, and -1, and deuterium splits ^{13}C resonances just as it splits proton signals.

(c)   The compound with one deuterium, H_2CDI, would show a triplet proton resonance with lines that have a 1:1:1 intensity ratio, because the deuterium can have three different values of spin ($+1$, 0, and -1). Because these are equally probable, they give rise to lines in the proton splitting pattern of equal intensity. The compound D_3CI has no proton NMR spectrum. For HCD_2I, the two deuteriums have the following combinations of spin values:

Total spin $= +2$: $+1, +1$ (one way)
Total spin $= +1$: $0, +1$ and $+1, 0$ (two ways)
Total spin $=\ \ 0$: $0, 0$; $+1, -1$; and $-1, +1$ (three ways)
Total spin $= -1$: $0, -1$ and $-1, 0$ (two ways)
Total spin $= -2$: $-1, -1$ (one way)

Each value of the total deuterium spin makes a different contribution to the position of the proton resonance; thus, there are five lines, with relative intensities proportional to relative probabilities of deuterium spin (that is, to the number of different ways that each spin can be achieved): 1:2:3:2:1.

Mass spectrometry could be used to easily determine the identity of each sample, as the samples each differ in molecular mass. IR could also be used: only D_3CI will show a C—D stretching absorption and no C—H stretching absorption in the IR spectrum.

13.51   The IR spectrum and the D_2O shake results indicate that compound A is an alcohol, and the absence of a resonance in the δ 3–4 region indicates that the alcohol is tertiary. Treatment of compound A with H_2SO_4 yields a compound B, which, from its δ 5.1 resonance in its NMR spectrum, is evidently an alkene. Evidently, the reaction with H_2SO_4 is a dehydration. If so, the molecular mass of 84 for compound B means that compound A has the molecular mass of compound B plus 18 (the molecular mass of H_2O). Hence, the molecular mass of A is $84 + 18 = 102$.

Now let's turn to the integral for compound A. The integrals (from high to low shift) are in the approximate ratio 1:4:6:3. Let's adopt the hypothesis that the O—H proton accounts for $1H$. If so, then compound A contains

14 hydrogens, and its formula is $C_6H_{14}O$, which gives the molecular mass of 102 hypothesized above. Why couldn't the alcohol have a formula $C_7H_{14}O$? Because this would give the wrong molecular mass.

In the NMR of compound A, the singlet at δ 1.2 integrates for $6H$. This can only be two methyl groups. The triplet just below δ 1.0 integrates for $3H$, and, from its splitting, corresponds to a CH_3CH_2— group. However, the methylene protons in this group must be split by other protons because they are not a simple quartet. The following structure would account for the complex splitting at δ 1.4, as well as the other facts:

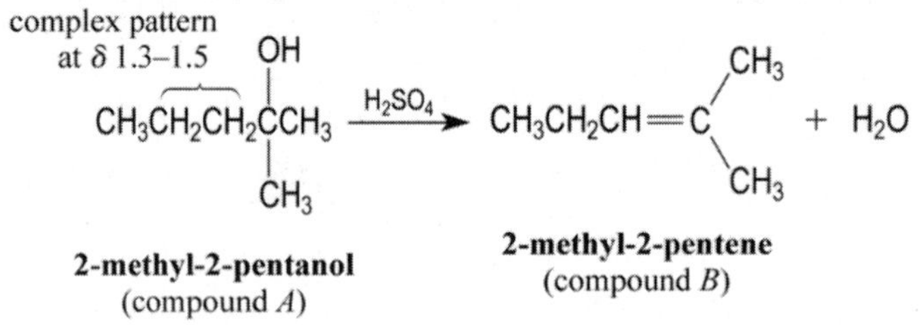

complex pattern
at δ 1.3–1.5

$$CH_3\overbrace{CH_2CH_2}C\underset{\underset{CH_3}{|}}{\overset{\overset{OH}{|}}{C}}CH_3 \xrightarrow{H_2SO_4} CH_3CH_2CH{=}C\overset{CH_3}{\underset{CH_3}{\diagdown}} + H_2O$$

2-methyl-2-pentanol
(compound A)

2-methyl-2-pentene
(compound B)

The dehydration would then give compound B shown above. In the NMR of this compound we expect a single vinylic proton split into a triplet, and this is what is observed at δ 5.1. (Some additional, very small, allylic splitting is also observed.) The NMR spectrum of compound B is rationalized as follows:

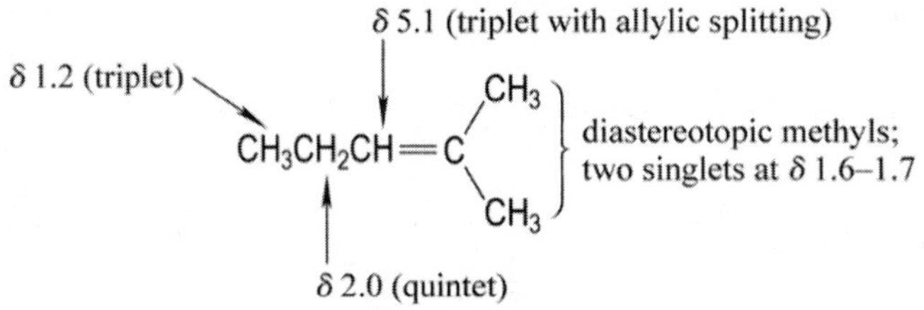

δ 5.1 (triplet with allylic splitting)

δ 1.2 (triplet)

$$CH_3CH_2CH{=}C\overset{CH_3}{\underset{CH_3}{\diagdown}}$$

diastereotopic methyls;
two singlets at δ 1.6–1.7

δ 2.0 (quintet)

13.52   (a)   The IR absorbance at 3200-3600 cm^{-1} suggests an –OH group is present, and there is a strong absorbance at 1090 cm^{-1} which suggests C–O single bonds, present either in alcohols or ethers. The NMR shows a resonance that disappears with a D_2O shake, confirming that an –OH group is present. Additionally from this spectrum, a six-hydrogen singlet suggests two methyl groups attached to a carbon that does not possess any protons. The other three-hydrogen singlet suggests another isolated methyl group, perhaps as part of an ether. With three of the six carbons and nine of the 14 hydrogens accounted for as belonging to methyl groups, you could start trying various structures to see what fits the rest of the data. The compound below fits the spectral data given:

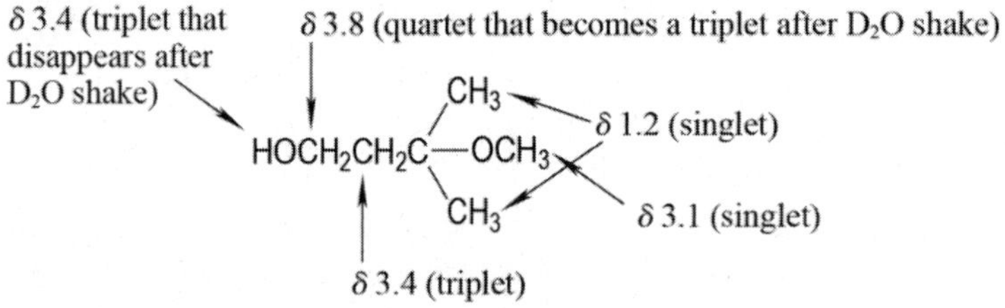

δ 3.4 (triplet that
disappears after
D_2O shake)

δ 3.8 (quartet that becomes a triplet after D_2O shake)

$$HOCH_2CH_2C\overset{CH_3}{\underset{CH_3}{\diagup}}{-}OCH_3$$

δ 1.2 (singlet)

δ 3.1 (singlet)

δ 3.4 (triplet)

The quartet at δ 3.8 is split by both the –OH proton and the adjacent CH_2 protons. After the D_2O shake, the –OH becomes an –OD and the signal at δ 3.8 is the only split by the CH_2 protons, into a triplet.

(b)   First, by calculating the unsaturation number (Sec. 4.3), it can be determined that the compound has either a double bond or a ring. The sharp absorbance on the IR spectrum at 1658 cm^{-1} suggests a carbon-carbon double bond (as to the absorbances at 3085 and 875 cm^{-1}, which correspond to protons attached to carbons of carbon-carbon double bonds). Both the broad stretch in the IR at 3200-3600 cm^{-1} and the signal in the NMR spectrum that disappears with a D_2O shake confirm an –OH group. Further inspection of the NMR spectrum shows a 3H singlet that tells us there is an isolated CH_3 group. The signals at δ 4.8 and 4.9 correspond to protons attached to a carbon that is part of a double bond. At this point, we have enough information to start trying various structures to see what fits the rest of the data. The only alkene alcohol that fits the data is:

Here is a related structure that is close but doesn't quite fit the data. Why not?

13.53   The formula, the IR spectrum, and the D_2O shake indicate that compound A is a five-carbon alcohol with one degree of unsaturation. This could be a ring or a double bond, but the NMR spectrum clearly indicates two vinylic protons near δ 5.5, a fact that suggests a double bond as the source of unsaturation. Moreover, the IR shows weak CAC stretch and a strong absorption at 965 cm^{-1}, which indicates a *trans*-alkene. The ^{13}C NMR resonances at δ 125.5 and δ 135.5, each with one attached hydrogen, support the diagnosis of an internal alkene. The alcohol is secondary because there is one proton in the α-oxygen region (δ 4.2) and one attached hydrogen for the corresponding carbon at δ 68.8; the latter proton is shifted a little more than normal for this type of proton; hence, this proton is also allylic. Furthermore, this α-proton is adjacent to four other protons. Two methyl groups are adjacent to single hydrogens. One of these methyls is allylic (δ 1.7) and the other is not (δ 1.2). To summarize, we have the following structural elements, which can only be assembled to give *trans*-3-buten-2-ol as the structure. The chirality of the proposed structure is consistent with the fact that compound A can be resolved into enantiomers.

trans-3-buten-2-ol

13.54   The mass spectrum indicates that the compound contains a single bromine (double molecular ion). The NMR spectrum indicates an ethoxy group (CH_3CH_2O—); this comes from the typical triplet–quartet pattern in which the —CH_2— group is at a typical α-oxygen chemical shift and shows splitting with *only* the CH_3. The spectrum indicates two vinylic hydrogens that split each other with a very small splitting, which, however, is larger than a geminal splitting. The IR spectrum shows no evidence for a trans alkene, nor does it show the characteristic absorption for a ACH$_2$ group. Hence, the vinylic hydrogens are cis. (The IR absorption at 694 cm^{-1} could be the C—H bending absorption for a cis alkene.) The ethoxy group counts for 29 mass units; the ^{79}Br for 79 mass units; two vinylic CH groups would complete the mass of 150. The structure can be only

(Z)-1-bromo-2-ethoxyethene

13.55 The sample of 2,5-hexanediol is a mixture of diastereomers, the racemate and the meso compound. Each of these stereoisomers has a ^{13}C NMR spectrum that consists of three resonances, but the two spectra are different, if only slightly so.

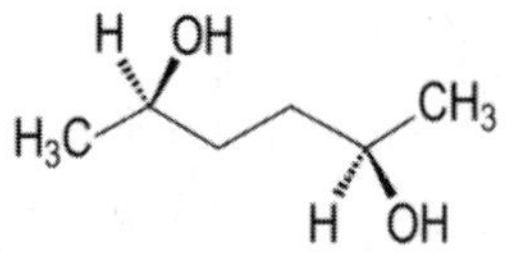

(±)-2,5-hexanediol
(the 2R,5R enantiomer is shown)

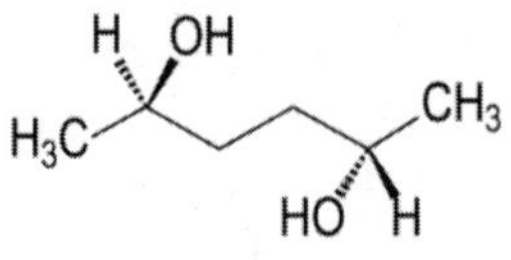

meso-2,5-hexanediol

13.56 (a) As always, we first draw the structure:

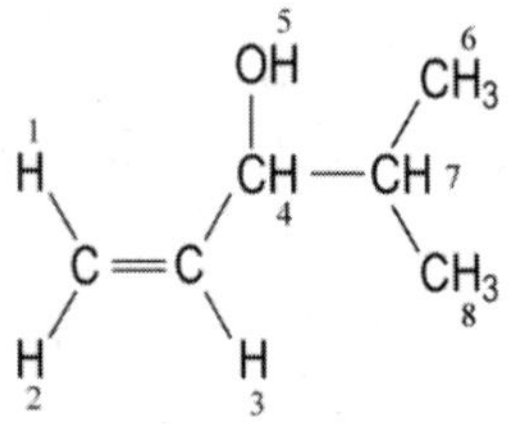

4-methyl-1-penten-3-ol

The proton NMR spectrum of 4-methyl-1-penten-3-ol should contain eight sets of absorptions because the compound contains eight chemically nonequivalent sets of protons (numbered 1–8 in the preceding structure). Notice that protons 1 and 2 are diastereotopic and therefore chemically nonequivalent, as are protons 6 and 8.

(b) Because all carbons are chemically nonequivalent, 4-methyl-1-penten-3-ol should have six carbon resonances in its ^{13}C NMR spectrum. (Notice that the carbons of methyl groups 6 and 8 in the structure shown in part (a), like the protons of these groups, are diastereotopic.)

13.57 (a) The NMR spectrum of a very dry sample would show splitting between protons 1 and 2. Hence, the resonance of proton 1 should be a triplet; the resonance of protons 2 should be either a quartet (if $J_{12} = J_{23}$) or a doublet of triplets (if $J_{12} \neq J_{23}$), where the J's are coupling constants. In a wet sample, the splitting associated with proton 1 is obliterated. In that case, the resonance of proton 1 should be a singlet; the resonance of protons 2, which is split only by protons 3, should be a triplet. The resonance of protons 3 should be a triplet in both the wet and the dry sample.

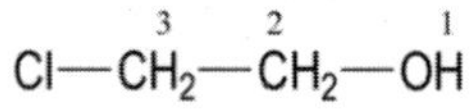

$$\overset{3}{Cl}-\overset{}{CH_2}-\overset{2}{CH_2}-\overset{1}{OH}$$

2-chloroethanol

(b) Following a D_2O shake, the resonance of proton 1 should disappear; and the resonance of protons 2 should be the same as it is in the wet sample—that is, a triplet.

13.58 The relevant protons are coded in the discussion as follows:

vitamin D₃

The two doublets centered at δ 6.2 are assigned to the diene protons 1 and 2. These protons are split by each other, but not significantly by any other protons. (Coupling constants depend on dihedral angle, and the large splitting evidently suggests an anti relationship about the single bond.) The two resonances centered at δ 4.9 are assigned to the geminal alkene protons 3 and 4. These two protons are diastereotopic, and are therefore chemically nonequivalent. Their splitting, although not expanded, is evidently the small splitting typical of geminal alkene protons (Table 13.3, text. p. 647). The multiplet just below δ 4 is assigned to the α-alcohol methane proton 6. We are given no information about a D_2O shake, so the proton 5 is not identified; it is probably within the complex pattern between δ 1 and δ 3. Within the inset on the right, the doublet integrating for three protons probably corresponds to a methyl group. The only CH_3—CH group are the methyl protons 8. In the inset, the doublet of doublets corresponds to six protons, which corresponds to the two diastereotopic methyl groups 9 and 10, each split by the adjacent methine hydrogen and having slightly different chemical shifts. The resonance at δ 0.5 is due to an unsplit methyl group and must correspond to protons 7.

13.59   The unsaturation number = 1. There are no alkene absorptions in the IR; therefore, we postulate the presence of a ring. The presence of an alcohol is indicated by the IR data, and the triplet at δ 3.5 must be an OH proton because it is lost as the result of a D_2O shake. Furthermore, it must be part of a —CH_2OH group because of its triplet splitting. The protons at δ 3.7 must be the α-protons, because their splitting collapses after a D_2O shake; furthermore, these protons are attached to a quaternary carbon, because they are not split further. This quaternary carbon must also bear a methyl group, because the 3H resonance at δ 1.3 is unsplit. We therefore have the following partial structure:

$$-\overset{\overset{\displaystyle CH_3}{|}}{\underset{|}{C}}-CH_2-OH$$

We are missing two carbons and an oxygen; recall that there must be a ring. The two doublets between δ 4.4 and δ 4.6 must be two CH_2 groups, and the remaining oxygen must complete a four-membered ring. Consider the following two structures:

$$\begin{array}{cc}
\overset{\overset{\displaystyle CH_3}{|}}{H_2C-\underset{|}{C}-CH_2-OH} & \overset{\overset{\displaystyle CH_3}{|}}{H_2C-\underset{|}{C}-CH_2-OH} \\
\underset{O-CH_2}{|\quad\quad\;} & \underset{H_2C-O}{|\quad\;\;} \\
A & B
\end{array}$$

In structure B, the two CH_2 groups would be at least triplets, not doublets (and their splitting would likely be more complex, because the two protons on each carbon are diastereotopic). Furthermore, the chemical shifts of both CH_2 sets indicate that they are both α to an oxygen; in structure B, only one CH_2 meets this criterion. Therefore, structure B is ruled out, leaving structure A as the only reasonable candidate.

The splitting of the CH_2 protons in A is rationalized as follows.

If you find an error in this manual, please visit http://people.pharmacy.purdue.edu/~loudonm/teaching/

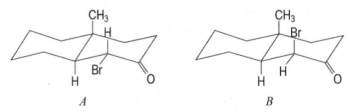

The two H^a protons are enantiotopic and therefore chemically equivalent; the same is true for the two H^b protons. Protons H^a and H^b split each other, and each is split into a doublet by its neighbor. Presumably H^a is the proton at greatest chemical shift because of its proximity to the OH group.

13.60   Both stereoisomers are based on a *trans*-decalin-like structure. You could determine which of the stereoisomers is which by inspecting the indicated protons. In compound *A*, the protons are anti to one another, and therefore would have a coupling constant between 8–15 Hz (see Sec. 13.5A and Fig. 13.9 on text p. 636). In compound *B*, they are gauche to one other and would have a coupling constant between 2–5 Hz.

Note that due to the nearby sp^2-hybridized carbon (the one that's double bonded to the oxygen), the two protons that we've focused on may not be exactly in gauche or anti orientations relative to each other. However, our analysis holds: we can inspect these two signals and expect a larger coupling constant between the protons in compound *A* because the angle between them is greater.

13.61   The ^{17}O NMR resonance of ^{17}O-water is split into a triplet by the attached two protons when a small amount of water is dissolved in CCl_4. The splitting pattern suggests that, in the presence of acid, the ^{17}O resonance is split by *three* protons. Evidently, this splitting is due to the hydronium ion, H_3O^+.

13.62   Shielding decreases chemical shift. Because a "naked" proton is completely *unshielded* by electrons, its chemical shift should be very large, and it should feel the full effect of the applied field. Because the range of proton chemical shifts within organic compounds extends to about 11–12 ppm (Fig. 13.4 on text p. 621), we expect the chemical shift of a "naked" proton to be even greater than this. Hence, the first answer, $\delta > 8$, is the correct one.

13.63   Note that the ^{13}C resonances from *unenriched* compounds will be much weaker than those from the isotopically enriched compounds and can be ignored. The proton-decoupled ^{13}C NMR spectrum will be a composite spectrum that consists of the spectra of the individual species present. These species, and their relative abundances, will be as follows: (*C = ^{13}C):

CH_3—CH_2—Br	relative probability = (0.5)(0.5) = 0.25; two doublets
CH_3—*CH_2—Br	relative probability = (0.5)(0.5) = 0.25; one singlet
*CH_3—CH_2—Br	relative probability = (0.5)(0.5) = 0.25; one singlet
CH_3—CH_2—Br	relative probability = (0.5)(0.5) = 0.25; no ^{13}C spectrum

Each resonance will show an apparent "triplet" pattern: the line from the singly-enriched species will be in the center of a doublet for the doubly-enriched species. The singly-enriched species are singlets because there is so little ^{13}C resulting from the natural abundance of ^{13}C at each carbon that it can be ignored. Each line of the doublet will have half the total intensity of the doublet, and therefore half the intensity of the singlet. A diagram of the spectrum is as follows:

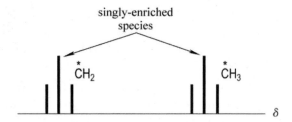

13.64   (a)   The resonance of the unpaired electron is split into four lines by the three neighboring protons according to the same $n + 1$ rule that applies to splitting between nuclei.

(b)   Use Eq. 13.17 on text p. 656 with $\gamma_n = \gamma_{electron} = 17.6 \times 10^6$ radians gauss^{-1} sec^{-1}, and $\mathbf{B}_0 = 3400$ gauss. We obtain

$$\nu_{electron} = \frac{\gamma_{electron}}{2\pi}\mathbf{B}_0 = \frac{17.6 \times 10^6 \text{ rad gauss}^{-1} \text{ sec}^{-1}}{2\pi \text{ rad}}(3400 \text{ gauss}) = 9.5 \times 10^9 \text{ sec}^{-1} \approx 10^{10} \text{ Hz}$$

Fig. 12.2 on text p. 572 shows that a frequency of 10^{10} Hz is in the microwave region of the electromagnetic spectrum. Indeed, ESR instruments employ microwave radiation to detect the magnetic resonance of electrons.

13.65   (a)   At 25 °C the conformations equilibrate too rapidly to be separately observed in the NMR spectrum; a single resonance is observed that is the "time-average" resonance of all conformations. At low temperature, however, the rates of the internal rotations are considerably reduced, and certain conformations can be separately observed:

enantiomeric conformations have
identical NMR spectra

The first of these conformations is the conformational diastereomer of the other two and hence has a different NMR spectrum. The other two conformations are enantiomeric and therefore have identical NMR spectra. Consequently, two resonances are observed: one for the first conformation, and the second for the other two.

(b)   The two lines at low temperature have different intensities because the diastereomers are present in different amounts. The intensity of each resonance is proportional to the amount of the conformation of which it is characteristic. *A priori,* if the conformations are equally probable, the resonance of the conformations with the gauche methyl groups should be twice as strong as that of the conformation with anti methyls. The actual intensities will likely differ from this 2:1 ratio in accordance with the relative stabilities of the conformations.

13.66   (a)   From Fig. 13.9, text p. 636, you know anti protons have large coupling constants and gauche protons have smaller coupling constants. When H^a is in the axial position (conformation E), it is split by adjacent protons that are anti and gauche. These protons would split H^a by different coupling constants. When H^a is in the equatorial position (conformation A), it is split by adjacent protons that are both gauche. These protons would split H^a by coupling constants that are about equal. Inspection of Fig. FP13.66 shows a resonance that is split into a triplet of triplets, or, a splitting pattern that would result from splitting by protons with different coupling constants, one large and one small. (This situation is similar to the one described by Fig. 13.10, text p. 638.) So, the resonance shown in FP13.66 belongs to proton H^a in the axial position, or conformation E.

(b) The other resonance, not shown, would come from H^a in the equatorial position (conformation A), and the adjacent splitting protons are both gauche. As described in part (a), the coupling constants coming from both would be small, and probably about equal. Thus, you might predict a quintet for this resonance, with small coupling constants between 3–4 Hz.

(c) One conformation is present in 3.39 times greater concentration than the other. The equilibrium constant, K_{eq}, is the ratio of the two conformations, 3.39:1, or 3.39. Using this value, the temperature, T, of 193 K (remember to convert Celsius to kelvins), and the gas constant of 8.314×10^{-3} kJ mol^{-1} K^{-1}, we can solve for ΔG.

$$\Delta G = -2.3RT \log K_{eq}$$

$$= -2.3(8.134 \times 10^{-3} \text{kJ mol}^{-1} \text{ K}^{-1})(193 \text{ K}) \log 3.39$$

$$= -1.96 \text{ kJ mol}^{-1}$$

The sign of ΔG is inconsequential; the difference in energy between the two conformations is what is significant. You know from previous discussions of the conformations of monosubstituted cyclohexanes (Sec. 7.3, on p. 282) that the larger substituent prefers to be in the equatorial position to minimize 1,3-diaxial interactions. So, the smaller resonance (not shown, relative integral = 1) corresponds to H^a in the axial position, and thus the iodine in the equatorial position, conformation E. So, conformation E is present at a 3.39 times greater conformation that A.

13.67 Number the carbons and view the Newman projection about the C1–C2 bond, with carbon-2 nearer the observer:

enantiomeric conformations have
identical ^{13}C NMR spectra

Assuming the spectrum is proton-decoupled, we would expect to see three resonances—one for each of the carbons—at room temperature. On cooling the sample, we would see two sets of three resonances. One set is due to the two enantiomeric conformations, and the other set is due to the remaining conformation. The largest chemical-shift change should be in the resonance for carbon-3, because its proximity to the bromine changes between the conformations. Smaller changes would be anticipated for carbons 1 and 2.

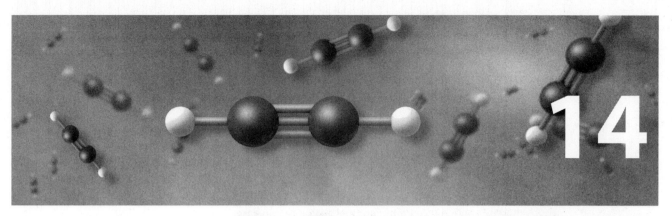

The Chemistry of Alkynes

STUDY GUIDE LINKS

14.1 Functional Group Preparations

Section 14.5 in the text discusses two ways to prepare aldehydes and ketones. At this point, you should be asking yourself, "What *other* ways do I know to make aldehydes and ketones?" Recall, for example, that oxidation of alcohols is an important method for preparation of aldehydes and ketones. Keep a separate review list that groups reactions by the *type of product formed*. Thus, you'll have a list of reactions for preparing aldehydes and ketones, a list for preparing alcohols, etc. You should write out each reaction in detail, noting the limitations on each. For example, what are the limitations of alkyne hydration as a method of ketone preparation? (Can *any* ketone be prepared with this method? Look at Problems 14.9–14.10 on text p. 693.) You can check how complete your list is by comparing it to the abbreviated list in Appendix V (p. A-11) of the text.

Use your list when working problems in synthesis. When your synthesis requires an aldehyde or ketone, consult your list; review the reactions if they start to look unfamiliar. Remember that putting together such lists keeps your studying *active* rather than passive.

14.2 Ammonia, Solvated Electrons, and Amide Ion

Sections 14.6 and 14.7 of the text have discussed several different uses of ammonia. The purpose of this Study Guide Link is to review and contrast these so that there is no confusion.

The first reagent is *liquid ammonia.* Ammonia can be condensed at –33 °C to a liquid. Once condensed, ammonia is easy to maintain in the liquid state because of its relatively high heat capacity. Liquid ammonia should not be confused with *aqueous* ammonia, which is a solution of ammonia in water. ("Household ammonia" is an aqueous ammonia solution.) If an alkali metal (such as sodium) is dissolved in liquid ammonia, a blue solution is obtained that is an excellent source of electrons—that is, an excellent reducing agent. This solution contains "solvated electrons"—electrons solvated by the hydrogens of the ammonia. The solution of solvated electrons is used in the reduction of alkynes to *trans*-alkenes (Sec. 14.6B). Sodium in liquid ammonia can be used to reduce other types of organic compounds as well, although these methods are not discussed explicitly in the text. Such reactions are called "dissolving metal reductions."

If a trace of Fe^{3+} ion is added to a solution of sodium in liquid ammonia, a *reaction* occurs between the sodium and the ammonia:

$$2\ Na\!\cdot\ +\ 2\ \ddot{N}H_3\ \xrightarrow{\ FeCl_3\ }\ 2\ Na^+\ {}^-\ddot{N}H_2\ +\ H_2$$

sodium amide
(sodamide)

Just as sodium reacts with water to give hydrogen gas and sodium hydroxide, sodium also reacts with ammonia to give hydrogen gas and sodium amide. The difference is that the reaction of sodium with water requires no catalyst, but the reaction with ammonia does. It is this solution—sodium amide in liquid ammonia—that is used in forming acetylenic anions from alkynes. Solid sodium amide can be purchased as a gray powder. Solutions of sodium amide can also be formed by adding this material to liquid ammonia or other solvents.

One other point: Do not confuse the basicity of amide ion with that of ammonia. Amide ion is the *conjugate base* of ammonia, which has a pK_a of about 35. Ammonia is also a base, but is the conjugate base of the *ammonium ion,* $^+NH_4$, which has a pK_a of 9.2. Most people normally think of ammonia as a base, but in the present context, ammonia is an *acid*, and the amide ion is its conjugate base. (See the discussion of amphoteric substances in Sec. 3.4B on text p. 99, and Problem 3.9(a) on the same page.)

 # REACTION REVIEW

I. ADDITION REACTIONS OF ALKYNES

A. ADDITION OF HYDROGEN HALIDES AND HALOGENS

1. Additions of halogens or hydrogen halides to the triple bond are somewhat slower in most cases than the same reactions of comparably substituted alkenes.
   a. The first addition to an alkyne gives a substituted alkene.
   b. The second addition of a hydrogen halide or a halogen to an alkyne is usually slower than the first because a halo alkene is less reactive than an alkyne.
   c. It is possible to isolate the substituted alkene resulting from addition of only one equivalent of halogen or hydrogen halide.

$-C\equiv C-\ \xrightarrow{\ Br_2\ }\ \overset{Br}{\underset{/}{\diagdown}}C\!=\!C\overset{/}{\underset{\diagdown}{}}\ \xrightarrow{\ Br_2\ }\ -\overset{\overset{Br}{|}}{\underset{\underset{Br}{|}}{C}}-\overset{\overset{Br}{|}}{\underset{\underset{Br}{|}}{C}}-$

2. HBr can be added to the triple bond in a regioselective addition that is analogous to the addition of HBr to alkenes.
   a. The bromine adds to the carbon of the triple bond that bears the alkyl substituent group.
   b. The regioselectivity is reversed in the presence of peroxides because free-radical intermediates are involved.

$$R-C\equiv C-H \xrightarrow{\text{HBr}} \begin{array}{c} \text{Br} \quad\quad \text{H} \\ \diagdown \diagup \\ C=C \\ \diagup \diagdown \\ R \text{H} \end{array}$$

$$R-C\equiv C-H \xrightarrow[\text{peroxides}]{\text{HBr}} \begin{array}{c} \text{H} \quad\quad \text{Br} \\ \diagdown \diagup \\ C=C \\ \diagup \diagdown \\ R \text{H} \end{array}$$

B. HYDRATION OF ALKYNES

1. The addition of water to the triple bond of an alkyne is catalyzed by strong acid or, more commonly, by a combination of dilute acid and mercuric ion, Hg^{2+}.

$$-C\equiv C- \;+\; H_2O \xrightarrow{H_2SO_4,\; HgSO_4} \begin{array}{c} \text{O} \\ \parallel \\ -CH_2-C- \end{array}$$

an alkyne   a ketone

a. The addition of water to a triple bond is called hydration.

b. Enols (alcohols containing an —OH group on a carbon of a double bond) are formed as intermediates in the hydration of alkynes.

    i. Most enols are unstable and are rapidly converted into the corresponding ketones or aldehydes.

    ii. Most aldehydes and ketones are in equilibrium with small amounts of the corresponding enols.

    iii. Because most enols are unstable, any synthesis designed to give an enol usually gives instead the corresponding aldehyde or ketone.

$$\begin{array}{c} \diagdown \text{OH} \\ C=C \\ \diagup \diagdown \end{array} \;\rightleftharpoons\; \begin{array}{c} \text{H} \text{O} \\ | \parallel \\ -C-C \\ | \diagdown \end{array}$$

an enol   an aldehyde
or ketone

2. The mechanism of alkyne hydration is very similar to that of acid-catalyzed hydration of alkenes.

a. In the first part the mechanism, the triple bond is protonated producing a vinylic cation. Then the —OH group of water adds to the carbon of the triple bond that bears the alkyl substituent.

b. The second part of the mechanism is the conversion of the enol into the ketone which is a rapid, acid-catalyzed process.

    i. Protonation of the double bond by aqueous acid gives another resonance-stabilized carbocation, which is also the conjugate acid of a ketone.

ii. In a Brønsted acid–base reaction, a β-proton is removed from the carbocation by a solvent water molecule to give the ketone product.

3. The hydration of alkynes is a useful way to prepare ketones provided that the starting material is a 1-alkyne or a symmetrical alkyne; most other alkynes give difficult-to-separate mixtures of isomers.

C. HYDROBORATION–OXIDATION OF ALKYNES

1. The hydroboration of alkynes is analogous to the same reaction of alkenes.
2. When alkynes react with diborane (B_2H_6), the elements of BH_3 are added to the triple bond.
   a. Oxidation of the organoborane with alkaline hydrogen peroxide yields the corresponding "alcohol," an enol.
   b. The enol reacts further to give the corresponding aldehyde or ketone.

3. The reaction conditions can be controlled so that only one addition takes place provided that the alkyne is not a 1-alkyne.
4. The hydroboration of 1-alkynes can be stopped after a single addition provided that organoborane containing highly branched groups is used instead of BH_3.
   a. One reagent developed for this purpose is called disiamylborane.
   b. Boron adds to the unbranched carbon atom of the triple bond, and hydrogen adds to the branched carbon.
   c. The disiamylborane molecule is so large and highly branched that only one equivalent can react with a 1-alkyne.

D. CATALYTIC HYDROGENATION OF ALKYNES

1. Alkynes undergo catalytic hydrogenation.
   a. Addition of one molar equivalent of H_2 to an alkyne gives an alkene.
   b. Addition of a second molar equivalent of H_2 gives an alkane.

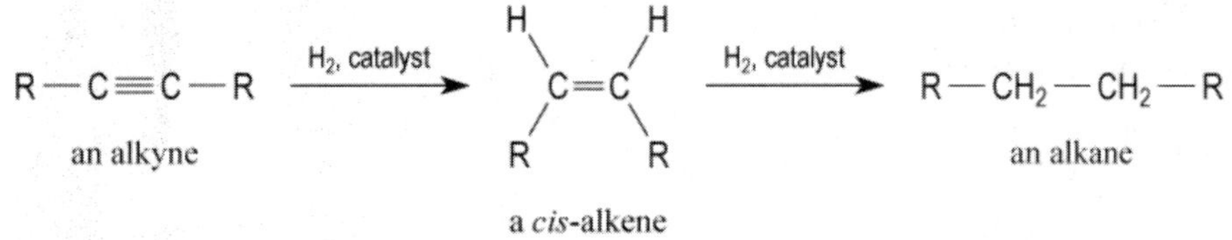

a *cis*-alkene

2. Hydrogenation of an alkyne may be stopped at the alkene stage if the reaction mixture contains a catalyst poison.

  a. A catalyst poison is a compound that disrupts the action of a catalyst. Some poisons can selectively block the hydrogenation of alkenes without preventing the hydrogenation of alkynes to alkenes.

  b. Useful catalyst poisons are salts of Pb^{2+} and certain nitrogen compounds such as pyridine, quinoline, or other amines.

  c. The Lindlar catalyst [Pd/CaCO$_3$ that has been washed with Pb(OAc)$_2$] is a commonly used poisoned catalyst.

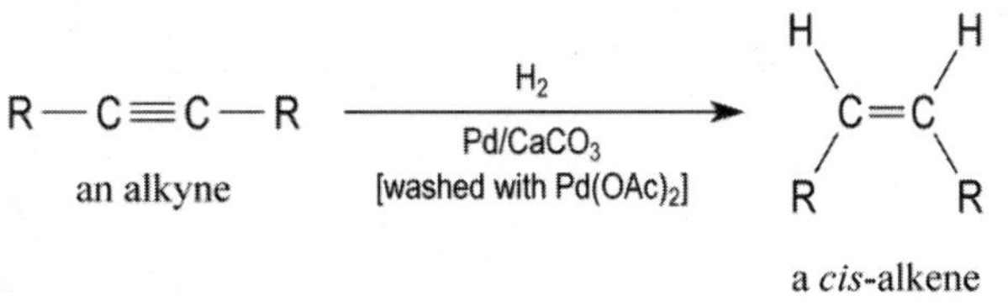

a *cis*-alkene

3. Hydrogenation of alkynes is a stereoselective syn-addition. When carried out with a poisoned catalyst, it is one of the best methods for preparing *cis*-alkenes.

E. REDUCTION OF ALKYNES WITH ALKALI METALS IN LIQUID AMMONIA

  1. Reaction of an alkyne with a solution of an alkali metal (usually sodium) in liquid ammonia gives an alkene. If the alkyne is not a 1-alkyne, the product is a *trans*-alkene. (1-Alkynes do not give alkenes; see Problem 14.40, text p. 710.)

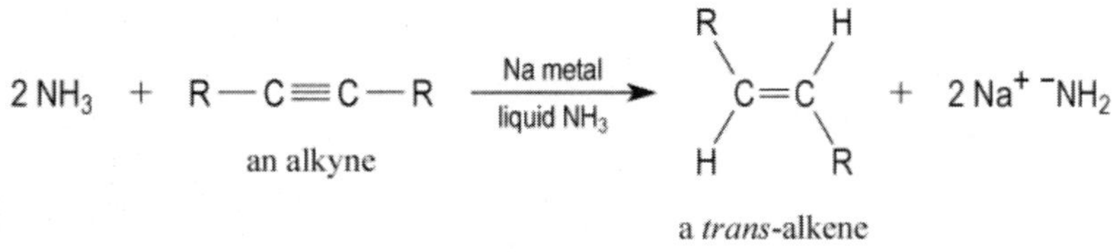

a *trans*-alkene

  2. The mechanism of this reaction involves a two-electron reduction:

  a. The deep blue solution of alkali metals dissolved in pure liquid ammonia is a source of electrons complexed to ammonia (solvated electrons).

solvated electron

  b. The addition of an electron to the triple bond results in a species (a radical anion) that has both an unpaired electron and a negative charge.

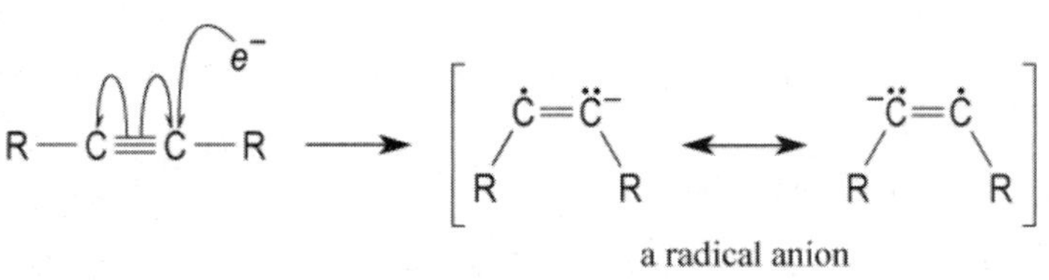

a radical anion

c. The radical anion is such a strong base that it removes a proton from ammonia to give a vinylic radical (a radical in which the unpaired electron is associated with one carbon of a double bond).

a radical anion   a vinylic radical

d. The resulting vinylic radical rapidly undergoes inversion.
   i. The equilibrium between the cis and trans radicals favors the trans radical for steric reasons.
   ii. Because there is much more of the trans radical, the ultimate product of the reaction is the *trans*-alkene.

a cis vinylic radical   a trans vinylic radical

e. The vinylic radical accepts an electron to form a vinylic anion.

a trans vinylic radical   a vinylic anion

f. This anion is more basic than the solvent and removes a proton from ammonia to complete the addition.

II. OTHER REACTIONS OF ALKYNES

A. FORMATION OF ACETYLENIC ANIONS

1. 1-Alkynes ("terminal acetylenes") are the most acidic of the common aliphatic hydrocarbons. (The pK_a of a 1-alkyne is about 25, whereas pK_a of an alkene is about 42, and the pK_a of an alkane is about 55.)

2. The conjugate bases of 1-alkynes, called acetylenic anions, are typically formed by the reaction of a 1-alkyne with a strong base such as sodium amide.

sodium amide
(sodamide)   an acetylenic anion

3. Closely related to acetylenic anions are acetylenic Grignard reagents, which are formed from more basic Grignard reagents by transmetallation.

$$R-C\equiv C-H \ + \ CH_3CH_2-MgBr \ \longrightarrow \ R-C\equiv C-MgBr \ + \ CH_3CH_3$$

an acetylenic
Grignard reagent

B. ACETYLENIC ANIONS AS NUCLEOPHILES

1. Acetylenic anions are strong bases; they are much stronger than hydroxide or alkoxides.

2. Acetylenic anions can be used as nucleophiles in S_N2 reactions to prepare other alkynes.

a. When an alkyl halide or sulfonate is used in this reaction, it must be primary or unbranched secondary.

$$R-C\equiv C{:}^- \ {}^+Na \ + \ I-CH_2CH_3 \ \longrightarrow \ R-C\equiv C-CH_2CH_3 \ + \ Na^+ \ {}^-I$$

b. This is another method of carbon–carbon bond formation.

3. Acetylenic anions and acetylenic Grignard reagents can also react with ethylene oxide to form acetylenic alcohols.

SOLUTIONS TO PROBLEMS

Solutions to In-Text Problems

14.1 (a) The *sp* hybridization of the carbon atoms of the triple bond requires that these two atoms and the two attached carbon atoms lie on a straight line. With a bond length of 1.2 Å for the triple bond and bond lengths of 1.47 Å for the attached carbon–carbon single bonds, the distance to be bridged by two remaining carbons of cyclohexyne is about 4.1 Å. This distance is too great to be connected by two carbon atoms at reasonable bond angles.

(b) Cyclodecyne is much more stable than cyclohexyne (and in fact can be isolated), because the distance of 4.1 Å referred to in part (a) can be bridged by six carbons with reasonable bond lengths and bond angles.

14.2

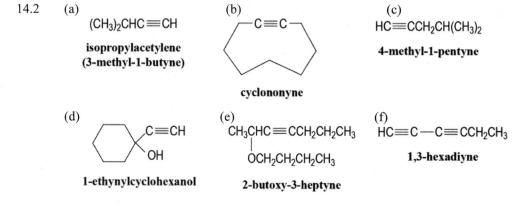

(a)
(CH₃)₂CHC≡CH

isopropylacetylene
(3-methyl-1-butyne)

(b)
C≡C

cyclononyne

(c)
HC≡CCH₂CH(CH₃)₂

4-methyl-1-pentyne

(d)
C≡CH
OH

1-ethynylcyclohexanol

(e)
CH₃CHC≡CCH₂CH₂CH₃
|
OCH₂CH₂CH₂CH₃

2-butoxy-3-heptyne

(f)
HC≡C—C≡CCH₂CH₃

1,3-hexadiyne

14.3 (a) 1-hexyne
(b) 5-decyne (common: dibutylacetylene)
(c) 2-methyl-3-pentyn-2-ol
(d) (*E*)-7-methoxy-3-propyl-5-hepten-1-yne
(e) 1-penten-4-yn-3-ol. The double bond receives numerical precedence over the triple bond when numbering is otherwise ambiguous.

14.4 A nine-proton singlet cries out, "*tert*-butyl group." The other resonance is in the position expected for an acetylenic proton. The 3300 cm⁻¹ absorption in the IR spectrum confirms the presence of an acetylenic C—H. *Tert*-butylacetylene (3,3-dimethyl-1-butyne, (CH₃)₃C—C'CH), which has a molecular mass of 82, is consistent with all the data.

14.5 (a) Because of its symmetry, 3-hexyne should have three chemically nonequivalent sets of carbons; in contrast, all six carbons of 2-hexyne are chemically nonequivalent. Thus, spectrum *A* is that of 2-hexyne, and spectrum *B* is that of 3-hexyne.

(b) Use the model compound 2-heptyne on text p. 687 to assist you.

δ 3.3   δ 79.1   δ 22.9   δ 14.6   δ 81.0   δ 12.7

H₃C—C≡C—CH₂—CH₂—CH₃   H₃C—CH₂—C≡C—CH₂—CH₃

    δ 75.4   δ 21.1   δ 13.6   δ 12.7   δ 81.0   δ 14.6

(Interchanging the δ 14.6 and δ 12.7 assignments in 3-hexyne is also a reasonable answer, although the assignments above are correct.)

14.6 The proton NMR spectrum of propyne should consist of two resonances. Evidently, the two happen to have the same chemical shift of δ 1.8, which is in fact a reasonable chemical shift for both acetylenic and propargylic protons.

14.7 Bromine addition to 3-hexyne could give two stereoisomers:

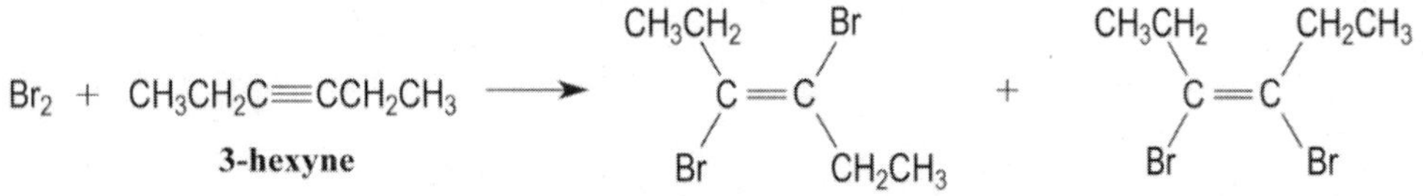

3-hexyne

(*E*)-3,4-dibromo-3-hexene (from *anti*-addition)

(*Z*)-3,4-dibromo-3-hexene (from *syn*-addition)

14.8 The product is (*Z*)-3-chloro-3-hexene. Notice that the product has *Z*-stereochemistry because of the relative group priorities, even though the addition is anti.

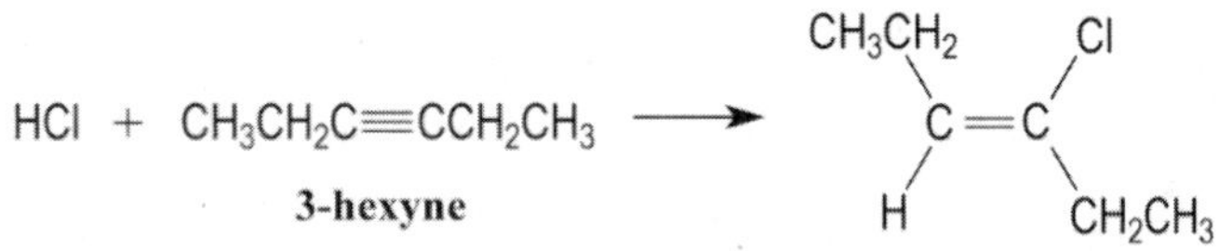

3-hexyne

(*Z*)-3-chloro-3-hexene (from *anti*-addition)

14.9 (a) HC≡$CCH_2CH_2CH_3$ **1-pentyne**

(b) $(CH_3)_3CC$≡CH **3,3-dimethyl-1-butyne**

(c) 5-Decyne would give the desired product. Notice that 4-decyne would give the desired product as a mixture with another product (what is it?).

$CH_3CH_2CH_2CH_2C$≡$CCH_2CH_2CH_2CH_3$

5-decyne

14.10 (a) The three-carbon alkyne propyne hydrates to give acetone, not propanal (the compound shown in the problem).

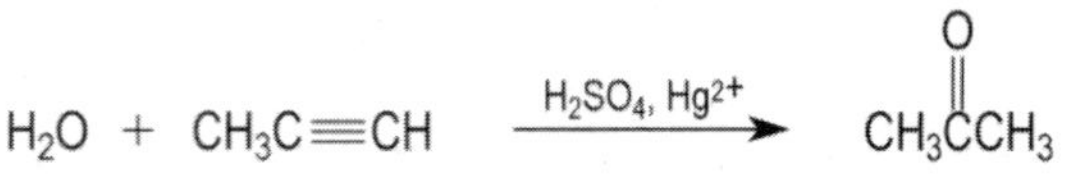

propyne **acetone**

(b) Hydration can only be used to prepare ketones that have at least two hydrogens on the carbon α (that is, adjacent) to the carbonyl carbon:

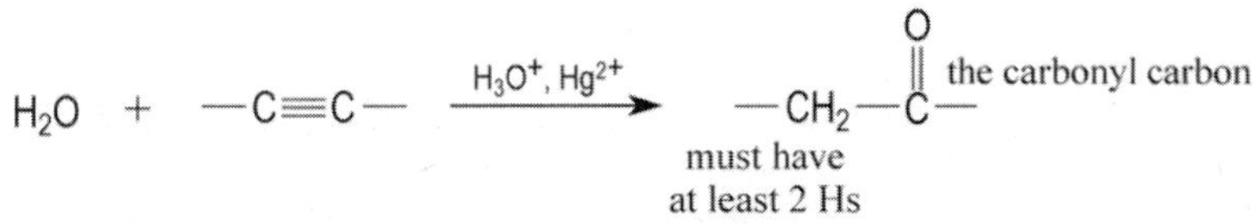

The ketone shown does not fulfill this requirement.

(c) The alkyne that would hydrate to this ketone (cyclohexanone) is cyclohexyne. Since cyclohexyne is too unstable to isolate, such a hydration reaction is impractical. (See the solution to Problem 14.1a for a discussion of the instability of cyclohexyne.)

14.11 Compounds *A* and *D* are enols. At equilibrium, they exist primarily as the ketones shown below.

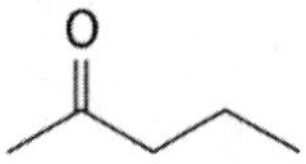

A

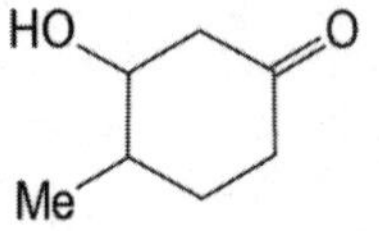

D

14.12 (a) Cyclohexylacetylene gives different products in the two reactions:

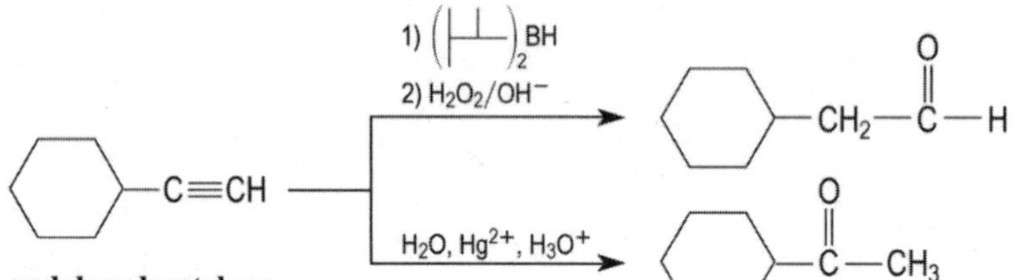

cyclohexylacetylene

(b) Because it is symmetrical, 2-butyne gives the same product in the two reactions:

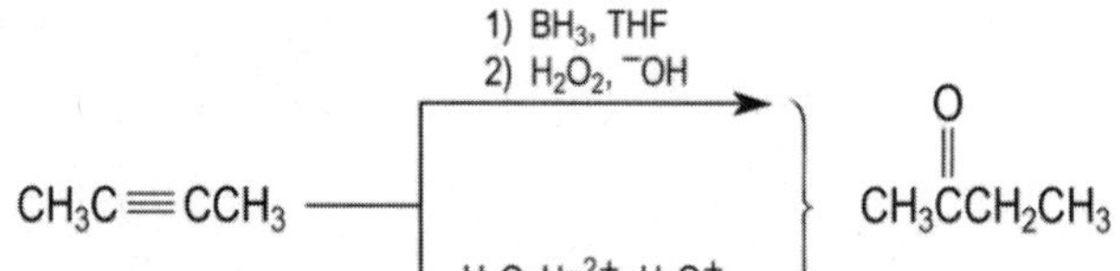

2-butyne

14.13 (a) The poisoned catalyst causes hydrogenation to stop at the alkene stage; the product is 1-octene, $CH_3(CH_2)_5CHACH_2$.

(b) The product is the alkane octane, $CH_3(CH_2)_6CH_3$.

(c) Because the molecule (as well as the product) contains its own catalyst poison, hydrogenation stops at the alkene. The product is

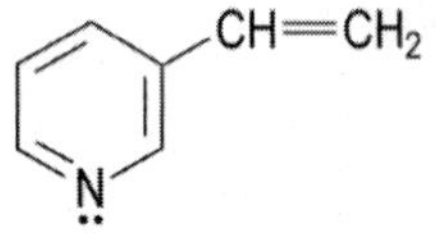

(d) The product is the alkene that results from syn-addition of D_2:

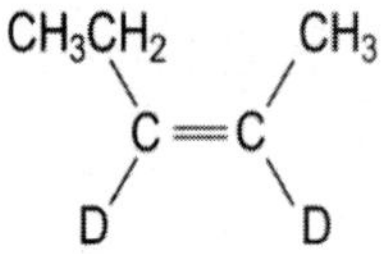

14.14 (a) The first reaction, an anti-addition, gives *trans*-3-hexene; and the second reaction, a syn-addition, gives (±)-hexane-3,4-d_2.

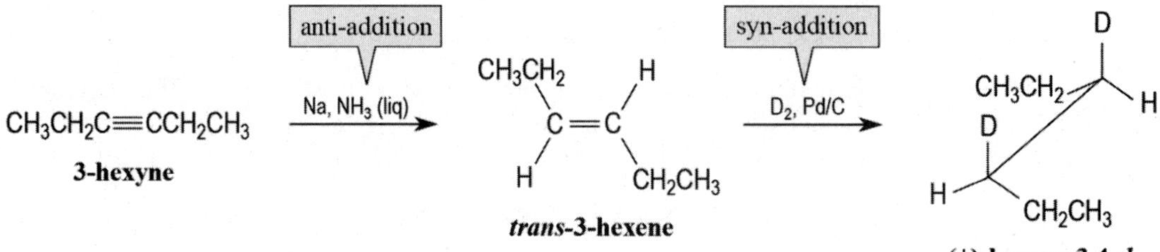

(b)   The first reaction is a syn-addition that gives *cis*-3-hexene; the second is another syn-addition that gives *meso*-hexane-3,4-d_2.

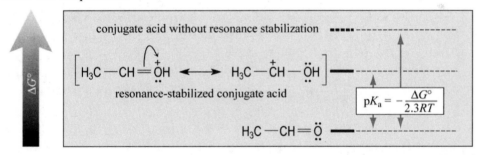

14.15   The conjugate acids are as follows:

$$H_3C-CH=\overset{+}{N}H_2 \qquad H_3C-C\equiv\overset{+}{N}H$$

$$AH^+ \qquad\qquad BH^+$$

The order of increasing acidity of the conjugate acids is $BH^+ > AH^+$. The order of increasing basicity of the compounds themselves is then $B < A$. This order follows from the principle developed in this section: the hydrogen with the greatest amount of *s* character in its C—H bond is most acidic. This problem demonstrates that it doesn't matter whether the atom to which the hydrogen is attached is charged or neutral, or whether it is a carbon or some other atom.

14.16   (a)   This is *not* the acidity order predicted by hybridization arguments. The proton in the O—H bond with more *s* character should be the most acidic if hybridization were the only effect.

(b)   The resonance structures of ion *B* and the energy diagram are shown below. The pK_a is negative because (in contrast to Fig. 3.3, text p. 116) the conjugate acid is at a higher free energy than the conjugate base. The free energy of the hypothetical conjugate acid without resonance stabilization is higher still. As this diagram shows, $-pK_a$ is smaller—that is, the pK_a is less negative—for the stabilized conjugate acid. In other words, resonance stabilization decreases its acidity. Evidently, its acidity is decreased so much that it is below that of species *A*.

(c)   The conjugate acids are stabilized by donation of hydrogen bonds to the solvent. Conjugate acid *A* can donate three hydrogen bonds, whereas conjugate acid *B* can only donate one. Such hydrogen bonding affords additional stabilization to acid *A*.

14.17   (a)   The pK_a of ammonia is about 35 (which indicates the basicity of the amide ion), and that of an alkyne acetylenic hydrogen is about 25 (Eq. 14.20, text p. 699). Use the method described in Eq. 3.26b, text p. 104, to find that the K_{eq} for Eq. 14.22 is about 10^{10}. With the pK_a of an alkane taken as 55, the K_{eq} for the reaction of the amide base with an alkane is about 10^{-20}.

(b)   As the calculation in part (a) shows, the reaction of sodium amide with an alkane is quite unfavorable. For that reason, this base cannot be used to form the conjugate-base anions of alkanes.

14.18   (a)   This is an S_N2 reaction of the anion with ethyl iodide to give 2-pentyne, $CH_3C'CCH_2CH_3$, plus sodium iodine, $Na^+\ I^-$.

(b)   The acetylenic anion reacts in an S_N2 reaction with butyl tosylate:

$$CH_3CH_2CH_2CH_2—OTs + Na^+ \;^-:C≡C—Ph \longrightarrow CH_3CH_2CH_2CH_2—C≡C—Ph + Na^+ \;^-OTs$$

butyl tosylate

(c)   The acetylenic Grignard reagent, like any other Grignard reagent (text Sec. 11.5C), opens ethylene oxide to give an alcohol:

$$CH_3C≡C—MgBr + \triangle\!\!\!\!O \longrightarrow CH_3C≡C—CH_2CH_2OMgBr \xrightarrow{H_3O^+} CH_3C≡C—CH_2CH_2OH$$

ethylene oxide

(d)   Sodium acetylide reacts with both alkyl halide groups to give a 1,8-nonadiyne, HC'C(CH$_2$)$_5$C'CH, plus two equivalents of sodium bromide, Na$^+$ Br$^-$.

14.19   Choke forgot the lessons of Sec. 9.5G: tertiary alkyl halides react with strong bases to give elimination products, not substitution products. Acetylenic anions are strong bases; thus, Choke's products were 2-methylpropene and propyne.

$$(CH_3)_3CBr + Na^+ \;^-:C≡CCH_3 \xrightarrow{\text{an E2 reaction}} (CH_3)_2C=CH_2 + HC≡CCH_3$$

tert-butyl bromide   **2-methylpropene**   **propyne**

14.20   The preparation involves the reaction of 3,3-dimethyl-1-butyne with sodium amide, then with methyl iodide.

$$(CH_3)_3CC≡CH \xrightarrow[\text{NH}_3\,(\text{liq})]{\text{NaNH}_2} (CH_3)_3CC≡C^{:-}\;Na^+ \xrightarrow{CH_3I} (CH_3)_3CC≡CCH_3$$

3,3-dimethyl-1-butyne   **4,4-dimethyl-2-pentyne**

14.21   Either alkyl group can be introduced in an S_N2 reaction of an acetylenic anion with an alkyl halide. The two preparations of 2-pentyne:

Method 1:

$$HC≡CCH_2CH_3 \xrightarrow[\text{NH}_3\,(\text{liq})]{\text{NaNH}_2} Na^+ \;^-:\!\bar{C}≡CCH_2CH_3 \xrightarrow[\text{iodomethane}]{H_3C—I} CH_3C≡CCH_2CH_3$$

1-butyne

Method 2:

$$CH_3C≡CH \xrightarrow[\text{NH}_3\,(\text{liq})]{\text{NaNH}_2} CH_3C≡C^{:-}\;Na^+ \xrightarrow[\text{iodoethane}]{CH_3CH_2—I} CH_3C≡CCH_2CH_3$$

propyne

14.22   As in the solution to Problem 14.21, either alkyl group can be introduced by the reaction with an alkyl halide. The alternative to Eq. 14.29 is to use the acetylenic anion derived from propyne followed by alkylation with 1-bromobutane.

$$CH_3C≡CH \xrightarrow[\text{NH}_3\,(\text{liq})]{\text{NaNH}_2} CH_3C≡C^{:-}\;Na^+ \xrightarrow[\text{1-bromobutane}]{CH_3CH_2CH_2CH_2—Br} CH_3C≡CCH_2CH_2CH_2CH_3$$

propyne   **2-heptyne**

However, the reaction shown in Eq. 14.29 is probably superior because CH$_3$Br is the more reactive alkyl halide.

14.23   (a)   A synthesis of 1-nonanol from compounds with five or fewer carbons:

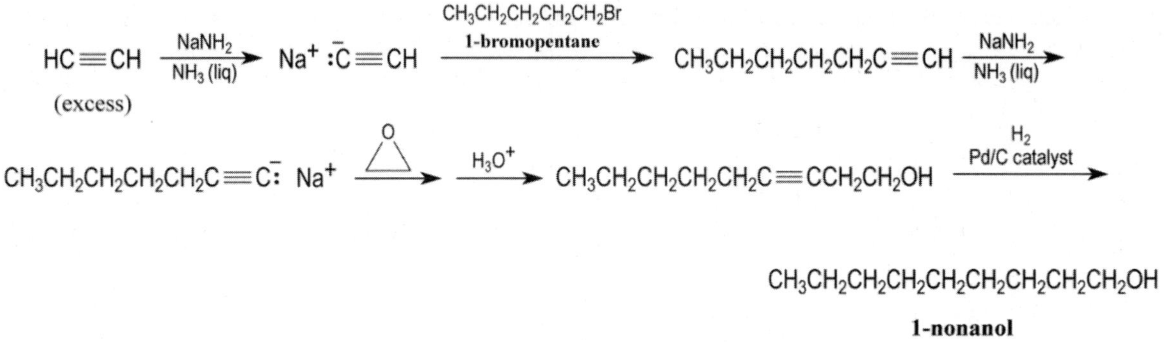

$HC≡CH$ $\xrightarrow[NH_3\,(liq)]{NaNH_2}$ $Na^+ :\overset{..}{C}≡CH$ $\xrightarrow[\text{1-bromopentane}]{CH_3CH_2CH_2CH_2CH_2Br}$ $CH_3CH_2CH_2CH_2CH_2C≡CH$ $\xrightarrow[NH_3\,(liq)]{NaNH_2}$

(excess)

$CH_3CH_2CH_2CH_2CH_2C≡\overset{..}{C}{:}^-\ Na^+$ $\xrightarrow{\triangle}$ $\xrightarrow{H_3O^+}$ $CH_3CH_2CH_2CH_2CH_2C≡CCH_2CH_2OH$ $\xrightarrow[\text{Pd/C catalyst}]{H_2}$

$CH_3CH_2CH_2CH_2CH_2CH_2CH_2CH_2CH_2OH$

1-nonanol

(b) A synthesis of 2-undecanone from compounds containing five or fewer carbons:

$CH_3(CH_2)_7CH_2OH$ $\xrightarrow{HBr}$ $CH_3(CH_2)_7CH_2Br$ $\xrightarrow[\text{[prepared in part (a)]}]{Na^+ :\overset{..}{C}≡CH}$ $CH_3(CH_2)_7CH_2—C≡CH$ $\xrightarrow[H_3O^+,\ Hg^{2+}]{H_2O}$

1-nonanol
[prepared in part (a)]

$CH_3(CH_2)_7CH_2—\overset{\overset{O}{\|}}{C}—CH_3$

2-undecanone

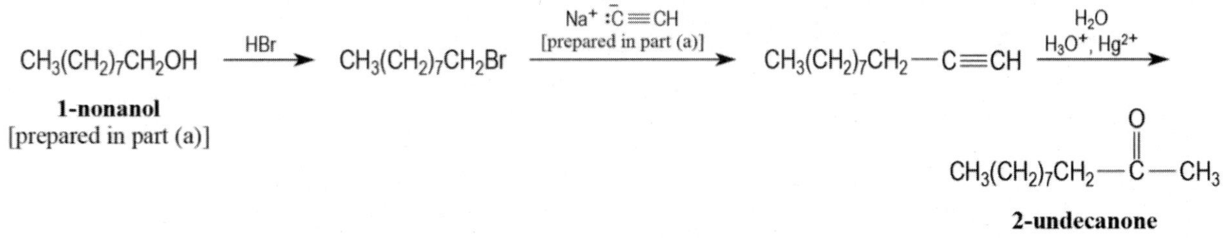

(c) A synthesis of *trans*-2-heptene from compounds containing five or fewer carbons:

$CH_3CH_2CH_2CH_2Br$ $\xrightarrow[\text{[prepared in part (a)]}]{Na^+ :\overset{..}{C}≡CH}$ $CH_3CH_2CH_2CH_2—C≡CH$ $\xrightarrow[NH_3\,(liq)]{NaNH_2}$ $CH_3CH_2CH_2CH_2C≡\overset{..}{C}{:}^-\ Na^+$ $\xrightarrow{CH_3Br}$

1-bromobutane

$CH_3CH_2CH_2CH_2C≡CCH_3$ $\xrightarrow[NH_3\,(liq.)]{Na}$

trans-2-heptene

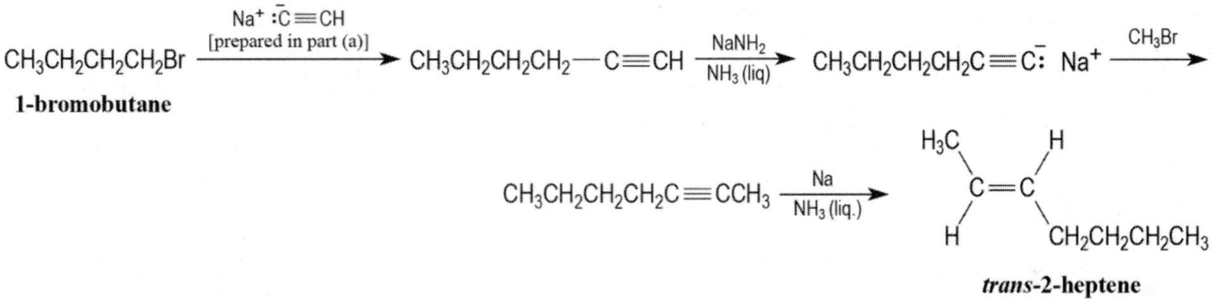

14.24 (a) When a *cis*-alkene is called for, catalytic hydrogenation of an alkyne is the preferred route. The acetylenic anion used in the first step is prepared by treating 1-butyne, $CH_3CH_2C'CH$, with $NaNH_2$ in liquid ammonia.

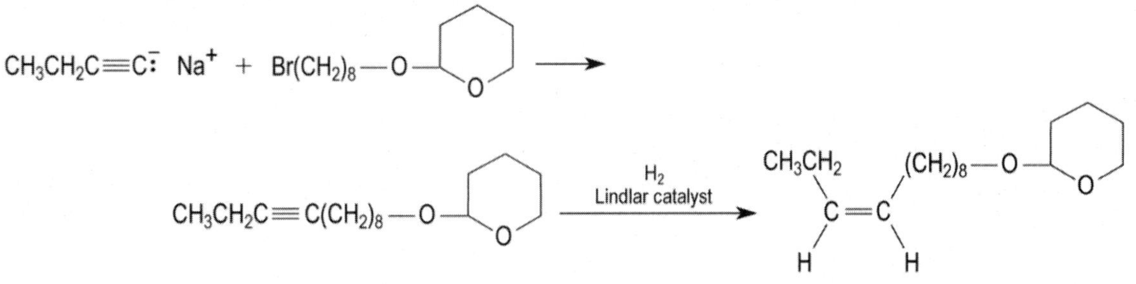

$CH_3CH_2C≡\overset{..}{C}{:}^-\ Na^+$ + $Br(CH_2)_8—O$ $\longrightarrow$

$CH_3CH_2C≡C(CH_2)_8—O$ $\xrightarrow[\text{Lindlar catalyst}]{H_2}$

(b) Because the required acetylene is already in hand from the part (a), treating it with Na in liquid ammonia will achieve the desired result:

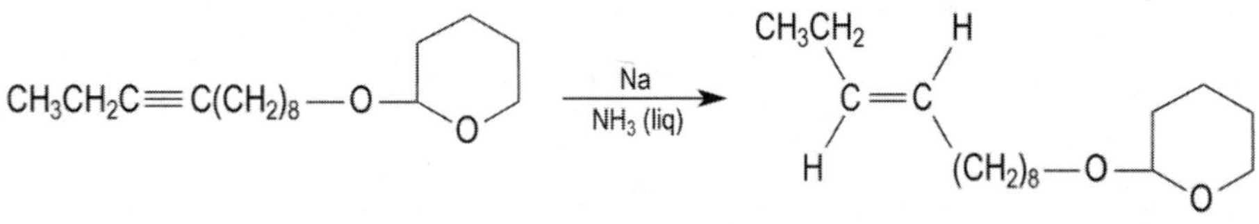

$CH_3CH_2C≡C(CH_2)_8—O$ $\xrightarrow[NH_3\,(liq)]{Na}$

14.25 Catalytic hydrogenation with a Lindlar catalyst should bring about hydrogenation of the alkyne to a cis-alkene group without affecting the existing alkene.

Solutions to Additional Problems

14.26 (a)

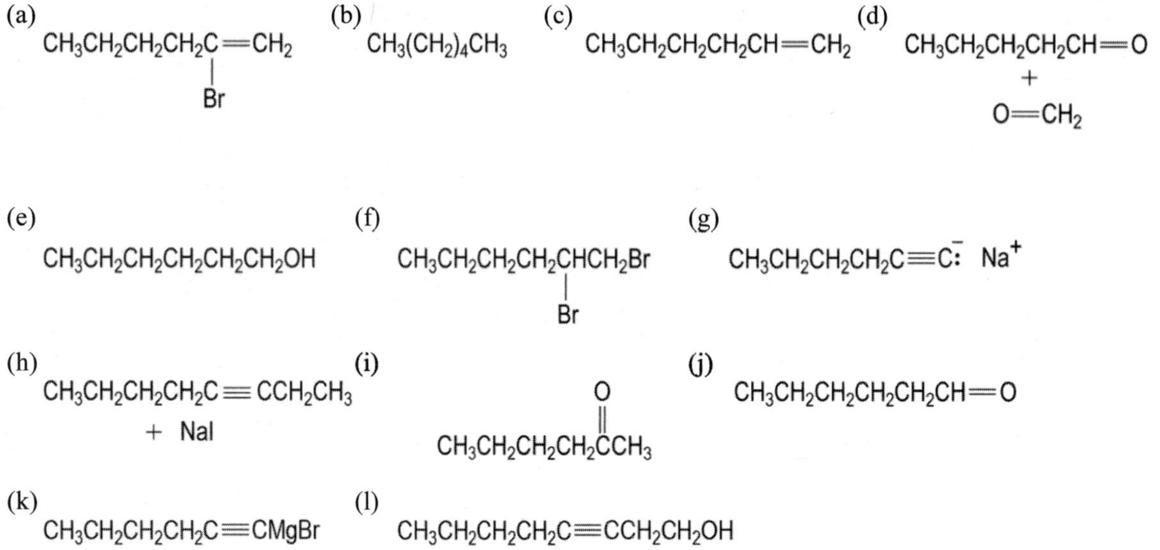

$$CH_3CH_2CH_2CH_2C{=}CH_2$$
$$\underset{Br}{|}$$

(b) $CH_3(CH_2)_4CH_3$

(c) $CH_3CH_2CH_2CH_2CH{=}CH_2$

(d) $CH_3CH_2CH_2CH_2CH{=}O$
$+$
$O{=}CH_2$

(e) $CH_3CH_2CH_2CH_2CH_2CH_2OH$

(f) $CH_3CH_2CH_2CH_2\underset{Br}{\underset{|}{C}}HCH_2Br$

(g) $CH_3CH_2CH_2CH_2C{\equiv}\overset{..}{C}{:}^-\ Na^+$

(h) $CH_3CH_2CH_2CH_2C{\equiv}CCH_2CH_3$
$+\ NaI$

(i)
$$CH_3CH_2CH_2CH_2\overset{\overset{O}{\|}}{C}CH_3$$

(j) $CH_3CH_2CH_2CH_2CH_2CH{=}O$

(k)
$CH_3CH_2CH_2CH_2C{\equiv}CMgBr$

(l) $CH_3CH_2CH_2CH_2C{\equiv}CCH_2CH_2OH$

14.27 (a) $CH_3CH_2CH_2CH_2CH_2CH_2CH_2CH_3$

(b)
$$\underset{\underset{H}{|}}{CH_3CH_2CH_2}\ \ \ \ \underset{\underset{H}{|}}{CH_2CH_2CH_3}$$
$$C{=}C$$

(c)
$$CH_3CH_2CH_2\overset{\overset{O}{\|}}{C}OH$$
(2 equivalents)

(d)
$$\underset{\underset{H}{|}}{CH_3CH_2CH_2}\ \ \ \ \underset{\underset{CH_2CH_2CH_3}{|}}{H}$$
$$C{=}C$$

(e)
$$CH_3CH_2CH_2\overset{\overset{O}{\|}}{C}CH_2CH_2CH_2CH_3$$

(f)
$$CH_3CH_2CH_2\overset{\overset{O}{\|}}{C}CH_2CH_2CH_2CH_3$$

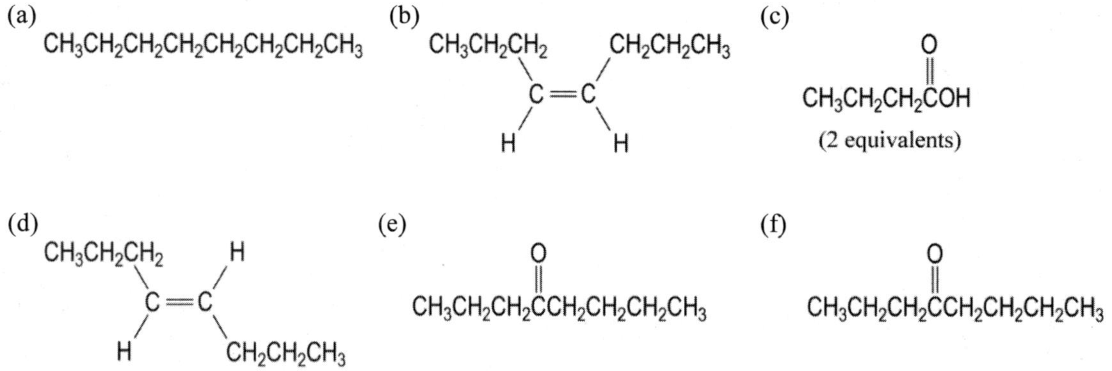

14.28 (a) 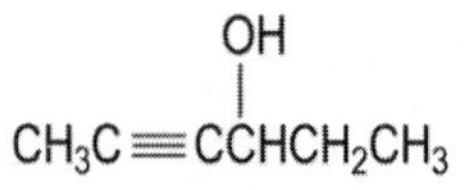
$$CH_3C{\equiv}C\underset{\underset{OH}{|}}{C}HCH_2CH_3$$
"2-hexyn-4-ol"
4-hexyn-3-ol

(c) 
$HC{\equiv}CH{-}CH{=}CH_2$
"1-butyn-3-ene"
1-buten-3-yne

(b) 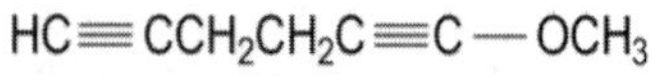
$HC{\equiv}CCH_2CH_2C{\equiv}C{-}OCH_3$
"6-methoxy-1,5-hexadiyne"
1-methoxy-1,5-hexadiyne

(d) 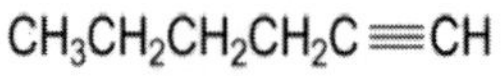
$CH_3CH_2CH_2CH_2C{\equiv}CH$
"5-hexyne"
1-hexyne

If you find an error in this manual, please visit http://people.pharmacy.purdue.edu/~loudonm/teaching/

14.29   (a)

—C≡CH

cyclopropylacetylene

(c)

$CH_3CH_2C≡CCH_2CH_3$

3-hexyne

(d)

$CH_3CH=CHCH_2C≡CH$ and

cis and trans

H_3C —C≡CH

cis and trans

(b)   Either the cis or the trans isomer of the cyclopropane derivative in part (d) is a satisfactory answer. Another is

$$CH_3CH_2CHC≡CH$$
with CH_3 substituent

14.30   (a)   Bonds with greater *s* character are shorter. (See text Sec.14.1.) Therefore, the order of increasing C—H bond length is

acetylene < ethylene < ethane

(b)   Hybridization affects the lengths of C—C bonds in the same way that it affects the lengths of C—H bonds. Therefore, the lengths of C—C bonds increase in the following order:

propyne < propene < propane

14.31   (a)   The order of increasing basicity is $F^- < CH_3CH_2O^- < HC'C^-$. The reason is the element effect operating across a row of the periodic table: the more electronegative the atom, the less basic it is. (Note that the element effect on acidity is more important than the effect of hybridization.)

(b)   Because alkynes are more acidic than alkenes, which are more acidic than alkanes, the acetylenic anion is less basic than the vinylic anion, which is less basic than the alkyl anion. Thus, the basicity order is

$$CH_3(CH_2)_3C≡C\overset{-}{:} \ < \ CH_3(CH_2)_3CH=\overset{..}{\underset{}{C}}H^- \ < \ CH_3(CH_2)_3\overset{..}{C}H_2^-$$

14.32   (a)   The alkyne should give off a gas (methane or ethane, respectively) when treated with CH_3MgBr or C_2H_5MgBr. (See Eqs. 14.23–14.24 on text p. 699.) The alkene is not acidic enough to react in this way.

(b)   Only the 1-alkyne should react with C_2H_5MgBr to release a gas (ethane).

(c)   The alkyne should react rapidly with Br_2 in an inert solvent, and should therefore rapidly decolorize such a bromine solution; the alkane will not react.

(d)   Forget about chemical tests; propyne is a gas and 1-decyne is a liquid at room temperature.

⚠ How would you know this? You know that *propane* is a gas, right? (It is used instead of natural gas in rural areas for heating or in barbecue grills as a fuel for cooking.) You also know that the presence of a double or triple bond has little effect on the physical properties of hydrocarbons. Thus, propyne is a gas also.

14.33   In some syntheses, more than one acceptable route is possible.

(a)

$HC≡CH$ $\xrightarrow[\text{liq. NH}_3]{\text{NaNH}_2}$ $Na^+ \ \overset{-}{:}C≡CH$ $\xrightarrow{\text{CH}_3\text{CH}_2\text{Br}}$ $CH_3CH_2C≡CH$ $\xrightarrow[\text{liq. NH}_3]{\text{NaNH}_2}$ $CH_3CH_2C≡C:^- \ Na^+$ $\xrightarrow{\text{BrCH}_2\text{CH}_3}$
(excess)

$CH_3CH_2C≡CCH_2CH_3$ $\xrightarrow[\text{Pd/C catalyst}]{D_2}$ $CH_3CH_2CD_2CD_2CH_2CH_3$

3-hexyne

(b)

$$HC \equiv CH \xrightarrow[\text{liq. NH}_3]{\text{NaNH}_2} HC \equiv C:^- \, Na^+ \xrightarrow{CH_3(CH_2)_3Br} CH_3(CH_2)_3C \equiv CH \xrightarrow[\text{Lindlar catalyst}]{H_2} CH_3(CH_2)_3CH = CH_2$$

(excess)   **1-hexyne**   **1-hexene**

(c)

$$CH_3CH_2C \equiv CCH_2CH_3 \xrightarrow[\text{Lindlar catalyst}]{H_2}$$

(with alkene intermediate, CH_3CH_2 and CH_2CH_3 on C=C with H, H)

$$\xrightarrow[\text{2) } H_2O_2, \, ^-OH]{\text{1) BH}_3, \text{ THF}} CH_3CH_2CHCH_2CH_2CH_3 \text{ (OH)}$$

3-hexyne
from part (a)   **3-hexanol**

Oxymercuration–reduction would work equally well in the last step because the alkene is symmetrical and reaction regioselectivity is not an issue.

(d)   1-Hexyne was prepared in part (b).

(e)

$$Na^+ \, :C \equiv CH \xrightarrow{CH_3CH_2CH_2CH_2CH_2Br} CH_3CH_2CH_2CH_2CH_2C \equiv CH \xrightarrow[\text{liq. NH}_3]{\text{NaNH}_2} \xrightarrow[\text{2) } H_3O^+]{\text{1) epoxide}}$$

$$CH_3CH_2CH_2CH_2CH_2C \equiv C-CH_2CH_2-OH \xrightarrow[\text{Pd/C catalyst}]{H_2} CH_3(CH_2)_7CH_2OH \xrightarrow[\text{2) } H_3O^+]{\text{1) KMnO}_4/\text{NaOH}} CH_3(CH_2)_7CO_2H$$

(f)

$$Na^+ \, :C \equiv CH \xrightarrow{(CH_3)_2CH(CH_2)_2Br} (CH_3)_2CH(CH_2)_2C \equiv CH \xrightarrow[\text{2) } H_2O_2, \, ^-OH]{\text{1) } [(CH_3)_2CHCH]_2BH \; (CH_3)} (CH_3)_2CH(CH_2)_3CH = O$$

from part (a)

(g)

$$CH_3CH_2C \equiv C:^- \, Na^+ \xrightarrow{CH_3Br} CH_3CH_2C \equiv CCH_3 \xrightarrow[\text{Lindlar catalyst}]{H_2}$$

from part (a)

(with product alkene CH_3CH_2 and CH_3 on C=C with H, H)

cis-2-pentene

(h)

$$Na^+ \, :C \equiv CH \xrightarrow{CH_3(CH_2)_3Br} CH_3(CH_2)_3C \equiv CH \xrightarrow[\text{liq. NH}_3]{\text{NaNH}_2} CH_3(CH_2)_3C \equiv C:^- \, Na^+ \xrightarrow{Br(CH_2)_3CH_3}$$

from part (f)

$$CH_3(CH_2)_3C \equiv C(CH_2)_3CH_3 \xrightarrow[\text{NH}_3 \text{ (liq.)}]{\text{Na metal}}$$

(with product alkene $CH_3(CH_2)_3$ and H, H and $(CH_2)_3CH_3$ on C=C)

trans-3-decene

(i)

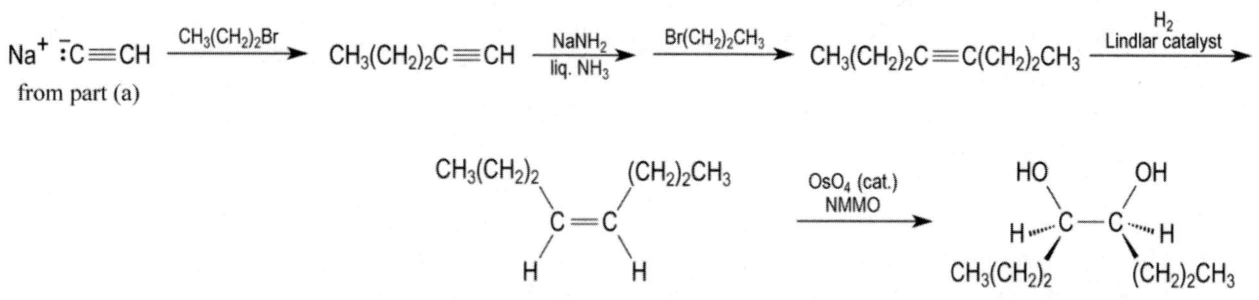

Na$^+$ $\bar{:}$C≡CH $\xrightarrow{\text{CH}_3(\text{CH}_2)_2\text{Br}}$ CH$_3$(CH$_2$)$_2$C≡CH $\xrightarrow[\text{liq. NH}_3]{\text{NaNH}_2}$ $\xrightarrow{\text{Br(CH}_2)_2\text{CH}_3}$ CH$_3$(CH$_2$)$_2$C≡C(CH$_2$)$_2$CH$_3$ $\xrightarrow[\text{Lindlar catalyst}]{\text{H}_2}$

from part (a)

CH$_3$(CH$_2$)$_2$ ⟍ ⟋ (CH$_2$)$_2$CH$_3$
          C=C
       H ⟋   ⟍ H

$\xrightarrow[\text{NMMO}]{\text{OsO}_4 \text{ (cat.)}}$

HO   OH
  \   /
H····C —— C····H
  /   \
CH$_3$(CH$_2$)$_2$ (CH$_2$)$_2$CH$_3$

meso-4,5-octanediol

An alternate route would be to form the *trans*-alkene from the alkyne by reduction with Na in liquid ammonia; then form the epoxide from the alkene using *meta*-chloroperbenzoic acid (text p. 488); and then allow the epoxide to react with water under acidic conditions to give the *meso*-diol.

(j)

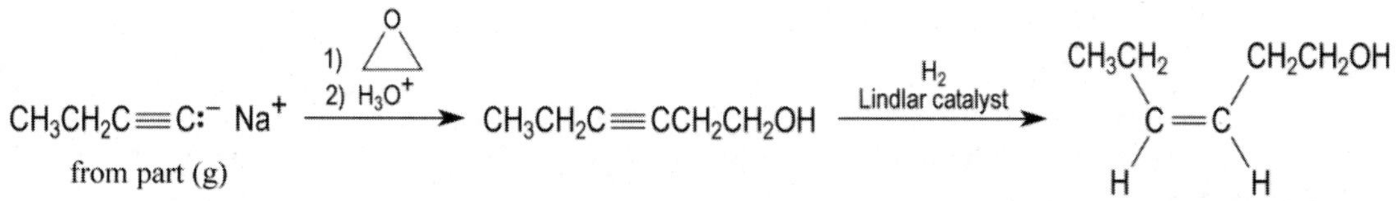

CH$_3$CH$_2$C≡C:$^-$ Na$^+$ $\xrightarrow[\text{2) H}_3\text{O}^+]{\text{1) } \triangle\text{O}}$ CH$_3$CH$_2$C≡CCH$_2$CH$_2$OH $\xrightarrow[\text{Lindlar catalyst}]{\text{H}_2}$

from part (g)

CH$_3$CH$_2$ ⟍ ⟋ CH$_2$CH$_2$OH
          C=C
       H ⟋   ⟍ H

(Z)-3-hexen-1-ol

14.34 (a)

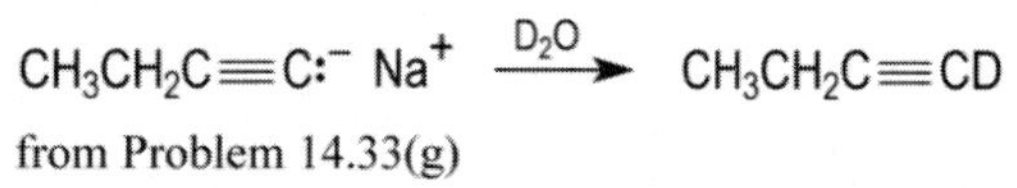

CH$_3$CH$_2$C≡C:$^-$ Na$^+$ $\xrightarrow{\text{D}_2\text{O}}$ CH$_3$CH$_2$C≡CD

from Problem 14.33(g)

(b)

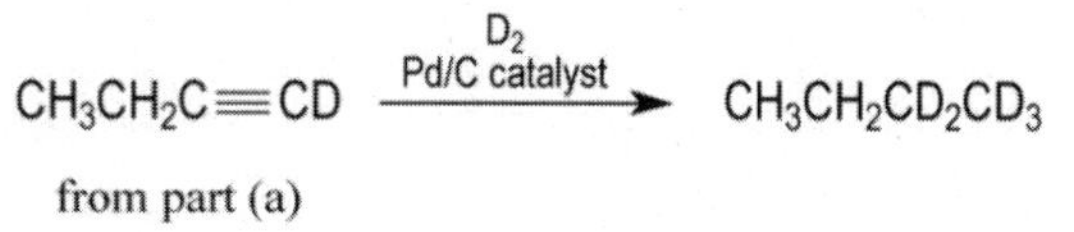

CH$_3$CH$_2$C≡CD $\xrightarrow[\text{Pd/C catalyst}]{\text{D}_2}$ CH$_3$CH$_2$CD$_2$CD$_3$

from part (a)

(c)

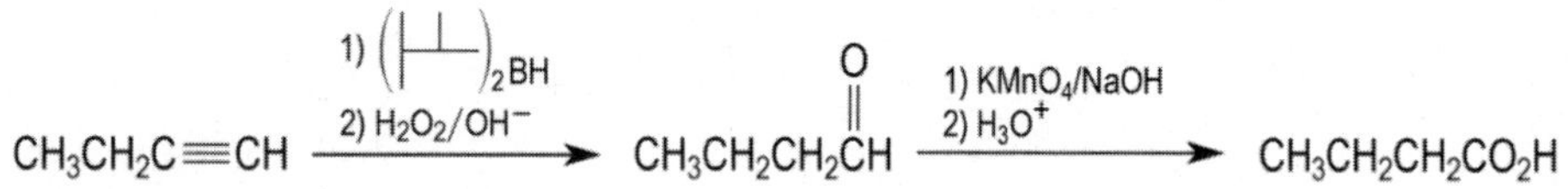

CH$_3$CH$_2$C≡CH $\xrightarrow[\text{2) H}_2\text{O}_2/\text{OH}^-]{\text{1) }(\text{⊢⊣})_2\text{BH}}$ CH$_3$CH$_2$CH$_2$CH(=O) $\xrightarrow[\text{2) H}_3\text{O}^+]{\text{1) KMnO}_4/\text{NaOH}}$ CH$_3$CH$_2$CH$_2$CO$_2$H

An alternate, albeit somewhat longer, route:

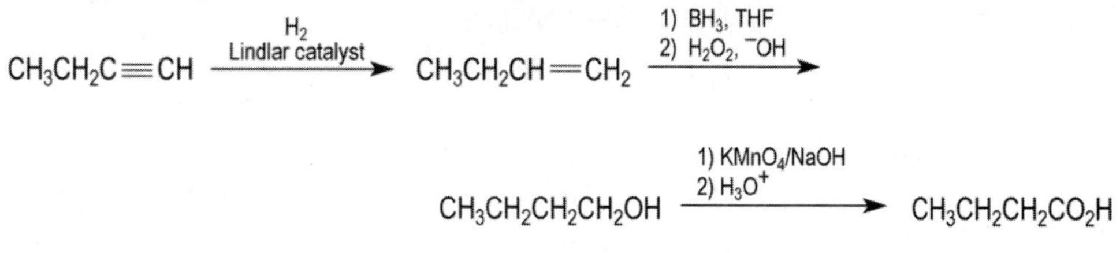

CH$_3$CH$_2$C≡CH $\xrightarrow[\text{Lindlar catalyst}]{\text{H}_2}$ CH$_3$CH$_2$CH=CH$_2$ $\xrightarrow[\text{2) H}_2\text{O}_2, \, ^-\text{OH}]{\text{1) BH}_3, \text{ THF}}$

CH$_3$CH$_2$CH$_2$CH$_2$OH $\xrightarrow[\text{2) H}_3\text{O}^+]{\text{1) KMnO}_4/\text{NaOH}}$ CH$_3$CH$_2$CH$_2$CO$_2$H

(d)

$$CH_3CH_2CH_2CH_2OH \xrightarrow{\text{HBr, H}_2\text{SO}_4} CH_3CH_2CH_2CH_2Br$$

from part (c)

$$CH_3CH_2CH_2CH_2OH \xrightarrow{\text{NaH}} CH_3CH_2CH_2CH_2O^- \ Na^+$$

$$\longrightarrow CH_3CH_2CH_2CH_2OCH_2CH_2CH_2CH_3$$

1-butoxybutane

(e)

$$CH_3CH_2C{\equiv}C{:}^- \ Na^+ \xrightarrow{\text{CH}_3\text{CH}_2\text{CH}_2\text{CH}_2\text{Br}} CH_3CH_2C{\equiv}CCH_2CH_2CH_2CH_3 \xrightarrow[\text{Lindlar catalyst}]{\text{H}_2}$$

from Problem 14.34(a)   **3-octyne**

a structure showing:

$$\underset{H}{\overset{CH_3CH_2}{\diagdown}} C = C \underset{H}{\overset{CH_2CH_2CH_2CH_3}{\diagup}} \xrightarrow[\text{Pd/C catalyst}]{\text{D}_2}$$

$$\overset{D \quad\quad D}{H\cdots\overset{|}{\underset{CH_3CH_2}{C}} - \overset{|}{\underset{CH_2CH_2CH_2CH_3}{C}}\cdots H}$$

(3R,4S)-3,4-dideuteriooctane
+ an equal amount of its enantiomer

The addition of D_2 and the addition of H_2 could be interchanged (with the catalysts remaining the same—that is, the Lindlar catalyst must be used first).

 In assigning configurations, notice that a CH_3CH_2CHD- group receives priority over a $-CH_2CH_2CH_2CH_3$ group.

(f)

$$CH_3CH_2C{\equiv}CCH_2CH_2CH_2CH_3 \xrightarrow[\text{Pd/C catalyst}]{\text{H}_2} CH_3CH_2CH_2CH_2CH_2CH_2CH_2CH_3$$

3-octyne   **octane**
from Problem 14.34(e)

(g)

$$CH_3CH_2C{\equiv}CH \xrightarrow{\text{H}_2\text{O, H}_3\text{O}^+, \text{Hg}^{2+}} CH_3CH_2\overset{O}{\overset{\|}{C}}CH_3$$

14.35 (a) Two constitutionally isomeric enols are possible, one of which can exist as *E* and *Z* stereoisomers.

$$\underset{H_3C}{\overset{H}{\diagdown}} C = C \underset{CH(CH_3)_2}{\overset{OH}{\diagup}} \qquad \underset{H_3C}{\overset{H}{\diagdown}} C = C \underset{OH}{\overset{CH(CH_3)_2}{\diagup}} \qquad \underset{CH_3CH_2}{\overset{HO}{\diagdown}} C = C \underset{CH_3}{\overset{CH_3}{\diagup}}$$

$$\underbrace{\hphantom{CH_3C = C CH_3CH_2 CH_2CH_3}}_{E \text{ and } Z \text{ stereoisomers}}$$

(b) Hydration would not be a good preparative method for this compound. The only alkyne that might give this compound as a product is 4-methyl-2-pentyne; but, because both carbons of its triple bond have one alkyl substituent, that alkyne would give a mixture of two constitutionally isomeric products of which the desired compound is only one component. These have very similar physical properties and are very difficult to separate.

$$H_2O \ + \ CH_3C \equiv CCH(CH_3)_2 \xrightarrow{H_3O^+, \ Hg^{2+}} CH_3CH_2 \overset{\overset{\displaystyle O}{\|}}{-C-} CH(CH_3)_2 \ + \ H_3C \overset{\overset{\displaystyle O}{\|}}{-C-} CH_2CH(CH_3)_2$$

4-methyl-2-pentyne

14.36   First look for the telltale C—H and C'C stretching absorptions; these are present in the spectrum of compound *B* near 3300 cm^{-1} and 2100 cm^{-1}. Hence, compound *B* is 1-hexyne. Compound *A* has no trace of C'C absorption and, more importantly, it has a 1640 cm^{-1} CAC stretching absorption as well as 910 cm^{-1} and 990 cm^{-1} C—H bending absorptions, all of which are consistent with a —CHACH$_2$ group. Therefore, compound *A* is 3-methyl-1,4-pentadiene. By elimination, compound *C* must be 2-hexyne.

$$\underset{\textbf{3-methyl-1,4-pentadiene}}{\overset{\overset{\displaystyle CH_3}{|}}{H_2C=CHCHCH=CH_2}} \qquad \underset{\textbf{1-hexyne}}{CH_3CH_2CH_2CH_2C \equiv CH} \qquad \underset{\textbf{2-hexyne}}{CH_3C \equiv CCH_2CH_2CH_3}$$

compound *A*   compound *B*   compound *C*

14.37   Because muscalure is a *cis*-alkene, we can prepare it by hydrogenation of an alkyne. The required alkyne is unbranched and can be prepared by a series of alkylations involving acetylenic anions and alkyl bromides. One of the required alkyl bromides is 1-bromooctane, which is prepared as follows. (For the preparation of sodium acetylide, Na$^+$ $^-$C'CH, see the solution to Problem 14.33(a).

$$CH_3CH_2C \equiv CH \xrightarrow[\text{Lindlar catalyst}]{H_2} CH_3CH_2CH=CH_2 \xrightarrow[\text{peroxides}]{HBr} CH_3CH_2CH_2CH_2Br \xrightarrow{Na^+ \ ^-:C \equiv CH}$$

$$CH_3(CH_2)_3C \equiv CH \xrightarrow[\text{Lindlar catalyst}]{H_2} CH_3(CH_2)_3CH=CH_2 \xrightarrow[\text{peroxides}]{HBr} CH_3(CH_2)_5Br \xrightarrow{Na^+ \ ^-:C \equiv CH}$$

$$CH_3(CH_2)_5C \equiv CH \xrightarrow[\text{Lindlar catalyst}]{H_2} CH_3(CH_2)_5CH=CH_2 \xrightarrow[\text{peroxides}]{HBr} CH_3(CH_2)_7Br$$

As you can see from this example, the carbon chain of an alkyne can be extended two carbons at a time using the following sequence (referred to below as the *alkyne n + 2 sequence*):

$$CH_3(CH_2)_nC \equiv CH \xrightarrow[\text{Lindlar catalyst}]{H_2} CH_3(CH_2)_nCH=CH_2 \xrightarrow[\text{peroxides}]{HBr} CH_3(CH_2)_{n+2}Br \xrightarrow{Na^+ \ ^-:C \equiv CH}$$

$$CH_3(CH_2)_{n+2}C \equiv CH$$

Consequently, 1-bromotridecane, CH$_3$(CH$_2$)$_{12}$Br, can be prepared by applying four successive "alkyne *n* + 2 sequences" to 1-pentyne (*n* = 2), and finishing with the hydrogenation and HBr-addition steps:

$$CH_3(CH_2)_2C \equiv CH \xrightarrow[\text{sequences}]{\overset{\text{four}}{\text{"alkene } n + 2\text{"}}} CH_3(CH_2)_{10}C \equiv CH \xrightarrow[\text{Lindlar catalyst}]{H_2}$$

1-pentyne

$$CH_3(CH_2)_{10}CH=CH_2 \xrightarrow[\text{peroxides}]{HBr} CH_3(CH_2)_{12}Br$$

1-bromotridecane

To complete the synthesis of muscalure, carry out two successive alkylations of acetylenic anions with the alkyl bromides prepared above, and then hydrogenate the resulting product.

$$Na^+ \ \overset{..}{:}C\equiv CH \xrightarrow{CH_3(CH_2)_7Br} CH_3(CH_2)_7C\equiv CH \xrightarrow[NH_3 \text{ (liq.)}]{NaNH_2} CH_3(CH_2)_7C\equiv C\overset{..}{:} \ Na^+ \xrightarrow{Br(CH_2)_{12}CH_3}$$

$$CH_3(CH_2)_7C\equiv C(CH_2)_{12}CH_3 \xrightarrow[\text{Lindlar catalyst}]{H_2}$$

muscalure

14.38   This problem is similar to the previous one. Disparlure is a *cis*-epoxide, which can be prepared by epoxidation of a *cis*-alkene. (Recall that this reaction proceeds with retention of stereochemistry; Sec. 11.3A.) The *cis*-alkene can be prepared by hydrogenation of an alkyne.

First prepare 1-bromodecane from 1-bromooctane, which was prepared in the solution to Problem 14.37.

$$CH_3(CH_2)_7Br \xrightarrow{Na^+ \ ^-C\equiv CH} CH_3(CH_2)_7C\equiv CH \xrightarrow[\text{Lindlar catalyst}]{H_2}$$

$$CH_3(CH_2)_7CH=CH_2 \xrightarrow[\text{peroxides}]{HBr} CH_3(CH_2)_9Br$$

1-bromodecane

Next, alkylate sodium acetylide with 1-bromo-3-methylbutane (isoamyl bromide). Apply the alkyne *n* + 2 sequence described in the previous solution, alkylate the conjugate base of the resulting alkyne with 1-bromo-decane, hydrogenate, and form the epoxide to complete the synthesis.

$$Br(CH_2)_2CH(CH_3)_2 \xrightarrow{Na^+ \ HC\equiv C^{-}} HC\equiv C(CH_2)_2CH(CH_3)_2 \xrightarrow[\text{sequence}]{\text{"alkene } n+2\text{"}} HC\equiv C(CH_2)_4CH(CH_3)_2 \xrightarrow[NH_3 \text{ (liq.)}]{NaNH_2}$$

$$^-{:}C\equiv C(CH_2)_4CH(CH_3)_2 \xrightarrow{CH_3(CH_2)_9Br} CH_3(CH_2)_9C\equiv C(CH_2)_4CH(CH_3)_2 \xrightarrow[\text{Lindlar catalyst}]{H_2}$$

$$\xrightarrow{mCPBA} CH_3(CH_2)_9\overset{}{\text{ⁿⁿ}}\overset{O}{\underset{H}{C}}-\underset{H}{C}\text{ⁿⁿ}(CH_2)_4CH(CH_3)_2$$

disparlure

14.39   (a)   In reaction (1), the "anionic carbon" of ethylmagnesium bromide acts as a base towards the C—H bond of ethynylmagnesium bromide. (The reacting C—Mg bond of the Grignard reagent is shown as ionic to stress its polar character. This is a convenient fiction that helps us to remember that the Grignard reagent is a strong carbon base.)

According to Eq. (2) in the problem, an excess of acetylene reacts with the BrMgC'CMgBr product of Eq. (1) to regenerate ethynylmagnesium bromide. Thus, the formation of BrMgC'CMgBr, the undesired by-product, is avoided by "recycling" it in Eq. (2) back to the desired product HC'CMgBr.

(b)   The "anionic carbon" of one Grignard reagent acts as a base toward the C—H bond of the other.

An excess of acetylene, by Le Châtelier's principle, drives the equilibrium to the left.

(c)   Precipitation of BrMgC'CMgBr would pull both reactions (1) and (2) to the right, thus depleting the desired product HC'CMgBr and forming more of the undesired one. Using a solvent that avoids this precipitation avoids these side-reactions.

14.40   (a)   The radical anion forms in this case just as it does with internal alkynes. (See Eq. 14.18a on text p. 696.) This radical anion is a strong base, and a 1-alkyne is acidic enough to react with it. (Internal alkynes, of course, lack this acidic hydrogen.) Recall that 1-alkynes have a pK_a of 25, whereas ammonia, the solvent, has a pK_a of about 32. Ammonia is the only acid available in the reaction of internal alkynes, but in the reaction of 1-alkynes, the alkyne itself is the strongest acid present. This is the source of the side reaction.

$$\text{Na}^+ \ R\overset{\cdot}{C}\!\!=\!\!\overset{\cdot\cdot}{C}H \quad H\!\!-\!\!C\!\!\equiv\!\!CR \longrightarrow R\overset{\cdot}{C}\!\!=\!\!CH_2 \ + \ \text{Na}^+ \ \overset{\cdot\cdot}{:}C\!\!\equiv\!\!CR$$

radical anion   vinylic radical   acetylenic anion

The vinylic radical is reduced as in the case of internal alkynes to give a vinylic anion:

$$R\overset{\cdot}{C}\!\!=\!\!CH_2 \ + \ \overset{\cdot}{N}a \longrightarrow R\overset{\cdot\cdot}{C}\!\!=\!\!CH_2 \ \text{Na}^+$$

vinylic radical   vinylic anion

Here we have another strong base; the pK_a of a vinylic proton is about 42. This anion therefore consumes yet another molecule of 1-alkyne to give the alkene.

$$R\overset{\cdot\cdot}{C}\!\!=\!\!CH_2 \ \text{Na}^+ \quad H\!\!-\!\!C\!\!\equiv\!\!CR \longrightarrow R\overset{\underset{|}{H}}{C}\!\!=\!\!CH_2 \ + \ \text{Na}^+ \ \overset{\cdot\cdot}{:}C\!\!\equiv\!\!CR$$

vinylic anion   alkene   acetylenic anion

Hence, we form *two* acetylenic anions for every one alkene formed; the acetylenic anions do not react further.

(b)   Presumably, the side reactions above could be avoided if a proton source were present that is a stronger acid than the acetylide ion. Ammonium sulfate (that is, $^+$NH$_4$ ion) was used for this purpose; the pK_a of $^+$NH$_4$ is 9.2. In the mechanism of part (a), simply substitute this acid everywhere a 1–alkyne is acting as an acid. With this ploy, all of the alkyne is reduced.

14.41   The order of increasing acidity is $C < A < B$. The acidity of carboxylic acids is increased when electronegative atoms or groups are nearby, as a result of the inductive effect. The alkyne group is more electronegative than the alkene, which is more electronegative than the alkane. The alkyne group, followed by the alkene group, draw electron density towards them and weaken the O–H bond, making for successively stronger acids. Additionally, the electron withdrawing ability of those groups stabilize the conjugate bases of those acids.

14.42   (a)   The IR spectrum suggests a 1-alkyne. In the NMR spectrum, the δ 3.31 singlet indicates a methoxy group, and the six-proton δ 2.41 singlet suggests two equivalent propargylic methyl groups bound to a quaternary carbon. The compound is 3-methoxy-3-methyl-1-butyne.

$$\text{HC}\!\!\equiv\!\!\text{CCCH}_3 \overset{\overset{\displaystyle CH_3}{|}}{\underset{\underset{\displaystyle OCH_3}{|}}{}}$$

3-methoxy-3-methyl-1-butyne

(b)   The IR data suggest a 1-alkyne, and this diagnosis is confirmed by the formation of a gas (ethane) when the compound is treated with C$_2$H$_5$MgBr. The three-proton δ 3.41 singlet suggests a methoxy group. The structure is 3-methoxypropyne, HC'CCH$_2$OCH$_3$.

(c) The IR data suggest that the compound is both a 1-alkyne and an alcohol. The δ 4.6 resonance for 1H indicates that the alcohol is secondary. Its splitting also shows that this α-hydrogen is adjacent to a methyl group, and its chemical shift shows that other functionality, probably the C'CH group, is adjacent. The **dry/wet contrast** also indicates that this is the resonance of an alcohol α-proton. A small splitting of the α-hydrogen by the acetylenic hydrogen would account for the secondary splitting of each quartet line into a doublet. (Recall that vinylic, and by extension, acetylenic, splitting can occur over four bonds.) The spectrum is consistent with only the following structure.

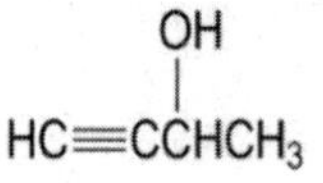

$$\underset{\textbf{3-butyn-2-ol}}{\overset{\overset{\displaystyle OH}{|}}{HC\equiv CCHCH_3}}$$

The methyl doublet at δ 1.5, the slightly split acetylenic-proton resonance at δ 2.5, and the D_2O-exchangeable resonance at δ 6.5 are all consistent with this structure.

(d) The IR data suggest a 1-alkyne. The presence of a methoxy group is indicated by the three-proton singlet at δ 3.79. There is one additional degree of unsaturation; because three carbons are already accounted for, the unsaturation must be in the remaining two. Consequently, there is a double bond. Cis stereochemistry is suggested by the small 6-Hz coupling constant between the vinylic protons. (See Table 13.3 on text p. 647.) The splitting suggests that one of the vinylic protons (δ 4.52) is also coupled to an acetylenic proton. The chemical shift of the other vinylic proton (δ 6.38) suggests that it is α to the methoxy oxygen. All of these data conspire to give the following structure:

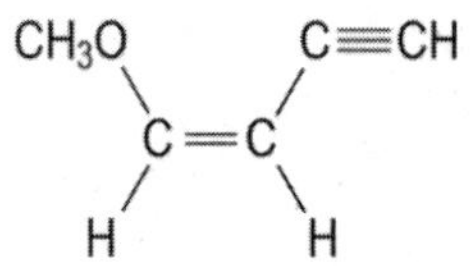

$$\underset{\textbf{(Z)-1-methoxy-1-buten-3-yne}}{\overset{\displaystyle CH_3O \qquad\qquad C\equiv CH}{\underset{\displaystyle H \qquad\qquad H}{\diagdown C = C \diagup}}}$$

14.43 (a) The IR absorptions indicate a carbon-carbon triple bond (2100 cm^{-1}) and a 1-alkyne (3300 cm^{-1}). The CMR data show that there are only four nonequivalent sets of carbons. The compound is 3,3-dimethyl-1-butyne, $(CH_3)_3CC'CH$, also known as *tert*-butylacetylene. (See also the solution to Problem 14.4.)

(b) Assuming that all resonances for chemically nonequivalent carbons are separately observable, 1-hexyne, $CH_3CH_2CH_2CH_2C'CH$, should have a CMR spectrum consisting of six resonances; 4-methyl-2-pentyne, $(CH_3)_2CHC'CCH_3$, should have a CMR spectrum consisting of five, because two of the methyl groups are chemically equivalent.

14.44 (a) This mechanism consists of two successive Brønsted acid–base reactions.

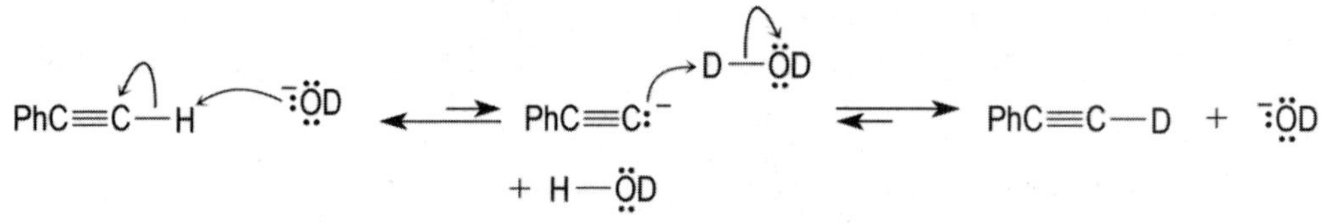

$$PhC\equiv C-H \quad \overset{..}{:\overset{..}{O}D} \quad \longleftrightarrow \quad PhC\equiv C:^- \quad \longleftrightarrow \quad PhC\equiv C-D \;+\; :\overset{..}{\overset{..}{O}}D$$
$$+\; H-\overset{..}{\overset{..}{O}}D$$

(b) A bromohydrin is formed; however, because the bromohydrin is also an enol, it is converted into the α-bromo ketone. (See Eqs. 14.6a–b on text p. 691 for the mechanism of the enol-to-ketone conversion.)

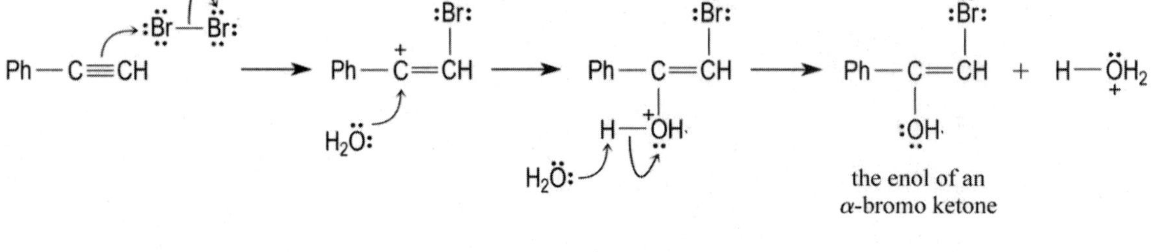

the enol of an
α-bromo ketone

A carbocation rather than a bromonium ion is shown above as the reactive intermediate in bromine addition, although a bromonium ion is a possible alternative. Because of its carbon–carbon double bond, a bromonium ion derived from bromine addition to an alkyne would be more strained than a bromonium ion derived from bromine addition to an alkene, and might be unstable enough to open to a carbocation.

14.45 The ozonolysis results define compound *B* as 1,5-hexadiene. $H_2CACHCH_2CH_2CHACH_2$. Because compound *B* is produced by hydrogenation of compound *A*, and because *two equivalents* of H_2 must be added to compound *A* (C_6H_6) to give compound *B* (C_6H_{10}), compound *A* must be 1,5-hexadiyne, $HC'CCH_2CH_2C'CH$.

14.46 Compound *A* has four degrees of unsaturation. Because the product of hydrogenation is butylcyclohexane, one degree of unsaturation is accounted for by a 6-membered ring. Two others are accounted for by the triple bond; remember, it was given that compound *A* is an alkyne. The triple bond cannot be within the ring (why?), nor can it be at the end of the four-carbon chain because of the absence of a reaction of *A* with ethylmagnesium bromide. Hydrogenation of *A* under conditions that hydrogenate triple bonds to double bonds but leave double bonds unaffected gives an alkene which, on ozonolysis, loses *two* carbons. The fact that a *tricarboxylic* acid is formed on ozonolysis shows that the double bond of *A* is within the ring. It is now possible to write several preliminary structures for *A*:

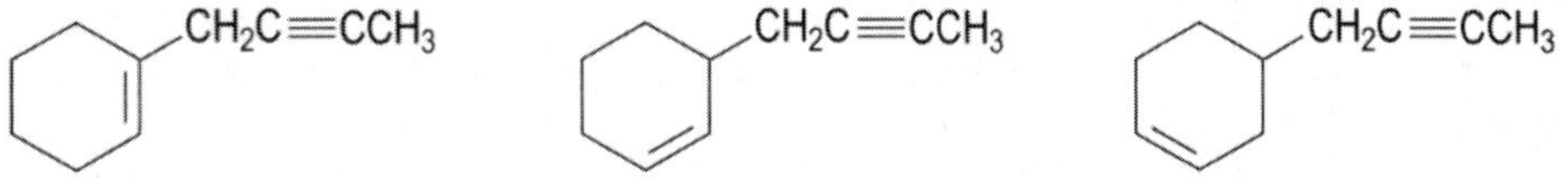

A1 *A2* *A3*

Compound *A1* is ruled out because it is not chiral; recall that compound *A* is optically active. To distinguish between *A2* and *A3*, imagine the results of hydrogenation over a poisoned catalyst followed by ozonolysis:

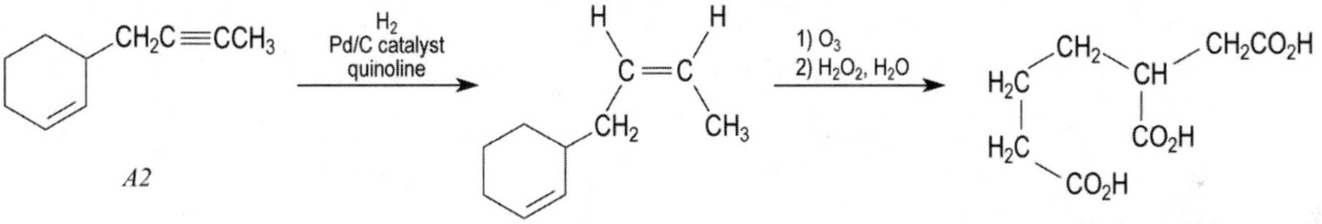

A2

chiral; would be optically active if *A2* were optically active

+ CH_3CO_2H

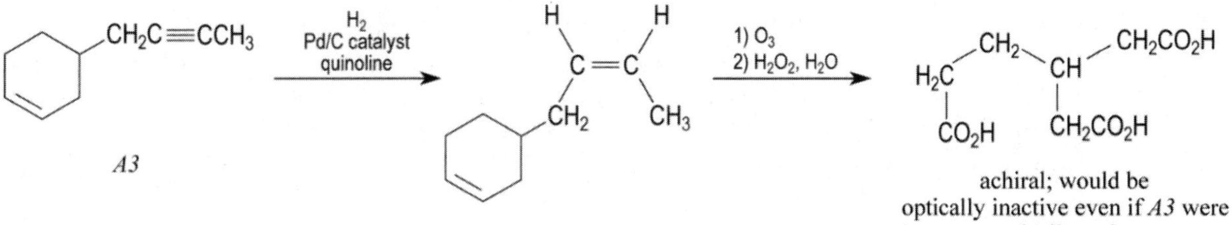

A3

achiral; would be optically inactive even if *A3* were optically active

+ CH_3CO_2H

The observation of optical activity in the ozonolysis product rules out *A3* and is consistent only with *A2* as the structure of compound *A*.

compound $A = A2$
3-(2-butynyl)cyclohexene

The data do not determine which enantiomer compound A is.

14.47 (a) The Grignard reagent converts 1-hexyne into the acetylenic Grignard reagent; see Eq. 14.23 on text p. 699. Protonolysis of the Grignard reagent by D_2O gives $CH_3CH_2CH_2CH_2C'CD$ as the final product.

(b) Butyllithium is a strong base; like a Grignard reagent, it converts a 1-alkyne into a lithium acetylide. Lithium acetylides, like acetylenic Grignard reagents, react as if they were acetylenic anions. Consequently, the lithium acetylide, as suggested by the hint, reacts as a nucleophile, displacing chloride ion from $(CH_3)_3Si$—Cl.

(c) 1-Octyne is converted into its conjugate-base acetylide ion with $NaNH_2$. This ion is alkylated by diethyl sulfate to give 3-undecyne.

(d) The lithium acetylide (as an acetylide ion) serves as a nucleophile in an S_N2 reaction with the alkyl halide; the better leaving group (chloride) is expelled. The product is 7-fluoro-1-heptyne, $F(CH_2)_5C'CH$, plus Li^+Cl^-.

(e) Bromine adds to the double bond; both bromines are subsequently eliminated in consecutive $NaNH_2$-promoted E2 reactions to give phenylacetylene. (Remember, $^-NH_2$ is a strong base and, like many strong bases, can promote E2 reactions.)

Two equivalents of $NaNH_2$ are required in the last step because one equivalent is consumed by reaction with phenylacetylene, PhC'CH, which is the product of the second elimination; this reaction gives the conjugate-base acetylenic anion, PhC'C:⁻. This anion is converted back into phenylacetylene by addition of acid (or water) at the conclusion of the reaction.

(f) As suggested by the hint, chloroform and a strong base give dichloromethylene, a carbene, which adds to the triple bond to give a cyclopropene. The formation of dichlorocarbene is shown in Eqs. 9.76a–b, text p. 435.

 If you find an error in this manual, please visit http://people.pharmacy.purdue.edu/~loudonm/teaching/

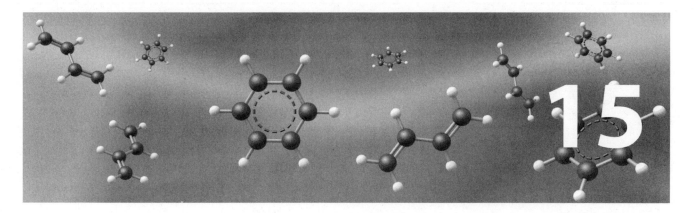

15

Dienes, Resonance, and Aromaticity

STUDY GUIDE LINKS

15.1 A Terminology Review

The purpose of this Study Guide Link is to review and to distinguish between the following terms used to describe addition reactions:

1. cycloaddition reactions
2. concerted reactions
3. pericyclic reactions

These terms in some cases may seem synonymous, but they are in fact different, and have very precise meanings. A *cycloaddition* is an addition that results in ring formation. Ozone addition and the Diels–Alder reaction are two examples of cycloadditions. The term *cycloaddition* succinctly describes the *outcome* of a reaction, but says nothing about the *mechanism* of the reaction. The terms *concerted* and *pericyclic*, in contrast, are descriptions of *mechanism*. Reactions that occur by mechanisms involving a single step—that is, mechanisms involving no reactive intermediates—are referred to as *concerted reactions*. Concerted reactions that occur by a cyclic flow of electrons are called *pericyclic reactions* (pericyclic = "around the circle"). All pericyclic reactions are concerted, but some concerted reactions are not pericyclic. Thus, hydroboration of alkenes (Sec. 5.4B of the text) is a concerted pericyclic reaction (although it is not a cycloaddition because a ring is not formed). The S_N2 reaction (Sec. 9.4 of the text) is concerted, but it is not pericyclic because the flow of electrons in the mechanism is not cyclic. The Diels–Alder reaction is a concerted cycloaddition, and it is also a pericyclic reaction. (Chapter 28 is devoted to a more detailed study of pericyclic reactions.)

FURTHER EXPLORATIONS

15.1   More on UV Spectroscopy

Why are peaks in UV spectra so broad? We are led to ask this question because the molecular orbital theory of UV spectroscopy discussed in the text implies that a $\pi \rightarrow \pi^*$ transition should occur at a very precise value of the energy, that is, at $E = \Delta\epsilon$.

The answer lies in the *rate* at which electronic transitions take place. It turns out that the motions of electrons are thousands of times faster than the motions of nuclei. As you learned in Chapter 12, the nuclei in molecules are vibrating constantly. When energy is absorbed from UV light by a molecule, this act of absorption "catches" the molecule in a variety of vibrational states. That is, some bonds are stretched, some are compressed, and all possible combinations of these states for all possible bonds can occur. Because the electronic transition occurs so much faster than bond vibrations, the "molecule" that absorbs UV energy is actually a complex mixture of molecules in various vibrational states. These states differ in energy, as do the excited states into which they are converted. These energy differences are reflected in a broadening of the energy at which the UV transition is observed. In gas-phase UV spectra, one can in many cases see very narrow peaks associated with the different vibrational states. (Spectroscopists call these peaks "vibrational fine structure.") Interaction of the absorbing molecules with solvent further blurs these vibrational energies, so that the absorption observed in solution is something of an "envelope" of all of these states. Thus, UV absorptions in solution are typically very broad.

15.2   The π Molecular Orbitals of Benzene

Only the benzene bonding π molecular orbital of lowest energy is shown in Fig. 15.16b on text p. 761. Benzene has five other π molecular orbitals, and they are shown in Fig. SG15.1 (following page).

As you can see, there are three bonding and three antibonding orbitals—six $2p$ orbitals, six MOs. Two of the bonding MOs—π_2 and π_3—have the same energy. Orbitals of the same energy are said to be *degenerate*. Because these MOs have the same energy, they have the same number of nodes (one). Two of the antibonding MOs—π_4^* and π_5^*—are also degenerate. Because there are six π electrons, there are just enough electrons to fill the three bonding MOs. All three of these MOs, because they are bonding, are lower in energy than an isolated $2p$ orbital, whose energy is indicated by the dashed line. Hence, benzene is stabilized by molecular-orbital formation.

In Sec. 15.7D, you will learn to calculate the energies of these MOs. When you do, you will find that the energy of π_1 is *particularly* low. This filled, low-energy MO, which encircles the entire molecule with π-electron density above and below the ring, is particularly important for the stability of benzene. This is the MO that is shown in Fig. 15.16b of the text. (Undoubtedly, the fact that the bond angles in benzene are exactly right for sp^2-hybridized carbons contributes to the molecule's stability as well.)

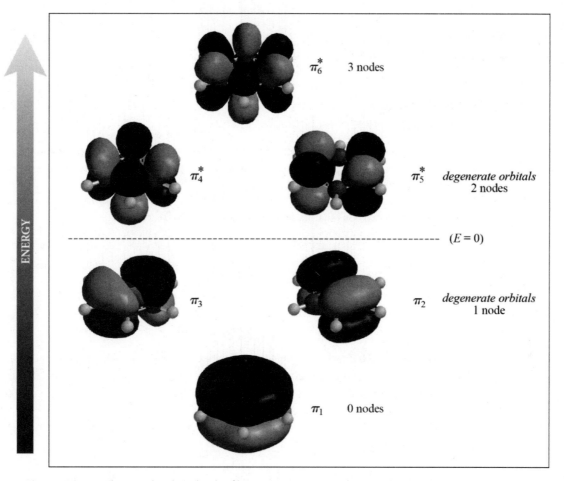

Figure SG15.1 The π molecular orbitals of benzene superimposed on a ball-and-stick model. Wave peaks are black, and wave troughs are gray. The $E = 0$ line is the energy of an electron in an isolated $2p$ orbital.

REACTION REVIEW

I. ADDITION OF HYDROGEN HALIDES TO CONJUGATED DIENES

A. 1,2- AND 1,4-ADDITIONS

1. Conjugated dienes react with hydrogen halides to give two types of addition product.
   a. The major product results from 1,2-addition. In a 1,2-addition, the hydrogen and halogen add to adjacent carbons.
   b. The minor product results from 1,4-addition, or conjugate addition. In a 1,4-addition, the hydrogen and halogen add to carbons that have a 1,4-relationship.

1,2-addition product   1,4-addition product

2. The 1,2-addition mechanism is analogous to the addition of HX with an ordinary alkene.
3. The 1,4-addition, or conjugate-addition, mechanism involves:
   a. protonation of a double bond to give an allylic carbocation in which the positive charge is not localized, but is instead shared by two different carbons.
   b. formation of two constitutional isomers by a Lewis acid–base association reaction of the halide ion with the carbocation at either of its electron-deficient carbons.

1,2-addition product   1,4-addition product

4. The product distribution in hydrogen-halide addition to a conjugated diene is determined by the relative rates of the product-determining steps.
   a. Formation of the 1,2-addition product is faster but reversible.
   b. Formation of the 1,4-addition product is slower but virtually irreversible.

II. FORMATION OF DIENE POLYMERS

1. The polymerization of 1,3-butadiene gives polybutadiene, an important type of synthetic rubber used in the manufacture of tires.

$$n\ H_2C=CH-CH=CH_2 \xrightarrow[\text{initiator}]{\text{free-radical}} -\!\!\left(CH_2-CH=CH-CH_2\right)_{\!\!n}\!\!-$$

1,3-butadiene   **polybutadiene**

a. Polybutadiene is referred to as a diene polymer because it comes from polymerization of a diene monomer.

b. Polybutadiene has only one double bond per unit, because one double bond is lost through the addition that takes place in the polymerization process.

2. Polymerization of dienes is a free-radical reaction. Although the polymerization product is largely the result of 1,4-addition, a small amount of 1,2-addition can occur as well.

3. Styrene–butadiene rubber, SBR, is an example of a copolymer, a polymer produced by the simultaneous polymerization of two or more monomers, in this case, styrene ($PhCH$=CH_2) and 1,3-butadiene.

III. THE DIELS–ALDER REACTION

A. INTRODUCTION

1. Many conjugated dienes undergo addition reactions, called Diels–Alder reactions, with certain alkenes or alkynes.

a. The conjugated diene component of this reaction is referred to simply as the diene.

b. The alkene component of this reaction is referred to as the dienophile.

2. Some of the dienophiles that react most readily in the Diels–Alder reaction bear substituent groups such as esters (—CO_2R), nitriles (—CN), or certain other unsaturated, electronegative groups that are conjugated with the alkene double bond. (Some alkynes also serve as dienophiles.)

3. When the diene is cyclic, bicyclic products are obtained in the Diels–Alder reaction.

4. The Diels–Alder reaction is

a. a cycloaddition reaction (an addition reaction that results in the formation of a ring).

b. a pericyclic reaction (a concerted reaction that involves a cyclic flow of electrons).

c. a 1,4-addition, or conjugate addition, reaction (an addition that occurs across the outer carbons of the diene).

d. a stereoselective *syn*-addition to both diene and dienophile.

B. EFFECT OF DIENE CONFORMATION ON THE DIELS–ALDER REACTION

1. In the Diels–Alder transition state, the diene component is in the *s*-cis conformation.

a. Dienes "locked" into *s*-trans conformations are unreactive in Diels–Alder reactions.

b. Dienes "locked" into *s*-cis conformations are unusually reactive and in many cases much more reactive than corresponding noncyclic dienes.

1,3-cyclopentadiene

"locked" into the *s*-cis diene
conformation by its structure

more reactive in Diels–Alder reactions

(2E,4E)-2,4-hexadiene

prefers the *s*-trans conformation

less reactive in Diels–Alder reactions

2. The Diels–Alder reactions of a *cis*-diene are much slower than the corresponding reactions of its *trans*-diene stereoisomer, which does not have the destabilizing repulsion in its *s*-cis conformation.

cis-diene

less reactive in Diels–Alder reactions

van der Waals repulsion

trans-diene

more reactive in Diels–Alder reactions

no methyl–methyl repulsion

   a. Severe van der Waals repulsion destabilizes the *s*-cis conformation of the diene.
   b. Severe van der Waals repulsion destabilizes the transition states for the Diels–Alder reactions of the *s*-cis conformation of the diene.

C. STEREOCHEMISTRY OF THE DIELS–ALDER REACTION

1. When a diene and dienophile react in a Diels–Alder reaction, they approach each other in parallel planes.
   a. This type of approach allows the π-electron clouds of the two components to overlap and form the bonds of the product.
   b. Each component adds to the other at one face.
      i. The diene undergoes a *syn*-addition to the dienophile.
      ii. The dienophile undergoes a 1,4-*syn*-addition to the diene.
   c. Groups that are cis in the alkene starting material are also cis in the product.
   d. *Syn*-addition to the diene is apparent if the terminal carbons of the diene unit are stereocenters.
2. A *syn*-addition requires that in the Diels–Alder product—
   a. the two inner substituents of the diene always have a cis relationship.
   b. the two outer substituents of the diene also always have a cis relationship.
   c. an inner substituent on one carbon of the diene is always trans to an outer substituent on the other.
   d. Cis substituents in the dienophile are cis in the product; trans substituents in the dienophile are trans in the product.

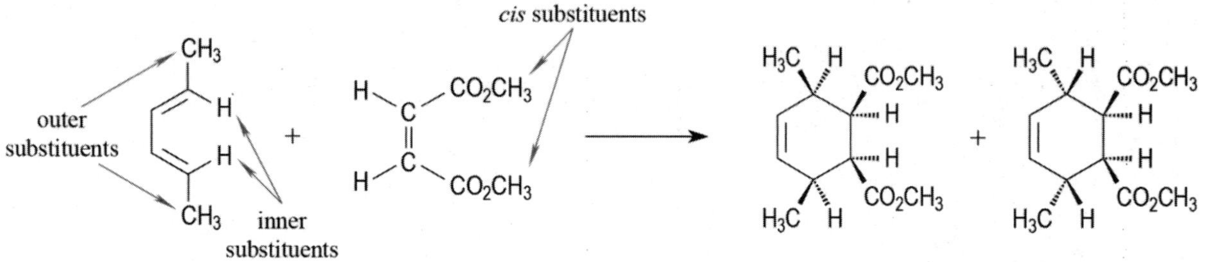

cis substituents

outer substituents

inner substituents

3. For a given diene and dienophile, two diastereomeric *syn*-addition products are possible and are abbreviated with the terms endo and exo.

a. In the endo product, the dienophile substituents are cis to the outer diene substituents.

endo-addition

CH₃ ←— outer diene substituent
CO₂CH₃ ←— dienophile substituent } cis

CO₂CH₃ ←— dienophile substituent
CH₃ ←— outer diene substituent } cis

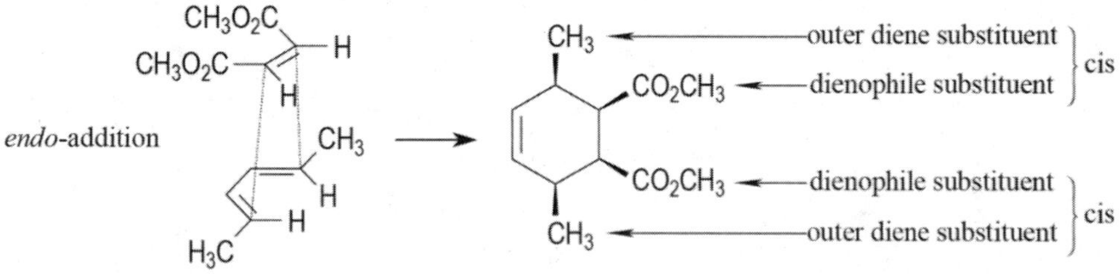

b. In the exo product, the dienophile substituents are trans to the outer diene substituents.

exo-addition

CH₃ ←— outer diene substituent
CO₂CH₃ ←— dienophile substituent } trans

CO₂CH₃ ←— dienophile substituent
CH₃ ←— outer diene substituent } trans

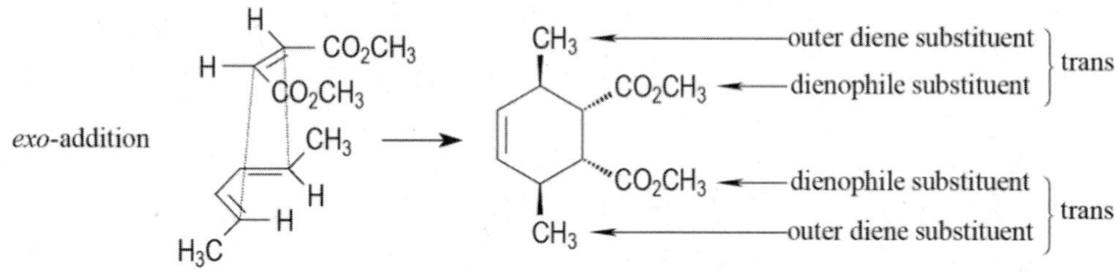

c. In many cases, the endo mode of addition is favored over the exo mode, particularly when cyclic dienes are used.

SOLUTIONS TO PROBLEMS

Solutions to In-Text Problems

15.1   (a)   1,3,5-Hexatriene (either cis or trans) has six MOs; three are bonding, and three are antibonding. These MOs are shown in Fig. SG15.2.

        (b)   Each MO of successively higher energy has an additional planar node.

        (c)   The nodes are shown in Fig. SG15.2. The nodes are symmetrically arranged about the center of the molecule. (The nodes of all MOs are shown.)

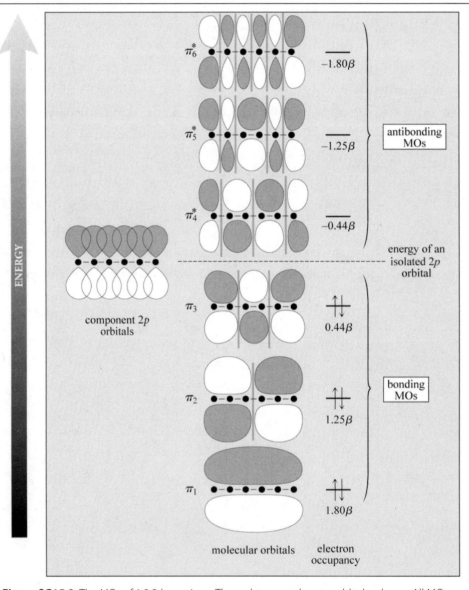

Figure SG15.2 The MOs of 1,3,5-hexatriene. The carbons are shown as black spheres. All MOs are projected onto the plane of the page. The nodes are the gray vertical lines.

15.2   The delocalization energy is the energy of each MO times the number of electrons in that MO, minus the same energy for three ethylenes:

$$\text{delocalization energy} = 2(1.80\beta) + 2(1.25\beta) + 2(0.44\beta) - (3)(2)(1.00\beta) = 0.98\beta$$

15.3   The conformations of (2E,4E)-2,4-hexadiene:

s-cis   s-trans

The conformations of (2E,4Z)-2,4-hexadiene:

greater
van der Waals
repulsions

s-cis   s-trans

(2E,4E)-2,4-hexadiene has the greater proportion of s-cis conformation because the van der Waals repulsions between inner groups are smaller in the (2E,4E)-stereoisomer. The reason is that the inner groups in the (2E,4E)-stereoisomer are both hydrogens, whereas, in the (2E,4Z)-stereoisomer, one of these groups is a methyl group, which is considerably larger than hydrogen. It follows then that (2E,4E)-2,4-hexadiene would also have the greater proportion of the gauche conformation. The (2E,4Z)-2,4-hexadiene isomer would still have van der Waals repulsions between the methyl group and the hydrogen, even when they are 60° apart.

15.4   First, draw the structures:

(E)-1,3-pentadiene   1,4-pentadiene   (E)-1,3-hexadiene   (E)-1,4-hexadiene

(E)-1,3-pentadiene has a double bond with one substituent and a double bond with two substituents (counting the carbon of other double bond), whereas 1,4-pentadiene has two double bonds with a single substituent. Hence, the energy difference between these two pentadiene isomers reflects not only the effect of conjugation but also the effect of additional double-bond substituents. *Both* effects stabilize the conjugated isomer. The two hexadiene isomers contain identical numbers of substituents; hence, the energy difference between these two isomers reflects only the effect of conjugation. Because the comparison of pentadiene isomers contains two effects that operate in the same direction, the energy difference between these isomers is expected to be greater, as observed.

15.5   (a)   The two enantiomers of an allene:

(b)   Because enantiomers have specific rotations of equal magnitudes and opposite signs, the other enantiomer has a specific rotation of +30.7°.

15.6   (a)   The energy of light is given by Eq. 12.3 on text p. 571 with the value of h, in kJ s mol^{-1}, given in Eq. 12.5 on text p. 571.

$$E = \frac{hc}{\lambda} = \frac{(3.99 \times 10^{-13} \text{ kJ s mol}^{-1})(3.00 \times 10^8 \text{ m s}^{-1})}{450 \times 10^{-9} \text{ m}} = 266 \text{ kJ mol}^{-1}$$

(The corresponding value in kcal mol^{-1} is 63.6.)

(b)   The calculation is identical to that in part (a) with different numbers:

$$E = \frac{hc}{\lambda} = \frac{(3.99 \times 10^{-13} \text{ kJ s mol}^{-1})(3.00 \times 10^8 \text{ m s}^{-1})}{250 \times 10^{-9} \text{ m}} = 479 \text{ kJ mol}^{-1}$$

(The corresponding value in kcal mol^{-1} is 114.)

15.7   (a)   Use Eq. 15.1 on text p. 720. An absorbance $A = 1.0$ means that $(I_0/I) = 10$; that is, one-tenth, or 10%, of the incident radiation is transmitted. When $A = 0$, all of the incident radiation is transmitted.

(b)   Use Eq. 15.1 on text p. 720 with $(I_0/I) = 2$. Thus, $A = \log(2) = 0.30$.

15.8   The piece with greater absorbance transmits less of the incident radiation. Therefore, the thick piece of glass has greater absorbance.

15.9   (a)   Apply Beer's law (Eq. 15.2 on text p. 721), with $\epsilon = 10{,}750$ absorbance units cm^{-1} L mol^{-1}, $l = 1.0$ cm, and, from Fig. 15.5 on text p. 720, $A = 0.800$ absorbance units. The concentration of isoprene is therefore

$$\frac{0.800 \text{ absorbance units}}{(10{,}750 \text{ absorbance units cm}^{-1} \text{ L mol}^{-1})(1 \text{ cm})} = 7.44 \times 10^{-5} \text{ mol L}^{-1}$$

(b)   The absorbance of the isoprene sample in Fig. 15.5 on text p. 720 at 235 nm is 0.225. With the concentration determined from part (a), Beer's law gives $A = 0.225$ absorbance units $= (\epsilon)(7.44 \times 10^{-5}$ mol $L^{-1})(1$ cm$)$, or $\epsilon = 3.02 \times 10^3$ absorbance units L mol^{-1} cm^{-1}. Another way to determine the extinction coefficient at a different wavelength is based on the fact that the ratio of absorbances at different wavelengths equals the ratio of the extinction coefficients. Hence, the extinction coefficient at 235 nm is

$$\epsilon_{235} = \epsilon_{225} \frac{A_{235}}{A_{225}} = (10{,}750 \text{ absorbance units L mol}^{-1} \text{ cm}^{-1}) \left(\frac{0.225 \text{ absorbance units}}{0.800 \text{ absorbance units}} \right)$$

or $\epsilon_{235} = 3.02 \times 10^3 = 3023$ absorbance units L mol^{-1} cm^{-1}.

15.10   (a)   The two alkyl substituents contribute +10 nm to the base λ_{max} of 217 for a predicted λ_{max} value of 227 nm.

(b)   The —CH_2— groups of the ring attached to the double bond are alkyl substituents. (If this point is not clear, see Study Guide Link 9.2 on p. 226 of this manual.) For three alkyl substituents the predicted λ_{max} is $(217 + 15)$ nm = 232 nm. (This calculation assumes that the molecule exists predominantly in the s-trans conformation.)

15.11   At pH = 9, fluorescein has the greater quantum yield, because the peak height of fluorescence is nearly as intense as the absorption. At pH = 5, the fluorescence is a smaller fraction of the absorbance.

15.12   Compound C is likely to emit yellow fluorescence. Of all the compounds given, it is the most highly conjugated and the most rigid, due to its fused aromatic ring system.

15.13   (a)   (b)

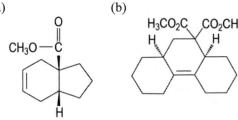

15.14   (a)   An analysis like that employed in Study Problem 15.1 (text p. 733) suggests two possibilities.

A   *B*

Pair *A* is preferred because, in many cases, the most reactive dienophiles are those with conjugated electronegative substituents. But if your answer was pair *B*, you have analyzed the problem correctly.

(b)   There are two possibilities:

A   *B*

Pair *A* is the more practical alternative because of the availability of starting materials, and especially because dienophiles with electron-withdrawing substituents are more reactive. But if your answer was pair *B*, you have analyzed the problem correctly.

15.15   (a)   With 1,3-butadiene as the diene and ethylene as the dienophile, the product would be cyclohexene.

1,3-butadiene   **ethylene**   **cyclohexene**

(b)   Cyclohexene, the product from the reaction of ethylene and 1,3-butadiene in part (a), is not observed because 1,3-butadiene reacts with itself to give 4-vinylcyclohexene:

4-vinylcyclohexene

(c)   Using an excess of ethylene (many more equivalents relative to 1,3-butadiene) could favor the formation of cyclohexane. The chances of 1,3-butadiene colliding with a molecule of ethylene would become much higher.

15.16   (a)   The dienophile can be oriented in two different ways relative to the diene.

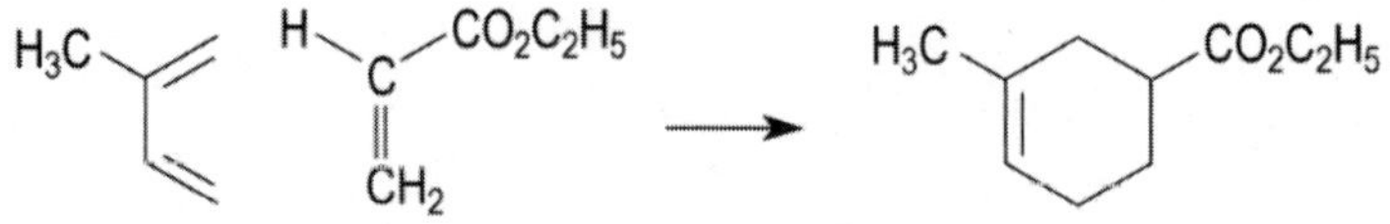

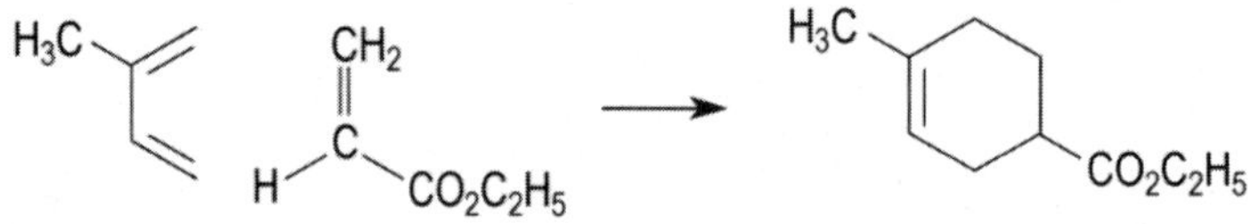

(b)   As in part (a), two possible orientations of the diene and dienophile lead to the following two possible constitutional isomers:

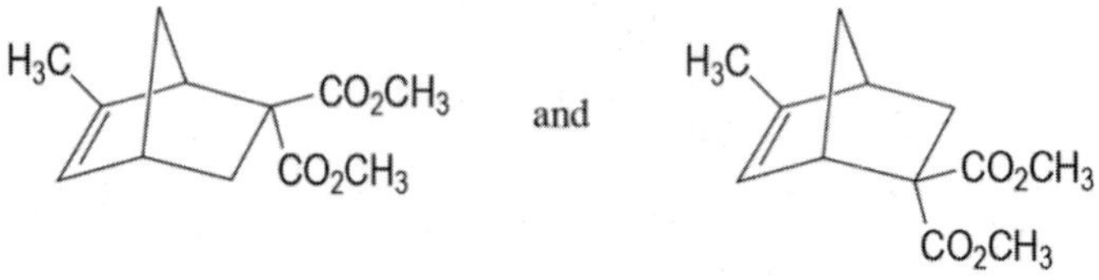

15.17   The Diels–Alder reaction requires that the diene assume an *s*-cis conformation in the transition state. In this conformation, the (2*E*,4*Z*)-diene has a methyl–hydrogen van der Waals repulsion that is absent in the (2*E*,4*E*)-diene. (This point was also used in the solution to Problem 15.3.) This repulsion raises the energy of the Diels–Alder transition state of this diene, and therefore reduces the rate of its reaction. Since the (2*E*,4*Z*)-diene reacts much more slowly than the (2*E*,4*E*)-diene, it follows that the (2*E*,4*E*)-diene is consumed in the reaction, and the (2*E*,4*Z*)-diene remains unreacted.

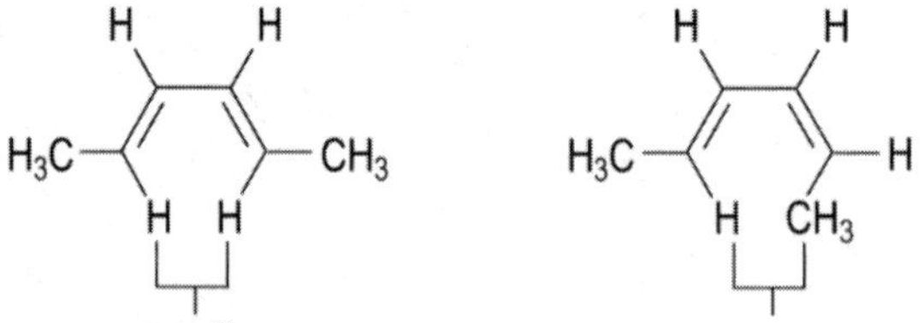

smaller
van der Waals repulsions

s-cis conformation of
the 2*E*,4*E* diene

greater
van der Waals repulsions

s-cis conformation of
the 2*E*,4*Z* diene

15.18   The triene contains two diene units with one double bond common to both. The dienophile reacts with the diene unit that is locked in an *s*-cis conformation.

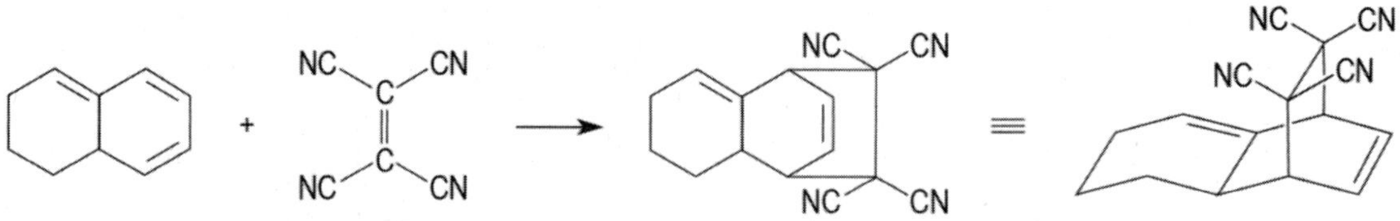

tetracyanoethylene
(TCNE)

15.19   (a)   The two acetoxy groups are cis in the product because they are both outer substituents in the *s*-cis conformation. (The racemate is formed; only one enantiomer is shown below.)

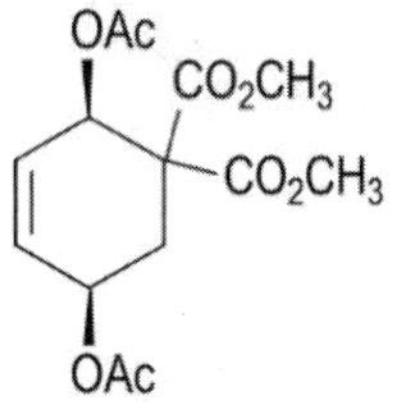

(b)   The two products correspond to the two possibilities in Eq. 15.15 on text p. 739. They result from addition of the diene at either of the two faces of the alkene (or the alkene at either of the two faces of the diene).

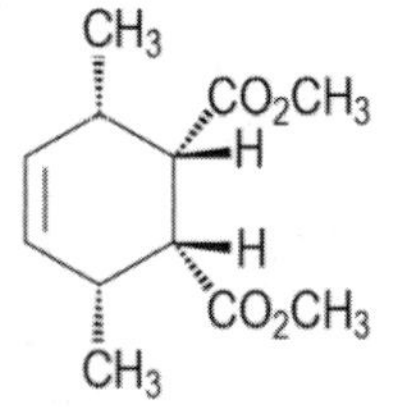

endo stereoisomers 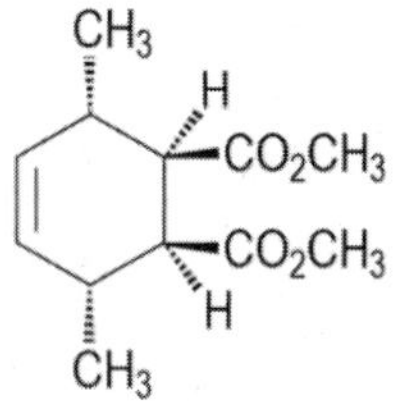   exo stereoisomers

Notice that both stereoisomers result from syn-addition. In other words, there are *two* possible modes of syn-addition.

(c)   The "inner" groups of the diene are "tied together" as the —CH_2— group of the five-membered ring.

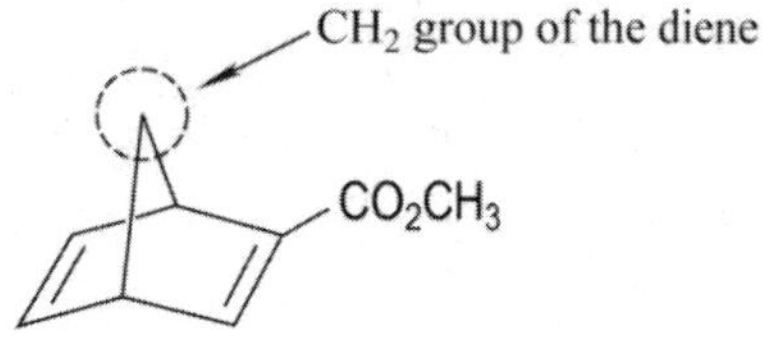

15.20   (a)

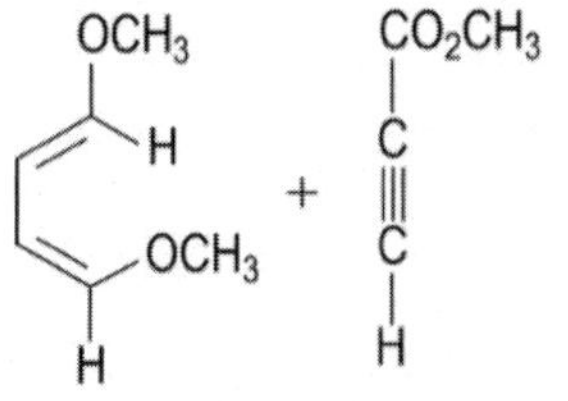

(c)

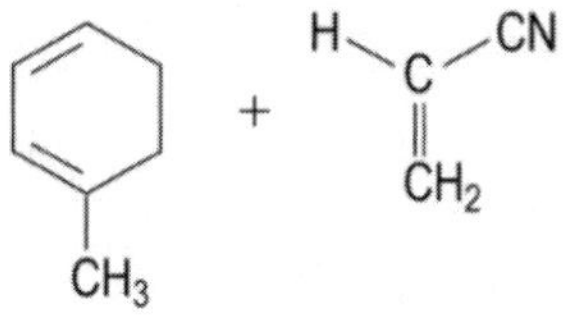

(b)

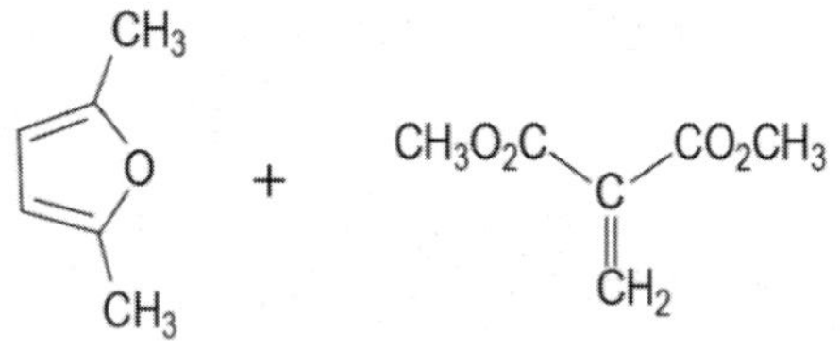

(d)

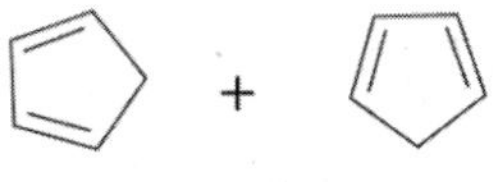

(that is, the reaction of two 1,3-cyclopentadiene molecules)

15.21   (a)   As shown by Eq. 15.15 on text p. 739, the *endo* diastereomer is the one in which the dienophile substituents (which, in this case, are the arms of the five-membered ring) are cis to the outer substituents when the diene is drawn in the *s*-cis conformation.

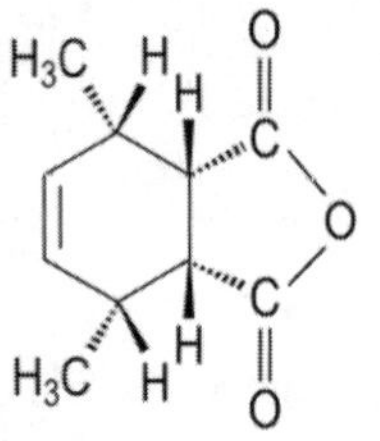
H3C
H
H
O
C
O
C
H
H3C
H
O

(b) Here are the diagrams. Notice that the product is formed initially in a boatlike conformation that subsequently changes to its more stable conformation. Be sure to make models if you have difficulty seeing these relationships.

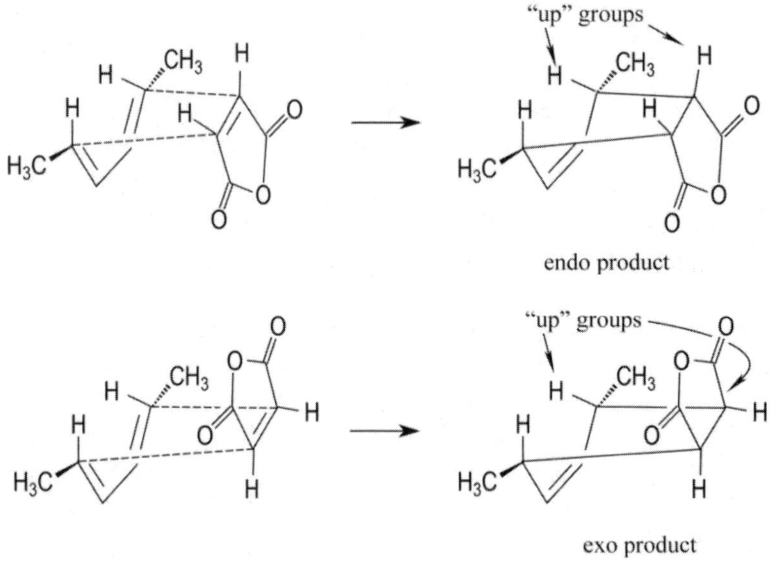
"up" groups
CH3
H3C
endo product
exo product

15.22 Use a method like the one in the solution to Problem 15.21:

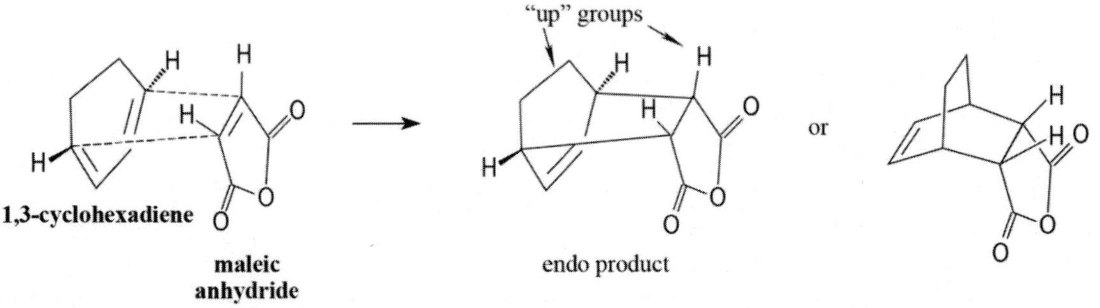
"up" groups
1,3-cyclohexadiene
maleic anhydride
endo product
or

15.23 (a) The carbocation intermediate in the addition of HCl has two sites of electron deficiency, either of which can undergo a Lewis acid–base association reaction with the chloride ion.

(b)   The products are a mixture of constitutionally isomeric ethers because the carbocation intermediate in the S_N1 reaction has two sites of electron deficiency that can react by a Lewis acid–base association reaction with the nucleophile ethanol. The details for the formation of 3-ethoxy-3-methylcyclohexene are shown below; you should provide the additional details of the curved-arrow mechanism for the formation of 3-ethoxy-1-methylcyclohexene.

3-ethoxy-1-methyl-cyclohexene

3-ethoxy-3-methyl-cyclohexene

 It is common (and useful) to show the derivation of different products by showing the separate "reactions" of different resonance structures. Be sure you remember, however, that the resonance structures are different structures used to depict *one* species. Thus, in the structures above, the carbocation is *one* species that reacts in two different ways, *not* a mixture of different species that undergo separate reactions.

15.24   (a)   An allylic carbocation intermediate undergoes a Lewis acid–base association reaction with bromide ion at the two sites of electron deficiency.

$$CH_3CH{=}CH{-}CH_2{-}\overset{+}{\ddot{O}}H_2 \longrightarrow \left[CH_3CH{=}CH{-}\overset{+}{C}H_2 \longleftrightarrow CH_3\overset{+}{C}H{-}CH{=}CH_2 \right] + \ddot{O}H_2$$

$$:\!\ddot{Br}\!:^-$$

$$(A) \qquad :\!\ddot{Br}\!:^- \qquad (B)$$

$$CH_3CH{=}CH{-}CH_2{-}\ddot{Br}\!: \; + \; CH_3CH{-}CH{=}CH_2$$

$$A \qquad\qquad\qquad :\!\ddot{Br}\!:$$
$$B$$

(b) This is an S_N1 reaction; the two products result respectively from the Lewis acid–base association reaction of water at the two electron-deficient carbons of the carbocation intermediate. (Acetone is a co-solvent that helps to dissolve the alkyl chloride in the aqueous solution.)

$$(CH_3)_2C{=}CH{-}CH{-}CH_3 \xrightarrow{\;OH_2\;} \left[(CH_3)_2C{=}CH{-}\overset{+}{C}H{-}CH_3 \longleftrightarrow (CH_3)_2\overset{+}{C}{-}CH{=}CH{-}CH_3 \right] \; :\!\ddot{C}l\!:^-$$

$$:\!\ddot{C}l\!: \qquad\qquad\qquad :\!\ddot{O}H_2 \qquad\qquad\qquad :\!\ddot{O}H_2$$

$$(CH_3)_2C{=}CH{-}CH{-}CH_3 \qquad + \qquad (CH_3)_2C{-}CH{=}CH{-}CH_3$$

$$\overset{+}{:}\!\ddot{O}H \qquad\qquad :\!\ddot{C}l\!:^- \quad \overset{+}{:}\!\ddot{O}H$$

$$H_2\ddot{O}\!: \longrightarrow H \qquad\qquad H_2\ddot{O}\!: \longrightarrow H$$

$$H_3\overset{..+}{O} \; :\!\ddot{C}l\!:^- \; + \; (CH_3)_2C{=}CH{-}CH{-}CH_3 \; + \; (CH_3)_2C{-}CH{=}CH{-}CH_3$$

$$\underbrace{\qquad\qquad}_{\text{HCl in water}} \qquad\qquad :\!\ddot{O}H \qquad\qquad\qquad :\!\ddot{O}H$$

15.25 (a) The allyl anion has four π electrons. Two occupy π_1, and two occupy π_2.

 (b) The negative charge is caused by the fourth electron, which occupies π_2. Because π_2 has a node through the central carbon, the negative charge is delocalized onto the terminal carbons.

 (c) The resonance structures of the allyl anion give the same conclusion; the unshared electron pair (and thus the negative charge) is delocalized to the two terminal carbons.

$$\left[H_2\overset{}{C}{=}CH{-}\overset{..-}{C}H_2 \longleftrightarrow H_2\overset{..-}{C}{-}CH{=}CH_2 \right]$$

allyl anion

15.26 The two constitutional isomers result respectively from "normal" addition (1,2-addition, product A) and conjugate addition (product B). The conjugate-addition product B is the major one at equilibrium because it has an internal double bond (that is, the double bond with the greater number of alkyl substituents). Although the stereochemistry of product B was not requested, it turns out that it is the trans isomer.

$$H_2C{=}CH{-}CH{=}CH_2 + Br_2 \longrightarrow H_2C{-}CH{-}CH{=}CH_2 \; + \; H_2C{-}CH{=}CH{-}CH_2$$

1,3-butadiene $\qquad\qquad\qquad Br \;\; Br \qquad\qquad Br \qquad\qquad\qquad Br$

$$A \qquad\qquad\qquad B$$

15.27 (a) The two electrons that make up the pi bond are shown explicitly here to help you track their movement.

(b)

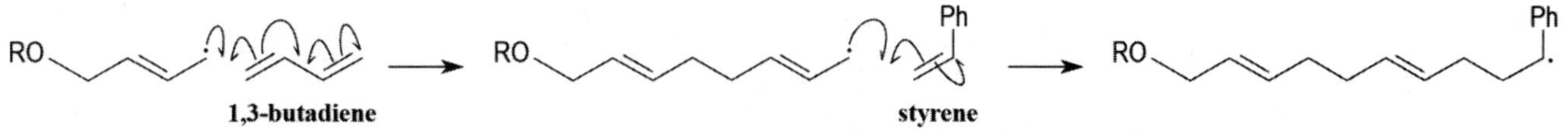

(c)   Before each bond forming reaction, the radical is able to relax into the most stable isomer. The less stable *cis*-isomer will rotate along one of its bonds to become the more stable *trans*-isomer, as shown below. Once the a reaction happens with the next diene, the stereochemistry of the *trans*-double bond is set.

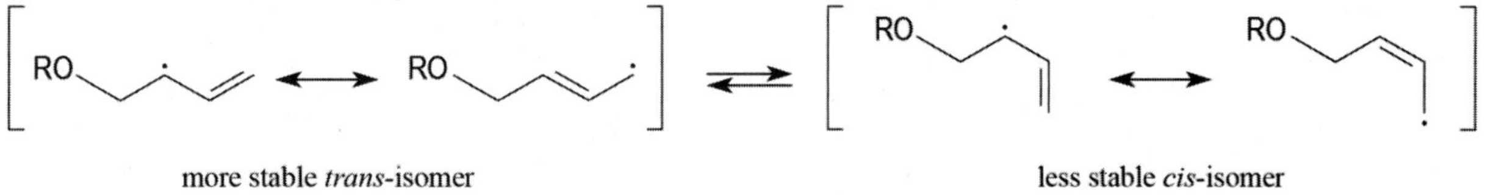

more stable *trans*-isomer   less stable *cis*-isomer

15.28   **(a)**   The second structure is more important because it has an octet on every atom.

(b)   The second structure is more **important because every atom has an octet**; however, the first structure has some importance because it **assigns positive charge to the more electro**positive atom, carbon.

(c)   The second structure is more important because it places electron deficiency (and positive charge) on the carbon with the greater number of alkyl substituents.

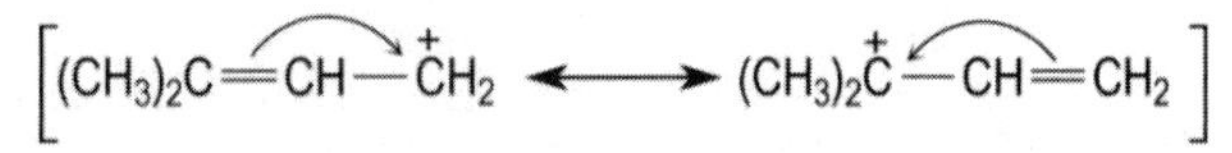

(d)   The second structure has no importance whatever, because it is geometrically impossible; it requires a bond angle of 180° at nitrogen. (See Guideline 4, text p. 753.)

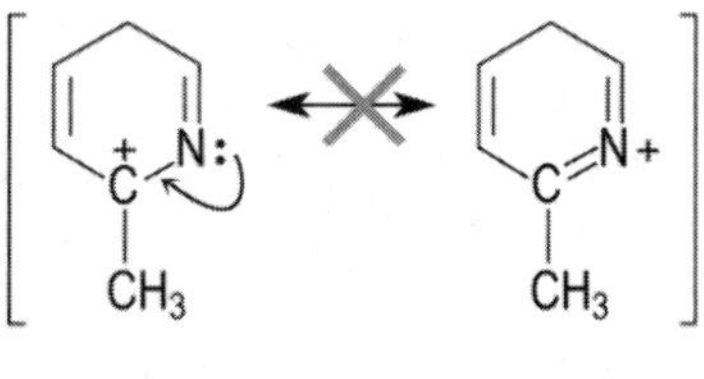

15.29   The electron-deficient carbon has an empty $2p$ orbital that overlaps with a filled $2p$ orbital on the oxygen. The other oxygen electron pair in an sp^2 orbital.

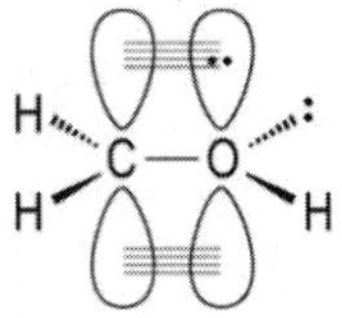

15.30 (a)

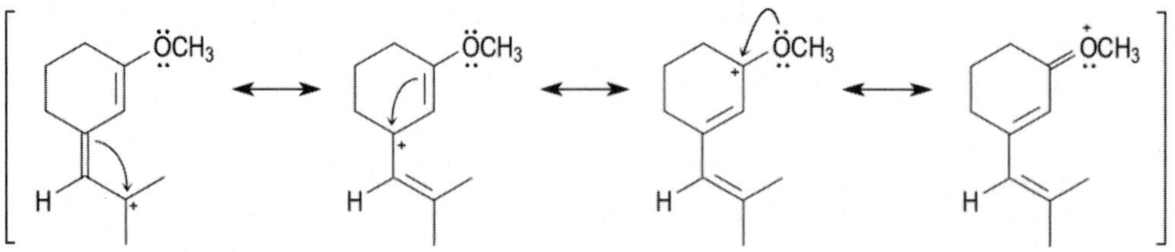

(b) The dashed lines represent bonds in between a full double bond and a single bond. The partial positive charges are not necessarily equal, but they exit on all of the atoms that have a +1 formal charge in one of the resonance structures in part (a). The resonance hybrid is a more accurate description of the electron distribution than any one of the resonance structures in part (a).

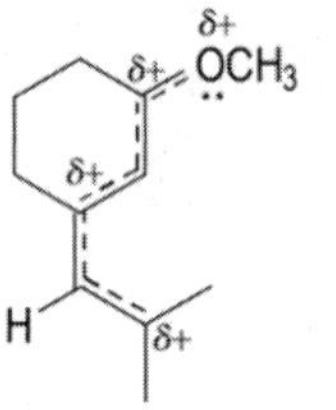

15.31 (a) The ion on the right is more stable; it has two important resonance structures (shown below); the one on the left has only the one structure shown in the problem.

(b) The radical on the left has three important resonance structures, whereas the radical on the right has two. The resonance structures for the more stable radical are shown here. (You should draw the resonance structure of the less stable radical.) In the more stable radical, the unpaired electron can be delocalized by both double bonds, whereas, in the less stable radical, the unpaired electron is delocalized by only one double bond.

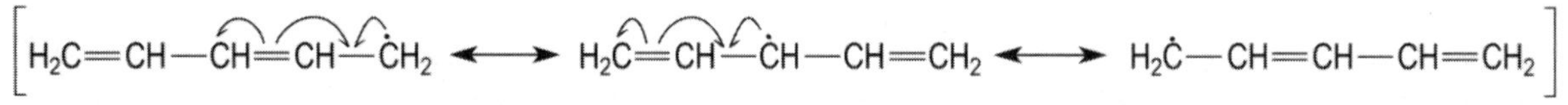

15.32 According to Hammond's postulate (Sec. 4.8D of the text), the reaction that involves the more stable carbocation intermediate will be faster. The carbocation intermediate involved in the solvolysis of compound B is resonance-stabilized; the structures of this ion are shown below. The intermediate involved in the solvolysis of A is not resonance-stabilized. Hence, the ion derived from B is *much* more stable than the one derived from A, and the solvolysis of B is therefore *much* faster than the solvolysis of A.

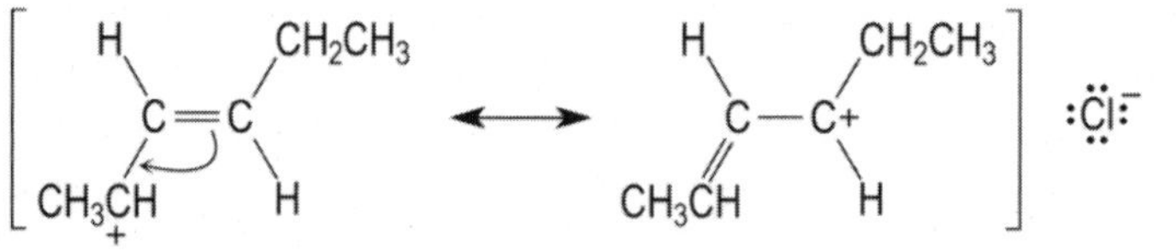

carbocation intermediate involved
in the solvolysis of compound B

15.33 (a) In the reactions below, HA is any acid that can protonate the anion.

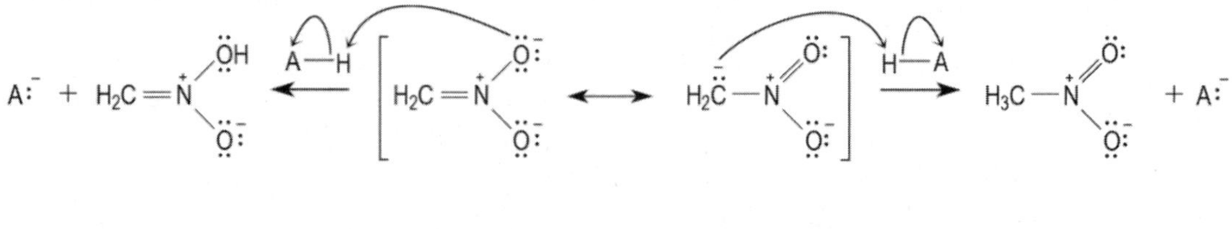

(b)   Compound *B* in problem 15.32 can give two different constitutional isomers when reacting with water via an S_N1 reaction. The intermediate formed after solvolysis is shown in the solution to problem 15.32. The resonance-stabilized carbocation can react with water to form two isomers. (The "–H$^+$" notation means that the additional step that involves removal of a proton after water has added to the carbocation is not shown, but understood.)

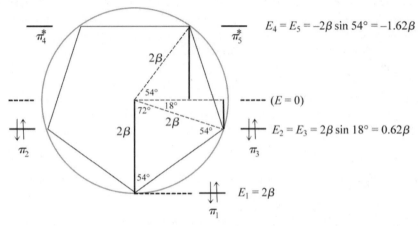

15.34   (a)   Follow the procedure in Study Problem 15.4 on text p. 764.

Step 1: Inscribe a regular pentagon within a circle of radius 2β with one vertex in the vertical position. (This is shown in the construction in the following step.)

Step 2: Place an energy level at each vertex. This gives five energy levels for the five MOs. Notice that two of the antibonding MOs, π_4^* and π_5^*, are degenerate (have the same energy), and two of the bonding MOs, π_2 and π_3, are also degenerate.

$$\pi_4^* \qquad \pi_5^* \qquad E_4 = E_5 = -2\beta \sin 54° = -1.62\beta$$

$$(E = 0)$$

$$2\beta \qquad \pi_2 \qquad \pi_3 \qquad E_2 = E_3 = 2\beta \sin 18° = 0.62\beta$$

$$\pi_1 \qquad E_1 = 2\beta$$

Step 3: Calculate the energies as multiples of β. Remember that the radius of the circle, and therefore E_1, is 2β. (Remember also that β is a negative number.) Some useful trigonometry is that the internal angle in a pentagon is 108°; the bisected internal angle is 54°; and, from the fact that the sum of the three angles of a triangle must be 180°, the angle between the lines that connect successive vertices to the center must be 72°. These angles are labeled in the construction.

Step 4: Add the π electrons to the MOs. This anion contains 6 π electrons distributed in three bonding MOs: two electrons in π_1, the MO at 2.0β, and two electrons in each of the two (degenerate) MOs, π_2 and π_3, at 0.62β. The antibonding MOs π_4^* and π_5^* at –1.62β are unoccupied.

(b)   The Frost circle construction for the cyclopropenyl cation is as follows:

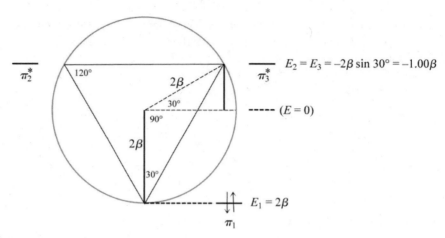

This cation contains two π electrons, which both occupy the bonding MO π_1. From Fig. 15.14, text p. 744, the π-electron energy of the allyl cation is $2(1.41\beta) = 2.82\beta$. The energy of cyclopropenyl is 4.0β. (Remember, β is a negative number.) The extra stabilization, 1.18β, is the contribution of aromaticity. (This does not take into account the destabilizing contribution of ring strain.)

15.35   Such a compound contains 10 π electrons and 10 MOs; 5 are bonding, and 5 are antibonding. One of the bonding π MOs lies at 2β; one of the antibonding π MOs lies at -2β. The four other bonding MOs are degenerate pairs—that is, $\pi_2 = \pi_3$, and $\pi_4 = \pi_5$—at energies < 0, and the four other antibonding MOs also lie in degenerate pairs at energies > 0. Because 10 is a $4n + 2$ number, there are just enough π electrons to fully occupy the bonding MOs.

15.36   Given that furan is aromatic, it can have $4n + 2\,\pi$-electrons (six electrons) if one of the electron pairs on oxygen is in a $2p$ orbital that is part of the π-electron system of the ring. The other electron pair on oxygen is in an sp^2 orbital whose axis is in the plane of the ring. (The electronic structure of furan is drawn in Fig. 26.2 on text p. 1330.)

15.37   An aromatic compound *cannot* have a single unpaired electron as part of its π-electron system, because the number of π-electrons required for aromaticity, $4n + 2$, must be an even number; a single unpaired electron would result in an odd number of electrons.

  However, a free radical could certainly be aromatic if the unpaired electron were not part of the π-electron system. An example is the *phenyl radical,* which could be formed conceptually by abstraction of a hydrogen atom from benzene:

phenyl radical

However, in such a radical, the unpaired electron itself does not contribute to the aromatic stability because it is not part of the $4n + 2$ π-electron system.

15.38   (a)   Thiophene is aromatic. One electron pair on sulfur is part of the aromatic π-electron system; the other is not (as in the oxygen of furan; see the solution to Problem 15.36.)

  (b)   This ion contains $4n$, not $4n + 2,\,\pi$ electrons, and is therefore not aromatic.

  (c)   This ion is not aromatic because it is not a *continuous* cycle of $2p$ orbitals. The carbon at the bottom has no p orbitals; it is a saturated carbon atom.

  (d)   Isoxazole is aromatic. Each double bond contributes two electrons to the π-electron system. One electron pair on the oxygen is also part of the π-electron system, but the other electron pair on the oxygen is not.

(See the solution to Problem 15.36.) The electron pair on the nitrogen is vinylic and, like the electron pair in pyridine, it is not part of the π-electron system.

(e)   This ion is aromatic. Each double bond contributes two electrons to the π-electron system. The cationic carbon bears an empty $2p$ orbital that contributes no electrons but is part of the π-electron system. One electron pair on the oxygen is part of the π-electron system, and the other is not, as in furan (Problem 15.36).

(f)   This compound is not aromatic because it has $4n$ rather than $4n + 2$ π electrons. The empty p orbital on boron, although part of the π-electron system, contributes no electrons.

15.39   Because compounds A and B are antiaromatic, they distort as much as possible to *avoid* overlap of the p orbitals involved in the two double bonds. (See the discussion of cyclobutadiene Sec. 15.7E in the text.) The single bonds are much longer than the double bonds, and the two types of carbon–carbon bonds are not equivalent. Because the deuteriums bridge a double bond in A and a single bond in B, and because the two types of carbon–carbon bonds are different, A and B are different compounds and are therefore in principle capable of independent existence. In contrast, all carbon–carbon bonds in benzene are equivalent—each is half single bond, half double bond. Consequently, all molecules with deuteriums on adjacent positions are identical and indistinguishable. To summarize:

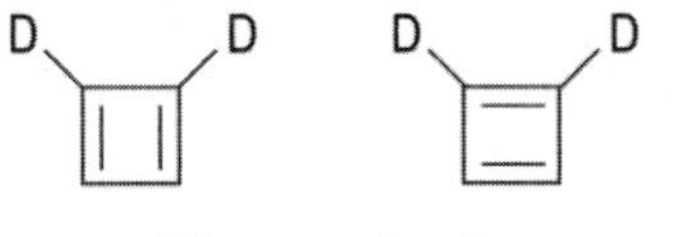

different molecules

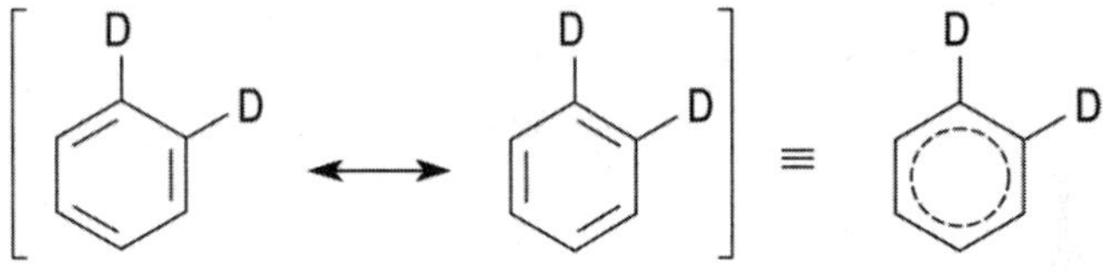

the same molecule

15.40   Compounds (b) and (f) contain $4n$ π electrons and are in principle antiaromatic.

15.41   The positive end of the acetylcholine ion would orient itself inside the barrel of the molecule, much like a ring sliding on a finger. The four faces of the benzene rings would point towards the positive charge, similar to the interaction shown in Fig. 15.20a on p. 775.

Solutions to Additional Problems

15.42   (a)   The structure on the right is somewhat more important because it is a tertiary free radical, whereas the structure on the left is a primary free radical.

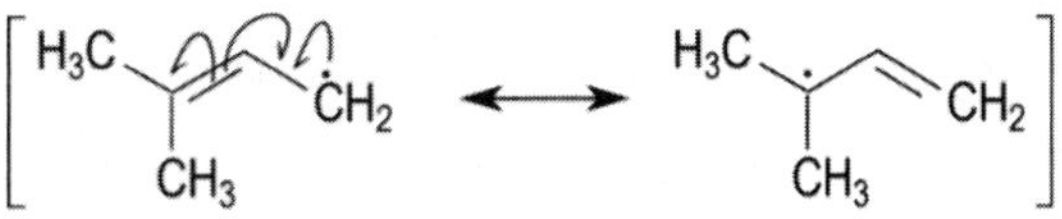

(b)   The structures are identical; they are of equal importance.

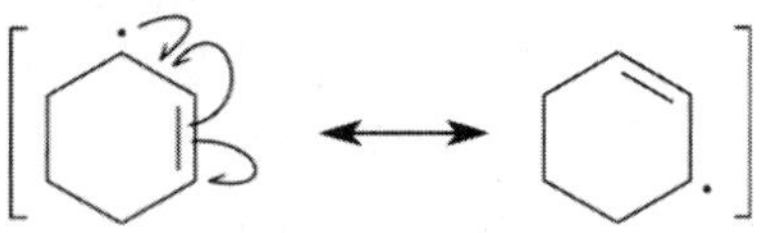

(c)   The middle structure is somewhat more important because the electronegative atom (oxygen) bears the negative charge and the double bonds are conjugated.

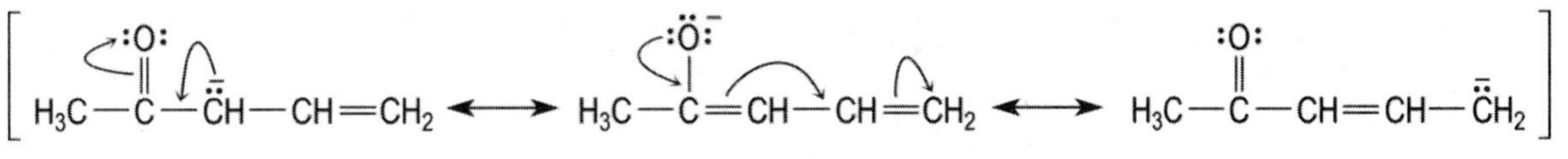

(d)   The fourth and fifth structures are most important because they are aromatic and because they have the negative charge on the most electronegative atom.

(e)   The first and second structures are both secondary carbocations, but the first and third structures have conjugated double bonds; in the third structure, the double bonds have the greatest number of alkyl substituents. All structures are important; the first two structures have somewhat greater importance because of carbocation stability.

(f)   The second structure is somewhat more important because it places electron deficiency (and positive charge) on a secondary carbon. (The first and third structures are identical, although only one of the carbons is written out explicitly.)

15.43   First, draw the structure of the starting material:

trans-1,3-pentadiene

(a)

$$CH_3CH—CH—CH=CH_2$$ + (H$_3$C, H, CH—CH$_2$Br, Br) + $CH_3CHCH=CHCH_2Br$

1,2-addition products

(cis and trans)
1,4-addition product

(b)   Within each pair, the first product is derived from the more stable carbocation intermediate (why?), and is therefore likely to be formed in greater amount. Notice that one of the 1,2-addition products is identical to one of the 1,4-addition products—that is, this compound can be formed by either process.

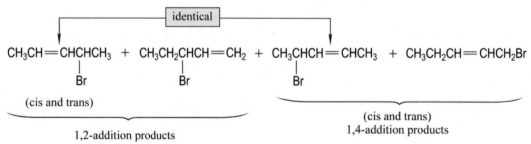

identical

$CH_3CH=CHCHCH_3$ + $CH_3CH_2CHCH=CH_2$ + $CH_3CHCH=CHCH_3$ + $CH_3CH_2CH=CHCH_2Br$

(cis and trans)

1,2-addition products

(cis and trans)
1,4-addition products

(c) $CH_3CH_2CH_2CH_2CH_3$

(d)   Hydration follows the same pattern as HBr addition; notice that, as in part (b), one of the 1,2-addition products is identical to one of the 1,4-addition products.

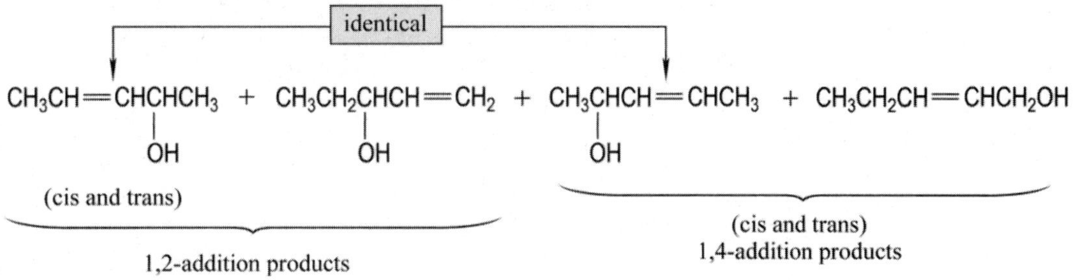

(e)   No reaction occurs.

(f)   A Diels–Alder reaction takes place, and the two possible adducts are formed. (Typically, the endo adduct predominates.)

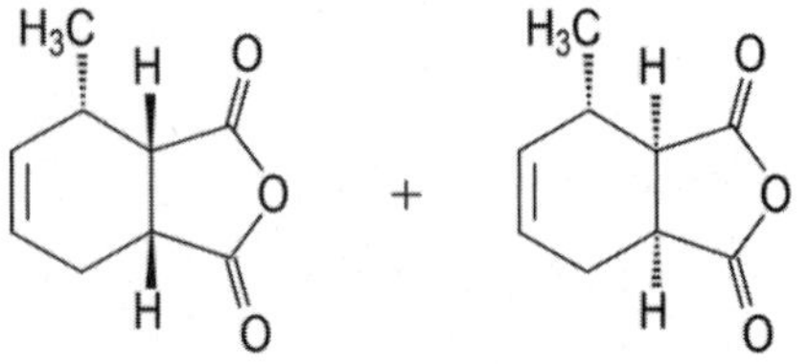

endo   exo

15.44   1,3-Cyclohexadiene gives 3-bromocyclohexene by either 1,2- or 1,4-addition of HBr.

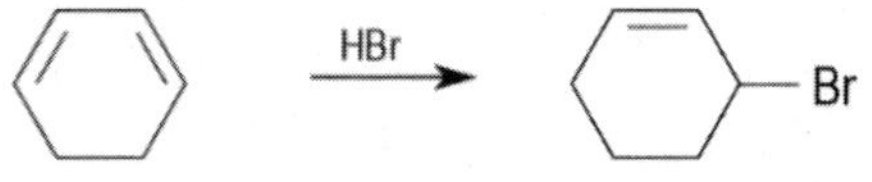

1,3-cyclohexadiene   **3-bromocyclohexane**

15.45   (a)   Since the perpendicular relationship is necessary for chirality of a cumulene, allenes, as well as other cumulenes with an even number of cumulated double bonds, can be chiral. Indeed, 2,3-heptadiene is a chiral allene, and therefore exists as a pair of enantiomers that can in principle be separated by an enantiomeric resolution.

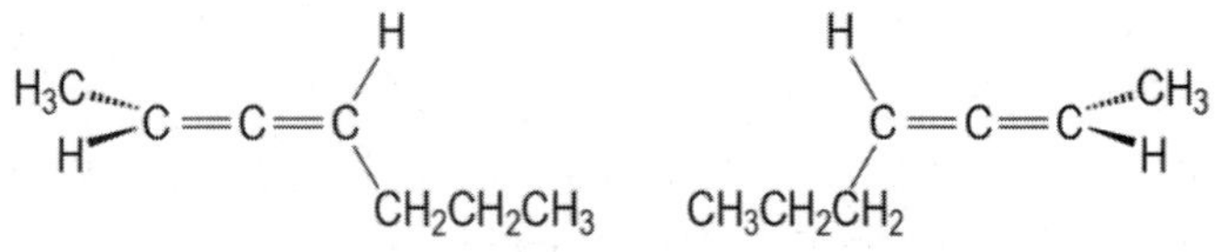

enantiomeric 2,3-heptadienes

(b)   The answers to both (a) and (b) follow from the geometry of cumulenes. Each additional cumulated double bond results in a 90° twist of one end of a molecule with respect to the other. Thus, in alkenes, the atoms connected to the double bond are in the same plane; in allenes, the atoms connected to the ends of the cumulated double bonds are in *perpendicular* planes; in cumulenes with three contiguous double bonds, the atoms attached to the terminal double bonds are in the same plane, as they are in alkenes. These can exist as *E* and *Z* stereoisomers, which are diastereomers:

Z isomer   *E* isomer

the diastereomeric 2,3,4-heptatrienes

15.46   (a)   This species contains six π electrons: two from the double bond, two from the anion, and two from the oxygen. Because $6 = 4n + 2$ for $n = 1$, this species is aromatic.

(b)   This radical has three π electrons: two from the double bond and one from the unpaired electron. It is therefore not aromatic.

(c)   Each double and triple bond contributes a pair of π electrons for a total of 18; therefore this compound is likely to be aromatic. Notice that the second pair of π electrons in a triple bond is in a π orbital that is in the plane of the ring; therefore, only the electrons in the π orbital that is perpendicular to the plane of the ring contribute to the aromatic π-electron system.

(d)   This anion has ten π electrons: two from each of the four double bonds and two from the anionic carbon. It is likely to be aromatic.

15.47   The species with $4n + 2$ π electrons are likely to be planar because the planar species are aromatic. Thus, tropylium ion and cyclooctatetraenyl dianion are planar species. Oxepin, with eight π electrons, avoids planarity and, like cyclooctatetraene, thus avoids antiaromaticity.

15.48   Severe van der Waals repulsions between the inner hydrogens (shown in the following structure) force the compound out of planarity. These van der Waals repulsions are so great in the planar conformation that not even aromaticity can compensate for the resulting destabilization.

15.49   Remember that increasing heat of formations correspond to decreasing stabilities.

(a)   *Heats of formation:* 1 < 2 < 4 << 3. *Reasons:* Compound (1) is more stable than compound (2) because a molecule with conjugated double bonds is more stable than an isomer with isolated double bonds. Compound (2) is more stable than compound (4) because trans alkenes are more stable than cis alkenes. Compound (3) is least stable because an allene is the least stable type of diene.

(b)   *Heats of formation:* 3 < 1 < 2. *Reasons:* Compound (3) is most stable because it is aromatic. Compound (1) is more stable than compound (2) because conjugated alkenes are more stable than isomeric alkynes.

(c)   *Heats of formation:* 2 < 3 < 1. *Reasons:* Compound (2) is most stable because it is aromatic. Compound (1) is least stable because molecules with cumulated double bonds are less stable than isomeric molecules with conjugated double bonds.

15.50   The principles are that conjugation increases the λ_{max} by 40–50 nm per double bond; each alkyl substituent increases λ_{max} by 5 nm; and an *s*-cis conformation of the diene unit increases the λ_{max} and decreases the absorption intensity relative to that of an analogous *s*-trans diene.

(a)   1,4-Cyclohexadiene is not conjugated and therefore has no UV absorption in the useful region of the spectrum; 1,3-cyclohexadiene, an *s*-cis diene, has absorption at 256 nm (see text page 725). The compound with the UV spectrum is therefore the 1,3-diene.

1,4-hexadiene   **1,3-hexadiene**
has no UV spectrum   conjugated; has a UV spectrum

(b)   The first compound has three alkyl substituents on conjugated double bonds; the second compound has two. Hence, the λ_{max} of the first compound should be about 5 nm greater than the λ_{max} of the second.

(c)   The second compound has four alkyl substituents on conjugated double bonds; the first compound has two. (The CH_2 group of the ring counts as a substituent on both double bonds.) Hence, the λ_{max} of the second compound should be about 10 nm greater than the λ_{max} of the first.

(d)   The first compound has three conjugated double bonds, whereas the second compound has two; one double bond is not conjugated with the other two. Hence, the first compound should have a considerably greater λ_{max} than the second.

(e)   Although the first compound is shown in the *s*-trans conformation, it doesn't exist in that conformation because of severe van der Waals repulsions between the *tert*-butyl groups. It can avoid these repulsions by internally rotating into its *s*-cis conformation. Although the *s*-cis conformation has a repulsion between the inner hydrogen and a *tert*-butyl group, this is much less severe than the repulsions between two *tert*-butyl groups in the *s*-trans conformation. Nevertheless, it is likely that the *s*-cis conformation will be distorted from planarity to relieve this repulsion. This will significantly lower the intensity of its UV absorption.

s-trans conformation   *s*-cis conformation

The *s*-trans conformation of the second diene, however, is more stable than the *s*-cis conformation; therefore, the second diene exists mostly in its *s*-trans conformation. Because *s*-cis dienes have considerably greater λ_{max} values than related *s*-trans dienes, the spectrum of the first diene should have a greater λ_{max} value, but a much lower intensity, than the spectrum of the second diene.

15.51   Compounds *B* and *C* are highly conjugated, and would thus absorb in the UV-A and UV-B range and could be used as sunscreens. Compound *A* contains only a single benzene ring and would not be an effective sunscreen because it would absorb at a smaller wavelength.

15.52   First, use Beer's law to determine the concentration of the diene:

$$A = 0.356 \text{ absorbance units} = \epsilon c l = (10{,}750) \text{ absorbance units L mol}^{-1} \text{ cm}^{-1} (c)(1.0 \text{ cm})$$

from which we calculate $c = 3.31 \times 10^{-5}$ mol L^{-1}.

Since the entire sample was diluted to one liter, the entire sample contains 3.31×10^{-5} mole of isoprene. (Any hydrogenation product lacks conjugated double bonds and does not have UV absorption.) The amount of isoprene present is

$$(1.00 \text{ L})(3.31 \times 10^{-5} \text{ mol L}^{-1})(68.12 \text{ g mol}^{-1}) = 2.25 \times 10^{-3} \text{ g} = 2.25 \text{ mg}$$

Because the mass of the entire sample is 75 mg, the mass percent of isoprene is $100(2.25 \text{ mg}/75.0 \text{ mg}) = 3.00\%$. That is, 3% by mass of the isolated material is unreacted isoprene.

15.53   The color of β-carotene is due to its chromophore of extensively conjugated double bonds. Catalytic hydrogenation would result in addition of hydrogen to these double bonds; hence, catalytic hydrogenation of a β-carotene sample would convert it from a red-orange conjugated alkene into a colorless alkane.

15.54   In order to observe fluorescence, a compound must first absorb light to excite electrons to a higher energy state. Without sunlight, there wouldn't be sufficient energy available to excite those electrons.

15.55   The energy difference between the two waves can be calculated using Eq. 15.3. The 10⁹ term at the end of the solution is to convert the difference in reciprocal wavelengths from nm to m.

$$\Delta E = \frac{hc}{\lambda_1} - \frac{hc}{\lambda_2} = hc\left(\frac{1}{\lambda_1} - \frac{1}{\lambda_2}\right) = (3.99\times10^{-13}\text{kJ mol}^{-1}\text{ s})(3\times10^8\text{ m s}^{-1})\left(\frac{1}{488} - \frac{1}{510}\right)$$

$$= (12\times10^{-5}\text{ kJ mol}^{-1})(8.8\times10^{-5}\text{ nm}^{-1})(10^9\text{ nm m}^{-1}) = 10.6\text{ kJ mol}^{-1}$$

15.56   (a)   Other resonance structures are possible, derived from movement of just one or two curved arrows at a time. Those shown represent the most important structures because they demonstrate varying degrees of aromaticity.

right ring is aromatic   left ring is aromatic   all rings are aromatic

(b)   The explanation for part (a) applies to this part as well.

15.57   The carbon skeleton of compound *A*, from the hydrogenation data, is related to that of hexane; that is, it has an unbranched chain of six carbons. The two degrees of unsaturation indicate that the compound is either a diene or an alkyne. The IR data suggests that the compound is not an alkyne; the partial hydrogenation data and the observation of optical activity indicate that the compound is an allene. The only possible chiral allene is 2,3-hexadiene, H₃C—CHACACH—CH₂CH₃. Indeed, partial hydrogenation of this diene would give a mixture of *cis*-2-hexene and *cis*-3-hexene.

15.58   Although mycomycin has no asymmetric carbons, it does contain stereocenters: the outer carbons of the allene unit. Evidently, mycomycin is one enantiomer of this chiral allene. The chirality of certain allenes is discussed in Sec. 15.1C.)

15.59   The positive end of the lysine side chain would be oriented towards the face of the aromatic ring of the phenylalanine residue, similar to the interaction shown in Fig. 15.20(a) on p. 775.

15.60   (a)   The dissociation constant, K_d, can be calculated by using a form of Eq. 3.33 on p. 110. The dissociation constant for *N*-methylquinolinium ion with molecule *A* is

$$K_d = e^{-\left(\frac{\Delta G_d^\circ}{RT}\right)} = e^{-\left(\frac{28.9 \text{ kJ mol}^{-1}}{(0.008314 \text{ kJ mol}^{-1} \text{ K}^{-1})(273 \text{ K})}\right)} = 2.96 \times 10^{-6}$$

The dissociation constant for 4-methylquinoline with molecule A is 5.66×10^{-5} (or 56.6×10^{-6}, for comparative purposes). This is a larger dissociation constant, meaning that at equilibrium there is a larger proportion of the dissociated species (molecule A and 4-methylquinoline) than there is of the complexed pair, relative to N-methylquinolinium ion and molecule A and its complexed pair. Restated, N-methylquinolinium ion binds more strongly to molecule A than N-methylquinoline does.

(b) Since N-methylquinolium ion is positively charged, it is attracted to the aromatic rings of the barrel-shaped molecule A more strongly than neutral 4-methylquinoline.

(c) In addition to the pi-cation interaction described in part (b), there are also electrostatic attractions between the negatively charged groups in molecule B and the positively charged N-methylquinolium ion.

15.61   In order for a diene to react in a Diels–Alder reaction, it must be able to assume an s-cis conformation in the transition state.

van der Waals repulsions are much more severe when R = *tert*-butyl than when R = H

When R = *tert*-butyl, the van der Waals repulsions in the s-cis conformation are much more severe than when R = methyl. Consequently, when R = *tert*-butyl, the transition state of any Diels–Alder reaction is destabilized by these van der Waals repulsions to a greater extent than when R = methyl. Evidently, the destabilization is so great that the Diels–Alder reaction with maleic anhydride does not occur at all when R = *tert*-butyl.

15.62   The analysis in Eqs. 15.10a–b on text p. 737 shows that the s-cis conformations of dienes with cis double bonds are destabilized by van der Waals repulsions, whereas the s-cis conformations of dienes with trans double bonds do not suffer the same repulsions.

van der Waals
repulsions

s-cis conformation
of an all-cis diene

s-cis conformation
of an all-trans diene

These van der Waals repulsions also destabilize the transition states of Diels–Alder reactions, which require s-cis conformations of the reacting dienes. Because the diene in the problem undergoes the Diels–Alder reaction under mild conditions, it is probably the all-trans diene.

15.63   The product of such a Diels–Alder reaction would be most interesting:

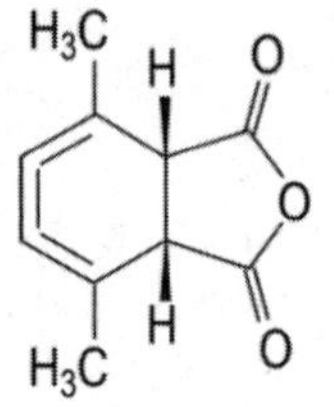

This is a cyclic triene containing three cumulated double bonds within the ring. Such an adduct could never form, because the requirement for linear geometry in the cumulated double bonds of the product would result in a compound with immense angle strain. A related explanation is that the $2p$ orbitals of maleic anhydride and those of the alkyne must overlap as shown in Figure 15.13 on text p. 736. Because the alkyne is constrained to linear geometry, the transition state for such a reaction would either be very strained (if the alkyne units were bent) or would contain extremely poor electronic overlap (if the alkyne units remained linear).

15.64   The answer to this problem lies in the *s*-cis conformations of the three dienes, which are as follows:

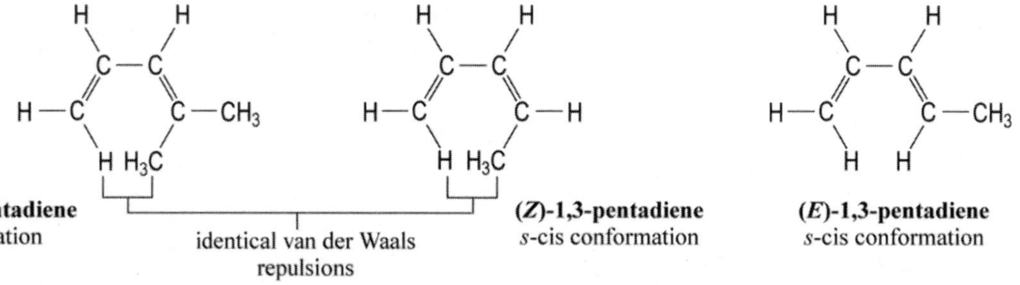

4-methyl-1,3-pentadiene   **(Z)-1,3-pentadiene**   **(E)-1,3-pentadiene**
s-cis conformation   *s*-cis conformation   *s*-cis conformation

identical van der Waals
repulsions

The reactivities of these three dienes in the Diels–Alder reactions correlates nicely with the accessibility of their *s*-cis conformations. (Remember, the transition state of the Diels–Alder reaction involves the *s*-cis conformation of the diene.) The van der Waals repulsions shown above destabilize the *s*-cis conformations of both 4-methyl-1,3-pentadiene and (*Z*)-1,3-pentadiene to the same extent because the interacting groups are identical. Hence, their reactivity is reduced by about the same amount relative to that of (*E*)-1,3-pentadiene, whose *s*-cis conformation has no such repulsions.

15.65   (a)   First, draw the structures of the two carbocations that would be produced by the alternative modes of protonation:

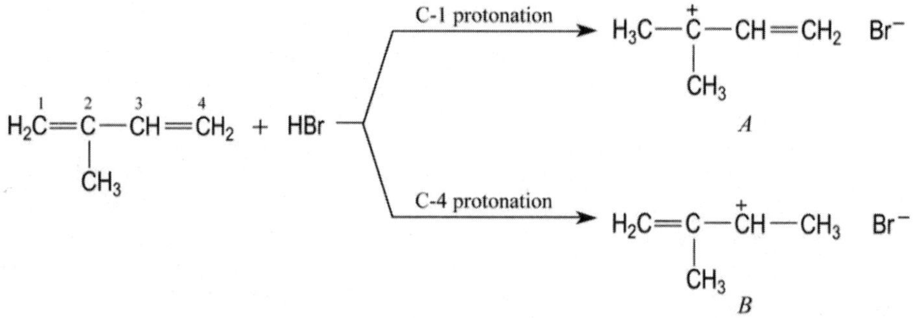

        Although both carbocations are allylic, carbocation *A* has a *tertiary* carbocation resonance structure, shown above, while neither resonance form of carbocation *B* is tertiary, Hence, carbocation *A*, from protonation at carbon-1, is the more stable carbocation.

   (b)   Because carbocation *A* is the more stable ion, it is the carbocation intermediate involved in HBr addition, and the products are thus derived from it. The two products come from the Lewis acid–base association reaction of bromide ion with the (electron-deficient) carbons that bear positive charge, which we deduce from the resonance structures of the ion.

resonance structures of carbocation *A*   1,2-addition product   1,4-addition product

(c)   Protonation at carbon-1 of *trans*-1,3,5-hexatriene gives the carbocation with the greatest number of resonance structures. (Verify this point by examining the carbocations obtained from protonation at other carbons.)

$$\text{H}_2\text{C} \xrightarrow[\text{protonation at C-1}]{\text{HBr}}$$

The three possible products are derived from the Lewis acid–base association reaction of Br⁻ at the positively charged carbons.

(d)   The kinetically controlled product is the 1,2-addition product in each case; this is labeled explicitly in part (b), and is compound *A* in part (c). The reason is given in the text; the 1,2-product results from collapse of the initially formed ion pair at the nearest site of positive charge. (See Eq. 15.27 on text p. 747.) In part (b), the thermodynamically controlled product is the 1,4-addition product, because it contains the double bond with the greater number of branches. The thermodynamically controlled product in part (c) is compound *C*, a 1,6-addition product, because it derives special stability from the conjugated arrangement of its double bonds and the greater amount of alkyl substitution at the double bond (relative to *A*).

15.66   (a)   The 1,2- and 1,4-addition products formed in the reaction of 1,3-pentadiene and HCl are identical if we assume that all double bonds retain their *E* stereochemistry.

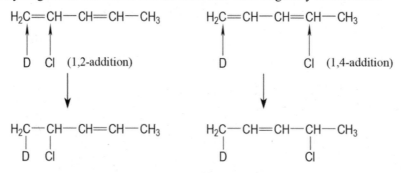

(b)   There can be no preference for 1,2- versus 1,4-addition on the basis of the product stability, because, as shown in part (a), the products are the same. The problem is that we can't tell one mode of addition from the other! The use of D—Cl, however, solves this problem. The use of deuterium allows us to distinguish between the hydrogen that has added and the ones that were originally in the diene.

The two products are different. The isotope has a negligible effect on product stability.

(c)   According to the text, the product of 1,2-addition is preferred, and it is shown in the solution to part (b). In fact, this result was observed experimentally. The actual ratio is about 70:30 in favor of 1,2-addition. (How would we tell one product from the other?) This experiment, reported in 1979 by J. E. Nordlander of Case Western Reserve University, established clearly that product stability has nothing to do with the kinetic preference. The arguments in the text are the only reasonable alternative.

15.67 (a)   Unless the reaction is allowed to run for a very long time, this is a kinetically controlled reaction, because the ratio of the two products is different from the ratio at equilibrium.

(b)   Compound *C* has a double bond with four substituents; compound *B* has a double bond with three substituents. Because substitution at the double bond enhances alkene stability, *C* is more stable and hence would be favored at equilibrium.

(c)   The reaction occurs through the following carbocation intermediate by a typical E1 process (see Sec. 10.2).

The formation of the product involves loss of a proton from this intermediate. Alkene *B* can form in two ways, namely, by abstraction of either one of two protons. Alkene *C* can form in only one way. Hence, the formation of alkene *B* is favored on statistical grounds. It is interesting that, in Fig. P15.67, the ratio of products at zero time is indeed about 2:1.

15.68   (a)   The transformations involved in the reaction between 1,3-cyclopentadiene and maleic anhydride:

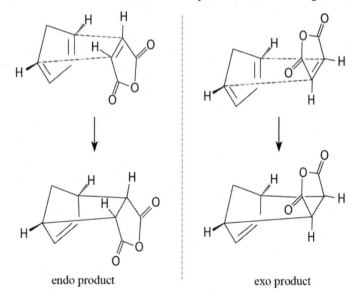

1,3-cyclopentadiene   maleic
                           anhydride

endo product

exo product

(b)   Because the initially formed product distribution is markedly different from the equilibrium distribution, the reaction is kinetically controlled.

(c)   At low temperature, the formation of product is kinetically controlled. The problem tells us that the endo stereoisomer is the major one formed. The transition state for this process is shown on the left side of the following diagram. The transition state for the other process is shown at right.

endo product   exo product

It has been suggested that kinetic control in the Diels–Alder reaction is due to "maximum accumulation of unsaturation," or "secondary orbital interactions." You can see that in only the endo transition state, the 2p orbitals of the anhydride carbonyl groups and some of the 2p orbitals of the diene unit are "face-to-face."

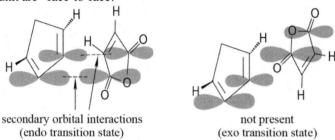

secondary orbital interactions   not present
(endo transition state)   (exo transition state)

The interaction between these orbitals evidently stabilizes the transition state. The endo product is formed almost exclusively. However, when the two products are allowed to come to equilibrium, there is little difference in their stabilities.

15.69   (a)   Product *A* is the kinetic product because it is formed more rapidly.

      (b)   The relative stabilities of the two products can be deduced from their chair conformations. The 1,3-diaxial interactions of the axial bromines with the axial methyl group (on the top) and the axial hydrogen (on the bottom) cause compound *A* to be less stable.

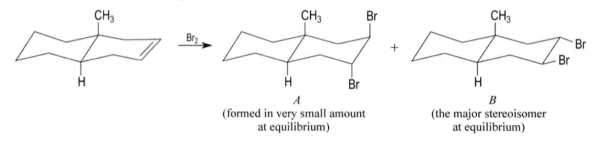

A
(formed in very small amount
at equilibrium)

B
(the major stereoisomer
at equilibrium)

      (c)   Two possible bromonium ions can be formed, which we'll call *X* and *Y*.

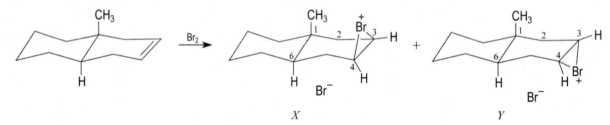

X

Y

The carbons are numbered for reference. We'll come back later to the issue of which is preferred.

    Here's the key point: The nucleophilic reaction of Br⁻ with the bromonium ion *must result in a product conformation in which the two Br atoms are anti!* This is a consequence of the geometry of backside substitution. Such a conformation can be reached by the reaction of Br⁻ at either carbon-3 or carbon-4 in each of the bromonium ions. We now analyze each of these reaction modes in turn. Backside substitution of Br⁻ at carbon-3 of *X* gives a *boat conformation of compound B*. This conformation can then relax to *B* itself.

     If you find an error in this manual, please visit http://people.pharmacy.purdue.edu/~loudonm/teaching/

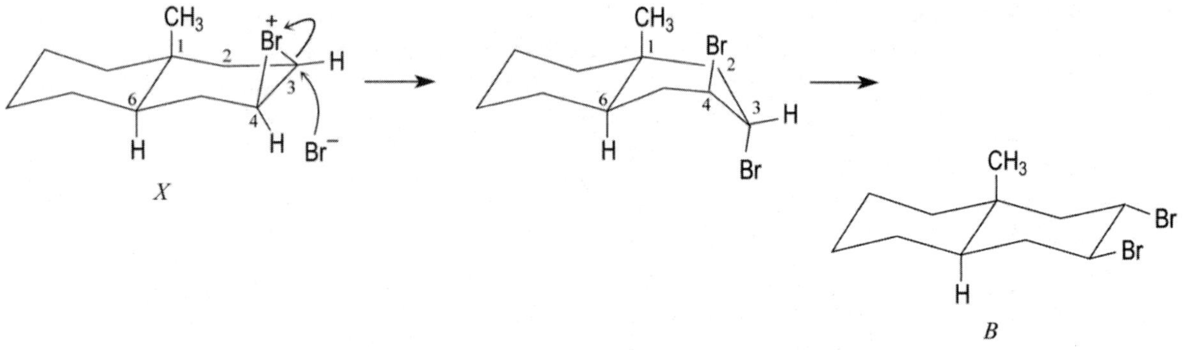

As the nucleophilic reaction occurs, carbon-3 swings down and the three groups bound to it swing up—inversion of configuration. If you have trouble seeing why this reaction should generate a boat, look at the *reverse* reaction. Start with compound *B* and imagine formation of a bromonium ion from *B* by the reverse reaction. To form this ion, the two bromines must achieve an anti conformation—that is, a dihedral angle of 180°. (Just run this backwards to see this.) Only by forming a boat can backside substitution occur in the reverse direction; therefore, it's also the only way backside substitution can occur in the forward direction. The forward and reverse of a reaction must occur by the same mechanism! (Principle of microscopic reversibility; text p. 175.)

The nucleophilic reaction of Br⁻ at carbon-4 of ion *X*, however, gives the required anti arrangement of the two bromines in a *chair conformation:*

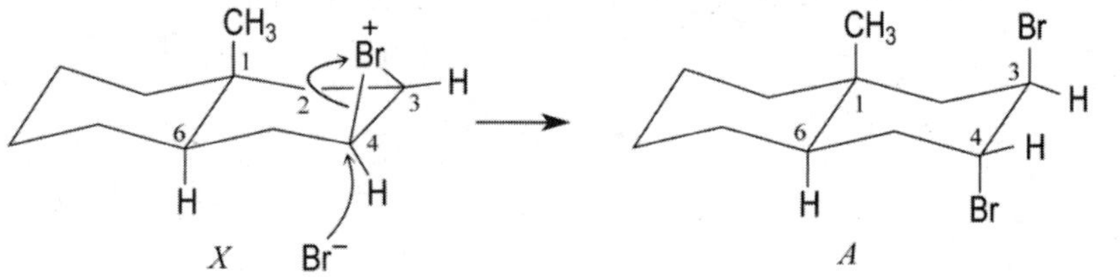

This gives compound *A* directly. Therefore, bromonium ion *X* is faced with two non-optimal choices: it can open to give a boat, or it can open to give the diaxial form *B*. Because boats have a very high energy, the latter option is the least energetic alternative and is therefore the one observed.

Now let's consider ion *Y* in the same way. The analysis is the same, except that *the reaction of Br⁻ at carbon-4 gives a boat conformation of* B, *and reaction of Br⁻ at carbon-3 gives* A. Here's the reaction at carbon-4 to give a boat:

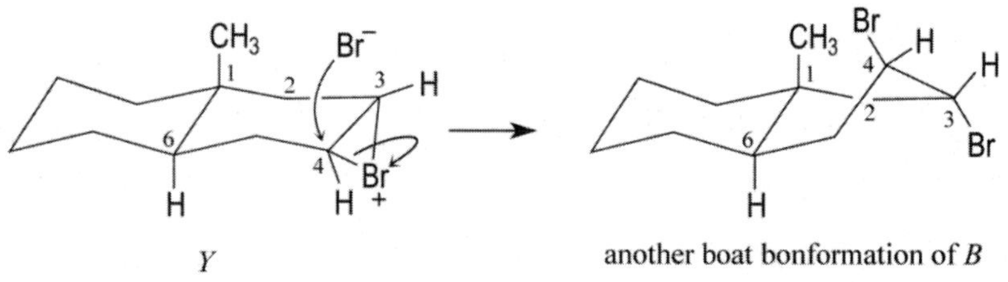

(We leave it to you to show the reaction of Br⁻ at carbon-3, which gives isomer *A* in the chair conformation.) Therefore ion *Y* is faced with the same two nonoptimal choices as ion *X*, and the result is the same. In other words, the less stable product is formed more rapidly. The more stable product is formed slowly because of the high-energy boat intermediate required.

Thus, the relative stability of the two ions *X* and *Y* is irrelevant. *Y* might be a little more stable because it avoids the 1,3-diaxial interaction with the methyl group. But the outcome is the same no matter which ion forms. In other words, the outcome is determined by the relative rates of the two product-determining steps in the reaction of either bromonium ion.

(d)   Either bromonium ion *X* or bromonium ion *Y* can form from compound *A*; this process is merely the reverse of the formation of the dibromides from the bromonium ion.

We've shown previously how bromide can react with either of these bromonium ions to give the more stable product via boat conformations. Although these reactions occur slowly, once they occur, they form the more stable product, thus providing a route to bring the two compounds into equilibrium.

15.70   Given that the addition occurs to the triple bond, the addition should occur regiospecifically to give the product in which the chlorine ends up on the internal carbon of the triple bond.

$$HC{\equiv}C{-}CH{=}CH_2 \ + \ HCl \ \longrightarrow \ H_2C{=}\overset{\overset{\displaystyle Cl}{|}}{C}{-}CH{=}CH_2$$

vinylacetylene
(1-buten-3-yne)   **chloroprene**

To deduce the structure of neoprene, follow the pattern for 1,4-addition polymerization shown in the solution to problem 15.27 in this manual, and Sec. 5.7 in the text.

$$\left(\!CH_2{-}\overset{\overset{\displaystyle Cl}{|}}{C}{=}CH{-}CH_2\!\right)_{\!n}$$

neoprene

15.71   Assume that *A* and *B* are 1,2- and 1,4-addition products.

$$H_2C{=}CH{-}CH{=}CH_2 \ \xrightarrow{\ Cl_2\ } \ H_2C{-}\underset{\underset{\displaystyle Cl}{|}}{\overset{}{C}}H{-}\underset{\underset{\displaystyle Cl}{|}}{\overset{}{C}}H{-}CH{=}CH_2 \ + \ ClCH_2{-}CH{=}CH{-}Cl$$

1,4-addition product

1,2-addition product

Which is *A* and which is *B*? And what is the stereochemistry of the 1,4-addition product? To answer these questions, note that further addition of Cl_2 to each gives a 1,2,3,4-tetrachlorobutane. Assume that addition of Cl_2 is an *anti*-addition. Because the 1,4-addition product has two stereocenters, *anti*-addition of Cl_2 will give the meso tetrachloride if the alkene is trans, and will give the racemic tetrachloride if the alkene is cis. (If this point is not clear, review Sec. 7.8C.) Because compound *B* gives a single tetrachloride *C* with meso stereochemistry, it follows, then, that the 1,4-addition product is compound *B*, and that it has trans stereochemistry. The 1,2-addition product, then, is compound *A*. Addition of Cl_2 to compound *A* gives both diastereomers of the tetrachloride. To summarize:

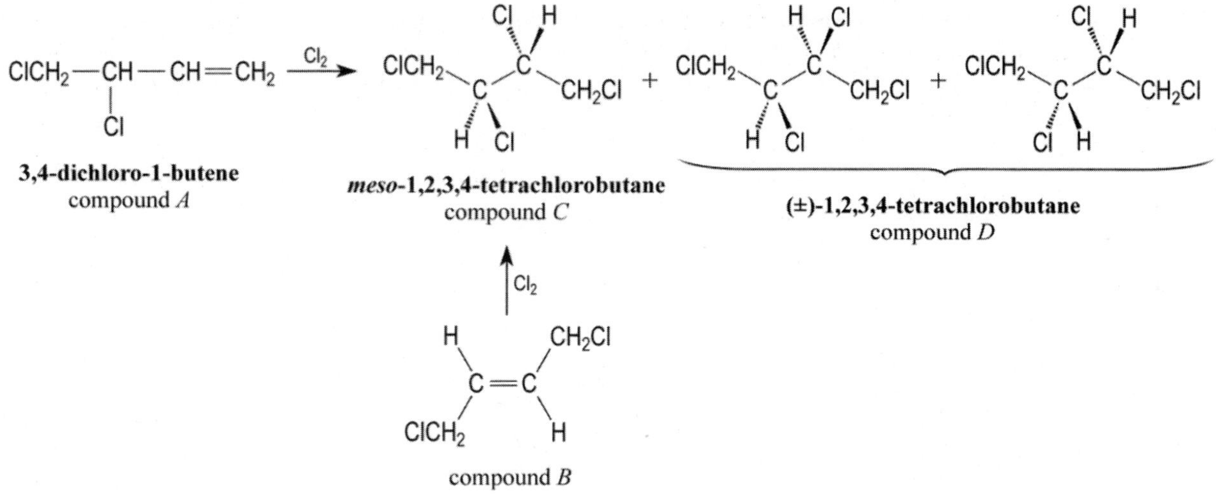

$ClCH_2-CH-CH=CH_2 \xrightarrow{Cl_2}$

3,4-dichloro-1-butene
compound *A*

meso-1,2,3,4-tetrachlorobutane
compound *C*

(±)-1,2,3,4-tetrachlorobutane
compound *D*

compound *B*

15.72   Reaction of KH with cyclopentadiene forms the potassium salt of the cyclopentadienyl anion (structure *X* below) plus dihydrogen, H_2. In this anion, all carbons, except for the isotope, are equivalent by resonance. The isotope makes no detectable difference in the relative importance of the resonance contributors—that is, except for the position of the isotope, all resonance contributors are identical. Hence, protonation of this anion by H_2O occurs at each carbon with equal (20%) probability. Protonation at carbons 2 and 5 gives the same product; and protonation at carbons 3 and 4 gives the same product. Hence, each of these products is obtained in (2 × 20%) = 40% yield.

X

from proton at—   C-1   C-2 and C-5   C-3 and C-4

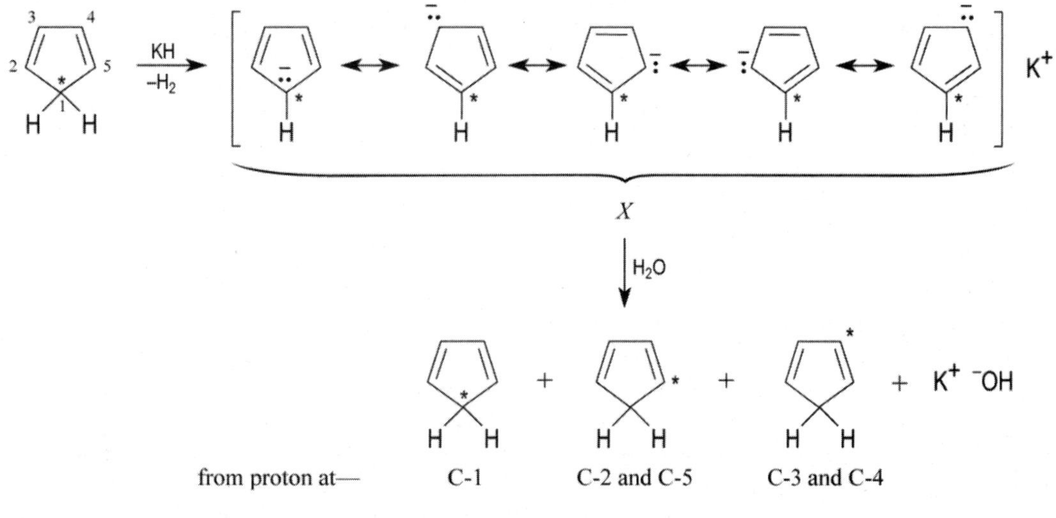

15.73   Borazole is very stable because it is aromatic; each nitrogen contributes two π electrons and each boron contributes zero electrons, for a total of six π electrons in the aromatic system. The resemblance of borazole to benzene is more obvious from its other two resonance structures:

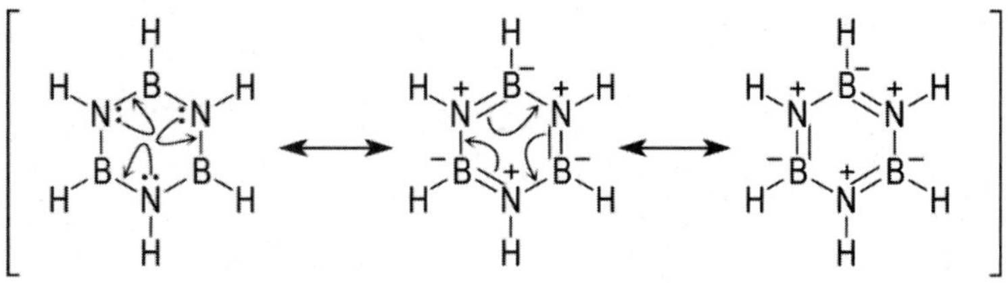

15.74   The conjugate-base anions of *both* compounds are aromatic. However, pyrrole itself is aromatic, whereas 1,3-cyclopentadiene is not. Therefore upon ionization, 1,3-cyclopentadiene *gains aromaticity,* whereas pyrrole does not. An amine is typically 23 pK_a units more acidic (32 versus 55) than an alkane; the gain of aromaticity in the ionization of the hydrocarbon makes it 2 pK_a units *more* acidic than the amine. The net difference of 25 pK_a units corresponds to $(2.3RT)(\Delta pK_a) = 5.71 \times 25 = 143$ kJ mol^{-1} for the gain of aromaticity. (See Eq. 3.34, text p. 110.) This energetic advantage of aromaticity is remarkably similar to the resonance energy of benzene (Sec. 15.7C of the text).

15.75 (a) The conjugate-base enolate ion is stabilized by the polar (electron-withdrawing) effect of the nearby carbonyl bond dipole as well as by resonance:

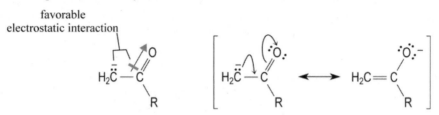

resonance stabilization of the ion

Recall (Sec. 3.6, text p. 116) that stabilization of a conjugate base increases the acidity of its conjugate acid.

(b) While the ionization of the molecule at either position would benefit from the polar effect of the nearby carbonyl group, only the conjugate-base enolate ion resulting from the ionization of hydrogen H^b would be enhanced by the resonance effect. The enolate ion resulting from the ionization of H^a is not stabilized by resonance, because resonance is geometrically impossible. The orbital interaction required for resonance interaction requires an impossible twisting of the molecule. (Guideline 4, text p. 753.) To put it another way, the resonance structure violates Bredt's rule:

the conjugate-base anion
from the loss of proton H^a

Therefore, the ionization of H^b would have the lower pK_a; that is, H^b is more acidic.

15.76 Compound *A* should react much more rapidly because the carbocation intermediate (shown below) has three important resonance contributors. In particular, an unshared pair of electrons on the oxygen can be delocalized in this cation; in the solvolysis of the other compound, the unshared pairs on oxygen have no resonance interaction with the positive charge in the carbocation.

carbocation intermediate involved in the solvolysis of *A*

By Hammond's postulate, the reaction involving the more stable carbocation intermediate is faster.

15.77 (a) The carbocation is stabilized through resonance by delocalizing the charge through the aromatic ring:

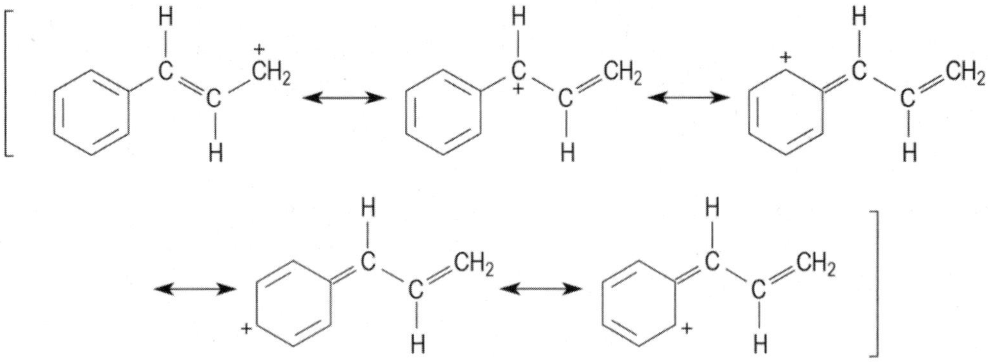

(b)   The two products below are the only two formed that keep the aromatic ring intact.

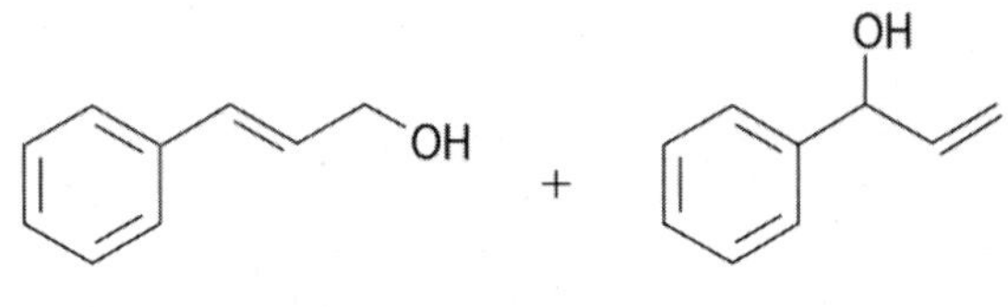

these are the only two products formed

Any other product would not be aromatic and would be much less stable.

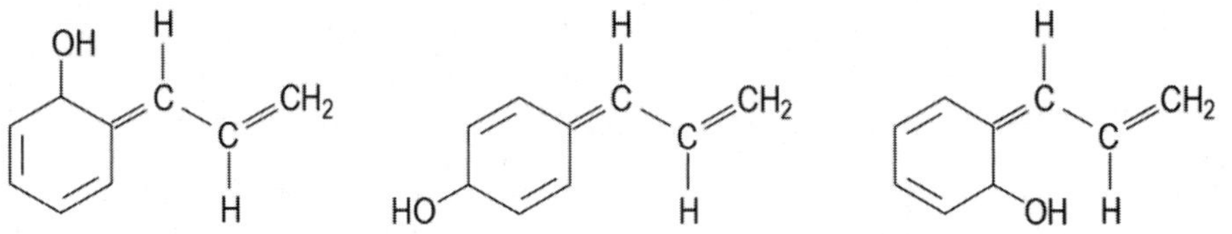

these products are not formed

(c)   The first product shown in part (b) is more highly conjugated than the second, and thus would absorb at a longer wavelength in the UV spectrum.

(d)   Isomer *A*, below, would give the same set of resonance stabilized carbocation intermediates, shown in part (a), as *trans*-cinnamyl chloride.

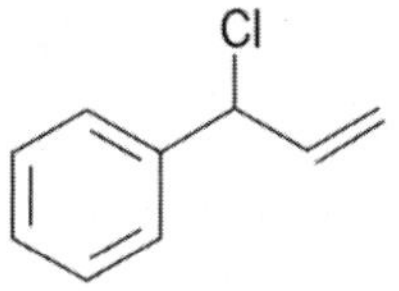

A

15.78   The carbocation intermediate in the S_N1 solvolysis of compound *A* is resonance-stabilized, but the carbocation intermediate in the solvolysis of compound *B* is not. Because the relative energies of transition states reflect the relative energies of the corresponding carbocation intermediates (Hammond's postulate), the solvolysis of *A*, which gives the more stable carbocation, is faster.

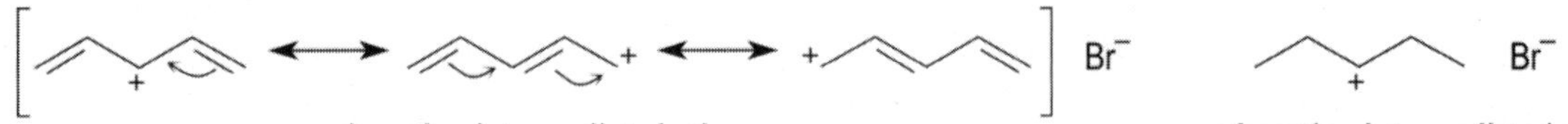

carbocation intermediate in the
solvolysis of compound *A*

carbocation intermediate in the
solvolysis of compound *B*

If compound *C* gives a doubly allylic carbocation, why should it be solvolytically inert? The answer is that the cation has a continuous cycle of π electrons. Because this cation contains $4n$ ($n = 1$) π electrons, it is *antiaromatic* and therefore very unstable. In fact, it is so unstable that it does not form.

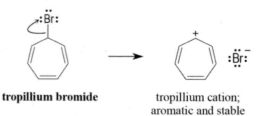

antiaromatic
and unstable

15.79   This compound behaves like a salt because it *is* a salt. Tropylium bromide ionizes readily to bromide ion and
the *tropylium cation,* which, because it has a continuous cycle of six ($4n + 2$, $n = 1$) π electrons, is aromatic,
very stable, and very easily formed.

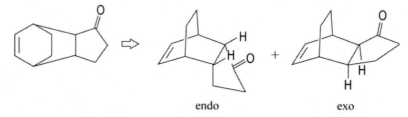

tropillium bromide   tropillium cation;
aromatic and stable

15.80   (a)   The triple bond hydrogenates, but the phenyl rings resist hydrogenation because they have aromatic
character. The product of syn-addition is the cis alkene.

Ph   Ph
\   /
C = C
/   \
H   H

***cis*-stilbene**
(***cis*-1,2-diphenylethylene)**

(b)   This is an ordinary Diels–Alder reaction; because the diene is cyclic, a bicyclic product is formed. Two
diastereomers are possible. (Each is formed as the racemate, only one enantiomer of which is shown
below.)

endo   exo

(c)   This is a Diels–Alder reaction with the alkyne as the dienophile. The phenyl rings are cis because, in the
s-cis conformation of the diene, they are both "outer" groups. (See text Eqs. 15.12a–b, text p. 738.)

(d)   The conjugated-diene part of the triene undergoes a Diels–Alder reaction with a double bond of the benzoquinone. Because there are two double bonds in the quinone, a second Diels–Alter reaction also occurs. Two constitutional isomers result from the second addition because the diene can add in two different orientations.

product of first
Diels–Alter reaction

Moreover, each of these constitutional isomers can exist as two diastereomers:

(Which of *these* are chiral and can therefore exist as enantiomers, and which are meso compounds?)

(e)   The maleic anhydride reacts with the diene unit that is "locked" into an *s*-cis conformation. The diastereomer of the product is formed that results from addition to the face of the steroid opposite to the angular methyl groups, which block the β-face of the molecule. (See Sec. 7.6D of the text.) In this mode of addition, van der Waals repulsions between maleic anhydride and the angular methyl groups are avoided.

(f)   Because the product has the same number of carbons as the starting material, the reaction is intramolecular. In fact, it is an intramolecular Diels–Alter reaction.

The bridgehead double bond in this product does not violate Bredt's rule (Sec. 7.6C) because it is part of a large ring.

(g)   The product is *nickelocene,* the nickel analog of ferrocene. (See Eq. 15.49, text p. 769.)

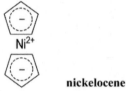

nickelocene

15.81   Applying the reverse of the mechanism in Fig. P15.87(a) to the Diels–Alder adduct in Fig. P15.87(b) to give 1,3-diphenylisobenzofuran and a dienophile leads to the conclusion that this dienophile must be *cyclopentyne.*

cyclopentyne

1,3-diphenylisobenzofuran

Cyclopentyne is formed by a β-elimination of $MgBr_2$ from the Grignard reagent formed from 1,2-dibromocyclopentene. Cyclopentyne is particularly unstable because it contains a triple bond in a small (five-membered) ring; the bond angles of the *sp*-hybridized carbons deviate severely from 180°. It helps to understand the elimination by thinking of the Grignard reagent as a carbanion:

$MgBr_2$

cyclopentyne
(reacts with 1,3-diphenyl-
isobenzofuran)

15.82   Follow the procedure used in solving Problem 15.81: mentally reverse the Diels–Alder reaction so that maleic anhydride is obtained; the other fragment obtained is compound *X,* which thus has the following structure:

reverse
Diels–Alder

**maleic
anhydride**

s-cis conformation

s-trans conformation

compound *X*

The formation of *X* results from ionization of the starting chloride to a tertiary allylic carbocation *A* followed by ring opening of *A* to a secondary carbocation *B*, which is captured by the nucleophile water. (Why should the tertiary allylic carbocation *A* rearrange to a secondary carbocation *B*?)

   If you find an error in this manual, please visit http://people.pharmacy.purdue.edu/~loudonm/teaching/

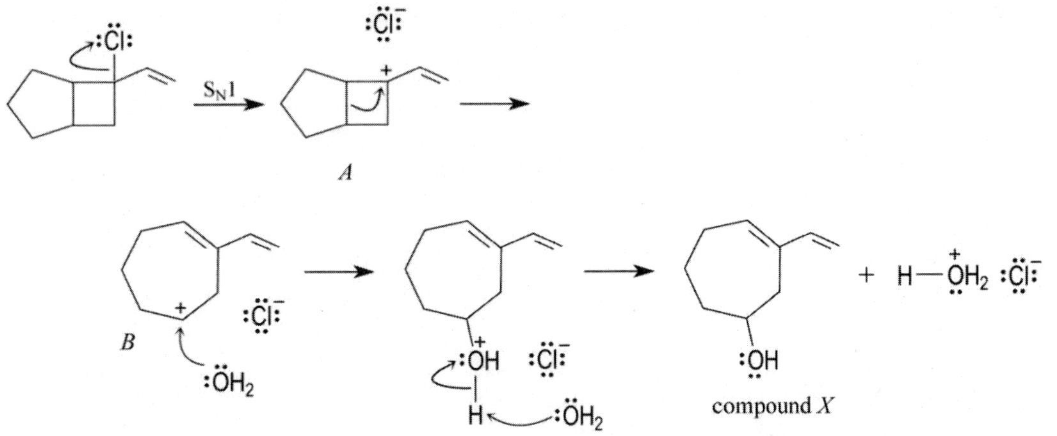

A

B

compound X

15.83   The structure of spiropentadiene is shown below; it undergoes a Diels–Alder reaction with two molar equivalents of 1,3-cyclopentadiene. Spiropentadiene is unstable because of its great ring strain. Although cyclopropane rings are retained in the product, they contain less ring strain than cyclopropene rings. (Why?)

spiropentadiene

15.84   **(a)**   The mechanism below is shown beginning with the protonated alcohol, which is formed under the strongly acidic conditions.

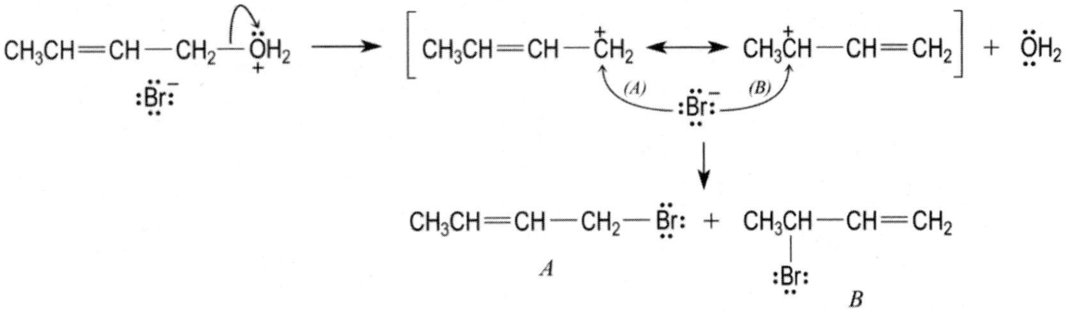

(b)   This reaction involves a rearrangement of carbocation *X1* to a more stable carbocation *X2* followed by loss of a β-proton from the rearranged ion. (Only the part of the molecule involved in the reaction is shown.)

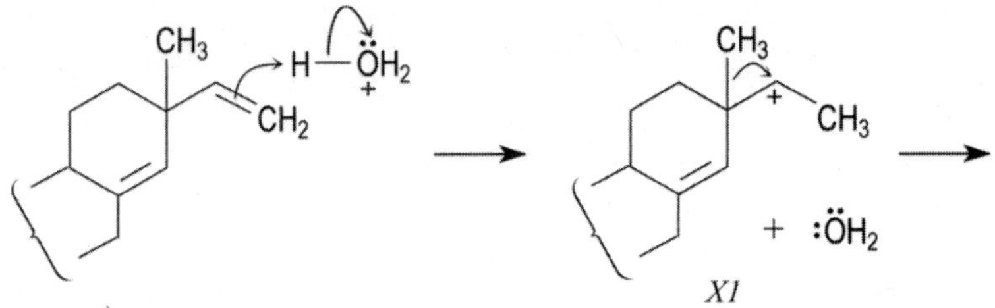

dextropimaric acid

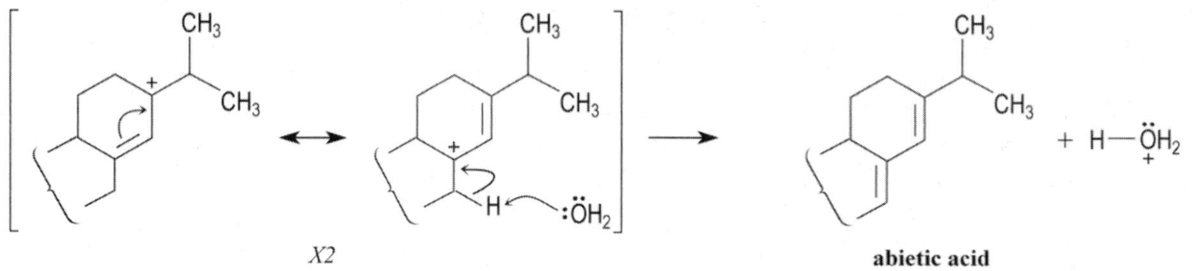

$X2$ **abietic acid**

(c)   This is a reverse Diels–Alder reaction followed by another Diels–Alder reaction.

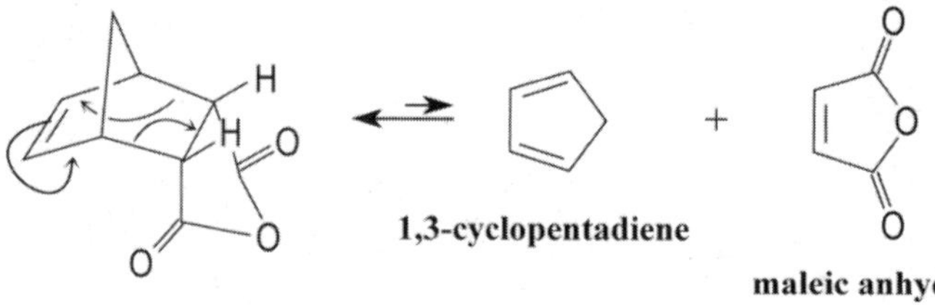

1,3-cyclopentadiene

maleic anhydride

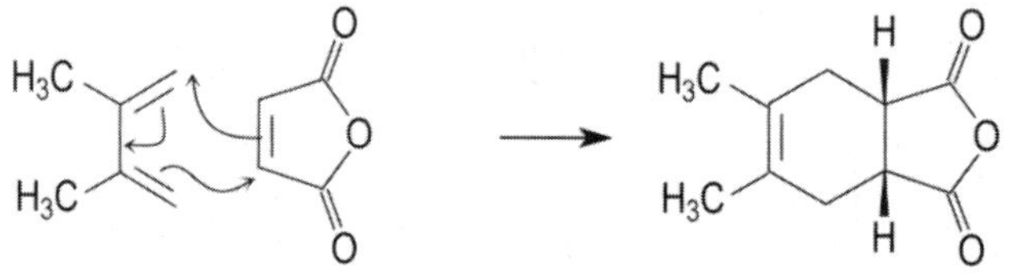

maleic anhydride   **(endo product)**

(d)   The product X is a Diels–Alder adduct between α-phellandrene and the alkyne:

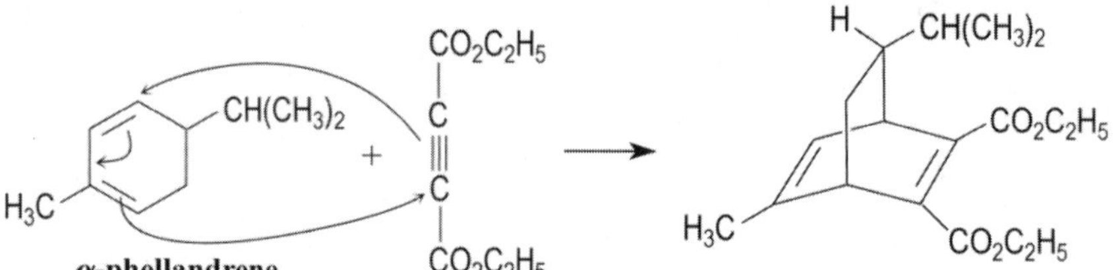

α-phellandrene X

(Because the absolute configuration of α-phellandrene is not specified, it could be either of the two enantiomers, or the racemate, and the same is true of the Diels–Alder adduct X.)

     Compound X undergoes a reverse Diels–Alder in a different manner to lose 3-methyl-1-butene and the aromatic compound diethyl 4-methylphthalate. The driving force for this reaction is, first, the formation of a volatile product (the alkene), and, second, the aromatic stability of the ultimate product.

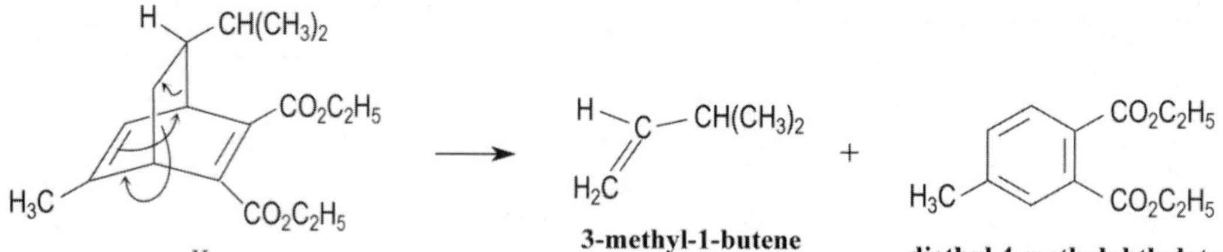

X **3-methyl-1-butene**   **diethyl 4-methylphthalate**

15.85   (a)   The exocyclic double bond is protonated in the manner shown below because this process gives the more stable carbocation—the carbocation with the greatest number of resonance structures. (The strong acid is represented by H—X.)

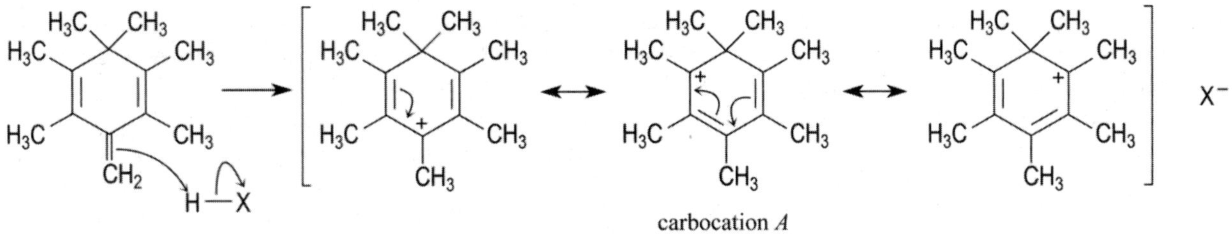

carbocation *A*

(b)   The protons nearest the positive charge have the greatest chemical shifts. These follow from the resonance structures above. The assignment of the methyl protons at δ 2.82 is clear from their relative integral; this methyl is the only one not equivalent to another methyl by symmetry. The two non-vinylic methyls have the smallest chemical shift. To summarize:

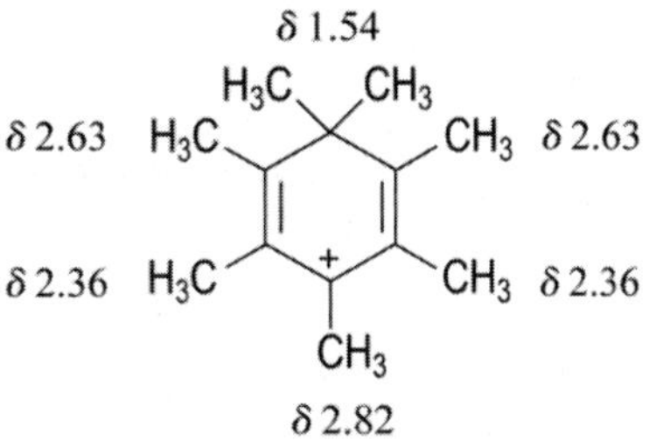

(c)   The signals coalesce when the temperature is raised because a chemical process takes place that makes the methyl groups equivalent over time. This reaction is nicknamed the "methyl walk."

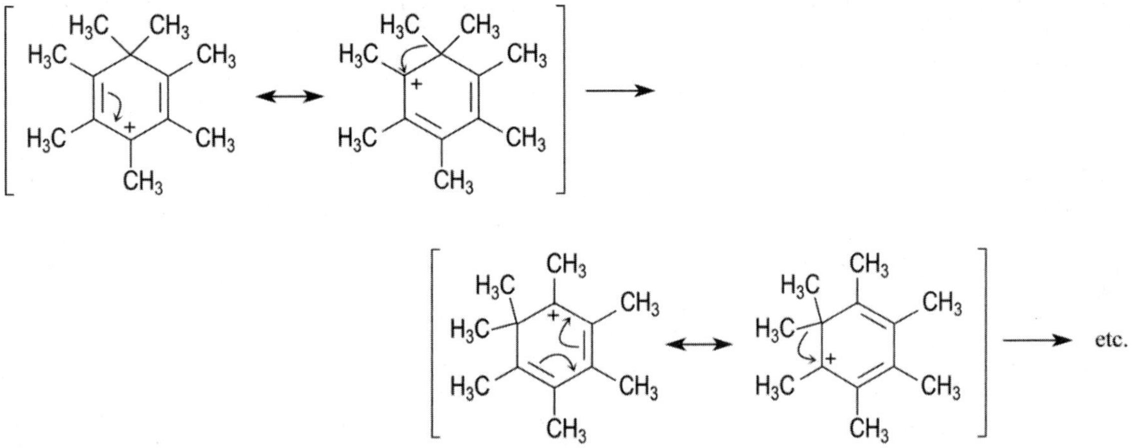

15.86   Remember that benzene rings have resonance structures in which the double bonds are completely delocalized. Let's draw [4]phenylene using the other resonance structures for the benzene rings:

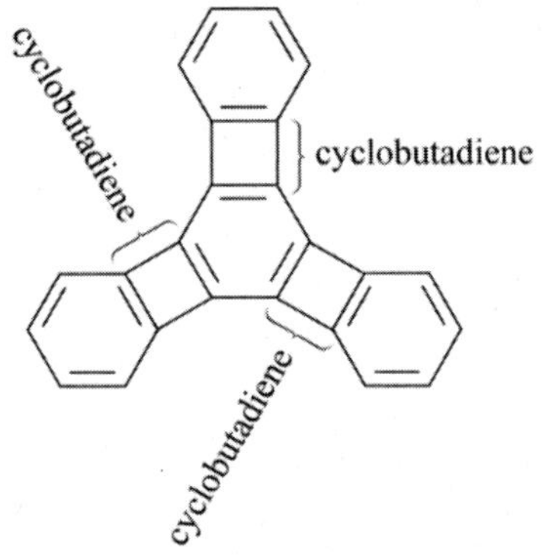

This structure shows that, in effect, that [4]phenylene incorporates three cyclobutadiene rings. Because these are antiaromatic, they are *very* unstable. What this means is that all of the benzene rings have no resonance stabilization, because when their π bonds are delocalized, cyclobutadienes result.

Hydrogenation solves this problem. When the central benzene ring is hydrogenated, the π bonds in the remaining benzene rings can be delocalized. (Cyclobutenes, unlike cyclobutadienes, are not continuous cycles of *p* orbitals and are *not* antiaromatic.) In other words, hydrogenation of the central benzene ring is especially favorable because it is accompanied by restoration of the aromaticity of all three outer benzene rings.

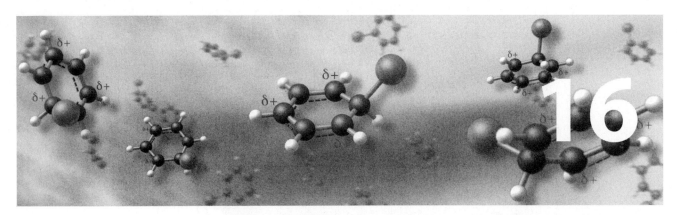

The Chemistry of Benzene
and Its Derivatives

STUDY GUIDE LINKS

16.1   NMR of Para-Substituted Benzene Derivatives

Be sure to realize that the NMR spectrum of a para-disubstituted benzene derivative shows the "two leaning doublet" pattern only when the two para substituents are different. When the substituents are the same, all ring protons are equivalent, and the resonance for the ring protons is a singlet.

The splitting pattern for para-disubstituted benzene derivatives is a little more subtle than might first appear, and can be rationalized as follows. Each ring proton has an ortho relationship to one nonequivalent proton and a para relationship to the other nonequivalent proton. Consider, for example, p-bromoethylbenzene, the structure of which is shown below, and the NMR spectrum of which is shown in Fig. 16.3 on text p. 796. In this structure, H^a is ortho to one H^b and para to the other H^b. Actually, both ortho and para splittings are present, but the para splittings are typically very small and not readily discernible in the spectrum. The major splitting observed is the one between adjacent protons—about 8 Hz in Fig. 16.3, a value within the range for ortho splitting given in Table 16.1 on text p. 795. Thus, to a useful approximation, the resonance of each proton is split into a doublet by its neighboring ortho proton. Since there are two chemically different types of protons—H^a and H^b—the spectrum consists of two such doublets.

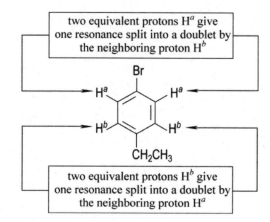

16.2   Different Sources of the Same Reactive Intermediate

The example of Eq. 16.21a–b on text pp. 807–808 is a further illustration of an idea that has been discussed in an earlier chapter (see Study Guide Link 11.2): *Different starting materials can serve as a source of the same reactive intermediate.*

In the Friedel–Crafts alkylation, the reactive intermediate is a carbocation. In this section of the text, you've learned that carbocation intermediates can be generated from alkyl halides and Lewis acids. But you've also learned that carbocations are formed when alkenes are subjected to strongly acidic conditions:

carbocation intermediate

The carbocation intermediates formed in such reactions can serve as electrophiles in electrophilic substitution just as carbocations generated in other ways can. Similarly, you've also learned that secondary and tertiary alcohols react with strong acids to generate carbocation intermediates (Sec. 10.2). These carbocations are transformed into alkene dehydration products if the reaction conditions are designed to remove the alkene from the reaction mixture. But if the alkene is not removed, and an aromatic hydrocarbon is present, these carbocation intermediates from alcohol dehydration can also serve as electrophiles in aromatic substitution. Eqs. 16.15a–b on text p. 808 illustrate this idea.

16.3   Reaction Conditions and Reaction Rate

What do we mean by "harsh" or "mild" reaction conditions? In electrophilic aromatic substitution, harsh conditions include high temperature, strong Lewis acids, high concentrations of reagents, and/or the use of reagents that generate high concentrations of electrophiles. Mild conditions include lower temperature, weaker Lewis acids (or no Lewis acids), and lower concentrations of reagents.

Why is it that such conditions affect reaction rate? Remember, to say that an aromatic compound is highly activated means that it *reacts relatively rapidly* under a given set of conditions. If it is deactivated, it *reacts much more slowly* under the same conditions. Recall (Sec. 4.8B) that reaction rates increase with increasing temperature. Recall also, from your study of rate laws (Sec. 9.3B), that many reaction rates increase with increasing reagent concentration. Thus, raising the temperature or the reagent concentration, or using a reagent that generates a high concentration of an electrophile, increases the reaction rate. Such strategies are necessary in order to get a highly *deactivated* (unreactive) compound to react at a convenient rate. However, use of harsh reaction conditions on a highly *activated* (reactive) compound in many cases leads to a greater degree of substitution than desired (or to other side reactions); hence, for highly activated compounds, milder conditions are used, such as weaker Lewis acids (or no Lewis acids), lower temperatures, or lower concentrations of reagents.

The examples in the text illustrate in a practical sense what is meant by "harsh" or "mild" conditions. It is *not* important to memorize the exact conditions for each reaction. It *is* important for you to understand why certain compounds are more activated toward substitution than others, and to understand conceptually that the success of reaction can depend on the rational choice of reaction conditions.

REACTION REVIEW

I. ELECTROPHILIC AROMATIC SUBSTITUTION REACTIONS OF BENZENE

A. HALOGENATION OF BENZENE

1. Benzene reacts with bromine in the presence of a Lewis acid catalyst ($FeBr_3$) to yield a product in which one bromine is substituted for a ring hydrogen.

$$Br_2 + \bigcirc \xrightarrow{FeBr_3} \bigcirc\!\!-Br + HBr$$

2. An analogous chlorination reaction using Cl_2 and $FeCl_3$ gives chlorobenzene.

3. Halogenation of benzene differs from the halogenation of alkenes by the type of product obtained and the reaction conditions.
   a. Alkenes give addition products and react spontaneously under mild conditions.
   b. Benzene gives substitution products and requires a Lewis acid as well as relatively severe conditions.

4. The mechanism of the bromination of benzene consists of the following steps:
   a. The first step is formation of a complex between Br_2 and the Lewis acid $FeBr_3$.

$$:\!\overset{..}{\underset{..}{Br}}\!-\!\overset{..}{\underset{..}{Br}}: + \ FeBr_3 \ \rightleftharpoons \ \overset{\delta^+}{:\!\overset{..}{\underset{..}{Br}}}\!----\!\overset{\delta^-}{\overset{..}{\underset{..}{Br}}}\!----\overset{\delta^+}{FeBr_3}$$

   i. This complexation makes one of the bromines a much better leaving group.

   ii. This complex of Br_2 and $FeBr_3$ reacts as if it contained the electron-deficient species $:\!\overset{..}{\underset{..}{Br}}\!^+$.

   b. In the second step, this complex reacts with the π electrons of the benzene ring.

$$\bigcirc + \ :\!\overset{..}{\underset{..}{Br}}\!:^{\delta^+} \ \rightleftharpoons \ \bigcirc\!\!\!<^{H}_{Br} \ + \ FeBr_4^{\ -}$$

resonance-stabilized
carbocation intermediate

    i. This step disrupts the aromatic stabilization of the benzene ring (thus the need for the harsh conditions).

    ii. The carbocation intermediate forms only because it is resonance stabilized.

  c. The reaction is completed when a bromide ion acts as a base to remove the ring proton and give the products; by losing a β-proton, the carbocation can form a stable aromatic compound.

B. NITRATION OF BENZENE

1. Benzene reacts with concentrated nitric acid, usually in the presence of a sulfuric acid catalyst, to form nitrobenzene.

2. In this reaction, called nitration, the nitro group, —NO_2, is introduced into the benzene ring by electrophilic substitution; the electrophile is the nitronium ion, $^+NO_2$.

$$H_2SO_4 \ + \ HNO_3 \ \longrightarrow \ {}^+NO_2 \ HSO_4^- \ + \ H_2O$$

C. SULFONATION OF BENZENE

1. Benzene reacts with a solution of sulfur trioxide in H_2SO_4, called fuming sulfuric acid, to form benzenesulfonic acid.

2. In this reaction, called sulfonation, the sulfonic acid group, —SO_3H, is introduced into the benzene ring by electrophilic substitution; the electrophile is the neutral compound sulfur trioxide, SO_3.

3. Sulfonic acids such as benzenesulfonic acid are rather strong acids.

4. Sulfonation, unlike many electrophilic aromatic substitution reactions, is reversible; the sulfonic acid group is replaced by a hydrogen when sulfonic acids are heated with steam.

D. FRIEDEL–CRAFTS ALKYLATION OF BENZENE

    1. The reaction of an alkyl halide with benzene in the presence of a Lewis acid gives an alkylbenzene; this type of reaction is called Friedel–Crafts alkylation.

    2. The electrophile in a Friedel–Crafts reaction is formed by complexation of the Lewis acid $AlCl_3$ with the halogen of an alkyl halide.
        a. Either the alkyl halide–Lewis acid complex, or the carbocation derived from it, can serve as the electrophile in a Friedel–Crafts reaction.
        b. Rearrangements of alkyl groups are observed in some Friedel–Crafts alkylations if the carbocation intermediate is prone to rearrangement.

    3. A catalytic amount (much less than one equivalent) of the $AlCl_3$ catalyst can be used in this reaction.
    4. The alkylbenzene products are more reactive than benzene itself.
        a. The product can undergo further alkylation, and mixtures of products alkylated to different extents are observed.
        b. A monoalkylation product can be obtained in good yield if a large excess of the starting material is used; this strategy is practical only if the starting material is cheap, and if it can be readily separated from the product.
    5. Alkenes or alcohols with an acid catalyst can also be used as the alkylating agents in Friedel–Crafts alkylation reactions.

    6. Friedel–Crafts alkylation is generally not useful on compounds that are more deactivated than benzene itself.

E. FRIEDEL–CRAFTS ACYLATION OF BENZENE

    1. When benzene reacts with an acid chloride in the presence of a Lewis acid catalyst such as aluminum trichloride, $AlCl_3$, a ketone is formed.

a. This reaction is an example of a Friedel–Crafts acylation.

b. An acyl group, typically derived from an acid chloride, is introduced into an aromatic ring in the presence of a Lewis acid catalyst.

2. The electrophile in the Friedel–Crafts acylation reaction is a carbocation called an acylium ion.

a. This ion is formed when the acid chloride reacts with the Lewis acid $AlCl_3$.

an acylium ion

b. The substitution occurs when the benzene π electrons react with the acylium ion, and a proton is lost from the resulting carbocation intermediate.

3. The ketone product of the Friedel–Crafts acylation reacts with the Lewis acid catalyst to form a complex that is catalytically inactive; this has two consequences:

a. Slightly more than one equivalent of the catalyst must be used:

  i. one equivalent to react with the product.

  ii. an additional catalytic amount to ensure the presence of catalyst throughout the reaction.

b. The complex must be destroyed before the ketone product can be isolated; this is usually accomplished by pouring the reaction mixture into ice water.

4. Because the ketone products of acylation are much less reactive than the benzene starting material, acylation occurs only once.

5. Friedel–Crafts acylation does not occur on a benzene ring substituted solely with one or more meta-directing groups. Thus, for example, nitrobenzene does not undergo Friedel–Crafts acylation.

6. The Friedel–Crafts acylation occurs intramolecularly when acylation results in the formation of a five- or six-membered ring.

a. The intramolecular process is much faster than reaction with the acylium ion on the phenyl ring of another molecule.

b. This type of reaction can only occur at an adjacent ortho position.

7. The Friedel–Crafts acylation reaction is important for two reasons.

a. It is an excellent method for the synthesis of aromatic ketones.

b. It is another method for the formation of carbon–carbon bonds (see a partial list on text p. 810 or a complete list in Appendix VI, text page A-13xr).

II. ADDITION REACTIONS OF BENZENE DERIVATIVES

A. HYDROGENATION OF BENZENE DERIVATIVES

1. Because of its aromatic stability, the benzene ring is resistant to conditions used to hydrogenate ordinary double bonds.

2. Aromatic rings can be hydrogenated under extreme conditions of temperature and/or pressure.

3. Catalytic hydrogenation of benzene derivatives gives the corresponding cyclohexanes, and cannot be stopped at the cyclohexadiene or cyclohexene stages.

4. Hydrogenation of the first double bond of benzene is an endothermic (thermodynamically unfavorable) reaction and requires energy (heat or pressure) to take place.

5. Once hydrogenation of the first double bond has taken place, the aromatic character of the benzene ring is lost; consequently, hydrogenation of the remaining double bonds under the harsh reaction conditions is instantaneous.

TABLE SG16.1	Electrophilic Aromatic Substitution Summary	
Electrophile	**Reaction Name**	**Product**
$Cl—\overset{+}{Cl}----\overset{-}{Fe}Cl_3$	chlorination	chlorobenzene
$Br—\overset{+}{Br}----\overset{-}{Fe}Br_3$	bromination	bromobenzene
SO_3 in H_2SO_4	sulfonation	benzenesulfonic acid
$^+NO_2$	nitration	nitrobenzene
$R—C\equiv\overset{+}{\overset{..}{O}}\ \ ^-AlCl_4$	Friedel–Crafts acylation	an aryl ketone
$R—Cl----AlCl_3$ or $R^+\ ^-AlCl_4$	Friedel–Crafts alkylation	an alkylbenzene

SOLUTIONS TO PROBLEMS

Solutions to In-Text Problems

16.1 (a) *m*-chloroethylbenzene, or 1-chloro-3-ethylbenzene
(b) *o*-diethylbenzene or 1,2-diethylbenzene
(c) *p*-nitrostyrene or 1-nitro-4-vinylbenzene
(d) 2,4-dichlorophenol
(e) 2-bromo-1-chloro-5-fluoro-4-iodo-3-nitrobenzene
(f) benzylbenzene or (phenylmethyl)benzene (also commonly called diphenylmethane)

16.2 (a)

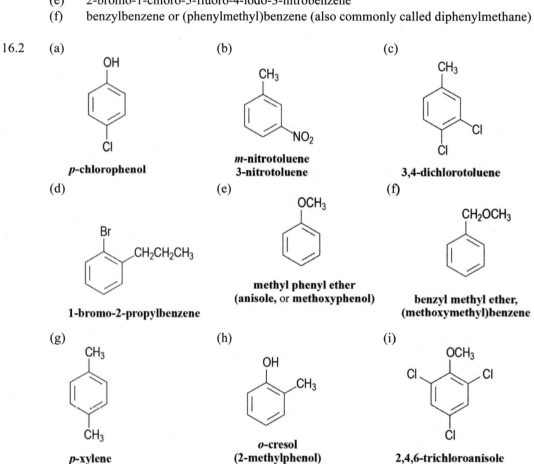

(a) **p-chlorophenol**

(b) **m-nitrotoluene**
3-nitrotoluene

(c) **3,4-dichlorotoluene**

(d) **1-bromo-2-propylbenzene**

(e) **methyl phenyl ether**
(anisole, or methoxyphenol)

(f) **benzyl methyl ether,**
(methoxymethyl)benzene

(g) **p-xylene**

(h) **o-cresol**
(2-methylphenol)

(i) **2,4,6-trichloroanisole**

16.3 Add about 25 °C per carbon relative to toluene (110.6 °C; see text p. 792):

(a) ethylbenzene: 136 °C (actual: 136 °C)   (c) *p*-xylene: 136 °C (actual: 138 °C)

(b) propylbenzene: 161 °C (actual: 159 °C)

16.4 The aromatic compound has NMR absorptions with greater chemical shift in each case because of the ring current (Fig. 16.1, text p. 794).

(a) Because thiophene is aromatic, its NMR absorptions occur at greater chemical shift than those of divinyl sulfide.

(b) The protons of benzene have considerably greater chemical shift because benzene is aromatic and 1,4-cyclohexadiene is not.

16.5 (a) This hydrocarbon is a continuous cycle of 18 π electrons. Because all atoms are trigonal and all bond angles are 120°, this compound can be planar. It meets the $4n + 2$ criterion for aromaticity for $n = 4$.

(b) The ring current brings about the expected increase in the local field at protons *outside* the ring, but, from Fig. 16.2 on text p. 794, it should have exactly the opposite effect on the protons *inside* the ring: These should experience a smaller local field. The resonance at δ 9.28 is that of the outer protons, and the resonance at $\delta(-2.99)$ is that of the inner protons. Because the local field at the outer protons is so large, the resonance frequency, and thus the chemical shift, of these protons is large. Because the local field at the inner protons is so small, the resonance frequency, and thus the chemical shift, of these protons is small. The relative integral is in agreement with this assignment.

16.6 (a) Mesitylene should give an NMR spectrum containing only two singlets in the integral ratio 3:1 for the methyl groups and ring protons, respectively. *p*-Ethyltoluene should contain a methyl triplet at smallest chemical shift (around δ 1.0), a methyl singlet at the next smallest chemical shift (around δ 2.3), and a methylene quartet at somewhat greater chemical shift. There are two possibilities for its ring protons. The most general one is a typical para-substitution pattern (apparent pair of doublets) for the ring protons in the δ 6.5–7.5 region. However, because the methyl and ethyl groups are so similar in their electronic characteristics, another possibility is that the nonequivalence of the ring protons is so slight as not to be apparent. In fact, this is the case; the ring protons are a singlet.

mesitylene
two singlets in its NMR spectrum

p-ethyltoluene
a methyl singlet, a methyl triplet, a methylene quartet,
and aromatic absorption in its NMR spectrum

(b) Among other features, the NMR spectrum of 1-bromo-4-ethylbenzene has a typical ethyl quartet and a typical para-substitution pattern for the ring protons, as shown in Fig. 16.3, text p. 796, whereas the spectrum of (2-bromoethyl)benzene should show a pair of triplets for the methylene protons and a complex pattern for the ring protons. If this isn't enough to distinguish the two compounds, the integral of the ring protons relative to the integral of the remaining protons is different in the two compounds.

1-bromo-4-ethylbenzene   **(2-bromoethyl)benzene**

16.7 (a) The signal for an exchangeable proton at δ 3.76 indicates that the compound contains an —OH group. It also contains aromatic absorptions with a splitting pattern that suggests para substitution. The splitting patterns of the δ 1.27 and δ 4.60 resonances indicate the partial structure H_3C—CH, and the chemical shift of the CH is consistent with its being α to both the ring and the OH group. The compound is

1-(4-methylphenyl)ethanol

(b)   The IR spectrum indicates the presence of an OH group, and the chemical shift of the broad NMR resonance (δ 6.0) suggests that this could be a phenol. The splitting patterns of the δ 1.17 and δ 2.58 resonances show that the compound also contains an ethyl group, and the splitting pattern of the ring protons shows that the compound is a para-disubstituted benzene derivative. The compound is *p*-ethylphenol.

$$CH_3CH_2 - \langle \rangle - OH$$

p-ethylphenol
(4-ethylphenol)

16.8   The following two methyl ethers with the formula $C_7H_6OCl_2$ would show five lines in their proton-decoupled ^{13}C NMR spectra.

1,3-dichloro-2-methoxybenzene   **1,3-dichloro-5-methoxybenzene**

16.9   Count the resonances. Mesitylene has three resonances; isopropylbenzene has six.

mesitylene   **isopropylbenzene**

16.10 (a)   The two benzene rings are conjugated in compound *A*; because there are more π bonds in conjugation, the UV absorption occurs at longer wavelength than it does in ethylbenzene.

       (b)   Compound *B* is like two connected molecules of compound *C* in the same sense that compound *A* is like two connected ethylbenzene molecules. It might seem that the absorption of compound *B* should occur at longer wavelength than that of compound *C* for the reason given in part (a). That this is *not* the case means that the two rings in compound *B* are not conjugated. They are not conjugated because the rings are not coplanar; and they are not coplanar because the ortho methyl groups on the two rings have severe van der Waals repulsions that are relieved when the molecule adopts a nonplanar conformation:

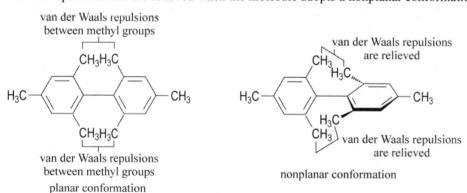

16.11   Because styrene has a double bond in conjugation with the ring and ethylbenzene does not, styrene has a greater λ_{max} in its UV spectrum.

16.12   Apply the steps shown in Eqs. 16.6–16.7 on text pp. 800–801 to the para position of bromobenzene.

16.13   The protonation of SO_3 by H_2SO_4:

The mechanism of sulfonation with the conjugate acid of SO_3 as the electrophile:

16.14   Apply the sulfonation mechanism shown in Eq. 16.13 on text p. 804 to the para position of toluene. Any of the resonance structures can be deprotonated by the base to regenerate aromaticity in the ring. Other resonance structures, not shown, exist with the negative charge delocalized on to each of the oxygens attached to the sulfur.

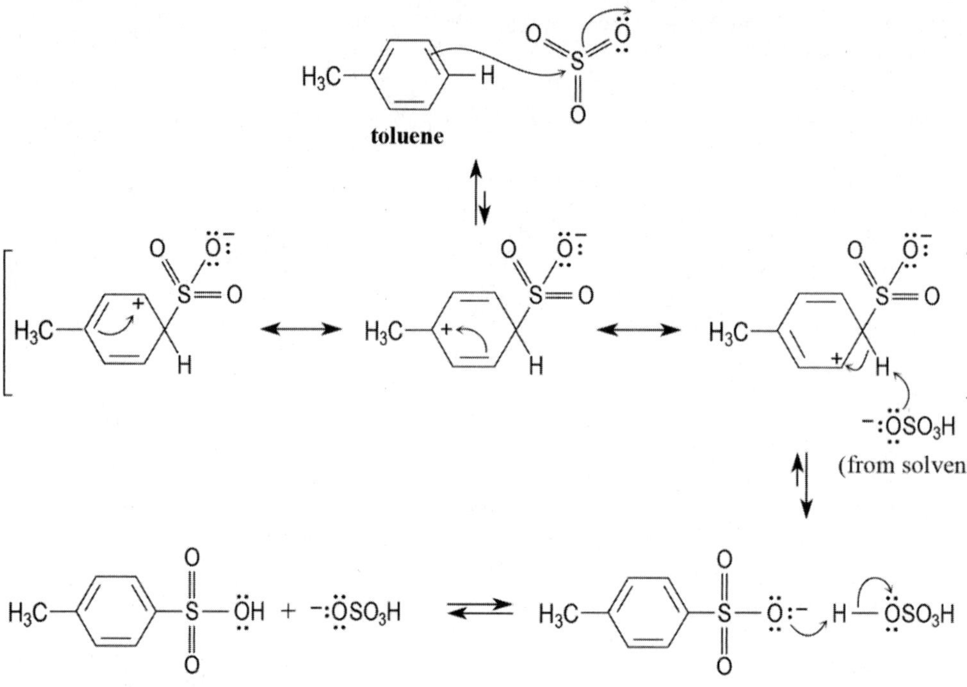

toluene

−:ÖSO₃H
(from solvent)

**p-toluenesulfonic acid
(p-TsOH)**

16.15 (a) As discussed in Study Guide Link 16.2, carbocations generated in a variety of ways can be used as electrophiles in Friedel–Crafts reactions. In this case, the carbocation produced by the protonation of cyclohexene is the electrophile.

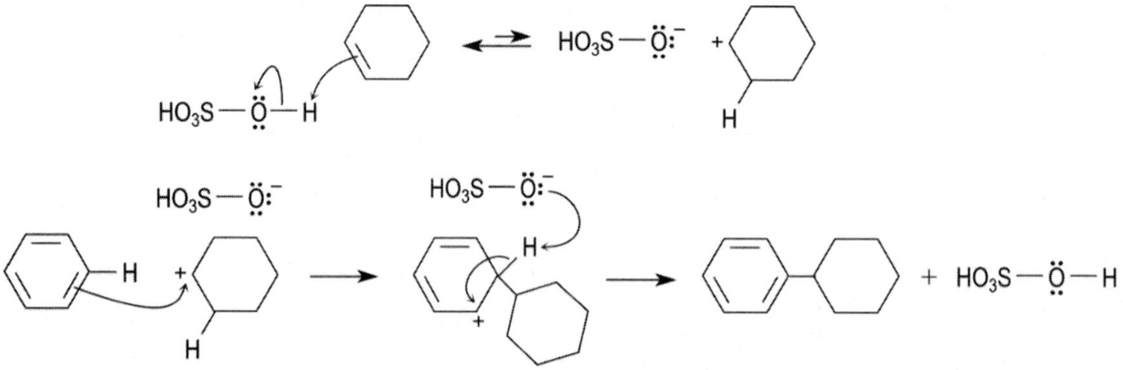

(b) The same product is obtained because the same carbocation electrophile is involved. It is generated from cyclohexanol by protonation and loss of water. The carbocation then reacts with benzene as in part (a).

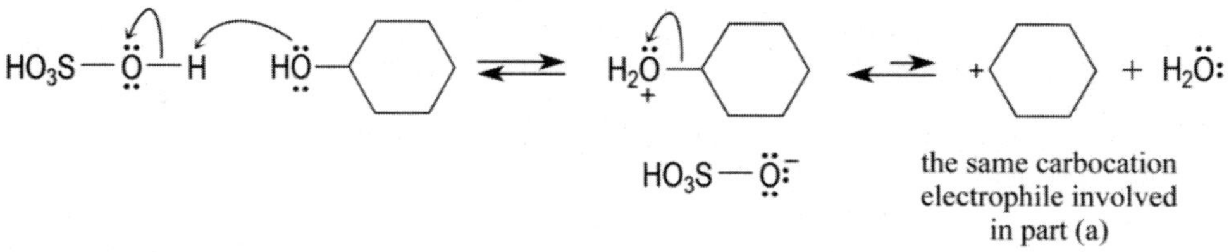

HO₃S—Ö:⁻

the same carbocation
electrophile involved
in part (a)

16.16 The product is *tert*-butylbenzene. The role of the Lewis acid BF₃ is to promote the ionization of HF. The mechanism of the reaction is as follows:

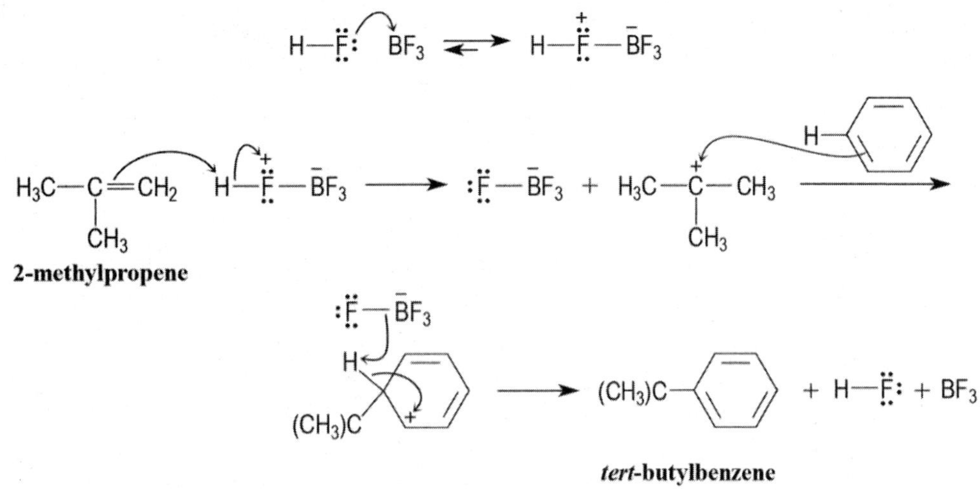

2-methylpropene

tert-butylbenzene

16.17   The hint and the fact that the product has the same number of carbons as the starting material suggest an intramolecular Friedel–Crafts alkylation.

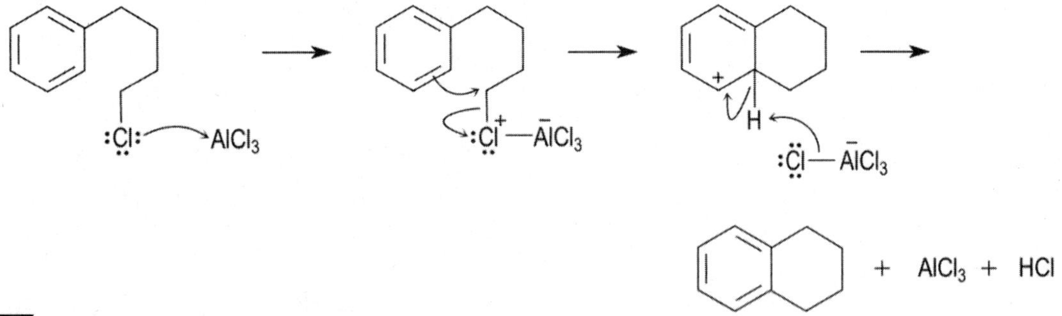

Remember that intramolecular reactions involving the formation of five- and six-membered rings are particularly rapid. This is another example of neighboring-group participation; see Sec. 11.8 of the text.

16.18   (a)   (b)

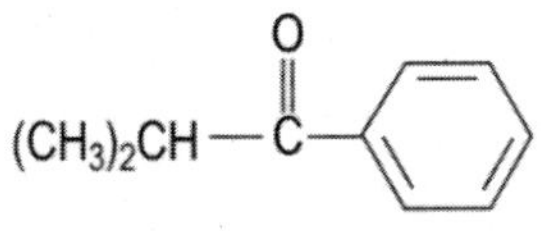

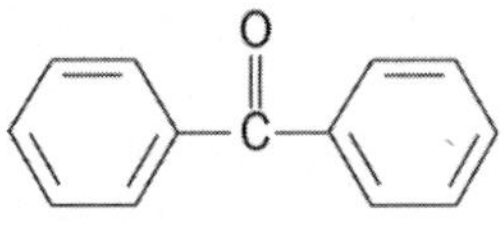

isobutyrophenone   **benzophenone**

16.19   The two possible Friedel–Crafts reactions:

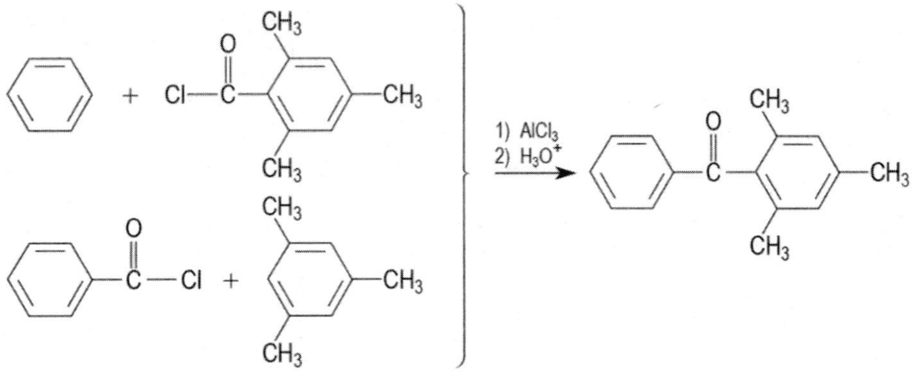

16.20 That the product contains the same number of carbons as the starting material suggests an intramolecular Friedel–Crafts reaction. The mechanism below begins with the acylium ion. (For the formation of this ion from the acid chloride, see Eq. 16.23 on text p. 808).

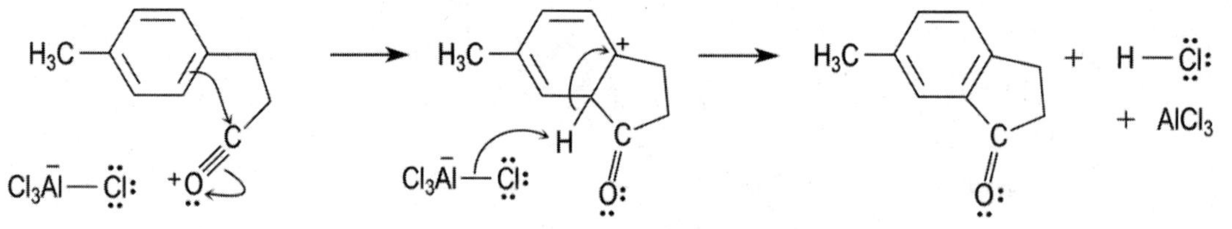

an acylium
ion

16.21 (a) Table 16.2 indicates that the methoxy group is an ortho, para-directing group. (The product is mostly the para isomer.)

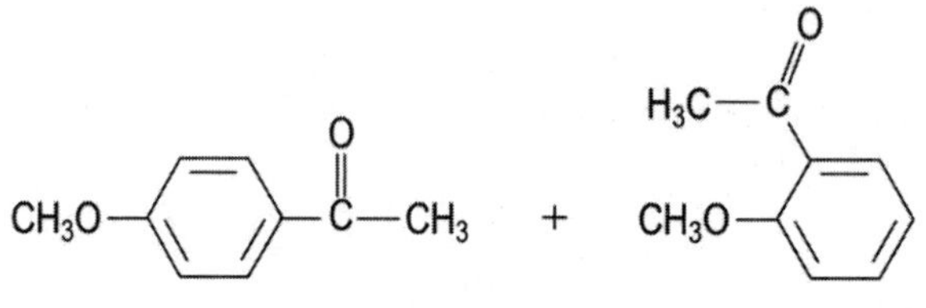

p-methoxyacetophenone   **o-methoxyacetophenone**

(b) Table 16.2 indicates that alkyl groups are ortho, para-directing groups, and the ethyl group is a typical alkyl group:

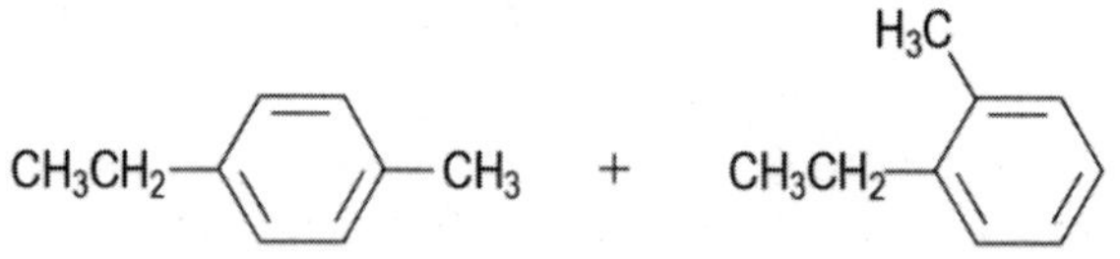

1-ethyl-4-methylbenzene   **1-ethyl-2-methylbenzene**
(p-ethyltoluene)   **(o-ethyltoluene)**

16.22 Let E^+ be a general electrophile. The four resonance structures of the carbocation intermediate that results from reaction of E^+ at the position ortho to the methoxy group of anisole are as follows:

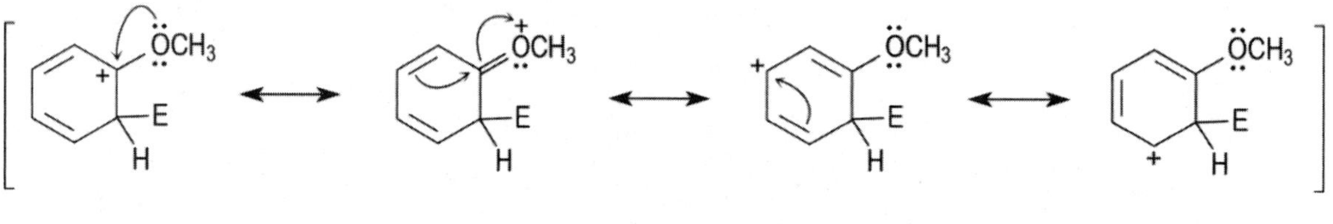

16.23 (a) Because substitution occurs para to the phenyl group, the phenyl group is evidently an ortho, para-directing substituent.

(b) The following resonance structures show that the electrons of the phenyl substituent can be used to stabilize the carbocation intermediate when substitution occurs at the para position. The electrons of the phenyl substituent cannot be delocalized in this way when substitution occurs at the meta position.

16.24   Notice that, except for alkyl groups, all ortho, para-directing groups have electron pairs on atoms adjacent to the benzene ring.

(a)   This substituent, like the methoxy group, is an ortho, para-directing group, a point that is confirmed by the fourth entry of Table 16.2 on text p. 812.

p-bromoacetanilide, or
N-(4-bromophenyl)acetamide   **o-bromoacetanilide**, or
N-(2-bromophenyl)acetamide

(b)   The three strong carbon-fluorine bond dipoles result in substantial partial positive charge on the carbon of the CF_3 group; consequently, this is a meta-directing group.

1-bromo-3-trifluoromethylbenzene

(c)   The carbocation intermediate involved in meta substitution has a greater separation between the positive charge of the carbocation and the positive charge of the substituent than does the carbocation intermediate involved in ortho, para substitution. (The argument is similar to that used with electrophilic substitution reactions of nitrobenzene on text pp. 814–815.) Therefore, the substituent is a meta-directing group.

(3-bromophenyl)triethylammonium ion

(d)   The *tert*-butyl group, like all other alkyl groups, is an ortho, para-directing substituent.

Br—⟨benzene⟩—C(CH₃)₃   +   ⟨benzene with Br⟩—C(CH₃)₃

1-bromo-4-*tert*-butylbenzene   **1-bromo-2-*tert*-butylbenzene**

(e)   The —O⁻ group has both unshared electron pairs and a negative charge with which to stabilize the carbocation intermediate involved in ortho, para substitution. Consequently, this group is an ortho, para-directing group.

Br—⟨benzene⟩—Ö:⁻   +   ⟨benzene with Br⟩—Ö:⁻

4-bromophenoxide ion   **2-bromophenoxide ion**

 Because of the negative charge on the oxygen, the "carbocation intermediate" is not a cation at all, but rather a neutral compound. Moreover, the —O⁻ group is so strongly activating that substitution does not stop at the monobromo derivatives but continues until all ortho and para positions have been filled. (See Eq. 18.87 on text p. 926.)

16.25   (a)   The reaction-free energy profiles for electrophilic substitution of benzene, chlorobenzene at the para position, and chlorobenzene at the meta position are shown in Fig. SG16.1. Notice that chlorobenzene is less reactive (that is, has a greater standard free energy of activation) than benzene because chlorine is a deactivating substituent. Notice also that para-substitution reactions on chlorobenzene are faster than meta-substitution reactions because chlorine is an ortho, para-directing group.

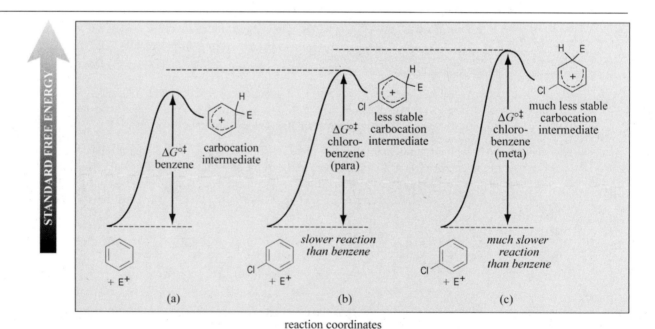

Figure SG16.1 Reaction-free energy profiles to accompany the solution to Problem 16.25(a).

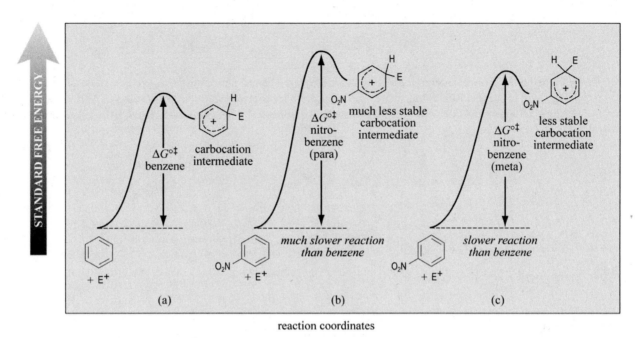

reaction coordinates

Figure SG16.2 Reaction-free energy profiles to accompany the solution to Problem 16.25(b).

(b)   The reaction-free energy profiles for electrophilic substitution of benzene, nitrobenzene at the meta position, and nitrobenzene at the para position are shown in Fig. SG16.2. Notice that nitrobenzene is less reactive than benzene because the nitro group is a deactivating substituent. Notice also that meta-substitution reactions on nitrobenzene are faster than para-substitution reactions because the nitro group is a meta-directing group.

16.26   Although both the CH_3O— and the CH_3S— groups are activating, ortho, para-directing groups, the orbitals on sulfur containing the unshared electron pairs, like those on chlorine, are derived from quantum level 3. Consequently, the overlap of sulfur orbitals with the carbon orbitals of the ring is poorer than the overlap of the oxygen orbitals. (See Fig. 16.7 on text page 819; the orbitals on sulfur are analogous to those on chlorine in the figure.) In other words, the electron-donating resonance interaction of the sulfur is weaker than that of the oxygen. Although the electron-withdrawing (and therefore rate-retarding) polar effect of oxygen is much greater than that of sulfur, the resonance interaction of oxygen is so much more powerful that it is the dominant effect.

16.27   Bromination of *N,N*-dimethylaniline is faster because nitrogen has an unshared electron pair that can stabilize the carbocation intermediate by resonance. As in the case of oxygen, the electron-withdrawing polar effect of nitrogen is much less important than its electron-donating resonance effect.

16.28   In this reaction, iodide ion, I^-, is oxidized by hydrogen peroxide, H_2O_2. Every oxidation is accompanied by a reduction, so the hydrogen peroxide is reduced to water.

$$I^- \longrightarrow I{-}OH$$

$$H_2O_2 \longrightarrow H_2O$$

The balanced half reactions are shown below. The reactions were balanced according to the method shown in Sec. 10.6A.

$$2\,H_2O + 2\,I^- \longrightarrow 2\,I{-}OH + 2\,H^+ + 2\,e^-$$

$$2\,H^+ + 2\,e^- + H_2O_2 \longrightarrow H_2O + H_2O$$

16.29   Tyrosine (*A*) would undergo iodination more rapidly. The aromatic ring of *B* is deactivated, or less electron-rich, due to the electron-withdrawing ability of the very electronegative fluorine atoms. Additionally, the carbocation intermediate would be inductively destabilized by the fluorines, resulting in a slower reaction.

aromatic ring is less nucleophilic   carbocation intermediate is destabilized
due to electron-withdrawing fluorines   by adjacent positve charges

16.30   (a)   Each substituent is an ortho, para-directing group. Two products satisfy the directing effects of both groups.

2-bromo-4-methyl-   4-bromo-2-methyl-
benzenesulfonic acid   benzenesulfonic acid

(b)   As in part (a), each substituent is an ortho, para-directing group. Two products satisfy the directing effects of both groups.

2-bromo-1-nitro-   4-bromo-1-nitro-
4-iodobenzene   2-iodobenzene

16.31   (a)   The order of increasingly harsh reaction conditions is *m*-xylene < benzene < *p*-dichlorobenzene. The reason is that the methyl groups of *m*-xylene are activating groups and the chlorines of *p*-dichlorobenzene are deactivating groups.

(b)   The order is anisole < toluene < chlorobenzene. Chlorobenzene requires the harshest conditions because chlorine is a deactivating group. Anisole requires the mildest conditions because the methoxy group is more activating than the methyl group of toluene. (See Table 16.2 on text p. 812.)

16.32   Because both groups are *meta*-directing groups, it might seem that either could be introduced first. However, the Friedel–Crafts acylation cannot be carried out on nitrobenzene because, as discussed on text p. 815, the nitro group is too deactivating. Consequently, the acetyl group must be introduced first.

benzene   **acetophenone**   ***m*-nitroacetophenone**

16.33   (a)   Hydrogenate cyclohexylbenzene, which, in turn, is prepared as shown in Eqs. 16.21a–b on text pp. 807–808. Note that chlorocyclohexane and AlCl$_3$, or cyclohexanol and either H$_2$SO$_4$ or H$_3$PO$_4$, may be used as the source of the electrophile instead of cyclohexene; see Problem 16.15(b), text p. 808 and Study Guide Link 16.2.

cyclohexylbenzene   **cyclohexylcyclohexane**

(b)   Hydrogenate *tert*-butylbenzene, which, in turn, is prepared by Friedel–Crafts alkylation as shown in Eq. 16.18 on text p. 807 or by the reaction shown in the solution to Problem 16.16.

***tert*-butylbenzene**   ***tert*-butylcyclohexane**

16.34   The nucleophilic nitrogen adds to the less sterically-hindered side of the epoxide, and from the top face, as drawn. A proton transfer (likely mediated by water) follows, yielding the neutral, covalently-modified DNA molecule.

Solutions to Additional Problems

16.35   (a)   No reaction.

(b)

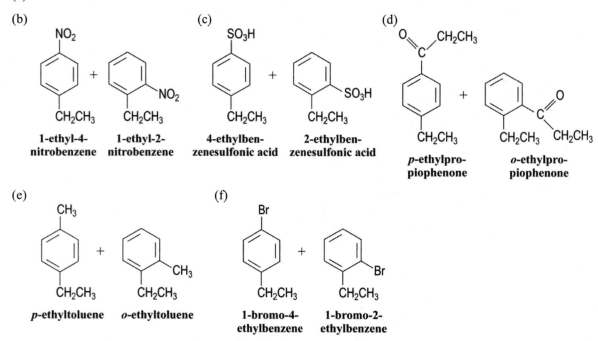

1-ethyl-4-nitrobenzene   1-ethyl-2-nitrobenzene

(c)   4-ethylbenzenesulfonic acid   2-ethylbenzenesulfonic acid

(d)   *p*-ethylpropiophenone   *o*-ethylpropiophenone

(e)

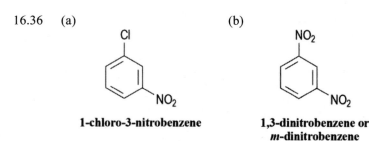

p-ethyltoluene   *o*-ethyltoluene

(f)   1-bromo-4-ethylbenzene   1-bromo-2-ethylbenzene

In part (e), ethylbenzene must be present in large excess for the monosubstitution products to be obtained. If ethylbenzene is not present in excess, products of dimethylation, trimethylation, etc., will also be formed; see Eq. 16.19 on text p. 807 and the discussion at the bottom of text p. 825.

16.36   (a)

1-chloro-3-nitrobenzene

(b)

**1,3-dinitrobenzene or
m-dinitrobenzene**

(c)   No electrophilic aromatic substitution reaction takes place. Friedel–Crafts acylation does not take place on any benzene derivative less reactive than the halobenzenes. In fact, nitrobenzene can be used as an inert solvent for Friedel–Crafts acylation.

16.37   A compound containing a benzene ring must have at least six carbon atoms and four degrees of unsaturation. Compound *A*, with three degrees of unsaturation, cannot contain a benzene ring; compound *C* has only five carbons and therefore cannot contain a benzene ring. Compounds *B* and *D* have more than six carbons and have five and four degrees of unsaturation, respectively, and therefore could contain a benzene ring.

16.38   (a)   Assume that the carbon-chlorine bond dipoles dominate the dipole moment. In the following diagram, the small gray arrows are the C—Cl bond dipoles and the darker arrows are the resultant molecular dipole moments. This diagram shows that the order of increasing dipole moment is *p*-dichlorobenzene ($\mu = 0$) < *m*-dichlorobenzene < *o*-dichlorobenzene.

(b)   Boiling points are increased by increasing attractions between molecules. To the extent that molecular attractions are dominated by attractions between molecular dipole moments, the order of increasing boiling point is predicted to be the same as the order of increasing dipole moment. (In fact, the boiling point of the ortho isomer (180 °C) is 6–7 °C higher than the boiling points of the meta and para isomers, which are within one degree of each other.)

16.39   Only compound *A* should have a proton NMR spectrum containing one six-proton singlet in the alkyl region (about δ 1.5). Only compound *B* should have a proton NMR spectrum that contains two closely spaced singlets in the integral ratio 2:1 in the benzylic proton region of the spectrum (about δ 2.3). Only compound *C* should have a proton NMR spectrum that contains two well-separated singlets in the integral ratio 3:1, the larger in the benzylic proton region and the smaller at about δ 4.5. All three compounds are also readily distinguished by the relative integrals of their aromatic proton absorptions.

16.40   Ethylbenzene has a three-proton triplet and, at somewhat greater chemical shift, in the benzylic proton region, a two-proton quartet. *p*-Xylene has a six-proton singlet in the benzylic region. Styrene, Ph—CH=CH$_2$, has no protons in the benzylic region.

16.41   (a)   The cyclopentadienyl anion is aromatic; its five resonance structures show that all carbons (and hydrogens) are chemically equivalent. Hence, the proton NMR spectrum of the ion consists of one singlet.

hybrid structure

(b)   Fig. 16.2 on text p. 794 shows that aromatic protons located in the plane of the ring and outside of the ring experience an augmented local field and thus a greater chemical shift. However, in the region above and below the ring, the induced field has the opposite direction, and consequently protons located in this region experience a reduced local field and thus a smaller chemical shift. Such is the case with the methyl group in the problem; the local field at this group is so small that its resonance occurs at nearly 1.7 ppm *smaller* chemical shift than that of TMS.

16.42   The electron density of the unshared electron pairs on the oxygen of the methoxy group is delocalized to the ortho and para positions on the ring, but not to the meta position:

Thus, there is more electron density at the ortho proton H^a than at the meta proton H^b. Since greater electron density results in a reduced local field $\mathbf{B}_0$ (see Sec. 13.3A), the chemical shift of proton H^a is smaller.

16.43 In each synthesis that involves substitution on a benzene derivative that contains an ortho, para-directing group, only the product resulting from para substitution is shown.

(a)

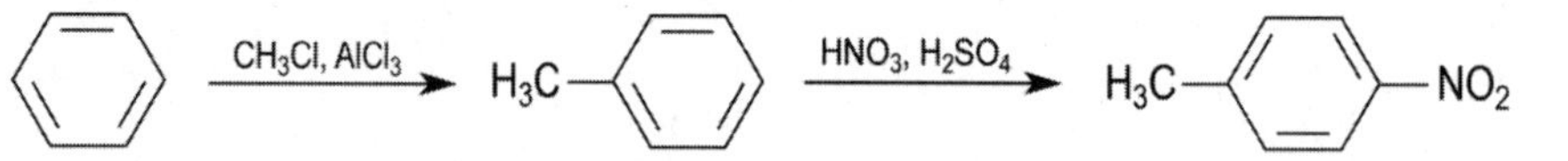

benzene (excess) **toluene** ***p*-nitrotoluene**

(b)

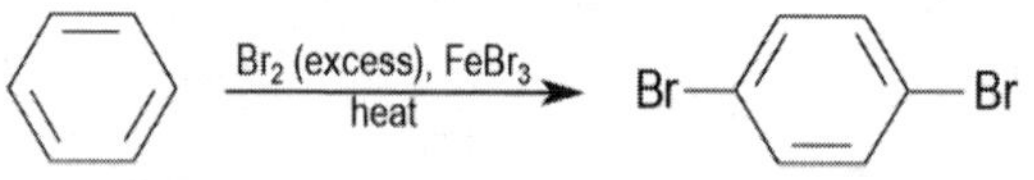

benzene ***p*-dibromobenzene**

(c)

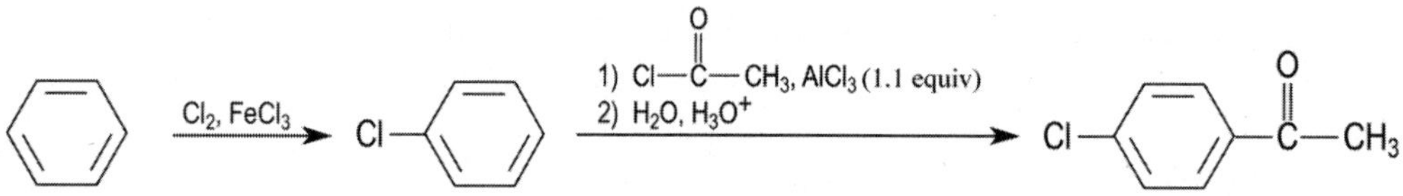

benzene **chlorobenzene** ***p*-chloroacetophenone**

(d)

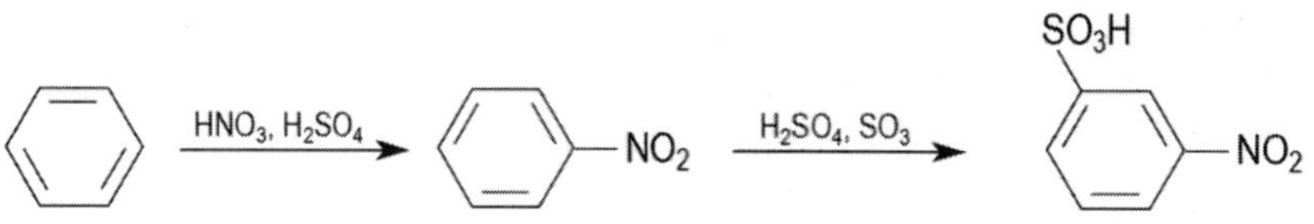

benzene **nitrobenzene** ***m*-nitrobenzenesulfonic acid**

(e)

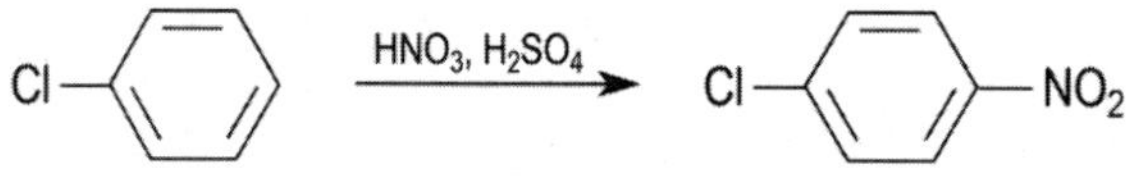

chlorobenzene ***p*-nitrochlorobenzene**

(f)

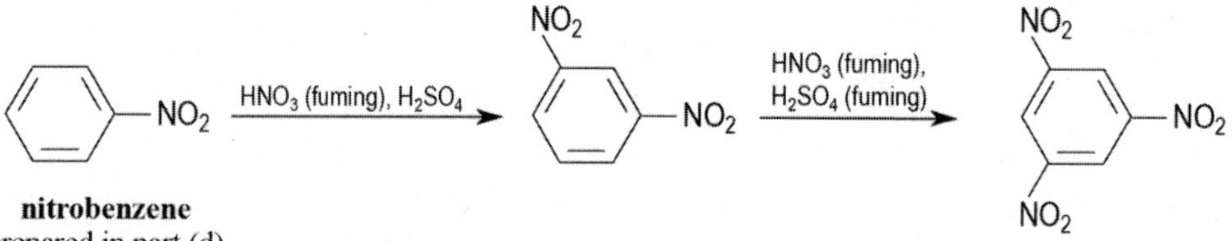

nitrobenzene prepared in part (d) **1,3,5-trinitrobenzene**

(g)

H$_3$C—⟨ ⟩—NO$_2$ $\xrightarrow{\text{Br}_2 \text{ (excess), FeBr}_3}$ 2,6-dibromo-1-nitrotoluene (with Br, H$_3$C, NO$_2$, Br substituents)

p-nitrotoluene
prepared in part (a)

2,6-dibromo-1-nitrotoluene

(h)

CH$_3$ / NO$_2$ ring $\xrightarrow[\text{heat}]{\text{Br}_2 \text{ (excess), FeBr}_3}$ CH$_3$, Br, NO$_2$, Br ring

o-nitrotoluene
by-product of the
synthesis in part (a)

2,4-dibromo-6-nitrotoluene

Nitration of toluene actually gives more ortho isomer than para isomer of toluene, and the two nitrotoluene isomers are readily separated by fractional distillation. (See text Sec. 18.5A, subsection: *The Ortho, Para Ratio*.) *o*-Nitrotoluene is a good starting material for a number of ortho-substituted benzene derivatives.

(i)

benzene $\xrightarrow{\text{C}_2\text{H}_5\text{Cl, AlCl}_3}$ ⟨ ⟩—CH$_2$CH$_3$ $\xrightarrow[\text{2) H}_3\text{O}^+\text{, H}_2\text{O}]{\text{1) CH}_3\overset{\text{O}}{\overset{\|}{\text{CCl}}}\text{, AlCl}_3}$ CH$_3$\overset{\text{O}}{\overset{\|}{\text{C}}}—⟨ ⟩—CH$_2$CH$_3$ $\xrightarrow{\text{HNO}_3\text{, H}_2\text{SO}_4}$

benzene
(excess)

ethylbenzene

p-ethylacetophenone

CH$_3$\overset{\text{O}}{\overset{\|}{\text{C}}}—⟨ NO$_2$ ⟩—CH$_2$CH$_3$

4-ethyl-3-nitroacetophenone

(j)

⟨ ⟩ + ⟨ ⟩ $\xrightarrow{\text{H}_2\text{SO}_4}$ cyclopentylbenzene

benzene   **cyclopentene**   **cyclopentylbenzene**
(excess)

Cyclopentanol may be substituted for cyclopentene in this synthesis, or chlorocyclopentane and AlCl$_3$ catalyst may be used instead of cyclopentene and H$_2$SO$_4$. Note that each of these possible starting materials serves as a source of the same carbocation, the cyclopentyl cation.

16.44 (a) In mesitylene, the three methyl groups all direct nitration to the same position; that is, they activate the same position of the ring. In 1,2,4-trimethylbenzene, only two of the three methyl groups activate the same ring position. Both trimethylbenzenes are more activated than toluene. Hence, the order is

toluene < 1,2,4-trimethylbenzene < mesitylene

(b)   The reactivity order follows from the relative activating effects of the substituents. (See the last column of Table 16.2 on text p. 812.)

nitrobenzene < chlorobenzene < benzene

(c)   In *p*-chloroanisole, the chlorine and methoxy group direct to different positions; that is, the chlorine strongly deactivates the positions that the methoxy group activate, and the methoxy group deactivates the positions that are least deactivated by the chlorine. In the meta isomer, the directing effects of the two substituents do not conflict. In anisole, the deactivating chloro group is absent altogether. The order is

p-chloroanisole < *m*-chloroanisole < anisole

(d)   The reactivity order follows from the relative activating effects of the substituents. (See the last column of Table 16.2 on text p. 812.)

p-bromoacetophenone < acetophenone < *p*-methoxyacetophenone

16.45 (a)   The boron on benzeneboronic acid has an empty $2p$ orbital and is not capable of stabilizing a carbocation by resonance. It would be difficult to introduce electron deficiency into a compound that is already electron deficient; hence, this compound undergoes substitution much more slowly than benzene itself. Moreover, the boron–oxygen bond dipoles place substantial positive charge on boron. The boronic acid group thus has the characteristics of a meta-directing group, and the principal mononitration product is *m*-nitrobenzeneboronic acid.

(b)   For the reasons given in the solution to Problem 16.24(c), the $—^+\text{N}(\text{CH}_3)_3$ group is a meta-directing group. This compound reacts more slowly than benzene, and the product is the *meta*-nitro derivative.

(c)   One benzene ring serves as an ortho, para-directing substituent on the other. (The directing effect of the phenyl substituent is explained in the solution to Problem 16.23.) Because the phenyl substituent can stabilize the carbocation intermediate in nitration, it is an activating group. Hence, this compound is more reactive than benzene, and the product is the *para*-nitro derivative (plus perhaps a small amount of the *ortho*-nitro isomer).

(d)   Only the positions marked with an asterisk (*) are activated by *both* the phenyl and methoxy groups. (You should show that a carbocation intermediate formed in the other ring at *any* position cannot be resonance-stabilized by the unshared pair of the methoxy group.) Because of this activation, nitration at the asterisked positions is more rapid than nitration of benzene itself; the products are as follows:

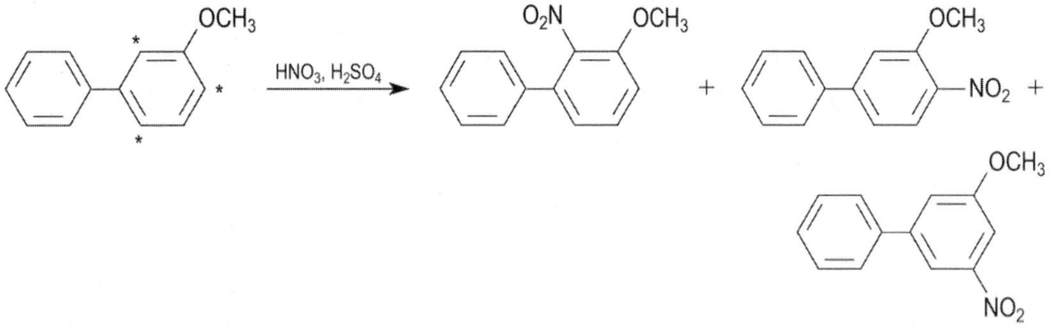

16.46   The reactivity order is $A < B < D < C$. Compound C is most reactive because the substituent has an unshared electron pair that can be used to stabilize the intermediate carbocation by resonance. Compounds B and D have alkyl substituents, which stabilize carbocations; however, the alkyl group of compound B contains a positively charged group that would interact unfavorably with a carbocation, offsetting the stabilizing effect of the alkyl carbon. Compound A has a positively charged, electronegative substituent attached directly to the ring that would interact most unfavorably with the carbocation. (See the solution to Problem 16.45b.) Compounds C and D undergo bromination at the ortho and para positions; compound A undergoes bromination at the meta position; and the position of substitution in compound B depends on the balance of the stabilizing effect of the alkyl group and the destabilizing effect of the positive charge. (In fact, this compound brominates in the ortho and para positions.)

16.47   First draw the resonance structure of compound B that has an electronic octet around phosphorus:

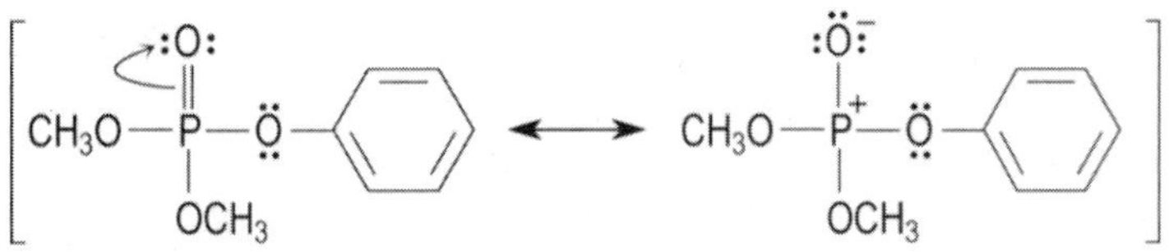

This is an important resonance structure because of its adherence to the octet rule. With this structure in mind, consider then the carbocation intermediate in nitration:

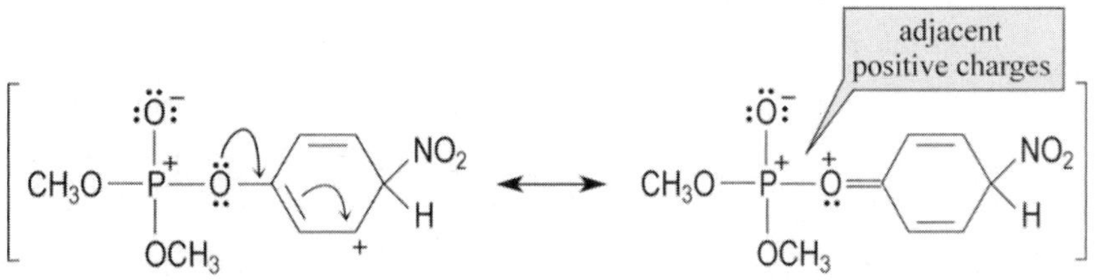

For an oxygen unshared pair to exert a stabilizing resonance effect on the carbocation, a destabilizing interaction must be simultaneously introduced as a result of the adjacent positive charges on oxygen and phosphorus. Hence, this resonance interaction is relatively unimportant—or much less important than in compound A. Because the resonance interaction of its electron pairs is weak or nonexistent, the oxygen acts more like an electronegative substituent *without* unshared pairs. Such a substituent (like the nitrogen in a nitro group) should be a deactivating, meta-directing group, as observed.

   Phenyl acetate has a dipole resonance structure for the C=O group as well, but such a structure does not have an octet on carbon. Furthermore, it has no other oxygens bound to the carbon to exert an electron-withdrawing polar effect. Therefore, the resonance structure that involves stabilization of the carbocation intermediate by the oxygen is relatively more important—so much so, that the para position is activated.

16.48   The two possibilities are the acylation of anisole by benzoyl chloride (pair A), or the acylation of benzene by p-methoxybenzoyl chloride (pair B). Because the methoxy group activates electrophilic substitution at the *ortho* and *para* positions, the Friedel–Crafts reaction of pair A should occur under the milder conditions.

| benzoyl chloride | anisole | *p*-methoxybenzoyl chloride | benzene |

A B

(Both reactions would work, however; and the use of pair B avoids the possibility of *ortho* substitution, although, with the Friedel–Crafts reaction, this does not generally occur to a great extent.)

16.49   The unsaturation number of compounds A and B is 5; hence, both compounds could be aromatic. Compound C is a hydrocarbon; because hydrogenation of C gives a compound (compound D) with the molecular formula C_9H_{10}, then the formula of compound C could be C_9H_8. If so, then the reaction of compounds A and B with acid is a dehydration. If compounds A and B are alcohols, their —OH groups must be on adjacent carbons in order for them to dehydrate to the same alkene. Compound D, the hydrogenation product, has an unsaturation number of 5; nitration confirms that it and, by deduction, the other compounds, are aromatic. Since compound D can contain no ordinary double bonds (they would have been hydrogenated), it must contain a ring *in addition to* the aromatic ring. The structures of the compounds that meet all these criteria are as follows:

A B C D

Note that compound D, because of its symmetry, would give only two nitration products:

D

The nitration results rule out isomers such as the following:

would give four
mononitration products;
therefore, cannot be D.

If the compound at left
cannot be D,
this cannot be A.

16.50   Any compound must have an unsaturation number of 6 and must contain a benzene ring.

***p*-divinylbenzene**

structures such as this without a benzene ring
would hydrogenate to *cis*-1,4-diethylcyclohexane

Structures such as the one on the right that do not contain a benzene ring do not meet the criterion, because they would undergo hydrogenation of the ring double bonds.

16.51   The λ_{max} of styrene is due to the conjugation of the alkene π bond with the p orbitals of the benzene ring. (Only one ring p orbital is shown below). Both π bonds of phenylacetylene cannot be simultaneously conjugated with the $2p$ orbitals of the benzene ring. (The nonoverlapping p orbitals are shown in gray below.) Hence, the λ_{max} of phenylacetylene is also due to the conjugation of a *single* π bond with the benzene ring. Consequently, the structural features of both compounds responsible for their UV spectra are identical.

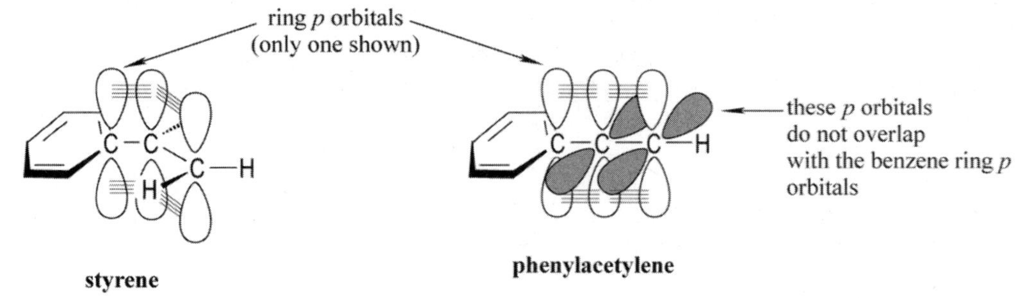

16.52   For this product to form, benzene must react with the electrophile shown in the following equation:

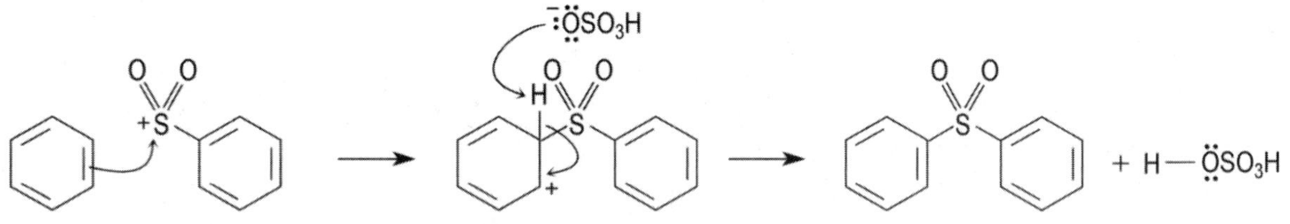

The electrophile, in turn, is formed from benzenesulfonic acid by the following mechanism:

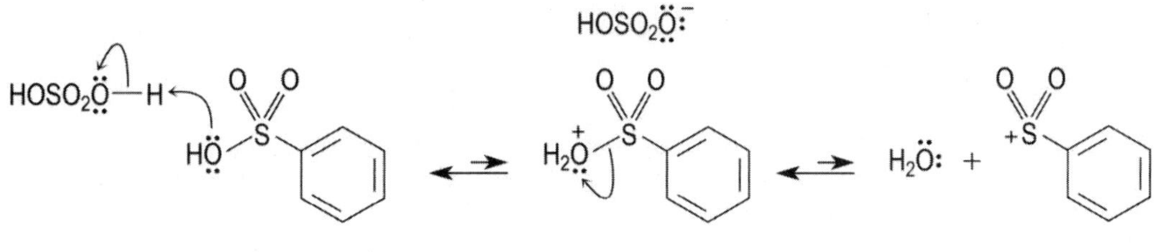

16.53   Run the electrophilic aromatic substitution reaction in reverse.

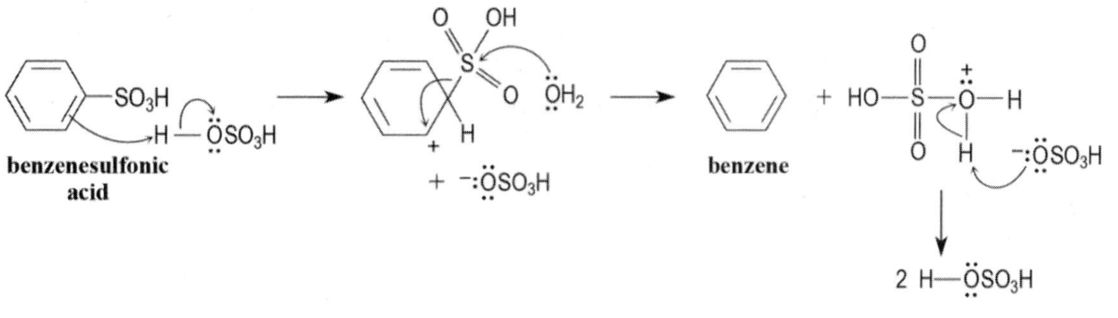

There are several variations on this mechanism. For example, SO₃H could be lost from the carbocation intermediate to give protonated SO_3 (that is, $^+SO_3H$; see Problem 16.13 on text p. 805) which could react with water to give SO_3 (sulfur trioxide) and H_3O^+. Sulfur trioxide reacts vigorously with water to give H_2SO_4 (sulfuric acid). The important aspect of the mechanism is the protonation of the ring and loss of a species which would serve as an electrophile in the reverse reaction.

16.54   The product has an unsaturation number of 6, which means that, relative to the starting material, the product has an additional ring or double bond. The fact that it does not react with Br_2 suggests that a new ring has been formed. In fact, formation of a carbocation by protonation of the alcohol and loss of water leads to an electrophilic aromatic substitution reaction that forms a new six-membered ring.

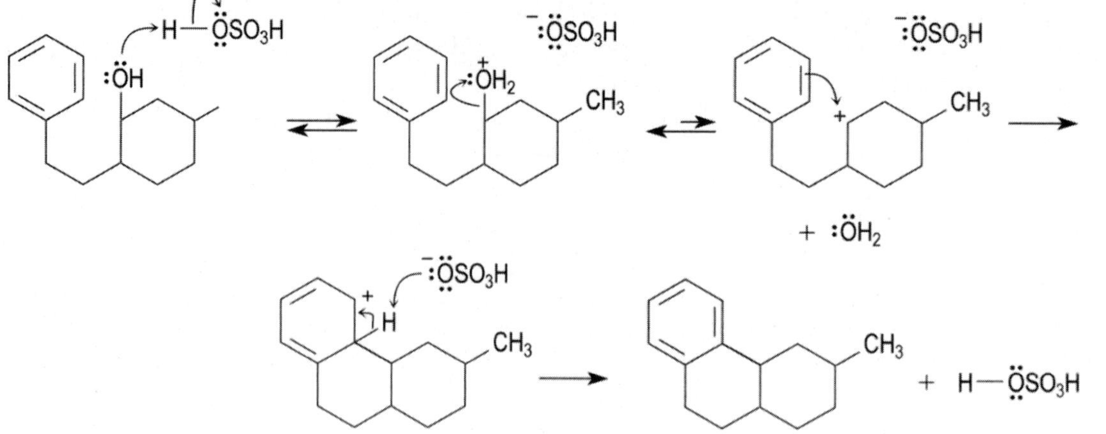

16.55   (a) and (b)

Generation of the electrophile: The electrophile is the carbocation generated by protonation of the alcohol oxygen and loss of water.

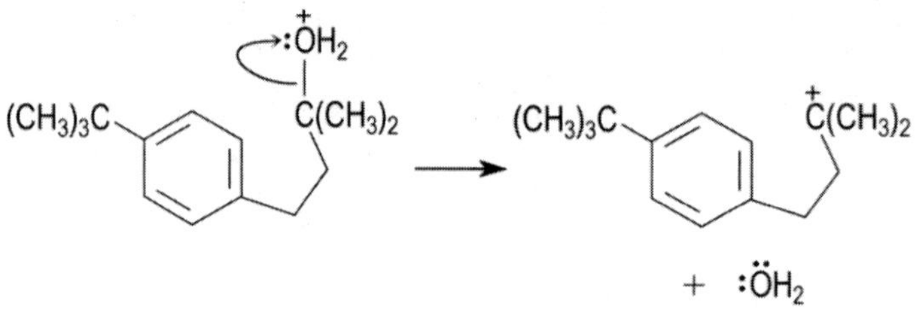

A Lewis acid–base association reaction of the benzene π electrons with the electrophile to generate another carbocation:

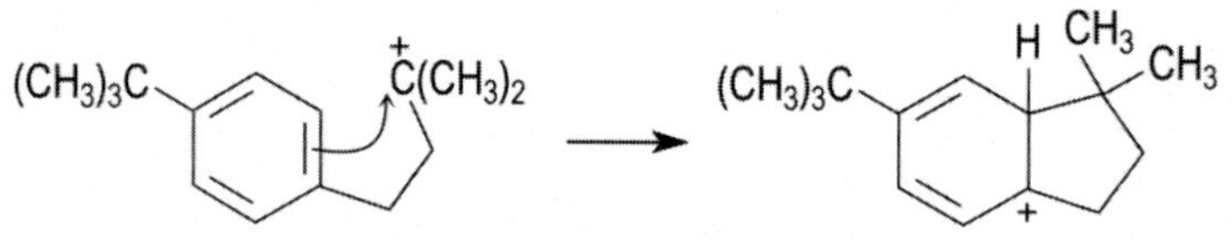

Loss of a β-proton to the Brønsted base H_2O to form the new aromatic compound:

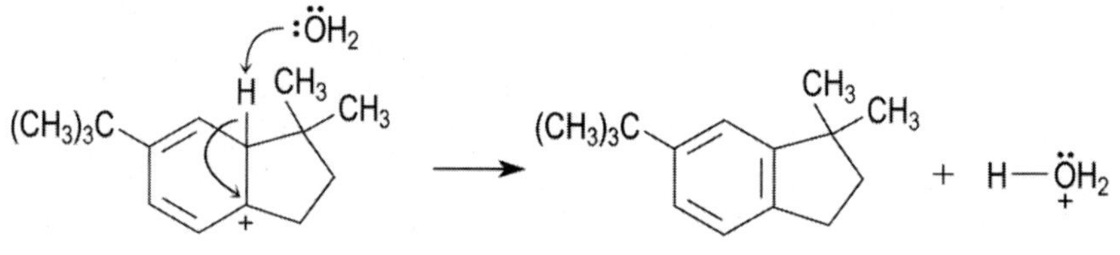

A

(c)   The aromatic ring of compound A has three alkyl substituents. Two of them—the *tert*-butyl group and one of the ring bonds—direct substitution by their electronic effects to the positions indicated by the asterisk:

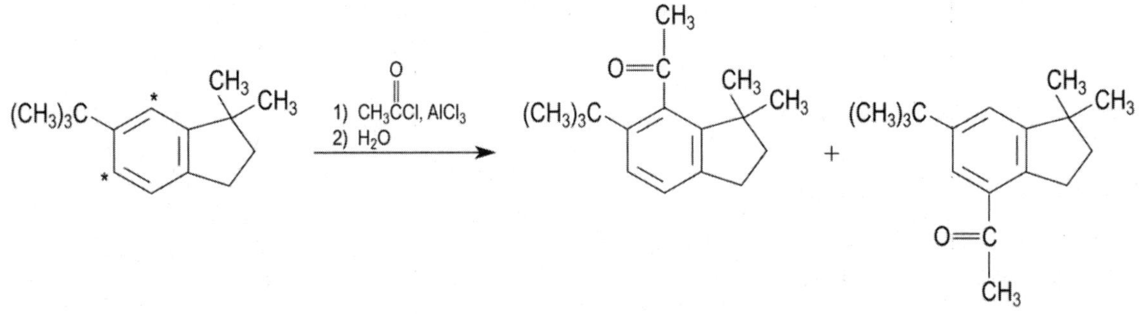

Electrophilic substitution might have occurred at either of these positions to give either or both of the compounds shown in the foregoing equation. Both of these positions, however, are ortho to the very large *tert*-butyl group, and one of them is ortho to *two* highly branched groups. For steric reasons substitution cannot occur at these very congested positions. Hence, it occurs at the remaining position. Note that ring position meta to alkyl substituents are not deactivated; they are simply less activated than positions that are ortho and para to alkyl substituents. Furthermore, the remaining ring position is activated by one alkyl substituent.

16.56   Protonation of the double bond of styrene by the acid gives a carbocation A, which serves as an electrophile to react with the double bond of a second molecule of styrene to give carbocation B, which loses a proton to give alkene X.

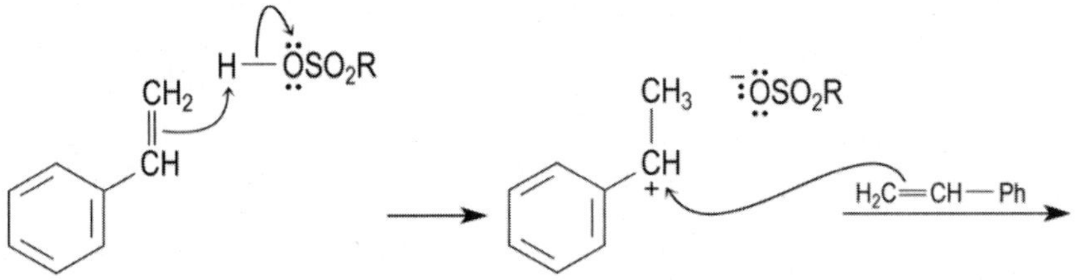

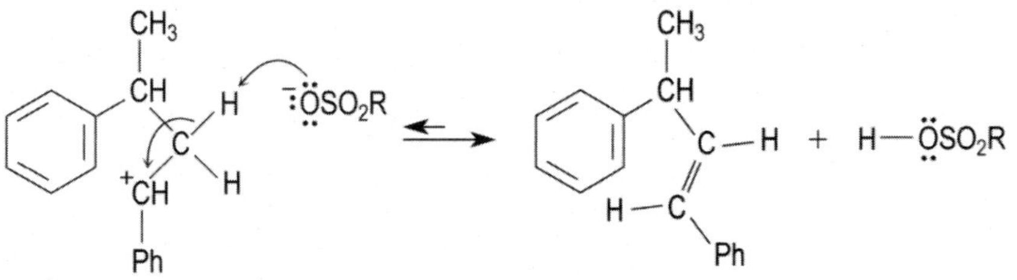

Carbocation A reacts with the double bond of styrene rather than the ring because the ring, being aromatic, is more resistant to electrophilic substitution than the "ordinary" double bond of styrene.

　　Carbocation B, formed by reprotonation of X, undergoes an internal (intramolecular) electrophilic aromatic substitution. This reaction is accelerated over an ordinary electrophilic aromatic substitution because it is intramolecular (proximity effect; Sec. 11.8).

Y and *Z*
(each is a racemate)

B

Carbocation *B* is chiral, but because it is formed from achiral starting materials, it is the racemate. When each stereoisomer of the racemate reacts, a new stereocenter is formed. Both enantiomers of *B* give both possible configurations of the stereocenter. Therefore, two diastereomers (*Y* and *Z*) are formed, each of which is a racemate.

16.57   Equation 16.15c referred to in the problem shows that metal electrophiles, by forming complexes with halide ions, can make them better leaving groups in the same way that protonation of an alcohol makes the OH a better leaving group. (Think of the metal cation as a "fat proton.") Thus, the silver ion assists the ionization of the bromide group of 2-bromooctane by complex formation. Carbocation formation by an S_N1–E1 mechanism is followed by the usual combination of S_N1 substitution and E1 elimination products.

16.58   The formula of compound *B* would be useful in solving this problem. Consider the integral in the NMR spectrum of compound *B*. The entire spectrum integrates for ten hydrogens, and we can hypothesize that no carbons are lost on the reaction with sodium ethoxide. Hence, the formula of compound *B* is thus C_9H_{10}. Thus, compound *A* loses the elements of HBr when it reacts with a base—clearly an elimination reaction—and adds one molar equivalent of H_2 on catalytic hydrogenation. Therefore, compound *B* is an alkene formed by an E2 reaction from compound *A*. The NMR of compound *B* shows that it has five aromatic hydrogens; hence, the remaining unsaturation of compound *B* is due to a monosubstituted benzene ring. The resonances for the vinylic hydrogens in the NMR spectrum centered at δ 5.1 integrate for two protons. These two protons are chemically nonequivalent, and they show no large splittings typical of a *cis*- or *trans*-alkene. Hence, the vinylic protons are in the part structure H_2C=. The chemical shift and integration of the δ 2.1 resonance suggests a vinylic methyl group. The methyl group cannot be benzylic because this would require one fewer aromatic hydrogen and one more vinylic hydrogen. The structure of *B* is determined and the structure of *C*, its hydrogenation product, follows:

B
isopropenylbenzene
(α-methylstyrene)

C
isopropylbenzene

Two possible alkyl halides can be considered for *A*. The chirality of *A* allows us to decide between them.

CH₃
|
C—CH₃
|
Br

CH₂Br
|
CH—CH₃

A
(2-bromo-1-methylethyl)benzene

16.59   (a)   The IR spectrum indicates the presence of an aromatic ring. The singlet at δ 3.72 suggests a methoxy group, and the apparent pair of doublets indicates a para-disubstituted benzene ring. The M + 2 peak in the mass spectrum with about one-third the intensity of the M = 142 peak suggests the presence of chlorine. A chlorine, a —OCH₃ group, and a para-disubstituted benzene ring fully account for the mass of 142. The compound *p*-chloroanisole (1-chloro-4-methoxybenzene) is consistent with all the data.

Cl—⟨benzene ring⟩—OCH₃

p-chloroanisole
(1-chloro-4-methoxybenzene)

When interpreting mass spectra, don't forget that you must use exact masses for isotopes. Thus, mass M = 142 corresponds to the compound containing ³⁵Cl, and the mass of the M + 2 peak corresponds to the compound containing ³⁷Cl.

(b)   The singlet at δ 3.8 suggests a methyl ether. The doublet of doublets centered at δ 7 suggest a *para*-disubstituted benzene ring. The complex pattern centered at δ 6.1 corresponds to two vinylic hydrogens, and the UV spectrum suggests a vinylic group conjugated with the benzene ring. The chemical shift of the δ 1.9 resonance suggests a vinylic methyl group, and its splitting of 6.6 Hz shows that this methyl group is coupled to one of the vinylic hydrogens. The 15.7 Hz splitting within the vinylic proton resonance shows that these protons are trans; the 965 cm⁻¹ IR absorption confirms this deduction. All of these data conspire to suggest that compound *B* is (*E*)-4-methoxy-1-(1-propenyl)benzene, known more commonly as *trans-p*-methoxy-β-methylstyrene.

H
|
C—CH₃
‖
CH₃O—⟨benzene ring⟩—C
|
H

(E)-4-methoxy-1-(1-propenyl)benzene
(trans-p-methoxy-β-methylstyrene)

16.60   (a)   *p*-Dibromobenzene can give only one mononitro derivative; hence, it must be compound *A*. *o*-Dibromobenzene can give two mononitro derivatives, and is therefore compound *B*. *m*-Dibromobenzene is compound *C*.

Br ... Br →[nitration] Br ... NO₂ ... Br

A

Br ... Br →[nitration] Br ... Br ... NO₂   +   O₂N ... Br ... Br

B

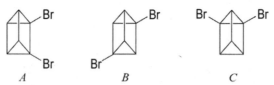

(b) Answer this question by deciding in how many different ways a single nitro group may be substituted for a hydrogen in each isomer. Because isomer separation methods (such as crystallization) based on conventional physical properties were used to differentiate isomers, enantiomeric differences between products were not evident; that is, only constitutional isomers or diastereomers are considered to be different compounds.

(c) The following two products derived from nitration of compound *C* were probably formed in smallest amount. In the formation of compound *C1*, the nitro group and the two bromines are involved in severe van der Waals repulsions; and the formation of compound *C2* satisfies the directing effect of neither bromine substituent.

(d) The three dibromobenzene isomers would be differentiated most readily by the number of resonances in their proton-decoupled CMR spectra. Compound *A,* the para isomer, has two resonances; compound *B,* the ortho isomer, has three resonances; and compound *C*, the meta isomer, has four resonances. The number of resonances in each case, of course, corresponds to the number of chemically nonequivalent sets of carbons.

16.61 (a) Each alkyl halide group reacts with benzene in a separate Friedel–Crafts alkylation reaction.

(b) The formula indicates that successive electrophilic aromatic substitution reactions have occurred; the product is triphenylmethane, Ph_3CH.

(c) The reaction is an intramolecular Friedel–Crafts alkylation that forms a ring. Note that the oxygen activates the substitution.

(d) Comparison of the formula of the product to that of naphthalene shows that one molar equivalent of the acylating agent has been introduced. Since there is no chlorine in the product, the reaction must involve a double acylation of the naphthalene ring by both ends of the acid chloride. The only way that this can occur with the formation of rings of reasonable size is for the acylation to occur across ortho positions of one benzene ring, or across the peri positions, which are the two positions on either side of the ring junction. These three possibilities account for the three products:

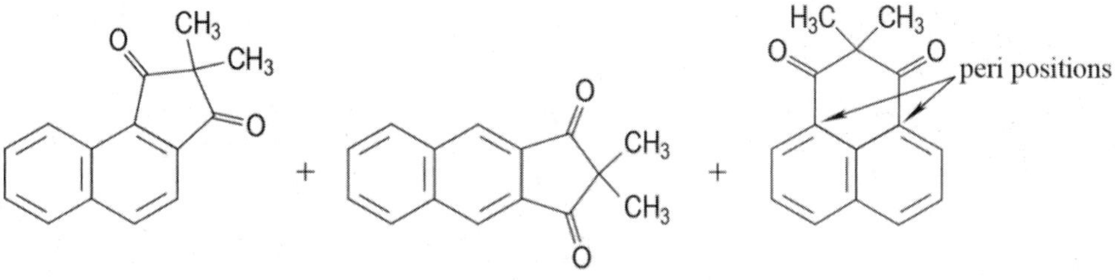

(e)   Similar to the reaction mechanism shown in Eq. 16.37a and 16.37b, hypoiodous acid will install an iodine ortho to an OH group on a benzene ring.

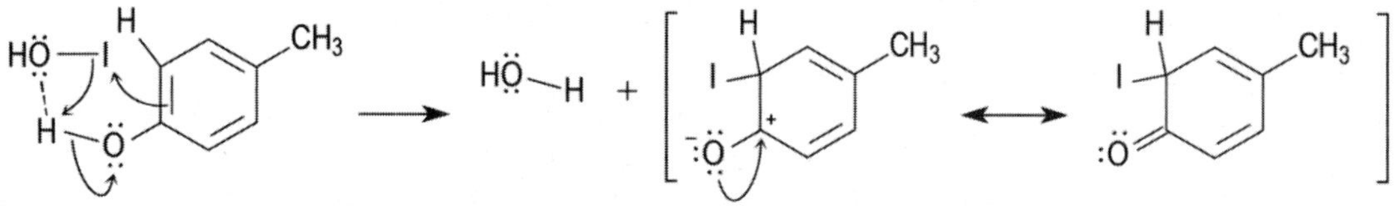

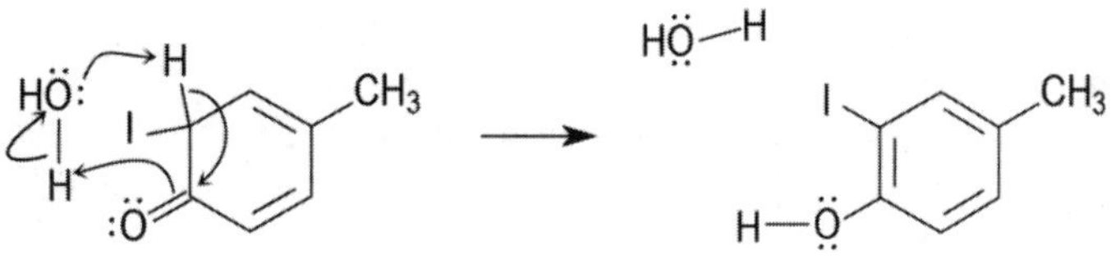

Since there is a large excess of I–OH, the other ortho position may be iodinated as well.

(f)   The cyclohexyl group, an alkyl group, directs nitration to the ortho and para positions of the benzene ring.

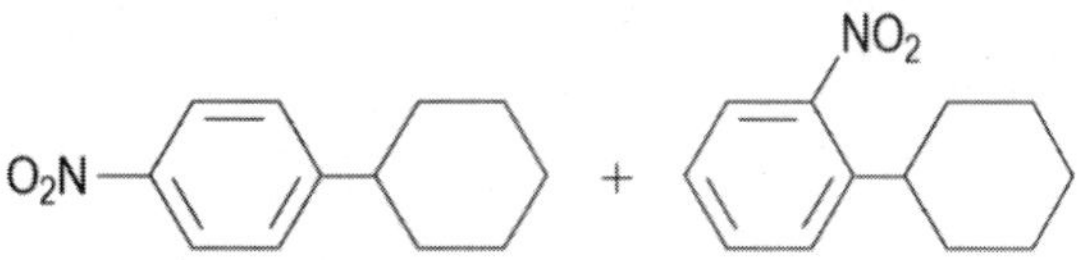

(mostly)

(g)   Ferrocene, like other aromatic compounds, undergoes a Friedel–Crafts acylation reactions.

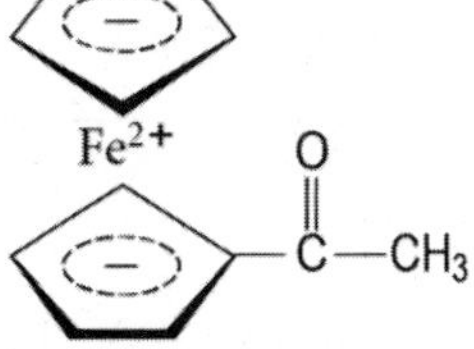

(h)   The nitro group is directed by both substituents to the position ortho to the methoxy group, and the bromine in the second reaction is directed to the other position ortho to the methoxy group.

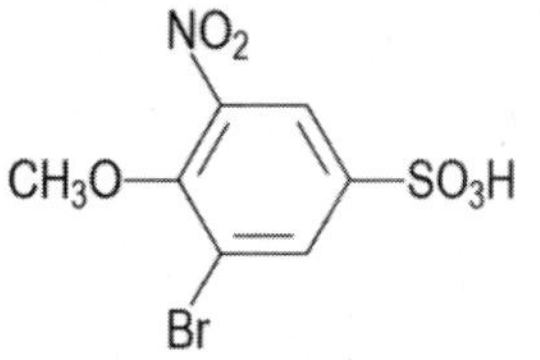

3-bromo-4-methoxy-5-nitrobenzenesulfonic acid

16.62   (a)   Because carbon-4 is para to the methoxy group, the carbocation intermediate involved in nitration at this carbon can be stabilized by electron delocalization from the oxygen, as it is in nitration at the para position of anisole. (See the colored structure in Eq. 16.29 on text p. 812.) Hence, the methoxy group activates substitution at carbon-4, and nitration at carbon-4 of 1-methoxynaphthalene is faster than nitration of naphthalene.

      (b)   When nitration occurs at carbon-5, the unshared electrons of the oxygen can be used to stabilize the carbocation intermediate by resonance; consequently, nitration at carbon-5 of 1-methoxynaphthalene is faster than nitration of naphthalene itself.

      (c)   Nitration at carbon-6 is analogous to nitration at a meta position of anisole; see Eq. 16.30 on text p. 813. In this case, the oxygen electrons *cannot* be delocalized to stabilize the carbocation. The only effect of the methoxy group in this case is its rate-retarding polar effect. Therefore, nitration at carbon-6 of 1-methoxynaphthalene is slower than nitration of naphthalene.

16.63   Friedel–Crafts acylation (or any other electrophilic substitution reaction) at carbon-2 of furan gives a carbocation intermediate with more resonance structures than acylation at carbon-3. Consequently, the carbocation intermediate from acylation at carbon-2 is more stable, and, by Hammond's postulate, acylation at carbon-2 is faster.

The carbocation intermediate from acylation at carbon-2:

three resonance structures

The carbocation intermediate from acylation at carbon-3:

two resonance structures

16.64   The balance between protonation on oxygen and protonation on a carbon of the ring depends on the relative basicities of an aromatic "double bond" and the oxygen of an aromatic ether. The fact that anisole protonates on oxygen shows that an ether oxygen is more basic than the aromatic π-electron system:

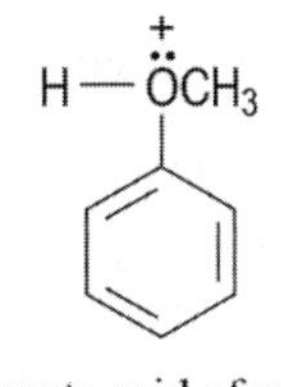

$$H-\overset{+}{\underset{}{\ddot{O}}}CH_3$$

conjugate acid of anisole

That 1,3,5-trimethoxybenzene protonates on carbon, then, means that the resulting conjugate-acid cation is stabilized sufficiently that the aromatic carbons have enhanced basicity. The stability of this cation is due to resonance stabilization involving all three oxygens:

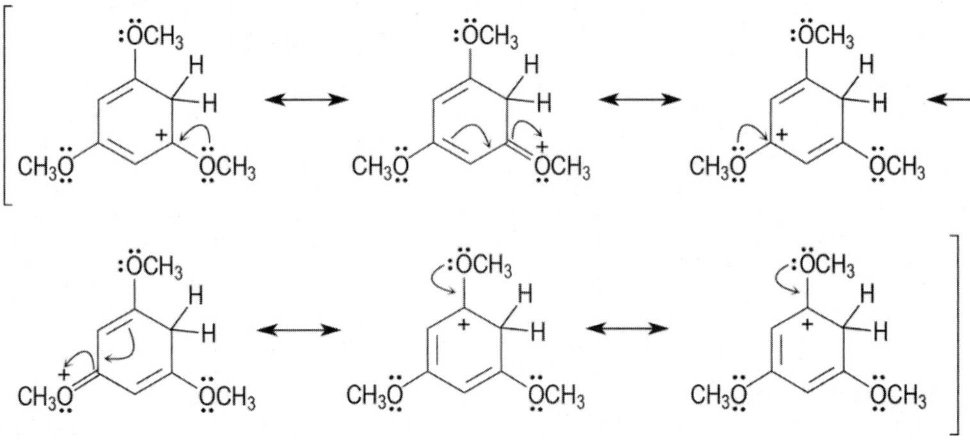

16.65   (a)   Furan derivatives, like 1,3-cyclopentadiene derivatives, are in effect conjugated dienes locked into *s*-cis conformations. Such dienes are reactive in Diels–Alder reactions. The product *A* results from such a reaction.

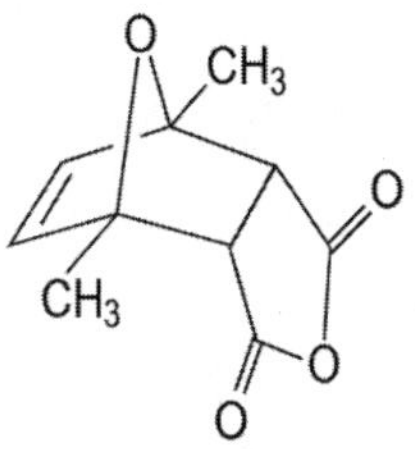

compound *A*

(b)   Evidently, the acidic conditions promote loss of the bridging ether oxygen as water. The mechanistic steps involved are protonation, carbocation formation, loss of a proton, and then repetition of a similar sequence. The driving force for this reaction is the formation of the aromatic ring.

16.66   In this reaction a *tert*-butyl cation is lost rather than a proton from the carbocation intermediate. The
electrophile, a nitronium ion $^+NO_2$, is generated by the mechanism shown in Eqs. 16.11a–d on text pp. 803–804.

16.67   At the higher temperature, the two ortho methyl groups (labeled *a* in the problem) are chemically equivalent on
the NMR time scale because rotation of the isopropyl group is rapid; the resonance of these groups occurs at
δ 2.25. However, at the lower temperature, rotation of the isopropyl group is slow, and, as a result, the two
ortho methyl groups are no longer chemically equivalent. Hence, the resonances of the ring methyl groups occur
at different chemical shifts at the lower temperature. Evidently, one ortho methyl group is wedged between the
two isopropyl methyls, and the other is in the same plane as the isopropyl C—H group, as shown by the
structure in the problem.

16.68   (a)   The two terminal rings of hexahelicene cannot lie in the same plane because, if they did, their hydrogens
would have severe van der Waals repulsions. Consequently, the molecule is somewhat bent out of plane

so that one of the terminal rings lies over the other. This causes the molecule to trace one turn of a helix, which is a chiral object. This optical rotation, by the way, is *huge*!

(b)   Normally the two rings of a biphenyl derivative lie in the same plane to maximize conjugation, that is, to maximize overlap of their π-electron systems. In this case, however, if the rings were to lie in the same plane, the very large sulfonic acid (— SO_3H) group on one ring would have severe van der Waals repulsions with an ortho hydrogen of the other ring. Consequently, the molecule adopts the conformation shown in the problem, in which the planes of the two rings are perpendicular. However, this conformation is chiral and is thus capable of showing optical activity. Evidently, rotation about the central carbon–carbon bond is slow enough that the molecule can be resolved into isolable enantiomeric conformations! When the molecule is heated, the internal rotation "reaction" (like all reactions) is accelerated, and the optically active molecule racemizes. This situation is also discussed in Further Exploration 6.2 on page 150–1 of this manual; a similar situation is the subject of solution 16.10(b) earlier in this chapter.

Allylic and Benzylic Reactivity

STUDY GUIDE LINKS

17.1   Synthetic Equivalence

The benzylic oxidation provides an excellent way to introduce a carboxylic acid group into a benzene ring. (None of the electrophilic aromatic substitution reactions in Chapter 16 provide a way to introduce this group directly.) Suppose, for example, we wish to prepare *p*-nitrobenzoic acid from benzene. The desired acid can be formed by benzylic oxidation of any *p*-alkylnitrobenzene with benzylic hydrogens—say, *p*-ethylnitrobenzene:

The *p*-ethylnitrobenzene comes, in turn, from nitration of ethylbenzene:

And the ethylbenzene comes from Friedel–Crafts alkylation of benzene (Eq. 16.20, text p. 807).

If we wanted to prepare *m*-nitrobenzoic acid, we would *first* oxidize ethylbenzene to benzoic acid, and then nitrate the benzoic acid, thus taking advantage of the *meta*-directing effect of the carboxylic acid group:

ethylbenzene   benzoic acid   *m*-nitrobenzoic acid

In these schemes, an alkyl group—specifically, the ethyl group—has been utilized as the *synthetic equivalent* of a carboxylic acid group. That is, the ethyl group is something that is *easily converted into* a carboxylic acid group. (Some people use the word *synthon* to mean the same thing as *synthetic equivalent*.) A person who is skilled in synthesis tends to see various functional groups in terms of their synthetic equivalents. For example, when such a person sees a carboxylic acid group on a benzene ring, an alkyl group pops into mind. For an aldehyde, a primary alcohol comes to mind. (What would you think of as a synthetic equivalent of a *cis*-alkene?) Of course, a group may have more than one synthetic equivalent, and the appropriate synthetic equivalency will generally depend on the exact situation at hand.

By coupling the Friedel–Crafts alkylation with benzylic oxidation, a connection has been made that was not explicitly discussed in the text. As you improve in your mastery of organic chemistry, this is the sort of connection that you should begin to make on your own. A good student will begin to think of compounds "out of context," that is, in ways not directly related to the text material. For example, a student will see a structure and ask, "How could I synthesize this?" even though its synthesis might not be the context in which the structure is encountered. Professional chemists tend to think this way. A structure or a reaction presented in one context will trigger the imagination to think of a wholly new context in which it might be useful. This is one way that new ideas are born. Awareness of this "intellectual triggering" process is one of the best reasons that scientists study the professional literature. And this works not only in chemistry; it works in fields as diverse as physics and medicine.

FURTHER EXPLORATIONS

17.1 Addition versus Substitution with Bromine

The purpose of this Further Exploration is to explain the effect of low bromine concentration in promoting free-radical substitution. The explanation of this effect lies in the rate laws for addition and substitution. The rate law for *ionic addition* (Sec. 5.2A) at low bromine concentrations is

$$\text{rate of ionic addition} = k_{\text{addition}}[\text{alkene}][\text{Br}_2]$$

Although it is not discussed in the text, addition of Br_2 can also take place by a free-radical chain mechanism. At low Br_2 concentrations, the rate of free-radical addition follows the rate law

$$\text{rate of free-radical addition} = k_{\text{fr-addition}}[\text{alkene}][\text{Br}_2]^{3/2}$$

The rather strange-looking $3/2$-order in bromine, roughly speaking, is due to the involvement of a bromine atom ("half of a bromine molecule") produced in the initiation step together with the involvement of a bromine molecule in the rate-limiting propagation step—one and one-half bromine molecules total.

Finally, the rate of free-radical substitution at low Br_2 concentrations has the rate law

$$\text{rate of free-radical substitution} = k_{\text{substitution}}[\text{alkene}][\text{Br}_2]^{1/2}$$

The $1/2$-order dependence in bromine is due to the involvement of a bromine atom that is produced in the initiation step.

A comparison of these rate laws reveals that the substitution reaction has the least dependence on bromine concentration. Hence, the rate of substitution not only *increases less rapidly* with *increasing* bromine concentration than the other reactions, but also *decreases less rapidly* with *decreasing* bromine concentration. At very low bromine concentration, then, free-radical substitution has the greatest rate, and is therefore the observed reaction.

The success of *N*-bromosuccinimide (NBS) in CCl₄ as a reagent for substitution of allylic and benzylic hydrogens by bromine is due to the fact that it provides an experimentally convenient way to maintain a low bromine concentration. It wasn't planned that way; NBS in CCl₄ for many years was known as a good reagent for carrying out allylic and benzylic brominations with no competing additions. Careful research later developed the rationale presented here and in the text.

 # REACTION REVIEW

I. ALLYLIC AND BENZYLIC SUBSTITUTION REACTIONS

A. ALLYLIC AND BENZYLIC BROMINATION WITH NBS

1. *N*-Bromosuccinimide (NBS) can be used to bring about substitution of allylic and benzylic hydrogens with bromine.

N-bromosuccinimide
(NBS)

2. The mechanism of allylic or benzylic bromination with NBS entails the following steps:

   a. The initiation step is the formation of a bromine atom by homolytic cleavage of the N—Br bond in NBS itself.

   b. The ensuing substitution reaction has three propagation steps:

      i. The bromine atom abstracts an allylic or a benzylic hydrogen.

      ii. The HBr thus formed reacts with the NBS in the second propagation step (by an ionic mechanism) to produce a Br₂ molecule.

iii. The last propagation step is the reaction of this bromine molecule with the radical formed in the first propagation step.

c. The Br_2 concentration remains low because it can be generated no faster than an HBr molecule and an allylic radical are generated.

d. The low solubility of NBS in CCl_4 and the consequent low $[Br_2]$ is crucial to the success of allylic bromination with NBS.

3. Because the unpaired electron of an unsymmetrical allylic radical is shared by two chemically nonequivalent carbons, such a radical can react to give two constitutionally isomeric products.

B. BENZYLIC BROMINATION

1. Benzylic hydrogens can be substituted at allylic positions with Br_2 in the presence of light.

2. This is a free-radical chain reaction that involves the following propagation steps:

a. abstraction of a benzylic hydrogen by a bromine atom in preference to a nonbenzylic hydrogen.

i. This propagation step gives rise to the selectivity for substitution of the benzylic hydrogen.

ii. The reason for this selectivity is that the benzylic radical which is formed has greater stability than nonbenzylic radicals.

b. reaction of the benzylic radical with a molecule of bromine to generate a molecule of product along with another bromine atom, which can then react with another benzylic hydrogen.

3. Allylic bromination also occurs under these conditions but is generally not used because of the competition with bromine addition to the double bond; NBS bromination is the preferred method for carrying out allylic bromination.

II. BENZYLIC OXIDATION REACTIONS

A. ALLYLIC AND BENZYLIC ALCOHOLS

1. Allylic and benzylic alcohols are oxidized selectively by a suspension of activated manganese(IV) dioxide, MnO_2, because they react much more rapidly than "ordinary" alcohols. (See Eqs. 17.34a–c, text p. 852 for the mechanism.)

    a. Primary allylic alcohols and primary benzylic alcohols are oxidized to aldehydes.

| a primary allyl alcohol | an aldehyde | a primary benzyl alcohol | an aldehyde |

    b. Secondary allylic alcohols and secondary benzylic alcohols are oxidized to ketones.

| a secondary allyl alcohol | an ketone | a secondary benzyl alcohol | an ketone |

    c. "Activated MnO_2" is obtained by the oxidation–reduction reaction of potassium permanganate, $KMnO_4$, with an Mn^{2+} salt such as $MnSO_4$ under alkaline or acidic conditions followed by thorough drying.

$$2 H_2O + 3 MnSO_4 + 2 KMnO_4 \xrightarrow{H_2O} 5 MnO_2 + 2 H_2SO_4 + K_2SO_4$$

manganese dioxide

B. OXIDATION AT THE ALLYLIC OR BENZYLIC POSITION BY CYTOCHROME P450

1. The enzyme cytochrome P450 (CyP450) can catalyze the replacement of allylic or benzylic protons with hydroxyl groups. (Less commonly, other protons can also be replaced.) It is likely that a radical mechanism is operating (see Sec. 17.5B).

C. SIDE-CHAIN OXIDATION OF ALKYLBENZENES

1. Treatment of alkylbenzene derivatives with strong oxidizing agents under vigorous conditions converts the alkyl side chain into a carboxylic acid group.

a. The benzene ring is left intact.

b. The alkyl side chain, regardless of length, is converted into a carboxylic acid group.

c. The conditions for this side-chain oxidation are generally vigorous: heat, high concentrations of oxidant, and/or long reaction times.

d. Common oxidants are Na_2CrO_7 (sodium dichromate), CrO_3 (chromium trioxide), $KMnO_4$ (potassium permanganate), or O_2 and special catalysts.

2. Oxidation of alkyl side chains requires the presence of a benzylic hydrogen; benzene derivatives with no benzylic hydrogens are resistant to side-chain oxidation.

SOLUTIONS TO PROBLEMS

Solutions to In-Text Problems

17.1   The allylic carbons are indicated with an asterisk (*).

(a)   (b)

17.2   The benzylic carbons are indicated with an asterisk (*).

(a)   (b)

17.3   (a)   The order of increasing S_N1 reactivity is (2) < (1) < (3). The rates depend on the stabilities of the respective carbocation intermediates. One resonance structure for the carbocation intermediate in the solvolysis of compound (3) is a tertiary carbocation; the resonance structures of the carbocation intermediates derived from compounds (1) and (2) are secondary carbocations.

carbocation intermediate in the solvolysis of compound (3)

Consequently, the carbocation intermediate derived from compound (3) is more stable, relative to the starting alkyl halide, than the intermediates in the solvolysis reactions of the other compounds. The intermediate in the solvolysis of compound (2) is least stable because of the electron-withdrawing polar effect of the oxygen. Note that an oxygen in the meta position cannot be involved in resonance; consequently, only its rate-retarding polar effect operates.

Note that the rates of S_N1 solvolysis reactions parallel the rates of electrophilic aromatic substitution reactions. Just as a methoxy group activates an aromatic substitution reaction at an ortho or para position, it also activates a reaction involving a benzylic cation at an ortho or para position. Just as a methoxy group *deactivates* aromatic substitution at a meta position, it also deactivates a reaction involving a benzylic cation at a meta position.

(b)   The reactivity order is (2) < (3) < (1). The S_N1 reaction of compound (2) is slowest because the polar effect of the *meta*-chloro substituent destabilizes the intermediate carbocation. The reaction of compound (3) is faster because the resonance effect of the *para*-chloro group partially offsets its polar effect.

carbocation intermediate in the solvolysis of compound (3)

Compound (1) reacts most rapidly because the carbocation intermediate is not destabilized by the deactivating polar effect of a chloro substituent, which outweighs its resonance effect.

17.4   The alkyl halide is the one that reacts to give the same carbocation intermediate, that is, a carbocation with the same resonance structures:

the same carbocation intermediate
that is involved in text Eq. 17.7

17.5   The carbocation formed when trityl chloride ionizes, the *trityl cation* (Ph_3C^+), is stabilized by delocalization of electrons from *all three* phenyl rings. This carbocation has more resonance structures than the carbocations formed from the other alkyl halides in the table, and is thus so stable that the transition state leading to its formation also has very low energy; consequently, it is formed very rapidly.

trityl chloride
(triphenylmethyl chloride)

trityl cation
(triphenylmethyl cation)

17.6   The number of products depends on (1) whether all of benzylic or allylic positions are equivalent, and (2) whether the resonance structures of the free-radical intermediate are identical.

(a)   All allylic positions of cyclohexene are chemically equivalent, and the two resonance structures are identical. Hence, only one allylic bromination product is possible.

cyclohexene
(starting material)

free-radical intermediate
(identical resonance structures)

3-bromocyclohexene
product

(b)   The two allylic positions are nonequivalent, but only one has allylic hydrogens; hence, only one can react. The two resonance structures of the resulting free radical are nonequivalent; hence, two different products can be formed.

3,3-dimethylcyclohexene
(starting material)

free-radical intermediate

6-bromo-3,3-dimethylcyclohexene

3-bromo-4,4-dimethylcyclohexene

products

(c)   *Trans*-2-pentene (middle, below) has two distinguishable allylic positions. In each of the allylic radicals that results from hydrogen abstraction at the two positions, the unpaired electron is delocalized to two different carbons. (See Study Problem 17.1.) The abstraction of H^a leads to two products, but abstraction of H^b leads to one product, because the resonance structures of the intermediate free radical are identical (assuming that the double bond retains its trans stereochemistry).

(E)-1-bromo-2-pentene
+

3-bromo-1-pentene

products from abstraction of H^a

free-radical intermediate
from abstraction of H^a

trans-2-pentene
(starting material)

free-radical intermediate
from abstraction of H^b
(equivalent resonance structures)

(E)-4-bromo-2-pentene

product from abstraction of H^b

(d)   A benzylic hydrogen is abstracted rather than a hydrogen of the *tert*-butyl group because a more stable benzylic free-radical intermediate is obtained.

4-*tert*-butyltoluene

free-radical intermediate

1-bromomethyl-4-*tert*-butylbenzene

(e)   A benzylic hydrogen is abstracted from the isopropyl group rather than a hydrogen of the two methyl groups because a more stable benzylic free-radical intermediate is obtained.

1-isopropyl-4-nitrobenzene

free-radical intermediate

1-(1-bromo-1-methylethyl)-4-nitrobenzene

17.7   (a)   The initially formed Grignard reagent (middle, below) undergoes a rapid allylic rearrangement; each Grignard reagent in the equilibrium can react with D_2O.

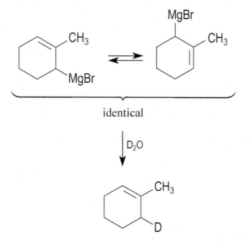

allyl Grignard reagent in equilibrium

(b)   Because the two Grignard reagents in rapid equilibrium are identical, only one product is obtained:

identical

17.8   (a)   Although this alkyl halide has two chemically nonequivalent types of β-hydrogens, an allylic hydrogen is more acidic than a nonallylic one; hence, the product is the conjugated diene derived from elimination of the allylic hydrogen and the bromine.

H ◄——— this β-proton is abstracted
by the base because it
is allylic

K⁺ ⁻OC(CH₃)₃ →

1,3-cyclohexadiene

(b)   The benzylic proton is abstracted; β-elimination gives a vinylic ether.

OCH₃
|
C═CH₂

(1-methoxyvinyl)benzene
(α-methoxystyrene)

17.9   Reaction with concentrated HBr can involve an allylic carbocation intermediate; consequently, a mixture of products, *A* and *B*, could be formed:

trans-2-buten-1-ol

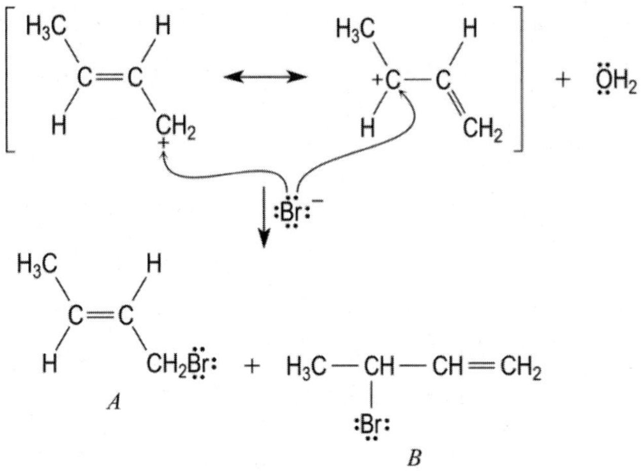

Reaction of the tosylate with NaBr in acetone, in contrast, involves an S_N2 reaction; consequently, there is no reactive intermediate, and only product A is formed.

17.10   (a)   (b)

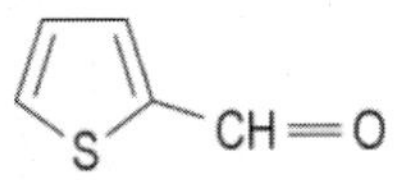

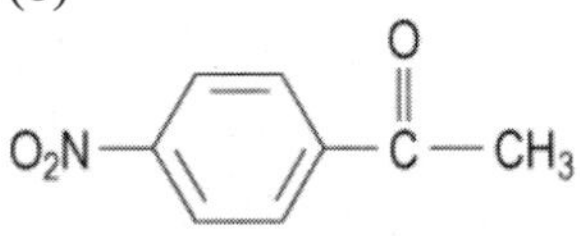

(c) 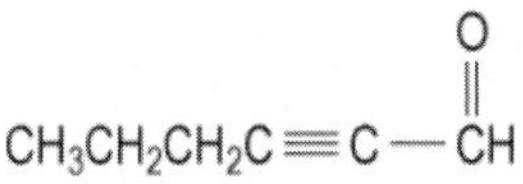   (d)

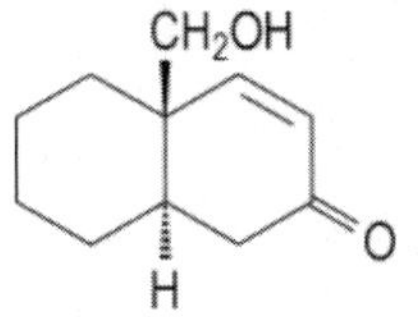

$$CH_3CH_2CH_2C \equiv C - \overset{\overset{O}{\|}}{C}H$$

17.11   (a) 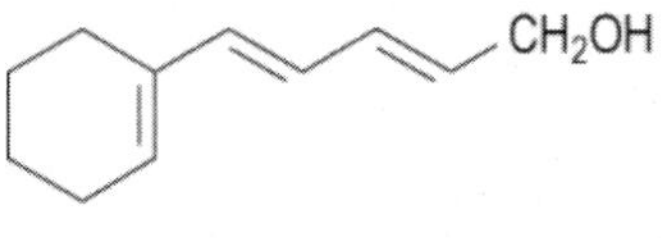   (b) 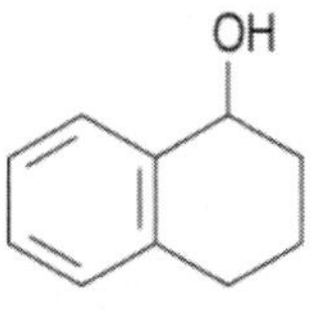   (c)

17.12   PCC oxidizes all primary alcohols to aldehydes and all secondary alcohols to ketones; MnO_2 oxidizes only allylic or benzylic alcohols (primary alcohols to aldehydes and secondary alcohols to ketones).

(a)   Both methods give the same product.

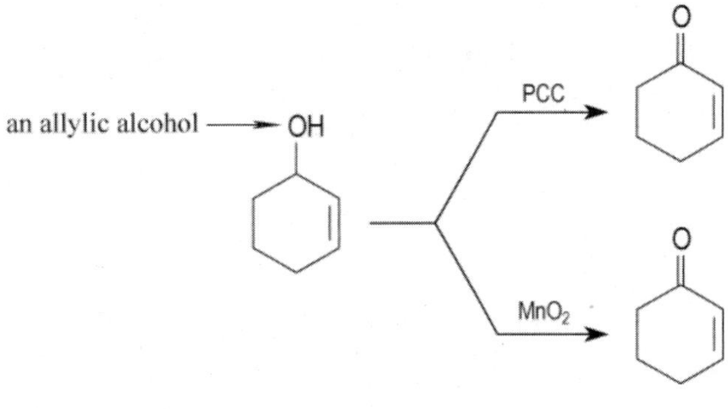

an allylic alcohol ⟶ OH

PCC ⟶

MnO_2 ⟶

(b)

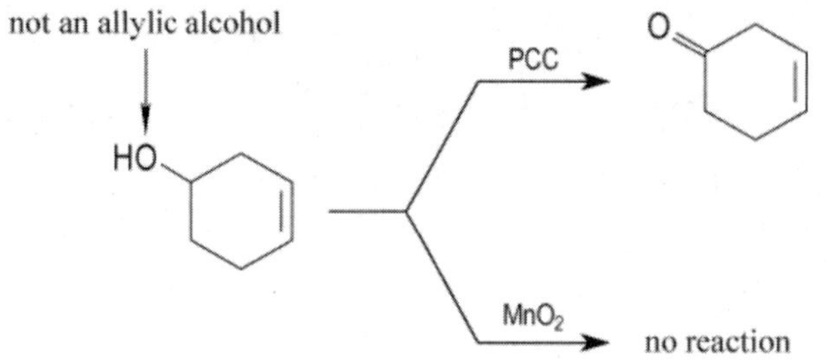

(c)   There is no reaction in either case.

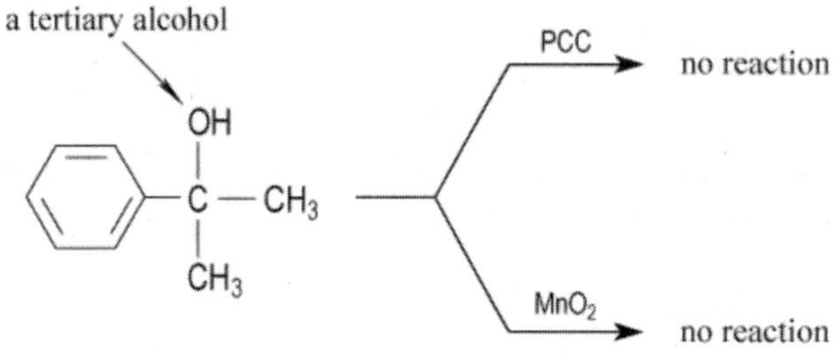

(d)

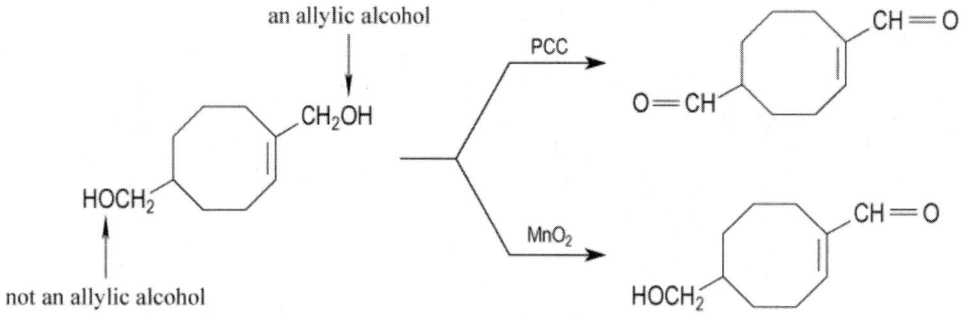

17.13   Oxidation by CyP450 occurs at the benzylic position.
(a)   (b)

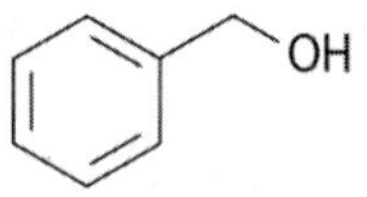

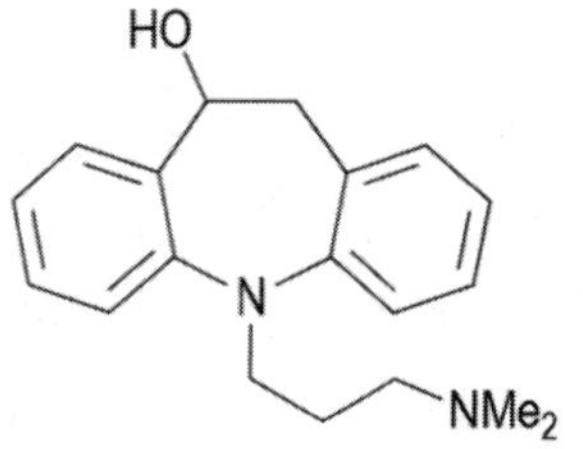

17.14   Remember that once the benzylic hydrogen (or in this case, deuterium) is removed, resonance can put the unpaired electron on either of two carbons. Due to the other deuteriums present in this compound, oxidation at either of the two positions will lead to different products.

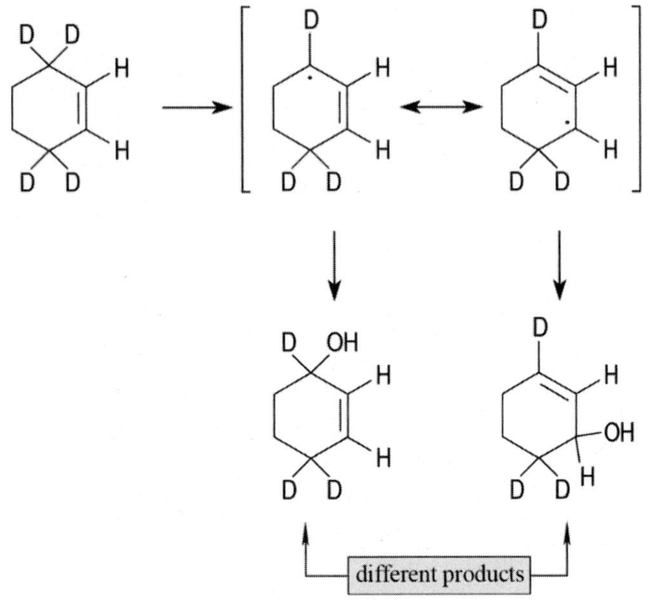

different products

17.15 (a)

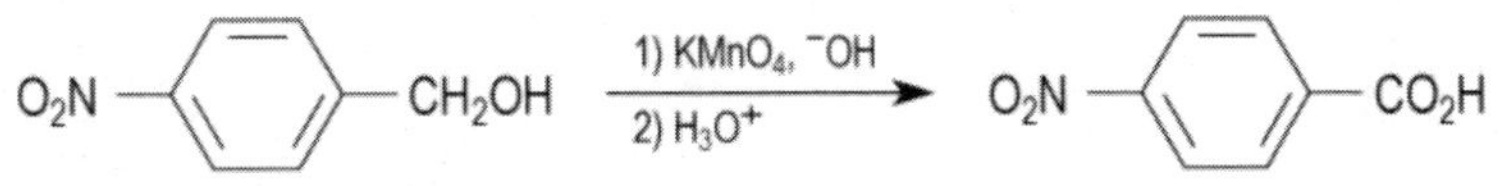

p-nitrobenzyl alcohol   *p*-nitrobenzoic acid

(b)   The butyl group is oxidized because it has α-hydrogens; the *tert*-butyl group is not affected.

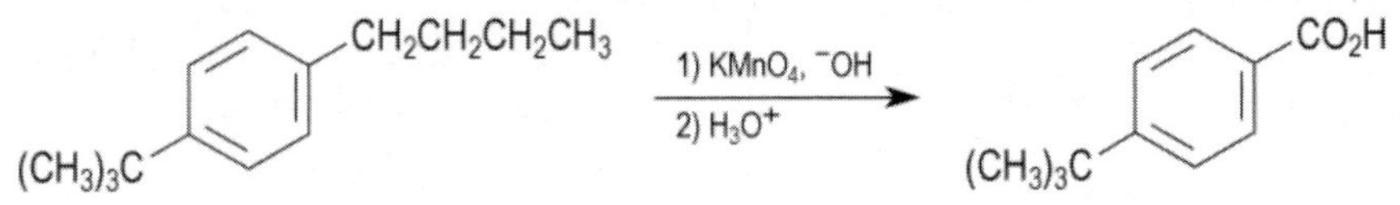

1-butyl-4-*tert*-butylbenzene   4-*tert*-butylbenzoic acid

17.16 (a)   The oxidation product shows that compound *A* is an *ortho*-disubstituted dialkylbenzene. Only compound *A = o*-xylene is consistent with this analysis and with the formula.

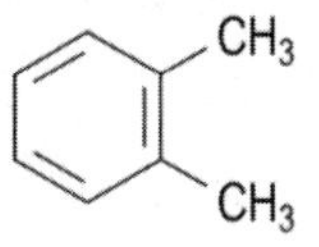

o-xylene

(b)   Because one carbon is lost as a result of the oxidation, and because the benzene ring accounts for all four degrees of unsaturation, compound *B* must be ethylbenzene.

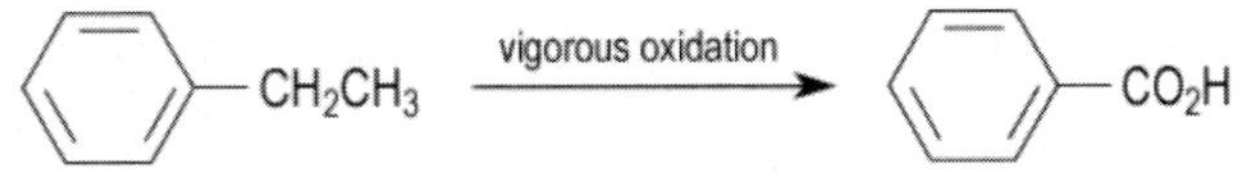

ethylbenzene   benzoic acid

17.17 (a)   Vitamin A is a diterpene because it contains four isoprene skeletons, which are shown as heavy bonds.

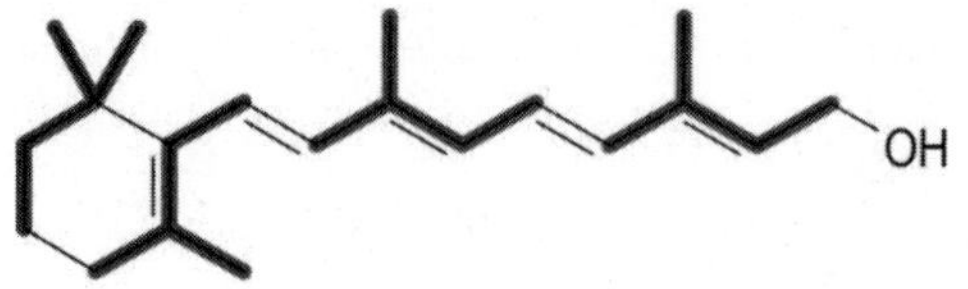

vitamin A

(b)   Caryophyllene is a sesquiterpene because it contains three isoprene skeletons, which are shown as heavy bonds.

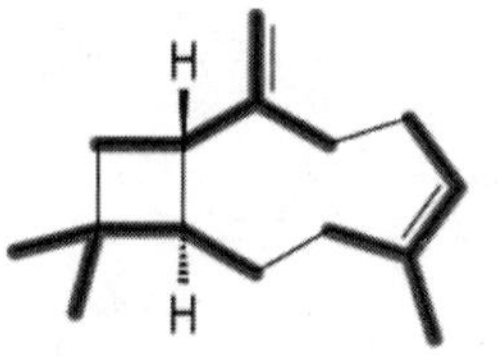

caryophyllene

17.18   (a)   A biosynthetic mechanism for limonene is as follows: (B: = a base.)

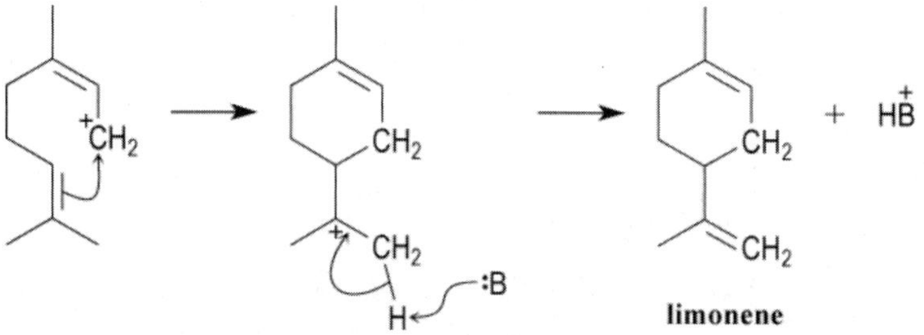

limonene

(b)   Ionization of geranyl pyrophosphate is followed by reaction with the pyrophosphate anion on the other electron-deficient carbon of the resonance-stabilized carbocation; rotation about a single bond is followed by ionization of pyrophosphate to give the desired carbocation.

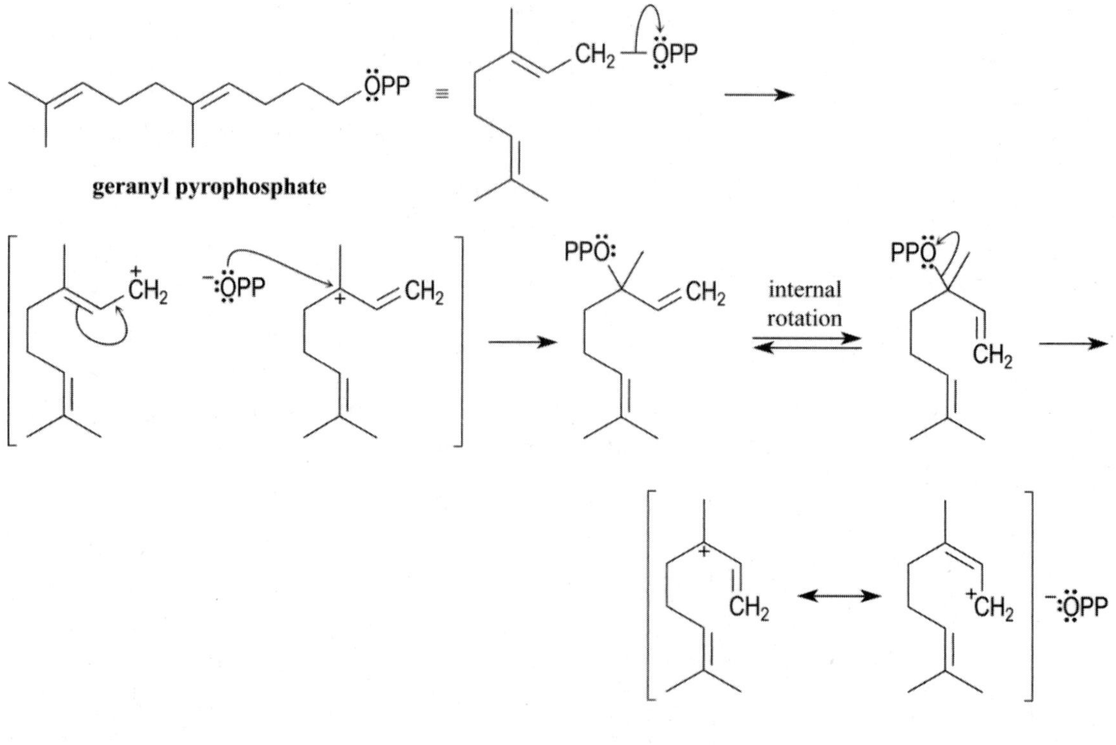

geranyl pyrophosphate

internal
rotation

17.19   (a)   Geranyl pyrophosphate is converted into farnesyl pyrophosphate by a mechanism exactly analogous to the one shown in Eqs. 17.44 on text p. 863. Then farnesyl pyrophosphate hydrolyzes to farnesol; see text Eq. 17.45.

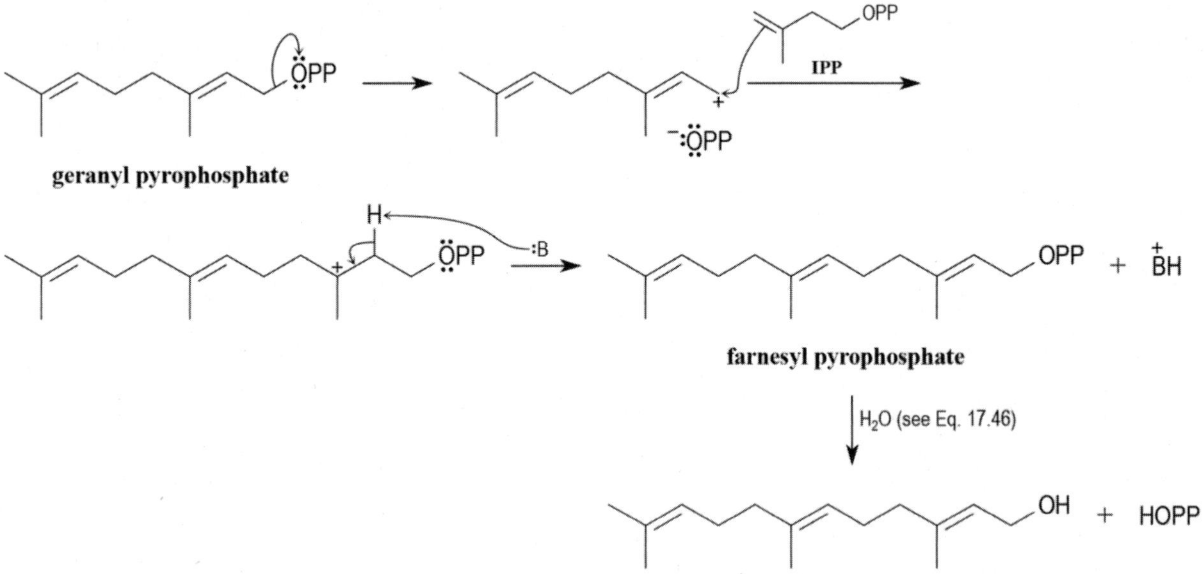

geranyl pyrophosphate

farnesyl pyrophosphate

H_2O (see Eq. 17.46)

farnesol

(b)   To see the new bond connections that have to be made, draw α-pinene in a planar projection. The projection on the left shows the molecule as it is in the text; turning this projection clockwise 120° in the plane of the page yields the second projection; and turning this projection over yields the third projection, which can be related more easily to the carbocation in the previous problem.

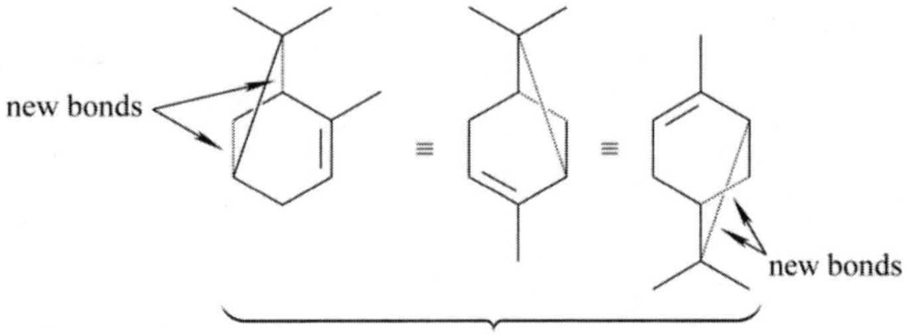

α-pinene

Here is the mechanism:

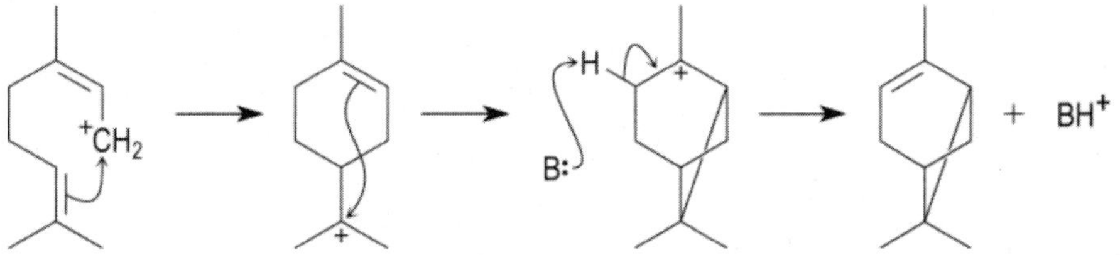

17.20   (a)

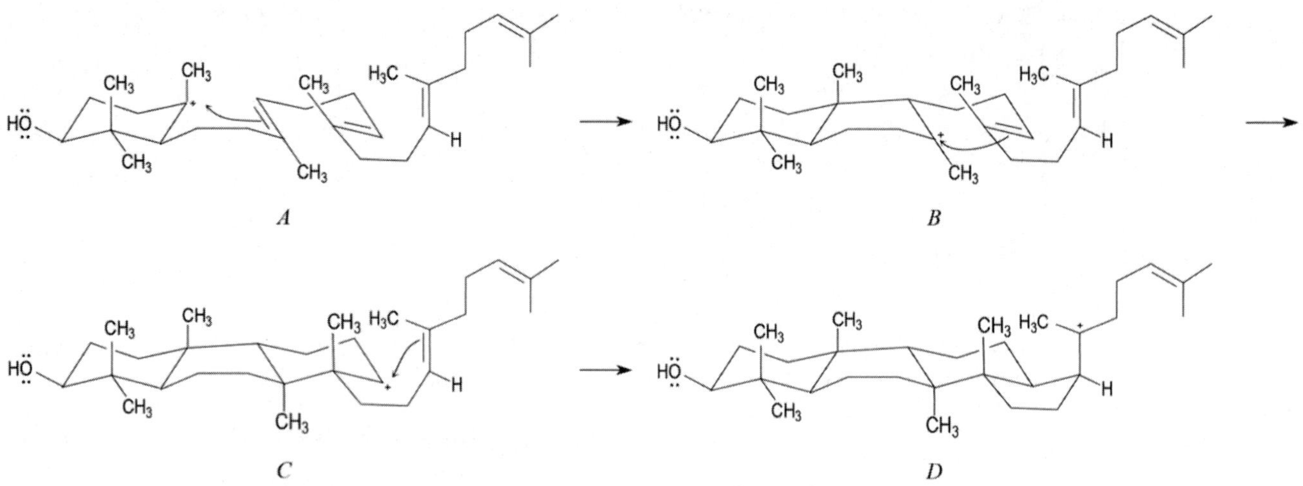

A → *B* →

C → *D*

(b)   If intermediate *D* is formed as part of a stepwise process, as shown in part (a), it would require that carbocation *C*, a secondary carbocation, exist as a discrete intermediate. However, it's energetically unfavorable that a secondary carbocation would form from a tertiary carbocation (*B*). Thus, a concerted mechanism is more likely, where all bonds involved break and form simultaneously.

17.21

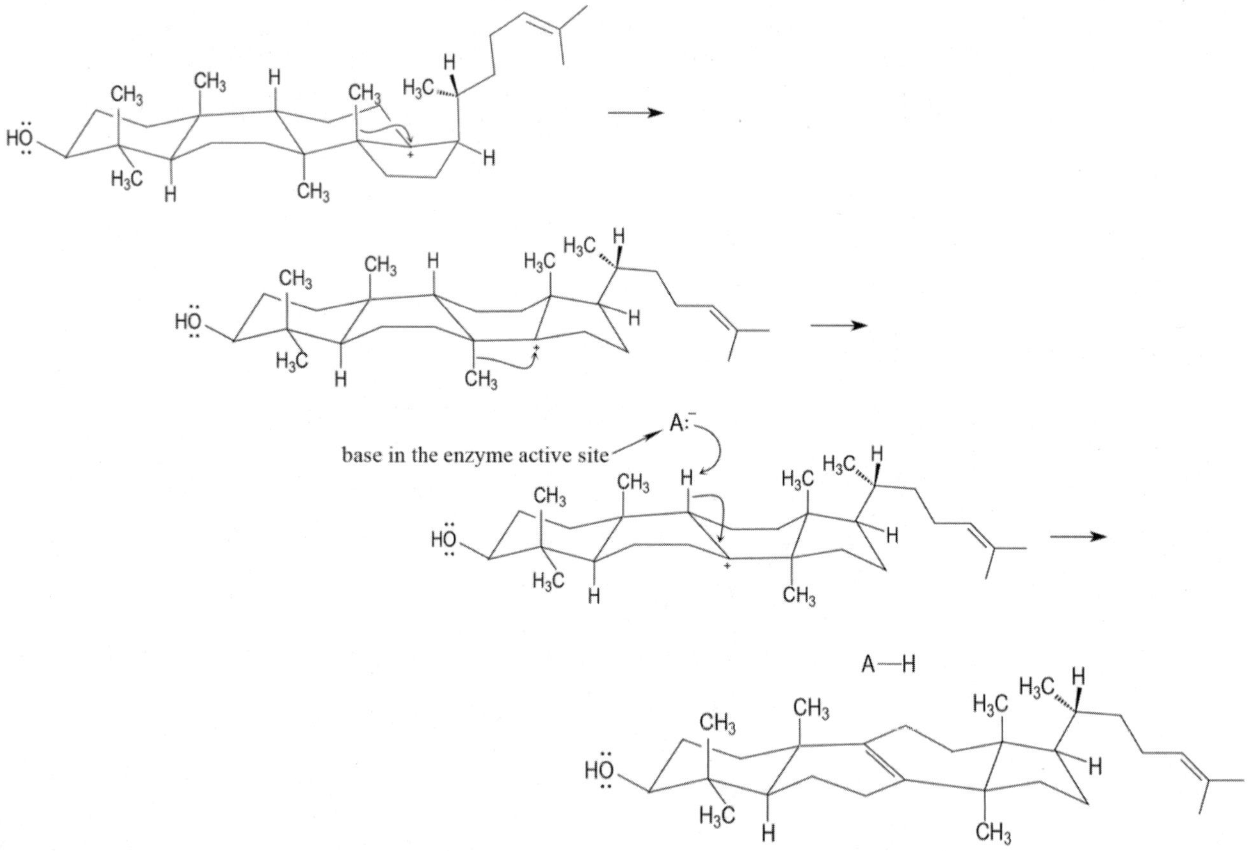

base in the enzyme active site

lanosterol

Solutions to Additional Problems

17.22   The structure of the starting material is:

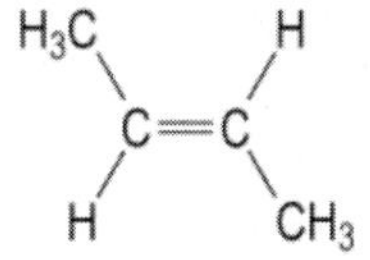

***trans*-2-butene**

(a)

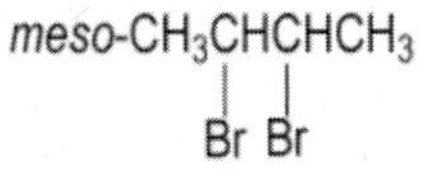

meso-CH$_3$CHCHCH$_3$
         | |
         Br Br

***meso*-2,3-dibromobutane**

(b)

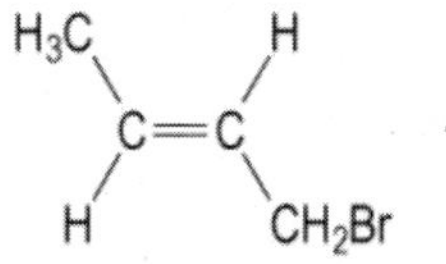

+   CH$_3$CHCH=CH$_2$

(*E*)-1-bromo-2-butene 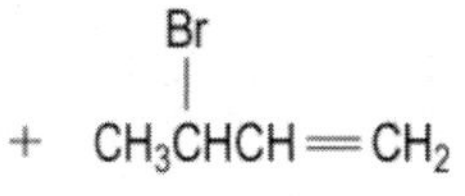   **3-bromo-1-butene**

(c)

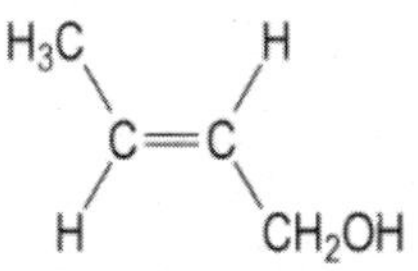

+   CH$_3$CHCH=CH$_2$

(*E*)-3-buten-ol 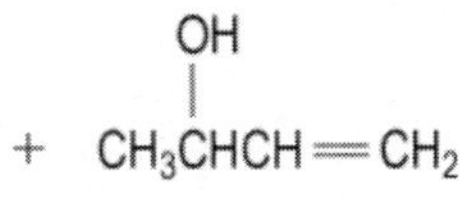   **1-buten-3-ol**

(d)

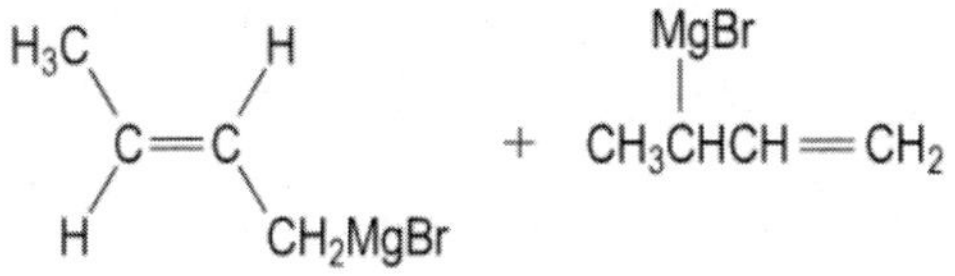

+   CH$_3$CHCH=CH$_2$

(e)

+   CH$_3$CHCH=CH$_2$

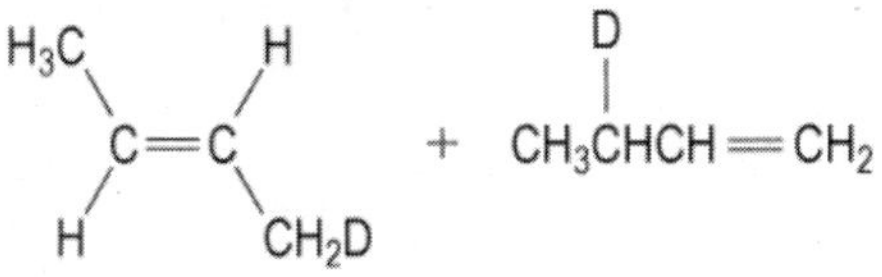

17.23   The structure of the starting material is:

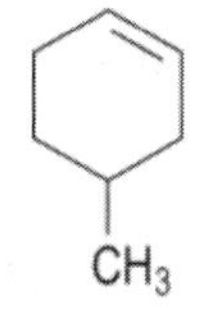

4-methylcyclohexene

(a)

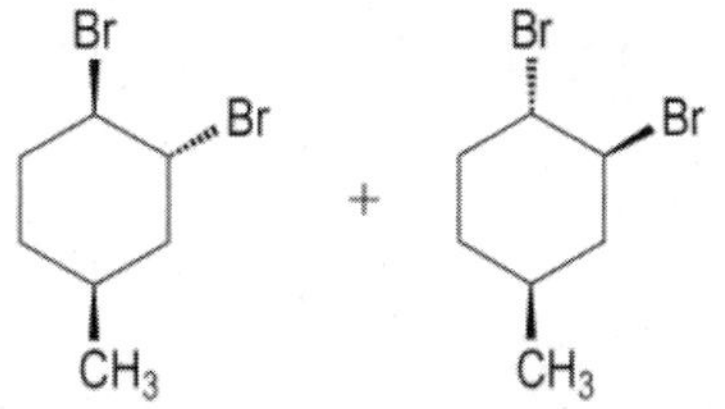

**(1*R*,2*R*,4*S*)-
1,2-dibromo-4-
methylcyclohexane**   **(1*S*,2*S*,4*S*)-
1,2-dibromo-4-
methylcyclohexane**

(and their enantiomers)

(b)

3-bromo-4-methyl-cyclohexene
(two diastereomers and their enantiomers)

3-bromo-5-methyl-cyclohexene
(two diastereomers and their enantiomers)

3-bromo-6-methyl-cyclohexene
(two diastereomers and their enantiomers)

(c)

6-methyl-3-cyclohexenol
(two diastereomers and their enantiomers)

4-methyl-2-cyclohexenol
(two diastereomers and their enantiomers)

4-methyl-2-cyclohexenol
(two diastereomers and their enantiomers)

(d)   As in the previous parts, all four stereoisomers of each compound are formed.

> ⚠ Don't forget that Grignard reagents undergo a very rapid allylic rearrangement (Eq. 17.23, text p. 847).

(e)   As in the previous part, all four stereoisomers of each compound are formed.

17.24   Compounds (a), (d), and (e) are terpenes. The isoprene skeletons are shown with heavy bonds.

(a)

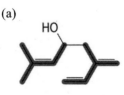

ipsdienol

(b)   Compound (b) is not a terpene.

(c)   Modhephene is believed to be formed from a terpene by a skeletal rearrangement, but its terpene origin is not obvious from its structure.

(d)

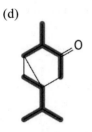

β-thujone

(e)

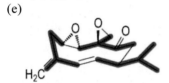

periplanone B

(f)   Compound (f) is not a terpene.

17.25   (a)   Zoapatanol is a diterpene; the isoprene units are shown with heavy bonds.

(b)

17.26   In reaction (1), the acidic conditions promote protonation of the alcohol and ionization to a carbocation, which reacts in a Lewis acid–base association reaction with ethanol at the two nonequivalent electron-deficient carbons. (The mechanism, which is discussed in Sec. 11.2C in the text, is outlined below; fill in the curved-arrow notation.)

In reaction (2), the carbon-oxygen bond is never broken; consequently, no allylic rearrangement can occur. This is a Williamson ether synthesis that involves formation of the alkoxide anion, which, in turn, is alkylated by ethyl iodide. (See Sec. 11.2A in the text; Problem 11.58 on text pp. 562 illustrates the same idea.)

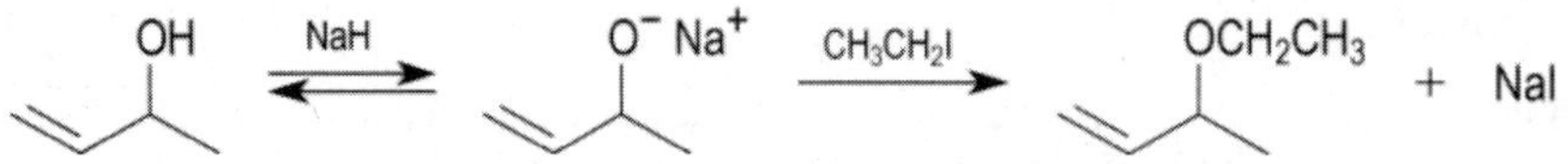

$$\text{(structure: but-3-en-2-ol, OH)} \xrightarrow{\text{NaH}} \text{(O}^-\text{Na}^+) \xrightarrow{\text{CH}_3\text{CH}_2\text{I}} \text{(OCH}_2\text{CH}_3) + \text{NaI}$$

17.27 (a) The allylic-rearrangement product B [(E)-1-bromo-2-butene] could be formed by ionization to a carbocation and bromide ion followed by reaction with the bromide ion on the other electron-deficient carbon.

(b) The curved-arrow notation is shown in the following scheme.

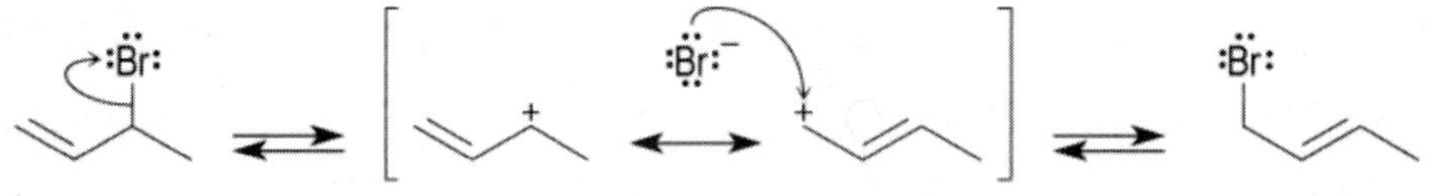

$$\text{3-bromo-1-butene} \rightleftharpoons \left[\text{allylic cation resonance structures} \right] \rightleftharpoons \text{(E)-1-bromo-2-butene}$$

3-bromo-1-butene
(compound A)

(E)-1-bromo-2-butene
(compound B)

(c) Compound B, the rearrangement product, is favored at equilibrium because it has the double bond with the greater number of alkyl branches.

17.28 (a)

$$\text{PhCH}_3 \xrightarrow[\text{CCl}_4]{\text{NBS, peroxides}} \text{PhCH}_2\text{Br} \xrightarrow[\text{CH}_3\text{OH}]{\text{Na}^+ \ ^-\text{OCH}_3} \text{PhCH}_2\text{OCH}_3$$

benzyl methyl ether

(b)

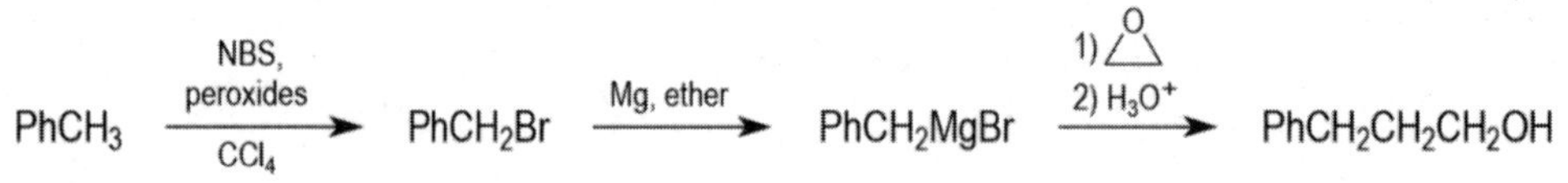

$$\text{PhCH}_3 \xrightarrow[\text{CCl}_4]{\text{NBS, peroxides}} \text{PhCH}_2\text{Br} \xrightarrow{\text{Mg, ether}} \text{PhCH}_2\text{MgBr} \xrightarrow[\text{2) H}_3\text{O}^+]{\text{1) epoxide}} \text{PhCH}_2\text{CH}_2\text{CH}_2\text{OH}$$

toluene

3-phenyl-1-propanol

(c) In the first set of reactions, note that allyl bromide is a particularly reactive alkylating agent.

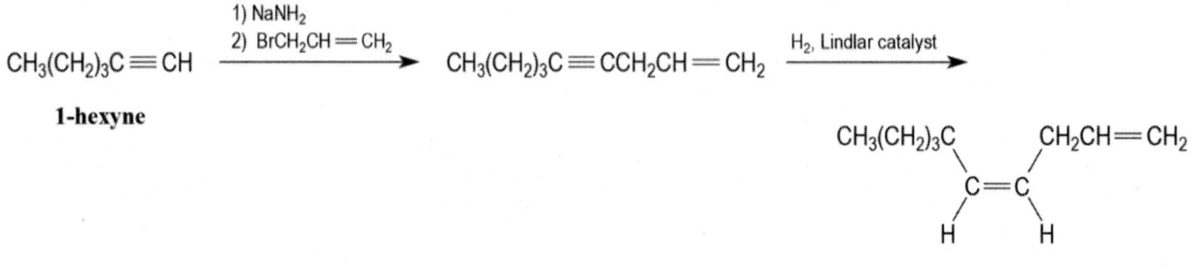

$$\text{CH}_3(\text{CH}_2)_3\text{C}\equiv\text{CH} \xrightarrow[\text{2) BrCH}_2\text{CH}=\text{CH}_2]{\text{1) NaNH}_2} \text{CH}_3(\text{CH}_2)_3\text{C}\equiv\text{CCH}_2\text{CH}=\text{CH}_2 \xrightarrow{\text{H}_2, \text{ Lindlar catalyst}}$$

1-hexyne

$$\begin{array}{c}\text{CH}_3(\text{CH}_2)_3\text{C} \quad\quad \text{CH}_2\text{CH}=\text{CH}_2 \\ \text{C}=\text{C} \\ \text{H} \quad\quad\quad \text{H}\end{array}$$

(Z)-1,4-nonadiene

(d)

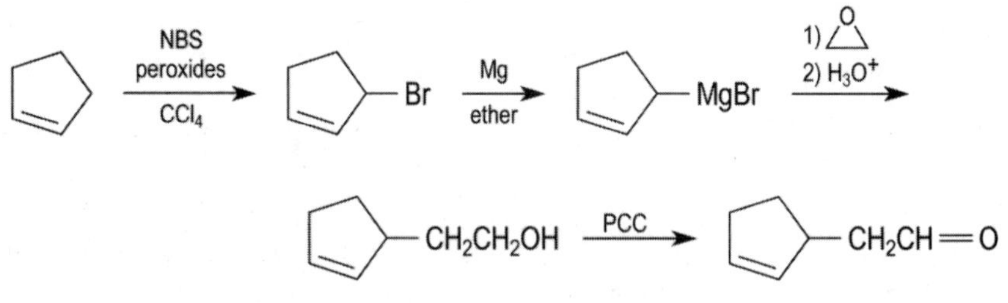

$$\text{(cyclopentene)} \xrightarrow[\text{CCl}_4]{\text{NBS peroxides}} \text{(cyclopentene)}-\text{Br} \xrightarrow{\text{Mg, ether}} \text{(cyclopentene)}-\text{MgBr} \xrightarrow[\text{2) H}_3\text{O}^+]{\text{1) epoxide}}$$

$$\text{(cyclopentene)}-\text{CH}_2\text{CH}_2\text{OH} \xrightarrow{\text{PCC}} \text{(cyclopentene)}-\text{CH}_2\text{CH}=\text{O}$$

(e)

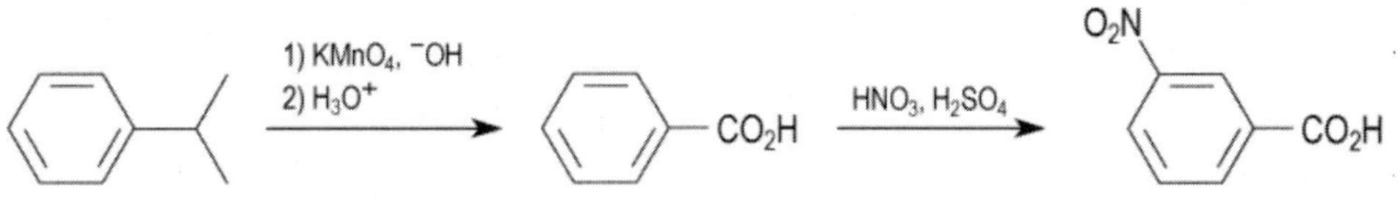

cumene   **m-nitrobenzoic acid**

(f)

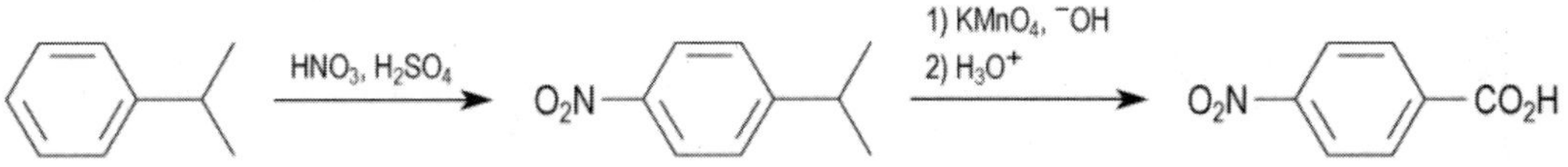

cumene   **p-nitrobenzoic acid**

Compare the solutions to parts (e) and (f) and notice how reversing the sequence of the oxidation and nitration steps brings the directing effects of different substituents into play during nitration.

(g)

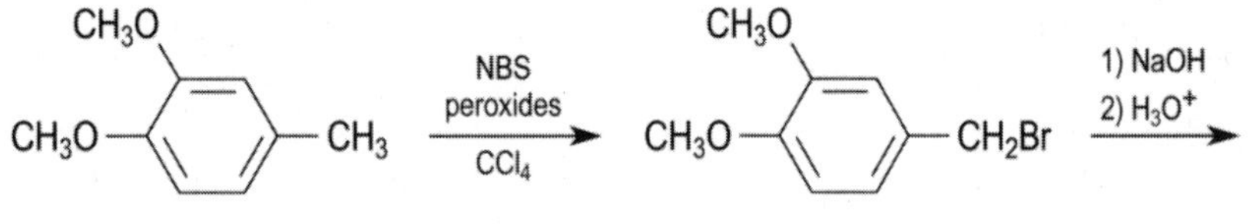

1,2-dimethoxy-4-methylbenzene

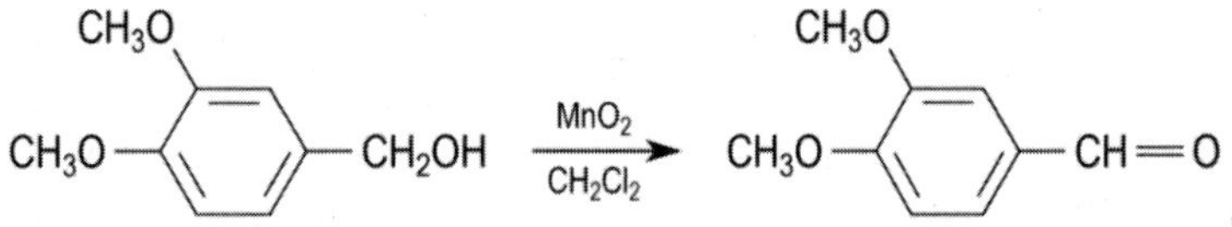

3,4-dimethoxybenzaldehyde

(h)

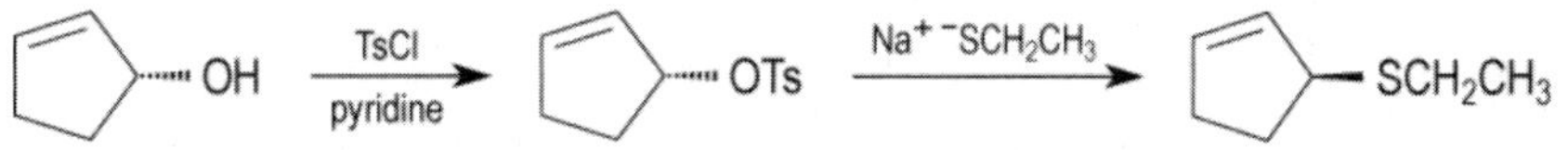

17.29   The compounds that give the most stable carbocation intermediates are the ones that undergo the most rapid solvolysis. This problem deals with the effect of substituent on the stability of the carbocation intermediate. The key is to analyze the balance of resonance and polar substituent effects just as you would for electrophilic aromatic substitution. The order of increasing reactivity is (4) < (1) < (3) < (2). Thus, compound (2) reacts most rapidly because the carbocation intermediate is stabilized by the *electron-donating resonance effect* of the *p*-methoxy substituent:

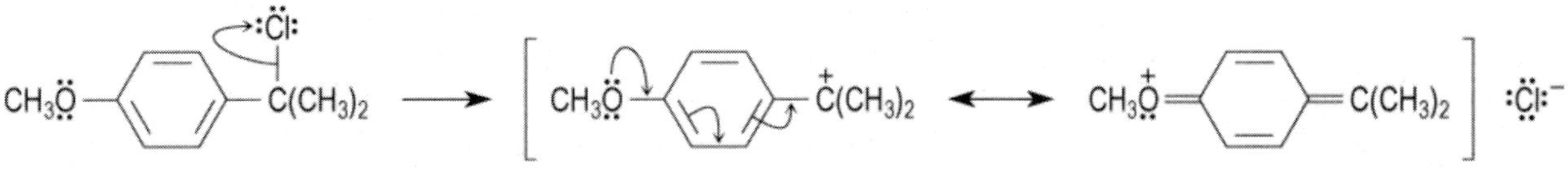

As in electrophilic substitution, the resonance effect of the *p*-methoxy group strongly outweighs its electron-withdrawing polar effect. In compound (3), there is a similar resonance effect; however, the polar effects of halogen substituents outweigh their resonance effects. Consequently, compound (3) reacts more slowly. The nitro group exerts no resonance effect in the carbocation intermediates derived from compounds (1) and (4); the question is then whether its polar effect is stronger from the meta or para position. As in electrophilic aromatic substitution, a *para*-nitro group destabilizes a carbocation intermediate more than a *meta*-nitro group because, in a *para*-nitro carbocation, positive charge is on adjacent atoms:

positive charge is on
adjacent atoms

In the *meta*-nitro carbocation, positive charge does *not* reside on adjacent atoms. Consequently, the *meta*-nitro carbocation is more stable (or perhaps we should say *less unstable*) than the *para*-nitro carbocation, and *m*-nitro-*tert*-cumyl chloride solvolyzes more rapidly than *p*-nitro-*tert*-cumyl chloride.

17.30   Acid-catalyzed dehydration involves a carbocation intermediate; see Sec. 10.2 of the text. The problem, then, really asks, "Which compound gives the most stable carbocation intermediate?" Compounds *B* and *C* both involve a *p*-methoxy substituent; the carbocation-stabilizing effect of this substituent in a similar situation is shown in Eq. 17.6, text p. 839. In the dehydration of *B*, the carbocation intermediate, a tertiary carbocation, is more stable than the carbocation in the dehydration of *C*, a secondary carbocation. The *p*-nitro carbocation is the least stable because the nitro group is an electron withdrawing, carbocation-destabilizing group, just as it is in electrophilic aromatic substitution (and for the same reasons; see the discussion on text p. 814–815). Therefore, the order of increasing reactivity is *A* < *C* < *B*.

17.31   The question is, essentially, "Which —OH group, when protonated and lost as water (Sec. 10.2A), gives the more stable carbocation?" Loss of the OH on the left gives a carbocation that is both tertiary and doubly benzylic. Loss of the OH on the right gives a carbocation that is secondary and singly benzylic. The choice is clear; the —OH group on the left is lost.

17.32   The solution to this problem, like the previous three solutions, hinges on an analysis of the relative stabilities of the carbocation intermediates involved in the S_N1 reactions of the two compounds. The carbocation intermediate in the solvolysis of compound *A* is resonance-stabilized:

If you find an error in this manual, please visit http://people.pharmacy.purdue.edu/~loudonm/teaching/

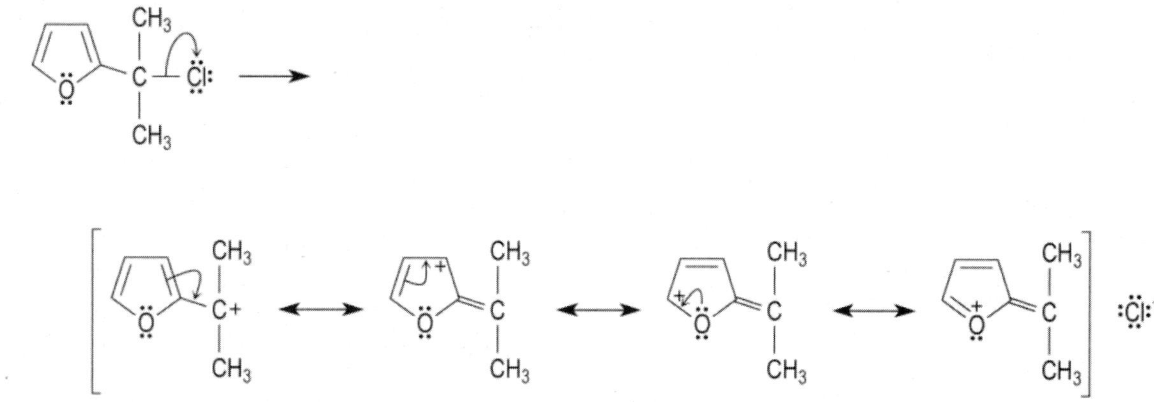

The carbocation intermediate involved in the solvolysis of compound *B* is not resonance-stabilized, and in fact is somewhat destabilized by the electron-withdrawing polar effect of the oxygen. The greater stability of the carbocation derived from compound *A* results in a greater solvolysis rate.

17.33 The fact that benzoic acid is obtained by chromic acid oxidation shows that all compounds contain a monosubstituted benzene ring. The NBS reaction is a benzylic bromination, and the alcohol produced by solvolysis of the resulting bromide must be tertiary, since it cannot be oxidized with CrO_3 and pyridine. The structures of compound *A*, *B*, and *C* are therefore as follows:

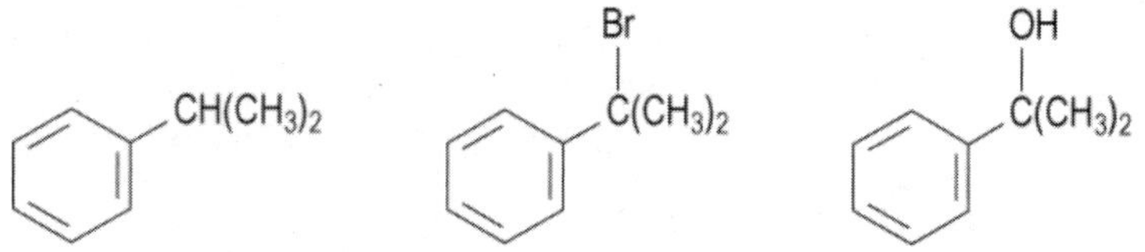

17.34 Compound *A* has an unsaturation number of 5. Because it ultimately affords phthalic acid, it is an ortho-disubstituted benzene derivative. The enantiomeric resolution and oxidation results show that compound *C* is a chiral secondary alcohol, and the α-carbon of this alcohol is the asymmetric carbon, because conversion to a ketone destroys its chirality. Compound *C*, in turn, arises from solvolysis of *B*, a product of either allylic or benzylic bromination. Because *A* contains a benzene ring, the formation of compound *B* is probably a benzylic bromination. A by-product of the formation of compound *C* is compound *D*, presumably an alkene, because it can be hydrogenated back to compound *A*. The only way to accommodate all of these data is for the additional degree of unsaturation in compound *A* to be due to a ring. Structures of all compounds that fit these data are as follows:

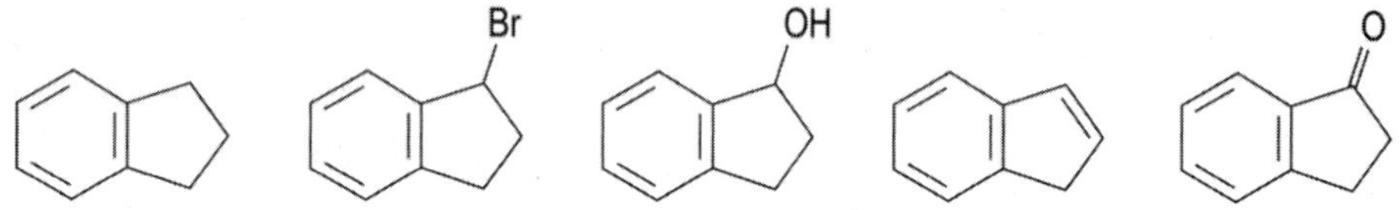

The following structure for compound *A* is ruled out because it would give a *tertiary* alkyl halide, which, in turn, would give a *tertiary* alcohol on solvolysis; a tertiary alcohol could not be oxidized to a ketone.

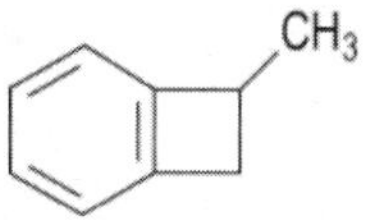

17.35 Alkyl halides with allylic or benzylic β-hydrogens undergo more rapid E2 reaction because the allylic or benzylic hydrogens are more acidic than ordinary hydrogens (See Sec. 17.3B in the text.) Because compound *A*

has allylic β-hydrogens and compound *B* does not, the E2 reaction of compound *A* is faster, and this reaction proceeds in the following manner:

In the presence of excess base this reaction does not stop here but would undoubtedly give the following *anion* as a product. Can you see why? Can you give a curved-arrow mechanism for this transformation?

17.36   The greater λ_{max} value indicates the formation of a species with more extensive conjugation. Protonation of the hydroxy group and loss of water gives this species, which is

benzyl cation
$\lambda_{max} = 442$ nm

17.37   First, examine each structure and determine if it is optically active or not. Then, since MnO_2 oxidizes allylic alcohols and benzylic alcohols selectively, identify the structures containing one or more allyl alcohol groups. Then draw the corresponding structures that would result from oxidation by MnO_2.

A   mirror plane
(optically inactive) $\xrightarrow{MnO_2}$ no reaction

B   allylic alcohol $\xrightarrow{MnO_2}$ optically inactive

C   allylic alcohol $\xrightarrow{MnO_2}$ optically active

D   mirror plane
(meso and optically inactive)   allylic alcohols $\xrightarrow{MnO_2}$ optically inactive

E   allylic alcohols $\xrightarrow{MnO_2}$ optically inactive

(a)   Structures B and E are optically active compounds and are oxidized by MnO_2 to optically inactive compounds.

(b)   Structure C is an optically active compound that is oxidized by MnO_2 to an optically active compound.

(c)   Structure D is an optically inactive compound that is to an optically inactive compound.

(d)   Structure A is a compound that is not oxidized by MnO_2.

17.38

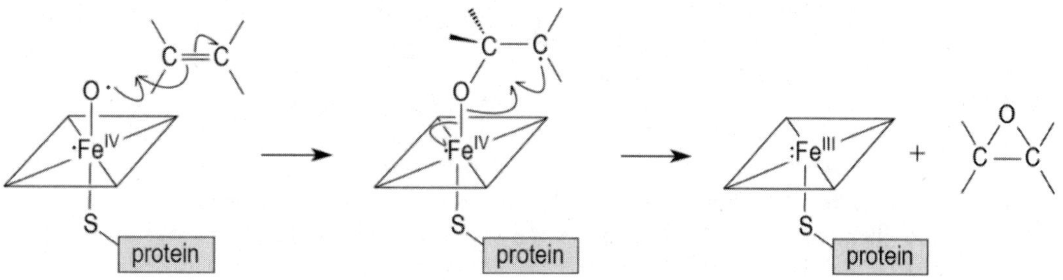

CyP450

17.39   (a)

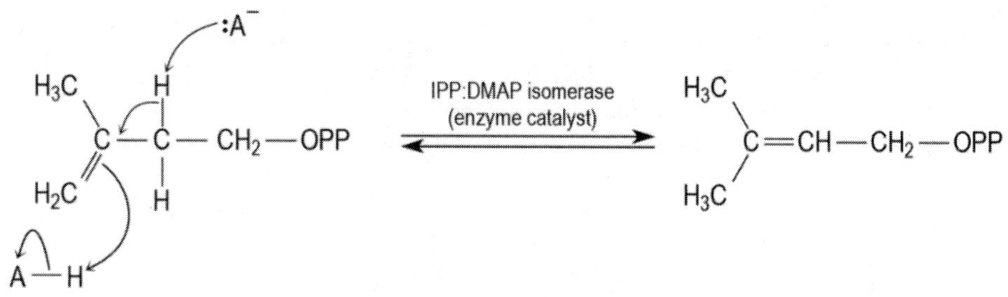

(b)   In an S_N1-like mechanism, the pyrophosphate leaves producing a resonance-stabilized allylic carbocation which is attacked by water and deprotonated to yield linalool.

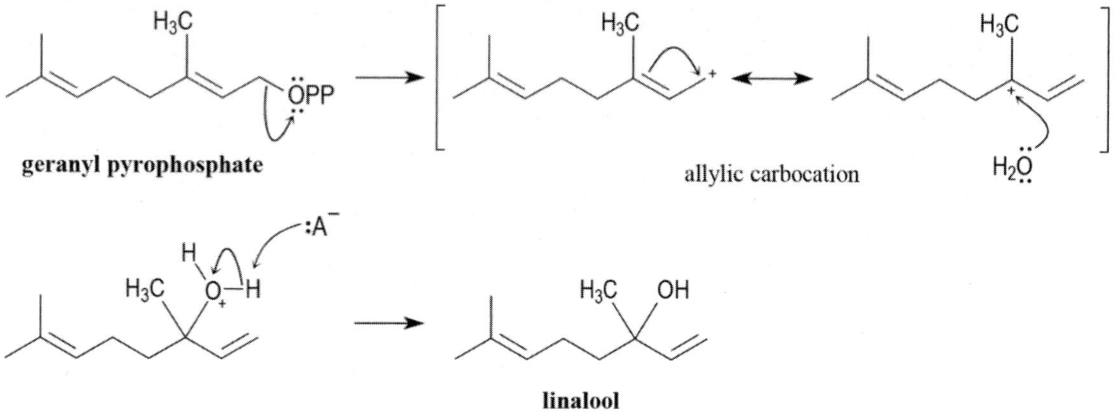

geranyl pyrophosphate   allylic carbocation

linalool

(c)   There are two potential places where the epoxide could form: at either the trisubstituted alkene or at the monosubstituted alkene. Inspecting compound *B* shows you that the monosubstituted double bond still exists intact, so the epoxide must have formed at the trisubstituted alkene (compound *A*). Compound *A* then undergoes an intramolecular cyclization, and proton transfers yield product *B*.

17.40 (a)   A deuterium-labeled compound, such as the one below, could be used to establish that a reaction took place via inversion. If the product had the opposite configuration as the starting material, then it could be deduced that the mechanism proceeded by inversion, or an S_N2 mechanism.

(b)   As noted in Sec. 10.9B, alcohol dehydrogenase catalyzes the removal of the pro-*R* hydrogen of ethanol. It can be deduced that the pro-*R* hydrogen of geraniol produced in part (a) would also have its pro-*R* hydrogen (in this case, the deuterium) removed by the enzyme. Thus, if the geraniol product in part (a) was oxidized by the enzyme the aldehyde produced would contain deuterium.

(Alternatively, if you chose to begin with the *S*-deuterium-labeled gernayl phosphate in part (a), then the product geraniol would be *R*, and the product of oxidation by alcohol dehydrogenase would *not* contain deuterium.)

17.41 (a)

(b)   When trying to solve problems like these, especially with structures containing rings, it's always a good idea to number carbons when trying to figure out the nature of the transformation that occurred. In this case, the numbers are arbitrary and have nothing to do with nomenclature. Since we need to take this linear molecule and form one containing two rings, we should begin with an intramolecular reaction. The

–OPP group is a good leaving group, so use one of the alkene double bonds as a nucleophile. Here, we're joining C-2 with C-7, as numbered below, and forming a six-membered ring.

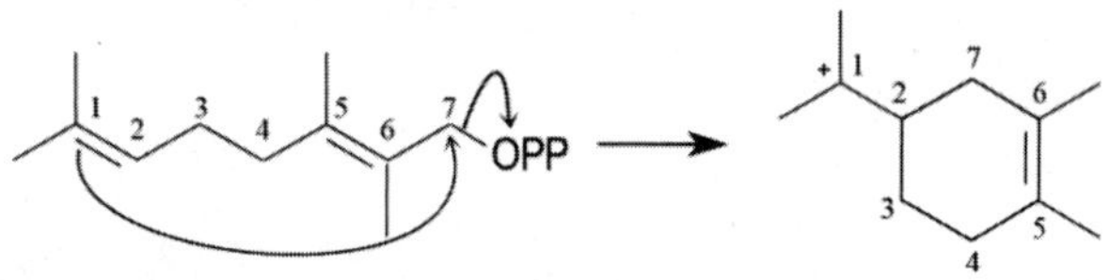

The intermediate formed as a result of this intramolecular reaction leaves a positive charge behind on C-1. Since a second ring must be formed, we can form a bond between C-5 and C-1. This leaves a carbocation at C-6, which can then react with water and then be deprotonated to yield the product.

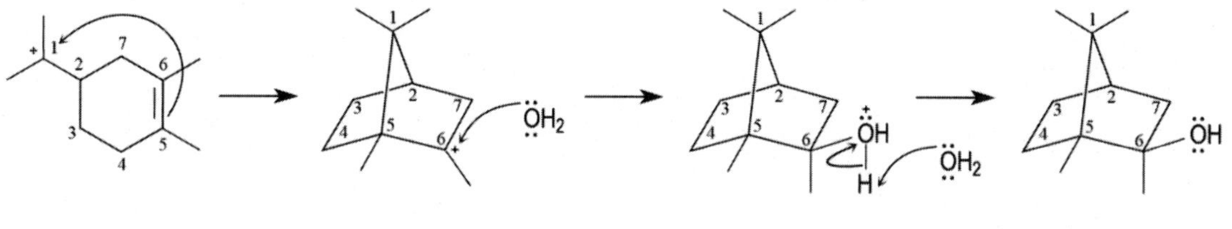

17.42   First analyze the relationship of the isoprene skeletons. Then use steps like the ones shown in Eqs. 17.44–17.46, text pp. 863, to assemble the parts from IPP and DMAP. Start with farnesyl pyrophosphate, the biosynthesis of which is shown in the solution to Problem 17.19(a), text p. 864. Note that **B:** = a base.

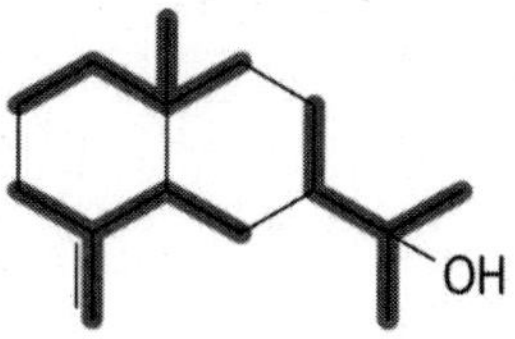

eudesmol

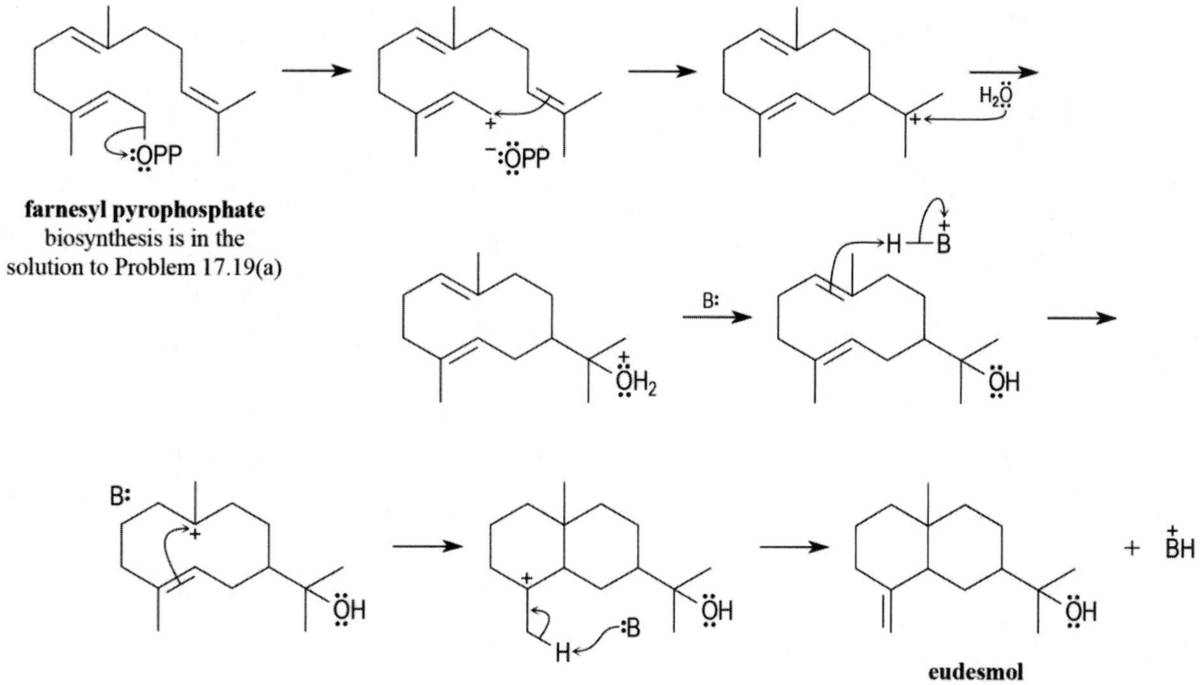

farnesyl pyrophosphate
biosynthesis is in the
solution to Problem 17.19(a)

eudesmol

A different arrangement of isoprene units in eudesmol can also be envisioned:

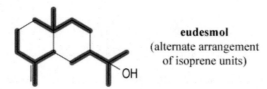

eudesmol
(alternate arrangement
of isoprene units)

A biosynthetic scheme based on this arrangement would be an equally correct answer. An isotope-labeling experiment would be required to distinguish between the two arrangements.

17.43 Compound A can be protonated at the "upper" double bond to give a tertiary carbocation X, which serves as the electrophile in a Lewis acid–base association reaction with the "lower" double bond to give a carbocation Y. Loss of acetone gives intermediate B.

The conversion of a tertiary carbocation X to a secondary carbocation Y is driven by the replacement of a π bond by a σ bond, and ultimately by the loss of the very stable molecule acetone.

Compound B protonates to give carbocation Z, which rearranges to a tertiary carbocation. This carbocation undergoes a Lewis acid–base association with water to give geosmin after loss of a proton.

17.44 (a) Although the conjugate-base anion of 1,4-pentadiene is doubly allylic and resonance-stabilized, the conjugate-base anion of 1,3-cyclopentadiene is in addition aromatic. (See text pp. 767–768 for a discussion of this case.) Consequently, much less energy is required for the ionization of 1,3-

cyclopentadiene, and its pK_a is therefore much lower. (The pK_a difference between these two compounds is estimated to be 10–15 units.)

(b)   The carbocation intermediate in the solvolysis of 3-bromo-1,4-pentadiene is doubly allylic and therefore resonance-stabilized. Consequently, formation of this ion is relatively fast. Although the carbocation intermediate that results from the ionization of 5-bromo-1,3-cyclopentadiene *appears* to be doubly allylic and resonance-stabilized, it is in fact a cyclic system of 4n π electrons, and is therefore *antiaromatic* and particularly *unstable*. Consequently, this ion does not form. (Problem 15.67 on text p. 783 also deals with this case.)

17.45   (a)   Two allylic Grignard reagents are in equilibrium, and, each reacts with D_2O.

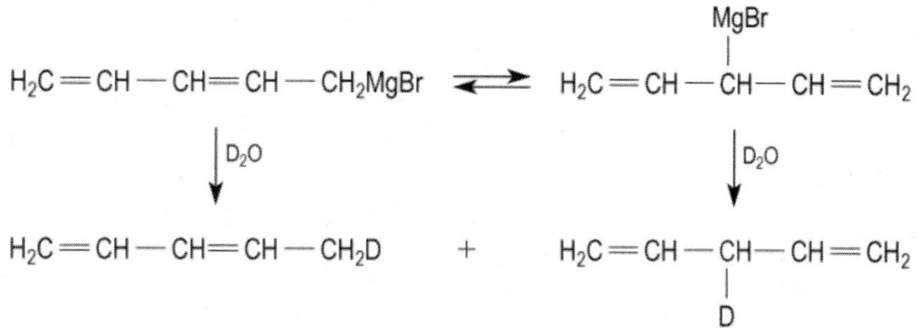

(Note in this and the following part that a third product resulting from allylic rearrangement of the Grignard reagent is identical to the first.)

(b)   Two allylic Grignard reagents are in equilibrium, and each reacts with ethylene oxide:

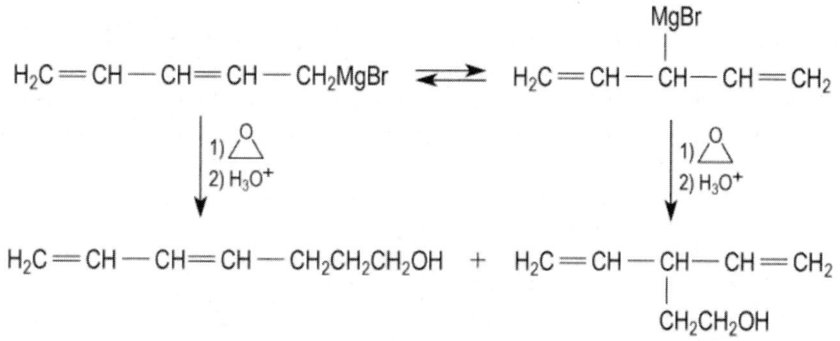

(c)   This is an S_N2 reaction of the ethanethiolate anion; because there is no reactive intermediate, only one product is formed:

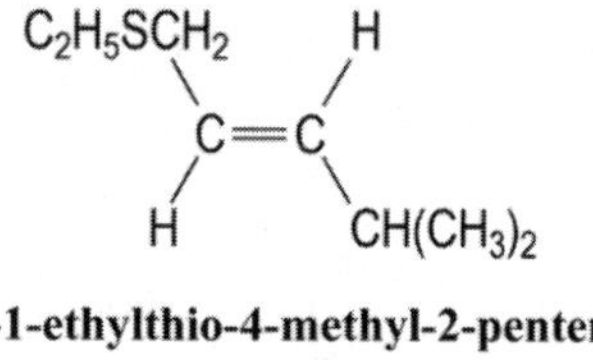

(*E*)-1-ethylthio-4-methyl-2-pentene

(d)   In the absence of a good nucleophile, the allylic bromide undergoes a slow S_N1 ionization, and ethanol attacks the two electron-deficient carbons of the resonance-stabilized carbocation to give two constitutionally isomeric ethyl ethers.

(e) The unpaired electron in the free-radical intermediate is shared between two chemically nonequivalent carbons; two corresponding products are formed.

(f) The observed product is derived from substitution of the tertiary benzylic hydrogen. This substitution involves a tertiary benzylic free-radical intermediate, whereas substitution of a *p*-methyl hydrogen would involve a less stable primary benzylic free-radical intermediate.

(g) The dibromide formed in the bromine-addition reaction undergoes two successive E2 reactions to give naphthalene, an aromatic compound. Both eliminations involve benzylic protons, and the final one is driven by aromaticity of the product.

naphthalene

(h) This product results from protonation of the alkene at the terminal carbon, because this gives the most stable carbocation intermediate. In this example, the 1,2-addition product not only is the more rapidly formed (kinetic) product but also is the more stable (thermodynamic) product. It is more stable thermodynamically because it is the only product that has a double bond conjugated with the benzene ring.

(i) All benzylic carbons—even the ones in the ring—are oxidized to carboxylic acid groups.

1,2,3-benzenetricarboxylic acid
(hemimellitic acid)

17.46 (a) Propargylic Grignard reagents, like allylic Grignard reagents, are an equilibrium mixture of two constitutional isomers. Each reacts with H_2O.

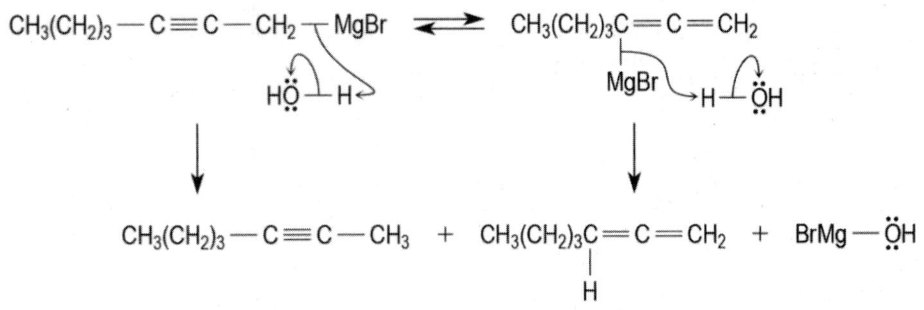

$$CH_3(CH_2)_3 - C \equiv C - CH_2 - MgBr \rightleftharpoons CH_3(CH_2)_3C = C = CH_2$$

$$CH_3(CH_2)_3 - C \equiv C - CH_3 \ + \ CH_3(CH_2)_3C = C = CH_2 \ + \ BrMg - \ddot{O}H$$

$$\underset{H}{\overset{|}{}}$$

(b)   Protonation of the triple bond gives a resonance-stabilized vinylic cation which is attacked intramolecularly by the oxygen of the —OH group. (H—OTs = *p*-toluenesulfonic acid; see text p. 443 for its structure.)

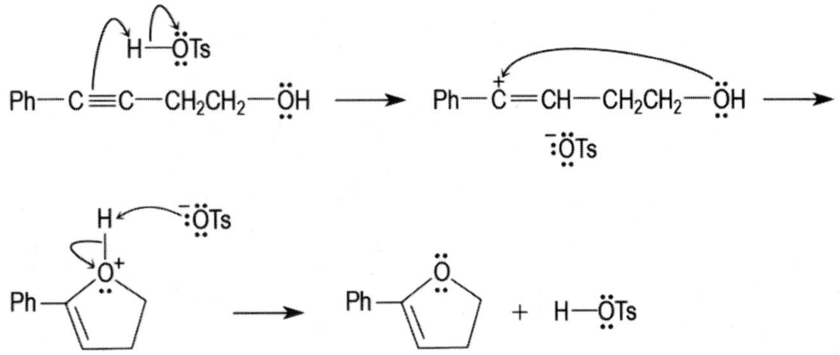

$$Ph - C \equiv C - CH_2CH_2 - \ddot{O}H \longrightarrow Ph - \overset{+}{C} = CH - CH_2CH_2 - \ddot{O}H \longrightarrow$$

$$:\ddot{O}Ts$$

(c)   A hydrogen on the central carbon is more acidic than an acetylenic hydrogen because the conjugate-base anion resulting from removal of the central hydrogen is both allylic and propargylic, and is therefore doubly resonance-stabilized. The conjugate-base anion is alkylated by allyl bromide.

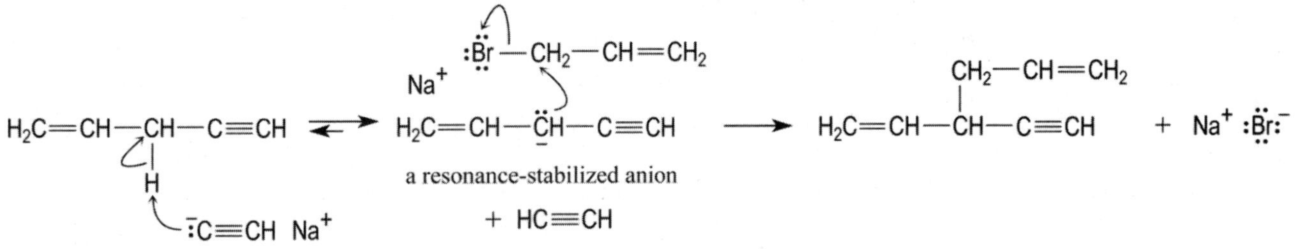

$$H_2C = CH - CH - C \equiv CH \rightleftharpoons H_2C = CH - \ddot{C}H - C \equiv CH \longrightarrow H_2C = CH - CH - C \equiv CH \ + \ Na^+ \ :\ddot{B}r:^-$$

a resonance-stabilized anion

+ HC≡CH

In the resonance structures of the anion intermediate, the negative charge is delocalized to two other carbons. (Draw these structures.) While the mechanism above shows why the indicated product is reasonable, it does not explain why products derived from the other possible resonance structures are not observed (or reported).

(d) Protonation of the alcohol and loss of water give an allylic carbocation that can react with ethanol at either of two electron-deficient carbons to give a mixture of two constitutionally isomeric ethyl ethers. The following mechanism begins with the protonated alcohol.

(e) The formation of a cyclopropane suggests the intermediacy of the following carbene:

$$(CH_3)_2C = C = C:$$

This carbene is formed by loss of chloride ion from an acetylenic ion. The carbene then adds to cyclohexene to give the cyclopropane. (See Eq. 9.75, text p. 434.)

17.47 (a) Formation of a resonance-stabilized propargylic anion (that is, an anion that is "allylic" to a triple bond) is followed by protonation at the internal carbon. Formation of another resonance-stabilized anion followed by protonation gives the 1-alkyne, which is the most acidic species in the reaction mixture. Consequently, its conjugate-base anion is formed irreversibly, and all equilibria are thus pulled toward this anion. The net result is migration of the triple bond to the end of the carbon chain.

H←⁻:B

$H_2C \xrightarrow{} C \equiv CCH_2CH_2CH_3$ ⇌ $\left[H_2\ddot{C} - C \equiv CCH_2CH_2CH_3 \longleftrightarrow H_2C = C = \ddot{C}CH_2CH_2CH_3 \right]$ ⇌

B─H←

H←⁻:B

$HC = C = CHCH_2CH_2CH_3$ ⇌ $\left[H\ddot{C} = C = CHCH_2CH_2CH_3 \longleftrightarrow HC \equiv C - \ddot{C}HCH_2CH_2CH_3 \right]$ ⇌

B─H←

$B:\longrightarrow H \overset{\frown}{-} C \equiv C - CH_2CH_2CH_2CH_3$ ⇌ $^-:C \equiv C - CH_2CH_2CH_2CH_3$ + B─H

Addition of water to the reaction mixture gives the 1-alkyne. Thus, the reaction shown here followed by addition of water results in the migration of a triple bond from an interior position to the 1-position of a carbon chain provided there are no branches. It has been shown that the triple bond migrates with equal frequency to both ends of the chain.

You might ask why a 1-alkyne is more acidic than a propargylic hydrogen when the conjugate-base anion of the latter is resonance-stabilized, and an acetylenic anion is not. This is a reflection of the effect of hybridization on acidity discussed in Sec. 14.7A of the text. A *doubly allylic* hydrogen is more acidic than an acetylenic hydrogen, but a singly allylic hydrogen is not.

(b)   The question is whether the triple bond migrates to the end of the carbon chain nearer to the methyl branch or to the end of the chain farther from the methyl branch. Once we consider the mechanism shown in part (a), the answer becomes clear. The migration of the triple bond occurs *away* from the methyl branch, because the mechanism of the reaction requires a stepwise migration of the triple bond, and a triple bond cannot form at a carbon that bears a branch because a carbon have no more than four bonds.

triple bond cannot form
between these two carbons

$CH_3CH_2\overset{\frown}{CH} - C \equiv C - CH_2CH_2CH_3$ $\xrightarrow[\substack{\text{zipper reaction} \\ \text{[see part (a)]}}]{B:^-}$ $CH_3CH_2CHCH_2CH_2CH_2 - C \equiv C:^-$
$\qquad\qquad |$ $\qquad\qquad\qquad\qquad\qquad\qquad\qquad\qquad\qquad\qquad\qquad |$
$\qquad\quad CH_3$ $\qquad\qquad\qquad\qquad\qquad\qquad\qquad\qquad\qquad\qquad\qquad\quad CH_3$

3-methyl-4-octyne

17.48   This reaction, like the one in the previous problem, is a "zipper" reaction brought about by a series of acid-base equilibria. In this case, however, the equilibrium is not driven by ionization of the product, because all C—H bonds are less acidic than *tert*-butyl alcohol. Instead, the equilibrium lies to the right because the product is more stable than the starting material. This, in turn, is so for two reasons: (1) because the double bonds in the products are conjugated with the benzene ring and the double bond in the starting material is not; and (2) because the double bond in the product has more branches than the double bond in the starting material.

double bond is *not* conjugated and has one branch

double bond *is* conjugated and has two branches

17.49 The equilibrium lies to the right because the double bond has four alkyl substituents whereas, in the starting material, it has three. Recall that alkyl substitution at double bonds is a stabilizing effect (Sec. 4.5B in the text). The mechanism involves simply protonation of the double bond to give the benzylic cation and loss of a proton to give the product.

17.50 **(a)** Solvolysis reactions are faster when the carbocation intermediate is stabilized. The only difference between compounds *A* and *B* is the presence of the nearby double bond in *A*, so we can conclude that this feature must stabilize the carbocation. If we consider resonance, then intermediate *A* has additional resonance structures that intermediate *B* does not, making *A* more stable. A more stable intermediate results in a lower activation energy, and a faster rate (see Sec. 9.6B and Sec. 17.1)

resonance stabilized intermediate

no resonance stabilization

(b) The retention of stereochemistry seen in reaction *A* can be attributed to the structure of the carbocation shown in the answer to part (a). One side of the carbocation is blocked by the presence of the pi electrons of the double bond. The electrons interact with the left-pointing lobe (as drawn) of the *p*-orbital, leaving only the right-pointing lobe available to accept the incoming nucleophile.

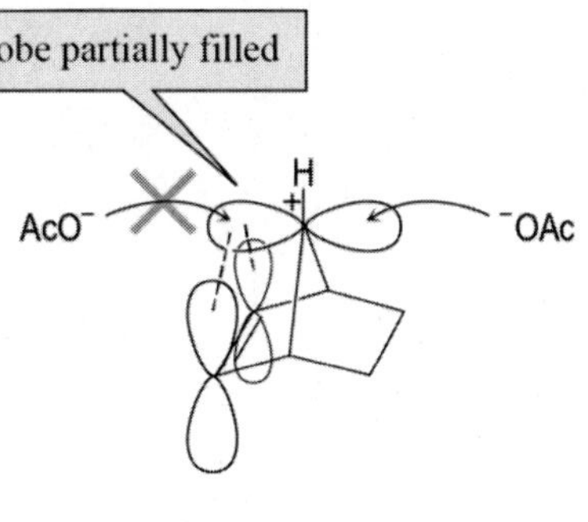

lobe partially filled

17.51   (a)

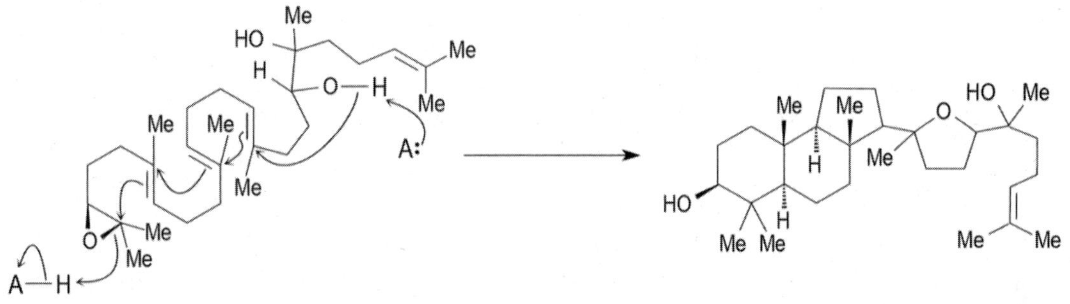

(b)

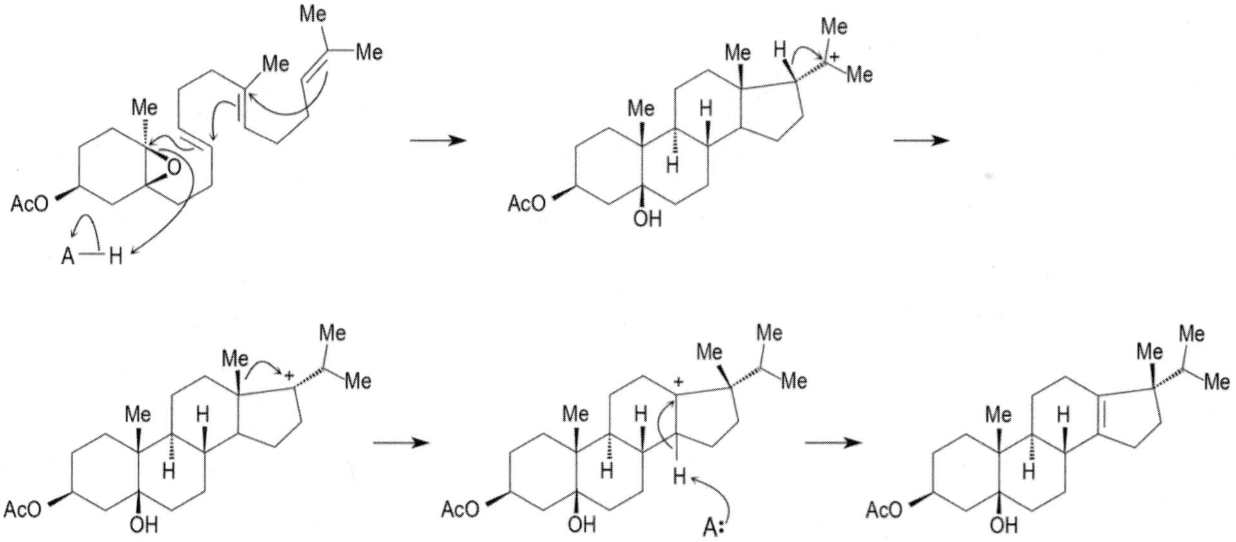

17.52   (a)   CyP450 oxidizes compounds at their benzylic positions (see Sec. 17.5B). Compound *A* is:

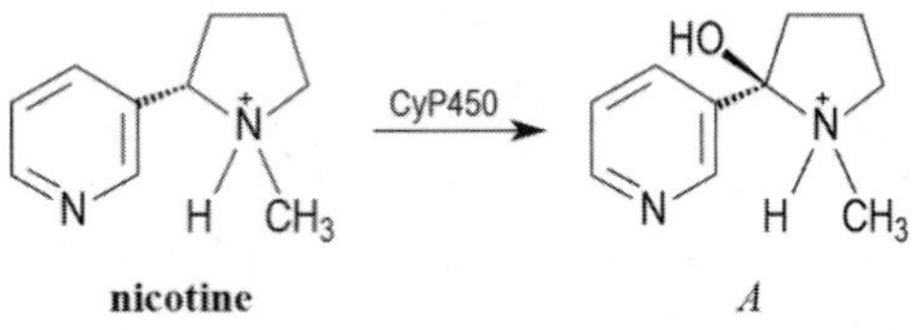

nicotine   *A*

(b)

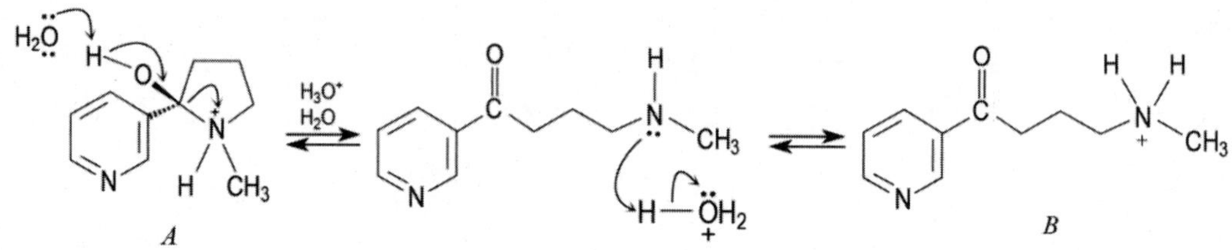

17.53   From the structure of the Diels–Alder product C, compound B must be the conjugated diene 2-chloro-1,3-butadiene. Because compound B is an allylic rearrangement product of compound A, compound A must be the allene 4-chloro-1,2-butadiene. (Note that 1-chloro-1,3-butadiene would not give the same Diels-Alder product.)

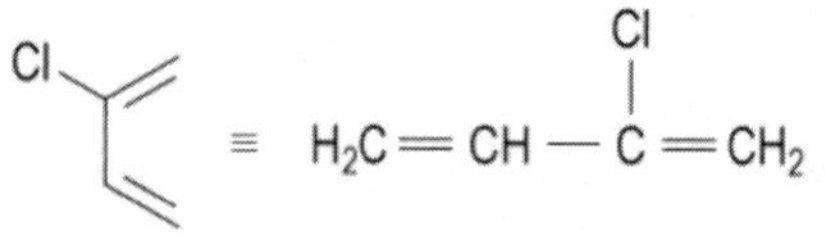

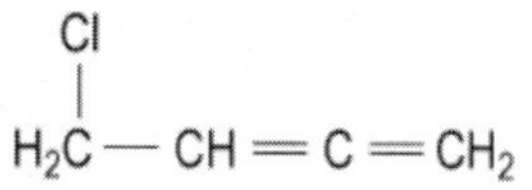

17.54   (a)   The stability of the trityl radical is due to delocalization of the unpaired electron into all three benzene rings.

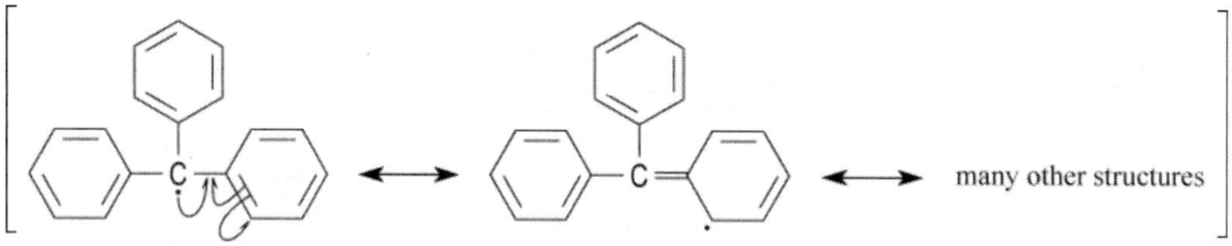

(b)   Hexaphenylethane could be formed by the recombination of two trityl radicals:

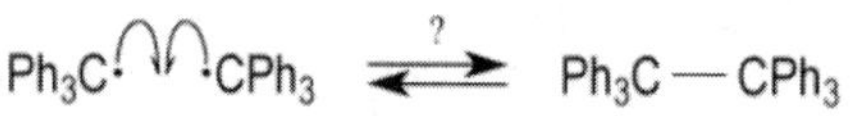

hexaphenylethane

Hexaphenylethane is destabilized by van der Waals repulsions between gauche pairs of phenyl groups.

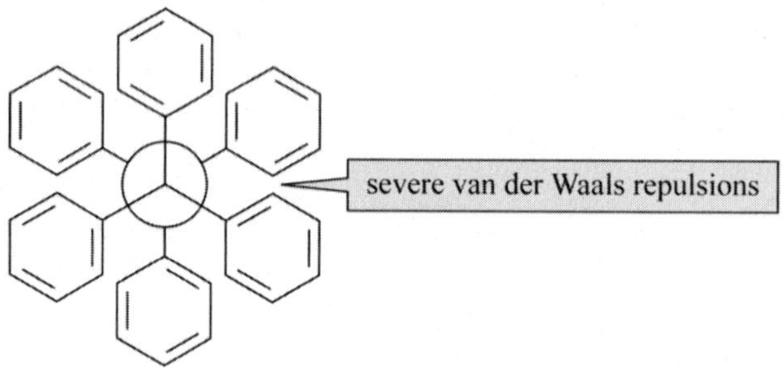

severe van der Waals repulsions

This destabilization is evidently so great, and the drive for recombination so powerful, that aromatic stability in one of the benzene rings can be sacrificed in forming the dimer shown in part (c).

(c)   The dimer of the trityl radical is formed by the following mechanism:

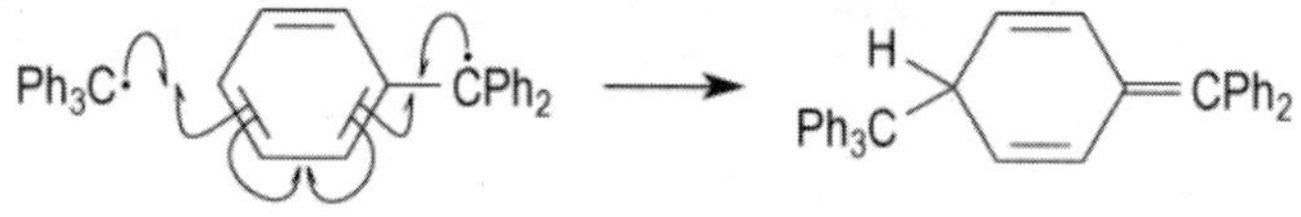

17.55   (a)   The triphenylmethyl anion (trityl anion) is stabilized by resonance interaction with all three benzene rings:

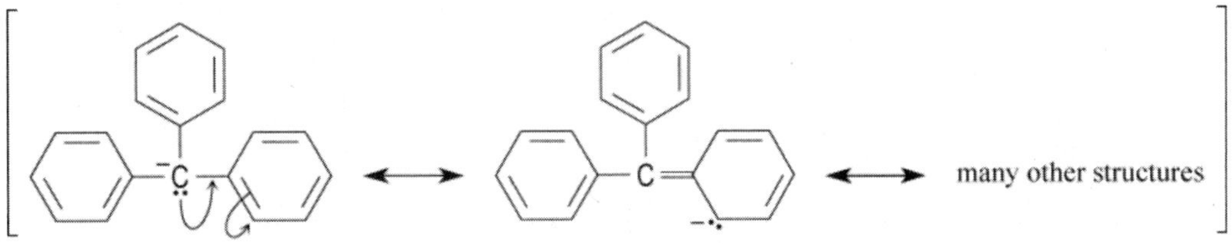

(b)   The resonance interaction shown in part (a) is optimized if the three rings are coplanar; when the rings are coplanar, overlap is optimum between the $2p$ orbitals of the central carbon and those of the rings. However, when the rings are coplanar, there are significant van der Waals repulsions between hydrogens on different rings:

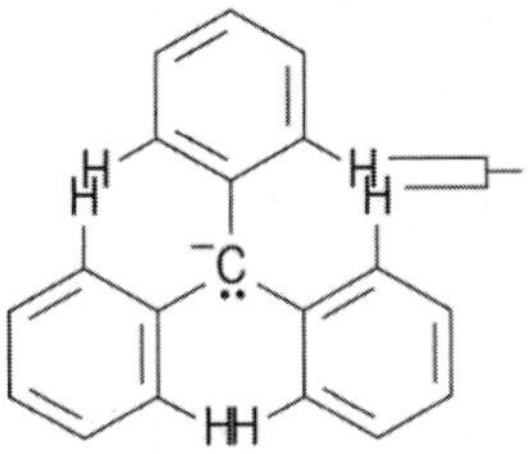

These van der Waals repulsions force the rings in the triphenylmethyl anion to twist significantly away from coplanarity, with a resulting cost in orbital overlap. (The molecule resembles a molecular "propeller.") In other words, in the triphenylmethyl anion, conjugation is not as effective as it would be if the rings were all coplanar. In fluoradene, the hydrogens that are the source of these repulsions are replaced by bonds that constrain the rings to coplanarity. The resulting increase in orbital overlap brings about a concomitant increase in stabilization of the anion. As a result, the pK_a of the conjugate acid is lowered significantly relative to that of triphenylmethane. The conjugate-base anion of fluoradene is really the "ultimate benzylic anion."

17.56   Section 7.8C shows that anti stereochemistry is one of the major pieces of evidence that a bromonium ion is involved in bromine addition to alkenes; similar conclusions hold for chlorine addition. The text also shows that a mechanism involving carbocation intermediates predicts mixed syn- and anti-addition (Eq. 7.38, text p. 312).

This problem deals with the *competition* between the two mechanisms: one involving a cyclic *chloronium ion* intermediate and the other involving a carbocation intermediate. The loss of stereoselectivity is evidence for the involvement of a carbocation intermediate. When the carbocation intermediate is resonance-stabilized (last two entries in the table), it is stable enough to compete energetically with the chloronium ion as a reactive intermediate. When R = p-methoxyphenyl, the carbocation intermediate is further resonance-stabilized by the para substituent. (See, for example, Eq. 17.6 on text p. 839.) In this case, the carbocation intermediate is stable enough, and the carbocation mechanism thus important enough, that nearly complete loss of stereoselectivity is observed.

17.57   Compound *A* is a benzylic bromination product, and it appears that compounds *B* and *C* are S_N1 products, each derived respectively from reaction of one of the nucleophiles present with a carbocation intermediate derived from *A*. Recall that S_N1 reactions are generally accompanied by E1 reactions; hence, it is reasonable to suppose that MPTP is such a product, namely, an alkene. The molecular formula of MPTP, $C_{12}H_{15}N$, indicates an unsaturation number of 6. Because five unsaturations are accounted for by the two rings in the starting material, the formula is consistent with the hypothesis that MPTP is an alkene.

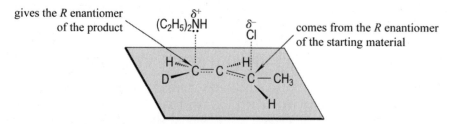

MPTP

17.58 (a) Fact (1) indicates that the transition state involves a molecule of $(C_2H_5)_2NH$ and a molecule of alkyl halide. Fact (2) indicates that the indicated starting materials undergo the reaction shown; had this not been established, one could have postulated that one of the starting alkyl halides is converted into the other by the reaction conditions, and that the observed product could have originated from only one of the two starting compounds. Fact (3) establishes that the observed product is the actual product of the reaction, and is not derived from a subsequent reaction of its allylic isomer.

Now to the mechanisms. In the reaction of the second alkyl chloride an allylic rearrangement has occurred. Such allylic rearrangements generally suggest the involvement of carbocation intermediates, which have two resonance forms, and can be attacked at two different carbons. However, the observation of second-order kinetics rules out an S_N1 reaction. Hence, the reaction is evidently a direct substitution at the *allylic carbon*, thus:

$$(C_2H_5)_2\ddot{N}H \quad H_2C{=}CH{-}CH{-}CH_3 \longrightarrow (C_2H_5)_2\overset{+}{N}H{-}CH_2{-}CH{=}CH{-}CH_3 \quad + \quad :\ddot{C}l:^-$$

$$\underset{:\ddot{C}l:}{|}$$

This concerted allylic substitution mechanism has been termed the S_N2' *mechanism*. The other isomer reacts by the conventional S_N2 mechanism, in which the nucleophile reacts at the carbon bearing the halide.

(b) The large size of the nucleophile dictates that its attack occurs at the unbranched carbon to avoid van der Waals repulsions. In one case, this is the allylic carbon; in the other, it is the α-carbon.

17.59 Imagine, in the transition state of the S_N2' reaction, that the three carbons that either start or finish the reaction as part of a double bond, as well as the atoms attached to these carbons, lie in a common plane with the nucleophile and the leaving group either above or below this plane. Placing all groups with their stereochemical configurations corresponding to those of the reactant and the first product shows that approach of the nucleophile and departure of the leaving group occur from the *same face* of the plane. Hence, the major product is formed by a syn-substitution, that is, a *frontside* substitution.

Note that the product is a *trans*-alkene with R stereochemistry at the asymmetric carbon. The minor product, a *cis*-alkene with S stereochemistry, also comes from a syn-substitution in which reaction of the nucleophile and loss of the leaving group both occur from the lower face of the same plane. In order for this to happen, the transition state must adopt a conformation that leads to the *cis*-alkene. (You should demonstrate this to yourself with a diagram like the one above.) Because *trans*-alkenes are more stable than *cis*-alkenes, the "transoid" transition state—the one shown above—is more stable and accounts for 95% of the reaction. However, the point is that *both* products result from syn-substitution. You should show that if the nucleophile and leaving group were situated on opposite faces, the *R* enantiomer of the product would have a cis double bond and the *S* enantiomer of the product would have a trans double bond, a result not consistent with the observations given in the problem. The syn stereochemistry observed in the S_N2' reaction contrasts with the stereochemistry of the S_N2 reaction, which is an anti-substitution; that is, attack of the nucleophile and loss of the leaving group occur from opposite sides of the molecule. (See Fig. 9.2 on text p. 395; note the relationship of the nucleophile and the leaving group with respect to the plane shown in color.

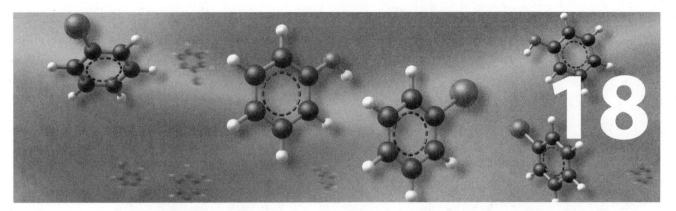

18

The Chemistry of Aryl Halides, Vinylic Halides, and Phenols
Transition-Metal Catalysis

STUDY GUIDE LINKS

18.1   Contrast of Aromatic Substitution Reactions

The name "aromatic substitution" may suggest some connection to the reactions you studied in Chapter 16, such as nitration, halogenation, sulfonation, etc. Other than the fact that a group is replaced on an aromatic ring, *nucleophilic aromatic substitution and electrophilic aromatic substitution have little in common.* In *electrophilic* aromatic substitution, an electrophile, or Lewis acid, reacts with the benzene π electrons, and a *proton* is lost from the benzene ring. In *nucleophilic* aromatic substitution, a *nucleophile,* or Lewis base, reacts at a carbon of the benzene ring and a *halide* is lost from the benzene ring.

Electrophilic aromatic substitution involves introduction of *electron deficiency* (and positive charge) into the benzene ring. Nitro groups and other electron-attracting groups *destabilize* this charge and thus retard the reaction. Nucleophilic aromatic substitution involves introduction of *more electrons* (and negative charge) into the benzene ring. Nitro groups and other electron-attracting groups *stabilize* this charge and thus accelerate this type of reaction.

Notice that *substitution,* like *addition,* is a type of reaction that can occur by a variety of mechanisms. The word *substitution* by itself conveys no information about mechanism. Thus, a substitution can occur by a concerted (S_N2) mechanism, an ionization (S_N1) mechanism, a two-step nucleophilic mechanism (as in nucleophilic aromatic substitution), a two-step electrophilic mechanism (as in electrophilic aromatic substitution), a free-radical mechanism, or (as shown in Secs. 18.5–18.6 of the text) other mechanisms. The mechanism that operates depends on the reactants and the reaction conditions. Students sometimes forget this point and assume that the word *substitution* means "S_N2 reaction," because that is the first substitution reaction they learn.

18.2   The Cumene Hydroperoxide Rearrangement

The formation of phenol and acetone (Eq. 18.103 on text p. 933) is an acid-catalyzed rearrangement of cumene hydroperoxide. This Study Guide Link discusses the mechanism of this reaction.

In the first step of the mechanism, an oxygen of the hydroperoxide is protonated:

We'll now treat the rearrangement as if it were a stepwise reaction. Loss of water gives an *electron-deficient oxygen,* the oxygen analog of a carbocation. Rearrangement occurs to give a much more stable resonance-stabilized carbocation. (The electron-deficient oxygen structure is so unstable that loss of water and the rearrangement undoubtedly take place in a concerted manner.)

a resonance-stabilized carbocation

This rearrangement is very similar to the rearrangement that occurs in the oxidation of organoboranes, the mechanism of which was discussed in Further Exploration 5.1 in this manual.

The Lewis acid–base association reaction of water with the carbocation gives a type of alcohol known as a hemiacetal.

a hemiacetal

You should now be able to complete the acid-catalyzed conversion of the hemiacetal into phenol and acetone. Start by protonating the oxygen of the PhO— group. What has to happen next to form phenol? Finish the mechanism to give acetone.

FURTHER EXPLORATIONS

18.1   The Fries Rearrangement

There's more to the Friedel–Crafts reactions of phenols than meets the eye. It turns out that phenols react with acid chlorides to give phenyl esters; this is a reaction you'll study in Chapter 21.

phenol   a phenyl ester

(Acid chlorides are *electrophiles* in this reaction just as they are in the Friedel–Crafts reaction. The phenol oxygen, like the aromatic ring, is a source of electrons because of its unshared pairs.) In the presence of AlCl₃—the catalyst used in Friedel–Crafts reactions—phenyl esters rearrange to ketones:

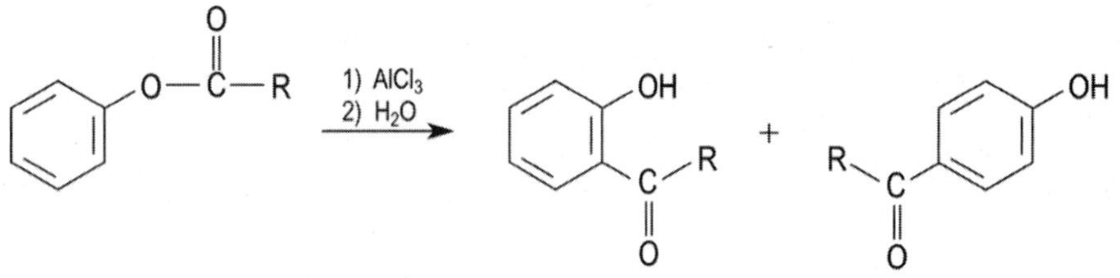

This reaction, known as the *Fries rearrangement,* can be clearly demonstrated because phenyl esters, which can be prepared as pure compounds in the absence of AlCl₃, can be shown to rearrange when treated with AlCl₃. The products of the Fries rearrangement, as you can see, are exactly what would be expected from a direct Friedel–Crafts reaction. Hence, the "Friedel–Crafts reaction" of a phenol almost surely involves ester formation followed by the Fries rearrangement, or a *combination* of the Fries rearrangement and a direct Friedel–Crafts reaction. In any event, the *end result* is as if a Friedel-Crafts reaction had taken place directly.

It is interesting that the "Friedel–Crafts reactions" of phenols in many cases give significant amounts of *ortho*-substituted products. These can be separated from the *para* isomers by application of the principles of hydrogen bonding. The *ortho* isomers contain *intramolecular* hydrogen bonds:

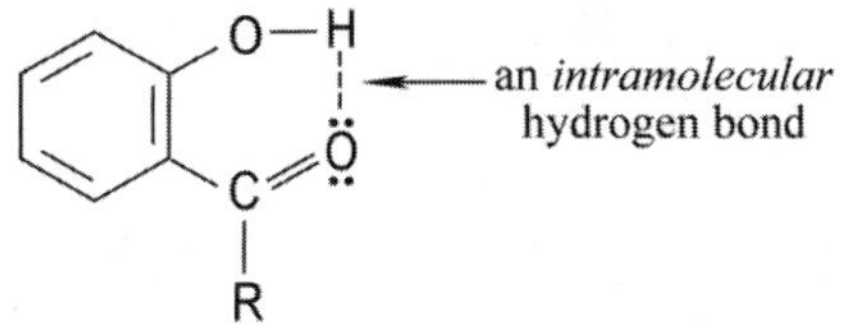

Such intramolecular hydrogen bonding is not possible for the isomeric *para*-substituted phenol. Because the hydrogen-bonding capability of the phenolic —OH group is satisfied within the *ortho*-substituted molecule, this —OH group forms hydrogen bonds less effectively to acceptor atoms of *other molecules*—for example, solvents or chromatography supports such as silica gel— than the —OH group of the *para*-substituted isomer. This is another manifestation of the *proximity effect* (Sec. 11.8)—in this case, one that enhances an intramolecular chemical equilibrium relative to an analogous intermolecular equilibrium. Hence, the *ortho*-substituted isomer is generally much less soluble in solvents that accept hydrogen bonds, and moves through a silica gel chromatography column more rapidly, than the *para*-substituted isomer. Thus, the great difference in hydrogen-bonding properties of the two isomers makes separation a relatively simple matter. Exactly the same principles can be utilized in designing separations of *o*- and *p*-nitrophenol.

REACTION REVIEW

I. REACTIONS OF ARYL AND VINYL HALIDES

A. ELIMINATION REACTIONS OF VINYLIC HALIDES

1. Base-promoted β-elimination reactions of vinylic halides do occur and can be useful in the synthesis of alkynes.

$$Ph-CH=\overset{\overset{\displaystyle Br}{|}}{C}-Ph \quad \xrightarrow[(CH_3)_3COH]{(CH_3)_3CO^- K^+} \quad Ph-CH\equiv C-Ph \ + \ K^+ Br^- \ + \ (CH_3)_3COH$$

B. NUCLEOPHILIC AROMATIC SUBSTITUTION

1. Aryl halides that have one or more nitro groups ortho or para to the halogen undergo nucleophilic substitution reactions under relatively mild conditions.

trifluralin
(a pre-emergence herbicide)

2. The mechanism involves a resonance-stabilized anionic intermediate called a Meisenheimer complex. (See discussion of mechanism and reactivity of a Meisenheimer complex in Sec. 18.4 of the text.)

3. The reaction is faster when there are more nitro groups ortho and para to the halogen leaving group.
   a. Ortho and para nitro groups accelerate the reaction because the rate-limiting transition state resembles the Meisenheimer complex, and the nitro groups stabilize this complex by resonance.
   b. Other substituents that can provide resonance stabilization to the Meisenheimer complex can also activate nucleophilic aromatic substitution.

4. The effect of the halogen on the rate is quite different from that in the S_N1 or S_N2 reaction of alkyl halides.

$$Ar—F \ \gg \ Ar—Cl \ \sim \ Ar—Br \ \sim \ Ar—I$$

C. THE HECK REACTION

1. In the Heck reaction, an alkene is coupled to an aryl bromide, aryl iodide, or aryl triflate under the influence of a Pd(0) catalyst.

$$Ar—Br \ + \ \overset{\diagdown}{\underset{\diagup}{C}}=\overset{\diagup}{\underset{\diagdown}{C}} \ + \ N(C_2H_5)_3 \ \xrightarrow[CH_3CN]{PdL_4} \ \overset{Ar}{\underset{\diagup}{\overset{\diagdown}{C}}}=\overset{\diagup}{\underset{\diagdown}{C}} \ + \ Br^- \overset{+}{N}(C_2H_5)_3 \qquad L = P-\left(\text{aryl}\right)_3$$

   a. The actual catalytically active species is PdL_2, formed by two ligand dissociations.

 If you find an error in this manual, please visit http://people.pharmacy.purdue.edu/~loudonm/teaching/

b. The PdL$_2$ thus generated enters into the catalytic cycle, which consists of the following steps:

   i. oxidative addition of the aryl halide

   ii. ligand substitution of the alkene for one of the catalyst ligands

   iii. 1,2-insertion of the alkene in which the aryl group migrates to carbon-2 of the alkene

   iv. β-elimination of the newly formed aryl substituted alkene followed by ligand association

   v. ligand dissociation of the aryl substituted alkene

   vi. ligand association of a catalyst ligand

   vii. reductive elimination of hydrogen halide, which is neutralized by the basic triethylamine present

$$HBr + N(C_2H_5)_3 \longrightarrow Br^- \overset{+}{N}(C_2H_5)_3$$

   c. Pd(II) catalysts can be used with aryl iodides, but it is thought that the Pd(II) is reduced to Pd(0) during the reaction.

  2. With cyclic alkenes, β-elimination occurs at the β-carbon to which the aryl group was not added owing to the stereochemical constraints of the reaction mechanism.

  3. When the Heck reaction is applied to unsymmetrically substituted alkenes, two products are in principle possible, because insertion might occur at either of the alkene carbons.

   a. When R = phenyl, CO_2R (ester), CN, or other relatively electronegative group, the aryl halide tends to react at the unsubstituted carbon, to usually give the E (trans) stereoisomer.

   b. When R = alkyl, mixtures of products are often observed.

  4. The Heck reaction is another example of a reaction that can be used to form carbon–carbon bonds.

D. THE SUZUKI COUPLING

  1. A Suzuki coupling (also known as a Suzuki reaction or a Suzuki–Miyaura coupling reaction) joins an aryl or vinylic boronic acid to an aryl or vinylic iodide or bromide.

   a. The coupling is catalyzed by a Pd(0) catalyst (Pd(PPh$_3$)$_4$ or Pd(OAc)$_2$/PPh$_3$, as in the Heck reaction).

   b. The reaction occurs in the presence of a base, typically aqueous NaOH or Na$_2$CO$_3$.

   c. The reaction may be used to prepare:

     i. biaryls—compounds in which two aryl rings are connected by a σ bond

     ii. aryl-substituted alkenes

     iii. conjugated alkenes

d. Coupling occurs with retention of alkene stereochemistry.

e. Suzuki coupling avoids issues of regiochemistry that can sometimes occur with the Heck reaction.

2. The necessary boronic acids may be purchased commercially or prepared in two ways:

a. the reaction of a Grignard or organolithium reagent with trimethyl borate followed by aqueous acid (see Eq. 18.50, text p. 905).

b. the hydroboration of 1-alkynes with catecholborane or disiamylborane (see Sec. 14.5B of the text).

E. ALKENE METATHESIS

1. In an alkene metathesis, the groups at each end of the double bonds are interchanged.

2. A number of transition-metal catalysts have been developed for alkene metathesis. Some are based on tungsten and molybdenum, which tend to be air-sensitive and less tolerant to other functional groups. The most common laboratory catalysts are based on ruthenium ruthenium (*e.g.*, G1 and G2), which can be easily handled in the presence of alcohols, phenols, and other groups.

3. In many cases, the catalysts bring some or all of the possible alkenes into equilibrium. For an alkene metathesis reaction of two unsymmetrical alkenes, ten alkenes (not counting stereoisomers) can be formed. The production of the desired product can be increased by:

a. using an excess of one of the reactants

b. utilizing Le Châtelier's principle (*e.g.*, where the by-product is a gas and bubbles out of the reaction mixture).

c. exploiting the fact that different alkenes undergo metathesis at greatly different rates. Alkene metathesis is very sensitive to the steric environment of the alkene double bonds.

4. One of the most important applications of alkene metathesis is for closing rings, even medium- and large-sized rings.

II. OTHER EXAMPLES OF TRANSITION-METAL CATALYZED REACTIONS

A. ZIEGLER–NATTA CATALYST

1. One of the most important transition-metal catalysts in commerce is a catalyst formed from $TiCl_3$ and $(CH_3CH_2)_2AlCl$, called the Ziegler–Natta catalyst.

2. This catalyst brings about the polymerization of ethylene and alkenes at 25 °C and 1 atm pressure to give high-density polyethylene.

3. One possible mechanism involves a two-step iterative process:

   a. a ligand association

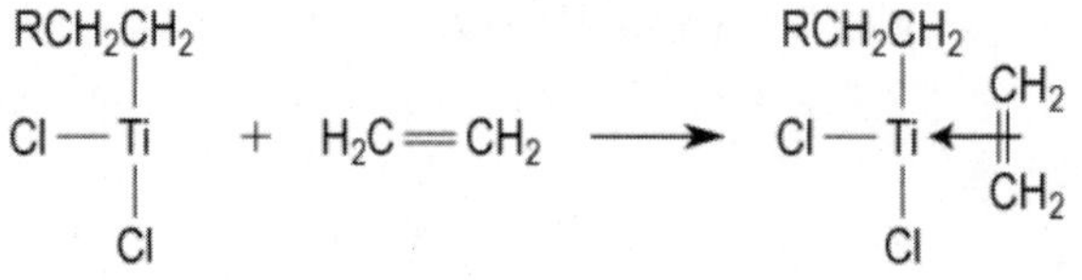

   b. a 1,2-ligand insertion.

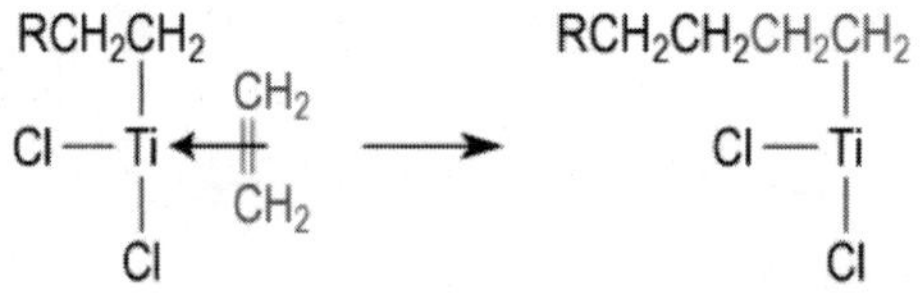

B. HYDROFORMYLATION

1. Hydroformylation is another commercially important process that involves a transition-metal catalyst, in this case, tetracarbonylhydridocobalt(I) catalyst, sometimes called the oxo process. (See Problem 18.29 on text p. 913.)

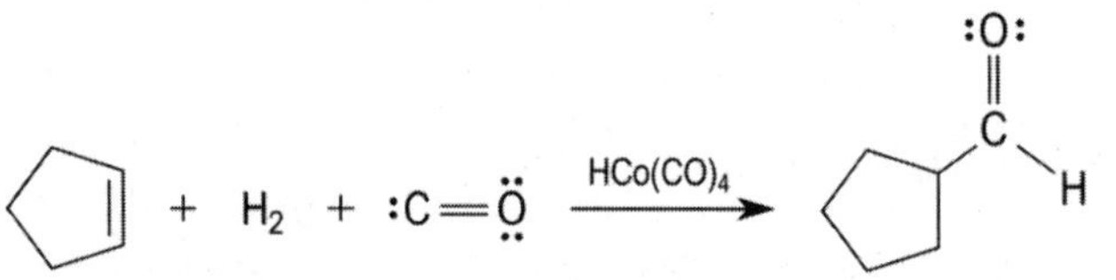

C. CATALYTIC HYDROGENATION

1. An important transition-metal catalyzed reaction is the homogeneous catalytic hydrogenation of akenes using a soluble rhodium(I) catalyst called Wilkinson's catalyst, $ClRh(PPh_3)_3$. (See Problem 18.15, text p. 901.)

2. Catalytic hydrogenation is an extremely important reaction that occurs over carbon-supported transition metals such as Ni, Pd, and Pt. The mechanism of catalytic hydrogenation is not definitively known.

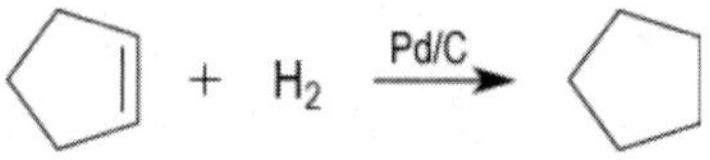

III. REACTIONS OF PHENOLS AND PHENOLATES

A. IONIZATION OF PHENOLS

1. Phenols are weakly acidic; the pK_a of phenol itself is 9.95.

2. Phenols are acidic enough to be converted completely into their conjugate-base phenolate ions (phenoxide ions) by one equivalent of NaOH.

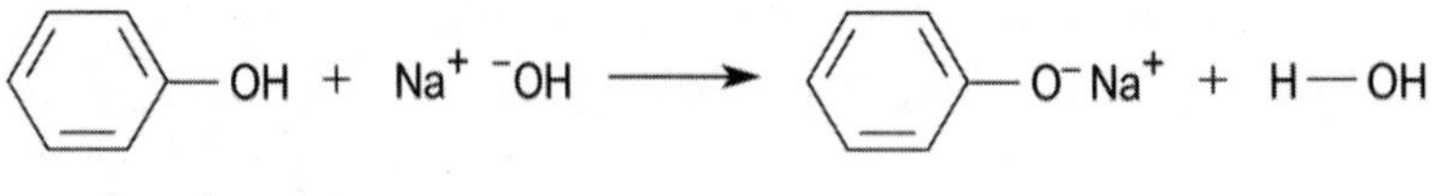

phenol
$pK_a = 9.95$

B. USE OF PHENOXIDES AS NUCLEOPHILES

1. Phenoxides can be used as nucleophiles.
2. Aryl ethers can be prepared by the reaction of a phenoxide anion and an alkyl halide (another example of the Williamson ether synthesis).

C. SUBSTITUTION AT THE ARYL–OXYGEN BOND: THE STILLE REACTION

1. Pd(0) catalysts can catalyze substitution at the aryl–oxygen bond if the phenolic OH group is converted into a trifluoromethanesulfonate ester, which is a very good leaving group.
   a. Aryl trifluoromethanesulfonate esters are prepared from phenols and trifluoromethanesulfonic anhydride (triflic anhydride).

   b. Trifluoromethanesulfonate esters are nicknamed triflates; the triflate group is often abbreviated —OTf.
   c. The triflate group is one of the best leaving groups because it is an extremely weak base.
2. Aryl triflates react readily with organotin derivatives in the presence of Pd(0) catalysts to give coupling products.

   a. The organotin compounds used in the Stille reaction are either commercially available or readily prepared from Grignard Reagents and commercially available trialkyltin chlorides.

   b. In general, vinylic groups and aryl groups are transferred from the tin more rapidly than alkyl groups.
   c. If a tetraalkylstannane is used, alkyl groups can also be transferred.
3. The mechanism of the Stille reaction probably involves the following steps:
   a. an oxidative addition of the aryl triflate to the 14-electron Pd(PPh₃)₂

   b. a ligand substitution by excess chloride ion (as LiCl) to the resulting unstable complex

c. a second ligand substitution in which the aryl or alkyl groups on the organotin compound (which resembles a carbanion and is nucleophilic) substitutes for the chloride on the Pd.

d. a reductive elimination to give the product and regenerate the catalyst.

4. The Stille reaction can be carried out with aryl halides instead of aryl triflates.
5. The transferred alkyl group does not rearrange in the Stille reaction.
6. This is another method that can be used to form carbon–carbon bonds.

D. ELECTROPHILIC AROMATIC SUBSTITUTION REACTIONS OF PHENOLS

1. Phenols undergo electrophilic aromatic substitution reactions.
2. Phenols react rapidly with bromine.
   a. Phenol can be halogenated once under mild conditions that do not affect benzene itself.

   b. Phenol reacts with Br_2 in H_2O (bromine water) to give 2,4,6-tribromophenol.

   i. Bromine reacts with water to give protonated hypobromous acid, a more potent electrophile than bromine itself.

   ii. In aqueous solutions near neutrality, phenol partially ionizes to its conjugate-base phenoxide anion; this anion is very reactive and brominates instantly, thereby pulling the phenol–phenolate equilibrium to the right.

   iii. Phenoxide ion is much more reactive than phenol because the reactive intermediate is not a carbocation, but is instead a more stable neutral molecule.

(also ionizes)

c. In strongly acidic solution, in which formation of the phenolate anion is suppressed, bromination can be stopped at the 2,4-dibromophenol stage.

3. Phenol is very reactive in other electrophilic substitution reactions.
    a. Phenol can be nitrated once under mild conditions.
    b. Direct nitration is not the preferred method for synthesis of di- and trinitrophenol, because the concentrated HNO_3 required for multiple nitrations is an oxidizing agent. Instead, nucleophilic aromatic substitution is used. (See Sec. 18.9 in the text for a discussion of mechanism.)

c. The great reactivity of phenol in electrophilic aromatic substitution does not extend to the Friedel–Crafts acylation reaction.
    i. Friedel–Crafts acylation of phenol occurs slowly, but may be carried out successfully at elevated temperatures; the ring is acylated only once.
    ii. Phenols are less reactive because they react rapidly with the $AlCl_3$ catalyst.

E. OXIDATION OF PHENOLS TO QUINONES
    1. *p*-Hydroxyphenols (hydroquinones), *o*-hydroxyphenols (catechols), and phenols with an unsubstituted position ortho or para to the hydroxy group are oxidized to quinones.

2. The oxidation of hydroquinone and its derivatives to the corresponding *p*-benzoquinone can also be carried out reversibly in an electrochemical cell.

F. INDUSTRIAL PREPARATION AND USE OF PHENOL

1. Phenol and acetone are prepared from a single starting material, cumene, which in turn comes from benzene and propene, two compounds obtained from petroleum.

2. The production of phenol and acetone is a two-step process.

   a. Cumene undergoes an autoxidation with molecular oxygen to form cumene hydroperoxide.

   cumene   cumene hydroperoxide

   b. The cumene hydroperoxide undergoes an acid-catalyzed rearrangement that yields both acetone and phenol.

   cumene hydroperoxide   phenol   acetone

SOLUTIONS TO PROBLEMS

Solutions to In-Text Problems

18.1 (a) *p*-Bromotoluene reacts most slowly (that is, not at all) because it is an aryl halide; aryl halides do not undergo S_N2 reactions. Benzyl bromide is a benzylic halide and therefore reacts most rapidly in an S_N2 reaction (see text Sec. 17.4, text p. 850).

Br—⟨ ⟩—CH₃   ≪   ⟨ ⟩—CH₂CH₂CH₂Br   ≪   ⟨ ⟩—CH₂Br

***p*-bromotoluene**   **(3-bromopropyl)benzene**   **benzyl bromide**

(b) 1-Bromocyclohexene, a vinylic halide, does not react by the S_N2 mechanism; 1-(bromomethyl)cyclohexene, an allylic halide, reacts most rapidly. (See text Sec. 17.4, text p. 850.)

⟨ ⟩—Br   ≪   ⟨ ⟩—Br   <   ⟨ ⟩—CH₂Br

1-bromocyclohexane   **bromocyclohexane**   **1-(bromomethyl)cyclohexane**

18.2 Assume that E2 reactions of *vinylic* halides follow the same trends as E2 reactions of *alkyl* halides. Therefore, because an anti-elimination is faster than the corresponding syn-elimination (Sec. 9.5E, text pp. 409–11), the reaction of compound *B* is faster than that of compound *A*; and the reaction of compound *A* is faster than that of compound *C* because bromide is a better leaving group than chloride. Because eliminations involving benzylic β-hydrogens are faster than other eliminations, compounds *A*–*C*, in which the β-hydrogens are benzylic, react more rapidly than compound *D*, in which the β-hydrogens are not benzylic. Furthermore, the elimination reaction of compound *D* gives a triple bond that is conjugated with only one benzene ring (and thus has only one phenyl branch; see structures that follow), whereas all of the others give triple bonds that are conjugated with two benzene rings (and thus have two phenyl branches). Hence, the desired order is *D* < *C* < *A* < *B*.

The product of the elimination reactions of compounds *A*–*C* is diphenylacetylene, Ph—C≡C—Ph; the product of the elimination reaction of compound *D* is phenylacetylene, Ph—C≡C—H.

18.3 (a) The order of increasing S_N1–E1 reactivity is *A* ≪ *B* < *C*. The benzylic halide C reacts fastest because its ionization gives a resonance-stabilized benzylic carbocation; and the vinylic halide *A* reacts slowest because vinylic halides are nearly inert in S_N1 reactions for the reasons given in Sec. 18.3 of the text.

(b) The reactivity order is *B* ≪ *C* < *A*. The reaction of compound *B* is slowest because vinylic halides are virtually inert in S_N1 reactions; and the reaction of compound *A* is fastest because its ionization gives a resonance-stabilized allylic carbocation.

18.4 (a) The nitrogen of the amine acts as a nucleophile in a nucleophilic aromatic substitution reaction.

O₂N—⟨ ⟩—$\overset{+}{N}$H₂Et Cl⁻   ⇌   O₂N—⟨ ⟩—N̈HEt + HCl
        | |
       NO₂   NO₂

The nitro groups reduce the basicity of the product to the point that it is deprotonated under the reaction conditions. (Can you give a reason why?) Because you have not yet studied the basicity of aromatic amines, however, either answer should be considered correct.

(b)   The product results from nucleophilic aromatic substitution by the thiolate group:

$$\text{(structure: benzene ring with } \ddot{S}\text{—Bu group ortho to NO}_2\text{)} + F^-$$

(c)   No reaction, because the compound contains no substituent that can activate nucleophilic aromatic substitution by stabilizing the anionic intermediate (the Meisenheimer complex).

18.5   (a)   The second compound, *p*-fluoronitrobenzene, reacts most rapidly because only in the reaction of this compound is the intermediate Meisenheimer complex stabilized by resonance interaction of an unshared electron pair with the nitro substituent.

(b)   The first compound should react more rapidly because the Meisenheimer complex is stabilized by the resonance interaction of both substituents; the complex in the case of the second compound is stabilized by resonance involving only a nitro substituent.

18.6   The bond to one allylic carbon is an X-type bond. That is, if we break the metal–carbon bond, a carbon anion is formed. The double bond is an L-type ligand. Breaking its coordination gives a neutral alkene.

As with the case of the cyclopentadienyl anion discussed immediately preceding the problem in the text, this is a bookkeeping fiction; the L- and X-type character is actually spread out over the allyl system:

It also happens that allyl can bond to a metal in some cases as an X-type ligand; that is, the π electrons in some cases aren't involved in coordination at all. (See Footnote ** in Table 18.1, text p. 891.) However, we know that this situation is not relevant to the present problem, as the problem tells us to be concerned with LX-type coordination.

18.7   (a)   For each electron pair moved out of a metal–oxygen bond onto an attached oxygen, the oxygen takes on a negative charge. Hence, by definition, each bond to oxygen counts as one X-type interaction. The number of X-type bonds to Mn is 7; and, because the overall charge is on oxygen, the charge on the metal, Q_M in Eq. 18.23, text p. 891, is 0. Therefore, the oxidation state is +7.

The following is an equivalent structure for permanganate in which the negative charge is on Mn:

$$O = \overset{\overset{\displaystyle O}{\|}}{\underset{\underset{\displaystyle O}{\|}}{Mn^-}} = O$$

In this case the oxidation state is $+8 - 1 = +7$; obviously, since resonance structures are different representations of the same compound, we get the same result from either structure.

(b)   The PPh_3 ligands are L-type ligands; hence, there are no X-type ligands, and, because the charge on Pd is 0, the oxidation state of Pd is 0.

You might be wondering about the prefix *tetrakis* in the name of this complex. The prefixes *bis*, *tris*, and *tetrakis* are used as numerical prefixes instead of *di*, *tri*, and *tetra* when the group that is enumerated itself contains multiple substituents. Thus, the ligand triphenylphosphine has three phenyl groups on the phosphorus (thus the prefix *tri* in the name of this ligand). There are four triphenylphosphine ligands—thus the prefix *tetrakis*.

(c)   The Cp ligand (Table 18.1) can be viewed as an L_2X ligand for bookkeeping purposes. Hence, with two X-type bonds and no charge on the iron, the oxidation state of Fe is +2.

18.8   With one X-type ligand (Cl) and zero charge, oxidation state of rhodium (Rh) in the starting material is +1. In the product, two new X-type bonds have been added; hence, the oxidation state is +3. The overall reaction is an oxidation. (This is an example of an *oxidative addition*, an important type of reaction that is considered later in this section.)

18.9   (a)   Consulting Fig. 18.3, text p. 889, we see that neutral tungsten (W) has 6 valence electrons. Add two electrons for the charge. The ligands are L-type ligands. Hence, this is a d^8 complex. Alternatively, the oxidation state of tungsten is, by Eq. 18.23, text p. 891, $0 - 2 = -2$. We then apply Eq. 18.25, text p. 892: n = valence electron count in neutral W (6) – the oxidation state (–2) = $6 - (-2) = 8$.

(b)   Pd has ten valence electrons in the neutral atom. There are no charges and no X-type ligands in the complex; hence, this is a d^{10} complex. Using Eq. 18.25, text p. 892, with an oxidation state of 0, we get the same answer.

(c)   We've seen this complex before, in Problem 18.8. We calculated in the solution to that problem that the oxidation state of rhodium (Rh) is +3. Fig. 18.3 shows that neutral Rh has 9 valence electrons. Hence, application of Eq. 18.25, text p. 892, shows that this is a d^6 complex. Alternately, we start with 9 electrons and remove 3 for the X-type ligands to get $9 - 3 = 6$ for the value of n.

18.10   (a)   For Rh, start with the number of electrons in the neutral atom (9); subtract charges on the metal (0); add the number of X-type ligands (3); and add twice the number of L-type ligands ($2 \times 3 = 6$) to get $9 + 3 + 6 = 18$ electrons. Alternatively, because we know $n = 6$ from the solution to Problem 18.9(c), Eq. 18.26 on p. 893 can be applied to obtain the electron count. This is an 18-electron complex.

(b)   For Tc, start with the number of electrons in the neutral atom (7); subtract charges on the metal (1); add the number of X-type ligands (0); and add twice the number of L-type ligands ($2 \times 6 = 12$) to get $7 - 1 + 12 = 18$ electrons.

18.11   Neutral iron (Fe) has 8 electrons. Because CO is an L-type ligand, it is counted twice in the electron count. We simply solve for x in $8 + 2x = 18$ and obtain $x = 5$. $Fe(CO)_5$, or pentacarbonyliron(0), is in fact a stable complex that can be purchased commercially.

18.12   Vanadium (V) in $V(CO)_5$ has an electron count of $5 + (2 \times 6) = 17$. If the tendency of metals is to have a total of 18 electrons, then V should have a significant driving force to obtain an extra electron, that is, to be reduced. Indeed, the total electron count in $[V(CO)_6]^-$ is 18, and this anion is relatively stable.

18.13   (a)   Cyanide (^-CN) is an X-type ligand. If we strip the four cyanide ligands from the Zn (zinc), a Zn^{2+} ion remains. From Fig. 18.3 on text p. 889, Zn has 12 valence electrons; hence, Zn^{2+} has 10. This is exactly the number needed to fill all of the $3d$ orbitals with two electrons each. Four valence orbitals—the $4s$ and the $4p$ orbitals—are left empty, and these are hybridized to form more directed orbitals. (This situation is exactly like carbon hybridization in methane, except that we are using orbitals from period 4.) Hybridizing one $4s$ and three $4p$ orbitals gives four sp^3 hybrid orbitals, which, as we know from methane, are directed to the corners of a regular tetrahedron. Each of these empty orbitals accepts a pair of electrons from a cyanide ligand. Thus, the $Zn(CN)_4]^{2-}$ complex is tetrahedral (the gray dashed lines show the tetrahedron; the black lines are bonds).

(b)   Triphenylphosphine (PPh_3) is an L-type ligand. If we strip the four PPh_3 ligands from the Pd (palladium), a Pd (0) atom remains. From Fig. 18.3 on text p. 889, Pd has 10 valence electrons. This is exactly the number needed to fill all of the $4d$ orbitals with two electrons each. Four valence orbitals—the $5s$ and the $5p$ orbitals—are left empty, and these are hybridized to form more directed orbitals. (This situation is exactly like carbon hybridization in methane, except that we are using orbitals from period 5.) Hybridize of one $5s$ and three $5p$ orbitals gives four sp^3 hybrid orbitals, which, as we know from methane, are directed to the corners of a regular tetrahedron. Each of these empty orbitals accepts a pair of electrons from a PPh_3 ligand. Thus, the $Pd(PPh_3)_4$ complex is tetrahedral (the gray dashed lines show the tetrahedron; the black lines are bonds).

Unhybridized Pd (0) (d^{10}):

18.14   The student has neglected to understand that when an X-type ligand reacts as a nucleophile with a metal to displace an L-type ligand, the charge changes by one unit, because electrons from X-type ligands are assumed to be divided between the metal and the ligand. In other words, charge is not balanced in this equation. If we put a negative charge on the Pd, as required by this reaction, the oxidation state of Pd is unchanged.

18.15   (a)   *Oxidative addition of H_2 to the catalyst:*

oxidation state: +1
electron count: $16e^-$

oxidation state: +3
electron count: $18e^-$

Ligand substitution of one PPh_3 by the alkene:

oxidation state: +3
electron count: $18e^-$

oxidation state: +3
electron count: $18e^-$

1,2-Insertion of the alkene into an Rh—H bond and addition of the previously expelled PPh_3:

oxidation state: +3
electron count: $18e^-$

Note: empty orbital on Rh because H departed with its two electrons

oxidation state: +3
electron count: $16e^-$

oxidation state: +3
electron count: $18e^-$

Reductive elimination of the product to regenerate the catalyst:

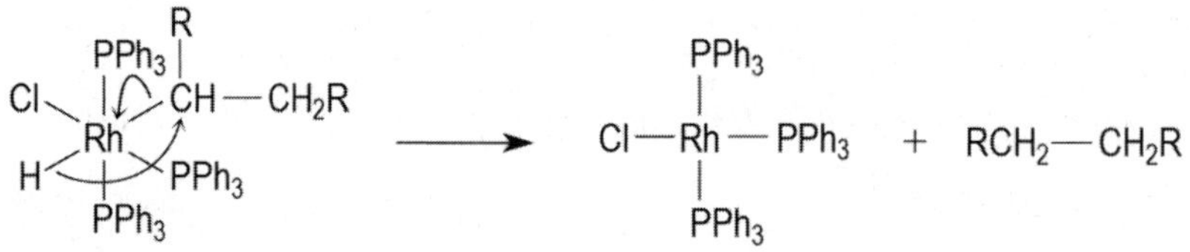

oxidation state: +3   oxidation state: +1
electron count: 18e^- electron count: 16e^-

(b)   The 1,2-insertion is a syn-addition:

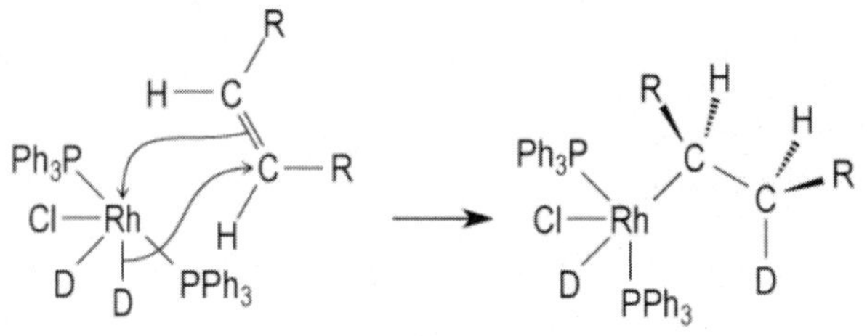

And the reductive elimination is also syn:

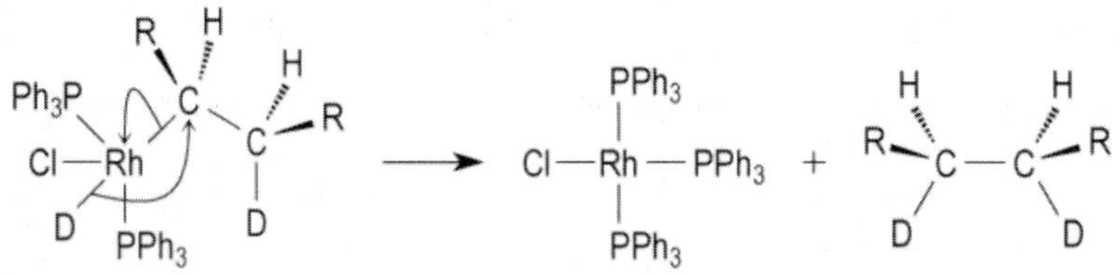

Because we started with a symmetrical *cis*-alkene, the product is the meso stereoisomer.

18.16   The steps in Eq. 18.43b, text p. 902, of the text are numbered for reference.

Step 1:

Fundamental process: oxidative addition
Oxidation state of Pd starting catalyst (PdL$_2$): 0
Electron count of Pd in the starting catalyst: 14e^-. (Note that neutral Pd is a 10-electron atom.)
Oxidation state of Pd in the product: +2
Electron count of Pd in the product: 16e^-

Step 2: (From here on, the reactant has the same properties as the product of the previous step.)

Fundamental process: ligand substitution
Oxidation state of Pd in the product: +2
Electron count of Pd in the product: 16e^-

Step 3:

Fundamental process: 1,2-ligand insertion
Oxidation state of Pd in the product: +2
Electron count of Pd in the product: 14e^-

Step 4:

Fundamental process: β-elimination
Oxidation state of Pd in the product: +2
Electron count of Pd in the product: 16e^-

Step 5:

Fundamental process: ligand dissociation
Oxidation state of Pd in the product: +2
Electron count of Pd in the product: $14e^-$

Step 6:

Fundamental process: ligand association
Oxidation state of Pd in the product: +2
Electron count of Pd in the product: $16e^-$

(Steps 5 and 6 together result in a ligand substitution.)

Step 7:

Fundamental process: reductive elimination
Oxidation state of Pd in the product: 0
Electron count of Pd in the product: $14e^-$

(The catalyst is regenerated in this step.)

18.17   The two **products result from coupling at the two chemically nonequivalent carbons of the alkene:**

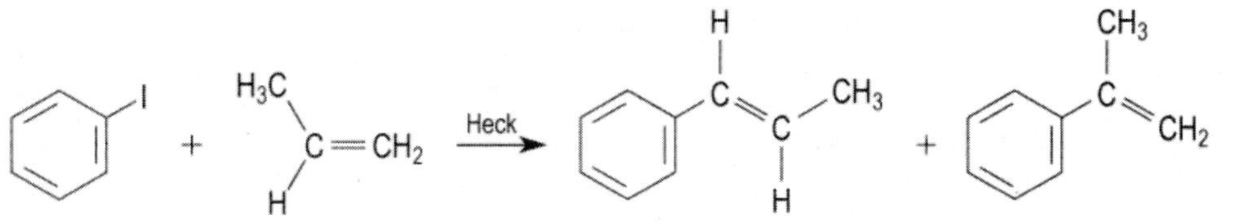

iodobenzene   propene   **(1-propenyl)benzene**   **isopropenylbenzene**
                                        (from coupling at C-1)   (from coupling at C-2)

That is, in step 3 of the mechanism described in Eq. 18.43b, text p. 902, and in the previous solution, 1,2-ligand insertion can occur to either carbon of the double bond. Because these two carbons are nonequivalent, two products are obtained.

18.18   Either aryl substituent could originate from the aryl halide or from the alkene. Remember, if the alkene substituent is aryl (as in these cases), substitution occurs mainly at the less branched carbon.

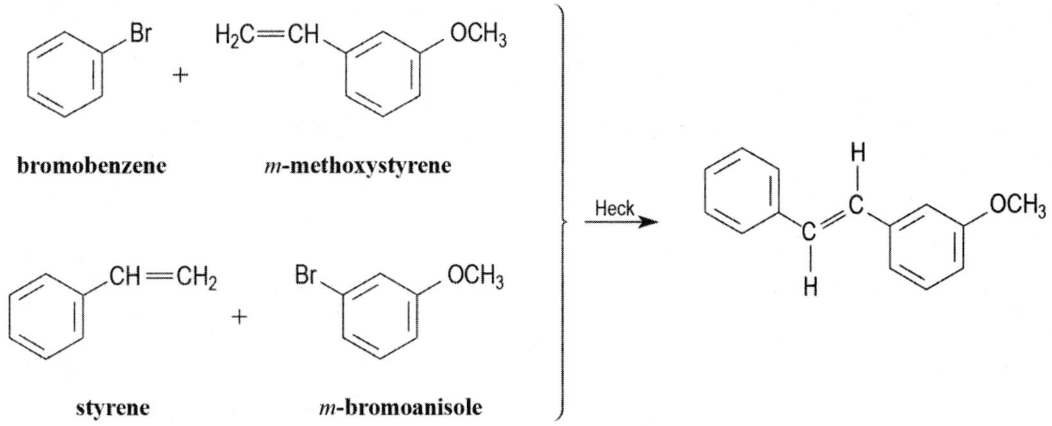

bromobenzene   *m*-methoxystyrene

styrene   *m*-bromoanisole

18.19   The environment around the Pd following the initial oxidative addition is rather congested with the large PPh$_3$ ligands, the aryl group, and the halide. Ligand replacement by the alkene is more favorable for alkenes containing fewer bulky substituents. As a consequence of the Heck reaction itself, the product is always more branched than the starting material, and hence, the starting alkene will "win" the competition for ligand replacement over the product alkene more often than not. If we assume that the subsequent rate constants for substitution are not very different for the two alkenes, the higher concentration of the catalyst-starting alkene complex leads to a greater reaction rate for this alkene. (Remember, rate = k[complex], where k = the rate constant. Given equal k values, the complex present in higher concentration wins the rate competition.)

18.20   This is essentially like the cyclohexene case given in Eq. 18.44 (text p. 902) and subsequent discussion.

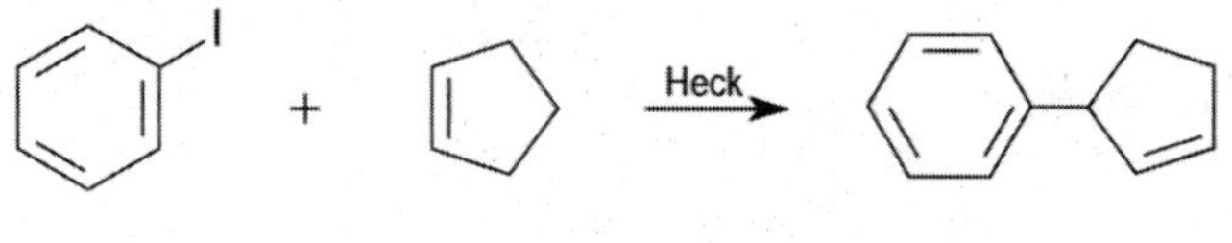

iodobenzene   **cyclopentene**   **(2-cyclopentenyl)benzene**

18.21   As illustrated in Eq. 18.49, text p. 904, Suzuki coupling occurs with retention of the alkene stereochemistry.

(a)   (b) 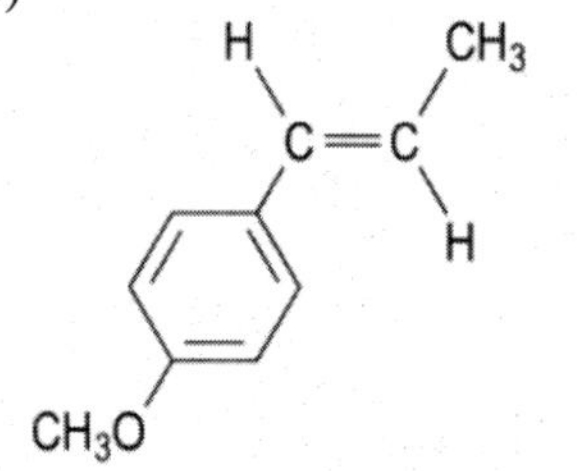   (c)

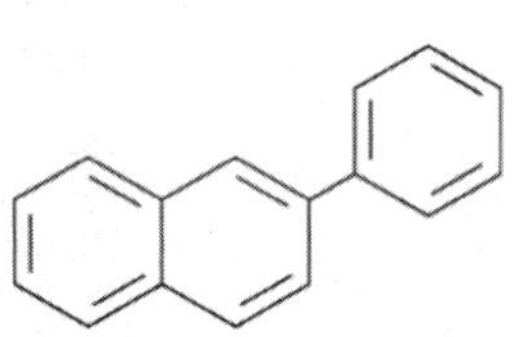

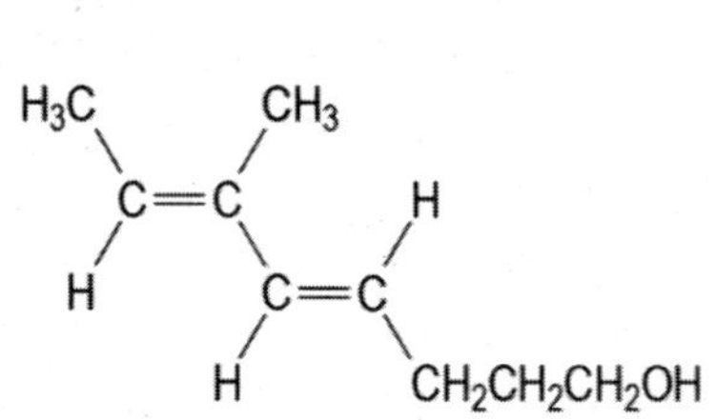

18.22   The starting materials for a Suzuki coupling can be determined by mentally "splitting" the aryl–aryl bond to form an aryl bromide with one part and an aryl boronic acid with the other part.

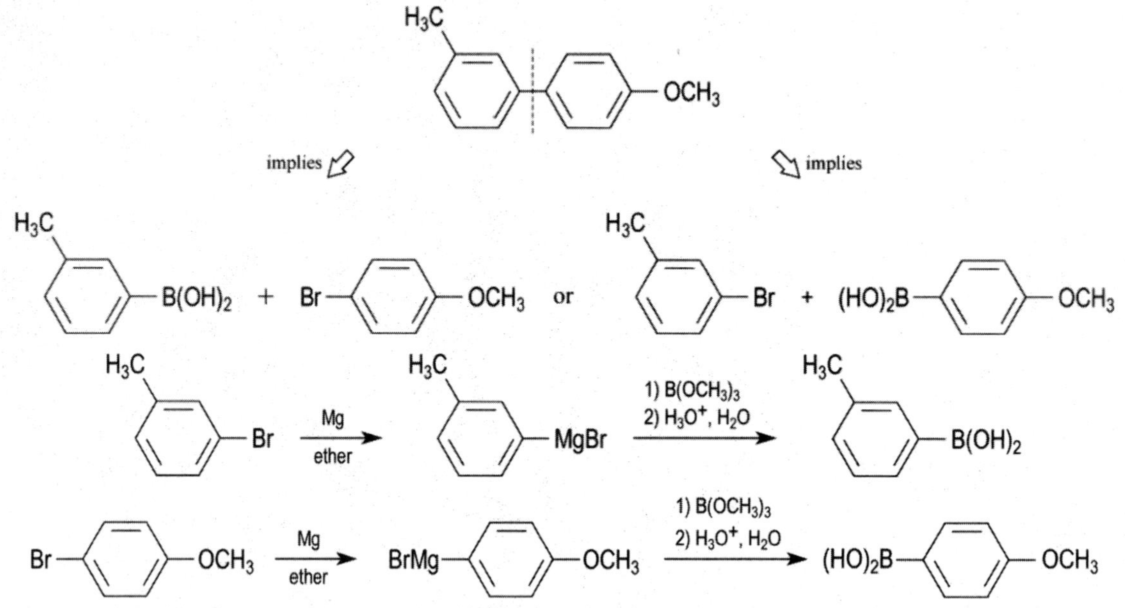

Sequence #1:

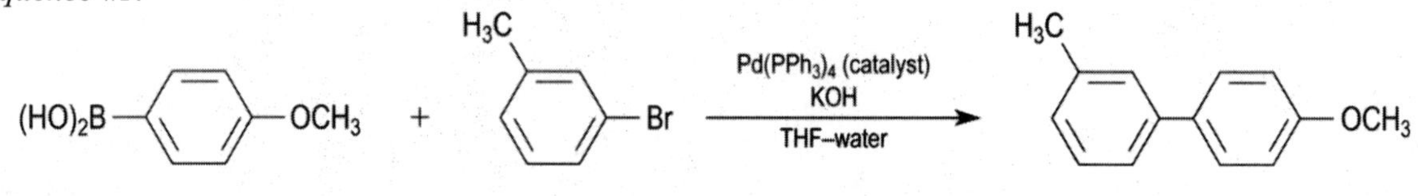

Sequence #2:

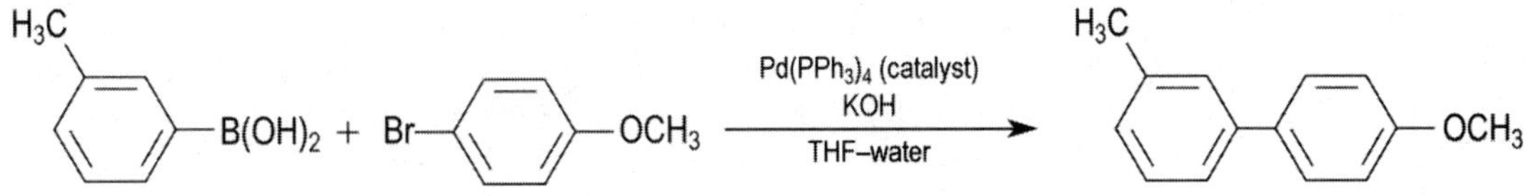

18.23 Start by breaking the compound at the bond between the aryl ring and the alkene; then place a bromine atom (Br) on one of the cleaved fragments and a boronic acid [B(OH)$_2$] or catecholborane (see Eq. 18.51 text p. 905) on the other:

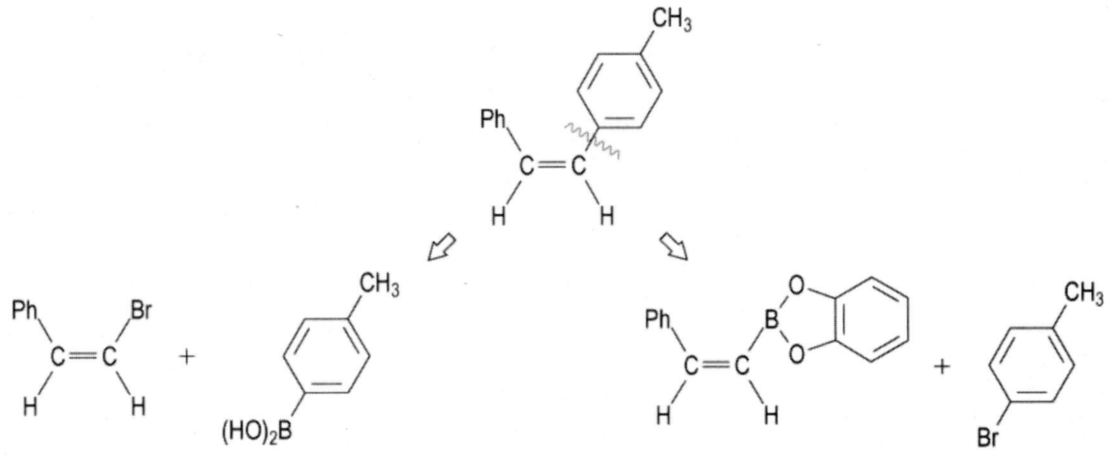

18.24 This reaction involves three sequences of similar steps: a Brønsted acid–base reaction with H$_3$O$^+$; a Lewis acid–base dissociation of methanol from boron; and a Lewis acid–base association of water (from solvent) with boron. Proton transfers are mediated by solvent, even when they're not shown explicitly.

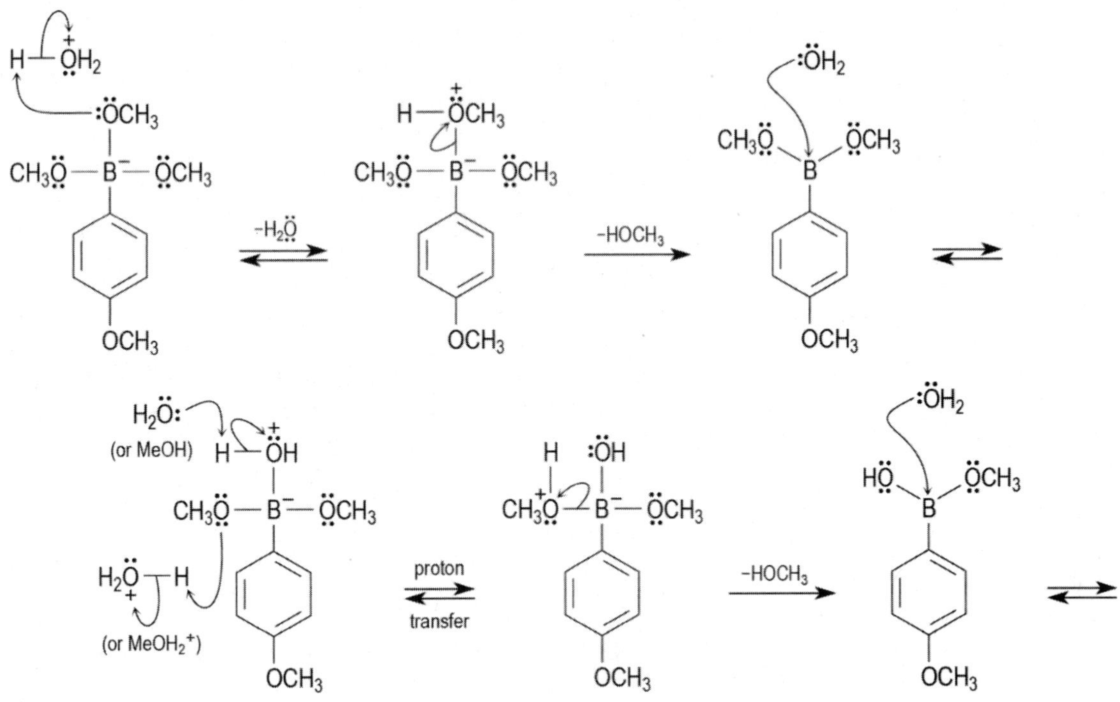

    If you find an error in this manual, please visit http://people.pharmacy.purdue.edu/~loudonm/teaching/

18.25   First, mentally break the two alkenes at the double bond to produce four halves: *A*, bearing the R^1 group; *B* bearing the R^2 group; *C* bearing the R^3 group; and *D* bearing the R^4 group. Now systematically combine one half with another to form trans alkenes—don't forget that one half can combine with another half of the same kind and that *A* + *B* will give the same product as *B* + *A*.

$$A = A \ + \ A = B \ + \ A = C \ + \ A = D \ +$$
$$B = B \ + \ B = C \ + \ B = D \ +$$
$$C = C \ + \ C = D \ +$$
$$D = D$$

18.26   (a)   (b)   (c)

18.27   (a)

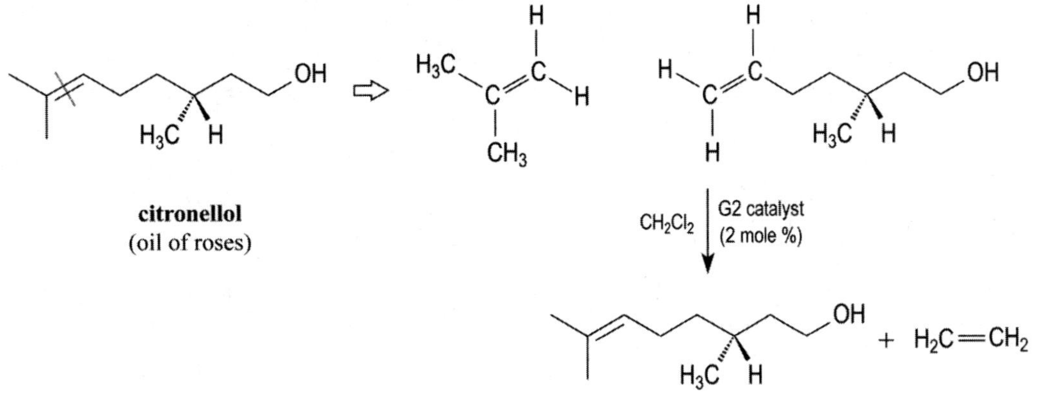

citronellol
(oil of roses)

(b)

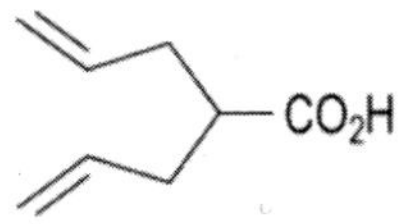

(c)

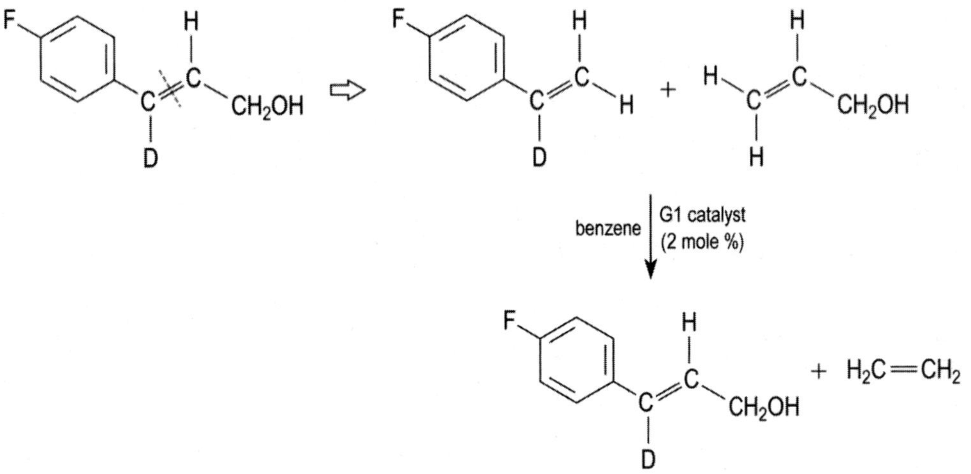

18.28

18.29 A mechanistic outline is as follows.

The classifications of the various steps are:

(1) Substitution of one L-type ligand for another
(2) 1,2-Ligand insertion
(3) Association with L-type ligand
(4) 1,1-Ligand insertion
(5) Oxidative addition
(6) Reductive elimination
(7) Association with L-type ligand

18.30 *(a)* 2,4-Dinitrophenol is more acidic because its conjugate-base anion has more important resonance structures. In the 2,4-dinitrophenoxide anion, both nitro substituents are involved in resonance stabilization; in 2,5-dinitrophenoxide anion, only the 2-nitro group is involved in resonance stabilization.

(The 5-nitro group is not involved for the reasons shown in the discussion of Eq. 18.69 on text p. 915.) The actual pK_a values are

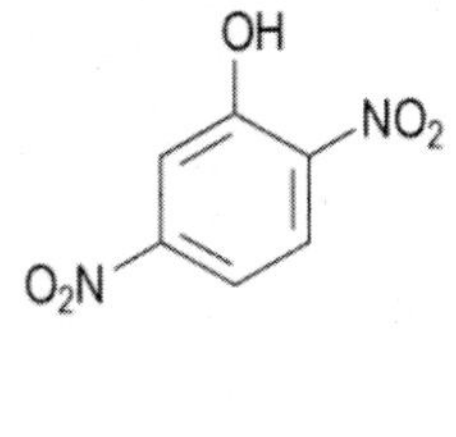

2,5-dinitrophenol
$pK_a = 5.22$

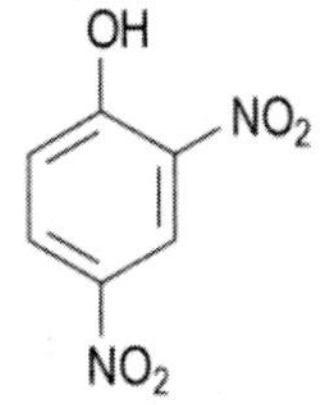

2,4-dinitrophenol
$pK_a = 4.11$

(b)   *Meta*-chlorophenol is more acidic because the conjugate-base anion, *m*-chlorophenoxide, is stabilized by the electron-withdrawing polar effect of the chloro substituent. The actual pK_a values are

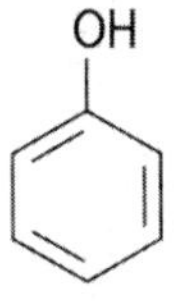

phenol
$pK_a = 9.95$

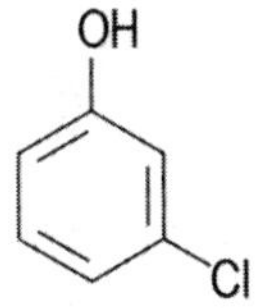

_m_-chlorophenol
$pK_a = 9.02$

(c)   The *para*-isomer of the conjugate-base phenoxide anion has more important resonance structures. The key resonance structure is the one below, in which the substituent group participates in the resonance stabilization. (The actual pK_a values are 7.7 and 8.0 for the *para*- and *meta*-isomers, respectively.)

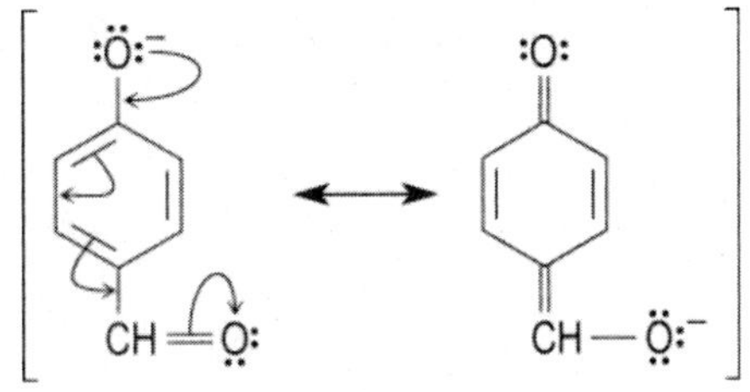

18.31   (a)

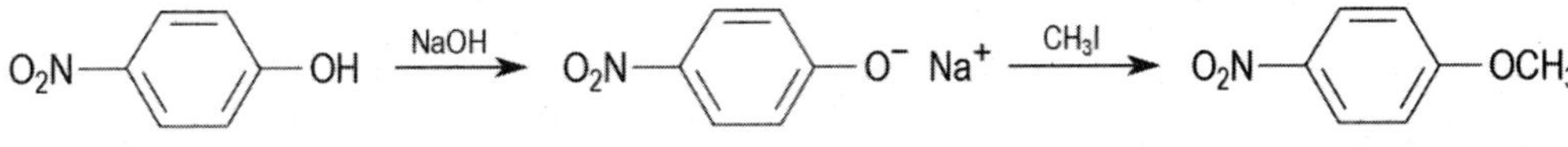

4-nitrophenol
(_p_-nitrophenol)

(_p_-nitroanisole)

(b)

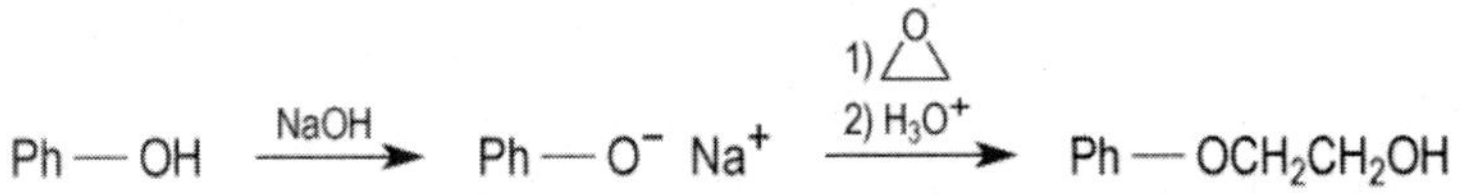

$$Ph-OH \xrightarrow{NaOH} Ph-O^- \; Na^+ \xrightarrow[\text{2) } H_3O^+]{\text{1) } \triangle} Ph-OCH_2CH_2OH$$

2-phenoxyethanol

18.32   The solubility of most phenols in base due to their −1 charge when ionized. Their conjugate bases are ionic compounds, and ionic compounds have greater solubility than nonionic ones. The compound shown in the problem is cationic by virtue of its positively charged nitrogen; ionization of the OH group converts it into a neutral compound. Neutral species are less soluble in water than compounds with a net charge.

Compounds that contain both a positive and a negative charge, such as the conjugate base of the compound shown in the problem, are called *zwitterions* ("double-ions"), or *inner salts*. These types of compounds are discussed in Chapter 27.

18.33   (a)

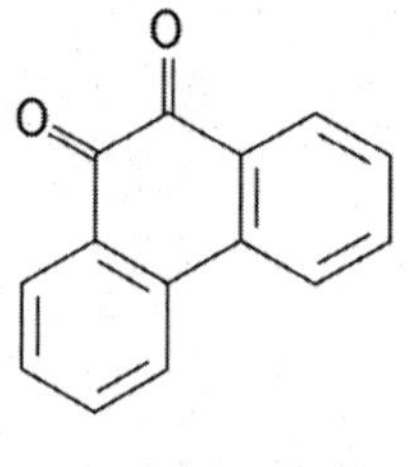

9,10-phenanthraquinone
(a *o*-quinone)

   (b)

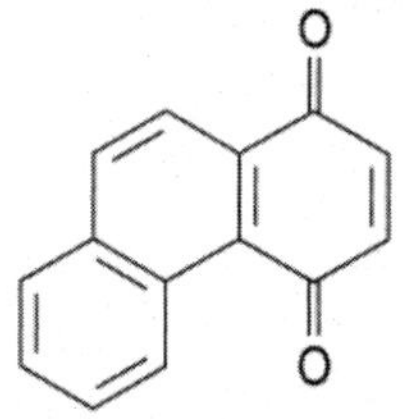

1,4-phenanthraquinone
(a *p*-quinone)

18.34   (a)   A *para*-quinone is formed as in Eq. 18.72, text p. 918; the nitro group is unaffected.

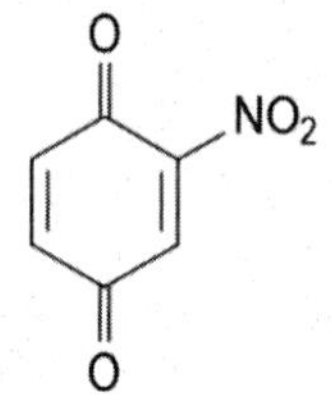

   (b)   The *para*-quinone is formed because it is more stable than the alternative *ortho*-quinone.

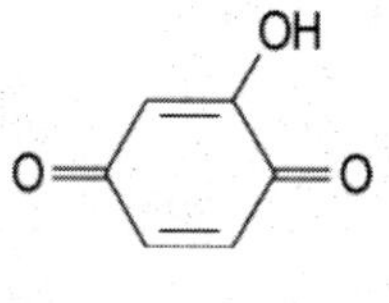

18.35   (a)   Let the long alkyl chain in vitamin E (structure on text p. 921) be abbreviated as —alkyl. The most stable radical is formed by abstraction of the phenolic hydrogen atom

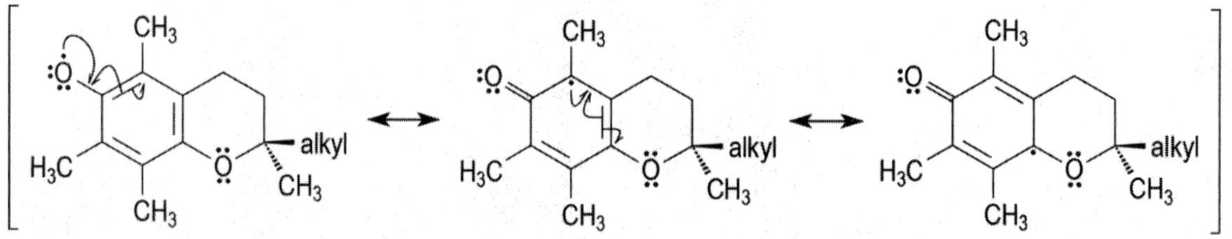

Another possible resonance structure that can derived by the fishhook notation is structure *B*:

The oxygen with the unpaired electron in this structure violates the octet rule—it has nine electrons. In molecular orbital terms, this structure would require the occupancy of an antibonding orbital and thus would reduce the stability of the radical. Hence, this resonance structure is not important.

(b)   As in part (a), the phenolic hydrogen atom is abstracted because a resonance-stabilized radical is formed.

18.36   (a)

(b)

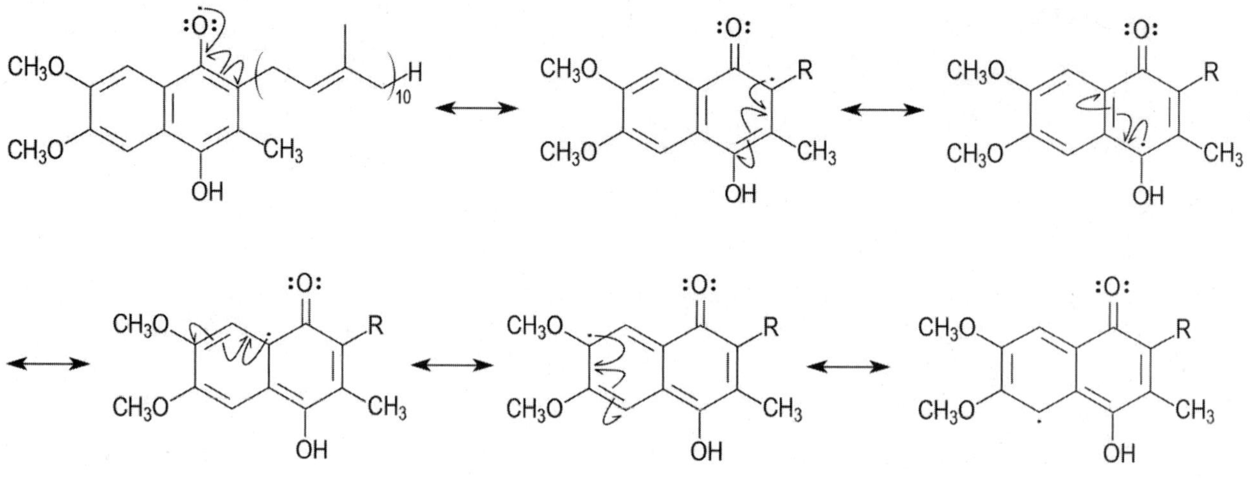

18.37 (a)

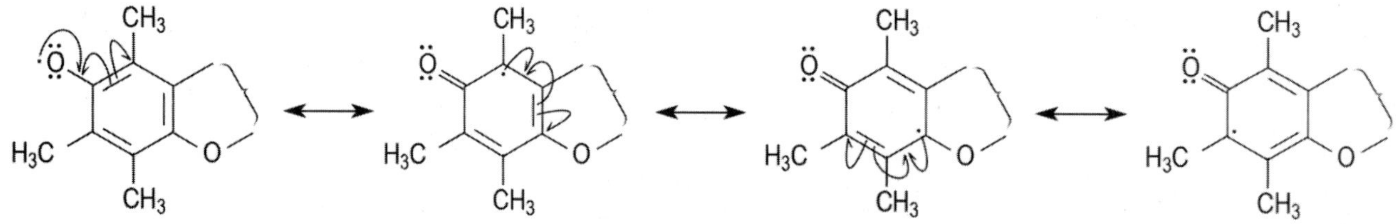

(b) You can use the second resonance structure shown in part (a) to react with the radical R•.

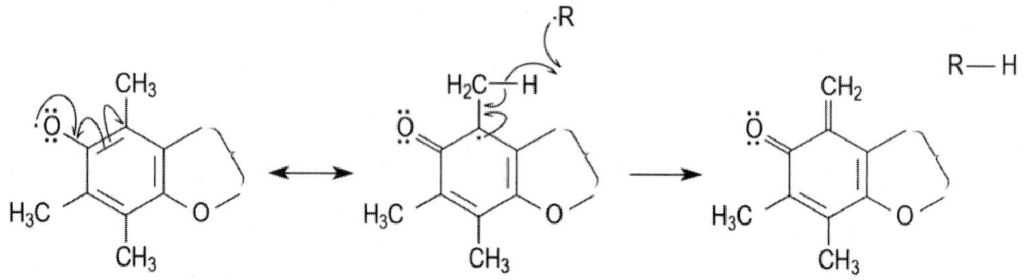

18.38 The long, hydrophobic hydrocarbon chain of ubiquinone (Eq. 18.78, text p. 921,) anchors it in the cellular hydrophobic membrane of the mitochondria organelle (see Sec. 8.7A).

18.39 (a)

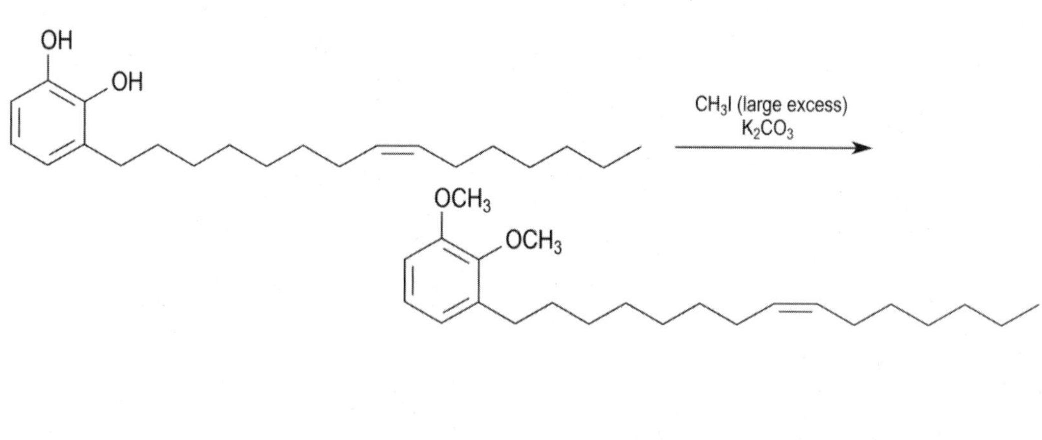

(b)   Unlike the phenolic groups of urushiol, the methyl ether groups would not be oxidized to a quinone and would thus not react with a thiol group of a protein to form the "foreign" protein sensed by the immune system. An allergic skin response like that provoked by urushiol would probably not occur.

18.40   (a)   (b)   (c)

OH   CH$_3$ OH   OH   OH O
                                              O$_2$N   C—CH$_3$
Br   Cl   +   Cl
              NO$_2$ Br

18.41   The electrophile is the *tert*-butyl cation, which is formed by protonation of the alcohol and the Lewis acid–base dissociation of water.

H—$\overset{..}{\underset{..}{O}}SO_3$H $^-$:$\overset{..}{\underset{..}{O}}SO_3$H

(CH$_3$)$_3$C—$\overset{..}{\underset{..}{O}}$H   ⟶   (CH$_3$)$_3$C—$\overset{+}{\underset{..}{O}}H_2$ ⟶   (CH$_3$)$_3$C+   + $\overset{..}{\underset{..}{O}}H_2$

tert-butyl cation

$^-$:$\overset{..}{\underset{..}{O}}SO_3$H

(CH$_3$)$_3$C+   ⟨benzene⟩—OH   ⟶   (CH$_3$)$_3$C—⟨benzene⟩—OH   ⟶   (CH$_3$)$_3$C—⟨benzene⟩—OH

+ H—$\overset{..}{\underset{..}{O}}SO_3$H

18.42   (a)   Neopentyl phenyl ether does not readily cleave with hot concentrated HBr. Such a reaction would require either an S$_N$1 or S$_N$2 reaction of bromide ion at a phenyl–oxygen bond, which does not occur, or an S$_N$1 or S$_N$2 reaction at the primary carbon of a neopentyl group. The S$_N$1 reaction does not occur because it would require the formation of a primary carbocation. (Such a reaction could be forced with very long reaction times and heat, and would occur with rearrangement of the neopentyl group.) The transition state of an S$_N$2 reaction at the primary carbon of a neopentyl group is so beset by van der Waals repulsions that such a reaction does not take place; see Fig. 9.4(b) on text p. 397.

In contrast, benzyl methyl ether can cleave by both S$_N$1 and S$_N$2 mechanisms at the benzylic carbon, or by an S$_N$2 mechanism at the methyl carbon. Here is the cleavage by an S$_N$2 mechanism involving the nucleophilic reaction of Br$^-$ at the methyl carbon of the protonated ether:

:$\overset{..}{\underset{..}{Br}}$:$^-$

H

PhCH$_2$—$\overset{+}{\underset{..}{O}}$—CH$_3$ ⟶   PhCH$_2$—$\overset{..}{\underset{..}{O}}$H   +   :$\overset{..}{\underset{..}{Br}}$—CH$_3$

benzyl methyl ether   **benzyl alcohol   methyl bromide**
(after protonation by HBr)

And here is the S_N2 mechanism at the benzylic carbon (Sec. 17.4 in the text):

benzyl methyl ether **benzyl bromide** **methyl alcohol**
(after protonation by HBr)

An S_N1 mechanism at the benzylic carbon is also possible that parallels the mechanism shown in part (a): protonation of the ether, formation of a benzylic cation, and a Lewis acid–base reaction of Br^- with this carbocation. The products are the same as those derived from the S_N2 reaction immediately previous. *p*-Methoxytoluene can cleave at the methyl–oxygen bond by an S_N2 mechanism.

p-methoxytoluene **p-cresol** **methyl bromide**
(after protonation by HBr)

(b) Diphenyl ether does not cleave with hot, concentrated HBr, because such a cleavage would require either an S_N1 reaction or an S_N2 reaction at a phenyl–oxygen bond; as this section of the text shows, such reactions do not occur. In contrast, *tert*-butyl phenyl ether cleaves by an S_N1 mechanism involving protonation of the oxygen and loss of phenol to form a *tert*-butyl cation, which undergoes a Lewis acid–base association reaction with bromide ion to form *tert*-butyl bromide.

tert-butyl phenyl ether **tert-butyl bromide**

18.43 The product is phenylacetylene, $Ph-C\equiv CH$, along with the usual by-products, $Li^+ \ ^-OTf$ and $ClSi(CH_3)_3$.

18.44 The triflate derivative of *p*-nitrophenol (prepared from *p*-nitrophenol and triflic anhydride; Eq. 18.97, text p. 930) and the (*Z*)-stereoisomer of the appropriate trimethylstannyl derivative would be required.

p-nitrophenyl triflate

18.45 The alcohol *A* can form a relatively stable tertiary benzylic carbocation under the acidic conditions of the reaction. This serves as a Friedel–Crafts alkylating agent for phenol. (Note the similarity to the mechanism shown in the solution to Problem 18.42, text p. 929.)

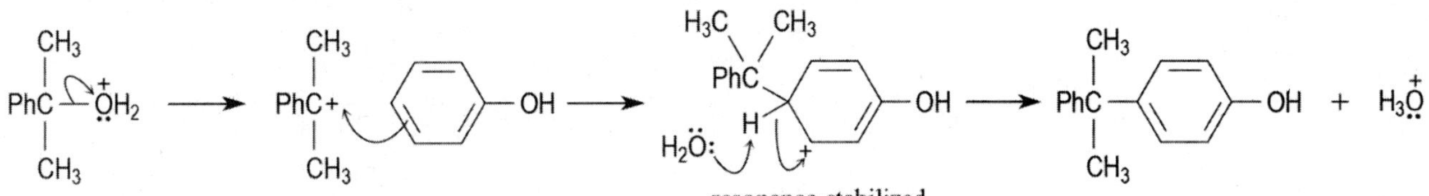

conjugate acid of *A*   +   :ÖH₂

resonance-stabilized
carbocation intermediate

B

Solutions to Additional Problems

18.46 (a)   No reaction
(b)   No reaction
(c)   No reaction
(d)   (e)   (f)   (g)   (h)

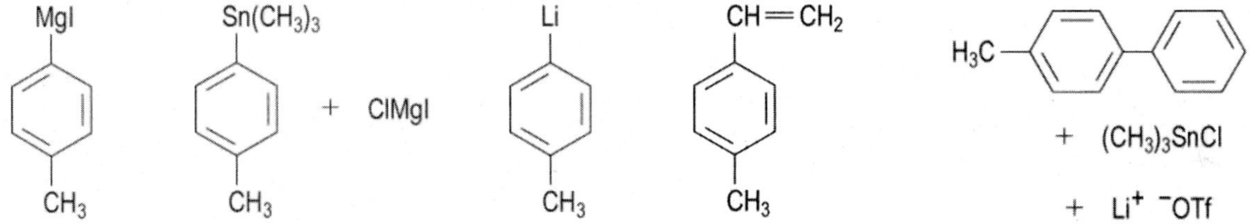

(i)   (j)   (k)

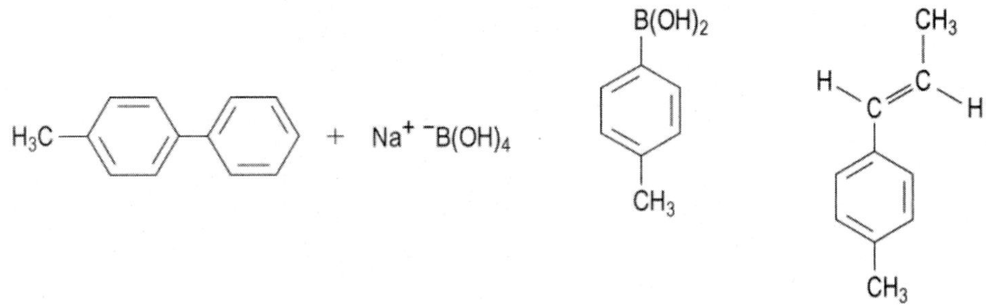

18.47 (a)   (b)

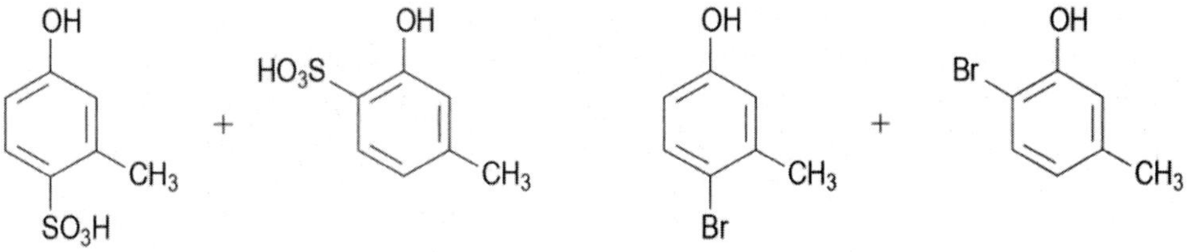

(c)   Both benzylic bromination and ring bromination take place; see Eq. 18.84, text p. 925.

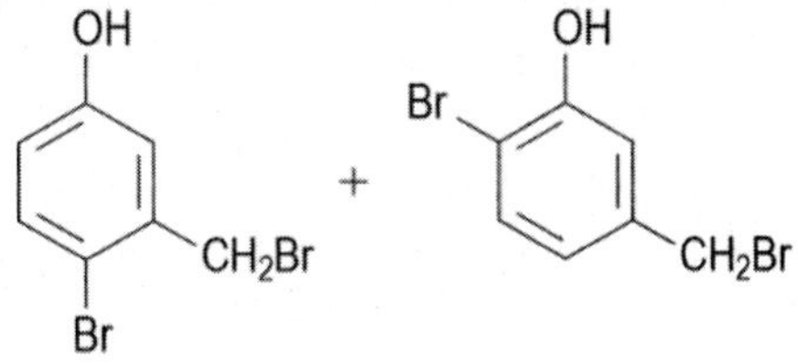

(d)   (e)   (f)

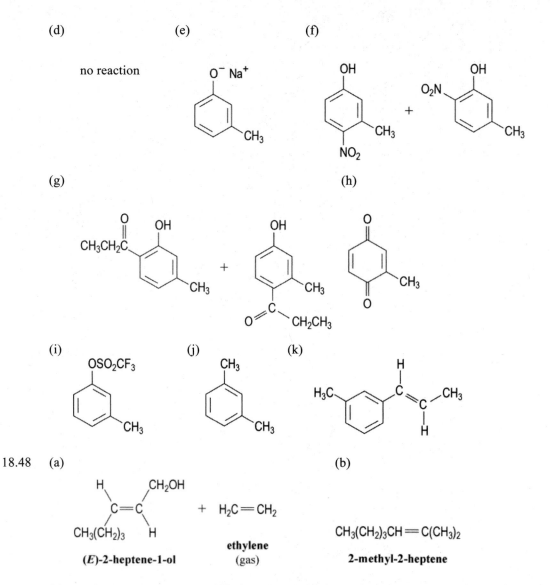

no reaction

18.48   (a)   (b)

$CH_3(CH_2)_3CH=C(CH_3)_2$

(E)-2-heptene-1-ol   **ethylene**   **2-methyl-2-heptene**
                                (gas)

18.49   (a)   Mercaptans are more acidic than phenols; and aryl mercaptans (thiophenols) are more acidic than alkyl mercaptans for the same reason that phenols are more acidic than alcohols.   The acidity order is

cyclohexanol   cyclohexyl mercaptan   benzenethiol

$\longleftarrow$————————————— pK_a ———————————

———————————— acidity ————————————$\longrightarrow$

(b)   Phenol is most acidic because its conjugate-base anion is stabilized by both the polar and resonance effects of the phenyl group. The conjugate-base anion of benzyl alcohol is stabilized by the polar effect of the phenyl group. The conjugate-base anion of cyclohexanol has none of these stabilizing contributions. The acidity order is

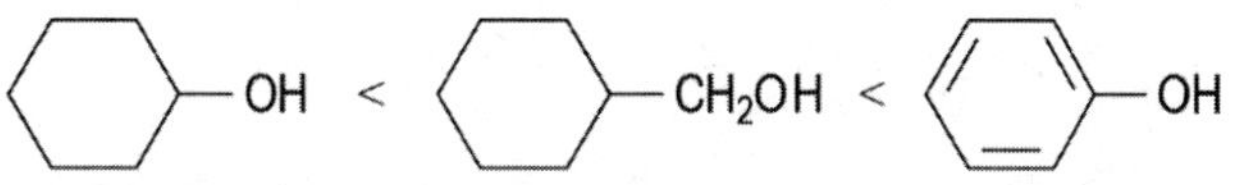

cyclohexanol   **benzyl alcohol**   **phenol**

pK_a

acidity

(c)   The *p*-nitro substituent stabilizes the *p*-nitrophenolate ion by both its resonance effect (see text Eq. 18.68 on p. 915) and its polar effect; the *p*-chloro substituent stabilizes *p*-chlorophenolate ion only by its polar effect. The third compound is nitric acid, a strong acid. Its conjugate-base anion, the nitrate ion, is stabilized both by the polar effect of the nitro group, which is much closer to the anionic oxygen than it is in the *p*-nitrophenolate ion, and by a resonance effect:

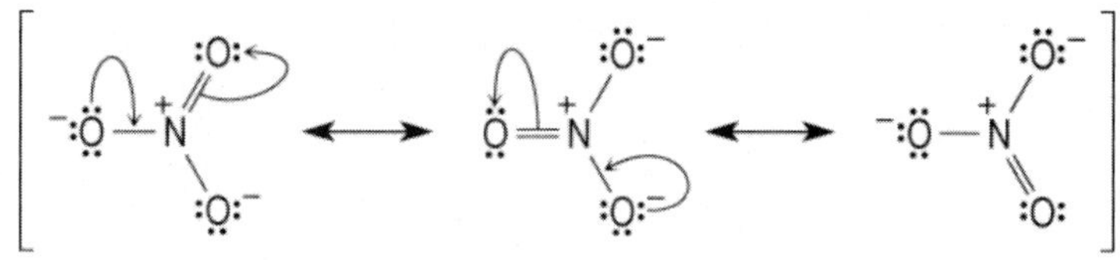

nitrite ion

Here is the acidity order:

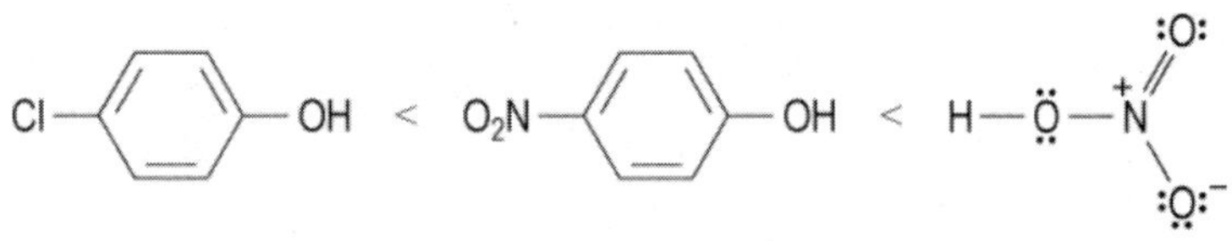

p-chlorophenol   **p-nitrophenol**   **nitric acid**

pK_a

acidity

(d)   4-Nitrophenol is more acidic than phenol (see text p. 915), and benzenethiols are more acidic than phenols (element effect.) The acidity order is

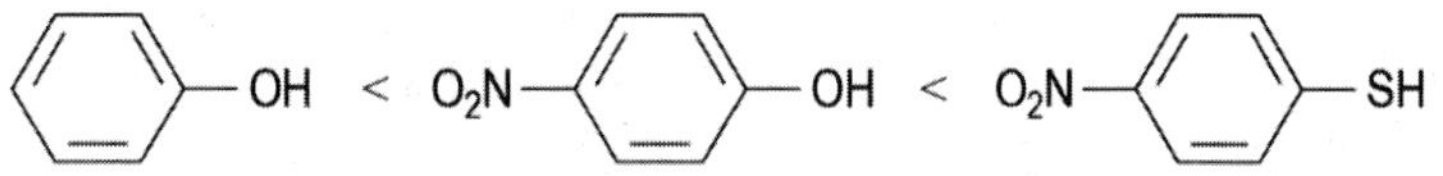

phenol   **4-nitrophenol**   **4-nitrobenzenethiol**

pK_a

acidity

(e)   The *para*-aryl substituents all stabilize the respective conjugate-base alkoxide ions by both a resonance and a polar effect. The nitro group of the *meta*-nitrophenyl substituent exerts an additional stabilizing polar effect; and the nitro group of the *para*-nitrophenyl substituent stabilizes the conjugate-base phenoxide ion by both a polar effect and a resonance effect:

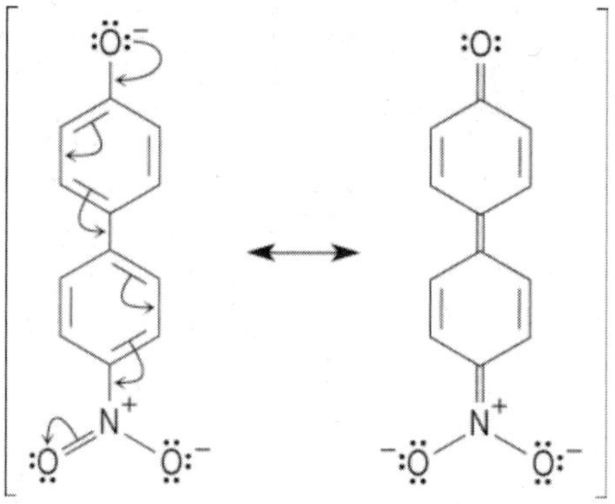

$$C \;<\; A \;<\; D \;<\; B$$

$$\longleftarrow \quad pK_a \quad \longrightarrow$$
$$\longleftarrow \text{ acidity } \longrightarrow$$

18.50   The enol *A* is more acidic (by about five pK_a units). The reason is that its conjugate-base anion is stabilized by resonance as shown below, whereas the conjugate-base anion of compound *B* is not resonance-stabilized.

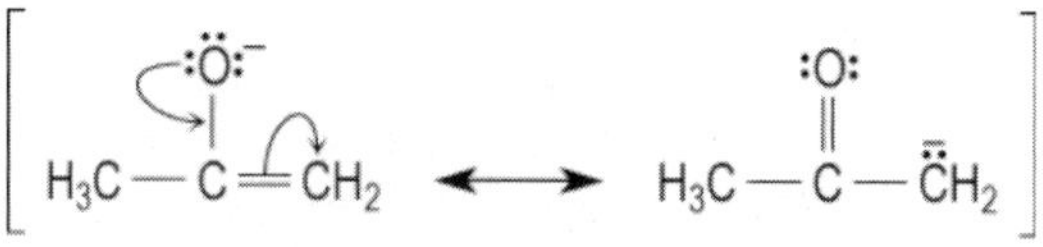

conjugate base of compound *A*

18.51   (a)

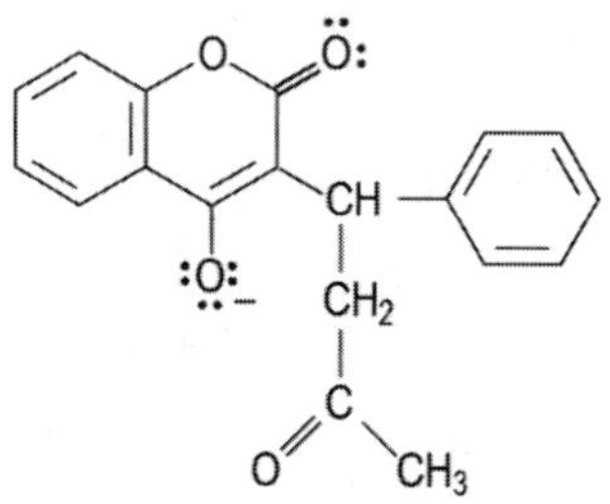

(b)   The resonance structures below show that the negative charge is delocalized over the two oxygens and the α-carbon.

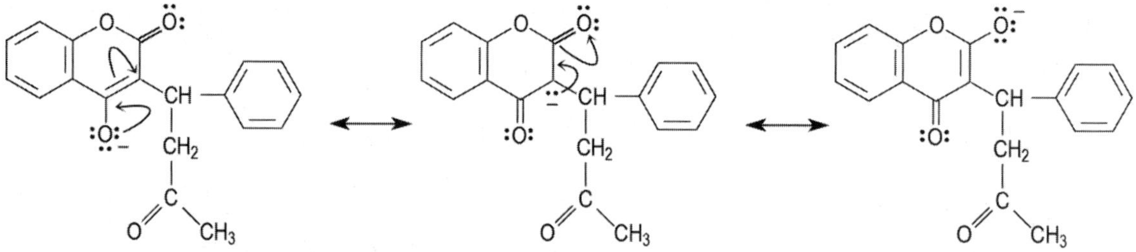

(c)   The conjugate base form of Warfarin predominates at pH = 7.4. When the pH of a solution is more than two units greater than the pK_a of a particular proton, we can say that the proton is >99% dissociated, as stated by rule 3 on text p. 107. Additionally, you can solve for the ratio of the dissociated to associated forms by using the Henderson-Hasselbalch equation, Eq. 3.29e on text p. 106.

18.52   Hydrogenation of phenol *A* yields the corresponding cyclohexanol. Compound *B* reacts readily with HBr to give the benzylic bromide; the corresponding Grignard reagent undergoes protonolysis to give *p*-xylene. The ether group in compound *C* is cleaved by HBr to give bromoethane, C_2H_5Br (which is volatile under the high-temperature reaction conditions), and *m*-cresol.

(a)

OH

3,5-dimethylphenol
(compound *A*)

(b)

H_3C—⬡—CH_2OH

***p*-methylbenzyl alcohol**
(compound *B*)

(c)

OC_2H_5

***m*-ethoxytoluene**
(compound *C*)

18.53   (a)   Cyclohexanol is ionized only to a very small extent in aqueous NaOH solution; phenol ionizes almost completely.

⬡—OH $\underset{\text{NaOH}}{\rightleftarrows}$ ⬡—O⁻ Na⁺   ⬡—OH $\underset{\text{NaOH}}{\rightleftarrows}$ ⬡—O⁻ Na⁺

(b)   Both compounds react rapidly with NaH to give H_2 and their conjugate bases: sodium cyclohexanolate in the case of cyclohexanol, and sodium phenolate (sodium phenoxide) in the case of phenol.

⬡—O⁻ Na⁺   ⬡—O⁻ Na⁺

sodium cyclohexanolate   **sodium phenolate**

(c)   Both compounds are converted into their triflate derivatives.

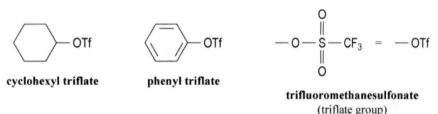

⬡—OTf   ⬡—OTf $-O-\overset{\displaystyle O}{\underset{\displaystyle O}{\overset{||}{\underset{||}{S}}}}-CF_3$ = —OTf

cyclohexyl triflate   **phenyl triflate**   **trifluoromethanesulfonate**
(triflate group)

(d)   Only cyclohexanol reacts with concentrated HBr; because the C—O bond in phenol cannot be broken by S_N1 or S_N2 reactions, phenol is inert, except for a small degree of protonation of the phenol oxygen. The product in the case of cyclohexanol is bromocyclohexane.

⬡—Br   **bromocylcohexane**

(e)   The phenol is brominated on the ring to give the following products; the alcohol is unaffected.

2-bromophenol   **4-bromophenol**

(f)   Cyclohexanol is oxidized to cyclohexanone, whereas phenol is oxidized to *p*-benzoquinone.

cyclohexanone   ***p*-benzoquinone**

(g)   Cyclohexanol undergoes dehydration to cyclohexene; phenol undergoes sulfonation, an electrophilic aromatic substitution reaction.

cyclohexene   **2-hydroxy-**   **4-hydroxy-**
                    benzenesulfonic   **benzenesulfonic**
                    acid   **acid**

18.54   (a)   The second compound reacts in a nucleophilic aromatic substitution reaction. The first compound does not react in this way because it is not activated toward nucleophilic aromatic substitution; the nitro group is not attached to the ring.

In the interest of total accuracy we point out that the first compound actually does react, but in a different way. A hydrogen of the —CH₂— group is acidic enough to ionize in alcoholic KOH; can you see why? But it does not liberate F⁻.

(b)   The last compound, diphenyl ether, cannot be prepared by a Williamson ether synthesis, because such a synthesis would require an S_N2 reaction to occur at the carbon of a benzene ring. Notice that the first ether, tetrahydrofuran, can be prepared by an intramolecular Williamson synthesis from 4-bromo-1-butanol, Br—(CH₂)₄—OH. The second ether, ethoxybenzene, can be prepared by a reaction of sodium phenolate, PhO⁻ Na⁺, with ethyl iodide.

(c)   The first compound undergoes solvolysis by an S_N1 mechanism to give a mixture of an alcohol, an ether, an alkene, and HBr (thus the acidic solution). The second compound is inert, because aryl halides do not undergo solvolysis under neutral conditions.

(d)   Diphenyl ether will not cleave with HI because such a cleavage would require an S_N1 or S_N2 reaction at the carbon of a benzene ring, and such reactions do not occur. Phenyl cyclohexyl ether, however, would cleave to phenol and iodocyclohexane:

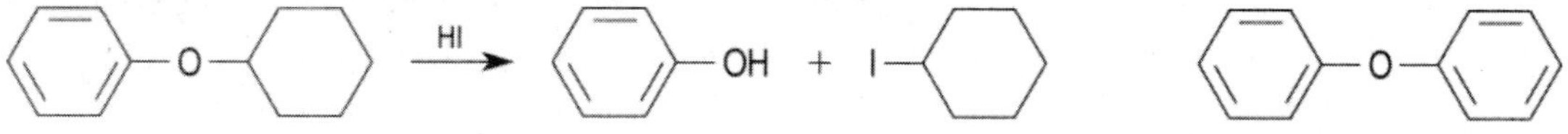

cyclohexyl phenyl ether

diphenyl ether
(inert to HI)

18.55   The phenol group is unaffected in both compounds (except for a small amount of protonation of the phenol oxygen); the alcohol and ether groups react.

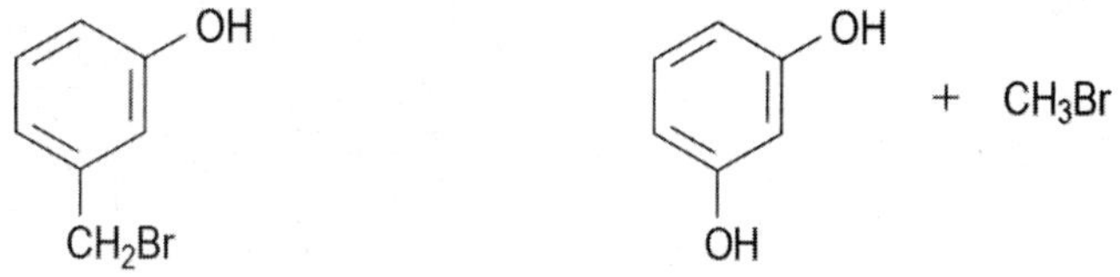

18.56   Potassium benzenethiolate, a base, reacts with the tertiary alkyl halide in an E2 reaction:

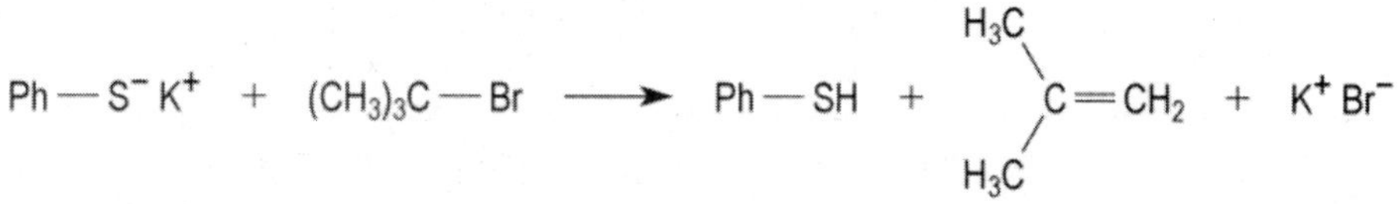

$$Ph - S^- K^+ \ + \ (CH_3)_3C - Br \ \longrightarrow \ Ph - SH \ + \ \underset{H_3C}{\overset{H_3C}{}} C = CH_2 \ + \ K^+ Br^-$$

18.57   Sodium ethoxide converts the thiol completely into its conjugate-base thiolate ion.

(a)   The thiolate ion is alkylated. (Remember from Sec. 17.4 of the text, that allyl bromide is a particularly reactive alkylating agent.)

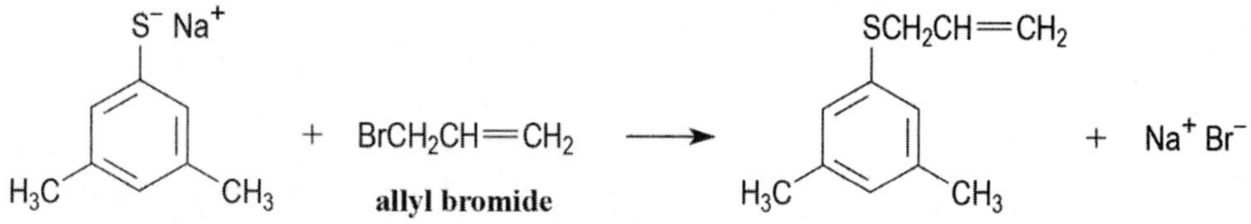

allyl bromide

product of thiol ionization

(b)   No further reaction of the thiolate occurs because bromobenzene, an aryl halide, is inert to nucleophilic substitution and elimination reactions. (Sec. 18.1 of the text.)

18.58   (a)   The reaction of phenol as a base with the acid H_2SO_4:

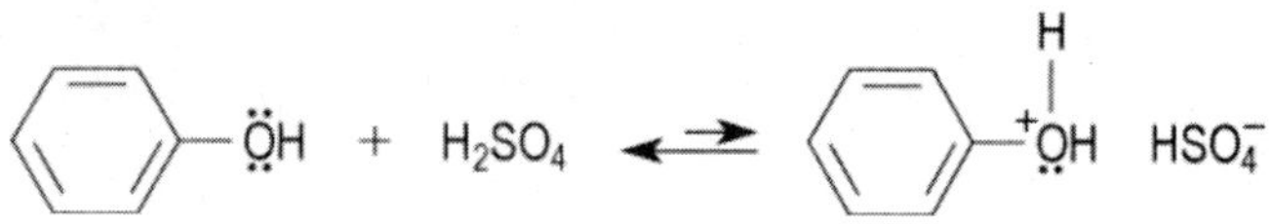

(b)   First, consider polar effects. The phenyl group has an *electron-withdrawing* polar effect; this destabilizes the positively-charged conjugate acid of phenol. Such an effect is absent in the conjugate acid of the alcohol cyclohexanol because its ring has no double bonds. Therefore, the polar effect reduces the basicity of the phenol relative to the alcohol. Second, consider the resonance effect of the benzene ring. The source of this effect is the delocalization of an oxygen unshared electron pair into the ring:

resonance structures of phenol

Such an effect is absent in the conjugate acid of the phenol because it would place *two* positive charges on the oxygen, an electronegative atom.

conjugate acid of phenol

Therefore, resonance stabilizes phenol relative to its conjugate acid and therefore reduces its basicity. (Another way to think of this situation is to imagine that delocalization of the oxygen unshared electron pair makes it "less available" for protonation.) Such a resonance effect is not present in the alcohol. The conclusion is that a phenol is much less basic than an alcohol because of both the polar effect and the resonance effect of the ring. Therefore, the alcohol is the stronger base.

18.59   Because NaOH can ionize a phenol, particularly a relatively acidic phenol such as *p*-nitrophenol, which has a pK_a of 7.21, the spectrum *B* (and the corresponding yellow color) is due to the conjugate base of the phenol. The UV absorption reflects conjugation of the extra unshared pair with the ring and its delocalization into the *p*-nitro group, as shown in Eq. 18.68, text p. 915. Addition of acid lowers the pH so that the un-ionized phenol is regenerated, and the UV spectrum of the phenoxide disappears. To summarize:

18.60   It is the *un-ionized* form of vanillin that has the typical odor. In NaOH solution, the phenol group of vanillin ionizes to its conjugate-base phenoxide ion; because vanillin is no longer present, and because ionic compounds such as the conjugate-base phenoxide are not volatile, the odor disappears. (A compound has to be volatile—that is, it must have a significant vapor pressure—to enter the gas phase and thus reach the nostrils.) Acidification of the solution brings about protonation of the phenoxide and regeneration of vanillin and, hence, the characteristic odor.

neutral compound:
volatile, has vanilla odor

ionic compound:
nonvolatile, has no odor

18.61   The two principles involved in the solution to this problem are, first, that phenoxide ions are more soluble in water than are their conjugate-acid phenols; and second, that an acid is ionized when the pH of the solution is well above the pK_a of the acid. The second principle can be demonstrated in two ways. Either use the rule of thumb that when the pH is more than two units greater than the pK_a of the acid the acid is >99% dissociated (point 3 on text p. 107), or use Eq. 3.31a (or Eq. 3.31b) to find the percent of the acid in the ionized form (or un-ionized form) at each given pH.(text p. 101). These formulas can be solved to find precisely the ratio of $[A^-]/[HA]$.

Condition (1) corresponds to a pH of about 1; neither phenol is ionized (much less than 1%), and therefore neither is soluble in this solution. Under condition (2), some fraction of 2,4-dinitrophenol is ionized (about 44% by Eq. 3.31a), but under condition (3), which is a pH value almost three units higher than the pK_a of the phenol, 2,4-dinitrophenol is >99% ionized. Because this pH value is more than three units below the pK_a of p-cresol, this phenol is not ionized at pH = 7. Therefore, 2,4-dinitrophenol should be relatively soluble at this pH because it is ionized, and p-cresol remains insoluble in the aqueous buffer and can be recovered from the ether layer. This is the best condition to separate the two phenols by extraction. Condition (4) corresponds to a pH of about 13; both phenols are ionized at this pH, and both should be relatively soluble (as their conjugate-base phenoxides) in the aqueous layer.

18.62   (a)   Because the S_N2 reaction requires approach of the nucleophile from the backside of the C—Br bond, and because this would require approach of the nucleophile along the axis of the triple bond, the S_N2 reaction is impossible.

(b)   1-Haloalkynes cannot undergo an S_N1 reaction. Such a reaction would require that a carbocation be formed at an sp-hybridized carbon. Yet carbocations optimally require sp^2 hybridization, which, in turn, requires trigonal planar geometry. An alkyne carbon cannot achieve this geometry because its three bonds are connected to the same atom. Hence, the sp^2 hybridization required for carbocation formation is impossible.

18.63   (a)

4-hydroxy-3-methoxycinnamic acid
(ferulic acid)

(b)   No. The phenolic group is meta to the alkene and carboxylic acid groups and the radical that is formed from proton abstraction cannot delocalize to the alkene and carboxylic acid groups.

3-hydroxy-4-methoxycinnamic acid

single electron skips over this carbon

18.64

18.65   (a)   The oxidation state is the number of valence electrons in the neutral atom (9) less the charge (+1) less the number of bonds to X-type ligands (2). (The nitrogens that have three bonds *in addition to* the metal–nitrogen bond are L-type ligands; the other nitrogen is an X-type ligand. The second X-type ligand is the cyano (—CN) group.) The oxidation state of Co is +3.

(b)   The number n is the valence electrons in the neutral atom (9) less the oxidation state (+3). Hence, Co in this complex is a d^6 metal.

(c)   The electron count is the number of electrons in the neutral atom (9) less the charge (+1) plus the number of X-type ligands (2) plus twice the number of L-type ligands (4). The electron count is therefore $18e^-$. This complex thus follows the eighteen-electron rule.

18.66   (a)   The number n is the valence electrons in the neutral atom (8) less the number of X-type ligands (5: two nitrogens, sulfur, and count oxygen twice since it is doubly bonded). Hence, Fe in this complex is a d^3 metal.

(b)   The electron count is n (3) plus twice the number of all ligands (7). The electron count is therefore $17e^-$.

18.67   By loss of a proton from the hydroxy group, the carbocation intermediate becomes the neutral compound that precipitates. (Formation of the electrophile is shown in Eq. 18.86, text p. 925.)

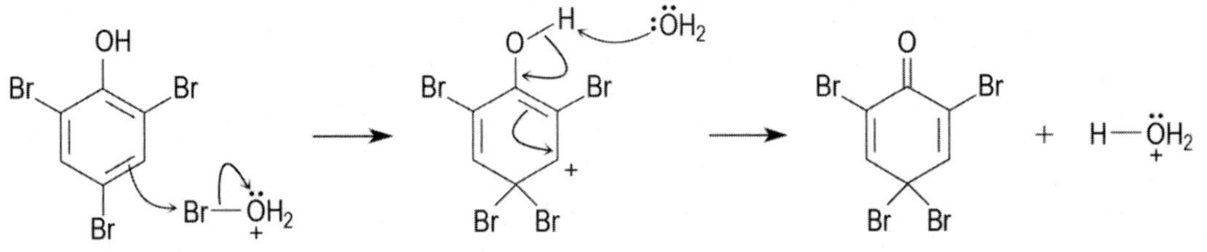

18.68   Solvolysis of the benzylic bromide results in a carbocation intermediate that can lose a proton to give the bracketed intermediate, an example of a *quinone methide*, which reacts with hydroxide ion as shown in the following equation. (See also Problem 18.85, text p. 943, for the formation of a similar type of intermediate.)

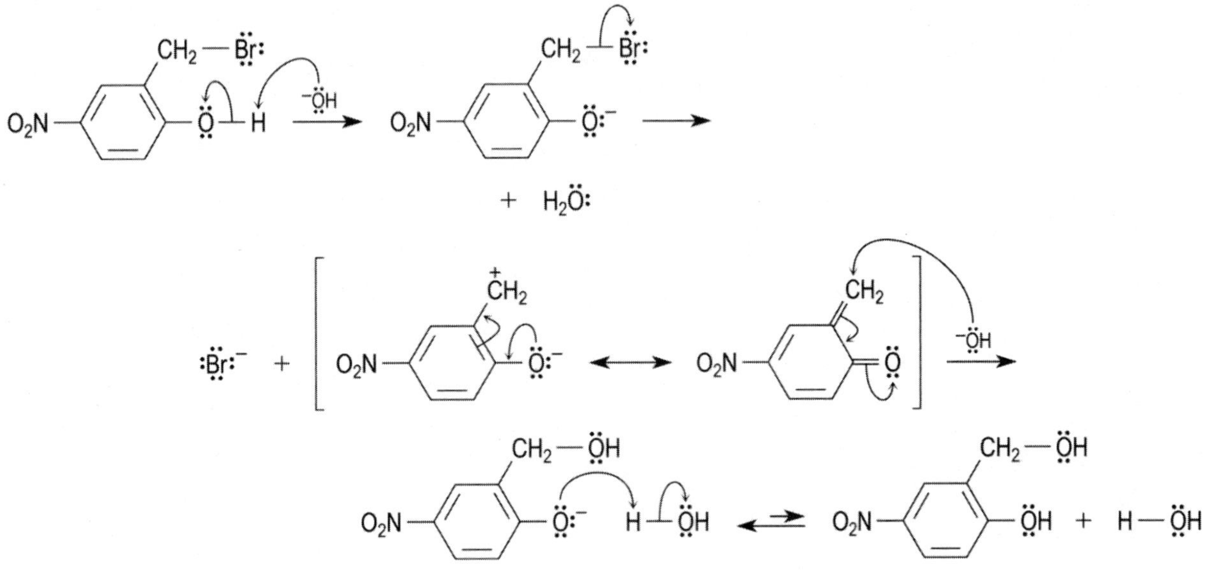

18.69   (a)

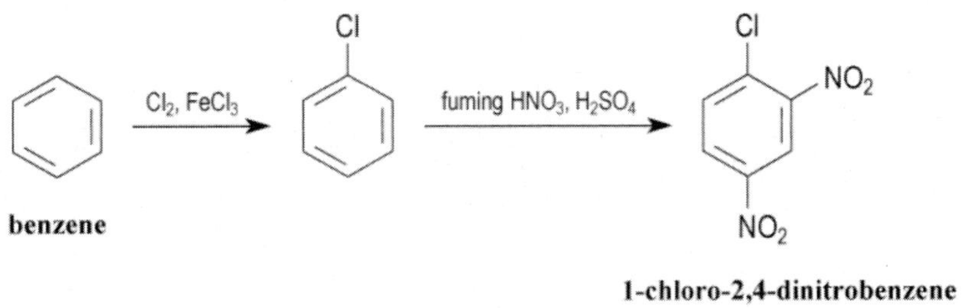

1-chloro-2,4-dinitrobenzene

(b)

1-chloro-3,5-dinitrobenzene

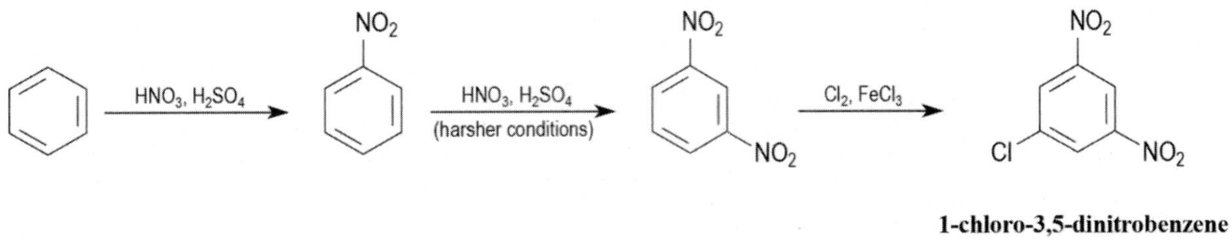

(c)

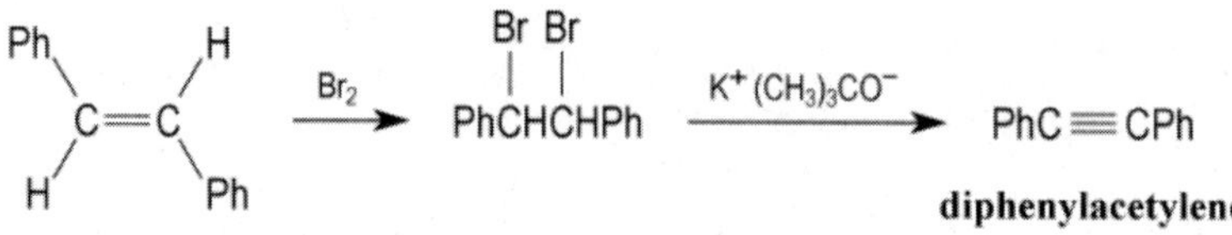

diphenylacetylene

(d)

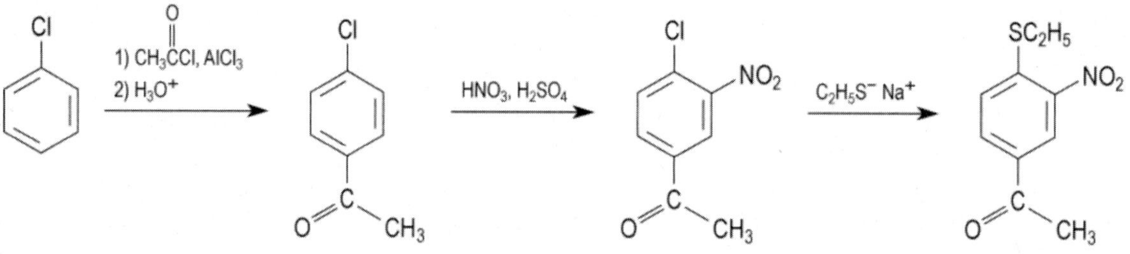

(e)

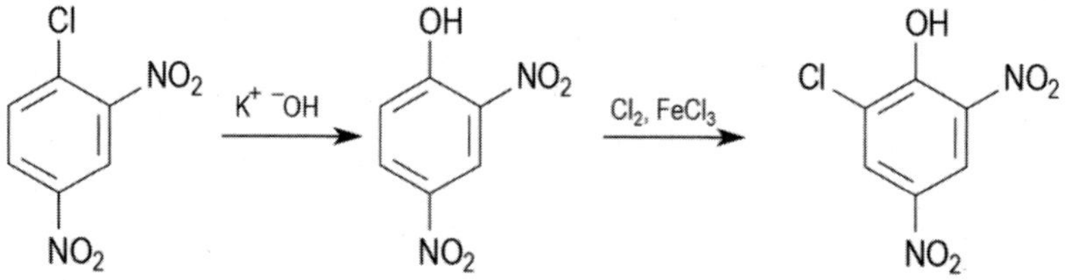

prepared in part (a) **2-chloro-4,6-dinitrophenol**

(f) See Eq. 18.94, text p. 928. Other sources of the *tert*-butyl cation electrophile could also be used, such as 2-methylpropene and acid. *Tert*-butyl chloride and $AlCl_3$ would probably not work well because of the tendency of phenols to form complexes with $AlCl_3$.

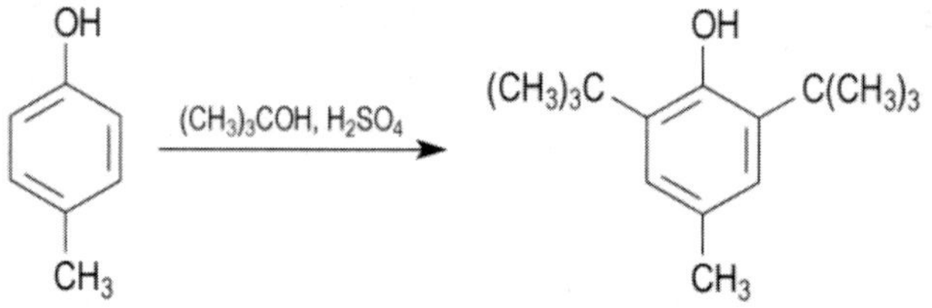

***p*-cresol** **2,6-di-*tert*-butyl-4-methylphenol (BHT)**

(g)

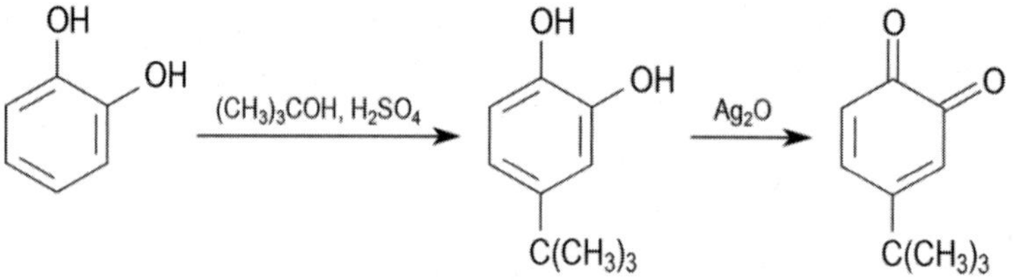

(h)

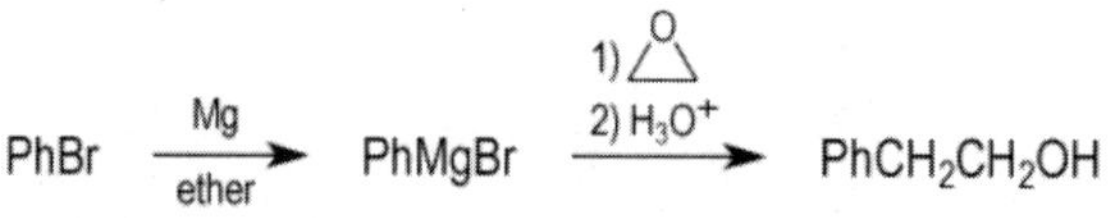

bromobenzene **2-phenyl-1-ethanol**

(i)

(j)   First, prepare bromobenzene from benzene.

Next, prepare 1-bromo-3-nitrobenzene from benzene.

Then carry out a Heck reaction of either aryl halide with ethylene. We'll use 1-bromo-3-nitrobenzene.

Finally, carry out another Heck reaction using *m*-nitrostyrene and bromobenzene (using somewhat harsher conditions).

(k)   First, prepare the Grignard reagent from the aryl bromide, and use it to prepare the corresponding aryl boronic acid. Couple the resulting aryl boronic acid with (E)-1-bromo-2-phenylethene in the presence of a Pd(0) catalyst and aqueous base (Suzuki coupling).

Alternatively, couple the aryl bromide with (E)-1-catecholboranyl-2-phenylethene in the presence of a Pd(0) catalyst, sodium ethoxide, and a non-aqueous solvent (Suzuki coupling)—water would hydrolyze the catecholboranyl reagent.

(l)   First, nitrate bromobenzene to produce 1-bromo-4-nitrobenzene.

Then carry out a Heck reaction with ethylene.

Finally, react *p*-nitrostyrene with *cis*-2-buten-1,4-diol in an alkene metathesis.

(m)   First, prepare the Grignard reagent from the alkyne, and use it to prepare the corresponding stannane.

$$PhC{\equiv}C{-}H + CH_3MgBr \longrightarrow PhC{\equiv}C{-}MgBr + CH_4$$

$$PhC{\equiv}C{-}MgBr + ClSn(CH_3)_3 \longrightarrow PhC{\equiv}C{-}Sn(CH_3)_3 + ClMgBr$$

Next, prepare the triflate of the phenol.

$$F_3C-\overset{O}{\underset{O}{\overset{\|}{\underset{\|}{S}}}}-O-\overset{O}{\underset{O}{\overset{\|}{\underset{\|}{S}}}}-CF_3$$

HO—〈benzene ring〉—OCH$_3$ $\xrightarrow[\text{pyridine}]{\text{triflic anhydride}}$ TfO—〈benzene ring〉—OCH$_3$

4-methoxyphenol

Finally, couple the triflate and the stannane in a Stille reaction.

PhC≡C—Sn(CH$_3$)$_3$ + TfO—〈benzene ring〉—OCH$_3$ $\xrightarrow[\text{LiCl}]{\text{Pd(PPh}_3)_4}$ PhC≡C—〈benzene ring〉—OCH$_3$

(n) The indicated compound may be obtained by a Suzuki coupling. First, prepare a vinylic boronic acid derivative by reacting 1-hexyne with catecholborane (see Eq. 18.51, text p. 905).

CH$_3$(CH$_2$)$_3$C≡CH + H—B〈catechol〉 $\xrightarrow{\text{THF}}$ vinylic boronate product

1-hexyne

catecholborane

CH$_3$CH$_2$CH$_2$CH$_2$

Next, prepare the aryl halide. Brominate resorcinol under mild conditions to produce a monobrominated product. Then carry out a Williamson ether synthesis with an excess of ethyl iodide.

〈resorcinol with two OH〉 $\xrightarrow[\text{CCl}_4]{\text{Br}_2}$ 〈bromoresorcinol with Br and two OH〉 $\xrightarrow[\text{2) C}_2\text{H}_5\text{I}]{\text{1) NaOH, H}_2\text{O}}$ 〈aryl bromide with Br, OC$_2$H$_5$, OC$_2$H$_5$〉

Finally, couple the vinylic boronic acid derivative with the aryl halide in a Suzuki reaction.

vinylic boronate + aryl bromide $\xrightarrow[\text{H}_2\text{O/THF}]{\substack{\text{Pd(PPh}_3)_4 \\ \text{(4.5 mole \%)} \\ \text{KOH}}}$ product

CH$_3$CH$_2$CH$_2$CH$_2$ ... OC$_2$H$_5$ ... OC$_2$H$_5$

Another route to this compound involves alkene metathesis between 1-hexene and 2,4-diethoxystyrene. Show how each of these compounds can be synthesized.

(o) Start with (2-cyclohexenyl)benzene, which is prepared from iodobenzene and cyclohexene by a Heck reaction as shown in Eq. 18.44, text p. 902.

〈(2-cyclohexenyl)benzene〉 $\xrightarrow{\text{m-chloroperoxybenzoic acid (Cl-C}_6\text{H}_4\text{-CO}_3\text{H)}}$ 〈epoxide product〉

(2-cyclohexenyl)benzene

It is reasonable to suppose that epoxidation will occur at the face of the ring opposite to that occupied by the bulky phenyl group. (As noted in the problem, the product is a racemate because the reagents are achiral.)

18.70 (a) Fluoride ion is displaced in a nucleophilic aromatic substitution reaction by ethanethiolate ion.

$$O_2N-\langle\text{benzene}\rangle-SC_2H_5 \ + \ F^-$$

(b) The product is formed by a nucleophilic aromatic substitution reaction in which chloride is displaced by hydroxide ion. The product is ionized by $^-$OH present in the reaction mixture; addition of acid (H_3O^+) forms the neutral phenol.

(c) The NaOH ionizes the phenol. The resulting phenolate ion is methylated by dimethyl sulfate to form the methyl ether, which is then hydrogenated to the corresponding cyclohexane.

cis-1-ethyl-4-methoxycyclohexane

(d) No reaction occurs under the mild conditions.
(e) This is a Suzuki coupling reaction in which pyridine derivatives are used instead of benzene derivatives.

(f)

$+ \ Na^+ \ ^-OTf \ + \ B(OH)_3$

(g)   Bromine is introduced at the ring position that is activated by two hydroxy groups.

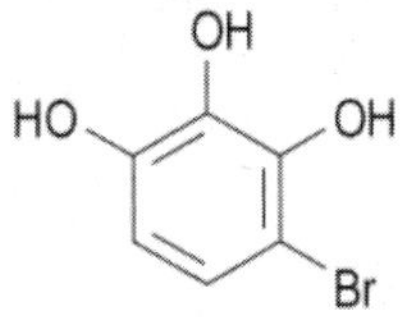

(h)   This is a Stille reaction that results in a coupling between the stannane and the triflate derivative of the naphthol.

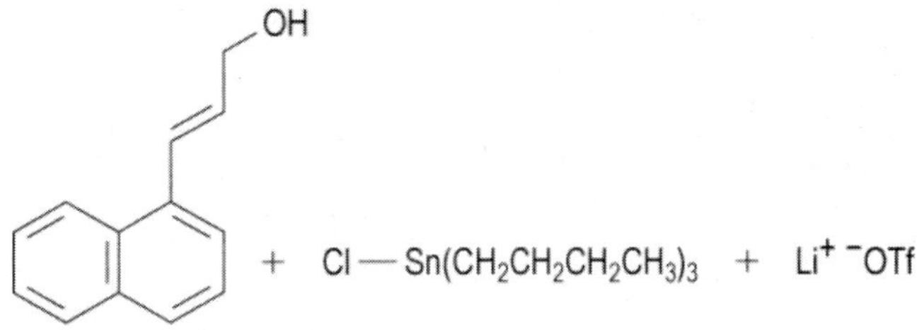

(i)   This is a Heck reaction between a *vinylic* halide and an alkene. Vinylic halides react in many cases like aryl halides in Pd-mediated coupling reactions. The coupling with cyclic alkenes result in coupling at an allylic position, as in Eqs. 18.43 and 18.44 on text p. 846–847.

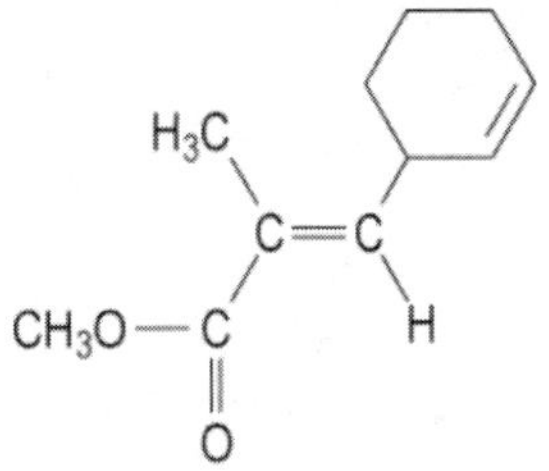

(j)

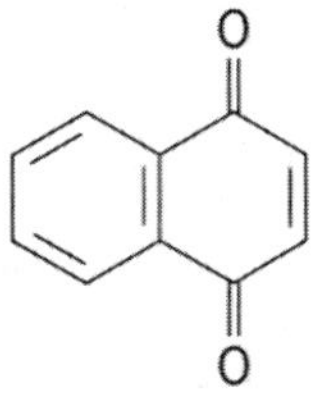

1,4-naphthoquinone

(k)   Oxidation occurs to give the quinone.

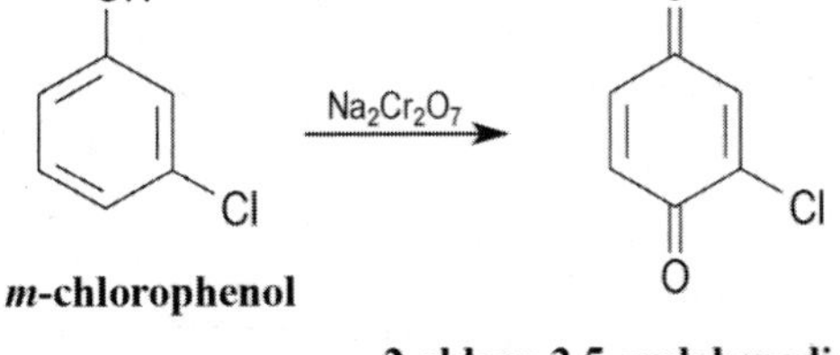

m-chlorophenol

2-chloro-2,5-cyclohexadien-1,4-dione

(l)   The methanesulfonate derivative of 2,4-dinitrophenol undergoes a nucleophilic substitution reaction in which the methanesulfonate group is displaced by methoxide.

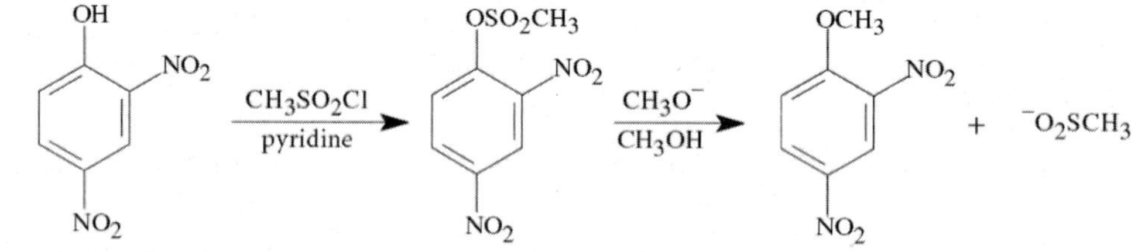

2,4-dinitrophenol   1-methoxy-2,4-dinitrobenzene

(m)   Nucleophilic aromatic substitution occurs at the position that is *ortho* and *para* to the nitro groups, because this involves the more stable anionic intermediate (Meisenheimer complex).

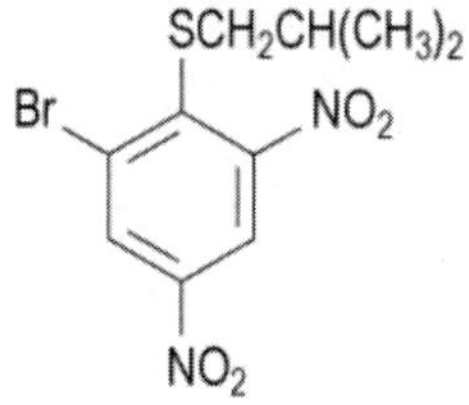

(n)   The methyl ethers are cleaved but the phenyl ethers are not. [See the solution to Problem 18.54(d).]

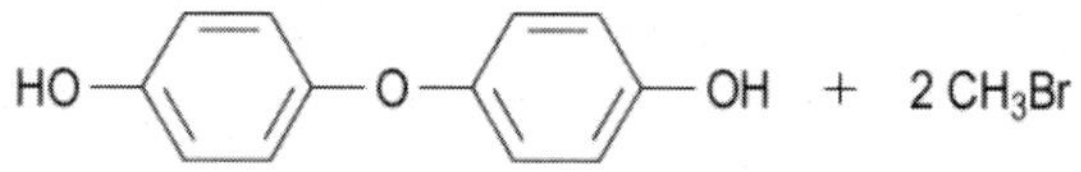

$$HO{-}\bigcirc{-}O{-}\bigcirc{-}OH \ + \ 2\,CH_3Br$$

(o)   Alkene metathesis produces the more stable trans alkene; asymmetric epoxidation (see Sec. 11.11 in the text) produces the epoxide shown (see Eq. 11.81a, text p. 556, for an equivalent reaction).

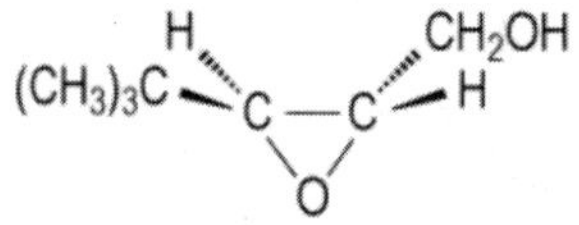

(p)   The intramolecular alkene metathesis forms a 12-membered cyclic alkene. Recall that 8- or larger membered cyclic alkenes can accommodate a trans double bond (see Sec. 7.6C in the text). Thus compounds *X* and *Y* are cis/trans isomers.

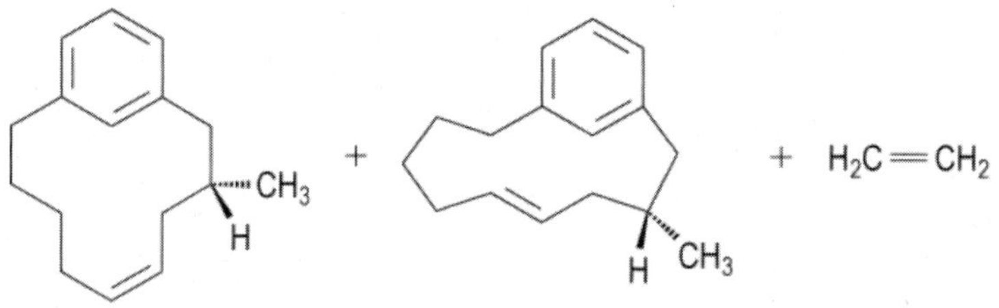

compounds *X* and *Y*

Since the only difference between compound *X* or compound *Y* is the stereochemistry at the alkene double bond, catalytic hydrogenation of either compound *X* or compound *Y* produces compound *Z*.

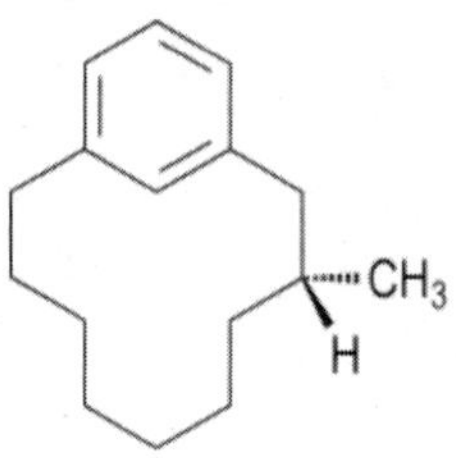

compound *Z*

(q)   The large bulky 9-BBN converts the alkene to an organoborane (see Section 5.4B in the text) in which the boron adds regioselectively to the terminal carbon. Subsequent reaction with the vinylic bromide in the presence of a palladium catalyst and aqueous sodium carbonate couples to alkyl borane and the vinyl bromide. In this reaction, a new stereocenter (*) is formed. Neglecting the stereochemistry at this center for the moment, the result of the reaction is as follows:

Note that the *cis*-1,5-cyclooctanediol is produced when 9-BBN is hydrolyzed; 9-BBN is formed by reacting borane with 1,5-cyclooctadiene.

Now to the stereochemistry of the new stereocenter. There can be no doubt that the 9-BBN reacts with the face of the double bond at the side opposite the angular methyl group. However, there are two possible conformations about the single bond between the ring and the vinylic carbon at which such a reaction is possible; each conformation gives a different stereoisomer.

conformation *A*

conformation *B*

The product from conformation *B* has much more steric interference with the angular methyl group because of the large group attached to boron. Therefore, the product is derived from conformation *A*. It has *R* stereochemistry because introduction of the OH group changes priorities at the stereocenter.

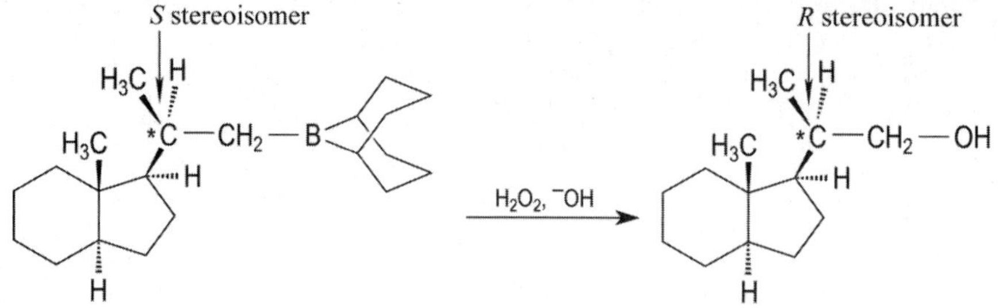

18.71   (a)   The simplest way to visualize the polymer resulting from the ROMP reaction is to break the alkene at the C=C bond and then draw the result with a set of parentheses. Finally, mentally break the section between the fragment ends, and then place them on either side of a C=C bond.

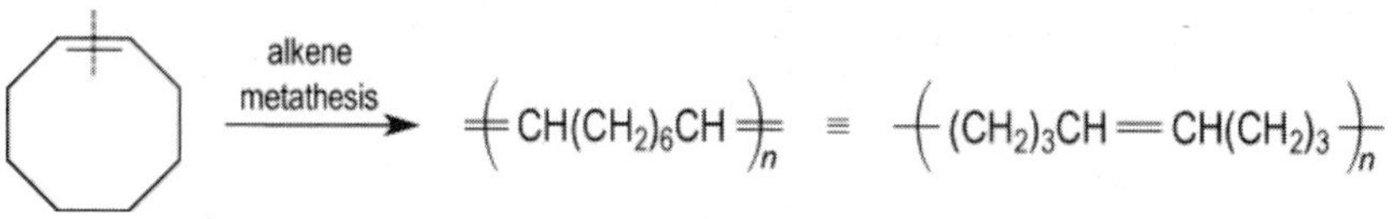

(b)

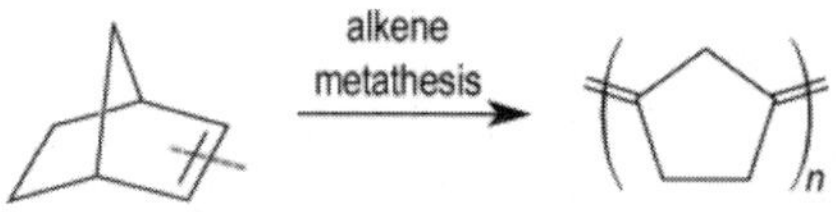

18.72   (a)   In the Diels–Alder reaction of 1,3-cyclopentadiene to yield *endo*-dicyclopentadiene, one of the 1,2-cyclopentadiene molecules as the "diene" and the other acts as the "dienophile."

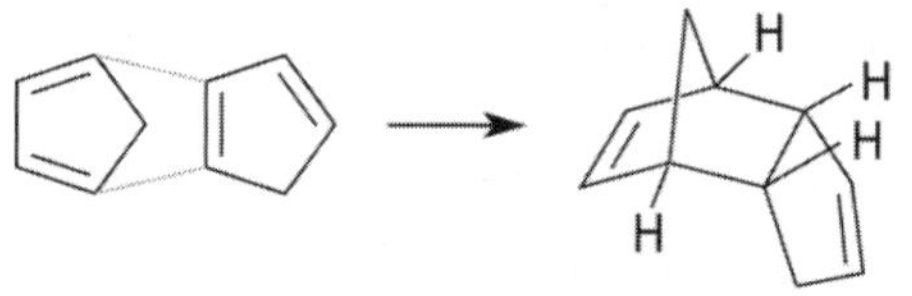

endo-dicyclopentadiene

(b)   The ROMP reaction using *endo*-dicyclopentadiene produces a cyclopentane ring bearing four alkene "ends" that can recombine is many different ways producing a 3-dimensional network. Consider the difference in strength between a thread and a piece of cloth (2-dimensional lattice). A 3-dimensional lattice is magnitudes stronger and more rigid than a 1- or 2-dimensional structure.

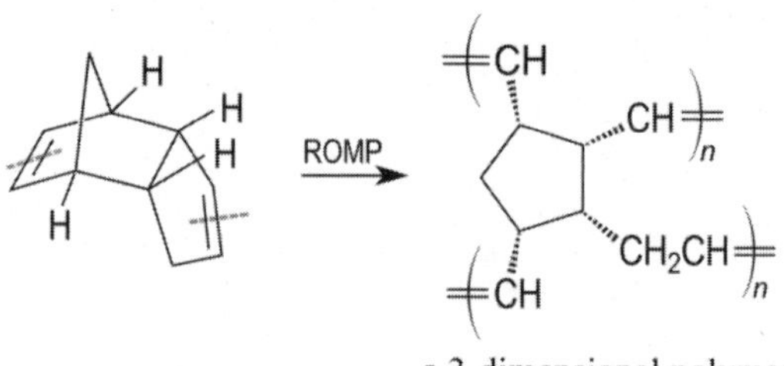

a 3-dimensional polymer

This structure is simplified, because each double bond is connected to a double bond at *any* of the possible positions in another monomer.

18.73   The basic principle needed to understand the results is that elimination is most rapid when it occurs with anti stereochemistry. In the first reaction, anti-elimination leads to the observed product. In the second reaction, formation of the alkyne requires a slower syn-elimination; hence, another process can compete, namely, elimination of a methyl hydrogen and the bromine to form the allene. In the first reaction, elimination of the methyl hydrogen to form the allene is a slower process because allenes are not so stable as alkynes. The transition state for elimination is destabilized by its allene-like character. (The methyl hydrogens are approximately as acidic as the vinylic hydrogens because they are allylic.) It is also possible that the allene is the only product formed in the second reaction and that the alkyne is formed by a base-catalyzed isomerization of the allene:

18.74   Because vinylic halides do not undergo S_N2 or S_N1 reactions, this substitution must occur by an unusual mechanism. The key to this solution is to recognize that the leaving-group effect and the effect of a *para*-nitro substituent are much the same as they are in nucleophilic aromatic substitution. This reaction, in fact, occurs by a mechanism that is much like the mechanism of nucleophilic aromatic substitution, except that it takes place at a vinylic carbon. The *para*-nitro group accelerates the reaction by affording additional resonance stabilization of the anionic intermediate *A*.

anionic intermediate *A*

anion *A*

18.75   The spectrum is consistent with the formation of an anionic intermediate—a stable Meisenheimer complex. (The dotted lines symbolize resonance delocalization of the negative charge.) If you draw out the resonance structures for this ion, you will see that charge is delocalized to the carbons bearing the nitro groups, and it can also be delocalized into the nitro groups. However, charge is *not* delocalized to the other carbons. Hence, the protons on these carbons do not show the smaller chemical shift that would be expected if there were high electron density on these carbons.

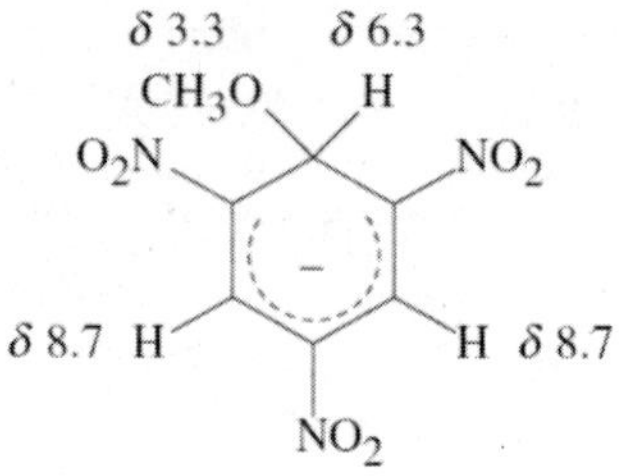

18.76 The reaction that forms compound X is a *hydroalumination*, and it is completely analogous to hydroboration. The structure of X is

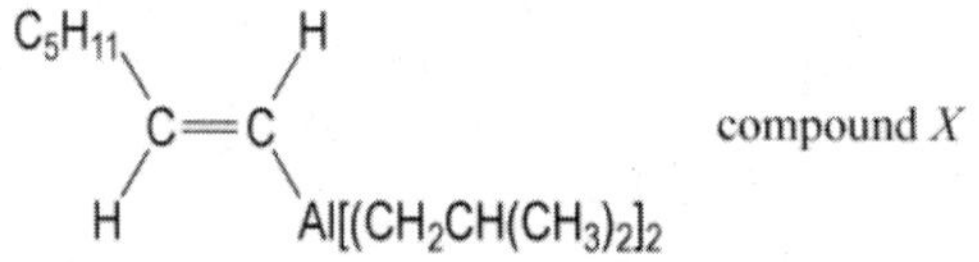

compound X

The second reaction is very much like a Stille coupling. We expect the carbon–aluminum bond to behave much like other carbon–metal bonds, for example, the carbon–tin bond in a stannane. And the vinylic iodide group is much like a triflate group in the sense that it is a good leaving group. The structure of the coupling product Y is

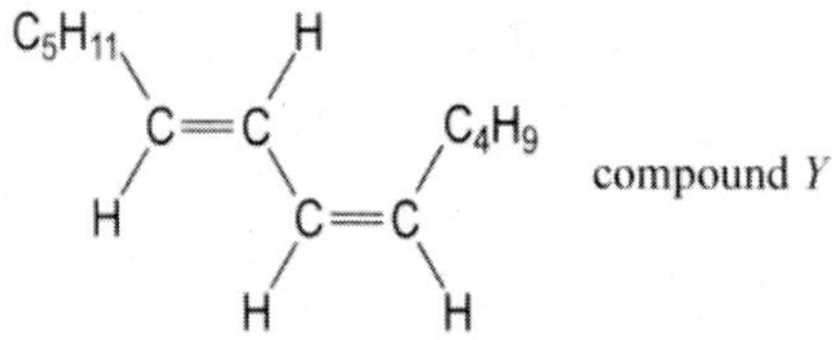

compound Y

The by-product of the reaction is the dialkyliodoalane, $I—Al[CH_2CH(CH_3)_2]_2$.

18.77 If PPh_3 reacts with a chlorine, an additional electron pair can be moved to the Pd.

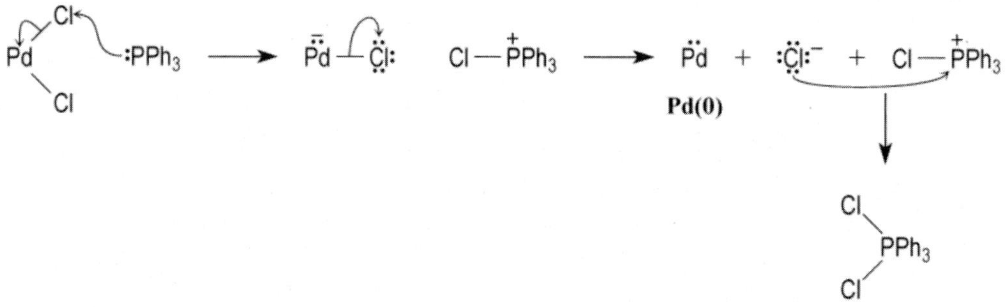

18.78 **(a)** This is a nucleophilic aromatic substitution reaction. Notice that the cyano group ($:N\equiv C—$) can stabilize the intermediate A by resonance.

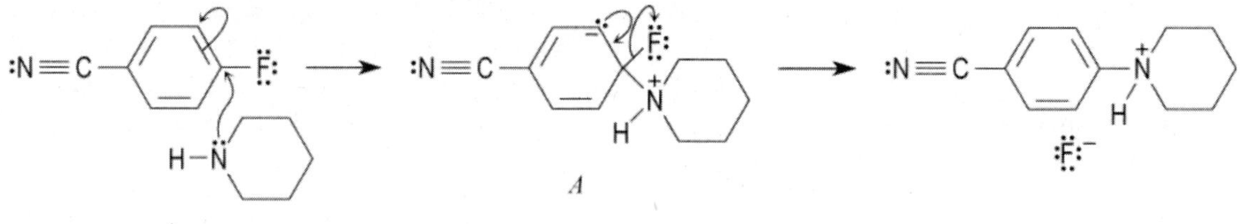

(b) Protonation of 2-methylpropene gives the *tert*-butyl cation, which serves as the electrophile in an electrophilic aromatic substitution reaction.

18.79 The actual catalytic species is probably Pd(PPh₃)₂, as shown by Eq. 18.43a on text p. 902. The likely steps include an oxidative addition, followed by a ligand substitution, and finally a reductive elimination to yield the product.

18.80 (a)

(b) As noted on text pp. 401–3, ¹⁸F-containing compounds are useful in positron emission tomography, or PET.

18.81 In 1-chloro-4-nitrobenzene, the dipole contributions of the carbon–chlorine bond and the nitro group are oriented in opposite directions; consequently, the net dipole moment is smaller than it is in nitrobenzene, in which only a dipole contribution of the nitro group is present.

The analysis of *p*-nitroanisole is similar, except that there is an additional effect: the very powerful electron-donating resonance effect of the methoxy group. Delocalization of an unshared electron pair from the oxygen of the methoxy group into the nitro group creates separation of charge that results in a large dipole moment. A similar resonance effect is present in 1-chloro-4-nitrobenzene, but, because the resonance effect of a chloro group is much weaker than that of a methoxy group (Fig. 16.7, text p. 819), it is not sufficient to offset the opposing contribution of the chlorine–carbon bond dipole.

This resonance structure has
a large separation of charge, and
hence a significant contribution
to the dipole moment.

18.82   Because the glycol starting material is also a chlorohydrin, and because the phenoxide salt is a base, epoxide formation takes place rapidly, accompanied by the formation of NaCl. (See Sec. 11.3B in the text.) It is the epoxide that actually alkylates the phenol more slowly. The hydroxide ion produced in the last step maintains the phenol in its ionized form.

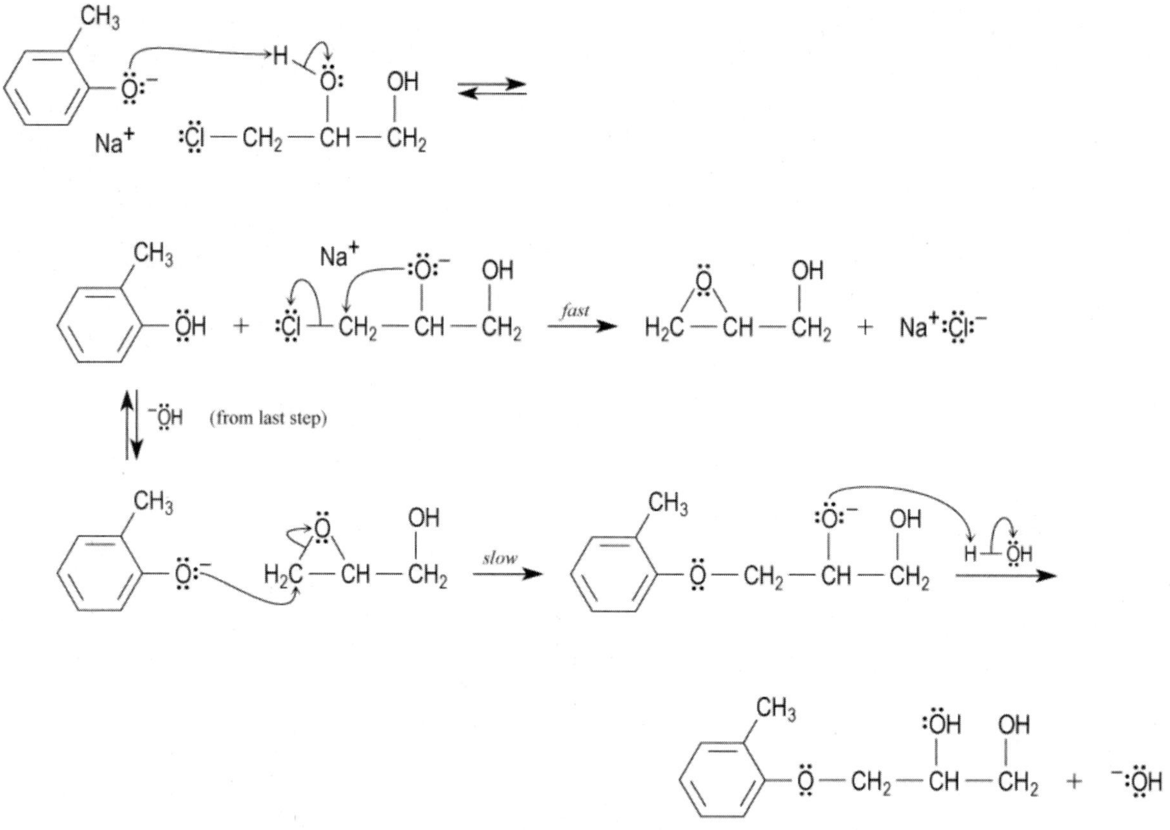

mephenesin

18.83   The only way to accommodate these results is for the labeled carbon and an adjacent carbon to become equivalent at some point. The hint suggests a β-elimination to give an alkyne called *benzyne*.

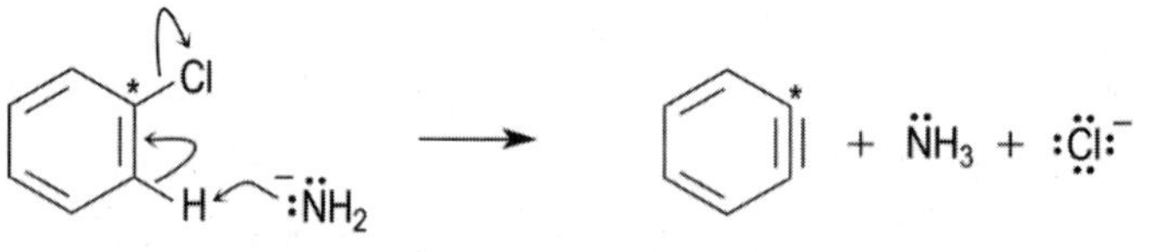

benzyne

Because benzyne is highly strained (why?), it undergoes an unusual reaction: a nucleophilic reaction of the $^-NH_2$ at a carbon of the triple bond to give a carbanion, which is protonated to give the aniline product. Because benzyne is symmetrical except for the label, the reactions of the $^-NH_2$ at the two carbons of the triple bond are equally likely. Consequently, the products are formed in identical amounts.

   If you find an error in this manual, please visit http://people.pharmacy.purdue.edu/~loudonm/teaching/

18.84   Deduce the structure of the "very interesting intermediate" by mentally imagining a "reverse Diels–Alder" reaction of triptycene that yields anthracene and the intermediate, which is benzyne:

benzyne

"reverse
Diels-Alder"

triptycene   **anthracene**

The Grignard reagent has carbanion character, and this "carbanion" is a strong base. Elimination of the weaker base fluoride gives benzyne:

The Grignard reagent has carbanion character, and this "carbanion" is a strong base. Elimination of the weaker base fluoride gives benzyne:

Because alkynes require linear geometry, it is difficult to incorporate then into six-membered rings. Therefore, benzyne is highly *strained* and, although it is a neutral molecule, it is very unstable. (Benzyne is about 205 kJ mol^{-1} (49 kcal mol^{-1}) more unstable than an ordinary alkyne.) Indeed, benzyne has been too reactive to isolate except at temperatures near absolute zero. (See also the solution to Problem 18.83 in this manual.)

18.85   The six-proton singlet in the NMR spectrum indicates that the two methyl groups ortho to the oxygen have been retained; this resonance cannot be due to one ortho methyl and one para methyl, because these would not be chemically equivalent. The two-proton singlet at δ 5.49 indicates two alkene protons, and the two-proton singlet at δ 6.76 is accounted for by the two ring protons, which evidently remain. The total number of protons accounted for by the NMR spectrum are two fewer than are present in the starting material. The conditions are much like those for oxidation of a phenol to a quinone. The structure below for compound *A* fits the data; the singlets are all broad because of very slight splitting over more than three bonds.

δ 1.90 H$_3$C   CH$_3$ δ 1.90

δ 6.76 H   H δ 6.76

δ 5.49 H   H δ 5.49

compound *A*

18.86 The two reactions are intramolecular Friedel–Crafts acylations. The first equivalent of $AlBr_3$, a strong Lewis acid, forms a complex with the acid chloride group to generate the electrophile, an acylium ion, as discussed in Sec. 16.4F of the text, and is also consumed by formation of a complex with the product. As the lower reaction in the problem shows, the acylium ion reacts at a position para to a methoxy group, because a methoxy group is both ortho, para-directing and strongly activating. The question, then, is why reaction occurs at the ring that does *not* have methoxy substituents when more Lewis acid is present. The answer is that the second and third equivalents of Lewis acid form complexes with the methoxy groups:

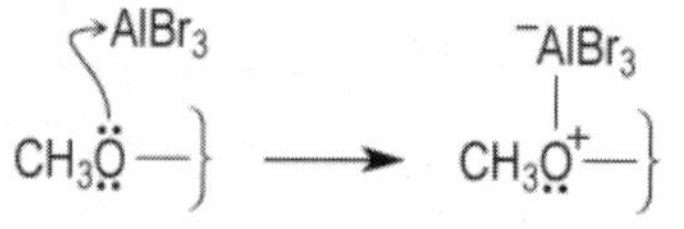

A complexed methoxy group cannot donate an unshared electron pair by resonance, because its unshared pair is already "donated" to the aluminum. Consequently, the methoxy group, through this complexation, is transformed into an electronegative group that is incapable of a resonance effect. An electronegative group that cannot donate electrons by resonance is a *deactivating group*, and a ring containing such a group (or two such groups) undergoes substitution more slowly than a ring without such substituents. In other words, acylation of the ring that does not contain such deactivating substituents is the faster acylation.

18.87 To begin, examine the starting material and the product. Note that left chain (heavy dotted line) in the starting material becomes the "top" of the three 6-membered rings, the middle chain becomes the "bottom" of the three 6-membered rings, and the right chain (heavy solid gray line) becomes the 5-membered ring. Label the four rings A, B, C, and D for reference. Also note that the sp- and sp^2-hybridized carbons (denoted by the diamonds) in the starting material correspond to the sp^2-hybridized carbons in the product (denoted by the diamonds). Simplify the starting material, using a series of R groups, to make drawing the mechanism easier.

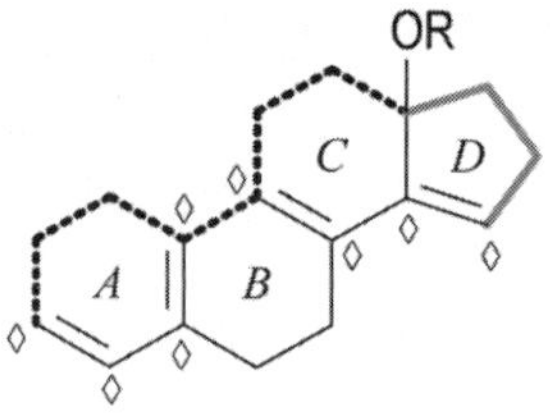

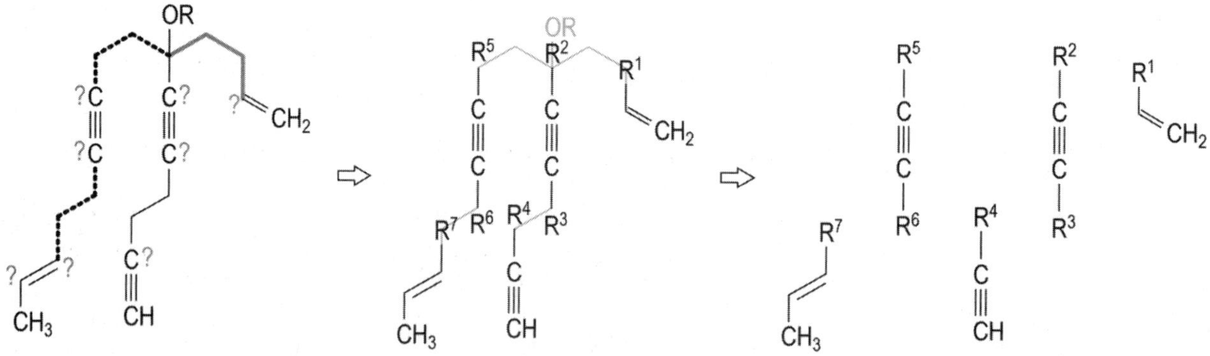

Starting from the right side of the starting material, the first step is loss of the $P(Cy)_3$ ligand by ligand dissociation. Ruthenium accepts a pair of electrons from the alkene in $R^1CH{=}CH_2$. A cycloaddition occurs forming a metallacycle (a ligand insertion). The metallacycle undergoes a cycloreversion to form a new alkene, which dissociates from the complex leaving the catalyst "primed" with the starting material.

Grubbs G1 catalyst

Next, ruthenium accepts a pair of electrons from the alkyne in $R^2C{\equiv}CR^3$. A cycloaddition occurs forming a metallacycle (a ligand insertion). The metallacycle undergoes a cycloreversion to form a new alkene, which dissociates from the complex leaving the catalyst "primed" with the next alkene group. Note that R^1 and R^2 in the newly formed alkene are actually the cyclopentene (*D* ring) and the top portion of the product.

Next, ruthenium accepts a pair of electrons from the alkyne in $R^6C{\equiv}C{-}$. A cycloaddition occurs forming a metallacycle (a ligand insertion). The metallacycle undergoes a cycloreversion to form a cyclohexene (the *C* ring), which dissociates from the complex leaving the catalyst "primed" with the next alkene group.

Next, ruthenium accepts a pair of electrons from the alkyne in HC≡C— of the terminal end of the middle chain of the starting material. A cycloaddition occurs forming a metallacycle (a ligand insertion). The metallacycle undergoes a cycloreversion to form another cyclohexene (the *B* ring), which dissociates from the complex leaving the catalyst "primed" with the next alkene group.

Lastly, ruthenium accepts a pair of electrons from the alkene in R^6 of the left chain of the starting material. A cycloaddition occurs forming a metallacycle (a ligand insertion). The metallacycle undergoes a cycloreversion to form another cyclohexene (the *A* ring), which dissociates from the complex leaving the catalyst "primed" with the next alkene group.

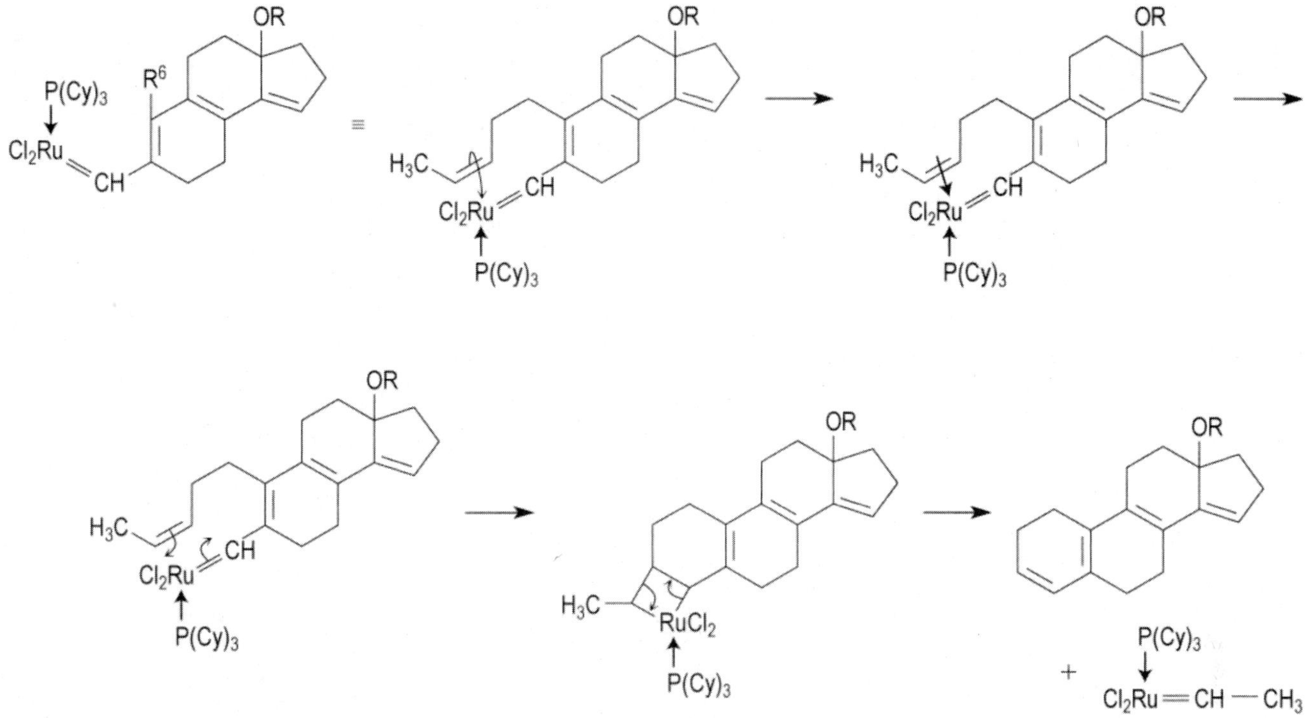

Repetition of this cycle gives the by-product alkene propene in the first step.

18.88   (a)   We consider only the cross-metathesis products. The pairwise mechanism predicts a 1:1 reaction between the two alkenes and thus produces only one product.

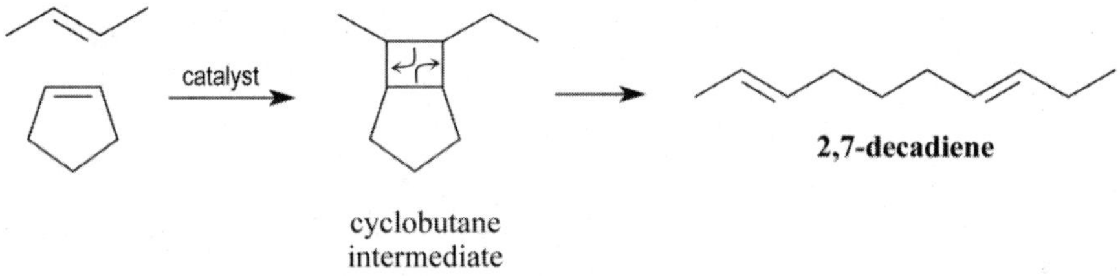

cyclobutane
intermediate

2,7-decadiene

In contrast, the metathesis of cyclopentane with (E)-2-pentene produces three products in a ratio of 1:2:1. Cyclopentene can produce only one "fragment" (fragment A), a "double-headed" fragment, while (E)-2-pentene produces two (fragments B and C). Thus, fragment A can combine with two fragment Bs, two fragment Cs, or one of each (in two different combinations): $B + A + B$ (2,7-nonadiene), $C + A + C$ (3,8-undecadiene), and $B + A + C$ and $C + A + B$ (2,7-decadiene). Note that $B + A + C$ and $C + A + B$ produce the same product and thus twice as much of that product is formed.

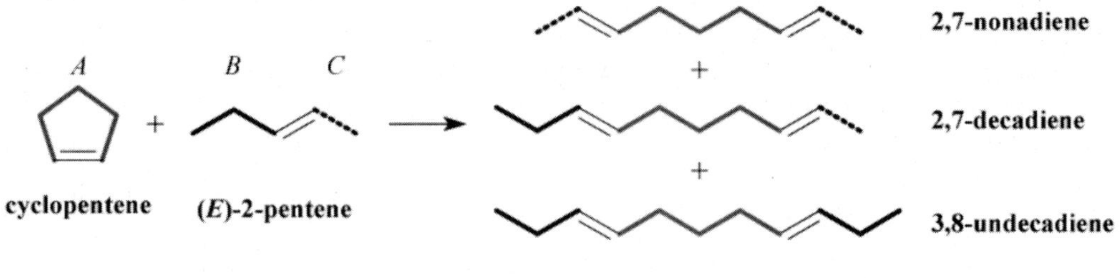

(b)   The pairwise mechanism can only give ethylene and ethylene-d_4, because at no point in the mechanism does one of the CH_2 (or CD_2) groups become detached from its parent molecule:

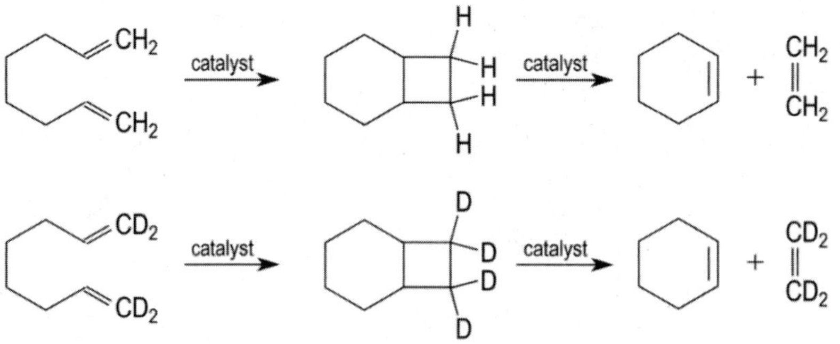

To see the results of the metallacycle mechanism, note that $M{=}CH_2$ and $M{=}CD_2$ (where M = the ruthenium and its ligands) are formed in the first catalytic cycle. (We leave it to you to fill in the details.)

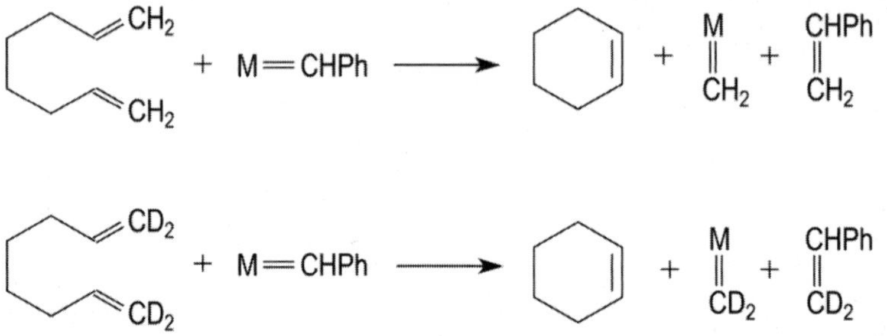

Now, ethylene-d_2 can form in *two ways:* by the reaction of $M{=}CH_2$ with the d_4-diene starting material, or by the reaction of $M{=}CD_2$ with the undeuterated diene:

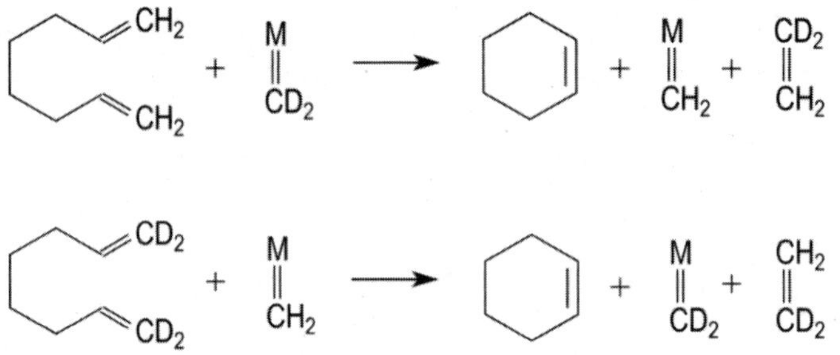

However, ethylene-d_4 can form in only one way—the reaction of $M{=}CD_2$ with the d_4-diene starting material; and ethylene-d_0 can form in only one way—the reaction of $M{=}CH_2$ with the undeuterated diene starting material. Consequently, the formation of ethylene-d_2 enjoys a two-fold statistical advantage, and the ratio of the three ethylenes is therefore $H_2C{=}CH_2 : H_2C{=}CD_2 : D_2C{=}CD_2 = 1 : 2 : 1$. The result of this very elegant experiment was that the three ethylenes were formed in this statistical ratio.

18.89 (a) The vinylic boronic acid reacts with bromine via a bromonium ion to give the dibromo product. Rotation around the C—C bond positions the boronic acid anti to the bromine on the adjacent carbon. Hydroxide ion (a Brønsted base) reacts with the boron (a Brønsted acid) to form borate, and excellent leaving group. Anti-elimination of the borate and bromide affords the vinyl bromide; the overall reaction thus occurs with *inversion* of configuration.

(b)   Iodine adds to the vinylic boronic acid to form an iodonium ion, to which hydroxide ion adds to form an iodohydrin (see Sec. 5.2B of the text). Rotation around the C—C bond positions the boronic acid anti to the hydroxy group on the adjacent carbon. The hydroxy group (a Brønsted base) reacts with the boron (a Brønsted acid) to form a cyclic borate that breaks down by a syn-elimination to afford the vinylic iodide with overall *retention* of configuration.

18.90   First, recognize that the bombykol carbon skeleton is comprised of only the 10-undecyne-1-ol and 1-pentyne carbon skeletons, thus a "coupling" of the two starting materials would give the desired product. Suzuki coupling involves a vinyl halide and a vinyl boronic acid. But which starting material should be converted to the boronic acid and which should be converted to the vinyl halide? Recall that reaction of a borohydride, such as catecholborane (see Eq. 18.51, text p. 905), with an alkyne produces a trans alkene boronic acid product via syn-addition of the borohydride. Inspecting the stereochemistry of bombykol, the hydroxynonyl group [—(CH$_2$)$_9$OH] is trans to the rest of the molecule while the propyl group is cis. Thus, reaction of the boronic acid formed by the addition of catecholborane to 1-pentyne followed by bromine addition in dichloromethane and the resulting product treated with aqueous sodium hydroxide [see Problem 18.89(a), text p. 943] produces a *cis*-vinylic bromide. Coupling the *trans*-boronic acid with the *cis*-vinylic bromide with a palladium catalyst in the presence of sodium ethoxide will produce bombykol.

18.91 The hydride of potassium tri(isopropoxy)borohydride forms a Lewis acid–base association product with the boron. This positions the hydride for backside substitution at the vinylic carbon, which occurs with inversion of configuration.

This mechanism would appear to violate the principle developed on text p. 881—that backside substitution does not occur at a vinylic carbon. However such a mechanism is reasonable in this case, because, first, the hydride is small and suffers little steric repulsion with the other hydrogen on the same side of the double bond; and second, the reaction should be accelerated because it is intramolecular (Sec. 11.8). This acceleration counterbalances the normally slow substitution at a vinylic carbon.

18.92 Whether the nucleophile occurs in biological systems or in the laboratory, nucleophilic substitution does not occur on aryl derivatives (see Sec. 18.1 of the text) unless the ring is activated by electron-withdrawing groups (such as carbonyl groups or nitro groups) that stabilize the Meisenheimer complex in nucleophilic aromatic substitution. (See Sec. 18.4 of the text.) Furthermore, both fluoride and cyanide are fairly strong bases (see Table 3.1, text p. 102) and are therefore very poor leaving groups.

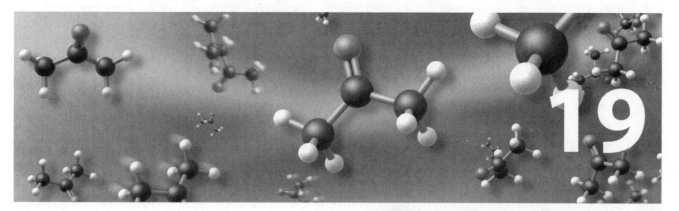

19

The Chemistry of Aldehydes and Ketones
Carbonyl-Addition Reactions

STUDY GUIDE LINKS

 ## 19.1 Lewis Acid Catalysis

Equation 16.15c on text p. 806 focused on Lewis acid interactions with leaving groups. What you should notice about both Grignard reagents and lithium aluminum hydride is the important role of Lewis acid catalysis in promoting reactions of these reagents with carbonyl compounds (as well as with epoxides; Eq. 11.41, text p. 530). Each reagent has its own "built-in" Lewis acid catalyst. As explained in the text, the Lewis acid catalyst in $LiAlH_4$ reductions is the lithium ion, Li^+; and the Lewis acid catalyst in Grignard reactions is the magnesium of the Grignard reagent.

Why not use proton (Brønsted) acids as catalysts in these reactions? Don't forget that both $LiAlH_4$ and Grignard reagents react instantaneously with protons of even weak acids such as water and alcohols. Consequently, the use of proton acids is not an option, because such acids destroy the reagents. Only *after* the addition of these reagents takes place can a proton source be added to the reaction mixture to replace the Lewis acid (Li^+ or ^+MgBr) with a hydrogen.

What you should notice here is that the Lewis acid serves the role of a "substitute proton"—or "fat proton," as one of the author's colleagues likes to call it—and fulfills *exactly* the same catalytic role that a proton would fulfill if a proton *could* be used.

 ## 19.2 Reactions That Form Carbon–Carbon Bonds

It's important to pay special attention to the reactions that can be used to form carbon–carbon bonds, for these are the reactions that are used to build up carbon skeletons. A complete list of these reactions, in the order in which they occur in the text, can be found in Appendix VI of the text, p. A-13. The three reactions for forming carbon–carbon bonds you either have encountered, or will encounter, in this chapter are cyanohydrin formation (Sec. 19.7), reactions of Grignard and related reagents with aldehydes and ketones (Sec. 19.9), and the Wittig alkene synthesis (Sec. 19.13). How many others can you list? Can you give examples of each? Do you understand the limitations of each, if any, and the reasons for these limitations?

19.3   Alcohol Synthesis

Study Guide Link 19.2 suggested making special note of reactions that form carbon–carbon bonds. Another approach to reviewing reactions is to classify them by the types of products they give. For example, the reactions of Grignard reagents with aldehydes and ketones, as you've seen, give alcohols as products. The other reaction type you've studied in this chapter that can be used to prepare alcohols is the hydride reduction of aldehydes and ketones (Sec. 19.8). Both reactions are methods for alcohol synthesis of *major* importance. What's more, a great variety of other compounds can be prepared from alcohols. Thus, these methods are important for the preparation of other types of organic compounds as well.

A complete list of alcohol syntheses, as well as the methods used to prepare every other major functional group, is found in Appendix V of the text on p. A-10. The reactions listed are given in the order that they are presented in the text. How many of these can you name without looking at the list? What are the limitations of each? Do you understand the reasons for the limitations?

19.4   Hemiacetal Protonation

You may ask why, in Eq. 19.51b, text p. 980, the hemiacetal is protonated on the —OH oxygen rather than on the —OCH$_3$ oxygen. This is a reasonable question because ethers and alcohols have similar basicities. The answer is that protonation on the —OCH$_3$ oxygen *does* occur, as does loss of methanol from the resulting protonated species. However, these steps are merely the reverse of hemiacetal formation and lead back to aldehyde or ketone and the starting alcohol. Under conditions of excess alcohol and removal of water, the equilibrium shifts to favor the alcohol.

When writing mechanisms, we usually do not write out reasonable steps that are not important in the formation of the product of interest.

19.5   Mechanism of Carbinolamine Formation

You are asked to write the mechanism for acid-catalyzed carbinolamine formation because it is another example of *carbonyl addition*. It is important to understand this reaction mechanistically because you'll see it repeatedly in organic chemistry and biochemistry. If you *really* want to make progress, take these suggestions seriously.

Are you having trouble getting started? In any carbonyl addition, *identify the nucleophile.* The nucleophilic atom is the nitrogen of the amine. *The nucleophile reacts at the carbonyl carbon* because electrons can flow onto the electronegative oxygen. (See section 19.7A in the text.)

$$
\begin{array}{ccc}
:O: & & :\ddot{O}:^{-} \\
\parallel & & | \\
R{-}\overset{}{C}{-}R & \longleftrightarrow & R{-}\overset{}{C}{-}R \\
| & & | \\
H_2\ddot{N}{-}R & & \underset{+}{H_2N{-}R}
\end{array}
$$

(SG19.1)

Perhaps you added a proton to the carbonyl oxygen first, and *then* let the amine react as a nucleophile. *This is a reasonable step for a beginning student to take,* because it follows the mechanism for acid-catalyzed hydration.

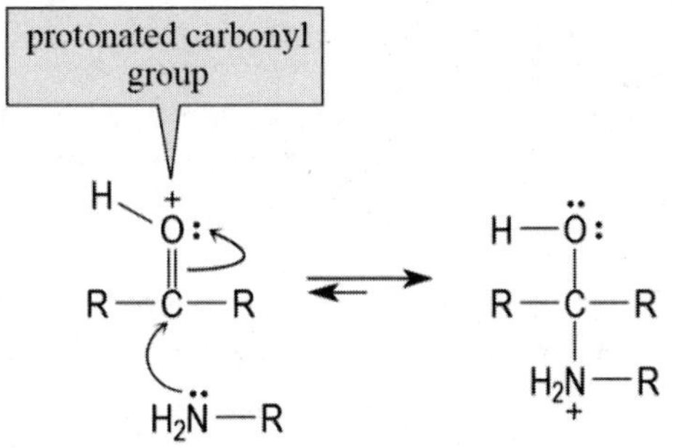

protonated carbonyl group

$$
\begin{array}{ccc}
\overset{H}{\underset{|}{\overset{+}{O}}}\!:\curvearrowright & & H-\ddot{O}: \\
R-\overset{|}{\underset{|}{C}}-R & \rightleftharpoons & R-\overset{|}{\underset{|}{C}}-R \\
\overset{|}{\underset{H_2\ddot{N}-R}{}} & & \overset{|}{\underset{\overset{+}{H_2N}-R}{}}
\end{array}
$$

(SG19.2)

Which is correct? For purposes of *your understanding,* either mechanism represents a reasonable first step. However, let's think about this issue in a little more detail. Remember that carbonyl oxygens are *very weak bases* and amine nitrogens are *fairly strong bases*—about like ammonia. If the acid is strong enough to provide a significant concentration of the protonated carbonyl compound, then surely under the same conditions the amine nitrogen would be completely protonated to an ammonium ion (see Eq. 19.61 on text p. 985) and would no longer be nucleophilic. Hence, the first mechanism—a nucleophilic reaction of the neutral amine with the neutral carbonyl compound (Eq. SG19.1)—is more likely to be correct.

Once the reaction of the carbonyl compound with the amine has occurred, a proton is transferred *from* H_3O^+ *to* the carbonyl oxygen and *from* the nitrogen to H_2O. Use *two separate steps* for these transfers. It doesn't matter which you write first. Notice that the oxygen, because of its negative charge, is an alkoxide, which is a rather strong base, and is efficiently protonated by rather low concentrations of H_3O^+.

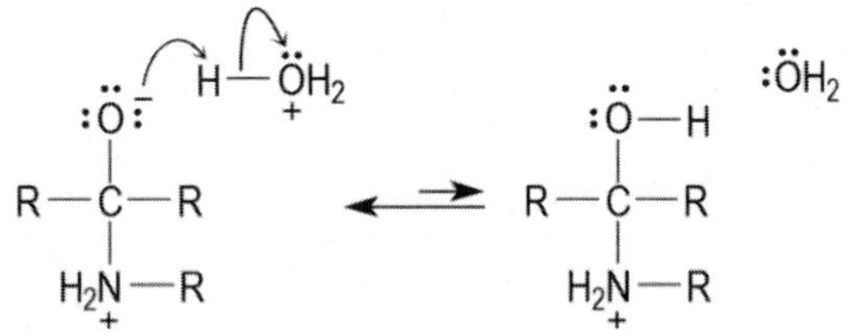

$$
\begin{array}{ccc}
:\!\overset{\frown}{\ddot{O}}:\!{}^{-}\ \overset{\frown}{H}-\overset{+}{\ddot{O}}H_2 & & :\ddot{O}-H \qquad :\ddot{O}H_2 \\
R-\overset{|}{\underset{|}{C}}-R & \rightleftharpoons & R-\overset{|}{\underset{|}{C}}-R \\
\overset{|}{\underset{\overset{+}{H_2N}-R}{}} & & \overset{|}{\underset{\overset{+}{H_2N}-R}{}}
\end{array}
$$

(SG19.3)

The final step is deprotonation of the nitrogen. Because the nitrogen of the carbinolamine is a base, this final step is an equilibrium. If the acid concentration is high enough, a significant amount of the protonated carbinolamine could be present.

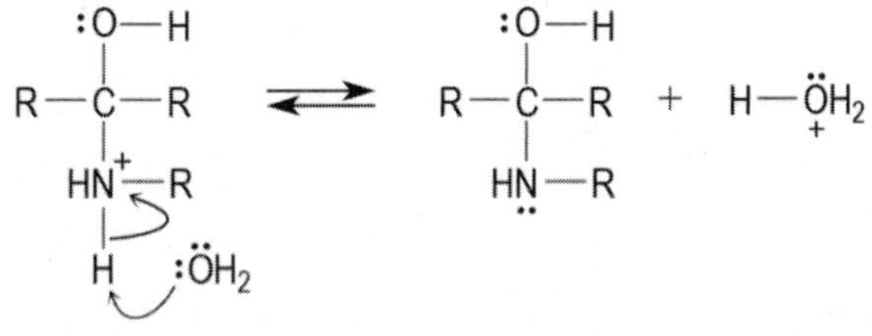

$$
\begin{array}{ccc}
:\ddot{O}-H & & :\ddot{O}-H \\
R-\overset{|}{\underset{|}{C}}-R & \rightleftharpoons & R-\overset{|}{\underset{|}{C}}-R \ +\ H-\overset{+}{\ddot{O}}H_2 \\
\overset{|}{\underset{\overset{+}{HN}-R}{}} & & \overset{|}{\underset{\ddot{H}N-R}{}} \\
\overset{|}{\underset{H\ \ :\ddot{O}H_2}{}} &
\end{array}
$$

(SG19.4)

Notice that when H_3O^+ is the acid used in the mechanism, then its conjugate base H_2O must be used as the base—don't make the mistake of using a strong base like HO^- in this step.

As illustrated here, the mechanism, including the proton-transfer steps, is written *one step at a time.* Students are often tempted to try to show everything in one step. This is not correct because simultaneous collisions of more than two molecules are highly improbable. More important is that consolidating several steps into one can lead to confusion. Remember to write mechanisms *one step at a time.*

Also remember again that all mechanisms involve elementary steps that can only be Lewis acid–base associations, Lewis acid-base dissociations, or electron–pair displacements, of which Brønsted acid–base reactions are a special case. (See Sec. 3.4A of the text.)

19.6   Dehydration of Carbinolamines

You should also try to write the mechanism for dehydration of a carbinolamine in acid, using the dehydration of alcohols as your guide. (See Sec. 10.2 of the text.) When you do this, notice that the carbocation intermediate is resonance-stabilized. This is one reason why the dehydration is so rapid.

Dehydration of carbinolamines, *unlike* dehydration of ordinary alcohols, can also occur in base. In this mechanism, an anion intermediate may be involved:

Notice that hydroxide ion, ⁻OH, acts as a leaving group in the second step of the mechanism. Hydroxide cannot act as a leaving group in the dehydrations of ordinary alcohols because the carbon–oxygen bond is strong and because hydroxide ion is too basic. In this case, however, the formation of the strong C=N double bond (which is considerably stronger than a C=C double bond) provides an additional driving force for the reaction. In addition, the extra electron pair on the nitrogen anion is very basic (its conjugate-acid pK_a is about 25) and thus provides a strong electronic "push" to expel the leaving group.

 19.7   Mechanism of the Wolff–Kishner Reaction

The text indicates that the mechanism of the Wolff–Kishner reaction is a series of acid–base reactions. Here they are. Can you supply the curved-arrow notation?

Research has suggested that the neutral diazene intermediate may not actually exist—the proton transfers that precede and follow it are concerted (occur together), possibly involving multiple solvent molecules. However, that's a detail.

Notice that the last intermediate is a carbanion. Carbanions, with conjugate-acid pK_a values in the 55–60 range, are generally too unstable to exist. If this one does really exist, it is probably because a very stable molecule (dinitrogen) is produced along with it. That is, the stability of dinitrogen offsets the instability of the carbanion.

FURTHER EXPLORATIONS

 ## 19.1   IR Absorptions of Cyclic Ketones

The increase of the infrared carbonyl absorption frequency with decreasing ring size is a very useful trend that can be used to diagnose the presence of small rings. The main reason for this trend is a coupling phenomenon. In the presentation of IR spectroscopy in Chapter 12, IR absorption is treated as a phenomenon associated with isolated bonds. Even though this is a highly useful approximation, the effects of nearby bonds in some cases cannot be ignored, and the carbonyl stretching frequencies of the cyclic ketones is one of those cases.

Imagine two extreme situations. In the first, the carbonyl group forms an angle of 180° with adjacent C—C bonds (as in ketene, $H_2C\!=\!C\!=\!O$). In such a situation, because the vibrations of both bonds occur along the same line, expansion of the carbon–oxygen bond that occurs during a vibration also compresses the carbon-carbon bond; and compression of the carbon-oxygen bond also expands the carbon–carbon bond. The two stretching vibrations are *coupled:*

$$H_2C \!=\!=\!= C \!=\!=\!= O \quad \xleftrightarrow{\substack{\text{expansion of C=O bond;}\\ \text{compression of C=C bond}}} \quad H_2C\!=\!C\!=\!=\!=O$$

$$\Big\updownarrow\substack{\text{compression of C=O bond;}\\ \text{expansion of C=C bond}}$$

$$H_2C \!=\!=\!=\!= C\!=\!O$$

The additional energy required to cause both bonds to vibrate at the same time is reflected in a higher vibrational frequency for both bonds.

Now consider a hypothetical situation in which the $C\!=\!O$ bond is perpendicular to adjacent bonds. In such a situation, the $C\!=\!O$ bond can stretch and compress without stretching the adjacent C—C bond:

$$H_2C-\overset{\overset{\displaystyle O}{\|}}{C}-CH_2 \quad \xleftrightarrow{\text{C=O bond stretch}} \quad H_2C\diagdown\underset{\displaystyle C}{}\diagup CH_2 \;\; \overset{\displaystyle O}{\|}$$

(Remember that bending of the C—C bond requires much less energy than stretching.) At this angle, the stretching vibrations of the two bonds are completely uncoupled. Hence, it requires much less energy to stretch the $C\!=\!O$ bond in this situation than when the two bonds are co-linear.

An intermediate situation occurs at bond angles between 180° and 90°: the stretching vibrations are partially coupled. The more closely the $C—C\!=\!O$ angle approaches 180°, the greater the coupling and the higher the vibration frequency. Notice that the bond angle in question is the one between the $C\!=\!O$ bond and the ring C—C bond, not the internal angle within the ring. This angle varies from 120° in cyclohexanone to greater values for the ketones with smaller rings. As the data in Eq. 19.4 on text p. 954 of the text show, the carbonyl stretching frequencies increase toward smaller ring sizes.

Another factor that contributes to carbonyl stretching frequencies is the relative strengths of the $C\!=\!O$ bonds. The $C\!=\!O$ bond in ketene involves an *sp*-hybridized carbon atom, whereas the $C\!=\!O$ bond in cyclohexanone involves an sp^2-hybridized carbon. The amount of *s* character in the $C\!=\!O$ σ bond is therefore greater in ketene, and the amount of *s* character increases from cyclohexanone through the smaller rings. As with C—H bonds (Eq. 14.27, text p. 700), the strengths of C=O bonds, and hence, their absorption frequencies, increase with increasing *s* character.

The coupling effect discussed above is believed to dominate the trend toward higher frequencies, with the bond–strength effect making a secondary contribution.

A similar trend can be seen with carbon–carbon double bond stretching frequencies:

C=C stretching frequencies: 1650 cm^{-1} 1672 cm^{-1} 1781 cm^{-1}

19.2   Use of Imines for Characterization of Aldehydes and Ketones

As noted in the text, another use for imines, which was more important before the advent of spectroscopy than it is now, is in the characterization of aldehydes and ketones. When a new compound was synthesized, it was typically characterized by conversion into two or more crystalline compounds called *derivatives*. These derivatives served as the basis for subsequent identification of the new compound when it was isolated from another source or from a different reaction. It was important to prepare derivatives because they eliminated the ambiguity that could arise if two compounds have very similar melting points or boiling points. It is relatively improbable that two compounds with the same melting or boiling points will give two crystalline derivatives with the same melting points.

Certain imines were frequently used as solid derivatives for characterization of aldehydes and ketones. These imines, and the amines from which they are derived, are listed in Table 19.3 on text p. 985. For example, the 2,4-DNP derivative of acetone is prepared by formation of an imine with 2,4-dintrophenylhydrazine:

$$H_3C \overset{O}{\underset{}{C}} CH_3 \ + \ H_2\ddot{N}-\ddot{N}H-\underset{NO_2}{\overset{NO_2}{\bigcirc}}-NO_2 \ \xrightarrow[\text{EtOH}]{\text{dilute } H_2SO_4} \ H_3C\overset{\ddot{N}-\ddot{N}H-}{\underset{}{C}}CH_3 \ + \ H_2O$$

acetone   **2,4-dinitrophenylhydrazine (2,4-DNP)**   a 2,4-dinitrophenylhydrazone (2,4-DNP derivative of acetone)   (precipitates)

To illustrate how such derivatives might have been used in structure verification, suppose that a chemist had isolated a liquid that could be either 6-methyl-2-cyclohexenone or 2-methyl-2-cyclohexenone. The boiling points of these compounds are too similar for an unambiguous identification. Yet the melting point of either a 2,4-DNP derivative or a semicarbazone (see Table 19.3) would quickly establish which compound had been isolated.

	68–70 °C (16 mm)	65–70 °C (16 mm)
boiling point	68–70 °C (16 mm)	65–70 °C (16 mm)
semicarbazone, mp	177–178 °C	207–208 °C
2,4-DNP derivative, mp	162–164 °C	207–208 °C

Today, the identity of the compound could be readily established by modern spectroscopic techniques (explain how). But it is still important to be familiar with the imine derivatives in Table 19.3 because references to the use of such derivatives are commonplace in the older chemical literature.

REACTION REVIEW

I. CARBONYL-ADDITION REACTIONS

A. HYDRATION OF ALDEHYDES AND KETONES

1. Aldehydes and ketones react reversibly with water; this reaction is called hydration.

a carbonyl hydrate

Acid-catalyzed hydration of aldehydes and ketones is reminiscent of acid-catalyzed hydration of alkenes, although hydration of aldehydes and ketones is much faster.
a. The first step is protonation of the carbonyl oxygen; the protonated carbonyl compound is a much stronger Lewis acid (electron acceptor) than an unprotonated carbonyl compound.
b. The electron-deficient carbon reacts with H_2O, a nucleophile.
c. Loss of a proton to the solvent (water) completes the reaction.

a carbonyl hydrate

2. Hydration of aldehydes and ketones also occurs in neutral and basic solution.

B. CYANOHYDRIN FORMATION

1. Hydrogen cyanide, HCN, reacts reversibly with aldehydes and ketones by a nucleophilic addition mechanism to give cyanohydrins.
a. A cyanide ion, formed by ionization of the weak acid HCN, reacts with the carbonyl group at the carbonyl carbon; the carbonyl oxygen becomes negatively charged.
b. The negatively charged oxygen is a relatively strong base, and is protonated by either water or HCN to complete the addition.

a cyanohydrin

2. Cyanohydrin formation favors the cyanohydrin addition product in the case of aldehydes and methyl ketones, but not in the case of aryl ketones.
3. Cyanohydrin formation is another method of forming carbon–carbon bonds.

C. REDUCTION OF ALDEHYDES AND KETONES TO ALCOHOLS

1. Aldehydes and ketones are reduced to alcohols with either lithium aluminum hydride ($LiAlH_4$) or sodium borohydride ($NaBH_4$), which serve as sources of nucleophilic hydride ion (H:⁻).
a. These reactions, which are not reversible, result in the net addition of the elements of H_2 across the C═O bond.
b. Reduction of an aldehyde gives a primary alcohol.
c. Reduction of a ketone gives a secondary alcohol.

2. $LiAlH_4$ and $NaBH_4$ reductions are generally referred to as hydride reductions and are examples of nucleophilic addition.
3. The reaction of $LiAlH_4$ with aldehydes and ketones involves the nucleophilic reaction of hydride (delivered from $^-AlH_4$) on the carbonyl carbon; all four hydride equivalents of $LiAlH_4$ are active in this reaction.
   a. A lithium ion coordinated to the carbonyl oxygen acts as a Lewis-acid catalyst.
   b. Hydride reacts with the carbonyl carbon to give an alkoxide addition compound.
   c. The alkoxide salt is converted by protonation in a separate step into the alcohol product; the proton source is water (or an aqueous solution of a weak acid such as $^+NH_4\ ^-Cl$).

   d. $LiAlH_4$ reacts violently with water and therefore must be used in dry solvents such as anhydrous ether or THF.
4. The reaction of $NaBH_4$ with aldehydes and ketones involves the nucleophilic reaction of hydride on the carbonyl carbon; all four hydride equivalents of $NaBH_4$ are active in this reaction.
   a. The sodium ion of $NaBH_4$ does not form as strong a bond to the carbonyl oxygen as the lithium ion.
   b. $NaBH_4$ reductions are carried out in protic solvents, such as alcohols; hydrogen bonding between the alcohol solvent and the carbonyl group serves as a weak acid catalysis that activates the carbonyl group.
   c. $NaBH_4$ reacts only slowly with alcohols, and it can be used in water if the solution is not acidic.

5. $LiAlH_4$ is a much more reactive reagent than $NaBH_4$.
   a. A number of functional groups react with $LiAlH_4$ but not $NaBH_4$:
      i. alkyl halides
      ii. alkyl tosylates
      iii. esters
      iv. nitro groups
   b. The greater selectivity and safety of $NaBH_4$ make it the preferred reagent in many applications.
6. Aldehydes and ketones can also be reduced to alcohols by catalytic hydrogenation.
   a. This reaction is analogous to the catalytic hydrogenation of an alkene.

   b. It is usually possible to use catalytic hydrogenation for the selective reduction of an alkene double bond in the presence of a carbonyl group; palladium catalysts are particularly effective for this purpose.

D. REACTIONS OF ALDEHYDES AND KETONES WITH GRIGNARD REAGENTS

1. The reaction of Grignard reagents with carbonyl groups is the most important application of Grignard reagents in organic chemistry; addition to aldehydes and ketones in an ether solvent, followed by protonolysis, gives alcohols. (See Appendix V, Section F, text page A-10, for a review of other syntheses of alcohols.)

an aldehyde
or ketone

an alcohol

2. The reaction of Grignard reagents with aldehydes and ketones is another example of carbonyl addition; the addition of Grignard reagents to aldehydes and ketones is not reversible.

3. The net effect of the Grignard reaction, followed by protonolysis, is addition of R—H (R = an alkyl or aryl group) across the C=O double bond; this addition is not reversible.

   a. Primary alcohols are synthesized by the addition of Grignard reagents to formaldehyde.

   b. Secondary alcohols are synthesized by the addition of Grignard reagents to aldehydes other than formaldehyde.

   c. Tertiary alcohols are synthesized by the addition of Grignard reagents to ketones.

4. The Grignard synthesis of a tertiary alcohol, or in some cases a secondary alcohol, can be extended to an alkene synthesis by dehydration of the alcohol with strong acid during the protonation step.

5. The Grignard reaction is an excellent method of carbon–carbon bond formation. (See Appendix VI, text page A-13, for a review of other reactions used to form carbon–carbon bonds.)

6. In the mechanism of the Grignard addition—

   a. the magnesium of the Grignard reagent, a Lewis acid, bonds to the carbonyl oxygen; this bonding makes the carbonyl carbon more electrophilic.

   b. the carbon group of the Grignard reagent reacts as a nucleophile at the carbonyl carbon to form a halomagnesium alkoxide.

   c. addition of dilute acid in a separate protonolysis step to the reaction mixture gives an alcohol.

7. The reactions of organolithium and sodium acetylide reagents with aldehydes and ketones are fundamentally similar to the Grignard reaction.

E. PREPARATION AND HYDROLYSIS OF ACETALS

1. When an aldehyde or ketone reacts with a large excess of an alcohol in the presence of a trace of strong acid, an acetal is formed.
   a. An acetal is a di-ether in which both ether oxygens are bound to the same carbon.
   b. Acetals are ethers of hydrates or *gem*-diols.

2. Two equivalents of alcohol are consumed in the formation of acetals, but one equivalent of a 1,2- or 1,3-diol can react to form a cyclic acetal, in which the acetal group is part of a five- or six-membered ring, respectively.

a cyclic acetal

3. Acetal formation is reversible and involves an acid-catalyzed carbonyl addition followed by a substitution that occurs by an S_N1 mechanism:
   a. The first step in the mechanism of acetal formation is acid-catalyzed addition of the alcohol to the carbonyl group to give a hemiacetal (a compound with an —OR and —OH group on the same carbon).
   b. The hemiacetal is converted into an acetal by substitution of the —OH group by another —OR group.

4. The reaction is driven to the right by applying Le Châtelier's principle in one or both of the following ways:
   a. use of excess alcohol as the solvent.
   b. removal of the water by-product.
5. Acetals in the presence of acid and excess water are transformed rapidly back into the corresponding carbonyl compounds and alcohols.

an acetal

a. This process is called acetal hydrolysis.
   b. By the principle of microscopic reversibility, the mechanism of acetal hydrolysis is the reverse of the mechanism of acetal formation.
   c. Acetal hydrolysis, like hemiacetal formation, is acid-catalyzed.
6. The interconversion of hemiacetals and aldehydes is catalyzed not only by acids, but by bases as well; however, the interconversion of hemiacetals and acetals is catalyzed only by acids. For this reason, the hydrolysis of acetals is catalyzed only by acids; acetals are stable in basic and neutral solutions.
7. Hemiacetals in most cases cannot be isolated because they react further to yield acetals or decompose to aldehydes or ketones plus water.
   a. Simple aldehydes form appreciable amounts of hemiacetals in alcohol solution, just as they form appreciable amounts of hydrates in water.
   b. Five- and six-membered cyclic hemiacetals form spontaneously from the corresponding hydroxy aldehydes, and most are stable, isolable compounds.

II. OXIDATION OF ALDEHYDES AND KETONES

A. OXIDATION OF ALDEHYDES

1. Aldehydes can be oxidized to carboxylic acids.

an aldehyde   a carboxylic acid

an aldehyde   an aldehyde hydrate   a carboxylic acid

2. Common oxidants, such as aqueous Cr(VI) reagents, nitric acid, or aqueous $KMnO_4/NaOH$, can be used in the oxidation of aldehydes to carboxylic acids; these oxidizing agents are the same ones used for oxidizing alcohols.
3. Some aldehyde oxidations begin as addition reactions.
   a. In the oxidation of aldehydes by Cr(VI) reagents, the hydrate, not the aldehyde, is actually the species oxidized.
   b. For this reason, some water should be present in solution in order for aldehyde oxidations with Cr(VI) to occur at a reasonable rate.

4. In the laboratory, aldehydes can be conveniently oxidized to carboxylic acids with Ag(I) reagents.
   a. Aldehydes that contain double bonds or alcohol —OH groups (functional groups that react with other oxidizing reagents) are oxidized by Ag_2O without oxidizing these other functional groups.
   b. If the silver ion is solubilized as its ammonia complex, $^+Ag(NH_3)_2$, oxidation of the aldehyde is accompanied by the deposition of a metallic silver mirror on the walls of the reaction vessel. (This observation can be used as a convenient test for aldehydes, known as the Tollens test.)

ammonium salt of
a carboxylic acid

5. Many aldehydes are oxidized by the oxygen in air upon standing for long periods of time; this process is another example of autoxidation.

B. OXIDATION OF KETONES

1. Ketones cannot be oxidized without breaking carbon–carbon bonds.
2. Ketones are resistant to mild oxidation with Cr(VI) reagents.

III. REACTIONS OF ALDEHYDES AND KETONES WITH AMINES

A. IMINE FORMATION WITH PRIMARY AMINES

1. Imines are prepared by the reaction of aldehydes or ketones with primary amines.
    a. Formation of imines is reversible and generally takes place with acid or base catalysis, or with heat.

$$\text{C=O} + \text{H}_2\text{NR} \underset{\text{H}_3\text{O}^+ \text{ or HO}^-}{\rightleftarrows} \text{C=N–R} + \text{H}_2\text{O}$$

an imine

    b. Imine formation is typically driven to completion in one or both of the following ways:
        i. Precipitation of the imine.
        ii. Removal of water.
2. Imine formation consists of a carbonyl-addition reaction followed by β-elimination.
    a. The first step of the mechanism is a nucleophilic addition to the carbonyl group in which the nucleophile (an amine) reacts with an aldehyde or ketone to give an unstable addition compound called a carbinolamine.
    b. The carbinolamine undergoes acid-catalyzed dehydration to form an imine.
    c. Dehydration of the carbinolamine is typically the rate-limiting step of imine formation.

$$\text{C=O} + \text{H}_2\text{NR} \xrightarrow{\text{addition}} \underset{\text{OH}}{\overset{\text{NHR}}{\text{C}}} \xrightarrow{\text{dehydration}} \text{C=N–R} + \text{H}_2\text{O}$$

a carbinolamine   an imine

3. Certain types of imine adducts sometimes find use as derivatives of aldehydes and ketones because they are in most cases solids with well-defined melting points. (See Table 19.3, text page 985, for the corresponding structures, and FE 19.2 in this manual.)
    a. Aldehydes or ketones + hydroxylamine → oximes.
    b. Aldehydes or ketones + hydrazine → hydrazones.
    c. Aldehydes or ketones + phenylhydrazine → phenylhydrazones.
    d. Aldehydes or ketones + 2,4-dinitrophenylhydrazine → 2,4-dinitrophenylhydrazones (2,4-DNP derivatives).
    e. Aldehydes or ketones + semicarbazide → semicarbazones.

B. ENAMINE FORMATION WITH SECONDARY AMINES

1. Formation of an enamine occurs when a secondary amine reacts with an aldehyde or ketone that has an α-hydrogen.

an aldehyde or ketone   a secondary   an enamine
with an α-hydrogen   amine

    a. Just as most aldehydes and ketones are more stable than their corresponding enols, most imines are more stable than their corresponding enamines.
    b. Because secondary amines cannot form imines, they form enamines instead.
2. Like imine formation, enamine formation is reversible and must be driven to completion by the removal of one of the reaction products (usually water).
    a. The mechanism of enamine formation begins like the mechanism of imine formation, as a nucleophilic addition to give a carbinolamine intermediate.
    b. Dehydration of a carbinolamine involves proton loss from an adjacent carbon.

a carbinolamine   an enamine

3. Enamines, like imines, revert to the corresponding carbonyl compounds and amines in aqueous acid.

C. REACTIONS OF ALDEHYDES AND KETONES WITH TERTIARY AMINES

   1. Tertiary amines do not react with aldehydes and ketones to form stable derivatives.

IV. CONVERSION OF KETONES AND ALDEHYDES INTO ALKANES AND ALKENES

A. REDUCTION OF CARBONYL GROUPS TO METHYLENE GROUPS

   1. A carbonyl group of an aldehyde or ketone can be reduced completely to a methylene ($-CH_2-$) group.
   2. One procedure for effecting this transformation involves heating the aldehyde or ketone with hydrazine (H_2N-NH_2) and strong base.

   a. This reaction, called the Wolff–Kishner reduction, typically utilizes ethylene glycol or similar compounds as co-solvents.
   b. The high boiling points of these solvents allow the reaction mixtures to reach the high temperatures required for the reduction to take place at a reasonable rate.
   3. The Wolff–Kishner reduction is an extension of imine formation; an intermediate in the reduction is a hydrazone (an imine of hydrazine).
   4. The Wolff–Kishner reduction takes place under strongly basic conditions.

   5. The same overall transformation can be achieved under acidic conditions by a reaction called the Clemmensen reduction, which involves reduction of an aldehyde or ketone with zinc amalgam (a solution of zinc metal in mercury) in the presence of HCl.

   6. The Wolff–Kishner and Clemmensen reactions are particularly useful for the introduction of alkyl substituents into benzene rings by the following sequence:
   a. Friedel–Crafts acylation to give an aryl ketone.
   b. Wolff–Kishner or Clemmensen reduction to yield the corresponding alkyl group.

B. THE WITTIG ALKENE SYNTHESIS

1. Another addition–elimination reaction, called the Wittig reaction, is an important method for preparing alkenes from aldehydes and ketones.

a Wittig reagent   an aldehyde   triphenylphosphine   an alkene
                     or ketone   oxide

2. The nucleophile in the Wittig reaction is a type of ylid.
    a. An ylid is any compound with opposite charges on adjacent covalently bound atoms, each of which has an electronic octet.
    b. Because phosphorus can accommodate more than eight valence electrons, a phosphorus ylid has an uncharged resonance structure.
    c. Although the structures of phosphorus ylids are sometimes written with phosphorus–carbon double bonds, the charged structures, in which each atom has an octet of electrons, are very important contributors.

an ylid

3. Preparation of phosphorus ylids:
    a. An alkyl halide reacts with triphenylphosphine (Ph_3P) in an S_N2 reaction to give a phosphonium salt. (Because the alkylation of triphenylphosphine is a typical S_N2 reaction, it is limited for the most part to methyl and primary alkyl halides.)
    b. The phosphonium salt is converted into its conjugate base, the ylid, by reaction with a strong base such as an organolithium reagent (commonly *n*-butyllithium, or *n*-BuLi, as in the example below).

4. The mechanism of the Wittig reaction:
    a. A nucleophile (the anionic carbon of the ylid) reacts with the carbonyl carbon.
       i. The anionic oxygen in the resulting species reacts with phosphorous to form an oxaphosphetane intermediate.
       ii. An oxaphosphetane is a saturated four-membered ring containing both oxygen and phosphorus as ring atoms.
    b. Under the usual reaction conditions, the oxaphosphetane spontaneously decomposes to the alkene and the by-product triphenylphosphine oxide.

an oxaphosphetane   triphenyl-
                                     phosphine oxide

5. The Wittig reaction is especially important because it gives alkenes in which the position of the alkene double bond is unambiguous.
    a. The reaction is thus completely regioselective.
    b. The reaction can be used for the preparation of alkenes that would be difficult to prepare by other reactions.
6. To plan the synthesis of an alkene by the Wittig reaction, consider the origin of each part of the product, and then reason deductively.
    a. One carbon of the alkene double bond originates from the alkyl halide used to prepare the ylid; the other is the carbonyl carbon of the aldehyde or ketone.
    b. In principle, two Wittig syntheses are possible for any given alkene.

c. Most Wittig syntheses are planned so that the most reactive alkyl halide (a methyl or primary alkyl halide) can be used as one of the starting materials. [See (3a) above.]

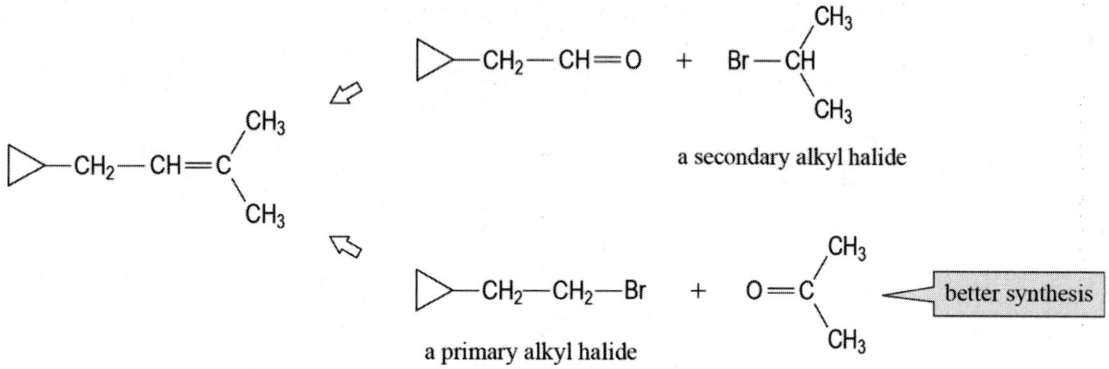

$$\triangleright\text{—CH}_2\text{—CH}{=}\text{O} \quad + \quad \text{Br—CH}\begin{smallmatrix}\text{CH}_3\\ \\ \text{CH}_3\end{smallmatrix}$$

a secondary alkyl halide

$$\triangleright\text{—CH}_2\text{—CH}{=}\text{C}\begin{smallmatrix}\text{CH}_3\\ \\ \text{CH}_3\end{smallmatrix}$$

$$\triangleright\text{—CH}_2\text{—CH}_2\text{—Br} \quad + \quad \text{O}{=}\text{C}\begin{smallmatrix}\text{CH}_3\\ \\ \text{CH}_3\end{smallmatrix} \quad \longleftarrow \text{ better synthesis}$$

a primary alkyl halide

d. The Wittig reaction in many cases gives mixtures of *E* and *Z* isomers, although certain modifications of the Wittig reaction that avoid this problem have been developed.

  If you find an error in this manual, please visit http://people.pharmacy.purdue.edu/~loudonm/teaching/

SOLUTIONS TO PROBLEMS

Solutions to In-Text Problems

19.1   (a)

(CH$_3$)$_2$CHCH=O

(b)

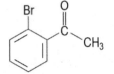

(c)

(d)

ClCH$_2$CH$_2$CH$_2$CH

(e)

(f)

(g)

(h)

19.2   (a)   2-propanone
       (b)   2,4-dimethyl-3-pentanone

(CH$_3$)$_2$CHCCH(CH$_3$)$_2$

**2,4-dimethyl-3-pentanone
(diisopropyl ketone)**

       (c)   3-allyl-2,4-pentanedione
       (d)   (*E*)-3-ethoxy-2-propenal
       (e)   3-methylcyclobutanecarbaldehyde
       (f)   4,4-dimethyl-2,5-cyclohexadienone

19.3   (a)   The aldehyde has a somewhat higher carbonyl stretching frequency, and has a C—H stretching
             absorption near 2700 cm^{-1} that is not present in the IR spectrum of the ketone.

CH$_3$CH$_2$CH$_2$CH$_2$CH$_2$CH=O
hexanal

cyclohexanone

       (b)   2-Cyclohexenone has a lower carbonyl stretching frequency because its two double bonds are
             conjugated.

2-cyclohexenone
has conjugated double bonds

3-cyclohexenone

       (c)   3-Buten-2-ol has both O—H and C=C stretching absorptions and 2-butanone does not; 2-butanone has a
             carbonyl stretching absorption, and 3-buten-2-ol does not.

$$\underset{\text{3-buten-2-ol}}{\text{CH}_3\text{CHCH}\!=\!\text{CH}_2}$$ (with OH on the CHOH carbon)

$$\underset{\text{2-butanone}}{\text{CH}_3\text{CCH}_2\text{CH}_3}$$ (with carbonyl O)

19.4   (a)   The IR spectrum suggests an aldehyde (carbonyl stretch at 1720 cm^{-1} and aldehyde C—H stretch at 2710 cm^{-1}, and the low-field absorption at δ 9.65 confirms the presence of the aldehyde hydrogen. The six-proton doublet at δ 1.2 and the septet at δ 2.4 indicate an isopropyl group $(\text{CH}_3)_2\text{CH}$—. Because the aldehyde carbon is the only remaining carbon, the structure is completely determined:

$$(\text{CH}_3)_2\text{CH}\!-\!\overset{\overset{\text{O}}{\|}}{\text{CH}}$$

isobutyraldehyde
(2-methylpropanal)

The CH of the isopropyl group and the aldehyde hydrogen split each other slightly.

   (b)   The compound is 2-butanone:

$$\text{CH}_3\overset{\overset{\text{O}}{\|}}{\text{C}}\text{CH}_2\text{CH}_3$$

2-butanone

   (c)   The high frequency of the carbonyl absorption suggests a strained ring. (See Eq. 19.4, text p. 954.) In fact, cyclobutanone matches the IR stretching frequency perfectly and the NMR fits as well:

δ 3.09 (triplet)

δ 2.01 (quintuplet)

δ 3.09 (triplet)

molecular mass = 70.1

cyclobutanone

   (d)   The molecular formula indicates an unsaturation number of 5. The IR spectrum suggests a conjugated carbonyl group and an aromatic ring, and these two features fully account for the unsaturation. The NMR indicates the following partial structures:

$$\text{CH}_3\text{CH}_2\!-\!\text{O}\!-\!\qquad -\overset{\overset{\text{O}}{\|}}{\text{C}}\text{CH}_3$$

and these define the structure as

$$\text{CH}_3\text{CH}_2\text{O}\!-\!\!\left\langle\!\!\bigcirc\!\!\right\rangle\!\!-\!\overset{\overset{\text{O}}{\|}}{\text{C}}\text{CH}_3$$

p-ethoxyacetophenone

19.5   The low intensities of the carbonyl carbon absorption at δ 212.6 and the α-carbon absorption at δ 44.2 show that these two carbons have no attached hydrogens; therefore this compound must have the following partial structure:

$$\text{C}\!-\!\overset{\overset{\text{O}}{\|}}{\text{C}}\!-\!\underset{\underset{\text{C}}{|}}{\overset{\overset{\text{C}}{|}}{\text{C}}}\!-\!\text{C}$$

Because all six carbons and their connectivities are accounted for, the only possible choice is to add the appropriate number of hydrogens. The compound is 3,3-dimethyl-2-butanone:

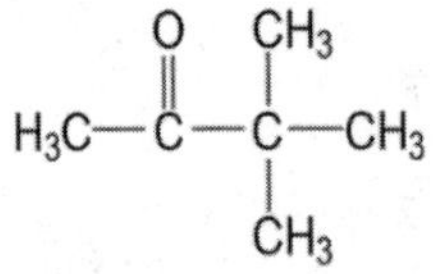

3,3-dimethyl-2-butanone

19.6 The structure and carbon NMR assignments of 2-ethylbutanal are shown below. The two methyl groups are chemically equivalent, and the two methylene groups are chemically equivalent; all carbons with different carbon NMR chemical shifts are chemically nonequivalent.

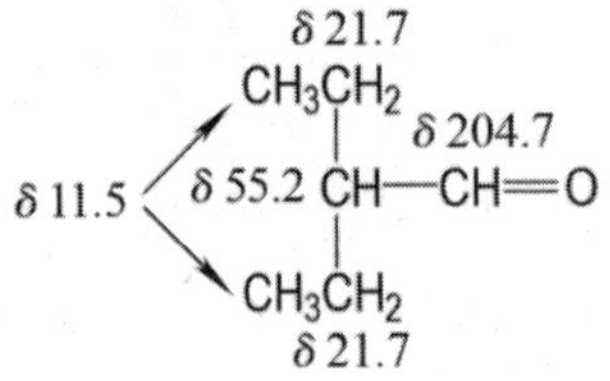

2-ethylbutanal

19.7 (a) The double bonds in 2-cyclohexenone are conjugated, but the double bonds in 3-cyclohexenone are not. Consequently, 2-cyclohexenone has the UV spectrum with the greater λ_{max}.

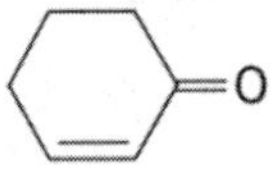

2-cyclohexenone
(conjugated; has UV spectrum
with the greater λ_{max})

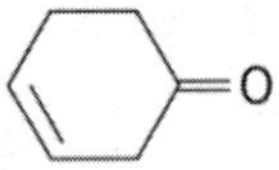

3-cyclohexenone

(b) The first compound has more conjugated double bonds and therefore will have the UV spectrum with the greater λ_{max} and larger extinction coefficient ϵ.

(c) In 1-phenyl-2-propanone, the aromatic ring is not conjugated with the carbonyl group; in p-methylacetophenone the aromatic ring and the carbonyl group are conjugated. Because of its additional conjugation, p-methylacetophenone has the UV spectrum with the greater λ_{max} and larger extinction coefficient ϵ.

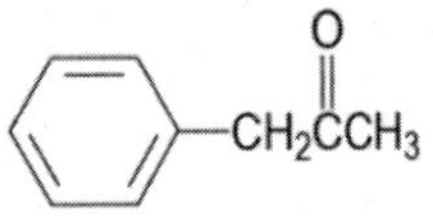

1-phenyl-2-propanone

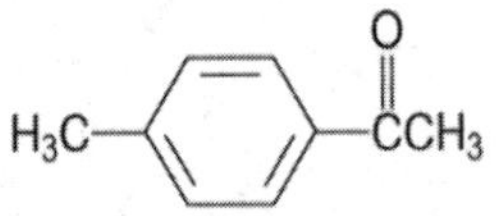

p-methylacetophenone
(has the UV spectrum with
the greater ϵ and λ_{max})

19.8 Addition of NaOH brings about the ionization of the phenolic —OH group of p-hydroxyacetophenone. Because the electron pair associated with the anion can be delocalized into the ring and into the carbonyl group, this conjugated species has a unique UV spectrum that is different from that of the un-ionized compound. Because p-methoxyacetophenone does not ionize in NaOH solution, its UV spectrum is not affected by addition of base.

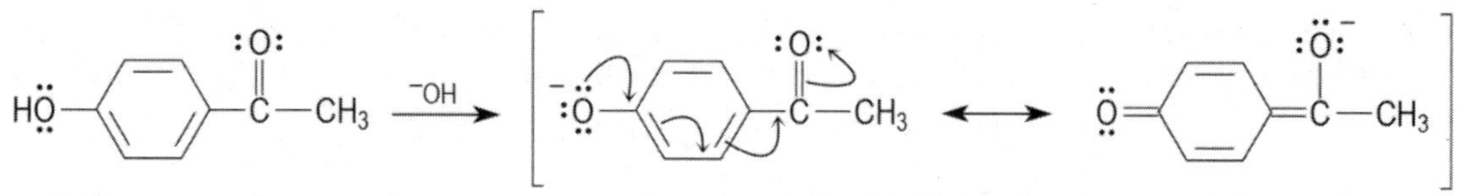

p-hydroxyacetophenone

electrons associated with negative charge are delocalized;
electron delocalization is reflected in the UV spectrum as a greater λ_{max}.

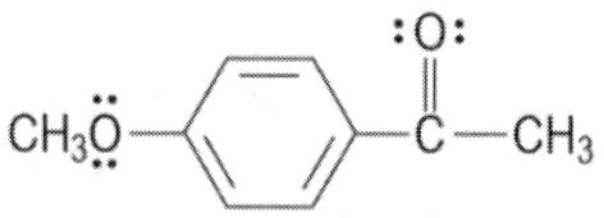

p-methoxyacetophenone
cannot ionize;
base has no effect on the UV spectrum

19.9   Compound *A*, vanillin, should have a $\pi \rightarrow \pi^*$ absorption at a greater λ_{max} when dissolved in NaOH solution because the resulting phenolate can delocalize into the carboxaldehyde group; the resulting phenolate from compound *B*, isovanillin, on the other hand, can only delocalize in the aromatic ring.

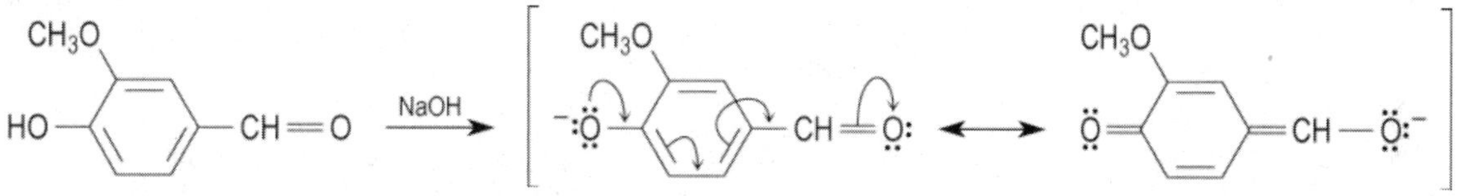

19.10   (a)   In both ketones, inductive cleavage of the molecular ion to give an acyl radical and a carbocation accounts for the $m/z = 57$ peak:

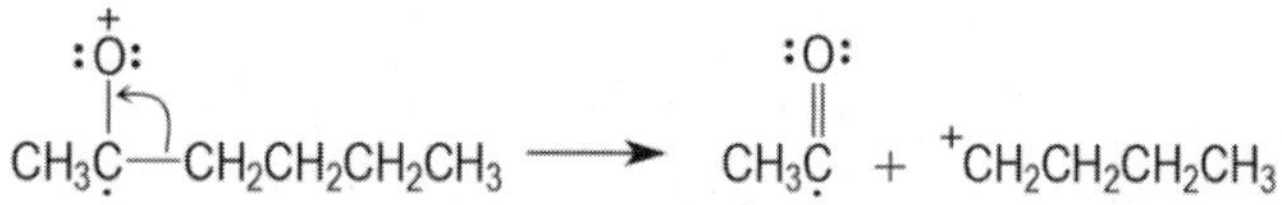

molecular ion of 2-hexanone

butyl cation
($m/z = 57$)

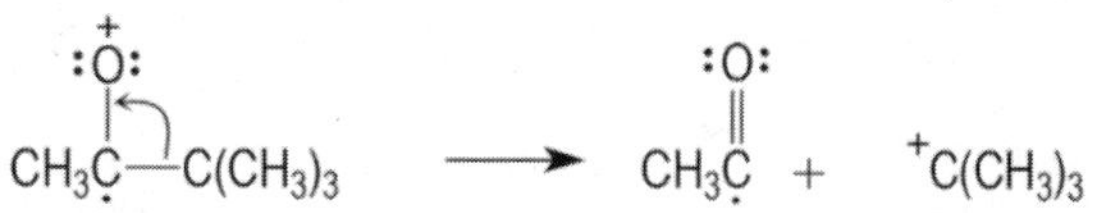

molecular ion of
3,3-dimethyl-2-butanone

***tert*-butyl cation**
($m/z = 57$)

Because the carbocation formed from 3,3-dimethyl-2-butanone is more stable (why?), more of this fragmentation mode is observed for 3,3-dimethyl-2-butanone; that is, the $m/z = 57$ peak is more abundant.

(b)   Its even mass suggests that the $m/z = 58$ fragment is an odd-electron ion. Such an ion can be readily produced by a McLafferty rearrangement from the molecular ion of 2-hexanone, but not from the molecular ion of 3,3-dimethyl-2-butanone:

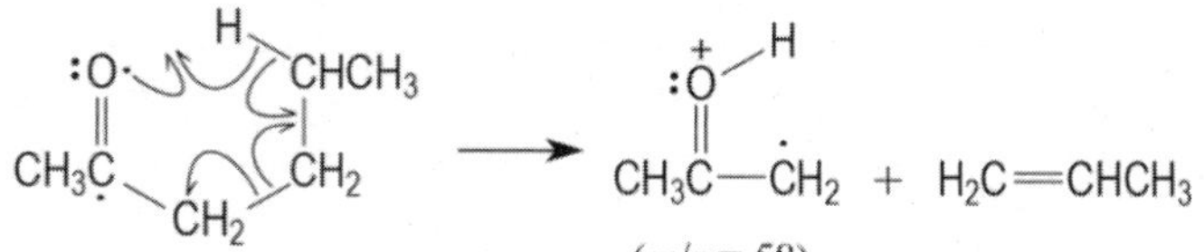

($m/z = 58$)

molecular ion of 2-hexanone

19.11   The mass spectrum of 2-heptanone should have major peaks at $m/z = 43$ (from α-cleavage), 71 (from inductive cleavage), and 58 (from McLafferty rearrangement). The mass spectrum of 3-heptanone should have a major peak resulting from both inductive cleavage and α-cleavage at $m/z = 57$ and a major peak resulting from McLafferty rearrangement at $m/z = 72$. Notice that the position of the even-mass, odd-electron ion is a major distinguishing feature. We leave it to you to draw out these fragmentations.

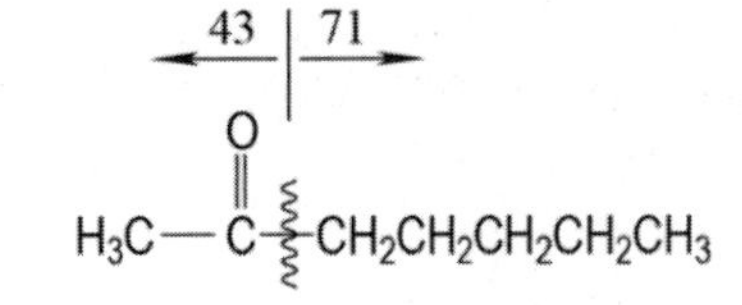

2-heptanone
McLafferty rearrangement at $m/z = 58$

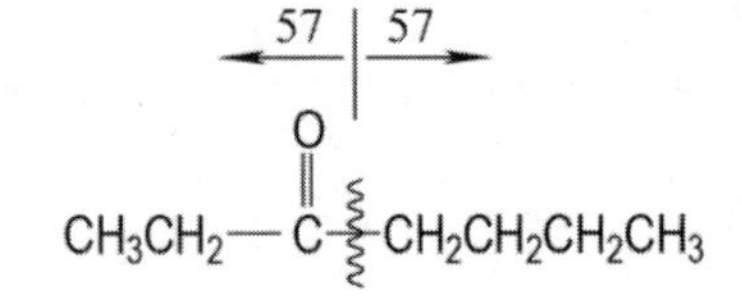

3-heptanone
McLafferty rearrangement at $m/z = 72$

19.12   (a)   The solvolysis mechanism of (chloromethoxy)methane in ethanol:

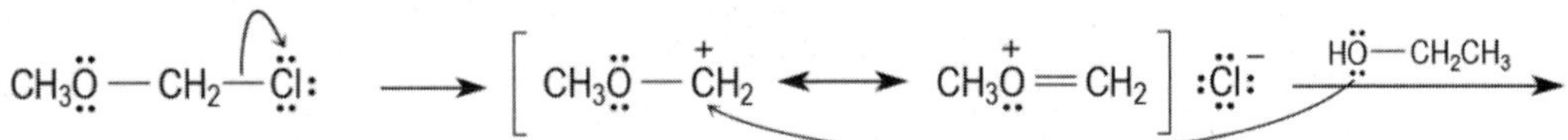

(chloromethoxy)methane

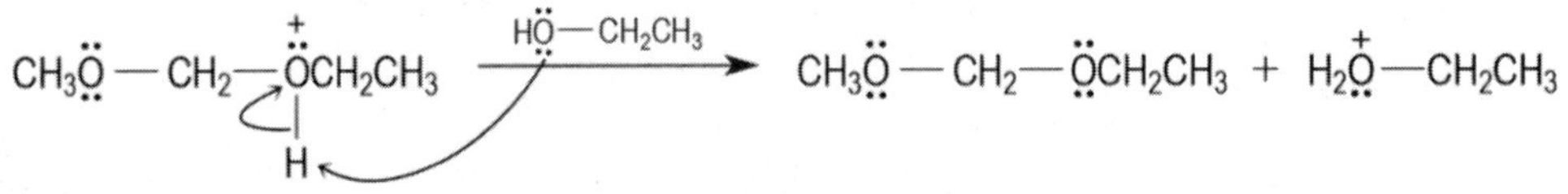

(b)   (Chloromethoxy)methane undergoes more rapid solvolysis than 1-chlorobutane because the carbocation intermediate is resonance-stabilized [as shown in the mechanism of part (a)] and is therefore more stable relative to starting material than the ordinary primary carbocation intermediate derived from 1-chlorobutane. (In fact, 1-chlorobutane probably solvolyzes by an S_N2 mechanism; the S_N1 reaction would no doubt be even slower than the reaction indicated in the problem.)

19.13   The products follow from a consideration of the mechanism in Study Problem 19.2 on text p. 961. The first steps of this mechanism consist of protonation of an —OH group and loss of water to form a carbocation intermediate.

(a)   When the two —OH groups are chemically nonequivalent, the —OH group lost is the one that gives the more stable carbocation intermediate—in this case, the tertiary —OH.

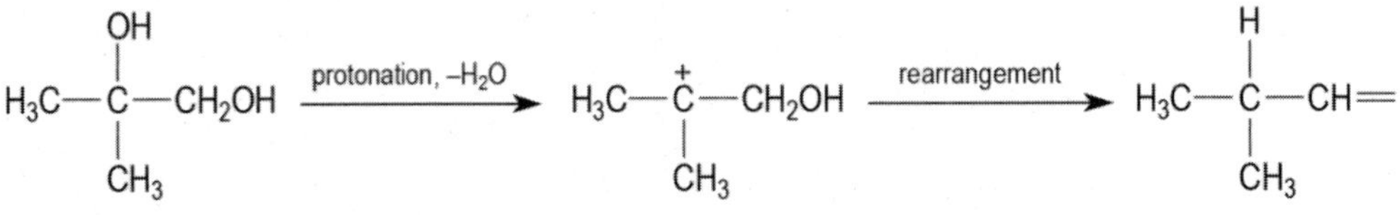

(b)   The reaction is exactly like the one shown in Study Problem 19.2 on text p. 961 with phenyl instead of methyl substituents. The product is the following ketone:

$$Ph_3C-\overset{\displaystyle O}{\overset{\|}{C}}-Ph$$

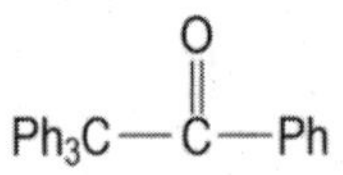

(c)   In this case, rearrangement involves a ring expansion. If reasoning through this type of rearrangement gives you difficulty, be sure to consult Further Exploration 4.4 (p. 78 of this manual) for assistance. The product is the following ketone:

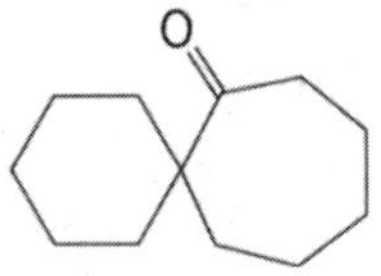

19.14 (a) Protonated *p*-methoxybenzaldehyde is essentially a benzylic carbocation; as you learned in Chapter 17, benzylic carbocations are stabilized by substituents that can donate electrons by resonance, and this cation is no exception; the resonance effect of the *p*-methoxy group stabilizes the cationic conjugate acid:

conjugate acid of *p*-methoxybenzaldehyde

Stabilization of the conjugate acid of any compound *increases* the compound's basicity. The *p*-nitro group *destabilizes* the corresponding cation that results from protonation of *p*-nitrobenzaldehyde. (See the solution to Problem 17.29 on pp. 529–530 of this manual, for a very similar case.) Destabilization of the conjugate acid of a compound *decreases* the compound's basicity.

(b) The conjugate acid of 3-buten-2-one has more important resonance structures than the conjugate acid of 2-butanone and is therefore more stable relative to unprotonated ketone than the conjugate acid of 2-butanone. Greater stability of the conjugate acid means that the ketone is more basic.

conjugate acid of 3-buten-2-one

19.15 (a) The mechanism of the hydroxide-catalyzed hydration of acetaldehyde:

(b) The curved-arrow mechanism for the decomposition of acetone cyanohydrin in aqueous hydroxide:

acetone cyanohydrin

The ketone–cyanohydrin equilibrium favors the ketone at high pH because hydroxide ion is a much stronger base than cyanide ion. (The pK_a of water is 15.7; the pK_a of hydrogen cyanide is 9.40; see Table 3.1, text p. 102.) The weaker base is the better leaving group.

19.16 (a) The acid-catalyzed addition of methanol involves mechanistic steps like those involved in acid-catalyzed hydration (Eqs. 19.20a–b on text pp. 966). The carbonyl group is protonated, and methanol reacts as a nucleophile at the carbonyl carbon.

The reaction mechanism with curved arrows, showing the acid-catalyzed addition of methanol to benzaldehyde, is displayed across the top of the page.

Remember that we typically invoke as a base the conjugate base of whatever acid is involved in the mechanism. Thus, if H_3O^+ is involved in a mechanism, its conjugate base (H_2O) acts as the base. We would not invoke both H_3O^+ and ^-OH in the same mechanism because a strong acid and a strong base cannot coexist in solution in significant concentrations.

(b) The methoxide-catalyzed addition of methanol to benzaldehyde is very similar to the hydroxide-catalyzed hydration mechanism (Problem 19.15a):

19.17 The data in Table 19.2, text p. 967, of the text show that hydration of benzaldehyde is less than 0.01 times as favorable as hydration of an unconjugated aliphatic aldehyde such as acetaldehyde. The assumption that the same principles apply to cyanohydrin formation leads to the prediction that propanal should have the greater proportion of cyanohydrin at equilibrium. The structure of this cyanohydrin is as follows:

$$CH_3CH_2CH\overset{\displaystyle OH}{\underset{\displaystyle C\equiv N}{|}}$$ **cyanohydrin derivative of propanal**

19.18 We use the same principles to predict reactivity that we use to predict relative equilibrium constants. The more prone a carbonyl compound is to form an addition product, the more reactive it is.

(a) The second compound, bromoacetone, is more reactive because the electronegative atom is closer to the carbonyl group.

(b) The first compound, 2,3-butanedione, is more reactive, because the partial positive charge on one carbonyl destabilizes the molecule by its repulsive interaction with the partial positive charge on the other. (See Problem 19.43, text p. 997, for a similar situation.)

(c) The first compound, *p*-nitrobenzaldehyde, is more reactive, because the *p*-nitro group raises the energy of the molecule by an unfavorable interaction of the positive charge on the nitrogen with the positive charge on the carbonyl carbon. In contrast, the *p*-methoxy group in the second compound, *p*-methoxybenzaldehyde, stabilizes the molecule by a resonance interaction with the positive charge on the carbonyl carbon.

some resonance structures of *p*-methoxybenzaldehyde

(d) The second compound, cyclopropanone, is more reactive for the following reason. A carbonyl compound is most stable when the bond angles at the carbonyl group can be close to 120°; however, in cyclopropanone they are constrained by the three-membered ring to be 60°. As a result, there is considerable strain in the carbonyl form. In the hydrate, the preferred bond angle is approximately tetrahedral (109.5°). Because this angle is closer to the 60° bond angle enforced by the ring than the bond angle in cyclopropanone is, there is less strain in the hydrate than there is in cyclopropanone. Hence, formation of the hydrate relieves some of the strain in cyclopropanone. In other words, the ketone is destabilized relative to the hydrate by strain; there is a driving force to form the hydrate. Because there is little or no strain in cyclopentanone, such effects do not come into play.

19.19   (a)   (b)   (c)

cyclopentanecarbaldehyde   **2-butanone**

An extra equivalent of LiAlH$_4$ would have to be used with the compound in (c), because the hydride would react vigorously with the tertiary alcohol to form the conjugate-base alkoxide. The tertiary alcohol would be re-formed when a weak acid is added to the reaction mixture. NaBH$_4$, on the other hand, would reduce the compound shown in (c) without this extra complication.

19.20   The tertiary alcohol *C* could not be synthesized by a hydride reduction, because only primary and secondary alcohols can be prepared by this method.

19.21   (a)   The zinc ion acts as a Lewis acid catalyst by bonding to the carbonyl oxygen, making the carbonyl carbon more electrophilic, as shown in the answer to 19.21(b). The role of the zinc ion is analogous to the role of the lithium ion in reductions with lithium aluminum hydride (see Eq. 19.28a).

         (b)

19.22   All three compounds below will react with hydrogen and a catalyst to give the product shown in the problem. Recall from section 7.8E that catalytic hydrogenation is a syn addition. This fact is important when considering how the third compound in this group reacts.

The following two compounds may appear to work, however each would produce a mixture of the desired product and a stereoisomer.

See if you can determine why it is that both stereoisomers below form when either of the two starting materials above are subjected to catalytic hydrogenation.

19.23 In each case, ethyl bromide, CH_3CH_2Br, reacts with Mg to give ethylmagnesium bromide, CH_3CH_2MgBr, which is then allowed to react as shown below.

(a)

benzaldehyde   1-phenyl-1-propanol

(b)

2-butanone

3-methyl-3-pentanol

(c)

ethylene oxide
(oxirane)   1-butanol

 Don't forget what you've already learned about the reactions of Grignard reagents; see Sec. 11.15C of the text.

(d)

formaldehyde   1-propanol   propanal

(e)

benzophenone

1,1-diphenyl-1-propanol   (1-phenyl-1-propenyl)benzene

(f)

1) CH$_3$CH$_2$MgBr
2) H$_3$O$^+$

cyclopentanone

1-ethyl-1-cyclopentanol

19.24 Any of the three alkyl groups bound to the α-carbon of the alcohol can in principle originate from the Grignard reagent.

Synthesis #1:

1) CH$_3$MgI
2) H$_3$O$^+$

3-hexanone

3-methyl-3-hexanol

Synthesis #2:

1) CH$_3$CH$_2$CH$_2$MgBr
2) H$_3$O$^+$

2-butanone

3-methyl-3-hexanol

Synthesis #3:

1) CH$_3$CH$_2$MgBr
2) H$_3$O$^+$

2-pentanone

3-methyl-3-hexanol

In synthesis #1, the Grignard reagent is prepared by the reaction of CH$_3$I (methyl iodide) and Mg in dry ether; in synthesis #2, the Grignard reagent is prepared from the similar reaction of CH$_3$CH$_2$CH$_2$—Br (propyl bromide) and Mg; and in synthesis #3, the Grignard reagent is prepared from ethyl bromide. Generally, it doesn't matter whether you chose alkyl chlorides, bromides, or iodides from which to synthesize Grignard reagents.

19.25 (a)

OCH$_2$CH$_3$

OCH$_2$CH$_3$

cyclopentanone diethyl acetal
(1,1-diethoxycyclopentane)

 Note that acetals are named as ethers in the IUPAC system. In the acetals derived from ketones in the common system were once called *ketals*. Thus, in older common nomenclature, the compound in the solution to part (a) would have been called *cyclopentanone diethyl ketal*.

(b)

OCH(CH$_3$)$_2$

CH$_3$CH$_2$CH$_2$CH

OCH(CH$_3$)$_2$

butyraldehyde diisopropyl acetal
(1,1-diisopropoxybutane)

19.26 In each case, a diol reacts with an aldehyde or ketone.

(a)

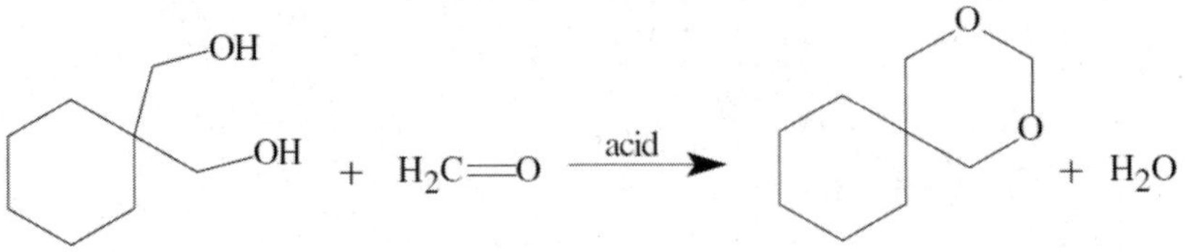

(b)

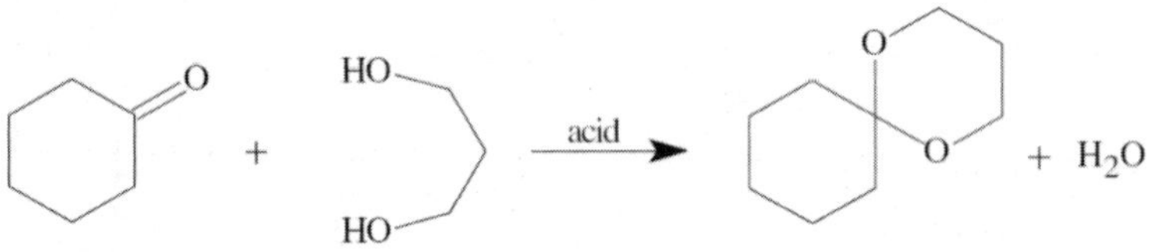

19.27   (a)   The formula indicates addition of two carbon atoms and it indicates one degree of unsaturation. If a diethyl acetal were formed and the ring were opened, the formula of the product would be $C_9H_{20}O_3$. The one degree of unsaturation suggests that the ring is intact. The product is a mixed acetal:

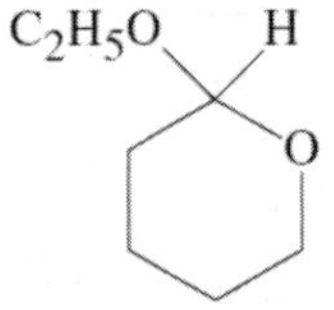

(b)   When five- and six-membered rings can be formed, diols generally react with aldehydes and ketones to give cyclic acetals, and this case is no exception:

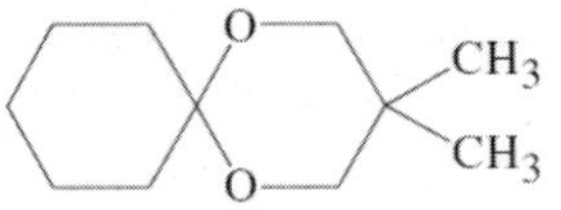

19.28   We protect the ketone as an acetal, and then prepare the Grignard reagent, which is allowed to react with acetaldehyde. The protonolysis step produces the alcohol and removes the acetal protecting group. (In the following synthesis, any common alcohol or ethylene glycol can be used instead of methanol.)

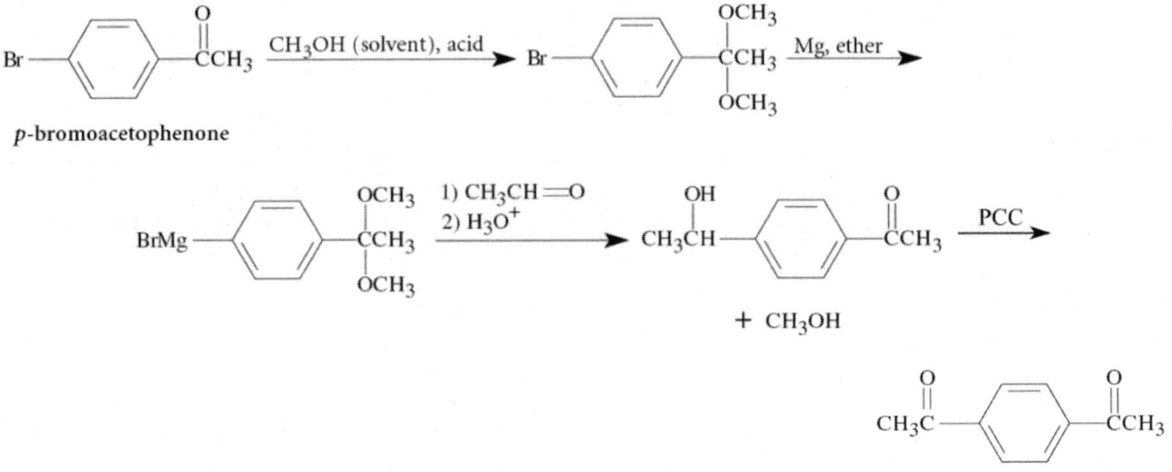

p-bromoacetophenone

Note that this compound could not be made by a Friedel–Crafts acylation of acetophenone. (Why?) The reaction sequence shown has maneuvered two meta-directing groups into a para relationship!

19.29   (a)   (b)

CH₃CH₂CH₂CH₂CHCH═NCH₂CH₃

$$CH_3CH_2CH_2CH_2CHCH{=}NCH_2CH_3$$
$$CH_3$$

19.30   First, the carbinolamine intermediate is formed. This intermediate then undergoes base-catalyzed dehydration to give the oxime.

cyclohexanone

carbinolamine intermediate

cyclohexanone oxime

 Note that it is equally appropriate to write the loss of water and formation of the carbon–nitrogen double bond as one step, thus avoiding the necessity of drawing resonance structures:

cyclohexanone oxime

19.31   Imine hydrolysis is the reverse of imine formation. Therefore, retrace the steps of imine formation as illustrated in the solution to Problem 19.30, starting with the imine and working backwards to the aldehyde and the amine.

PhCH═NC₂H₅   PhCH═NC₂H₅   PhCH─NC₂H₅

19.32

19.33   (a)   (b)

19.34   Any compound with a carbonyl group on any of the prospective alkyl carbons could in principle serve as a starting material. (The answer is restricted to compounds containing only one carbonyl group.)

4-isobutylbenzaldehyde   *p*-methylisobutyrophenone   2-methyl-3-(4-methylphenyl)propanal

19.35   Alkylate the phenol, carry out a Friedel–Crafts acylation, and then apply the Wolff–Kishner reaction. Alkylation of the phenol should precede the Friedel–Crafts reaction because of the sluggish reactivity of phenols in the Friedel–Crafts acylation reaction; see text Sec. 18.9 in the text. Note that Friedel–Crafts alkylation of the ring with $ClCH_2CH_2CH_3$ would give some rearrangement product in addition to the desired product; see Eq. 19.74a on text p. 990.

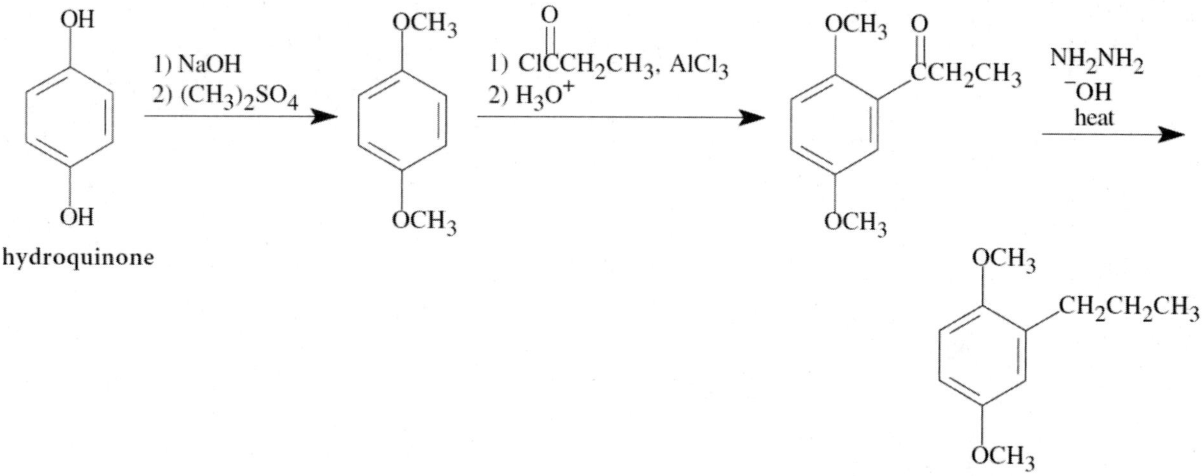

1,4-dimethoxy-2-propylbenzene

19.36 (a)

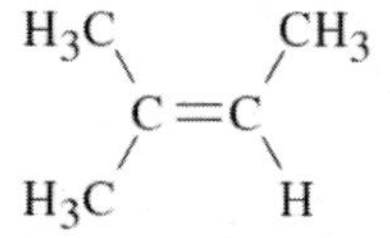

2-methyl-2-butene

(b)

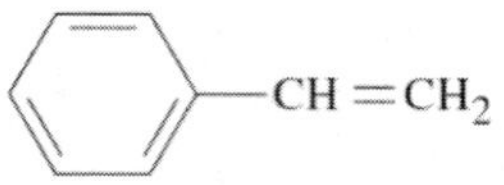

styrene

19.37 (a) Either "half" of the alkene can in principle be derived from an aldehyde.

The first synthesis:

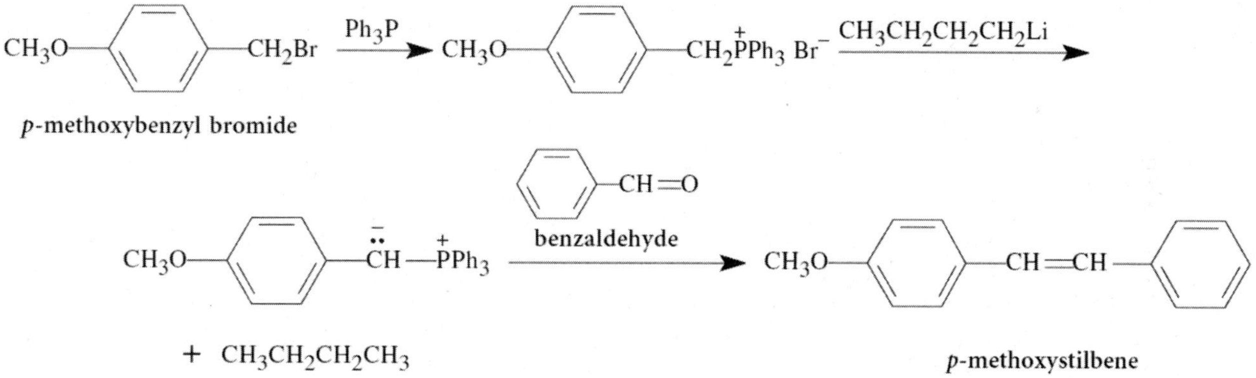

p-methoxybenzyl bromide

benzaldehyde

p-methoxystilbene

+ CH₃CH₂CH₂CH₃

The second synthesis:

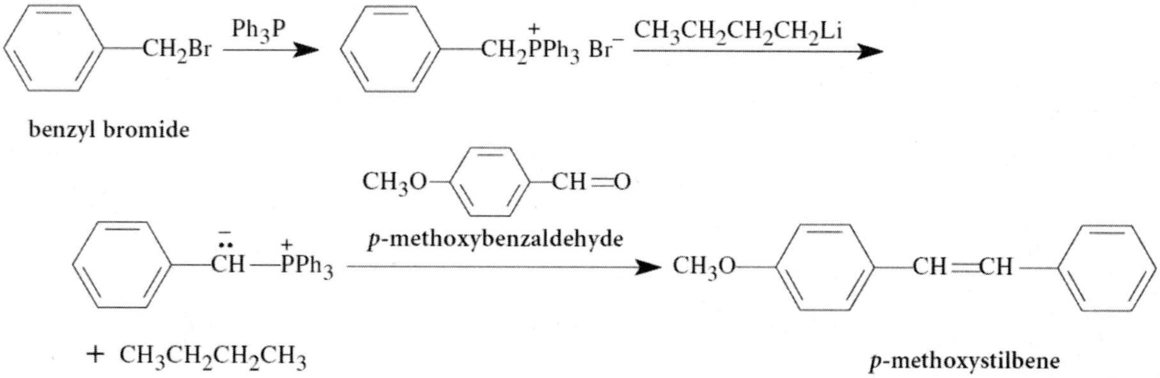

benzyl bromide

p-methoxybenzaldehyde

+ CH₃CH₂CH₂CH₃

p-methoxystilbene

(b)

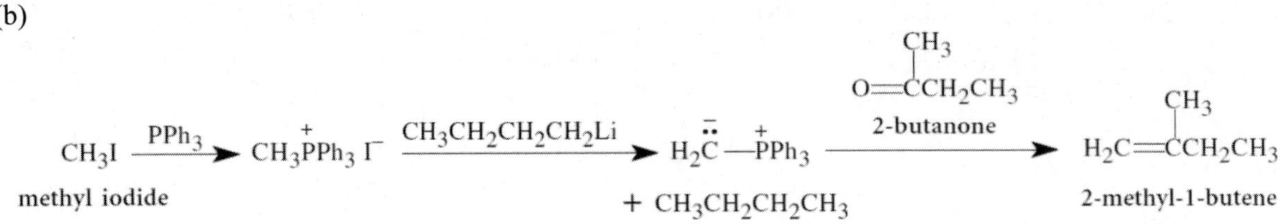

$$CH_3I \xrightarrow{PPh_3} CH_3\overset{+}{P}Ph_3 \ \overset{-}{I} \xrightarrow{CH_3CH_2CH_2CH_2Li} \overset{\cdot\cdot}{H_2C}\!-\!\overset{+}{P}Ph_3 \xrightarrow{2\text{-butanone}} H_2C\!=\!CCH_2CH_3$$

methyl iodide

$$+\ CH_3CH_2CH_2CH_3$$

O=CCH_2CH_3 (2-butanone) with CH_3

2-methyl-1-butene with CH_3

(c)

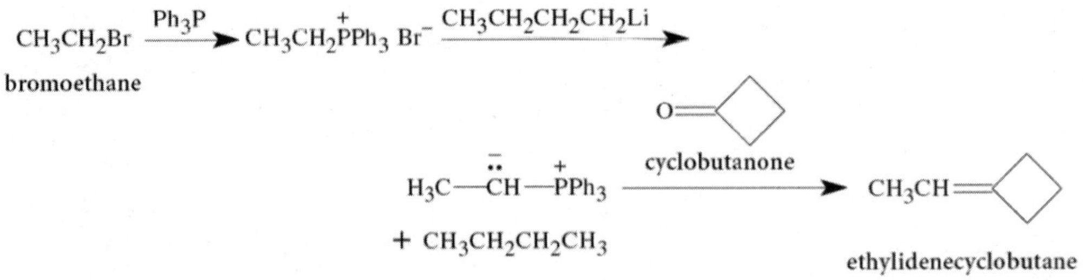

$$CH_3CH_2Br \xrightarrow{Ph_3P} CH_3CH_2\overset{+}{P}Ph_3 \ Br^- \xrightarrow{CH_3CH_2CH_2CH_2Li}$$

bromoethane

$$H_3C\!-\!\overset{\cdot\cdot}{C}H\!-\!\overset{+}{P}Ph_3 \xrightarrow{\text{cyclobutanone}} CH_3CH=\!\!\square$$

$$+\ CH_3CH_2CH_2CH_3$$

O=□ cyclobutanone

ethylidenecyclobutane

19.38   The starting material has five degrees of unsaturation, four of which are accounted for by a benzene ring. Consequently, the compound must be both an aldehyde and an alcohol:

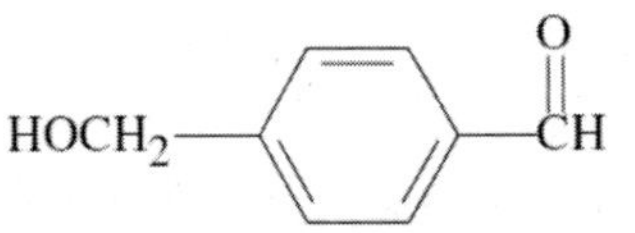

$$HOCH_2\!-\!\!\bigcirc\!\!-\!\overset{O}{\overset{\|}{C}}H$$ 4-(hydroxymethyl)benzaldehyde

19.39   Silver(I) oxide oxidizes the aldehyde selectively to the following carboxylic acid:

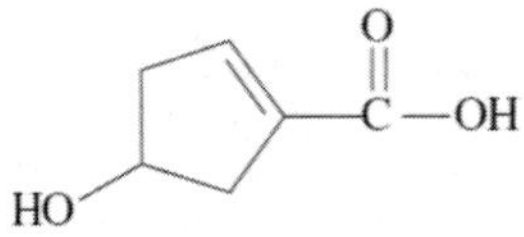

$$\underset{HO}{}\overset{O}{\overset{\|}{C}}\!-\!OH$$

4-hydroxy-1-cyclopentenecarboxylic acid

Solutions to Additional Problems

19.40   Only organic products are shown in the answers below.

(a)   Acetone is protonated on oxygen to give the conjugate acid shown below, but unless the acid is very strong, this reaction does not occur to a great extent.

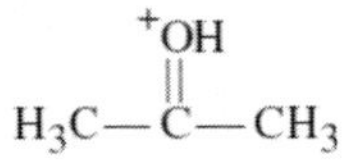

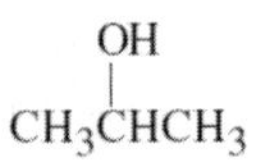

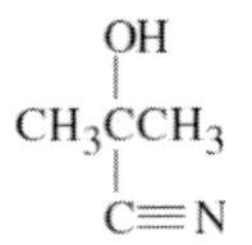

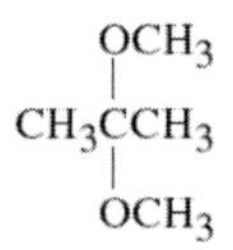

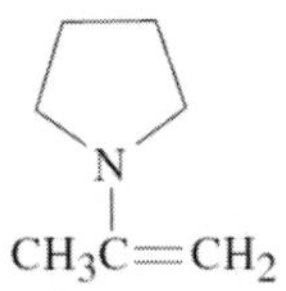

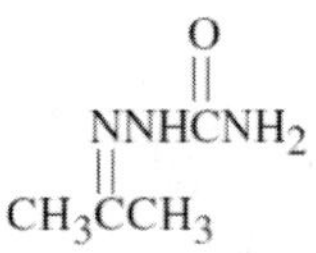

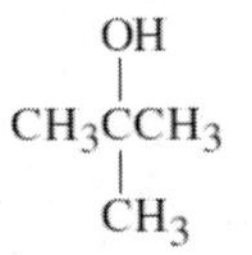

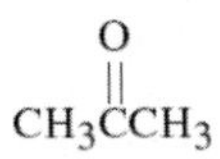

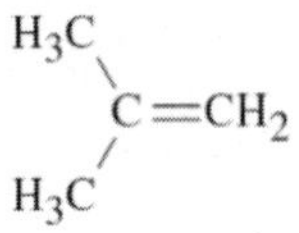

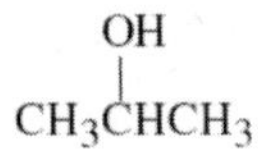

$$H_3C\!-\!\overset{+OH}{\overset{\|}{C}}\!-\!CH_3$$

(b)	(c)	(d)	(e)	(f)	(g)
$\underset{CH_3CHCH_3}{\overset{OH}{\|}}$	No reaction	$\underset{\underset{C\equiv N}{\|}}{\overset{OH}{\overset{\|}{CH_3CCH_3}}}$	$\underset{\underset{OCH_3}{\|}}{\overset{OCH_3}{\overset{\|}{CH_3CCH_3}}}$	pyrrolidine ring with $CH_3C\!=\!CH_2$	$\underset{CH_3CCH_3}{\overset{O}{\overset{\|}{}}}$ with NNHCNH_2

(h)	(i)	(j)	(k)	(l)	(m)
$\underset{\underset{CH_3}{\|}}{\overset{OH}{\overset{\|}{CH_3CCH_3}}}$	$\underset{CH_3CCH_3}{\overset{O}{\overset{\|}{}}}$	$\underset{H_3C}{\overset{H_3C}{}}\!\!C\!=\!CH_2$	$\underset{CH_3CHCH_3}{\overset{OH}{\|}}$	$(CH_3)_2C\!=\!CH_2$ $+\ Ph_3P\!=\!O$	$CH_3CH_2CH_3$

19.41   Only organic products are shown.

(a)   (b)   (c)   (d)

OH   OH   O   O
| | || ||
CH₃CH₂CH₂CHPh   CH₃CH₂CH₂CH₂   CH₃CH₂CH₂C—OH   CH₃CH₂CH₂C—OH

(e)   (f)   (g)   (h)

NOH   O   CH₃CH₂CH₂CH₃   CH₃CH₂CH₂CH=CHCH₃
|| ||
CH₃CH₂CH₂CH   CH₃CH₂CH₂C—OH   cis and trans

19.42   **(a)**   This is a simple addition reaction in which the sulfur of the bisulfite ion reacts as a nucleophile at the carbonyl carbon of the the aldehyde. In the last step of the mechanism, the initially formed addition product *A* ionizes because it is a fairly strong acid.

$$
\begin{array}{c}
\text{:O:}^- \\
\| \\
\text{RCH} \\
| \\
\text{:SO}_2\text{OH}
\end{array}
\;\rightleftharpoons\;
\begin{array}{c}
\text{:O:}^- \\
| \\
\text{RCH} \\
| \\
\text{SO}_2\text{OH}
\end{array}
\;\xrightarrow{\text{H—SO}_3^-}\;
\begin{array}{c}
\text{:OH} \\
| \\
\text{RCH} \\
| \\
\text{SO}_2\text{O—H} \;\;\text{:SO}_3^{2-}
\end{array}
\;\rightleftharpoons\;
\begin{array}{c}
\text{:OH} \\
| \\
\text{RCH} \;\; + \; \text{H—SO}_3^- \\
| \\
\text{SO}_2\text{O:}^-
\end{array}
$$

A

(b)   Acid, that is, H_3O^+, destroys sodium bisulfite by reacting with it and converting it into SO_2, a gas, and H_2O. Destruction of bisulfite in this manner eliminates a reactant in the equilibria shown above and pulls the reaction to the left, that is, toward free aldehyde. Hydroxide ion reacts with bisulfite and converts it into sulfite ion, $(OSO_2)^{2-}$; because bisulfite is removed as a reactant by this process as well, the equilibrium is also pulled to the left.

In the mechanisms presented thus far in this Study Guide and for the most part in the text, unshared electron pairs have been drawn in explicitly. Omitting these electron pairs can save some time and can eliminate a great deal of tedium in writing mechanisms; most chemists do not draw electron pairs explicitly. This shortcut is useful only if you realize that the unshared electron pairs which are not shown are understood to be present. In many of the mechanisms shown subsequently, electron pairs are omitted in the most common situations. (In unusual or uncommon situations, the electron pairs are retained.) Here are some typical situations in which electron pairs should not be necessary:

$$
H_2O \longrightarrow \overset{|}{\underset{|}{C}}{=}O \longrightarrow H_2\overset{+}{O}-\overset{|}{\underset{|}{C}}-O^- \quad \text{means} \quad H_2\overset{..}{O}: \longrightarrow \overset{|}{\underset{|}{C}}{=}\overset{..}{\underset{..}{O}} \longrightarrow H_2\overset{+}{O}-\overset{|}{\underset{|}{C}}-\overset{..}{\underset{..}{O}}:^-
$$

$$
(CH_3)_3C-Br \longrightarrow (CH_3)_3C^+ \;\; Br^- \quad \text{means} \quad (CH_3)_3C-\overset{..}{\underset{..}{Br}}: \longrightarrow (CH_3)_3C^+ \;\; :\overset{..}{\underset{..}{Br}}:^-
$$

$$
R_3N \quad \text{means} \quad R_3N:
$$

$$
R_3C^- \quad \text{means} \quad R_3\overset{-}{C}:
$$

Thus, an oxygen with three bonds and a positive charge has one unshared electron pair; an oxygen with one bond and a negative charge has three unshared pairs; an uncharged nitrogen with three bonds or a negatively charged carbon with three bonds has one unshared pair; and a halide ion has four unshared pairs.

The reason we do not have to show unshared valence electrons is that if the formal charge on an atom and the number of bonds to an atom are known, the number of unshared valence electrons are automatically known. It follows that the formal charge must be shown; otherwise, the valence electron count of an atom is undetermined.

Most of the mechanisms shown after this point will be written without showing the unshared pairs. If you ever become confused by the absence of electron pairs, *do not hesitate to draw them in explicitly*.

(c) The bisulfite addition product of 2-methylpentanal:

$$CH_3CH_2CH_2\overset{\overset{\displaystyle OH}{|}}{\underset{\underset{\displaystyle CH_3}{|}}{C}}H\overset{\overset{\displaystyle O}{\|}}{\underset{\underset{\displaystyle O}{\|}}{S}}\!-\!O^-\;Na^+$$

19.43 The central carbonyl group is hydrated for the following reason. The carbon of any carbonyl group bears a partial positive charge. The central carbonyl group is therefore bonded to two partially positive carbonyl carbons, and the positive charge on each of these carbons has an unfavorable repulsive interaction with the partial positive charge on the central carbon (see problem 19.18b for a similar situation). In addition, the other two carbonyl groups are conjugated with the benzene ring. As explained in the text, adjacent electronegative groups make carbonyl-addition reactions more favorable, and groups that are conjugated with the carbonyl group make carbonyl-addition reactions less favorable. It follows that hydration of the central carbonyl group is the more favorable process. The structure of the hydrate is as follows:

hydrate of ninhydrin

19.44 (a) Two diastereomeric alcohols are formed by the nucleophilic reaction of a hydride at the upper and lower face, respectively, of the carbonyl group followed by protonation of the resulting alkoxides.

from the nucleophilic reaction of hydride at the upper face of the carbonyl carbon

+

from the nucleophilic reaction of hydride at the lower face of the carbonyl carbon

(b) Both cis and trans isomers of the alkene $PhCH\!\!=\!\!CHCH_3$ [(1-propenyl)benzene, also known as β-methylstyrene] are formed.

19.45 A five-membered *cyclic acetal* is formed when the middle OH group and one of the end OH groups of glycerol react with hexanal under acid-catalyzed conditions. A six-membered *cyclic acetal* is formed when both of the end OH groups of glycerol react with hexanal under acid-catalyzed conditions. This results in two sets of two diastereomers.

The five-membered cyclic acetals are chiral; each diastereomer is a racemate. The six-membered cyclic acetals are not chiral.

19.46   (a)   Glycerol has three hydroxy groups. Two possible cyclic acetals can be formed; one (*A*) contains a six-membered ring, and the other (*B*) contains a five-membered ring.

(b)   Only compound *B* is chiral, and for this reason only compound *B* can be resolved into enantiomers; hence, compound *B* is the observed compound.

19.47   The two separable isomers are the diastereomeric acetals:

19.48   (a)   The reaction is formation of a hydrazone, a type of imine; see Table 19.3, text p. 985.

(b)   The reaction is a straightforward dimethyl acetal formation.

1,1-dimethoxy-4-methylcyclohexane

(c)   The formula shows that only one of the carbonyl groups is involved in acetal formation; the aldehyde carbonyl reacts selectively because aldehydes form addition compounds more rapidly than do ketones, and because addition reactions of aldehydes are thermodynamically more favorable than those of ketones.

1,1-dimethoxy-2-propanone

(d)   This reaction is internal acetal formation, in which the two alcohol groups form an acetal with the carbonyl group within the same molecule.

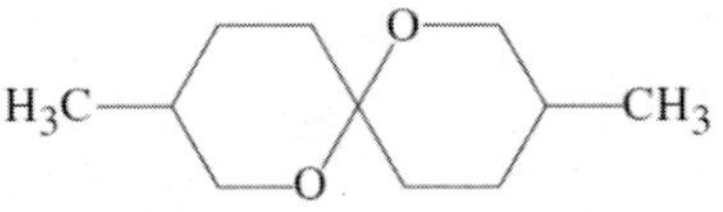

$$H_3C \cdots \text{(spiro bicyclic structure with two O)} \cdots CH_3$$

(e)   This is a Grignard addition to the ketone to give a tertiary alcohol that subsequently dehydrates under the acidic conditions to an alkene. Whether the dehydration occurs depends on the acid concentration and whether the conditions are designed to remove water.

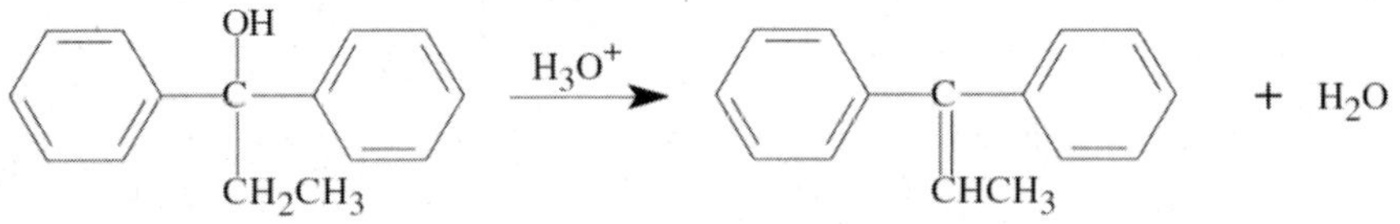

$$\text{1,1-diphenyl-1-propanol} \xrightarrow{H_3O^+} \text{diphenyl alkene} \;+\; H_2O$$

1,1-diphenyl-1-propanol

(f)   This is a Wittig reaction.

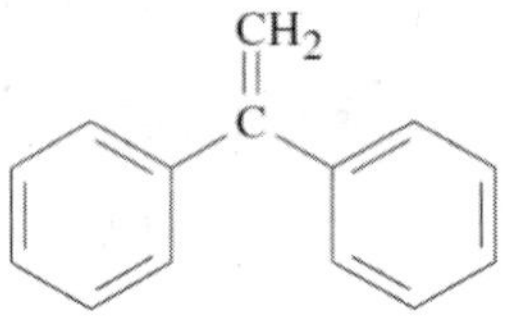

$$\text{1,1-diphenylethylene (CH}_2\text{=C(Ph)}_2\text{)}$$

(g)   A "double" Wittig reagent is formed from the two benzylic bromides, and each Wittig reagent reacts at a different carbonyl group of the same aldehyde molecule to form a diene.

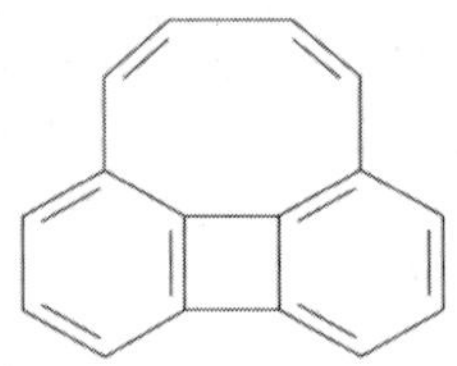

19.49 (a)   In this reaction, the carbonyl group is reduced to an alcohol, which is subsequently ionized by NaH. The resulting intermediate is essentially a Wittig oxaphosphetane intermediate, as the hint suggests; this intermediate decomposes to form an alkene.

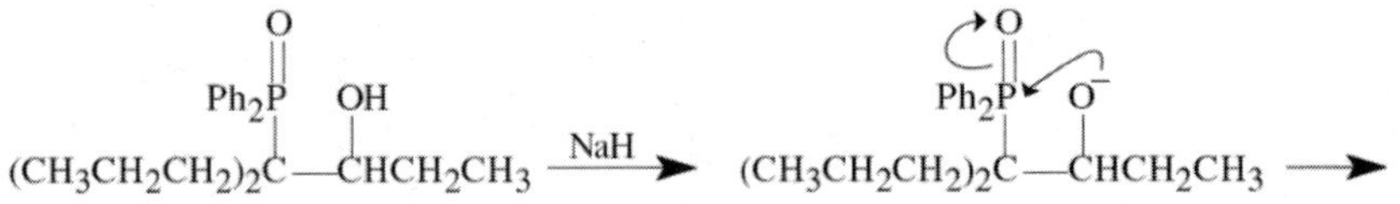

$$(CH_3CH_2CH_2)_2C(PPh_2)(OH)\!-\!CHCH_2CH_3 \xrightarrow{NaH} (CH_3CH_2CH_2)_2C(PPh_2)(O^-)\!-\!CHCH_2CH_3 \longrightarrow$$

product of NaBH4 reduction

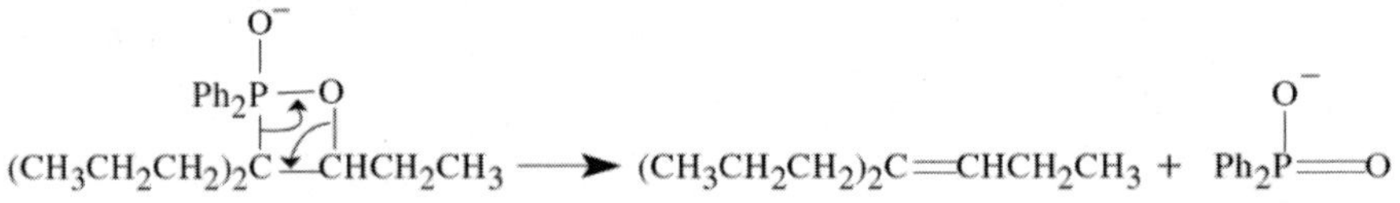

$$(CH_3CH_2CH_2)_2C\!-\!CHCH_2CH_3 \longrightarrow (CH_3CH_2CH_2)_2C\!=\!CHCH_2CH_3 \;+\; Ph_2P\!=\!O$$

an oxaphosphetane
(see Eq. 19.73a)

(b)   An initially formed imine reacts with sodium borohydride to form an amine. (Although this reaction is discussed in Sec. 23.7B, text p. 1199, the product follows directly from the hint.)

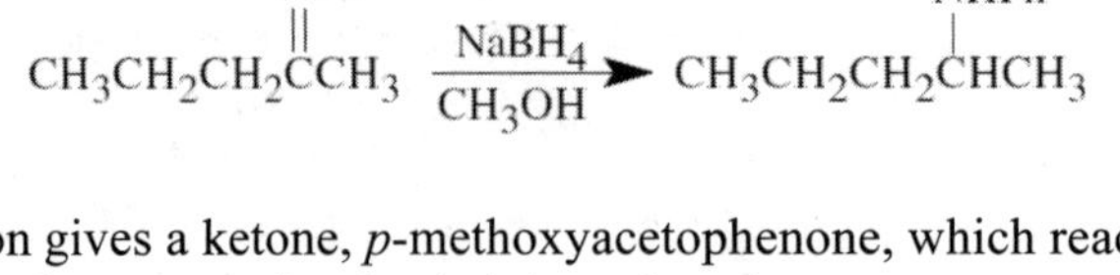

$$CH_3CH_2CH_2\overset{NPh}{\overset{\|}{C}}CH_3 \xrightarrow[CH_3OH]{NaBH_4} CH_3CH_2CH_2\overset{NHPh}{\overset{|}{C}H}CH_3$$

19.50 (a)   Friedel–Crafts acylation gives a ketone, *p*-methoxyacetophenone, which reacts in a Wolff–Kishner reduction of the carbonyl group to give 1-ethyl-4-methoxybenzene.

H_3CO—⟨benzene⟩—$\overset{\overset{O}{\|}}{C}$—$CH_3$ $\xrightarrow{\text{Wolff-Kishner}}$ CH_3O—⟨benzene⟩—CH_2CH_3

p-methoxyacetophenone
product of Friedel-Crafts acylation

1-ethyl-4-methoxybenzene
(*p*-ethylanisole)

(b)   Prepare the triflate of the phenol and subject it to a Stille reaction (Sec. 18.10B) with tetraethylstannane.

CH_3O—⟨benzene⟩—OH $\xrightarrow[\text{anhydride}]{\text{triflic}}$ CH_3O—⟨benzene⟩—OTf $\xrightarrow[\text{LiCl (excess)}]{\overset{(CH_3CH_2)_4Sn}{Pd(0)\ (cat.),\ PPh_3}}$

hydroquinone monomethyl ether
(*p*-methoxyphenol)

CH_3O—⟨benzene⟩—CH_2CH_3

19.51   The data indicate that compound *A* is a benzene derivative with an additional degree of unsaturation. Its reactivity in the Clemmensen reduction and the formation of a xylene suggests that it has two substituents on a benzene ring, one of which is a methyl group and one of which is an aldehyde. Only *p*-methylbenzaldehyde would give, after Clemmensen reduction, a compound (*p*-xylene) that in turn gives, because of its symmetry, one and only one monobromination product, 2-bromo-1,4-dimethylbenzene.

H_3C—⟨benzene⟩—CH=O $\xrightarrow{\text{Zn/Hg, HCl}}$ H_3C—⟨benzene⟩—CH_3 $\xrightarrow{\text{Br}_2,\ \text{Fe}}$ H_3C—⟨benzene⟩—CH_3 (with Br)

4--methylbenzaldehyde
(compound *A*)

p-xylene
(compound *A*)

2-bromo-1,4-dimethylbenzene

19.52   As always, bear in mind that there is often more than one acceptable synthesis that fits the parameters given.
(a)

$CH_3CH_2CH_2\overset{\overset{O}{\|}}{CH}$ $\xrightarrow[\text{2) H}_3\text{O}^+]{\text{1) PhMgBr}}$ $CH_3CH_2CH_2\overset{\overset{OH}{|}}{CH}$—$Ph$ $\xrightarrow{\text{PCC}}$ $CH_3CH_2CH_2\overset{\overset{O}{\|}}{C}$—$Ph$

butyraldehyde   **butyrophenone**

(b)

⟨cyclohexanone⟩=O $\xrightarrow[\text{CH}_3\text{OH}]{\text{NaBH}_4}$ ⟨cyclohexyl⟩—OH $\xrightarrow[\text{H}_2\text{SO}_4]{\text{conc. HBr}}$ ⟨cyclohexyl⟩—Br $\xrightarrow{\text{Mg, ether}}$

cyclohexanone

⟨cyclohexyl⟩—$MgBr$ $\xrightarrow[\text{2) H}_3\text{O}^+]{\text{1) CH}_3\overset{\overset{O}{\|}}{C}\text{CH}_3}$ ⟨cyclohexyl⟩—$\overset{\overset{OH}{|}}{C}(CH_3)_2$

2-cyclohexyl-2-propanol

(c)

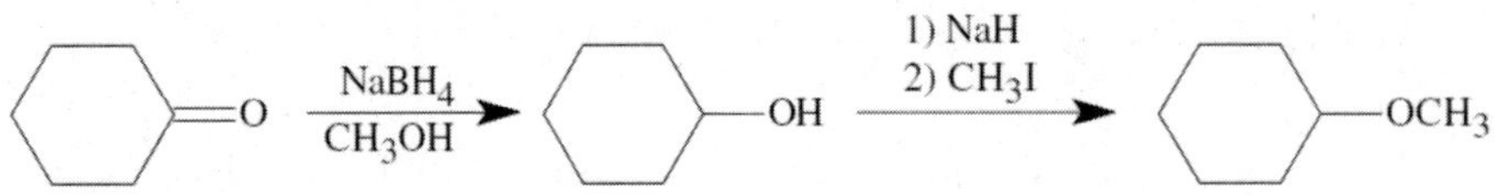

cyclohexanone   cyclohexyl methyl ether
(methoxycyclohexane)

(d)

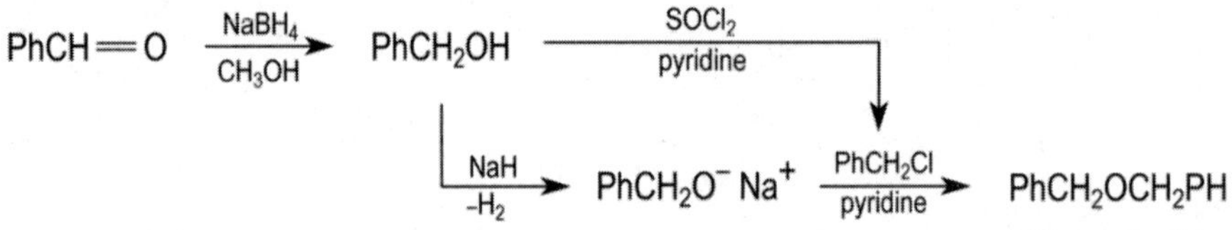

dibenzyl ether

(e)

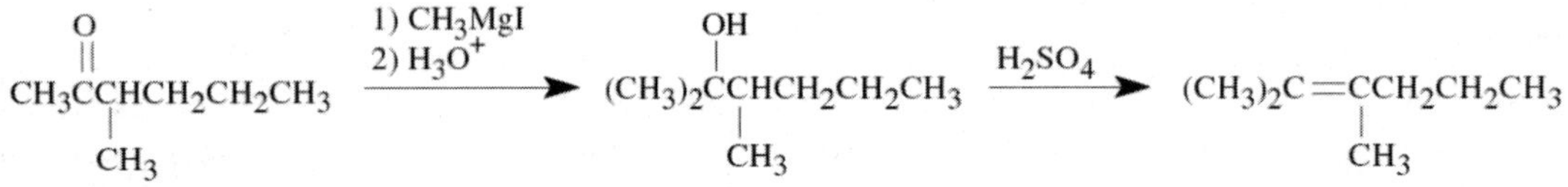

3-methyl-2-hexanone   **2,3-dimethyl-2-hexene**

(f)

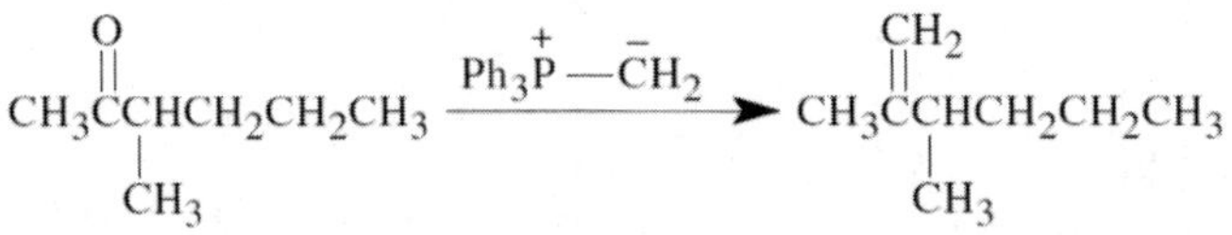

3-methyl-2-hexanone   **2,3-dimethyl-1-hexene**

(g)

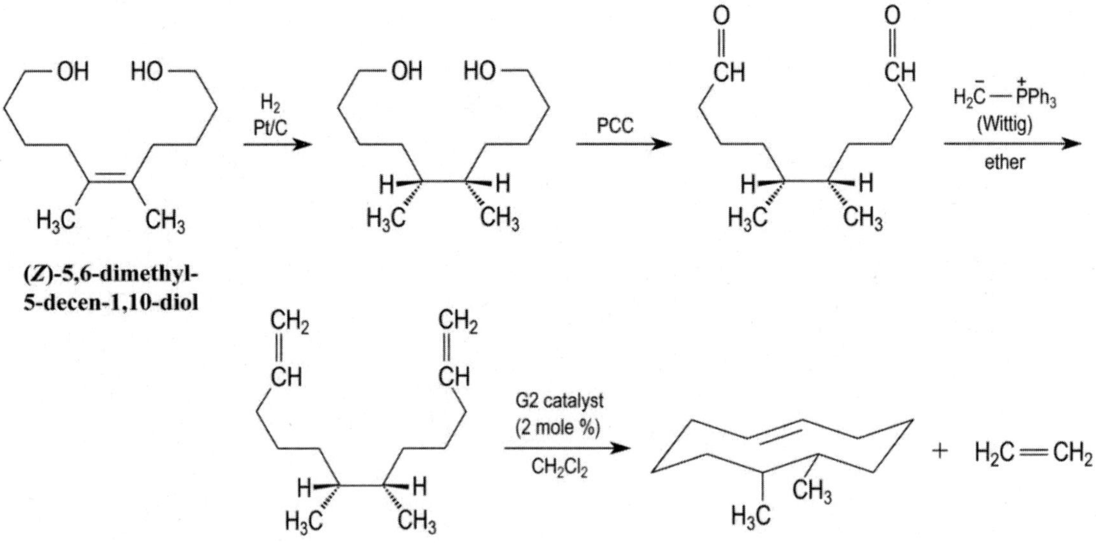

**(Z)-5,6-dimethyl-
5-decen-1,10-diol**

(h)

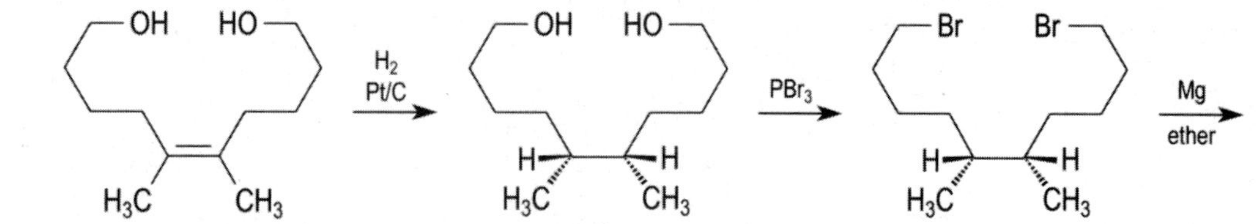

(Z)-5,6-dimethyl-
5-decen-1,10-diol

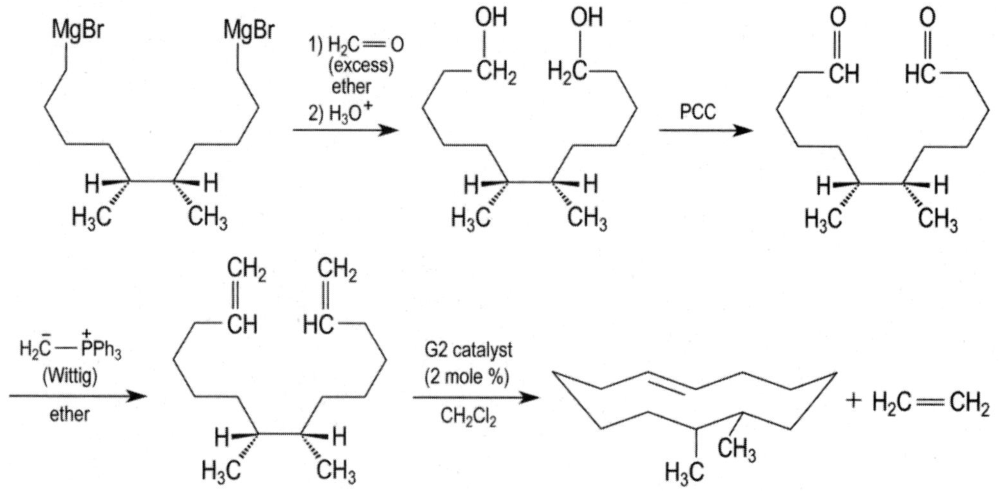

(i)

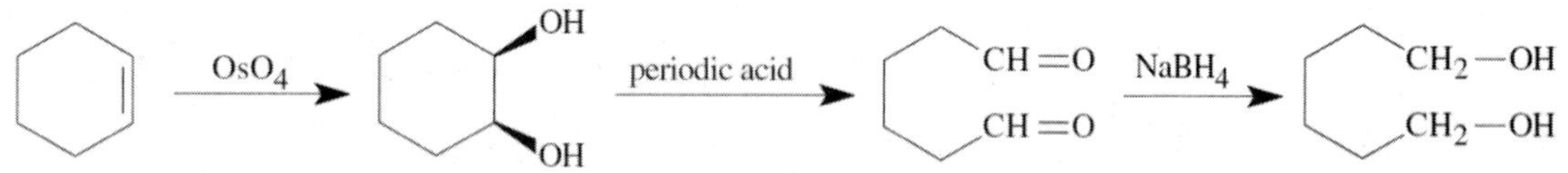

cyclohexene   1,6-hexanediol

(j)   In the following synthesis, the two acyl groups could be introduced in the opposite order.

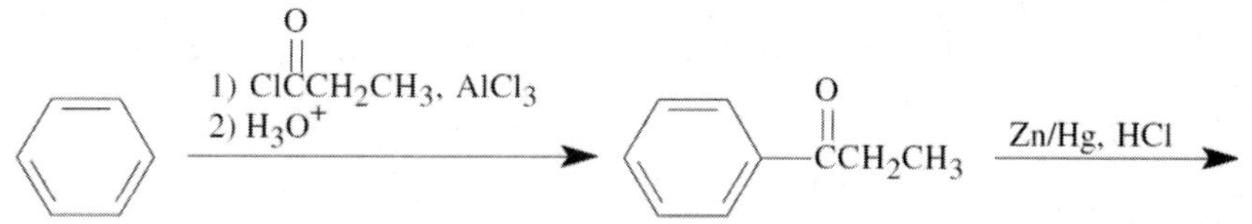

benzene

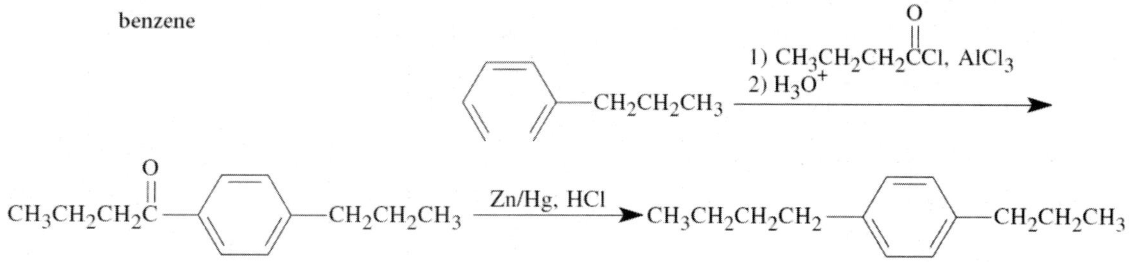

1-butyl-4-propylbenzene

(k)   First prepare the Grignard reagent benzylmagnesium bromide from benzaldehyde:

Then let this Grignard reagent react with benzaldehyde and oxidize the resulting alcohol:

(l) The first hint indicates that a protecting group must be used. The second hint is meant to call attention to the hemiacetal linkage which, because it is a five-membered *cyclic* hemiacetal, will form spontaneously from the corresponding aldehyde.

(m) A protecting group must be used in this synthesis.

If you proposed protonolysis of the Grignard reagent in D_3O^+ you have the right idea. However, use of D_3O^+ would actually result in the following product:

The reason is that acidic D_2O brings about the exchange of protons that are α to an aldehyde or ketone carbonyl group. This is a reaction you will learn about in Chapter 22. (See Eq. 22.6 on text p. 1108.) To avoid this exchange use D_2O for protonolysis of the Grignard reagent, which gives $Mg(OD)_2$ as a by-product. Because this by-product maintains basic conditions, the acetal is left intact. Hydrolysis of the acetal is then carried out with H_3O^+ in H_2O.

(n)

19.53   (a)   An Lewis acid–base dissociation of the bromide—essentially an S_N1 reaction—gives an α-hydroxy carbocation, which is, by resonance, also the conjugate acid of acetophenone. Deprotonation by bromide ion gives the ketone and HBr.

(b)   Part (a) of this problem shows that α-bromo alcohols break down spontaneously to aldehydes or ketones.

Osmium tetroxide reacts to form the glycol A, which is also an α-bromo alcohol. This breaks down to the aldehyde as implied by the foregoing reaction:

19.54   The molecular formula of compound *A* shows that two equivalents of methanol are added to the alkyne. The fact that the product hydrolyzes to acetophenone indicates that the two methoxy groups of compound *A* are on the same carbon, that is, that compound *A* is an acetal.

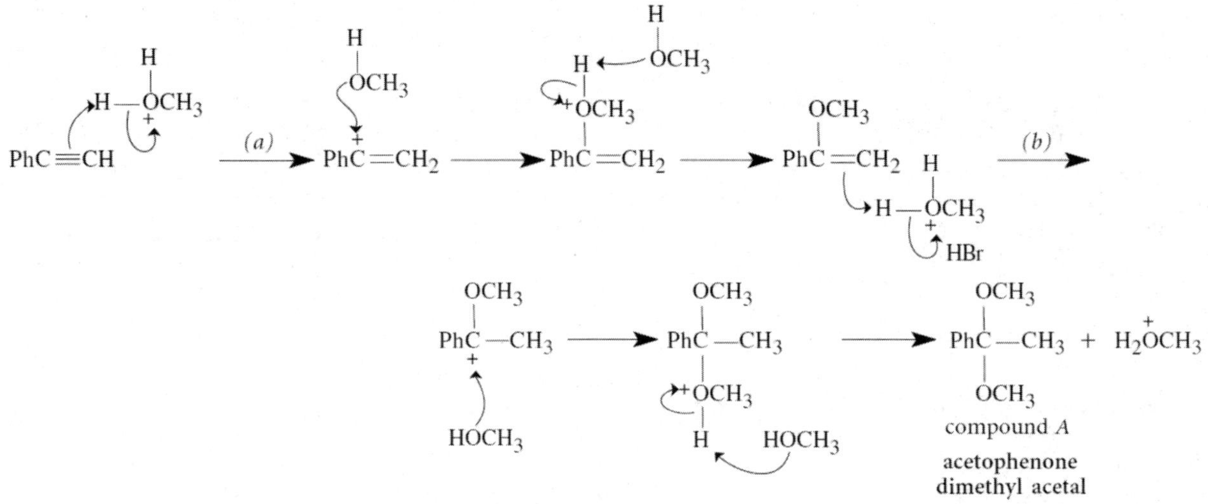

In the step labeled (*a*), the acid $CH_3\overset{+}{O}H_2$ is used to protonate the alkyne because it is the major acidic species present when H_2SO_4 is dissolved in methanol (just as H_3O^+ is the major acidic species present when H_2SO_4 is dissolved in water). Protonation occurs on the terminal carbon because it gives a carbocation that is benzylic and therefore resonance-stabilized. In the step labeled (*b*), protonation again occurs on the terminal carbon because the resulting carbocation is resonance-stabilized by electron donation from both the benzene ring and the neighboring oxygen.

19.55   (a)   A THP ether is a cyclic acetal. In aqueous acid, the cyclic acetal undergoes hydrolysis to give, besides the ROH, 5-hydroxypentanal, which is in equilibrium with the cyclic hemiacetal 2-hydroxypyran. The equilibrium between the two favors the cyclic hemiacetal. (See Eq. 19.49 on text p. 979.)

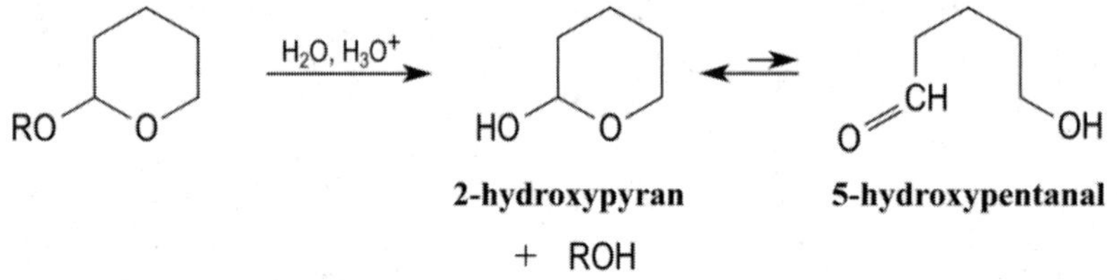

2-hydroxypyran   **5-hydroxypentanal**

+ ROH

(b)

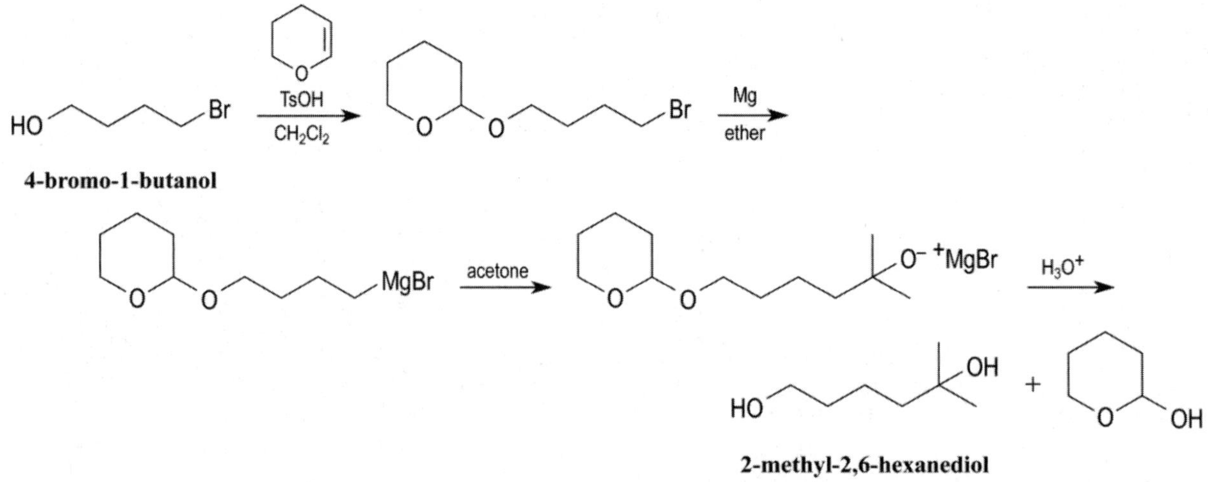

4-bromo-1-butanol

2-methyl-2,6-hexanediol

19.56   (a)

$$CH_3CH_2CH_2CH_2CH{=}O \quad \text{and} \quad H_2NNH\text{—}\langle\text{phenyl ring}\rangle\text{—}OCH_3$$

pentanal (valeraldehyde)

p-methoxyphenylhydrazine

(b)   Because a cyclic imine is formed, the reaction must be intramolecular; that is, the amine and the carbonyl group that react must be in the same molecule.

19.57   (a)

(b)   The NADPH is behind the plane created by the carbonyl group of the 3-ketobutanoyl thioester. The pro-(*R*) hydrogen is delivered to the back side of the carbonyl. In the transition state, the bond between the hydride and the carbonyl carbon begins to form, and the oxygen is pushed forward out of the plane, while developing a partial negative charge.

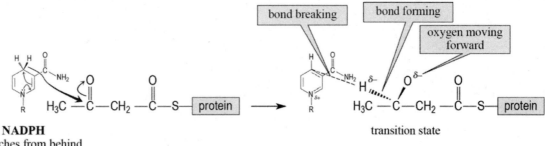

bond breaking

bond forming

oxygen moving forward

NADPH
(approaches from behind
the carbonyl group)

transition state

(c)   The –OH group of a tyrosine or serine residue (as shown) can stabilize the developing negative charge on the carbonyl oxygen through hydrogen bonding and lowers the energy of the transition state, speeding up the enzymatic reaction.

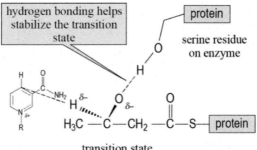

hydrogen bonding helps
stabilize the transition
state

protein

serine residue
on enzyme

transition state

19.58   Because compound *A* has one carbon more than the product *D*, a carbon is lost at some point; the most likely place is the benzylic oxidation that gives compound *D* itself. The structure of compound *D* shows that all compounds contain benzene rings with four substituent groups. That compound *A* can be regenerated by oxidation of optically active *B* suggests that *B* is an alcohol and *A* is a ketone. Compound *B* cannot be a primary alcohol because such an alcohol could not be optically active. (The chirality has to be associated with the alcohol functionality because optical activity is lost when the alcohol is oxidized.) It cannot be a tertiary alcohol because it could not be formed by LiAlH$_4$ reduction and could not be oxidized. Compound A has one degree of unsaturation in addition to the four accounted for by its benzene ring and the one accounted for by its carbonyl group; the only way to accommodate all the data is for compound A to contain a ring. The correct structures are as follows:

19.59   Compound A is an aldehyde because it is oxidized with Ag(I). Because there is no additional unsaturation, the remaining oxygen is accounted for by either an alcohol or an ether functional group. Because Clemmensen reduction of the CrO$_3$ oxidation product gives a compound without oxygens, the CrO$_3$ oxidation product of A must be a keto aldehyde or a dialdehyde. Because compound A is oxidized to a dicarboxylic acid by H$_2$CrO$_4$, it must contain a primary alcohol; hence, its CrO$_3$ oxidation product must be a dialdehyde. Compound A, then, is a chiral hydroxy aldehyde that is oxidized by CrO$_3$ to an achiral dialdehyde. The Clemmensen reduction product shows that all compounds have the carbon skeleton of 3-methylpentane. The compounds with this carbon skeleton that fit all the data are the following.

19.60   (a)   The mechanism below begins with the protonated aldehyde chloral, which serves as a carbocation electrophile in the ring alkylation of chlorobenzene. The resulting product, an alcohol, dehydrates under the acidic conditions to give another carbocation that alkylates a second chlorobenzene molecule and thus forms the product.

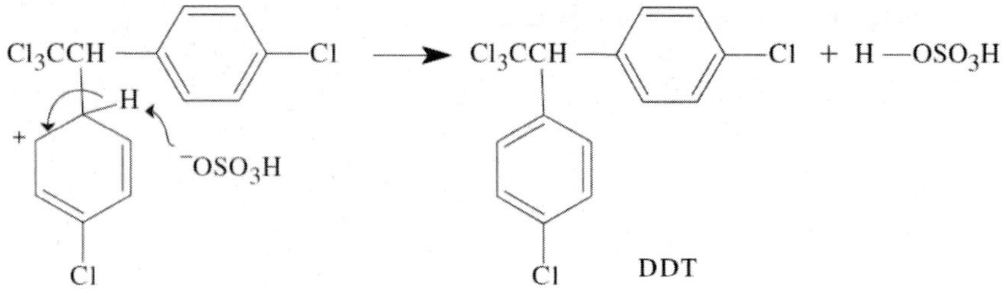

DDT

(b)

a resonance-stabilized
carbocation

(resonance stablized)

bisphenol A

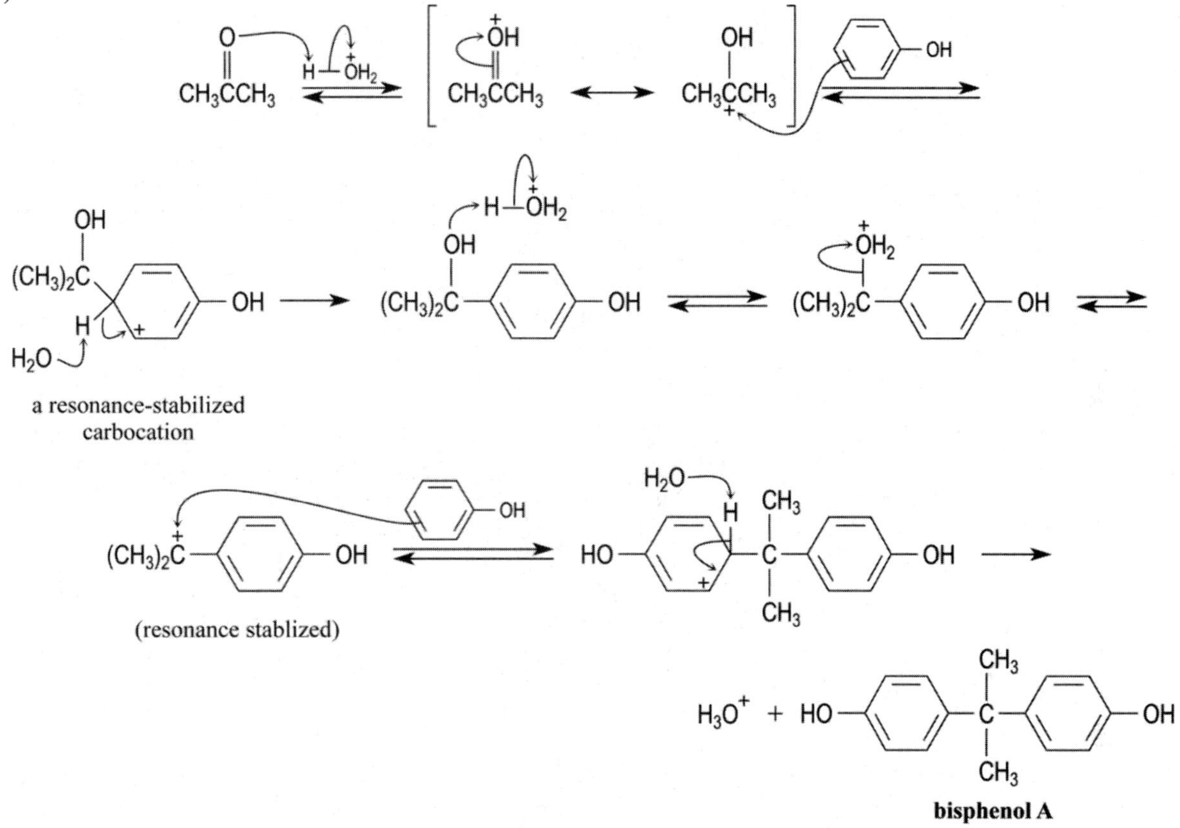

19.61   The first step is the formation of an imine, by the standard mechanism (see the solution to problem 19.30 in this manual). The imine carbon-nitrogen double bond can be thought of as a less reactive version of a carbonyl carbon-oxygen double bond; draw a resonance structure of the imine to convince yourself of the partial positive charge on the carbon. An intramolecular Friedel-Crafts "acylation" occurs, closing the ring. Regeneration of the aromatic ring by a base yields salsolinol.

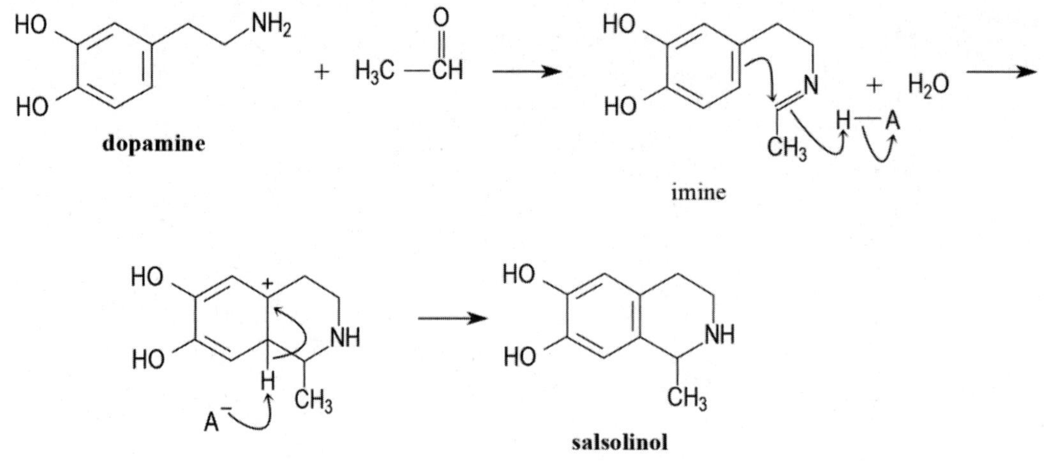

dopamine

imine

salsolinol

19.62 (a) In this case, LiAlD$_4$ serves as a source of nucleophilic isotopic hydride (deuteride); deuteride opens the epoxide with inversion of configuration.

(racemate)

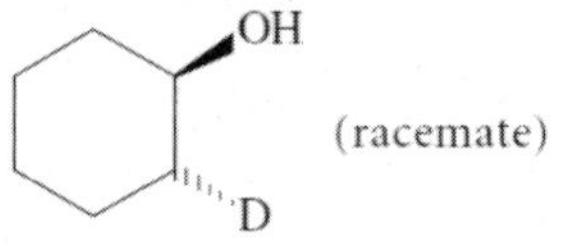

(b) Hydride ion, delivered from LiAlH$_4$, opens the epoxide with inversion of configuration. Because the epoxide carbons are enantiotopic and therefore chemically equivalent, the nucleophilic reaction of hydride at each of these carbons occurs at the same rate. As a result, the racemic product is formed. An equivalent view is that the starting epoxide is a meso compound. Hence, the chiral product must be a racemate.

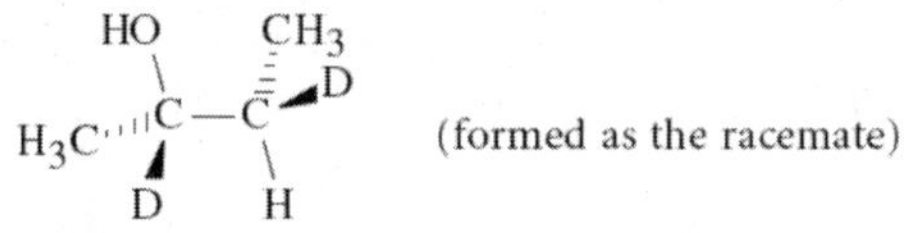

(formed as the racemate)

19.63 (a) Thumbs wants a Grignard reagent to react selectively with a ketone in the presence of an aldehyde. Because aldehydes are more reactive than ketones, the aldehyde, not the ketone, will react most rapidly.

(b) Thumbs is trying to form a Wittig reagent by first carrying out an S$_N$2 reaction of triphenylphosphine (Ph$_3$P) with neopentyl bromide. Because neopentyl bromide is virtually unreactive in S$_N$2 reactions, this reaction will not occur. (See Fig. 9.4(b) on text p. 397 and associated discussion.)

19.64 (a)

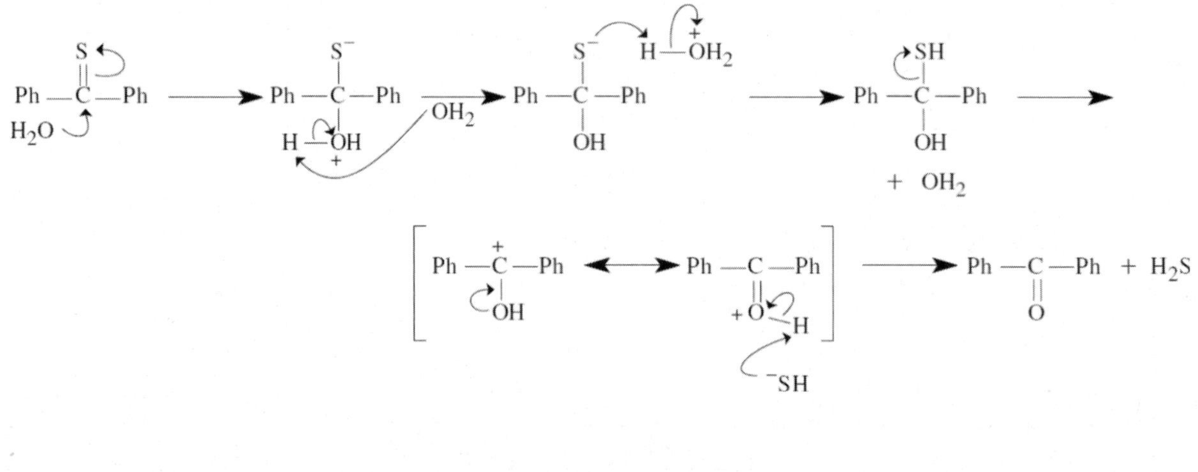

Compare loss of the —SH group, which does not require protonation, to loss of the —OH group in the acid-catalyzed decomposition of a hydrate, which does require protonation. Protonation of the —SH group is not necessary for it to serve as a leaving group for two reasons. First, the C—S bond is substantially weaker than the O—H bond; and second, ‾SH is a much weaker base than ‾OH. Furthermore, the —SH group itself is a much weaker base than the —OH group; consequently, much less —SH than —OH is protonated under neutral or dilute-acid conditions.

(b)   Opening of the epoxide by the phosphorus compound yields an oxaphosphetane, the same type of compound that serves as an intermediate in the Wittig reaction; this compound collapses to an alkene as it does in the Wittig reaction. (See also the solution to Problem 19.49a, text p. 998.)

$$(CH_3O)_3P: \quad H_2C\!-\!CHCH_3 \longrightarrow (CH_3O)_3\overset{+}{P} \quad O^- \quad H_2C\!-\!CHCH_3 \longrightarrow (CH_3)O)_3P\!-\!O \quad H_2C\!-\!CHCH_3 \longrightarrow$$

an oxaphosphetane

$$(CH_3O)_3P\!=\!O \ + \ H_2C\!=\!CHCH_3$$

(c)   The imine, formed by the mechanism shown in Eqs. 19.60a–b on text p. 984, is in equilibrium with a small amount of an enamine (just as an aldehyde is in equilibrium with a small amount of enol). The nitrogen of the enamine serves as a nucleophile that reacts with the second aldehyde carbonyl group intramolecularly to form the ring.

(d)   Protonation of the aldehyde gives an α-hydroxy carbocation that alkylates the double bond. The mechanism below starts with the protonated aldehyde.

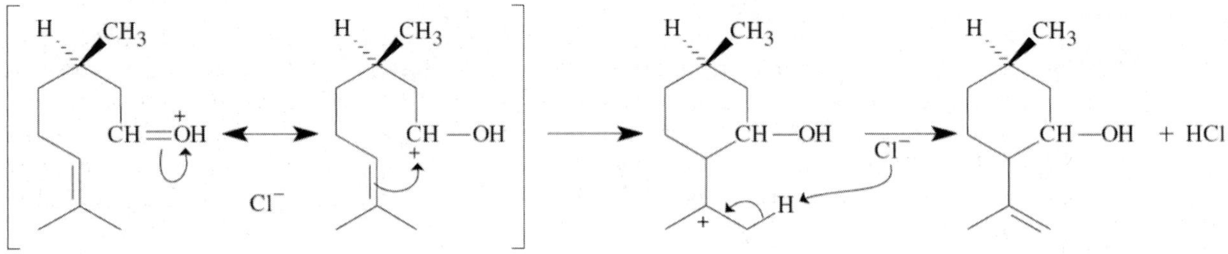

protonated aldehyde

(e)   The carbonyl oxygen serves as a nucleophile that reacts at the carbonyl carbon of a protonated acetaldehyde molecule.

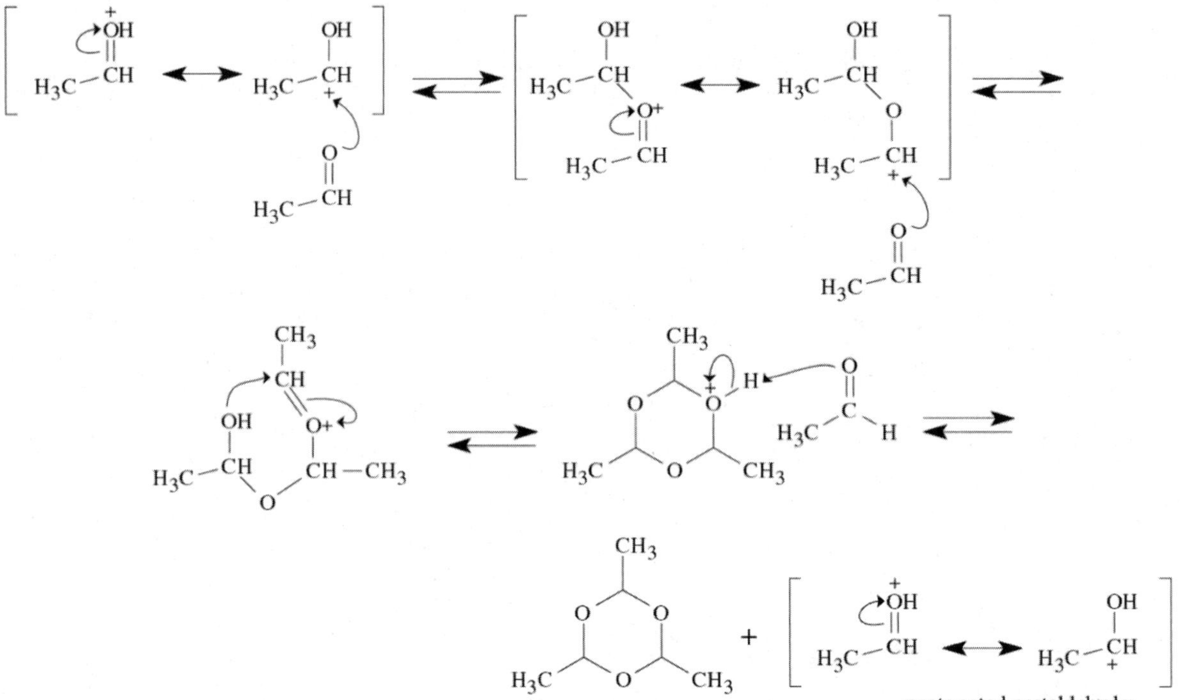

paraldehyde

protonated acetaldehyde;
reacts to give more paraldehyde

19.65   (a)   The synthesis of 4-methyl-3-heptanol:

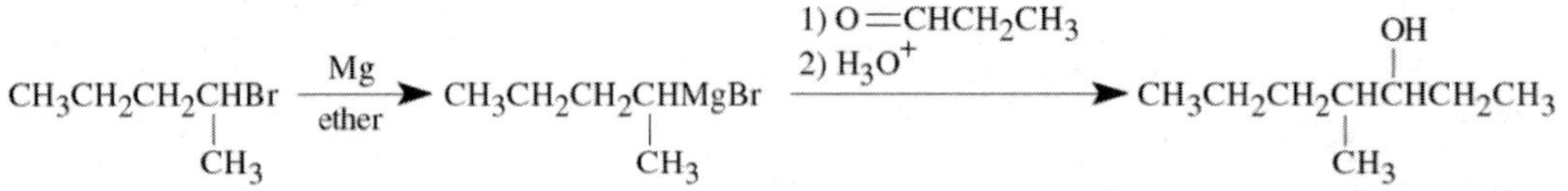

$$CH_3CH_2CH_2CHBr \xrightarrow[\text{ether}]{Mg} CH_3CH_2CH_2CHMgBr \xrightarrow[\text{2) } H_3O^+]{\text{1) } O=CHCH_2CH_3} CH_3CH_2CH_2CHCHCH_2CH_3$$

with CH_3 substituent on the CHBr and CHMgBr carbons, and OH / CH_3 substituents on the product:

4-methyl-3-heptanol

(b)   Because 4-methyl-3-heptanol contains two asymmetric carbons, it can exist as diastereomers. Even if the starting alkyl halide were enantiomerically pure, and even if the Grignard reagent could be prevented from racemizing, the product would likely be a mixture of diastereomers.

19.66   (a)   The NMR spectrum suggests a great degree of symmetry: four aromatic protons, all equivalent, and two methyl groups adjacent to a carbonyl group. The IR spectrum suggests the presence of a carbonyl group conjugated with the aromatic ring. The chemical shift of the ring protons is greater than usual because of the electronegative carbonyl groups. Because all four ring protons are chemically equivalent, they appear in the NMR spectrum as a singlet.

$$H_3C-\overset{\overset{\displaystyle O}{\|}}{C}-\underset{}{\text{⟨benzene⟩}}-\overset{\overset{\displaystyle O}{\|}}{C}-CH_3$$

p-diacetylbenzene

(b)   The NMR spectrum indicates the presence of a *tert*-butyl group and an aldehyde. The compound is

$(CH_3)_3C-CH=O$ 2,2-dimethylpropanal (pivalaldehyde)

(c)   The NMR indicates two vinylic hydrogens as well as an aldehyde adjacent to a CH. The IR absorption at 970 cm^{-1} shows that the vinylic hydrogens are trans. The double bond and the aldehyde account for all of the unsaturation. The UV absorption at 215 nm indicates that the alkene and aldehyde double bonds are conjugated. The allylic CH$_2$ absorption at δ 2.3 is essentially a quartet, which suggests three hydrogens on adjacent carbons, that is, a vinylic hydrogen (because the absorption is allylic) and two others. The high-field triplet integrates for three protons, and thus represents a methyl group; its splitting suggests an adjacent CH$_2$ group. The absorption at δ 5.2 imples a —CH$_2$— adjacent to five protons. The following structure is thus defined by the data:

$$\begin{array}{c} CH_3CH_2CH_2 \\ \diagdown \\ \diagup \\ H \end{array} C{=}C \begin{array}{c} H \\ \diagdown \\ \diagup \\ CH{=}O \end{array} \quad \textit{(E)}\text{-2-hexenal}$$

19.67   The double absorptions of equal intensity in the mass spectrum indicate the presence of one bromine. The IR absorptions indicate a ketone in which the carbonyl group is conjugated to a benzene ring (1678 cm^{-1}) as well as aromatic "double bonds" (1600 cm^{-1}). The NMR spectrum indicates a *para*-disubstituted benzene ring and a methyl group adjacent to a carbonyl group. The compound has structure *A* below (*p*-bromoacetophenone). This compound has the molecular mass (198 for ^{79}Br) indicated by the mass spectrum.

$$Br-\underset{}{\text{⟨benzene⟩}}-\overset{\overset{\displaystyle O}{\|}}{C}-CH_3 \qquad\qquad H_3C-\underset{}{\text{⟨benzene⟩}}-\overset{\overset{\displaystyle O}{\|}}{C}-Br$$

p-bromoacetophenone   *B*
A

If you proposed structure *B*, it is not a bad answer, because you probably don't know about the IR spectra of acid halides. However, the IR carbonyl absorption of such a compound occurs at much higher frequency; furthermore, the chemical shift of the methyl group is more consistent with its being adjacent to a carbonyl group.

19.68   (a)   The cyclic trimer can exist as a mixture of diastereomers:

$$\text{and}$$

β-form α-form
two diastereomers of the cyclic trimer of chloral

(You were not expected to specify which form is which.)

    (Note that paraldehyde in Problem 19.64, part (e), can also exist as diastereomers.)

(b)   The higher-melting form has the more symmetrical structure; in this case, the β-form. See Sec. 8.5D.

(c)   The α-form has two chemically nonequivalent sets of protons; hence, its proton NMR spectrum has two resonances in a 2:1 ratio. Because all protons in the β-form are chemically equivalent, this form has only one resonance in its NMR spectrum. Related conclusions apply to the carbon NMR spectra: the α-form has four lines in its spectrum, whereas the β-form has only two.

19.69   (a)   Subject the ketone benzyl phenyl ketone to an excess of isotopically enriched water in the presence of an acidic or basic catalyst. Several cycles of formation and decomposition of the hydrate will eventually "wash out" the isotope of lower atomic mass. (Note that *O = ^{18}O)

$$
\underset{\substack{\text{benzyl phenyl}\\ \text{ketone}}}{\overset{O}{\underset{\|}{PhCCH_2Ph}}} + \underset{\substack{\text{(large}\\ \text{excess)}}}{\overset{*}{H_2O}} \;\rightleftharpoons\; \underset{\substack{OH\\ *}}{\overset{OH}{\underset{|}{PhCCH_2Ph}}} \;\longrightarrow\; \overset{O^*}{\underset{\|}{PhCCH_2Ph}} + H_2O
$$

Note that the last step is effectively irreversible only because of the large excess of isotopically enriched water; the unenriched water is present at very low concentration at all times. The ratio of enriched to unenriched ketone will be essentially the same as the ratio of enriched to unenriched water at equilibrium. Once the ketone is formed, reduce it with $LiAlH_4$ or $NaBH_4$ to provide the desired product:

$$
\overset{O^*}{\underset{\|}{PhCCH_2Ph}} \quad\xrightarrow[\text{2) }H_3O^+]{\text{1) }LiAlH_4}\quad \overset{*}{\underset{|}{OH}}{PhCHCH_2Ph}
$$

Another reasonable synthesis is oxymercuration-reduction of *cis*-or *trans*-stilbene ("1,2-diphenylethylene") with isotopically enriched water used as the nucleophile in the oxymercuration step.

$$
\underset{\textit{cis- or trans-}\textbf{stilbene}}{PhCH{=\!=}CHPh} \quad\xrightarrow[\text{2) }NaBH_4]{\text{1) }Hg(OAc)_2/H_2O^*/THF}\quad \overset{*}{\underset{|}{OH}}{PhCHCH_2Ph}
$$

The S_N1 reaction of the corresponding alkyl bromide with isotopic water would not be a satisfactory synthesis because this particular S_N1 reaction gives mostly *trans*-stilbene (structure above), the elimination (E1) product. This alkene is the major product because it has an internal conjugated double bond and is therefore very stable.

19.70   (a)   In 1,2-cyclopentanedione the C=O bond dipoles are constrained by the geometry of the ring to be aligned at an angle of about 72°. This alignment provides a significant resultant dipole, as shown by diagram *A*:

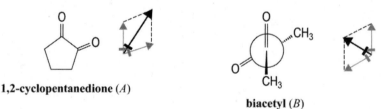

1,2-cyclopentanedione (*A*)   biacetyl (*B*)

In biacetyl (2,3-butanedione), rotation about the central carbon–carbon bond is possible. Evidently, the two bond dipoles are oriented so that their resultant is less than the resultant in 1,2-cyclopentanedione. Notice that if their dihedral angle were 180°, their resultant would be zero. Hence, the bond dipoles are aligned at an angle somewhere between 72° and 180°, as shown in the Newman projection of diagram *B* above. (The angle turns out to be about 135°.) Aligned bond dipoles cause electrostatic repulsion. By undergoing internal rotation, biacetyl avoids this destabilizing effect.

The alignment in biacetyl occurs because it is the alignment of lowest energy; there are two reasons for this reduced energy. First, the alignment reduces the electrostatic repulsion between the negative ends of the bond dipoles. Second, it prevents overlap between the π-electron systems of the two carbonyl groups that would occur at an angle of either 0° or 180°. This overlap is unfavorable because it creates a charge distribution in the molecule that involves an electron-deficient oxygen, as symbolized by the following resonance structures:

electron-deficient oxygen

The information given in the problem plus a simplifying assumption that the cyclopentane ring has the shape of a regular pentagon allows you to use the law of cosines to calculate the bond dipole of a C=O bond as well as the angle between the two carbonyl bond dipoles in biacetyl (given above as about 135°). Try it!

(b)   The $n \rightarrow \pi^*$ absorption is characteristic of the carbonyl group. This absorption disappears because a reaction occurs in which the carbonyl group is converted into another group that does not have this absorption. This reaction is addition of ethanethiol to give the sulfur analog of a hemiacetal:

$$H_3C-CH=O \; + \; CH_3CH_2SH \; \rightleftharpoons \; H_3C-CH-OH$$
$$\underset{\substack{\text{SCH}_2\text{CH}_3}}{}$$

has a carbonyl group; has $n \rightarrow \pi^*$ absorption

has no carbonyl group; has no $n \rightarrow \pi^*$ absorption

(c)   The Tollens test (oxidation with the ammonia complex of Ag$^+$) is a characteristic reaction of aldehydes. This compound reacts slowly because very little of it is in the aldehyde form; rather, the molecule exists largely as a cyclic hemiacetal.

reacts rapidly with the Tollens reagent

The molecule exists mostly in this cyclic hemiacetal form, which is not an aldehyde, and which therefore does not react with the Tollens reagent.

19.71   The molecular formula requires three degrees of unsaturation. The six-proton singlet suggests two chemically equivalent methyl groups. The IR indicates the presence of a carbonyl group. The two doublets at about δ 6 and δ 7.5 are vinylic hydrogens, one of which is significantly shifted by some electronegative group, no doubt the carbonyl group. If the vinylic hydrogens account for one double bond and a carbonyl group accounts for another, we are still missing one degree of unsaturation, which thus must be a ring. The two remaining protons are in the chemical-shift region for protons α to a carbonyl. The data leave only one possibility:

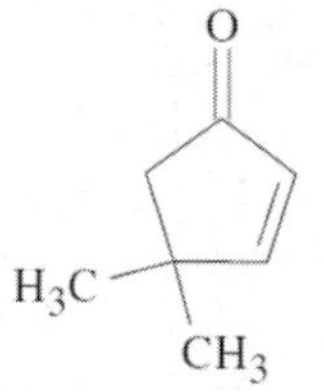

O

H₃C / CH₃   4,4-dimethyl-2-cyclopentenone

If we return to Eq. 19.4 on text p. 954, we see that cyclopentanones should have IR carbonyl absorptions at about 1745 cm^{-1}. However, conjugation reduces this frequency by 40–45 cm^{-1}, as shown by the trends in Eq. 19.2 on text p. 953. This means that the observed carbonyl frequency of 1703 cm^{-1} is exactly what we would expect for this compound. Furthermore, the small vinylic coupling constant is expected for two protons in a cis relationship (see Table 13.3 on text p. 647).

19.72   The formula indicates one degree of unsaturation which is accounted for by a ketone (IR absorption at 1710 cm^{-1}, CMR absorption at δ 204.9 with zero attached hydrogens). The remaining oxygens cannot be alcohols (absence of O—H stretch in the IR) or other carbonyls. The CMR-DEPT data indicate the presence of at least one methyl ether (δ 53.5) as well as a carbon bearing one hydrogen (δ 101.7) that could also be attached to two oxygens; thus, an acetal group is a possibility. The proton NMR data indicate a methyl group attached to a carbonyl. The only carbon unaccounted for is a methylene group with a CMR absorption at δ 47.2, which places it adjacent to the carbonyl and near, but not adjacent to, the acetal. All the data conspire to define the following structure:

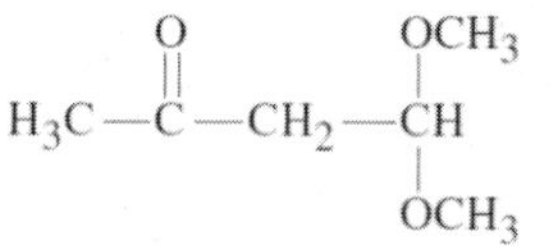

$$\underset{\text{OCH}_3}{\overset{\text{O}}{H_3C-\overset{\|}{C}-CH_2-\underset{\overset{|}{\text{OCH}_3}}{\overset{|}{CH}}}} \quad \text{4,4-dimethoxy-2-butanone}$$

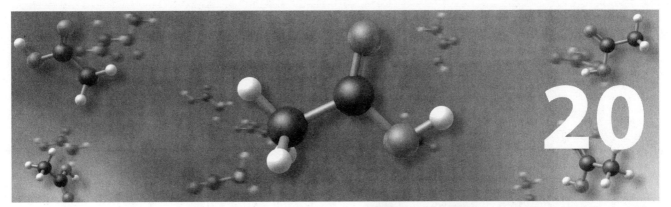

The Chemistry of Carboxylic Acids

STUDY GUIDE LINKS

 ### 20.1 Reactions of Bases with Carboxylic Acids

In Eq. 20.3 (text p. 1012), water is the base that reacts with a carboxylic acid, because water is the most abundant base present in the aqueous solution of an acid. However, you should not forget that carboxylic acids can in principle react with *any* base. (For example, write the reaction of a carboxylic acid with one equivalent of a Grignard reagent.)

Another point worth special note is that the reaction of strong bases not only with carboxylic acids, but also with alcohols and water, is very fast—so fast that a reaction occurs at virtually every encounter between an acid molecule and a base molecule. This becomes very important in situations in which proton transfer competes with other reactions for a limiting amount of a base. Suppose, for example, we want the aldehyde group in the following molecule to react with a Grignard reagent. The first equivalent of Grignard reagent reacts not with the aldehyde, but rather with the carboxylic acid, because proton transfer is such a fast reaction.

To achieve a reaction of the aldehyde, a second equivalent of the Grignard reagent must be added:

(The aldehyde carbonyl is much more reactive than the carboxylate carbonyl for reasons that are discussed in Chapter 21.)

In summary, proton transfers to strong bases from O—H groups in most cases are much faster than other reactions of strong bases.

20.2 Resonance Effect on Carboxylic Acid Acidity

The resonance stabilization of carboxylate ions was once considered the *major* effect responsible for the enhanced acidity of carboxylic acids. It is now believed to be less important than originally thought (although still significant) because *carboxylic acids are also resonance-stabilized.*

Figure 3.3 on text p. 116 shows that stabilizing a conjugate-base carboxylate decreases the pK_a of the corresponding carboxylic acid. It follows that stabilizing a carboxylic acid, that is, lowering its energy relative to that of its conjugate-base carboxylate, *increases* its pK_a. Hence, the resonance stabilization of an acid and the resonance stabilization of its carboxylate are opposing effects. However, *resonance is more important in the carboxylate ion* because the carboxylate resonance structures *disperse* or *delocalize* charge, whereas the carboxylic acid resonance structures *separate* charge. Structures that disperse a single type of charge are more important than those that separate opposite charges, because separating opposite charges requires energy, by the electrostatic law (Eq. 3.44, text p. 117). For this reason, a carboxylate ion is stabilized *more* by resonance than its conjugate-acid carboxylic acid. If the two were equally stabilized by resonance, the resonance effect would cancel when the free energies of the carboxylic acid and the carboxylate are compared, and it would have no effect on the pK_a. (The same point applies to the acidity of phenols; see Sec 18.7A in the text.)

20.3 Mechanism of Acid Chloride Formation

The formation of acid chlorides follows a mechanism that is fundamentally the same as the mechanism for esterification. In esterification, the carbonyl group is activated by protonation. In acid chloride formation, it is activated by reaction with the Lewis acid $SOCl_2$. A chloride ion is displaced from $SOCl_2$ in the process:

As a result of this reaction the carbonyl oxygen is more electrophilic. In other words, the Lewis acid $SOCl_2$ activates the carbonyl carbon toward nucleophilic attack in much the same sense that a proton does. (You've seen that other Lewis acids, such as the lithium ion of $LiAlH_4$ or the magnesium of a Grignard reagent, can also activate attack at a carbonyl carbon.) The nucleophile chloride ion produced in the step above reacts at the carbonyl carbon:

Notice the formation of a tetrahedral addition intermediate, as in esterification. *This is a common mechanistic thread that runs throughout carbonyl chemistry.* The —OSOCl group is an excellent leaving group and departs to give a protonated acid chloride. Loss of the —OSOCl group is irreversible, because it decomposes to SO_2 (a gas) and chloride ion:

Loss of a proton to chloride ion gives HCl (another gaseous by-product) and the acid chloride:

 ## 20.4   More on Synthetic Equivalents

Study Guide Link 17.1 (p. 509 of this manual) discussed the notion of a *synthetic equivalent*. Acid chlorides present another opportunity to think about this concept. Recall that acid chlorides are the acylating agents used in Friedel–Crafts acylation (Sec. 16.4F). Because acid chlorides are often prepared from carboxylic acids, it follows, then, that the acyl group of an aromatic ketone can originate from a carboxylic acid. Thus, a carboxylic acid or acid chloride can be viewed as the synthetic equivalent of the acyl group in an aromatic ketone.

 ## 20.5   Mechanism of Anhydride Formation

Working through the mechanism of anhydride formation will allow you to see again the patterns involved in carbonyl substitution reactions. Consider the mechanism for the formation of the anhydride of a general carboxylic acid, RCO_2H, by acetic anhydride. The acetic anhydride serves as a Lewis acid to activate the carbonyl group of an R—CO_2H molecule:

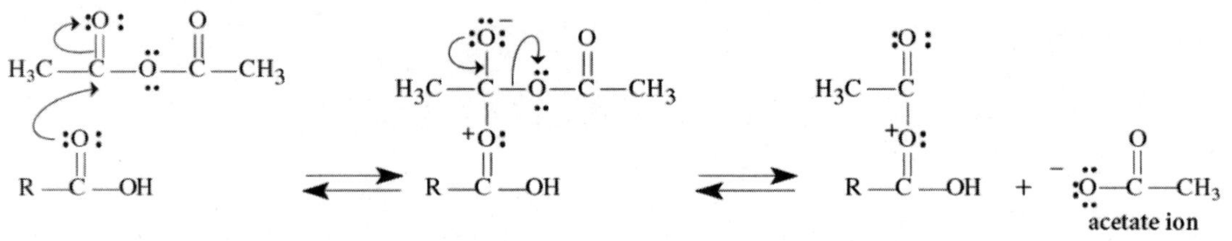

The activated carbonyl group reacts with another molecule of carboxylic acid and loses a proton to the acetate ion formed in the previous step to give a tetrahedral addition intermediate:

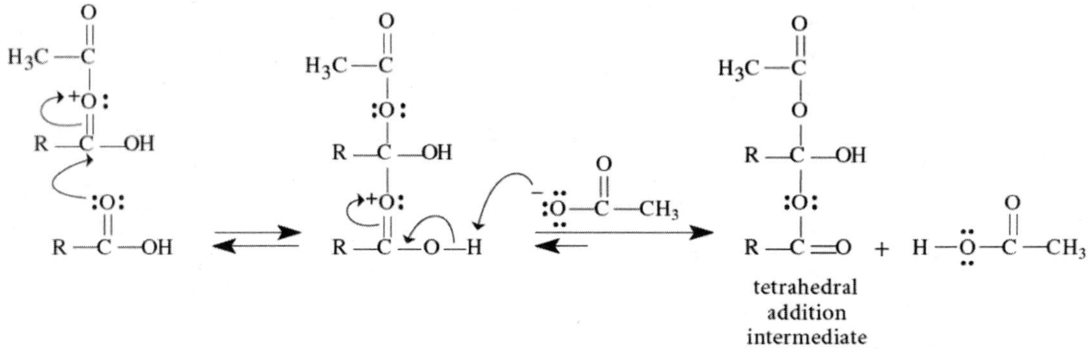

You should now complete the mechanism by showing, first, the loss of another acetate ion as a leaving group, and second, loss of a proton:

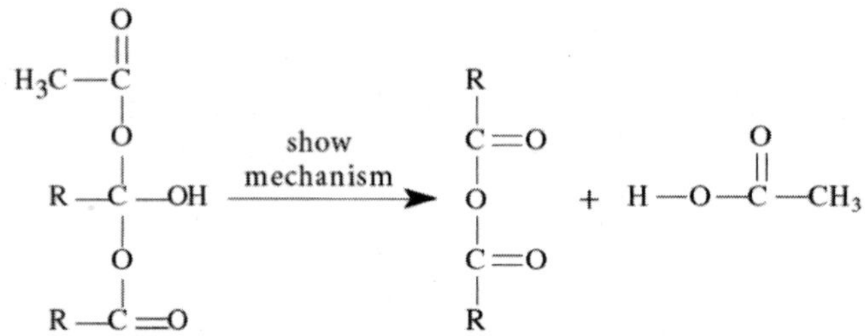

The mechanisms of anhydride formation with P_2O_5 and with $POCl_3$ are similar.

Anhydride formation is an equilibrium. When acetic anhydride is used as the reagent, this equilibrium is driven to the right by the use of a large excess of acetic anhydride (Le Châtelier's principle in operation). When cyclic anhydrides are being formed, the preference for intramolecular reactions that form five- and six-membered rings (Sec. 11.8) provides not only a kinetic advantage to the reaction—a greater reaction rate—but also an equilibrium advantage. Consequently, reactions that form five- and six-membered cyclic anhydrides in many cases require milder conditions.

FURTHER EXPLORATIONS

 ### 20.1   Chemical Shifts of Carbonyl Carbons

As pointed out in the text, the electronegativity of oxygen might lead one to expect the chemical shift for the carbonyl group of acetic acid to be *greater* than that of acetone. An additional puzzle is provided by the carbonyl chemical shift of the acetate ion, the conjugate base of acetic acid:

$$H_3C—\overset{\overset{O}{\|}}{C} \overset{\delta\,178}{\diagup} OH \qquad\qquad H_3C—\overset{\overset{O}{\|}}{C} \overset{\delta\,182}{\diagup} O^-$$

acetic acid   acetate ion

If chemical shifts in carbon NMR spectroscopy correlate with electronegativity, as they do in proton NMR, it seems that the acetate ion, because of its relatively electron-rich —O^-, should have a *smaller* carbonyl chemical shift than acetic acid. The data above show that this is not the case.

A related phenomenon is the chemical shift of the O—H proton, which was discussed in Sec. 13.7D on text p. 651–653. In the gas phase (in which hydrogen bonding and other solvent effects are absent), this proton has a chemical shift of $\delta\,0.8$, a much smaller value than would be expected for a proton directly attached to an electronegative atom.

These cases illustrate the point that *chemical shifts are not determined solely by electronegativity of nearby atoms.* You learned, for example, that induced circulation of π electrons in the presence of the applied field accounts for the significant chemical shifts of both aromatic and alkene carbons as well as the shifts of aromatic and alkene protons—shifts far greater than would be predicted on the basis of the electronegativity of an sp^2 hybridized carbon (Fig. 13.15 on text p. 646 and Fig. 16.2 on text p. 794). In the cases discussed here, induced electron circulation also accounts for the results, except that the circulation involved is that of the oxygen unshared electrons, and it occurs in such a way as to oppose, and thus reduce, the applied field in the region of the adjacent atoms; as a result, a lower frequency is required for resonance of these atoms. The result is a smaller chemical shift. The α-carbon atoms of acetone have no such contribution, because they have no unshared electrons. Hence, the carbonyl carbon chemical shifts of ketones in carbon NMR are greater than those of carboxylic acids. In acetate ion, the contribution of one unshared pair is partially offset by that of another, because the two orbitals are directed differently in space with respect to the applied field; hence, acetate absorbs at a chemical shift that is closer to the "normal" carbonyl value.

 ### 20.2   More on Surfactants

The word "surfactant" means "surface-active compound." To say that something is *surface active* means that is lowers the *surface tension* of a liquid. Without getting into a technical discussion of surface tension, we can say that it is the resistance of a surface to penetration by another object. Water has a particularly high surface tension. (If you've ever seen someone float a coin on the surface of a glass of water, you've witnessed the effects of surface tension.) The high surface tension of water is due to the tight network of strong hydrogen bonds at the water surface. Because water cannot form hydrogen bonds to air, water molecules form particularly strong hydrogen bonds to each other at the air–water interface.

When a surfactant is dissolved in water, the surfactant molecules tend to collect at the surface of the water with the hydrocarbon tails pointed toward the air–water interface, as shown in Fig. SG20.1. This orientation of surfactant molecules at the surface allows the solvent water molecules to interact with the polar head groups and places the nonpolar tails in contact with air, which is

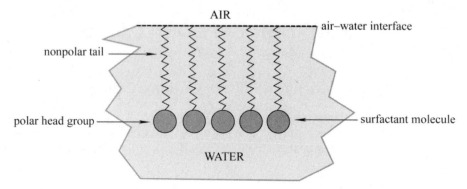

AIR
air–water interface
nonpolar tail
polar head group
surfactant molecule
WATER

Figure SG20.1 A diagram showing schematically the orientation of surfactant molecules at the interface of an aqueous solution and air. The nonpolar tails are oriented towards the interface.

composed mostly of other nonpolar molecules. In this situation the network of tight hydrogen bonds at the surface of pure water is no longer present. As a result, the surface tension of the solution is lowered. One perceptible manifestation of this phenomenon is aqueous solutions of surfactants (for example, soap solutions) are slippery to the touch.

20.3 Orthoesters

We can ask whether esters react further under the esterification conditions to form compounds analogous to acetals. (See Sec. 19.10A.)

$$R-\overset{\overset{\displaystyle O}{\|}}{C}-OCH_3 \;+\; 2\,CH_3OH \;\;\underset{acid}{\cancel{\longrightarrow}}\;\; R-\underset{\underset{\displaystyle OCH_3}{|}}{\overset{\overset{\displaystyle OCH_3}{|}}{C}}-OCH_3 \;+\; H_2O$$

an orthoester
(not observed)

Such compounds, called orthoesters, are indeed known. However, they *cannot* be formed in this manner because the equilibrium constants for addition to the carbonyl groups of acid derivatives are only about 10^{-8} times those for formation of aldehyde addition compounds. Such equilibria are simply too unfavorable to be driven to completion in the usual ways.

An ester has more resonance structures and is more stable relative to its addition products; addition reactions are less favorable.

An aldehyde or ketone has fewer resonance structures and is less stable relative to its addition products; addition reactions are more favorable.

Hence, the equilibrium constant for an addition to a ketone is more favorable than the equilibrium constant for the same addition to an ester. This point is illustrated in Fig. SG20.2.

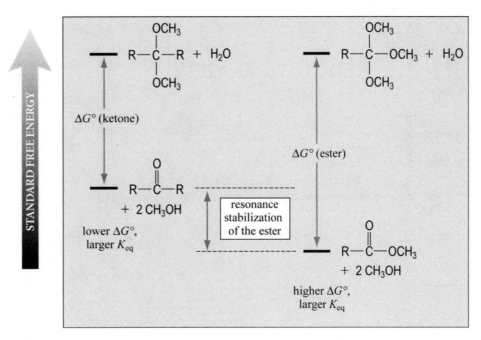

Figure SG20.2 The greater resonance stabilization of esters causes addition reactions of esters to have smaller equilibrium constants than the corresponding addition reactions of ketones.

 20.4   Mechanism of the LiAlH₄ Reduction of Carboxylic Acids

The mechanism of the LiAlH$_4$ reduction of carboxylic acids is presented in the text only in outline. Further Exploration fills in the details.

Reduction of the carboxylate ion by AlH$_3$ (step *a* in the following equation) is analogous to the addition of BH$_3$ to the double bond of an alkene. (BH$_3$ also reduces carboxylic acids to primary alcohols.)

Notice that a tetrahedral addition intermediate is formed in step *a*. Breakdown of this intermediate (step *b*) expels Li$^+$ H$_2$AlO$^-$, which also can act as a reducing agent. Further reduction of the aldehyde to the alcohol (Sec. 19.8A) can occur either by this species or by excess LiAlH$_4$ present in the reaction mixture.

REACTION REVIEW

I. SYNTHESIS AND REACTIVITY OF CARBOXYLIC ACIDS

A. SYNTHESIS OF CARBOXYLIC ACIDS—REVIEW

   1. Methods for preparing carboxylic acids introduced in previous chapters:
      a. Oxidation of primary alcohols (see Sec. 10.6B, text p. 480).

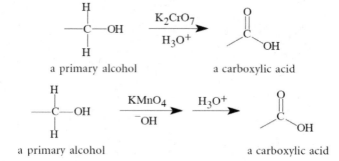

      b. Side-chain oxidation of alkylbenzenes (see Sec. 17.5C, text p. 480).

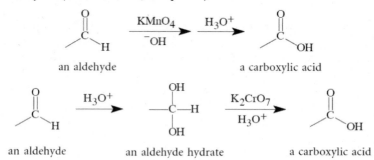

      c. Oxidation of aldehydes (see Sec. 19.14, text p. 994).

      d. Ozonolysis of alkenes followed by oxidative workup with H_2O_2 (see Sec. 5.5, text page 198).

B. SYNTHESIS OF CARBOXYLIC ACIDS WITH GRIGNARD REAGENTS

1. The reaction of a Grignard reagent with carbon dioxide, followed by protonolysis, gives a carboxylic acid.
   a. Addition of a Grignard reagent to carbon dioxide gives the halomagnesium salt of a carboxylic acid.
   b. When aqueous acid is added to the reaction mixture in a separate reaction step, the free carboxylic acid is formed.

2. The reaction of Grignard reagents with CO_2 is another method for the formation of carbon–carbon bonds. (See Appendix VI, text page A-13, for a review of other carbon–carbon bond-forming reactions.)

C. INTRODUCTION TO CARBOXYLIC ACID REACTIONS

1. Reactions of carboxylic acids can be categorized into four types:
   a. Reactions at the carbonyl group.
      i. The most typical reaction at the carbonyl group is substitution at the carbonyl carbon. In such a reaction, the —OH of the carboxy group is typically substituted by a nucleophilic group.
      ii. Reaction of the carbonyl oxygen with an electrophile (Lewis acid or Brønsted acid), that is, the reaction of the carbonyl oxygen as a base.
   b. Reactions at the carboxylate oxygen, such as reactions of the carboxylate oxygen as a nucleophile.
   c. Loss of the carboxy group as CO_2 (decarboxylation).
   d. Reactions involving the α-carbon.
2. Many substitution reactions at the carbonyl carbon are acid catalyzed; that is, the reactions of nucleophiles at the carbonyl carbon are catalyzed by the reactions of acids at the carbonyl oxygen.

II. REACTION OF CARBOXYLIC ACIDS AT THE CARBONYL GROUP

A. ACID-CATALYZED ESTERIFICATION OF CARBOXYLIC ACIDS

1. Esters are carboxylic acid derivatives in which the proton on the carboxylate oxygen in effect has been replaced by an alkyl or aryl group.
2. When a carboxylic acid is treated with a large excess of an alcohol in the presence of a strong acid catalyst, an ester is formed; this reaction is called acid-catalyzed esterification (or Fischer esterification).

an ester

   a. Acid-catalyzed esterification is a substitution of —OH at the carbonyl group of the acid by the —OR group of the alcohol; the —OH group leaves as water.
   b. The equilibrium constants for esterification with most primary alcohols are near one.
      i. The reaction is driven toward completion by applying Le Châtelier's principle.
      ii. Use of the reactant alcohol as the solvent, which ensures that the alcohol is present in large excess, drives the equilibrium toward the ester product.
   c. Acid-catalyzed esterification cannot be applied to the synthesis of esters from phenols or tertiary alcohols.
3. The mechanism of acid-catalyzed esterification serves as a model for the mechanisms of other acid-catalyzed reactions of carboxylic acids and their derivatives.
   a. The first step of the mechanism is protonation of the carboxyl oxygen.
      i. The catalyzing acid is the conjugate acid of the solvent, formed by reaction of the solvent with a strong acid such as H_2SO_4.
      ii. Protonation of a carbonyl oxygen makes the carbonyl carbon more electrophilic because the carbonyl oxygen becomes a better electron acceptor.

b. Reaction of the alcohol with the carbonyl carbon of this carbocation, followed by loss of a proton, gives a tetrahedral addition intermediate.

tetrahedral
addition intermediate

c. The tetrahedral addition intermediate, after protonation, loses water to give the conjugate acid of the ester.

d. Loss of a proton gives the ester product and regenerates the acid catalyst.

B. SYNTHESIS OF ACID CHLORIDES

1. Acid chlorides are carboxylic acid derivatives in which the —OH group has been replaced by —Cl.
   a. Acid chlorides are often prepared from the corresponding carboxylic acids.
   b. Two reagents used for this purpose are thionyl chloride, $SOCl_2$, and phosphorus pentachloride, PCl_5.

$$R-C(=O)-OH \xrightarrow[\text{or } PCl_5]{SOCl_2} R-C(=O)-Cl + HCl + SO_2 \text{ (or } POCl_3)$$

2. Acid chloride synthesis fits the general pattern of substitution reactions at a carbonyl group; in this case, —OH is substituted by —Cl.
3. Sulfonyl chlorides are the acid chlorides of sulfonic acids.
   a. Sulfonyl chlorides are prepared by treatment of sulfonic acids or their sodium salts with PCl_5.
   b. Aromatic sulfonyl chlorides can be prepared directly by the reaction of aromatic compounds with chlorosulfonic acid, $ClSO_3H$.
      i. This reaction is a variation of aromatic sulfonation, an electrophilic aromatic substitution reaction.
      ii. Chlorosulfonic acid, the acid chloride of sulfuric acid, acts as an electrophile in this reaction.
      iii. The sulfonic acid produced in the reaction is converted into the sulfonyl chloride by reaction with another equivalent of chlorosulfonic acid.

C. SYNTHESIS OF ANHYDRIDES

1. Carboxylic acid anhydrides are carboxylic acid derivatives in which the hydroxy group has been replaced by an acyloxy group.
   a. Anhydrides are prepared by treatment of carboxylic acids with strong dehydrating agents (usually P_2O_5).

an anhydride

b. Most anhydrides can themselves be used as reagents for the preparation of other anhydrides.
2. Some dicarboxylic acids react with acetic anhydride to form cyclic anhydrides—compounds in which the anhydride group is part of a ring.
   a. Cyclic anhydrides containing five- and six-membered rings are readily prepared from the corresponding dicarboxylic acids.
   b. Formation of cyclic anhydrides with five- or six-membered rings is so facile that in some cases it occurs on heating the dicarboxylic acid.

3. Anhydrides, like acid chlorides, are used in the synthesis of other carboxylic acid derivatives.

D. REDUCTION OF CARBOXYLIC ACIDS TO ALCOHOLS

1. When a carboxylic acid is treated with $LiAlH_4$, then with dilute acid, a primary alcohol is formed.
   a. Before the reduction itself takes place, $LiAlH_4$ reacts with the acidic hydrogen of the carboxylic acid to give the lithium salt of the carboxylic acid and one equivalent of hydrogen gas. (The lithium salt of the carboxylic acid is the species that is actually reduced.)

2. The reduction occurs in two stages:
   a. AlH_3 (formed from the reaction of $LiAlH_4$ with the acidic proton of the carboxylic acid) reduces the carboxylate ion to an aldehyde.
   b. The aldehyde is rapidly reduced further to give, after protonolysis, the primary alcohol. (Because the aldehyde is more reactive than the carboxylate salt, it cannot be isolated.)

3. The $LiAlH_4$ reduction of a carboxylic acid incorporates two different types of carbonyl reactions:
   a. A net substitution reaction at the carbonyl group to give an aldehyde intermediate.
   b. An addition to the aldehyde thus formed.
4. Sodium borohydride, $NaBH_4$, does not reduce carboxylic acids, although it does react with the acidic hydrogens of carboxylic acids.

III. REACTION OF CARBOXYLIC ACIDS AT THE OXYGEN OF THE —OH GROUP

A. ESTERIFICATION BY ALKYLATION

1. When carboxylic acids are treated with diazomethane in ether solution, they are rapidly converted into their methyl esters.

a. Protonation of diazomethane by the carboxylic acid gives the methyldiazonium ion; this ion has one of the best leaving groups, dinitrogen.

b. An S_N2 reaction of the methyldiazonium ion with the carboxylate oxygen results in the displacement of N_2 and formation of the ester.

2. The reaction of certain alkyl halides with carboxylate ions give esters.
   a. This is an S_N2 reaction in which the carboxylate ion, formed by a Brønsted acid–base reaction of the acid and a base (such as K_2CO_3), acts as the nucleophile that reacts with the alkyl halide.

$$R—CO_2H \xrightarrow[-KHCO_3]{K_2CO_3} R—CO_2^-\ K^+ \xrightarrow[-KBr]{PhCH_2Br} R—CO_2—CH_2Ph$$

   b. This reaction works best on alkyl halides that are especially reactive in the S_N2 reaction, such as methyl iodide and benzylic or allylic halides, because carboxylate ions are relatively poor nucleophiles.

IV. DECARBOXYLATION OF CARBOXYLIC ACIDS

1. The loss of carbon dioxide from a carboxylic acid is called decarboxylation.
2. Certain types of carboxylic acids are readily decarboxylated:
   a. β-keto acids (carboxylic acids with a keto group in the β-position).
   b. malonic acid derivatives.
   c. derivatives of carbonic acid.

3. β-Keto acids readily decarboxylate at room temperature in acidic solution.
   a. Decarboxylation of a β-keto acid involves an enol intermediate that is formed by an internal proton transfer from the carboxylic acid group to the carbonyl oxygen atom of the ketone.
   b. The enol is transformed spontaneously into the corresponding ketone.

enol intermediate

4. The acid form of the β-keto acid decarboxylates more readily than the conjugate-base carboxylate form because the latter has no acidic proton that can be donated to the β-carbonyl oxygen.
5. Malonic acid and its derivatives readily decarboxylate upon heating in acid solution. This reaction, which does not occur in base, bears a close resemblance to the decarboxylation of β-keto acids, because both types of acid have a carbonyl group β to the carboxy group.

$$\underset{\text{a malonic acid derivative}}{HO-\overset{\overset{O}{\|}}{\underset{\beta}{C}}-\overset{\alpha}{\underset{/\backslash}{C}}-\overset{\overset{O}{\|}}{C}-OH} \quad\xrightarrow{\text{heat}}\quad HO-\overset{\overset{O}{\|}}{C}-\underset{/\backslash}{C}-H \;+\; CO_2$$

6. Carbonic acid is unstable and decarboxylates spontaneously in acid solution to carbon dioxide and water.
   a. Carbonic acid derivatives in which only one carboxylate oxygen is involved in ester or amide formation also decarboxylate under acidic conditions.

$$R-\underset{\underset{H}{\overset{|}{N}}}{\overset{\overset{O}{\|}}{C}}-OH \quad\rightleftharpoons\quad R-NH_2 \;+\; CO_2 \quad\underset{\xleftarrow{\hspace{1cm}}}{\xrightarrow{H_3O^+}}\quad R-\overset{+}{N}H_3$$

   b. Under basic conditions, carbonic acid is converted into its salts bicarbonate and carbonate, which do not decarboxylate.
   c. Carbonic acid derivatives in which both carboxylic acid groups are involved in ester or amide formation are stable.
      i. Carbonate esters are diesters of carbonic acid.
      ii. Ureas are diamides of carbonic acid.
      iii. Phosgene is the acid chloride of carbonic acid.

SOLUTIONS TO PROBLEMS

Solutions to In-Text Problems

20.1 (a)   (b)   (c)   (d)

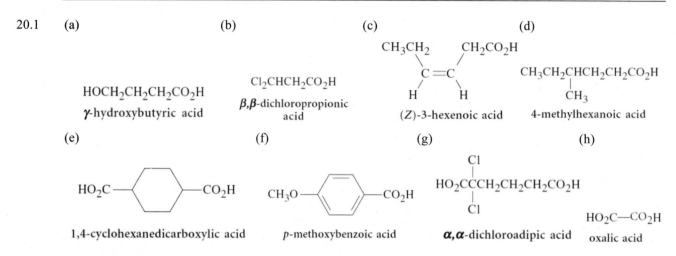

HOCH$_2$CH$_2$CH$_2$CO$_2$H
γ-hydroxybutyric acid

Cl$_2$CHCH$_2$CO$_2$H
β,β-dichloropropionic
acid

(*Z*)-3-hexenoic acid

CH$_3$CH$_2$CHCH$_2$CH$_2$CO$_2$H
           |
          CH$_3$
4-methylhexanoic acid

(e)   (f)   (g)   (h)

1,4-cyclohexanedicarboxylic acid   *p*-methoxybenzoic acid   *α,α*-dichloroadipic acid   oxalic acid

20.2 (a)   2,2-dimethylbutanoic acid (common: *α,α*-dimethylbutyric acid)
(b)   9-methyldecanoic acid (common: *ω*-methylcapric acid). Note that the term *ω* (omega, the last letter of the Greek alphabet) is used in common nomenclature for a branch at the end of a carbon chain.
(c)   3-hydroxy-4-oxocyclohexanecarboxylic acid
(d)   2,4-dichlorobenzoic acid
(e)   2-methylpropanedioic acid (common: methylmalonic acid)
(f)   cyclopropanecarboxylic acid

20.3   The dimer owes its stability to intermolecular hydrogen bonds. The solvent carbon tetrachloride has no donor or acceptor atoms to compete for these hydrogen bonds. However, the solvent water can compete as both a donor and an acceptor for hydrogen bonds. Hence, the dimer has less of an energetic advantage in water and, as a result, there is a lower concentration of dimer (in fact, essentially none) in aqueous solution and a higher concentration in CCl$_4$.

20.4   The IR spectrum indicates the presence of a carboxy group. A carboxy group accounts for 45 mass units; this leaves 43 mass units unaccounted for. The NMR spectrum indicates seven hydrogens in addition to the carboxy hydrogen at δ 10. A group with seven hydrogens and 43 mass units must be either a propyl or an isopropyl group; the splitting in the NMR spectrum indicates the latter. The compound is 2-methylpropanoic acid (isobutyric acid), (CH$_3$)$_2$CH—CO$_2$H.

20.5   The NMR data indicate a *para*-substituted benzoic acid derivative; given this deduction, the para substituent must be a chlorine.

p-chlorobenzoic acid
(4-chlorobenzoic acid)

20.6   First write the structures of the two compounds.

$$\underset{\substack{|\\CH_3}}{\overset{\substack{CH_3\\|}}{HO_2C-C-CH_2-CO_2H}}$$ $$HO_2CCH_2CH_2CH_2CH_2CO_2H$$

α,α-dimethylsuccinic acid   adipic acid

(a)   α,α-Dimethylsuccinic acid should have five resonances in its CMR spectrum because it has five chemically nonequivalent sets of carbons. Adipic acid, in contrast, should have three resonances because it has only three chemically nonequivalent sets of carbons.

(b)   The proton NMR spectrum of only α,α-dimethylsuccinic acid consists of three singlets; the proton NMR spectrum of adipic acid is more complex.

20.7   (a)   The first and second ionizations of succinic acid:

$$HO_2CCH_2CH_2CO_2H \underset{H_3O^+}{\overset{H_2O}{\rightleftharpoons}} {}^-O_2CCH_2CH_2CO_2H \underset{H_3O^+}{\overset{H_2O}{\rightleftharpoons}} {}^-O_2CCH_2CH_2CO_2^-$$

$$pK_a = 4.21 \qquad\qquad pK_a = 5.64$$

(b)   The first pK_a is lower because the electron-withdrawing polar effect of an un-ionized carboxy group stabilizes the anion formed by ionization of the other carboxy group. The second pK_a is higher because the negative charge of an ionized carboxy group destabilizes the anion formed by the second ionization. This destabilization is the result of the electrostatic repulsion of like charges.

20.8   Extract an ether solution of the two compounds with an aqueous solution of $NaHCO_3$, Na_2CO_3, or NaOH. The acid will ionize and its conjugate-base anion will dissolve in the aqueous layer as the sodium salt; *p*-bromotoluene will remain in the ether layer. After isolating the aqueous layer, acidify it with concentrated HCl; neutral *p*-bromobenzoic acid will precipitate.

20.9   (a)

cyclopentanol   cyclopentanecarboxylic acid

(b)

1-heptene

$$CH_3(CH_2)_5CH_2MgBr \xrightarrow[2)\ H_3O^+]{1)\ CO_2} CH_3(CH_2)_5CH_2CO_2H$$

octanoic acid

One could also use HBr in the presence of peroxides to form the alkyl bromide directly from the alkene. In general, this alternative scheme will always be applicable in the conversion of a 1-alkene into the corresponding primary alkyl halide.

20.10   (a)

3-methylhexanoic acid   ethyl 3-methylhexanoate

(b)   Both carboxy groups are esterified; the product is dipropyl adipate:

$$\text{HOC(CH}_2)_4\text{COH} \quad \xrightarrow[\text{CH}_3\text{CH}_2\text{CH}_2\text{OH}]{\text{H}_2\text{SO}_4\text{(cat.)}} \quad \text{CH}_3\text{CH}_2\text{CH}_2\text{OC(CH}_2)_4\text{COCH}_2\text{CH}_2\text{CH}_3}$$

adipic acid   dipropyl adipate

20.11   (a)   Follow the reverse of the steps shown in Eqs. 20.18a–d, text pp. 1020–1021, with R— = Ph—.

(b)   To favor ester hydrolysis rather than ester formation, use a large excess of water as solvent rather than an alcohol. By Le Châtelier's principle, this drives the carboxylic acid–ester equilibrium toward the carboxylic acid.

20.12   (a)   The mechanism of hydroxide-promoted ester hydrolysis:

Notice that under the basic conditions the carboxylic acid product is ionized in the last step. This drives base-promoted ester hydrolysis to the right.

The base-promoted hydrolysis of esters is called *saponification*, because soaps are made this way from the base-promoted hydrolysis of fats, which are nothing more than esters of the fatty acids. We'll study saponification in Sec. 21.7A, text p. 1061. However, if you were able to construct this mechanism, you're ahead of the game—you understand the reaction already!

(b)   As shown in the last step of the mechanism in the solution to part (a), methoxide reacts with the *acidic proton* of benzoic acid to form the benzoate ion much more rapidly than it reacts at the carbonyl carbon. (See Study Guide Link 20.1 on p. 659 of this manual for a discussion of this point.)

20.13   The structures of isobutyric and benzoic acids are given in Table 20.1, text p. 1006. The products are as follows:

(a)   (b)   (c)   (d)

$(\text{CH}_3)_2\text{CHC}-\text{OCH}_3$
methyl isobutyrate

$\text{CH}_3\text{O}-\text{C}-\text{CH}_2\text{CH}_2-\text{C}-\text{OCH}_3$
dimethyl succinate

$(\text{CH}_3)_2\text{CHC}-\text{OCH}_2\text{Ph}$
benzyl isobutyrate

$\text{Ph}-\text{C}-\text{OCH}_2\text{CH}=\text{CH}_2$
allyl benzoate

20.14   In this reaction the carboxylic acid is alkylated by the *tert*-butyl cation, which is formed by protonation of 2-methylpropene.

A variation on this mechanism proposed by many students is alkylation of the carboxylate oxygen rather than the carbonyl oxygen by the *tert*-butyl cation:

Although this appears to be a more direct route to the ester, the mechanism given above is more likely to be correct for the following reason. Just as a carboxylic acid is *protonated* on the carbonyl oxygen rather than the carboxylate oxygen (why? see Sec. 20.4B of the text), it should be *alkylated* on the carbonyl oxygen for the same reason.

20.15   (a)   (b)   (c)

2-methylbutanoyl chloride

p-methoxybenzoyl chloride
(anisoyl chloride)

1-propanesulfonyl chloride

20.16   (a)   (b)   (c)

ethanesulfonyl chloride

benzoyl chloride

p-toluenesulfonyl chloride
(tosyl chloride)

20.17   Study Guide Link 20.4 in this manual, points out that a carboxylic acid is the synthetic equivalent of an acyl group in an aromatic ketone. This concept is realized by applying the Friedel–Crafts acylation reaction to the corresponding acid chloride. This synthesis takes advantage of the fact that the methoxy group is an ortho, para-directing substituent in electrophilic aromatic substitution.

benzoic acid   benzoyl chloride   *p*-methoxybenzophenone

20.18   (a)   The corresponding anhydride is formed:

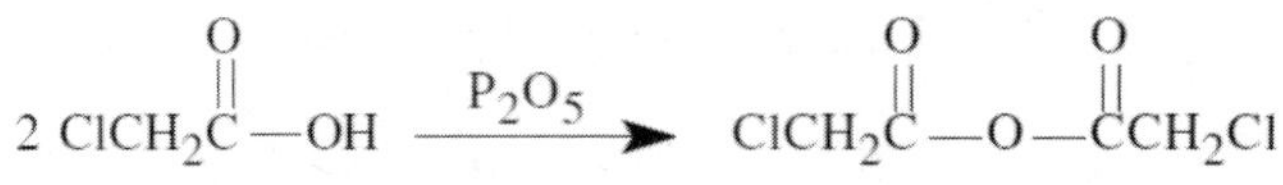

$$2 \ ClCH_2\overset{\overset{\displaystyle O}{\|}}{C}-OH \quad \xrightarrow{P_2O_5} \quad ClCH_2\overset{\overset{\displaystyle O}{\|}}{C}-O-\overset{\overset{\displaystyle O}{\|}}{C}CH_2Cl$$

chloroacetic acid   chloroacetic anhydride

(b)

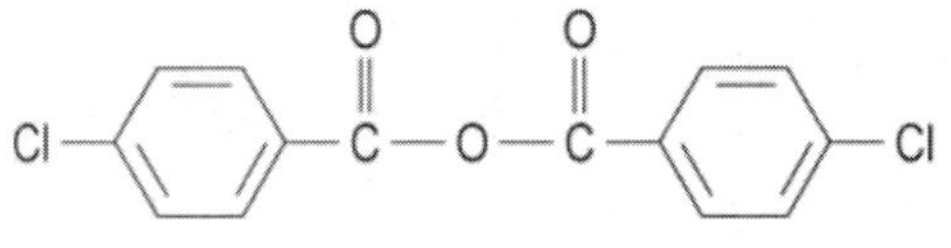

p-chlorobenzoic anhydride

20.19   (a)   Maleic anhydride forms a cyclic anhydride because its carboxy groups are cis. A cyclic anhydride of fumaric acid would require that a five-membered ring contain a trans double bond.

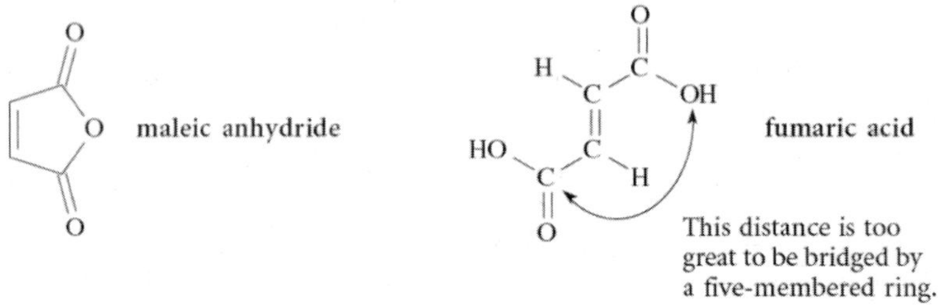

maleic anhydride   fumaric acid

This distance is too great to be bridged by a five-membered ring.

   (b)   On heating, 2,3-dimethylbutanedioic acid forms a cyclic anhydride containing a five-membered ring. Because a cyclic anhydride of a-methylmalonic acid would contain a very strained four-membered ring, it is not formed on heating.

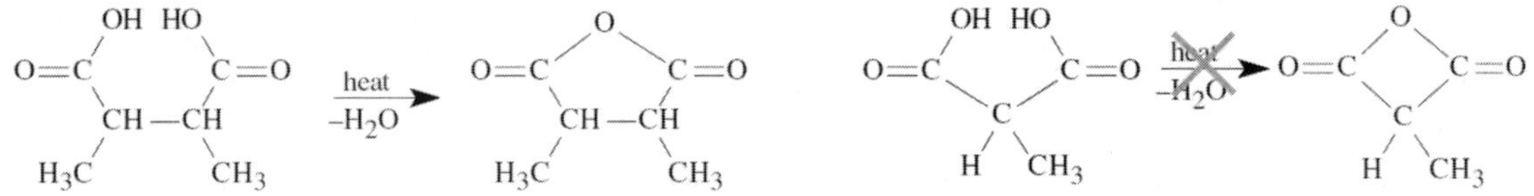

2,3-dimethylbutanedioic acid   2,3-dimethylbutanedioic anhydride   **α**-methylmalonic acid   **α**-methylmalonic anhydride

20.20   (a)   The compound must contain three oxygens and two degrees of unsaturation in addition to the benzene ring. *p*-Formylbenzoic acid meets the indicated criteria.

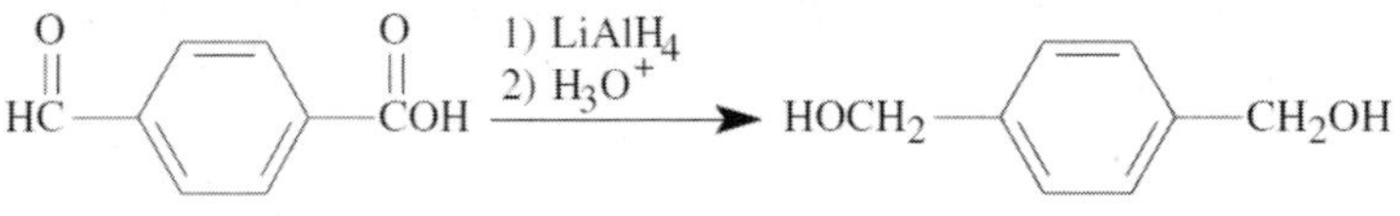

p-formylbenzoic acid

   (b)   1,4-Benzenedicarboxylic acid (terephthalic acid) is a compound with the formula $C_8H_6O_4$ that gives the indicated diol on treatment with $LiAlH_4$ followed by protonolysis.

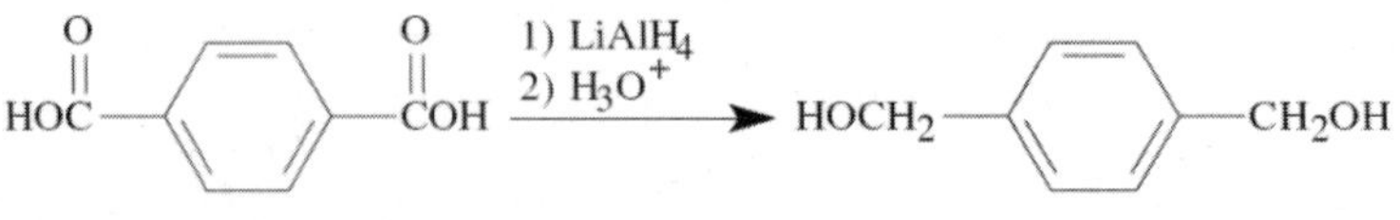

1,4-benzenedicarboxylic acid (terephthalic acid)

20.21   (a)   This synthesis requires the addition of one carbon. Follow the general scheme in Study Problem 20.2, text p. 1029.

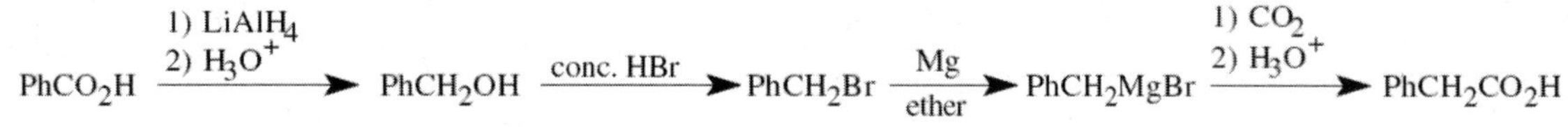

$$PhCO_2H \xrightarrow[\text{2) } H_3O^+]{\text{1) LiAlH}_4} PhCH_2OH \xrightarrow{\text{conc. HBr}} PhCH_2Br \xrightarrow[\text{ether}]{Mg} PhCH_2MgBr \xrightarrow[\text{2) } H_3O^+]{\text{1) CO}_2} PhCH_2CO_2H$$

benzoic acid

phenylacetic acid

(b)   This synthesis requires the addition of two carbons, which can be accomplished by the reaction of a Grignard reagent with oxirane (text Sec. 11.5C, text p. 529).

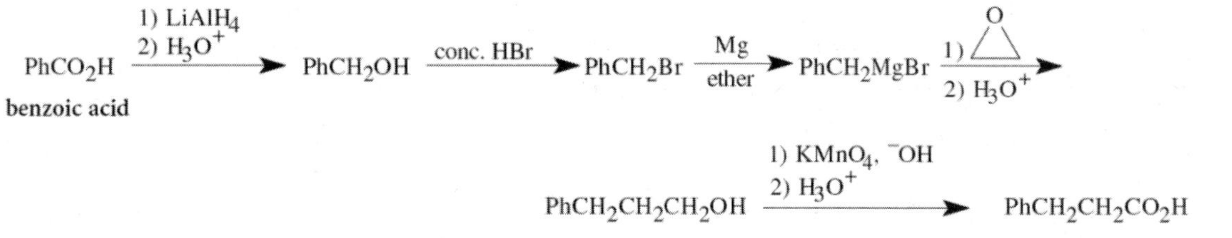

$$PhCO_2H \xrightarrow[\text{2) } H_3O^+]{\text{1) LiAlH}_4} PhCH_2OH \xrightarrow{\text{conc. HBr}} PhCH_2Br \xrightarrow[\text{ether}]{Mg} PhCH_2MgBr \xrightarrow[\text{2) } H_3O^+]{\text{1)}\ \triangle}$$

benzoic acid

$$PhCH_2CH_2CH_2OH \xrightarrow[\text{2) } H_3O^+]{\text{1) KMnO}_4,\ ^-OH} PhCH_2CH_2CO_2H$$

3-phenylpropanoic acid

20.22   Each compound decarboxylates; the net effect is the replacement of the carboxy group by a hydrogen. (In part (c), the amine product is protonated under the acidic conditions.)

(a)   (b)   (c)

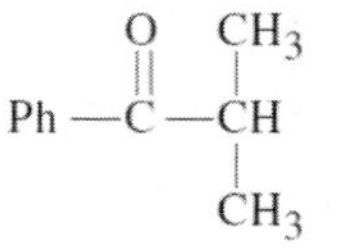

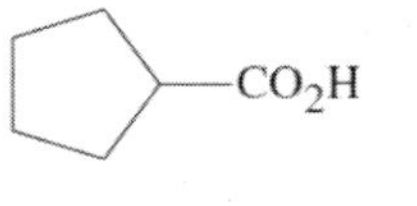

$$CH_3CH_2NH_2 \underset{}{\overset{H_3O^+}{\rightleftarrows}} CH_3CH_2\overset{+}{N}H_3$$

20.23   The following β-keto acids (and their enantiomers) will decarboxylate to give 2-methylcyclohexanone:

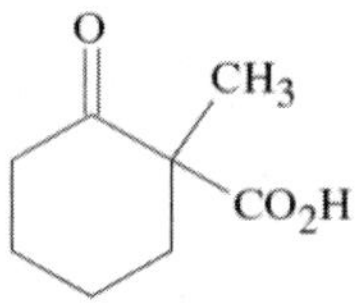

1-methyl-2-oxocyclo-
hexanecarboxylic acid

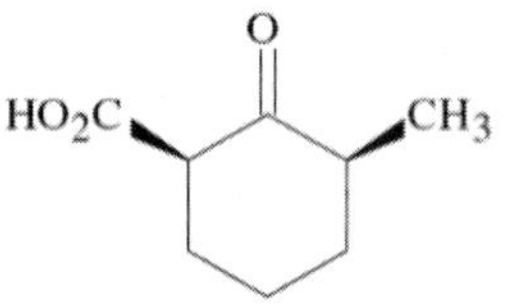

cis-3-methyl-2-oxocyclo-
hexanecarboxylic acid

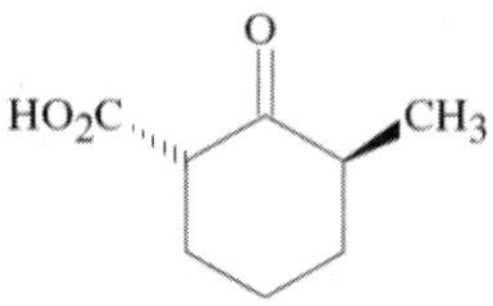

trans-3-methyl-2-oxocyclo-
hexanecarboxylic acid

20.24   Were this carboxylic acid to decarboxylate by the mechanism shown in Eq. 20.40 on p. 1030 of the text, the following enol would be formed as an intermediate:

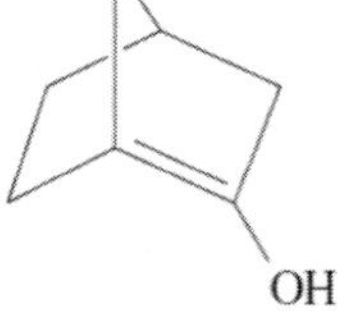

This enol is very unstable because it violates Bredt's rule (Sec. 7.6C, text pp. 297–8); that is, the double bond is twisted to such an extent that the 2*p* orbitals on each carbon cannot overlap.

20.25   (a)   The second resonance structure contains a carbene carbon—a very unstable electronic configuration for carbon that, while neutral, has less than eight electrons.

(b)   The second structure contains two *sp*-hybridized atoms (the carbon and nitrogen of the triple bond). The ideal bond angles for triple bonds are 180 degrees, which is impossible to obtain in a five-membered ring.

20.26   In the first step, the hydroxyl group of malic acid is oxidized by NAD⁺. An mechanism of an analogous enzymatic reaction mediated by NAD⁺, the oxidation of ethanol, is discussed in Sec. 10.8 in the text.

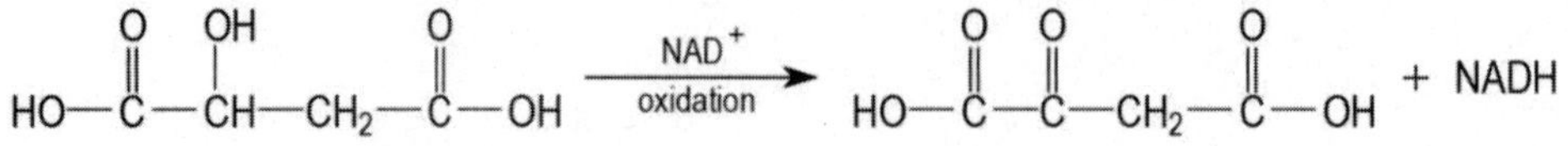

malic acid → oxalacetate + NADH

Oxalacetate is a β-keto carboxylic acid, which spontaneously decarboxylates (loses carbon dioxide) to an enol (see Sec. 20.11A in the text), which then isomerizes into a ketone, in this case pyruvate:

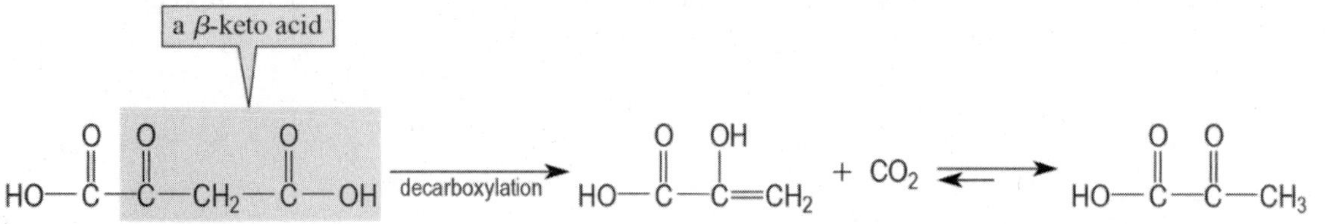

oxalacetate → an enol + CO₂ ⇌ pyruvate

Finally, the NADH that was generated in the first step serves as a reducing agent to reduce the pyruvate to lactic acid (see Sec. 19.8B in the text). In the process, NADH is oxidized back to NAD⁺, which can reenter the enzymatic cycle in step one, above.

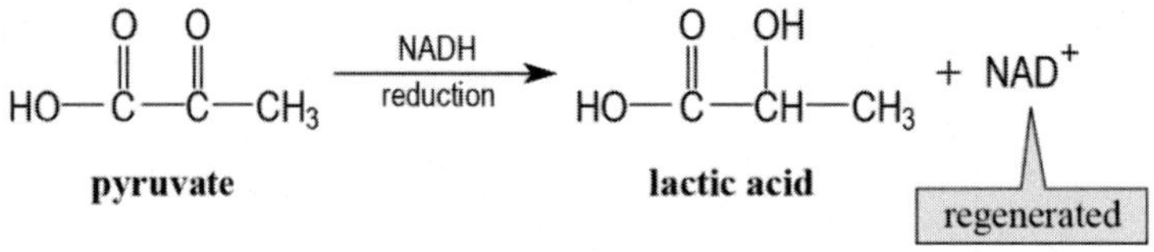

pyruvate → lactic acid + NAD⁺ (regenerated)

Solutions to Additional Problems

20.27 (a)

$$CH_3CH_2CH_2\overset{O}{\overset{\|}{C}}OCH_2CH_3$$

ethyl butyrate

(b)

$$CH_3CH_2CH_2\overset{O}{\overset{\|}{C}}O^- \ Na^+$$

sodium butyrate

(c)

$$CH_3CH_2CH_2CH_2OH$$

1-butanol

(d)

no reaction

(e)

$$CH_3CH_2CH_2\overset{O}{\overset{\|}{C}}Cl$$

butyryl chloride

(f)

$$CH_3CH_2CH_2\overset{O}{\overset{\|}{C}}OCH_3$$

methyl butyrate

(g)

$$(CH_3CH_2CH_2CH_2O)_2CHCH_3$$

acetaldehyde dibutyl acetal

(h)

$$CH_3CH_2CH_2\overset{O}{\overset{\|}{C}}Ph$$

butyrophenone

(i)

$$CH_3CH_2CH_2CH_2Ph$$

butylbenzene

20.28 (a)   (b)   (c)   (d)

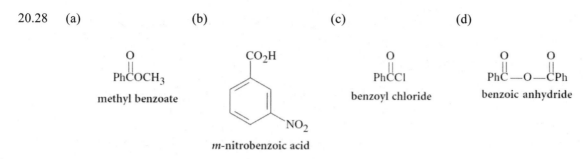

20.29 The dicarboxylic acids with the formula $C_6H_{10}O_4$:

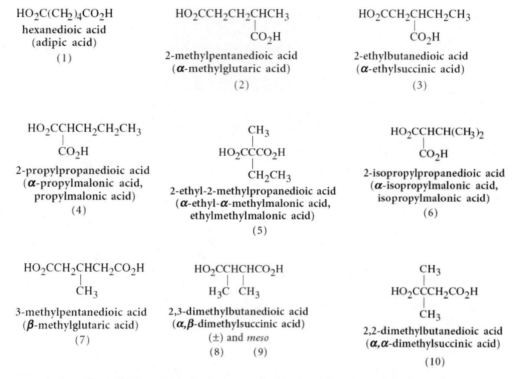

HO$_2$C(CH$_2$)$_4$CO$_2$H
hexanedioic acid
(adipic acid)
(1)

HO$_2$CCH$_2$CH$_2$CHCH$_3$
|
CO$_2$H
2-methylpentanedioic acid
(α-methylglutaric acid)
(2)

HO$_2$CCH$_2$CHCH$_2$CH$_3$
|
CO$_2$H
2-ethylbutanedioic acid
(α-ethylsuccinic acid)
(3)

HO$_2$CCHCH$_2$CH$_2$CH$_3$
|
CO$_2$H
2-propylpropanedioic acid
(α-propylmalonic acid,
propylmalonic acid)
(4)

CH$_3$
|
HO$_2$CCCO$_2$H
|
CH$_2$CH$_3$
2-ethyl-2-methylpropanedioic acid
(α-ethyl-α-methylmalonic acid,
ethylmethylmalonic acid)
(5)

HO$_2$CCHCH(CH$_3$)$_2$
|
CO$_2$H
2-isopropylpropanedioic acid
(α-isopropylmalonic acid,
isopropylmalonic acid)
(6)

HO$_2$CCH$_2$CHCH$_2$CO$_2$H
|
CH$_3$
3-methylpentanedioic acid
(β-methylglutaric acid)
(7)

HO$_2$CCHCHCO$_2$H
| |
H$_3$C CH$_3$
2,3-dimethylbutanedioic acid
(α,β-dimethylsuccinic acid)
($\pm$) and *meso*
(8)   (9)

CH$_3$
|
HO$_2$CCCH$_2$CO$_2$H
|
CH$_3$
2,2-dimethylbutanedioic acid
(α,α-dimethylsuccinic acid)
(10)

Compounds (2), (3), and (8) are chiral. Compounds (2), (3), (7), (8), (9), and (10) would readily form cyclic anhydrides on heating, because those anhydrides contain five- or six-membered rings. Compounds (4), (5), and (6) would readily decarboxylate on heating because they are malonic acid derivatives.

20.30 (a)   First calculate the number of moles of 0.1 M NaOH required for the neutralization:

$$\text{mol NaOH} = (0.100 \text{ mol NaOH L}^{-1})(8.61 \times 10^{-3} \text{ L}) = 8.61 \times 10^{-4}$$

The same molar amount of the carboxylic acid must be present because one mole of NaOH is required to neutralize one mole of a monocarboxylic acid. The molecular mass is

$$\text{molecular mass of the acid} = (100 \times 10^{-3} \text{ g}) \div (8.61 \times 10^{-4} \text{ mol}) = 116 \text{ g mol}^{-1}$$

(b)   First, calculate the number of millimoles of succinic acid (see Table 20.1, text p. 1006) in 100 mg of succinic acid:

$$\text{mmol succinic acid} = \frac{100 \text{ mg succinic acid}}{118.10 \text{ mg succinic acid (mmol succinic acid)}^{-1}} = 0.847$$

Using the facts that sodium hydroxide reacts with the carboxylic acid groups and that succinic acid has two dicarboxylic acid groups per molecule, calculate the millimoles of carboxylic acid groups present in the sample:

$$\text{mmol carboxylic acid} = \frac{2\ \text{mmol carboxylic acid}}{\text{mmol succinic acid}} \times 0.847\ \text{mmol succinic acid} = 1.69$$

Finally, calculate the volume 0.1 M NaOH required for the neutralization:

$$\text{mL of NaOH} = \frac{1\ \text{mL of NaOH}}{0.1\ \text{mmol of NaOH}} \times \frac{1\ \text{mmol of NaOH}}{1\ \text{mmol of carboxylic acid}} \times 1.69\ \text{mmol of carboxylic acid} = 16.9$$

20.31 In each case, the base reacts with acetic acid in a Brønsted acid–base reaction to form the conjugate base of acetic acid (acetate ion) and a by-product.

(a)

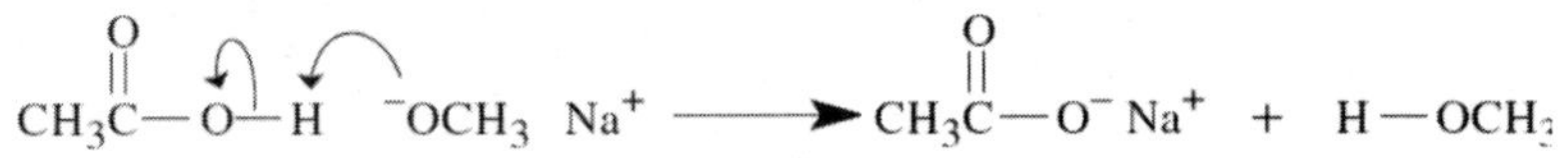

(b)

(c)

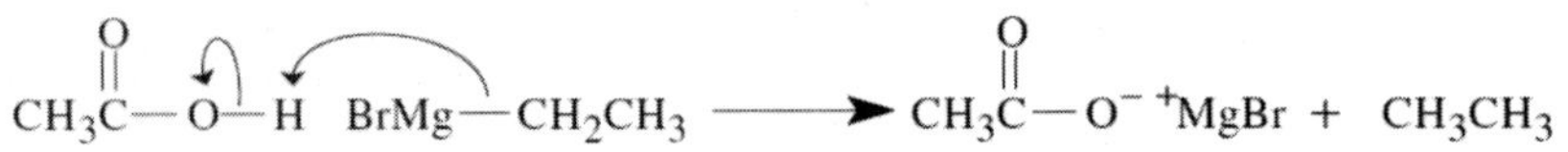

(d)

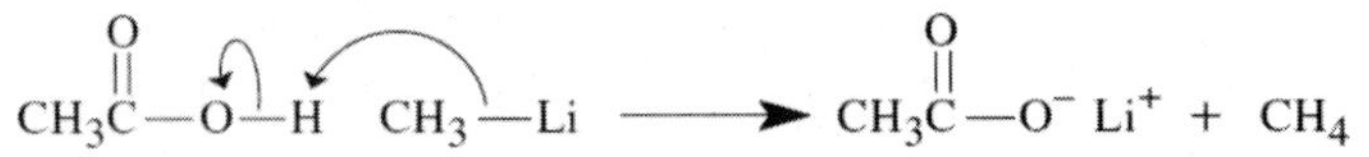

(e)

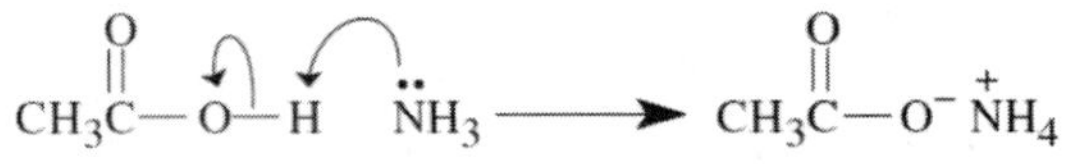

(f)

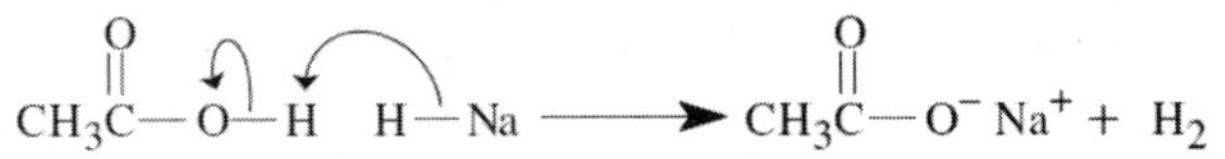

(g)

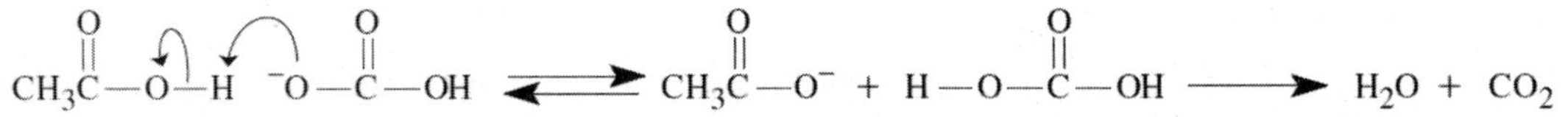

bicarbonate ion carbonic acid

20.32 Only enough NaOH is present to ionize one of the carboxy groups. The question, then, is which if any of the carboxy groups is more acidic. Because polar effects of substituents decrease with increasing distance of the substituents from the reaction center, the acid-strengthening polar effect of the chlorines has a greater effect on the carboxy group to which they are closer. Therefore, in the presence of one molar equivalent of NaOH, it is this carboxy group that is principally ionized.

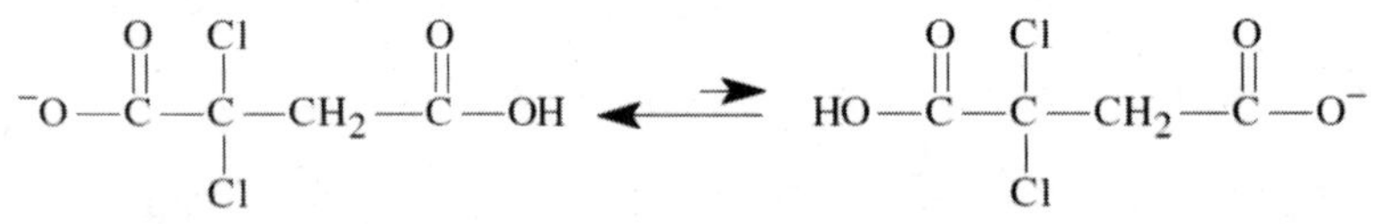

major species present little of this anion is present

20.33 As the chain length becomes larger the distance between the two carboxy groups becomes greater. Because polar effects decrease with distance, the polar effect is negligible when the chain length is great, and the two pK_a values are nearly equal. (However, they are not *exactly* equal; for an explanation, see the icon comment following the solution to Problem 3.51 on p. 65 of this manual.)

20.34   The trimethylammonium substituent in B has the greatest acid-strengthening polar effect because it carries a full positive charge. The methoxy substituent in A has a more modest acid-strengthening polar effect associated with the C=O bond dipole. The polar effect of a methyl group is negligible, and is slightly acid-weakening. Therefore, the order of increasing acidity (and therefore decreasing pK_a is $C < A < B$.

20.35   Carry out a calculation of pH using the dissociation-constant expression for each acid. (See Eq. 3.19, text p. 101.) Let the concentration of the undissociated acid be $0.1 - x$; the concentrations of both H_3O^+ and $CH_3CO_2^-$ are then x.

$$K_a = 10^{-4.76} = 1.74 \times 10^{-5} = \frac{[H_3O^+][CH_3CH_2O^-]}{[CH_3CO_2H]} = \frac{x^2}{0.1 - x}$$

Because the dissociation constant of acetic acid is small, use the approximation that x is small compared with the amount of acetic acid present.

$$(x^2)/0.1 = 1.74 \times 10^{-5}$$
$$x = [H_3O^+] = 1.32 \times 10^{-3}$$
$$-\log x = pH = 2.88$$

Thus, the pH is less than 3, and acetic acid therefore turns litmus paper red.

   An analogous calculation reveals that the pH of a 0.1 M solution of phenol ($pK_a = 9.95$) is 5.48. Since this pH is greater than 3, a 0.1 M solution of phenol ("carbolic acid") does not turn litmus paper red.

20.36   (a)   Consider the following equilibrium, in which AH is an acid, and S is the solvent:

$$AH + S \rightleftarrows A^- + \overset{+}{S}H$$

Suppose that AH is more acidic than ^+SH. Then this equilibrium lies to the right, and ^+SH, the conjugate acid of the solvent, is the major species present. That is, because AH cannot survive, the strongest acid that can exist is the conjugate acid of the solvent.

(b)   Use the result in part (a). When acetic acid is the solvent, the most acidic species that can exist is the *conjugate acid* of acetic acid (that is, protonated acetic acid; see structure in Eq. 20.8, text p. 1015, with R— = H_3C—), which has a pK_a of about –6 (Eq. 20.8). Table 3.1 on text p. 102 gives the pK_a of HBr as –8 to –9.5. Consequently, the HBr–acetic acid system has an effective pK_a of about –6. In aqueous solution, HBr is dissociated to give H_3O^+, the conjugate acid of water, which has a pK_a of –1.7; therefore the effective pK_a of aqueous HBr is –1.7. Hence, HBr in acetic acid is far more acidic than HBr in water because of the greater acidity of the conjugate acid of acetic acid, the solvent.

20.37   The proton and the π bond move back and forth between the two oxygens so rapidly that the two oxygens retain their identities for only a tiny fraction of a second. The mechanism varies with the situation. In a solvent (for example, water) in which the acid readily dissociates, even to a small extent, the two oxygens in the conjugate-base carboxylate ion are equivalent and are protonated with essentially equal frequency; these reactions are very fast. (In the equations that follow, one of the oxygens is marked with an asterisk (*) so that its fate can be traced throughout the reaction.)

In solvents in which the carboxylic acid molecule cannot readily dissociate, protons can jump back and forth within a carboxylic acid hydrogen-bonded dimer (text p. 1010):

20.38   (a)

$$(CH_3)_2CHCOH \xrightarrow[\text{H}_2\text{SO}_4 \text{ (catalyst)}]{\text{HOCH}_2\text{CH}_2\text{CH}_3 \text{ (solvent)}} (CH_3)_2CHCOCH_2CH_2CH_3$$

isobutyric acid
(2-methylpropanoic acid)

propyl isobutyrate

(b)

$$(CH_3)_2CHCOH \xrightarrow{\text{CH}_3\text{OH (solvent), H}_2\text{SO}_4 \text{ catalyst}} (CH_3)_2CHCOCH_3$$

methyl isobutyrate

(c)   Notice in this synthesis that two additional carbons must be added. Incorporate the strategy for a two-carbon chain extension of an alcohol shown in Study Problem 11.7 on text pp. 553–4.

$$(CH_3)_2CHCOH \xrightarrow[\text{2) H}_3\text{O}^+]{\text{1) LiAlH}_4} (CH_3)_2CHCH_2OH \xrightarrow[\text{H}_2\text{SO}_4]{\text{conc. HBr}} (CH_3)_2CHCH_2Br \xrightarrow[\text{ether}]{\text{Mg}} \xrightarrow[\text{2) H}_3\text{O}^+]{\text{1)} \triangle}$$

2-methylpropanoic acid

$$(CH_3)_2CHCH_2CH_2CH_2OH \xrightarrow[\text{2) H}_3\text{O}^+]{\text{1) KMnO}_4, \text{ }^-\text{OH}} (CH_3)_2CHCH_2CH_2COH$$

4-methylpentanoic acid

(d)

$$(CH_3)_2CHCOH \xrightarrow{\text{SOCl}_2} (CH_3)_2CHCCl \xrightarrow[\text{2) H}_3\text{O}^+]{\text{1) benzene, AlCl}_3} (CH_3)_2CHC-\langle \bigcirc \rangle$$

isobutyrophenone
(2-methyl-1-phenyl-1-propanone)

(e)   In part (e) and (f), CrO$_3$(pyridine)$_2$ is an alternative to pyridine chlorochromate (PCC). Either is correct.

$$(CH_3)_2CHCH_2OH \xrightarrow{\text{CrO}_3\text{(pyridine)}_2} (CH_3)_2CHCH{=}O \xrightarrow[\text{2) H}_3\text{O}^+]{\text{1) CH}_3\text{MgI}} (CH_3)_2CH\overset{\text{OH}}{\underset{}{C}}HCH_3 \xrightarrow{\text{CrO}_3\text{(pyridine)}_2}$$

prepared in part (c)

$$(CH_3)_2CHCCH_3$$

3-methyl-2-butanone

(f)

$$(CH_3)_2CHCH_2OH \xrightarrow{\text{CrO}_3\text{(pyridine)}_2} (CH_3)_2CHCH{=}O \xrightarrow[\text{(Wittig reaction)}]{\overset{-}{\text{CH}_2}-\overset{+}{\text{PPh}_3}} (CH_3)_2CHCH{=}CH_2$$

prepared in part (c)

3-methyl-1-butene

20.39   Convert each enantiomer of 4-methylhexanoic acid into 3-methylhexane by any process that does not involve breaking a bond to the asymmetric carbon. Because each enantiomer of 4-methylhexanoic acid has known absolute configuration, the configuration of the corresponding enantiomer of 3-methylhexane will be determined. Suppose, for example, that (S)-4-methylhexanoic acid gives (–)-3-methylhexane. Then the two molecules should have corresponding configurations, and the configuration of (–)-3-methylhexane would then be S:

$$\underset{\underset{\text{(S)-4-methylhexanoic acid}}{\begin{array}{c}\text{CH}_2\text{CH}_2\text{CO}_2\text{H}\\ \text{H}\cdots\text{C}\text{—CH}_3\\ \text{CH}_2\text{CH}_3\end{array}}}{} \longrightarrow \underset{\underset{\text{(S)-3-methylhexane}}{\begin{array}{c}\text{CH}_2\text{CH}_2\text{CH}_3\\ \text{H}\cdots\text{C}\text{—CH}_3\\ \text{CH}_2\text{CH}_3\end{array}}}{}$$

The following sequence of reactions would work for such a conversion:

$$\underset{\text{CH}_3}{\text{CH}_3\text{CH}_2\text{CHCH}_2\text{CH}_2\text{CO}_2\text{H}} \xrightarrow[\text{2) H}_3\text{O}^+]{\text{1) LiAlH}_4} \underset{\text{CH}_3}{\text{CH}_3\text{CH}_2\text{CHCH}_2\text{CH}_2\text{CH}_2\text{OH}} \xrightarrow[\text{H}_2\text{SO}_4]{\text{conc. HBr}}$$

$$\underset{\text{CH}_3}{\text{CH}_3\text{CH}_2\text{CHCH}_2\text{CH}_2\text{CH}_2\text{Br}} \xrightarrow[\text{ether}]{\text{Mg}} \underset{\text{CH}_3}{\text{CH}_3\text{CH}_2\text{CHCH}_2\text{CH}_2\text{CH}_2\text{MgBr}} \xrightarrow{\text{H}_3\text{O}^+} \underset{\text{CH}_3}{\text{CH}_3\text{CH}_2\text{CHCH}_2\text{CH}_2\text{CH}_3}$$

20.40 (a) The substitutive name of valproic acid is 2-propylpentanoic acid.

(b) The common name of valproic acid is α-propylvaleric acid.

(c) A number of syntheses are possible. Here are two:

Synthesis #1:

$$\underset{\text{CH}_3\text{CH}_2\text{CH}_2\text{CH}}{\overset{\text{O}}{\|}} \xrightarrow[\text{2) H}_3\text{O}^+]{\text{1) CH}_3\text{CH}_2\text{CH}_2\text{MgBr}} \underset{\underset{\text{4-heptanol}}{\text{CH}_3\text{CH}_2\text{CH}_2\text{CHCH}_2\text{CH}_2\text{CH}_3}}{\overset{\text{OH}}{|}} \xrightarrow{\text{CrO}_3(\text{pyridine})_2}$$

$$\underset{\text{CH}_3\text{CH}_2\text{CH}_2\text{CCH}_2\text{CH}_2\text{CH}_3}{\overset{\text{O}}{\|}} \xrightarrow[\text{(Wittig reaction)}]{\text{H}_2\text{C}\text{—PPh}_3} \underset{\text{CH}_3\text{CH}_2\text{CH}_2\text{CCH}_2\text{CH}_2\text{CH}_3}{\overset{\text{CH}_2}{\|}} \xrightarrow[\text{2) H}_2\text{O}_2/\text{NaOH}]{\text{1) BH}_3/\text{THF}}$$

$$\underset{\text{CH}_3\text{CH}_2\text{CH}_2\text{CHCH}_2\text{CH}_2\text{CH}_3}{\overset{\text{CH}_2\text{OH}}{|}} \xrightarrow[\text{2) H}_3\text{O}^+]{\text{1) KMnO}_4, \text{ }^-\text{OH}} \underset{\underset{\text{valproic acid}}{\text{CH}_3\text{CH}_2\text{CH}_2\text{CHCH}_2\text{CH}_2\text{CH}_3}}{\overset{\text{CO}_2\text{H}}{|}}$$

Synthesis #2:

$$\underset{\underset{\text{4-heptanol}}{\text{CH}_3\text{CH}_2\text{CH}_2\text{CHCH}_2\text{CH}_2\text{CH}_3}}{\overset{\text{OH}}{|}} \xrightarrow[\text{tosyl chloride}]{\text{H}_3\text{C}—\bigcirc—\text{SO}_2\text{Cl, pyridine}}$$

$$\underset{\text{CH}_3\text{CH}_2\text{CH}_2\text{CHCH}_2\text{CH}_2\text{CH}_3}{\overset{\text{OTs}}{|}} \xrightarrow{\text{NaBr in DMSO}} \underset{\text{CH}_3\text{CH}_2\text{CH}_2\text{CHCH}_2\text{CH}_2\text{CH}_3}{\overset{\text{Br}}{|}} \xrightarrow[\text{3) H}_3\text{O}^+]{\overset{\text{1) Mg, ether}}{\text{2) CO}_2}}$$

$$\underset{\underset{\text{valproic acid}}{\text{CH}_3\text{CH}_2\text{CH}_2\text{CHCH}_2\text{CH}_2\text{CH}_3}}{\overset{\text{CO}_2\text{H}}{|}}$$

20.41 To confirm the labeling pattern, oxidize phenylacetic acid vigorously with $KMnO_4$ or other oxidizing agent that will convert it into benzoic acid. (*C = ^{14}C.)

$$\bigcirc—\text{CH}_2\overset{*}{\text{C}}\text{O}_2\text{H} \xrightarrow[\text{2) H}_3\text{O}^+]{\text{1) KMnO}_4, \text{ }^-\text{OH, heat}} \bigcirc—\text{CO}_2\text{H} + \overset{*}{\text{C}}\text{O}_2$$

If all the radiolabel of phenylacetic acid is at the carbonyl carbon, the product benzoic acid should be completely devoid of radioactivity. However, if there is labeling elsewhere, the fraction of radioactivity located in carbons other than the carbonyl carbon is equal to the amount of radioactivity in the product benzoic acid divided by the total radioactivity in the starting phenylacetic acid.

20.42   Use a method that does not involve loss of oxygen from benzoic acid. Thus, acid-catalyzed esterification should not be used, because, as Eq. 20.18c on text p. 1021 shows, this method results in cleavage of the bond between the carbonyl carbon and the carboxylate oxygen. Esterification with either diazomethane or methyl iodide and K_2CO_3 would be the preferred method because, as the discussion in Sec. 20.8B, text p. 1023, shows, these methods do not involve loss of oxygen. (O* = ^{18}O.)

preferred method: Ph—C(=O*)—OH $\xrightarrow[\text{CH}_3\text{I/acetone/K}_2\text{CO}_3]{\text{CH}_2\text{N}_2, \text{ ether or}}$ Ph—C(=O*)—OCH$_3$ all isotopic oxygen is retained

inferior method: Ph—C(=O*)—OH $\xrightarrow[\text{H}_2\text{SO}_4 \text{ (catalyst)}]{\text{CH}_3\text{OH (solvent)}}$ Ph—C(=O*)—OCH$_3$ + H$_2$O   some isotopic oxygen is lost

20.43   Dissolve the mixture in a suitable solvent; since the company specializes in chlorinated organic compounds, methylene chloride might be readily available for the purpose. Extract with 5% $NaHCO_3$ solution. *p*-Chlorobenzoic acid is extracted as its conjugate-base anion into the aqueous layer. Acidification of the aqueous layer will give *p*-chlorobenzoic acid itself. Then extract the methylene chloride solution with 5% aqueous NaOH solution. *p*-Chlorophenol will be extracted into the aqueous layer as its conjugate-base , *p*-chlorophenolate anion; acidification will afford *p*-chlorophenol. The basis for this separation is Eq. 20.7 on text p. 1014. When the pK_a of an acid is much lower than the pH of the solution, the acid is converted into its conjugate base; that is, the ratio of $[RCO_2^-]$ to $[RCO_2H]$ is large. The pK_a of *p*-chlorobenzoic acid, which is about 4 (see Table 20.2 in the text), is much lower than the pH of aqueous sodium bicarbonate, which is about 8.5. Hence, *p*-chlorobenzoic acid is more than 99.99% ionized by the sodium bicarbonate solution. The pK_a of *p*-chlorophenol, 9.4, is higher than the pH of aqueous sodium bicarbonate, and thus this phenol is not appreciably ionized. However, the pK_a of the phenol is much lower than the pH of 0.1 *M* aqueous NaOH solution, which is about 13; hence, the phenol is ionized by, and thus extracted into, this solution.

Once the acid and the phenol are removed, evaporation of the methylene chloride gives a mixture of 4-chlorocyclohexanol and chlorocyclohexane. The boiling point of the alcohol is much higher because of hydrogen bonding; consequently, these two remaining substances can be separated by fractional distillation.

20.44   The essence of this solution is to determine the ionization state of penicillin-G at the different pH values. The principles involved are discussed in sec. 20.4A, described by Eq. 20.7. Because penicillin G is a carboxylic acid, its pK_a should be in the 3–5 range. Because the pH of blood, 7.4, is considerably higher than the pK_a of the drug, penicillin-G is ionized in blood. Because the pH of stomach acid is lower than the pK_a of penicillin-G, the penicillin is largely un-ionized in stomach acid. Because carboxylate ions are generally more soluble in aqueous solution than un-ionized carboxylic acids, penicillin-G is more soluble in blood than it is in stomach acid.

20.45   (a)   For this problem, let

Ar— = $(CH_3)_2N$—⟨ ⟩—

The color of crystal violet is due to the overlap of the empty $2p$ orbital of the positively charged carbon with the π orbitals of the benzene ring. The bleaching is due to the reaction of the carbocation with hydroxide to form the alcohol. In the alcohol, there is no longer an empty $2p$ orbital; hence, the conjugation decreases and the color disappears.

Ar$_3$C$^+$ $^-$OH $\longrightarrow$ Ar$_3$C—OH
crystal violet   (colorless)
(colored)

(b)   SDS, a typical detergent, forms micelles. Despite its charge, crystal violet is incorporated into the interior of the SDS micelles because the "greasy" benzene rings interact favorably with the "greasy" hydrocarbon tails of SDS. (Think of this as a solubility phenomenon; benzene is more soluble in hydrocarbons than it is in water.) In the interior of a micelle, crystal violet is no longer accessible to hydroxide ion. Because hydroxide is strongly solvated by water, hydroxide ion cannot easily enter the interior of the micelles, and therefore it cannot react with the entrapped crystal violet. The very slow reaction that does occur is probably due to hydroxide reacting with the very small amount of crystal violet in solution.

20.46   (a)   (b)   (c)

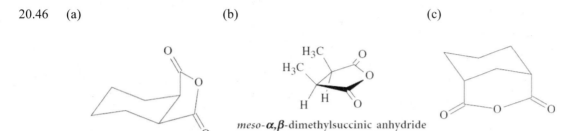

meso-α,β-dimethylsuccinic anhydride

20.47   Always remember that correct syntheses other than the one given may be possible. Be sure to ask your professor or teaching assistant if your synthesis differs from the one given and you are not sure whether it is correct.

(a)

$$CH_3CH_2CO_2H \xrightarrow[\text{2) H}_3\text{O}^+]{\text{1) LiAlH}_4} CH_3CH_2CH_2OH \xrightarrow[\text{H}_2\text{SO}_4]{\text{conc. HBr}} CH_3CH_2CH_2Br \xrightarrow[\text{ether}]{\text{Mg}}$$

propanoic acid

$$CH_3CH_2CH_2MgBr \xrightarrow[\text{2) H}_3\text{O}^+]{\text{1) CH}_3\text{CH}=\text{O}} CH_3CH_2CH_2\overset{\text{OH}}{\underset{}{\text{C}}}HCH_3$$

2-pentanol

(b)

$$H_2C=CHCH_2OH \xrightarrow{\text{H}_2\text{, cat.}} CH_3CH_2CH_2OH \xrightarrow[\text{2) H}_3\text{O}^+]{\text{1) KMnO}_4\text{, }^-\text{OH}} CH_3CH_2CO_2H$$

allyl alcohol

$$\downarrow \text{HBr}$$

$$H_2C=CHCH_2Br$$

$$\downarrow \text{K}_2\text{CO}_3\text{, acetone}$$

$$CH_3CH_3\overset{\text{O}}{\overset{\|}{\text{C}}}OCH_2CH=CH_2$$

    Acid-catalyzed esterification of propionic acid (prepared as shown above) with allyl alcohol could also be used, but this reaction would require an excess of allyl alcohol.

(c)

$$CH_3(CH_2)_3CO_2H \xrightarrow[\text{2) H}_3\text{O}^+]{\text{1) LiAlH}_4} CH_3(CH_2)_3CH_2OH \xrightarrow[\text{H}_2\text{SO}_4]{\text{conc. HBr}} CH_3(CH_2)_3CH_2Br \xrightarrow[\text{ether}]{\text{Mg}}$$

pentanoic acid

$$CH_3(CH_2)_3CH_2MgBr \xrightarrow[\text{2) H}_3\text{O}^+]{\text{1) (CH}_3)_2\text{C}=\text{O}} CH_3(CH_2)_3CH_2\overset{\text{OH}}{\underset{}{\text{C}}}(CH_3)_2 \xrightarrow{\text{HBr}}$$

$$CH_3(CH_2)_3CH_2\overset{\text{Br}}{\underset{}{\text{C}}}(CH_3)_2 \xrightarrow[\text{ether}]{\text{Mg}} CH_3(CH_2)_3CH_2\overset{\text{MgBr}}{\underset{}{\text{C}}}(CH_3)_2 \xrightarrow{\text{H}_3\text{O}^+} CH_3(CH_2)_3CH_2CH(CH_3)_2$$

2-methylheptane

(d)

toluene $\xrightarrow[\text{2) H}_3\text{O}^+]{\text{1) KMnO}_4,\ ^-\text{OH, heat}}$ (benzoic acid) $\xrightarrow{\text{HNO}_3,\ \text{H}_2\text{SO}_4}$ *m*-nitrobenzoic acid

(e)

$$\text{PhCO}_2\text{H} \xrightarrow[\text{2) H}_3\text{O}^+]{\text{1) LiAlH}_4} \text{PhCH}_2\text{OH} \xrightarrow{\text{conc. HBr}} \text{PhCH}_2\text{Br} \xrightarrow[\text{ether}]{\text{Mg}} \text{PhCH}_2\text{MgBr} \xrightarrow[\text{2) H}_2\text{SO}_4]{\text{1) O=CHCH}_2\text{Ph}}$$

benzoic acid

$$\text{PhCH=CHCHPh} \xrightarrow{\text{H}_2,\ \text{cat.}} \text{PhCH}_2\text{CH}_2\text{CH}_2\text{Ph}$$

(3-phenylpropyl)benzene

Although the rigorously correct substitutive name for this compound is (3-phenylpropyl)-benzene, most chemists would call the product in part (e) 1,3-diphenylpropane.

(f) First, reduce the ketone to a hydroxyl group to form a γ-hydroxy acid. These compounds spontaneously cyclize to form cyclic esters, or lactones. See Sec. 20.8A in the text for a discussion of the reaction mechanism of esterification from alcohols and carboxylic acids. Although the general reaction is catalyzed (sped up) with an acid catalyst, the cyclization in the reaction below, which results in the formation of a five-membered ring, requires no catalyst because it proceeds at an appreciable rate on its own.

(structures: γ-keto acid $\xrightarrow[\text{Pd/C catalyst}]{\text{H}_2}$ γ-hydroxy acid $\rightleftharpoons$ lactone + H_2O)

(g)

(norbornene structure) $\xrightarrow[\text{2) H}_2\text{O}_2,\ \text{H}_2\text{O}]{\text{1) O}_3}$ (diacid structure with CO$_2$H, CO$_2$H) $\equiv$ (cyclopentane structure with CO$_2$H, CO$_2$H)

cis-1,3-cyclopentanedicarboxylic acid

(h)

$$\text{CH}_3\overset{\text{O}}{\underset{\|}{\text{C}}}(\text{CH}_2)_3\text{Br} \xrightarrow[\text{HCl (catalyst)}]{\text{HOCH}_2\text{CH}_2\text{OH}}$$ (dioxolane with H$_3$C, (CH$_2$)$_3$Br) $\xrightarrow[\text{ether}]{\text{Mg}}$ (dioxolane with H$_3$C, (CH$_2$)$_3$MgBr) $\xrightarrow[\text{2) H}_3\text{O}^+]{\text{1) CO}_2}$ $\text{CH}_3\overset{\text{O}}{\underset{\|}{\text{C}}}(\text{CH}_2)_3\text{CO}_2\text{H}$

5-bromo-2-pentanone

5-oxohexanoic acid

(i)

$$\text{Ph}-\text{C}\equiv\text{C}-\text{H} \xrightarrow{\text{C}_2\text{H}_5\text{MgBr}} \text{Ph}-\text{C}\equiv\text{CMgBr} \xrightarrow[\text{2) H}_3\text{O}^+]{\text{1) CO}_2}$$

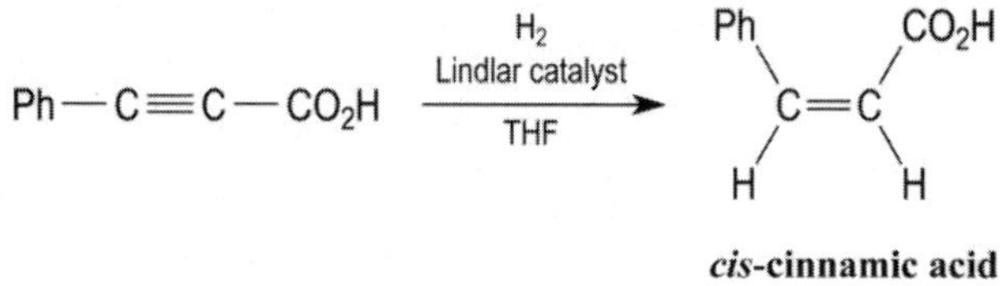
Ph—C≡C—CO2H
H2
Lindlar catalyst
THF
Ph
CO2H
C=C
H
H

***cis*-cinnamic acid**

20.48 (a) The two different compounds are the diastereomers that result respectively from loss of each of the two carboxy groups from compound *A*.

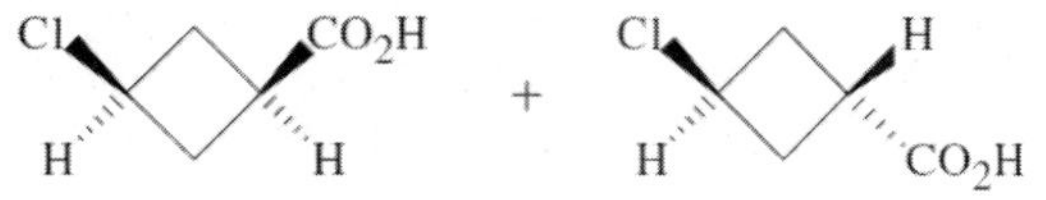

(b) The question is whether loss of the different carboxy groups gives rise to the same compound or to different compounds. Draw the two possible structures and determine whether they are different or identical. In fact, the two are identical; therefore, only one product is formed when compound *B* decarboxylates.

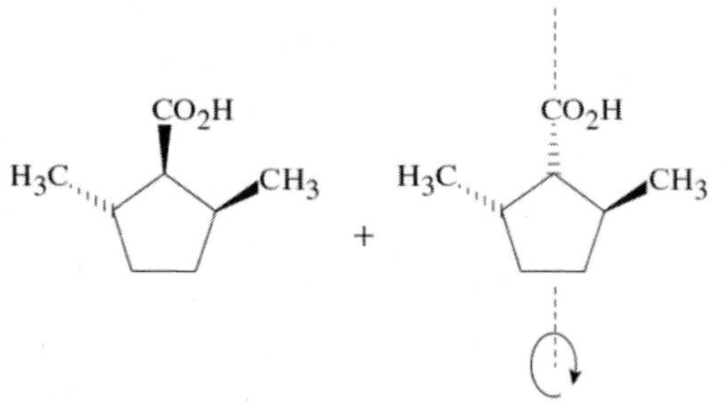

Rotation 180° about the dashed axis shows that these structure are congruent and are therefore identical.

20.49 (a) The first ionization of squaric acid has the lower pK_a, just as the first ionization of a dicarboxylic acid has a lower pK_a than the second. (For the reason, see the solution to Problem 20.7(b) on p. 487 of this manual.)

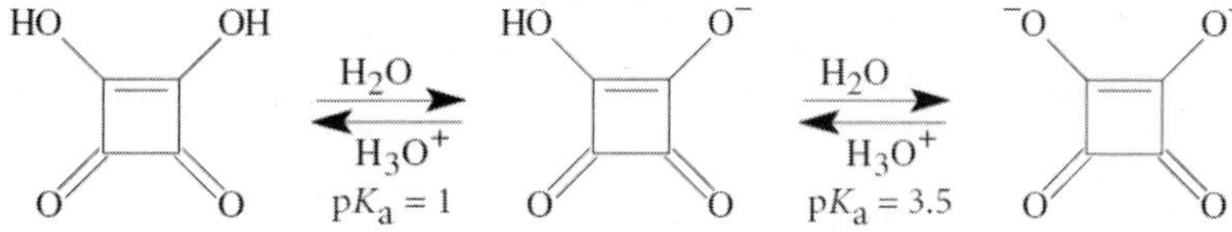
H2O
H3O+
pKa = 1
H2O
H3O+
pKa = 3.5

(b) One reason for the greater acidity of squaric acid is the electron-withdrawing polar effect of the two carbonyl groups, which stabilizes the anions. Another reason for the greater acidity of squaric acid is the resonance stabilization of the conjugate-base anions that results from interaction of the unshared pairs on each of the oxygens with a carbonyl group.

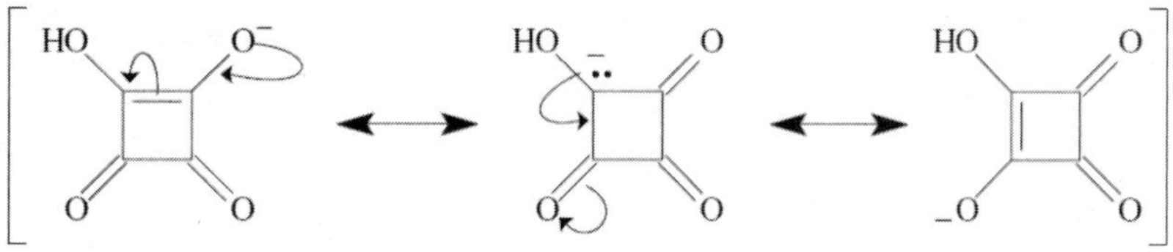

In the conjugate base of an enol, in contrast, the stabilizing polar effect of a carbonyl oxygen is not present, and the only resonance stabilization results from interaction of the electron pair with the double bond.

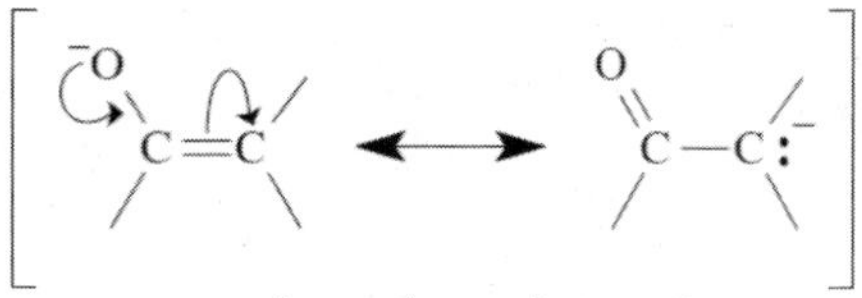

conjugate base of an enol

 If you find an error in this manual, please visit http://people.pharmacy.purdue.edu/~loudonm/teaching/

(c) The reactions of squaric acid with a number of reagents are analogous to the corresponding reactions of carboxylic acids. Thus, reaction with $SOCl_2$ results in the formation of the "di-acid chloride" *A*; and reaction with ethanol and an acid catalyst results in formation of the "di-ester" *B*.

20.50 (a) The reagent involved is essentially a diazomethane molecule in which the hydrogens of diazomethane have been replaced by phenyl groups. Just as diazomethane forms methyl esters, this reagent forms a diphenylmethyl ester.

diphenylmethyl 4-methylbenzoate

(b) The KOH converts benzoic acid into its conjugate-base benzoate anion, which is alkylated by benzyl chloride to give benzyl benzoate.

potassium benzoate   benzyl benzoate

(c) Ethylene glycol is a diol and the acid, terephthalic acid (1,4-benzenedicarboxylic acid), is a dicarboxylic acid. Each end of both the diacid and the diol can be esterified; the result is a polymeric ester, polyethylene terephthalate. You may be wearing some of this compound, which is known in the industrial world as *polyester*.

polyethylene terephthalate (polyester)

(d) This is an oxymercuration-reduction reaction in which acetic acid rather than water serves as the nucleophile that opens the mercurinium ion. (See the solution to Problem 5.35(c) on p. 136 of this manual.)

1-methylcyclohexyl acetate

(e) The conjugate-base carboxylate of propionic acid acts as the nucleophile in a ring-opening reaction of the epoxide. Because the conditions are basic, reaction occurs at the less branched carbon of the epoxide. Note that KOH reacts much more rapidly with the acidic hydrogen of the acid than it does with the epoxide; because only one equivalent of KOH is present, reaction of hydroxide ion with the epoxide is not a competing reaction.

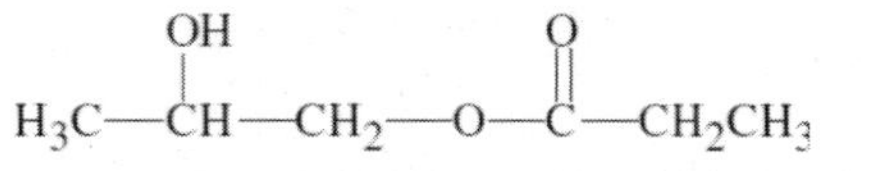

2-hydroxy-1-propyl propanoate
(*β*-hydroxypropyl propionate)

(f)   This is an intramolecular variation of the esterification shown in Eq. 20.22, text p. 1023. In this case, potassium carbonate converts the carboxylic acid into its conjugate-base potassium carboxylate, which is then intramolecularly alkylated to form the cyclic ester (lactone).

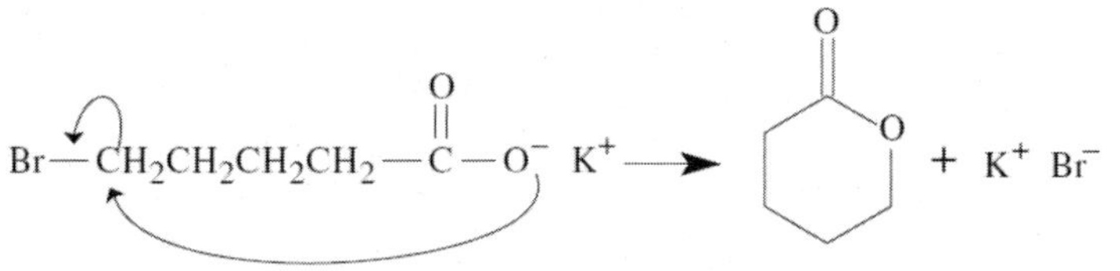

(g)   Potassium carbonate converts both the carboxy group and the phenol —OH groups into their respective conjugate-base anions; these are all alkylated by dimethyl sulfate.

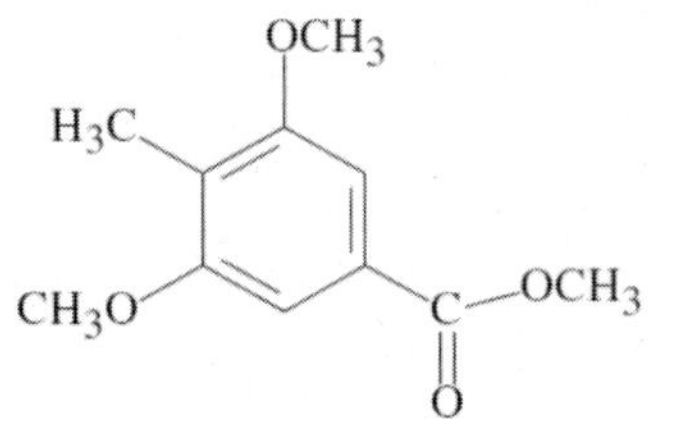

methyl 3,5-dimethoxy-4-methylbenzoate

(h)   Chlorosulfonation of chlorobenzene gives electrophilic aromatic substitution at the para position. (See Eq. 20.28a–b on text pp. 1025–6.)

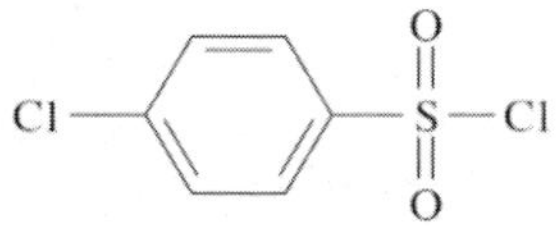

p-chlorobenzenesulfonyl chloride

20.51 (a)   The hint refers to the fact that organolithium reagents, like Grignard reagents, are strong bases and react instantaneously with carboxylic acids to give their conjugate-base lithium carboxylate salts:

A second equivalent of organolithium reagent undergoes a carbonyl addition with the carboxylate salt to give compound *A*, which is the conjugate-base di-anion of a ketone hydrate. Since compound *A* does not contain a good leaving group, it is stable until acid is added. Protonation forms the hydrate, which then decomposes to the ketone by the reverse of the mechanism shown for hydrate formation in Eqs. 19.20a–b on text p. 966.

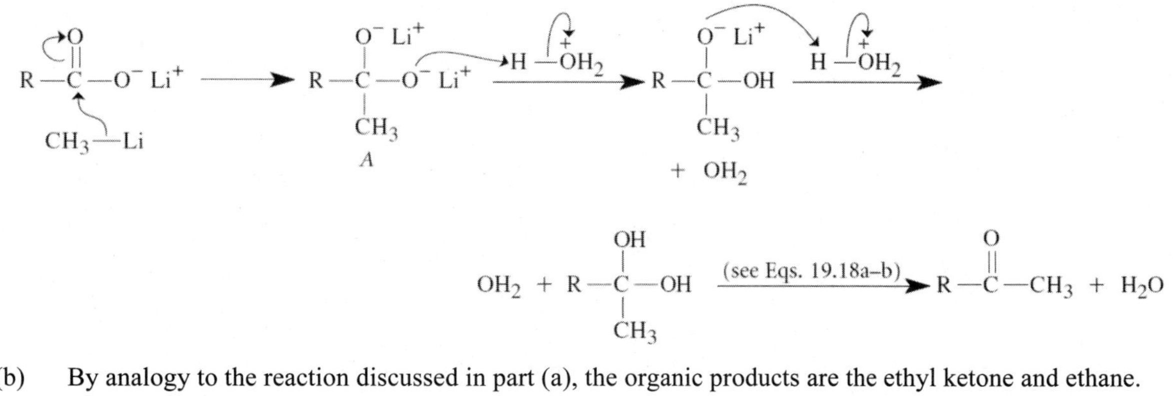

(b)   By analogy to the reaction discussed in part (a), the organic products are the ethyl ketone and ethane.

20.52 (a) Orthoesters have the same relationship to esters that acetals have to ketones, and the hydrolysis mechanisms of both orthoesters and acetals are virtually identical.

(b) The mechanism is much like that for acetal formation, except that the carboxy group rather than a second alcohol molecule reacts with the α-alkoxy carbocation intermediate. (Recall that many intramolecular reactions that form small rings are faster than related intermolecular reactions; see Sec. 11.8A–B, text p. 539.)

(c)   Sulfuric acid protonates the alkene to give a carbocation, which is reacts with carbon monoxide to give an acylium ion (Eq. 16.23, text p. 808). This ion, in turn, reacts with water to give, after appropriate proton transfers, the carboxylic acid.

(d)   The pattern is much like that in the solution to part (c), except that acetic acid rather than carbon monoxide reacts with the carbocation intermediate.

(e)   Acid-promoted opening of the epoxide gives a tertiary and benzylic carbocation, which decarboxylates to the enol. The enol then forms the aldehyde by a mechanism like that shown in Eqs. 14.6a–b on text p. 691.

The other C=O bond of the epoxide does not rupture instead because (1) the carbocation that would result is secondary, and (2) carbocations α to carbonyl groups are very unstable. (The latter point is considered in Chapter 22 in Eq. 22.37, on text p. 1118.)

(f)   The carbon of diazomethane acts as a nucleophile at the carbonyl carbon to form tetrahedral adduct *T*. This intermediate has two fates. To give the major product, we show loss of N_2 to give a primary carbocation, which then rearranges. In reality, the loss of dinitrogen and the rearrangement occur in a concerted manner, but the transition state undoubtedly has carbocation character.

Alternatively, the carbocation can collapse with the anionic oxygen or, more likely, an internal "S_N2" reaction can take place within *T* itself, to give the minor product.

(g)   The alkene reacts with iodine to form an iodium ion, analogous to bromonium ion formation discussed in Sec. 5.2A, text p. 184. The –OH group of the carboxylic acid opens the iodium ion in an intramolecular nucleophilic attack, analogous to halohydrin formation discussed in Sec. 5.2B, text p. 186.

20.53 (a)   The protonated imine nitrogen serves as an "electron sink" that accepts electrons from the departing carbon dioxide. What follows is the acid-mediated formation of a ketone from an imine, which is the reverse of imine formation from an amine and a ketone, discussed in Sec 19.11A, text p. 984. Note that the "proton transfer" steps are abbreviated in the mechanism below.

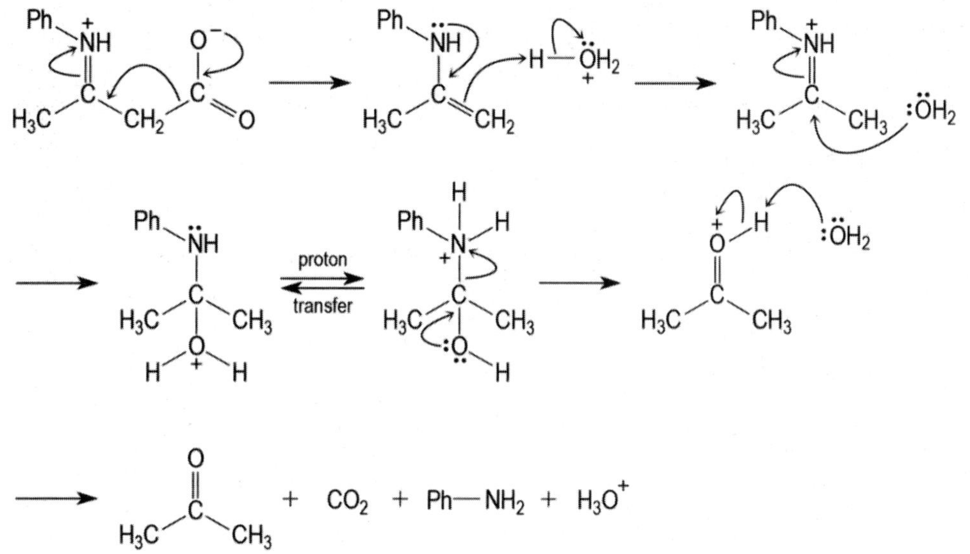

(b)   At low pH, the nitrogen is protonated, as in part (a), but the carboxylic acid is also protonated. This form of the compound won't decarboxylate.

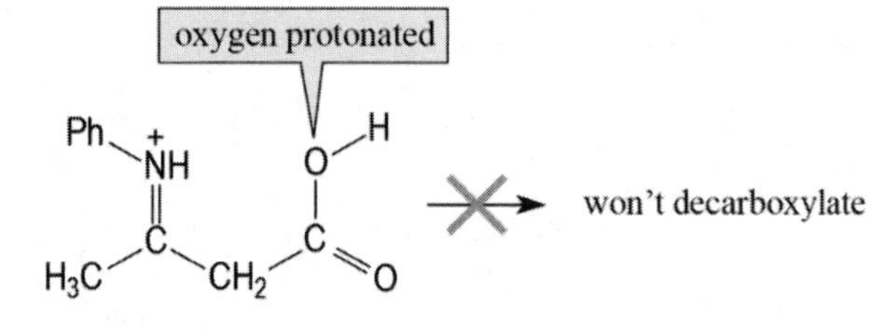

won't decarboxylate

At high pH, the carboxylic acid is deprotonated, but so is the nitrogen of the imine. Decarboxylation of this form would produce a very strong, unstable base.

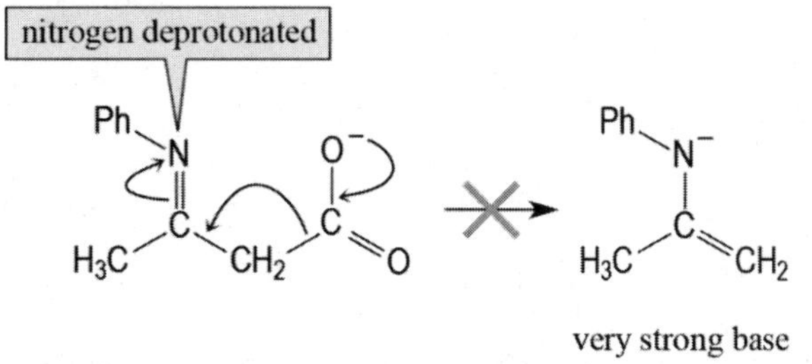

very strong base

(c)   The primary amine reacts with the ketone in acetoacetic acid to form an imine, as described in Sec. 19.11A, text p. 984. The β-imino carboxylic acid decarboxylates, and the catalyst amine is regenerated when the imine reacts with water via a mechanism that is the reverse of imine formation, as shown in part (a).

20.54 (a)

(b) The positively charged nitrogen acts as an electron sink necessary for decarboxylation.

(c) The driving force for this reaction is the regeneration of aromaticity. You can check your answer by showing the product of the reaction of lysine with PLP, which also produces imine *Y*, as stated in the problem.

$$X \qquad\qquad Y$$

20.55   (a)   The formation of phosphate suggests that it serves as a leaving group and that the reaction proceeds via solvolysis, as described in Sec. 9.6, text p. 418.

$$[X]$$

isopentenyl pyrophosphate

Because this is an E1-like mechanism, the stability of the carbocation intermediate has a direct effect on the rate of the reaction. The methyl group donates electron density through induction, stabilizing the electron-deficient carbon. The fluoromethyl group withdraws electron density due to the very electronegative fluorine atom, which destabilizes the electron-deficient carbon. Thus, the carbocation that results is much higher in energy and the reaction rate is slower.

carbocation stabilized; solvolysis faster

$$[X]$$

carbocation destabilized; solvolysis slower

(b)   The positively charged carbon is electron deficient and acts as an electron sink necessary for decarboxylation.

20.56   (a)   The even mass of the $m/z = 74$ ion shows that the ion is an odd-electron ion. McLafferty rearrangement is a common mechanism for formation of such ions in carbonyl compounds. (See Sec. 19.3E, text p. 958.)

H₃C ... HC ... H₂C ... CH ... CH₃ ... O ... C ... OH
molecular ion
→
H₃C ... HC ... H₂C ... CH ... OH ... CH₃
$m/z = 74$

(b)   The molecular mass of benzoic acid is 122; therefore, the $m/z = 105$ peak in its mass spectrum represents a mass loss of 17 units, which corresponds to loss of an —OH group. This loss can occur by an α-cleavage mechanism.

Ph —C— OH → Ph —C   +   •OH
molecular ion $m/z = 105$

The $m/z = 77$ peak corresponds to a phenyl cation, which can be lost by inductive cleavage at the other side of the carboxy group.

$\left[Ph-C-OH \longleftrightarrow Ph-C-OH \right]$ → Ph^+ +   :C—OH
molecular ion $m/z = 77$

 You may recall that aryl cations are very unstable. (See Fig. 18.2, text p. 884.) The high electron energies involved in mass spectroscopy enable the formation of such intermediates that would ordinarily not form in solution.

20.57   (a)   The presence of two exchangeable hydrogens suggests that compound *A* is a dicarboxylic acid. The carbon NMR chemical shifts suggest the presence of a methyl group, a carbonyl group, and a carbon bound to a carbonyl group. Because compound *A* is a dicarboxylic acid, each carbon NMR resonance must correspond to two carbons. Compound *A* is 2,3-dimethylbutanedioic acid; the fact that it can be resolved into enantiomers shows that it is the racemate rather than the *meso* diastereomer.

H₃C ... H ... H ... C—C—CO₂H ... HO₂C ... CH₃

(±)-2,3-dimethylbutanedioic acid
(compound *A*;
one of two enantiomers)

(b)   The *meso* stereoisomer isomer would have ^{1}H and ^{13}C NMR spectra that are almost identical to that of *A*. Because it is a diastereomer of *A*, it has a different melting point.

H ... H ... H₃C—C—C—CH₃ ... HO₂C ... CO₂H

***meso*-2,3-dimethylbutanedioic acid**

20.58   A hydrocarbon cannot have an odd mass; hence, the peak in the mass spectrum at $m/z = 67$ cannot be the molecular ion. Given that compound B is a dicarboxylic acid, compound A must contain a double bond within a ring. Calculate the number of moles of carboxy groups titrated as follows:

$$\text{moles of carboxy groups} = (13.7 \text{ mL})(0.100 \text{ mol L}^{-1})(1.00 \times 10^{-3} \text{ L}^{-1}) = 1.37 \times 10^{-3}$$

Noting that there are two carboxy groups per mole of B, we calculate the molecular mass of compound B as follows:

$$\text{molecular mass of } B = \frac{(100 \times 10^{-3})(2.00 \text{ moles of carboxy groups/mole of } B)}{1.37 \times 10^{-3} \text{ moles of carboxy groups}} = 146 \text{ g mol}^{-1}$$

From the following transformations, the atomic masses of the four oxygens introduced, and the molecular mass of compound B, the molecular mass of compound A can be deduced as follows:

A
molecular mass = x

B
molecular mass = $x + 64 = 146$

$x = 146 - 64 = 82$

If the molecular mass of compound A is 82, then the $m/z = 67$ peak in its mass spectrum corresponds to loss of 15 mass units, which corresponds to the loss of a methyl group. Evidently, compounds A and B contain a ring with a methyl branch. This branch is positioned such that there are three allylic hydrogens (from the NMR integration data) and it is also positioned such that dicarboxylic acid B is chiral. Compound A is 3-methylcyclopentene.

A
3-methylcyclopentene

B
2-methylpentanedioic acid

20.59   (a)   The 2300–3200 cm^{-1} and 1710 cm^{-1} absorptions clearly indicate a carboxylic acid, and this is confirmed by the δ 11.7 absorption in the NMR. The NMR shows five aromatic protons; hence, there is a monosubstituted benzene ring. The two mutually split triplets indicate a partial structure —CH_2CH_2—. We've accounted for all of the carbons and all of the unsaturation. However, what about the third oxygen? The chemical shift of the triplet at δ 4.1 shows that these protons must be adjacent to an oxygen. Hence, our part structure should be modified to —O—CH_2CH_2—. All that it left is to put the benzene ring at one end of the chain and the carboxy group at the other. This can be done in two ways:

A

B

Only structure A is a stable compound. (Why can't B exist? See Eq. 20.43a on text p. 1031.) Structure A is the correct answer.

(b)   The IR spectrum indicates the presence of a carboxylic acid, and the NMR indicates the presence of an ethoxy group (—OCH_2CH_3) and a *para*-disubstituted benzene ring. The compound is 4-ethoxybenzoic acid.

HO_2C —〈 〉— OCH_2CH_3 4-ethoxybenzoic acid

(c)   The IR spectrum indicates a carboxylic acid; the carbonyl absorption indicates that the carbonyl could be conjugated, and this is consistent with the UV absorption. The IR also shows *trans*-alkene absorption at 981 cm^{-1}. Let's now use the mass spectrum to see how much mass remains after subtracting the masses of the structural features we just deduced: 114 – 45 (carboxy group) – 13 (alkene CH) – 13 (alkene CH) = 43. A mass of 43 units is characteristic of a propyl group or an isopropyl group. A propyl group (CH$_3$CH$_2$CH$_2$—) would require a three-proton methyl triplet; this is observed at about δ 1. An isopropyl group would require a six-proton methyl doublet; this is not observed. The resonances at δ 1.5 and δ 2.1 could be the remaining protons of a propyl group. All the data are consistent with the following structure:

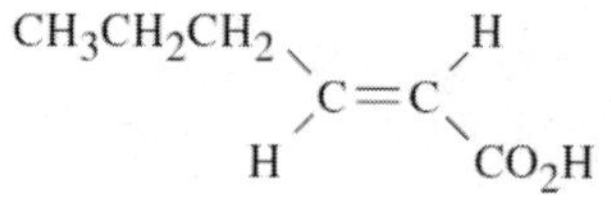

(*E*)-2-hexenoic acid

(You should account for the remaining splitting that was not discussed above.)

20.60   The titration data suggest that compound *A* is both a carboxylic acid and a phenol, and the IR spectrum shows both —OH and carbonyl absorptions. The change in the UV spectrum at high pH is also consistent with the presence of a group that ionizes at high pH and is conjugated with a benzene ring, namely, the phenolic —OH group. The resonances near δ 7 indicate a *para*-disubstituted benzene ring. The two triplets at δ 2.4 and δ 2.7 integrates for four hydrogens; and they indicate a partial structure —CH$_2$CH$_2$— . Add the masses of the groups known to be present: the *para*-disubstituted benzene ring (76 units), the phenol —OH group (17 units), the carboxy group (45 units), and the —CH$_2$CH$_2$— group (28 units), and we have accounted for the *m/z* = 166 peak in the mass spectrum as the molecular ion. Compound *A* is 3-(4-hydroxyphenyl)propanoic acid.

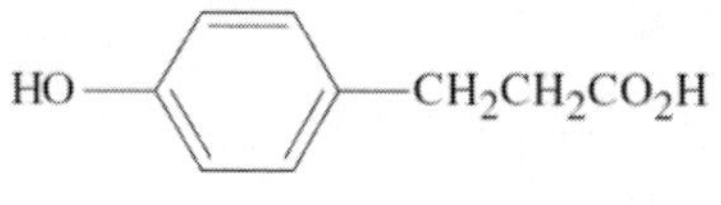

3-(4-hydroxyphenyl)propanoic acid
(compound *A*)

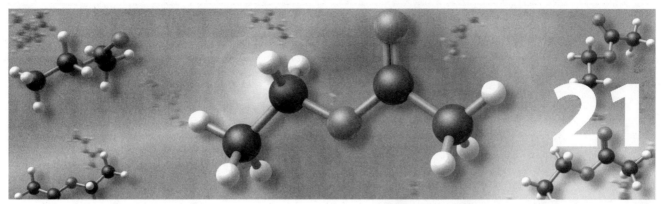

The Chemistry of Carboxylic Acid Derivatives

STUDY GUIDE LINKS

21.1 Solving Structure Problems Involving Nitrogen-Containing Compounds

Amides and nitriles are the first nitrogen-containing functional groups that are considered in the text. When you solve structure problems you should remember certain things about compounds that contain nitrogen. First, a compound that contains an odd number of nitrogen atoms has an *odd molecular mass*. (If you studied mass spectrometry (Chapter 12), you will appreciate that this means that the parent ion in the mass spectrum of such a compound will occur at an odd m/z value.) Second, the special formula given below (see also Eq. 4.7 on text p. 144) is used to calculate the unsaturation number U of a compound containing nitrogen. If C is the number of carbons, H is the number of hydrogens, X is the number of halogens, and N is the number of nitrogens, then U is given by

$$U = \frac{(2C+2)+N-(H+X)}{2}$$

The implication of this formula is that every nitrogen increases the number of hydrogens in a fully saturated molecule by one relative to the number of hydrogens in the corresponding hydrocarbon. Finally, when a compound contains nitrogen, you can't tell whether an ion in its mass spectrum is an odd-electron ion or an even-electron ion simply by its mass, because the number of nitrogens in the ion determines whether its mass is odd or even. When a compound contains zero or an even number of nitrogens, odd-electron ions have odd masses; when a compound contains an odd number of nitrogens, odd-electron ions have even masses.

21.2 Basicity of Nitriles

Let's place the basicity of nitriles within the context of other basicities that you've studied. Consider the following series:

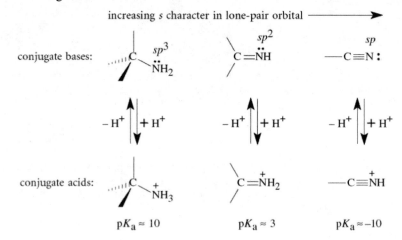

Notice that, as the amount of s character in the lone-pair orbital of the conjugate base increases, the pK_a of the conjugate acid decreases. Although the pK_a values are much different, this is exactly the same trend observed in the acidities of hydrocarbons (Sec. 14.7A, text pp. 698):

increasing s character in lone-pair orbital ⟶

conjugate bases:

$pK_a \approx 55$–60 $pK_a \approx 42$ $pK_a \approx 25$

In other words, just as an acetylenic anion, with its lone pair in an sp orbital, is the least basic hydrocarbon anion, a nitrile, also with a lone pair in an sp orbital, is the least basic among the analogous series of nitrogen compounds.

 21.3 Mechanism of Ester Hydrolysis

Some students try to combine the features of the acid-catalyzed hydrolysis and saponification reactions by using strong acid to protonate the carbonyl oxygen and hydroxide to act as the nucleophile.

Implicit in such a scheme is the incorrect notion that when the base is added to the acidic solution, the carbonyl oxygen of the ester would remain protonated and that all hydronium ions would not dissociate. However, both protonated esters and hydronium ions are strong acids. Remember: *the deprotonation of strong acids by strong bases is instantaneous.* (See Study Guide Link 20.1 on p. 659 of this manual.) The very first event that would occur on addition of base to the solution would be deprotonation of the protonated carbonyl oxygen and rapid neutralization of the acid (H_3O^+) in solution. The nucleophilic reaction of ^-OH on the ester carbonyl carbon is a much slower reaction. All of the hydronium ions in solution would react with the hydroxide ion before hydroxide could react to a significant extent at the carbonyl carbon.

Remember: in mechanisms, an acid and its conjugate base act in tandem. If H_3O^+ is the acid, H_2O should be used as the base (and the nucleophile). If ^-OH is the base (or nucleophile), H_2O should be used as the acid.

21.4   Another Look at the Friedel–Crafts Reaction

The Friedel–Crafts acylation reaction is first classified in the text as an *electrophilic aromatic substitution* reaction (Sec. 16.4F on text p. 808), because the reaction is first considered as a reaction of benzene derivatives. However, in terms of what happens to the acid chloride, the reaction can also be viewed as a nucleophilic acyl substitution reaction.

$$\langle\!\!\bigcirc\!\!\rangle\!\!-\!H \;+\; Cl-\overset{\overset{\displaystyle O}{\|}}{C}-CH_3 \;\xrightarrow[\text{2) } H_3O^+]{\text{1) } AlCl_3}\; \langle\!\!\bigcirc\!\!\rangle\!\!-\!\overset{\overset{\displaystyle O}{\|}}{C}-CH_3 \;+\; HCl$$

If the Friedel–Crafts acylation is considered to be a nucleophilic acyl substitution reaction, the nucleophilic electrons are the π electrons of the aromatic ring. Because benzene and its derivatives are very weak nucleophiles, a strong Lewis acid such as $AlCl_3$ is required to activate the acid chloride. Review again the mechanism of Friedel–Crafts acylation (Eq. 16.23 and 16.24 on text p. 809), paying particular attention to what happens to the acid chloride.

21.5   Esters and Nucleophiles

Notice that most esters of *carboxylic acids* react with nucleophiles at the carbonyl carbon—that is, they undergo nucleophilic acyl substitution reactions. The result is that the bond between the carbonyl group and the ester oxygen—the *acyl–oxygen bond*—is cleaved. (Further Exploration 21.2 discusses one exception to this generalization.)

$$\underset{\substack{\text{cleavage occurs at the}\\ \text{acyl-oxygen bond}}}{R-\overset{\overset{\displaystyle O}{\|}}{C}\!\!\begin{array}{c}\wr\\\wr\end{array}\!\!OCH_3} \;+\; {}^-OH \;\longrightarrow\; R-\overset{\overset{\displaystyle O}{\|}}{C}-O^- \;+\; CH_3OH$$

Esters of *sulfonic acids* react in a fundamentally different manner. These reactions are presented in Sec. 10.4A (text pp. 465–468), where the focus is on the sulfonate ester groups acting essentially as equivalents of halide leaving groups. As the discussion in the text shows, most sulfonate esters react with nucleophiles not by the reaction of the nucleophile at the sulfur, but by S_N2 reaction of the nucleophile on the *alkyl group* (if the alkyl group is primary or secondary); in such a reaction, the bond between the alkyl group and the ester oxygen is cleaved. (See Eq. 10.26 on text p. 467.)

$$\underset{\substack{\text{cleavage occurs at the}\\ \text{alkyl-oxygen bond}}}{R-\overset{\overset{\displaystyle O}{\|}}{\underset{\underset{\displaystyle O}{\|}}{S}}-O\!\!\begin{array}{c}\wr\\\wr\end{array}\!\!CH_3} \;+\; {}^-OH \;\xrightarrow{S_N2}\; R-\overset{\overset{\displaystyle O}{\|}}{\underset{\underset{\displaystyle O}{\|}}{S}}-O^- \;+\; CH_3OH$$

You may have noticed that we are returning with increasing frequency to earlier parts of the text and looking at reactions you've studied earlier in somewhat different contexts. If you willingly participate in this type of exercise, you'll gain a lot of insight about organic chemistry. In particular, you'll see how the parts begin to "fit together," that is, you'll begin to see the "forest" and not just the individual "trees."

FURTHER EXPLORATIONS

 ## 21.1   NMR Evidence for Internal Rotation in Amides

Internal rotation about the carbonyl-nitrogen bond of amides can be conveniently studied by NMR using the principles discussed in Sec. 13.8 on p. 653 of the text. Consider, for example, internal rotation in *N,N*-dimethylacetamide:

$H_3C-\overset{\overset{\displaystyle O}{\|}}{C}-\overset{|}{\underset{\underset{\displaystyle CH_3}{|}}{N}}-CH_3$ } these methyl groups are diastereotopic and hence chemically nonequivalent

The *N*-methyl groups in this compound are diastereotopic; one is cis to the carbonyl oxygen, and the other is trans. At low temperature, rotation about the C—N bond is slow enough that these methyl groups are observed as separate singlets in the NMR spectrum. However, when the temperature is raised, these signals broaden and, at about 60 °C, they coalesce into one singlet at a chemical shift that is the average of the two individual methyl chemical shifts. This is illustrated in Fig. SG21.1 on the following page.

 ## 21.2   Cleavage of Tertiary Esters and Carbonless Carbon Paper

The hydrolysis of tertiary esters has some noteworthy differences from the hydrolysis of other esters discussed in Sec. 21.7A. First, the saponification of tertiary esters is considerably slower than that of primary and secondary esters. For example, the saponification of *tert*-butyl acetate occurs at about 0.01 times the rate of the saponification of methyl acetate. The reason is that in the transition state for saponification, the methyl branches in the *tert*-butyl group are involved in van der Waals repulsions with the hydroxide nucleophile as it approaches the carbonyl carbon. A practical consequence of this rate difference is that a methyl or ethyl ester can generally be saponified without affecting a *tert*-butyl ester in the same molecule.

$(CH_3)_3CO-\overset{\overset{\displaystyle O}{\|}}{C}-\!\!\!\!\!\!\!\!\!\!\!\!\langle \rangle\!\!\!\!\!\!\!\!\!\!\!\!-\overset{\overset{\displaystyle O}{\|}}{C}-OCH_3 \xrightarrow[\text{(1 equiv.)}]{\text{NaOH}} (CH_3)_3CO-\overset{\overset{\displaystyle O}{\|}}{C}-\!\!\!\!\!\!\!\!\!\!\!\!\langle \rangle\!\!\!\!\!\!\!\!\!\!\!\!-\overset{\overset{\displaystyle O}{\|}}{C}-O^-\,Na^+ \ + \ HOCH_3$

tertiary ester   methyl ester is
is not hydrolyzed   hydrolyzed

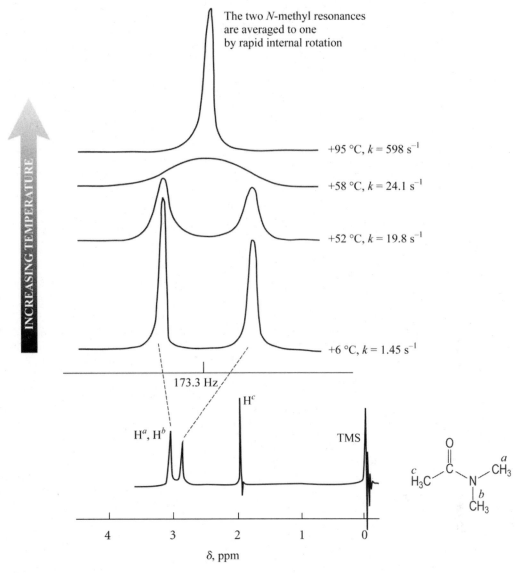

Figure SG21.1 The 60 MHz NMR spectrum of *N,N*-dimethylacetamide changes with increasing temperature as the rate of rotation about the carbonyl–nitrogen bond increases. The full spectrum of this compound is shown in the lower part of the figure, and the pair of singlets near δ 2.9 (173.3 Hz) is expanded in the upper part of the figure and shown as a function of temperature. The two *N*-methyl groups are individually observable at temperatures at or below 52° (lower two spectra) because internal rotation about the carbonyl–nitrogen bond is relatively slow at these temperatures. As the temperature is raised, the two resonances broaden and coalesce into one because the internal rotation becomes too rapid for the NMR experiment to resolve the individual lines. (See Fig. 13.21 on text p. 655 for a related phenomenon.) The *k* values are the first-order rate constants for internal rotation; a *k* of 1.45 s^{-1} corresponds to an approximate lifetime of 0.5 s for the individual conformations.

A second difference is that the acid-catalyzed hydrolysis of tertiary esters occurs by a mechanism that is completely different from that of primary and secondary esters. The first step in the hydrolysis mechanism is the same—protonation of the carbonyl oxygen. But there the similarity ends. The protonated ester dissociates by an S$_N$1 mechanism into a tertiary carbocation and the carboxylic acid.

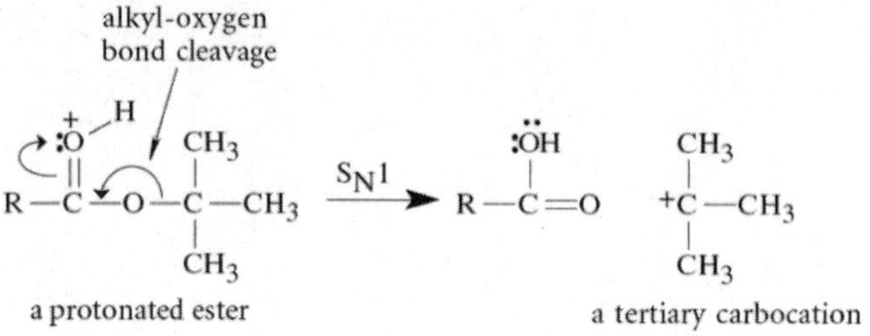

alkyl-oxygen
bond cleavage

a protonated ester → S_N1 → a tertiary carbocation

The Lewis acid–base association of water (or any other nucleophile that might be present) occurs at the tertiary carbocation.

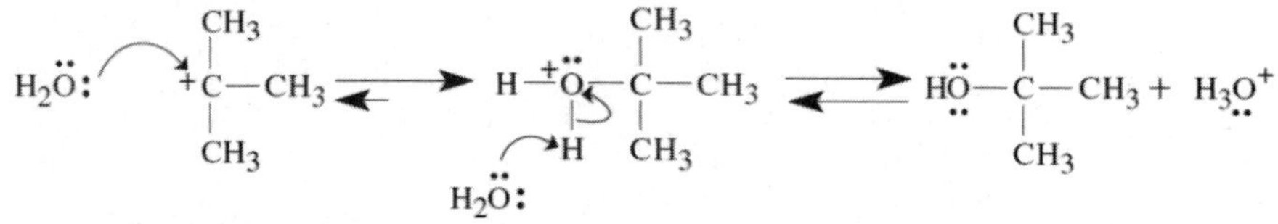

Notice that hydrolysis of tertiary esters by this mechanism involves breaking the *alkyl–oxygen bond*. In contrast, the mechanism of hydrolysis of primary and secondary esters involves breaking the *carbonyl–oxygen bond*. Consequently, the hydrolysis of tertiary esters is *not* a nucleophilic acyl substitution reaction. Rather, it is more like the S_N1 mechanism for substitution of a tertiary alcohol (see Eqs. 10.20b–c on text p. 463). Esters of primary and secondary alcohols in most cases do not cleave by an S_N1 mechanism because the carbocation intermediates that would be involved are much less stable.

"Carbonless carbon paper" (sometimes called NCR paper, for "No Carbon Required") is a very ingenious commercial application of tertiary ester cleavage as applied to lactones. The active component in "carbonless carbon paper" is the colorless lactone *A*, which is encapsulated within microscopic particles of a phenol–formaldehyde resin (Sec. 19.15 on text p. 995) and deposited on a paper base that also contains an acidic layer. When the particles are broken by the pressure of a pen or typewriter, the lactone is brought into contact with the acidic layer. This causes an S_N1-like opening of the lactone ring to give a relatively stable carbocation *B*. This carbocation has a blue-violet color.

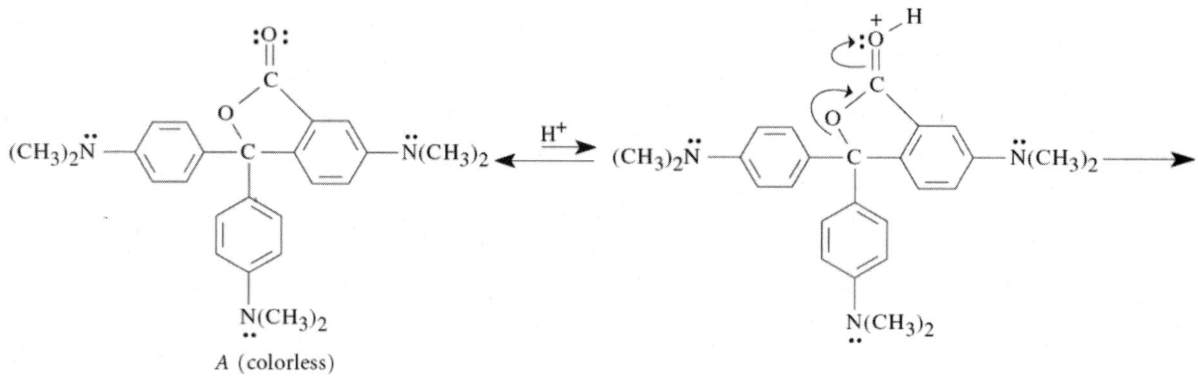

A (colorless)

B (blue-violet)   N(CH₃)₂

It takes some time for color to develop because the reaction is not instantaneous. As you may know from experience, the color in carbonless carbon paper intensifies with time.

21.3 Reaction of Tertiary Amines with Acid Chlorides

Tertiary amines react with acid chlorides to form *acylammonium salts.*

pyridine
(a tertiary amine)   an acylammonium salt

Acylammonium salts, because of their positive charge adjacent to the carbonyl carbon, are very reactive and are converted into amides by reaction with primary or secondary amines.

pyridinium chloride
(the HCl salt of pyridine)

If the acid chloride has acidic α-hydrogens, another possible reaction is the formation of a *ketene* by a β-elimination:

a ketene

Ketenes also react rapidly with primary and secondary amines to yield amides.

Despite these side reactions, then, tertiary amines can be used as catalysts for amide formation with primary and secondary amines because the products of these side reactions—acylammonium salts and ketenes—themselves react with amines to form amides.

REACTION REVIEW

I. HYDROLYSIS OF CARBOXYLIC ACID DERIVATIVES

A. HYDROLYSIS OF ACID CHLORIDES AND ANHYDRIDES

1. Acid chlorides and anhydrides react rapidly with water, even in the absence of acids or bases.
2. The hydrolysis reactions of acid chlorides and anhydrides are almost never used for the preparation of carboxylic acids because these derivatives are themselves usually prepared from acids.

$$R-\overset{\overset{\displaystyle O}{\|}}{C}-Cl \xrightarrow{H_2O} R-\overset{\overset{\displaystyle O}{\|}}{C}-OH + HCl$$

$$R-\overset{\overset{\displaystyle O}{\|}}{C}-O-\overset{\overset{\displaystyle O}{\|}}{C}-R \xrightarrow{H_2O} 2\,R-\overset{\overset{\displaystyle O}{\|}}{C}-OH$$

B. BASIC HYDROLYSIS (SAPONIFICATION) OF ESTERS AND LACTONES

1. Esters hydrolyze in aqueous hydroxide.

$$\underset{\text{an ester}}{R-\overset{\overset{\displaystyle O}{\|}}{C}-OR'} + {}^-OH \longrightarrow \underset{\text{a carboxylate anion}}{R-\overset{\overset{\displaystyle O}{\|}}{C}-O^-} + HOR'$$

2. This process is called saponification; the term saponification is sometimes used to refer to hydrolysis in base of any acid derivative.
3. The mechanism of ester saponification involves:
   a. The nucleophilic reaction of hydroxide anion at the carbonyl carbon to give a tetrahedral addition intermediate from which an alkoxide ion is expelled.
   b. The alkoxide ion thus formed reacts with the carboxylic acid to give the carboxylate salt and the alcohol.

4. The equilibrium in this reaction lies far to the right because the carboxylic acid is a much stronger acid than the liberated alcohol; saponification is effectively irreversible.
5. Many esters can be saponified with just one equivalent of ⁻OH, although an excess of ⁻OH is often used as a matter of convenience.
6. Saponification converts a lactone completely into the salt of the corresponding hydroxy acid; upon acidification, the hydroxy acid forms.

$$O=\!\!\overset{O}{\underset{}{\diagdown}}\!\!-CH_3 \xrightleftharpoons{HO^-} {}^-O-\overset{\overset{\displaystyle O}{\|}}{C}-CH_2-CH_2-\overset{\overset{\displaystyle OH}{|}}{C}H-CH_3 \xrightarrow{H_3O^+} HO-\overset{\overset{\displaystyle O}{\|}}{C}-CH_2-CH_2-\overset{\overset{\displaystyle OH}{|}}{C}H-CH_3$$

   a. If a hydroxy acid is allowed to stand in acidic solution, it comes to equilibrium with the corresponding lactone. This reaction is an acid-catalyzed, intramolecular esterification.
   b. Lactones containing five- and six-membered rings are favored at equilibrium over their corresponding hydroxy acids; those with ring sizes smaller than five or larger than six are less stable than their corresponding hydroxy acids.
7. Ester and lactone saponifications are examples of acyl substitution.
   a. The mechanisms of these reactions are classified as nucleophilic acyl substitution mechanisms.

   b. In a nucleophilic acyl substitution reaction, the substituting group reacts at the carbonyl carbon as a nucleophile, and a leaving group (another base) departs, usually in a separate step.

C. ACID-CATALYZED HYDROLYSIS OF ESTERS

1. Esters can be hydrolyzed to carboxylic acids in aqueous solutions of strong acids; in most cases this reaction is slower than base-promoted hydrolysis and must be carried out with an excess of water (in which esters are insoluble).

2. By the principle of microscopic reversibility, the mechanism of acid-catalyzed hydrolysis is the exact reverse of acid-catalyzed esterification.

   a. The ester is first protonated by the acid catalyst; protonation makes the carbonyl carbon more electrophilic by making the carbonyl oxygen a better acceptor of electrons.

   b. Water, as the nucleophile, reacts as a nucleophile at the carbonyl carbon and then loses a proton to give the tetrahedral intermediate.

   c. Protonation of the leaving oxygen converts it into a better leaving group.

   d. Loss of the protonated leaving group gives a protonated carboxylic acid, from which a proton is removed to give the carboxylic acid itself.

3. Saponification, followed by acidification, is a much more convenient method for hydrolysis of most esters than acid-catalyzed hydrolysis because—

   a. it is faster.

   b. it is irreversible.

   c. it can be carried out not only in water but also in a variety of solvents, even in alcohols.

4. Ester hydrolysis is another example of nucleophilic acyl substitution.

D. HYDROLYSIS OF AMIDES

1. Amides can be hydrolyzed to carboxylic acids and ammonia or amines by heating them in acidic or basic solution.

2. In acid, protonation of the ammonia or amine by-product drives the hydrolysis equilibrium to completion; the amine can be isolated, if desired, by addition of base to the reaction mixture following hydrolysis.

3. Hydrolysis of amides in base is analogous to saponification of esters; the reaction is driven to completion by formation of the carboxylic acid salt.

4. The conditions of both acid- and base-promoted amide hydrolysis are considerably more severe than the corresponding reactions of esters.

5. The mechanisms of amide hydrolysis are typical nucleophilic acyl substitution mechanisms.

E. HYDROLYSIS OF NITRILES

1. Nitriles are hydrolyzed to carboxylic acids and ammonia by heating them in acidic or basic solution.

2. The conditions of nitrile hydrolysis are considerably more severe than the corresponding reactions of esters and amides.

3. The mechanism of nitrile hydrolysis in acidic solution involves:
   a. protonation of the nitrogen, which makes the nitrile carbon more electrophilic.
   b. reaction of the nucleophile water at the nitrile carbon and loss of a proton, which give an intermediate called an imidic acid, the nitrogen analog of an enol.
   c. conversion of the unstable imidic acid under the reaction conditions into an amide.
   d. hydrolysis of the amide to a carboxylic acid and ammonium ion. (See Eqs. 21.20a–c on text p. 1066 for the detailed mechanism.)

4. In base, the nitrile group reacts with basic nucleophiles at the nitrile carbon and, as a result, the electronegative nitrogen assumes a negative charge.
   a. Proton transfer gives an imidic acid, which ionizes in base.
   b. The imidic acid reacts further to give the corresponding amide, which in turn, hydrolyzes under the reaction conditions to the carboxylate salt of the corresponding carboxylic acid. (See Eqs. 21.21a–b on text p. 1067 for the detailed mechanism.)

5. The hydrolysis of nitriles is a useful way to prepare carboxylic acids because nitriles, unlike many other carboxylic acid derivatives, are generally synthesized from compounds other than the acids themselves, such as alkyl halides and sulfonates.

II. REACTIONS OF CARBOXYLIC ACID DERIVATIVES WITH ALCOHOLS

A. REACTION OF ACID CHLORIDES AND ANHYDRIDES WITH ALCOHOLS AND PHENOLS

1. Esters are formed rapidly when acid chlorides react with alcohols or phenols, usually in the presence of a tertiary amine such as pyridine, or a related base.

2. Esters of tertiary alcohols and phenols, which cannot be prepared by acid-catalyzed esterification, can be prepared by this method.
3. Anhydrides react with alcohols and phenols in much the same way as acid chlorides. Cyclic anhydrides react with alcohols and phenols to give half-esters.

a cyclic anhydride   a half-ester

4. Sulfonate esters are prepared by the analogous reactions of sulfonyl chlorides with alcohols.

B. REACTION OF ESTERS WITH ALCOHOLS

1. When an ester reacts with an alcohol under acidic conditions, or with an alkoxide under basic conditions, a new ester is formed.

2. This type of reaction, called transesterification, typically has an equilibrium constant near unity.
3. The reaction is driven to completion by the use of an excess of the displacing alcohol or by removal of a relatively volatile alcohol by-product as it is formed.

III. REACTIONS OF CARBOXYLIC ACID DERIVATIVES WITH AMINES

A. REACTION OF ACID CHLORIDES AND ANHYDRIDES WITH AMMONIA AND AMINES

1. Acid chlorides react rapidly and irreversibly with ammonia or amines by a nucleophilic acyl substitution reaction mechanism to give amides.
   a. Reaction with ammonia yields a primary amide.
   b. Reaction with a primary amine yields a secondary amide.
   c. Reaction with a secondary amine yields a tertiary amide.

2. The amine reacts as a nucleophile at the carbonyl group to form a tetrahedral intermediate, which expels chloride ion; a proton-transfer step yields the amide.
   a. An important aspect of amide formation is the proton transfer in the last step of the mechanism.
   b. Unless another base is added to the reaction mixture, the starting amine acts as the base in this step; if the only base present is the amine nucleophile, then at least two equivalents must be used:
      i. one equivalent as the nucleophile.

ii. one equivalent as the base in the final proton-transfer step.

3. In the Schotten–Baumann technique for amide formation, the reaction is run with a water-insoluble acid chloride and an amine in a separate layer over an aqueous solution of NaOH. The NaOH neutralizes HCl produced as a by-product of the reaction.

4. A tertiary amine such as pyridine can be used as the second amine equivalent. The presence of a tertiary amine does not interfere with amide formation because a tertiary amine itself cannot form an amide.

5. Anhydrides react with amines in much the same way as acid chlorides.

a. Half-amides of dicarboxylic acids are produced in analogous reactions of amines and cyclic anhydrides.

b. These compounds can be cyclized to imides by treatment with dehydrating agents, or in some cases, simply by heating.

a cyclic anhydride   a half-amide   a cyclic imide

B. REACTION OF ESTERS WITH AMINES

1. The reaction of an ester with ammonia or amines yields an amide.

2. The reaction of esters with hydroxylamine (NH_2OH) gives *N*-hydroxyamides; these compounds are known as hydroxamic acids.

a. This chemistry forms the basis for the hydroxamate test, used mostly for esters.

b. The hydroxamic acid products are easily recognized because they form highly colored complexes with ferric ion.

a hydroxamic acid

IV. REDUCTION OF CARBOXYLIC ACID DERIVATIVES

A. REDUCTION OF ACID CHLORIDES TO ALDEHYDES AND ALCOHOLS

1. Acid chlorides can be reduced to aldehydes by either of two procedures:

a. Hydrogenation over a catalyst that has been deactivated, or poisoned, with an amine, such as quinoline, that has been heated with sulfur.

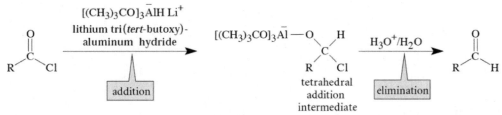

    i. This reaction is called the Rosenmund reduction.

    ii. The poisoning of the catalyst prevents further reduction of the aldehyde product.

  b. Reaction at low temperature with lithium tri(*tert*-butoxy)aluminum hydride.

    i. The hydride reagent used in this reduction is derived by the replacement of three hydrogens of LiAlH$_4$ by *tert*-butoxy groups; this reagent is much less reactive, and is more selective, than LiAlH$_4$.

    ii. Because acid chlorides are more reactive than aldehydes toward nucleophiles, the reagent reacts preferentially with the acid chloride reactant rather than with the product aldehyde.

  2. Acid chlorides (and anhydrides) react with LiAlH$_4$ to give primary alcohols.

B. REDUCTION OF ESTERS TO PRIMARY ALCOHOLS

  1. Esters can be reduced to alcohols with LiAlH$_4$. This reaction involves a nucleophilic acyl substitution reaction followed by a carbonyl-addition reaction.

  a. The active nucleophile in LiAlH$_4$ reductions is the hydride ion, H:$^-$, which replaces alkoxide at the carbonyl group of the ester to give an aldehyde.

  b. The aldehyde cannot be isolated because it reacts rapidly with LiAlH$_4$ to give, after protonolysis, the alcohol.

  c. Two alcohols are formed:

    i. One is derived from the acyl group of the ester

    ii. One is derived from the alkoxy group (usually methanol or ethanol, which is discarded).

  2. Sodium borohydride reacts very sluggishly or not at all with most esters.

C. REDUCTION OF AMIDES TO AMINES

  1. Amides are reduced to amines with LiAlH$_4$.

  a. Primary amines are obtained from primary amides.

  b. Secondary amines are obtained from secondary amides.

  c. Tertiary amines are obtained from tertiary amides.

  2. The mechanism of the reaction of a secondary amide with lithium aluminum hydride involves—

  a. formation of the lithium salt of the amide by reaction of the amide N—H proton with an equivalent of hydride, a strong base.

  b. reaction of the lithium salt of the amide, a Lewis base, with AlH$_3$, a Lewis acid.

  c. delivery of hydride to the C=N double bond.

d. loss of ⁻OAlH₂, a fairly good leaving group (better than the nitrogen group), to form an imine.

e. reduction of the imine and protonolysis in a separate step to yield the product amine. (See Eqs. 21.55a–e on text pp. 1081–1082 for the detailed mechanism.)

    i. An excess of acid will convert the amine, which is a base, into its conjugate-acid ammonium ion.
    ii. Hydroxide is required to neutralize this ammonium salt and thus give the neutral amine.

3. The reductions of primary and tertiary amides involve somewhat different mechanisms, but they too involve loss of oxygen rather than nitrogen as a leaving group.

D. REDUCTION OF NITRILES TO AMINES

1. Nitriles are reduced to primary amines by reaction with LiAlH₄, followed by protonolysis in a separate step.

$$R-C\equiv N \xrightarrow[\substack{2)\ H_3O^+/H_2O \\ 3)\ HO^-}]{1)\ LiAlH_4} R-CH_2-NH_2$$

a. An excess of acid will convert the amine, which is a base, into its conjugate-acid ammonium ion.
b. Hydroxide is required to neutralize this ammonium salt and thus give the neutral amine.

2. Nitriles are also reduced to primary amines by catalytic hydrogenation. An intermediate in the reaction is the imine, which is not isolated but is hydrogenated to the amine product.

$$R-C\equiv N \xrightarrow{H_2,\ (cat.)} R-CH=NH \xrightarrow{H_2,\ (cat.)} R-CH_2-NH_2$$

V. REACTIONS OF CARBOXYLIC ACID DERIVATIVES WITH ORGANOMETALLIC REAGENTS

A. REACTION OF ACID CHLORIDES WITH LITHIUM DIALKYLCUPRATES

1. The reaction of lithium dialkylcuprates with acid chlorides gives ketones in excellent yield; lithium dialkylcuprates typically react with acid chlorides and aldehydes, very slowly with ketones, and not at all with esters.

2. Lithium dialkylcuprates react much like Grignard or lithium reagents, but are less reactive; a lithium dialkylcuprate can be considered conceptually as an alkyl anion complexed with copper.
a. Lithium dialkylcuprate reagents are prepared by the reaction of two equivalents of an organolithium reagent with one equivalent of a cuprous halide such as CuBr.

$$RLi + CuI \longrightarrow LiI + CuR \xrightarrow{RLi} LiCuR_2$$

b. The first equivalent forms an alkylcopper compound; the driving force for this reaction is the preference of lithium, the more electronegative metal, to exist as an ion (Li⁺).
c. The copper of an alkylcopper reagent is a Lewis acid and reacts accordingly with "alkyl anion" from a second equivalent of the organolithium reagent; the product of this reaction is a lithium dialkylcuprate.

B. REACTION OF ESTERS WITH GRIGNARD REAGENTS

1. In the reaction of esters with Grignard reagents, tertiary alcohols are formed after protonolysis. (Secondary alcohols are formed from esters of formic acid after protonolysis.)
   a. Two equivalents of organometallic reagent react per mole of ester.
   b. A second alcohol (derived from the alkoxy group, usually methanol or ethanol) is produced in the reaction and is typically discarded.

2. This reaction is a nucleophilic acyl substitution followed by an addition.
   a. A ketone is formed in the substitution step along with an alkoxide leaving group.
   b. The ketone intermediate is not isolated because ketones are more reactive than esters toward nucleophilic reagents.
   c. The ketone reacts with a second equivalent of the Grignard reagent to form a magnesium alkoxide, which, after protonolysis, gives the alcohol.
3. This reaction is a very important method for the synthesis of alcohols in which at least two of the groups on the α-carbon of the alcohol product are identical.

VI. REACTIONS OF CARBOXYLIC ACID DERIVATIVES WITH OTHER NUCLEOPHILES

A. REACTION OF ACID CHLORIDES WITH CARBOXYLATE SALTS

1. Even though carboxylate salts are weak nucleophiles, acid chlorides are reactive enough to react with carboxylate salts to give anhydrides.
2. The reaction of acid chlorides with carboxylate salts can be used to prepare mixed anhydrides.

VII. SYNTHESIS OF CARBOXYLIC ACID DERIVATIVES—REVIEW

A. SYNTHESIS OF ACID CHLORIDES

1. Reaction of carboxylic acids with $SOCl_2$ or PCl_5 (see Sec. 20.9A, text p. 1024).

B. SYNTHESIS OF ANHYDRIDES

1. Reaction of carboxylic acids with dehydrating agents (see Sec. 20.9B, text p. 1026).

2. Reaction of acid chlorides with carboxylate salts (see Sec. 21.8A, text p. 1071).

C. SYNTHESIS OF ESTERS

    1. Acid-catalyzed esterification of carboxylic acids (see Sec. 20.8A, text p. 1019).

$$R\text{—}CO_2H \; + \; HOR' \; \underset{}{\overset{H_3O^+}{\rightleftharpoons}} \; R\text{—}CO_2R' \; + \; H_2O$$

    2. Alkylation of carboxylic acids or carboxylates (see Sec. 20.8B, text page 1023).

$$R\text{—}CO_2H \; + \; CH_2N_2 \; \longrightarrow$$

$$R\text{—}CO_2H \; + \; H_3C\text{—}I \; \xrightarrow{K_2CO_3}$$

$$\left.\right\} \quad \underset{O}{\overset{O}{R\text{—}\overset{\|}{C}\text{—}OCH_3}}$$

    3. Reaction of acid chlorides and anhydrides with alcohols or phenols (see Sec. 21.8A, text page 1071).

$$\underset{}{\overset{O}{\underset{R\text{—}\overset{\|}{C}\text{—}Cl}{}}} \; + \; HOR' \; \xrightarrow{\text{base}} \; \underset{}{\overset{O}{\underset{R\text{—}\overset{\|}{C}\text{—}OR'}{}}}$$

    4. Transesterification of other esters (see Sec. 21.8C, text page 1075).

$$R\text{—}CO_2R' \; + \; HOR'' \; \rightleftharpoons \; R\text{—}CO_2R'' \; + \; HOR'$$

D. SYNTHESIS OF AMIDES

    1. Reaction of acid chlorides, anhydrides, or esters with amines (see Sec. 21.8, text p. 1071).

$$\underset{}{\overset{O}{\underset{R\text{—}\overset{\|}{C}\text{—}X}{}}} \; + \; HNR'R'' \; \longrightarrow \; \underset{}{\overset{O}{\underset{R\text{—}\overset{\|}{C}\text{—}NR'R''}{}}} \qquad \text{—}X = \text{—}Cl, \; \underset{}{\overset{O}{\underset{\text{—}O\text{—}\overset{\|}{C}\text{—}R,}{}}} \; \text{or} \; \text{—}OR$$

    2. When X = halide or carboxylate, a base or a second equivalent of an amine is required to neutralize the HX generated as a byproduct. In many cases, a tertiary amine such as pyridine or triethylamine is used.

E. SYNTHESIS OF NITRILES

    1. S_N2 reaction of cyanide ion with alkyl halides or sulfonate esters. In this reaction, primary or unbranched secondary alkyl halides or sulfonate esters are required (see Sec. 9.2, text p. 387).

$$R\text{—}CH_2\text{—}X \; + \; {}^-C\equiv N \; \longrightarrow \; R\text{—}CH_2\text{—}C\equiv N \; + \; X^- \qquad X = \text{halide ion, } {}^-OTs$$

    2. Cyanohydrin formation (see Table 9.1, text p. 385).

$$\underset{}{\overset{O}{\underset{R\text{—}\overset{\|}{C}\text{—}R'}{}}} \; + \; H\text{—}C\equiv N \; \longrightarrow \; \underset{\underset{C\equiv N}{|}}{\overset{\overset{OH}{|}}{R\text{—}\overset{}{C}\text{—}R'}}$$

SOLUTIONS TO PROBLEMS

Solutions to In-Text Problems

21.1   (a)

$N\equiv CCH_2CH_2CH_2CH_2CO_2H$

5-cyanopentanoic acid

(b)

$CH_3CH_2CH_2CH_2\overset{\displaystyle O}{\overset{\|}{C}}OCH(CH_3)_2$

isopropyl valerate

(c)

$CH_3CH_2O-\overset{\displaystyle O}{\overset{\|}{C}}-CH_2-\overset{\displaystyle O}{\overset{\|}{C}}-OCH_3$

ethyl methyl malonate

(d)

$CH_3\overset{\displaystyle O}{\overset{\|}{C}}O-$⟨cyclohexyl⟩

cyclohexyl acetate

(e)

$H\overset{\displaystyle O}{\overset{\|}{C}}N(CH_3)_2$

N,N-dimethylformamide

(f)

γ-valerolactone

CH_3

(g)

glutarimide

(h)

$(CH_3)_2\overset{\displaystyle O}{\overset{\|}{C}}\overset{}{C}Cl$
|
Cl

α-chloroisobutyryl
chloride

(i)

$HO_2CCH_2CHCH_2CH_2CO_2H$

$O=\overset{}{C}\overset{}{\diagdown}OCH_2CH_3$

3-ethoxycarbonylhexanedioic acid

21.2   (a)   butanenitrile (common: butyronitrile)
     (b)   Cyclopropanecarbonyl chloride
     (c)   isopentyl 3-methylbutanoate (*common*: isoamyl isovalerate)

 The *isoamyl* group is the same as an isopentyl or 3-methylbutyl group:

$H_3C\diagdown$
$\quad CHCH_2CH_2—$ isoamyl group
$H_3C\diagup$

     (d)   N,N-dimethylbenzamide
     (e)   (Z)-2-Butenedioic anhydride (common name more often used: maleic anhydride)
     (f)   Ethyl 3-oxopentanoate (common: ethyl β-ketovalerate)
     (g)   1-Methyl-3-butenyl propanoate. (Notice the numbering from the point of attachment to the carboxylate oxygen.)

21.3   (a)   It is the E conformation.

(b)

Z conformation

21.4   Any lactam containing a relatively small ring (fewer than eight members) must exist in an *E* conformation. An example is δ-valerolactam. A small-ring lactam cannot exist in a *Z* conformation, because such a conformation would result in a transoid bond within the ring, a very strained and therefore unstable arrangement. (This is the same reason that small-ring (*E*)-cycloalkenes are unstable.) If you try to construct a model of a small-ring lactam with its amide bond in a *Z* conformation, the strain of such a conformation should be obvious.

δ-valerolactam

amide bond in the *Z* conformation

This distance requires
at least four more carbons.

21.5   As shown by the data above the problem, a carboxylic acid has a higher boiling point than an ester because it can both donate and accept hydrogen bonds within its liquid state; hydrogen bonding does not occur in the ester. Consequently, pentanoic acid (valeric acid) has a higher boiling point than methyl butanoate. Here are the actual data:

$$CH_3CH_2CH_2CH_2CO_2H \qquad CH_3CH_2CH_2CO_2CH_3$$

pentanoic acid   methyl butanoate
(valeric acid)   (methyl butyrate)
bp 186 °C   bp 98 °C

21.6   (a)   2-Butanone should have the higher dipole moment, because, other things being equal, molecules with higher dipole moments have higher boiling points.

        (b)   The dipole-moment data require that esters, like amides, exist predominantly in the *Z* conformation. In this conformation of an ester, the resultant of the bond dipoles of the two carbon–oxygen single bonds opposes the bond dipole of the C=O bond; the overall dipole moment of the ester is thus reduced. In the ketone, the C=O bond dipole is not opposed by any other significant bond dipole.

If the ester were in an *E* conformation, the resultant bond dipoles of the two carbon–oxygen single bonds would *augment* the bond dipole of the C=O bond. In this case, the dipole moment of the ester would be larger than that of the ketone.

The experimental facts bear out these predictions. γ-Butyrolactone, an "ester" that is locked into an *E* conformation, has a greater dipole moment than 2-butanone, which has a greater dipole moment than methyl acetate, an ester which exists mostly in the *Z* conformation.

	γ-butyrolactone	2-butanone	methyl acetate
dipole moments:	4.1D	2.8 D	1.5 D

21.7   (a)   The carbonyl absorption of the ester occurs at higher frequency, and only the carboxylic acid has the characteristic strong, broad O—H stretching absorption in 2400–3600 cm^{-1} region.

       (b)   In the proton NMR spectra, the amide has a single methyl resonance that is a doublet, whereas the nitrile has two methyl resonances that are singlets. In addition, the integral of the aromatic proton absorptions in the nitrile is smaller in proportion to the integral of the other absorptions because there are fewer aromatic protons.

N-methylbenzamide   **2,4-dimethylbenzonitrile**

       (c)   In methyl propionate, the singlet for the methyl of the CH$_3$O— group has a greater chemical shift (near δ 4) than the methyl singlet in ethyl acetate (near δ 2). In ethyl acetate, the quartet for the —CH$_2$— group has a greater chemical shift (near δ 4) than the —CH$_2$— quartet in methyl propionate (near δ 2).

ethyl acetate   methyl propionate

       (d)   In *N*-methylpropanamide, the *N*-methyl group is a doublet at about δ 3. *N*-Ethylacetamide has no doublet resonances. In *N*-methylpropanamide, the α-protons are a quartet near δ 2.5. In *N*-ethylacetamide, the α-protons are a singlet at δ 2. The NMR spectrum of *N*-methylpropanamide has no singlets.

N-methylpropanamide   *N*-ethylacetamide

       (e)   Ethyl isobutyrate, (CH$_3$)$_2$CHCO$_2$CH$_2$CH$_3$, has five CMR resonances, whereas ethyl butyrate, CH$_3$CH$_2$CH$_2$CO$_2$CH$_2$CH$_3$, has six.

21.8   That the compound is an amide is confirmed by the N—H absorption at 3300 cm^{-1} and by the carbonyl absorption at 1650 cm^{-1}. The compound is *N*-ethylacetamide; the —CH$_2$— resonance of the ethyl group is split by both the adjacent methyl protons and the N—H proton.

H$_3$C—C—NHCH$_2$CH$_3$ *N*-ethylacetamide

21.9   (a)   The first ester is more basic because its conjugate acid is stabilized not only by resonance interaction with the ester oxygen, but also by resonance interaction with the double bond; that is, the conjugate acid of the first ester has one more important resonance structure than the conjugate acid of the second. (See Problem 19.14(b), text p. 962, and its solution for a similar situation.)

$$CH_3CH = CH - \overset{\overset{+}{O}H}{\underset{\|}{C}} - OCH_3 \longleftrightarrow CH_3\overset{+}{C}H - CH = \overset{OH}{\underset{\|}{C}} - OCH_3$$

resonance interaction of the protonated carbonyl
group with the double bond

(b)　The first ester is more basic because its conjugate acid is stabilized by resonance interaction of the *p*-methoxy oxygen.

$$CH_3\overset{..}{\underset{..}{O}} - \langle \rangle - \overset{\overset{..}{:O:}}{\underset{\|}{C}} - OC_2H_5 \longleftrightarrow CH_3\overset{..}{O} = \langle \rangle = \overset{\overset{..}{:O:^-}}{C} - OC_2H_5$$

The meta position of the methoxy group in the second ester does not allow a similar resonance interaction; consequently, only its electron-withdrawing polar effect operates. This polar effect (Sec 3.6C, text p. 115) destabilizes the conjugate acid of the second ester. (See Problem 19.14a on text p. 962 and its solution on p. 629 of this manual for a similar situation.)

(c)　The second ester is more basic because its conjugate acid is stabilized through the polar effect (Sec 3.6C, text p. 115) of the attached electron-donating methyl group. The first ester is less basic because its conjugate acid is destabilized by the polar effect of the attached electron-withdrawing trifluoromethyl group.

destabilized

stabilized

$$\left[F_3C - \overset{\overset{+}{O}H}{\underset{\|}{C}} - OEt \longleftrightarrow F_3C - \overset{OH}{\underset{\overset{+}{|}}{C}} - OEt \right] \qquad \left[H_3C - \overset{\overset{+}{O}H}{\underset{\|}{C}} - OEt \longleftrightarrow H_3C - \overset{OH}{\underset{\overset{+}{|}}{C}} - OEt \right]$$

stronger acid　　　　　　　　　　　　　　　　weaker acid

21.10　(a)　The mechanism of hydrolysis in acidic solution is much like that for the hydrolysis of an ester.

Ph —C—NHCH₃

(b)　The mechanism of the base-promoted hydrolysis of *N*-methylbenzamide is essentially a saponification mechanism analogous to that shown for esters in Eq. 21.9a–b, text pp. 1061–2.

tetrahedral
addition
intermediate

21.11 (a) The hydrolysis products consist of the conjugate base of the carboxylic acid and an amine:

isobutyrate ion

pyrrolidine

(b) The hydrolysis products consist of the conjugate base of a carboxylic acid and an amine, both of which are a part of the same molecule:

21.12 *Resonance effects*: The application of the resonance effect for predicting the relative rates of acid-catalyzed ester and amide hydrolyses is exactly the same as it is for the corresponding base-promoted hydrolyses. (See Eq. 21.26, text p. 1069 and the discussion in Sec. 21.7E of the text.)

Leaving-group basicities: In the acid-catalyzed hydrolysis of an ester, the leaving group is an alcohol; in the acid-catalyzed hydrolysis of an amide, the leaving group is ammonia or an amine. Alcohols are much weaker bases (conjugate-acid $pK_a = -2$ to -3) than amines or ammonia (conjugate-acid $pK_a = 9$–10). Because weaker bases are better leaving groups, esters should hydrolyze more rapidly, and they do.

21.13 First, consider the stability of the starting materials. Acetyl fluoride is stabilized by resonance, due to effective overlap of the $2p$ orbitals of fluorine and the carbonyl carbon. Acetyl chloride is not stabilized as effectively, due to poor overlap of the $3p$ chlorine orbital and $2p$ orbital of carbon (Fig. 16.7, text p. 819). Thus, acetyl fluoride is more stable.

unimportant structure

Next, consider the stabilities of the intermediates. The tetrahedral intermediates of acid fluorides are stabilized than the intermediates of acid chlorides are, due to the greater effective positive charge on the carbonyl carbon attached to the fluorine.

partial charge on carbon larger, stabilizes charge on oxygen better

partial charge on carbon smaller, stabilizes charge on oxygen worse

If the stability of the starting material was the dominant factor in determining the rate of the reaction, the acid chloride would react faster, due to it being less stable. If the stability of the intermediate was the dominant factor, then the acid fluoride would react faster, because its intermediate (and, by Hammond's Postulate, its transition state leading to the intermediate) would be lower in energy. Since we know from the question that the acid chloride reacts faster, it must be the stabilities of the starting materials that determine the rate of the reaction.

21.14 (a) Because esters hydrolyze much more rapidly than nitriles, one equivalent of base brings about the selective hydrolysis of the ester.

$$N\equiv C-CH_2-\overset{\overset{\displaystyle O}{\|}}{C}-O^- \ + \ HOCH_3$$

(b)   This is an ordinary ester saponification followed by acidification.

$$F-\langle\ \rangle-\overset{\overset{\displaystyle O}{\|}}{C}-OH \ + \ HOCH_3$$

(c)   Hydrolysis of one amide gives carbamic acid, which spontaneously decarboxylates; see Eqs. 20.42a–b on text p. 1031.

$$H_2N-\overset{\overset{\displaystyle O}{\|}}{C}-NH_2 \ + \ H_2O \ \xrightarrow{H_3O^+,\ heat} \ \overset{+}{N}H_4 + H_2N-\overset{\overset{\displaystyle O}{\|}}{C}-OH \ \xrightarrow{H_3O^+} \ \overset{+}{N}H_4 + CO_2$$

carbamic acid

21.15   First, prepare the acid chloride from hexanoic acid:

$$CH_3(CH_2)_4\overset{\overset{\displaystyle O}{\|}}{C}-OH \ \xrightarrow{SOCl_2} \ CH_3(CH_2)_4\overset{\overset{\displaystyle O}{\|}}{C}-Cl$$

hexanoic acid   hexanoyl chloride

(a)   To prepare the ester, allow the acid chloride to react with ethanol:

$$CH_3(CH_2)_4\overset{\overset{\displaystyle O}{\|}}{C}-Cl \ \xrightarrow{HOCH_2CH_3} \ CH_3(CH_2)_4\overset{\overset{\displaystyle O}{\|}}{C}-OCH_2CH_3$$

hexanoyl chloride   ethyl hexanoate

(b)   To prepare the secondary amide, allow the acid chloride to react with the appropriate amine:

$$CH_3(CH_2)_4\overset{\overset{\displaystyle O}{\|}}{C}-Cl \ \xrightarrow{H_2NCH_3\ (excess)} \ CH_3(CH_2)_4\overset{\overset{\displaystyle O}{\|}}{C}-NHCH_3$$

hexanoyl chloride   *N*-methylhexanamide

21.16   (a)   (b)   (c)

$$CH_3CH_2\overset{\overset{\displaystyle O}{\|}}{C}-N(CH_3)_2$$
$$(CH_3)_2\overset{+}{N}H_2\ Cl^-$$

$$\triangleright\!-\!\overset{\overset{\displaystyle O}{\|}}{C}-O-\!\triangleleft$$

$$PhCH_2\overset{\overset{\displaystyle O}{\|}}{C}-SCH_2CH_3$$

(d)   (e)   (f)

$$CH_3CH_2CH_2\overset{\overset{\displaystyle O}{\|}}{C}-O-\overset{\overset{\displaystyle O}{\|}}{C}CH_3$$

$$Cl-\overset{\overset{\displaystyle O}{\|}}{C}-OCH_3$$

$$CH_3O-\overset{\overset{\displaystyle O}{\|}}{C}-OCH_3$$

(g)   (h)

$$\overset{\overset{\displaystyle O}{\|}}{\underset{O\diagdown\quad\diagup O}{}}$$

$$+\ 2\ C_2H_5OH$$

$$\langle\ \rangle\overset{\overset{\displaystyle O}{\|}}{\underset{\overset{\displaystyle C}{\|}}{\underset{\displaystyle O}{}}}\!\!\begin{array}{l}C-OCH_3\\[6pt]C-OH\end{array}$$

21.17   Because sulfonate esters behave like alkyl halides, cleavage in *A* occurs at the alkyl–oxygen bond. (See Study Guide Link 21.5 on p. 701 of this manual for a discussion.) Consequently, the oxygen isotope ends up in the product alcohol. Cleavage in *B* occurs at the acyl–oxygen bond; therefore, the isotope ends up in the product acetate ion. The products are therefore as follows: (*O = ^{18}O):

products from *A*: products from *B*:

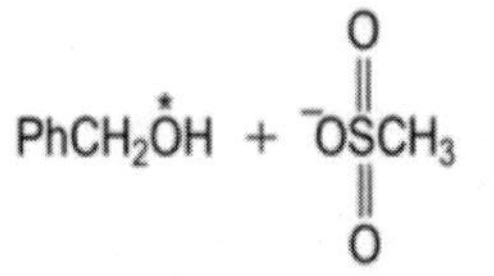

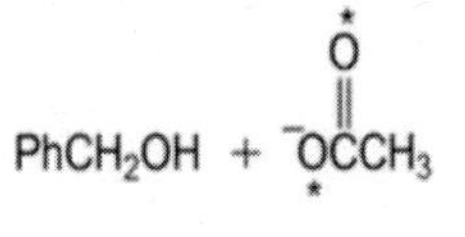

In the products from *B*, the isotope is distributed equally between the two carboxylate oxygens, because these oxygens are equivalent by resonance.

21.18 (a)

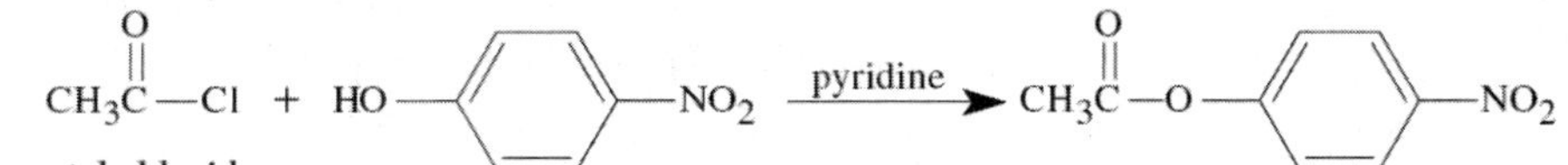

acetyl chloride; *p*-nitrophenol; *p*-nitrophenyl acetate

(b)

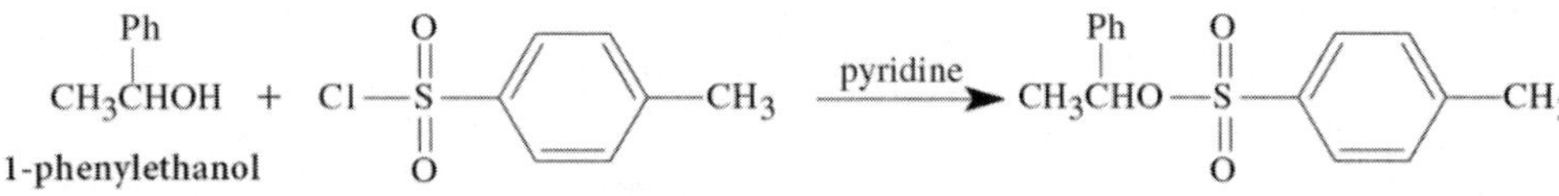

1-phenylethanol; *p*-toluenesulfonyl chloride; 1-phenylethyl tosylate

(c)

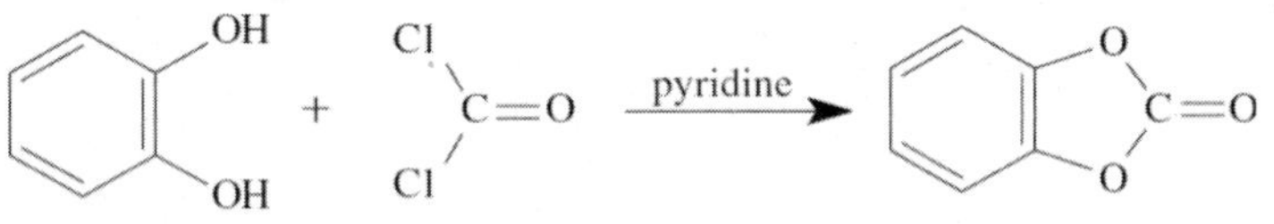

catechol; phosgene

(d)

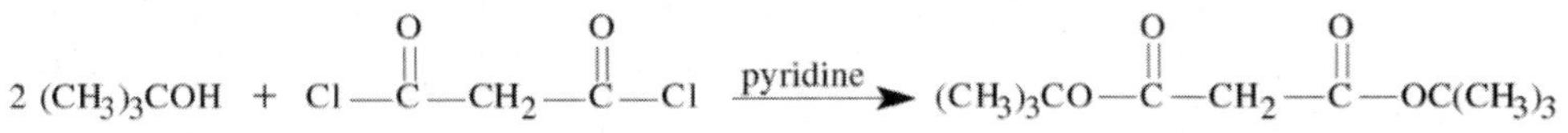

tert-butyl alcohol; malonyl dichloride; di-*tert*-butyl malonate

21.19 (a) There are two cyclic amide (or lactam) functional groups present in the compound. Since one lactam is contained in a four-membered ring, it is considerably more reactive than the other due to the ring strain that exists. So, one equivalent of sodium methoxide will react preferentially at the more reactive lactam.

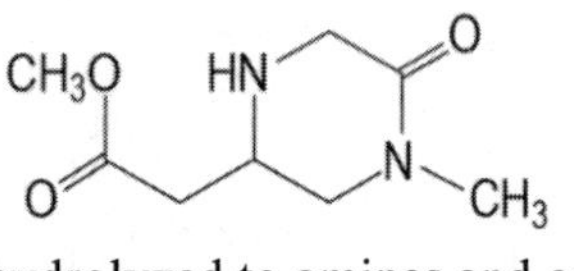

(b) In this case, both lactams would be hydrolyzed to amines and carboxylic acids. Shown below is the neutral form of the product. In strong acid, the amines would be protonated.

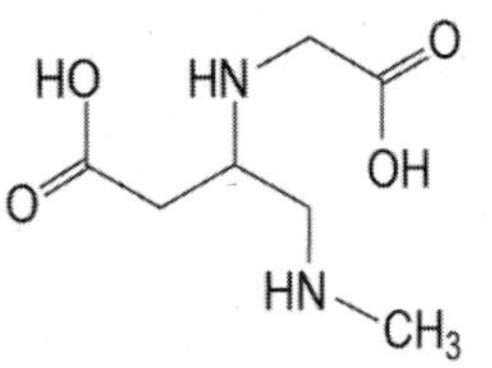

21.20 (a) Carry out one of the reductions in Sec. 21.9D:

$$PhC(=O)-Cl \xrightarrow[\text{2) } H_3O^+]{\text{1) } Li^+ \; H\bar{A}l[OC(CH_3)_3]_3} PhC(=O)-H$$

benzoyl chloride   benzaldehyde

(b) Reduce the tertiary amide with LiAlH₄.

$$PhC(=O)-Cl \; + \; HN\langle \longrightarrow PhC(=O)-N\langle \xrightarrow[\text{2) } H_3O^+]{\text{1) } LiAlH_4} PhCH_2N\langle$$

benzoyl chloride   pyrrolidine   N-benzylpyrrolidine
                     (excess)

21.21 (a) $PhCH_2CH_2NH_2$

(b) Although the ester reacts most rapidly, the nitrile also reacts because an excess of LiAlH₄ is present. The product is the amino alcohol $HOCH_2CH_2CH_2NH_2$ plus the by-product ethanol, C_2H_5OH.

(c) Both esters are reduced.

$$2 \; CH_3CH_2OH \quad + \quad Ph\overset{\displaystyle OH}{\underset{\displaystyle |}{C}}HCH_2OH$$

ethanol   1-phenyl-1,2-ethanediol

(from reduction
of both the acetoxy
group and the ethyl ester)

21.22 Among the compounds that would give $(CH_3)_2CHCH_2CH_2CH_2NH_2$ as the result of LiAlH₄ reduction are the following:

$$(CH_3)_2CHCH_2CH_2C\equiv N \qquad (CH_3)_2CHCH_2CH_2C(=O)-NH_2$$

4-methylpentanenitrile   4-methylpentanamide

21.23 (a) As shown in Eq. 21.58, text p. 1083, the reduction involves an imine intermediate. This intermediate *A* can react with a molecule of amine product to form the nitrogen analog of an acetal, which then can lose ammonia to form a different imine *B*, which in turn is reduced to the by-product. (You should be able to provide the curved-arrow notation for the reactions involved in the *A* ⇌ *B* equilibrium in the overall process outlined below.)

$$RC\equiv N \xrightarrow{H_2, \text{ cat.}} RCH=NH \underset{A}{\overset{H_2NCH_2R}{\rightleftarrows}} R\overset{|}{\underset{|}{C}}H-NH_2 \rightleftarrows NH_3 + R\overset{|}{\underset{|}{C}}H \xrightarrow{H_2, \text{ cat.}} R\overset{|}{\underset{|}{C}}H_2$$

with NHCH₂R under RCH—NH₂ group, NCH₂R (labeled *B*) and NHCH₂R.

$$RCH=NH \xrightarrow{H_2, \text{ cat.}} RCH_2NH_2$$

(b) By Le Châtelier's principle, ammonia drives the equilibria in the above equation away from imine *B* and back toward imine *A*, which is eventually reduced.

21.24 (a)

$$PhCO_2H \xrightarrow{C_2H_5OH, \; H_2SO_4} PhCO_2CH_2CH_3 \xrightarrow[\text{2) } H_3O^+]{\text{1) } PhMgBr \text{ (excess)}} Ph_3C-OH$$

benzoic acid   triphenylmethanol

(b)

$$CH_3CH_2CH_2COH \xrightarrow{PCl_5} CH_3CH_2CH_2CCl \xrightarrow[\text{2) }H_3O^+]{\text{1) Li}^+ \text{ }^-\text{Cu(CH}_3)_2} CH_3CH_2CH_2CCH_3 \xrightarrow[\text{2) }H_3O^+]{\text{1) CH}_3CH_2MgI}$$

butyric acid

$$CH_3CH_2CH_2\overset{\overset{\displaystyle OH}{|}}{\underset{\underset{\displaystyle CH_3}{|}}{C}}CH_2CH_3$$

3-methyl-3-hexanol

(c)   First, the nitrile is hydrolyzed to isobutyric acid, which is converted into its acid chloride:

$$(CH_3)_2CHC\equiv N \xrightarrow{H_2O, H_3O^+, \text{ heat}} (CH_3)_2CHC-OH \xrightarrow{SOCl_2} (CH_3)_2CHC-Cl$$

isobutyronitrile   isobutyric acid   isobutyryl chloride

One synthesis involves reaction of the acid chloride with lithium dimethylcuprate:

$$(CH_3)_2CHC-Cl \xrightarrow[\text{2) }H_3O^+]{\text{1) Li}^+ \text{ }^-\text{Cu(CH}_3)_2} (CH_3)_2CHC-CH_3 \xrightarrow[\text{2) }H_3O^+]{\text{1) CH}_3MgI} (CH_3)_2CHC-CH_3$$
$$\text{with } CH_3$$

2,3-dimethyl-2-butanol

The second synthesis involves the Grignard reaction of an ester (see Eq. 21.63, text p. 1086):

$$(CH_3)_2CHC-OH \xrightarrow[\text{H}_2SO_4]{C_2H_5OH \text{ (solvent)}} (CH_3)_2CHC-OC_2H_5 \xrightarrow[\text{2) }H_3O^+]{\text{1) CH}_3MgI \text{ (excess)}} (CH_3)_2CHC-CH_3$$
$$\text{with } CH_3$$

2,3-dimethyl-2-butanol

  Although you have not studied this reaction explicitly, a more direct route to the alcohol is the reaction of the acid chloride with excess CH_3MgI followed by protonolysis.

$$(CH_3)_2CHC-Cl \xrightarrow[\text{2) }H_3O^+]{\text{1) CH}_3MgI \text{ (excess)}} (CH_3)_2CHC-CH_3$$
$$\text{with } CH_3$$

Can you see why acid chlorides should react the same way as esters in this reaction?

(d)

$$CH_3CH_2COH \xrightarrow{PCl_5} CH_3CH_2CCl \xrightarrow[\text{2) }H_3O^+]{\text{1) Li}^+ \text{ }^-\text{Cu(C}_2H_5)_2} CH_3CH_2CCH_2CH_3$$

propionic acid   **3-pentanone**

21.25   (a)   The reaction of a Grignard reagent with ethyl formate gives a secondary alcohol in which the two alkyl groups at the α-carbon are identical.

$$CH_3CH_2OCH + 2 \text{ RMgBr} \longrightarrow \xrightarrow{H_3O^+} R-\overset{\overset{\displaystyle OH}{|}}{C}H-R + CH_3CH_2OH$$

ethyl formate

(b)   The reaction of a Grignard reagent with ethyl formate gives a secondary alcohol in which the two alkyl groups at the α-carbon are identical.

$$\underset{\text{ethyl formate}}{\text{CH}_3\text{CH}_2\text{O}\overset{\overset{\displaystyle O}{\|}}{\text{C}}\text{H}} + 2\ \text{CH}_3\text{CH}_2\text{MgBr} \xrightarrow{\text{H}_3\text{O}^+} \underset{\text{3-pentanol}}{\text{CH}_3\text{CH}_2\overset{\overset{\displaystyle OH}{|}}{\text{C}}\text{HCH}_2\text{CH}_3} + \text{CH}_3\text{CH}_2\text{OH}$$

21.26 (a) Because only the acid chloride reacts, the product is the following keto nitrile.

$$\text{N}\equiv\text{C(CH}_2)_{10}\overset{\overset{\displaystyle O}{\|}}{\text{C}}\text{CH}_3 \qquad \text{12-oxotridecanenitrile}$$

(b) Because only the acid chloride reacts, the product is the following keto ester:

$$\text{CH}_3(\text{CH}_2)_3\text{O}-\overset{\overset{\displaystyle O}{\|}}{\text{C}}(\text{CH}_2)_4\overset{\overset{\displaystyle O}{\|}}{\text{C}}-\text{CH}_3 \qquad \text{butyl 6-oxoheptanoate}$$

21.27 The first method involves carbonation of the corresponding Grignard reagent; the second involves hydrolysis of the nitrile.

$$\underset{\substack{\text{1-bromo-4-methylpentane}}}{(\text{CH}_3)_2\text{CH(CH}_2)_3\text{Br}} \Big\langle \begin{array}{l} \xrightarrow[\text{ether}]{\text{Mg}}\ \xrightarrow[\substack{2)\ \text{H}_3\text{O}^+}]{1)\ \text{CO}_2}\ \underset{\text{5-methylhexanoic acid}}{(\text{CH}_3)_2\text{CH(CH}_2)_3\text{CO}_2\text{H}} \xleftarrow{\substack{\text{H}_2\text{O, H}_3\text{O}^+,\ \text{heat}}} \\[2em] \xrightarrow[\text{DMSO}]{\text{Na}^+\ {}^-\text{CN}}\ (\text{CH}_3)_2\text{CH(CH}_2)_3\text{C}\equiv\text{N} \end{array}$$

21.28 The required nitrile is 2-hydroxypropanenitrile. This is a cyanohydrin, and this cyanohydrin can be prepared by addition of cyanide to acetaldehyde.

$$\underset{\text{acetaldehyde}}{\text{CH}_3\overset{\overset{\displaystyle O}{\|}}{\text{C}}\text{H}} \xrightarrow{\text{NaCN, H}_2\text{O}} \underset{\substack{\text{2-hydroxypropanenitrile}}}{\text{CH}_3\overset{\overset{\displaystyle OH}{|}}{\underset{\underset{\displaystyle \text{C}\equiv\text{N}}{|}}{\text{C}}}\text{H}} \xrightarrow{\text{H}_2\text{O, H}_3\text{O}^+,\ \text{heat}} \underset{\substack{\text{2-hydroxypropanoic acid}}}{\text{CH}_3\overset{\overset{\displaystyle OH}{|}}{\underset{\underset{\displaystyle \text{CO}_2\text{H}}{|}}{\text{C}}}\text{H}}$$

 Notice that the alcohol does *not* dehydrate under the rather severe conditions of nitrile hydrolysis. The reason is that the carbocation intermediate that would be involved is destabilized by both the cyano group and the carboxy group of the product. Can you think of a reason why?

21.29 Amides are more resistant to base-promoted hydrolysis than esters; the same is true of *polymeric* amides and esters. Nylon is more resistant to hydrolysis than polyester.

21.30 Process (*a*) is catalytic hydrogenation. Because furan is aromatic, high pressure and/or heat might be required. Process (*b*) is ether cleavage with HCl and ZnCl$_2$ (or other acidic catalyst) and heat. Process (*c*) consists of the S$_\text{N}$2 reactions of cyanide ion with the dichloride. Process (*d*) is catalytic hydrogenation or LiAlH4 reduction followed by protonolysis. Process (*e*) is nitrile hydrolysis with aqueous acid and heat. Finally, process (*f*) is to mix the amine and the carboxylic acid and heat (Eq. 21.72, text p. 1090), or to form the di-acid chloride and allow it to react with the amine.

21.31 First, water opens the lactam to an amino acid *A*. The amino group of this compound serves as a nucleophile to open another lactam molecule; and the resulting amino group thus liberated repeats the process, thus growing the polymer chain:

$$\text{proton transfers} \qquad \xrightarrow{} \underset{A}{\text{HO}\overset{\overset{\displaystyle O}{\|}}{\text{C}}(\text{CH}_2)_5\text{NH}_2}$$

The top portion shows a reaction mechanism for nylon-6 formation from ε-caprolactam:

A ... NH$_2$(CH$_2$)$_5$CO$_2$H ... proton transfers ...

$$HO\overset{O}{\underset{||}{C}}(CH_2)_5NH-\overset{O}{\underset{||}{C}}(CH_2)_5NH_2 \xrightarrow{\text{similar reactions}} \text{nylon-6}$$

this amino group reacts with
another ε-caprolactam molecule

Solutions to Additional Problems

21.32   (a)

PhCO$_2$H

+ CH$_3$CH$_2$OH

(b)

PhCO$_2^-$ Na$^+$

+ CH$_3$CH$_2$OH

(c)

$$Ph\overset{O}{\underset{||}{C}}-NH_2$$

+ CH$_3$CH$_2$OH

(d)

PhCH$_2$OH

+ CH$_3$CH$_2$OH

(e)

$$\underset{\underset{CH_2CH_2CH_3}{|}}{\overset{\overset{OH}{|}}{PhCCH_2CH_2CH_3}}$$

+ CH$_3$CH$_2$OH

(f)

$$\underset{\underset{CH_2CH_2CH_3}{|}}{\overset{\overset{O}{||}}{PhCCH_2CH_2CH_3}} \; + \; \text{(pyridinium)} \; \overset{+}{N}H \; Cl^-$$

with OCCH$_3$ group

(g)

$$\underset{\underset{CH_2CH_2CH_3}{|}}{\overset{\overset{O\;\;\;\;O}{\diagdown\!S\!\diagup}}{PhCCH_2CH_2CH_3}} \;\; Ph$$

(h)   In this reaction, ethoxide displaces ethoxide, but the reaction is "invisible" because the product is the same as the starting material. Thus, there is no *net* reaction.

21.33   (a)

CH$_3$CH$_2$CO$_2$H

(b)

$$CH_3CH_2\overset{O}{\underset{||}{C}}SCH_2CH_3$$

$$+ \; \text{(pyridinium)} \; \overset{+}{N}H \; Cl^-$$

(c)

$$CH_3CH_2\overset{O}{\underset{||}{C}}OC(CH_3)_3$$

$$+ \; \text{(pyridinium)} \; \overset{+}{N}H \; Cl^-$$

(d)

$$CH_3CH_2\overset{O}{\underset{||}{C}}CH_3$$

(e)

$$CH_3CH_2\overset{O}{\underset{||}{C}}H$$

(f)

$$CH_3CH_2\overset{O}{\underset{||}{C}}-\!\!\!\!\bigcirc\!\!\!\!-CH_3$$

(g)

$$CH_3CH_2\overset{O}{\underset{||}{C}}NHCH(CH_3)_2$$

$$+ \;\; Cl^- \; H_2\overset{+}{N}CH(CH_3)_2$$

(h)

$$CH_3CH_2\overset{O\;\;O}{\underset{||\;\;||}{C}OC}Ph$$

(i)

$$CH_3CH_2\overset{O}{\underset{||}{C}}O-\!\!\!\!\bigcirc\!\!\!\!-CH_3$$

$$+ \; \text{(pyridinium)} \; \overset{+}{N}H \; Cl^-$$

21.34   (a)   (b)   (c)   (d)   (e)

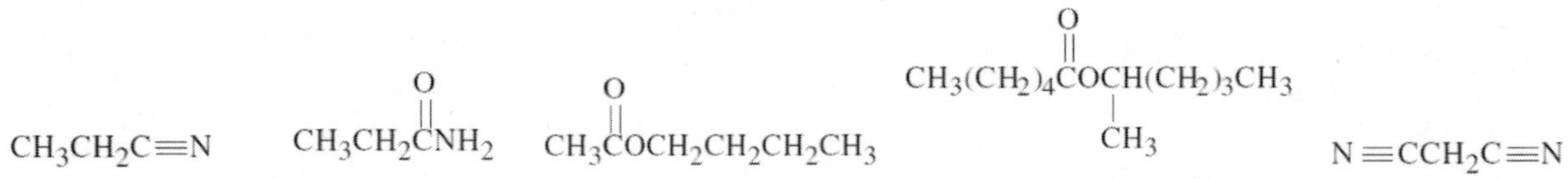

$$CH_3(CH_2)_4COCH(CH_2)_3CH_3$$
$$|$$
$$CH_3$$

$CH_3CH_2C \equiv N$ $CH_3CH_2\overset{O}{\overset{||}{C}}NH_2$ $CH_3\overset{O}{\overset{||}{C}}OCH_2CH_2CH_2CH_3$ $N \equiv CCH_2C \equiv N$

propionitrile   propionamide   butyl acetate   1-methylpentyl hexanoate   malononitrile

Malononitrile (propanedinitrile) in part (e) hydrolyzes in aqueous acid to malonic acid and ammonium ion, and the malonic acid decarboxylates to acetic acid and carbon dioxide. (See Sec. 20.11 in the text.)

21.35   The easiest way to work this problem is to start with the structure and work backwards. Because carbonation of a Grignard reagent gives a carboxylic acid, compound *D* must be the corresponding alkyl bromide. Compound *E* must be the hydrocarbon that gives compound *D* on benzylic bromination. The Wolff–Kishner reaction produces compound *E* from compound *B*, which therefore must be a ketone. Reduction of ketone *B* with sodium borohydride gives alcohol *C*, which is converted into alkyl bromide *D*. Finally, the structure of compound *B* shows that reaction *A* must be a Friedel–Crafts acylation. In summary:

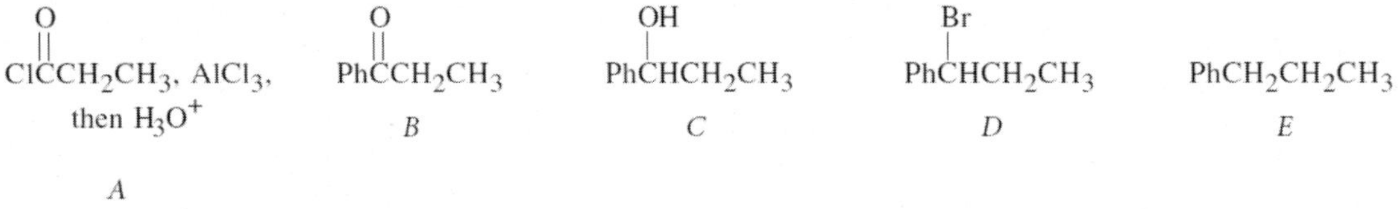

$Cl\overset{O}{\overset{||}{C}}CH_2CH_3$, $AlCl_3$, $Ph\overset{O}{\overset{||}{C}}CH_2CH_3$ $Ph\overset{OH}{\overset{|}{C}}HCH_2CH_3$ $Ph\overset{Br}{\overset{|}{C}}HCH_2CH_3$ $PhCH_2CH_2CH_3$
     then H_3O^+ *B*   *C*   *D*   *E*

    A

21.36   The product is (*R*)-(−)-1-phenyl-1,2-ethanediol. Notice that the asymmetric carbon is unaffected by these transformations.

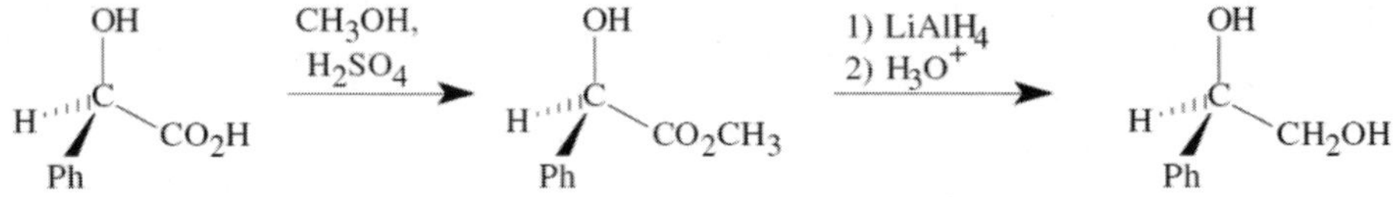

(*R*)-(−)-mandelic acid $\xrightarrow{\substack{CH_3OH, \\ H_2SO_4}}$ (with CO_2CH_3) $\xrightarrow{\substack{1)\ LiAlH_4 \\ 2)\ H_3O^+}}$ (*R*)-(−)-1-phenyl-1,2-ethanediol
(a 1,2-glycol; reacts with periodic acid)

21.37   Both the ketone and the carboxylic acid are reduced with $LiAlH_4$; although the carboxylic acid is ionized by $NaBH_4$ (but regenerated on addition of acid), only the ketone is reduced.

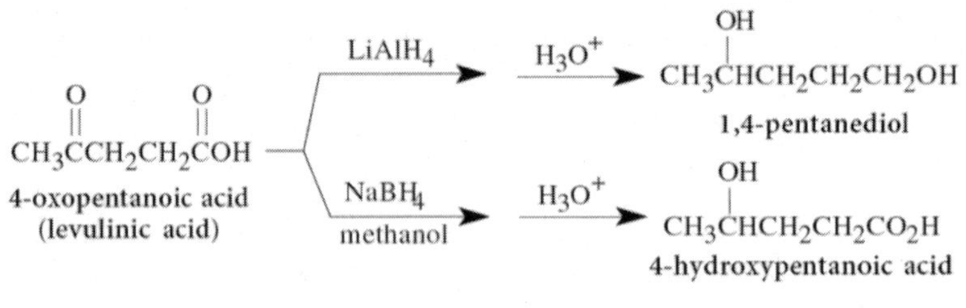

$CH_3\overset{O}{\overset{||}{C}}CH_2CH_2\overset{O}{\overset{||}{C}}OH$
4-oxopentanoic acid
(levulinic acid)

$\xrightarrow{LiAlH_4}$ $\xrightarrow{H_3O^+}$ $CH_3\overset{OH}{\overset{|}{C}}HCH_2CH_2CH_2OH$
1,4-pentanediol

$\xrightarrow[\text{methanol}]{NaBH_4}$ $\xrightarrow{H_3O^+}$ $CH_3\overset{OH}{\overset{|}{C}}HCH_2CH_2CO_2H$
4-hydroxypentanoic acid

In acidic solution, 4-hydroxypentanoic acid is in equilibrium with the corresponding lactone, which is favored in the equilibrium. (See Eq. 21.12, text p. 1064.)

21.38   (a)   Compound *A* has the formula $C_{11}H_{13}NO$. The formula of compound *B* indicates that it differs from *A* by the addition of the elements of water. The fact that compound *C* is a lactone with an oxygen at carbon-4 and an additional hydrogen at carbon-3 suggests that addition of water to the double bond of compound *A*, that is, alkene hydration, occurs metabolically to give an alcohol *B*. The lactone *C* is formed by cyclization of compound *B*, that is, an intramolecular esterification. The structure of compound *B* is given in the mechanism of part (b).

(b)   A curved-arrow mechanism for the formation of compound *C* from compound *B*:

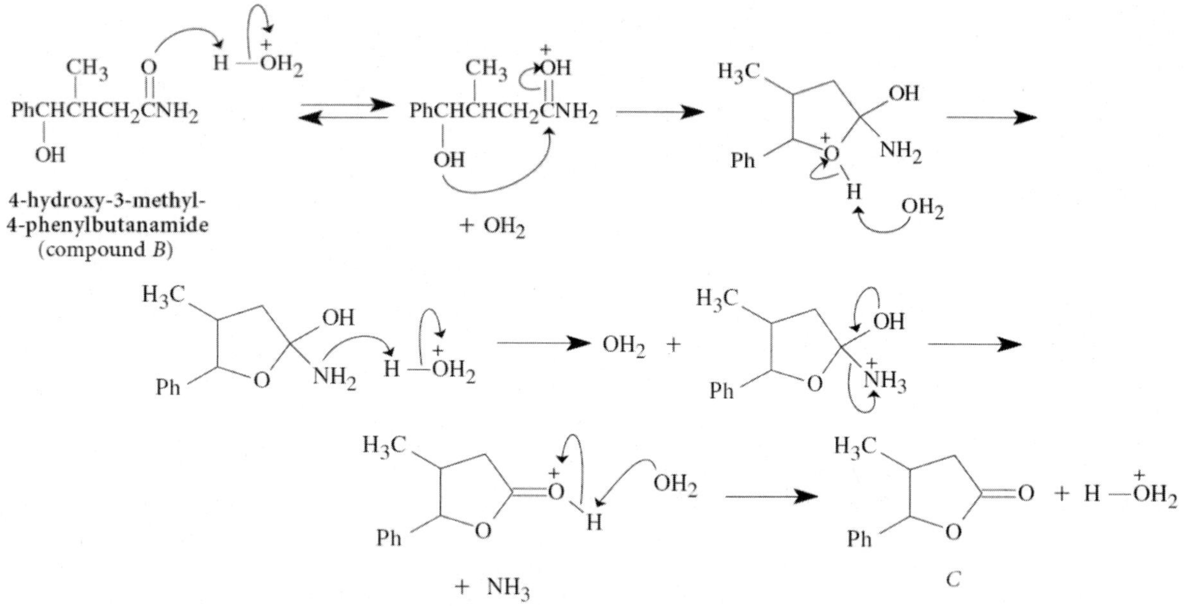

4-hydroxy-3-methyl-
4-phenylbutanamide
(compound *B*)

+ OH₂

+ NH₃
(protonated under
the acidic conditions)

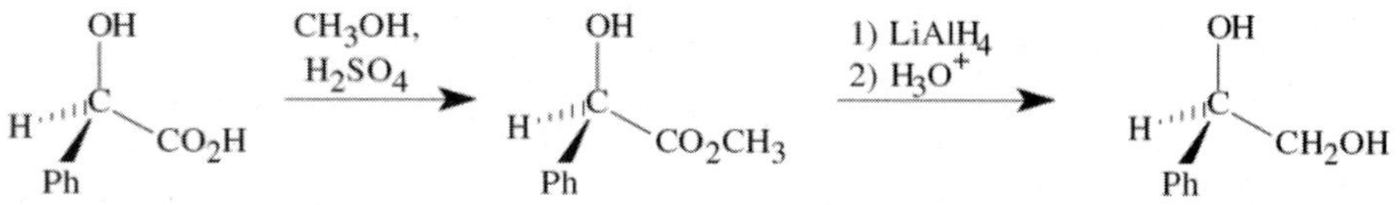

C

(*R*)-(−)-mandelic acid

(*R*)-(−)-1-phenyl-1,2-ethanediol
(a 1,2-glycol; reacts with periodic acid)

21.39 (a) In aqueous base the ester groups are saponified to give glycerol (1,2,3-propanetriol) and three equivalents of sodium oleate, the sodium salt of oleic acid.

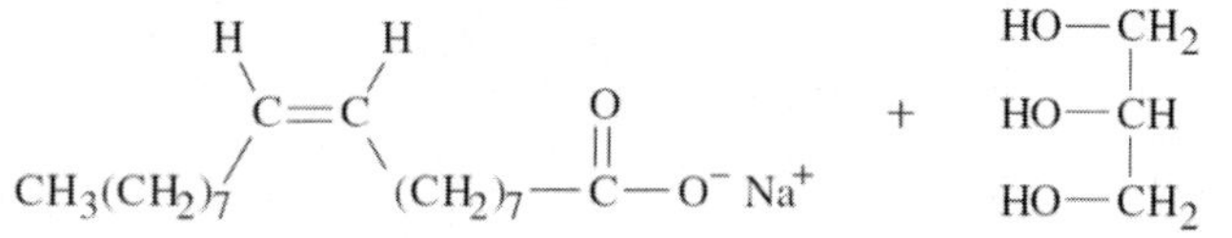

sodium oleate

glycerol

(b) The structure of glyceryl tristearate:

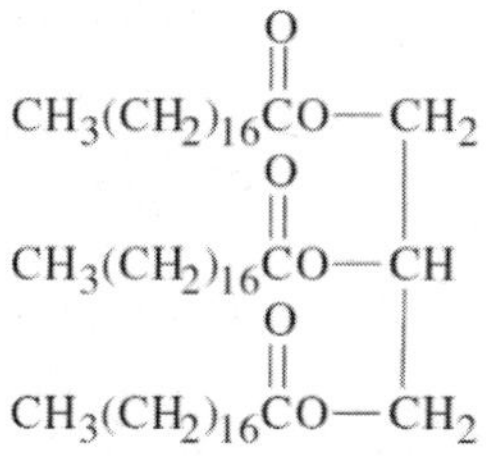

glyceryl tristearate

Because unsaturated fats are oils and saturated fats are solids, glyceryl trioleate is a liquid, whereas the product of hydrogenation is a solid. In fact, glyceryl trioleate is a major component of olive oil, whereas glyceryl tristearate is a major component of lard.

21.40 (a)

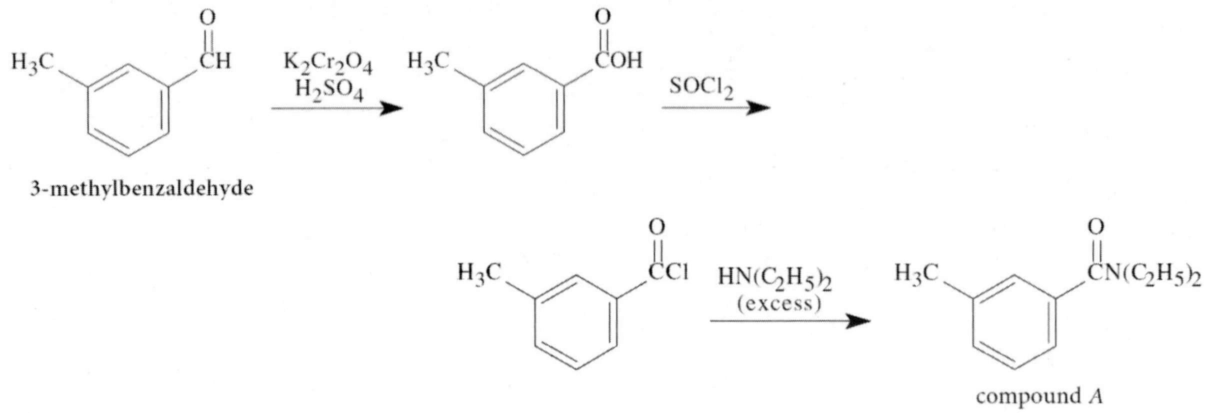

3-methylbenzaldehyde

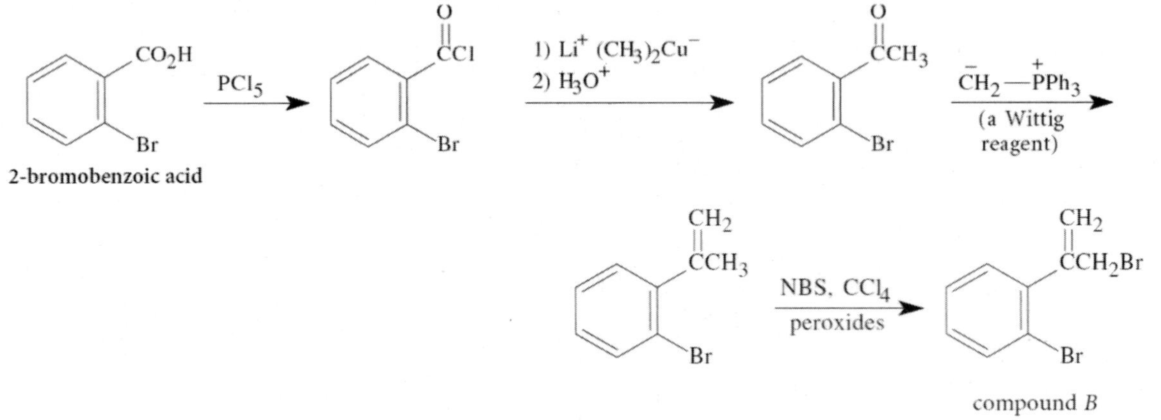

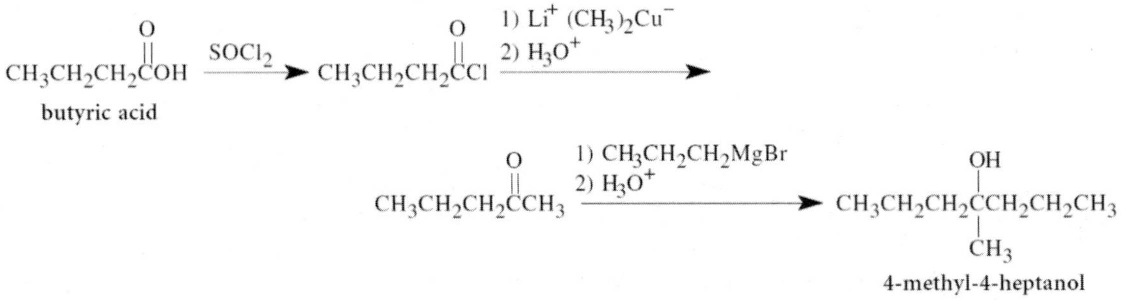

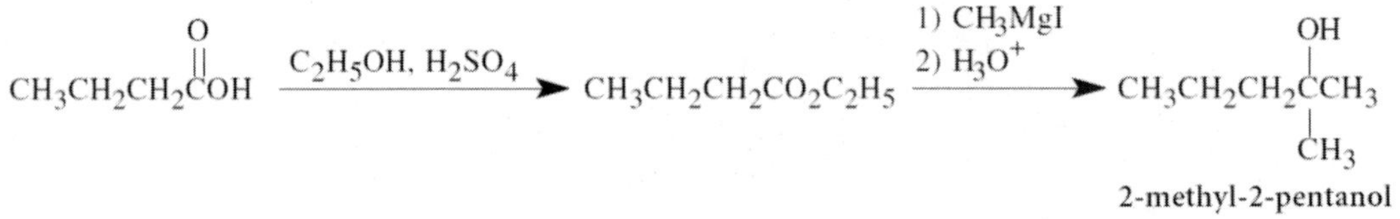

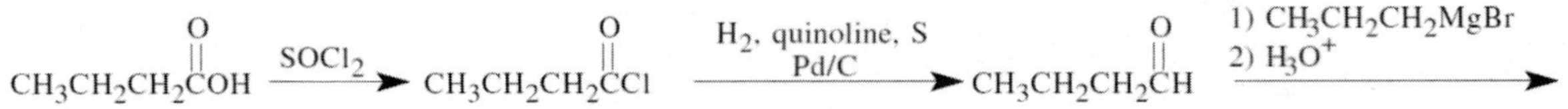

compound A

(b)

2-bromobenzoic acid

compound B

21.41 (a)

butyric acid

4-methyl-4-heptanol

(b)

2-methyl-2-pentanol

(c)

OH
|
$CH_3CH_2CH_2CHCH_2CH_2CH_3$

4-heptanol

(d)

$$CH_3CH_2CH_2\overset{O}{\underset{\|}{C}}OH \xrightarrow[2) \ H_3O^+]{1) \ LiAlH_4} CH_3CH_2CH_2CH_2OH \xrightarrow[H_2SO_4]{conc. \ HBr} CH_3CH_2CH_2CH_2Br \xrightarrow[ether]{Mg}$$

$$CH_3CH_2CH_2CH_2MgBr \xrightarrow[2) \ H_3O^+]{1) \ \triangle O} CH_3CH_2CH_2CH_2CH_2CH_2OH \xrightarrow[2) \ H_3O^+]{1) \ KMnO_4, \ ^-OH}$$

$$CH_3CH_2CH_2CH_2CH_2CO_2H \xrightarrow{SOCl_2} CH_3CH_2CH_2CH_2CH_2\overset{O}{\underset{\|}{C}}Cl \xrightarrow{NH_3 \ (excess)}$$

$$CH_3CH_2CH_2CH_2CH_2\overset{O}{\underset{\|}{C}}NH_2 \xrightarrow[2) \ H_2O]{1) \ LiAlH_4} CH_3CH_2CH_2CH_2CH_2CH_2NH_2$$

1-hexanamine (hexylamine)

(e)

$$CH_3CH_2CH_2\overset{O}{\underset{\|}{C}}OH \xrightarrow[2) \ H_3O^+]{1) \ LiAlH_4} CH_3CH_2CH_2CH_2OH \xrightarrow[H_2SO_4]{conc. \ HBr} CH_3CH_2CH_2CH_2Br \xrightarrow[ether]{Mg}$$

$$CH_3CH_2CH_2CH_2MgBr \xrightarrow[2) \ H_3O^+]{1) \ CO_2} CH_3CH_2CH_2CH_2CO_2H \xrightarrow{SOCl_2} CH_3CH_2CH_2CH_2\overset{O}{\underset{\|}{C}}Cl \xrightarrow{NH_3 \ (excess)}$$

$$CH_3CH_2CH_2CH_2\overset{O}{\underset{\|}{C}}NH_2 \xrightarrow[2) \ H_2O]{1) \ LiAlH_4} CH_3CH_2CH_2CH_2CH_2NH_2$$

1-pentanamine (pentylamine)

(f)

$$CH_3CH_2CH_2\overset{O}{\underset{\|}{C}}OH \xrightarrow{SOCl_2} CH_3CH_2CH_2\overset{O}{\underset{\|}{C}}Cl \xrightarrow[(excess)]{NH_3} CH_3CH_2CH_2\overset{O}{\underset{\|}{C}}NH_2 \xrightarrow[2) \ H_2O]{1) \ LiAlH_4} CH_3CH_2CH_2CH_2NH_2$$

**1-butanamine
(butylamine)**

21.42 **(a)**   First of all, two sets of constitutional isomers can be obtained from the reaction of 1-phenylethanol at either of the nonequivalent carbonyl groups of the anhydride:

α-phenylglutaric
anhydride   1-phenylethanol   *A*   *B*

Each of these products has two asymmetric carbon stereocenters and can thus exist as four stereoisomers. All stereoisomers of *A* can in principle be separated from all stereoisomers of *B* because constitutional isomers have different properties. The four stereoisomers of *A* can be separated into two pairs of enantiomers; the same is true of *B*. Consequently, four compounds can in principle be obtained from this reaction mixture without enantiomeric resolution.

(b)   The same two constitutional isomers would be obtained, and each would be obtained as a pair of diastereomers. However, only one enantiomer of each diastereomer (the one with the *S* configuration at the carbon derived from the alcohol) would be formed.

> If the diastereomeric esters from part (b) were separated from each other and saponified, the *α*-phenylglutaric acid formed would be enantiomerically pure. The process described here would therefore be useful as an enantiomeric resolution of *α*-phenylglutaric acid (or any other chiral dicarboxylic acid from which a cyclic anhydride can be readily formed).

21.43   Because the two carbonyl groups of the anhydride are not equivalent, two different products are possible; each corresponds to reaction with the ethanol at a different carbonyl group. Because ethyl propionate has a higher molecular mass than ethyl acetate, ethyl propionate should have the higher boiling point. Since it is given that the ester with the lower boiling point is formed in greater amount, then ethyl acetate is the major product.

$$CH_3\overset{\overset{\displaystyle O}{\|}}{C}-O-\overset{\overset{\displaystyle O}{\|}}{C}CH_2CH_3 \quad + \quad CH_3CH_2OH \quad \longrightarrow$$
acetic propionic anhydride

$$CH_3\overset{\overset{\displaystyle O}{\|}}{C}-OCH_2CH_3 \quad + \quad HO-\overset{\overset{\displaystyle O}{\|}}{C}CH_2CH_3 \quad + \quad CH_3CH_2O-\overset{\overset{\displaystyle O}{\|}}{C}CH_2CH_3 \quad + \quad CH_3\overset{\overset{\displaystyle O}{\|}}{C}-OH$$

ethyl acetate · ethyl propionate

major products

21.44   Carboxylate salts are less reactive than esters in nucleophilic acyl substitution reactions for two reasons. First, resonance stabilization of carboxylate salts is particularly important because carboxylates have two identical, important resonance structures; see Eq. 20.4 on text p. 1013. Resonance stabilization of esters is less important because one resonance structure separates charge; see the structures in Eq. 21.26, text p. 1069. Because resonance stabilization reduces reactivity, carboxylates are less reactive than esters.

The second and major reason that carboxylate salts are less reactive is that they contain no leaving group. The oxygen anion of the carboxylate, in order to serve as a leaving group, would have to depart as an oxide ion, O_2^-. This very basic leaving group is unprecedented in the usual reactions of organic chemistry.

> Reactions such as $LiAlH_4$ reduction that occur on carboxylate salts might at first glance appear to involve oxide ions as leaving groups. In such reactions, however, the "oxygen anion" is bonded to another species that allows the oxygen to serve as a leaving group. (See the details in Further Exploration 20.4 on p. 665 of this manual.)

21.45   (a)   Thiol esters are less stabilized by resonance than esters because the resonance interaction of sulfur unshared electron pairs with a carbonyl group is less effective than the resonance interaction of oxygen unshared electron pairs. The reason for this difference is that the orbitals on sulfur are derived from quantum level 3, and such orbitals overlap poorly with the π orbitals of a carbonyl group, which are derived from quantum level 2. (See Fig. 16.7 on text p. 819 and the accompanying discussion of this point.) Thus, resonance arguments suggest that a thiol ester should hydrolyze more rapidly than an ester.

(b)   Sulfur is considerably less electronegative than oxygen. Therefore, the polar stabilization of the thioester transition state is less than the polar stabilization of the ester transition state by oxygen. This argument suggests that thioesters should be less reactive than esters.

> Notice that the answers to parts (a) and (b) seem to contradict each other. The result is that the two effects tend to cancel, and the reactivity of thioesters and oxygen esters is about the same, which is in accord with experimental observation.

21.46   (a)   Synthesis of nylon-4,6 requires mixing 1,4-butanediamine and adipoyl dichloride.

$$H_2N(CH_2)_4NH_2 \; + \; \overset{O}{\underset{\parallel}{Cl}C}(CH_2)_4\overset{O}{\underset{\parallel}{C}}Cl \longrightarrow \text{nylon-4,6}$$

1,4-butanediamine   adipoyl dichloride

(Alternatively, 1,4-butanediamine can be heated with adipic acid; see Eq. 21.72 on text p. 1090.) The preparations of these materials are as follows:

$$\underset{\text{succinic acid}}{\overset{O\quad\quad O}{HO\overset{\parallel}{C}CH_2CH_2\overset{\parallel}{C}OH}} \xrightarrow{SOCl_2} \overset{O\quad\quad O}{Cl\overset{\parallel}{C}CH_2CH_2\overset{\parallel}{C}Cl} \xrightarrow{NH_3 \text{ (excess)}} \overset{O\quad\quad O}{H_2N\overset{\parallel}{C}CH_2CH_2\overset{\parallel}{C}NH_2} \xrightarrow[2)\ H_2O]{1)\ LiAlH_4} \underset{\text{1,4-butanediamine}}{H_2N(CH_2)_4NH_2}$$

$$\underset{\text{adipic acid}}{\overset{O\quad\quad O}{HO\overset{\parallel}{C}(CH_2)_4\overset{\parallel}{C}OH}} \xrightarrow{SOCl_2} \underset{\text{adipoyl dichloride}}{\overset{O\quad\quad O}{Cl\overset{\parallel}{C}(CH_2)_4\overset{\parallel}{C}Cl}}$$

 Why do you think heating the dicarboxylic acid with the diamine is preferred industrially to reaction of the diamine with the dicarboxylic acid dichloride? (There are at least two reasons.)

(b)   The preparation of this polyester requires heating 1,4-butanediol with glutaric acid.

(Alternatively, the di-acid chloride of glutaric acid could be mixed with 1,4-butanediol.)

$$\underset{\text{succinic acid}}{\overset{O\quad\quad O}{HO\overset{\parallel}{C}CH_2CH_2\overset{\parallel}{C}OH}} \xrightarrow[2)\ H_3O^+]{1)\ LiAlH_4} \underset{\text{1,4-butanediol}}{HO(CH_2)_4OH} \xrightarrow[\text{heat}]{\overset{\overset{O\quad\quad O}{HO\overset{\parallel}{C}(CH_2)_3\overset{\parallel}{C}OH}}{\text{glutaric acid}}} \left(\!\!\!\overset{}{O(CH_2)_4O}-\overset{O\quad\quad O}{\overset{\parallel}{C}(CH_2)_3\overset{\parallel}{C}}\!\!\!\right)_{\!\!n}$$

21.47   (a)

$$\left[\overset{O}{\underset{\parallel}{C}}-\!\!\!\left\langle\right\rangle\!\!\!-\overset{O}{\underset{\parallel}{C}}-\underset{H}{N}-\!\!\!\left\langle\right\rangle\!\!\!-\underset{H}{N}\right]_n$$

(b)

21.48   The IR spectrum of compound *A* indicates that it is an anhydride, and probably a cyclic anhydride containing a five-membered ring. (Compare the carbonyl absorptions of compound *A* with those in Table 21.3 on text p. 1054; the 1050 cm^{-1} absorption is a C—O stretching absorption). Addition of methanol to a cyclic anhydride should give a methyl half-ester, compound *B*. Indeed, the IR spectrum of compound *B* indicates the presence of both a carboxylic acid and an ester group. Subtracting the elements of methanol (CH_4O) from the formula of compound *B* gives the formula of compound *A*, $C_4H_4O_3$. Compound *A* is succinic anhydride.

Two points about the NMR spectra are worth noting. First, the methylene protons of compound *B* give a complex resonance at δ 2.7 because they are chemically nonequivalent but have very similar chemical shifts. This is the condition under which complex, non-first-order splitting can occur. (See text Sec. 13.5B, pp. 641–643.) That these protons should have similar chemical shifts is a reasonable observation because both sets of protons are in very similar electronic environments: both are on carbons that are α to carbonyl groups. The second point is that the proton of the carboxy group of compound *B* is not observed because the spectrum is taken in D_2O. Recall (Sec. 13.7D, text pp. 651–653) that O—H protons are rapidly exchanged for deuterium in when D_2O is present, and that deuterium nuclei are "silent" in proton NMR.

succinic anhydride	methyl hydrogen succinate
(compound *A*)	(compound *B*)

21.49   The high-frequency IR absorption and the reaction with water to give a dicarboxylic acid suggest that this compound is the anhydride of α,α-dimethylmalonic acid. The formation of a half ester in the reaction with methanol also supports the hypothesis that this compound is an anhydride.

α,α-dimethylmalonic anhydride

In most cases cyclic anhydrides show two carbonyl stretching absorptions. Can you think of a reason why this particular anhydride shows only one? (*Hint:* See Study Problem 12.2 on text p. 584.) Is the missing carbonyl absorption at lower or higher frequency than the one that is observed at 1820 cm^{-1}? (*Hint:* See Table 21.3, text p. 1054, and extrapolate from the absorptions of six-and five-membered cyclic anhydrides.)

21.50   (a)   Because phenols are very reactive in electrophilic substitution reactions, bromine in CCl_4 will effect bromination of the phenol ring as well as addition to the alkene double bond. (See Eq. 18.68 on text p. 915.)

$$CH_2NHC(CH_2)_4CHCHCH(CH_3)_2$$

bromine addition occurs with anti stereochemistry to give the enantiomeric (*S,R*) and (*R,S*) stereoisomers in equal amounts.

(b)   Dilute aqueous NaOH will bring about ionization of the phenol to its conjugate-base phenolate ion. The amide is unaffected, because amide hydrolysis requires heat and strong base.

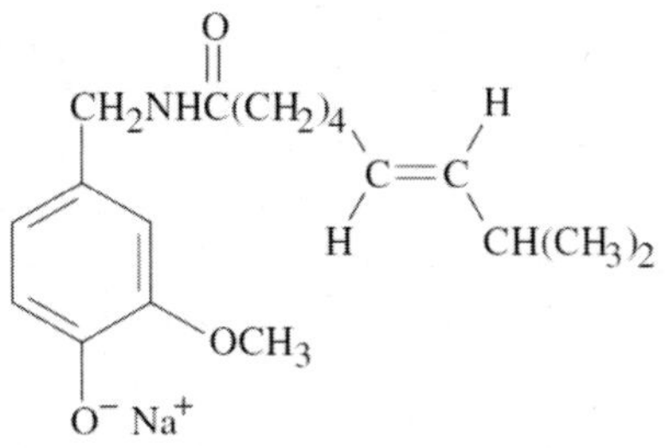

The structure shows: $CH_2NHC(=O)(CH_2)_4$ connected to a $C=C$ double bond with H substituents and $CH(CH_3)_2$ group, on a benzene ring with OCH_3 and $O^- Na^+$ substituents.

(c)   No significant reaction occurs with dilute aqueous HCl. (The carbonyl oxygen of the amide is protonated to a small extent.)

(d)   Catalytic hydrogenation brings about addition of hydrogen to the alkene double bond.

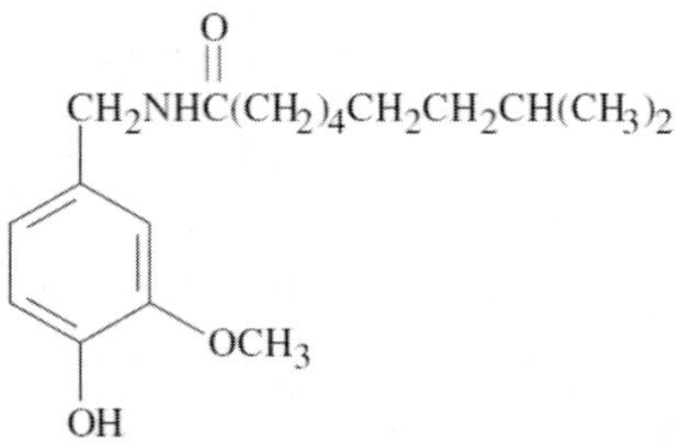

The structure shows: $CH_2NHC(=O)(CH_2)_4CH_2CH_2CH(CH_3)_2$ on a benzene ring with OCH_3 and OH substituents.

Depending on the conditions, hydrogenation of the *single bond* between the benzylic CH_2 and the nitrogen can also occur; such a reaction is called *hydrogenolysis*. Because you have not studied this reaction, it is ignored.

(e)   Heating the product of (d) in 6 *M* HCl brings about amide hydrolysis.

(It is also conceivable that some cleavage of the methyl ether will occur to give methyl chloride and the diphenol.)

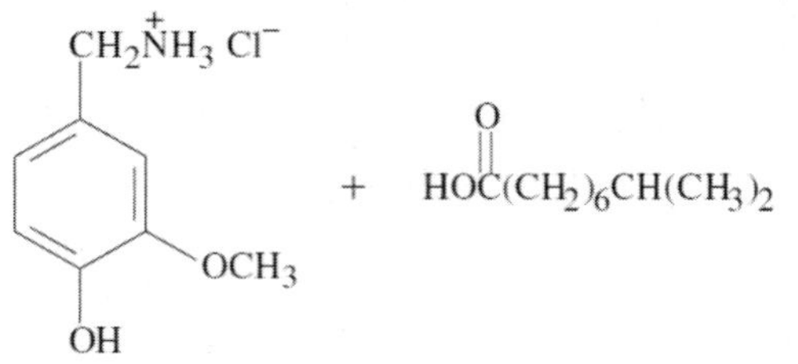

The structure shows: $CH_2\overset{+}{N}H_3\ Cl^-$ on a benzene ring with OCH_3 and OH substituents, plus $HOC(=O)(CH_2)_6CH(CH_3)_2$.

(f)   The phenolate ion formed in part (b) is alkylated to give a second methyl ether group.

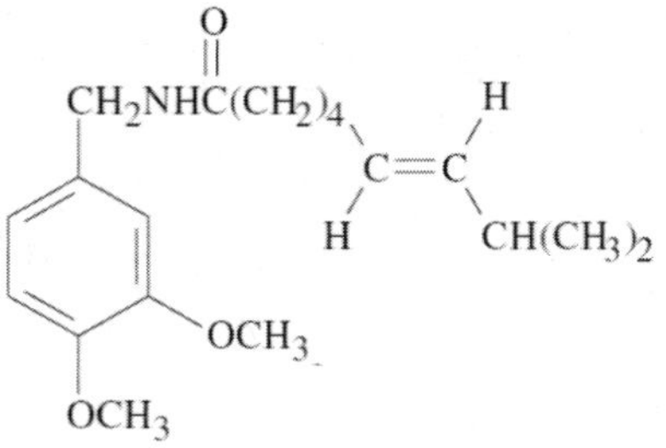

The structure shows: $CH_2NHC(=O)(CH_2)_4$ connected to a $C=C$ double bond with H substituents and $CH(CH_3)_2$ group, on a benzene ring with OCH_3 and OCH_3 substituents.

(g)   On heating, the same amide hydrolysis observed in part (e) will take place, and the methyl ether will cleave.

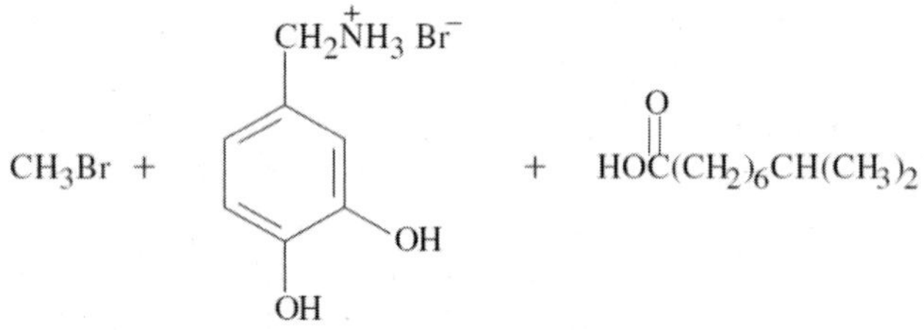

The structure shows: CH_3Br + a benzene ring with $CH_2\overset{+}{N}H_3\ Br^-$, OH and OH substituents, plus $HOC(=O)(CH_2)_6CH(CH_3)_2$.

21.51   Use the titration data to calculate the molecular mass of the ester. The amount of 1 *M* NaOH solution consumed in the saponification reaction is (15.00 mL – 5.30 mL) = 9.70 mL. Make the provisional assumption that there is one ester group per molecule of compound *A*. Then one mole of NaOH is consumed per mole of ester saponified, and the molecular mass of compound *A* is then

$$\text{molecular mass of } A = \frac{2.00 \text{ g}}{(9.70 \times 10^{-3} \text{ L})(1.00 \text{ mol L}^{-1})} = 206 \text{ g mol}^{-1}$$

Next, try to deduce a structure from the data in the problem. The structure of compound *C*, the alcohol that results from the saponification of compound *A*, can be determined; it is the alcohol that can be oxidized to acetophenone. Therefore, compound *C* is 1-phenylethanol.

1-phenylethanol
(compound *C*)   acetophenone

The molecular mass of compound *C* is 122 g mol^{-1}. Now, from the overall transformation of ester hydrolysis, the molecular mass of ester *A* must equal the sum of the molecular masses of compounds *B* and *C* minus the molecular mass of H_2O, that is,

$$\text{molecular mass of } A = 206 \text{ g mol}^{-1} = \text{molecular mass of } B + 122 \text{ g mol}^{-1} - 18 \text{ g mol}^{-1}$$

or

$$\text{molecular mass of } B = 206 \text{ g mol}^{-1} + 18 \text{ g mol}^{-1} - 122 \text{ g mol}^{-1} = 102 \text{ g mol}^{-1}$$

Compound *B* is a carboxylic acid; its carboxy group accounts for 45 molecular mass units. Therefore, the remainder of the carboxylic acid has a mass of 57 units, which is the mass of a butyl group. Because acid *B* is optically active, it is chiral; therefore the butyl group must contain an asymmetric carbon stereocenter. Only a *sec*-butyl group meets this criterion. The structures of compounds *B* and *A* are therefore

2-methylbutanoic acid
(compound *B*)

1-phenylethyl 2-methylbutanoate
(compound *A*)

As noted in the problem, the absolute configurations of the asymmetric carbons cannot be determined from the data. Hence, compounds *B* and *C* are each one of two possible enantiomers, and compound *A* is the particular stereoisomer with the corresponding absolute configurations of its asymmetric carbons. Compound *D* is the racemate of compound *C*, and compound *E* is the diastereomer of compound *A* with the following configurations:

compound *E*

21.52   (a)   This is a reprise of Problem 10.65 on text p. 509, which is answered on p. 297 of this manual. The very weak acid HCN is not significantly dissociated. Consequently, the solution is insufficiently acidic to effect protonation of the —OH group of the alcohol. This protonation is necessary to convert this group into a good leaving group. Furthermore, there is virtually no cyanide ion ($^-$C≡N) present, and hence virtually no nucleophile to displace the —OH group.

    (b)   First of all, an excess of an alcohol is generally required to drive acid-catalyzed esterification to completion. However, even if some of the adipic acid is converted into its ester, there is no reason why this monoester would not be essentially as reactive as adipic acid itself, and the reaction mixture would ultimately contain a mixture of adipic acid, its monomethyl ester, and its dimethyl ester. The yield of the desired monoester would be poor, and it would have to be separated from both the di-ester and the unreacted adipic acid.

(c) There is no reason why acetic acid should form an anhydride only with benzoic acid, and vice-versa; substantial amounts of acetic anhydride and benzoic anhydride should also be obtained. If all the reactions were totally random, the desired unsymmetrical anhydride would be formed in a maximum of 50% yield.

(d) The hydroxide ion reacts *much* more rapidly with the O—H proton of the phenol than it does with the ester. (See Study Guide Link 20.1 on p. 659 of this manual for a discussion of this point.) Consequently, the one equivalent of hydroxide is consumed by this reaction, and no base is left to saponify the ester.

(e) Several functional groups in the β-lactam molecule are more reactive toward acid hydrolysis than the amide indicated. Because esters are more reactive than amides, the acetate ester will undoubtedly also hydrolyze under the reaction conditions. The β-lactam ring itself is an amide, and because hydrolysis of this amide relieves substantial ring strain, this amide should also hydrolyze more rapidly. Once the β-lactam hydrolyzes, the resulting enamine also should hydrolyze readily in acid, and the N—C—S linkage is also unstable toward hydrolysis in the same sense that an acetal is unstable. With so many faster competing processes, Klutz has no hope for a selective reaction.

21.53 Determine the structure of compound B and reason backwards. Compound B is a diol because it gives a diacetate derivative C. Because compound B is the diol that can be oxidized to β-methylglutaric acid, compound B is 3-methyl-1,5-pentanediol.

Compound A can be neither a hydroxy acid, a dicarboxylic acid, an oxo acid, nor an ester of these, because it contains only two oxygens. If it has only one carboxy group (either as a carboxylic acid or an ester) then it must contain either a ring or a carbon–carbon double bond. The structure of the LiAlH$_4$ reduction product B rules out a carbon–carbon double bond; hence, compound A contains a ring. Therefore, compound A is a lactone, and X is the sodium salt of the corresponding hydroxy acid, which is soluble in aqueous base. Compound A appears to dissolve when it is converted into the water-soluble compound X.

21.54 (a) The "alcohol" formed as one of the saponification products is an enol, which spontaneously reverts to the corresponding ketone. (See Eq. 14.5c on text p. 690.)

$$CH_3\overset{\overset{\displaystyle O}{\|}}{C}O^- \ Na^+ \ + \ HOC{=\!\!=}CH_2 \ \longrightarrow \ O{=\!\!=}C-CH_3$$

$$\underset{\text{Ph}}{|} \qquad\qquad \underset{\text{Ph}}{|}$$

an enol

initially formed
saponification products

(b)   Transesterification of the formate ester occurs.

$$CH_3\overset{\overset{\displaystyle O}{\|}}{C}CH_2OH \ + \ CH_3O\overset{\overset{\displaystyle O}{\|}}{C}H$$

(c)   The formula shows that the product includes the carbonyl group of phosgene (the acid chloride) minus two HCl molecules. Because the product has an unsaturation number of 7, a ring must be formed. Because both amines and carboxylic acids can react with acid chlorides, the product, called *isatoic anhydride*, is simultaneously a cyclic anhydride and a cyclic amide (lactam).

isatoic anhydride

(d)   Both amino groups are acylated by the excess of the acid chloride.

$$Ph-\overset{\overset{\displaystyle O}{\|}}{C}-NH-NH-\overset{\overset{\displaystyle O}{\|}}{C}-Ph$$

(e)   Both the amide and the nitrile are hydrolyzed.

$$PhCO_2H \ + \ H_3\overset{+}{N}(CH_2)_5CO_2H \ + \ \overset{+}{N}H_4$$

(f)   The cyclic amide is reduced to a cyclic amine.

(g)   The Grignard reactions of lactones are much like those of esters, except that the alcohol displaced by the nucleophilic acyl substitution reaction remains as part of the same molecule.

$$\underset{\underset{\displaystyle CH_3}{|}}{\overset{\overset{\displaystyle OH \qquad\quad OH}{|\qquad\quad |}}{CH_3CCH_2CH_2CH(CH_2)_4CH_3}}$$

(h)   Diethyl carbonate reacts first to give an ester, then a ketone, and then a tertiary alcohol (after protonolysis) in which all three groups attached to the α-carbon are the same.

$$C_2H_5O-\overset{\overset{\displaystyle O}{\|}}{C}-OC_2H_5 \ \xrightarrow{R-MgBr} \ R-\overset{\overset{\displaystyle O}{\|}}{C}-OC_2H_5 \ \xrightarrow{R-MgBr} \ R-\overset{\overset{\displaystyle O}{\|}}{C}-R \ \xrightarrow{R-MgBr} \ \xrightarrow{H_3O^+} \ R-\overset{\overset{\displaystyle OH}{|}}{\underset{\underset{\displaystyle R}{|}}{C}}-R$$

$$+ \ BrMg^+ \ ^-OC_2H_5 \qquad\qquad + \ BrMg^+ \ ^-OC_2H_5$$

From this pattern, the product of the reaction given is 3-ethyl-3-pentanol (R = ethyl in the preceding tertiary alcohol structure).

$$CH_3CH_2-\overset{\overset{\displaystyle OH}{|}}{\underset{\underset{\displaystyle CH_2CH_3}{|}}{C}}-CH_2CH_3 \qquad \text{3-ethyl-3-pentanol}$$

(i)   Lithium aluminum hydride reacts with acid chlorides in the same way that it reacts with esters, only faster. The product is $(CH_3)_3CCH_2OH$ (2,2-dimethyl-1-propanol).

(j)   Because aldehydes are much more reactive than esters, only the aldehyde reacts when only one equivalent of the Grignard reagent is used.

$$\underset{\text{methyl 5-hydroxy-5-phenylpentanoate}}{CH_3O\overset{\overset{\displaystyle O}{\|}}{C}(CH_2)_3\overset{\overset{\displaystyle OH}{|}}{C}HPh}$$

(k)   Transesterification occurs to give three equivalents of methyl stearate and one of glycerol.

$$\underset{\text{methyl stearate}}{3\ CH_3(CH_2)_{16}CO_2CH_3}\quad +\quad \underset{\text{glycerol}}{\overset{\overset{\displaystyle OH\quad OH\quad OH}{|\qquad |\qquad |}}{CH_2-CH-CH_2}}$$

(l)   The ester group undergoes intramolecular aminolysis to form a γ-lactam. Such intramolecular nucleophilic substitution reactions are very common when five- and six-membered rings can be formed. This is another manifestation of the greater rate of intramolecular reactions (Sec. 11.8 of the text).

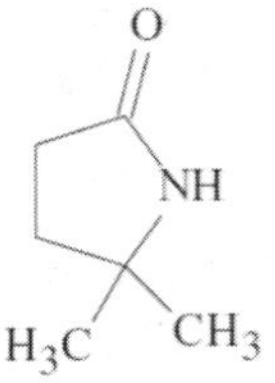

$+\quad CH_3OH$

21.55   (a)   The odd molecular mass indicates the presence of an odd number of nitrogens, and the hydroxamate test and the IR carbonyl absorption at 1733 cm^{-1} indicate the presence of an ester. The IR absorption at 2237 cm^{-1} indicates a nitrile. The triplet–quartet pattern in the NMR spectrum clearly indicates that compound *B* is an ethyl ester. Subtracting the masses of all the atoms accounted for leaves 14 mass units, the mass of a CH_2 group. Compound *A* is ethyl cyanoacetate.

$$N\equiv C-CH_2-\overset{\overset{\displaystyle O}{\|}}{C}-OCH_2CH_3$$
ethyl cyanoacetate (compound *A*)

(b)   From its IR absorption, compound *B* is an ester. The formula indicates an unsaturation number of 1, which is completely accounted for by the ester carbonyl group. The chemical shift and splitting of the δ 4.0 resonance indicates the part-structure —CH_2CH_2O—. The hint in the figure caption indicates that the δ 1.6 resonance is the other CH_2 of this part-structure; because this consists of six lines, then a more complete part-structure for this group is $CH_3CH_2CH_2O$—. The triplet at δ 0.95 is consistent with the methyl group of this part structure. This leaves an ethyl group unaccounted for; from its chemical shift, the CH_2 of the ethyl group is α to a carbonyl group. Compound *B* is propyl propionate.

$$CH_3CH_2-\overset{\overset{\displaystyle O}{\|}}{C}-OCH_2CH_2CH_3$$
propyl propionate
(compound *B*)

(c)   The odd molecular mass indicates the presence of nitrogen. The IR spectrum indicates the presence of both a nitrile and an alcohol. The broad, D_2O-exchangeable resonance at δ 3.4 in the NMR spectrum confirms the presence of an O—H group, and the pair of triplets indicates a —CH_2CH_2— group. Compound *C* is 3-hydroxypropanenitrile (β-hydroxypropionitrile).

$$HO-CH_2CH_2C\equiv N\quad \text{3-hydroxypropanenitrile}$$
(compound *C*)

(d)   The IR absorption indicates the presence of an ester. The doubled peaks in the mass spectrum indicate the presence of a single bromine. Consequently, compound *D* contains the elements CO_2 (from the ester) and Br. The remaining mass could correspond to C_4H_9. Therefore, adopt $C_5H_9O_2Br$ as a provisional formula. In the NMR spectrum, the triplet and the doublet at low chemical shift cannot be splitting each other, because their integrations are not consistent with such a situation. That is, a mutually split triplet (which indicates two adjacent hydrogens) and doublet (which indicates one adjacent hydrogen) must have an integral ratio of 1:2, not 1:1 as observed. Evidently, the two absorptions at δ 4.23 and δ 4.37 are the resonances that are split by the two triplets at lower chemical shift. From their chemical shifts, some of these protons are probably α to the Br and some are probably α to the oxygen. If the ester is an ethyl ester, then the high-field triplet is the resonance of the methyl group, and, from its integration, the doublet is also due to a methyl group on a carbon adjacent to a CH. These deductions conspire to identify compound *D* as ethyl 2-bromopropanoate.

$$CH_3CH - \overset{\overset{\displaystyle O}{\|}}{C} - OCH_2CH_3$$
$$|$$
$$Br$$

ethyl 2-bromopropanoate
(compound *D*)

(e) The mass spectrum shows an odd mass, which indicates the presence of nitrogen. The IR spectrum could indicate the presence of a nitrile, although, if so, the C≡N absorption is about 50 cm^{-1} lower in frequency than the C≡N absorption of an ordinary nitrile. The IR spectrum also indicates a trans alkene. The UV spectrum indicates extensive conjugation. This shift of the nitrile IR absorption is similar to that observed with conjugated carbonyl groups, and it suggests that the nitrile group is conjugated with the double bond. The NMR shows resonances for 7 protons. The large splitting in the resonances at δ 5.85 and δ 7.35 confirms the presence of the trans alkene. Evidently, the apparent singlet at δ 7.4 corresponds to five protons, that is, to a monosubstituted benzene ring. Compound *E* is (*E*)-3-phenylpropenenitrile.

$$\begin{array}{ccc} H & & C{\equiv}N \\ \backslash & & / \\ & C{=}C & \\ / & & \backslash \\ Ph & & H \end{array}$$

(*E*)-3-phenylpropenenitrile
(compound *E*)

(f) The unsaturation number is 5, and the NMR resonances in the δ 6.8–7.4 region indicate a para-substituted benzene ring. The broad resonance at δ 7.6 could be an amide N—H, and the IR absorption at 1659 cm^{-1} is confirmatory. The triplet–quartet pattern at δ 1.4 and δ 4.0 could indicate an ethyl ester, but we reject this idea because there is no ester carbonyl absorption. Hence, this quartet is evidently due to a *para*-CH$_3$CH$_2$O— substituent on the ring. The singlet at δ 2.1 is at a chemical shift consistent possibly with an *N*-methyl group or with protons α to a carbonyl group. Here are two possible structures:

$$CH_3CH_2O-\!\!\!\bigcirc\!\!\!-\overset{\overset{\displaystyle O}{\|}}{C}-NHCH_3 \qquad CH_3CH_2O-\!\!\!\bigcirc\!\!\!-NH-\overset{\overset{\displaystyle O}{\|}}{C}-CH_3$$

4-ethoxy-*N*-methylbenzamide　　　　　　**N-4-ethoxyphenylacetamide**
(*p*-ethoxyacetanilide)
(compound *F*)

The absence of splitting in the δ 2.1 resonance rules out the first structure, for which we would expect a doublet for the *N*-methyl group. Hence, the second structure is the correct one.

(g) The IR absorptions indicate the presence of a primary amide; the odd mass is consistent with the presence of nitrogen. The weak CMR resonances are probably due to carbons that do not bear hydrogens. The δ 180.5 resonance is the amide carbonyl group; the resonance at δ 38.0 is a quaternary carbon, which, from its chemical shift, could be α to the carbonyl carbon. Subtracting the mass of the carboxamide group from the mass of the molecule gives the remaining mass as 57 units, which is the mass of a C$_4$H$_9$ group. Only a *tert*-butyl group has the symmetry required to fit the CMR data. Compound *G* is 2,2-dimethylpropanamide (pivalamide).

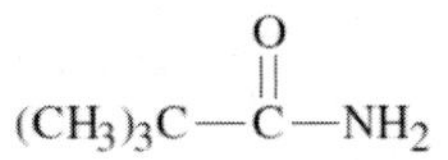

$$(CH_3)_3C—\overset{\overset{\displaystyle O}{\|}}{C}—NH_2$$

2,2-dimethylpropanamide (pivalamide)
(compound *G*)

21.56   (a)   PCC can be used as an oxidizing agent instead of CrO$_3$(pyridine)$_2$.

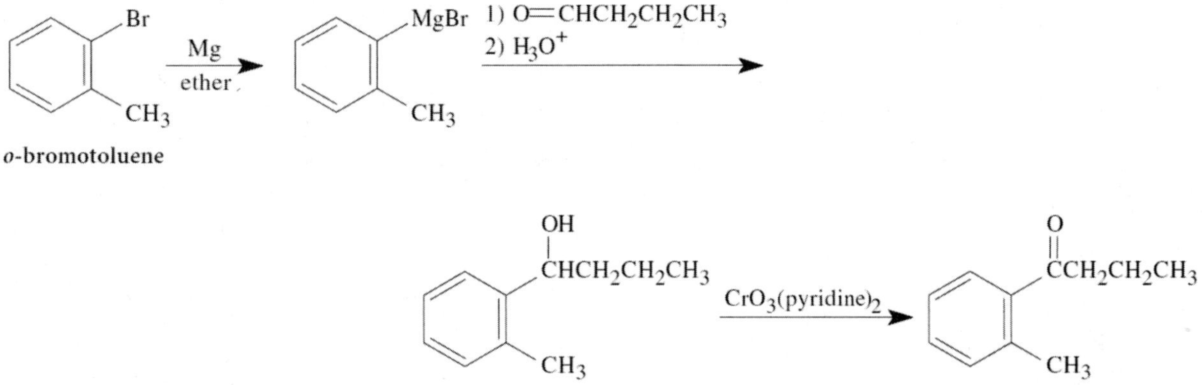

o-bromotoluene

o-methylbutyrophenone

(b)

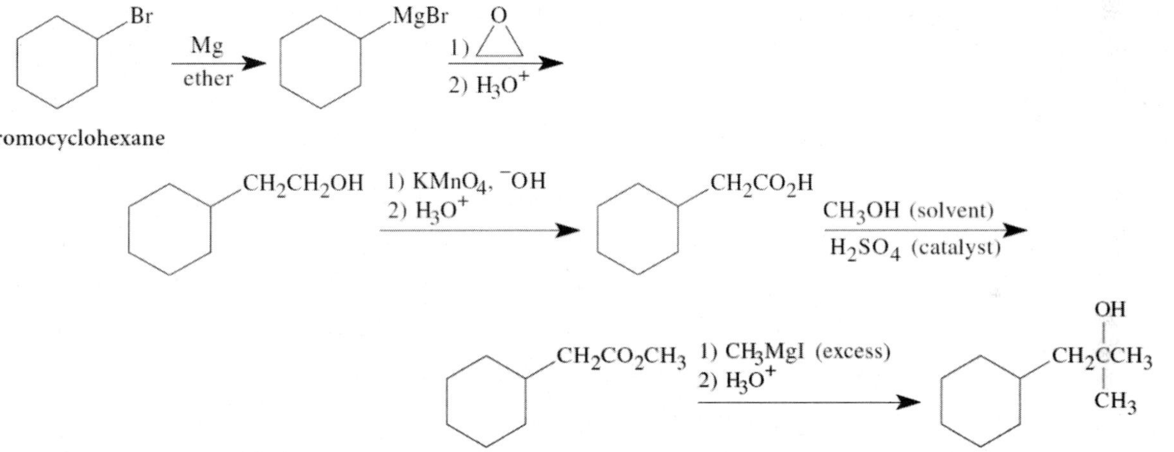

bromocyclohexane

1-cyclohexyl-2-methyl-
2-propanol

(c)

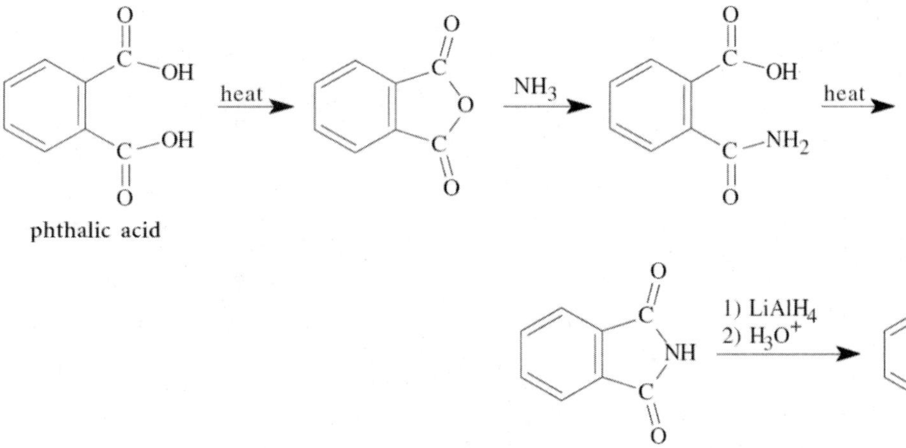

phthalic acid

(d)

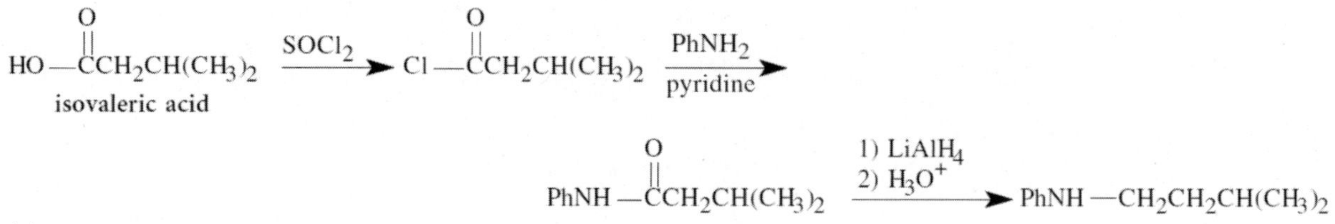

$$HO-\underset{O}{\overset{O}{C}}CH_2CH(CH_3)_2 \xrightarrow{SOCl_2} Cl-\underset{O}{\overset{O}{C}}CH_2CH(CH_3)_2 \xrightarrow[\text{pyridine}]{PhNH_2}$$

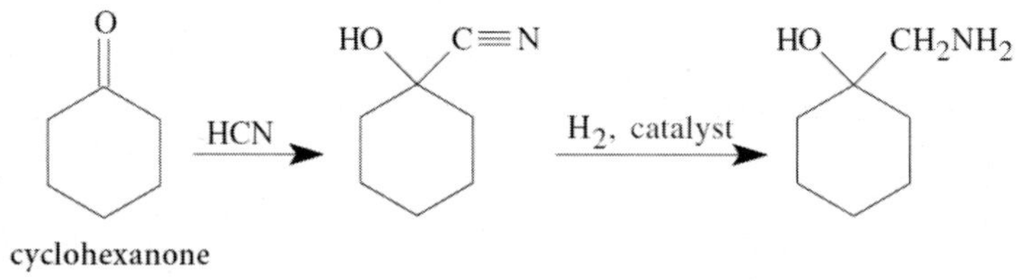

isovaleric acid

$$PhNH-\underset{O}{\overset{O}{C}}CH_2CH(CH_3)_2 \xrightarrow[\text{2) H}_3O^+]{\text{1) LiAlH}_4} PhNH-CH_2CH_2CH(CH_3)_2$$

(e)

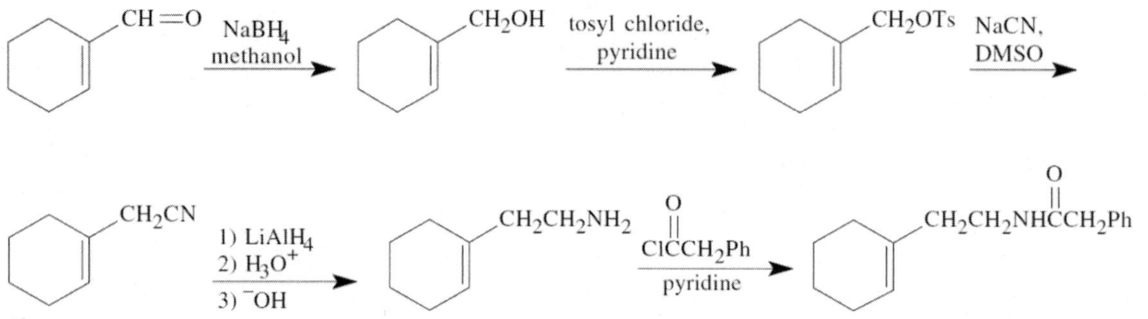

CH=O $\xrightarrow[\text{methanol}]{NaBH_4}$ CH₂OH $\xrightarrow[\text{pyridine}]{\text{tosyl chloride,}}$ CH₂OTs $\xrightarrow[\text{DMSO}]{\text{NaCN,}}$

CH₂CN $\xrightarrow[\substack{\text{2) H}_3O^+ \\ \text{3) }^-OH}]{\text{1) LiAlH}_4}$ CH₂CH₂NH₂ $\xrightarrow[\text{pyridine}]{\overset{O}{\overset{\|}{Cl}CCH_2Ph}}$ CH₂CH₂NHĈCH₂Ph

(f)

$\xrightarrow{HCN}$ (HO)(C≡N) $\xrightarrow{H_2, \text{ catalyst}}$ (HO)(CH₂NH₂)

cyclohexanone

(g)

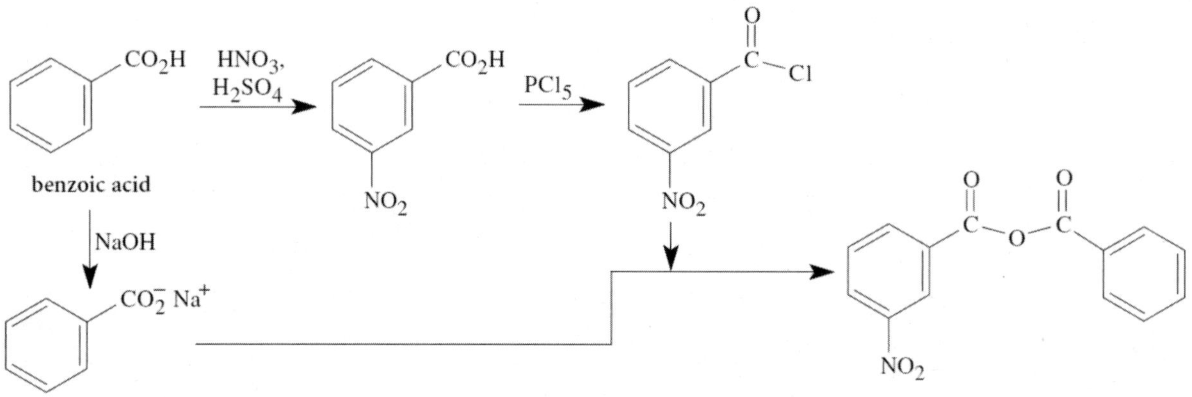

CO₂H $\xrightarrow[\text{H}_2SO_4]{HNO_3,}$ CO₂H (NO₂) $\xrightarrow{PCl_5}$ $\overset{O}{\overset{\|}{C}}$-Cl (NO₂)

benzoic acid

$\downarrow$ NaOH

CO₂⁻ Na⁺

$$\overset{O}{\overset{\|}{C}}-O-\overset{O}{\overset{\|}{C}}$$ (NO₂)

(h)

$$HO-\underset{O}{\overset{O}{C}}-CH_2CH_2CH_2-\underset{O}{\overset{O}{C}}-OC_2H_5 \xrightarrow{SOCl_2} Cl-\underset{O}{\overset{O}{C}}-CH_2CH_2CH_2-\underset{O}{\overset{O}{C}}-OC_2H_5 \xrightarrow[\text{(2) H}_3O^+]{\text{(1) (CH}_3)_2\text{CuLi}}$$

$$H_3C-\underset{O}{\overset{O}{C}}-CH_2CH_2CH_2-\underset{O}{\overset{O}{C}}-OC_2H_5 \xrightarrow[\text{CH}_3OH]{NaBH_4} H_3C-\underset{OH}{\overset{OH}{CH}}-CH_2CH_2CH_2-\underset{O}{\overset{O}{C}}-OC_2H_5 \xrightarrow[\text{H}_2O]{H_3O^+}$$

+ HOC₂H₅

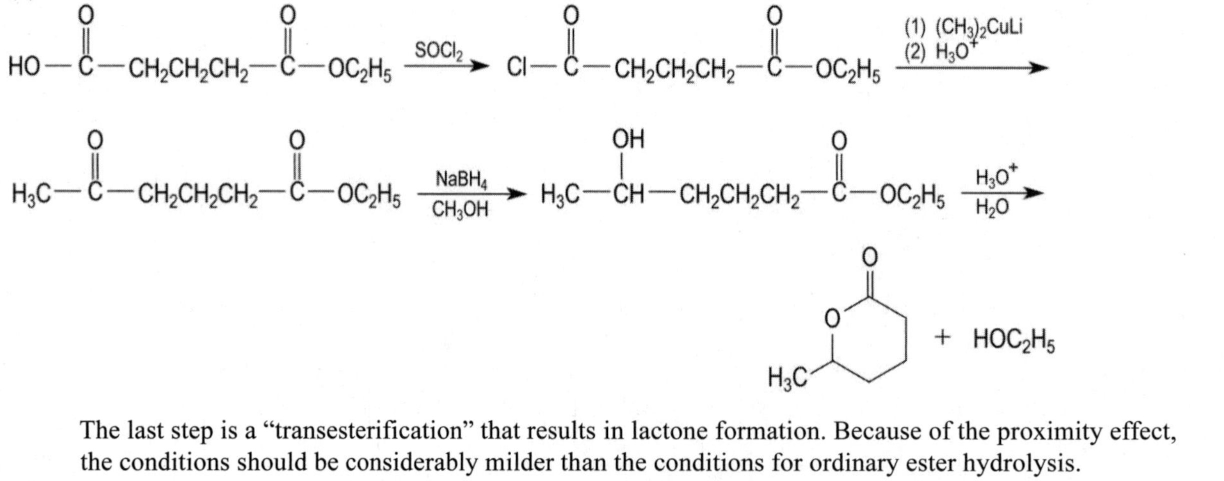

The last step is a "transesterification" that results in lactone formation. Because of the proximity effect, the conditions should be considerably milder than the conditions for ordinary ester hydrolysis.

(i)

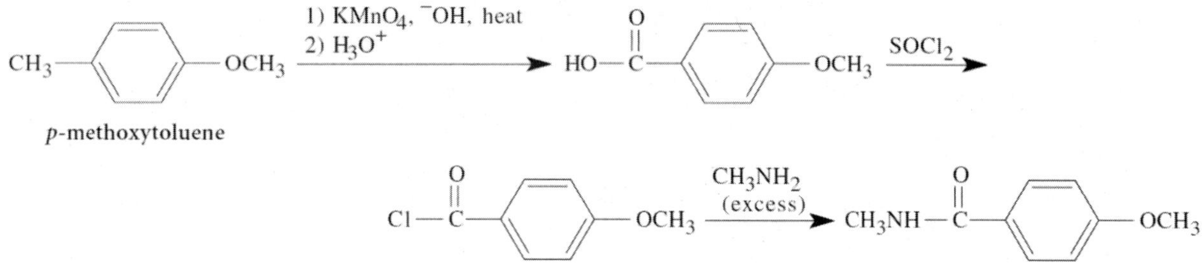

(j)

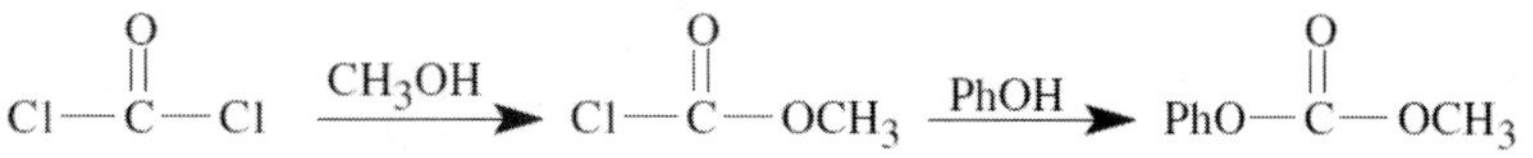

phosgene
(excess)

An excess of phosgene is used in the first step to ensure that only one chlorine is displaced. It is probably best to use methanol as the first nucleophile because phenyl esters are fairly reactive and could be transesterified by methanol if the order were reversed.

(k)

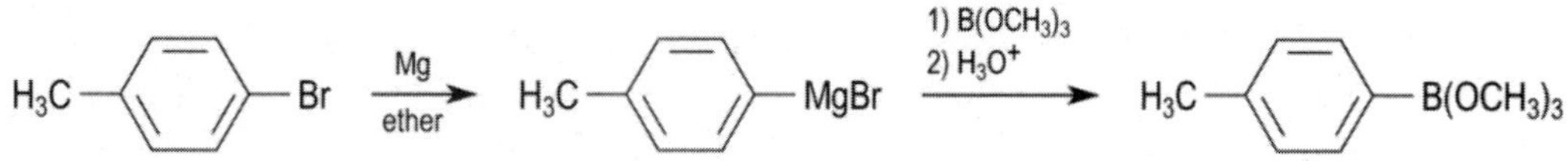

***p*-bromotoluene**

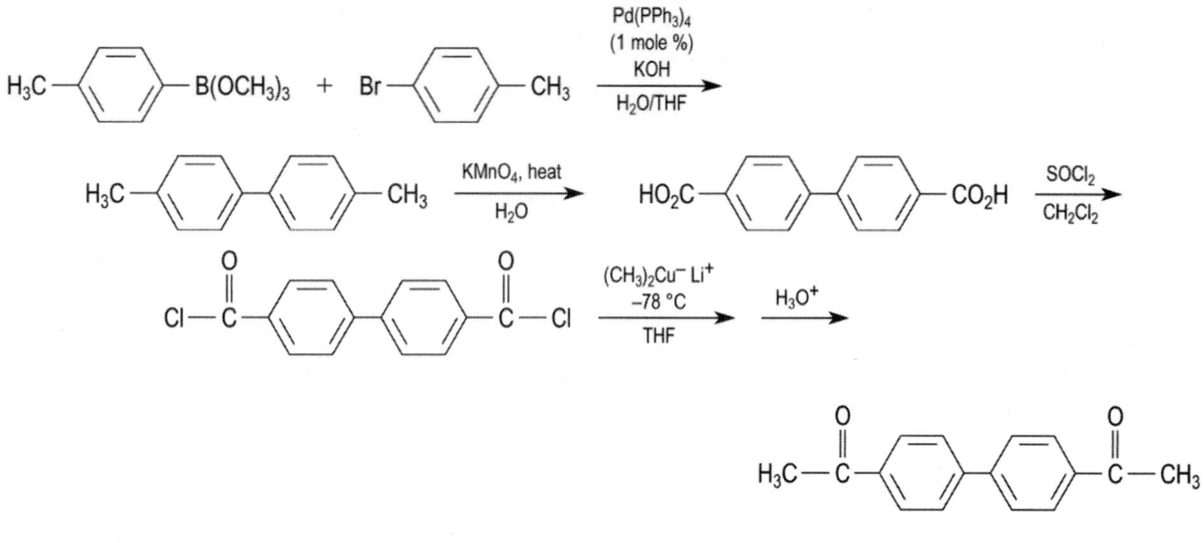

21.57 (a) The reversible hydration of carbon dioxide to carbonic acid is an acid-catalyzed addition of water that is mechanistically identical to the hydration of aldehydes and ketones shown in Eq. 19.29a–b, text p. 966.

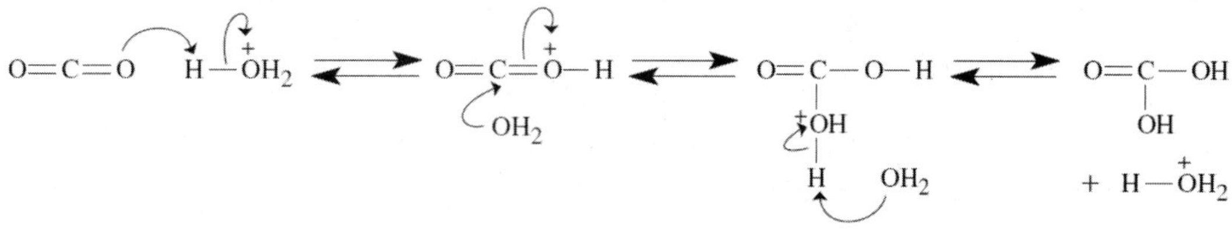

(b) The first step forms a Grignard reagent. The carbon-magnesium bond is highly polarized towards carbon, and reacts to form an alkene and displace the carboxylate leaving group.

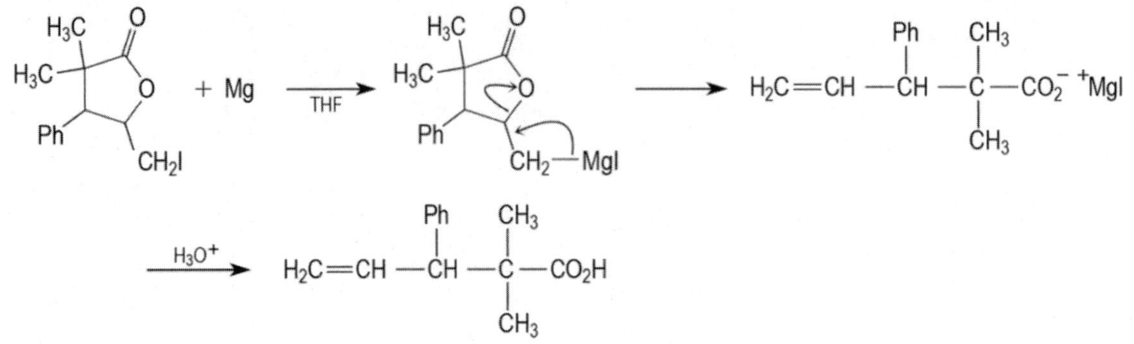

(c)   Protonation of the carbonyl oxygen produces a good leaving group which is displaced by bromide ion. The carboxy group is then esterified by the usual mechanism (Eqs. 20.18a–c, text pp. 1020–1).

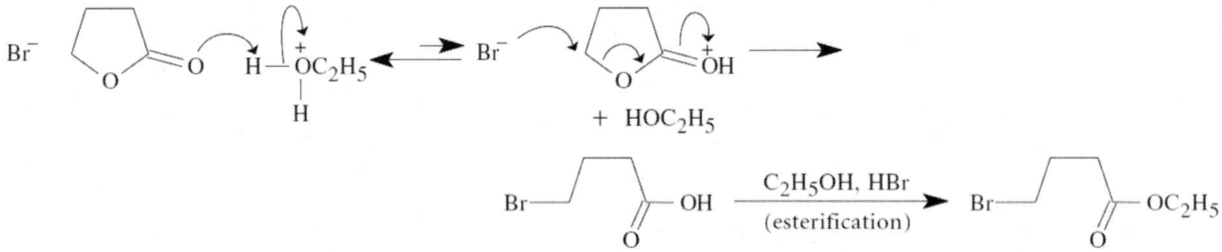

An alternative reasonable mechanism involves transesterification of the lactone by ethanol to give a hydroxy ester, followed by conversion of the alcohol to a bromide by an acid-catalyzed S_N2 reaction. (See Eqs. 10.21a–b, text p. 463.) However, it is likely that the mechanism shown above is the correct one, because the equilibrium between lactone and hydroxy ester is likely to strongly favor the lactone, and because a protonated carboxy group is an excellent leaving group.

(d)   A mercurinium ion forms by the standard mechanism (Sec. 5.4A), generating an equivalent of trifluoroacetate. The carbonyl electrons react with the alkene, which opens the mercurinium ion in a series of steps that are likely concerted, giving compound A. Sodium borohydride then exchanges the mercury for hydrogen.

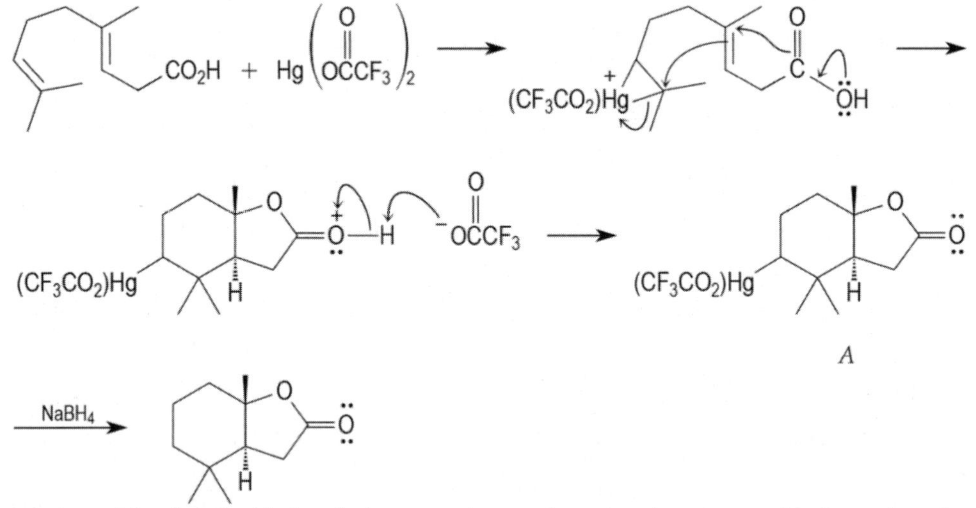

(e)   Protonation of the ring double bond gives a tertiary carbocation that reacts with the carbonyl oxygen of the ester to give the lactone.

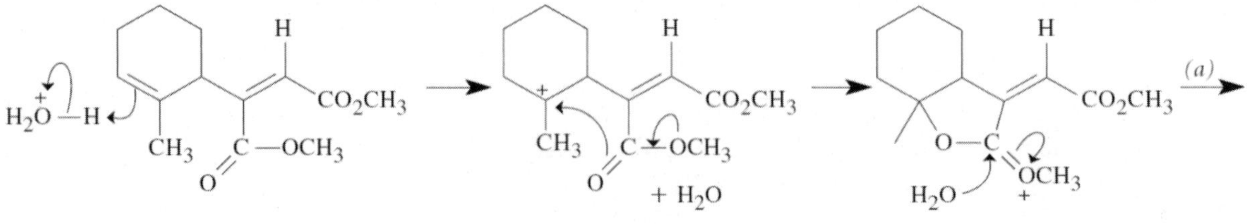

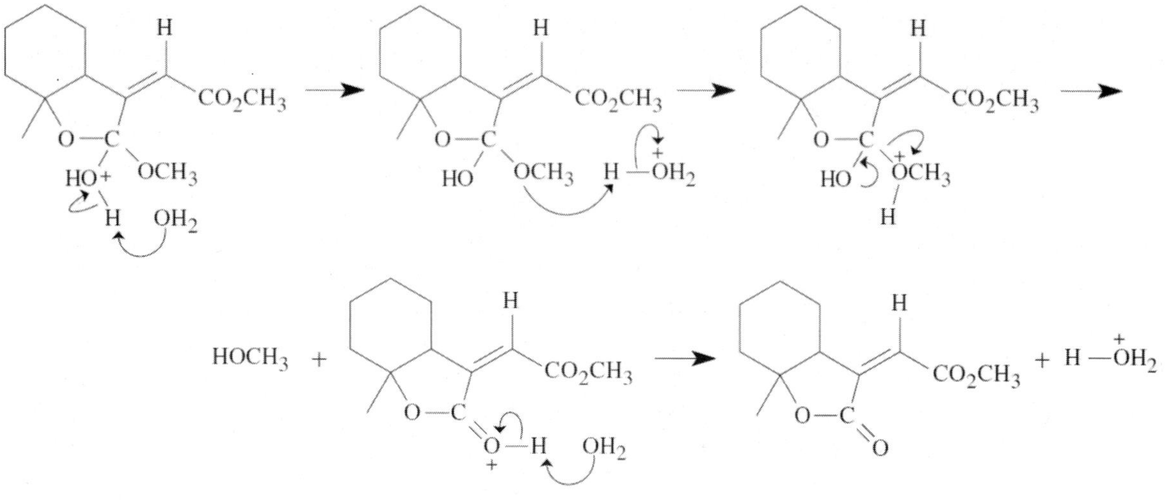

As an alternative to step *(a)* and the subsequent steps in the foregoing mechanism, water could react with the methyl group:

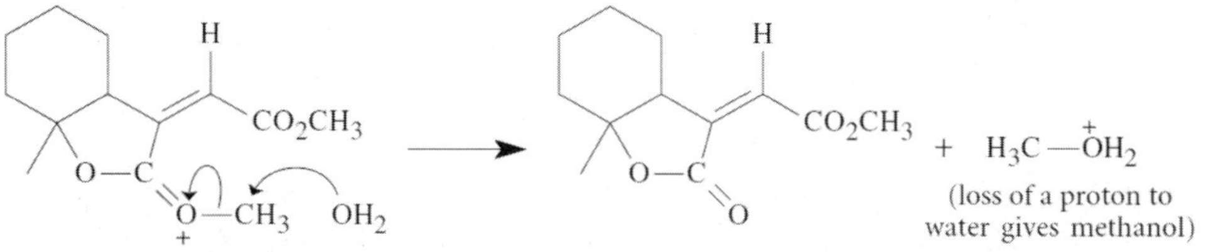

(loss of a proton to water gives methanol)

How might carrying out the reaction in ^{18}O-water enable one to distinguish between these two mechanisms?

21.58   Addition of the Grignard reagent to the carbon–nitrogen triple bond is expected by analogy with carbonyl-group reactions. The product *A* is an imine.

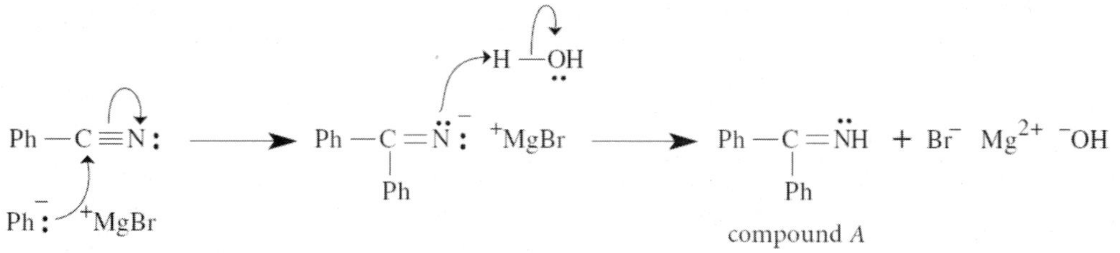

compound *A*

This imine hydrolyzes in acidic solution to benzophenone:

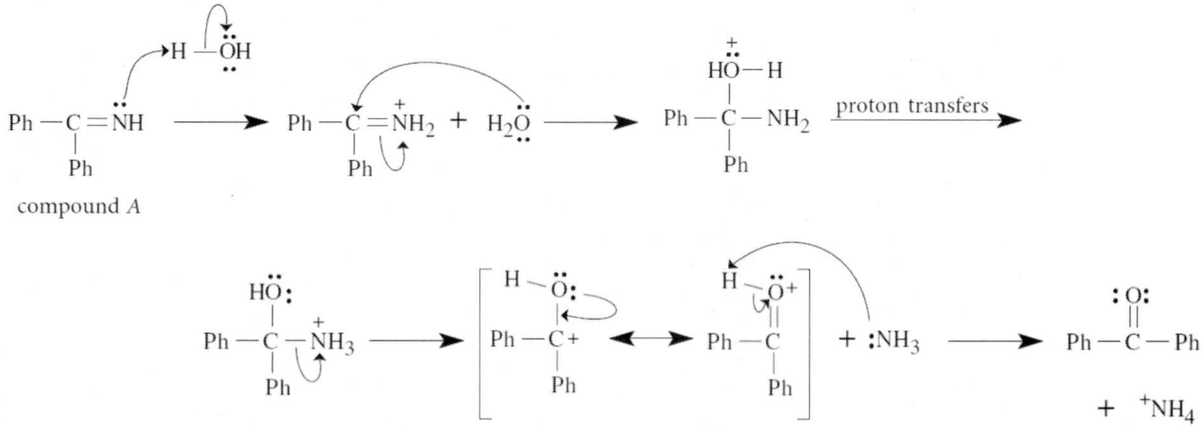

compound *A*

21.59 Compound *A* is an *O*-substituted oxime, which is formed by the mechanism summarized in Eqs. 19.60a–b on text p. 984. Compound *A* undergoes a base-promoted *β*-elimination in base to give the nitrile. (In the following mechanism, the 2,4-dinitrophenyl group is abbreviated Ar.)

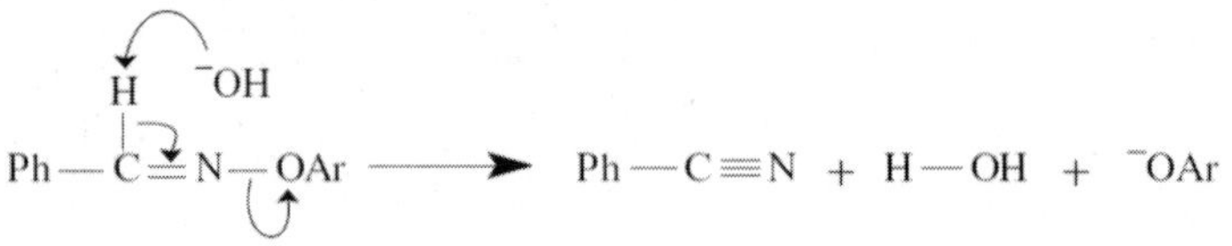
H
⁻OH
Ph—C=N—OAr
Ph—C≡N + H—OH + ⁻OAr
compound *A*

21.60 (a)

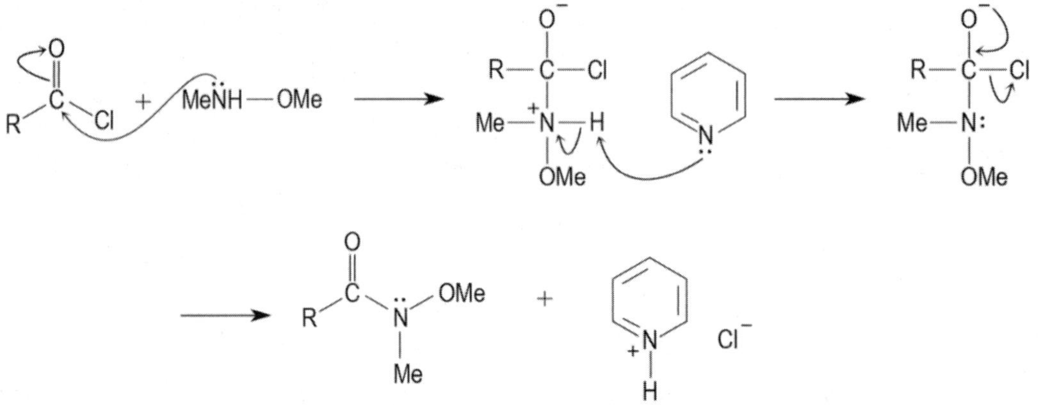
MeNH—OMe
OMe
Weinreb amide **pyridinium chloride**

(b) The first reason that this intermediate doesn't break down is because the leaving group is especially unstable.

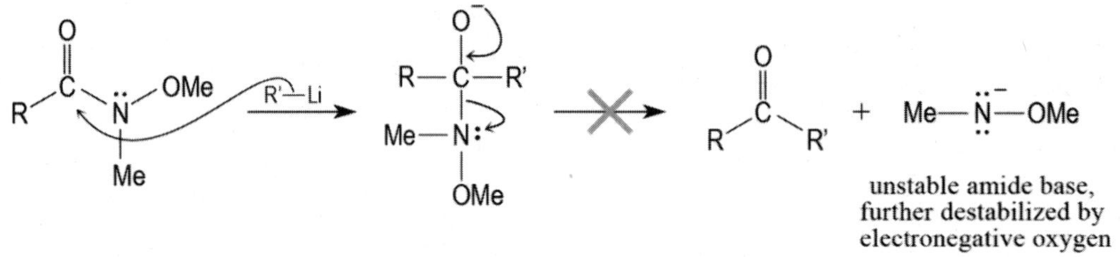
R'—Li
Me—N—OMe
unstable amide base,
further destabilized by
electronegative oxygen

Additionally, the lithium ion (or magnesium, if this were a Grignard reaction) complexes with the intermediate, stabilizing it, and helping it resist decomposition.

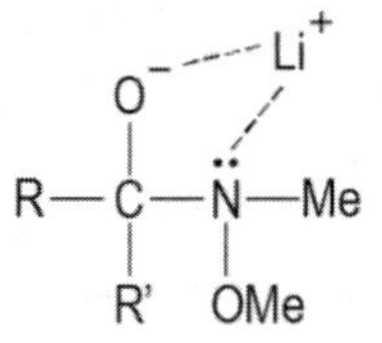

(c)

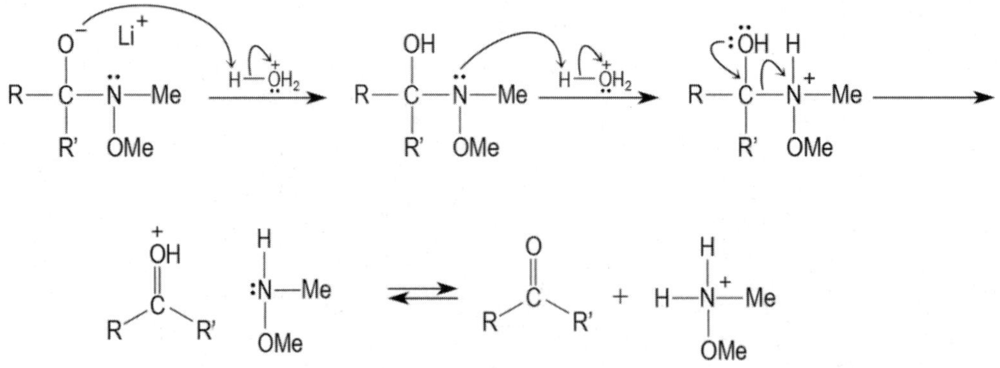

(d)

(e)   When the acid chloride reacts with a Grignard reagent, the reaction doesn't stop after just one addition; the ketone continues to react with the Grignard, which is usually present in excess.

21.61   Compound *A* is an oxime (see Table 19.3 on text p. 985). It is formed by the mechanism for imine formation summarized in Eqs. 19.60a–b on text p. 984. Protonation of the —OH group and loss of water *(a)* gives an electron-deficient nitrogen, a very unstable species. Subsequent rearrangement *(b)* gives a carbocation that is somewhat resonance stabilized. (We say "somewhat" because incorporation of a triple bond within a seven-membered ring introduces considerable strain.) This cation reacts with water, and, after several proton-transfer steps, ε-caprolactam is formed.

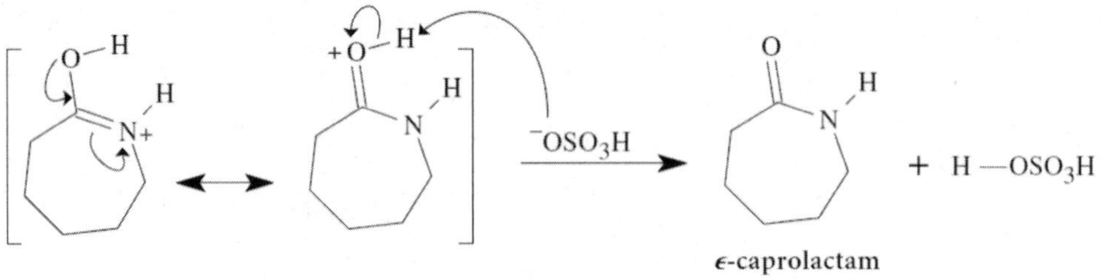

ϵ-caprolactam

Steps *(a)* and *(b)* occur in a single concerted step because a species containing an electron-deficient nitrogen is very unstable. However, two steps are shown above to demonstrate the driving force for the rearrangement.

21.62 (a) Because the carbonyl group has trigonal-planar geometry, it must undergo internal rotation about the bond to the ring so that its plane is perpendicular to the plane of the benzene ring in order to avoid significant van der Waals repulsions with the ortho methyl groups. This occurs even though conjugation is lost.

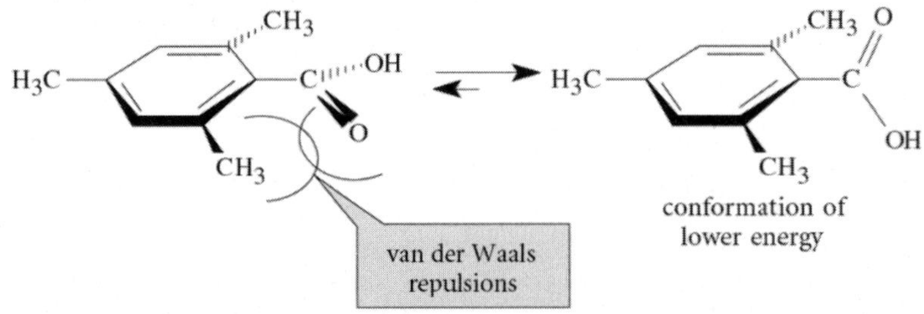

conformation of
lower energy

(b) Remember that nucleophiles approach a carbonyl carbon from above or below the plane of the carbonyl group. (See Fig. 21.4 on text p. 1063.) In the "perpendicular" conformation shown on the right, the path of a nucleophile to the carbonyl carbon is blocked on both sides by the ortho methyl groups. Because ester hydrolysis occurs by the nucleophilic reaction of a water molecule at the carbonyl carbon, this mechanism, and thus the hydrolysis reaction, cannot occur.

(c) For the same reason that ester *hydrolysis* cannot occur, ester *formation* also cannot occur by any mechanism that involves the reaction of a nucleophile at the carbonyl carbon. Because acid-catalyzed esterification involves the nucleophilic reaction of an alcohol molecule at the carbon of a protonated carbonyl group, this reaction does not take place. However, esterification with diazomethane can occur because the carboxylate oxygen reacts as a nucleophile with protonated diazomethane (see Eq. 20.21b on text p. 1023). The reaction is less susceptible to van der Waals repulsions because it takes place at an atom that is more remote from the ortho methyl groups.

21.63 (a) Phthalic anhydride can be formed by the nucleophilic reaction of a carboxy group at the carbonyl carbon of the protonated amide. (Note that the amide carbonyl oxygen is about 10^5 times as basic as the carbonyl oxygen of the carboxylic acid; see Eqs. 21.4a–b on text p. 1058.)

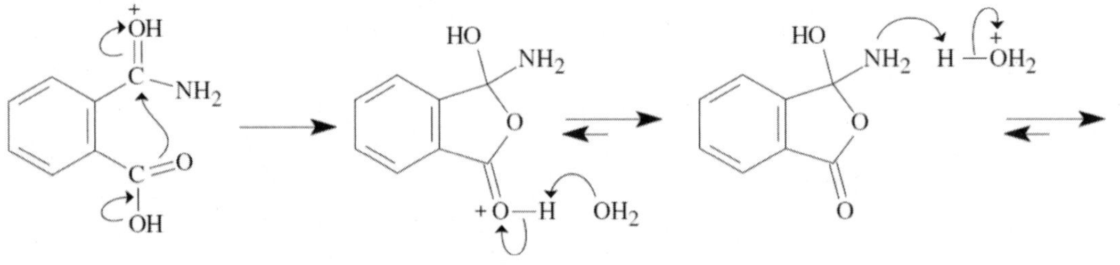

phthalic anhydride

(b)   The phthalic anhydride is labeled with ^{13}C on only one of its two carbonyl carbons. However, the two carbonyl groups, except for isotopic differences, are chemically indistinguishable and therefore react at essentially equal rates with isotopic water:

| from the reaction of isotopic water at carbonyl (*a*) | from the reaction of isotopic water at carbonyl (*b*) |

(c)   Formation of the anhydride intermediate must be faster than hydrolysis by the usual mechanism; otherwise it would not occur. Anhydride formation is an *intramolecular* reaction, whereas the ordinary hydrolysis mechanism is *intermolecular*. Intramolecular reactions that involve the formation of five-membered rings are usually much faster than their intermolecular counterparts. (The reasons for the advantage of intramolecular reactions are discussed in Sec. 11.8, text pp. 539–547.) Although the anhydride is formed very rapidly, it is also hydrolyzed rapidly because anhydrides are generally very reactive toward hydrolysis. Thus, the overall hydrolysis is accelerated because an anhydride intermediate is formed rapidly and then destroyed rapidly.

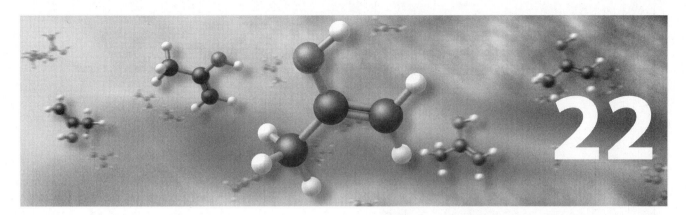

The Chemistry of Enolate Ions, Enols, and α,β-Unsaturated Carbonyl Compounds

STUDY GUIDE LINKS

22.1   Ionization versus Nucleophilic Reaction at the Carbonyl Carbon

One of the things you learned about Brønsted bases is that they can usually act as nucleophiles, and that nucleophiles react at the carbon atoms of carbonyl groups. Yet in Eqs. 22.3 and 22.4 of the text (p. 1104), bases are shown removing protons. You might be wondering, "What is the major reaction: nucleophilic reaction at the carbonyl carbon, or removal of an α-hydrogen to give an enolate ion?" The answer is that both can occur. This section of the text focuses on the proton removal, but later sections of this chapter show the interplay between both types of reactions. Nevertheless, it might help you to see just one example here.

Suppose an ester is treated with aqueous NaOH. Recall (Sec. 21.7A) that NaOH saponifies esters. Saponification results from a nucleophilic reaction of hydroxide ion at the carbonyl carbon of an ester. However, while saponification is taking place, hydroxide ion removes the α-protons of a few ester molecules to form some enolate ions. This reaction has no consequences for saponification because saponification occurs much more rapidly than the reactions of enolate ions (many of which are discussed in this chapter). For this reason, it was not important to consider enolate-ion formation in the discussion of saponification.

The point of this chapter is to focus on situations in which enolate ion formation is significant, and ultimately on what happens when *both* proton removal and nucleophilic reaction at a carbonyl carbon can occur.

22.2   Kinetic versus Thermodynamic Stability of Enols

When we think of unstable organic compounds we often think of molecules containing strained rings or molecules containing van der Waals repulsions. The structures of enols have no such destabilizing features. Yet enols in most cases cannot exist for significant lengths of time, while other less stable molecules can be put in bottles and stored for years. Why this seeming contradiction?

The essence of this problem is the meaning of the word *stability* as applied to chemical substances. *Thermodynamic stability* is a term used to describe the relative energy contents of molecules. In general, thermodynamic stability has very little to do with whether a compound can be isolated. *Kinetic stability* is a term used to describe how rapidly molecules revert to more stable molecules. In order for a molecule to be *kinetically unstable*, a mechanism must exist for its conversion into a more stable molecule. For a compound to be kinetically unstable, it must be thermodynamically unstable, but the converse need not be true. Some *thermodynamically unstable compounds* are easy to isolate because they are *kinetically stable*. For example, almost all hydrocarbons in the presence of oxygen are thermodynamically unstable relative to CO_2 and H_2O by hundreds of kJ mol^{-1}; yet hydrocarbons do not rapidly revert to CO_2 and H_2O because there is no readily available mechanism by which this conversion can be accomplished. If we supply a flame, however, such a conversion mechanism becomes available—we call it *combustion*—and we see very graphically how unstable hydrocarbons really are!

Examples of kinetic stability and instability in carbonyl chemistry are provided by acetals (Sec. 19.10, text pp. 978–983) and carbonyl hydrates (Sec. 19.7B, text pp. 967–969). The hydrate of cyclohexanone is both thermodynamically and kinetically unstable with respect to the ketone and water:

HO   OH   O

(cyclohexane ring) ⇌ **fast** → (cyclohexanone ring)   + H_2O

the hydrate   cyclohexanone
of cyclohexanone

Yet replacing the O—H hydrogens with alkyl groups gives compounds called *acetals* that are thermodynamically unstable, but *kinetically stable*, in neutral and basic solution.

CH$_3$O   OCH$_3$ O

H_2O + (cyclohexane ring) ⇌ **too slow to measure** → (cyclohexanone ring)   + 2 CH$_3$OH

cyclohexanone   cyclohexanone
dimethyl acetal

(1,1-dimethoxy-
cyclohexane)

The base-catalyzed mechanism of hydrate decomposition requires the presence of an O—H hydrogen, which acetals lack. Hence, the base-catalyzed pathway of hydrate decomposition cannot operate with acetals. Therefore, acetals, although thermodynamically unstable relative to their corresponding carbonyl compounds, are kinetically stable in basic solution. This is why acetals are good protecting groups for carbonyl groups under basic or neutral conditions. Under acidic conditions, however, acetals are kinetically unstable because an acid-catalyzed mechanism exists for their conversion into ketones.

The thermodynamic stabilities of most enols are not much less than the stabilities of their corresponding carbonyl isomers—typically, about 30–40 kJ mol^{-1} (7–10 kcal mol^{-1}). What makes enols difficult to isolate is their *kinetic* instabilities. Simple paths, or mechanisms, exist for their rapid transformation into carbonyl compounds in the presence of acids or bases; these mechanisms are shown in Eqs. 22.17a–b in the text (p. 1112). Enols are kinetically unstable molecules because the conditions under which they are converted into carbonyl compounds—the presence of acids and/or bases—are so prevalent. It's hard to generate an enol without an acid or base around. In fact, enols themselves are weak acids, and can catalyze their own slow conversion into ketones. Enols simply don't stand much of a chance for survival.

In the mid-1970s, chemists figured out ways to generate simple enols in solution in the absence of other acids and bases and, indeed, found that they could be isolated and observed for significant periods of time.

22.3   Dehydration of β-Hydroxy Carbonyl Compounds

At first sight, the dehydration of β-hydroxy carbonyl compounds would appear to be an ordinary acid-catalyzed dehydration as discussed in Sec. 10.2 of the text. Recall that, in the mechanism of such a dehydration, protonation of the hydroxy group is followed by a Lewis acid–base dissociation to form a carbocation and water:

Loss of a C—H proton from carbocation *A* gives the α,β-unsaturated ketone. However, a different mechanism—and, as it turns out, the correct one—is possible. This mechanism involves dehydration not of the ketone, but instead of the *enol*.

an enol form

The dehydration involves protonation of the —OH group, formation of carbocation *B*, and loss of a proton from this carbocation:

The reason the enol mechanism is favored is that *carbocation B is much more stable than carbocation A.* Can you see why? Draw resonance structure(s) for carbocation *B*. Is carbocation *A* resonance-stabilized? What is the *polar effect* of the carbonyl group on the stability of carbocation *A*?

22.4   Understanding Condensation Reactions

Condensations such as the aldol condensation and its variants, as well as the other reactions you'll encounter in Chapter 22, present a particular challenge to the beginning student, because the products are somewhat more complex than the products of many typical organic reactions, and because more than one functional-group transformation is involved (an addition followed by a dehydration). The way to develop an understanding of these reactions is to write the mechanism of each new reaction you see. First, identify the sources of the atoms in the product. (See suggestions 1–3 in Problem 4.65, text p. 179; the discussion in Sec. 22.4D of the text should also be useful.) Use the base-catalyzed reaction of acetaldehyde (Eqs. 22.40a–b, text p. 1120, and Eq. 22.44, text p. 1121) and the acid-catalyzed reaction of acetone (Eqs. 22.46a–c, text pp. 1122–1123) as models. If you still have difficulty, *seek assistance!* You should have taken this approach with the crossed-aldol products in Study Problem 22.1 on text p. 1124. Be sure you understand the reactions shown in Eqs. 22.47 and 22.48 on text pp. 1124–1125. If you work through these examples, the reaction in Eq. 22.53 on text p. 1128 should be clear. The only thing different in this case is that the condensation occurs internally. Which carbons of the starting material are joined in the product? What is the structure of the enol intermediate involved? Use Eq. 22.46c on text p. 1123 to help you understand the formation of the carbon–carbon bond.

Just as in a sport, or in music, or in any other worthwhile endeavor, improving your skill requires practice. If you are concerned or discouraged that these transformations are not immediately obvious, don't be! They are more likely to become easier to understand, however, if you'll take the approach described here.

22.5 Variants of the Aldol and Claisen Condensations

Many variations of the aldol and Claisen condensations exist. The text has considered a few: the Claisen–Schmidt, Reformatsky, and Knoevenagel variations of the aldol condensation, and the Dieckmann variation of the Claisen condensation. Possibly you'll hear of the Perkin condensation, the Stobbe condensation, or other variations that have no name. To cover all these in a text of reasonable (or even unreasonable) size is neither possible nor even necessary. For if you master the principles of reactivity contained in Chapters 21 and 22, understanding variations of these condensation reactions will only be a matter of re-applying the principles. In fact, mastering these principles might even allow you to invent your own condensation reaction!

22.6 Further Analysis of the Claisen Condensation

If you imagine the Claisen condensation on an ester of general structure A, you'll see that saponification and decarboxylation *must* yield a *symmetrical ketone* of general structure C. (A symmetrical ketone is a ketone in which the two groups attached to the carbonyl carbon are the same.)

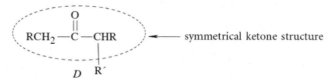

(Be sure to go through this case in detail to verify the foregoing statement.) Alkylation of the Claisen product B can lead to a variety of *substituted derivatives* of C. For example, treatment of B with NaOEt, then a general alkyl halide R'—I, would give, after saponification, acidification, and decarboxylation, a general derivative D.

$$RCH_2 - \overset{\overset{\textstyle O}{\|}}{C} - \underset{\underset{\textstyle R'}{|}}{\overset{}{C}}HR \qquad \longleftarrow \text{ symmetrical ketone structure}$$

$$D$$

Notice the symmetrical ketone structure "lurking" as the circled part of D.

When you are asked to prepare a ketone by the Claisen condensation, look for such symmetrical-ketone "substructures" within the target molecule. Problems 22.40 and 22.41 (text p. 1152) give you practice in applying this strategy.

22.7   Synthetic Equivalents in Conjugate Addition

Study Guide Links 17.1 (p. 507 of this manual) and 20.4 (p. 661 of this manual) discussed the notion of *synthetic equivalents* (and the corresponding property, *synthetic equivalence*), that is, thinking of one group in terms of other groups from which it might be formed. This notion is particularly useful in planning conjugate additions. In Study Problem 22.6 on text p. 1164, for example, the di(ethoxycarbonyl)methyl anion ($^-$:CH(CO$_2$Et)$_2$, the conjugate base of diethyl malonate,) is used as the *synthetic equivalent* of a carboxymethyl anion ($^-$:CH$_2$CO$_2$H), which itself cannot exist. (Why can't this anion exist?).

What are some other synthetic equivalents that are useful in conjugate-addition reactions?

$$\text{:CN is equivalent to} \quad \overset{\displaystyle O}{\underset{\displaystyle}{^-\text{:C}}}\text{—OH} \qquad\qquad \text{(see Eq. 22.99 on text p. 1157)}$$

$$\text{:CN is also equivalent to :CH}_2\text{—NH}_2 \qquad\qquad \text{(see Sec. 21.9C on text p. 1082)}$$

$$\underset{\displaystyle \text{CO}_2\text{Et}}{\overset{\displaystyle O}{^-\text{:CH—C—R}}} \text{ is equivalent to } \overset{\displaystyle O}{^-\text{:CH}_2\text{—C—R}} \qquad \text{(see Eq. 22.87 on text p. 1150)}$$

(You should show how each group can be converted into the other, for example, how a —C≡N group can be converted into a —CH$_2$NH$_2$ group.)

When you can start thinking in terms of synthetic equivalents you really have started to master the art of organic synthesis.

FURTHER EXPLORATIONS

 ## 22.1   Malonic Ester Alkylation

In Eq. 22.77b of the text (p. 1146), a *secondary* alkyl halide is used to alkylate the enolate ion of malonic ester. When you studied the S_N2 reaction, you learned that a side reaction is E2 elimination, and that this side reaction is particularly prevalent with secondary and tertiary alkyl halides. Because the product of Eq. 22.64b is formed in high yield, elimination is evidently not a problem in this case, and one might ask why.

Recall (Sec. 9.5G) that, although most strong bases are good nucleophiles, a number of relatively weak bases are also good nucleophiles, and that branched or secondary alkyl halides undergo S_N2 reactions with such nucleophiles with little interference from E2 reactions. (Note particularly the example in Eq. 9.48 on text p. 416.) The enolate ion of diethyl malonate is in this category. Chemists have found that delocalized (that is, resonance-stabilized) carbanions such as enolate ions are particularly good nucleophiles (for reasons that are not completely understood). Consequently, the S_N2 reaction is particularly rapid for these anions, and the E2 reaction does not interfere unless highly branched secondary or tertiary alkyl halides are used.

 ## 22.2   Alkylation of Enolate Ions Derived from Ketones

The acetoacetic ester synthesis requires the presence of the ester group of a β-keto ester; this group is then "thrown away" by hydrolysis and decarboxylation at the end of the synthesis and replaced by a hydrogen. For the same reasons that carboxylic acids with quaternary α-carbons cannot be prepared by the malonic ester synthesis, ketones with quaternary α-carbons cannot be prepared by the acetoacetic ester synthesis. For example, suppose we want to prepare 2,2-dimethylcyclohexanone:

quaternary carbon

2,2-dimethylcyclohexanone

Although 2-methylcyclohexanone is easily prepared by a Dieckman condensation–alkylation sequence, introduction of the second methyl group is not possible in such a reaction scheme.

Why not, then, resort to the strategy used in Sec. 22.8B of the text? 2-Methylcyclohexanone could be alkylated *directly* by forming its enolate ion with a strong base such as LDA or LCHIA and treatment of that enolate with methyl iodide:

Can you see the problem with this approach? With an ester, only one enolate ion is possible. But with a ketone, in general *two* enolate ions are possible:

The problem, then, is how to form one enolate and not the other, and to keep the two enolate ions from equilibrating once the desired one is formed.

But the problems don't stop here. Suppose a way could be found to form the desired enolate ion. An additional problem is that enolate ions can react in more than one way. The enolate ions of ketones are not only nucleophilic at carbon; they are also nucleophilic at *oxygen*:

Alkylation at oxygen gives a vinyl ether—sometimes called an *enol ether,* because conceptually it is derived by replacement of the —OH hydrogen of an enol with an alkyl group. In general, both products are formed. (Alkylation of oxygen is not usually observed with esters because enols of esters—and enol ethers of esters—are *much* less stable relative to their parent esters than enols of ketones are; see Eq. 22.13 on text p. 1110.)

Direct alkylation of ketones, then, is a much more complex problem than alkylation of esters. Nevertheless, chemists have discovered ways to control, in many cases, which enolate ion is formed, and to control which alkylation product is produced. (The selective formation of enolates was discussed in Section 22.4C, text p. 1123.) These solutions, although beyond the scope of an introductory text, are conceptually not difficult to understand if you understand the principles of enolate-ion chemistry.

22.3   Conjugate Addition of Organocuprate Reagents

Although the mechanism presented in the text is conceptually useful, it does not explain the effect of copper in promoting conjugate addition. Because carbonyl addition rather than conjugate addition is observed in the *absence* of copper, it is evident that copper has a special effect. A substantial body of evidence has shown that conjugate-addition reactions of lithium organocuprate reagents involve free-radical intermediates, and that conjugate addition is a consequence of a free-radical mechanism.

REACTION REVIEW

I. α-HALOGENATION OF CARBONYL COMPOUNDS

A. ACID-CATALYZED α-HALOGENATION OF ALDEHYDES AND KETONES

1. Halogenation of an aldehyde or ketone in acidic solution usually results in the replacement of one α-hydrogen by halogen.

   a. Enols are reactive intermediates in these reactions.

   b. Enols add only one halogen atom.

2. The rate law implies that even though the reaction is a halogenation, the rate is independent of the halogen concentration.

$$\text{rate} = k[\text{ketone}][H_3O^+]$$

   a. Halogens are not involved in the transition state for the rate-limiting step of the reaction.

   b. Enol formation is the rate-limiting process in acid-catalyzed halogenation of aldehydes and ketones.

3. Introduction of a second halogen is much slower than introduction of the first halogen.

B. HALOGENATION OF ALDEHYDES AND KETONES IN BASE; THE HALOFORM REACTION

1. Halogenation of aldehydes and ketones with α-hydrogens also occurs in base; all α-hydrogens are substituted by halogen.

2. When the aldehyde or ketone starting material is either acetaldehyde or a methyl ketone, the trihalomethyl carbonyl compound is not stable under the reaction conditions, and it reacts further to give, after acidification of the reaction mixture, a carboxylic acid and a haloform.

   a. This reaction is called the haloform reaction.

   b. A carbon–carbon bond is broken.

3. The mechanism of the haloform reaction involves the formation of an enolate ion as a reactive intermediate.

   a. The enolate ion reacts as a nucleophile with halogen to give an α-halo carbonyl compound.

   b. A dihalo and trihalo carbonyl compound are formed more rapidly in successive halogenations.

   c. A carbon–carbon bond is broken when the trihalo carbonyl compound undergoes a nucleophilic acyl substitution reaction.

      i. The leaving group in this reaction is a trihalomethyl anion, which is much less basic than ordinary carbanions.

      ii. The trihalomethyl anion reacts irreversibly with the carboxylic acid by-product to drive the overall haloform reaction to completion.

      iii. The carboxylic acid itself can be isolated by acidifying the reaction mixture.

4. The haloform reaction can be used to prepare carboxylic acids from readily available methyl ketones.
5. The haloform reaction is used as a qualitative test for methyl ketones, called the iodoform test.
   a. A compound of unknown structure is mixed with alkaline I_2.
   b. A yellow precipitate of iodoform (HCI_3) is taken as evidence for a methyl ketone (or acetaldehyde).
   c. Alcohols of the structure $R-\overset{\overset{\displaystyle OH}{|}}{C}H-CH_3$ also give a positive iodoform test because they are oxidized to methyl ketones (or to acetaldehyde, in the case of ethanol) by the basic iodine solution.

C. α-BROMINATION OF CARBOXYLIC ACIDS: THE HELL–VOLHARD–ZELINSKY REACTION

1. A bromine is substituted for an α-hydrogen when a carboxylic acid is treated with Br_2 and a catalytic amount of red phosphorus or PBr_3; this reaction is called the Hell–Volhard–Zelinsky (HVZ) reaction.
2. The first stage in the mechanism of this reaction is the conversion of the carboxylic acid into a small amount of acid bromide by the catalyst PBr_3.
   a. The acid bromide enolizes in the presence of acid; this reaction is similar to the acid-catalyzed bromination of ketones.
   b. The enol of the acid bromide is the species that actually brominates.

   i. When a small amount of PBr_3 catalyst is used, the α-bromo acid bromide reacts with the carboxylic acid to form more acid bromide, which is then brominated.
   ii. If one full equivalent of PBr_3 catalyst is used, the α-bromo acid bromide is the reaction product.
   iii. The reaction mixture can be treated with an alcohol to give an α-bromo ester.

D. REACTIONS OF α-HALO CARBONYL COMPOUNDS

1. Most α-halo carbonyl compounds are very reactive in S_N2 reactions, and can be used to prepare other α-substituted carbonyl compounds.
2. In the case of α-halo ketones, nucleophiles used in these reactions must not be too basic.
   a. Stronger bases promote enolate-ion formation; and the enolate ions of α-halo ketones undergo other reactions.
   b. More basic nucleophiles can be used with α-halo acids because, under basic conditions, α-halo acids are ionized to form their carboxylate conjugate-base anions, and these do not undergo the side-reactions that α-halo ketones undergo.
3. α-Halo carbonyl compounds react so slowly by the S_N1 mechanism that this reaction is not useful; reactions that require the formation of carbocations alpha to carbonyl groups generally do not occur.

II. CONDENSATIONS INVOLVING ENOLATE IONS AND ENOLS OF ALDEHYDES AND KETONES

A. BASE-CATALYZED ALDOL ADDITION AND CONDENSATION REACTIONS

1. In aqueous base, aldehydes and ketones with α-hydrogens undergo a reaction called the aldol addition.

$$2 \ \underset{H}{\overset{O}{-\!\!C\!\!-\!\!C\!\!-\!\!R}} \ \underset{\xleftarrow{\hspace{1cm}}}{\xrightarrow{H_2O, \ ^-OH}} \ \underset{H \quad R}{\overset{OH \quad O}{-\!\!C\!\!-\!\!C\!\!-\!\!C\!\!-\!\!C\!\!-\!\!R}}$$

an aldehyde or ketone   a β-hydroxy aldehyde or ketone

  a. The term aldol is both a trivial name for 3-hydroxybutanal and a generic name for β-hydroxy aldehydes.
  b. The aldol addition is a very important and general reaction of aldehydes and ketones that have α-hydrogens.
  c. The aldol addition is another nucleophilic addition to a carbonyl group.
2. The base-catalyzed aldol addition involves an enolate ion as an intermediate; the enolate ion adds to a second molecule of aldehyde or ketone.

$$^-OH \quad \underset{CH_2-C-H}{\overset{H \quad O}{}} \ \xrightleftharpoons{} \ H_2O \ + \ \left[\ \underset{}{\overset{O}{H_2\bar{C}-C-H}} \ \longleftrightarrow \ \underset{}{\overset{O^-}{H_2C=C-H}} \ \right]$$

an enolate ion

$$\underset{}{\overset{O}{H_3C-C-H}} \ \underset{}{\overset{O}{H_2\bar{C}-C-H}} \ \xrightleftharpoons{} \ \underset{H}{\overset{HO-H \quad O^-}{H_3C-C-CH_2-C-H}} \ \xrightleftharpoons{} \ \underset{H \quad aldol}{\overset{HO^- \quad OH \quad O}{H_3C-C-CH_2-C-H}}$$

3. The aldol addition is reversible.
  a. The equilibrium for the aldol addition is much more favorable for aldehydes than for ketones.
  b. For example, the aldol-addition product of acetone is obtained only if an apparatus is used that allows the product to be removed from the base catalyst as it is formed.
4. Under severe conditions (higher base concentration and/or heat), the product of aldol addition undergoes a dehydration reaction. (The sequence of reactions consisting of the aldol addition followed by dehydration is called the aldol condensation; a condensation is a reaction in which two molecules combine to form a larger molecule with the elimination of a small molecule, in many cases water.)

$$2 \ R-CH_2-\overset{O}{\overset{\|}{C}}-H \ \underset{\substack{addition \\ via \ enolate}}{\xrightleftharpoons{^-OH}} \ R-CH_2-\underset{R}{\overset{OH \quad O}{CH-CH-C-H}} \ \underset{dehydration}{\xrightarrow{^-OH}} \ R-CH_2-\underset{R}{\overset{O}{CH=C-C-H}}$$

  a. The dehydration part of the aldol condensation is catalyzed by base, and occurs in two distinct steps through a carbanion intermediate.
  b. Base-catalyzed dehydration reactions of alcohols are unusual; β-hydroxy aldehydes and β-hydroxy ketones undergo base-catalyzed dehydration because—
    i. the α-hydrogen is relatively acidic.
    ii. the product is conjugated and therefore particularly stable.
5. The product of an aldol addition–dehydration sequence is an α,β-unsaturated carbonyl compound.
6. The aldol condensation is an important method for the preparation of certain α,β-unsaturated carbonyl compounds. Whether the aldol addition product or the condensation product is formed depends on the reaction conditions, which must be determined on a case-by-case basis. (See Section 22.4B, text p. 1122.)

B. ACID-CATALYZED ALDOL CONDENSATION
1. Aldol condensations are also catalyzed by acid.
  a. Acid-catalyzed aldol condensations generally give α,β-unsaturated carbonyl compounds as products.
  b. The addition products usually cannot be isolated.
2. Both the conjugate acid and the enol isomer of the aldehyde or ketone are key reactive intermediates. (See Eqs. 22.46a–b on text page 1122–1123 for a detailed mechanism.)
3. The α-hydroxy aldehyde or ketone formed in the reaction spontaneously undergoes acid-catalyzed dehydration to give an α,β-unsaturated carbonyl compound; this dehydration drives the aldol condensation to completion.

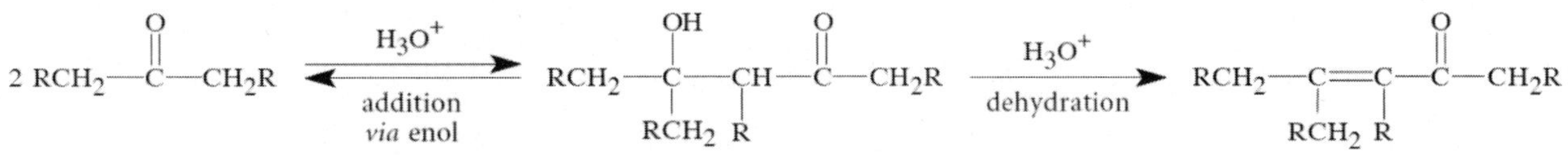

C. SPECIAL TYPES OF ALDOL CONDENSATION

1. When two different carbonyl compounds are used, a crossed aldol condensation occurs; in many cases, the result is a difficult-to-separate mixture.

2. Under the usual conditions (aqueous or alcoholic acid or base), useful crossed aldol condensations as a practical matter are limited to situations in which a ketone with α-hydrogens is condensed with an aldehyde without α-hydrogens.

3. An important example of this type of crossed aldol condensation is the Claisen–Schmidt condensation.

    a. A ketone with α-hydrogens is condensed with an aromatic aldehyde that has no α-hydrogens.

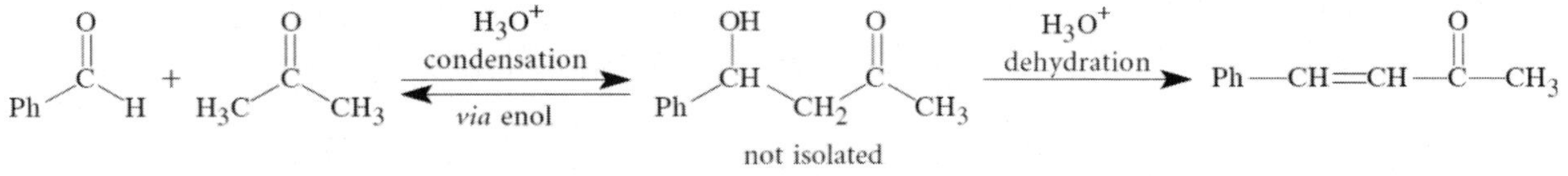

    b. The Claisen–Schmidt condensation occurs instead of other competing processes for three reasons:

        i. Because the aldehyde has no α-hydrogens, it cannot act as the enolate or enol component of the aldol condensation.

        ii. Addition of the ketone enolate or enol to a ketone occurs more slowly than addition to an aldehyde; aldehydes are more reactive than ketones.

        iii. The aldol addition reaction is reversible; thus, to the extent that this side reaction occurs, it is reversed under the reaction conditions.

    c. The Claisen–Schmidt condensation, like other aldol condensations, can be catalyzed by either acid or base.

4. When a molecule contains more than one aldehyde or ketone group, an intramolecular reaction (a reaction within the same molecule) is possible; the resulting formation of a ring is particularly favorable when five- and six-membered rings can be formed.

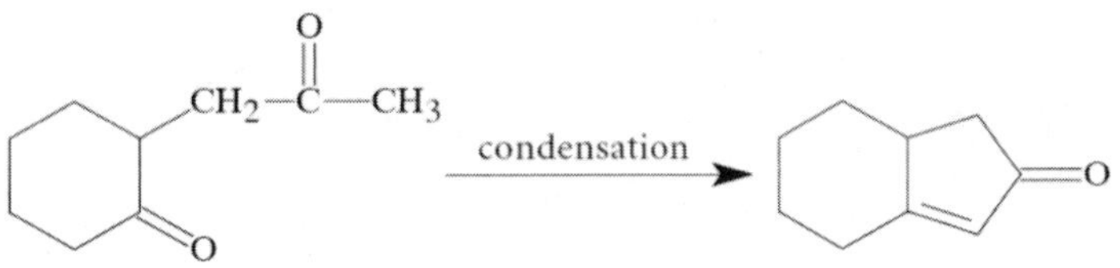

5. Directed aldol reactions involve pre-forming one enolate ion at cold temperatures, then adding the second ketone or aldehyde, thus promoting only the desired reaction.

    a. A strong, non-nucleophilic amide base such as LDA is used to generate the first enolate ion selectively (see Reactions IV.C.1. in this section).

    b. A different aldehyde (or ketone) is added to the solution containing the enolate ion, and only one aldol reaction occurs.

    c. Generally, the resulting ketone is less reactive than the starting aldehyde, so side reactions are not favored.

III. CONDENSATION REACTIONS INVOLVING ESTER ENOLATE IONS

A. CLAISEN CONDENSATION

1. The Claisen condensation is used for the formation of β-dicarbonyl compounds (compounds with two carbonyl groups in a β-relationship), particularly—

    a. β-keto esters (compounds in which a ketone carbonyl group is β to an ester carbonyl group)

    b. β-diketones (See Sec. 22.6D, text p. 1138.)

$$\text{EtO}^- + 2\ \text{RCH}_2\overset{\overset{\displaystyle O}{\|}}{C}-\text{OEt} \longrightarrow \text{RCH}_2\overset{\overset{\displaystyle O}{\|}}{C}-\overset{\overset{\displaystyle }{\underset{\underset{\displaystyle R}{|}}{\bar{C}}}}{}-\overset{\overset{\displaystyle O}{\|}}{C}-\text{OEt} \xrightarrow{\text{H}_3\text{O}^+} \text{RCH}_2\overset{\overset{\displaystyle O}{\|}}{C}-\overset{\underset{\underset{\displaystyle R}{|}}{CH}}{}-\overset{\overset{\displaystyle O}{\|}}{C}-\text{OEt}$$

an ester

\+ EtOH

a β-keto ester

$$\text{EtO}^- + \text{RCH}_2\overset{\overset{\displaystyle O}{\|}}{C}\text{R}' + -\overset{\overset{\displaystyle O}{\|}}{C}-\text{OEt} \longrightarrow -\overset{\overset{\displaystyle O}{\|}}{C}-\overset{\underset{\underset{\displaystyle R}{|}}{\bar{C}}}{}-\overset{\overset{\displaystyle O}{\|}}{C}\text{R}' \xrightarrow{\text{H}_3\text{O}^+} -\overset{\overset{\displaystyle O}{\|}}{C}-\overset{\underset{\underset{\displaystyle R}{|}}{CH}}{}-\overset{\overset{\displaystyle O}{\|}}{C}\text{R}'$$

a ketone   an ester

\+ EtOH

a β-diketone

2. The mechanism of the Claisen condensation:

a. An enolate ion is formed by the reaction of an ester with ethoxide as the base. (Although the ester enolate ion is formed in low concentration, it is a strong base and good nucleophile.)

$$\text{EtO}^- \overset{H}{\curvearrowright}\ \text{R}-\overset{\overset{\displaystyle O}{\|}}{CH}-\overset{\overset{\displaystyle O}{\|}}{C}-\text{OEt} \rightleftharpoons \left[\text{R}-\overset{\overset{\displaystyle O}{\|}}{\bar{CH}}-\overset{\overset{\displaystyle O}{\|}}{C}-\text{OEt} \longleftrightarrow \text{R}-CH=\overset{\overset{\displaystyle \bar{O}}{|}}{C}-\text{OEt} \right]$$

b. The ester enolate undergoes a nucleophilic acyl substitution reaction with a second molecule of ester.

$$\text{R}-CH_2-\overset{\overset{\displaystyle O}{\|}}{C}-\text{OEt} + \text{R}-\overset{\overset{\displaystyle O}{\|}}{\bar{CH}}-\overset{\overset{\displaystyle O}{\|}}{C}-\text{OEt} \rightleftharpoons \text{R}-CH_2-\overset{\underset{\underset{\displaystyle \text{OEt R}}{|}}{\overset{\displaystyle \bar{O}}{|}}}{C}-CH-\overset{\overset{\displaystyle O}{\|}}{C}-\text{OEt} \rightleftharpoons$$

$$\text{EtO}^- + \text{R}-CH_2-\overset{\overset{\displaystyle O}{\|}}{C}-\overset{\underset{\underset{\displaystyle R}{|}}{CH}}{}-\overset{\overset{\displaystyle O}{\|}}{C}-\text{OEt}$$

c. The overall equilibrium favors the reactants; that is, all β-keto esters are less stable than the esters from which they are derived.

d. The Claisen condensation has to be driven to completion by applying Le Châtelier's principle.

   i. The most common technique is to use one equivalent of alkoxide catalyst.

   ii. The hydrogens adjacent to both carbonyl groups in the β-keto ester product are especially acidic ($pK_a \approx 10$–11); the alkoxide (conjugate acid $pK_a = 16$) removes one of these hydrogens to form quantitatively the conjugate base of the product.

$$\text{EtO}^- \quad \text{R}-CH_2-\overset{\overset{\displaystyle O}{\|}}{C}-\overset{\underset{\underset{\displaystyle R}{|}}{\overset{\displaystyle H}{|}}}{C}-\overset{\overset{\displaystyle O}{\|}}{C}-\text{OEt} \rightleftharpoons \text{R}-CH_2-\overset{\overset{\displaystyle O}{\|}}{C}-\overset{\underset{\underset{\displaystyle R}{|}}{\bar{C}}}{}-\overset{\overset{\displaystyle O}{\|}}{C}-\text{OEt} + \text{EtOH}$$

$pK_a = 10$–11 $pK_a = 16$

e. The un-ionized β-keto ester product is formed when acid is added subsequently to the reaction mixture.

$$\text{R}-CH_2-\overset{\overset{\displaystyle O}{\|}}{C}-\overset{\underset{\underset{\displaystyle R}{|}}{\bar{C}}}{}-\overset{\overset{\displaystyle O}{\|}}{C}-\text{OEt} \xrightarrow{\text{H}_3\text{O}^+} \text{R}-CH_2-\overset{\overset{\displaystyle O}{\|}}{C}-\overset{\underset{\underset{\displaystyle R}{|}}{CH}}{}-\overset{\overset{\displaystyle O}{\|}}{C}-\text{OEt}$$

3. Attempts to condense an ester that has only one α-hydrogen result in little or no product, because the desired condensation product has no α-hydrogens acidic enough to react completely with the alkoxide.

B. DIECKMANN CONDENSATION

1. Intramolecular Claisen condensations take place readily when five- or six-membered rings can be formed.

2. The intramolecular Claisen condensation reaction is called the Dieckmann condensation.

3. As with the Claisen condensation, one equivalent of base is necessary; the α-proton must be removed from the initially formed product in order for the reaction to be driven to completion.

C. CROSSED CLAISEN CONDENSATION

1. The Claisen condensation of two different esters is called a crossed Claisen condensation; the crossed Claisen condensation of two esters that both have α-hydrogens gives a mixture of compounds that are typically difficult to separate.

2. Crossed Claisen condensations are useful if one ester is especially reactive or has no α-hydrogens. Formate esters and diethyl carbonate (a less reactive ester that is used in excess) are frequently used in this type of reaction.

ethyl formate (G = H)
diethyl carbonate (G = EtO)

3. Another type of crossed Claisen condensation is the reaction of ketones with esters.

a. In this type of reaction the enolate ion of a ketone reacts as a nucleophile at the carbonyl group of an ester.

i. The equilibrium for the aldol addition of two ketones favors the reactants, whereas the Claisen condensation is irreversible because one equivalent of base is used to form the enolate ion of the product.

ii. Ketones are far more acidic than esters; the enolate ion of the ketone is formed in much greater concentration than the enolate ion of the ester.

b. The product is a β-dicarbonyl compound, which is especially acidic and is ionized completely by the one equivalent of base; ionization makes this reaction irreversible.

IV. ALKYLATION OF ESTER ENOLATE IONS

A. MALONIC ESTER SYNTHESIS

1. Diethyl malonate (malonic ester), like many other β-dicarbonyl compounds, has unusually acidic α-hydrogens ($pK_a = 12.9$).

a. Its conjugate-base enolate ion can be formed completely with alkoxide bases such as sodium ethoxide, whose conjugate acid, ethanol, has a pK_a of about 16.

b. The conjugate-base anion of diethyl malonate is nucleophilic, and it reacts with alkyl halides and sulfonate esters (RX, R´X in the following equation) in typical S_N2 reactions.

c. This reaction can be used to introduce alkyl groups at the α-position of malonic ester.

diethyl malonate
$pK_a = 12.9$

$+$ EtOH $+$ X$^-$ $+$ EtOH $+$ X$^-$

2. The malonic ester synthesis can be extended to the preparation of carboxylic acids. (See Eqs. 22.77a–c, text p. 1146.)

   a. Saponification of the diester and acidification of the resulting solution give a substituted malonic acid derivative.

   b. Heating any malonic acid derivative causes it to decarboxylate.

   c. The result of alkylation, saponification, and decarboxylation is a carboxylic acid that conceptually is a substituted acetic acid—an acetic acid molecule with one or two alkyl substituents on its α-carbon.

3. The overall sequence of ionization, alkylation, saponification, and decarboxylation starting from diethyl malonate is called the malonic ester synthesis.

   a. The alkylation step of the malonic ester synthesis results in the formation of one or two new carbon–carbon bonds.

   b. The anion of malonic acid can be alkylated twice in two successive reactions with different alkyl halides (if desired) to give, after hydrolysis and decarboxylation, a disubstituted acetic acid.

B. ACETOACETIC ESTER SYNTHESIS

1. β-Keto esters (pK_a = 10–11), which are substantially more acidic than ordinary esters (pK_a = 25), are completely ionized by alkoxide bases, which have conjugate-acid pK_a values near 16.

   a. The resulting enolate ions can be alkylated by alkyl halides or sulfonate esters.

   b. Dialkylation of β-keto esters is also possible.

   c. The alkylated derivatives of ethyl acetoacetate can be hydrolyzed and decarboxylated to give ketones.

2. The alkylation of ethyl acetoacetate followed by saponification, protonation, and decarboxylation to give a ketone is called the acetoacetic ester synthesis and involves the construction of one or two new carbon–carbon bonds.

3. Whether a target ketone can be prepared by the acetoacetic ester synthesis can be determined by mentally reversing the synthesis (see Eq. 22.87, text p. 1150):

   a. Replace an α-hydrogen on the target ketone with a —CO_2Et group.

   b. Determine whether the resulting β-keto ester can be prepared by a Claisen condensation or from other β-keto esters by alkylation or di-alkylation with appropriate alkyl halides.

C. DIRECT ALKYLATION OF ENOLATE IONS DERIVED FROM MONOESTERS

1. A family of very strong, highly branched nitrogen bases can be used to form stable enolate ions rapidly at −78° from esters.

   a. The nitrogen bases themselves are generated from the corresponding amines and butyllithium at −78° in tetrahydrofuran (THF).

$$CH_3CH_2CH_2CH_2-Li \ + \ i\text{-}Pr_2NH \xrightarrow{-78°} i\text{-}Pr_2N^- \ Li^+ \ + \ CH_3CH_2CH_2CH_3$$

$$pK_a = 32\text{–}35 \qquad\qquad pK_a = 55\text{–}60$$

b. These amide bases are strong enough to convert esters rapidly and irreversibly into their conjugate-base enolate ions.

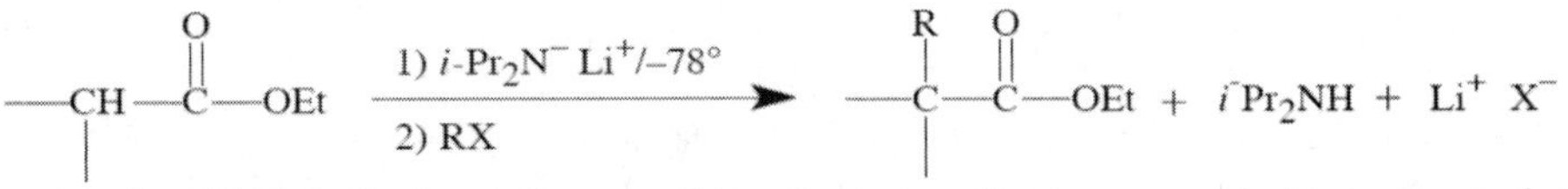

i. The ester is added to the base; the reaction of esters with strong amide bases is much faster, even at $-78°$, than the Claisen condensation; the enolate is formed rapidly.

ii. Attack of the amide base on the carbonyl carbon of the ester, a reaction that competes with proton removal, is retarded by van der Waals repulsions between groups on the carbonyl compound and the large branched groups on the amide base.

iii. This type of van der Waals repulsions has been termed F-strain, or "front-strain."

c. The ester enolate anions formed with these bases can be alkylated with alkyl halides.

d. Esters with quaternary α-carbon atoms can be prepared with this method.

2. This method of ester alkylation is considerably more expensive than the malonic ester synthesis and requires special inert-atmosphere techniques.

D. ALDOL REACTIONS OF ESTER ENOLATES

1. Ester enolates can serve as nucleophiles in aldol reactions.

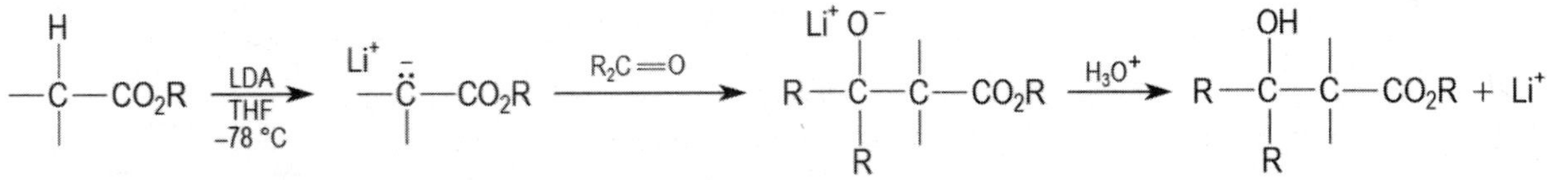

2. The Reformatsky reaction utilizes an organozinc compound as the nucleophile, which behaves analogously to a Grignard reagent, however it is less reactive.

a. The organozinc compound (Reformatsky reagent) does not necessarily have to be pre-formed in a second step as shown below. It can be formed in the presence of the aldehyde or ketone.

b. Organozinc compounds are less reactive than Grignard reagents, therefore they can be formed from esters.

c. The actual structure of the Reformatsky reagent more complex than is shown, however it behaves as if it were as simple as depicted.

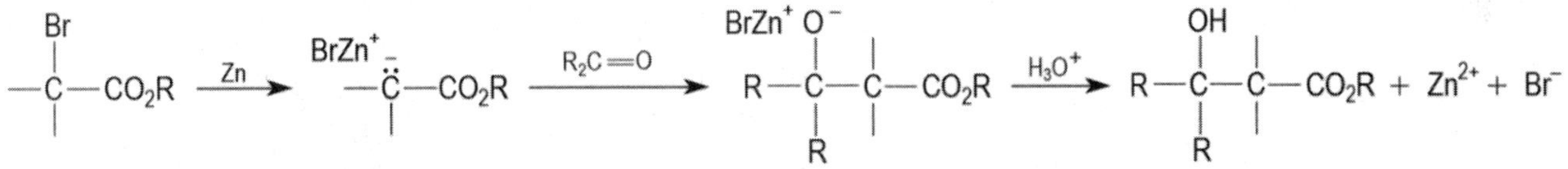

Reformatsky
reagent

3. The Knoevenagel reaction is an aldol reaction that utilizes the enolate of a malonic ester or acetoacetic ester as the nucleophile. The products of these reactions are α,β-unsaturated carbonyl compounds.

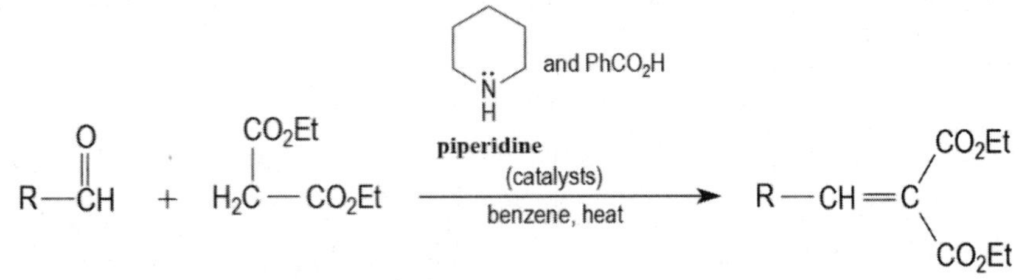

a. First, the base piperidine deprotonates the malonic ester.

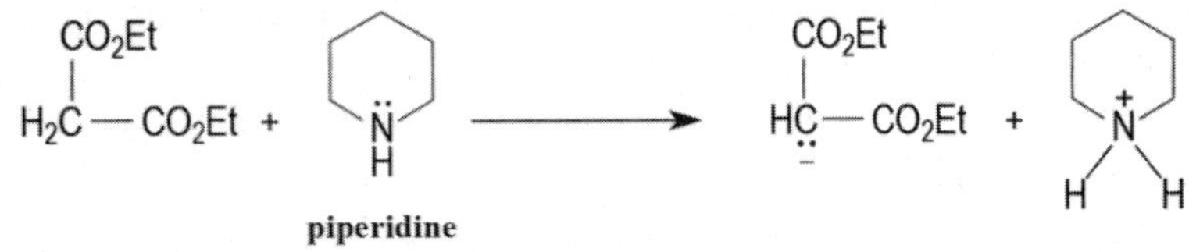

piperidine

b. Then, the aldol reaction proceeds as normal, and the addition product is protonated by the piperidinium cation.

c. Finally, the addition product, which is not isolated, is deprotonated by the piperidine, which drives the reaction to completion through a dehydration reaction. The hydroxide by-product then reacts with the piperidinium ion to regenerate the catalyst.

V. CONJUGATE-ADDITION REACTIONS

A. CONJUGATE ADDITION TO α,β-UNSATURATED CARBONYL COMPOUNDS

1. Addition of a nucleophile (Nuc⁻ in the following equation) to the double bond of an α,β-unsaturated carbonyl compound is an example of conjugate addition (1,4-addition); nucleophilic conjugate addition has no parallel in the reactions of simple conjugated dienes.

2. Nucleophilic addition to the carbon–carbon double bonds of α,β-unsaturated aldehydes, ketones, esters, and nitriles is a rather general reaction that can be observed with a variety of nucleophiles.

   a. The addition of a nucleophile to acrylonitrile (H_2C=CH—C≡N) is a useful reaction called cyanoethylation.

Nuc—H + H_2C=CH—CN →(base)→ Nuc—CH_2—CH_2—CN

nucleophile   acrylonitrile   cyanoethylated nucleophile

   b. Quinones are α,β-unsaturated carbonyl compounds, and they also undergo similar addition reactions.
   c. When cyanide is the nucleophile, a new carbon–carbon bond is formed; the nitrile group can then be converted into a carboxylic acid group by acid hydrolysis.

3. Nucleophilic addition to the double bond in an α,β-unsaturated carbonyl compound occurs because it gives a resonance-stabilized enolate ion intermediate.

4. Acid-catalyzed conjugate additions to α,β-unsaturated carbonyl compounds are also known.

B. CONJUGATE ADDITION OF ENOLATE IONS: MICHAEL ADDITION

1. Enolate ions derived from malonic ester derivatives, β-keto esters, and the like undergo conjugate-addition reactions with α,β-unsaturated carbonyl compounds.

a. Conjugate additions of carbanions to α,β-unsaturated carbonyl compounds are called Michael additions.

b. Many Michael additions can originate from either of two pairs of reactants.

c. To maximize conjugate addition, choose the pair of reactants with the less basic enolate ion.

2. In one useful variation of the Michael addition, called the Robinson annulation, the immediate product of the reaction can be subjected to an aldol condensation that closes a ring. (An annulation is a ring-forming reaction.)

VI. REACTIONS OF α,β-UNSATURATED CARBONYL COMPOUNDS WITH OTHER NUCLEOPHILES

A. REDUCTION OF α,β-UNSATURATED CARBONYL COMPOUNDS

1. The carbonyl group of an α,β-unsaturated aldehyde or ketone is reduced to an alcohol with lithium aluminum hydride.

a. This transformation involves the nucleophilic reaction of hydride at the carbonyl carbon and is therefore a carbonyl addition.

b. Carbonyl addition is not only faster than conjugate addition but, in this case, also irreversible; reduction of the carbonyl group with LiAlH$_4$ is a kinetically controlled reaction.

2. Many α,β-unsaturated carbonyl compounds are reduced by NaBH$_4$ to give mixtures of both carbonyl-addition products and conjugate-addition products.

3. The carbon–carbon double bond of an α,β-unsaturated carbonyl compound can in most cases be reduced selectively by catalytic hydrogenation.

B. ADDITION OF ORGANOLITHIUM REAGENTS TO THE CARBONYL GROUP

1. Organolithium reagents react with α,β-unsaturated carbonyl compounds to yield products of carbonyl addition.

2. The product is the result of kinetic control, since carbonyl addition is more rapid than conjugate addition and it is also irreversible.

C. CONJUGATE ADDITION OF LITHIUM DIALKYLCUPRATE REAGENTS

1. Lithium dialkylcuprate reagents give exclusively products of conjugate addition when they react with α,β-unsaturated esters and ketones.

2. Although conjugate addition of lithium dialkylcuprates actually involves a radical mechanism (see Further Exploration 22.3 on p. 755 of this manual), the reaction can be conceptualized mechanistically to be similar to other conjugate additions.

   a. A nucleophilic reaction of an "alkyl anion" of the dialkylcuprate on the double bond gives a resonance-stabilized enolate ion.

   b. Protonation of the enolate ion by the addition of water gives the conjugate-addition product.

3. Conjugate addition also occurs with Grignard reagents in the presence of added CuCl.

SOLUTIONS TO PROBLEMS

Solutions to In-Text Problems

22.1   (a)   The acidic hydrogens are the α-hydrogens on the carbons between the two carbonyl groups. (—OEt = ethoxy group = —OCH$_2$CH$_3$.)

acidic hydrogens

$$\underset{\text{diethyl malonate}}{\text{EtOC—CH}_2\text{—COEt}}$$

The reason that these hydrogens are particularly acidic is that their conjugate-base enolate ions are stabilized by the polar effects and resonance effects of two carbonyl groups, whereas the conjugate-base enolate ion of an ordinary ester is stabilized by the corresponding effects of only one carbonyl group. The resonance structures of the conjugate-base enolate ion of diethyl malonate are as follows

$$\left[\text{EtO—C—CH—C—OEt} \longleftrightarrow \text{EtO—C—CH=C—OEt} \longleftrightarrow \text{EtO—C=CH—C—OEt} \right]$$

resonance structures for the conjugate-base enolate ion of diethyl malonate

(b)   The acidic hydrogens are the α-hydrogens on the carbons between the two carbonyl groups. (—OEt = ethoxy group = —OCH$_2$CH$_3$.)

acidic hydrogens

$$\underset{\text{ethyl acetoacetate}}{\text{CH}_3\text{C—CH}_2\text{—COEt}}$$

The reason that these hydrogens are particularly acidic is that the conjugate-base enolate ion is stabilized by the polar effects and resonance effects of two carbonyl groups, whereas the conjugate-base enolate ion of an ordinary ester is stabilized by the corresponding effects of only one carbonyl group. The resonance structures of the conjugate-base enolate ion of ethyl acetoacetate are as follows:

$$\left[\text{CH}_3\text{—C—CH—C—OEt} \longleftrightarrow \text{CH}_3\text{—C—CH=C—OEt} \longleftrightarrow \text{CH}_3\text{—C=CH—C—OEt} \right]$$

22.2   The imide of succinic acid (succinimide) is more acidic, because its conjugate-base anion is stabilized by the polar effects and resonance effects of two adjacent carbonyl groups. The conjugate-base anion of succinic acid diamide is stabilized by resonance interaction with only one carbonyl group.

conjugate-base anion of succinimide

$$\underset{\text{conjugate-base anion of succinic acid diamide}}{\text{HN—C—CH}_2\text{CH}_2\text{—C—NH}_2}$$

The resonance effect in the conjugate base of succinimide far outweighs the statistical effect of two amide groups in succinic acid diamide.

22.3   A mechanism for replacement of one hydrogen in the reaction of Eq. 22.6, text p. 1108, is shown in the following equation. (The mechanisms for replacement of the others are identical.) Only the α-hydrogens are replaced because the enolate ion is the only carbanion stable enough to be formed. The carbanion intermediates required in a similar mechanism for the replacement of the hydrogens other than the α-hydrogens are not resonance-stabilized.

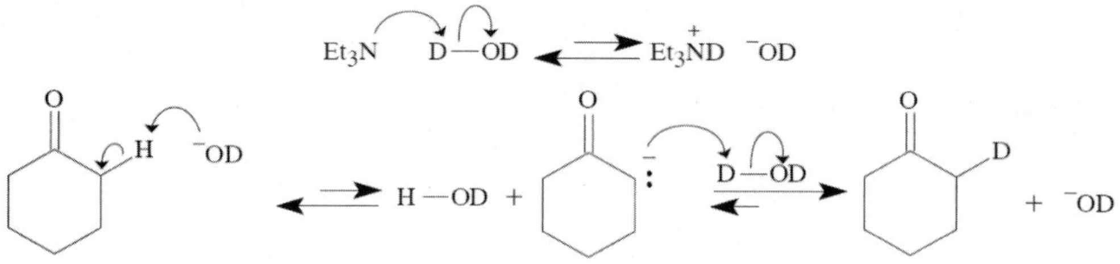

resonance-stabilized
enolate ion

22.4   The proton NMR spectrum would change as follows:

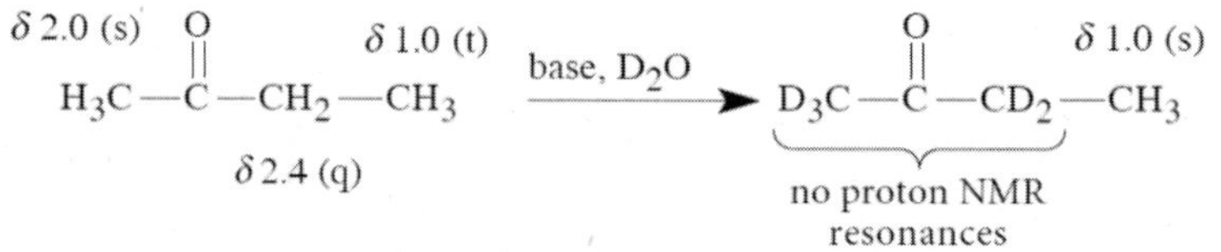

That is, in basic D_2O, the resonances for the α-protons disappear and the triplet at δ 1.0 becomes a singlet.

22.5   Exchange does not occur because the orbital containing the unshared electron pair on the α-carbon of the enolate ion is perpendicular to the π-electron system of the carbonyl group; therefore, the enolate ion is not stabilized by resonance. Another way to view this situation is suggested by the hint: the resonance structure of the enolate ion violates Bredt's rule (Sec. 7.6C). This situation is discussed in Guideline 4 for resonance structures, Sec. 15.6B, text p. 753, and in the figure at the top of p. 1106.

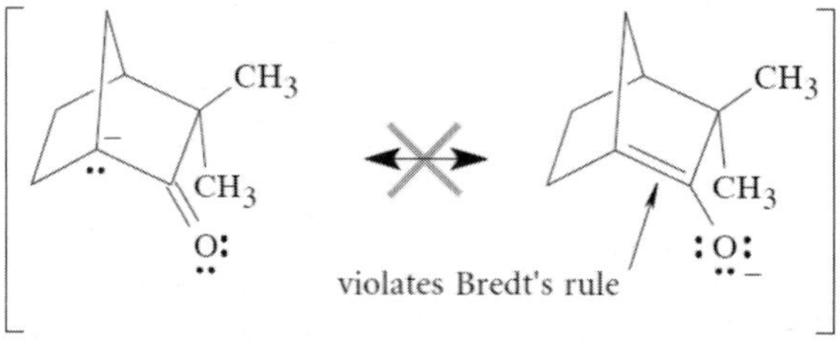

violates Bredt's rule

22.6   (a)   All α-hydrogens are exchanged for deuterium. (The methyl hydrogens of the *tert*-butyl group are not α-hydrogens.)

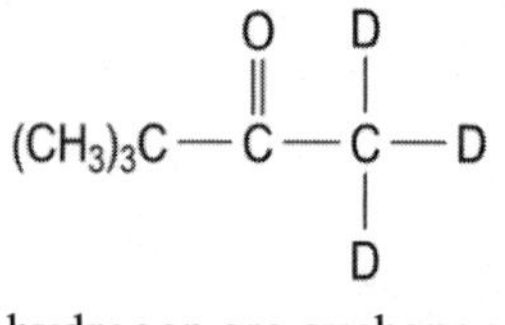

(b)   All α-hydrogens as well as the O—H hydrogen are exchanged for deuterium:

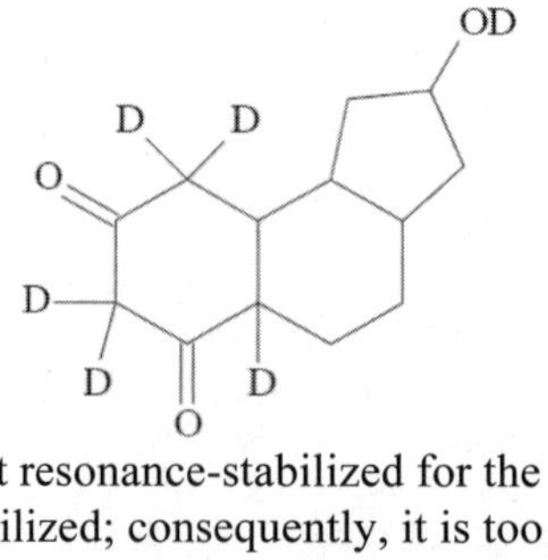

(c)   The only possible enolate ion is not resonance-stabilized for the same reason that the enolate ion in Problem 22.5 is not resonance-stabilized; consequently, it is too unstable to form. Because deuterium exchange of the α-hydrogens would involve an enolate ion intermediate, deuterium exchange cannot occur.

22.7   (a)

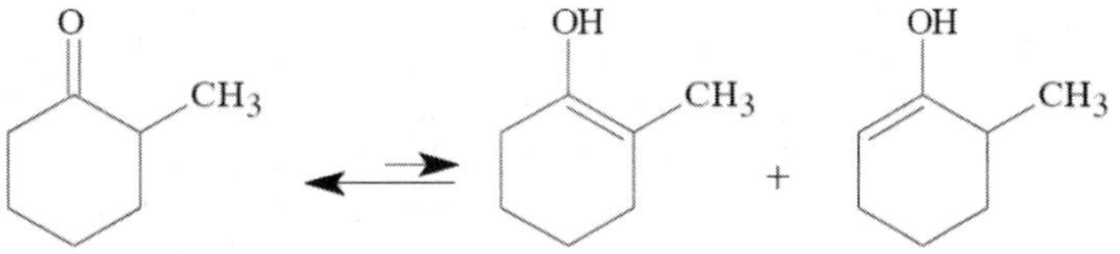

2-methylcyclohexanone   enol forms of 2-methylcyclohexanone

(b)

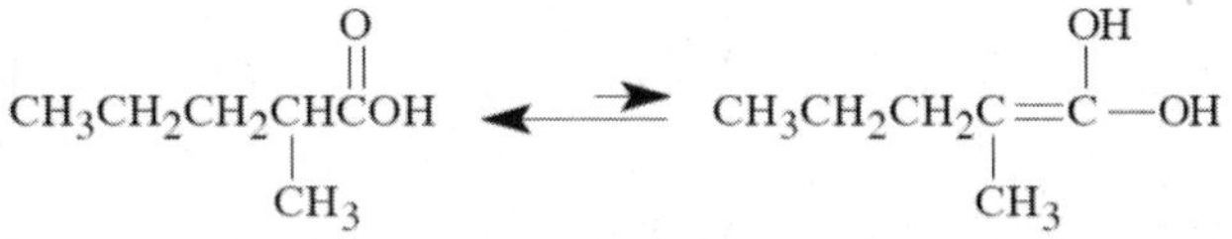

2-methylpentanoic acid   enol form of 2-methylpentanoic acid

(c)   Benzaldehyde, PhCH=O, has no enol forms because it has no α-hydrogens.

(d)

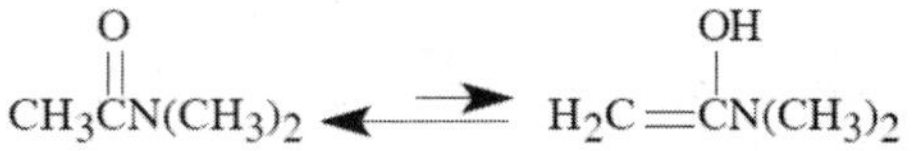

N,N-dimethylacetamide   enol form of
                                              N,N-dimethylacetamide

(e)   *N,N*-Dimethylformamide has no enol forms because it has no α-hydrogens.

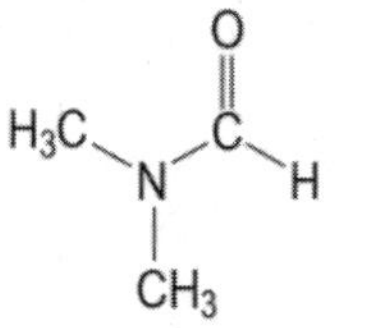

N,N-dimethylformamide

22.8   The enol forms of 2-butanone.

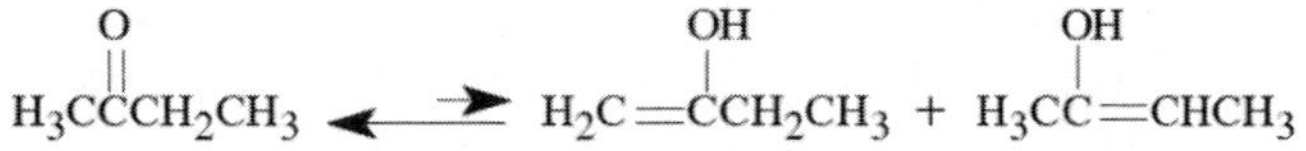

2-butanone   least stable   (E) and (Z)
                                              most stable
                        enol forms of 2-butanone

We know that double bonds are stabilized by alkyl substitution (see Sec. 4.5B, text pp. 149–151). Hence the enol on the left in the previous equation is the least stable.

22.9   (a)   The "enol" form of a nitro compound:

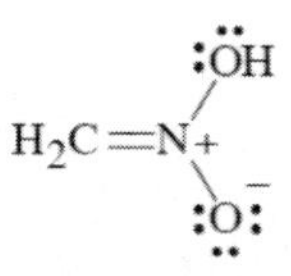

"enol" form of nitromethane
(an *aci*-nitro compound)

(b)   The "enol" form of an amide that has a carbon–nitrogen double bond is called an *imidic* acid; see Eq. 21.20b, text p. 1066. Notice that this particular amide cannot enolize toward the α-carbon because it has no α-hydrogens on that carbon.

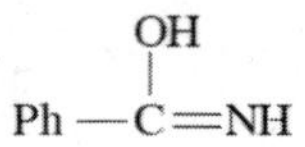

$$\text{OH}$$
$$\text{Ph} - \overset{|}{\text{C}} = \text{NH}$$

"enol" form of benzamide
(an imidic acid)

22.10 (a) One source of stabilization of the enols of β-diketones is their intramolecular hydrogen bonds; see the structure preceding Problem 22.7 on text p. 1111. In aqueous solution, the solvent can serve as a hydrogen-bond donor to the carbonyl oxygen; hence the energetic advantage of the enol that results from intramolecular hydrogen bonding is reduced in aqueous solution. In a solvent that cannot donate hydrogen bonds, such as hexane, there are no external hydrogen bonds, and the internal hydrogen bond thus provides an energetic advantage to the enol.

(b) As the structure above Problem 22.7 on text p. 1111 shows, the enol form of 2,4-pentanedione has a conjugated π-electron system involving a carbonyl double bond and a carbon–carbon double bond. This conjugated system in the enol is the source of the UV spectrum. The absorption is stronger in hexane because there is a greater proportion of the enol form in hexane for the reasons discussed in the solution to part (a) of this problem.

22.11 (a) The enol is formed by the mechanism shown in Eq. 22.17b on text p. 1112. The enol is achiral, and it can be protonated at either face of the double bond at the same rate to give the racemate of the protonated ketone.

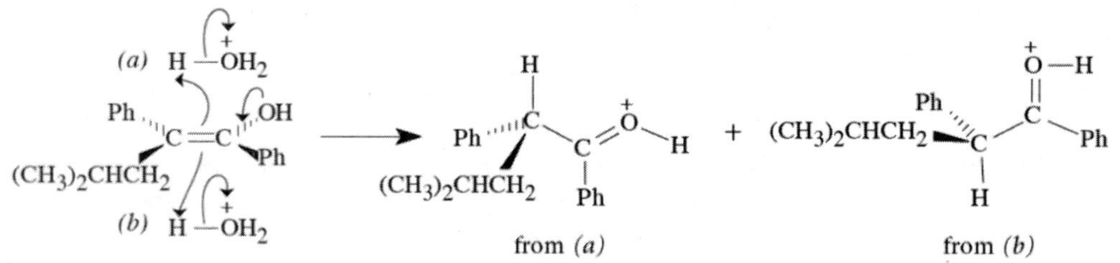

from (a)   from (b)

(b) The enol is formed by the mechanism shown in Eq. 22.17b on text p. 1112, except that D_3O^+ is the acid. This results in the "washout" of one α-hydrogen into the large excess of deuterated solvent.

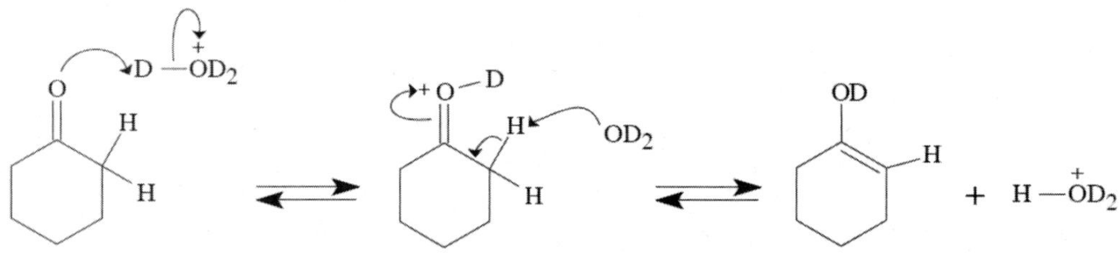

Protonation of the double bond by D_3O^+ gives a deuterium at the α-position. Replacement of one α-hydrogen by deuterium is shown; the mechanism for replacement of the other α-hydrogens is identical.

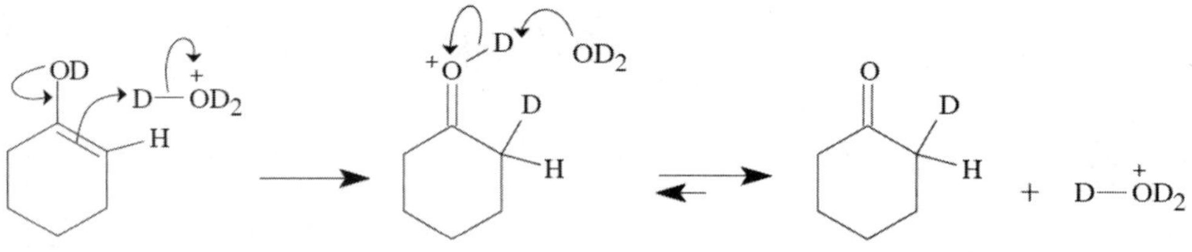

22.12 (a) The reaction free-energy diagram is given in Fig. SG22.1.
(b) The reaction free-energy diagram is given in Fig. SG22.2. Notice that the first two steps are identical to those shown in Fig. SG22.1 in part (a), because the process is identical.

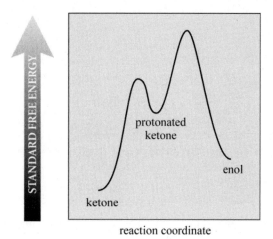

Figure SG22.1 Reaction free-energy diagram for acid-catalyzed enolization of a ketone to accompany the solution to Problem 22.12(a).

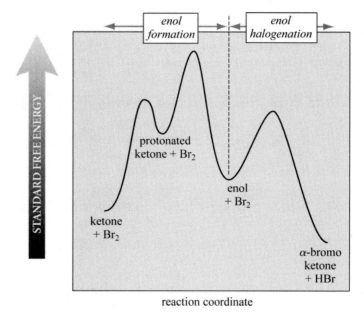

Figure SG22.2 Reaction free-energy diagram to accompany the solution to Problem 22.12(b). Notice that the rate-limiting steps for enol formation and overall halogenation are the same.

22.13 (a) Because enolization is the process responsible for acid-catalyzed racemization, and because the rate-limiting process in acid-catalyzed ketone halogenation is enolization, the rates of the two processes are the same—the rate of enolization.

(b) Because the rate of ketone halogenation is independent of the halogen concentration, the rates of halogenation of the same ketone with two different halogens are also independent of halogen concentration and therefore independent of the identity of the halogen itself. In fact, the rate in both cases is the rate of enolization, as in part (a). The two processes are compared at the same acid concentration because the enolization process is acid-catalyzed (Eq. 22.24, text p. 1114).

22.14 (a)   (c)

In part (b), benzophenone does not halogenate because it cannot form an enolate ion (it has no α-hydrogens), and the aromatic rings are deactivated by the carbonyl group toward electrophilic halogenation.

In part (c), the alcohol is first oxidized to the corresponding ketone, acetophenone, which then undergoes the haloform reaction.

22.15 2,5-Hexanedione gives succinic acid and iodoform when subjected to the iodoform reaction.

After iodination of the methyl group occurs once, subsequent iodinations are faster. Evidently, iodination of a —CH_2— group is either slower or it is reversible.

22.16 (a)   The α-bromo acid bromide reacts with ethanol to give the α-bromo ester.

> In part (a), neutral ethanol is not basic enough to act as a nucleophile and displace the α-bromine.

(b)   The α-bromo acid bromide is formed first. Then ammonia reacts as a nucleophile with it in both an acyl substitution reaction to give the amide, and in an S_N2 reaction at the α-carbon to give the α-amino amide of the α-amino acid alanine.

22.17 (a)   The base pyridine displaces the α-bromine in an S_N2 reaction.

(b)   The base acetate ion displaces the α-bromine to give a compound that is an example of a *phenacyl ester*.

22.18 One equivalent of NaOH is consumed by its reaction with the carboxylic acid. The second equivalent forms the 2,4-dichlorophenolate ion, which is the nucleophile in the ensuing S_N2 reaction.

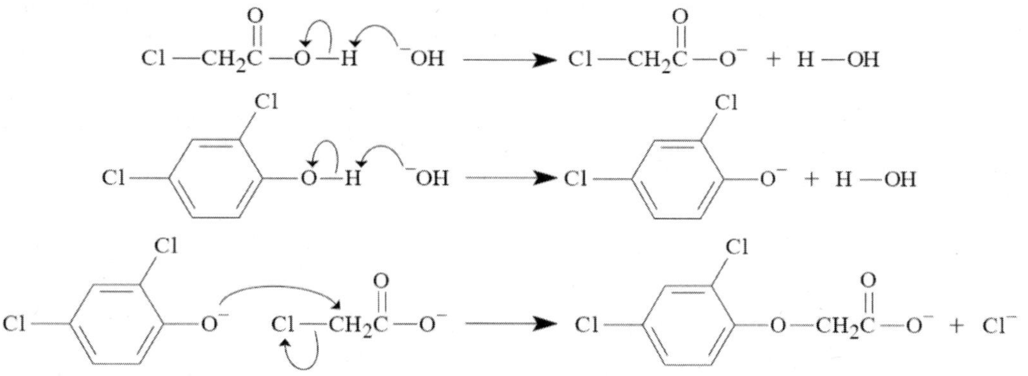

The final step of Eq. 22.34 (protonation of the carboxylate to give the carboxylic acid) does not take place until acid is added to the reaction mixture; under the basic conditions of the substitution reaction the product is ionized.

22.19 (a) The aldol addition reaction of phenylacetaldehyde:

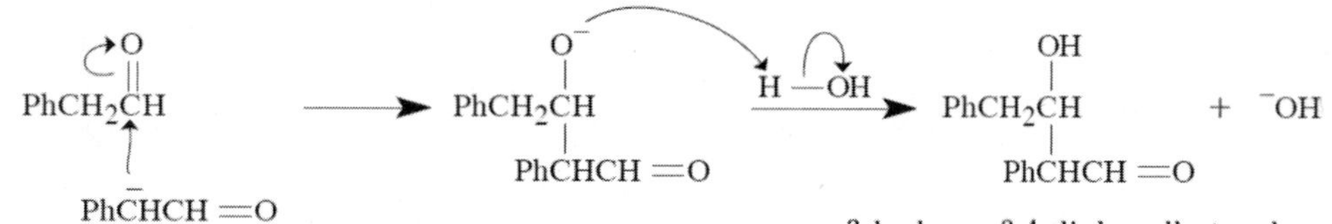

enolate ion of phenylacetaldehyde

3-hydroxy-2,4-diphenylbutanal

(b) The aldol addition reaction of propionaldehyde:

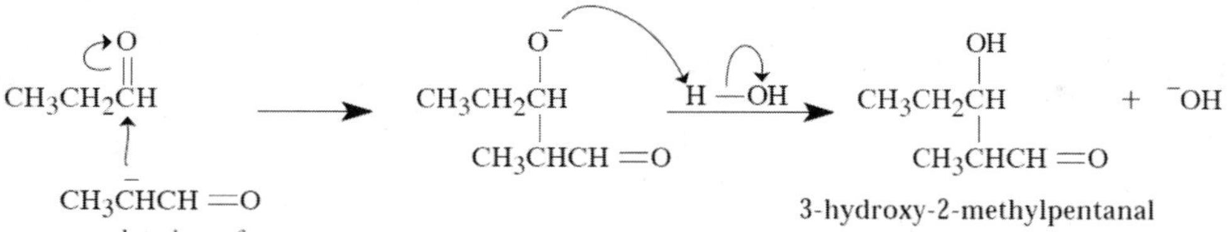

enolate ion of propionaldehyde

3-hydroxy-2-methylpentanal

22.20 (a)

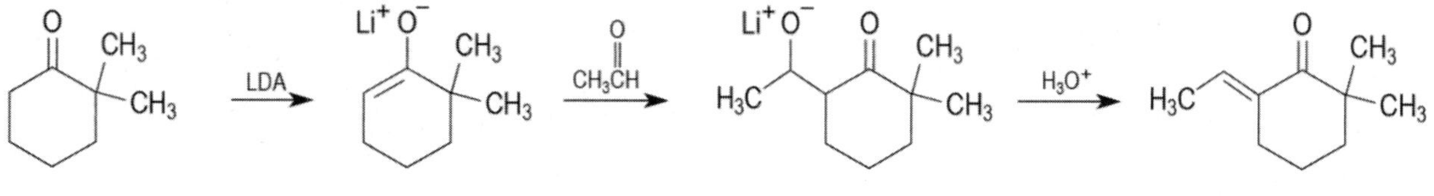

2,2-dimethylcyclohexanone simplified enolate structure aldol addition product aldol dehydration product

(b) The dehydration product can exist as either cis or trans isomers. Diastereomers have different energies and are therefore formed in different amounts (see Sec. 7.7B).

22.21 (a) The problem states that equal molar amounts of each component are used. (Otherwise two moles of the aldehyde might react—one mole at each set of ketone α-hydrogens.) The enolate ion of the ketone reacts with the aldehyde because aldehyde carbonyl groups are more reactive in additions than ketone carbonyl groups.

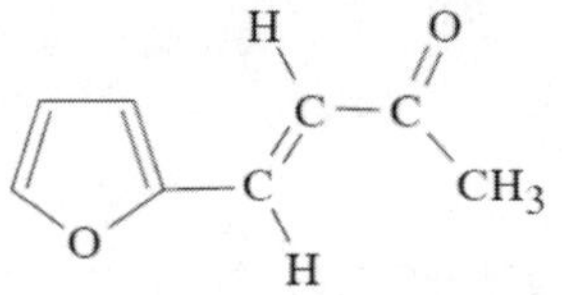

(b) As in part (a), the enolate ion of the ketone adds to the aldehyde carbonyl group to give product *A* because aldehyde carbonyl groups are more reactive than ketone carbonyl groups. It is also likely that aldol condensation of two molecules of hexanal will occur to give product *B*.

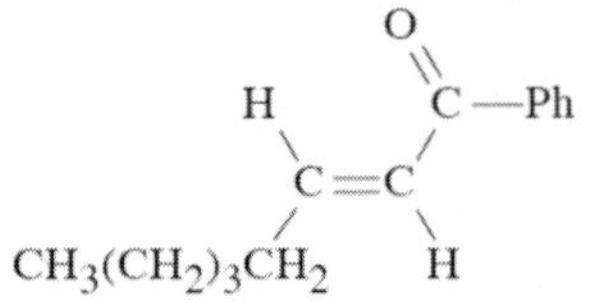

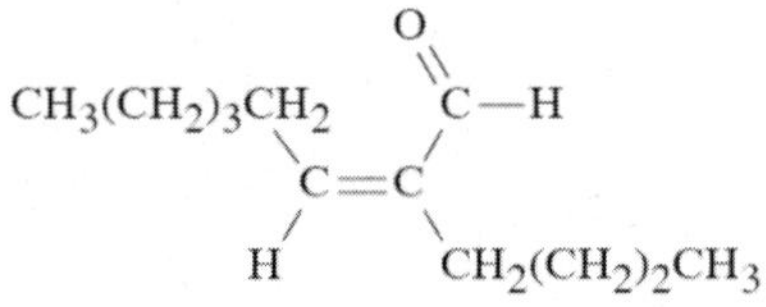

(c) This is a directed aldol condensation. The ketone enolate is generated at low temperatures and the aldehyde is added after. This process minimizes the potential for forming by-products.

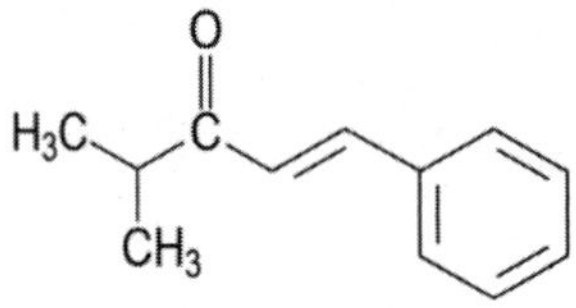

(d) This is an intramolecular aldol condensation.

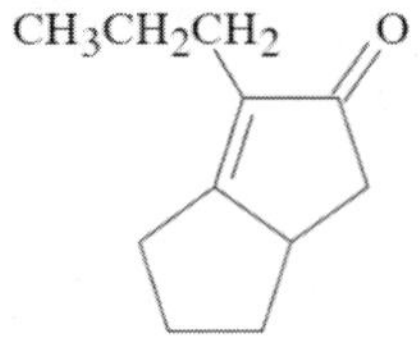

22.22 (a) In this case, the enolate ion of a symmetrical ketone, 3-pentanone, would have to react with *p*-methoxybenzaldehyde. There is only one possible enolate ion, and the aldehyde carbonyl group is much more reactive than the ketone carbonyl group. Hence, the reaction is reasonable.

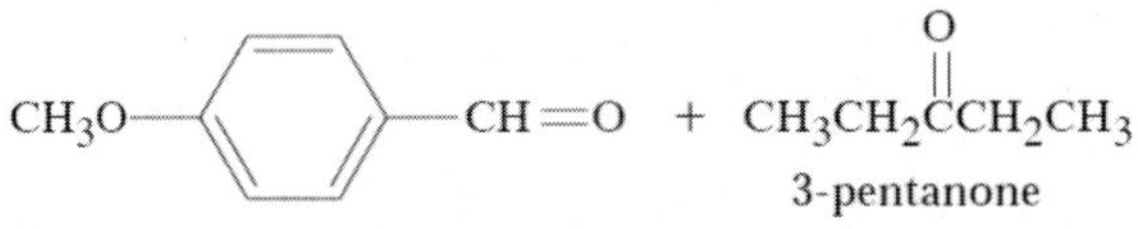

p-methoxybenzaldehyde

(b) The product shown would require the following starting materials. However, since the enolate ion required for the synthesis is only one of two that could form, at least two isomeric products are possible of which the desired product is only one. The desired product would therefore be formed as one component of a mixture of isomers.

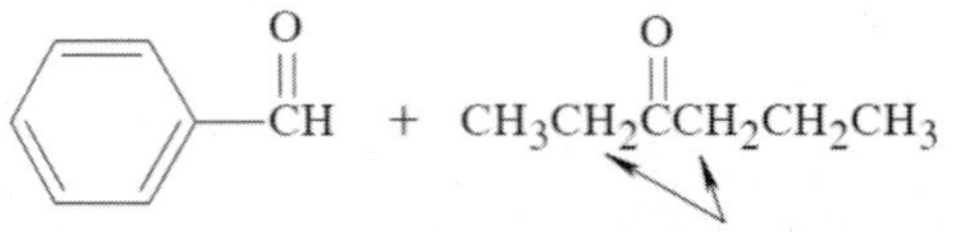

an enolate ion could
form at either carbon

(c) This product would require the intramolecular aldol condensation of the following keto aldehyde. In this case, the enolate ion formed at the α-carbon of the aldehyde would have to react with the ketone carbonyl group. Because the aldehyde carbonyl group is more reactive, and because its intramolecular reaction with an enolate ion of the ketone would also give a six-membered ring, synthesis of the compound shown in the problem is not reasonable. The following reaction would likely occur instead:

This compound is required
for the product given.

This compound
would be formed instead.

(d)   This product is the result of an intramolecular reaction of a ketone enolate and an aldehyde carbonyl group; hence, its synthesis from the following starting material is reasonable.

(e)   A double aldol condensation of the following starting materials would give the desired product. The reaction is particularly favorable because the product is highly conjugated.

(f)   This is a reaction of two equivalents of benzaldehyde with one of acetone. Because an aldehyde without α-hydrogens is involved, the reaction is a reasonable example of the Claisen–Schmidt condensation.

$$2 \ PhCH=O \ + \ CH_3\overset{O}{\overset{||}{C}}CH_3$$

(g)   This is a directed aldol reaction, therefore the enolate is generated first, at low temperatures, and the aldehyde is added afterwards.

aldol addition product   aldol dehydration product

(h)   Because the diketone starting material is symmetrical, the product shown is the only possible one, and thus the synthesis is reasonable.

22.23   First, break the double bond as shown in Eq. 22.54 on text p. 1129 to reveal the possible starting material.

The product shown in the problem results from the reaction of an enol formed at carbon a with carbonyl group B. Reaction of the enol formed at carbon b with carbonyl B would require formation of a strained four-membered ring (product bB below) and is thus not likely. Reaction of an enol formed at carbon c with carbonyl A also would require the formation of a strained four-membered ring (product cA below). Reaction of an enol formed at carbon d with carbonyl A would give bicyclic compound dA:

product bB product cA product dA

Compound dA is a modest violation of Bredt's rule (Sec. 7.6C, text pp. 297–8; build a model to verify that it is strained). The aldol *addition* product could form, thus avoiding dehydration to the strained alkene, but aldol additions of ketones are highly reversible. The product shown in the problem is thus the only one that can readily be driven to completion by dehydration.

22.24 (a)   Pyruvate forms an enamine with the lysine residue of the enzyme in an analogous process to the one shown in Eqs. 22.56a–b on text p. 1131. When (R)-glyceraldehyde-3-phosphate (G3P) enters the enzyme's active site, the α-carbon of the "enolate" reacts with the carbonyl of G3P, in an aldol-like addition reaction. Acids (HA) and their conjugate bases (A^-) within the active site protonate and deprotonate the intermediates as needed (see Eqs. 22.56c–d on text p. 1132).

Once the carbon-carbon bond is formed as a result of the aldol addition reaction, the imine is cleaved by the usual mechanism (see Eq. 22.56e for an analogous reaction, and Study Guide Links 19.5 and 19.6 on pp. 608–610 of this manual for details of the reaction mechanism.)

Finally, the KDPG cyclizes forming a hemiacetal, similar to the reaction shown in Eq. 22.57 on p. 1132 of the text. The mechanism is shown below, and detailed in Sec. 19.10A. Numbering carbons is helpful whenever you're writing a reaction mechanism that involves cyclization. Additionally, you can be sure that you've preserved the stereochemistry at each step by assigning R or S at each asymmetric carbon.

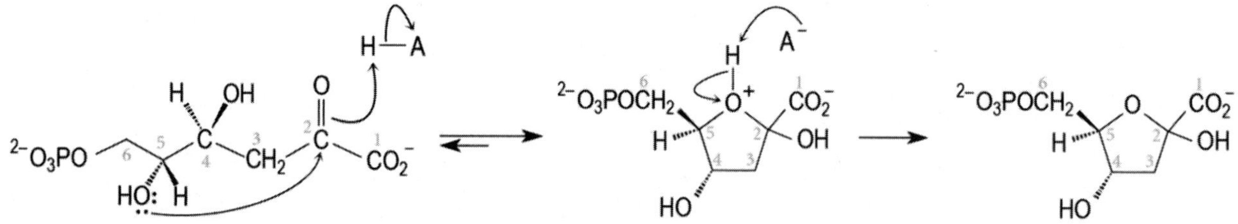

(b)   Diastereomeric forms of the cyclic hemiacetal of KDPG:

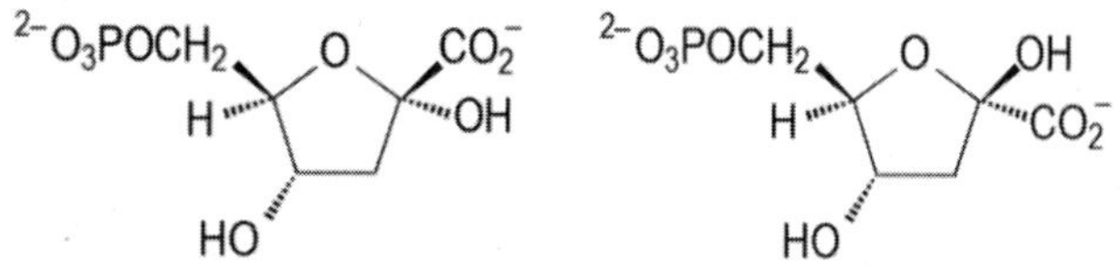

22.25   (a)   First, form an enamine (compound X) by the standard mechanism (see Sec. 19.11, text p. 984) at the carbonyl carbon marked with the asterisks.

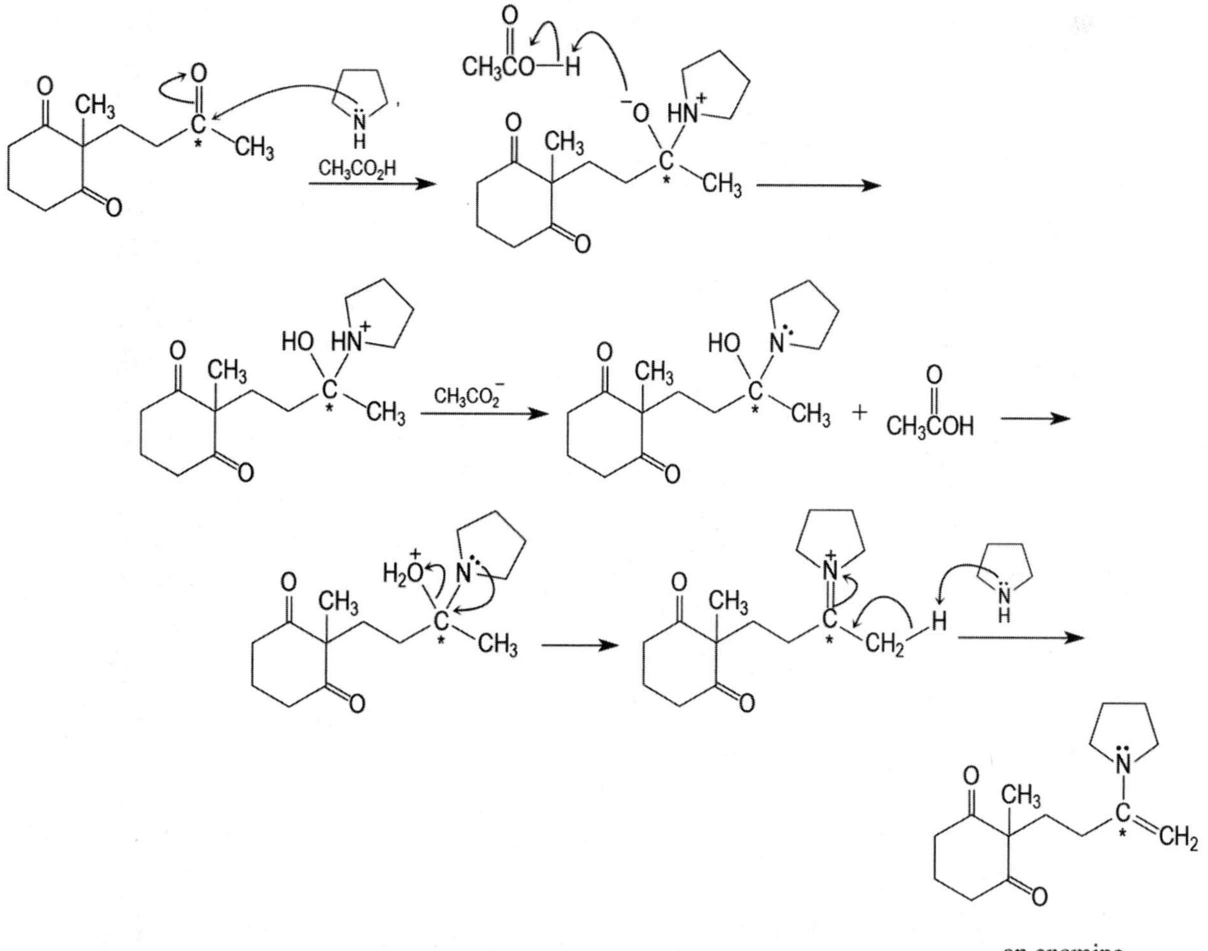

an enamine

(b)   Next, use one of the resonance structures of the enamine to do an intramolecular aldol addition reaction (Sec. 22.4A). After proton transfers, the enamine is regenerated.

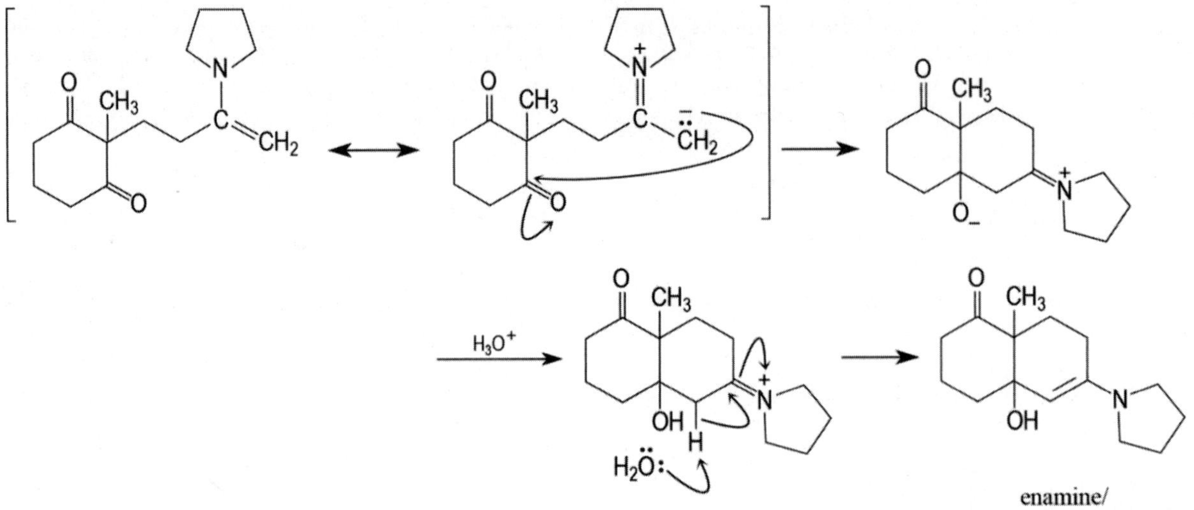

enamine/
aldol addition product

(c)   Treatment with acid hydrolyzes the enamine back to the ketone (Sec. 19.11, text p. 984). Continued treatment with acid will dehydrate the β-hydroxyketone (Sec. 22.4B) yielding an α,β-unsaturated ketone, or the aldol condensation product.

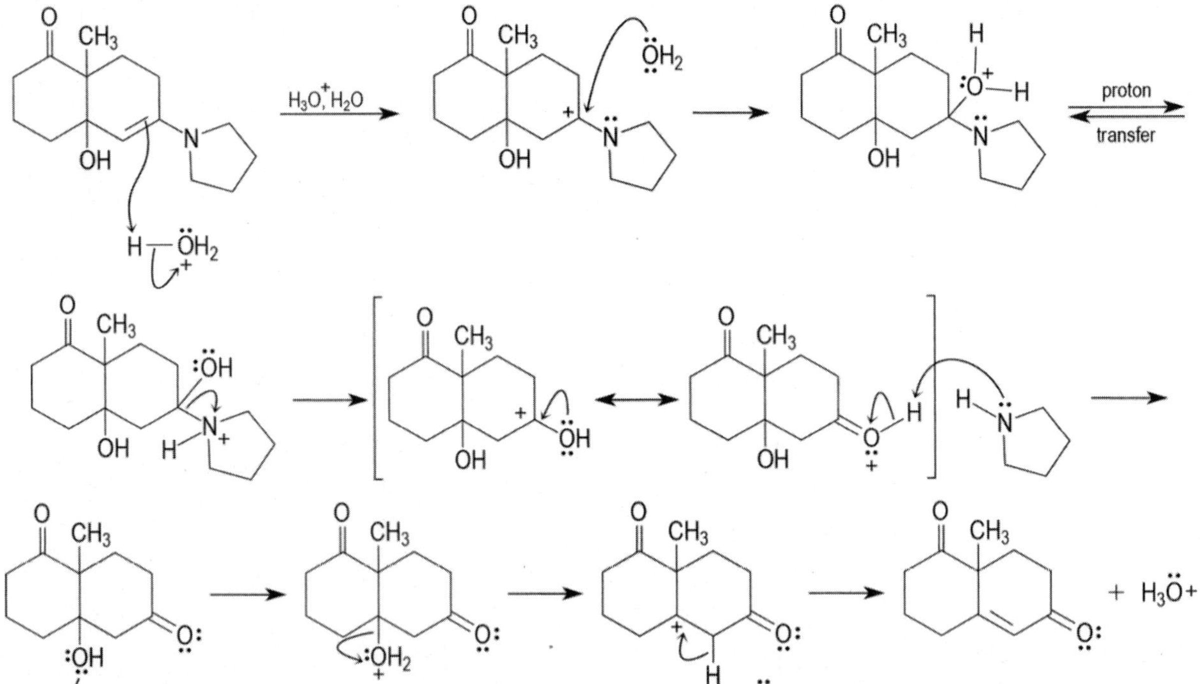

α,β-unsaturated ketone
(aldol condensation product)

22.26   (a)

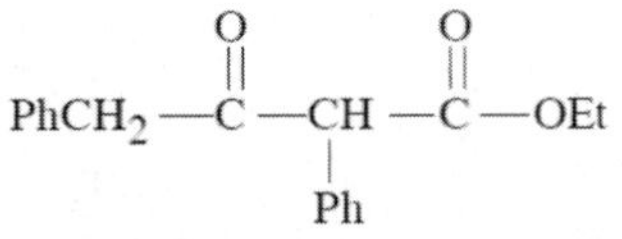

ethyl 2,4-diphenyl-3-oxobutyrate

(b)

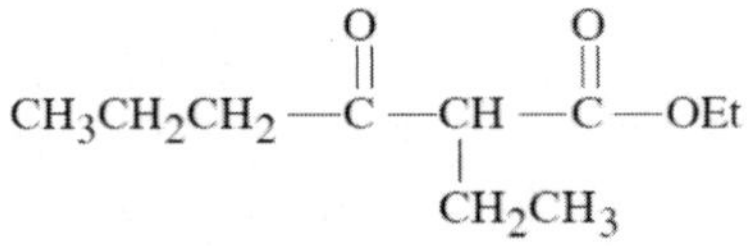

ethyl 2-ethyl-3-oxohexanoate

22.27   Hydroxide ion would *not* be a good base for the Claisen condensation because it would saponify ethyl acetate and convert it into acetate ion. The role of the ethoxy group in ethyl acetate is to serve as a leaving group; acetate ion lacks this leaving group and therefore will not react in the Claisen condensation.

22.28   (a, b) An ordinary Claisen condensation is driven to completion by ionization of an α-hydrogen more acidic than ethanol, the conjugate acid of the base ethoxide. Compound *A* lacks such an acidic α-hydrogen. Consequently, ethoxide reacts with the ketone carbonyl group and converts it into *B*. This reaction is driven to completion by ionization of the acidic α-hydrogen of compound *B*. Compound *B* itself is formed on acidification of the reaction mixture.

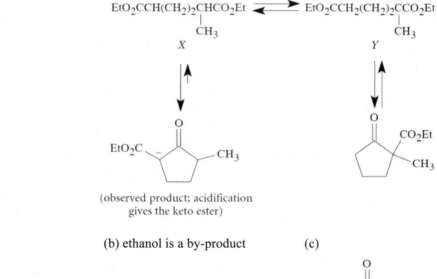

(c)   Use reasoning similar to that used in solving part (a). Two possible enolate ions, *X* and *Y*, can form; the product derived from ion *X* is ultimately observed because only this product is formed irreversibly by ionization.

(observed product; acidification gives the keto ester)

22.29   (a)   (b) ethanol is a by-product   (c)

$$EtO\overset{O}{\underset{}{C}}CH_2\overset{O}{\underset{}{C}}CMe_3$$

$$Ph\overset{O}{\underset{}{C}}CH_2\overset{O}{\underset{}{C}}Ph$$

In all three parts, ethanol is a by-product. In part (c), the carbanion formed at carbon *a* leads to the product. Hydrogens *b* are more acidic than hydrogens *a*, but a carbanion formed at carbon *b* cannot lead to a product containing a ring of reasonable size. Hence, the two anions are in an equilibrium that is driven by ionization of the product (Le Châtelier's principle).

$$\underset{a}{\overset{O}{\parallel}}\quad\underset{b}{\overset{Me}{\mid}}$$
$$CH_3CCH_2CCH(CO_2Et)_2$$
$$\mid$$
$$Me$$

22.30 (a) Apply the analysis shown in Eq. 22.69 on text p. 1136. The two possible sets of starting materials are as follows:

$$CH_3\overset{\overset{\displaystyle O}{\|}}{C}OEt \ + \ CH_3CH_2\overset{\overset{\displaystyle O}{\|}}{C}CH_2CH_3 \quad and \quad CH_3\overset{\overset{\displaystyle O}{\|}}{C}CH_2CH_3 \ + \ EtO\overset{\overset{\displaystyle O}{\|}}{C}CH_2CH_3$$

A   B

Set *A* could work provided that an excess of the ester is used. The analysis is similar to the analysis of Eq. 22.67 on text p. 1135. Because the ketone is symmetrical, only one enolate ion is possible. Set *B* would give a mixture of products because the ketone could form two different conjugate-base enolate ions, and there is no reason for one to be strongly preferred over the other.

(b) The two possible sets of starting materials are as follows:

$$CH_3CH_2\overset{\overset{\displaystyle O}{\|}}{C}CH_2CH_3 \ + \ EtO\overset{\overset{\displaystyle O}{\|}}{C}OEt \quad and \quad 2 \ CH_3CH_2\overset{\overset{\displaystyle O}{\|}}{C}OEt$$

A   B

Either set would work. In set *A*, a large excess of diethyl carbonate must be used. (See Eq. 22.65, text p. 1137, and the discussion that follows.) However, set *B* would be more convenient because only one ester is required as a starting material and because we would not have to separate the product from an excess of starting material, as in set *A*.

(c) Apply the analysis shown in Eq. 22.68 on text p. 1139. The two possible sets of starting materials are as follows:

A   B

Neither set of starting materials is a good choice. Because the desired product has no α-hydrogen more acidic than ethanol, it cannot be formed in a Claisen condensation because the equilibrium would be unfavorable. In fact, *A could* react in a Claisen condensation, but a different product would be formed. (Can you give its structure?) Reactants *B* are also not a good choice for the same reason: the desired product is not acidic enough to be ionized by ethoxide. Furthermore, in this case a different product would be formed.

(d) The two possible sets of starting materials are as follows:

A   B

Set *A* consists of two esters with α-hydrogens; four different products are possible. Furthermore, the α-hydrogens of ethyl phenylacetate are considerably more acidic than those of ethyl valerate (why?); therefore, the major product is likely to be derived from the enolate ion formed at this position, but this is not the desired product. In set *B*, there are two sets of α-hydrogens, but the hydrogens that are α to the phenyl ring are considerably more acidic (why?). Unfortunately, the desired product is *not* derived from the enolate ion formed at this position. Hence, both sets of starting materials are unsatisfactory. In summary, the desired product cannot be made by a Claisen condensation.

22.31   (a)

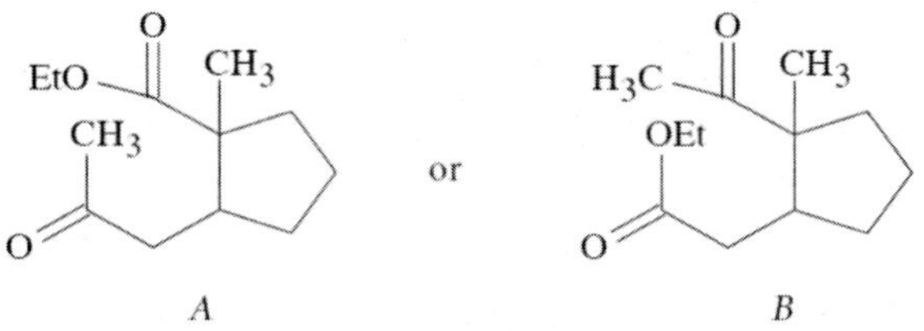

A   or   B

Either answer is satisfactory, but perhaps compound *B* would be a better choice because, in compound *A*, reaction of the ketone enolate as a nucleophile at the ester carbonyl carbon might be somewhat retarded by van der Waals repulsions with the α-methyl group.

(b)

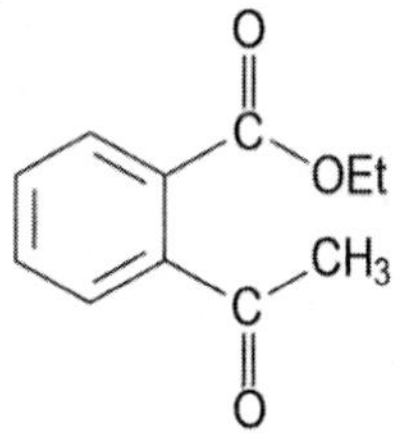

ethyl 2-acetylbenzoate

22.32   Follow the general process outlined in Eq. 22.75, text p. 1144, for every two carbons you want to add to the chain. First, the hexanoyl-ACP is transferred to the thiol of the SYNTH–SH part of the protein, and the ACP–SH is recharged with another malonyl group from malonyl-CoA.

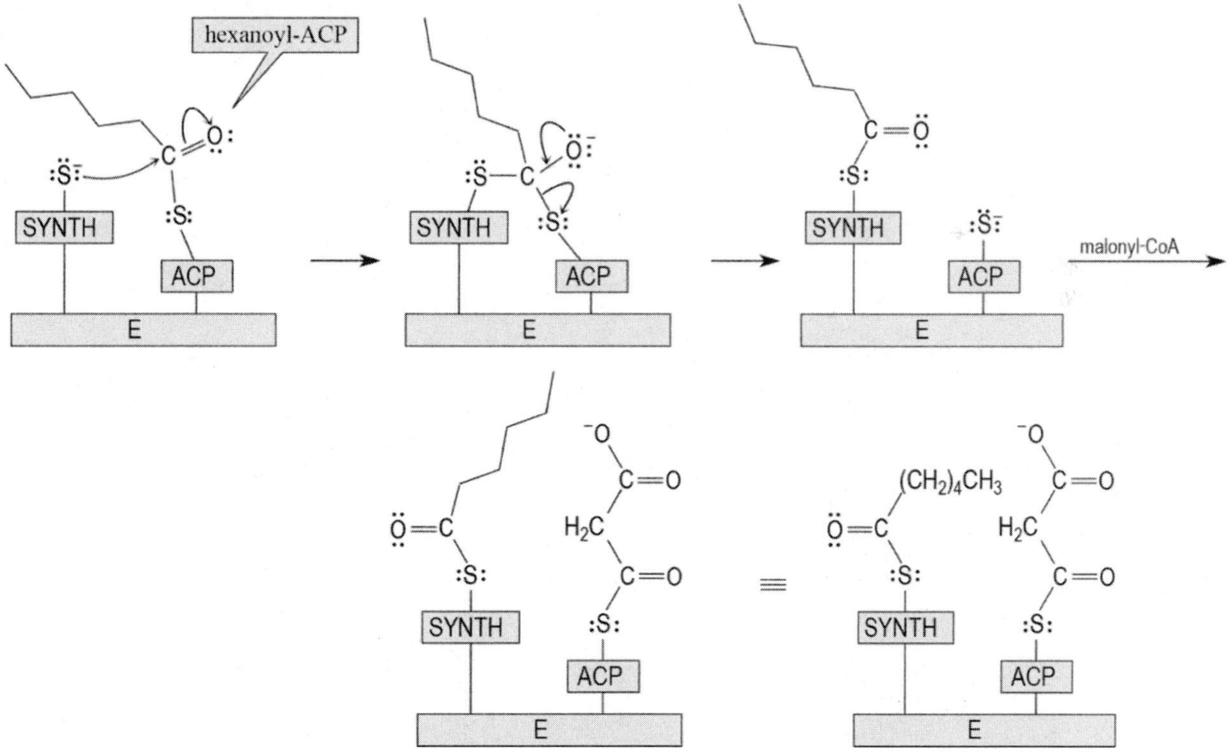

Then, the malonyl group loses carbon dioxide, yielding an enolate (see Eq. 23.73b in the text). The enolate reacts at the adjacent thioester, extending the chain by two more carbons.

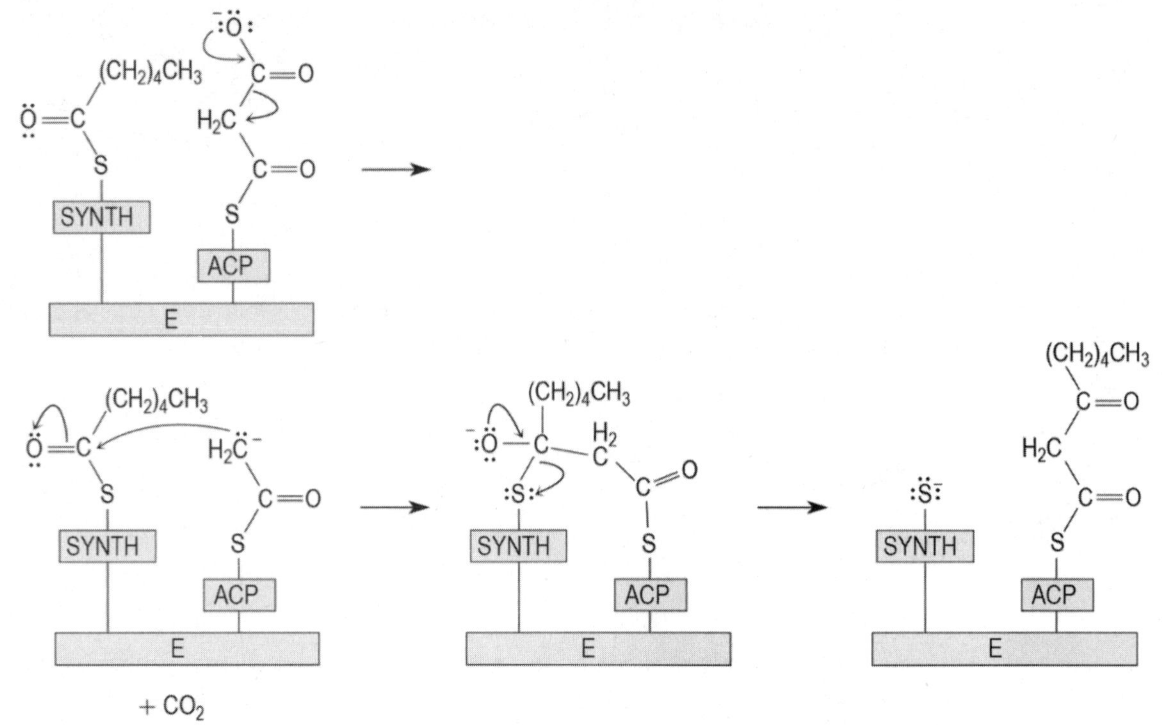

$+ CO_2$

Finally, the SACP-bound chain undergoes reduction, dehydration, and double bond reduction to yield octanoyl-ACP.

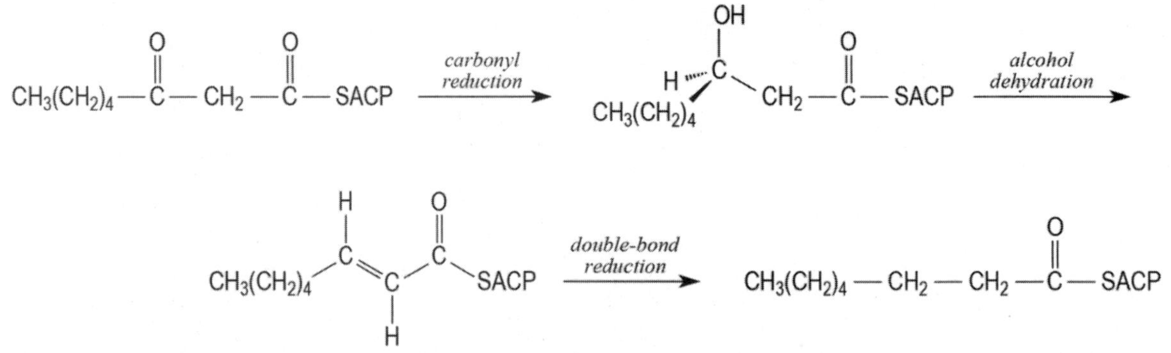

(The entire process can be repeated over and over again, adding two carbons to the growing chain at a time, until it is cleaved from the enzyme as shown in Eq. 22.76 in the text.)

22.33 Acetyl CoA is the ultimate source of carbon atoms in fatty acid biosynthesis, as noted in the text (Eqs. 22.71 – 22.76) and in the solution to problem 22.32 in this manual. If the carbonyl carbon of acetyl CoA is labeled, then every other carbon in a long chain fatty acid, like palmitic acid, will carry the label.

palmitate
(conjugate-base anion of **palmitic acid**)

22.34 (a)

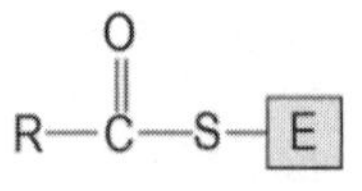

(b) H–A and A⁻ are acids and bases available inside the active site of the enzyme that assist in generating *X* and the acetylated CoA.

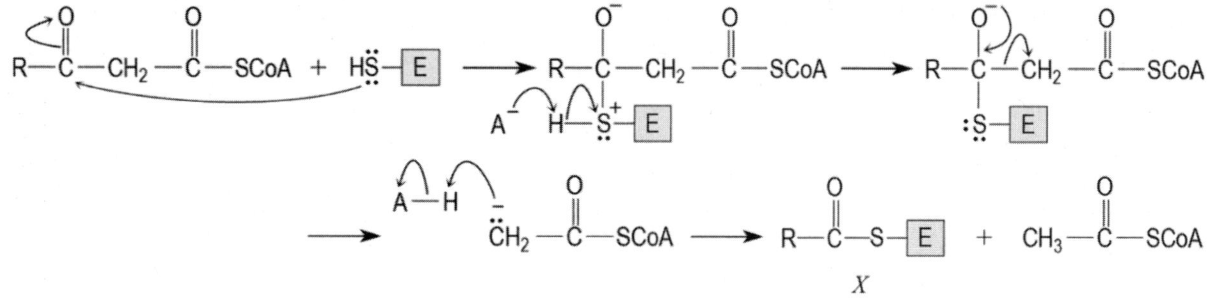

Compound *X* then reacts with CoA–SH to form the acetylated product and to regenerate the free enzyme.

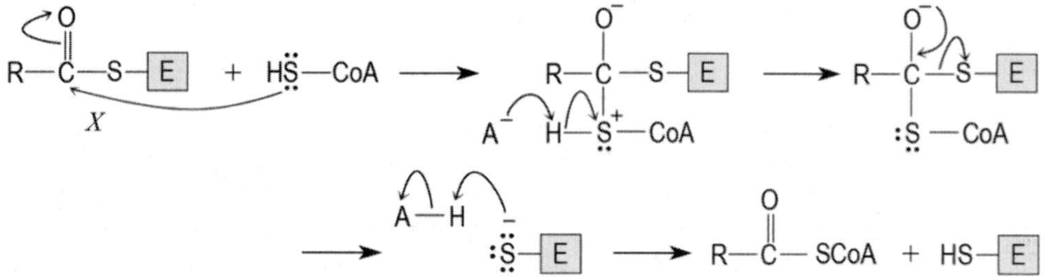

22.35 (a) 3-Phenylpropanoic acid can be prepared by a malonic ester synthesis:

$$CH_2(CO_2Et)_2 \xrightarrow[\text{benzyl chloride}]{\text{1) NaOEt, EtOH; 2) PhCH}_2\text{Cl}} PhCH_2CH(CO_2Et)_2 \xrightarrow[-\text{EtOH},\ -CO_2]{H_3O^+,\ \text{heat}} PhCH_2CH_2CO_2H$$

3-phenylpropanoic acid

Recall that benzyl chloride, $PhCH_2Cl$, is exceptionally reactive in S_N2 reactions. (See Eq. 17.29a–b, text p. 850.)

(b) 2-Ethylbutanoic acid can be prepared by a malonic ester synthesis:

$$CH_2(CO_2Et)_2 \xrightarrow{\text{1) NaOEt, EtOH; 2) EtBr}} EtCH(CO_2Et)_2 \xrightarrow{\text{1) NaOEt, EtOH; 2) EtBr}} Et_2C(CO_2Et)_2 \xrightarrow[-\text{EtOH},\ -CO_2]{H_3O^+,\ \text{heat}} Et_2CHCO_2H$$

2-ethylbutanoic acid

(c) 3,3-Dimethylbutanoic acid cannot be prepared by a malonic ester synthesis because the requisite alkyl halide, *tert*-butyl bromide, is virtually unreactive in S_N2 reactions. (See Sec. 9.4D in the text.)

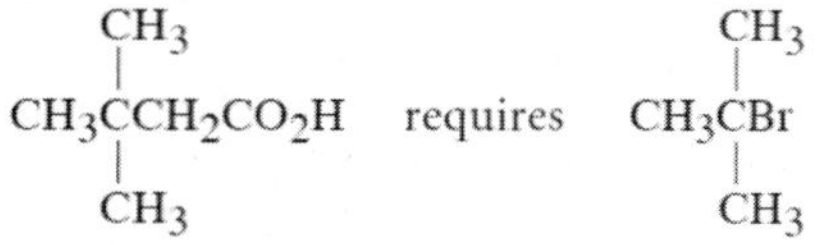

3,3-dimethylbutanoic acid requires *tert*-butyl bromide

22.36 On the assumption that hydrolysis of the ester groups and decarboxylation of one carboxy group have occurred, the unsaturation number (2) of the product calculated from the molecular formula indicates that a ring has been formed. The formation of cyclobutanecarboxylic acid by the following reaction sequence accounts for the results:

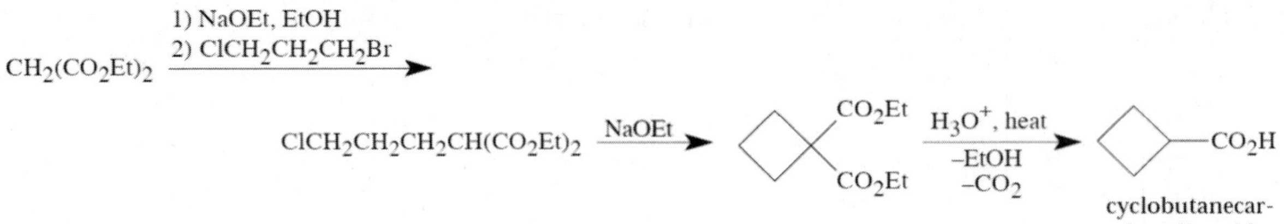

$$CH_2(CO_2Et)_2 \xrightarrow{\begin{array}{c}\text{1) NaOEt, EtOH}\\\text{2) ClCH}_2\text{CH}_2\text{CH}_2\text{Br}\end{array}}$$

$$ClCH_2CH_2CH_2CH(CO_2Et)_2 \xrightarrow{\text{NaOEt}} \quad \xrightarrow[\substack{-EtOH \\ -CO_2}]{H_3O^+, \text{ heat}} \quad -CO_2H$$

cyclobutanecar-
boxylic acid

The ring closure occurs by an internal "S_N2" reaction in which the enolate ion is the nucleophile:

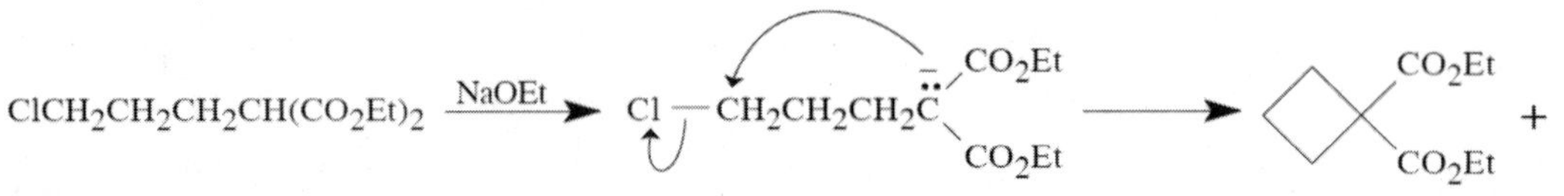

$$ClCH_2CH_2CH_2CH(CO_2Et)_2 \xrightarrow{\text{NaOEt}} \quad + \ Cl^-$$

22.37   (a)   Recall Sec. 18.1, text p. 880, where the four reasons for the lack of reactivity of aryl halides under S_N2 conditions is discussed. Firstly, the relatively high energy of the transition state required for the conversion of an sp^2-hybridized carbon into an sp-hybridized carbon (about 21 kJ mol^{-1} or 5 kcal mol^{-1}) that must occur in an S_N2 reaction at an aryl carbon. Secondly, the nucleophile, in this case the conjugate-base enolate of diethyl malonate, must approach the carbon–bromine bond in the plane of the phenyl ring, which would result in significant van der Waals repulsions (a steric effect) of both the nucleophile and the phenyl ring. Thirdly, the nucleophile must also approach the carbon–bromine bond through the plane of the phenyl ring, which is impossible. And fourthly, because the carbon at which substitution occurs would have to undergo stereochemical inversion, the reaction would necessarily yield a benzene derivative containing a twisted and highly strained double bond.

   (b)   Review the fundamental reactions of transition-metal complexes (Sec. 18.5E, text p. 896).
   Let L = P(t-Bu)$_3$.

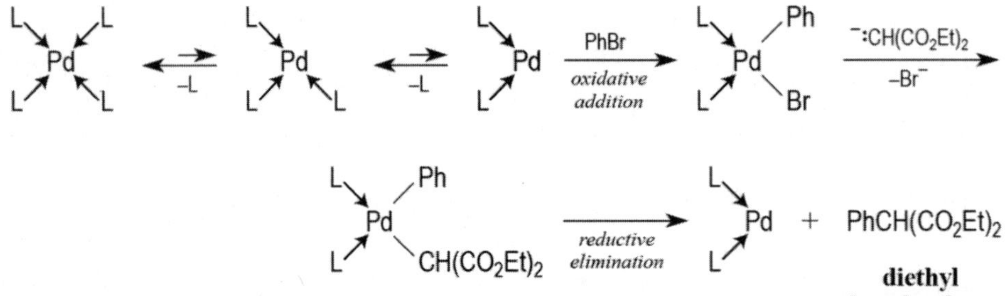

diethyl
phenylmalonate

22.38   Because part (c) involves formation of a carboxylic acid with a quaternary α-carbon, and because such compounds cannot be prepared by a malonic ester synthesis, we'll use our one allowed reaction of amide bases to form this compound. All of the other compounds can be prepared by the malonic ester synthesis, as follows:

   (a)   Note that allylic halides such as allyl bromide (H$_2$C=CHCH$_2$Br) are very reactive in S_N2 reactions. (See Eq. 17.28a–b, text p. 850.)

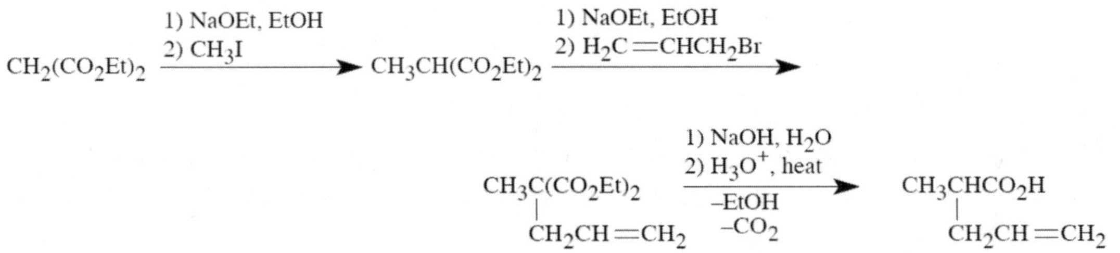

$$CH_2(CO_2Et)_2 \xrightarrow{\begin{array}{c}\text{1) NaOEt, EtOH}\\\text{2) CH}_3\text{I}\end{array}} CH_3CH(CO_2Et)_2 \xrightarrow{\begin{array}{c}\text{1) NaOEt, EtOH}\\\text{2) H}_2\text{C}=\text{CHCH}_2\text{Br}\end{array}}$$

$$\underset{\underset{CH_2CH=CH_2}{|}}{CH_3C(CO_2Et)_2} \xrightarrow[\substack{-EtOH \\ -CO_2}]{\begin{array}{c}\text{1) NaOH, H}_2\text{O}\\\text{2) H}_3\text{O}^+, \text{ heat}\end{array}} \underset{\underset{CH_2CH=CH_2}{|}}{CH_3CHCO_2H}$$

2-methyl-4-pentenoic
acid

(b)

$$CH_2(CO_2Et)_2 \xrightarrow[\text{2) PrBr}]{\text{1) NaOEt, EtOH}} PrCH(CO_2Et)_2 \xrightarrow[\text{2) PrBr}]{\text{1) NaOEt, EtOH}} Pr_2C(CO_2Et)_2 \xrightarrow[\substack{-EtOH \\ -CO_2}]{H_3O^+,\ heat} Pr_2CHCO_2H$$

valproic
acid
(2-propylpen-
tanoic acid)

(See also the solution to Problem 20.40, text p. 1037, on p. 683 of this manual.)

(c)

$$CH_2(CO_2Et)_2 \xrightarrow[\text{2) EtBr}]{\text{1) NaOEt, EtOH}} EtCH(CO_2Et)_2 \xrightarrow[\text{2) EtBr}]{\text{1) NaOEt, EtOH}} Et_2C(CO_2Et)_2 \xrightarrow[\substack{-EtOH \\ -CO_2}]{H_3O^+,\ heat}$$

$$Et_2CHCO_2H \xrightarrow[\text{H}_2\text{SO}_4\ (catalyst)]{\text{EtOH (solvent)}} Et_2CHCO_2Et \xrightarrow[\text{2) H}_2C{=}CHCH_2Br]{\text{1) LDA}} \underset{\substack{| \\ CH_2CH{=}CH_2}}{Et_2CCO_2Et}$$

2,2-diethyl-4-pen-
tenoic acid

22.39   This is a transition-metal catalyzed reaction similar to the one shown in the solution to Problem 22.37(b).

$$Ph{-}\underset{\underset{CH_3}{|}}{\overset{\overset{CH_3}{|}}{C}}{-}\overset{\overset{O}{\|}}{C}{-}OEt \quad \textbf{ethyl 2-methyl-2-phenylpropanoate}$$

22.40   (a)

$$\underset{\text{ethyl acetoacetate}}{CH_3\overset{\overset{O}{\|}}{C}CH_2CO_2Et} \xrightarrow[\substack{\text{2) BrCH}_2\text{CH(CH}_3)_2 \\ \text{isobutyl bromide}}]{\text{1) NaOEt, EtOH}} \underset{\underset{CH_2CH(CH_3)_2}{|}}{CH_3\overset{\overset{O}{\|}}{C}CHCO_2Et} \xrightarrow[\text{2) H}_3O^+,\ -CO_2]{\text{1) NaOH}} \underset{\text{5-methyl-2-hexanone}}{CH_3\overset{\overset{O}{\|}}{C}CH_2CH_2CH(CH_3)_2}$$

(b)

$$\underset{\text{ethyl acetoacetate}}{CH_3\overset{\overset{O}{\|}}{C}CH_2CO_2Et} \xrightarrow[\substack{\text{2) BrCH}_2\text{Ph} \\ \text{benzyl bromide}}]{\text{1) NaOEt, EtOH}} \underset{\underset{CH_2Ph}{|}}{CH_3\overset{\overset{O}{\|}}{C}CHCO_2Et} \xrightarrow[\text{2) H}_3O^+,\ -CO_2]{\text{1) NaOH}} \underset{\text{4-phenyl-2-butanone}}{CH_3\overset{\overset{O}{\|}}{C}CH_2CH_2Ph}$$

22.41   (a)   Follow the hint in Study Guide Link 22.6 on p. 752 of this manual.

$$\underset{}{CH_3\overset{\overset{CO_2Et}{|}}{CH}{-}\overset{\overset{O}{\|}}{C}{-}CH_2CH_3} \xrightarrow[\text{2) PhCH}_2Br]{\text{1) NaOEt, EtOH}} \underset{\underset{CH_2Ph}{|}}{CH_3\overset{\overset{CO_2Et}{|}}{C}{-}\overset{\overset{O}{\|}}{C}{-}CH_2CH_3} \xrightarrow[\text{2) H}_3O^+,\ -CO_2]{\text{1) NaOH}} \underset{\underset{CH_2Ph}{|}}{CH_3\overset{\overset{O}{\|}}{CH}CCH_2CH_3}$$

The starting β-keto ester is prepared by a Claisen condensation of ethyl propionate:

$$\underset{\text{ethyl propionate}}{2\ EtO\overset{\overset{O}{\|}}{C}CH_2CH_3} \xrightarrow[\text{2) H}_3O^+]{\text{1) NaOEt, EtOH}} \underset{}{CH_3\overset{\overset{CO_2Et}{|}}{CH}{-}\overset{\overset{O}{\|}}{C}CH_2CH_3}$$

(b)

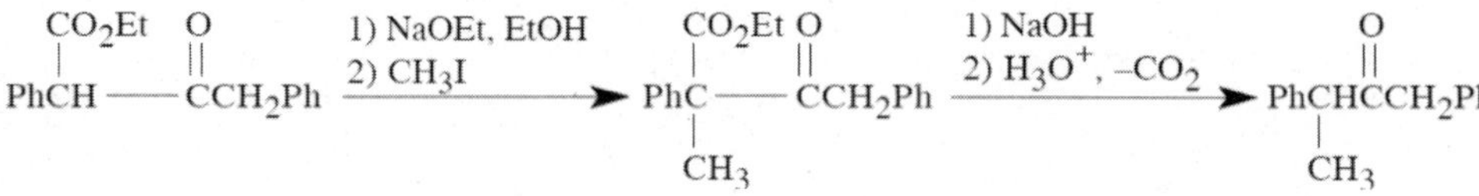

The starting β-keto ester is prepared by a Claisen condensation of ethyl phenylacetate:

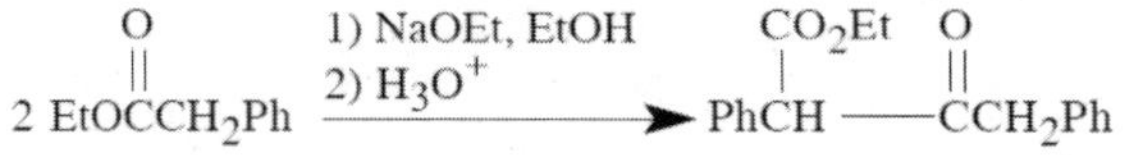

22.42 Compound A is the enolate ion of diethyl malonate. Reaction of the enolate ion as a nucleophile at the methylene carbon of the epoxide gives the alkoxide B, which, in turn, undergoes an intramolecular transesterification to give a lactone.

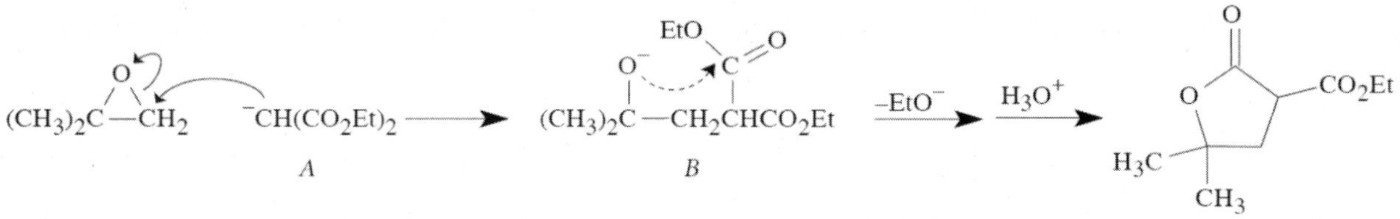

22.43 Once the enolate has been generated, it can react with either enantiomer of 2-phenylpropanal (reaction with the R enantiomer is shown). The products are diastereomers, formed in unequal amounts. The enantiomer of each product shown is also formed.

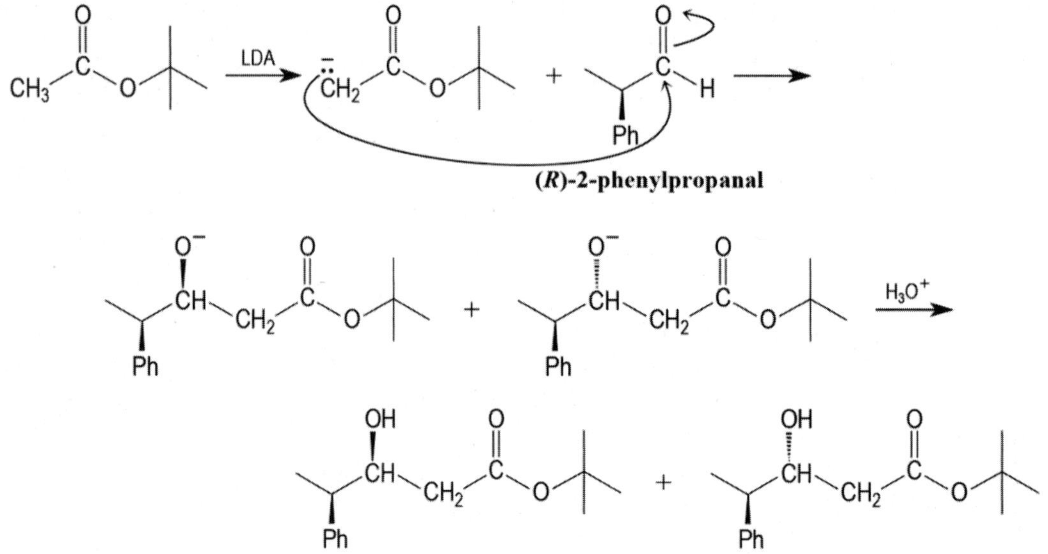

(R)-2-phenylpropanal

22.44 (a) Treatment of the a-bromoester with Zn yields a Reformatsky reagent (text Sec. 22.8D), an ester enolate. An aldol reaction with the ketone gives the zinc alkoxide, A. Treatment of A with acid would yield an aldol addition product.

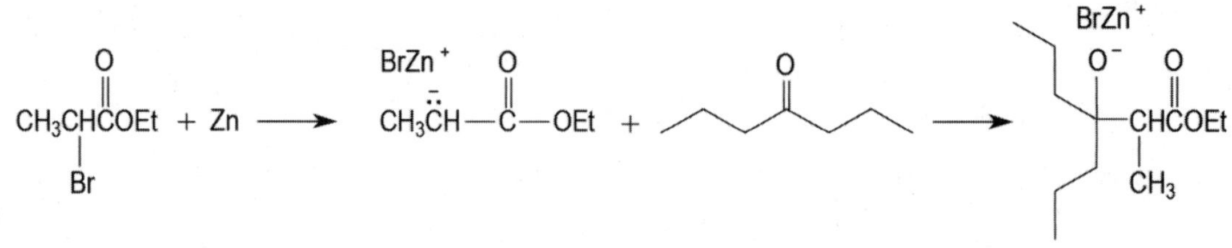

A

(b)   Compound *B* is an aldol condensation product, formed by a Knoevenagel reaction (text Sec. 22.8D). The mechanism is analogous to that shown in Eqs. 22.93a–b, text p. 11.54. Compound *C* is a carboxylic acid, as the problem states, formed by decarboxylation of the diester (text Sec. 20.11).

22.45   A transthioesterification transfers the acetyl group from acetyl-CoA onto the thiol of the enzyme.

An aldol addition reaction links the acetolacetyl-CoA to the enzyme. Finally, hydrolysis of the thioester releases the HMG-CoA from the enzyme.

22.46   (a)   (b)

methyl methacrylate
(c)   (d)

In part (d), saponification occurs instead of conjugate addition because saponification is an *irreversible* nucleophilic acyl substitution reaction.

When considering acid-catalyzed additions to α,β-unsaturated carbonyl compounds, pay careful attention to the mechanism in Eq. 22.106 on text p. 1160. In part (c), it is tempting to add HBr to the double bond to give the tertiary carbocation A; however, such a mechanism would lead to a product B that is different from the product observed:

Remember that carbocations α to carbonyl groups are quite unstable and are almost never formed; see Eq. 22.38 on text p. 1119. In contrast, the mechanism in Eq. 22.106 on text p. 1160 involves a ketone that is protonated on the carbonyl oxygen; this is a resonance-stabilized species in which positive charge is shared at the terminal carbon. Reaction of bromide with this carbon leads to the observed product.

22.47 (a) The product results from two successive conjugate additions. (Proton transfers are shown for simplicity as intramolecular processes, but they could be mediated by other acids and bases in solution.)

(b) This is a conjugate addition of the benzenethiolate anion to the carbon–carbon double bond. The benzenethiolate nucleophile, PhS⁻, is formed by the reaction of benzenethiol, PhSH, with ⁻OEt. A mixture of diastereomers is formed because the final protonation step can occur from the same face as the methyl group or from the face opposite the methyl group.

(mixture of cis and trans)

(c) The conjugate-addition product has no carbon–carbon double bond. Internal rotation within this adduct, followed by reversal of the addition, gives the isomerized product.

22.48 (a) The nucleophile must be a species that can be converted into a carboxymethyl (HO_2CCH_2-) group. Because the nucleophile ends up as a "substituted acetic acid," the conjugate-base enolate ion of diethyl malonate can serve as species X. Addition of X to the ester gives compound A. All ester groups of A are hydrolyzed and one of the resulting carboxy groups is decarboxylated when A is heated in aqueous acid.

(b)   The key is to recognize that the desired product results from the addition of *two equivalents* of acrylonitrile, which end up as the two —CH$_2$CH$_2$CO$_2$H groups of the product. The conjugate-base anion of ethyl acetoacetate would be a suitable nucleophile *Y*. Reaction of *Y* with one equivalent of acrylonitrile would give compound *B*. Ionization of *B* gives anion *C*, which reacts with a second equivalent of acrylonitrile to give compound *D*. The nitrile groups are hydrolyzed to carboxy groups and the ester group is hydrolyzed and decarboxylated to give the final product when compound *D* is heated in acid.

22.49   (a)   The first part of the mechanism is a Michael addition followed by loss of formic acid to give, following protonation, intermediate *A*.

Compound *A* then undergoes an intramolecular aldol condensation reaction.

(b) A Michael addition of the diethyl malonate conjugate-base anion gives compound *B*, which undergoes an intramolecular Claisen (Dieckmann) condensation. This Dieckmann condensation, like all Claisen-type condensations, is driven to completion by ionization of the product; the un-ionized product is formed when acid is added. These final steps are not shown in the mechanism below. Note that the tertiary hydrogen of compound *B* is the most acidic hydrogen, but although the anion resulting from removal of this proton is formed in the reaction mixture, it is depleted by the reaction that gives the product.

B (redrawn)

(ionizes under the basic conditions, and is regenerated when acid is added.)

In the foregoing mechanisms many of the steps are reversible and should rigorously be shown with equilibrium arrows. However, the focus of the problem is on the mechanism of the forward reaction and not on the reversibility of the reaction. In such cases, forward arrows only are shown.

22.50 (a)

NADPH

***trans*-crotonyl-ACP**

butyryl-ACP

NADP⁺

(b) Since the hydride (H⁻) was added to the β-carbon, the reaction is a conjugate addition.

(c) To solve this problem, first determine the structure of (2*S*,3*R*)-butyryl-ACP-2,3-d_2.

Next, rotate along the C2–C3 bond into the eclipsed conformation. Then, rotate the the structure so that the deuteriums are in the plane of the page. "Remove" the two deuteriums and analyze the resulting alkene. Since the alkene is trans, it confirms that the deuteriums were added syn.

trans-**crotonyl-ACP**

 Had the alkene that resulted from this analysis been cis, then we would have deduced that the addition of deuteriums was anti. This technique will always work for these types of problems.

22.51   (a)

$$CH_3CH{=}CHCOEt \xrightarrow{H_2,\ catalyst} CH_3CH_2CH_2COEt$$

ethyl 2-butenoate   ethyl butanoate

(b)

$$CH_3CH{=}CHCOEt \xrightarrow[\text{2) }H_3O^+]{\text{1) LiAlH}_4} CH_3CH{=}CHCH_2OH\ +\ EtOH$$

ethyl 2-butenoate   2-buten-1-ol

22.52   (a)

$$(CH_3)_2C{=}CHCCH_3 \xrightarrow[\text{2) }H_3O^+]{\text{1) Li}^+\text{(CH}_3)_2\text{Cu}^-} (CH_3)_3CCH_2CCH_3$$

mesityl oxide   4,4-dimethyl-2-pentanone
(4-methyl-3-penten-2-one)

(b)

$$(CH_3)_2C{=}CHCCH_3 \xrightarrow[\text{2) dil. }H_3O^+]{\text{1) CH}_3\text{Li}} (CH_3)_2C{=}CHCCH_3$$

mesityl oxide   with OH and CH₃ substituents
(4-methyl-3-penten-2-one)

2,4-dimethyl-2-penten-1-ol

(c)

$$(CH_3)_2C{=}CHCCH_3 \xrightarrow[\text{2) }H_3O^+]{\text{1) Li}^+\text{(C}_2\text{H}_5)_2\text{Cu}^-} C_2H_5CCH_2CCH_3 \xrightarrow[\text{2) }H_2SO_4]{\text{1) CH}_3\text{MgI}} C_2H_5CCH{=}C(CH_3)_2$$

2,4,4-trimethyl-2-hexene

Notice that the conjugate addition must *precede* the carbonyl addition. (If the carbonyl addition were carried out first, the product would not be an α,β-unsaturated carbonyl compound, and conjugate addition could not occur.)

22.53 (a) The question in this problem is what type of conjugate addition occurs (1,4 or 1,6). The 1,6-addition occurs because it gives the enolate ion intermediate with the greater number of resonance structures and hence, the more stable enolate ion.

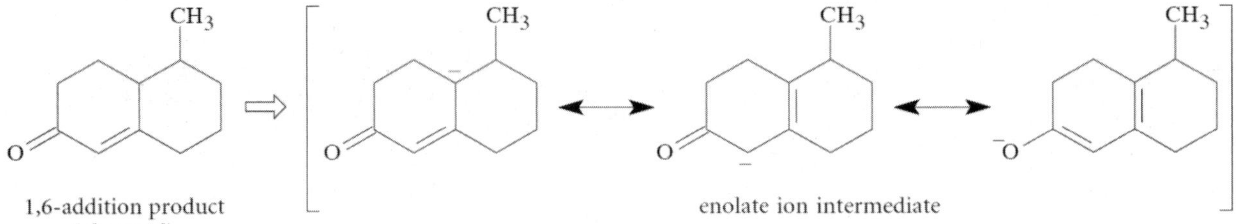

1,6-addition product
(observed)

enolate ion intermediate

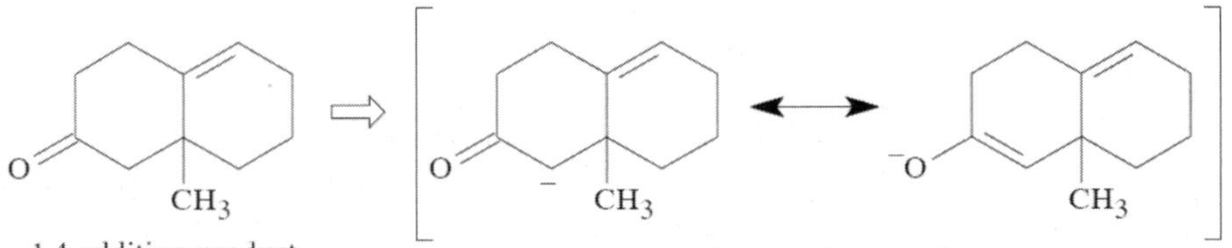

1,4-addition product

enolate ion intermediate

(b) Conjugate addition also occurs with α,β-unsaturated compounds in which a carbon–carbon triple bond is part of the conjugated π-electron system. The product is $(CH_3)_2C{=}CHCO_2Me$ (methyl 3-methyl-2-butenoate).

22.54 (a) An analysis similar to that used in Study Problem 22.7 on text p. 1170 reveals that either a methyl group or an ethyl group can be added in the conjugate addition. (Addition of an ethyl group is illustrated here.)

$$CH_3CC{=}CHCH_3 \xrightarrow[\text{2) } H_3O^+]{\text{1) } Li^+ (C_2H_5)_2Cu^-} CH_3CCHCHC_2H_5$$

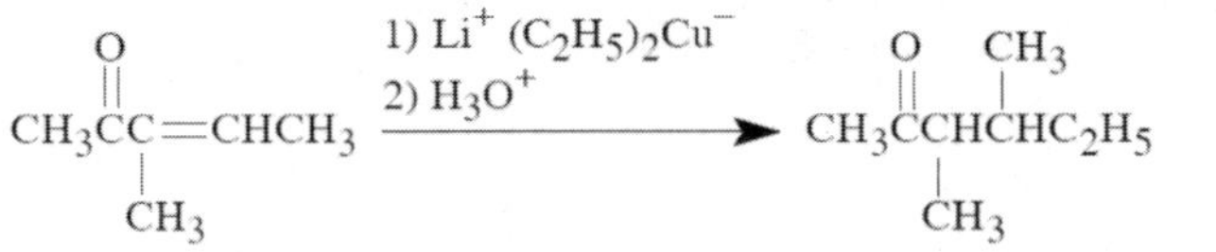

3-methyl-3-penten-2-one   3,4-dimethyl-2-hexanone

(b) An analysis similar to that used in Study Problem 22.7, text p. 1170, reveals that either a methyl group or an ethyl group can be added in the conjugate addition. (Addition of a methyl group is illustrated here.)

$$CH_3CC{=}CHC_2H_5 \xrightarrow[\text{2) } H_3O^+]{\text{1) } Li^+ (CH_3)_2Cu^-} CH_3CCHCHC_2H_5$$

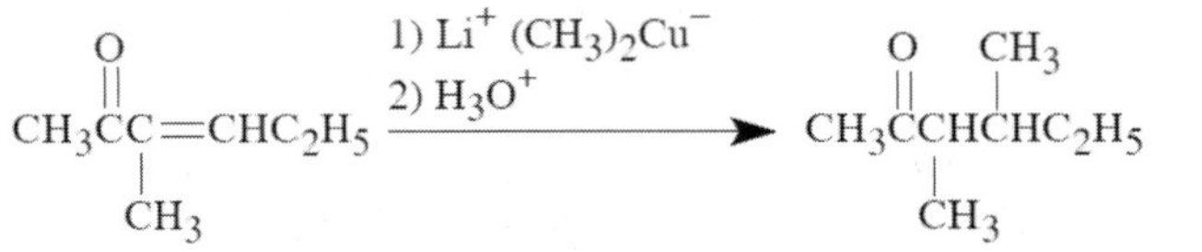

3-methyl-3-hexen-2-one   3,4-dimethyl-2-hexanone

(c) An analysis along the lines of Study Problem 22.7, text p. 1170, reveals the following two possibilities:

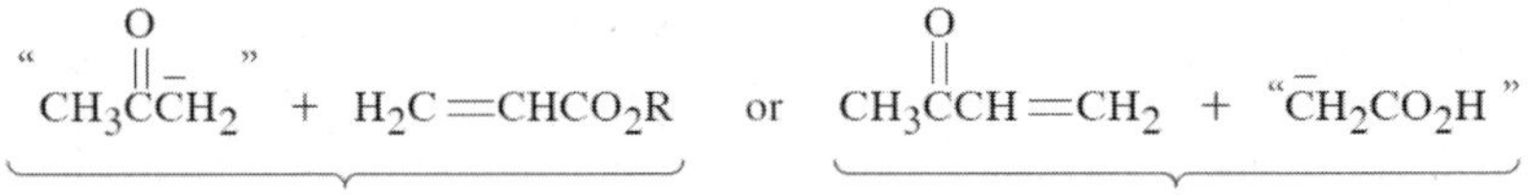

"$CH_3CCH_2^-$" + $H_2C{=}CHCO_2R$ or $CH_3CCH{=}CH_2$ + "$^-CH_2CO_2H$"

(a)   (b)

A practical synthetic equivalent to the anion in (a) is the conjugate base of ethyl acetoacetate:

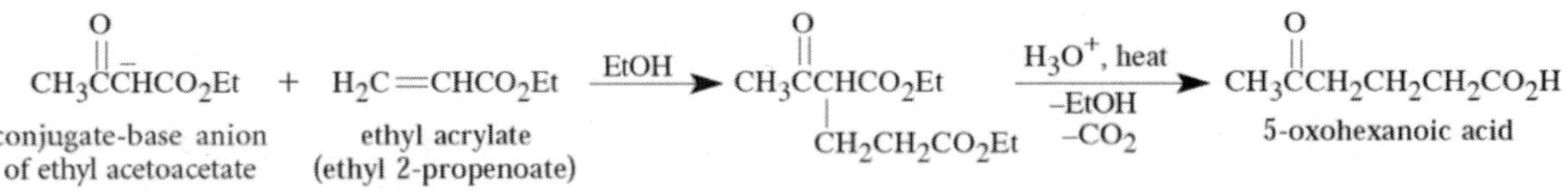

$$CH_3C\overset{-}{C}HCO_2Et + H_2C{=}CHCO_2Et \xrightarrow{\text{EtOH}} CH_3CCHCO_2Et \xrightarrow[\text{–EtOH } -CO_2]{H_3O^+,\text{ heat}} CH_3CCH_2CH_2CH_2CO_2H$$

conjugate-base anion   ethyl acrylate $CH_2CH_2CO_2Et$ 5-oxohexanoic acid
of ethyl acetoacetate   (ethyl 2-propenoate)

The practical synthetic equivalent to the anion in (b) is the conjugate base of diethyl malonate:

$$CH_3\overset{O}{\overset{\|}{C}}CH=CH_2 \; + \; {}^-CH(CO_2Et)_2 \xrightarrow{EtOH} CH_3\overset{O}{\overset{\|}{C}}CH_2CH_2CH(CO_2Et)_2 \xrightarrow[\substack{-EtOH \\ -CO_2}]{H_3O^+,\, heat} CH_3\overset{O}{\overset{\|}{C}}CH_2CH_2CH_2CO_2H$$

methyl vinyl ketone   conjugate-base   5-oxohexanoic acid
(3-buten-2-one)   anion of
diethyl malonate

(d)   An analysis similar to that in Study Problem 22.7, text p. 1170, suggests the following possibilities:

$$\underbrace{CH_3\overset{O}{\overset{\|}{C}}{}^- \; + \; H_2C=CHCO_2R}_{(a)} \quad or \quad \underbrace{CH_3\overset{O}{\overset{\|}{C}}CH=CH_2 \; + \; \overset{O}{\overset{\|}{C}}OH}_{(b)}$$

A practical equivalent for the anion in (a) might be an acetylide anion, $^-C\!\equiv\!CH$. The resulting acetylene could then be hydrated to the desired ketone. Unfortunately, acetylenic cuprate reagents do not work in conjugate additions because the acetylenic ligands do not transfer from the copper. Nevertheless, if you came up with this possibility, you are analyzing the problem with considerable sophistication.
     Anion (b) does have a simple practical equivalent: the cyanide ion:

$$CH_3\overset{O}{\overset{\|}{C}}CH=CH_2 \; + \; {}^-C\!\equiv\!N \xrightarrow[\substack{trace\ of \\ acid}]{EtOH} CH_3\overset{O}{\overset{\|}{C}}CH_2CH_2C\!\equiv\!N \xrightarrow{H_3O^+,\, heat} CH_3\overset{O}{\overset{\|}{C}}CH_2CH_2\overset{O}{\overset{\|}{C}}OH$$

methyl vinyl ketone   levulinic acid
(3-buten-2-one)   (4-oxopentanoic acid)

Solutions to Additional Problems

22.55   The structure of the starting material is

$$H_2C=CH\overset{O}{\overset{\|}{C}}CH_3$$

3-buten-2-one

(a)   (b)   (c)   (d)

$$\underset{(e)}{BrCH_2CH_2\overset{O}{\overset{\|}{C}}CH_3} \qquad \underset{(f)}{CH_3CH_2\overset{O}{\overset{\|}{C}}CH_3} \qquad \underset{(g)}{H_2C=CH\overset{OH}{\overset{|}{C}}HCH_3} \qquad \underset{(h)}{N\!\equiv\!CCH_2CH_2\overset{O}{\overset{\|}{C}}CH_3}$$

$$CH_3CH_2CH_2CH_2\overset{O}{\overset{\|}{C}}CH_3 \qquad (EtO_2C)_2CHCH_2CH_2\overset{O}{\overset{\|}{C}}CH_3$$

In part (h), the α,β-unsaturated ketone serves as a Diels–Alder dienophile.

22.56   The structure of the starting material is

$$\underset{H_3C}{\overset{H}{}}\!\!\!\diagdown C=C \diagup\!\!\!\overset{CO_2Et}{\underset{H}{}}$$

ethyl *trans*-2-butenoate

(a)

$$HO_2CCHCH_2CO_2H$$
$$\overset{|}{CH_3}$$

(b)

$$(CH_3)_2NCHCH_2CO_2Et$$
$$\overset{|}{CH_3}$$

(c)

$$\underset{H_3C}{\overset{H}{>}}C=C\underset{H}{\overset{CO_2^-}{<}} \quad + \quad EtOH$$

(d)

$$\underset{H_3C}{\overset{H}{>}}C=C\underset{H}{\overset{\overset{\displaystyle OH}{|}}{\overset{\displaystyle C(CH_3)_2}{<}}} \quad + \quad EtOH$$

(e)

$$CH_3CH_2CH_2CO_2Et$$

(f)

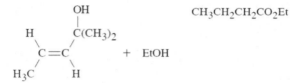

(both compounds are racemates)

22.57   The structure of the starting material is

$$\underset{\text{(isobutyraldehyde structure)}}{(CH_3)_2CH-\overset{\overset{\displaystyle O}{\|}}{C}-H}$$

In any answer where the condensation product (α,β-unsaturated carbonyl compound) is shown, the addition product (β-hydroxyketone) would also be an acceptable answer.

(a)

(b)

condensation product cannot
be formed (no α-protons exist
after the aldol addition)

(c)

by a Reformatsky reaction
(Sec. 22.8D)

(d)

by a Knoevenagel reaction
(Sec. 22.8D)

(e)

22.58   (a)

$$CH_3CH_2\overset{\overset{\displaystyle O}{\|}}{\underset{\overset{|}{CH_3}}{C}}... $$

$$CH_3CH_2CHCCH_3$$
$$\overset{|}{CH_3}$$

(either enantiomer)

(b)

$$CH_3CH_2\overset{\overset{\displaystyle O}{\|}}{C}CH_2CH_2CH_3$$

(two of several possibilities)

(c)

(d)

$$CH_3CH_2CHCH_2CH{=}O$$
$$|$$
$$CH_3$$

(Each could be any one of several stereoisomers—
six possible stereoisomers for the middle structure
and four possible stereoisomers for the other two.)

22.59   (a)   (b)   (c)

$$\overset{O}{\overset{\|}{CH_3CCH_2CH_3}}$$ $$\overset{O}{\overset{\|}{(CH_3)_2CHCH}} \quad or \quad \overset{O}{\overset{\|}{CH_3CH_2CH_2CH}}$$ $$\overset{OH}{\overset{|}{H_2C{=}CHCHCH_3}}$$

For part (c), recall (Eq. 22.28, text p. 1116) that alcohols of the form

$$\overset{OH}{\overset{|}{{-}CH{-}CH_3}}$$

give a positive haloform text, because the basic iodine solution used in the iodoform reaction oxidizes secondary alcohols. A secondary alcohol of this form is thus oxidized to a methyl ketone, which then gives the iodoform test.

22.60   (a)   This compound is an "ynol," the acetylenic analog of an enol. It is spontaneously converted into its isomeric ketene derivative, $CH_3CH{=}C{=}O$.

(b)   This compound, 1,3,5-cyclohexanetrione, exists as its "triene-triol" isomer phloroglucinol because the latter is aromatic.

1,3,5-benzenetriol
(phloroglucinol)

(c)   This compound is a hemiacetal, and it spontaneously decomposes to acetaldehyde and an enol *A*, which, in turn, spontaneously forms propionaldehyde.

$$\overset{OH}{\overset{|}{CH_3CHOCH{=}CHCH_3}} \longrightarrow \overset{O}{\overset{\|}{CH_3CH}} + HOCH{=}CHCH_3 \longrightarrow O{=}CHCH_2CH_3$$
$$\qquad\qquad\qquad\qquad\qquad acetaldehyde \qquad A \qquad\qquad\qquad propionaldehyde$$

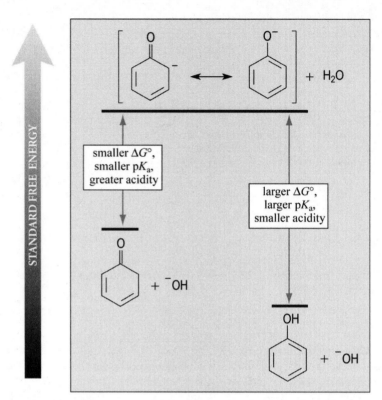

Figure SG22.3 Diagram for the solution to Problem 22.61(b), which shows the effect of relative stability on the acidity of two isomers. Because the conjugate acids of the two isomers are identical, the more stable isomer is less acidic. The phenol isomer is more stable because it is aromatic.

22.61 (a) The conjugate bases of the two compounds, because they are resonance structures, are identical.

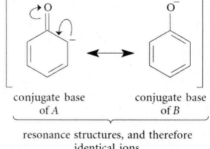

(b) Because compound B (phenol) is aromatic, it is more stable, and thus more energy is required to convert it into the conjugate base shown in part (a). From Eq. 3.34 on text p. 110, $pK_a = \Delta G_a°/2.3RT$. Thus, compound B has the greater pK_a (it is less acidic). This conclusion follows is demonstrated from Fig. SG22.3.

22.62 (a) Removal of the O—H proton from A and a C—H proton from B gives conjugate-base anions that are resonance structures and therefore identical. (B:⁻ = a general base.)

(b) Because the ketone is more stable than its enol, more energy must be expended for it to ionize. Consequently, the ketone is less acidic. The logic is similar to that used in the solution to Problem 22.61(b).

22.63 (a)

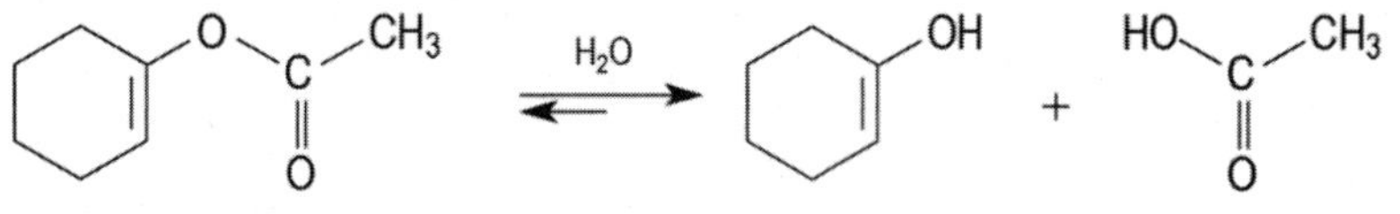

A

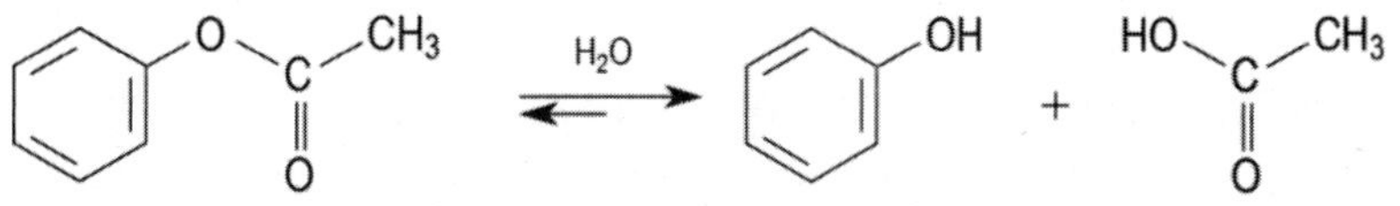

B

(b) Reaction *A* is much more favorable because the alcohol that forms after hydrolysis is an enol. Enols are much more stable in the keto form. Thus, the hydrolysis is essentially irreversible.

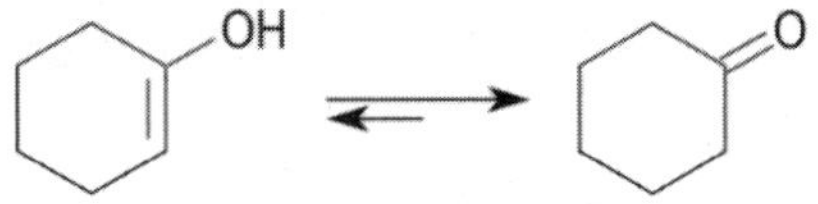

more stable
keto form

22.64 (a) The second compound is most acidic because its conjugate-base anion (structure below) has the greatest number of resonance structures and is therefore most stable. (Draw these structures!) This anion is stabilized by resonance interaction with two carbonyl groups and a benzene ring. In the first compound, the conjugate-base anion lacks the resonance interaction with a benzene ring; and in the third compound, the conjugate-base anion at the central carbon also lacks the resonance interaction with a benzene ring, and the conjugate-base anion at the benzylic carbon lacks the resonance interaction with a second carbonyl group.

(b) The first compound is more acidic because the conjugate-base anion (see following structure) has the greater number of resonance structures. It is stabilized by resonance interaction with both the carbon–carbon double bond and the carbonyl group. The two possible enolate ions of the second compound lack the resonance interaction with a carbon–carbon double bond.

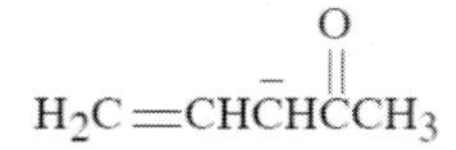

conjugate-base anion
of the most acidic compound

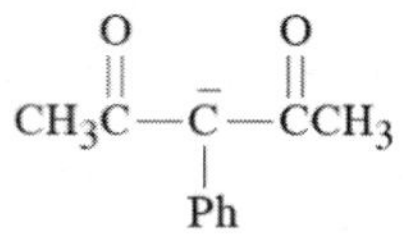

conjugate-base anion
of the most acidic compound

22.65 The order of increasing acidity (decreasing pK_a) is as follows. [Approximate pK_a values are in brackets.]

$$(3)\ [42] < (4) \approx (5)\ [25] < (1)[15] < (6)\ [10] < (2)\ [4.5]$$

Toluene is much less acidic than phenol, and amides are much less acidic than carboxylic acids (element effect). The pK_a values of 1-alkynes and esters are very similar (about 25). The remaining rankings require knowledge of approximate pK_a values. The point of this problem is that it is very important to know the typical pK_a values of organic compounds because so much of organic reactivity is based on acid–base principles.

22.66 (a) Bromination can occur at either of the two α-carbons; compounds *B* and *C* are diastereomers. (Each of the three compounds can also exist as enantiomers.)

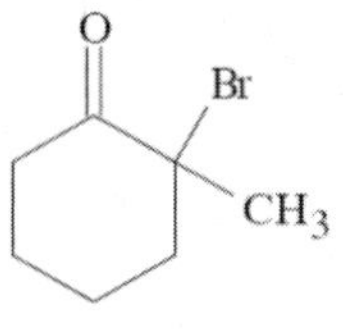

A

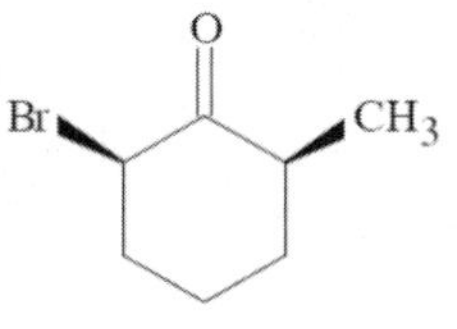

B

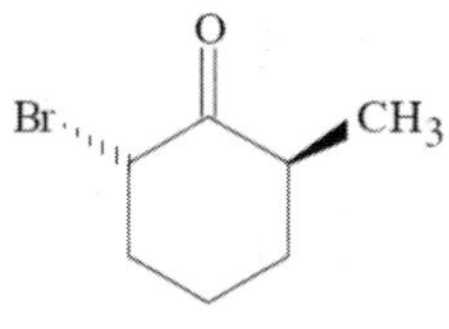

C

(b)   The enol leading to *A* is the more stable one because it has the greater number of alkyl substituents on the double bond. Hence, compound *A* is the α-bromo derivative that is formed.

enol intermediate
involved in the
formation of *A*
(more stable enol)

enol intermediate
involved in the
formation of
compounds *B* and *C*

22.67   The dibromo derivative is a mixture of diastereomers: the ($\pm$)-diastereomer (that is, the racemate) and the meso diastereomer.

($\pm$)-1,3-dibromo-1,3-diphenyl-2-propanone

meso-**1,3-dibromo-1,3-diphenyl-2-propanone**

22.68   The discussion on text pp. 1113–4 explains that enols are intermediates in the acid-catalyzed bromination of aldehydes and ketones, and that enol formation is the rate-limiting process. As explained in Sec. 20.11A, an enol intermediate is also involved in the decarboxylation of a β-keto acid. Thus, the bromination reaction described in the problem evidently involves an enol intermediate, and formation of the enol is the rate-limiting process. Hence, the rate of the bromination reaction is zero-order in bromine.

rate-limiting
process

$$CH_3CCH_2COH \longrightarrow CO_2 + CH_3C{=}CH_2 \xrightarrow{Br_2} CH_3CCH_2Br + HBr$$

22.69   It is reasonable to hypothesize that the carbon atom of the iodoform must originate from the central carbon of the β-diketone:

source of iodoform

$$Ph-C-CH_2-C-Ph$$

source of benzoic acid

Base-promoted iodination occurs at the central carbon by the usual mechanism involving enolate ions to give the diiodo ketone *A*. Hydroxide then displaces an α-diiodo enolate anion *B* in a nucleophilic acyl substitution reaction; this resonance-stabilized anion is a fairly good leaving group. Anion *B* is then iodinated, and the resulting triiodo ketone *C* undergoes the final steps of the haloform reaction illustrated in Eq. 22.27e on text p. 1116 to give iodoform and benzoate anion. The two equivalents of benzoate anion produced in this sequence give benzoic acid when the reaction mixture is acidified.

$$Ph-C-CH_2-C-Ph \xrightarrow{I_2, \text{ base}} Ph-C-CI_2-C-Ph \xrightarrow{^-OH}$$

A

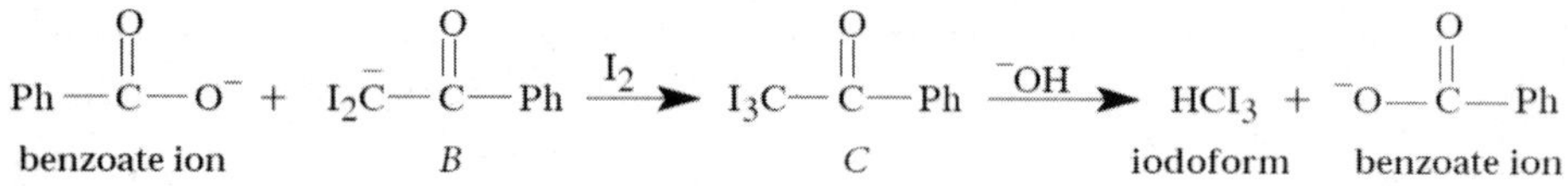

$$Ph-\overset{\overset{\displaystyle O}{\|}}{C}-O^- \ + \ I_2\overset{..}{C}-\overset{\overset{\displaystyle O}{\|}}{C}-Ph \ \xrightarrow{\ I_2\ } \ I_3C-\overset{\overset{\displaystyle O}{\|}}{C}-Ph \ \xrightarrow{\ ^-OH\ } \ HCI_3 \ + \ ^-O-\overset{\overset{\displaystyle O}{\|}}{C}-Ph$$

benzoate ion   *B*   *C*   iodoform   benzoate ion

22.70   **(a)**   The α-hydrogens and the N—H hydrogen are acidic enough to be replaced. These hydrogens are shown as deuteriums in the following structure.

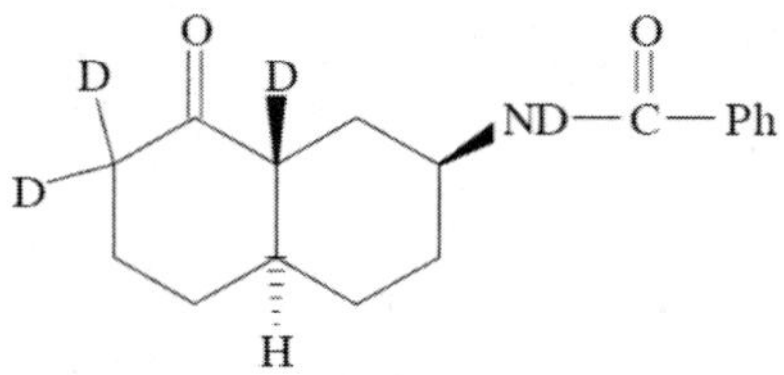

The stereochemistry of the deuterium at the ring junction could be a mixture of "up" and "down" (that is, trans and cis), because the planar carbanion intermediate could be protonated from either face. (See the solution to Problem 22.11a on p. 770 of this manual.) However, because the *trans*-decalin derivatives are more stable than *cis*-decalin derivatives, and because the reaction conditions provide a path for equilibration of the two isomers, the stereoisomer shown above should be the predominant one.

(b)   The two α-hydrogens are acidic enough to be replaced. These hydrogens are shown as deuteriums in the following structure.

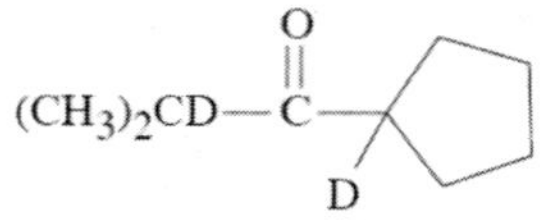

22.71   **(a)**   The isomerization favors compound *B* because trans-decalin derivatives are more stable than *cis*-decalin derivatives. (See Sec. 7.6B, text pp. 295–7.) The isomerization mechanism involves formation of an enolate ion intermediate, which can be protonated on either face to give *A* or *B*, respectively.

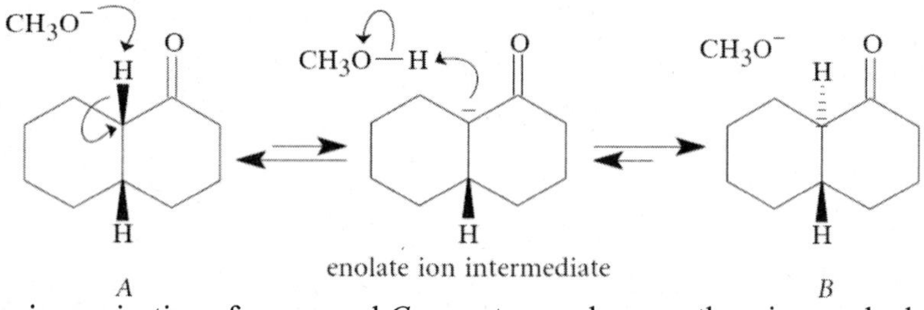

enolate ion intermediate

A   *B*

(b)   The analogous isomerization of compound *C* cannot occur because there is no α-hydrogen at the ring junction. Only the removal of an acidic hydrogen at a ring junction could lead to isomerization, and the hydrogen at the other carbon of the ring junction is not acidic enough to ionize under the conditions given.

22.72

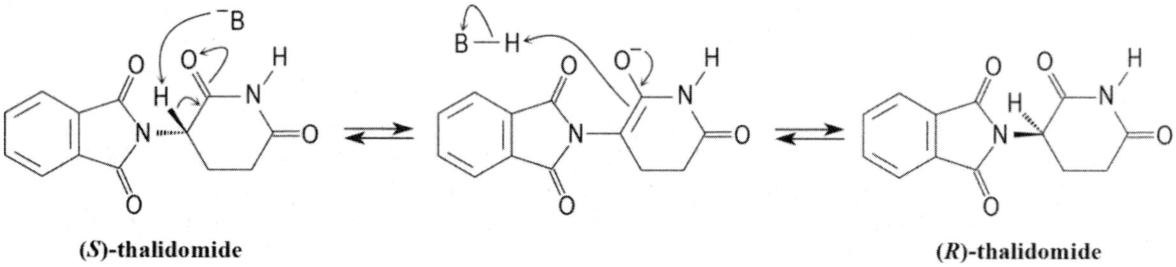

(S)-thalidomide   **(R)-thalidomide**

22.73   **(a)**   The equilibrium favors the α,β-unsaturated isomer because it is conjugated. Conjugation is a stabilizing effect because of the additional π-electron overlap that is possible in conjugated compounds. (See Sec. 15.1A, text pp. 713–716.)

(b)   The mechanism in aqueous base involves a resonance-stabilized enolate ion intermediate, which can be protonated on either the α-carbon or the γ-carbon to give the respective unsaturated ketones.

(c)   The mechanism below commences with the dienol 1,3-cyclohexadienol, which is formed by the usual acid-catalyzed enolization mechanism (Eq. 22.17b on text p. 1112). Protonation of this dienol gives a resonance-stabilized carbocation, which is the same as the protonated α,β-unsaturated ketone. Loss of the O—H proton gives the α,β-unsaturated ketone itself.

1,3-cyclohexadienol

(d)   The equilibrium constant for conversion of 4-methyl-3-cyclohexenone into 4-methyl-2-cyclohexenone should be smaller, because the additional alkyl substituent on the β,γ-double bond tends to offset the stabilizing effect of conjugation in the α,β-unsaturated derivative.

4-methyl-3-cyclohexenone   4-methyl-2-cyclohexenone

22.74   (a)   The exchange of H^a occurs by the usual enolate-ion mechanism.

Removal of either hydrogen H^c or H^d gives an anion that has resonance structures which show that it is an enolate ion. Hence, these hydrogens are acidic enough to undergo exchange. The exchange of H^c is shown explicitly below; you should show the exchange of H^d, which occurs by essentially the same mechanism.

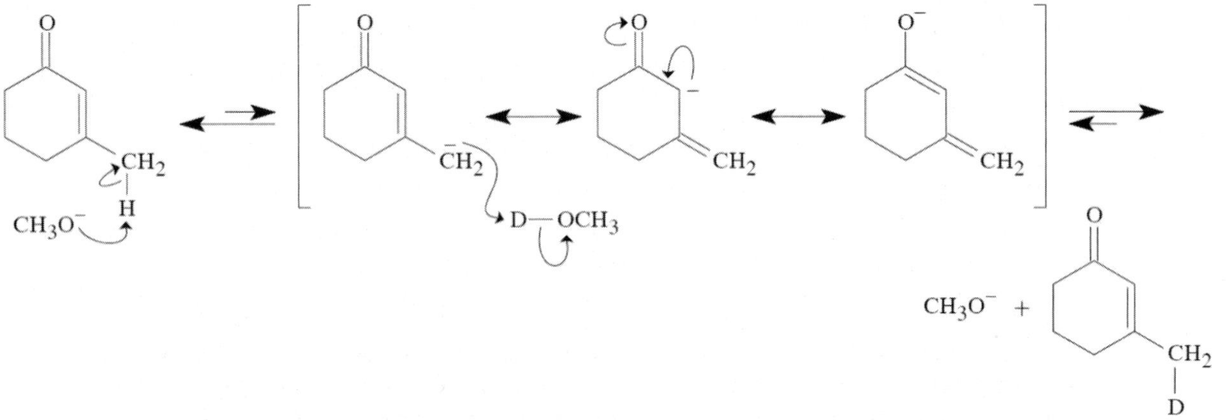

(b)   Although H^b is an α-proton, it is not acidic because the anion that would result from its removal is not resonance-stabilized. The reason is that the orbital containing the unshared electron pair in this anion is perpendicular to the π-electron system of the carbonyl group and cannot overlap with it. An equivalent resonance argument is that if the overlap which lies at the basis of resonance were to occur, the nuclei would have to move and a cumulated double bond and the attendant large amount of strain would be introduced into the six-membered ring.

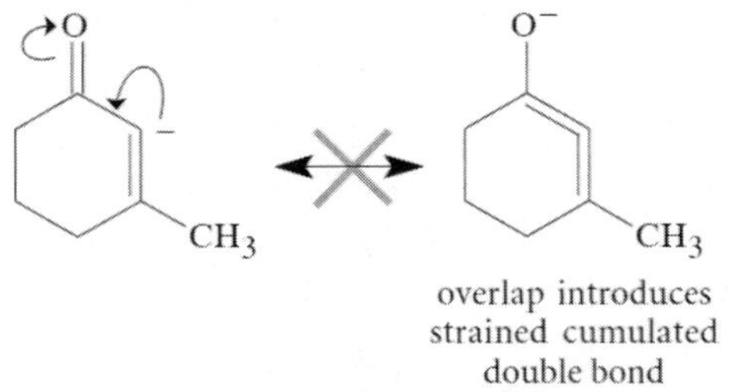

overlap introduces
strained cumulated
double bond

(c)   Hydrogen H^b is not acidic in the α,β-unsaturated ketone. However, as shown in the Problem 22.73 (text p. 1174), the α,β-unsaturated ketone readily isomerizes to a β,γ-unsaturated ketone. Ionization of H^b in this compound gives a resonance-stabilized anion, as shown in the solution to Problem 22.73b, p. 800 of this manual. Hence, H^b exchanges, and the β,γ-unsaturated ketone isomerizes back to the α,β-unsaturated ketone.

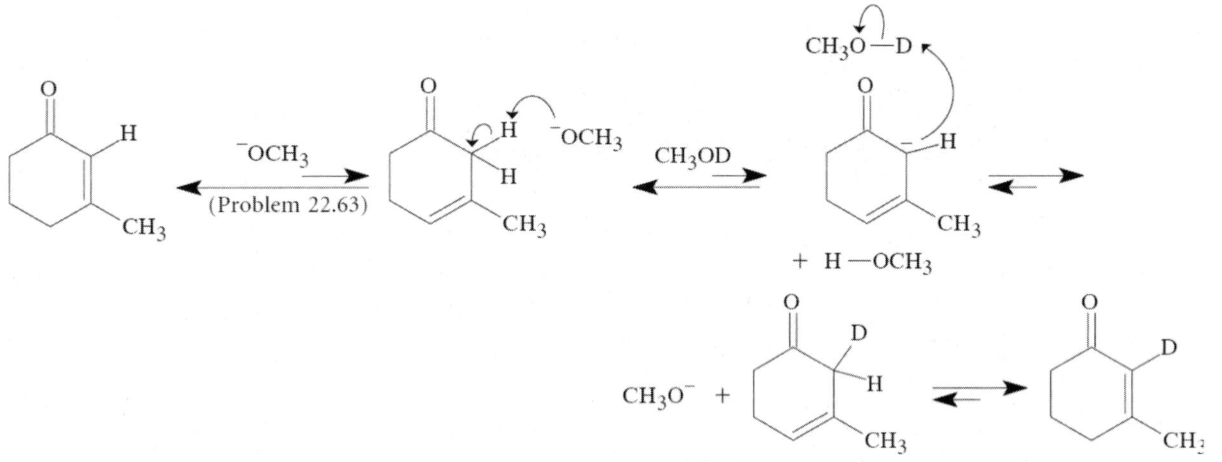

(d)   The protons on carbons indicated by arrows or the one attached to the oxygen would be exchanged for deuterium under these conditions.

22.75 (a) Although ethers are usually more basic than carbonyl oxygens, in compound *A* the carbonyl oxygen is protonated because the resulting carbocation has $4n + 2$ π electrons; that is, it is aromatic. Remember that increasing the stability of the conjugate acid in an acid–base equilibrium decreases its acidity (or increases the basicity of the conjugate base). The aromaticity of its conjugate acid thus explains why compound *A* is more basic than compound *B*. Compound *B* is protonated on the ether oxygen and has the basicity typical of an ether.

conjugate acid of compound A
(an aromatic species)

conjugate acid
of compound B

The relative basicities of these two compounds are deceptive because they involve protonation on different oxygens. The conjugate-acid pK_a of a ketone is typically –6. Hence, aromaticity actually raises the pK_a of the conjugate acid of *A* by about 6 units.

(b) Tropone is unusually basic because its conjugate acid is an aromatic carbocation.

conjugate acid of tropone
(an aromatic α-hydroxy carbocation)

The consequences of aromaticity on the pK_a of a conjugate acid were explored in part (a), and these apply equally here.

22.76 (a) As shown in Fig. SG22.2 on page 771 of this manual, the more stable a species is relative to its conjugate base the less acidic it is. Consequently, resonance stabilization of an α,β-unsaturated carboxylic acid decreases its acidity, assuming that the resonance structures for the conjugate base are less important.

(b) The pK_a of the last compound, butanoic acid, is the reference value—the pK_a to be expected for a four-carbon carboxylic acid containing no carbon–carbon double bonds. In the first compound, 3-butenoic acid, the carbon-carbon double bond is not conjugated with the carbonyl group. Hence, the difference between its pK_a value and that of butanoic acid is due to the electron-withdrawing polar effect of the carbon–carbon double bond. If the polar effect were the only effect operating in the middle compound, *trans*-2-butenoic acid (crotonic acid), it should be the most acidic compound in the series, because the polar effect of any group on acidity increases when the group, in this case a carbon–carbon double bond, is closer to the site of ionization. Because *trans*-2-butenoic acid is in fact *less* acidic than 3-butenoic acid, some acid-weakening effect must also be operating, and this is the resonance effect discussed in part (a).

22.77　(a)　The α-hydrogen of any nitro compound is particularly acidic because the conjugate-base anion is stabilized both by resonance and by the polar effect of the nitro group (note the positive charge next to the negative charge of the carbanion).

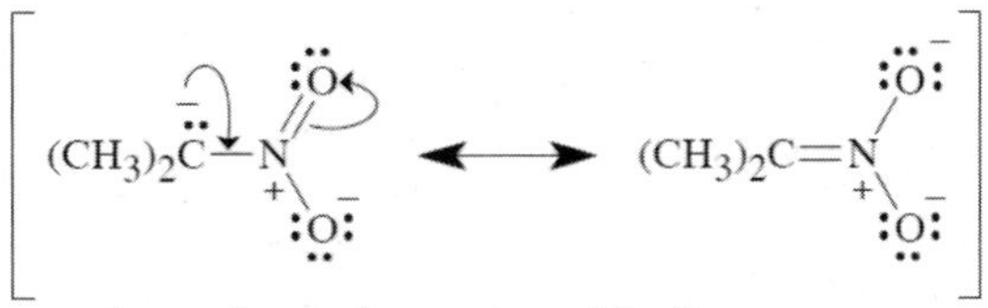

the conjugate-base anion of 2-nitropropane

(b)　To answer this question, ask where besides the anionic carbon there is negative charge and a pair of electrons that can be protonated. As the resonance structures above show, the negative charge of the carbanion is shared by the oxygens of the nitro group. Protonation of either oxygen gives the nitro analog of an enol, which is called an *aci*-nitro compound. This is the isomer that is requested in the problem. (See also Problem 22.9a, text p. 1112, and its solution on p. 769 of this manual.)

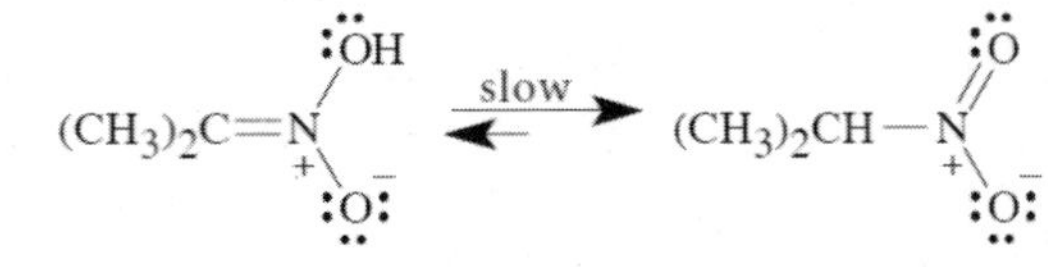

an *aci*-nitro compound

(c)　Because 2-nitropropane is much more acidic than ethanol, it is completely converted into its anion by sodium ethoxide. This anion, like many other "enolate ions," undergoes a Michael addition to ethyl acrylate to give ethyl 4-methyl-4-nitropentanoate.

$$(CH_3)_2CCH_2CH_2CO_2Et$$
$$|$$
$$NO_2$$

ethyl 4-methyl-4-nitropentanoate

22.78　Intermediate *A* is the anionic product of a crossed Claisen condensation. This anion undergoes a Michael addition to the α,β-unsaturated ketone, and the product of that reaction undergoes an intramolecular aldol condensation to give compound *B*. Hydroxide ion effects a reverse Claisen condensation that removes the formyl group. The steps are outlined below; you should provide the mechanistic details using the curved-arrow notation.

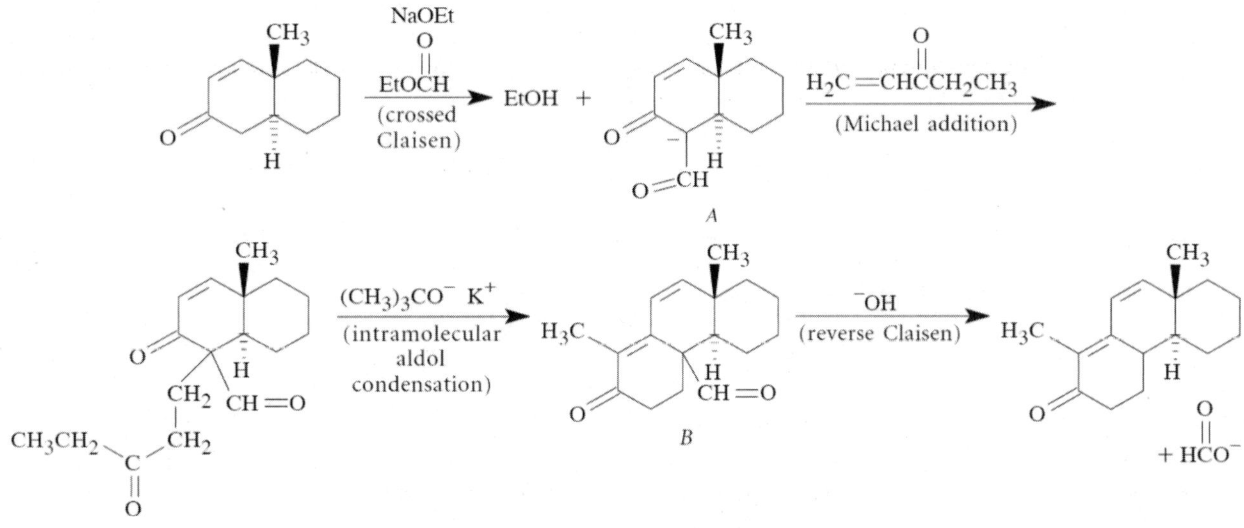

22.79　(a)　One full equivalent of base is required in the Claisen or Dieckmann condensation to ionize the product. Without this ionization, the condensation equilibrium is unfavorable. (See Sec. 22.6A, text pp. 1133–6.)

(b)   Sodium phenoxide, the conjugate base of phenol, is about 10^{-5} times as basic as sodium ethoxide. Thus, sodium phenoxide is not basic enough to ionize completely the Claisen condensation product. Without such an ionization, the condensation equilibrium is unfavorable.

(c)   The equilibrium for the Claisen condensation is rather unfavorable and requires the final ionization of the product to proceed to completion. Under acidic conditions, no base is available to cause this ionization. In addition, an acid-catalyzed Claisen condensation would require the enol of an ester, because enols are the α-carbon nucleophiles under acidic conditions. As shown in Eq. 22.13, p. 1110, the formation of enols from esters is *much* less favorable than the formation of enols from aldehydes or ketones. The much lower concentration of reactive intermediate would translate into a much reduced rate. In summary, both the rate of the reaction is very low, and its equilibrium constant is highly unfavorable; hence, no reaction is observed.

22.80   (a)   An alkylation of this type requires a base that will rapidly and completely convert the ester into its conjugate-base enolate anion so that the enolate ion is not present simultaneously with the ester. In the presence of sodium ethoxide, however, small amounts of the enolate ion are present together with large amounts of the un-ionized ester; consequently, the Claisen condensation will occur as the major reaction. In all likelihood methyl iodide will alkylate the anionic product of the Claisen condensation to give the following compound:

$$\underset{\substack{|\\ CH_3}}{CH_3CH_2\overset{\overset{\displaystyle O}{\|}}{C}-\overset{\overset{\displaystyle CH_3}{|}}{C}-\overset{\overset{\displaystyle O}{\|}}{C}OEt}$$

(b)   The conjugate-base anion of diethyl malonate would be formed with sodium ethoxide, but it would not react with bromobenzene because the latter compound, an aryl halide, does not undergo S_N2 reactions. (See Sec. 18.1, text p. 880.)

(c)   Alcohol dehydration involves a carbocation intermediate, and carbocations α to carbonyl groups are particularly unstable. The desired dehydration will not occur. The more likely reaction would be polymerization of the ester by reaction of the hydroxy group of one molecule with the carbonyl of the other.

(d)   The first step will effect α-halogenation of the carboxylic acid. However, an attempt to form the Grignard reagent will fail, because the Grignard reagent will be destroyed instantly by a protonolysis reaction with the α-bromo acid. Because the Grignard reagent does not survive, its reaction with acetaldehyde cannot occur.

(e)   This is a crossed aldol condensation in which the conjugate-base enolate ion of the ketone reacts with the aldehyde. A mixture of isomeric products will result because there are two possible conjugate-base enolate ions of the ketone that can react. Moreover, a very significant, if not predominant, competing reaction will be the aldol condensation of acetaldehyde itself.

(f)   In addition to bromination of the benzene ring, the acidic conditions will also promote α-bromination at the methyl group. Lewis acids can catalyze this reaction just as Brønsted acids can. In addition, HBr is a by-product of ring halogenation, and this can also catalyze α-halogenation.

22.81   (a)   The gas is H_2 and the species *A* is the sodium salt of the conjugate-base enolate ion of 2,4-pentanedione. As the resonance structures indicate, the negative charge, and therefore the nucleophilic character, of this anion is shared by both the anionic carbon and the oxygens.

$$\left[\underset{A}{CH_3\overset{\overset{\displaystyle O}{\|}}{C}-\overset{-}{C}H-\overset{\overset{\displaystyle O}{\|}}{C}CH_3 \longleftrightarrow CH_3\overset{\overset{\displaystyle O}{\|}}{C}-CH=\overset{\overset{\displaystyle O^-}{|}}{C}CH_3 \longleftrightarrow CH_3\overset{\overset{\displaystyle O^-}{|}}{C}=CH-\overset{\overset{\displaystyle O}{\|}}{C}CH_3} \right] Na^+$$

(b)   The three species that are formed all result from reaction with a nucleophilic atom with methyl iodide in an S_N2 reaction. The products are the two stereoisomeric ethers *B* and *C* and the alkylated β-diketone *D*.

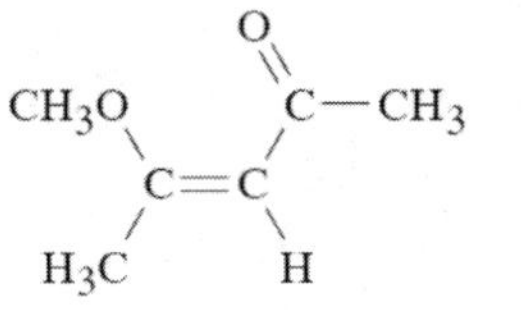

B

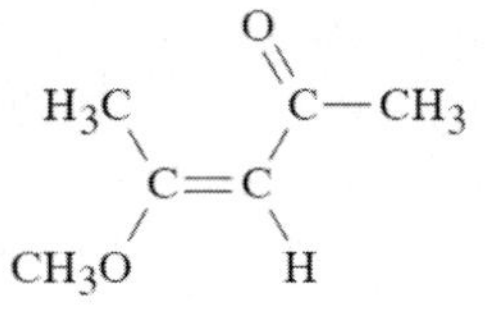

C

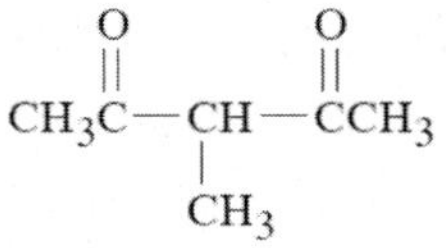

D

22.82 (a)

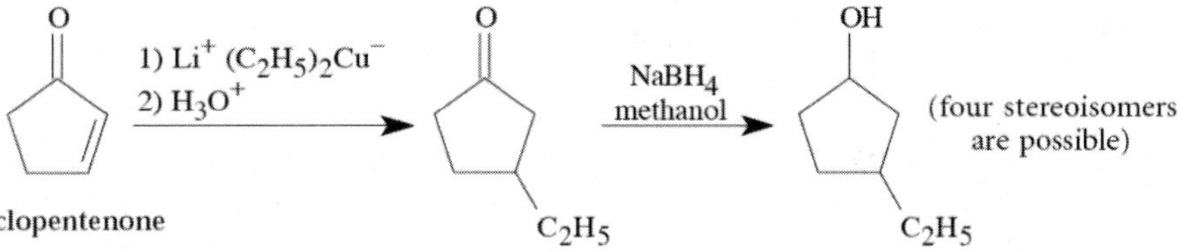

2-cyclopentenone

3-ethylcyclopentanol

(b)

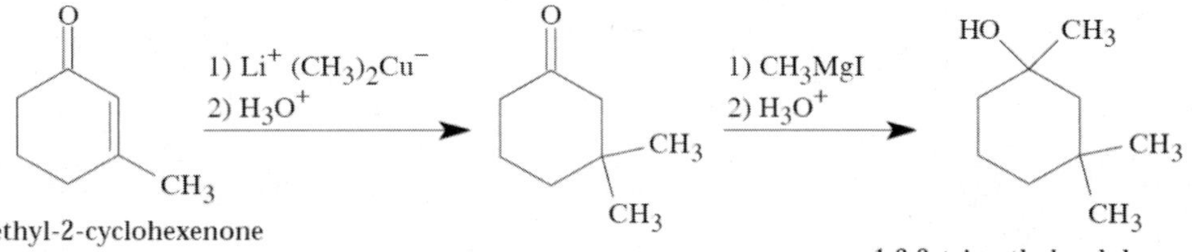

3-methyl-2-cyclohexenone

1,3,3-trimethylcyclohexanol

(c)

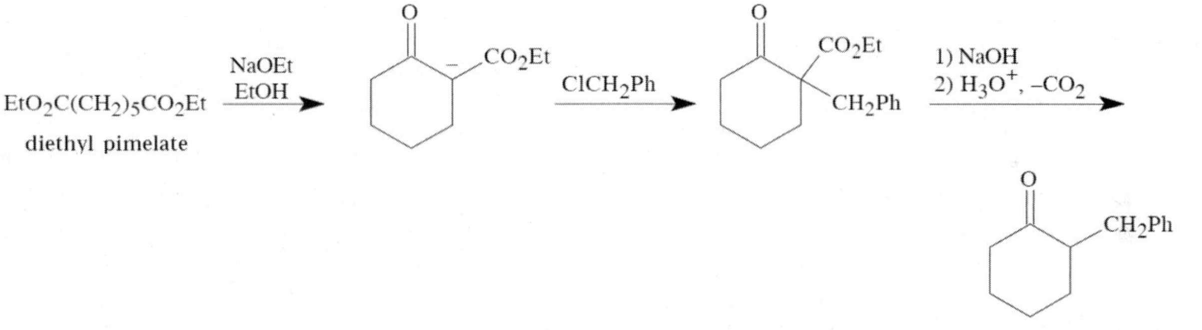

diethyl pimelate

2-benzylcyclohexanone

(d)

$EtO_2CCH_2CO_2Et$ (diethyl malonate) $\xrightarrow{\text{1) NaOEt, EtOH; 2) } CH_3I}$ $EtO_2CCH(CH_3)CO_2Et$ $\xrightarrow{\text{1) NaOEt, EtOH; 2) } CH_3I}$

$EtO_2CC(CH_3)_2CO_2Et$ $\xrightarrow{\text{1) } LiAlH_4\text{; 2) } H_3O^+}$ $HOCH_2C(CH_3)_2CH_2OH$

2,2-dimethyl-1,3-propanediol

(e)

$H_2C{=}CHCO_2Et$ (ethyl acrylate) $\xrightarrow{HCN}$ $N{\equiv}CCH_2CH_2CO_2Et$ $\xrightarrow{\text{1) } LiAlH_4\text{; 2) } H_3O^+\text{; 3) } ^-OH}$ $H_2NCH_2CH_2CH_2CH_2OH$

(f)

$$H_2C=CH-CN \xrightarrow{(CH_3CH_2)_2NH} (CH_3CH_2)_2NCH_2CH_2CN \xrightarrow[\substack{3) \; ^-OH}]{\substack{1) \; LiAlH_4 \\ 2) \; H_3O^+}} (CH_3CH_2)_2NCH_2CH_2CH_2NH_2$$

acrylonitrile

(g)

PhCH$_2$COH $\xrightarrow[\text{H}_2\text{SO}_4 \text{ (catalyst)}]{\text{EtOH (solvent)}}$ PhCH$_2$COEt $\xrightarrow[\text{NaOEt}]{\substack{\text{EtOCOEt} \\ \text{(excess)}}}$ PhCHCOEt | CO$_2$Et $\xrightarrow[\text{2) CH}_3\text{CH}_2\text{Br}]{\text{1) NaOEt, EtOH}}$

α-phenylacetic acid

PhC—COEt (with CH$_3$CH$_2$ and O above, CO$_2$Et below) $\xrightarrow[\substack{-\text{EtOH} \\ -\text{CO}_2}]{\text{H}_3\text{O}^+, \text{ heat}}$ CH$_3$CH$_2$CHCO$_2$H | Ph

2-phenylbutanoic acid

(h)

$$2 \text{ Ph}-\overset{O}{\underset{}{C}}-CH_3 \xrightarrow{H_2SO_4} Ph-\overset{O}{\underset{}{C}}-CH=\underset{Ph}{\underset{|}{C}}-CH_3 \xrightarrow{H_2, \text{ cat.}} Ph-\overset{O}{\underset{}{C}}-CH_2\underset{Ph}{\underset{|}{CH}}CH_3$$

acetophenone   1,3-diphenyl-1-butanone

(i)

HOCCH$_2$Ph $\xrightarrow{SOCl_2}$ ClCCH$_2$Ph $\xrightarrow[\text{2) H}_3\text{O}^+]{1)\ \bigcirc,\ AlCl_3}$ PhCCH$_2$Ph $\xrightarrow{D_2O, \text{ base}}$

phenylacetic acid

PhCCD$_2$Ph (O above C) $\xrightarrow[\text{2) dil. H}_3\text{O}^+]{\text{1) LiAlH}_4}$ PhCHCD$_2$Ph (OH above)

(j)

cyclohexanone

(k)

ClCCH$_3$ (O above) $\xrightarrow[\text{2) H}_3\text{O}^+]{\substack{\text{1) CH}_3\text{O}-\bigcirc, \\ \text{AlCl}_3}}$ [CH$_3$O–benzene ring–C(=O)–CH$_3$] $\xrightarrow{Cl_2, HCl}$ [CH$_3$O–benzene ring–C(=O)–CH$_2$Cl] $\xrightarrow{Na^+ \; ^-O-\bigcirc}$

acetyl chloride

(assume para is the major product, and can be separated from the ortho isomer)

[CH$_3$O–benzene ring–C(=O)–CH$_2$–O–phenyl]

(l)

$$(EtO_2C)_2CH_2 \xrightarrow[\text{2) CH}_3\text{CH}_2\text{CH}_2\text{Br}]{\text{1) NaOEt, EtOH}} (EtO_2C)_2CHCH_2CH_2CH_3 \xrightarrow[\text{2) CH}_3\text{I}]{\text{1) NaOEt, EtOH}}$$

diethyl malonate

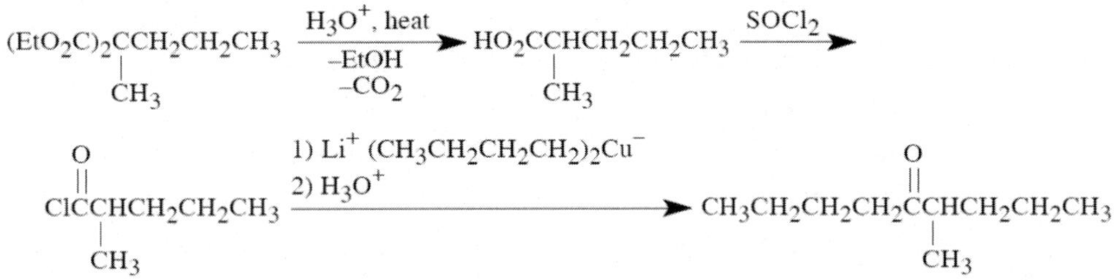

$$(EtO_2C)_2CCH_2CH_2CH_3 \xrightarrow[\substack{-EtOH \\ -CO_2}]{H_3O^+, \text{ heat}} HO_2CCHCH_2CH_2CH_3 \xrightarrow{SOCl_2}$$

with CH_3 substituents

$$ClCCHCH_2CH_2CH_3 \xrightarrow[\substack{1) \ Li^+ \ (CH_3CH_2CH_2CH_2)_2Cu^- \\ 2) \ H_3O^+}]{} CH_3CH_2CH_2CH_2CCHCH_2CH_2CH_3$$

4-methyl-5-decanone

(m)

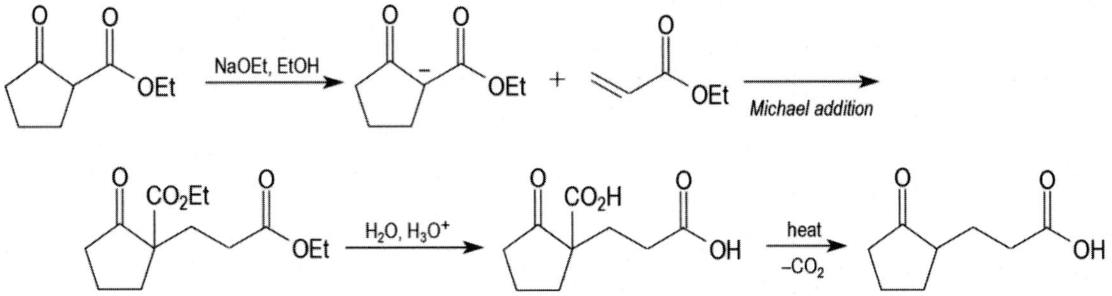

22.83   In the first step, the conjugate-base anion of diethyl malonate reacts with mesityl oxide, an α,β-unsaturated ketone, in a Michael addition.

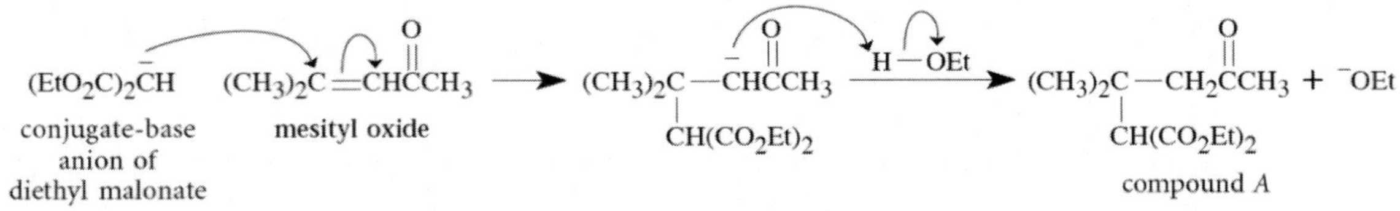

A crossed-Claisen condensation between the enolate ion of the methyl ketone and one of the ester groups then ensues.

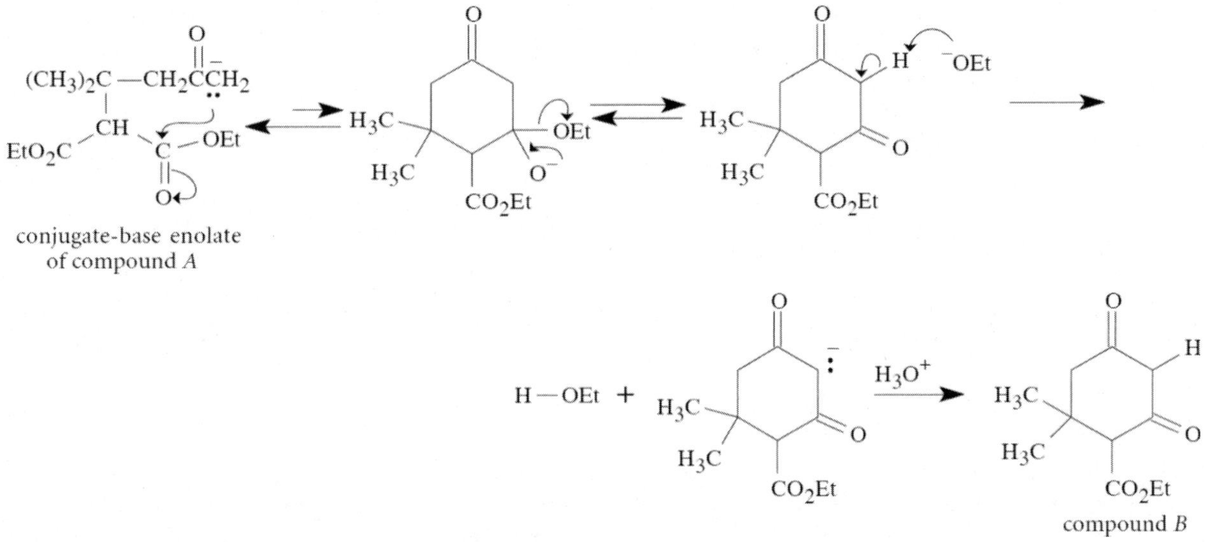

Dimedone results from the saponification and decarboxylation of the remaining ester group (a β-keto ester) in compound B.

Note that the acidic C—H ionizes again during the saponification. This ionization is what prevents a reverse Claisen condensation from occurring in which ⁻OH reacts wtih one of the carbonyls of the β-diketone.

22.84 This reaction is essentially a type of crossed-Claisen condensation in which the "enolate ion" is the conjugate-base anion of urea. A second reaction of the same type closes the ring. The following mechanism starts with the reaction of the ionized urea with one of the ester groups.

As in other Claisen-type condensation reactions, the product Veronal is ionized by the ethoxide base, and Veronal is regenerated when acid is added. (These final steps are not shown in the foregoing mechanism.)

22.85 The synthesis of pentothal involves, first, preparation of the substituted malonic ester derivative, and then condensing it with thiourea, the sulfur analog of urea.

22.86 Compound B is an aldehyde (it gives a positive Tollens test); therefore, compound A is a primary alcohol. Compound B is conjugated (it has a strong UV spectrum), and it has the same carbon skeleton as octanoic acid (an unbranched chain of eight carbons). Therefore, compound B is 2-octenal, and compound A is 2-octenol. (The data do not differentiate between the E and Z stereoisomers.) In order to form 2-octenol, lithium dibutylcuprate must undergo a *conjugate addition* to the epoxide:

$$CH_3CH_2CH_2CH_2CH_2CH=CHCH_2OH \xrightarrow{PCC} CH_3CH_2CH_2CH_2CH_2CH=CHCH=O$$

2-octenol (compound *A*)   2-octenal (compound *B*)

22.87   (a)   This is an aldol addition; dehydration cannot occur because the resulting alkene would violate Bredt's rule. The mechanism below begins with the enolate ion of the methyl ketone, which is formed when a proton is removed by potassium carbonate.

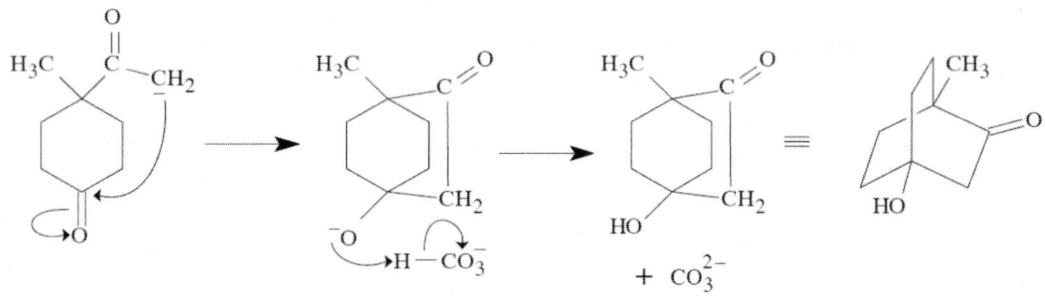

(b)   The mechanism shown below begins with an α-bromo enolate ion, formed by the removal of an α-proton by NaH. This ion displaces a bromine in an internal nucleophilic substitution reaction to form the ring. An E2 reaction of the resulting compound with NaH as the base gives the product. (Recall that E2 reactions are particularly rapid when the proton undergoing elimination is particularly acidic; see Sec. 17.3B, text pp. 848–9.)

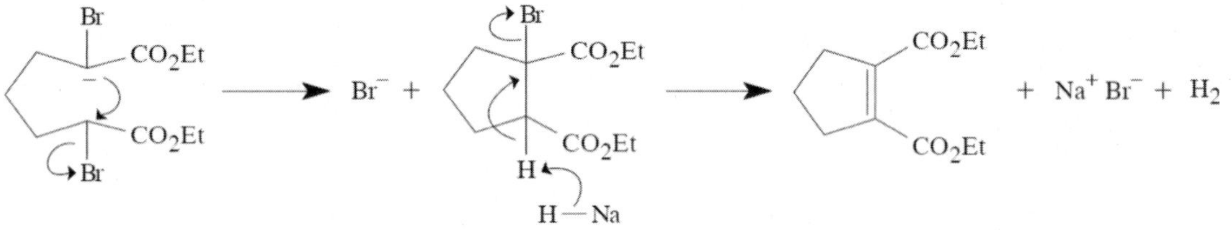

(c)   This is a double Claisen condensation followed by hydrolysis and decarboxylation of the ester groups. The mechanism below begins with the conjugate-base enolate ion of the diester.

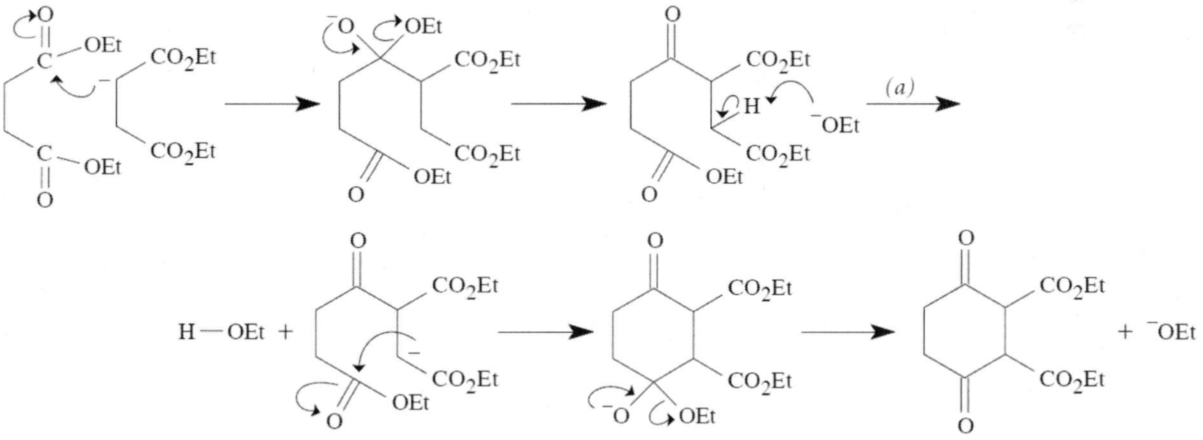

Compound *A* forms a conjugate-base enolate ion under the basic conditions but is regenerated when acid is added. Hydrolysis of the two ester groups and decarboxylation of the resulting β-keto acid groups give the product. In the step labeled *(a)*, the hydrogen that is removed is not the most acidic hydrogen in the molecule. (Which hydrogen is most acidic?) However, removal of the most acidic hydrogen does not lead to the observed product. Removal of the hydrogen shown, although occurring less frequently, leads rapidly to product.

Note also that the second condensation reaction could involve the reaction of a different enolate ion with a different ester group so that the β-keto ester shown on the right would be formed:

alternate enolate for
second condensation

Ester hydrolysis and decarboxylation of this compound would also give the observed product.

(d) This reaction is crossed aldol condensation followed by an intramolecular crossed aldol condensation followed by the same sequence with another molecule of dialdehyde. The following mechanism begins with the conjugate-base enolate ion of the diketone. Although the dehydration step is shown for brevity as a concerted process, it probably involves an enolate-ion intermediate. (See Eq. 22.44, text p. 1121.)

repeat the same
reaction sequence

repeat the same
reaction sequence

(e)

(f) A carbon–carbon double bond of a dienol on one ring reacts with the protonated carbonyl group on the other. The mechanism below begins with a protonated ketone.

(g)   This reaction is a double crossed-Claisen condensation followed by an intramolecular transesterification reaction in which the anionic oxygen of an enolate ion serves as the nucleophile. (The oxygen, rather than the anionic carbon, acts as the nucleophile because a six-membered ring is formed; reaction with the anionic carbon would give a strained four-membered ring.) The mechanism that follows begins with the conjugate-base enolate ion of acetone. Acid-catalyzed hydrolysis of the diester by the usual mechanism gives chelidonic acid.

diethyl ester of chelidonic acid

22.88   (a)   The conjugate-base enolate ion is alkylated by the alkyl halide. Then the conjugate-base enolate ion of the alkylation product is formed and reacts intramolecularly with the second alkyl halide group to form a ring. Hydrolysis and decarboxylation of the ester group give the following product.

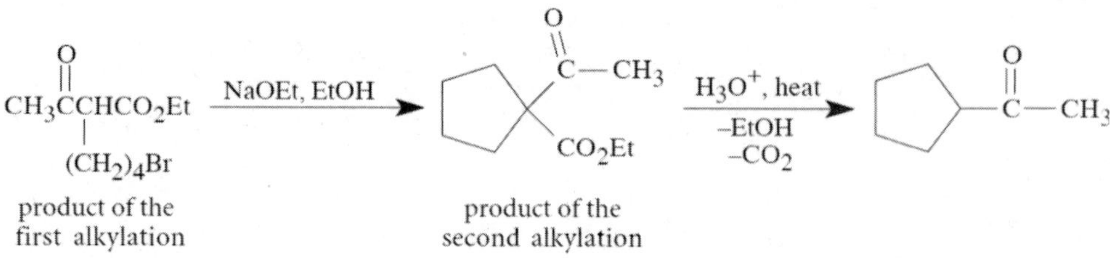

product of the
first alkylation

product of the
second alkylation

(b) The lactone, a cyclic ester, can form an enolate ion, and this ion is alkylated by methyl iodide.

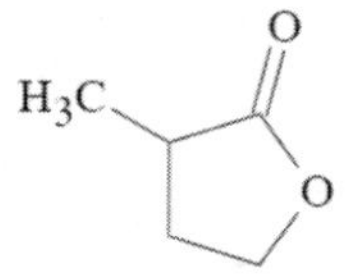

(c) This is a nucleophilic aromatic substitution reaction in which the nucleophile is an enolate ion. The ester groups are hydrolyzed, and one of the resulting carboxylic acids is decarboxylated, by the hot acidic conditions.

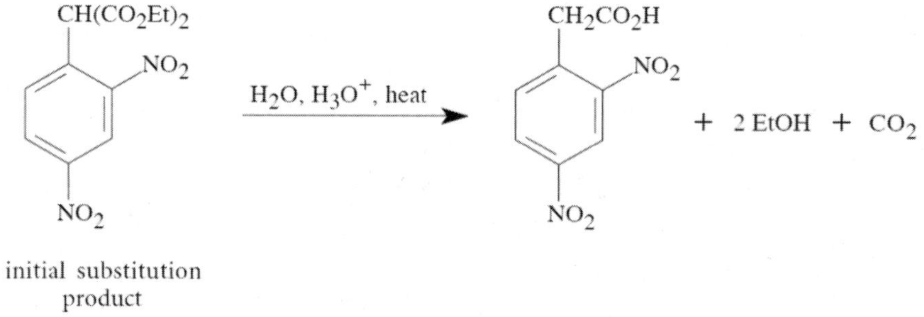

initial substitution
product

$+ \ 2\,\text{EtOH} \ + \ CO_2$

(d) Lithium aluminum hydride reduces the carbonyl group to an alcohol.

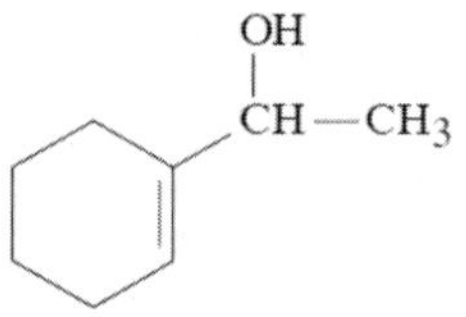

(e) This is a Michael addition in which the conjugate-base enolate ion of diethyl methylmalonate is the nucleophile.

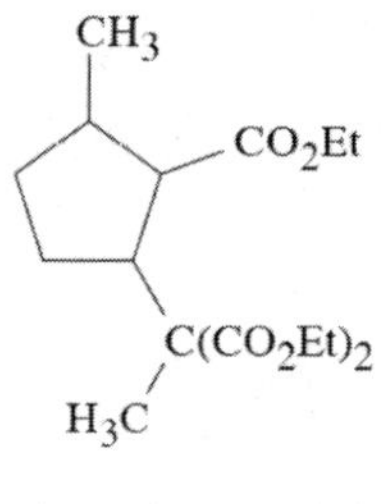

(f) First, an aldol condensation yields the product shown below. Conjugate addition of cyanide ion, followed by hydrolysis of the cyano group and the esters, and finally decarboxylation, yields the dicarboxylic acid.

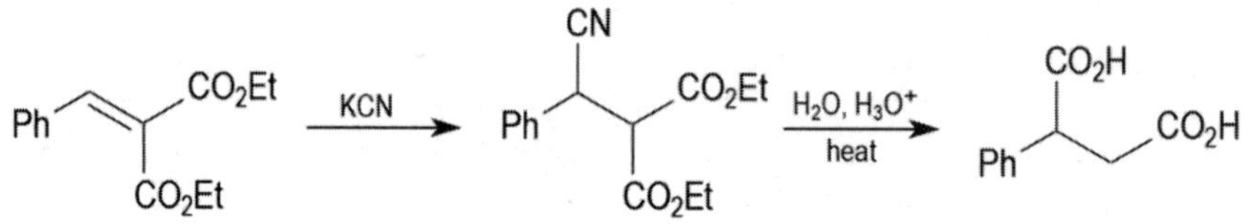

(g) The diethylmalonate displaces the chloride of the acid chloride. After hydrolysis of the carboxylic acids, the resulting β-keto diacid undergoes two equivalents of carboxylic acid to yield a ketone as a final product.

(h)   This is a Michael addition in which the conjugate-base enolate ion of acetoacetic ester is the nucleophile.

(i)   The magnesium organocuprate reagent undergoes conjugate addition; the acetal group hydrolyzes; and an acid-catalyzed intramolecular aldol condensation ensues.

(j)   This is a conjugate addition of the magnesium organocuprate reagent. Assume that the isopropenyl group (the group delivered from the organocuprate reagent) enters trans to the angular methyl group to avoid van der Waals repulsions.

(k)   Like part (j), this is a conjugate addition of the magnesium organocuprate reagent. Treatment with aqueous acid, followed by heat, hydrolyses the esters, and the compound decarboxylates.

(l)   The benzene π electrons serve as the nucleophile in an acid-catalyzed conjugate addition, which can also be viewed as a variation of Friedel–Crafts alkylation.

(m)   Under ordinary conditions, aryl halides do not react with the conjugate-base enolates; however, in the presence of a transition-metal catalyst, such as $Pd[P(t\text{-}Bu)_3]_4$, the reaction proceeds in excellent yield. Treatment with acid hydrolyzes the three ester groups to produce a malonic acid derivative, which under heat, decarboxylates to the product.

$$HO_2C-\!\!\!\langle\text{benzene}\rangle\!\!\!-CH_2CO_2H$$

22.89 (a) Use a conjugate addition of cyanide to an α,β-unsaturated ester, and then hydrolyze. (Either stereoisomer of the starting ester can be used.)

$$CH_3CH=CHCO_2Et \xrightarrow[\text{dil. HOAc}]{\text{NaCN}} N\equiv\overset{*}{C}CHCH_2CO_2Et \xrightarrow[\substack{-EtOH\\-NH_4^+}]{H_3O^+,\,heat} HO_2\overset{*}{C}CHCH_2CO_2H$$
(with CH_3 substituents below)

(b) Take the same approach as in part (a), but with a different ester.

$$EtO_2CC=CH_2 \xrightarrow[\text{dil. HOAc}]{\text{NaCN}} EtO_2\overset{*}{C}CHCH_2\overset{*}{C}\equiv N \xrightarrow[\substack{-EtOH\\-NH_4^+}]{H_3O^+,\,heat} HO_2\overset{*}{C}CHCH_2CO_2H$$
(with CH_3 substituents below)

22.90 (a) A conjugate addition of hydroxide to pulegone gives a β-hydroxy ketone, which then undergoes the reverse of an aldol addition to give acetone and 3-methylcyclohexanone.

$$H_3C-\overset{\displaystyle O}{\overset{\|}{C}}-CH_3 + \text{[3-methylcyclohexanone structure]} \longrightarrow \text{[product]} + {}^-OH$$

acetone

3-methylcyclohexanone

$$EtO_2CC=CH_2 \xrightarrow[\text{dil. HOAc}]{\text{NaCN}} EtO_2CCHCH_2\overset{*}{C}\equiv N \xrightarrow[\substack{-EtOH\\-NH_4^+}]{H_3O^+,\,heat} HO_2\overset{*}{C}CHCH_2CO_2H$$
(with CH_3 substituents below)

(b) The starting material is a β-hydroxy ketone. Ionization of the hydroxy group followed by the reverse of an aldol addition gives the product. The reaction is driven by relief of strain in the four-membered ring.

22.91 (a) A crossed Claisen condensation is followed by a reverse Claisen condensation, driven by expulsion of the more volatile ester, ethyl acetate, from the reaction mixture. The tertiary proton in compound A is very acidic and is removed to give a large amount of a conjugate-base anion (not shown) in equilibrium with a small amount of compound A; however, by Le Châtelier's principle, the reaction of A shown in the following mechanism eventually depletes this anion.

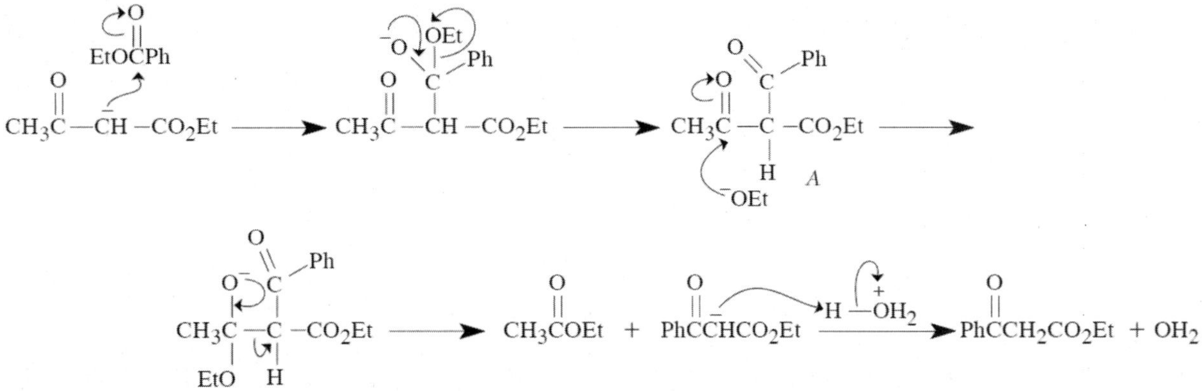

(The reaction mechanisms are shown as chemical structures.)

(b) The conjugate-base anion of diethyl malonate serves as the nucleophile in an epoxide-opening reaction. Normally, epoxide opening under basic conditions involves reaction with the nucleophile at the less branched carbon. In this case, however, reaction also occurs at the more branched carbon because it is benzylic and therefore activated toward nucleophilic substitution. (Only reaction at the benzylic carbon is shown below; reaction at the other carbon occurs by a similar mechanism.) The resulting alkoxide undergoes intramolecular transesterification to give a lactone; hydrolysis and decarboxylation of the remaining ester give the product. The lactone does not hydrolyze because the equilibrium favors lactone formation when five- and six-membered rings are involved; see Eqs. 21.12–21.13 on text p. 1110.

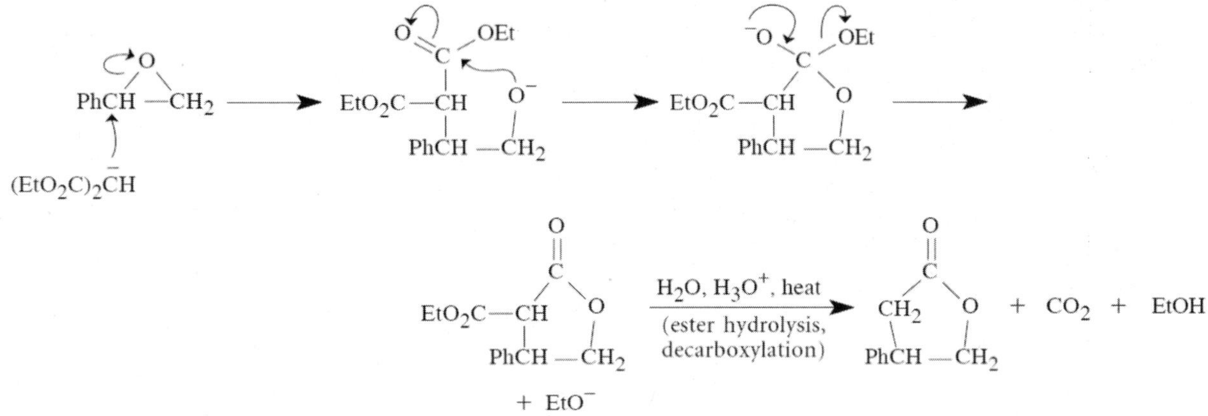

(c) This is an acid-catalyzed conjugate addition of water followed by elimination of phenol. The resulting enol is then spontaneously converted into the corresponding aldehyde.

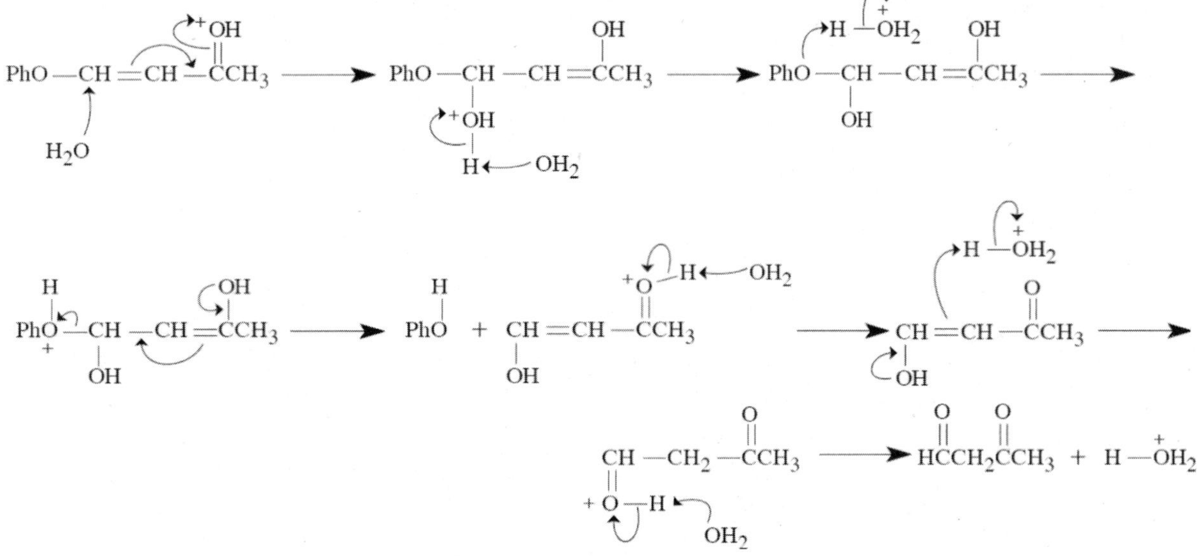

(d)   This mechanism is much like that in part (d), except that basic conditions are involved in all steps but the last.

(e)   Ionization of the benzylic hydrogen gives a carbanion that not only is allylic and benzylic but also is stabilized by the polar effect of the α-oxygen. The O—H hydrogen is also acidic, and removal of it occurs, but the resulting anion does not lead to a reaction. The equilibrium in the reaction is pulled to the right by formation of the ketone from the enol, and the overall reaction is favorable because aldehydes and ketones have a lower energy than their isomeric allylic alcohols. (See heats of formation in Eq. 22.110 on text p. 1161.) In addition, the ketone product in this case is conjugated with the benzene ring, whereas the double bond in the starting material is not. This provides an additional driving force for the reaction.

(f)   Conjugate addition with the nitrogen as a nucleophile gives compound A, which undergoes intramolecular transesterification. Alternatively, transesterification could occur first followed by conjugate addition.

(g) α-Cyano esters have acidities comparable to those of β-keto esters. (Why?) The conjugate-base enolate ion of the α-cyano ester adds to the aldehyde carbonyl group. Elimination of water gives an α,β-unsaturated nitrile, which undergoes conjugate addition with cyanide ion. Hydrolysis of the ester and decarboxylation of the resulting acid give the product.

$Ph(CH_2)_3CH$ (with C=O) ⟶ $Ph(CH_2)_3CH$ ... $EtO_2CCHC≡N$ / $EtO_2CCC≡N$ ⟶ $H—OH$ ⟶ $Ph(CH_2)_3CH$ (OH) $EtO_2C—C—C≡N$ ⟶ $Ph(CH_2)_3CH$ (OH) $EtO_2C—C—C≡N$ + $H—OH$ ⟶

^-OH + $Ph(CH_2)_3CH$, $EtO_2C—C—C≡N$ —^-CN/EtOH⟶ $Ph(CH_2)_3CH—C≡N$, $EtO_2C—C—C≡N$ (H—OEt) ⟶ $Ph(CH_2)_3CH—C≡N$, $EtO_2C—C—C≡N$ (H) —heat, H_2O, H_3O^+ / ester hydrolysis⟶ + EtO^-

$Ph(CH_2)_3CH—C≡N$, $O=C—C—C≡N$ (O—H) , $H—OH_2^+$ ⟶ $Ph(CH_2)_3CH—C≡N$, $O=C—C—C=NH^+$ (O—H) , $H_2O—H$ ⟶

$Ph(CH_2)_3CH—C≡N$, CO_2 + $H_2O—H^+$ $C=C=N—H^+—OH_2$ ⟶ H_2O + $H—OH_2^+$ + $Ph(CH_2)_3CH—C≡N$ ($CH_2C≡N$)

(h)   This reaction is a "conjugate addition" of cyanide ion on a cyclopropane. Notice that reaction at the least branched carbon of the cyclopropane occurs and that the immediate product of ring opening is a very stable anion. Heating in acid protonates this anion, hydrolyzes the ester and nitrile groups, and decarboxylates the resulting β-keto acid.

[cyclopropane-fused cyclopentanone with CO₂Me, attacked by ^-CN] ⟶ [cyclopentanone with $^-CO_2Me$ and CN] —H_2O, H_3O^+, heat (ester and nitrile hydrolysis followed by decarboxylation)⟶ [cyclopentanone with CH_2CO_2H] + CO_2 + $^+NH_4$ + MeOH

(i)   Conjugate addition of triphenylphosphine to the anhydride starting material followed by proton transfer to the initially formed anion gives an ylid *A*, which undergoes a Wittig reaction with the aldehyde butanal. Hydrolysis of the resulting anhydride gives the dicarboxylic acid.

$Ph_3P:$ [maleic anhydride] ⟶ Ph_3P^+ [anion] ⟶ Ph_3P^+ [ylid *A*] , $CH_3CH_2CH_2CH$ (C=O) ⟶

Ph_3P^+ [betaine with $^-O—CH$, $CH_2CH_2CH_3$] ⟶ [$CH_3CH_2CH_2CH=$ exocyclic anhydride] + $Ph_3P^+—O^-$ —H_2O, H_3O^+ / anhydride hydrolysis⟶ $CH_3CH_2CH_2CH=C$ (CH_2CO_2H , CO_2H)

22.92   (a)   The conjugate-base enolate ion of the α-bromo ester, formed by reaction of the α-bromo ester with potassium *tert*-butoxide, reacts with the carbonyl carbon of the aldehyde to give β-bromo alkoxide *A*, the conjugate base of a bromohydrin. As shown in Eqs. 11.20–11.21 on text p. 520, an alkoxide of this type readily cyclizes to an epoxide by an internal nucleophilic substitution mechanism.

A

(b) One of the two possible conjugate-base enolate ions undergoes an intramolecular nucleophilic substitution reaction to give the cyclopropane. Although the other enolate is undoubtedly also formed, its cyclization would lead to a different product. In addition, the enolate that leads to product is the more stable enolate (why?) In any case, the enolate ion shown evidently reacts more rapidly.

22.93 Conjugate addition of lithium dimethylcuprate to the α,β-unsaturated ketone can in principle give two diastereomers. Only product A and its Clemmensen reduction product C could be optically active. Compound B and its reduction product are meso compounds and are therefore achiral. Therefore, compound A, which has the (3S,5S) configuration, is the product of conjugate addition, and product C, which has the (1R,3R) configuration, is the Clemmensen reduction product.

22.94 The deuterium incorporation results show that the carbon–carbon double bond is protonated. This is exactly the way that enols are protonated (see the reverse of Eq. 22.17b on text p. 1112). As a result, a relatively stable α-alkoxy carbocation is formed, and this reacts with water to give a hemiacetal. The hemiacetal breaks down by the usual mechanism (see text Sec. 19.10A) to give acetaldehyde and the alcohol. Hydrolysis of vinyl ethers is much faster than hydrolysis of ordinary ethers because vinyl ether hydrolysis involves a relatively stable carbocation intermediate, an α-alkoxy carbocation.

   In the following mechanism, the asterisk indicates the position of the deuterium that is incorporated into the product if the reaction were run in D$_2$O. (All the other "hydrogens" of H$_2$O and H$_3$O$^+$ would be deuteriums as well, but these are not shown in the mechanism.)

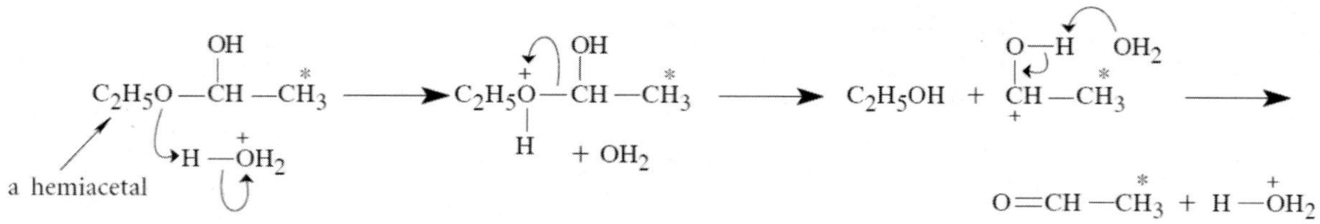

a hemiacetal

22.95   The Wittig reaction gives a vinylic ether, which, as shown in the previous problem, hydrolyzes rapidly to the corresponding aldehyde.

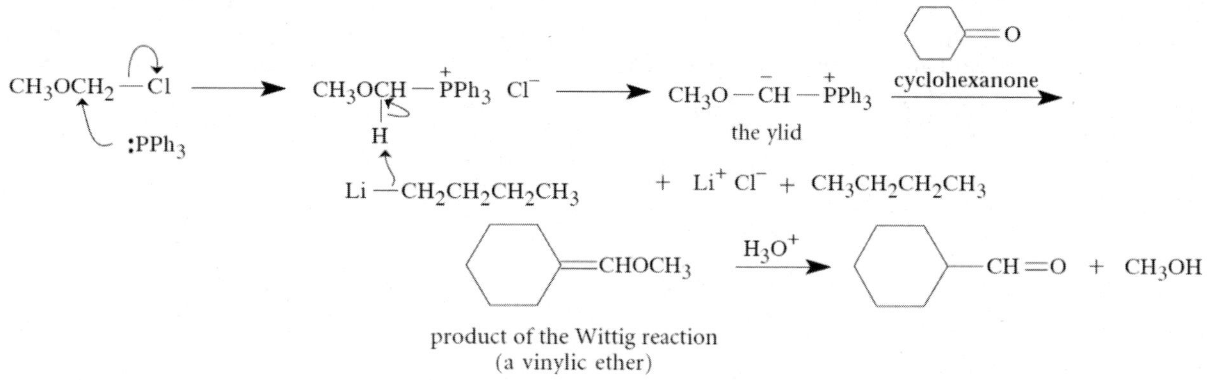

cyclohexanone

the ylid

$+ \; Li^+ \; Cl^- \; + \; CH_3CH_2CH_2CH_3$

product of the Wittig reaction
(a vinylic ether)

22.96   First, form the enolate of the acetyl-CoA with a base in the enzyme.

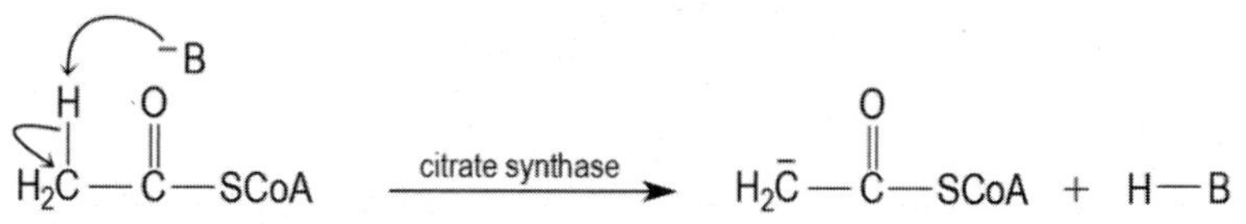

citrate synthase

Next, the enolate undergoes an aldol addition reaction with the ketone of the axalacetate.

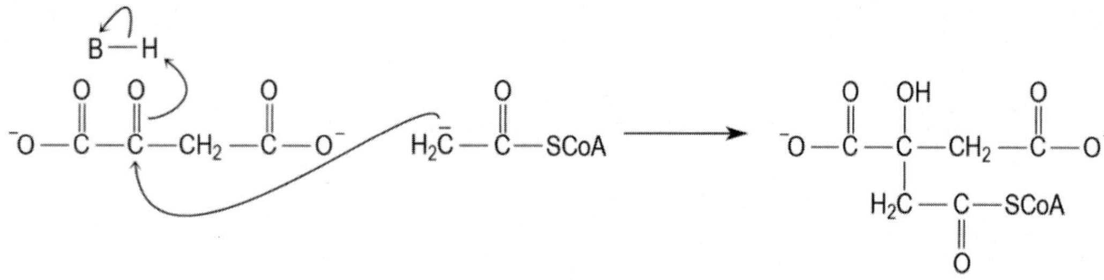

Finally, the coenzyme A is hydrolyzed off of the citrate, and the resulting carboxylic acid is deprotonated.

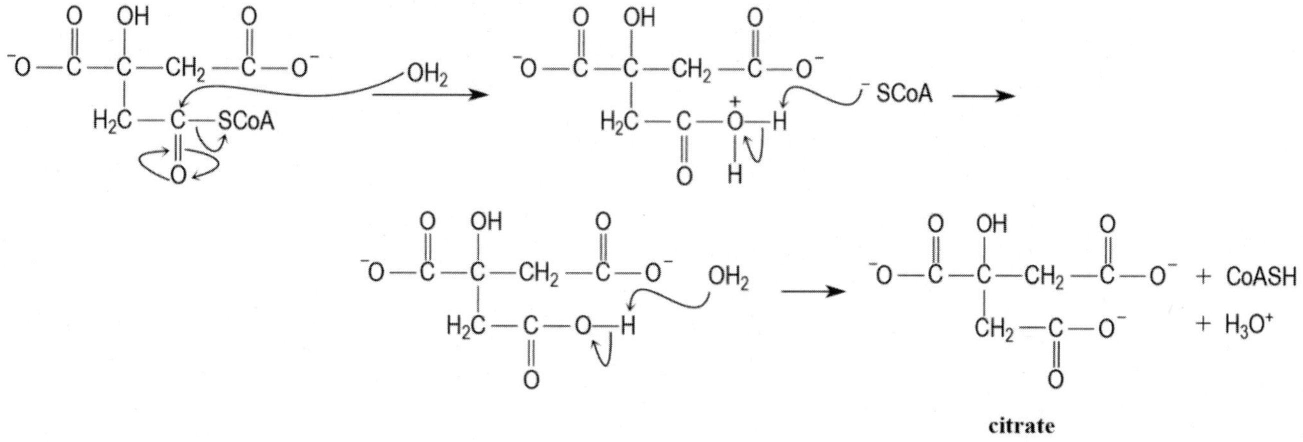

$+ \; CoASH$

$+ \; H_3O^+$

citrate

22.97 (a)   This reaction is analogous to an E2 reaction.

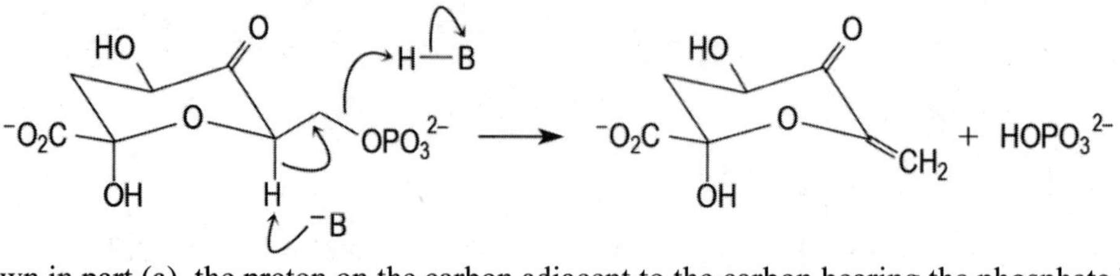

(b)   As shown in part (a), the proton on the carbon adjacent to the carbon bearing the phosphate leaving group needs to be removed by a base. In order to enhance the acidity of this proton, step 1 converts the other adjacent carbon, bearing an OH group, into a carbonyl group. As you know, protons on carbons α to a carbonyl are particularly acidic. After the proton has been removed in step 2, the carbonyl is converted back into an OH group, in step 3.

(c)   Reaction 4 also bears resemblance to an E2 reaction, or to the collapse of a tetrahedral intermediate, expelling a leaving group, as in nucleophilic acyl substitution reactions.

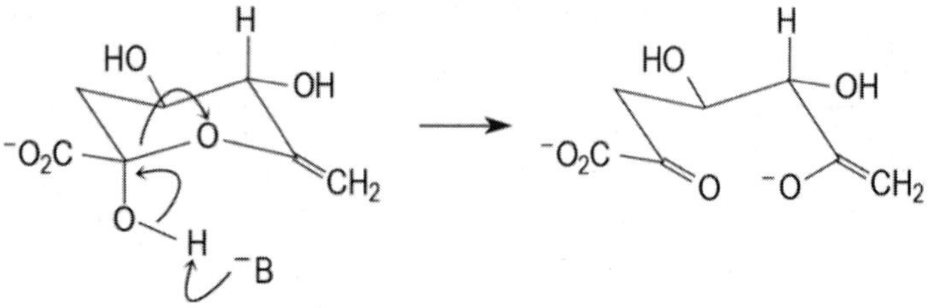

Reaction 5 is an aldol addition reaction.

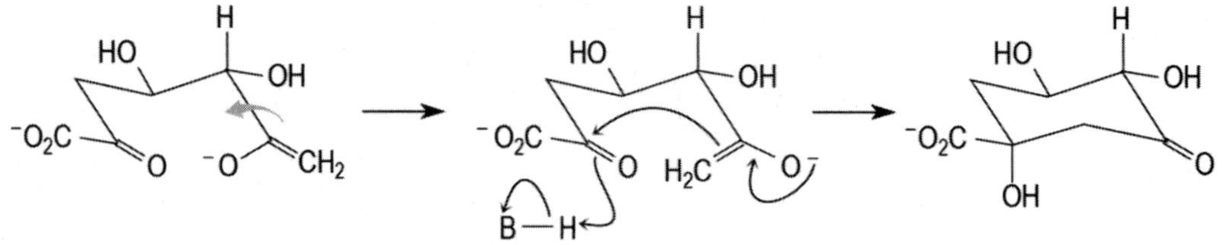

3-dehydroquinate

22.98   A mechanism accounting for the formation of A is shown below.

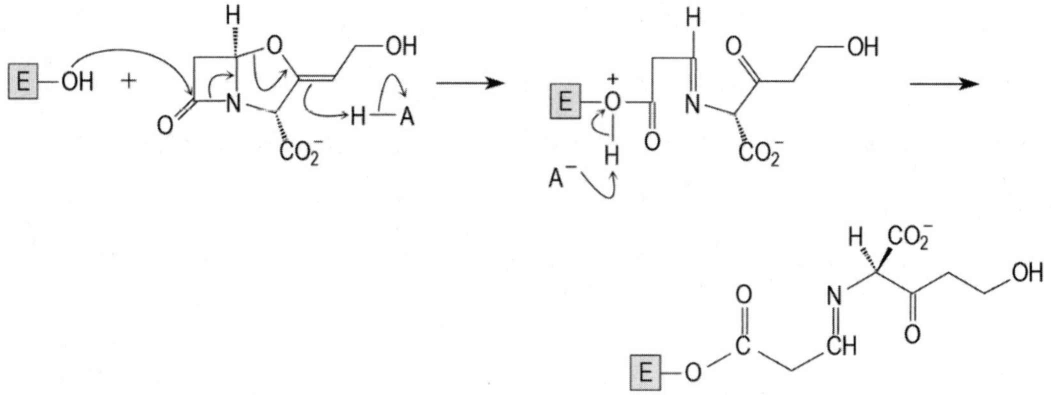

A

A mechanism showing the formation of *B*, and the structure of by-product *C*, is below.

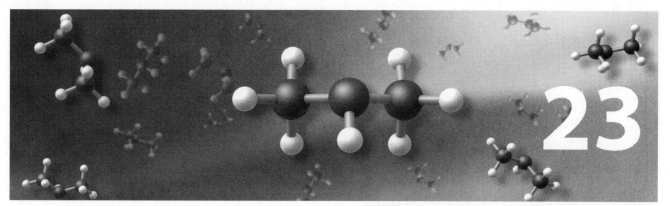

The Chemistry of Amines

STUDY GUIDE LINKS

23.1   Nitration of Aniline

Remember that rates of reactions are affected by both intrinsic reactivity (reflected in the magnitude of the rate constant) and concentrations of the species involved (Sec. 9.3B, text p. 389). Under the acidic conditions of nitration, the highly reactive free aniline is present in very small concentration; the much less reactive protonated aniline is present in much greater concentration.

$$\text{rate of formation of } para\text{-nitroaniline} = k_{para}[\text{PhNH}_2]$$

very large   very small

$$\text{rate of formation of } meta\text{-nitroaniline} = k_{meta}[\text{Ph}\overset{+}{\text{N}}\text{H}_3]$$

very small   very large

These two competing rates must be about the same, because comparable amounts of *meta-* and *para-*substituted products are formed in text Eq. 23.37 (text p. 1205).

23.2   Mechanism of the Curtius Rearrangement

Although the Curtius rearrangement is concerted, as shown in the text, it is helpful to *think* of the reaction in a stepwise fashion to see why the rearrangement occurs. There is a strong driving force for loss of dinitrogen, a very stable molecule:

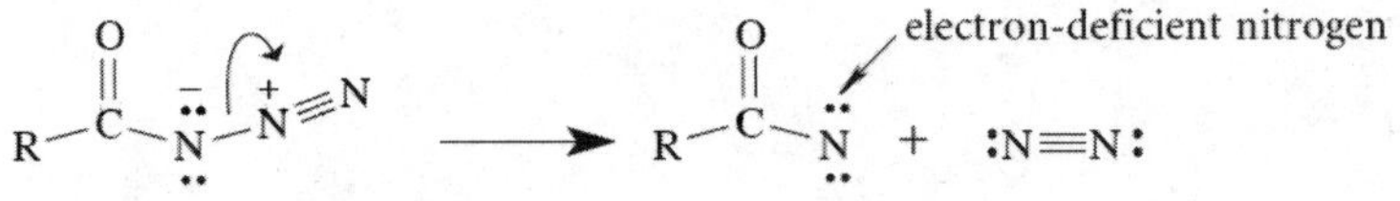

Loss of dinitrogen leaves behind an *electron-deficient* nitrogen; it is electron-deficient because it is short one electron pair of an octet.

Just as rearrangements of alkyl groups occur to the electron-deficient *carbons* of carbocations, so they also occur to electron-deficient nitrogens.

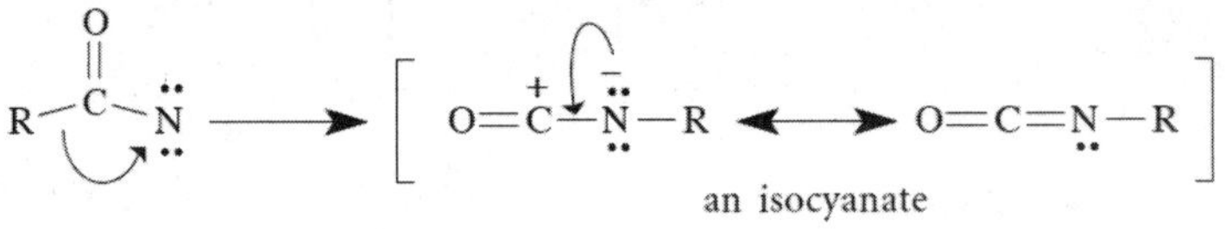

In other words, the Curtius rearrangement is driven by the very strong tendency of nitrogen to have a complete octet.

23.3   Formation and Decarboxylation of Carbamic Acids

The formation of carbamic acids by hydration of isocyanates is an acid-catalyzed addition to the C=N double bond:

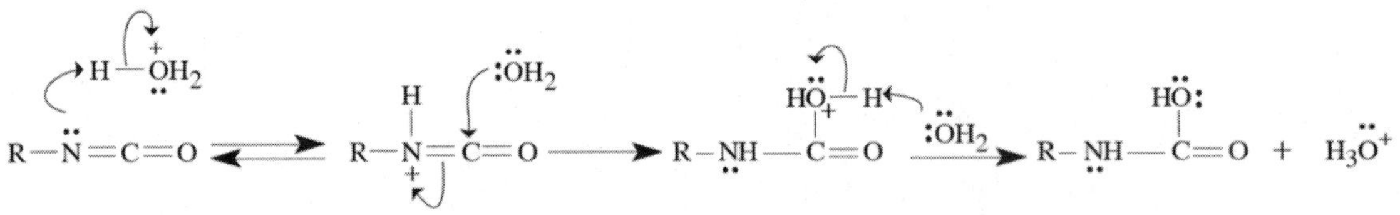

Hydration can also occur in base by the nucleophilic reaction of ⁻OH at the carbonyl carbon. (Write the mechanism of this reaction.) The hydration in basic solution gives the carbamate ion, the conjugate base of a carbamic acid. The hydration of isocyanates under basic conditions is important in the Hofmann rearrangement, which is also discussed in Sec. 23.11D of the text.

Decarboxylation is a concerted reaction in which the same water molecule acts as both an acid and base catalyst at the same time.

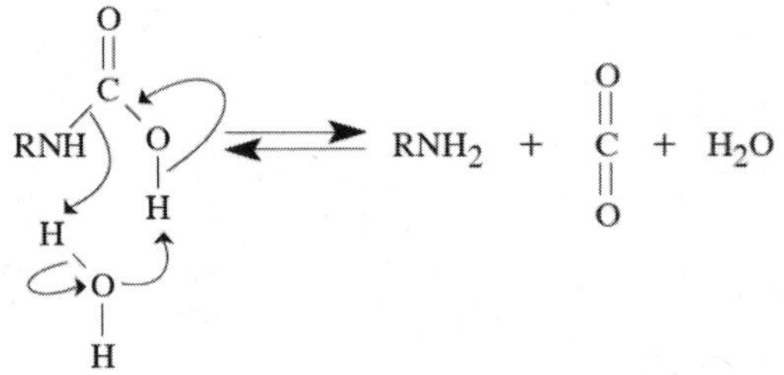

The decarboxylation reaction is reversible; that is, amines react readily with CO_2 to give carbamic acids. However, the decarboxylation reaction in acidic solution is driven to completion by two factors: (1) protonation of the product amine, and (2) loss of CO_2 from solution as a gas.

Decarboxylation is fastest in acidic solution, but also occurs in base:

This reaction is driven to completion by reaction of the hydroxide ion with CO_2 to form bicarbonate ion:

bicarbonate ion

If the solution is strongly basic, bicarbonate ion itself ionizes further to carbonate ion.

Decarboxylation of carbamates in basic solution is important in the Hofmann rearrangement.

FURTHER EXPLORATIONS

23.1   Alkyl Group Polarization in Ionization Reactions

The text points out that alkyl groups can stabilize *either* positive charge or negative charge, as evidenced by studies of gas-phase acidity of alcohols (Sec. 10.1D) and gas-phase basicity of amines (this chapter). The purpose of this section is to give you a more concrete feeling for the nature of these alkyl-group effects.

Consider first the stabilization of negative charge by alkyl groups. Let's consider the ionization reaction of *tert*-butyl alcohol:

$$(CH_3)_3C-OH \rightleftarrows (CH_3)_3C-O^- + H^+$$

tert-butyl alcohol   conjugate base of
                            tert-butyl alcohol

Using a quantum-mechanics program, we can examine the charge on each atom of the molecule before and after the reaction. These charges aren't *formal charges,* which, as you recall, are calculated by a bookkeeping procedure, but are the *actual* charges in the molecule that are present as a result of bond dipoles.

tert-butyl alcohol

conjugate base of
tert-butyl alcohol

(You can add the charges on the left to verify that they add to zero, which is the overall charge on the molecule; and you can add the charges on the right to verify that they add to –1. Remember that two of the three methyl groups are not shown). Notice that, in response to the greater negative charge on oxygen following ionization, the α-carbon (bold) takes on greater positive charge, and the positive charge on the hydrogens is reduced. This is another way of saying that *electrons flow away from the oxygen onto the hydrogens of the methyl groups.*

Now let's look at the protonation of trimethylamine in a similar way:

$$(CH_3)_3N\colon\ +\ H^+\ \underset{\longleftarrow}{\longrightarrow}\ (CH_3)_3\overset{+}{N}-H$$

trimethylamine   conjugate acid of
                              trimethylamine

Here is the result of the charge calculation:

trimethylamine   conjugate acid of
                              trimethylamine

Notice in this case that introduction of positive charge on the electronegative nitrogen causes the α-carbons (bold numbers) to become more *negative*; the additional negative charge is clearly coming from the hydrogens, which have become more positive.

These examples show graphically how the electron distribution in methyl groups can be polarized by nearby charges. Polarization can occur either way, and can therefore stabilize both positive and negative charges.

23.2   Structures of Amide Bases

Although the structures of lithium amide bases are written as ionic compounds for conceptual simplicity, research has shown that the structures of these species are considerably more complex. For example, lithium diisopropylamide in tetrahydrofuran (THF) has a symmetrical dimer structure (that is, it is a symmetrical aggregate of two units):

Each lithium is partially bonded to two nitrogens with "half bonds" (dashed lines in the structure above). The nitrogen–lithium bonds are covalent, although they have significant ionic character. Each lithium is strongly solvated by two solvent molecules (THF in this case). The degree to which these amide bases are aggregated into larger structures varies with their concentration and with the nature of the solvent.

23.3   Mechanism of Diazotization

Diazotization at first sight may seem to be a very strange reaction, but once you understand the reactivity of nitrous acid, diazotization should make more sense.

Nitrous acid in solution is in equilibrium with its anhydride, dinitrogen trioxide.

$$2 \text{ HNO}_2 \; \rightleftarrows \; \text{O}{=}\text{N}{-}\text{O}{-}\text{N}{=}\text{O} \; + \; \text{H}_2\text{O}$$
dinitrogen trioxide

Dinitrogen trioxide is the actual diazotizing agent under conditions of dilute acid; under conditions of very strong acid, the diazotizing agent is the nitrosyl cation, $^+$NO, formed by dehydration of nitrous acid. The dinitrogen trioxide mechanism is explored here.

The N=O double bonds of dinitrogen trioxide have a reactivity somewhat analogous to that of a carbonyl group. They react at nitrogen with nucleophiles such as amines with loss of nitrite.

(The pK_a of nitrous acid is 3.2; hence, the nitrite anion is a weak base, and a good leaving group.) The product of this sequence is a *nitrosamine*. Because the amine nitrogen loses only one hydrogen in forming a nitrosamine, secondary amines as well as primary amines form nitrosamines when they react with nitrous acid (Sec. 23.10C).

In the case of primary amines, nitrosamines react further under the acidic conditions. First, they undergo isomerization to give another intermediate called a diazoic acid. (You should draw the curved-arrow notation for this step, which is analogous to enol formation from a ketone.)

$$\text{RNH}{-}\text{N}{=}\text{O} \; \underset{}{\overset{\text{H}_2\text{O, H}_3\text{O}^+}{\rightleftarrows}} \; \text{R}{-}\text{N}{=}\text{N}{-}\text{OH}$$
a nitrosamine   a diazoic acid

Under the acidic conditions of diazotization, diazoic acids dehydrate to give diazonium ions:

a diazonium ion

REACTION REVIEW

I. AMINES AS BRØNSTED BASES AND ACIDS

A. BASICITY OF AMINES

1. Halogenation of an aldehyde or ketone in acidic solution usually results in the replacement of one α-hydrogen by halogen.

   a. Enols are reactive intermediates in these reactions.

   b. Enols add only one halogen atom.

$$\underset{\text{rate-limiting step}}{\overset{\overset{\displaystyle O}{\parallel}}{\underset{}{-C}}-CH- \;\underset{\longleftarrow}{\overset{H_3O^+}{\longrightarrow}}\; \underset{\text{enol}}{\overset{OH}{-C=C-}} \;\overset{Br_2}{\longrightarrow}\; \overset{\overset{\displaystyle O}{\parallel}\;\;\overset{\displaystyle Br}{\mid}}{-C-C-}\; + \; HBr}$$

2. The rate law implies that even though the reaction is a halogenation, the rate is independent of the halogen concentration.

$$\text{rate} = k[\text{ketone}][H_3O^+]$$

   a. Halogens are not involved in the transition state for the rate-limiting step of the reaction.

   b. Enol formation is the rate-limiting process in acid-catalyzed halogenation of aldehydes and ketones.

3. Introduction of a second halogen is much slower than introduction of the first halogen.

B. ACIDITY OF AMINES

1. Alkylamines have pK_a values in the 32–35 range.

2. Alkylamines react with organometallic reagents such as butyllithium to give amide bases. (The term "amide" should not be confused with the carboxylic acid derivative of the same name.)

$$\underset{\substack{\text{diisopropylamine} \\ pK_a \approx 35}}{\ddot{N}-H} + Li-CH_2CH_2CH_2CH_3 \longrightarrow \underset{\text{lithium diisopropylamide (LDA)}}{\ddot{N}{:}^-\ Li^+} = (i\text{-}Pr)_2\bar{N}\ Li^+ + CH_3CH_2CH_2CH_3$$

3. Lithium amides find frequent use as strong bases for forming the lithium enolates of carbonyl compounds. (See Sec. 22.8B, text p. 1148–9a, and Further Exploration 22.1 on p. 754 of this manual.)

II. SYNTHESIS OF AMINES; ALKYLATION REACTIONS

A. GENERAL

1. Amines are good nucleophiles (Lewis bases).

2. Three reactions of nucleophiles have been studied in previous sections:

   a. S_N2 reaction with alkyl halides, sulfonate esters, or epoxides (Sec. 9.1A, text p. 383; Sec. 9.4, text p. 391; Secs. 10.4A–C, text pp. 465–9; and Sec. 11.5, text p. 524).

   b. Addition to aldehydes, ketones, and α,β-unsaturated carbonyl compounds (Sec. 19.7, text p. 963; Sec. 19.11, text p. 984; and Sec. 22.9A, text p. 1156).

   c. Nucleophilic acyl substitution at the carbonyl groups of carboxylic acid derivatives (Sec. 21.8, text p. 1071).

B. DIRECT ALKYLATION OF AMINES

1. Treatment of ammonia or an amine with an alkyl halide or other alkylating agent (RX in the following equation) results in alkylation of the nitrogen. This is an example of an S_N2 reaction in which the amine acts as the nucleophile.

$$NH_3 \ + \ RX \ \longrightarrow \ RNH_2 \ + \ R_2NH \ + \ R_3N \ + \ R_4N^+ \ X^-$$

   a. The immediate product of the reaction is an alkylammonium ion.
   b. If this ammonium ion has N—H bonds, further alkylations can take place to give a complex product mixture.
   c. A mixture of products is formed because the alkylammonium ion produced initially can be deprotonated by the amine starting material.

2. Epoxides, as well as α,β-unsaturated carbonyl compounds and α,β-unsaturated nitriles, also react with amines or ammonia; multiple alkylation can occur with these alkylating agents as well.

3. In an alkylation reaction, the exact amount of each product obtained depends on the precise reaction conditions and on relative amounts of starting amine and alkyl halide; the utility of alkylation as a preparative method for amines is limited.

C. QUATERNIZATION OF AMINES

1. Amines can be converted into quaternary ammonium salts with excess alkyl halide or sulfonate ester (R—X) under forcing conditions.

   a. This process, called quaternization, is one of the most important synthetic applications of amine alkylation.
   b. The reaction is particularly useful when especially reactive alkyl halides are used.

2. Conversion of an amine into a quaternary ammonium salt with excess methyl iodide is called exhaustive methylation.

D. REDUCTIVE AMINATION

1. Primary and secondary amines form imines and enamines, respectively, when they react with aldehydes and ketones; in the presence of a reducing agent, imines and enamines are reduced to amines.

   a. Reduction of the C=N double bond is analogous to reduction of the C=O double bond.

b. The imine or enamine does not have to be isolated; because imines and enamines are reduced much more rapidly than carbonyl compounds, reduction of the carbonyl compound is not a competing reaction.

2. The process in which an amine reacts with a carbonyl compound and the resulting imine is reduced to an amine is called reductive amination. (It has also been referred to as reductive alkylation.)

   a. When the reduction step is carried out by hydride reducing agents such as $NaBH_4$, $NaBH_3CN$, or $NaBH(OAc)_3$, a proton source such as a protic solvent or dilute acid is required.

   b. An imminium ion is formed in solution by protonation and dehydration of the carbinolamine intermediate in the reductive amination reaction of a secondary amine and formaldehyde; this ion is rapidly and irreversibly reduced.

$$H_2C=O \xrightleftharpoons[\text{}]{R_2NH} R_2N-CH_2-OH \xrightleftharpoons[\text{}]{H_3O^+} R_2\overset{+}{N}=CH_2 \xrightarrow[\text{reduction}]{NaBH_3CN} R_2N-CH_3$$

$$\text{a carbinolamine} \qquad \text{an iminium ion}$$
$$+ \; H_2O$$

c. The reaction of an amine with an excess of formaldehyde is a useful way to introduce methyl groups to the level of a tertiary amine; quaternization does not occur in this reaction.

3. When the reducing agent is sodium cyanoborohydride ($NaBH_3CN$), the reaction is known as the Borch reaction.

$$\overset{\diagdown}{\underset{\diagup}{C}}=O \xrightarrow{RNH_2} \overset{\diagdown}{\underset{\diagup}{C}}=N^{\diagup R} \xrightarrow[\text{MeOH, HCl}]{NaBH_3CN} \overset{H}{\underset{|}{-C-}}NH-R$$

4. To determine the required starting materials for reductive amination:

   a. Starting with the desired compound, mentally reverse the reductive amination process by breaking one of the C=N bonds:

      i. On the nitrogen side, replace it with an N—H bond

      ii. On the carbon side, remove a hydrogen from the carbon and add a carbonyl oxygen.

   b. Repeat this process for each C=N bond and evaluate the best pathway.

III. OTHER SYNTHESES OF AMINES

A. GABRIEL SYNTHESIS OF PRIMARY AMINES

1. The alkylation of phthalimide anion followed by hydrolysis of the alkylated derivative to the primary amine is called the Gabriel synthesis.

phthalimide

a. The nitrogen in phthalimide has only one acidic hydrogen, and thus it can be alkylated only once.
  i. The conjugate base of phthalimide ($pK_a \approx 9$) is easily formed with KOH or NaOH.
  ii. The phthalimide anion can be alkylated by alkyl halides or sulfonate esters (primary or unbranched secondary, shown as RX in the preceding equation) in S_N2 reactions.
  iii. The *N*-alkylated phthalimide is converted into the free amine by amide hydrolysis in either strong acid or base.
b. Although *N*-alkylphthalimides also have a pair of unshared electrons on nitrogen, they do not alkylate further, because neutral imides are much less basic, and therefore less nucleophilic, than the phthalimide anion.
2. Multiple alkylation, which occurs in the direct alkylation of ammonia, does not occur in the Gabriel synthesis.

B. STAUDINGER REACTION FOR SYNTHESIS OF PRIMARY AMINES

1. Primary alkyl azides form phosphazides upon reaction with triphenylphosphine. Phosphazides spontaneously rearrange to iminophosphoranes, which can be hydrolyzed with water to yield primary amines.

an alkyl azide   a phospazide   an iminophosphorane

a. The alkyl azide starting materials are typically synthesized from alkyl halides or sulfonate esters by an S_N2 reaction, so relatively unhindered amines are produced by this method.
b. The rearrangement of the phosphazide to an iminophosphorane is shown in Eq. 23.56b on p. 1212 of the text, and is analogous to the rearrangement of oxaphosphetanes in the Wittig alkene synthesis (Eq. 19.78a, text p.991).
2. This method allows for the preparation of primary amines without multiple alkylations of the nitrogen.

C. REDUCTION OF NITRO COMPOUNDS

1. Nitro compounds can be reduced to amines under a variety of conditions.
a. The nitro group is usually reduced very easily by catalytic hydrogenation.
b. The reduction of aromatic nitro compounds to primary amines can also be accomplished with finely divided metal powders and HCl.

  i. Iron or tin powder is frequently used.
  ii. In this reaction the nitro compound is reduced at nitrogen, and the metal, which is oxidized to a metal ion, is the reducing agent.
2. Aromatic nitro compounds react with $LiAlH_4$, but the reduction products are azobenzenes, not amines.
3. Nitro groups do not react at all with $NaBH_4$ under the usual conditions.

D. AMINATION OF ARYL HALIDES AND ARYL TRIFLATES

1. Arylamines can be prepared by the direct amination of aryl chlorides and aryl bromides in the presence of a base and a Pd(0) catalyst.
a. Direct amination of aryl halides is sometimes called Buchwald–Hartwig amination.

b. The catalysts are typically of the form PdL_2, where L is a sterically demanding ligand, formed by mixing $Pd(OAc)_2$ and two equivalents of the ligand.

2. These amination reactions operate by more than one mechanism, but they are conceptually similar to the Heck (Sec. 18.6A), Suzuki (Sec. 18.6B), and Stille (18.10B) reactions.

   a. The amine starting material must lose a hydrogen.

     i. Secondary amines produce tertiary amine products.

     ii. Primary amines produce secondary amines that may react further unless the primary amine is itself an aryl amine or contains a large, highly branched alkyl group.

     iii. Nitro groups may be present.

E. HOFMANN REARRANGEMENT

1. Treatment of a primary amide with bromine in base gives rise to a rearrangement called the Hofmann rearrangement or Hofmann hypobromite reaction.

$$Br_2 + 2\ ^-OH + R{-}\overset{\overset{\displaystyle O}{\|}}{C}{-}NH_2 \longrightarrow R{-}NH_2 + CO_2 + 2\ Br^- + H_2O$$

2. The first step in the mechanism is ionization of the amide N—H; the resulting anion is then brominated.

   a. The *N*-bromoamide product is more acidic than the amide starting material and it ionizes.

     i. The *N*-bromo anion rearranges rapidly to an isocyanate. (See Eq. 23.78a–d, text p. 1178, for details.)

     ii. Because the Hofmann rearrangement is carried out in aqueous base, the isocyanate cannot be isolated.

   b. The isocyanate formed spontaneously hydrates to a carbamate ion, which then decarboxylates to the amine product under the strongly basic reaction conditions.

   c. When the reaction mixture is acidified, the carbamate decarboxylates to give the amine.

3. The Hofmann rearrangement takes place with retention of stereochemical configuration in the migrating alkyl group. (See the foregoing equation for an example.)

F. CURTIUS REARRANGEMENT

1. When an acyl azide is heated in an inert solvent such as benzene or toluene, it is transformed with loss of nitrogen into an isocyanate; this concerted reaction is called the Curtius rearrangement.

2. The overall transformation that occurs as a result of the Curtius rearrangement followed by hydration is the removal of the carbonyl carbon of the acyl azide as CO_2.

   a. The isocyanate product of the Curtius rearrangement can be transformed into a carbamic acid by hydration in either acid or base.

   b. Spontaneous decarboxylation of the carbamic acid gives the amine, which is protonated under the acidic conditions of the reaction.

   c. The free amine is obtained by neutralization.

3. The Curtius rearrangement takes place with retention of stereochemical configuration in the migrating alkyl group. (See the preceding equation.)

4. The Curtius reaction can be run under mild, neutral conditions.

   a. The isocyanate can be isolated if desired.

   b. Some acyl azides in the pure state can detonate without warning, and extreme caution is required in handling them.

5. An important use of the Curtius rearrangement is for the preparation of carbamic acid derivatives. (See summary in Eq. 23.73, text p. 1218.)

   a. Reaction of isocyanates with alcohols or phenols yields carbamate esters.

   b. Reaction of isocyanates with amines yields ureas.

6. Acyl azides can be prepared in two ways:

   a. The most straightforward method is the reaction of an acid chloride with sodium azide.

   b. An ester can be converted into an acyl derivative of hydrazine by aminolysis; the resulting amide, an acyl hydrazide, is then diazotized with nitrous acid to give the acyl azide.

G. AROMATIC SUBSTITUTION REACTIONS OF ANILINE DERIVATIVES

1. Aromatic amines can undergo electrophilic aromatic substitution reactions on the aromatic ring.

   a. The amino group is one of the most powerful ortho, para-directing groups in electrophilic aromatic substitution provided that it is not protonated under the reaction conditions.

  b. A protonated amino group does not have the unshared electron pair on nitrogen that gives rise through resonance to the activating, ortho, para-directing effect of a free amino group.

  c. Ammonium groups are meta-directing groups.

 2. Aniline can be nitrated regioselectively at the para position if the amino group is first protected from protonation by conversion into an acetamide derivative by acetylation.

H. SYNTHESIS OF AMINES: SUMMARY

 1. Methods of amine synthesis:

  a. Reduction of amides with LiAlH$_4$ (Sec. 21.9B, text p. 1180).

  b. Reduction of nitriles with LiAlH$_4$ (Sec. 21.9C, text p. 1182).

  c. Direct alkylation of amines (Sec. 23.7A, text p. 1198). This reaction is of limited utility, but it is useful for preparing quaternary ammonium salts.

  d. Reductive amination (Sec. 23.7B, text p. 1199).

  e. Aromatic substitution reactions of anilines (Sec. 23.9, text p. 1204).

  f. Gabriel synthesis of primary amines (Sec. 23.11A, text p. 1211).

  g. Staudinger reaction for synthesis of primary amines (Sec. 23.11A, text p. 1211).

  h. Reduction of nitro compounds (Sec. 23.11B, text p. 1213).

  i. Amination of aryl chlorides or aryl triflates in the presence of a Pd(0) catalyst (Sec. 23.11C, text p. 1214).

  j. Hofmann and Curtius rearrangements (Sec. 23.11D, text p. 1216).

 2. Methods c, d, e, and i represent methods of preparing amines from other amines. Because amides can be prepared from amines, method a can be placed in this category as well.

 3. Methods b, f, g, h, and j are limited to the preparation of primary amines.

 4. Methods a, b, d, h, and j can be used to obtain amines from other functional groups.

IV. SYNTHESES OF ORGANIC COMPOUNDS WITH AMINES

A. ACYLATION OF AMINES

 1. Amines can be converted into amides by acylation with acid chlorides, anhydrides or esters.

 2. The reaction of an amine with an acid chloride or anhydride requires either two equivalents of the amine or one equivalent of the amine and an additional equivalent of another base such as a tertiary amine or hydroxide ion.

B. HOFMANN ELIMINATION OF QUATERNARY AMMONIUM HYDROXIDES

 1. Alkenes can be formed from amines by a three-step process:

  a. exhaustive methylation

  b. conversion of the ammonium salt to the hydroxide

  c. Hofmann elimination.

2. The Hofmann elimination involves a quaternary ammonium hydroxide as the starting material; an amine acts as the leaving group.
   a. When a quaternary ammonium hydroxide is heated, a β-elimination reaction takes place to give an alkene, which distills from the reaction mixture.
   b. A quaternary ammonium hydroxide is formed by treating a quaternary ammonium salt with silver hydroxide (AgOH, formed from water and silver oxide, Ag_2O).
   c. The Hofmann elimination is conceptually analogous to the E2 reaction of alkyl halides, in which a proton and a halide ion are eliminated; in the Hofmann elimination, a proton and a tertiary amine are eliminated.
   d. Hofmann elimination generally occurs as an anti-elimination.
   e. The conditions of the Hofmann elimination are typically harsh.

C. DIAZOTIZATION; REACTIONS OF DIAZONIUM IONS

   1. Oxidation of amines generally occurs at the amino nitrogen; an important oxidation reaction of amines is called diazotization.

   a. Diazotization is the reaction of a primary amine with nitrous acid (HNO_2) to form a diazonium salt.
   b. A diazonium salt is a compound of the form R—$^+$N'≡N: X$^-$, in which X$^-$ is a typical anion.
   c. Both aliphatic and aromatic primary amines are readily diazotized.
   2. Diazonium ions incorporate one of the best leaving groups, molecular nitrogen.
   3. Aliphatic diazonium ions react immediately as they are formed by S_N1, E1, and/or S_N2 mechanisms to give substitution and elimination products along with nitrogen gas. (The rapid liberation of nitrogen gas on treatment with nitrous acid is a qualitative test for primary alkylamines.)
   4. Aromatic diazonium salts may be isolated and used in a variety of reactions; however, they are usually prepared in solution at 0–5 °C and used without isolation, because they lose nitrogen on heating and they are explosive in the dry state.

D. AROMATIC SUBSTITUTION WITH DIAZONIUM IONS

   1. Aryldiazonium ions react with aromatic compounds containing strongly activating substituent groups, such as amines and phenols, to give substituted azobenzenes.

   2. This is an electrophilic aromatic substitution reaction in which the terminal nitrogen of the diazonium ion is the electrophile. (See Eqs. 23.48a–b, text p. 1209, for a detailed mechanism.)
   3. The azobenzene derivatives formed in these reactions have extensive conjugated π-electron systems, and most of them are colored.
   4. Some of these compounds are used as dyes and indicators; as a class they are known as azo dyes. (An azo dye is a colored derivative of azobenzene.)

E. SUBSTITUTION REACTIONS OF ARYLDIAZONIUM SALTS

   1. Among the most important reactions of aryldiazonium salts are substitution reactions with cuprous halides; in these reactions the diazonium group is replaced by a halogen.

a. An analogous reaction occurs with CuCN.
   i. This reaction is another way of forming a carbon–carbon bond, in this case to an aromatic ring.
   ii. The resulting nitrile can be converted by hydrolysis into a carboxylic acid.
b. The reaction of an aryldiazonium ion with a cuprous salt is called the Sandmeyer reaction.
c. This reaction is an important method for the synthesis of aryl halides and nitriles.
2. Aryl iodides can also be made by the reaction of diazonium salts with potassium iodide; a cuprous salt is not required.

3. Diazonium salts can be hydrolyzed to phenols by heating them with water or by treating them with cuprous oxide (Cu_2O) and an excess of aqueous cupric nitrate [$Cu(NO_3)_2$] at room temperature.

4. The diazonium group is replaced by hydrogen when the diazonium salt is treated with hypophosphorous acid, H_3PO_2.

5. All of the diazonium salt reactions listed above are substitution reactions but none are S_N2 or S_N1 reactions because aromatic rings do not undergo substitution by these mechanisms.
   a. The Sandmeyer and related reactions occur by radical-like mechanisms mediated by the copper.
   b. The reaction of diazonium salts with KI probably occurs by a similar mechanism.
   c. The reaction of diazonium salts with H_3PO_2 has been shown definitely to be a free-radical chain reaction.
6. The substitution reactions of diazonium salts achieve ring-substitution patterns that cannot be obtained in other ways.

F. REACTIONS OF SECONDARY AND TERTIARY AMINES WITH NITROUS ACID

1. Secondary amines react with nitrous acid to yield *N*-nitrosoamines, usually called simply nitrosamines. Nitrosamines are known to be potent carcinogens.

$$R_2NH \ + \ HNO_2 \ \longrightarrow \ R_2N-N{=}O \qquad \text{a nitrosamine}$$

2. The nitrogen of tertiary amines does not react under the strongly acidic conditions used in diazotization reactions.

3. *N,N*-Disubstituted aromatic amines undergo electrophilic aromatic substitution on the benzene ring; the electrophile is the nitrosyl cation, which is generated from nitrous acid under acidic conditions.

SOLUTIONS TO PROBLEMS

Solutions to In-Text Problems

23.1   (a)   (b)   (c)

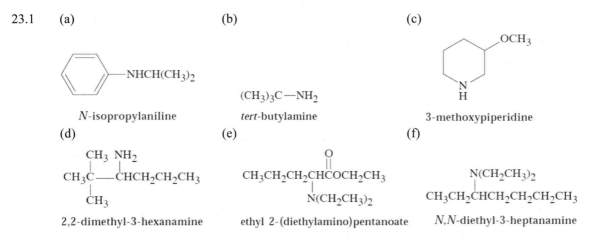

N-isopropylaniline   *tert*-butylamine   3-methoxypiperidine

(d)   (e)   (f)

2,2-dimethyl-3-hexanamine   ethyl 2-(diethylamino)pentanoate   *N*,*N*-diethyl-3-heptanamine

23.2   (a)   *N*-ethyl-*N*-methylisobutylamine (common), or *N*-ethyl-*N*,2-dimethyl-1-propanamine
       (b)   *N*,*N*-dimethyl-*p*-nitroaniline (common), or 1-dimethylamino-4-nitrobenzene
       (c)   dicyclohexylamine (common), or *N*-cyclohexylcyclohexanamine
       (d)   3-methylamino-1-pentanol. Note that the —OH group is the principal group.
       (e)   1-(2-chloroethyl)-3-propylpyrrolidine

23.3   (a)   The principle to apply is that the more double-bond character the carbon–nitrogen bond has, the shorter it
             is. The order of decreasing double-bond character, and hence increasing carbon–nitrogen bond length, is

                              p-nitroaniline < aniline < cyclohexylamine

             Because of resonance interaction of their nitrogen lone pairs with the benzene ring, aniline and
             p-nitroaniline have the most double-bond character; cyclohexylamine lacks this interaction and therefore
             has the least carbon–nitrogen double-bond character. In *p*-nitroaniline, the unshared electron pair on the
             amine nitrogen can be delocalized into the *p*-nitro group:

             This interaction is in addition to the resonance interaction of the unshared pair with the ring of the type
             shown in Eq. 23.3 on p. 1187 of the text. This additional delocalization increases the carbon–nitrogen
             double-bond character and decreases the carbon–nitrogen bond length.

       (b)   Apply the principle used in part (a). The order of increasing carbon–nitrogen bond length is

                                          A < *C* < *B*

             Compound *B* has a true carbon–nitrogen single bond, and hence has the longest carbon–nitrogen bond.
             Compound *A* has a true carbon–nitrogen double bond, and hence has the shortest carbon–nitrogen bond.
             The carbon–nitrogen bond of *C* has some double-bond character because of resonance interaction of the

nitrogen unshared pair with the double bond. (Draw the appropriate resonance structure.) Another way to compare the C—N bond lengths of compounds *B* and *C* is to note that the C—N bond in *B* is an sp^3–sp^3 single bond, whereas the C—N bond in *C* is an sp^2–sp^3 single bond, and to remember that σ bonds with greater *s* character are shorter. (See discussion in Secs. 4.B and 15.1A of the text)

23.4   The five-proton resonance at δ 7.18 in the NMR spectrum indicates the presence of a monosubstituted benzene ring (which accounts for 77 mass units), and the mutually split triplet–quartet pattern at δ 1.07 and δ 2.60 indicates an ethyl group (which accounts for 29 mass units). The IR spectrum indicates the presence of an amine; if it is a secondary amine, as suggested by the integral of the δ 0.91 resonance in the NMR spectrum, the NH accounts for 15 mass units, leaving 14 mass units unaccounted for. A CH_2 group could account for this remaining mass, and the two-proton singlet at δ 3.70 could correspond to such a resonance. On the assumption that the N—H proton is undergoing rapid exchange and does not show splitting with neighboring protons, *N*-ethylbenzylamine fits the data.

$$\text{C}_6\text{H}_5 - \text{CH}_2 - \text{NH} - \text{CH}_2\text{CH}_3 \qquad \textit{N-ethylbenzylamine}$$

The predicted chemical shift (Fig. 13.4, text p. 621) for the benzylic CH_2 group is greater than that of ordinary benzylic protons because these protons are α to a nitrogen; the chemical shift should also be greater than protons that are only α to a nitrogen. The chemical shift of δ 3.6 is quite reasonable. The $m/z = 120$ peak in the mass spectrum corresponds to loss of a methyl radical by α-cleavage; and the $m/z = 91$ peak corresponds to a benzyl cation, which is formed by inductive cleavage. (See Sec. 12.6D of the text for a discussion of these cleavage mechanisms.)

23.5   The NMR spectrum indicates the presence of only two phenyl hydrogens, and the singlet absorptions at δ 2.07 and δ 2.16 indicate the presence of chemically nonequivalent methyl groups in the ratio 2:1. Only compound (2), 2,4,6-trimethylaniline, fits the data. The two resonances at δ 2.07 and δ 2.16 correspond to the ortho and para methyl groups, respectively, and the δ 3.19 resonance to the —NH_2 protons of the amine.

23.6   (a)   2,2-Dimethyl-1-propanamine (neopentylamine, $(CH_3)_3CCH_2NH_2$) has a maximum of three resonances in its CMR spectrum. 2-Methyl-2-butanamine, $CH_3CH_2C(CH_3)_2NH_2$, should have four resonances in its CMR spectrum.

        (b)   *Trans*-1,4-cyclohexanediamine has only two nonequivalent sets of carbons and thus should have only two resonances in its CMR spectrum. *Trans*-1,2-cyclohexanediamine has three nonequivalent sets of carbons and therefore should have three resonances in its CMR spectrum. [Chemically equivalent (and in these cases, homotopic) carbons in each compound are indicated by the same symbol.]

trans-1,4-cyclohexanediamine
two nonequivalent sets of carbons

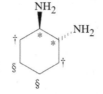

trans-1,2-cyclohexanediamine
three nonequivalent sets of carbons

23.7   (a)   The order of increasing basicity (and increasing conjugate-acid pK_a) is NH_3 (ammonia) < $PrNH_2$ (propylamine) < Pr_2NH (dipropylamine). As the text discussion in Sec. 23.5B, text pp. 1190–4, shows, within a series of amines of increasing alkyl substitution, basicity increases from ammonia to the secondary amine.

        (b)   The order of increasing basicity is

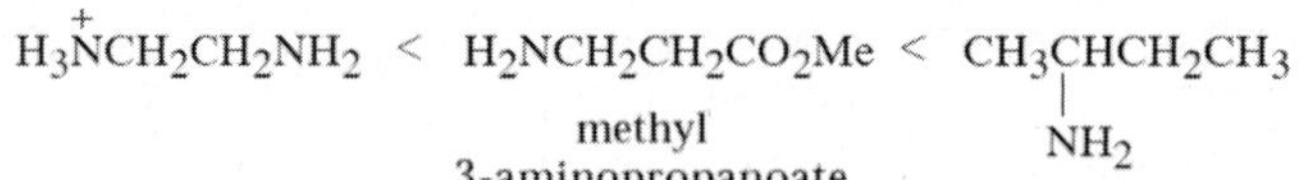

$$\overset{+}{H_3N}CH_2CH_2NH_2 \quad < \quad H_2NCH_2CH_2CO_2Me \quad < \quad CH_3\underset{\underset{NH_2}{|}}{C}HCH_2CH_3$$

methyl
3-aminopropanoate

sec-butylamine

(Note that *basicity* refers to the basicity of the uncharged amino group in the first compound, not the basicity of the *conjugate base* of the ammonium ion.) The electron-withdrawing polar effect of the partial positive charge on the carbonyl carbon of the ester reduces the basicity of its β-amino group, and the polar effect of the full positive charge in the ammonium ion reduces the basicity of its β-amino group even more. The polar effect of a substituent containing a fully charged atom is generally greater than the polar effect of a substituent containing a dipolar group.

(c) The order of increasing basicity is

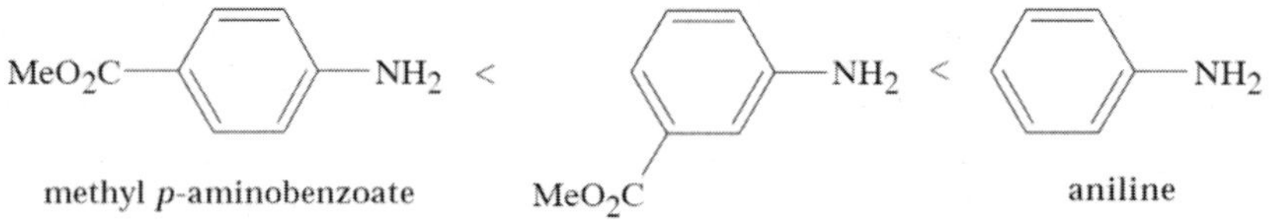

methyl *p*-aminobenzoate   methyl *m*-aminobenzoate   aniline

The methoxycarbonyl group (that is, the methyl ester group) decreases basicity in the first two derivatives by a polar effect. Furthermore, methyl *p*-aminobenzoate is stabilized by resonance interaction of the nitrogen unshared pair with the carbonyl group:

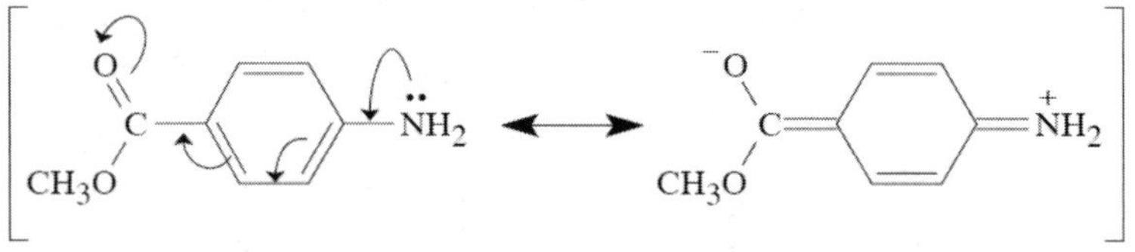

Because there is no unshared pair on nitrogen in the conjugate-acid ammonium ion, such resonance stabilization of the conjugate acid is not possible. As Fig. 23.2 on text p. 1191 shows, anything that stabilizes an amine relative to its conjugate-acid ammonium ion lowers its basicity. Note that resonance interaction between the carbonyl group and a *meta*-amino group is not possible.

(d) The order of increasing basicity is

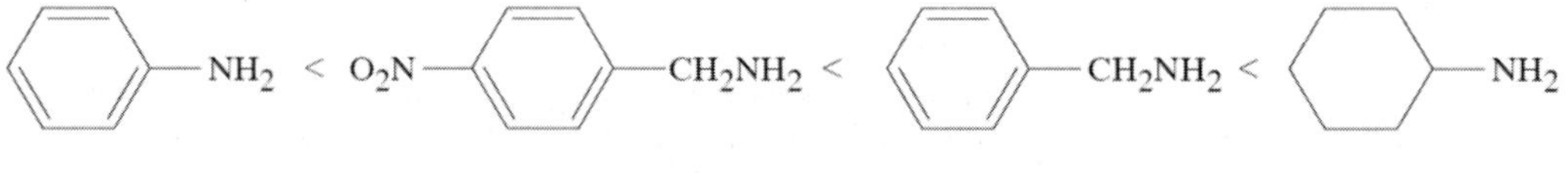

aniline   *p*-nitrobenzylamine   benzylamine   cyclohexylamine

The resonance and electron-withdrawing polar effect of the benzene ring severely reduces the nitrogen basicity of aniline relative to that of an alkylamine. The electron-withdrawing polar effect of the nitro group of *p*-nitrobenzylamine reduces its basicity relative to that of benzylamine. Finally, the electron-withdrawing polar effect of the benzene rings reduces the basicity of both benzylamines relative to cyclohexylamine, an ordinary alkylamine.

23.9   Dissolve the mixture in a low-boiling, water-insoluble organic solvent such as methylene chloride. Extract this solution with 5% sodium bicarbonate or 5% sodium hydroxide solution; *p*-chlorobenzoic acid will dissolve in the aqueous layer as its sodium salt. It can be recovered as the free acid by acidifying the aqueous extracts.

The methylene chloride solution now contains only *p*-chloroaniline and *p*-chlorotoluene. Extract the methylene chloride solution with 5% HCl solution. *p*-Chloroaniline will dissolve in the aqueous layer as its hydrochloride salt. It can be recovered as the free amine by neutralizing the aqueous layer with dilute NaOH.

p-Chlorotoluene remains in the methylene chloride layer; it can be isolated by drying the solution and evaporating the methylene chloride. (Note that the order of the acid and base extractions could be interchanged.)

23.10 Mix the racemic acid with one equivalent of the enantiomerically pure amine. (The *S* enantiomer of the amine is utilized in the equation below, but either *pure* enantiomer of the amine could be used.) The carboxylic acids will react with the amine to give a mixture of *diastereomeric* salts.

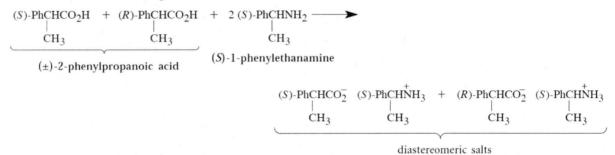

Because these salts have different properties, they can be separated by differential solubility, for example, by fractional crystallization from some solvent, probably an alcohol. After separating the salts, aqueous acid is added to each. The amine resolving agent dissolves in the aqueous acid solution (from which it can be recovered), leaving the enantiomerically pure free carboxylic acid as a solid that can be isolated by filtration and recrystallized.

$$(S)\text{-PhCHCO}_2^- \quad (S)\text{-PhCHNH}_3^+ \ + \ \text{HCl} \xrightarrow{\text{H}_2\text{O}} (S)\text{-PhCHCO}_2\text{H} \ + \ (S)\text{-PhCHNH}_3^+ \ \text{Cl}^-$$

$$\underset{\text{salt}}{\underbrace{\overset{|}{\text{CH}_3} \qquad\qquad \overset{|}{\text{CH}_3}}} \qquad\qquad\qquad \underset{\text{crystallizes}}{\overset{|}{\text{CH}_3}} \qquad \underset{\substack{\text{an ionic compound;} \\ \text{remains in solution}}}{\overset{|}{\text{CH}_3}}$$

Notice that not only is amine basicity important in forming the initial salt, but also it is important in separating the amine resolving agent from the resolved carboxylic acid.

23.11 (a)

$$(\text{CH}_3\text{CH}_2)_4\text{N}^+ \ \text{F}^-$$

tetraethylammonium fluoride

(b)

$$(\text{PhCH}_2)_2\overset{+}{\text{N}}(\text{CH}_3)_2 \ \text{Br}^-$$

dibenzyldimethylammonium bromide

23.12 Compound *A* has an asymmetric nitrogen stereocenter; therefore, it is chiral and can exist as two enantiomers. Compound *B* also contains an asymmetric nitrogen; however, compound *B* undergoes a Brønsted acid–base equilibrium by which it is in equilibrium with a very small amount of the free amine:

$$\underset{\substack{| \\ \text{C}_2\text{H}_5 \\ \\ \text{compound } B}}{\overset{\overset{\displaystyle \text{CH}_3}{|}}{\text{PhCH}_2\overset{+}{\text{N}}{-}\text{H}}} \ \text{Cl}^- \ \rightleftharpoons \ \underset{\substack{| \\ \text{C}_2\text{H}_5}}{\overset{\overset{\displaystyle \text{CH}_3}{|}}{\text{PhCH}_2\text{N}{:}}} \ + \ \text{H}{-}\text{Cl}$$

This equilibrium takes place so rapidly, and the minuscule amount of free amine undergoes nitrogen inversion so quickly, that the ammonium salt is rapidly racemized. Because compound *A* does not have a hydrogen that can be involved in an equilibrium of this type, it is not racemized.

23.13 (a) As detailed in Sec. 11.5A, nucleophiles can react with epoxides to form alcohols. Water serves to both deprotonate the protonated nitrogen after the initial reaction, and as an acid to protonate the resulting alkoxide oxygen.

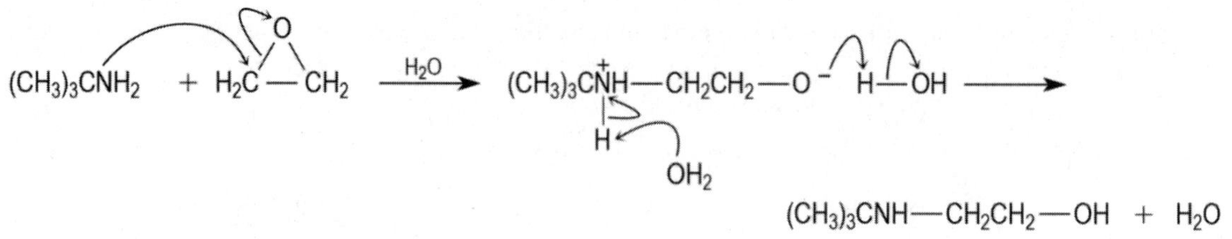

$$(CH_3)_3CNH_2 \ + \ H_2C-CH_2 \ \xrightarrow{H_2O} \ (CH_3)_3C\overset{+}{N}H-CH_2CH_2-O^- \ H-OH \ \longrightarrow$$

$$(CH_3)_3CNH-CH_2CH_2-OH \ + \ H_2O$$

Since the resulting product is also nucleophilic at nitrogen, a second addition can occur.

$$(CH_3)_3CNH-CH_2CH_2-OH \ \xrightarrow{H_2O} \ (CH_3)_3C\overset{+}{N}-CH_2CH_2-OH \ \longrightarrow$$

$$(CH_3)_3CN(CH_2CH_2OH)_2 \ + \ H_2O$$

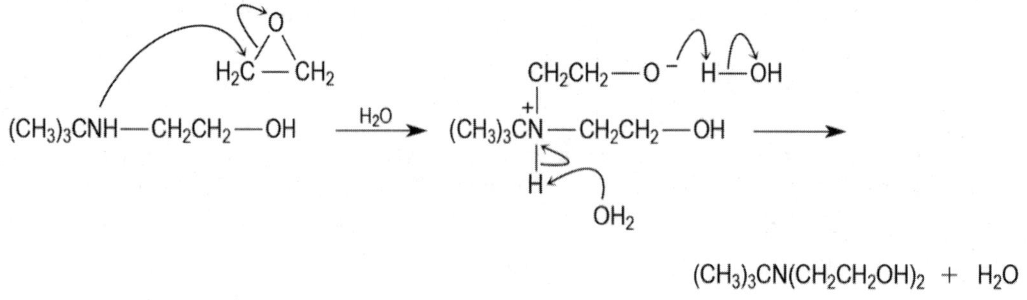

(b) As shown by Eqs. 22.101 and 22.102 on p. 1157 of the text, ammonia can act as a nucleophile in conjugate addition reactions.

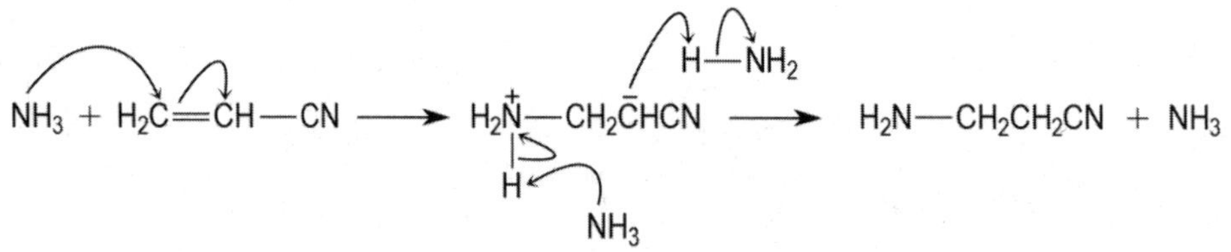

$$NH_3 \ + \ H_2C=CH-CN \ \longrightarrow \ H_2\overset{+}{N}-CH_2\bar{C}HCN \ \longrightarrow \ H_2N-CH_2CH_2CN \ + \ NH_3$$

Since the resulting product is also nucleophilic at nitrogen, a second addition can occur, leading to the dialkylated product formed in 57% yield as shown in Eq. 23.18. Write the mechanism, using the one shown above as a guide.

(c) A third alkylation doesn't occur in high yields most likely due to sterics, or the concentrations of the reagents. Increasing the concentration of acrylonitrile (2-propenenitrile) relative to ammonia would favor the formation of the trisubstituted compound.

23.14 The reaction of a primary amine and a ketone yields an imine, by the standard mechanism. This mechanism is shown in Study Guide Links 19.5 and 19.6, on pp. 608–610 of this manual.

23.15 Under the acidic conditions shown in Eqs. 23.23 (acetic acid) and 23.24 (HCl), the initial product would be pronated on the basic amine nitrogen. In order to isolate the neutral amine, base is added to remove the acidic proton.

23.16 The products of Eqs. 23.26 and 23.27 are both tertiary amines. Tertiary amines form unstable products with ketones and aldehydes, that quickly break back down into starting materials (see text p. 988).

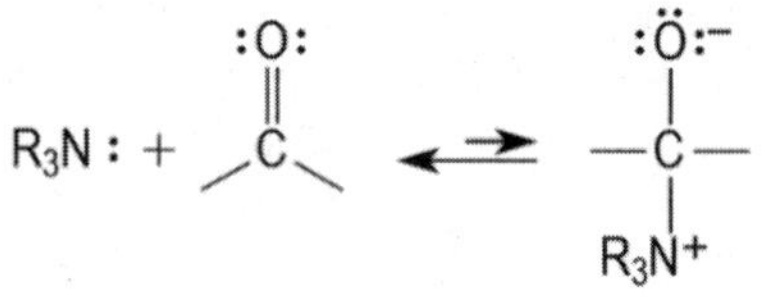

$$R_3N: \ + \ \underset{}{\overset{:O:}{\underset{\|}{C}}} \ \rightleftharpoons \ \underset{R_3N^+}{\overset{:\ddot{O}:^-}{\underset{|}{-C-}}}$$

Since no imine or enaime can form, there is no π-bond to reduce. Thus, quaternization cannot occur.

23.17 Either the ethyl group can originate from acetaldehyde, or the cyclohexyl group can originate from cyclohexanone.

23.18 The ethyl group could originate from acetaldehyde:

23.19 (a) Prepare the tertiary amine by reductive amination and alkylate it to give the ammonium salt.

An alternative synthesis would be to reductively aminate the same amine with formaldehyde to give trimethylamine and then alkylate this tertiary amine with benzyl bromide ($PhCH_2Br$).

(b) Exhaustively methylate benzylamine with a large excess of methyl bromide.

23.20 (a) Caleb has forgotten, if he ever knew, that aryl halides such as bromobenzene do not undergo S_N2 reactions (Sec. 18.1).

(b) A Pd(0) catalyst of the type PdL_2. This reaction is discussed in Sec. 23.11C, where L is:

23.21 The dimethylammonium ion can lose a proton to give dimethylamine, which can then be alkylated by methyl iodide.

23.22 (a)

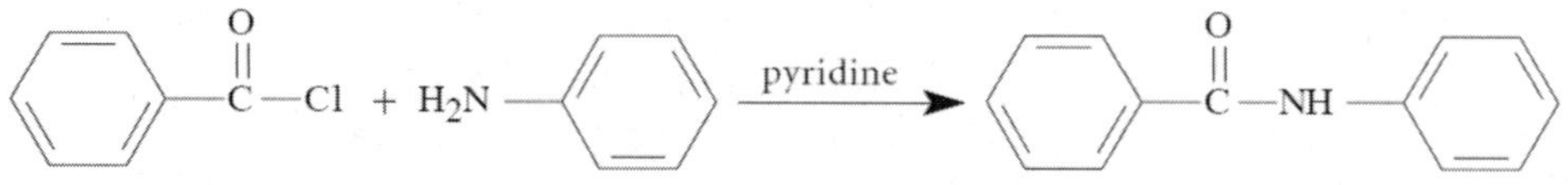

benzoyl chloride   aniline   *N*-phenylbenzamide

(b)

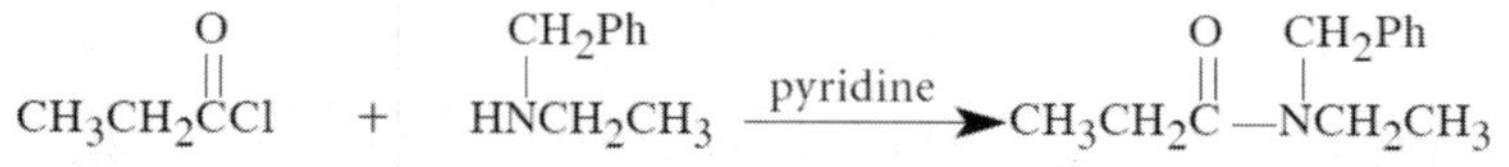

propanoyl chloride   benzylethylamine   *N*-benzyl-*N*-ethylpropanamide

23.23 (a) Draw the molecule in a conformation in which the trimethylammonium group is anti to the β-hydrogen, and examine the relative positions of the phenyl groups. In this part, the alkene product must have the *Z* configuration.

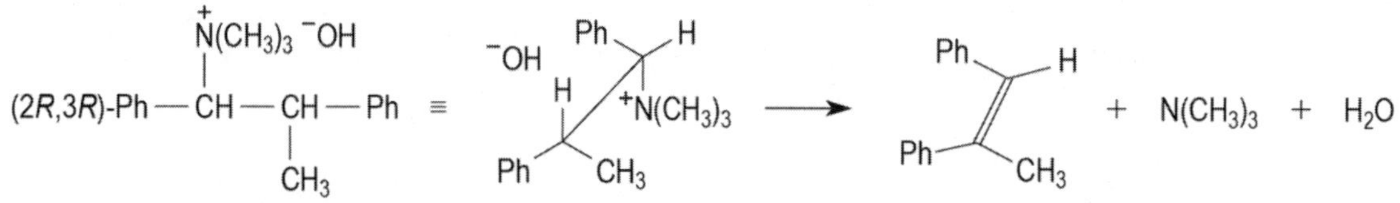

note that phenyl groups are gauche ⇨ phenyl groups are cis; therefore, alkene has *Z* configuration

(b) Draw the molecule in a conformation in which the trimethylammonium group is anti to the β-hydrogen, and examine the relative positions of the phenyl groups. In this part, the alkene product must have the *E* configuration.

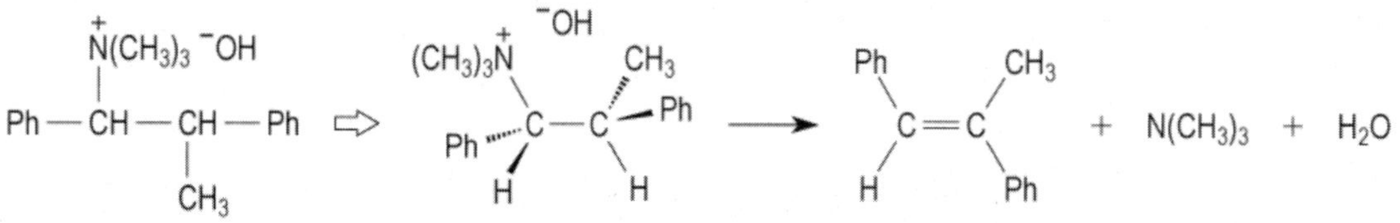

(2*R*,3*S*)

Note that pheny groups are anti ⇨ phenyl groups are cis; therefore, alkene has *E* configuration

23.24 (a) The product of the exhaustive methylation followed by heating with silver oxide is 3-methylcyclohexene, not 1-methylcyclohexene. Since the amino and methyl groups are in a trans orientation on the cyclohexane ring and Hofmann elimination generally occurs as an anti-elimination, the only hydrogen in an anti relationship to the amine group is the hydrogen on C6 of the ring.

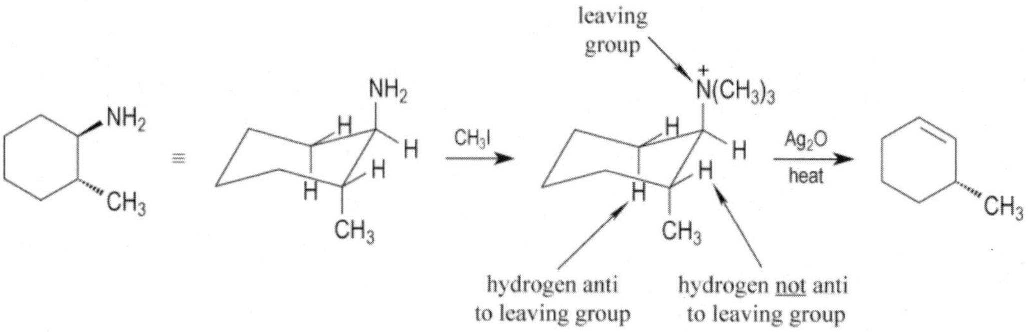

leaving group

hydrogen anti to leaving group   hydrogen not anti to leaving group

(b)   The hydrogen α to the carbonyl group is removed because it is considerably more acidic than the other β-hydrogens.

23.25   Coniine has one degree of unsaturation. Because the exhaustive methylation–Hofmann elimination sequence introduces a double bond, and because each of the products *A–C* contains only one double bond, it follows that coniine is a cyclic amine. Product *C* shows one point of attachment of the nitrogen in coniine, and shows that the coniine ring has a propyl side chain. Products *A* and *B* show the other point of attachment of the nitrogen: the alkene carbon that is common to both structures. Because five carbons intervene between points of attachment of the nitrogen, coniine must be a piperidine derivative (see text p. 1185):

2-propylpiperidine
(coniine)

Exhaustive methylation, Ag$_2$O treatment, and Hofmann elimination give three alkenes that result from the three possible points of β-elimination:

product resulting from exhaustive methylation
and Ag$_2$O treatment of coniine

In many problems, the text uses Ag$_2$O [silver(I) oxide] for the conversion of quaternary ammonium halides into quaternary ammonium hydroxides. Attempts to balance these reactions will show that a hydrogen is unaccounted for. The resolution of this problem is that silver(I) oxide is typically used in a *hydrated* form; hydrated Ag$_2$O is equivalent to AgOH. Consequently, you should assume in all problems involving Hofmann elimination that Ag$_2$O is operationally the same thing as AgOH.

23.26

23.27 Use an acetyl protecting group as outlined in Study Problem 23.4, text p. 1205.

If the acetyl protecting group were not used, the amine group on one molecule would react with a chlorosulfonyl group on the other and a polymer would be formed.

23.28 (a) Begin with *p*-nitroacetanilide, prepared as shown in Study Problem 23.4, text pp. 1205. Then convert it into 2,4-dinitroaniline. The acetyl protecting group is used to avoid protonation of the nitrogen, which would result in a significant amount of unwanted meta substitution. (See Eq. 23.37 on text p. 1205.)

(b) Begin with *p*-acetamidobenzenesulfonyl chloride, prepared as shown in the solution to Problem 23.27, p. 844 of this manual.

23.29   (a)

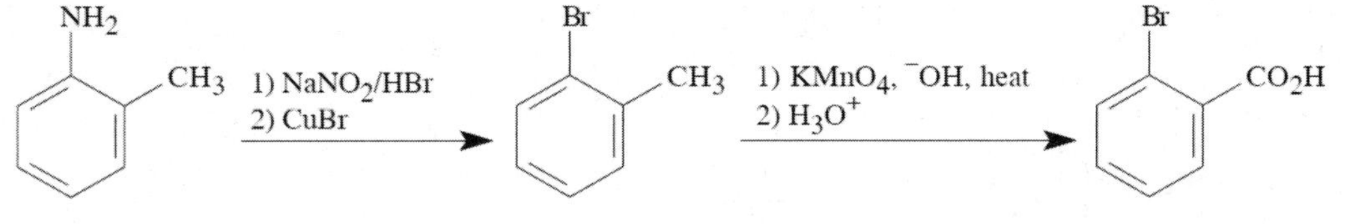

o-toluidine   2-bromobenzoic acid

(b)

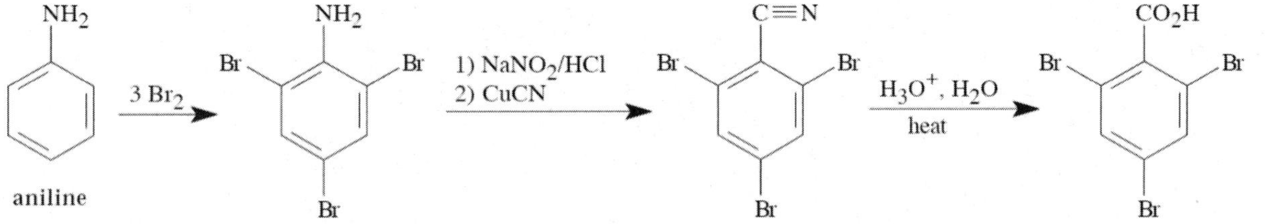

aniline

2,4,6-tribromobenzoic
acid

23.30   (a)   Because diazotization does not break the carbon–nitrogen bond, the diazonium ion intermediate, like the starting amine, has the *R* configuration. Hence, inversion of stereochemical configuration occurs in the reaction with water.

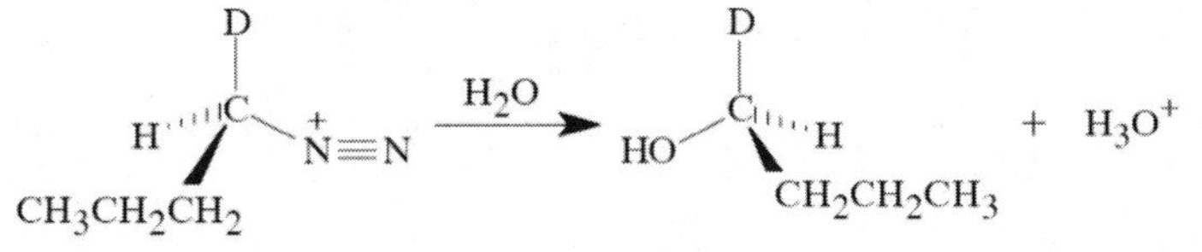

(b)   Because inversion of stereochemical configuration is observed, an S_N2 reaction in which water is the nucleophile and N_2 is the leaving group is likely to be involved. An S_N1 mechanism involving a carbocation intermediate would be expected to result in some racemization. (Why?)

23.31   (a)

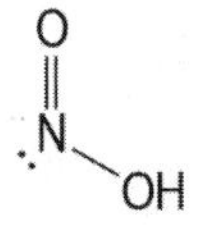

(b)

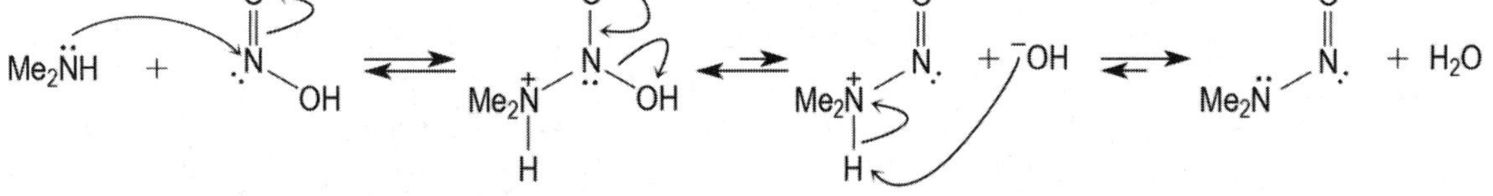

23.32   *p*-Aminobenzenesulfonic acid is prepared and diazotized to give the diazonium ion *A*. This synthesis begins with *p*-acetamidobenzenesulfonyl chloride, prepared as shown in the solution to Problem 23.27 on p. 844 of this manual. Formaldehyde is reductively aminated with aniline to give *N,N*-dimethylaniline, *B*. Compounds *A* and *B* are coupled to give methyl orange.

The reaction scheme shows:

p-acetamidobenzenesulfonyl chloride (see solution to Problem 23.22) reacting with H₃O⁺, H₂O, heat to give H₃N⁺—C₆H₄—SO₃H, then NaNO₂/H₂SO₄ to give the diazonium compound A (N≡N⁺—C₆H₄—SO₃H).

H_2N—C₆H₄ + $H_2C=O$ / $NaCNBH_3$ → $(CH_3)_2N$—C₆H₄ (**B**)

A + B → 1) H₂O 2) NaOH →

(CH₃)₂N—C₆H₄—N=N—C₆H₄—SO₃⁻ Na⁺

methyl orange

23.33 The following compounds would react to give FD & C Yellow No. 6:

Na⁺ ⁻O₃S—(naphthalene with OH) + N≡N⁺—C₆H₄—SO₃⁻ Na⁺ / Cl⁻ →

Na⁺ ⁻O₃S—(naphthalene)—N=N—C₆H₄—SO₃⁻ Na⁺ (with OH)   **FD & C yellow #6**

The position next to the ring junction (called the α-position) in a naphthalene ring is more reactive in electrophilic substitution than the ring position one carbon removed. Hence, of the two positions "ortho" to the phenolic —OH group, the α-position is the more reactive and gives the observed product.

23.34 (a) Protonation of the —OH oxygen and loss of water give the nitrosyl cation. (The chloride counter-ion is not shown.)

H_2O^+—H   HO—N=O → HO⁺(H)—N=O → H₂O + ⁺N=O

nitrosyl cation

(b) The loss of H₂O from protonated nitrous acid provides the nitrosyl cation, as shown in the solution to part (a) of this problem. This cation then serves as the electrophile in an electrophilic aromatic substitution reaction:

O=N⁺ + C₆H₅—N(CH₃)₂ → O=N—(C₆H₅ with H)⁺—N(CH₃)₂ + OH₂ →

nitrosyl cation   **resonance-stabilized carbocation intermediate**

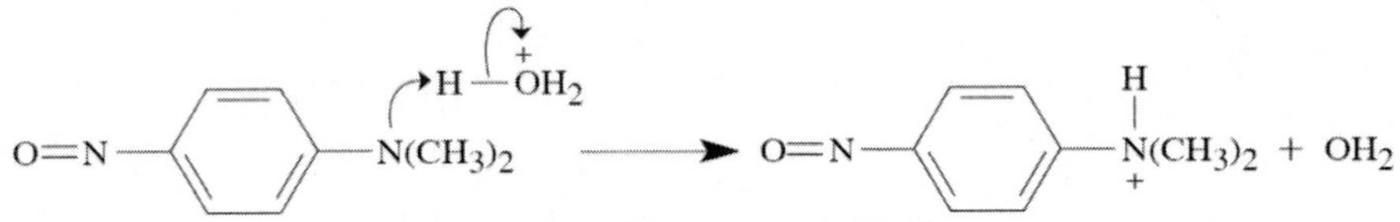

23.29 (a) In both the Gabriel synthesis and Staudinger reaction, only primary or unbranched secondary alkyl halides or sulfonates will react since each process involves an S_N2 mechanism.

(b) A possible alkyl halide starting material for this amine is 1-bromo-3-methylpentane.

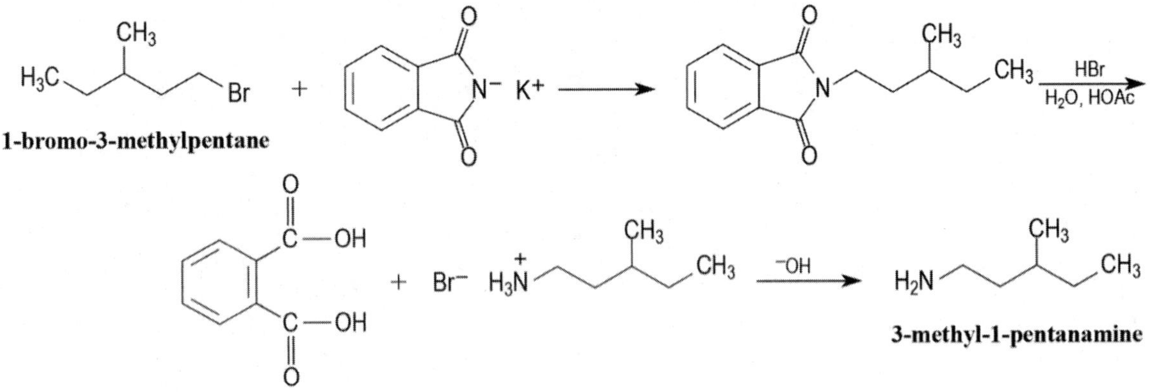

1-bromo-3-methylpentane

3-methyl-1-pentanamine

(c)

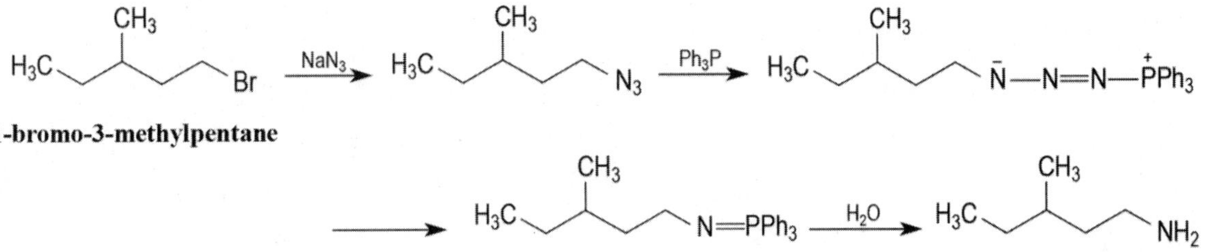

1-bromo-3-methylpentane

3-methyl-1-pentanamine

(d) The alkyl halide or sulfonate required to produce 2,2-dimethyl-1-propanamine is sterically hindered at the carbon on which the reaction would occur. *N*-Butylaniline is not a primary amine; only primary amines can be prepared by these methods. Therefore, 3-methyl-1-pentanamine is the only amine of the three that can be prepared by either method.

23.36 (a)

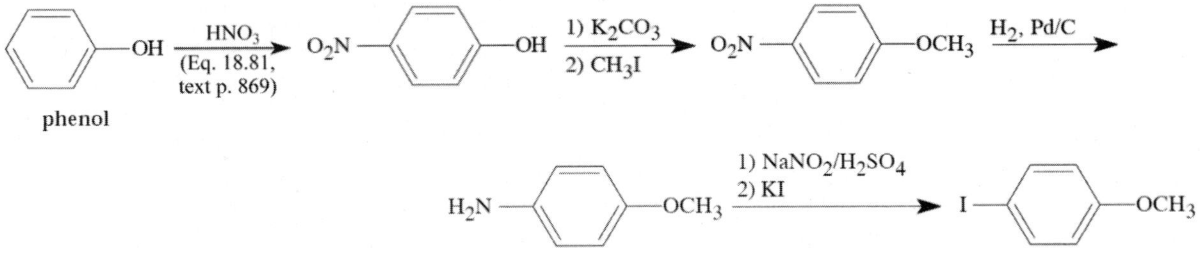

phenol

p-iodoanisole

Note that formation of the ether is carried out before reduction of the nitro group because, otherwise, the amino group could also be alkylated with methyl iodide.

(b)

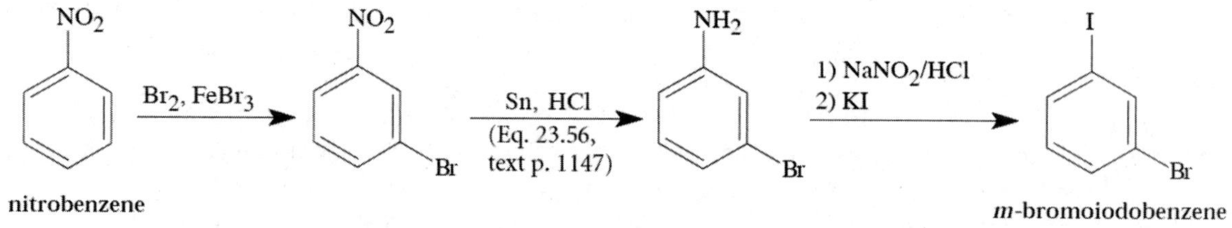
NO2
Br2, FeBr3
NO2
Br
Sn, HCl
(Eq. 23.56,
text p. 1147)
NH2
Br
1) NaNO2/HCl
2) KI
I
Br
nitrobenzene
m-bromoiodobenzene

23.37 (a)

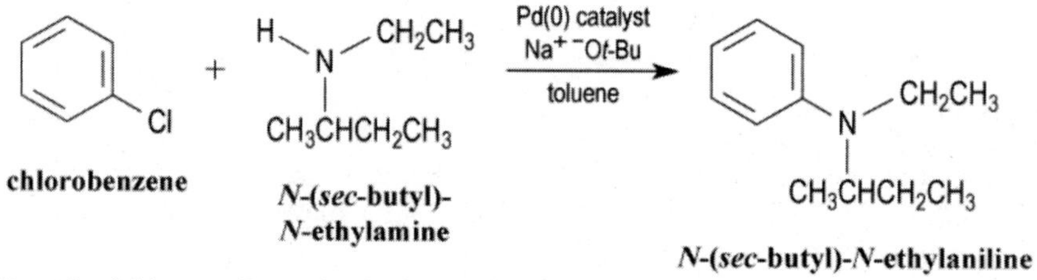
Cl
chlorobenzene
H
N
CH2CH3
CH3CHCH2CH3
N-(sec-butyl)-
N-ethylamine
Pd(0) catalyst
Na+ −Ot-Bu
toluene
N
CH2CH3
CH3CHCH2CH3
N-(sec-butyl)-N-ethylaniline

(b) (TfOTf = triflic anhydride; see Sec. 18.10B in the text for the structure.)

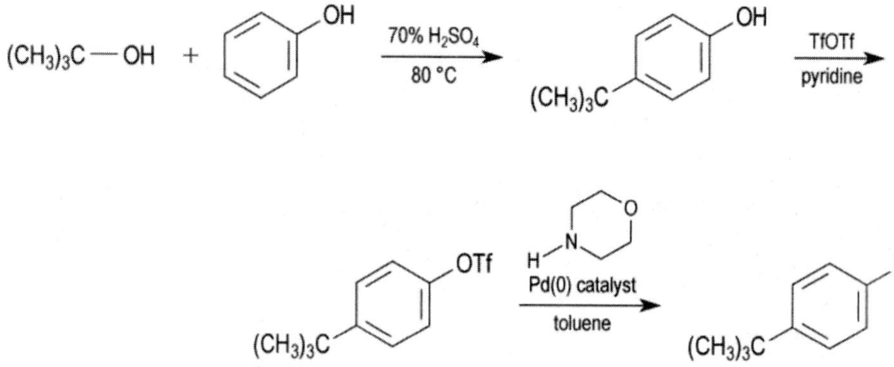
(CH3)3C—OH +
OH
70% H2SO4
80 °C
OH
(CH3)3C
TfOTf
pyridine
OTf
(CH3)3C
H
N
O
Pd(0) catalyst
toluene
N
O
(CH3)3C

(c)

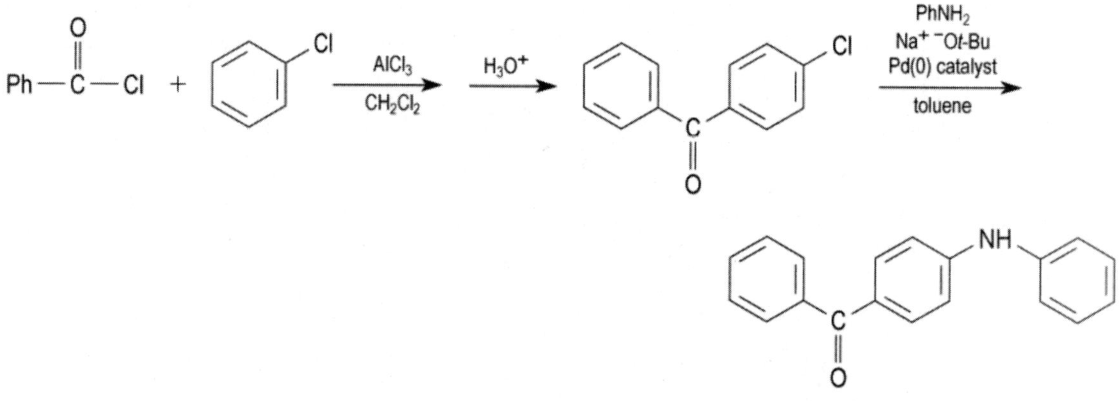
O
Ph—C—Cl +
Cl
AlCl3
CH2Cl2
H3O+
Cl
C
O
PhNH2
Na+ −Ot-Bu
Pd(0) catalyst
toluene
NH
C
O

23.38 (a) *Tert*-butylamine cannot be prepared by the Gabriel synthesis because it would require that the nitrogen of phthalimide be alkylated with *tert*-butyl bromide in an S_N2 reaction. Tertiary alkyl halides generally do not undergo S_N2 reactions but, under basic conditions, undergo E2 reactions instead.

(b) The Hofmann or Curtius rearrangement would be an ideal way to prepare *tert*-butylamine. (The Hofmann rearrangement is illustrated below.)

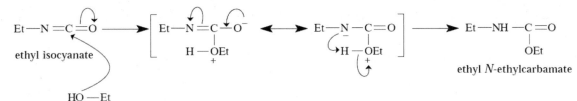

$$(CH_3)_3C-\overset{\overset{\displaystyle O}{\displaystyle \|}}{C}-NH_2 \quad \xrightarrow{\text{Br}_2, \text{ NaOH}} \quad (CH_3)_3C-NH_2$$

2,2-dimethylpropanamide   *tert*-butylamine
(pivalamide)

23.39 (a) The curved-arrow mechanism for the conversion of ethyl isocyanate into ethyl *N*-ethylcarbamate:

Et—N=C=O ⟶ [Et—N=C—O⁻ ⟷ Et—N—C=O] ⟶ Et—NH—C=O

ethyl isocyanate   H—OEt⁺   H—OEt⁺   OEt

HO—Et   ethyl *N*-ethylcarbamate

⚠ Note that the proton-transfer step, shown in the foregoing reaction for brevity as an intramolecular process, probably is intermolecular. Intramolecular proton transfers are possible, but they tend to take place fastest through 6, 7, and 8-membered rings. Nevertheless, we can abbreviate the more complex intermolecular processes as intramolecular processes, and we'll sometimes follow this practice for convenience.

(b) The curved-arrow mechanism for the reaction of ethylamine with ethyl isocyanate to give *N,N'*-diethylurea:

Et—N=C=O ⟶ [Et—N=C—O⁻ ⟷ Et—N—C=O] ⟶ Et—NH—C=O

H₂N—Et   H—NEt⁺   H—NEt⁺   HNEt

                        H   H   *N,N*-diethylurea

23.40 (a) In ethanol solvent, ethanol (or its conjugate base ethoxide) is the most abundant nucleophile present, and thus an ethyl carbamate is obtained as the product.

$$CH_3\underset{\underset{\displaystyle CH_3}{\displaystyle |}}{CH}-\overset{\overset{\displaystyle O}{\displaystyle \|}}{C}-NH_2 \quad \xrightarrow{\text{Br}_2, \text{ NaOEt}} \quad CH_3\underset{\underset{\displaystyle CH_3}{\displaystyle |}}{CH}-NH-\overset{\overset{\displaystyle O}{\displaystyle \|}}{C}-OEt$$

2-methylpropanamide   ethyl *N*-isopropylcarbamate

23.41 Pentyl isocyanate is an intermediate in the reaction. Because it is largely insoluble in water, it forms a separate phase. A small amount of pentyl isocyanate reacts with the aqueous NaOH to form pentylamine, which is then extracted into the isocyanate, with which it reacts to give the urea.

$$CH_3(CH_2)_4\overset{\overset{\displaystyle O}{\displaystyle \|}}{C}NH_2 \quad \xrightarrow[\text{H}_2\text{O}]{\text{Br}_2, \text{ NaOH}} \quad CH_3(CH_2)_4N=C=O \quad \xrightarrow{\text{H}_2\text{O, NaOH}} \quad CH_3(CH_2)_4NH_2 \ + \ CO_2$$

hexanamide   *A*

$$CH_3(CH_2)_4NH\overset{\overset{\displaystyle O}{\displaystyle \|}}{C}NH(CH_2)_4CH_3$$

N,N'-dipentylurea

B

23.42 **(a)**

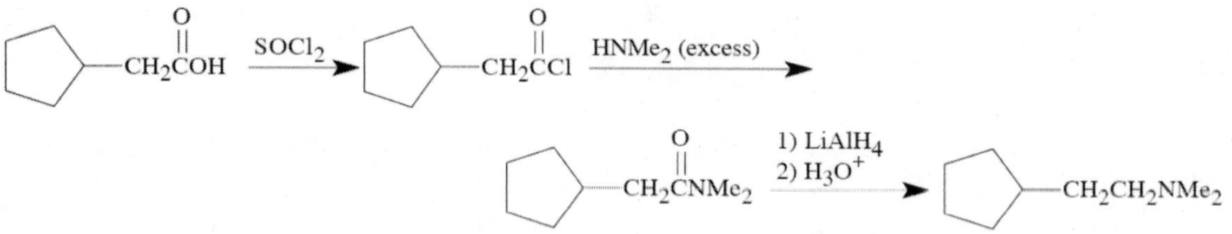

2-cyclopentyl-*N*,*N*-dimethylethanamine

(b)

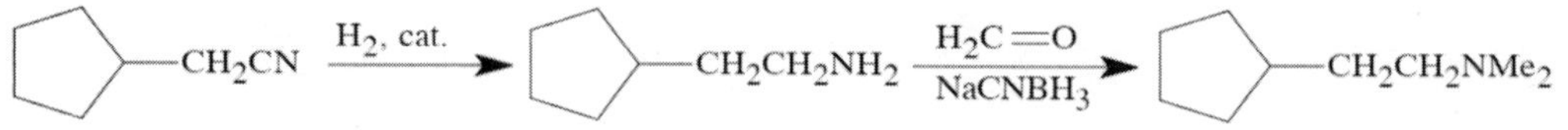

(c)

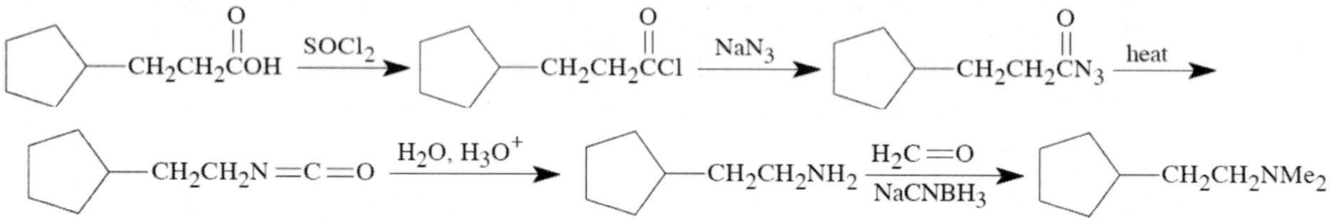

(d) One synthesis involves a reductive amination of the aldehyde:

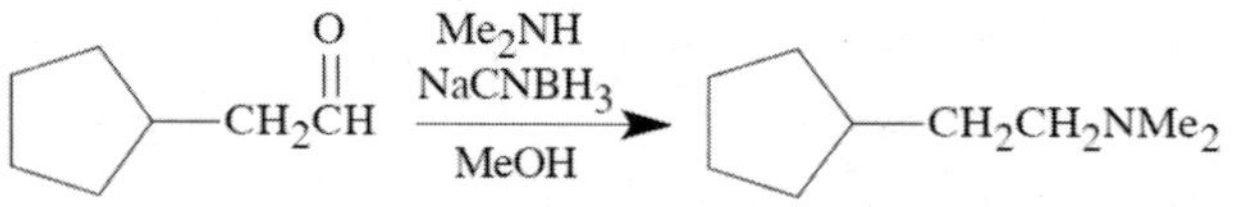

A second synthesis involves oxidation of the aldehyde to the carboxylic acid starting material of part (a) with aqueous H_2CrO_4, $KMnO_4$, or other common oxidant, and then proceeding as in part (a).

23.43 **(a)** In morphine, the nitrogen of the tertiary amine is the most basic atom. Therefore the conjugate acid of morphine has the following structure:

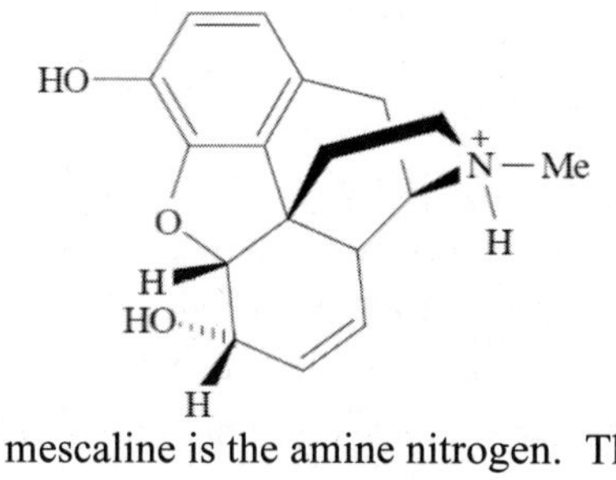

conjugate acid of morphine

(b) The most basic atom in mescaline is the amine nitrogen. Thus, the conjugate acid of mescaline has the following structure:

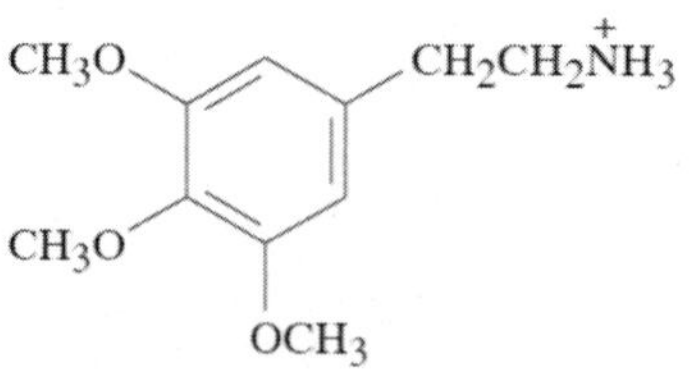

conjugate acid of mescaline

Solutions to Additional Problems

23.44   The structure of *p*-chloroaniline is

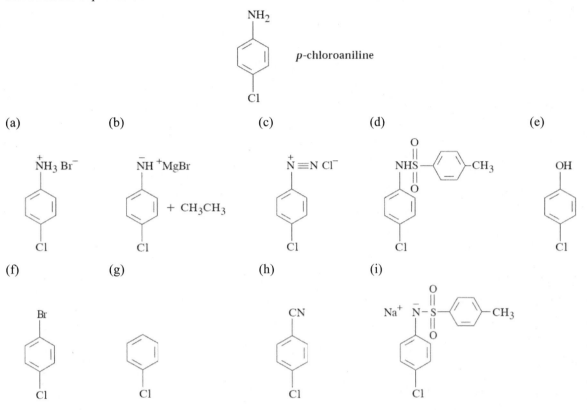

p-chloroaniline

(a)   (b)   (c)   (d)   (e)

(f)   (g)   (h)   (i)

Sulfonamides are acidic enough to be ionized by NaOH. How would you know this? An amide has a pK_a of about 16. We know that sulfonic acids are stronger than carboxylic acids by about 6 pK_a units; assuming the same difference hold for sulfonamides versus carboxamides, then a sulfonamide should have an N—H pK_a of about 16–6 = 10. This is about as acidic as a phenol, and acidic enough to be fully ionized by NaOH (which has a conjugate-acid pK_a = 15.7). Furthermore, the conjugate-base anion of this particular sulfonamide is resonance-stabilized by the phenyl group. We know that phenols are more acidic than alcohols by about 5 pK_a units; if the phenyl group affects the acidity of the sulfonamide in the same way, the pK_a of this compound could be as low as 5.

23.45   The structure of *N*-methylaniline is

N-methylaniline

(a)

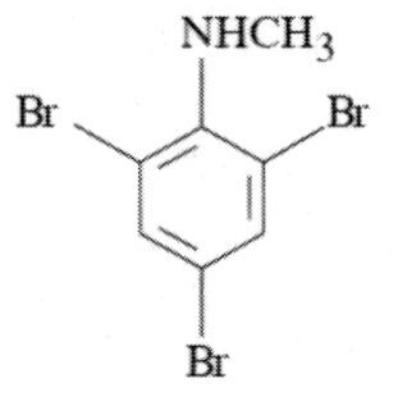

NHCH$_3$

Br, Br, Br (2,4,6-tribromo structure)

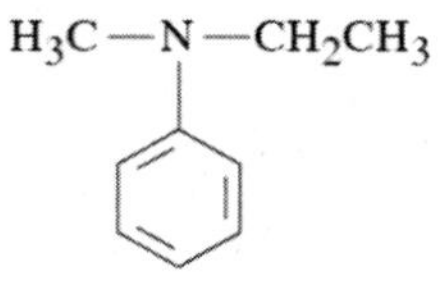

(b)

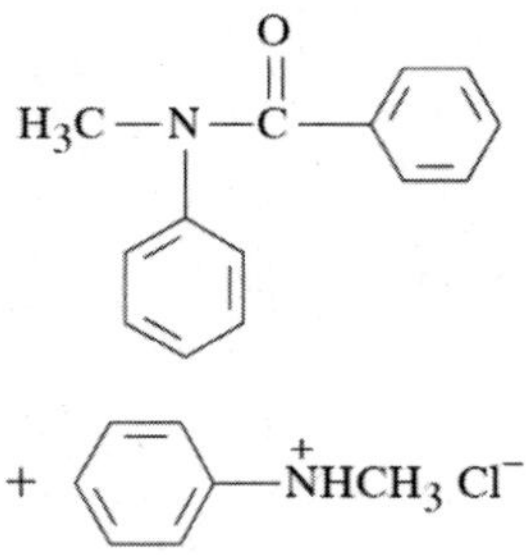

$H_3C-N-\overset{O}{\overset{||}{C}}-$Ph (with N bearing phenyl)

$+$ Ph$-\overset{+}{N}HCH_3$ Cl$^-$

(c)

$\overset{CH_2Ph}{\underset{}{H_3C-\overset{+}{N}-CH_2Ph}}$ ^-OH (with N bearing phenyl)

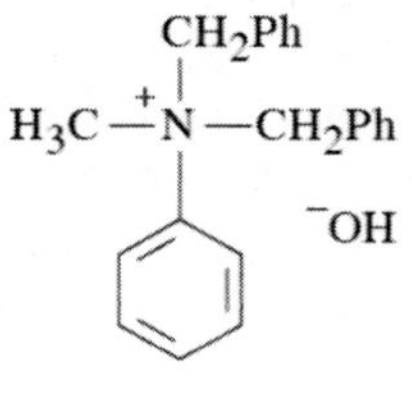

(d)

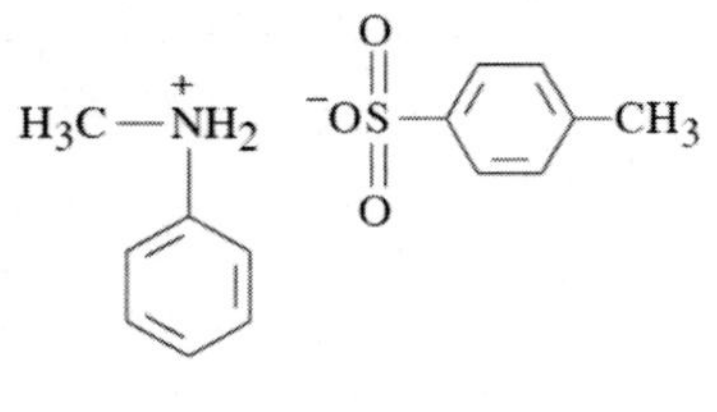

$H_3C-\overset{+}{N}H_2$ (with phenyl) $^-OS(=O)_2-$CH$_3$

(e)

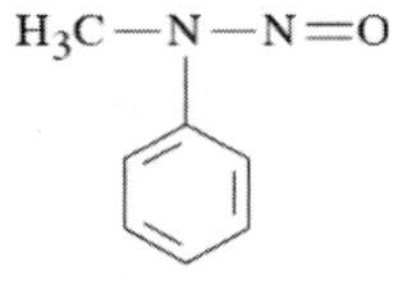

$H_3C-N-N=O$ (with phenyl on N)

(f)

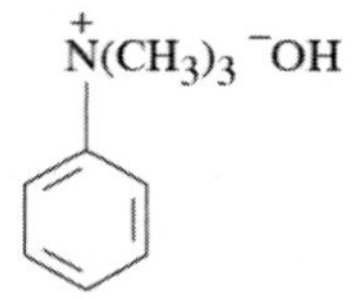

$\overset{+}{N}(CH_3)_3$ ^-OH (with phenyl)

(g)

$H_3C-N-CH_2CH_3$ (with phenyl on N)

(h)

Ph$_2$N$-$CH$_3$ (diphenyl methyl amine)

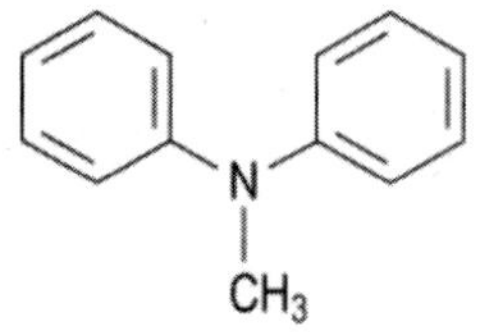

23.46 The structure of isopropylamine is $(CH_3)_2CH-NH_2$

(a)

$(CH_3)_2CH\overset{+}{N}H_3$ $^-OSO_3H$

(b)

no reaction

(c)

$(CH_3)_2CH\overset{..}{N}H$ Li$^+$

$+$ CH$_3$CH$_2$CH$_2$CH$_3$

(d)

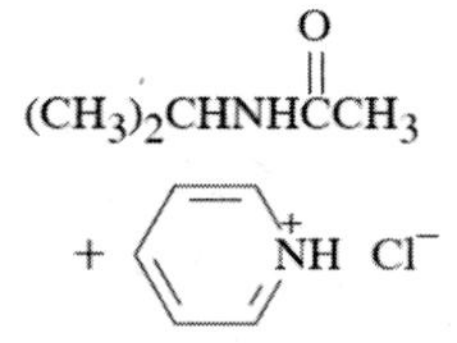

$(CH_3)_2CHNH\overset{O}{\overset{||}{C}}CH_3$

$+$ $\overset{+}{N}H$ Cl$^-$ (pyridinium)

(e)

$(CH_3)_2CHBr$

$+$ $(CH_3)_2CHOH$

$+$ $H_2C=CHCH_3$

(f)

$(CH_3)_2CHNHCH(CH_3)_2$

(g)

$(CH_3)_2CH\overset{+}{N}(CH_3)_3$ I$^-$

(h)

$(CH_3)_2CH\overset{+}{N}H_3$ ^-OCPh (with C=O)

(i)

$(CH_3)_2CHN(CH_3)_2$

(j)

H$_3$C, CH$_3$ (2,4-dimethyl aniline with N-isopropyl)

$\overset{|}{N}-CH(CH_3)_2$

H

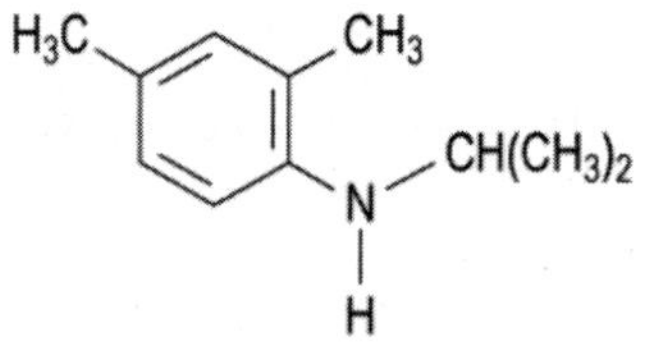

(k)

$H_2C=CHCH_3$

$+$ $N(CH_3)_3$

(l)

$(CH_3)_2CHNHCH_2CH_3$

 If you find an error in this manual, please visit http://people.pharmacy.purdue.edu/~loudonm/teaching/

23.47 (a)

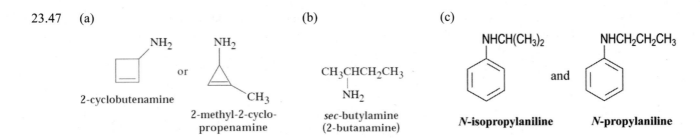

2-cyclobutenamine   or   2-methyl-2-cyclo-propenamine

(b) CH₃CHCH₂CH₃ with NH₂
sec-butylamine (2-butanamine)

(c) NHCH(CH₃)₂   and   NHCH₂CH₂CH₃
***N*-isopropylaniline   *N*-propylaniline**

(d)   The required compound could be either of the two imines $(CH_3)_2C=NCH_3$ or $(CH_3)_2CHN=CH_2$

23.48 (a)

$$CH_3CH_2CH_2CH_2CH_2CNCH_3$$ (with C=O)
N-methylhexanamide

$$CH_3CH_2CH_2CH_2CH_2CH_2CH_2CH_2NH_2$$
1-octanamine

$$CH_3CH_2CH_2CH_2CH_2CH_2N(CH_3)_2$$
N.N-dimethyl-1-hexanamine

N-Methylhexanamide is an amide, and it is therefore not significantly soluble in dilute HCl; the two amines are soluble as their conjugate-acid ammonium salts in dilute HCl solution. 1-Octanamine, like all primary alkylamines, gives off a gas (N_2) when diazotized with $NaNO_2$ in aqueous acid. *N,N*-Dimethyl-1-hexanamine does not give off a gas when treated with nitrous acid.

(b)

H_3C—⟨ ⟩—NH_2
p-methylaniline (*p*-toluidine)

⟨ ⟩—CH_2NH_2
benzylamine

H_3C—⟨ ⟩—OH
p-cresol

⟨ ⟩—OCH_3
anisole

Only the two amines, *p*-methylaniline and benzylamine, are soluble (as their conjugate-acid ammonium ions) in dilute HCl. Only the phenol *p*-cresol is soluble (as its conjugate-base phenolate) in dilute NaOH solution. Anisole is soluble in neither dilute acid nor dilute base. Only benzylamine gives off a gas (N_2) when diazotized at 0° with HNO_2. *p*-Methylaniline gives off N_2 when the corresponding diazonium salt is heated.

23.49 (a)   The amine nitrogen (but *not* the amide nitrogen) is the most basic atom in the molecule. Hence, labetalol hydrochloride is the ammonium ion that results from protonation of this nitrogen.

⟨ ⟩—$CH_2CH_2CHNHCH_2CH$—⟨ ⟩—OH · Cl^- labetalol hydrochloride
(with OH, C(=O)NH₂, H₃C, H groups shown)

(b)   When labetalol hydrochloride is treated with one molar equivalent of NaOH, a proton on the positively charged nitrogen is removed and neutral labetalol is formed.

(c)   Hot NaOH ionizes the phenol group and also brings about hydrolysis of the amide group to a carboxylate.

⟨ ⟩—$CH_2CH_2CHNHCH_2CH$—⟨ ⟩—$O^- Na^+$ + NH_3
(with OH, C(=O)O⁻Na⁺, CH₃ groups shown)

(d) Hot 6 *M* aqueous HCl solution results in protonation of the amine nitrogen (as in part (a)) and hydrolysis of the amide.

Depending on the severity of the conditions, the alcohol could dehydrate to give compound *A*, which is the conjugate acid of an enamine. This would hydrolyze to an amine and an aldehyde under the aqueous conditions.

compound *A*

(a resonance-stabilized carbocation)

23.50 (a) In 1 *M* HCl solution, cocaine is protonated on its amino nitrogen.

(b) Hot aqueous NaOH would bring about saponification of the ester groups:

(c) Concentrated, hot aqueous HCl would protonate the amine nitrogen (as in part (a)) and would bring about hydrolysis of the two ester groups.

H +N CH₃ Cl⁻

CO₂H

O
‖
+   HO—C—Ph   +   CH₃OH

H OH

23.51   Dissolve the mixture in a suitable low-boiling solvent such as ether or methylene chloride. Extract with 5% aqueous sodium bicarbonate solution. The carboxylic acid p-nitrobenzoic acid will dissolve in the aqueous solution as its sodium salt. Isolate this aqueous solution and acidify it with HCl to precipitate the carboxylic acid. Then extract the remaining organic solution with 5% aqueous NaOH solution. p-Chlorophenol will dissolve in the aqueous solution as its conjugate-base phenolate ion. Isolate this aqueous layer and acidify it to obtain the neutral phenol. Then extract the organic solution with 5% aqueous HCl solution. The amine aniline will dissolve in the aqueous layer as its conjugate-acid ammonium ion. Separate the aqueous layer and neutralize it with NaOH. Aniline will form a separate layer that can be separated, dried, and distilled. Finally, dry the remaining organic layer and remove the solvent to obtain nitrobenzene. (The acidic extraction could be carried out as the first step.)

23.52   The basicity of trifluralin is much less than that of N,N-diethylaniline. The electron-withdrawing polar effects of both the nitro substituents and the trifluoromethyl substituent destabilize the conjugate acid of trifluralin, and resonance interaction of the nitrogen unshared electron pair with the two nitro groups stabilizes trifluralin itself; this interaction is absent in the ammonium ion. (In fact, trifluralin is not basic enough to dissolve in dilute HCl.)

some resonance structures of trifluralin

23.53   (a)   Nitrous acid forms a diazonium ion from anthranilic acid. This diazonium ion undergoes coupling with N,N-dimethylaniline to give methyl red.

CO₂H

—N=N—⟨ ⟩—N(CH₃)₂   methyl red

(b)   In acidic solution, methyl red is protonated to give its conjugate acid. Protonation occurs on one of the diazo nitrogens because the resulting conjugate acid is resonance-stabilized.

ionizes at pH 2.3

CO₂H   CO₂H

+N=N—⟨ ⟩—N(CH₃)₂   ⟷   N—N=⟨ ⟩=+N(CH₃)₂
H   H

conjugate acid of methyl red

dissociates at pH 5.0

You may have noticed that the pK_a of the carboxy group is unusually low for the pK_a of a carboxylic acid. Can you explain why?

23.54 **(a)** First prepare *p*-nitroaniline from aniline as shown in Study Problem 23.4 on text p. 1205. Then diazotize this amine to give the diazonium salt. Couple the diazonium salt to salicylic acid to obtain alizarin yellow R.

p-nitroaniline

NaNO$_2$/HCl

salicylic acid

alizarin yellow R

(b) From the solution to the Problem (23.53 on p. 855 of this manual), or from the information in the problem, the pK_a of a protonated diazo group should be around 5. Consequently, at a pH value well above 5, the diazo nitrogens are unprotonated. The pK_a of the carboxy group should be around 4. Therefore, at pH = 9, the carboxy group should be ionized. The group that ionizes between 10 and 12 should affect the conjugated π-electron system because it affects the color. Since this group is not the diazo group, it must be the phenol. If the phenol has a pK_a of about 11, then at pH = 9 it is un-ionized.

structure of alizarin yellow R
at pH = 9

(c) At pH > 12, both the carboxy group and the phenol are ionized. (Show the resonance structures for the delocalization of the phenolate ion throughout the π-electron system of the rings, the diazo group, and the *p*-nitro group. This interaction is the basis for the color change.)

structure of alizarin yellow R
at pH > 12

23.55 **(a)** Amanda is evidently trying to run a Friedel–Crafts acylation on aniline. This cannot work as shown for two reasons. First, the aluminum trichloride, a Lewis acid, will react with the electron pair on the amino group, thus deactivating the ring. (This is much like its effect on phenol acylation; see Eq. 18.92 on text p. 927.) Second, a competing reaction is acylation of the amino group to form acetanilide, a reaction that will undoubtedly be catalyzed by the aluminum trichloride.

(b) Excess methyl iodide will form the quaternary salt $(CH_3)_3\overset{+}{C}N(CH_3)_3$ I⁻; the reaction will not stop at the secondary amine shown in the problem.

(c) Amanda is attempting to nitrate the ring and is relying on the activating, ortho, para-directing effect of the dimethylamino group to direct the substitution to the para position. The fallacy here is that under the very acidic conditions the dimethylamino group is protonated, and the protonated dimethylamino group is expected to be a meta director. (See Eq. 23.27 and related discussion on text p. 1205.)

(d) Although the aldehyde will indeed be reduced, so will the nitro group. (See Eq. 23.60 on text p. 1214.)

(e)   Amanda is attempting to displace water from the protonated alcohol with dimethylamine, which is a good nucleophile. Unfortunately, the acidic conditions required to convert the —OH group of the alcohol into a good leaving group also protonate the amine, which is much more basic than the alcohol. The protonated amine no longer has the unshared electron pair that is responsible for its nucleophilicity.

(f)   Amanda is attempting to carry out an S_N2 reaction on a tertiary alkyl halide. Unfortunately, tertiary alkyl halides do not undergo S_N2 reactions. If a reaction occurs at all, it will be the competing E2 process.

23.56   (a)

$$PhNH_2 \xrightarrow[\text{2) CuCN}]{\text{1) NaNO}_2\text{/HCl}} PhC{\equiv}N \xrightarrow{H_2, \text{ cat.}} PhCH_2NH_2$$
benzylamine

(b)

$$PhNH_2 \xrightarrow[\text{2) CuCN}]{\text{1) NaNO}_2\text{/HCl}} PhC{\equiv}N \xrightarrow{H_3O^+, H_2O, \text{ heat}} PhCOH \xrightarrow[\text{2) H}_3\text{O}^+]{\text{1) LiAlH}_4} PhCH_2OH$$
benzyl alcohol

(c)

$$PhC{\equiv}N \xrightarrow{H_3O^+, H_2O, \text{ heat}} PhCOH \xrightarrow[\text{2) H}_3\text{O}^+]{\text{1) LiAlH}_4} PhCH_2OH \xrightarrow{\text{conc. HBr}} PhCH_2Br \xrightarrow[\text{ether}]{Mg} \xrightarrow[\text{2) H}_3\text{O}^+]{\text{1) CO}_2}$$
prepared in part (a)

$$PhCH_2COH \xrightarrow{SOCl_2} PhCH_2CCl \xrightarrow{NH_3 \text{ (excess)}} PhCH_2CNH_2 \xrightarrow[\text{2) H}_2\text{O}]{\text{1) LiAlH}_4} PhCH_2CH_2NH_2$$
2-phenylethanamine

(d)

$$PhNH_2 + CH_3CCH_2CH_3 \xrightarrow[\text{CH}_3\text{OH}]{NaCNBH_3} PhNHCHCH_2CH_3$$
|
CH_3

N-phenyl-2-butanamine

(e)

Ph—NH_2 $\xrightarrow{Ac_2O}$ Ph—NHAc $\xrightarrow{Cl_2, \text{ FeCl}_3}$ Cl—Ph—NHAc $\xrightarrow[\text{2) NaOH (dilute)}]{\text{1) H}_3\text{O}^+, \text{H}_2\text{O, heat}}$

Cl—Ph—NH_2 $\xrightarrow[\text{2) CuCN}]{\text{1) NaNO}_2\text{/HCl}}$ Cl—Ph—C$\equiv$N $\xrightarrow[\text{H}_2\text{O}]{H_3O^+, \text{ heat}}$ Cl—Ph—CO_2H

p-chlorobenzoic acid

(f)

$$PhNH_2 \xrightarrow[\text{2) CuCl}]{\text{1) NaNO}_2\text{/HCl}} PhCl \xrightarrow[\text{toluene}]{\substack{\text{aniline} \\ \text{Pd(0) catalyst}}}$$
diphenylamine

23.57   The acetic anhydride treatment acetylates either the amino group or the hydroxy group. Because the initially formed product A can form an ethyl ether, its —OH group is not affected in the first reaction. Consequently, compound A is p-acetamidophenol and compound B is its ethyl ether. This is reasonable because, so long as the hydroxy group is not ionized, the amino group is the most basic group in the molecule, and is thus the more nucleophilic group. Compound A dissolves in base because the hydroxy proton can ionize.

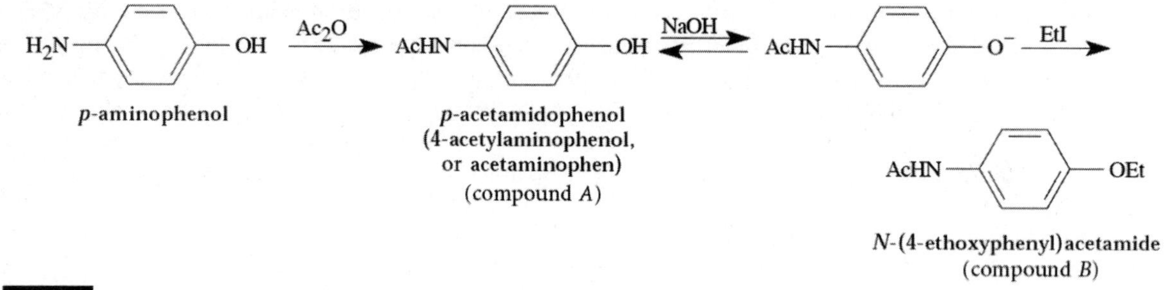
H2N OH
Ac2O
AcHN OH
NaOH
AcHN O−
EtI
p-aminophenol
p-acetamidophenol
(4-acetylaminophenol,
or acetaminophen)
(compound A)
AcHN OEt
N-(4-ethoxyphenyl)acetamide
(compound B)

In compound *A* the acidity of the amide N—H is likely to be comparable to that of the phenol (why?). Hence, the conjugate-base anion of the phenoxide is in equilibrium with the phenoxide ion. If both ions are indeed present, evidently the conjugate base phenolate is alkylated more rapidly. Alkylation of the oxygen then pulls the equilibrium between the anions towards the phenoxide:

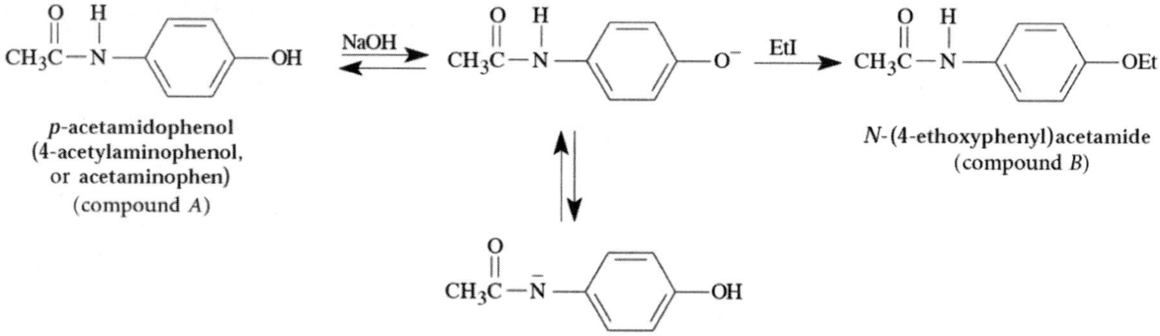
CH3C—N OH
NaOH
CH3C—N O−
EtI
CH3C—N OEt
p-acetamidophenol
(4-acetylaminophenol,
or acetaminophen)
(compound A)
N-(4-ethoxyphenyl)acetamide
(compound B)
CH3C—N OH

23.58 First identify compound *B* by completing the reaction sequence in Fig. P23.52, text p. 1160:

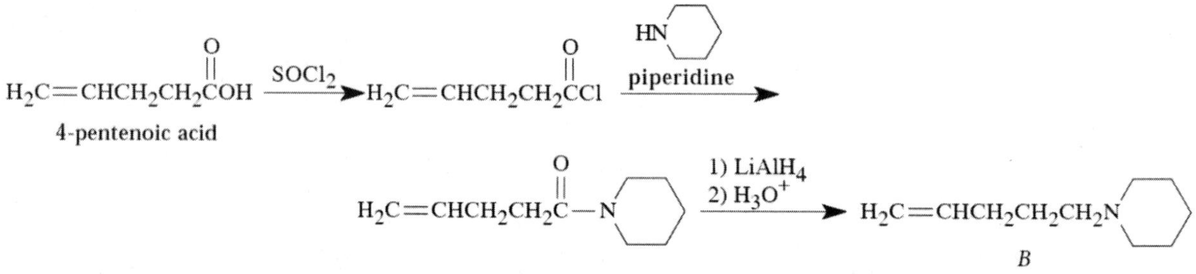
H2C=CHCH2CH2COH
SOCl2
H2C=CHCH2CH2CCl
HN
piperidine
4-pentenoic acid
H2C=CHCH2CH2C—N
1) LiAlH4
2) H3O+
H2C=CHCH2CH2CH2N
B

Compound *A*, because of its reaction with Ag^+ to give AgBr, appears to be an ionic compound. The formula of compound *B* shows that it must be derived from two molecules of 1,5-dibromopentane. Because the alkene *B* is obtained by an elimination reaction, it is reasonable to suppose that compound *A* is a quaternary ammonium salt. A reasonable structure of such a salt that could result from the reaction of two equivalents of 1,5-dibromopentane and one of ammonia is as follows. (This equation also shows the reaction of compound *A* with hydroxide ion to give compound *B*.)

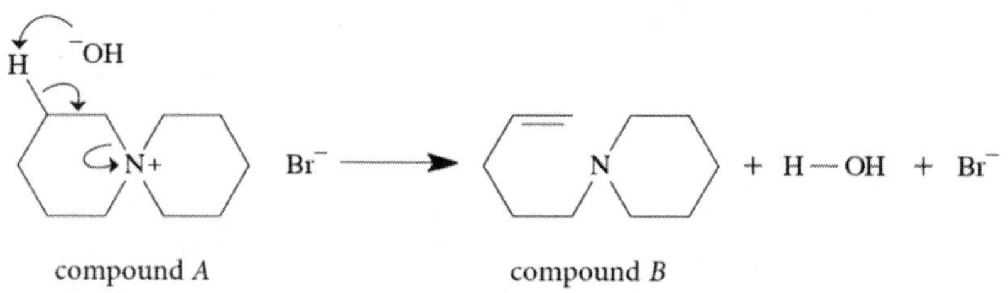
H −OH
N+
Br−
N
+ H—OH + Br−
compound A
compound B

The formation of compound *A* from 1,5-dibromopentane and ammonia thus must involve the following steps.

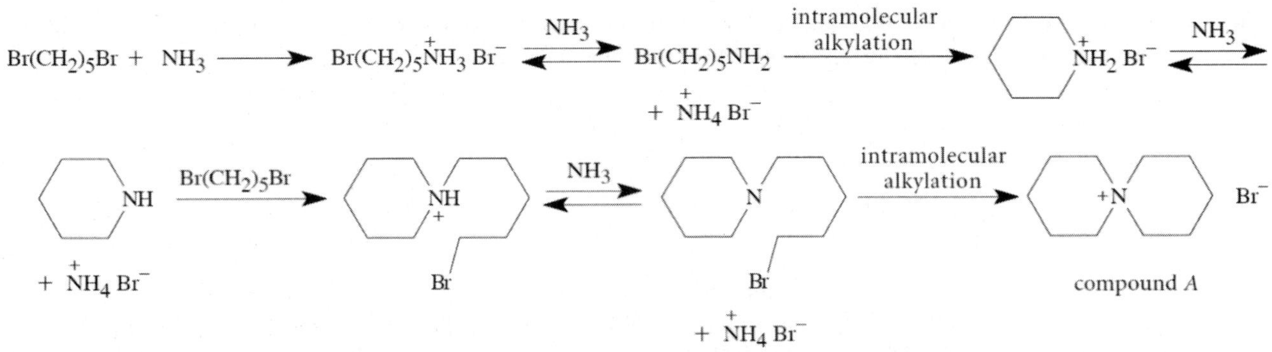

$$Br(CH_2)_5Br + NH_3 \longrightarrow Br(CH_2)_5\overset{+}{N}H_3 \ Br^- \underset{NH_3}{\rightleftharpoons} Br(CH_2)_5NH_2 \xrightarrow[\text{alkylation}]{\text{intramolecular}}$$

$$+ \ \overset{+}{N}H_4 \ Br^-$$

compound A

23.59   (a)   The barrier to internal rotation about the *N*-phenyl bond in *N*-methyl-*p*-nitroaniline is greater because there is more double bond character in this bond than there is in *N*-methylaniline. There is more double bond character because of the resonance interaction between the unshared electron pair on the amine nitrogen and the *p*-nitro group.

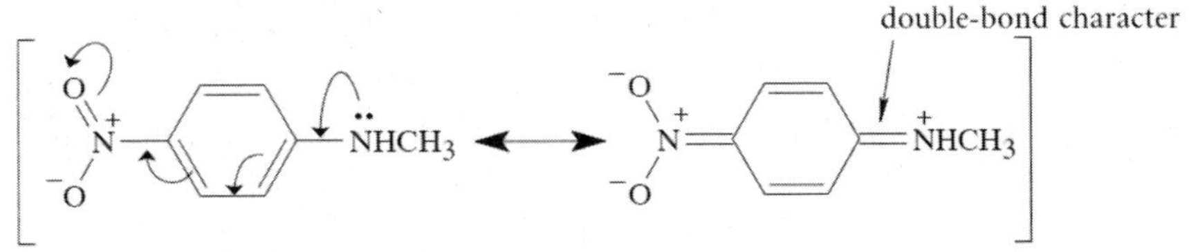

double-bond character

(b)   *Cis*-and *trans*-1,3-dimethylpyrrolidine rapidly interconvert because all that is required for this process is amine inversion, which is very rapid.

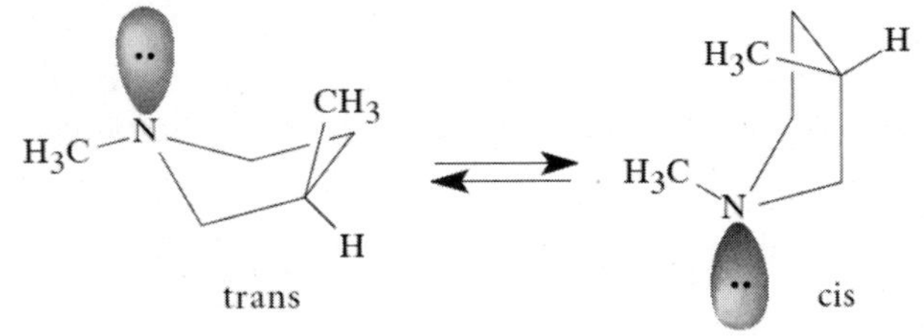

trans   cis

(c)   This compound is the nitrogen analog of an acetal and, by a mechanism similar to that for the hydrolysis of an acetal, decomposes to formaldehyde and two equivalents of methylamine in aqueous solution.

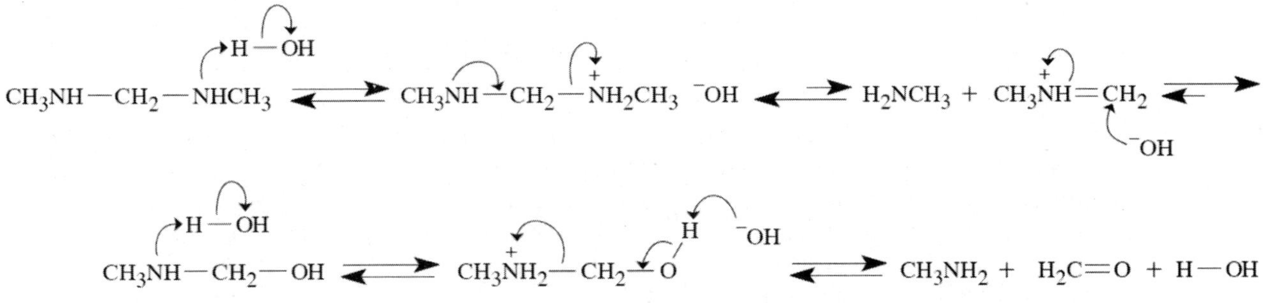

$$CH_3NH-CH_2-NHCH_3 \rightleftharpoons CH_3\overset{+}{N}H-CH_2-NH_2CH_3 \ ^-OH \rightleftharpoons H_2NCH_3 + CH_3\overset{+}{N}H=CH_2 \rightleftharpoons$$

$$^-OH$$

$$CH_3NH-CH_2-OH \rightleftharpoons CH_3\overset{+}{N}H_2-CH_2-O \rightleftharpoons CH_3NH_2 + H_2C=O + H-OH$$

(d)   This compound exists as an enamine for the same reason that *β*-diketones exist as enols: internal hydrogen bonding and conjugation stabilize the enamine form.

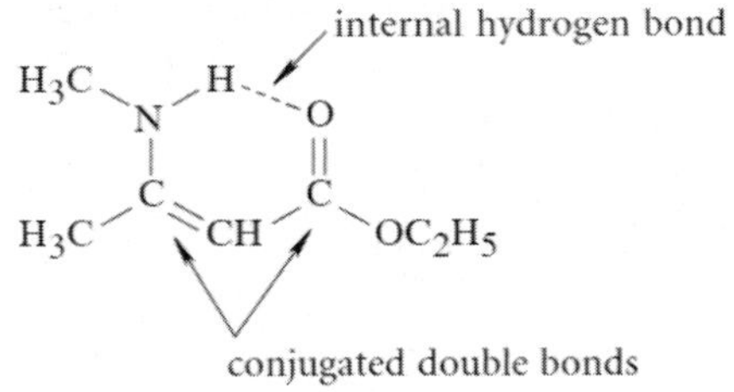

internal hydrogen bond

conjugated double bonds

(e)   The diazonium salt does not decompose to a carbocation because the carbocation has a $4n$ π-electron system and is therefore antiaromatic and very unstable.

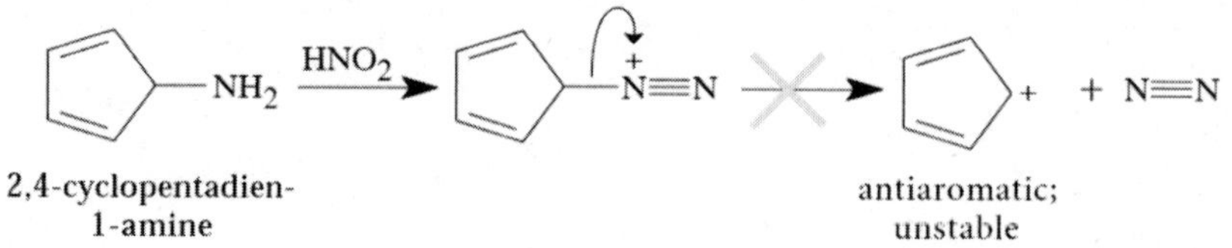

2,4-cyclopentadien-
1-amine

antiaromatic;
unstable

23.60   First, resolve 2-phenylbutanoic acid into enantiomers using an enantiomerically pure amine as the resolving agent. Then allow the appropriate enantiomer of the carboxylic acid to undergo the reactions shown in each part below. The key element in all these syntheses is that the Hofmann and Curtius rearrangements take place with retention of stereochemical configuration.

(a)

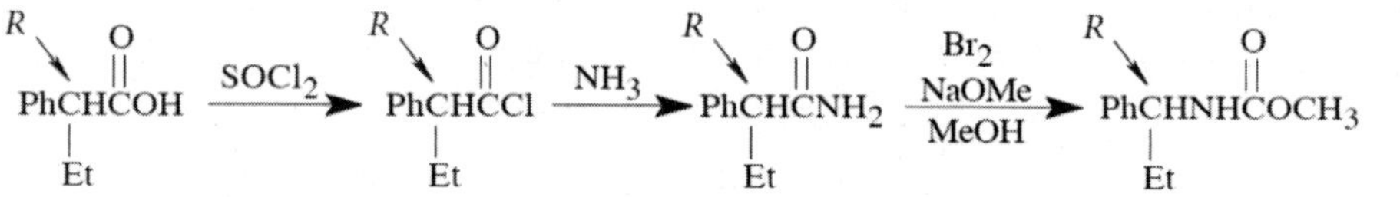

(b)   In this case, esterify the S enantiomer of the carboxylic acid that was obtained from the enantiomeric resolution.

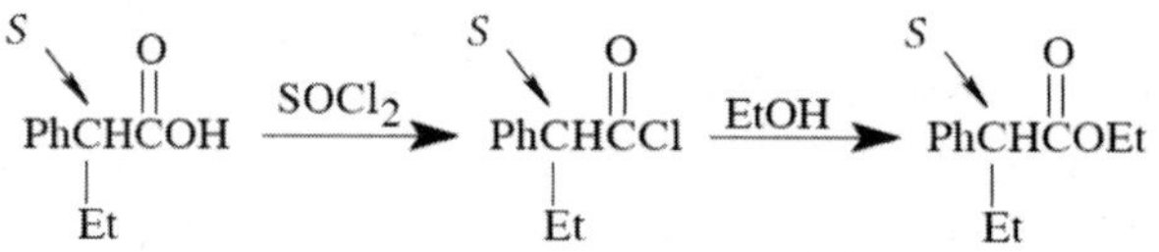

(c)   In this part, the isocyanate product from the Curtius rearrangement is divided in half, and part of it is converted into an amine. This amine is then allowed to react with the remaining isocyanate.

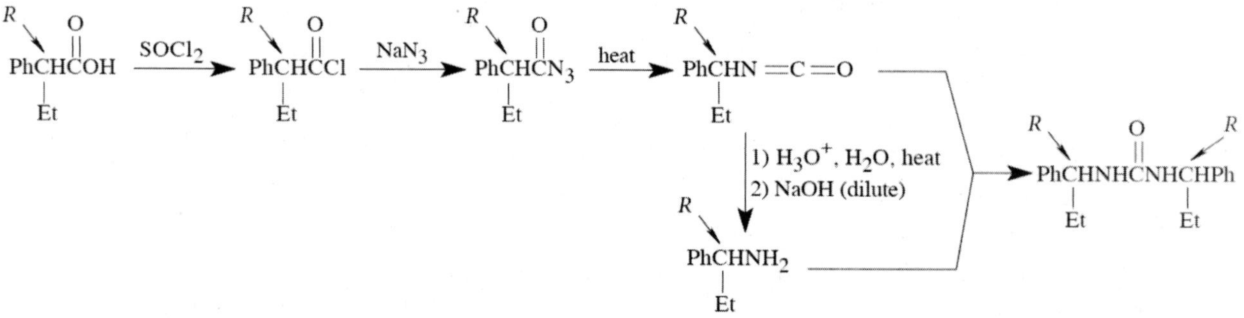

(d)   Use the same reactions in part (c), except prepare the (S)-amine from the (S)-carboxylic acid. Then let the (R)-isocyanate react with the (S)-amine to give the meso product. (The same product would be obtained from the reaction of the (S)-isocyanate with the (R)-amine.)

(e)   Use the same reactions in part (c) to prepare (R)-1-phenylpropanamine. Then couple the (R)-amine with ethyl p-chlorobenzoate in the presence of a Pd(0) catalyst and Na^+ ^-Ot-Bu (a Buchwald–Hartwig amination) followed by hydrolysis of the ester.

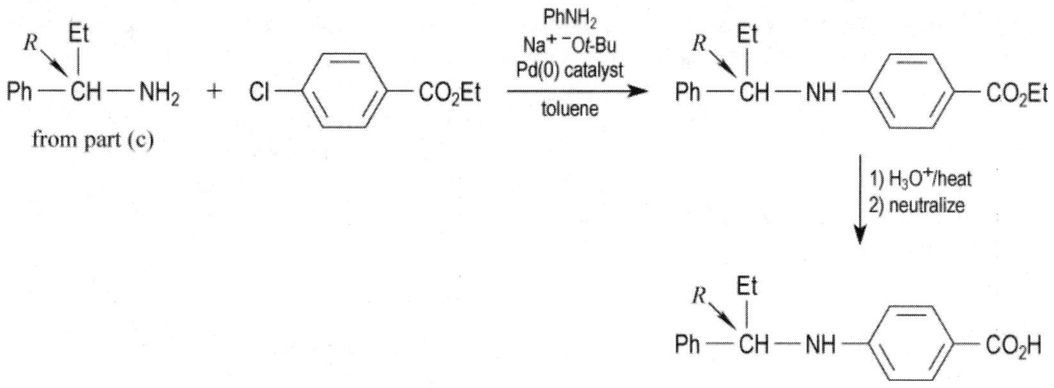

from part (c)

23.62 (a) Sodium benzenethiolate, $Na^+ PhS^-$, is a salt and thus soluble in water but not an organic solvent; 1-bromooctane is relatively nonpolar and thus insoluble in water. No reaction occurs initially because the two reactants cannot interact; however, with the addition of the phase-transfer catalyst tetrabutylammonium bromide, $(CH_3CH_2CH_2CH_2)_4N^+ Br^-$, the two reactants can interact to form octyl phenyl sulfide. (See text p. 1196.)

$$Ph—O—CH_2(CH_2)_6CH_3$$

octyl phenyl sulfide

(b) Phenylacetonitrile and 1,4-dibromobutane are not soluble in water, especially with an ionic compound like sodium hydroxide in the aqueous layer. Phenylacetonitrile and 1,4-dibromobutane are soluble in dichloromethane; the tetrabutylammonium bromide is a phase-transfer catalyst and the rapid stirring brings the hydroxide ion into the organic phase. The hydrogens of phenylacetonitrile that are α to both the phenyl ring and the nitrile group are acidic and may be removed by hydroxide ion. The resulting conjugate base reacts with the 1,4-dibromobutane in an S_N2 reaction:

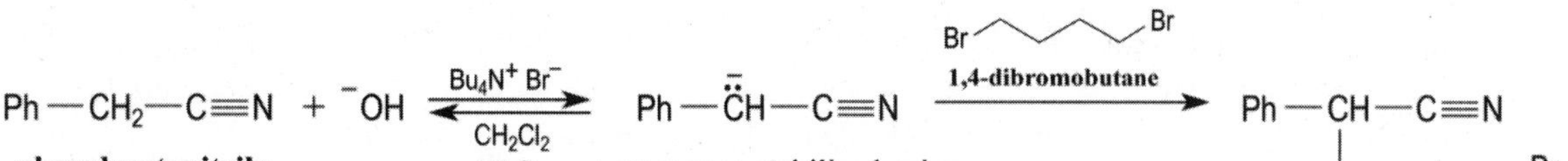

phenylacetonitrile a resonance-stabilized anion

The resulting product reacts with another equivalent of hydroxide to give the conjugate base, which undergoes an intramolecular S_N2 reaction to form the $C_{12}H_{13}N$ product in high yield.

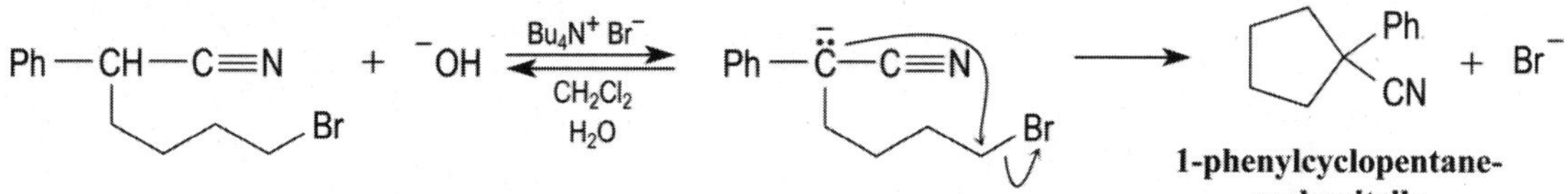

1-phenylcyclopentane-carbonitrile

(c) The reaction between morpholine and bromobenzene in the presence of aqueous NaOH and the catalyst $Pd[P(t\text{-Bu})_3]_2$ is a Buchwald–Hartwig amination reaction. The cetyltrimethylammonium bromide is a phase-transfer catalyst that brings the water-insoluble bromobenzene into contact with the reactants.

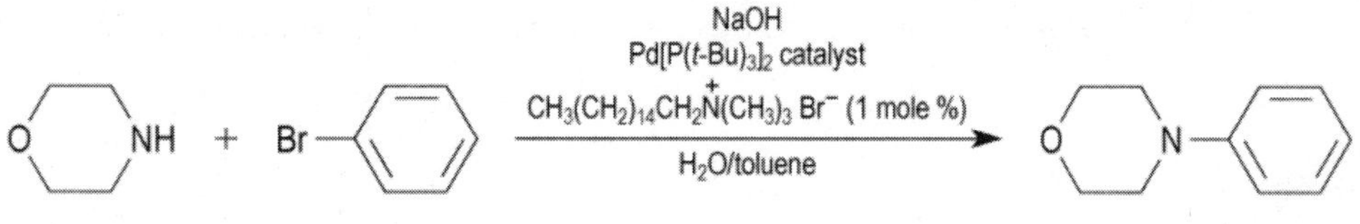

morpholine **bromobenzene** ***N*-phenylmorpholine**

23.63 Compound *A* appears to be an *N*-benzoyl amide of compound *B* because it is hydrolyzed to benzoic acid, and because it is re-formed when *B* is treated with benzoyl chloride. Compound *A* cannot be an ester because it contains only one oxygen. Compound *B* is therefore an achiral primary or secondary amine. It cannot be a primary amine because it does not liberate nitrogen gas after treatment with nitrous acid; thus, it must be secondary because a tertiary amine cannot form an amide. The formula of *B* is obtained by subtracting the formula of a benzoyl group (Ph—CO—, C_7H_5O) from the formula of *A* and adding a hydrogen. Thus, the formula of *B* is $C_{15}H_{23}N$. Compound *B* therefore contains five degrees of unsaturation. One degree of unsaturation is accounted for by a cyclohexane ring, reflected in the degradation product *D*, and the remaining four by a phenyl ring, reflected in the degradation product styrene, which is formed along with product *C*. There are only two structures for compound *B*, shown below as *B1* and *B2*, that would give the indicated degradation products:

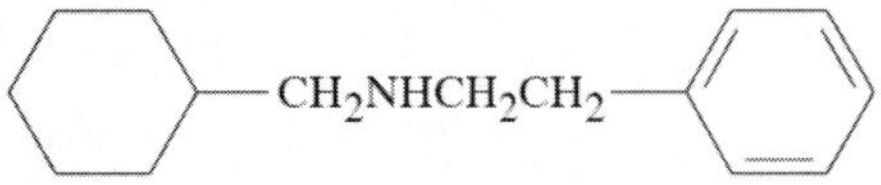

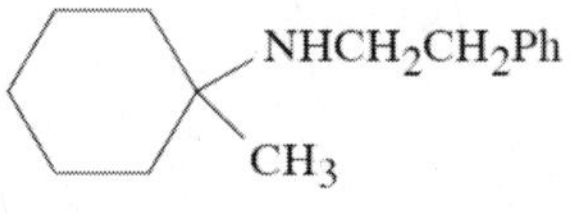

Although both are reasonable answers, structure *B1* is better because it would lead to *one* alkene *D*, whereas *B2* would lead to *two* different alkenes of which *D* is the major one. Taking *B1* as the correct structure for *B*, the identities of the unknown compounds are as follows:

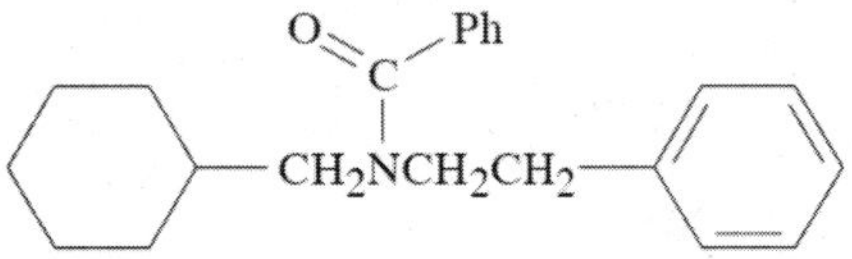

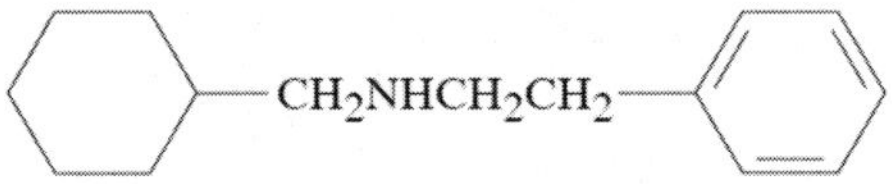

23.64   The oxidation data reveal that compound *A* contains a monosubstituted benzene ring, compounds *B* and *C* contain *para*-disubstituted benzene rings, and compound *C* additionally has an amino group directly attached to the ring. The pKa of compound C additionally confirms that it is an aniline derivative, and the fact that compound *C*, following diazotization, requires warming to evolve a gas (N_2) also indicates that it is an aniline derivative. Compound *A* is chiral, and furthermore, it must be the racemate, because it evidently forms diastereomeric salts with (+)-tartaric acid. Because six carbons are involved in a benzene ring, the problem is to determine how the remaining two carbons are arranged so that the resulting structures are consistent with the data. The following structures are the only reasonable possibilities:

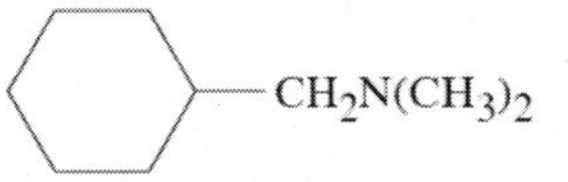

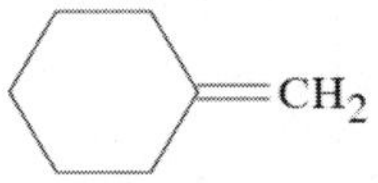

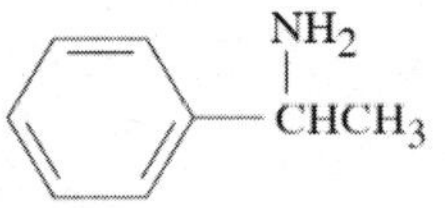

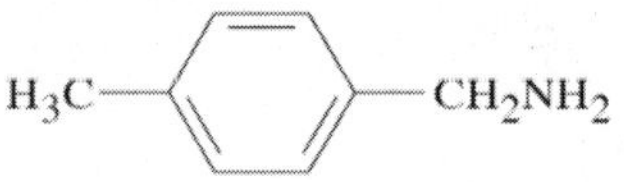

23.65   (a)   Since diazotization of R—NH₂ gives N_2 and products derived from R^+, then diazotization of H—NH₂ should give H^+, that is, H_3O^+, and N_2.

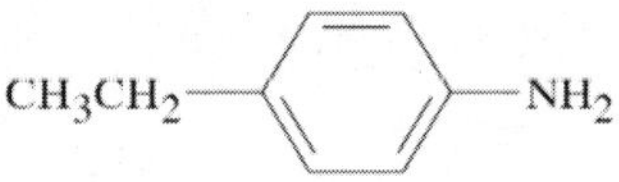

(b)   Exhaustive methylation is followed by Hofmann elimination in which a proton is removed from a β-carbon.

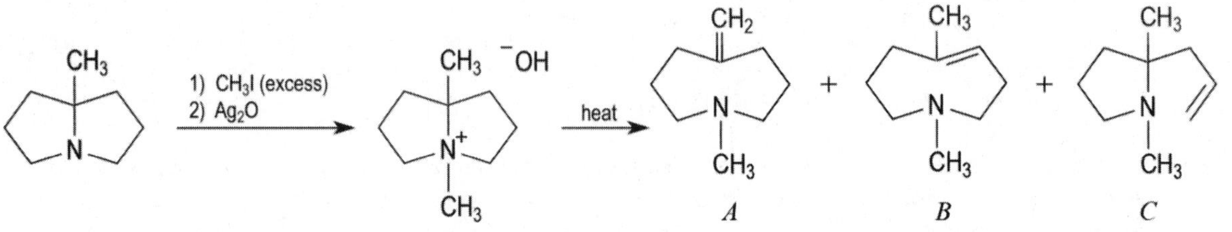

Although it is not discussed in the text, the regioselectivity of the Hofmann elimination tends to favor removal of a proton at the β-carbon with fewest alkyl substituents. Therefore, *A* is the predominant product.

(c)   The triamide resulting from treatment of the acid chloride is reduced with LiAlH$_4$ to give triamine *A*, which is exhaustively methylated and then subjected to Hofmann elimination to give the product, which is an isomer of benzene.

CH$_2$N(CH$_3$)$_2$

(CH$_3$)$_2$NHCH$_2$ CH$_2$N(CH$_3$)$_2$

A

1) CH$_3$I (excess)
2) Ag$_2$O, then heat

CH$_2$

H$_2$C   CH$_2$

+ 3 (CH$_3$)$_3$N

(d)   The nitrosyl cation acts as the electrophile in an electrophilic aromatic substitution reaction of phenol.

N=O

—OH   +   O=N— —OH

(e)   In this variation of the Sandmeyer reaction, copper(I) nitrite introduces a nitro group into the ring.

O$_2$N— —NO$_2$

p-Dinitrobenzene would be very difficult to prepare by electrophilic aromatic substitution; can you see why?

(f)   The amino group is alkylated twice by ethylene oxide to give CH$_3$CH$_2$CH$_2$CH$_2$N(CH$_2$CH$_2$OH)$_2$.

(g)   The amine reacts with the epoxide at the carbon with fewer alkyl substituents.

OH
|
(CH$_3$)$_2$C—CH$_2$NEt$_2$

(h)   The conjugate-base anion of phthalimide reacts at the carbon of the epoxide with no alkyl substituents to give compound *A*; hydrolysis of the amide groups gives the amine product.

O

OH
|
NCH$_2$CHCH$_3$

O

A

NaOH, H$_2$O, heat

O
||
C
   O$^-$

C
||
O
   O$^-$

+ H$_2$NCH$_2$CHCH$_3$
                    OH
                    |

(i)   The amine reacts with the epoxide with inversion of configuration. Because both starting materials are achiral (the epoxide is the meso stereoisomer), the product is the racemate; one enantiomer results from reaction at one carbon of the epoxide, and the other enantiomer results from reaction at the other carbon.

(j)   Vinylic halides are inert in S$_N$2 reactions, and allylic halides are very reactive. Because an excess of the amine is used, only one halide molecule reacts with each amine, and the allylic bromine is displaced.

$$H_2C{=}C{-}CH_2\overset{+}{N}H_2Et \;\; Br^-$$
$$|$$
$$Br$$

(k)   Catalytic hydrogenation reduces nitrobenzene to aniline, which then reductively aminates the aldehyde; because excess aldehyde is present, the resulting secondary amine then reductively aminates a second molecule of aldehyde. The final product is the tertiary amine *N,N*-dibutylaniline, PhN(CH$_2$CH$_2$CH$_2$CH$_3$)$_2$.

23.66   (a)

$$CH_3(CH_2)_3\overset{O}{\overset{||}{C}}OH \xrightarrow[\text{2) H}_3\text{O}^+]{\text{1) LiAlH}_4} CH_3(CH_2)_3CH_2OH \xrightarrow[\text{H}_2\text{SO}_4]{\text{conc. HBr}} CH_3(CH_2)_3CH_2Br \xrightarrow[\text{ether}]{\text{Mg}} \xrightarrow[\text{2) H}_3\text{O}^+]{\text{1) CO}_2}$$

pentanoic acid

$$CH_3(CH_2)_3CH_2\overset{O}{\overset{||}{C}}OH \xrightarrow{\text{SOCl}_2} CH_3(CH_2)_3CH_2\overset{O}{\overset{||}{C}}Cl \xrightarrow{\text{H}_2\text{NCH}_3 \text{ (excess)}} CH_3(CH_2)_3CH_2\overset{O}{\overset{||}{C}}NHCH_3 \xrightarrow[\text{3) NaOH}]{\substack{\text{1) LiAlH}_4 \\ \text{2) H}_3\text{O}^+}}$$

$$CH_3(CH_2)_3CH_2CH_2NHCH_3$$
N-methyl-1-hexanamine

(b)

$$CH_3(CH_2)_3\overset{O}{\overset{||}{C}}OH \xrightarrow{\text{SOCl}_2} CH_3(CH_2)_3\overset{O}{\overset{||}{C}}Cl \xrightarrow{\text{NH}_3 \text{ (excess)}} CH_3(CH_2)_3\overset{O}{\overset{||}{C}}NH_2 \xrightarrow[\text{3) NaOH}]{\substack{\text{1) LiAlH}_4 \\ \text{2) H}_3\text{O}^+}} CH_3(CH_2)_3CH_2NH_2$$

pentanoic acid   pentylamine

(c)

$$CH_3(CH_2)_3\overset{O}{\overset{||}{C}}OH \xrightarrow{\text{SOCl}_2} CH_3(CH_2)_3\overset{O}{\overset{||}{C}}Cl \xrightarrow{\text{HN(CH}_3)_2 \text{ (excess)}}$$

$$CH_3(CH_2)_3\overset{O}{\overset{||}{C}}N(CH_3)_2 \xrightarrow[\text{3) NaOH}]{\substack{\text{1) LiAlH}_4 \\ \text{2) H}_3\text{O}^+}} CH_3(CH_2)_3CH_2N(CH_3)_2$$
N,N-dimethyl-1-pentanamine

(d)

$$CH_3(CH_2)_3\overset{O}{\overset{||}{C}}NH_2 \xrightarrow[\text{H}_2\text{O}]{\text{Br}_2,\text{ NaOH}} CH_3(CH_2)_3NH_2$$

prepared in   butylamine
part (b)

(e)

$$CH_3(CH_2)_3CH_2\overset{O}{\overset{||}{C}}Cl \xrightarrow{\text{NH}_3 \text{ (excess)}} CH_3(CH_2)_3CH_2\overset{O}{\overset{||}{C}}NH_2 \xrightarrow[\text{3) NaOH}]{\substack{\text{1) LiAlH}_4 \\ \text{2) H}_3\text{O}^+}} CH_3(CH_2)_3CH_2CH_2NH_2$$

prepared in part (a)   hexylamine
(1-hexanamine)

(f)

$$O=CHCH_2CH_2CH_3 \xrightarrow[CH_3OH]{\substack{(CH_3)_2NH \\ NaCNBH_3}} (CH_3)_2NCH_2CH_2CH_2CH_3 \xrightarrow{PhCH_2Br} PhCH_2\overset{CH_3}{\underset{CH_3}{\overset{+}{N}}}-CH_2CH_2CH_2CH_3 \ Br^-$$

butyraldehyde

(g)

$$PhCH_3 \xrightarrow[\text{peroxides}]{NBS, CCl_4} PhCH_2Br \xrightarrow[\text{ether}]{Mg} \xrightarrow[2) H_3O^+]{1) \triangle\!\!\!\!O} PhCH_2CH_2CH_2OH \xrightarrow[2) H_3O^+]{1) KMnO_4, \ ^-OH}$$

toluene

$$PhCH_2CH_2\overset{O}{\overset{||}{C}}OH \xrightarrow{SOCl_2} PhCH_2CH_2\overset{O}{\overset{||}{C}}Cl \xrightarrow[\text{(excess)}]{H_2NEt} PhCH_2CH_2\overset{O}{\overset{||}{C}}NHEt \xrightarrow[\substack{2) H_3O^+ \\ 3) NaOH}]{1) LiAlH_4} PhCH_2CH_2CH_2NHEt$$

N-ethyl-3-phenyl-
1-propanamine

(h)

$$CH_2(CO_2Et)_2 \xrightarrow[2) CH_3CH_2CH_2Br]{1) NaOEt, EtOH} CH_3CH_2CH_2CH(CO_2Et)_2 \xrightarrow[2) CH_3I]{1) NaOEt, EtOH}$$

diethyl malonate

$$CH_3CH_2CH_2\underset{CH_3}{\overset{|}{C}}(CO_2Et)_2 \xrightarrow{H_3O^+, H_2O, \text{ heat}} CH_3CH_2CH_2\underset{CH_3}{\overset{|}{C}}H\overset{O}{\overset{||}{C}}OH \xrightarrow[2) NH_3]{1) SOCl_2}$$

$$CH_3CH_2CH_2\underset{CH_3}{\overset{|}{C}}H\overset{O}{\overset{||}{C}}NH_2 \xrightarrow[H_2O]{Br_2, NaOH} CH_3CH_2CH_2\underset{CH_3}{\overset{|}{C}}HNH_2$$

2-pentanamine

(i)

$$(CH_3)_2C=O \xrightarrow[2) H_3O^+]{1) LiAlH_4} (CH_3)_2CHOH \xrightarrow[H_2SO_4]{\text{conc. HBr}} (CH_3)_2CHBr \xrightarrow[\text{ether}]{Mg} \xrightarrow[2) H_3O^+]{1) \triangle\!\!\!\!O}$$

acetone

$$(CH_3)_2CHCH_2CH_2OH \xrightarrow[2) H_3O^+]{1) KMnO_4, \ ^-OH} (CH_3)_2CHCH_2\overset{O}{\overset{||}{C}}OH \xrightarrow[2) NH_3]{1) SOCl_2}$$

$$(CH_3)_2CHCH_2\overset{O}{\overset{||}{C}}NH_2 \xrightarrow[H_2O]{Br_2, NaOH} (CH_3)_2CHCH_2NH_2$$

isobutylamine

(j)

$$(CH_3)_2C=O \xrightarrow[2) H_3O^+]{1) LiAlH_4} (CH_3)_2CHOH \xrightarrow[H_2SO_4]{\text{conc. HBr}} (CH_3)_2CHBr \xrightarrow[\text{ether}]{Mg} \xrightarrow[2) H_3O^+]{1) \triangle\!\!\!\!O}$$

acetone

$$(CH_3)_2CHCH_2CH_2OH \xrightarrow[2) H_3O^+]{1) KMnO_4, \ ^-OH} (CH_3)_2CHCH_2\overset{O}{\overset{||}{C}}OH \xrightarrow[2) NH_3]{1) SOCl_2}$$

$$(CH_3)_2CHCH_2\overset{O}{\overset{||}{C}}NH_2 \xrightarrow[\substack{2) H_3O^+ \\ 3) NaOH}]{1) LiAlH_4} (CH_3)_2CHCH_2CH_2NH_2$$

isopentylamine

(k)

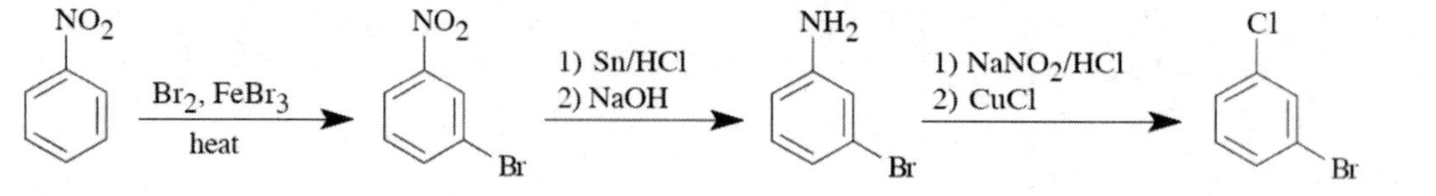

nitrobenzene *m*-chlorobromobenzene

(l)

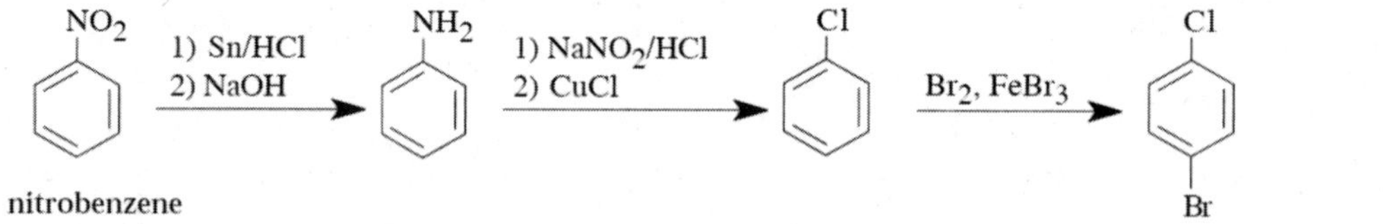

nitrobenzene *p*-chlorobromobenzene

(m) For the first reaction below, see Eq. 18.90 on text p. 927.

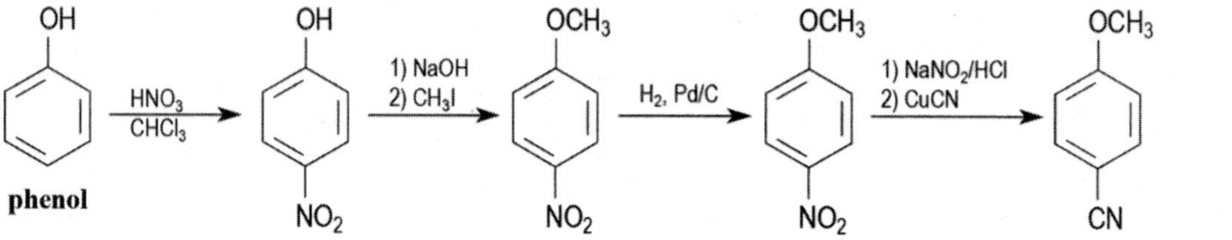

***p*-methoxybenzonitrile**

(n)

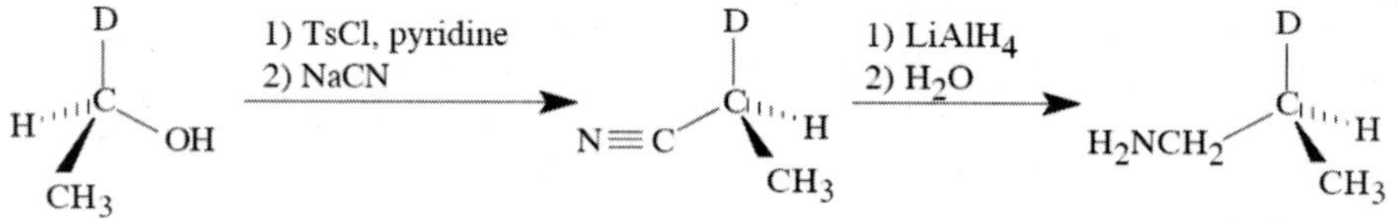

Note that the S_N2 reaction of cyanide ion proceeds with inversion of configuration.

(o)

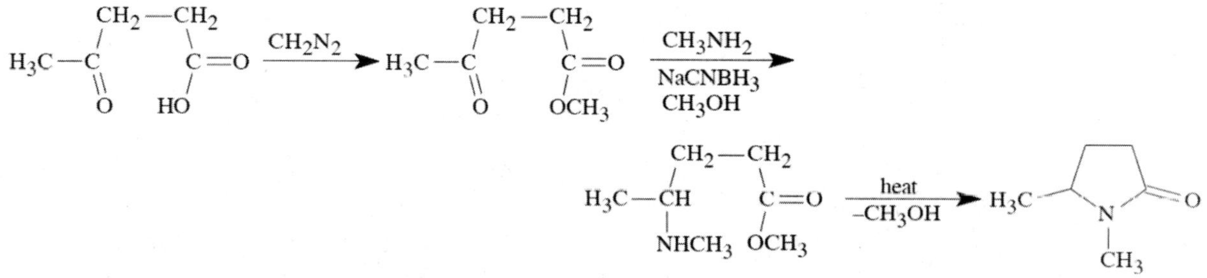

23.67 (a) Following the loss of an electron, the molecular ion of 2-methyl-2-heptanamine undergoes an α-cleavage to give $(CH_3)_2C{=}^{+}NH_2$, which has an *m*/*z* of 58. Following the loss of an electron, the molecular ion of *N*-ethyl-4-methylpentanamine undergoes an α-cleavage to give $CH_3CH{=}^{+}NHEt$, which has an *m*/*z* of 72.

2-methyl-2-heptanamine

$m/z = 58$

N-ethyl-4-methyl-2-pentanamine

$m/z = 72$

(b)   The α-cleavage of tributylamine produces a major peak at $m/z = 142$:

$m/z = 142$

23.68   At pH 1, methylamine exists almost completely as the methylammonium ion, CH_3—$^+NH_3$. The methyl quartet is expected for the presence of three neighboring nitrogen protons provided that chemical exchange is slow. As the pH is raised, more free methylamine is present and chemical exchange of the protons on the nitrogen can occur rapidly by the following mechanism. (The asterisk is used to differentiate the two nitrogens.)

$$CH_3\overset{+}{\underset{*}{N}}H_3 + CH_3NH_2 \rightleftharpoons CH_3\underset{*}{N}H_2 + CH_3\overset{+}{N}H_3$$

Evidently, this exchange is so fast on the NMR time scale that splitting is obliterated. (See Secs. 13.7D and 13.8, text pp. 651–6.)

23.69   The λ_{max} of aniline occurs at longer wavelength than that of benzene because of the conjugation of the aniline unshared electron pair with the π-electron system of the benzene ring. Recall that the more atoms are involved in a conjugated π-electron system, the greater is the intensity of the UV spectrum and the greater is the λ_{max} (Sec. 15.2C in the text). When aniline is protonated, the nitrogen no longer has an unshared pair, and the conjugated π-electron system is restricted to the phenyl ring, in which case the UV spectrum looks much as it does in benzene. In other words, protonated aniline, as far as the π-electron system and the UV spectrum are concerned, is essentially the same as benzene.

23.70   Assume that the —NH_2 or —NH protons of the amine rapidly exchange, so that they are not coupled to adjacent protons on carbon.

    Compound C should have the resonance at greatest chemical shift, because the proton α to the nitrogen is also benzylic and is a methine proton. This proton should be a triplet.

    Compound B should have the resonance at the next greatest chemical shift, because the protons α to the nitrogen are also benzylic, but are methylenes, which have smaller chemical shifts than methine protons by about 0.5–0.7 ppm. These α-protons should be a singlet. The relative integrations of the α-protons of B and C are different, as well.

Between compounds *A* and *D*, compound *A* has the resonance at smallest chemical shift—the methyl group, which should be a doublet. These methyl protons are neither benzylic nor α to a nitrogen; thus, they should have a resonance not far from δ 1.

All of the resonances of compound *D* are in the δ 2–3.5 range. This spectrum should consist of two triplets and a singlet (plus the NH proton). Other features of these spectra might be cited as well.

23.71 (a) With an unsaturation number of 4 and clear evidence for a para-substituted benzene ring (the four-proton pair of doublets centered at about δ 7), no other unsaturations are possible. The NMR indicates partial structures —OCH$_3$ and —CH$_2$CH$_2$—. With no other unsaturation, the nitrogen has to be an amine, and the broad, two-proton resonance near δ 1 suggests an —NH$_2$ group. The only structure that fits all the data is

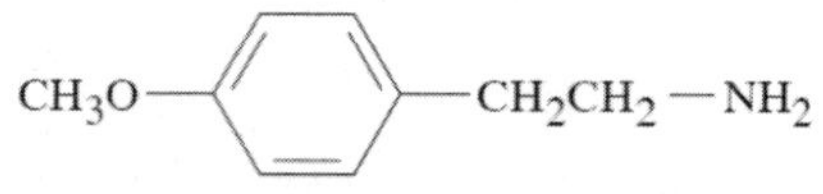

$$CH_3O—\langle \bigcirc \rangle—CH_2CH_2—NH_2$$

2-(4-methoxyphenyl)ethylamine
(compound *A*)

(b) The compound has no unsaturation and therefore must be an amine. The mutually split triplet–quartet pattern suggests an ethyl group, and the integration suggests two identical ethyl groups. If so, the δ 1.1 resonance has buried under it (as suggested by the hint) a broad resonance for two additional protons. Indeed, the chemical shift is reasonable for an —NH$_2$ group. Because there are two amine nitrogens, these have to be two chemically identical secondary amines. Two resonances have four identical protons; this suggests two chemical identical —CH$_2$— groups for each resonance. One of these resonances is a quartet and is part of the ethyl group discussed above; hence, there are two chemically equivalent ethyl groups. The other —CH$_2$— resonance is a singlet; hence this corresponds to two —CH$_2$— groups in identical environments that do not give rise to splitting. This molecule must have some internal symmetry. The structure is

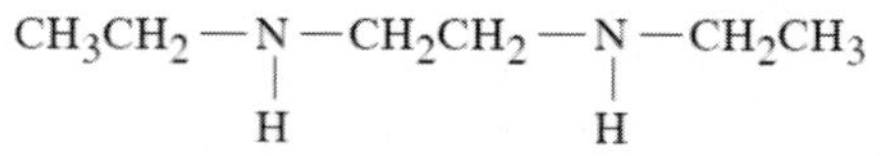

$$CH_3CH_2—\underset{\underset{H}{|}}{N}—CH_2CH_2—\underset{\underset{H}{|}}{N}—CH_2CH_3$$

N,N´-diethyl-1,2-ethanediamine
(compound *B*)

Although the two —CH$_2$— groups in the center of the molecule are adjacent, they do not show any splitting because they are chemically equivalent.

(c) This compound has only one degree of unsaturation, and both the NMR (δ 4.9) and the IR (898 cm^{-1}) suggest a vinylic methylene group (H$_2$C=) in which the protons are coupled to no other protons. The chemical shift of the three-proton singlet (δ 1.8) suggests an allylic methyl group. We have a two-proton singlet at δ 3.2 that is clearly α to the nitrogen, and an ethyl absorption in which the —CH$_2$— group is also α to the nitrogen. These data conspire to define the following structure:

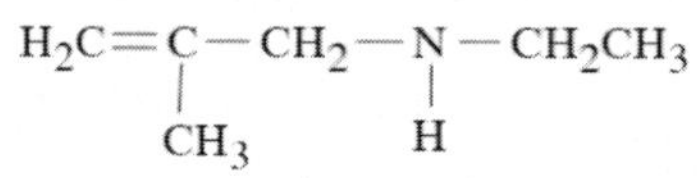

$$H_2C=\underset{\underset{CH_3}{|}}{C}—CH_2—\underset{\underset{H}{|}}{N}—CH_2CH_3$$

N-ethyl-2-methyl-2-propenamine
(compound *C*)

The absorptions of the two vinylic protons consist of two singlets near δ 4.8; geminal splitting is typically very small (Table 13.3, text p. 647) and evidently is not observed within the resolution of this spectrum.

23.72 The unsaturation number of this compound is 0. The instability of this compound in aqueous acid and the presence of two oxygens is consistent with the presence of an acetal. The six-proton singlet at δ 3.27 could be due to the methyl protons of a dimethyl acetal; the δ 53.2 resonance in the CMR spectrum could be due to the carbons of these methyl groups, and the δ 102.4 resonance could be due to the carbon of the acetal group, that is, —O—CH—O— . If so, the resonance in the proton NMR at δ 4.50 is due to the proton of this CH group, and its splitting indicates two adjacent protons. The resonance for these two protons is found at δ 2.45. This leaves a six-proton singlet at δ 2.30 unaccounted for; this could be due to the six methyl protons of a dimethylamino group. The compound is

$$(CH_3)_2N-CH_2-\underset{\underset{OCH_3}{|}}{\overset{\overset{OCH_3}{|}}{CH}}$$

dimethylaminoacetaldehyde dimethyl acetal
(2,2-dimethoxy-*N,N*-dimethylethanamine)

23.73 (a) The chemical data indicate the presence of a primary amine. The compound has one degree of unsaturation, and the chemical shifts in the CMR are too small for any carbon to be involved in a double bond. Hence, compound *A* contains a ring, and only CH and CH_2 groups are present. The presence of only three resonances means that two carbons are equivalent. Compound *A* is cyclobutylamine.

Cyclopropylmethanamine is ruled out by the fact that the carbon bound to one hydrogen has the largest chemical shift and therefore must be bound to the nitrogen.

cyclobutylamine   cyclopropylmethanamine
(compound *A*)   (ruled out)

(b) The IR data for this compound indicate a —CH=CH_2 group. The molecule is neither a primary amine (chemical data) nor a tertiary amine (it has an N—H). Hence it is a secondary amine. Two structures worth considering are *A* and *B*:

$$CH_3CH_2NHCH{=}CH_2 \qquad CH_3NHCH_2CH{=}CH_2$$

A   *B*

Because the CMR spectrum contains no resonances near δ 15, which would be expected for the methyl carbon of compound *A*, the structure must be *B*, *N*-methyl-2-propenamine. The chemical shifts of the carbon resonances are consistent with this structure.

23.74 (a) This reaction is related to the Hofmann rearrangement, on p. 1219 of the text.

(b) This reaction is related to the Hofmann rearrangement. Reaction with hydroxide on the carbonyl carbon of *N*-bromosuccinimide gives an *N*-bromo anion, which undergoes a Hofmann rearrangement to the corresponding isocyanate, which in turn, hydrolyzes and decarboxylates to the amine product.

 If you find an error in this manual, please visit http://people.pharmacy.purdue.edu/~loudonm/teaching/

(c)   This reaction involves a Curtius rearrangement (text p. 1217), followed by intramolecular formation of a cyclic urea (Eq. 23.73, text p. 1218).

23.75   Evidently, an anion is generated at the α-carbon of the ester, and this then adds to the aldehyde. Although one might be tempted to use the amine as a base to remove an α-hydrogen, vinylic hydrogens that are α to carbonyl groups are not particularly acidic, because the electron pair of the enolate ion that would form and the double bond cannot simultaneously be conjugated with the carbonyl. Rather, the amine acts as a nucleophile in a conjugate addition, which generates the required anion. Once the addition occurs, the amine is expelled again to regenerate the α,β-unsaturated product. (As in previous problems, proton transfers are shown as intramolecular processes for convenience.)

23.76   An intermediate in the formation of 1-pentanamine is the imine A. The other products arise from the fact that 1-pentanamine, B, can react with either the starting aldehyde, pentanal, or with imine A to give a new imine C. (Give the mechanism for this reaction.) Imine C can also undergo hydrogenation to the second product, dipentylamine, D. Dipentylamine can also react with other imine intermediates to form imminium ion E, which can undergo hydrogenation to form tripentylamine F, the tertiary amine by-product.

$$CH_3(CH_2)_3\overset{NH}{\overset{||}{CH}} + H_2NCH_2(CH_2)_3CH_3 \rightleftharpoons NH_3 + CH_3(CH_2)_3\overset{NCH_2(CH_2)_3CH_3}{\overset{||}{CH}} \xrightarrow{H_2, cat.} [CH_3(CH_2)_3CH_2]_2NH$$

A   B   C   D

$$CH_3(CH_2)_3\overset{NCH_2(CH_2)_3CH_3}{\overset{||}{CH}} + [CH_3(CH_2)_3CH_2]_2NH \rightleftharpoons H_2NCH_2(CH_2)_3CH_3 + CH_3(CH_2)_3\overset{\overset{+}{N}[CH_2(CH_2)_3CH_3]_2}{\overset{||}{CH}}$$

C   D   B   E

$$\xrightarrow{H_2, cat.} [CH_3(CH_2)_3CH_2]_3N$$

F

The yield of pentylamine (compound *B*) can be increased, and the amounts of the other by-products reduced, by including ammonia in the reaction mixture. Can you see why? (*Hint*: Apply Le Châtelier's principle.)

23.77   Both compound *A* and *C* undergo Hofmann elimination after exhaustive methylation followed by heating with Ag$_2$O. Compound *A* is converted to cyclooctatetraene (COT), compound *B*, and compound *C* is converted to "1,3,5-cyclohexatriene," which is actually benzene. COT contains $4n$ π electrons and is not aromatic; benzene contains $4n + 2$ π electrons and is aromatic (see discussion in Sec. 15.7B in the text). Under mild conditions, benzene does not react with bromine, while COT behaves as a typical cycloalkene and does react with bromine (see Eq. 15.43, text p. 759).

23.78   (a)   Amides, like amines, can be diazotized. The result is an acyldiazonium ion *A*, which rapidly hydrolyzes; remember that N$_2$ is a superb leaving group. This produces carbamic acid, which decomposes to ammonia and CO$_2$. Then ammonia is itself diazotized to give N$_2$ as shown in the solution to Problem 23.65(a) on p. 862 of this manual.

$$H_2N\overset{O}{\overset{||}{C}}NH_2 \xrightarrow{HNO_2} H_2N\overset{O}{\overset{||}{C}}\overset{+}{N}{\equiv}N \xrightarrow{H_2O} H_2N\overset{O}{\overset{||}{C}}OH \longrightarrow CO_2 + NH_3 \xrightarrow{HNO_2} N_2 + 2 H_2O$$

A   carbamic acid

(b)   The amine reacts intramolecularly with the carbonyl group to give a carbinolamine, which then reacts to form a cyclic imminium ion *A*, which is reduced by the sodium borohydride.

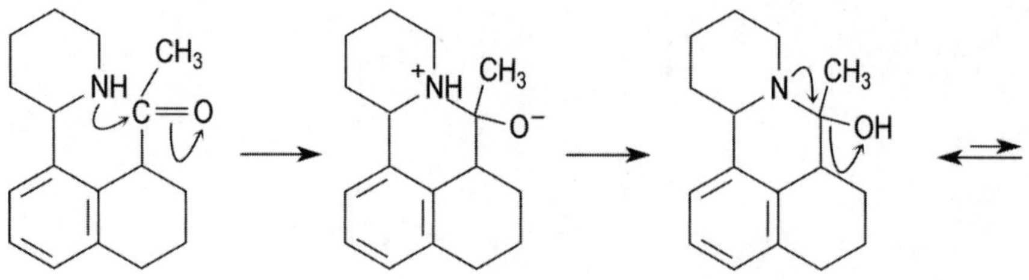

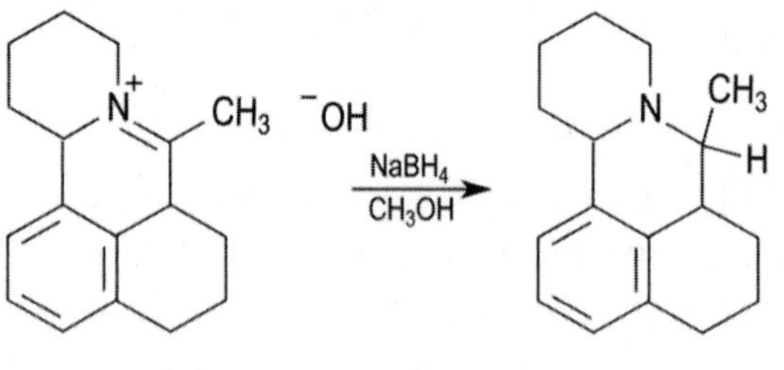

A

(c) Cyanide ion serves as a base to promote a β-elimination of trimethylamine and form an α,β-unsaturated ketone *A*. This undergoes conjugate addition of cyanide ion to give the product (Sec. 22.9A in the text).

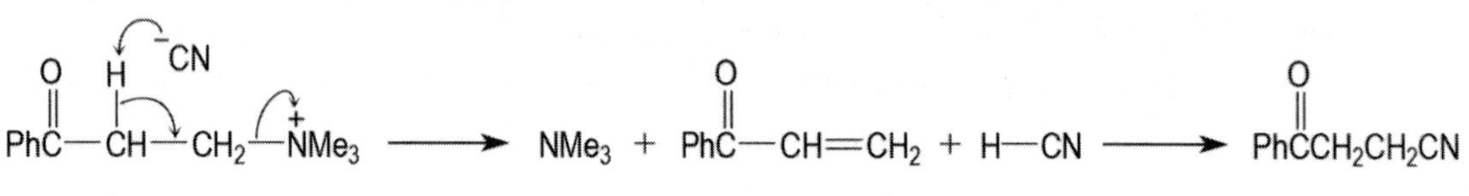

A

(d) The acyl azide *A* undergoes Curtius rearrangement to an enamine *B*, which hydrolyzes under the aqueous reaction conditions to the product ketone.

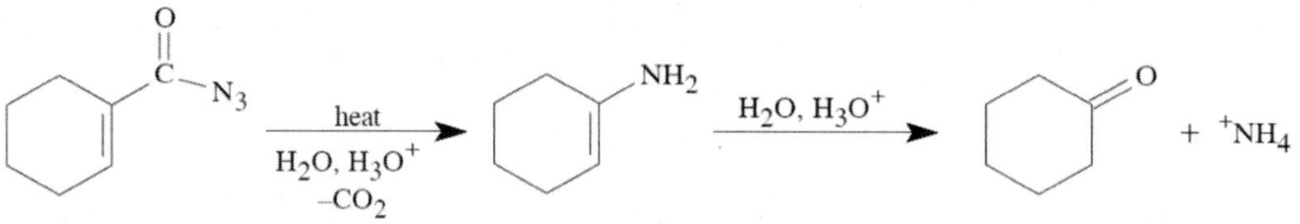

A   *B*

The mechanism of the Curtius rearrangement is shown on text p. 1217, and the hydrolysis of the enamine occurs as follows:

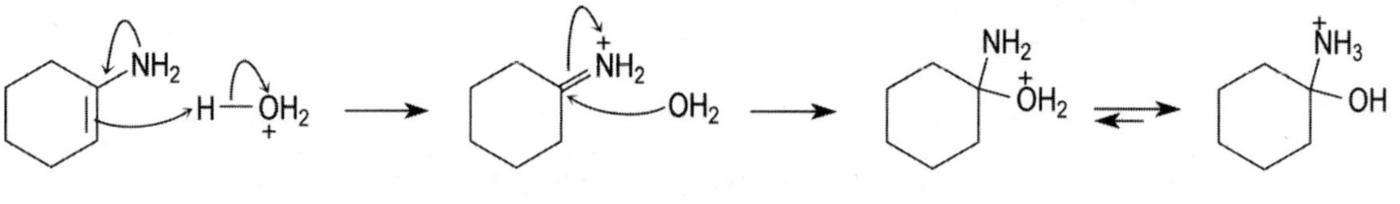

(a resonance-
stabilized cation)

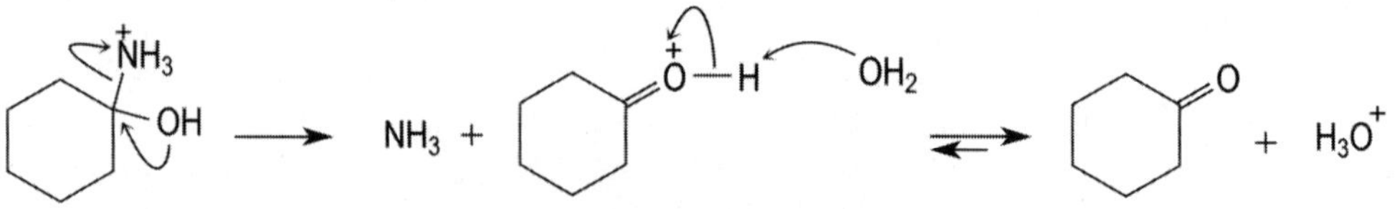

(a resonance-
stabilized cation)

(e) The diazonium ion *A* formed from the starting material loses nitrogen to give a carbocation *B*. This carbocation undergoes a rearrangement to the product. (See Study Problem 19.2 on text pp. 961–2 for a discussion of a closely related reaction.)

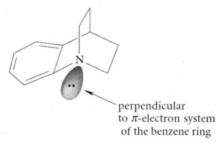

23.79 The conjugate-acid pK_a of amine C is a normal conjugate-acid pK_a for a tertiary amine. (See, for example, the pK_a of the conjugate acid of triethylamine in Table 23.1 on text p. 1190.) This observation shows that incorporation of an amine into a bicyclic structure has no significant effect on its conjugate-acid pK_a. The conjugate-acid pK_a of amine A is not unusual for an N,N-dialkylaniline. (See, for example, the pK_a of $PhN(CH_3)_2$ in Table 23.1.) Recall (text p. 1193) that the conjugate-acid pK_a values of aniline derivatives are considerably lower than those of tertiary alkylamines because of resonance interaction of the unshared electron pair on the nitrogen with the benzene ring. The conjugate-acid pK_a of compound B, then, has the unusual value that requires explanation. In this compound, the unshared pair on nitrogen *cannot* overlap with the π-electron system of the benzene ring because the π orbitals of the benzene ring and the orbital on the nitrogen are forced by the bicyclic structure out of coplanarity.

perpendicular
to π-electron system
of the benzene ring

Consequently, the basicity-lowering resonance effect is absent. For this reason, the conjugate-acid pK_a of compound B is higher than that of compound A. However, the basicity-lowering polar effect of the benzene ring remains; for this reason, the conjugate-acid pK_a of compound B is lower than that of compound C.

The pK_a data in the problem provide good estimates of the polar and resonance effects of a benzene ring on amine basicity. These data show that polar and resonance effects are about equally important.

23.80 The data show that the conjugate-acid pK_a values of a bicyclic tertiary amine C (which is the same as amine C in the previous problem) and a monocyclic tertiary amine B are not very different. Hence, the additional ring has no unusual effect on pK_a. The conjugate-acid pK_a of amide A is also fairly normal for an amide that is protonated on the carbonyl oxygen. (See Eq. 21.4b on text p. 1058.) The pK_a of amide D, then, is unusual. Normally, amides protonate on the carbonyl oxygen because the resulting carbocation is resonance-stabilized. In this case, however, the bicyclic structure of the ring prevents the overlap of the nitrogen unshared electron pair with the π-electron system of the carbonyl group. Hence, the electronic character of this nitrogen is more like that of an amine with a neighboring electron-withdrawing substituent (the carbonyl group) than it is like that of an amide. Thus, this amide behaves like an amine: it protonates on nitrogen. What about the hydrolysis rate? Recall that ordinary amides hydrolyze slowly because they are resonance-stabilized. (See Sec. 21.7E, text p. 1067.) When resonance stabilization is not present, as in the case of compound D, an amide hydrolyzes rapidly. In acidic solution, when the nitrogen is protonated, its hydrolysis rate is more like that of an acid chloride!

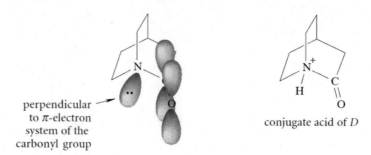

perpendicular
to π-electron
system of the
carbonyl group

conjugate acid of *D*

(See also the sidebar on text p. 1059 for a discussion of this case.)

23.81   **(a)**   The phosphine-containing probe compound reacts with the azide-containing protein as in a standard Staudinger reaction, as shown in Eqs. 23.56a–b on text p. 1212, releasing nitrogen as a by-product.

(b)   The negatively-charged nitrogen reacts with the ester in a nucleophilic acylation reaction to form an amide (see Sec. 21.8C for related reactions). The iminophosphorane is then hydrolyzed by water to give the probe-labeled protein.

(c)   Had the ester group not been present in such close proximity to the iminophosphorane nitrogen, the reaction in part (b) would not occur, and instead, water would simply hydrolyze the iminophosphorane into the triphenylphosphine-derivative and the amine linked to the protein, as in a standard Staudinger reaction (see Eq. 23.57 on text p. 1213).

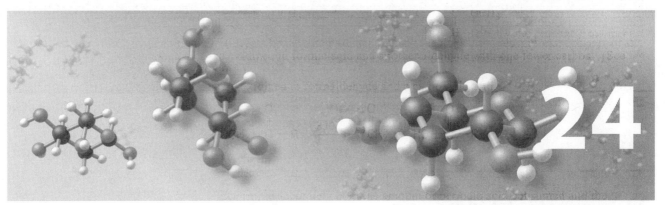

Carbohydrates

STUDY GUIDE LINKS

 ## 24.1 Acid Catalysis of Carbohydrate Reactions

Acid catalysis of glycoside formation involves protonation of the hydroxy group at the anomeric carbon (Eq. 24.28a on text p. 1255); acid catalysis of mutarotation involves protonation of the ring oxygen (Problem 24.9 on text p. 1251). You may be asking why these specific oxygens are protonated—why not others? The answer is that *any one* of the oxygens can indeed be protonated to small extents in acidic solution. However, *only* protonation at the ring oxygen leads to mutarotation, and *only* protonation of the —OH group at the anomeric carbon leads to glycoside formation. In other words, the various protonated forms are in rapid equilibrium and are present whether they are shown explicitly or not. Only certain of the protonated forms can react further, and these are the ones shown in the equations.

 ## 24.2 Configurations of Aldaric Acids

Consideration of the stereochemistry of aldaric acids reveals a subtle yet important aspect of the D,L configurational system. Certain aldaric acids can be derived from either a D or an L carbohydrate.

$$
\begin{array}{ccc}
\text{CO}_2\text{H} & & \text{CO}_2\text{H} \\
\text{H}\!-\!\!|\!-\!\text{OH} & & \text{HO}\!-\!\!|\!-\!\text{H} \\
\text{HO}\!-\!\!|\!-\!\text{H} & \xrightarrow{180°} & \text{HO}\!-\!\!|\!-\!\text{H} \\
\text{H}\!-\!\!|\!-\!\text{OH} & & \text{H}\!-\!\!|\!-\!\text{OH} \\
\text{H}\!-\!\!|\!-\!\text{OH} & & \text{HO}\!-\!\!|\!-\!\text{H} \\
\text{CO}_2\text{H} & & \text{CO}_2\text{H} \\
\end{array}
$$

D-glucaric acid (from D-glucose) or L-gularic acid (from L-gulose)
(the same compound)

In this case, whether the aldaric acid is classified as D or L depends on which carbon is specified as carbon-1. Because the two ends of the molecule are *constitutionally equivalent* (see text p. 488), the choice is completely arbitrary! The names D-glucaric acid and L-gularic acid are both correct names for this aldaric acid. The two names reflect the fact that this aldaric acid can be formed by nitric acid oxidation of either D-glucose or L-gulose.

This situation arises because the —OH groups on the endmost asymmetric carbons are on the same side of the Fischer projection. However, when these —OH groups are on opposite sides, the configuration is unambiguous. For example, D-tartaric acid also has constitutionally equivalent ends, but it has the D configuration no matter how it is turned.

D-tartaric acid

As you might expect, a similar situation arises in other derivatives with constitutionally equivalent ends, such as alditols. This issue is explored in Problem 24.21 on text p. 1260.

FURTHER EXPLORATIONS

 24.1   Nomenclature of Anomers

Long before the actual structures of the anomeric forms of many carbohydrates were known, the α-anomer of a D-carbohydrate was simply defined as "the more dextrorotatory of the two anomers." This convention, proposed by C. S. Hudson (1881–1952), an American carbohydrate chemist, is cumbersome because, in order to apply it, one has to know the optical rotations of the two anomers. When the structures of several anomers became known, it was found that in most cases the more dextrorotatory anomer of each D-carbohydrate has the configuration shown in the text for the α-anomer. Hudson's definition in terms of a physical property—optical rotation—was subsequently replaced by the structural definition used in the text.

24.2   More on the Fischer Proof

The Fischer proof of glucose stereochemistry has a number of interesting facets that are not discussed in the text. One of the most important experimental aspects of carbohydrate chemistry is that many carbohydrates are notoriously difficult to crystallize; they are frequently isolated as "syrups" and "gums." Because chemists in Fischer's day did not have available the sophisticated purification methods available today, they had to rely almost exclusively on crystallization as a purification method. Because carbohydrates were difficult to crystallize, they were very difficult to purify. Few chemists were interested in working on problems fraught with so many experimental difficulties; consequently, little progress had been made in the field of carbohydrate chemistry when Fischer began his work. In fact, Fischer himself, in some of his correspondence, deplored the dreadfully slow pace of carbohydrate research in his own laboratory resulting from the experimental difficulties of handling carbohydrates.

Two solutions to the problem of carbohydrate crystallization emerged from Fischer's long-standing interest in phenylhydrazine, $PhNHNH_2$. He had developed an indole synthesis (today called the *Fischer indole synthesis*) using phenylhydrazine as a starting material. When he allowed phenylhydrazine to react with aldohexoses, he found that very interesting compounds called *osazones* were obtained.

$$
\begin{array}{c}
\text{CH}{=}\text{O} \\
\text{H}\text{---}\text{OH} \\
\text{HO}\text{---}\text{H} \\
\text{H}\text{---}\text{OH} \\
\text{H}\text{---}\text{OH} \\
\text{CH}_2\text{OH}
\end{array}
\quad + \ 3\ \text{PhNHNH}_2 \quad \longrightarrow \quad
\begin{array}{c}
\text{CH}{=}\text{N}\text{---}\text{NHPh} \\
\text{C}{=}\text{N}\text{---}\text{NHPh} \\
\text{HO}\text{---}\text{H} \\
\text{H}\text{---}\text{OH} \\
\text{H}\text{---}\text{OH} \\
\text{CH}_2\text{OH}
\end{array}
\quad + \ \text{PhNH}_2 \ + \ \text{NH}_3 \ + \ \text{H}_2\text{O}
$$

(+)-glucose   glucosazone

This reaction is very reminiscent of phenylhydrazone formation (Table 19.3 on text p. 985). However, an unusual aspect of this reaction is that one equivalent of phenylhydrazine serves as an oxidizing agent, with the net result that a second substituted imine group is introduced. (For the mechanism of this reaction, see Loudon, *Organic Chemistry,* First edition, Addison–Wesley Publishing Co., 1984, pp. 1417–1418.) Osazones proved to be much more easily crystallized, and thus more easily characterized, than aldoses themselves. Furthermore, they served to confirm some of Fischer's stereochemical deductions. Notice that *neither carbon-1 nor carbon-2 are asymmetric in the osazone.* Consequently, isomeric aldoses that are epimeric only at carbon-2 give the same osazone. Fischer found that (+)-glucose and (+)-mannose give *different* phenylhydrazones (formed under milder conditions with phenylhydrazine), but give the same osazone. This fact, along with their simultaneous preparation in the Kiliani–Fischer synthesis from (–)-arabinose, showed that they are epimers at carbon-2.

Phenylhydrazine figured prominently in Fischer's research in a second way when Fischer found that the aminolysis reactions of aldonic acids (undoubtedly reacting as their lactones) with phenylhydrazine give phenylhydrazides, which, unlike the parent acids, are beautifully crystalline compounds.

$$
\begin{array}{c}
\text{CO}_2\text{H} \\
\text{H}\text{---}\text{OH} \\
\text{HO}\text{---}\text{H} \\
\text{H}\text{---}\text{OH} \\
\text{H}\text{---}\text{OH} \\
\text{CH}_2\text{OH}
\end{array}
\ \underset{\longleftarrow}{\overset{-\text{H}_2\text{O}}{\longrightarrow}}\
\begin{array}{c}
\text{C}{=}\text{O} \\
\text{H}\text{---}\text{OH} \\
\text{HO}\text{---}\text{H} \\
\text{H}\text{---}\text{O} \\
\text{H}\text{---}\text{OH} \\
\text{CH}_2\text{OH}
\end{array}
\ \overset{\text{PhNHNH}_2}{\underset{\substack{\text{(ester}\\ \text{aminolysis)}}}{\longrightarrow}}\
\begin{array}{c}
\text{C}{=}\text{O, NHNHPh} \\
\text{H}\text{---}\text{OH} \\
\text{HO}\text{---}\text{H} \\
\text{H}\text{---}\text{OH} \\
\text{H}\text{---}\text{OH} \\
\text{CH}_2\text{OH}
\end{array}
$$

D-gluconic acid   D-γ-gluconolactone   D-gluconic acid
                                                                 phenylhydrazide

One other interesting aspect of the Fischer proof has to do with the last step, which was made possible by a remarkable bit of serendipity. It is noted in the text that Fischer had prepared L-(+)-gulose in the course of his research; recall that it was the oxidation of D-glucose and L-gulose to the same aldaric acid that completed the Fischer proof. Equation 24.38 on text p. 1260 shows that D-glucaric acid can exist as a mixture of two different lactones (Step 4, text p. 1266). Fortuitously, Fischer's procedure for isolating the lactones gave mainly the 1,4-lactone. His synthesis of L-gulose began with this lactone.

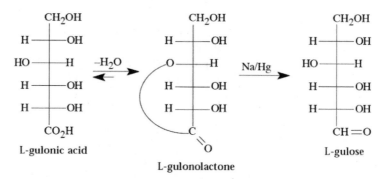

D-glucaric acid 1,4-lactone   D-glucuronic acid

L-gulonic acid   L-gulonolactone   L-gulose

"Na/Hg" is sodium amalgam, a solution of sodium in mercury, which was widely used as a reducing agent until lithium aluminum hydride and sodium borohydride were developed. The sodium–amalgam reductions shown above can be carried out in stages by controlling the conditions of each step. Recall that D-glucose is readily oxidized to D-glucaric acid-1,4 lactone with dilute nitric acid. Consequently, beginning with D-glucose, the overall sequence of glucose oxidation followed by the sodium amalgam reductions effects the *net interchange* of the CH=O and CH_2OH groups, and thus gives L-gulose from D-glucose. The serendipitous element was this: had Fischer isolated the 3,6-lactone of D-glucaric acid, it would have given back D-glucose when taken through this sequence of reactions, and the last step of the proof would not have been possible.

An intriguing account of Fischer's life and work, including details of the Fischer proof, was written by F. W. Lichtenthaler and published in *Angewandte Chemie International Edition in English,* vol. 31, No. 12, pp. 1541–1596 (1992). This journal is available in most chemistry libraries.

REACTION REVIEW

I. REACTIONS OF CARBOHYDRATES

A. BASE-CATALYZED ISOMERIZATION OF CARBOHYDRATES

1. In base, aldoses and ketoses rapidly equilibrate to mixtures of other aldose and ketoses; this transformation is an example of the Lobry de Bruyn–Alberda van Ekenstein reaction.
   a. An aldose can ionize to give a small amount of its enolate ion in base.
   b. Protonation of this enolate ion at one face gives back the aldose; protonation at the other face gives an epimer.

c. The enolate ion can also protonate on the oxygen to give a new enol, called an enediol, which contains a hydroxy group at both ends of a double bond.
d. The enediol is simultaneously the enol of two epimeric aldoses and an isomeric ketose, and the enediol is transformed into a mixture of these.

2. Several transformations of this type are important in metabolism.

B. ESTER AND ETHER DERIVATIVES OF CARBOHYDRATES

1. In the presence of concentrated base, carbohydrates are converted into ethers by reactive alkylating agents (Williamson ether synthesis).

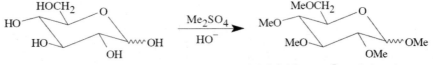

methyl 2,3,4,6-tetra-O-methyl-D-glucopyranoside

2. The alkoxy group at the anomeric carbon differs from other alkoxy groups in the alkylated carbohydrate.
   a. The alkoxy group at the anomeric carbon is part of a glycosidic linkage.
   b. Since it is an acetal, it can be hydrolyzed in aqueous acid under mild conditions; the other alkoxy groups are ordinary ethers and do not hydrolyze under these conditions.

2,3,4,6-tetra-*O*-methyl-D-glucopyranoside

3. The hydroxy groups of carbohydrates can be esterified; the resulting esters can be saponified in base or can be removed by transesterification with an alkoxide.

4. Ethers and esters are used as protecting groups in reactions involving carbohydrates; furthermore, they have broader solubility characteristics and greater volatility than the carbohydrates themselves.

C. OXIDATION AND REDUCTION REACTIONS OF CARBOHYDRATES

1. Treatment of an aldose with bromine water oxidizes aldoses to aldonic acids; this reaction is a useful test for aldoses. (Carbohydrates that can be oxidized by bromine water are called reducing sugars.)
   a. Aldonic acids exist in acidic solution as five-membered lactones called aldonolactones.

an aldonic acid   an aldolactone

   b. Glycosides are not oxidized by bromine water because the aldehyde carbonyl group is protected as an acetal. (Carbohydrates that cannot be oxidized by bromine water are called nonreducing sugars.)
2. Aldoses can also be oxidized with the Tollens reagent; the alkaline conditions of Tollens test, however, also promote the equilibration of ketoses and aldoses. For this reason, ketoses also give a positive Tollens test.
3. Both ends of an aldose are oxidized to carboxylic acid groups by dilute HNO_3, but the secondary alcohol groups are not affected; the oxidation product is an aldaric acid.
   a. Aldaric acids in acidic solution form lactones.

   b. Two different five-membered lactones are possible from a hexose or pentose, depending on which carboxylic acid group undergoes lactonization.

c. Under certain conditions, some aldaric acids derived from aldohexoses can be isolated as dilactones, in which both carboxylic acid groups are lactonized. (See Eq. 24.39 on text p. 1260.)

4. Many carbohydrates contain vicinal glycol units and are oxidized by periodic acid.

a. α-Hydroxy aldehydes are oxidized to formic acid and another aldehyde with one fewer carbon. (See Eq. 24.40 on text page 1261.)

b. Hydroxymethyl ketones are oxidized to formaldehyde and a carboxylic acid.

a hydroxymethyl
ketone

c. Because it is possible to determine accurately both the amount of periodic acid consumed and the amount of formic acid produced, periodic acid oxidation can be used to differentiate between pyranose and furanose structures for saccharide derivatives.

5. Aldoses and ketoses undergo many of the usual carbonyl reductions, such as catalytic hydrogenation and reaction with NaBH$_4$; an aldose is reduced to a primary alcohol called an alditol.

II. SYNTHESIS OF SUGARS FROM OTHER SUGARS

A. KILIANI–FISCHER SYNTHESIS; INCREASING THE LENGTH OF THE ALDOSE CHAIN

1. Addition of hydrogen cyanide to aldoses gives cyanohydrins; because the cyanohydrin product of such a reaction has an additional asymmetric carbon, it is formed as a mixture of two epimers.

a. These epimers are diastereomers and are typically formed in different amounts.

b. The mixture of cyanohydrins can be converted by catalytic hydrogenation into a mixture of imines, which are not isolated.

c. The imines would normally hydrogenate to amines, but the acidic aqueous reaction conditions intercept them and convert them into aldehydes, which are the aldose products.

2. The cyanohydrin formation–reduction sequence converts an aldose into two epimeric aldoses with one additional carbon; this process is known as the Kiliani–Fischer synthesis.

B. RUFF DEGRADATION; DECREASING THE LENGTH OF THE ALDOSE CHAIN

1. In the Ruff degradation, the calcium salt of an aldonic acid is oxidized with hydrogen peroxide in the presence of Fe^{3+}.

   a. An aldose is degraded to another aldose with one less carbon atom.

   b. The stereochemistry of the remaining groups remains the same.

SOLUTIONS TO PROBLEMS

Solutions to In-Text Problems

24.1   (a)   Many Fischer projections are possible for a compound with three asymmetric carbons. The two below are related by a 180° rotation in the plane of the page. If yours don't look like these, try to use the allowed manipulations of Fischer projections to see whether yours are equivalent. (There are eighteen possible correct Fischer projections!)

(b)   The following two of the nine possible Fischer projections of (*S*)-2-butanol are related by a cyclic permutation of the ethyl, —OH, and the methyl group.

24.2   (a)   The way to solve a problem of this sort is to manipulate one of the Fischer projections so that two groups in one projection are in the same positions as the corresponding groups in the other; then the relationship between the two molecules should be clear. Thus, manipulate the projection on the right so that the OH groups are in the vertical positions. It then becomes apparent that this structure and the one on the left in the problem are enantiomers.

(b)   Turning either projection 180° in the plane of the page establishes that it is identical to the other projection.

24.3   It is relatively simple to determine if a compound is meso using Fisher projections. First, perform a cyclic permutation at either end of the Fisher projection (see Rule 4, text p. 1236) to give similar groups at each end, if possible, as in *C* and *D* below. Then draw an imaginary horizontal line through the center of the Fisher projection. If the top half is a mirror image of the bottom half, the compound is meso; if the two halves are not mirror images, then the compound is not meso. Compound *A* has an aldehyde group on top and a primary alcohol group on the bottom, so it cannot be a meso compound. Compound *B* is a meso compound. This structure (as well as structure *D*) illustrates that the "mirror plane" can go through certain atoms. After a cyclic permutation at the top carbon of compound *C*, the OH group is on the right while the lower carbon has an OH group on the left; compound *C* is not a meso compound. After cyclic permutations at the top and bottom carbons of compound *D*, all three OH groups are on the same side; compound *D* is a meso compound.

	A		B		C		D
	CH=O		CH₂OH		CH₂OH		CH₂OH

Fischer projections A, B, C, D as shown.

A:
CH=O
H—OH
H—OH
H—OH
CH₂OH

B:
CH₂OH
HO—H
H—OH
HO—H
CH₂OH

C:
CH₂OH
H—OH
HO—H
HO—H
CH₂OH

D:
CH₂OH
H—OH
H—OH
H—OH
CH₂OH

A   *B*   *C*   *D*

24.4 (a) Use a cyclic permutation to transform the given Fischer projection into a standard form in which all of the backbone carbons are in a vertical line. This will show that the —OH group on carbon-5 is on the right. Thus, this aldose has the D-configuration.

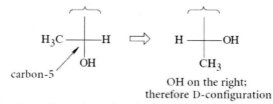

carbon-5

OH on the right;
therefore D-configuration

(b) An analysis of the carbon-3 configuration of D-glucose shows that carbon-3 has the *S* configuration. Consequently, the L-enantiomer of glucose has the *R* configuration at carbon-3.

  Remember that the configuration of an asymmetric carbon in a Fischer projection can be analyzed directly. To review, see discussion on text pp. 1237–8.

24.5 Use cyclic permutations to put all of the aldoses in the standard Fischer projection to see the relative locations of corresponding groups:

A:
CH=O
H—OH
HO—H
H—OH
HO—H
CH₂OH

B:
CH=O
H—OH
HO—H
H—OH
H—OH
CH₂OH

C:
CH=O
HO—H
H—OH
HO—H
HO—H
CH₂OH

A   *B*   *C*

Compounds *B* and *C* are enantiomers. Compound *A* is an epimer of *B* (at carbon-5) and a diastereomer of both *B* and *C*. (Remember that an epimer is a particular type of diastereomer.)

24.6 (a)

H₃C—CH—CH₂CH₂CH₂—CH=O   ⇌   ⇌
    |
    HO

(b)

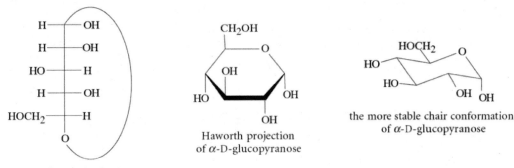

24.7   (a)   Recognize that α-D-glucopyranose is an anomer of β-D-glucopyranose, and convert the β-D-glucopyranose structures in Study Problem 24.2 on text p. 1245 to α-D-glucopyranose structures by inverting the configuration of carbon-1.

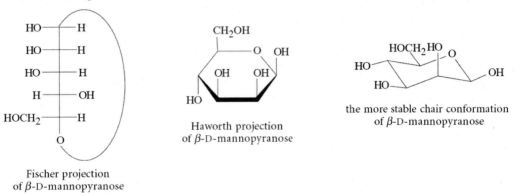

Fischer projection
of α-D-glucopyranose

Haworth projection
of α-D-glucopyranose

the more stable chair conformation
of α-D-glucopyranose

(b)   Recognize that D-mannose is epimeric to D-glucose at carbon-2, and thus convert the β-D-glucopyranose structures on text p. 1245 to β-D-mannopyranose structures by inverting the configuration of carbon-2.

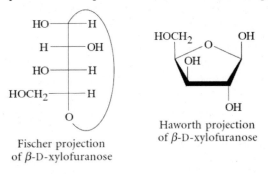

Fischer projection
of β-D-mannopyranose

Haworth projection
of β-D-mannopyranose

the more stable chair conformation
of β-D-mannopyranose

(c)   Start with the structure of D-xylose in Figure 24.3 (text p. 1240) and perform the usual manipulations. An easier solution is to recognize that D-xylose is epimeric to D-ribose at carbon-3. Thus, the structures for β-D-ribofuranose in Eq. 24.16 on text p. 1181 can be used if the configuration of carbon-3 is inverted.

Fischer projection
of β-D-xylofuranose

Haworth projection
of β-D-xylofuranose

(d)   Start with the Fischer projection for D-fructose from text p. 1242.

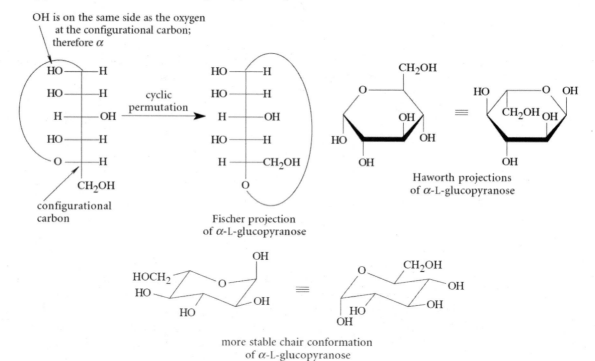

(e)   Draw the enantiomers of the corresponding structures for α-D-glucopyranose, which are shown on text pp. 1245–6 or in the answers to part (a) of this problem.

(f)   Start with the structures of α-D-glucopyranose in the solution for part (e) and make the bond to the anomeric carbon a "squiggly bond." (For the Fischer projection simply do not show the configuration. "Squiggly bonds" are not used in Fischer projections.)

24.8   (a)   Perform the reverse of the analysis in Study Problem 24.2 on text p. 1245 to reveal that this is α-D-ribose. But you can also look at Eq. 24.16 on text p. 1247 and see that the compound given is the α-anomer of β-D-ribose, which means that it is α-D-ribose.

(b)   This is the β-anomer of a D-hexopyranose that differs from D-glucose in its configuration at carbons 3 and 4; therefore, it is β-D-gulopyranose.

(c)   This is β-D-arabinose.

24.9 The following mechanism begins with the α-anomer protonated on the ring oxygen and ends with the β-anomer, also protonated on the ring oxygen.

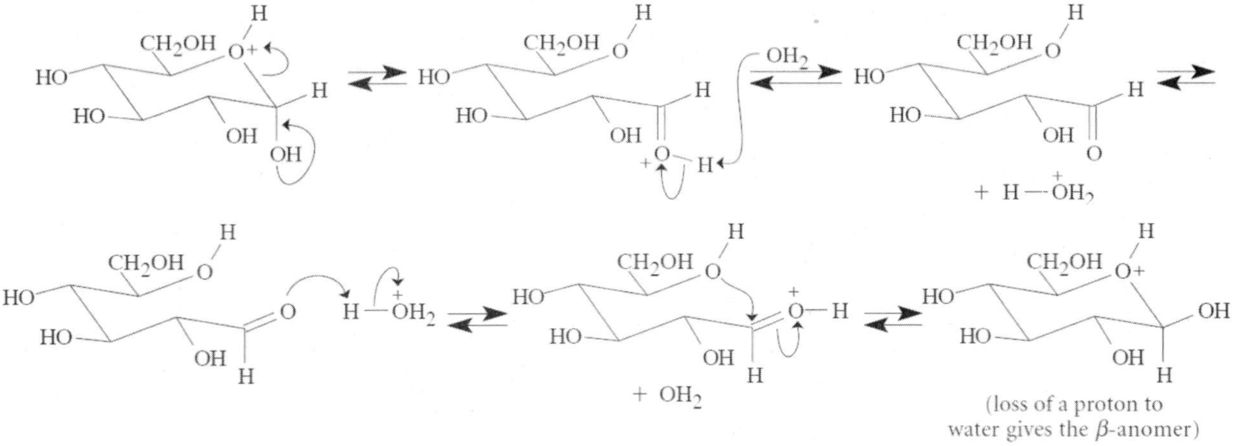

24.10 The mechanism for base-catalyzed mutarotation of glucose:

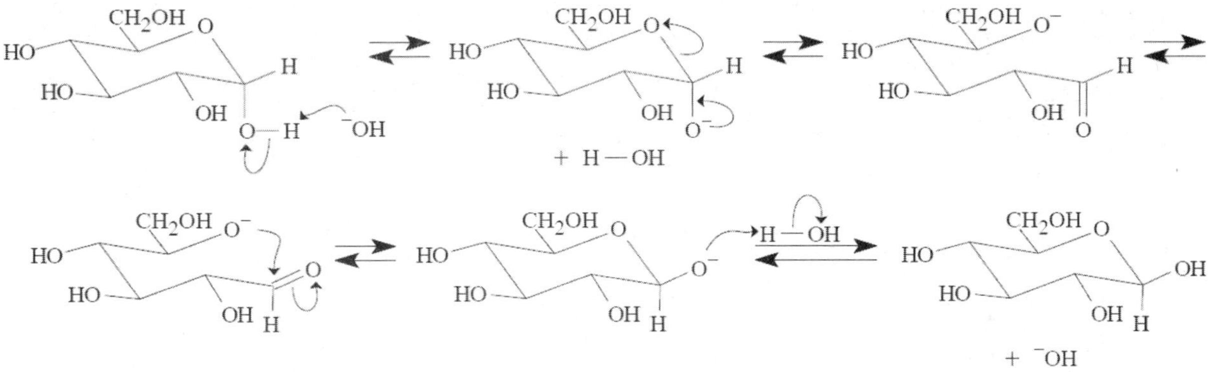

24.11 The β-pyranose forms of D-glucose and D-talose are as follows:

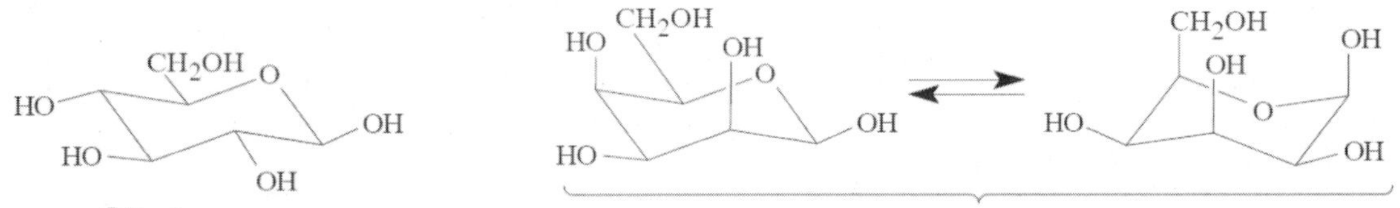

β-D-glucopyranose β-D-talopyranose

Both chair forms of the β-D-talopyranose molecule have a significantly greater number of 1,3-diaxial interactions than β-D-glucopyranose. These unfavorable interactions in β-D-talopyranose raise its energy and thus lower the relative amounts of these forms at equilibrium. Hence, talose contains a lower percentage of the β-pyranose form (and a lower percentage of the α-pyranose form for the same reason) and a higher percentage of the furanose and aldehyde forms at equilibrium.

24.12 (a) The easiest way to do this is to consider the relationship of D-allose to D-glucose and modify the conformational representation of β-D-glucopyranose (in which all of the ring substituents are equatorial) accordingly. Thus, Fig. 24.3 on text p. 1240 shows that D-allose is epimeric to D-glucose at carbon-3. Consequently, configurational inversion of the β-D-glucopyranose structure at carbon-3 gives the desired conformational representation of β-D-allopyranose.

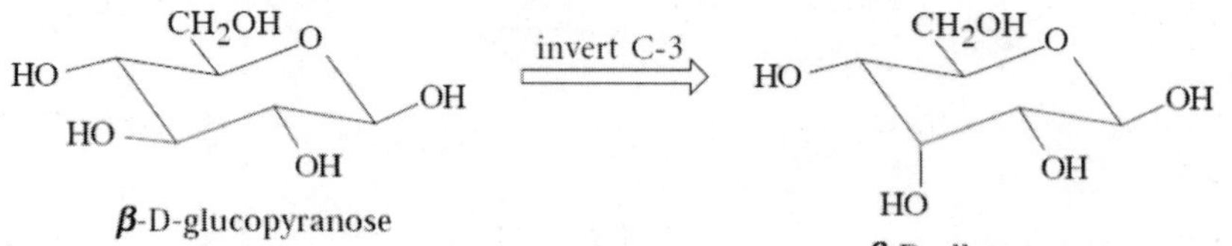

β-D-glucopyranose ⟹ (invert C-3) β-D-allopyranose

(b)   Follow the procedure in Study Problem 24.2 on text p. 1245. Note that in the second and third Fischer projections the configuration of carbon-5 cannot be represented without violating the Fischer-projection conventions. We must remember that this carbon has the R configuration, and this configuration must be shown explicitly in the conformational representation.

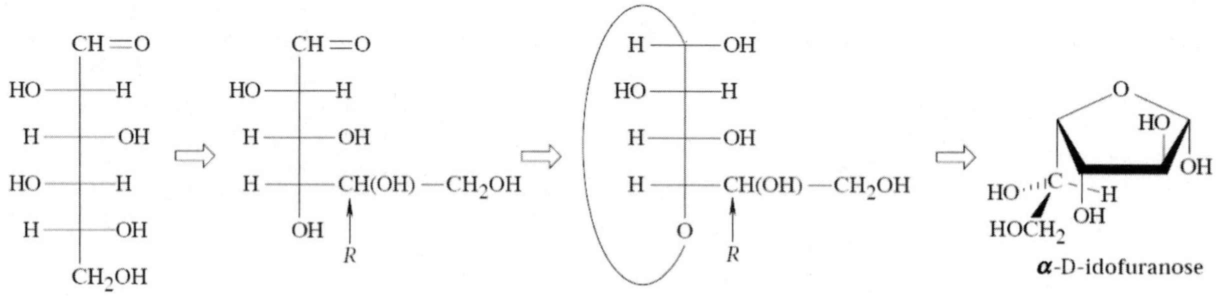

α-D-idofuranose

24.13   Each form of the sugar contributes its own optical rotation in proportion to the amount that is present. Let $N_i =$ the fraction of form i. (The dimensions of the specific rotations are abbreviated simply as degrees.)

$$N_{total} = 52.7° = N_\alpha(112.0°) + N_\beta(18.7°)$$

Since $N_\alpha + N_\beta = 1$, substitute $N_\beta = (1 - N_\alpha)$ and obtain

$$52.7° = N_\alpha(112.0°) + (1 - N_\alpha)(18.7°)$$

$$N_\alpha(112.0° - 18.7°) = 52.7° - 18.7° = 34.0°$$

$$N_\alpha = \frac{34.0°}{93.3°} = 0.36$$

$$N_\beta = 1 - N_\alpha = 0.64$$

Table 24.1 on text p. 1234 confirms that D-glucopyranose at equilibrium contains 36% of the α-form and 64% of the β-form.

24.14   (a)   D-Galactose would be transformed into the aldohexose that is epimeric at carbon-2, namely, D-talose. (See the structures in Fig. 24.3 on text p. 1240.) Also formed would be the ketose D-tagatose, which has the following structure:

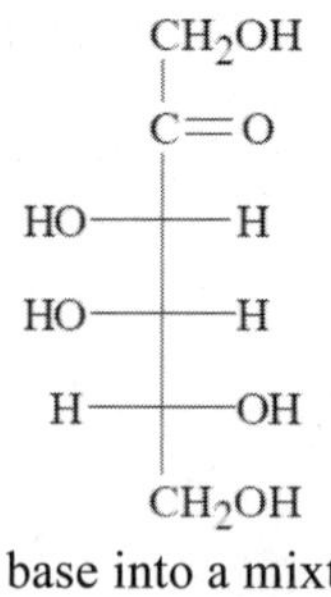

D-tagatose

(b)   D-Allose would be transformed by base into a mixture of D-altrose (structure in Fig. 24.3 on text p. 1240) and the ketose D-psicose.

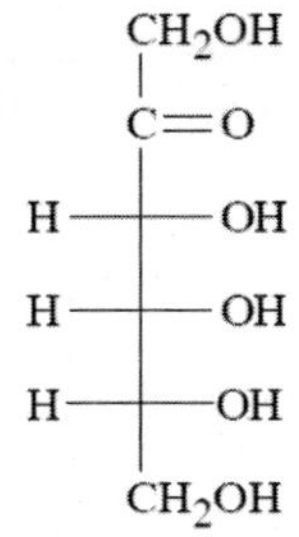

CH₂OH
|
C=O

H——OH

H——OH   D-psicose

H——OH

CH₂OH

24.15   (a)   The compound is the β-p-nitrophenyl glycoside of D-galactopyranose; its name is p-nitrophenyl β-D-galactopyranoside.

(b)   This compound is hydrolyzed in aqueous acid to p-nitrophenol and a (mutarotated) mixture of α-and β-D-galactopyranose.

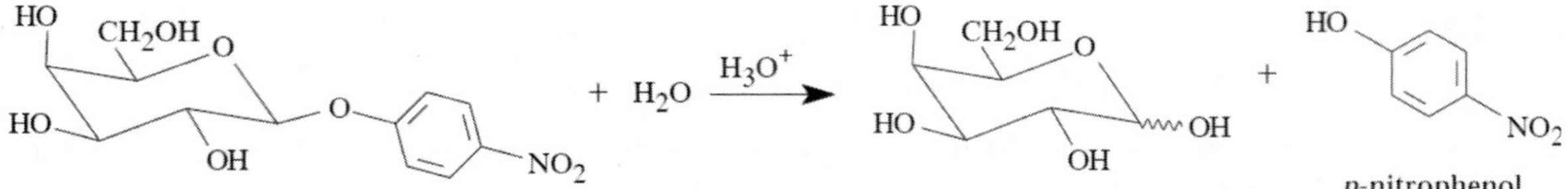

α- and β-D-galactopyranose   p-nitrophenol

24.16   Because naturally occurring glycosides generally have bonds between an alcohol or phenol oxygen and the anomeric carbon of a carbohydrate, it is reasonable to propose that the β-D-glycoside of vanillin and glucose has the following structure:

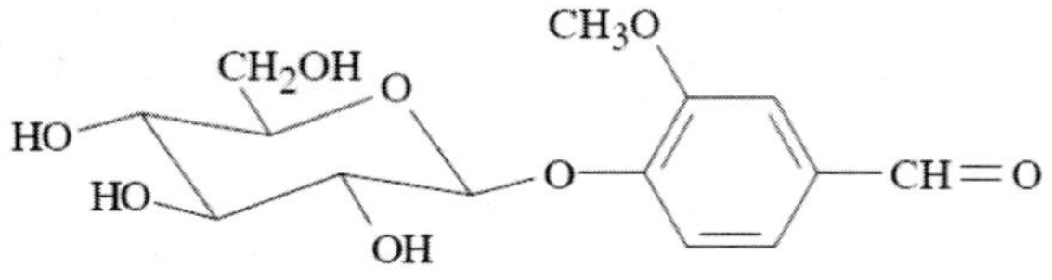

24.17   (a)   Start with the structure of β-D-fructofuranose on text p. 1249 and use it to draw the β-glycoside of methanol.

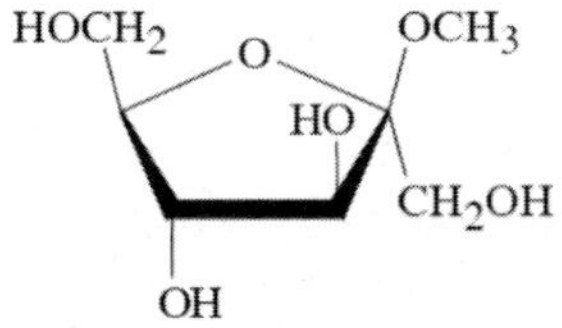

methyl β-D-fructofuranoside

(b)   Alkylate all of the hydroxy groups with benzyl chloride (PhCH₂Cl), a very reactive alkyl halide; then hydrolyze the benzyl glycoside. (Why is benzyl chloride very reactive? See Sec. 17.4, text p. 850.)

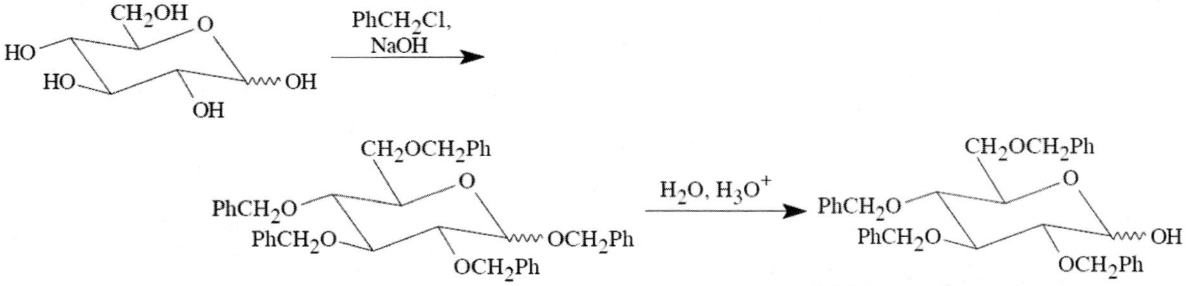

2,3,4,6-tetra-O-benzyl-D-glucopyranose

24.18   The protonation of an oxygen of the acetal group (the glycosidic oxygen in the example below) and loss of an alcohol (methanol in the example below) gives a *resonance-stabilized* carbocation intermediate. A similar mechanism applied to ordinary ethers gives carbocations that are not resonance-stabilized. Hammond's postulate implies that the hydrolysis reaction that involves the more stable carbocation intermediate should occur more rapidly.

glycoside protonated
on the C-1 oxygen

a resonance-stabilized carbocation

24.19 (a) Prepare the 1-*O*-ethyl derivative and use the Williamson synthesis to introduce the methyl groups.

EtOH,
HCl

$(CH_3)_2SO_4$,
NaOH

ethyl 2,3,4,6-tetra-*O*-methyl-
D-galactopyranoside

(b) Alkylate all of the hydroxy groups with benzyl chloride (PhCH₂Cl), a very reactive alkyl halide; then hydrolyze the benzyl glycoside. (Why is benzyl chloride very reactive? See Sec. 17.4, text p. 850.)

PhCH₂Cl,
NaOH

H_2O, H_3O^+

2,3,4,6-tetra-*O*-benzyl-D-glucopyranose

24.20 (a) (b) Notice that ribitol is a meso compound.

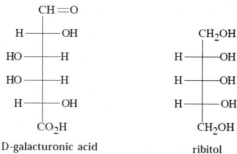

D-galacturonic acid

ribitol

24.21 The structures of D-glucaric and L-gularic acids are as follows. Turning either structure 180° in the plane of the page shows that the two structures are identical. (This identity was a key element in the Fischer proof of glucose stereochemistry, which is discussed in text Sec. 24.10A.)

CO₂H
H——OH
HO——H
H——OH
H——OH
CO₂H

D-glucaric acid
(derived from D-glucose)

CO₂H
HO——H
HO——H
H——OH
HO——H
CO₂H

L-gularic acid
(derived from L-gulose)

24.22   (a)   The structure of the aldaric acid derived from oxidation of D-galactose, and its 1,4-lactone:

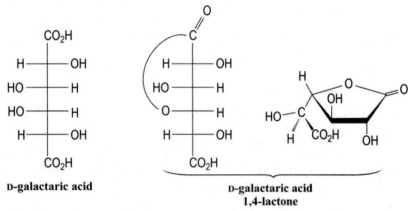

D-galactaric acid

**D-galactaric acid
1,4-lactone**

(b)   The structure of the aldaric acid derived from oxidation of D-mannose, and the structure of its 1,4-lactone:

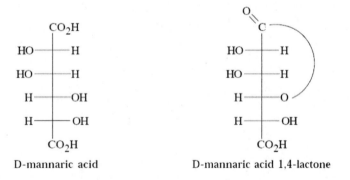

D-mannaric acid

D-mannaric acid 1,4-lactone

24.23   As is the case with carbohydrates, the primary alcohol groups are selectively oxidized by HNO₃.
(a)

HO''·····⬡——◀CO₂H

trans-3-hydroxycyclopen-
tane-1-carboxylic acid

(b)

HO₂C—CH₂CH₂—CO₂H

succinic acid

24.24   As Eq. 24.41a on text p. 1261 shows, periodate oxidation of a methyl pyranoside in the D-series gives a product containing carbons 1, 2, 4, 5, and 6 of the pyranoside. The secondary alcohols at carbons 2 and 4 are converted into aldehyde groups in all cases. In the oxidation products, only the carbons corresponding to carbons 1 and 5 in the pyranoside starting material are asymmetric. Since the configuration of carbon-5 determines whether the pyranoside has the D or the L configuration, and the configuration of carbon-1 determines whether the pyranoside is α or β, it follows that all methyl α-D-pyranosides give the same oxidation product with periodate.

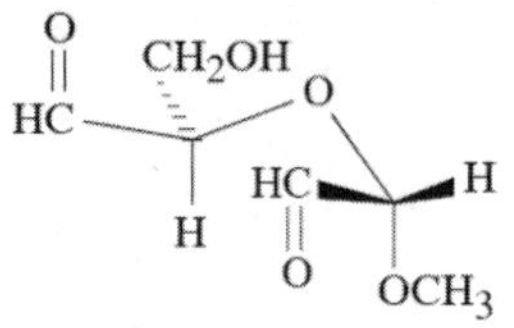

fragment obtained from
periodate oxidation of any
methyl α-D-pyranoside

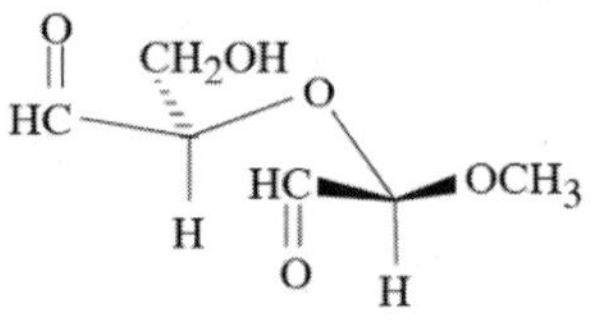

fragment obtained from
periodate oxidation of any
methyl β-D-pyranoside

(See the solution to Problem 24.53 on p. 905 of this manual for a more detailed consideration of the structures of the periodate oxidation products of carbohydrates.)

24.25   The discussion in the solution to Problem 24.24 (p. 892 of this manual) shows that the fragment obtained from oxidation of a β-D-pyranoside is the *diastereomer* of the fragment obtained from oxidation of an α-D-pyranoside. Diastereomers have different properties. Hence, these fragments, which can in principle be identified by their properties, including their optical rotations, can be used to determine whether the original pyranoside was α or β. This method works not only for the methyl galactopyranosides, but for *all* pyranosides. (Periodate oxidation was used to determine the anomeric configurations of many pyranosides before modern spectroscopic tools were available, and these results have withstood the scrutiny of modern analytical methods.)

24.26   The aldopentose *A* is D-ribose, and the aldohexose *B* is D-allose. The other hexose formed in the Kiliani–Fischer synthesis is D-altrose. (See Fig. 24.3 on text page 1240 for the structures of these aldoses. Note also in this figure the "Kiliani–Fischer relationships" among the aldoses.) Although D-xylose would also be oxidized to an optically inactive aldaric acid, its Kiliani–Fischer products, D-gulose and D-idose, would *both* be oxidized to optically active aldaric acids.

24.27   The aldopentose *A* is D-lyxose, and the aldohexose *B* is D-galactose. The other hexose *C* formed in the Kiliani–Fischer synthesis is D-talose. (See Fig. 24.3 on text page 1240 for the structures of these aldoses.) Although D-arabinose would also be oxidized to an optically active aldaric acid, its Kiliani–Fischer products, D-glucose and D-mannose, would both be oxidized to optically active aldaric acids.

24.28   Compound *A* is D-gulose, because it gives the same aldaric acid as L-glucose. The fact that some aldaric acids can be derived from two different aldoses is also addressed in the solution to 24.21 in this manual. The same point is discussed in Study Guide Link 24.2 on p. 877. In contrast, D-idose is the only aldohexose that can be oxidized to its aldaric acid.

24.29   Oxidize both (with dilute nitric acid) to their respective aldaric acids (which in this case are stereoisomeric tartaric acids). The compound that gives the optically active tartaric acid is D-threose; the compound that gives *meso*-tartaric acid (which is achiral and therefore optically inactive) is D-erythrose.

24.30   In such a case, the structure of D-(+)-glucose, indeed, the structures of *all* D-carbohydrates, would be the enantiomers of the structures shown in Fig. 24.3 on text p. 1240.

24.31   (a)   Aqueous HCl (1 *M*) would bring about the hydrolysis of lactobionic acid into one equivalent each of D-galactose and D-gluconic acid, which, under the acidic conditions, exists primarily as D-γ-gluconolactone (structure in Eq. 24.37 on text p. 1259).

         (b)   Dimethyl sulfate in the presence of NaOH methylates all of the hydroxy groups.

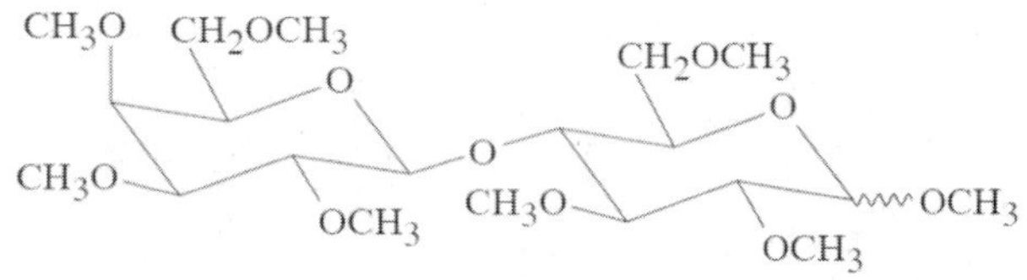

(c)   Hydrolysis of all acetal and hemiacetal groups (but not the other ethers) takes place.

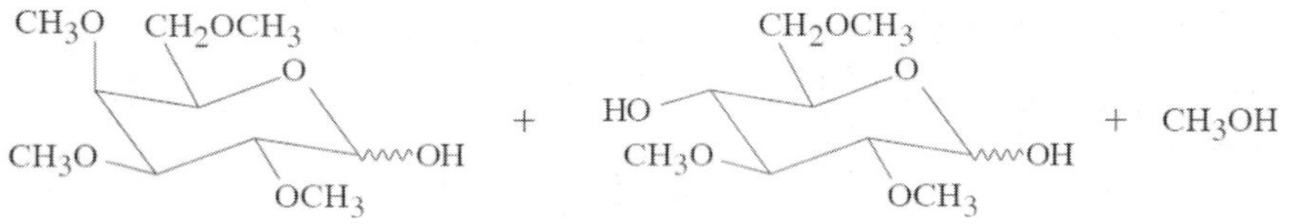

24.32   Because cellobiose is a dimer of D-glucose, it is hydrolyzed by aqueous acid into two molar equivalents of D-glucose.

24.33   The products would be a small amount of 2,3,4,6-tetra-*O*-methyl-D-glucopyranose (*A*, from the residue at the nonreducing end of the polymer) and mostly 2,3,6-tri-*O*-methyl-D-glucopyranose (*B*):

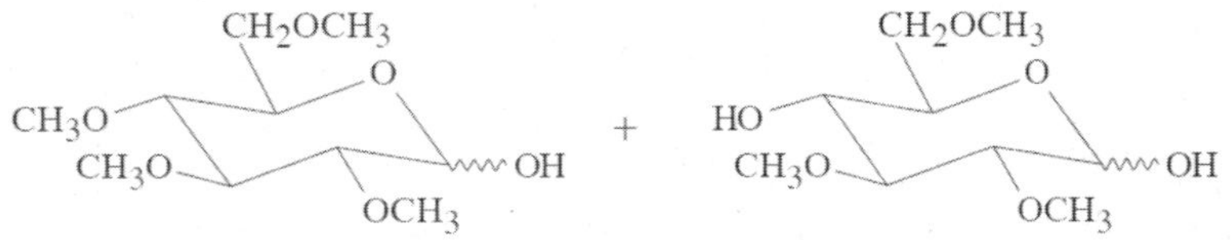

Solutions to Additional Problems

24.34   Note that carboxylic acids derived from carbohydrates usually exist as lactones, and aldoses exist primarily as cyclic hemiacetals. For simplicity they are represented below in their noncyclic forms.

(a)   (b)

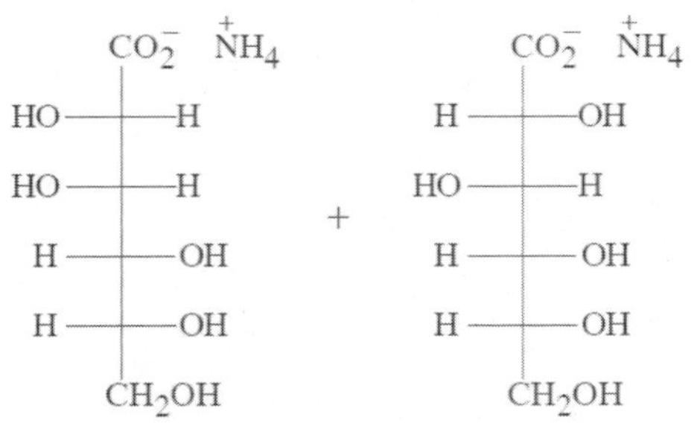

No reaction.

Note that in part (a), the basic conditions under which the Tollens reagent is used (aqueous ammonium hydroxide) promote the de Bruyn–van Ekenstein rearrangement (see Secs. 24.5 and 24.8A of the text). Thus, carbon-2 of D-mannose is epimerized.

(c)

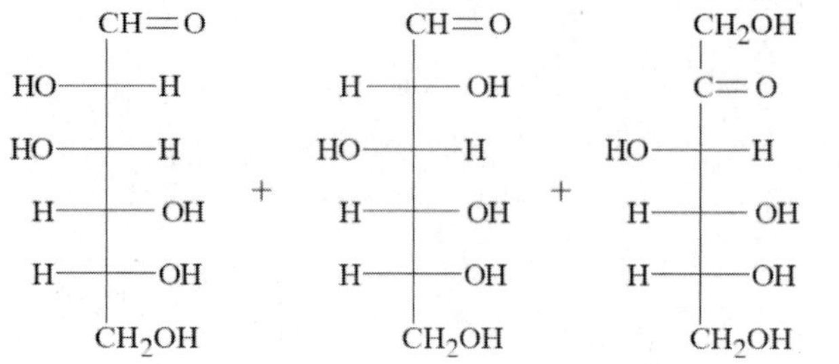

(d)

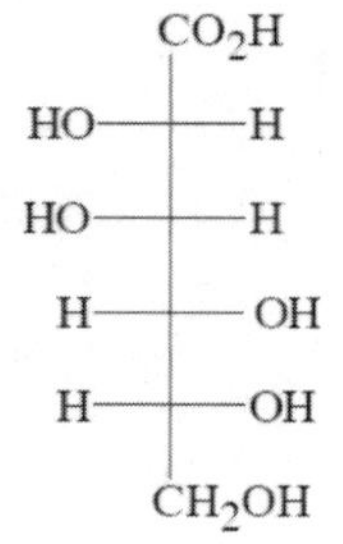

(e)

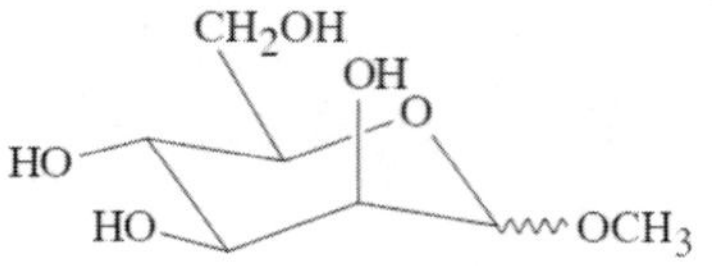

(f)

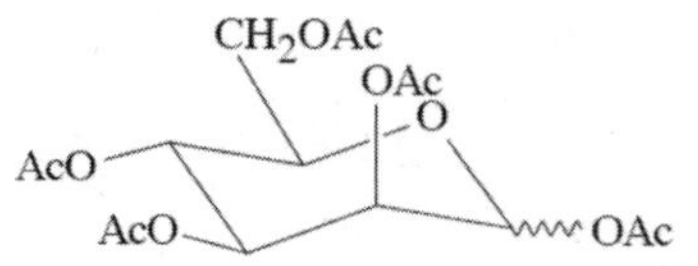

(g)

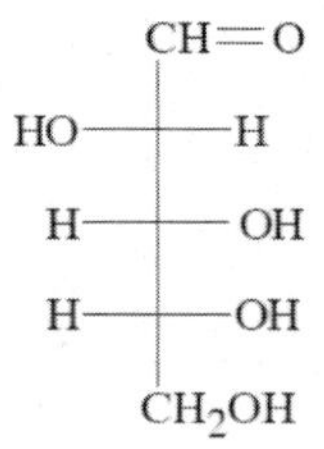

(h)

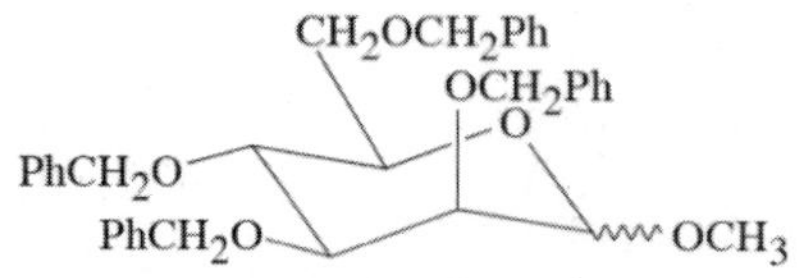

Reaction (g) is a Ruff degradation.

24.35   Note that carboxylic acids derived from carbohydrates usually exist as lactones, and aldoses exist primarily as cyclic hemiacetals. For simplicity they are represented below in their noncyclic forms.

(a)   (b) 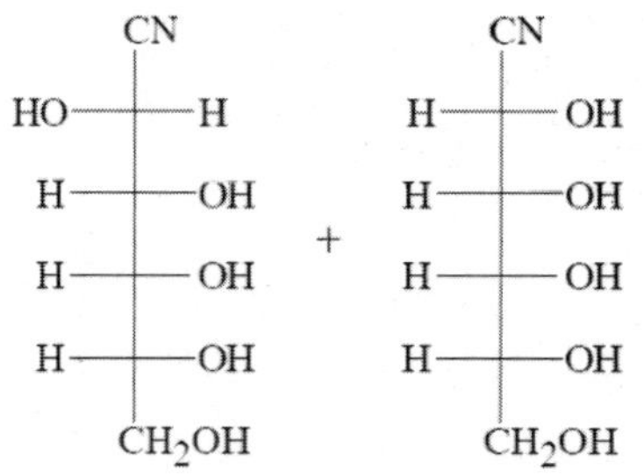   (c)

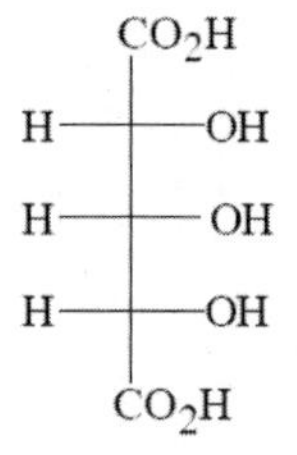

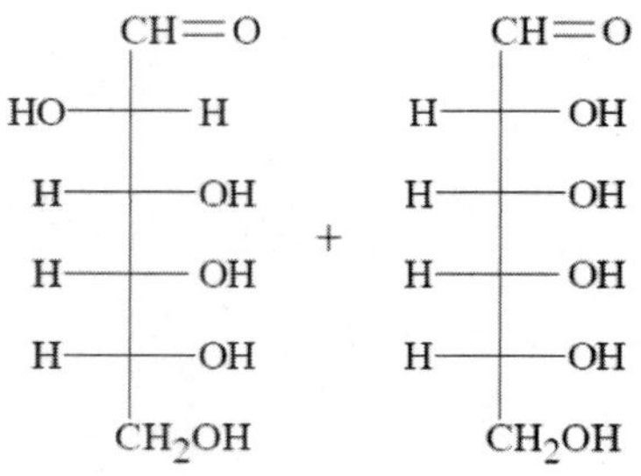

(d)

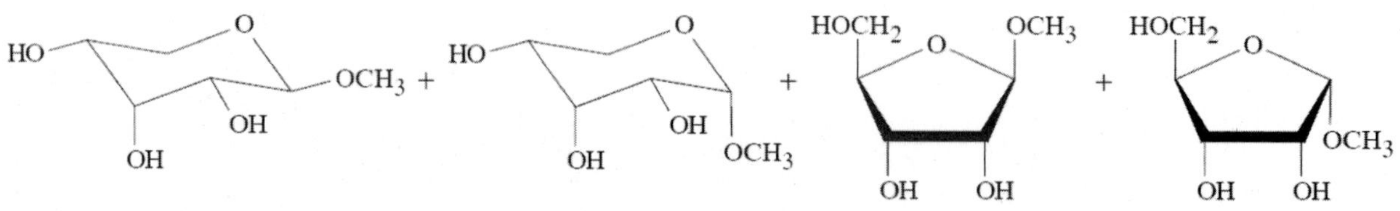

(e)

$$CH_3O \quad \cdots \quad O \quad OCH_3 \ + \ CH_3O \quad \cdots \quad O \quad + \quad CH_3OCH_2 \quad O \quad OCH_3 \quad CH_3OCH_2 \quad O \ +$$

24.36   (a)   (b)

α-D-talopyranose   propyl **β**-L-arabinopyranoside

 Remember that there are several valid ways to draw the chair conformations of pyranoses. Because the text probably has made you accustomed to seeing β-anomers with equatorial groups at carbon-1, you may be thinking that the chair conformation in the solution to part (b) shows the α-anomer. However, this *is* a correct representation of the β-anomer. Whether the β-anomer has an equatorial or axial group at the anomeric carbon depends on which chair conformation you choose to draw. The following equally valid representations of a different chair conformation of the same compound do have the group at the anomeric carbon in an equatorial position.

24.37   (a)   Turning this structure over 180° shows that it is the noncongruent mirror image of β-D-galactopyranose; therefore it is β-L-galactopyranose.

turn 180°

β-L-galactopyranose   β-D-galactopyranose

enantiomers

(b)   This pyranose is epimeric to β-D-glucopyranose at carbon-3. Consult Fig. 24.3 on text p. 1240 to discover that the aldohexose epimeric to glucose at carbon-3 is allose. Therefore, this compound is β-D-allopyranose.

(c)   First, notice that carbon-5 of this furanose has the *R* configuration. Since the D-hexoses have the *R* configuration at carbon-5 (the configurational carbon), and since the hemiacetal carbon bears a hydrogen, this must be a D-aldohexose. This can be converted into a Fischer projection as follows:

This shows that this compound is a form of D-glucose; it is β-D-glucofuranose.

24.38 (a) The D-series of the stereoisomeric 2-ketohexoses are shown below. (There is, of course, an enantiomeric L-series.) The problem didn't ask for the names, but they are given as part of our policy of extra value at no added cost.

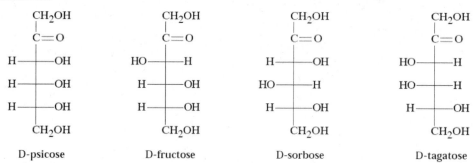

D-psicose   D-fructose   D-sorbose   D-tagatose

(b) An achiral ketopentose must be a meso compound:

$$\begin{array}{c} CH_2OH \\ H\!-\!\!-\!OH \\ C\!=\!O \\ H\!-\!\!-\!OH \\ CH_2OH \end{array}$$

(c) The structure of α-D-galactofuranose is derived from the Fischer projection as follows:

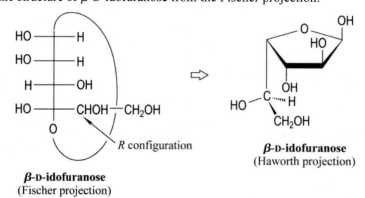

(d) Derivation of the structure of β-D-idofuranose from the Fischer projection:

This is identical to the structure in the solution to Problem 24.12b except for the configuration at the anomeric carbon.

24.39 (a) Diastereomers, epimers, and anomers. Remember that epimers and anomers are special types of diastereomers.
(b) Diastereomers and epimers
(c) Enantiomers
(d) Constitutional isomers

(e)   Constitutional isomers
(f)   The two names describe the same structure.

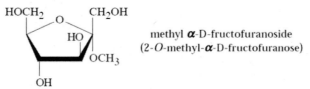

methyl α-D-fructofuranoside
(2-O-methyl-α-D-fructofuranose)

24.40   (a)   L-Sorbose is a hexose.
   (b)   L-Sorbose is a ketohexose, or hexulose.
   (c)   L-Sorbose is not a glycoside because it is not involved in acetal formation with an alcohol or a phenol.
   (d)   L-Sorbose is not an aldohexose; it is a ketohexose, as noted in the solution to part (b).
   (e)   This is another valid Fischer projection of L-sorbose obtained by a cyclic permutation at carbon-5.
   (f)   This compound is epimeric to L-sorbose at carbon-5, and thus it is not a form of L-sorbose. In fact, it is D-fructose.
   (g)   This is the α-furanose form of L-sorbose (α-L-sorbofuranose). The simplest way to see this is to work from the Haworth projection given in the problem to the Fischer projection:

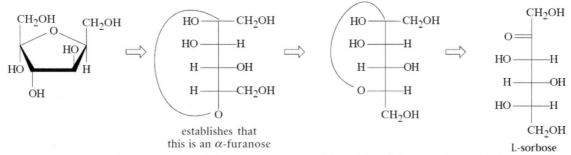

(h)   This structure is the enantiomer of the structure in part (g), and thus it is not a form of L-sorbose. (It is the α-furanose form of D-sorbose.)
   (i)   The relationship of this furanose to the one shown in part (g) is best discerned by turning either of the structures over 180°. This operation reveals that this structure differs in configuration from the one in part (g) at carbons 3 and 4. Hence, this is not a form of L-sorbose.
   (j)   This is an aldopyranose and therefore cannot be a form of L-sorbose. (It is β-D-glucopyranose.)
   (k)   This is the α-pyranose form of L-sorbose, α-L-sorbopyranose. As with part (g), the easiest way to see this is by working from the chair to the Haworth to the Fischer.

24.41   (a)   Raffinose is a nonreducing sugar because it has no hemiacetal groups; the saccharide residues at both ends are involved in acetal linkages.
   (b)   The glycoside linkages in raffinose:

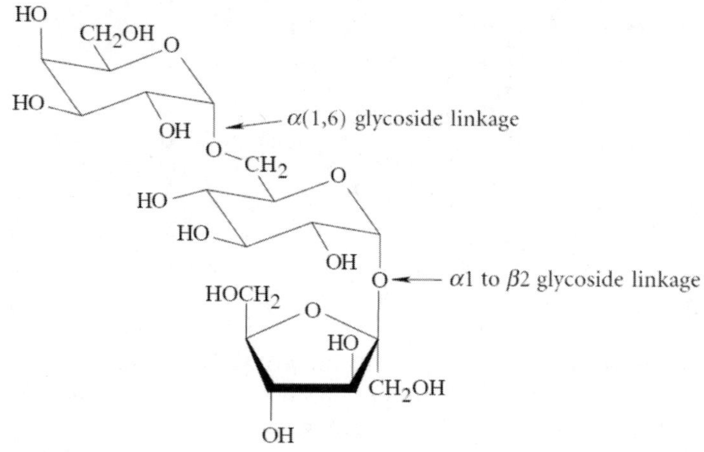

The type of linkage in the furanose residue can be determined systematically by using the Haworth projection to form the standard Fischer projection. (The glycosidic linkage is indicated as RO—.)

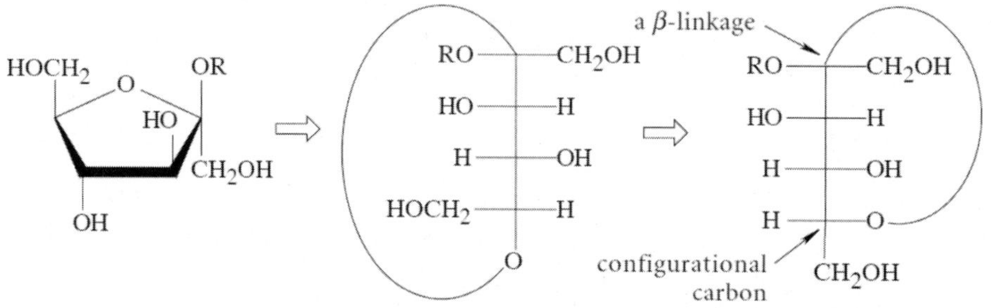

(c)   Raffinose is hydrolyzed into equal amounts of D-galactose, D-glucose, and D-fructose.

(d)   The products of methylation followed by hydrolysis are the following; each is a mixture of anomers.

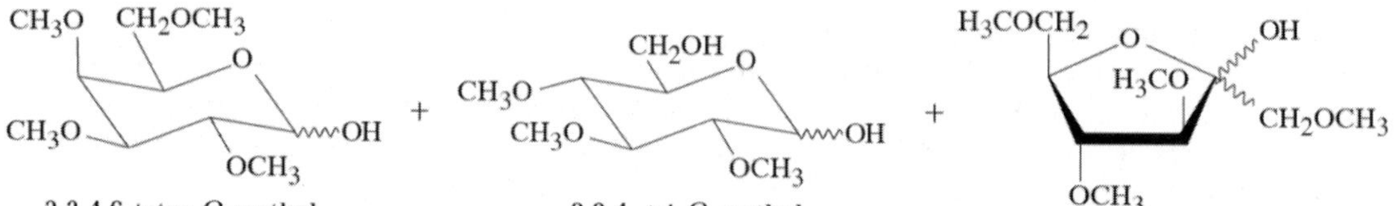

2,3,4,6-tetra-*O*-methyl-
D-galactopyranose

2,3,4,-tri-*O*-methyl-
D-glucopyranose

1,3,4,6-tetra-*O*-methyl-
D-fructofuranose

24.42   The structure of 3-*O*-β-D-glucopyranosyl-α-D-arabinofuranose:

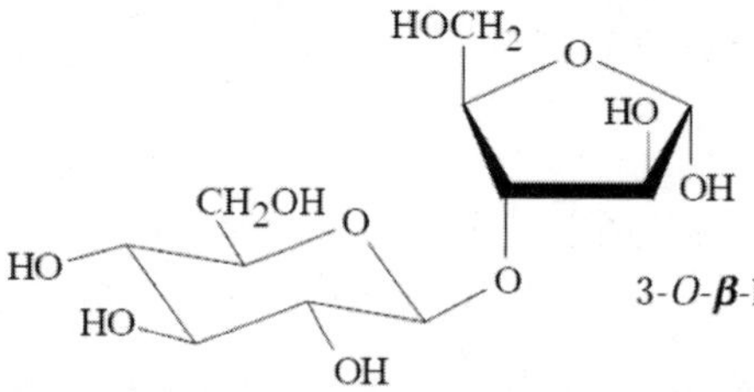

3-*O*-**β**-D-glucopyranosyl-**α**-D-arabinofuranose

24.43   (a)   Turn the chair conformation 180° about a vertical axis in the page to get the equivalent chair structure below. Then convert this to a Fischer projection as shown. These manipulations reveal that, when we draw the Fischer projection in standard form (carbon-6 vertical), the oxygen at carbon-5 is on the left. Hence, this enantiomer of fucose has the L configuration.

oxygen on left at C-5;
therefore L-configuration

(b)   Because the —OH at carbon-1 in this standard representation is on the same side as the oxygen at carbon-5, this is by definition the α-anomer.

(c)   Fucose is an aldose, because the hemiacetal carbon bears a hydrogen; thus, in the carbonyl form, there is also a hydrogen at carbon-1. Hence, the carbonyl form must be an aldehyde.

(d)   A Fischer projection of the aldehyde form of L-fucose:

L-fucose

 Fucose is classified as a *deoxy* carbohydrate; that is, it has one less —OH group than an ordinary aldohexose. (An aldohexose would have an —OH group at carbon-6.)

24.44   (a)   Because the osazones from the two carbohydrates are the same, and given that both are aldohexoses, the only stereochemical difference between them—a difference that is obliterated by osazone formation—is at carbon-2. Consequently, the two carbohydrates must be epimeric at carbon-2.

(b)   D-Ribose gives the same osazone as D-arabinose because D-ribose is epimeric to D-arabinose at carbon-2.

24.45   (a)   Just as the acetal units in carbohydrates undergo *hydrolysis* in water, they also undergo *methanolysis* in methanol. The result is a mixture of the methyl glycosides along with the by-product phenol.

phenyl **β**-D-glucopyranoside   →   methyl D-glucopyranoside
                                       (mixture of α and β)

(b)   Any 3- or 4-hydroxy aldehyde, like a carbohydrate, will form a cyclic acetal in acidic methanol.

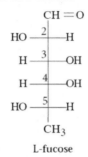

(c)   Osmium tetroxide forms *cis*-1,2-cyclohexanediol; periodic acid cleaves this diol to form the dialdehyde 1,6-hexanedial, $O{=}CHCH_2CH_2CH_2CH_2CH{=}O$; and this dialdehyde is reduced by sodium borohydride to 1,6-hexanediol, $HOCH_2CH_2CH_2CH_2CH_2CH_2OH$.

(d)   Cyclohexene reacts with OsO_4 to give *cis*-1,2-cyclohexanediol, which forms a cyclic acetal with acetone.

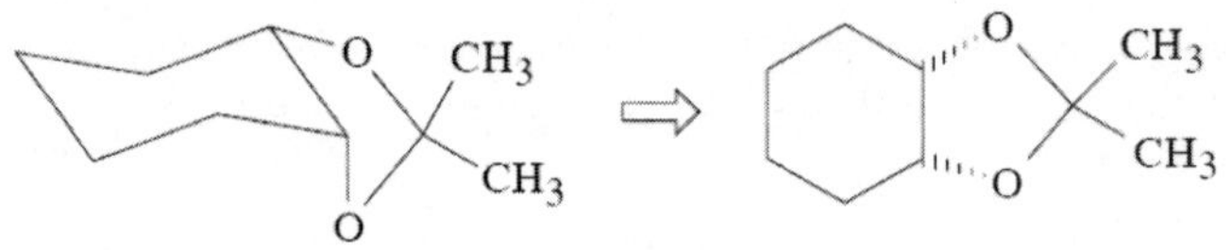

(e) All of the free hydroxy groups in sucrose are methylated. (For the structure of (+)-sucrose see text p. 1207.)

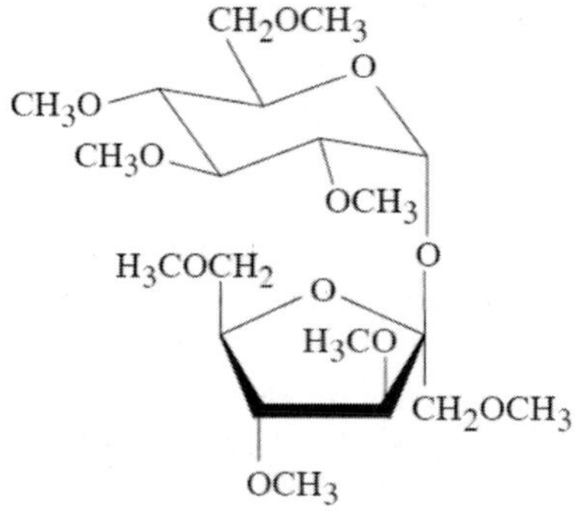

(f) Ethanolysis of the acetal linkage occurs in ethanol just as hydrolysis occurs in water. An ethyl pyranoside is also formed from the glucose residue at the reducing end of lactose.

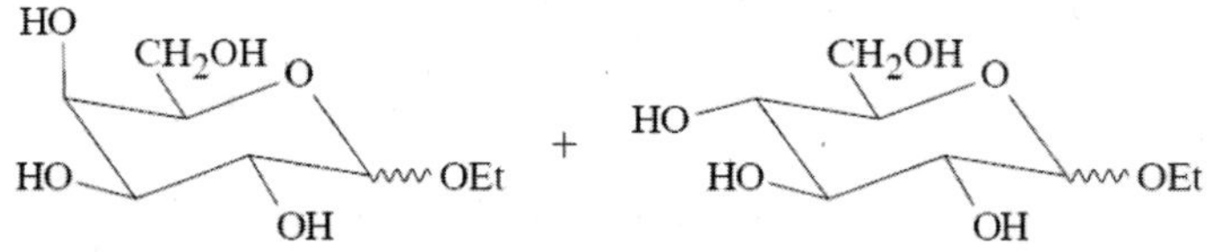

(g) This reaction is not unlike the one in part (a), in which one acetal is converted into another in methanol. Acid-catalyzed loss of methanol is followed by cyclization to give a mixture of the two methyl pyranosides.

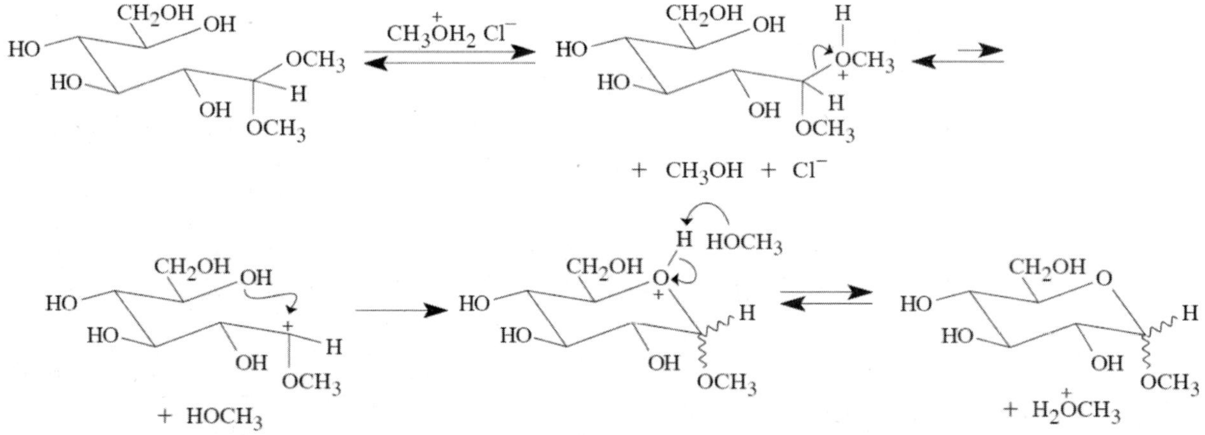

$$+ CH_3OH + Cl^-$$

(Note that $CH_3\overset{+}{O}H_2\ Cl^-$, the conjugate acid that results from protonation of the solvent by HCl, is the effective acid catalyst.)

24.46 (a) This is an isotopically labeled analog of glucose, and is prepared by the Kiliani–Fischer synthesis shown in Eqs. 24.44 and 24.45 on text pp. 1262–3, except that radioactive sodium cyanide (Na[14]CN) is used instead of ordinary NaCN. In this synthesis, ^{14}C-D-mannose will be a by-product.

(b) This compound is prepared in the same way as the compound in (a), except that ordinary (unlabeled) NaCN is used, and the reduction is carried out with tritium-enriched hydrogen ([3]H_2). In this synthesis, ^{3}H-D-mannose will be isolated as a by-product.

(c)   The synthesis of compound (c) requires removal of an *unlabeled* carbon from D-arabinose, replacing it with a labeled carbon, and then completion of an ordinary Kiliani–Fischer synthesis to give labeled mannose. (* = ^{14}C)

Each of the last two reactions also gives epimeric sugars—D-ribose and D-glucose, respectively—as by-products.

24.47   The results of the oxidation of *A* indicate that compound *A* is not a methyl pyranoside or furanoside, that is, the methyl ether is not at carbon-1. The Ruff degradation yields another reducing sugar. This means that the methyl ether cannot be at carbon-2. Only if the ether is at carbon-3 can one of the Kiliani–Fischer products be oxidized to an optically inactive (that is, a meso) compound:

24.48   (a)

D-(+)-galactose   D-lyxitol

(b)   D-Galactose, an aldohexose, exists predominantly in the pyranose form. The decarbonylation reaction only occurs when D-galactose is in the aldohexose form; the increased temperature (130 °C) is required to increase the rate at which the reactions in the pyranose $\rightleftarrows$ aldohexose occur. This strategy ensures that, once the small amount of the aldehyde form reacts with the organometallic reagent, the aldehyde form in the equilibrium is rapidly replenished.

CH=O
H——OH
HO——H
HO——H
H——OH
CH₂OH

D-galactose
(reacts with the
organometallic reagent)

⇌

β-D-galactopyranose
(does not react)

(c)   In the following mechanism, L = Ph₃P (the triphenylphosphino ligand).

Rh
rhodium (I)
a 16e⁻ complex

→ *ligand dissociation* →

Rh + L
rhodium (I)
a 14e⁻ complex

R—C—H + Rh → *ligand association* → R—C—H Rh → *oxidative addition* →

R—C Rh
H | Cl
L
rhodium (III)
a 16e⁻ complex

→ *alkyl migration (β-elimination of an alkyl group)* →

R—Rh CO
H | Cl
L

→ *reductive elimination* →

R—H + Rh CO
rhodium (I)
a 16e⁻ complex

By using special ligands and by altering the reaction conditions, investigators have been able to turn this into a catalytic process rather than one that is stoichiometric in rhodium.

24.49   Since the product of the decarbonylation reaction is optically inactive, compound *B* must be *meso*-(2*R*,3*S*)-butane-1,2,3,4-tetraol (erythritol) in which all of the —OH groups are on the right side in the Fisher projection, Thus, the aldopentose *A* is either D-ribose or D-arabinose (see Fig. 24.3, text p. 1240).

CH=O
?——?
H——OH
H——OH
CH₂OH

—(Ph₃P)₃RhCl→

CH₂OH
H——OH
H——OH
CH₂OH

erythritol
(a meso compound)
optically inactive

The products from the Kiliani–Fisher synthesis would then be either D-allose and D-altrose or D-glucose and D-mannose (see Fig. 24.3, text p. 1240). Since both of the products from the nitric acid oxidation are optically active, that means that D-allose could not be a product from the Kiliani–Fisher synthesis because it would produce an optically inactive aldaric acid. Thus, *A* is D-arabinose, *C* and *D* are D-glucose and D-mannose, and *E* and *F* are D-glucaric acid and D-mannaric acid. (See Eqs. 24.44–45, text pp. 1262–3, for the Kiliani–Fisher synthesis using D-arabinose as the starting aldose.)

```
        CH=O                        CH=O      CH=O      CO₂H      CO₂H
                                  H——OH    HO——H     H——OH    HO——H
    HO——H          CH₂OH       HO——H     HO——H     HO——H     HO——H
     H——OH      H——OH          H——OH     H——OH     H——OH     H——OH
     H——OH      H——OH          H——OH     H——OH     H——OH     H——OH
        CH₂OH       CH₂OH          CH₂OH     CH₂OH     CO₂H      CO₂H

         A            B              C         D          E          F
    D-arabinose    erythritol    D-glucose  D-mannose  D-glucaric  D-mannaric
                                                          acid        acid
```

24.50 The extract of mouse spleen evidently contains enzymes that catalyze the removal of a phosphate as well as the conversion of ribose into an optically inactive isomer. Ribitol and xylitol are the alditols derived from ribose and xylose, respectively, and therefore are epimeric at carbon-3. Compound *X* has one degree of unsaturation, and, because reduction gives alcohols that are epimeric at carbon-3, compound *X* is a 3-ketopentose. The structure for *X* shown in the following transformation fits the data:

```
        CH=O                           CH₂OH                    H    O
                                                                 \  //
     H——OH        extract of        H——OH        periodic         C
H₂O + H——OH       mouse spleen      C=O           acid            |
     H——OH       ───────────►       H——OH        ───────►        C=O  +  2 H₂C=O
        CH₂OPO₃²⁻                       CH₂OH                      |
                                                                 C
                                                                / \\
   ribose-5-phosphate               compound X              H      O

                                    +  HOPO₃²⁻
```

24.51 Compound *A* is the corresponding oxime (see Table 19.3 on text p. 985), and compound *B* results from acetylation of the hydroxy groups of the oxime. Compound *C* results from a sodium acetate-promoted β-elimination reaction. The conversion of compound *C* to D-arabinose is the result of a transesterification reaction, which removes all acetyl groups as methyl acetate, followed by the reverse of cyanohydrin formation.

Identity of compounds A and B:

```
        CH=N—OH                          CH=N—OAc
     H——OH                            H——OAc
    HO——H              A             AcO——H              B
     H——OH                            H——OAc
     H——OH                            H——OAc
        CH₂OH                            CH₂OAc
```

The mechanism for the conversion of compound B into compound C:

```
    AcO⁻   H
            ⏜
         C≡N—OAc                                C≡N
      H——OAc              ───────►   AcO—H  +  H——OAc

          B                                      C
```

The mechanism for the conversion of compound C into D-arabinose:

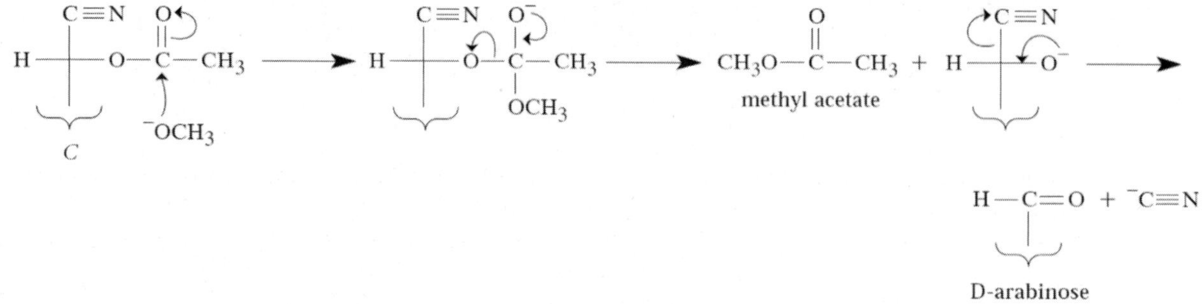

Note that the other acetate groups are removed in transesterification reactions, the mechanisms of which are identical to the first two steps of the foregoing mechanism.

24.52 Compound *A* is the lactone of the corresponding aldonic acid, and compound *B* is the amide that results from ester (lactone) ammonolysis. Chlorine and NaOH bring about a Hofmann rearrangement of amide *B* to give *C*, a carbinolamine, which, under the aqueous reaction conditions, breaks down spontaneously to ammonia and an aldehyde. (See Eq. 19.60a on text p. 984.) This aldehyde is the aldose that has one fewer carbon atom than the starting aldose. (This is illustrated below with D-glucose, but any aldose whose aldonic acid can form a lactone would undergo the same chemistry.)

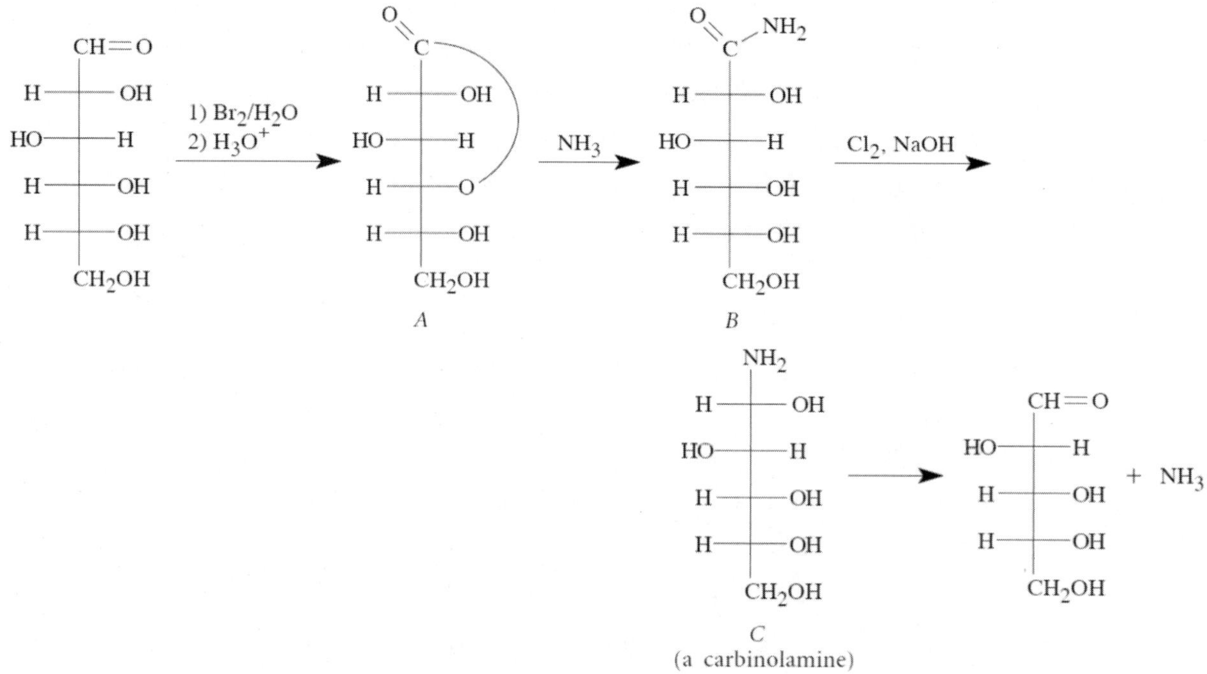

1) Br$_2$/H$_2$O
2) H$_3$O$^+$

NH$_3$

Cl$_2$, NaOH

+ NH$_3$

C
(a carbinolamine)

24.53 The product *X* might be reasonably expected to result from treatment of methyl α-L-rhamnopyranoside with periodic acid. (See the solution to Problem 24.23 on p. 647 of this manual.)

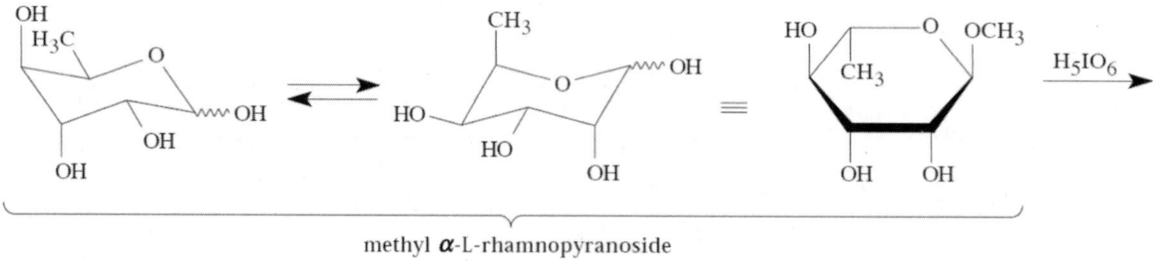

H$_5$IO$_6$

methyl **α**-L-rhamnopyranoside

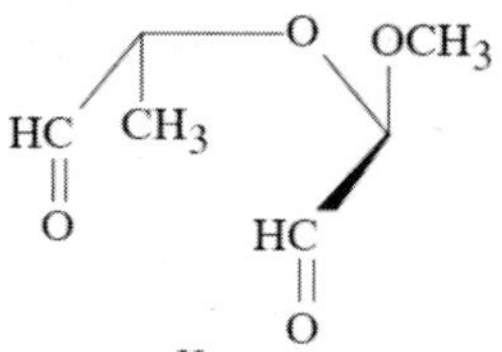

X

The formula of *X*, $C_6H_{10}O_4$, is short of the formula of actual product *A* by the elements of H_2O. Nevertheless, reduction of this product by $NaBH_4$ would give compound *C*. Evidently, compound *A* has two —OH groups, because methylation introduces two carbon atoms. Compound *A* results, first, from hydration of one of the carbonyl groups of *X*; an —OH group of the hydrate then reacts with the other carbonyl carbon to give a "double hemiacetal" *A*. The structure of *A* and its methylation product *B* are as follows:

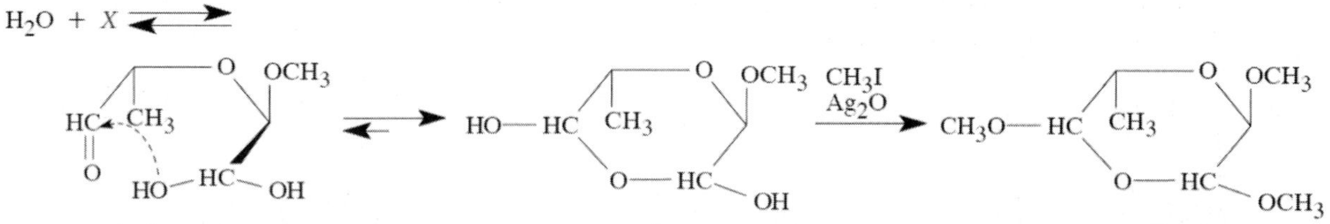

$$H_2O + X \rightleftharpoons$$

a hydrate of *X*   *A*   *B*

(Compound *B* is undoubtedly a mixture of stereoisomers.) Because compound *A* is a hemiacetal, it shows no carbonyl absorption. However, it is in equilibrium with a small amount of the dialdehyde *X*, which is reduced by $NaBH_4$ to give compound *C*. The reduction pulls the equilibrium between *A* and *X* toward *X* until all of this compound has been reduced.

24.54   The residue at the nonreducing ("left") end undergoes the following transformation:

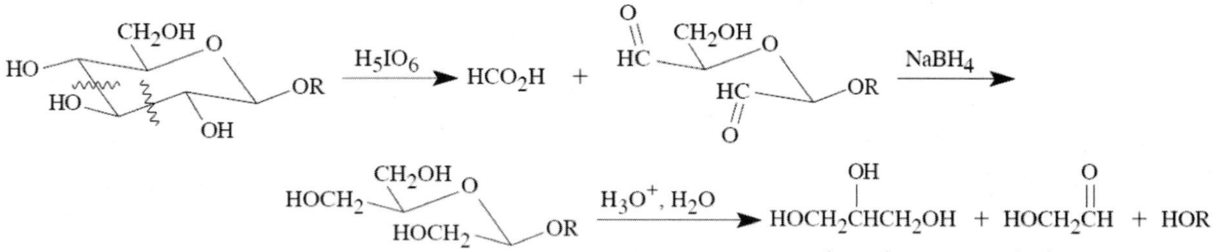

The internal residues undergo the following transformation:

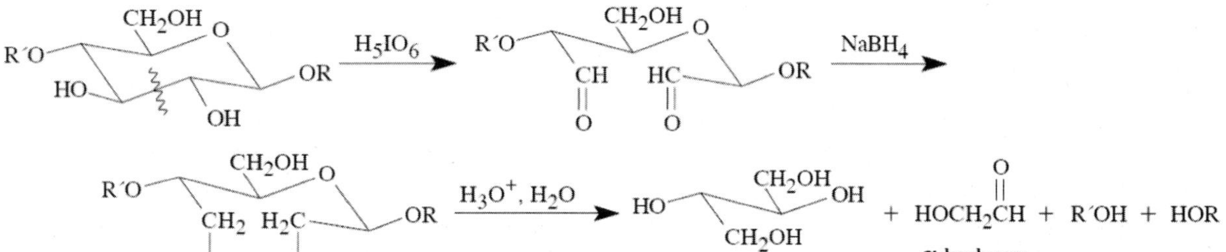

The residue at the reducing ("right") end gives erythritol, R'OH, and two equivalents of formic acid (HCO_2H). (Demonstrate this for yourself; see Eq. 24.40 on text p. 1261.) Erythritol therefore results from the cleavage reactions of all residues except for the one at the nonreducing end; and glycerol results from cleavage of that residue. Thus, glycerol is produced from one residue, and erythritol is produced from $n - 2 + 1 = n - 1$ residues. Since $n = 12$, eleven moles of erythritol per mole of glycerol are formed.

24.55 The hydrolysis of maltose by α-amylase shows that maltose contains two glucose residues connected by an α-glycosidic bond. Because maltose is a reducing sugar, a free hemiacetal group must be at carbon-1 of one of the glucose units. The 2,3,4,6-tetra-*O*-methyl-D-glucose must arise from the glucose residue at the nonreducing end. The question is which oxygen in the glucose residue at the reducing end is involved in the glycosidic linkage; the other product of methylation-hydrolysis provides evidence on this point. Because the oxygens at carbons 4 and 5 are not methylated in this product, one of these oxygens is involved in pyranoside or furanoside ring formation, and the other is involved in the glycosidic linkage. Two structures satisfy these requirements:

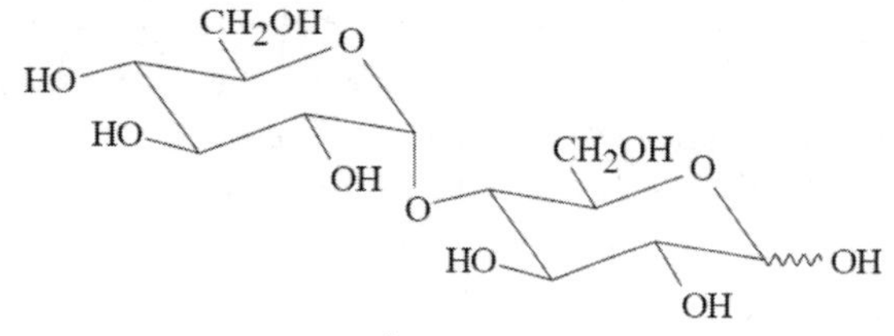
A

and

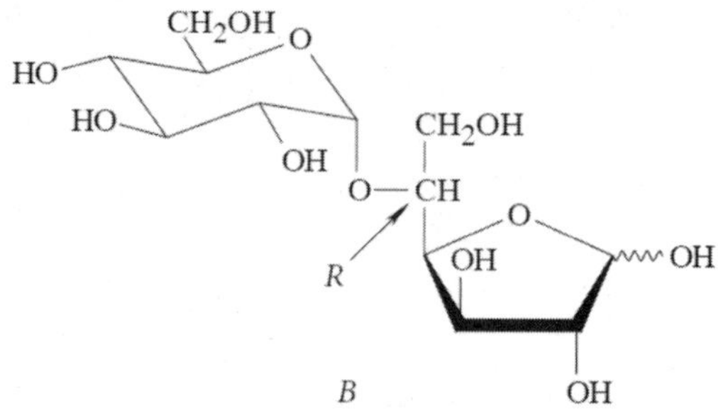
B

Methylation of maltobionic acid occurs under basic conditions; under these conditions any lactone present would be saponified. Hence, the additional oxygen that is methylated under these conditions must be the oxygen that was *within* the glycoside ring at the reducing end of maltose. In other words, oxidation and saponification expose this oxygen as an —OH group and thus make it available for methylation. Since the oxygen at carbon-5 is methylated, it must have been the oxygen within the saccharide ring. Since the oxygen at carbon-4 is *not* methylated, it must have been the oxygen involved in the glycosidic bond. Hence, structure *A* is the correct one for maltose. To summarize the chemistry using the correct structure:

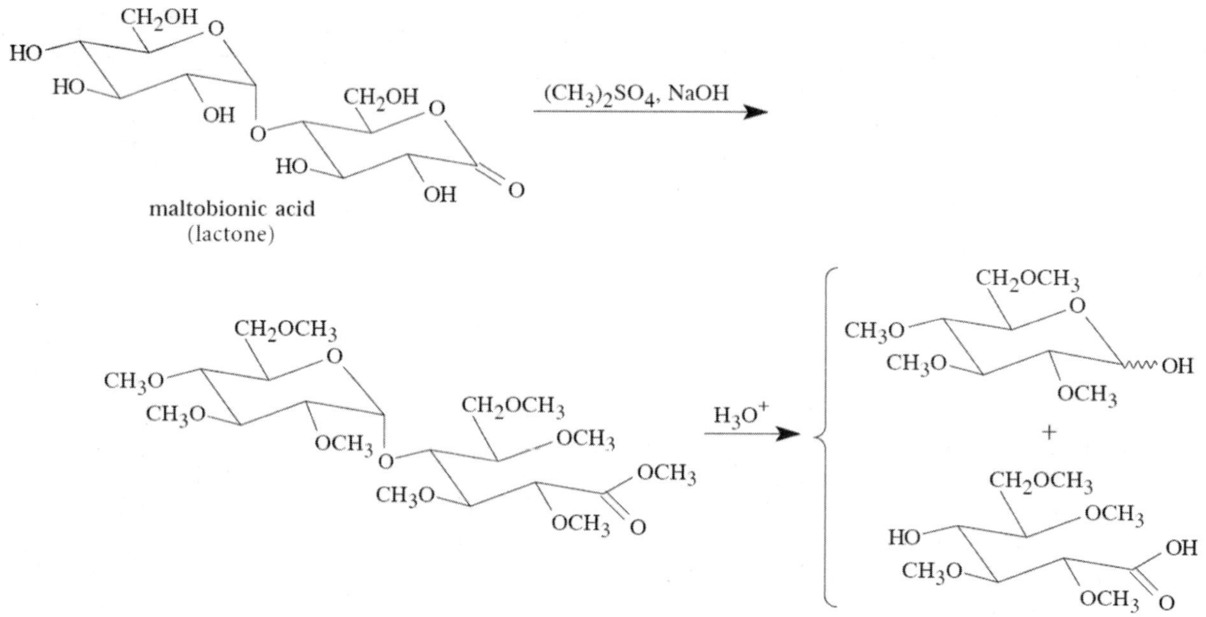

maltobionic acid (lactone)

24.56 The hydrolysis of planteose by almond emulsin establishes that an intact sucrose unit is connected to a galactose residue by an α-galactosyl glycosidic bond. The question, then, is which oxygen of the sucrose residue is involved in the glycosidic linkage to the galactose. Methylation followed by hydrolysis provides the answer. (It will be helpful to refer to the structure of sucrose on text p. 1271.) If the fructose residue of sucrose were *not* connected to the galactose, it would have four oxygens available for methylation: the oxygens at carbons 1, 3, 4, and 6. Because the oxygen at carbon-6 is not methylated, it must be the point of attachment of the galactose residue. Hence, the structure of planteose is as follows:

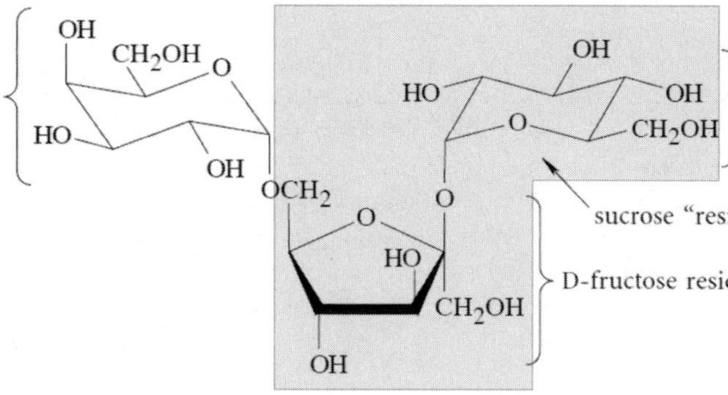

planteose

24.57 (a) The hydroxy groups of cellulose react with the anhydride to introduce succinate half-esters into the cellulose structure, as follows. (In these equations. HO—cellulose means any of the hydroxy groups of cellulose.)

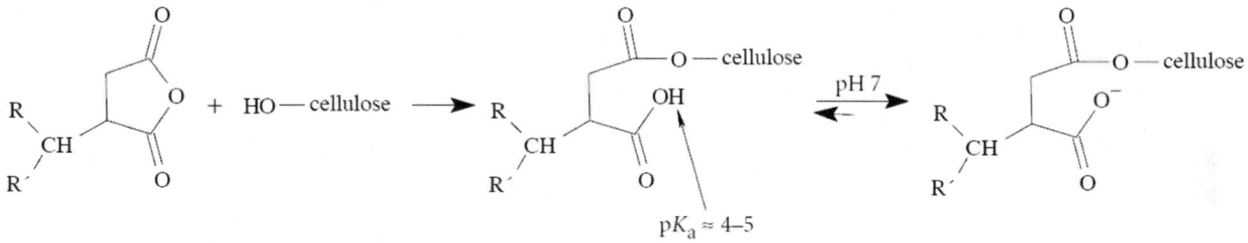

(b) Sizing introduces a large number of alkyl groups into the cellulose structure. These resist solvation by water for the same reasons that hydrocarbons are insoluble in water; they are "hydrophobic" groups. Hence, these groups tend to resist the incorporation of solvent water into the cellulose structure. Furthermore, many of the —OH groups in the cellulose, which are hydrogen-bond donors before the reaction, become ester oxygens as a result of the reaction and therefore can no longer donate hydrogen bonds.

(c) As the reaction in the solution to part (a) shows, the carboxylic acid groups introduced by the reaction are ionized by the pH 7 treatment because their pK_a values are well below the pH of the solution. When the paper is dried, these groups are left in their ionized state (with an appropriate counter-ion). The carboxylate groups are the slightly basic groups.

For example, when they come into contact with water, they react with water to produce a small amount of hydroxide ion:

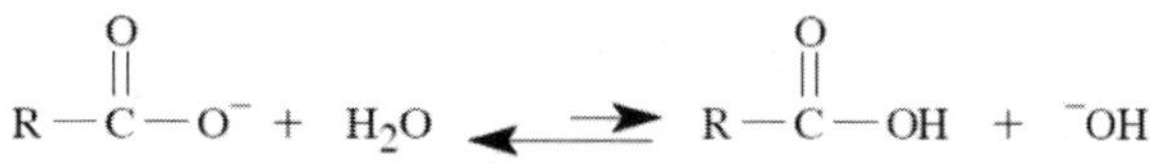

You can use the pK_a value for the carboxylic acid and the ion-product constant of water to calculate that the pH of an aqueous solution of a carboxylic acid salt should have a pH of around 9. Hence, when the "sized" paper comes into contact with acid, the acid is neutralized by the small amount of hydroxide ion present. So many carboxylate groups are present that the hydroxide ion is immediately replenished, by Le Châtelier's principle.

24.58 The reaction of an amine with an aldose gives a carbinolamine, which dehydrates to an iminium ion *A*. This ion is trapped by the hydroxy group at carbon-5 to give an aminoglycoside, which is shown as its conjugate acid *B*. (The *p*-methylphenyl group of the amine is abbreviated R.)

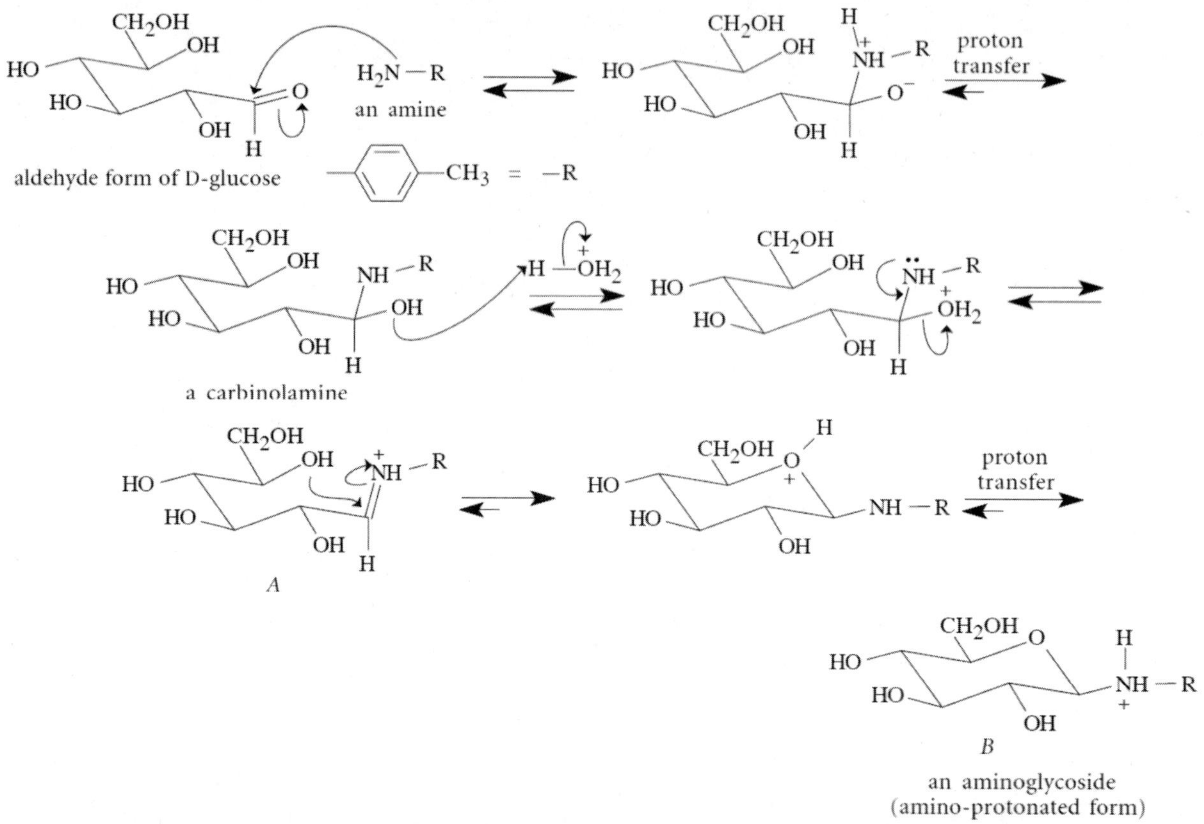
aldehyde form of D-glucose
an amine
proton transfer
a carbinolamine
A
proton transfer
B
an aminoglycoside
(amino-protonated form)

This reaction is called the *Maillard reaction*. From here on, only the part of D-glucose that is involved in the reaction is shown explicitly. The iminium ion *A* (which, by resonance, is also a carbocation) isomerizes to an enamine *C*, which is also an enol. Acid-catalyzed formation of the isomeric ketone completes the reaction. This isomerization is called the *Amadori rearrangement*.

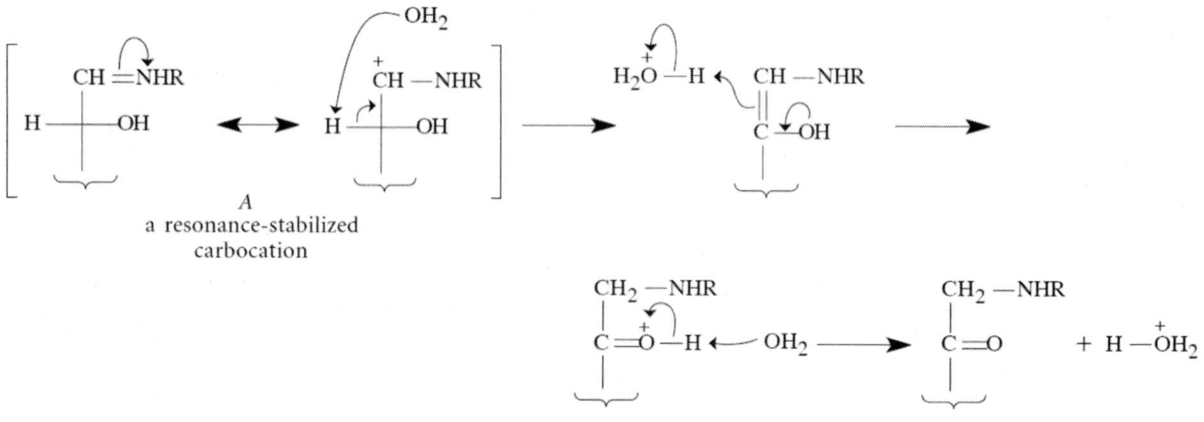
A
a resonance-stabilized carbocation

This reaction is similar to the Lobry de Bruyn–Alberda van Ekenstein rearrangement (text pp. 1251–2), except that imine and enamine intermediates are involved rather than aldehyde and enol intermediates. The driving force for the rearrangement is that the C=N double bond is lost and a stronger C=O double bond is formed.

24.59 This transformation is essentially a Lobry de Bruyn–Alberda van Ekenstein reaction, the detailed mechanism of which is shown in Eqs. 24.23–24.24b on text p. 1252. The aldehyde form of D-glucosamine is in equilibrium with a small amount of its enol *A*. This enol is also an enamine, which is in equilibrium with an imine *B*. Because imine formation is reversible in aqueous acid and base, the imine hydrolyzes to give the corresponding ketose, D-fructose, along with the by-product ammonia. Under the basic conditions, D-fructose is also in equilibrium with D-glucose and D-mannose, as shown on text p. 1251.

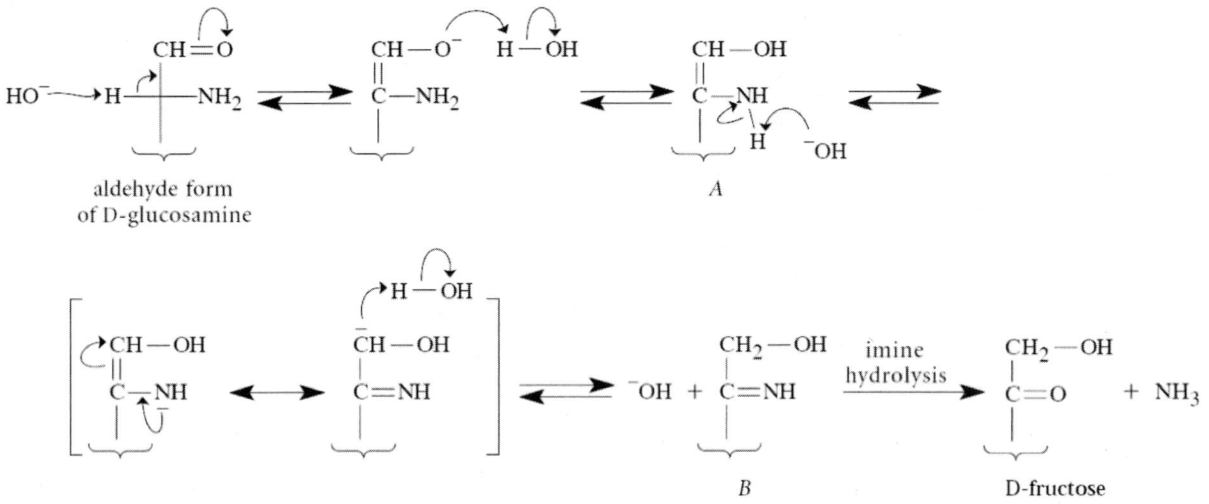

aldehyde form
of D-glucosamine

A

B   D-fructose

24.60   (a)   The acidic hydrogen is the one that ionizes to give a conjugate-base anion in which the negative charge is delocalized into the carbonyl group. This anion is stabilized by the electron-withdrawing polar effects of nearby oxygens and by the same type of resonance interactions that are present in a carboxylate ion. Consequently, ascorbic acid is about as acidic as a carboxylic acid.

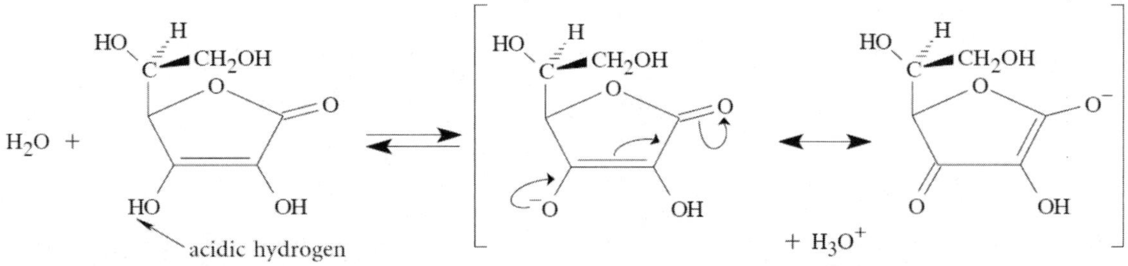

acidic hydrogen $+ H_3O^+$

   (b)   The reaction sequence with the missing structures added begins with compound A. The fact that carbon-2 of L-sorbitol (or carbon-5 of D-glucitol) becomes the acetal carbon as a result of the reaction labeled "C" in the problem shows that the —OH group on this carbon is the one that is oxidized; this carbon is therefore at the ketone level of oxidation. $KMnO_4$ oxidizes the —CH$_2$OH group to a carboxylate; because the other —OH groups are protected as acetals, they are stable under the basic conditions of permanganate oxidation. Hydrolysis of the oxidation product in acid removes the acetal protecting groups, protonates the carboxylate group, and promotes lactone formation.

      Can you think of one or more reasons why ascorbic acid should exist as an enediol rather than as an α-keto lactone? (See Sec. 22.2, and the discussion of carbonyl dipoles on text p. 919.)

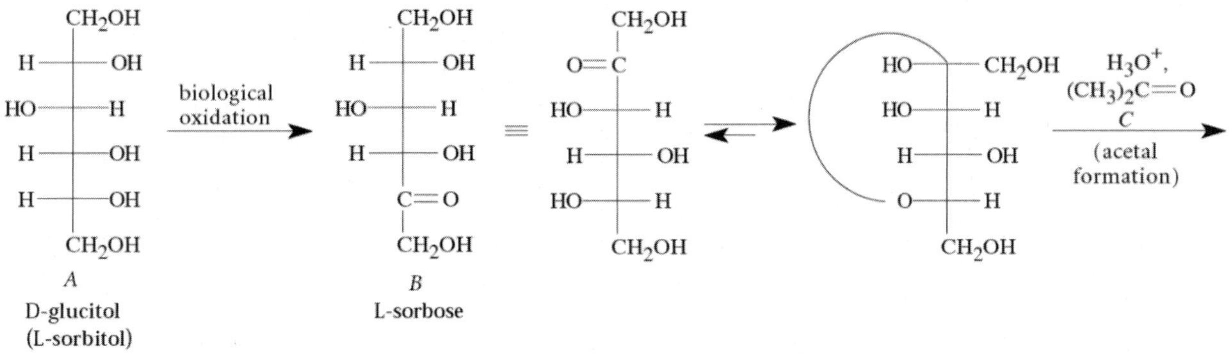

A   B
D-glucitol   L-sorbose
(L-sorbitol)

The reaction sequence shows a bicyclic acetal-protected structure being oxidized by $KMnO_4$, ^-OH to give a carboxylate (CO_2^-), then treated with H_3O^+, H_2O (lactone formation), giving intermediate products including $+ 2 (CH_3)_2C{=}O$ and, after enolization, L-ascorbic acid.

24.61 **(a)** The anhydro form of D-idose has the following chair conformation:

1,6-anhydro-D-idopyranose

(b) The issue is the relative stabilities of the anhydro and ordinary pyranose forms of D-idose and D-glucose. In the chair conformation of 1,6-anhydro-D-glucopyranose, all of the hydroxy groups are axial. The many resulting 1,3-diaxial interactions destabilize this form of glucose. In the more stable chair form of D-gluco-pyranose, in contrast, all of the hydroxy groups are equatorial. Thus, the pyranose form of D-glucose is much more stable than the 1,6-anhydro form, and D-glucose exists primarily as an ordinary pyranose. In 1,6-anhydro-D-idopyranose, all of the hydroxy groups are equatorial. In the more stable of the chair conformations of the ordinary pyranose form of D-idose, the —CH_2OH group is forced into an axial position. Formation of the ether bond between carbons 1 and 6 relieves one set of 1,3-diaxial interactions in this form of D-idose. Thus, the 1,6-anhydro form of D-idose is more stable than the pyranose form, and therefore D-idose exists mostly in the 1,6-anhydro form.

β-D-glucopyranose   1,6-anhydro-D-glucopyranose

β-D-idopyranose   1,6-anhydro-D-idopyranose

24.62   (a)   The resonance at δ 5.24 is from the α-anomer, and the resonance at δ 4.66 is from the β-anomer. We know this from the coupling constants. The Karplus curve indicates that the protons that have a dihedral angle of 180° (the C1 and C2 protons of the β-anomer) should have the larger coupling constant.

(b)   The integral of the β-anomer is 66.3% of the total, and the integral of the α-anomer is 34.7% of the total. These are the percentages of the two anomers. (See Table 24.1 of the text; compare this with the percentages of the two pyranose anomers of glucose.)

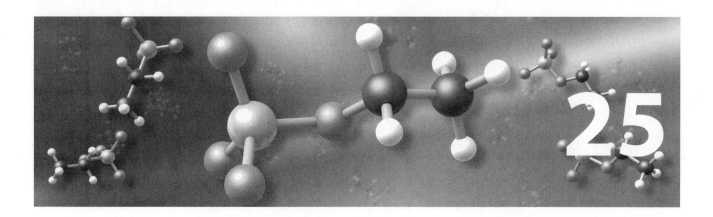

The Chemistry of Thioesters, Phosphate Esters, and Phosphate Anhydrides

FURTHER EXPLORATIONS

 ### 25.1   Pentacovalent Intermediates in Phosphate Ester Hydrolysis

The evidence for pentacovalent intermediates in phosphate ester hydrolysis hinges on three key points:

1. Compounds of pentacovalent phosphorus contain two types of bonds, called **apical** (or **axial**) bonds, and **equatorial** bonds. The apical bonds are somewhat longer; they have a bond angle of 180° with each other, and a bond angle of 90° with the equatorial bonds. The equatorial bonds have bond angles of 120° with each other and define a plane that contains the phosphorus at the center.

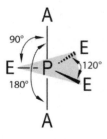

(The grey triangle indicates the plane of the equatorial bonds.)

2. In pentacovalent compounds derived from phosphate esters in which the phosphorus is part of a five-membered ring, one O—P ring bond must be apical and the other must be axial. The

reason is that the preferred O—P—O bond angle is close to 90°.

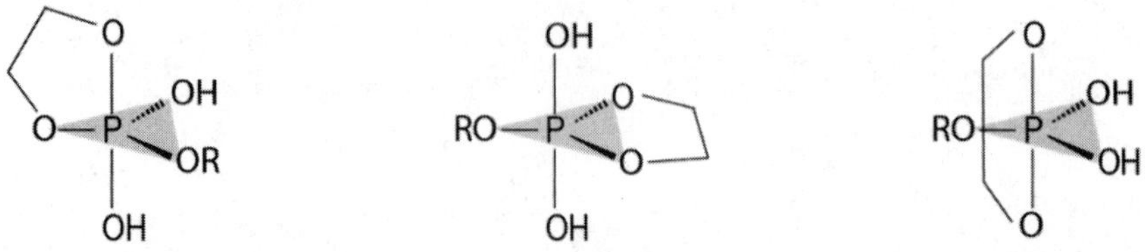

preferred geometry	strained	too strained to exist
O—P—O angle = 90°	O—P—O angle = 120°	O—P—O angle = 180°
(the ring bridges apical and equatorial positions)	(the ring bridges two equatorial positions)	(the ring bridges two axial positions)

If the ring bridges two equatorial positions, the O—P—O angle is 120° and the ring is strained. A molecule in which the ring bridges two apical positions has impossibly long bond lengths and is too strained to exist.

3. Leaving groups depart from apical positions. Because, in the reverse reaction, the roles of nucleophiles and leaving groups are reversed, nucleophiles must also enter from apical positions.

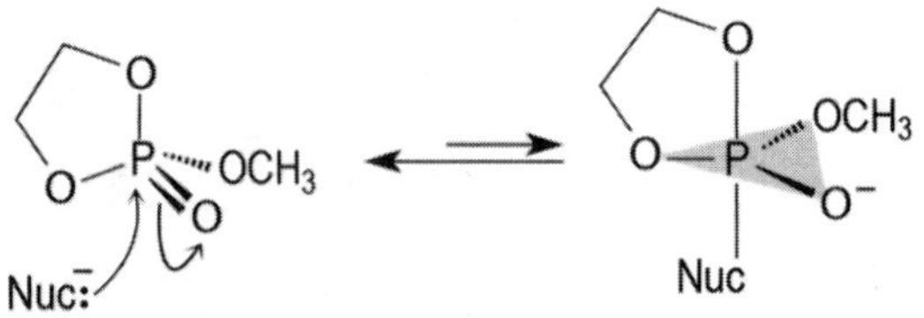

Thus, when a nucleophile reacts at the PAO bond in a tetrahedral phosphate ester, it reacts with phosphorus along a path that is approximately perpendicular to the P=O bond, and the new bond to the nucleophile is apical in the pentacovalent intermediate. The bond to the leaving group is also apical so that the leaving group departs from the pentacovalent intermediate along a path that is also perpendicular to the P=O bond that is formed. This situation is much the same as in carbonyl chemistry (text Fig. 21.4, p. 1063), and it is the result of the positioning of the antibonding molecular orbitals that serve as the destination for the nucleophile electrons or departure point for the leaving-group electrons.

Now we're ready to examine the evidence for pentacovalent intermediates. When ethylene methyl phosphate reacts with water at pH 5.5, two products are formed corresponding to loss of the two possible leaving groups:

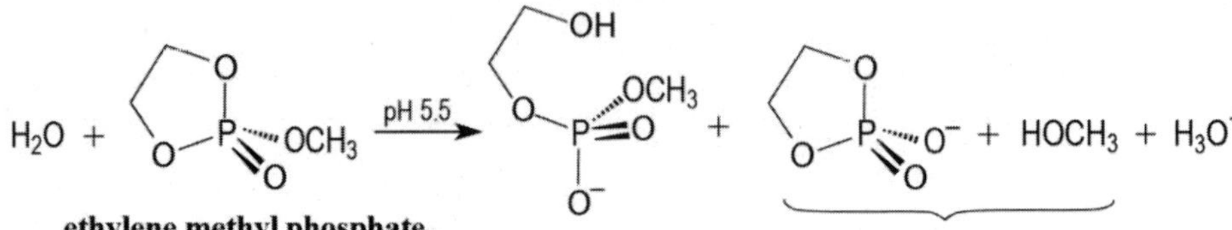

ethylene methyl phosphate

A
(70% of product)

B
(30% of product)

Let's consider the possibility of a pentacovalent intermediate. The first pentacovalent intermediate has to be the following, from points 2 and 3:

where the three forms of the intermediate *P1*, *P2*, and *P3* differ only in the position of a proton. Because one bond to a ring oxygen is apical, that oxygen can become a leaving group. This accounts for the "70% product."

Although this result is consistent with a pentacovalent intermediate, it could also occur in a concerted manner, with proton transfers occurring before, during, or after the substitution.

It is the "30% product" *B* that is the key to the argument for pentacovalent intermediates. For the OCH₃ group to be the leaving group, it must occupy an apical position in a pentacovalent intermediate. However, one apical position is occupied by the nucleophile and the other by a P—O ring bond. For the OCH₃ group to become apical, the molecule undergoes a very subtle internal reorganization called a **pseudorotation**. In a pseudorotation, two of the equatorial bonds lengthen, their bond angle to phosphorus opens to 180°, and they become apical: these would have to be the bond to the OCH₃ group (because it is to become the leaving group) and the formerly equatorial P—O ring bond; and two of the apical bonds shorten, their bond angles close to 120°, and they become equatorial: these would have to be formerly apical P—O ring bond, and the bond to the nucleophilic oxygen. The P—O⁻ bond (which was a PAO bond in the starting ester) remains equatorial; this bond is termed the *pivot bond*. Pseudorotation on *P2* gives *P4*, which, after proton transfers to form *P4´*, loses apical methanol to give product *B*.

Here is the logic: *there can't be a pseudorotation if there isn't an actual pentacovalent intermediate that lasts long enough for pseudorotation to take place.* Therefore, it seems that hydrolysis of this cyclic ester must involve a pentacovalent intermediate. Presumably the products *A* and *B* arise from two different pentacovalent intermediates (*P2* and *P2´*) that are related by a pseudorotation.

It is interesting that saponification of ethylene methyl phosphate in base (⁻OH) gives only product *A*. Therefore, there is no direct evidence for pentacovalent intermediates under these conditions.

 # REACTION REVIEW

I. REACTIONS OF THIOESTERS WITH NUCLEOPHILES

A. HYDROLYSIS OF THIOESTERS

1. Thioesters undergo saponification like esters.

$$R-\overset{\overset{\displaystyle O}{\|}}{C}-SR' + {}^-OH \longrightarrow R-\overset{\overset{\displaystyle O}{\|}}{C}-OH + {}^-SR' \longrightarrow R-\overset{\overset{\displaystyle O}{\|}}{C}-O^- + HSR'$$

$pK_a = 4.76 \qquad\qquad pK_a = 10.5$

B. REACTIONS OF THIOESTERS WITH OTHER NUCLEOPHILES

1. Thioesters, like esters, can react with other nucleophiles in acyl substitution reactions.

![Thioester acyl substitution mechanism: thioester + ⁻Nuc → addition → tetrahedral addition intermediate → elimination → acyl-Nuc product + ⁻SR′]

2. Thioesters can also be reduced to alcohols and thiols with standard laboratory reducing agents (NaBH₄ and LiAlH₄) and biological reducing agents like NADPH.

![Thioester reduction with LiAlH₄: thioester → addition → tetrahedral addition intermediate → elimination → aldehyde intermediate (not isolated) → 1) LiAlH₄ 2) H₃O⁺, H₂O → R—CH₂—OH; alcohol and thiol products formed; + Li⁺ ⁻SR′ → H₃O⁺, H₂O → R′—SH]

II. REACTIONS OF PHOSPHATE ESTERS

A. HYDROLYSIS OF PHOSPHATE ESTERS

1. Phosphate triesters can react with a nucleophile (such as hydroxide) at carbon, cleaving the C–O bond, in a reaction analogous to the S_N2 reaction of sulfonate esters (see Sec. 10.4A).

![Phosphate triester hydrolysis: HÖ⁻ attacks CH₃CH₂—Ö—P(=O)(OEt)—OEt with C—O cleavage → CH₃CH₂—ÖH + ⁻Ö—P(=O)(OEt)—OEt]

2. Phosphate triesters can also react with nucleophiles at phosphorous.

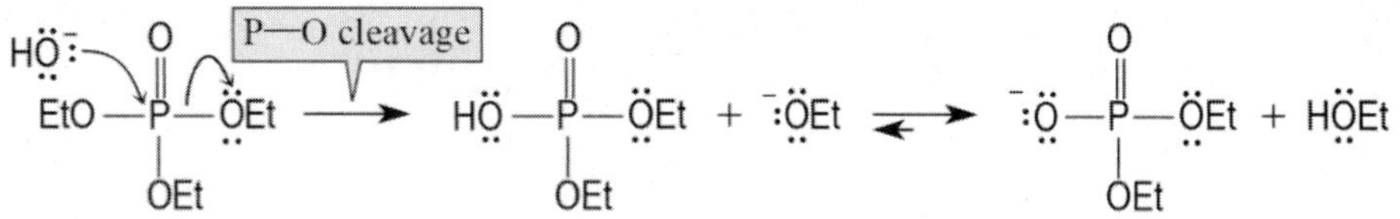
P—O cleavage

a. Two mechanisms are possible, and either is acceptable.

i. An addition-elimination reaction mechanism, analogous to carboxylate ester saponification, is one mode of hydrolysis. This mechanism proceeds through a pentacovalent *intermediate*.

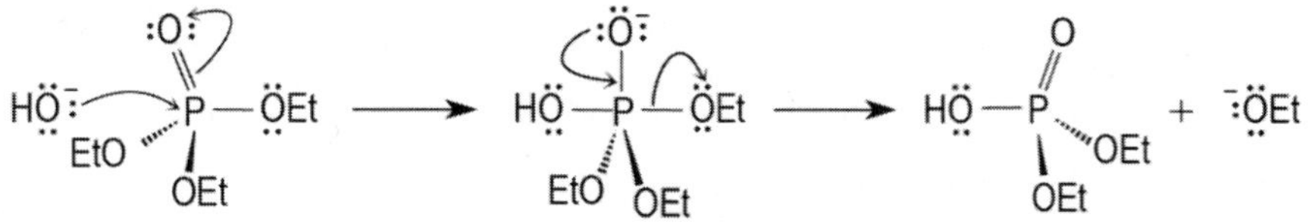
pentacovalent intermediate
(ionizes to give the product)

ii. A concerted reaction mechanism, analogous to an S_N2 reaction, is another mode of hydrolysis. This mechanism proceeds through a pentacovalent *transition state*.

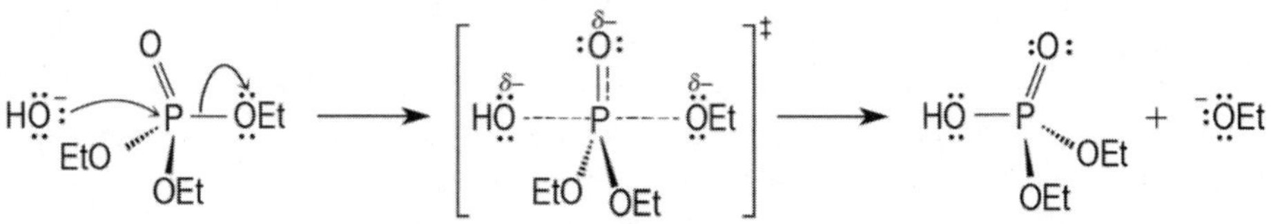
transition state
(ionizes to give the product)

3. Phosphate diesters and monoesters can also be hydrolyzed, but at a much slower rate than triesters due to the nearby repulsive negative charges (see Eq. 25.15, text p. 1298).

III. REACTIONS OF PHOSPHATE ANHYDRIDES

A. Hydrolysis

1. Phosphate anhydrides, such as adenosine triphosphate (ATP), can be hydrolyzed at either the α-, β-, or γ-phosphorus (see Sec. 25.7A).

B. Reactions of Acyl Phosphates with Nucleophiles

1. Acyl phosphates, like other carboxylic acid derivatives, react with nucleophiles in addition-elimination reactions.

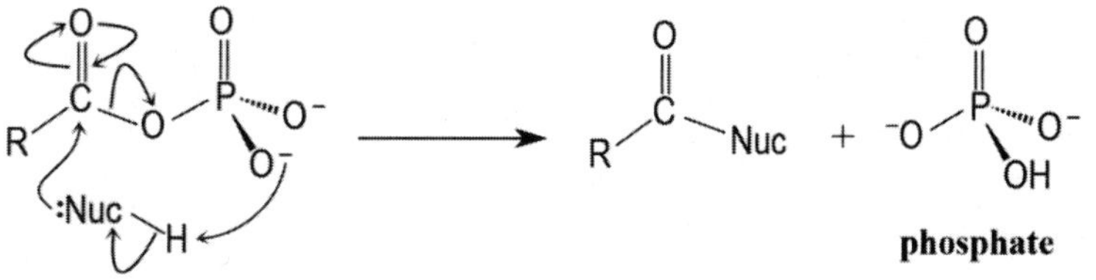
phosphate

C. Reaction at Carbon

1. Alkyl pyrophosphates can also react with nucleophiles at carbon, generating a pyrophosphate leaving group. Depending on the structure, these reactions can proceed through either an S_N2 or S_N1 mechanism.

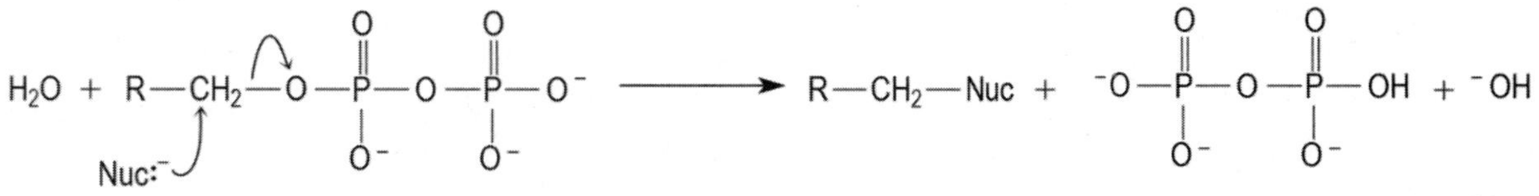

 If you find an error in this manual, please visit http://people.pharmacy.purdue.edu/~loudonm/teaching/

SOLUTIONS TO PROBLEMS

Solutions to In-Text Problems

25.1   (a)   (b)   (c)

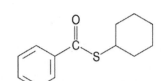

cyclohexyl thiobenzoate   *S*-isopropyl butanethioate   *S*-phenyl cyclohexanecarbothioate

25.2   Allyl butanethioate (common) and *S*-2-propenyl butanethioate (IUPAC)

25.3   To solve this problem, review the discussion on text p. 107, and utilize Eqs. 3.31a–b. There are two protons that can dissociate from methyl phosphate, and so there are three forms: the un-ionized form (charge = 0), the mono-anion form (charge = –1) and the di-anion form (charge = –2). Consider each proton's pK_a, one at a time. The most acidic proton has a pK_a of 2.3. Statement 3 on p. 107 says that when the pK_a is much lower than the pH, than we can treat that proton as effectively being 100% dissociated from the acid. Therefore, there is *essentially none* of the neutral form of methyl phosphate at pH = 7.4; it's either all mono-anion, all di-anion, or a mixture.

Next, consider the second proton, with a pK_a = 6.7. Statement 3 tells us that since this pK_a is less than the pH, it's going to be partly dissociated. If the pK_a was 1 unit less than the pH, it would be 90% dissociated, and if the pK_a was equal to the pH, it would be 0% dissociated. The difference, Δ, is equal to 0.7, so the ratio of the –2 form to the –1 form must be somewhere between 0–90%, but probably closer to 90%. To calculate the amount of the di-anion (the –2 form) exactly, use Eq. 3.31a. Solving for f_A (in this case, f_A = the fraction of di-anion relative to mono-ion) gives 0.83, or 83%. The rest (17%) must be the mono-anion (you can check your work by using Eq. 3.31b and solving for f_{AH}).

So, at this pH, there is approximately 83% of the di-anionic, 17% of the mono-anionic, and none of the neutral form of methyl phosphate.

25.4

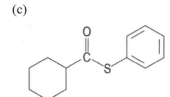

25.5

25.6   If the oxygen were sp^3-hybridized, then the C—O—P bond angle would be closer to 109.5°. It is closer to 120°, which is the ideal bond angle for sp^2-hybridized atoms. The significant positive charge on phosphorus in one resonance structure suggests that another important resonance structure would have the oxygen bearing the positive charge, and in this configuration, it is sp^2-hybridized.

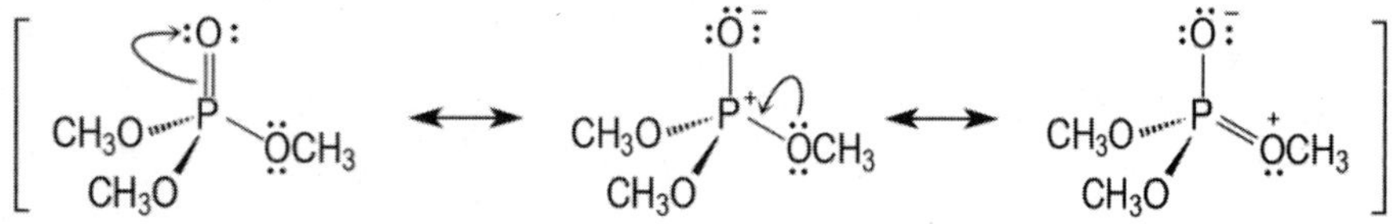

Other resonance structures exist with positive charges on the other oxygens as well.

25.7   Despite the central phosphorus being tetrahedral with four different groups attached, it cannot be isolated as optically active because this phosphate ester is actually a mixture of enantiomers in equilibrium. The proton can move rapidly between the two oxygens by acid–base reactions, shown below.

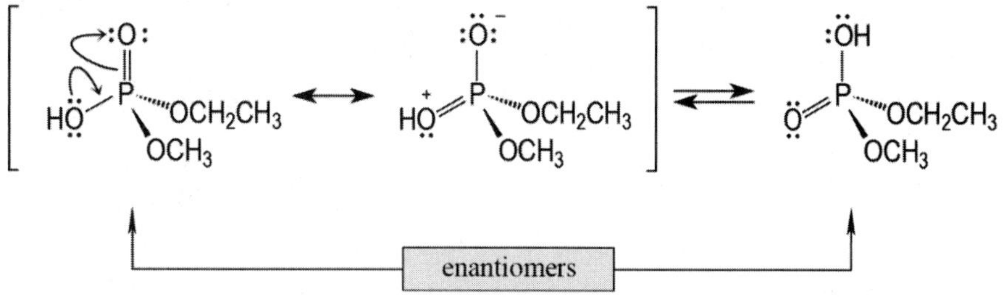

25.8   The CH$_2$ proton signal would be split into five lines, or a quintet, by the $n + 1$ rule: three hydrogens on the adjacent carbon and one phosphorous on the adjacent oxygen.

25.9   (a)

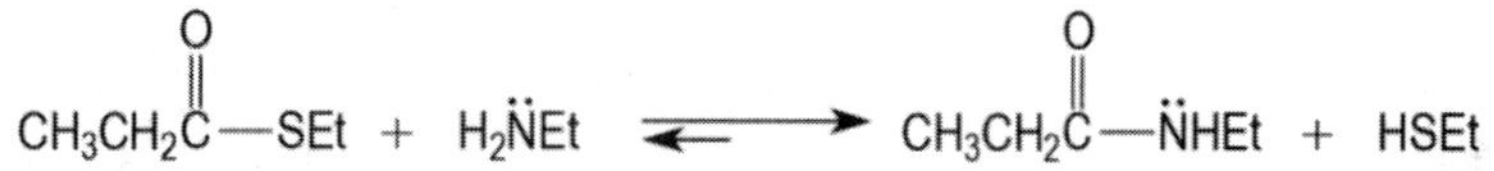

(b)   Under acidic conditions, the amine is protonated, H_3N^+Et, and no longer nucleophilic. There is no lone electron pair available, therefore, the reaction in part (a) cannot take place.

25.10

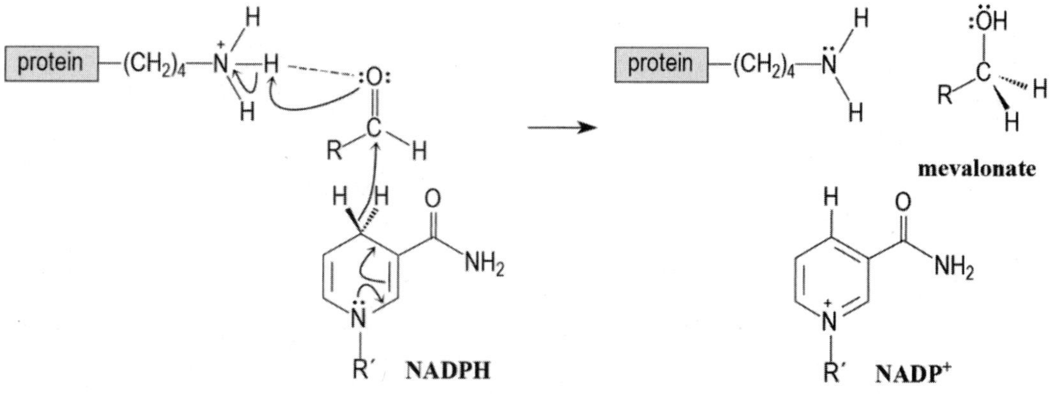

25.11   (a)   As noted on text p. 1302, there are only three isotopes of oxygen: ^{16}O, ^{17}O, and ^{18}O. This fact makes studying the stereochemistry of the reaction in Eq. 25.21 on text p. 1302 impossible when the nucleophile is water (in this case, water labeled with ^{17}O). The product would contain two ^{17}O atoms, and with the proton free to associate with any of the oxygens, regardless of the isotope present, the phosphorus would only contain three different attached groups, and would thus be achiral.

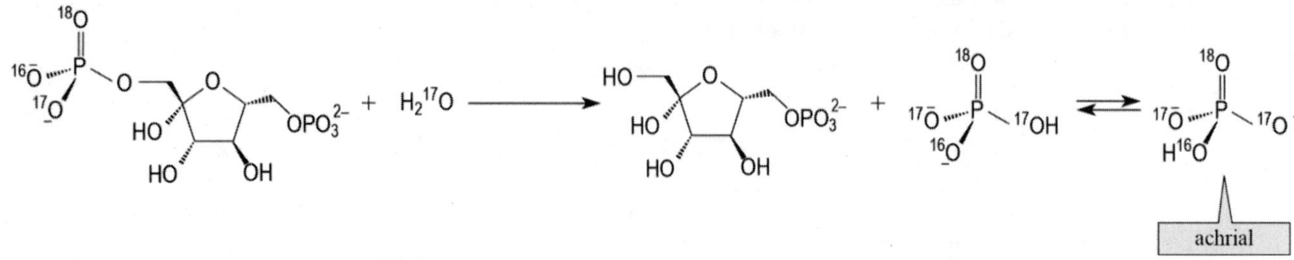

Sulfur effectively gives us a "fourth" isotope of oxygen, assuming [as stated in part (b)] that it doesn't change the overall mechanism of the reaction. With four "oxygens" (S, ^{16}O, ^{17}O, and ^{18}O), and thus four different groups attached to the tetrahedral phosphorous in the product, we can probe the stereochemistry of the reaction mechanism by comparing the R/S configuration of the starting material and of the product. This experiment is explored in part (b).

(b)   The mechanism of F16BP is described in Fig 25.5, text p. 1302. Assuming inversion of configuration, the product of the reaction and its stereochemistry is shown below.

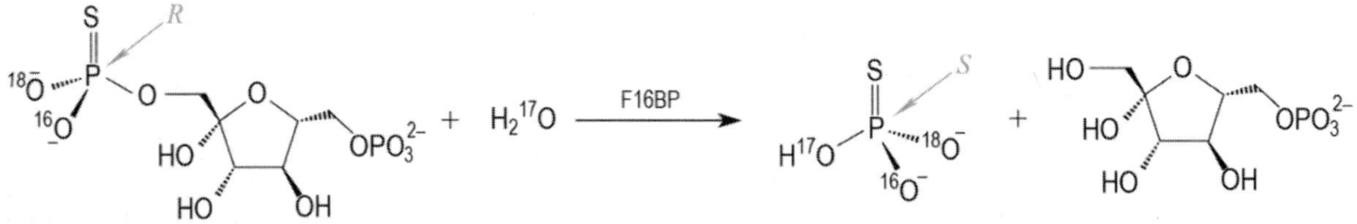

25.12   This mechanism utilizes a double inversion.

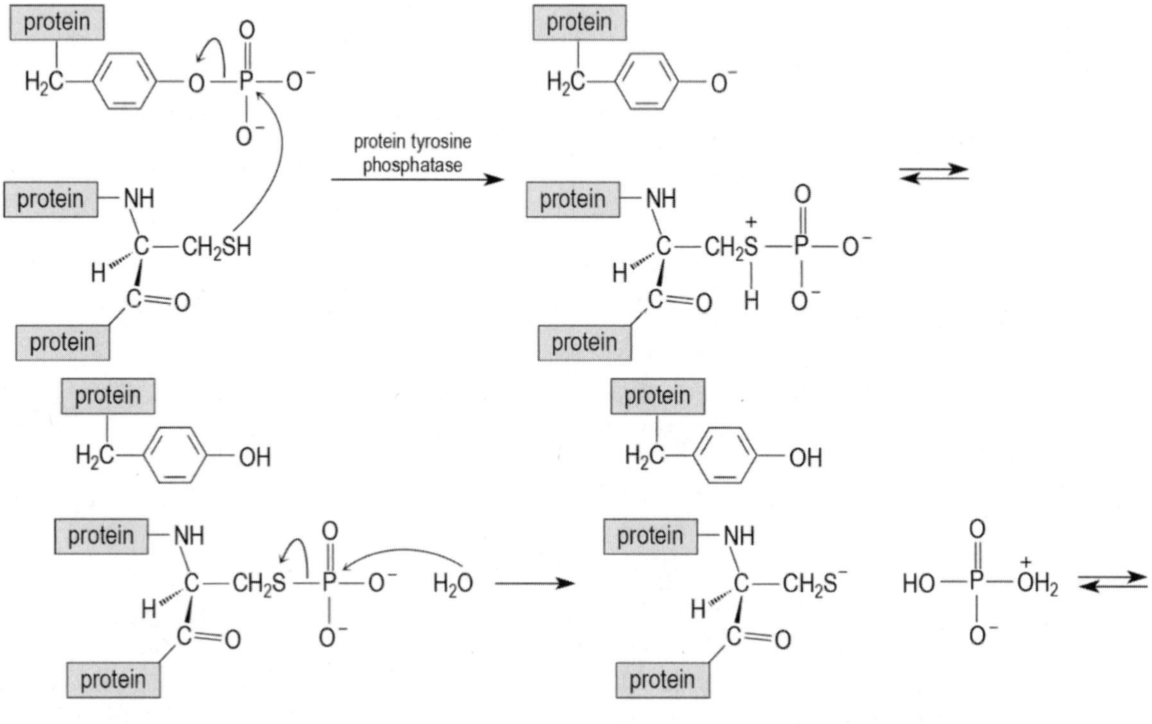

25.13   (a)   The glutamic acid residue provides the base (*1*) to deprotonate a water molecule, making hydroxide, which is more nucleophilic. The various magnesium ions (*2*) stabilize the negative charges that develop as a result of the reaction, as discussed in Eq. 25.23 and the accompanying text on p. 1305. The arginine and tyrosine residues provide hydrogen bonds (*3*) to the oxygens of the leaving group, stabilizing the ion. In total, all features in the active site of the enzyme serve to speed up the reaction by both minimizing or enhancing charges where needed, and aligning reactants.

(b)   From the reaction mechanism shown, inversion at the phosphorous would be observed.

25.14   (a)   First, the fatty acid is adenylated by ATP, as shown in Eq. 25.25c on text p. 1308.

Next, the fatty acyl adenylate, a mixed anhydride, reacts with the thiol group of coenzyme A, followed by a proton transfer step, to produce a fatty acyl coenzyme A. Notice that the mixed anhydride reacts at the carboxy group; in carboxylic-phosphoric anhydrides, the carboxy group is typically more reactive, as discussed in the following section of the text.

(b)   As described by Eq. 25.23–24 and the accompanying text on p. 1305–6, after pyrophosphate is generated in a reaction, inorganic phosphatase catalyzes its hydrolysis to yield two equivalents of phosphate. This reaction is necessary to prevent the reverse of the reaction shown in the second step described in part (a), and to drive the reaction forward.

25.15　(a)

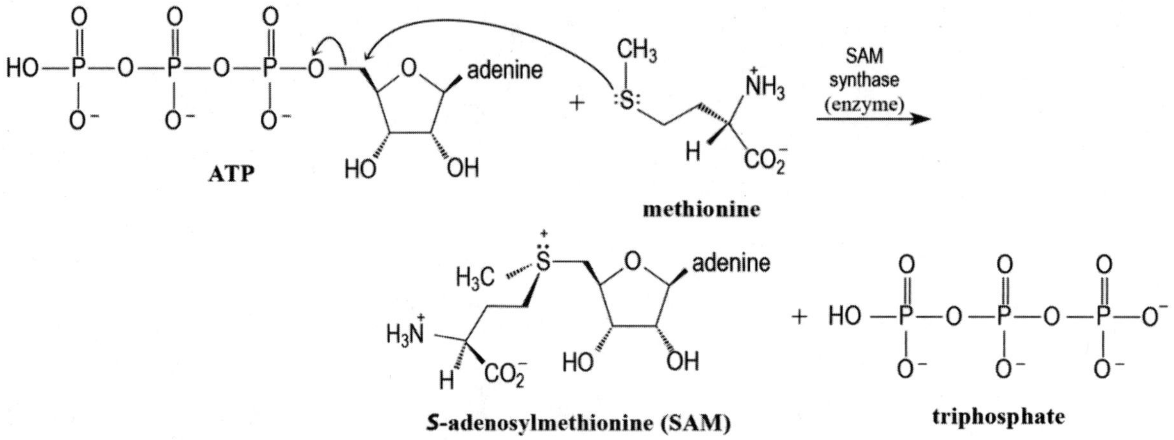

S-adenosylmethionine (SAM)　　　**triphosphate**

(b)　Since triphosphate is a leaving group that is highly negatively charged, the magnesium ions likely neutralize these charges as the S_N2 reaction is taking place.

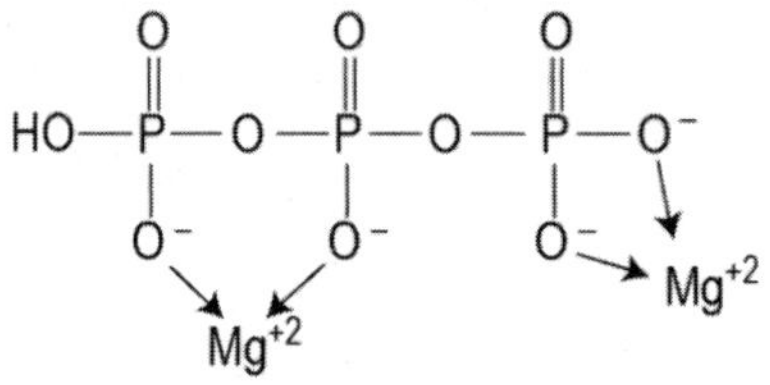

25.16　(a)

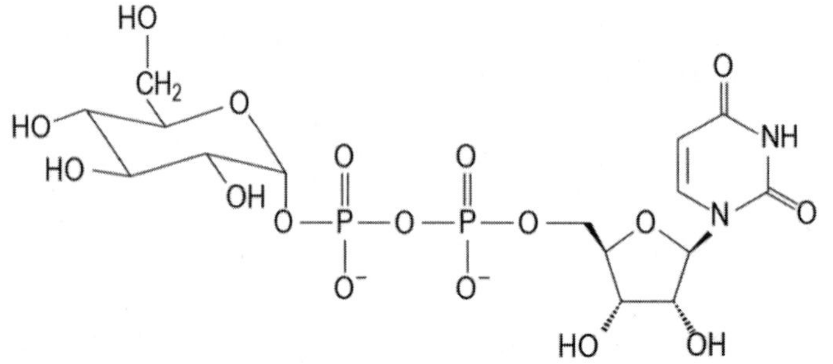

(b)　Examine the structure of sucrose, and imagine breaking the glycosidic bond, forming the individual monosaccharide pieces.

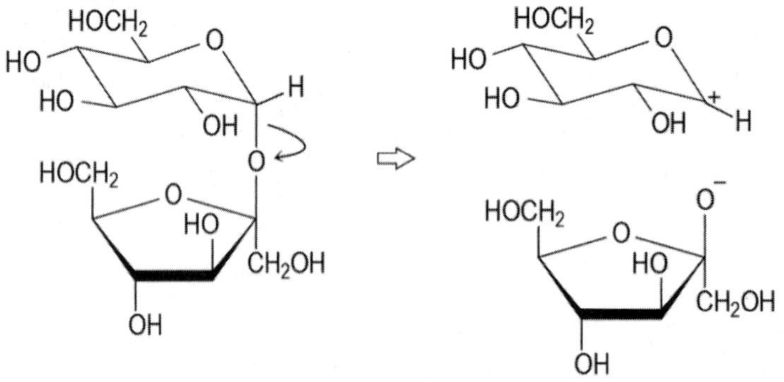

From this analysis, it appears that the fructose moiety must approach from the axial orientation relative to the glucose. In order to do so, the reaction mechanism must proceed through an S_N1 mechanism because the leaving group currently occupies the axial position. The carbocation is stabilized by resonance with the adjacent oxygen.

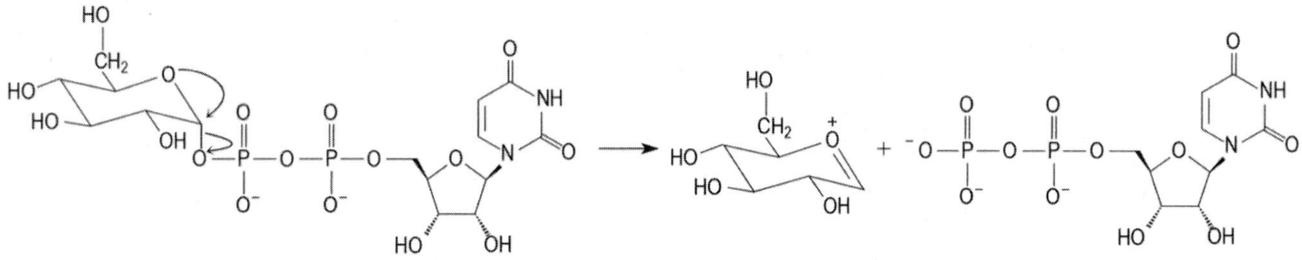

Fructose can then act as a nucleophile as shown below, forming an α-glycosidic bond with glucose. Deprotonation yields sucrose and UDP.

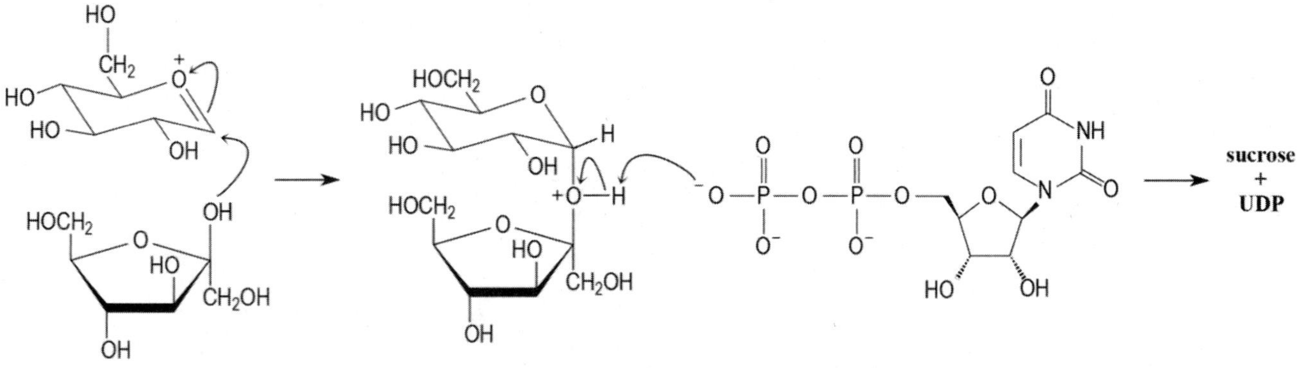

(c) These positively-charged residues likely stabilize the negatively-charged UDP leaving group.

25.17 (a) First, consider how Figs. 25.8 and 25.9 on text pp. 1316–7 were derived. See the solution depicted in Fig. SG25.1 on the next page.

(b) To solve this problem, use a method similar to the one given by Eq. 25.36 on text p. 1316. The problem gives values for the $\Delta G^{\circ\prime}$ of hydrolysis of acetylcholine and acetyl-CoA. You want to find $\Delta G^{\circ\prime}$ for the conversion of choline and acetyl-CoA to acetylcholine and HS–CoA. Since the two given reactions need to sum to the reaction under inspection, reverse the reaction for hydrolysis of acetylcholine, and flip the sign of $\Delta G^{\circ\prime}$ for that reaction. Canceling the reactants on either side of the equilibrium gives the reaction for the synthesis of acetylcholine. Summing the $\Delta G^{\circ\prime}$ values gives –6.3 kJ mol–1 for the overall reaction.

$$\text{acetyl CoA} + \cancel{H_2O} \rightleftharpoons \cancel{CH_3CO_2^-} + \text{HSCoA} + \cancel{H^+} \qquad \Delta G^{\circ\prime} = -31.4\ \text{kJ mol}^{-1}$$

$$\cancel{CH_3CO_2^-} + \text{choline} + \cancel{H^+} \rightleftharpoons \text{acetylcholine} + \cancel{H_2O} \qquad \Delta G^{\circ\prime} = +25.1\ \text{kJ mol}^{-1}$$

$$\text{acetyl CoA} + \text{choline} \rightleftharpoons \text{acetylcholine} + \text{HSCoA} \qquad \Delta G^{\circ\prime} = -6.3\ \text{kJ mol}^{-1}$$

(c) If $\Delta G^{\circ\prime} = -2.3RT \log K_{eq}$, and assuming a temperature of 37 °C (310 K), $K_{eq} = 11.5$.

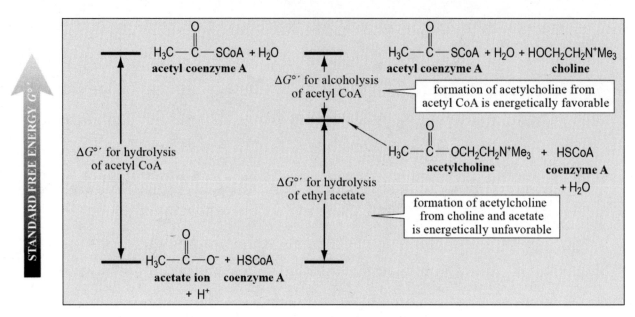

Figure SG25.1 A free-energy diagram to accompany the solution to problem 25.17(b)

25.18 (a) Use a technique similar to that shown in the solution to 25.17(a). It is given that phosphocreatine + ADP gives creatine + ATP, and 12.5 kJ mol^{-1} of energy is given off. As shown in Eq. 25.36 on text p. 1316, ATP + water gives ADP + phosphate, and 30.5 kJ mol^{-1} of energy is given off. Summing these two equations and cancelling, gives the equation for the hydrolysis of phosphocreatine:

$$\text{phosphocreatine} + \text{water} \rightleftharpoons \text{creatine} + \text{phosphate} + 43.0 \text{ kJ mol}^{-1}$$

Since a large amount of energy is given off, phosphocreatine is a very "high energy" compound.

(b) Phosphocreatine is such a high energy molecule because it contains a high energy bond—the phosphorous-nitrogen bond. Resonance structures show that each of these atoms bears some positive charge, which is unstable. Hydrolysis of this bond relieves this unfavorable interaction, and gives off energy.

two adjacent
positive charges

25.19 If the concentrations of ATP, ADP, and phosphate are all 0.001 M, then by Eq. 25.37 on text p. 1318, $\Delta G^{o\prime} = -48.29$ kJ mol^{-1} (see Study Problem 25.1 for a similar example).

25.20 As discussed on text p. 1306, and shown by Eq. 25.24, the conversion of pyrophosphate to phosphate by inorganic pyrophosphatase makes many reactions very favorable by coupling them to this irreversible reaction. If ADP were used as an adenosyl donor, then the product, pyrophosphate, would be hydrolyzed by SAM synthase, and there would be no pyrophosphate left for hydrolysis by inorganic pyrophosphatase. Essentially, the beneficial effects of hydrolyzing two high-energy phosphate bonds is lost, because the starting material was ADP instead of ATP. We now can see why even small concentrations of methionine can be completely converted into SAM–the high-energy releasing processes of conversion of ATP to SAM and triphosphate, followed by triphosphate to pyrophosphate and phosphate, followed finally by pyrophosphate to phosphate are very favorable processes.

Solutions to Additional Problems

25.21

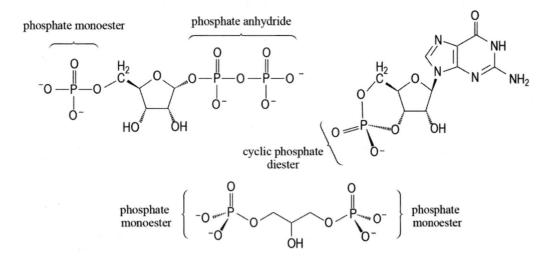

25.22   The anomeric proton is split by both the proton on the adjacent carbon, and the phosphorus on the adjacent oxygen. The coupling constant of 3.5 Hz comes from the hydrogen, since hydrogens with a gauche relationship are typically split by this amount (see Fig. 13.9, text p. 636). The other 7.3 Hz coupling constant must then come from the phosphorous. Since the coupling constants are inequivalent, a doublet of doublets is observed (Sec. 13.5 in the text).

25.23   (a)   The order of reactivity in hydrolysis, from least reactive to most, is $B < C < A$. In general, carboxylic esters are more reactive than phosphate esters (see Sec. 25.6A in the text), so A is the most reactive. Of the two phosphate esters, consider the transition states. Compound B has two negative charges to begin with, and another is brought in by the nucleophile, making the transition state very unstable. Compound C has one less negative charge in the transition state, so it reacts faster than B but slower than A (see Eq. 25.15 on text p. 1298).

(b)   Carboxylic acid derivatives' reactivities increase in the general order: amide < ester < anhydride < acid chloride (see Sec. 21.7 of the text). It follows that the phosphorous analogs of these compounds would react in the same order: $C < B < A < D$.

25.24   (a)   (b)   (c)

25.25   When base (triethylamine) is present, the product is an amide because the amine end of 2-aminoethanol reacts at the carbonyl of the thioester to yield an amide, C.

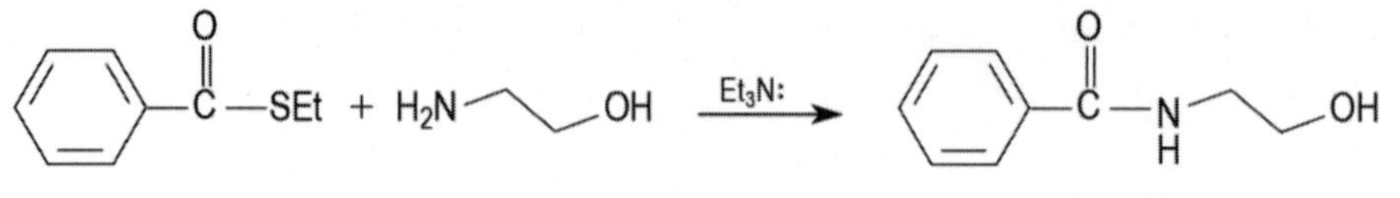

amide *C*

When acid (*p*-toluenesulfonic acid) is present, the amine is protonated, and non-nucleophilic. In that case, the alcohol end of the 2-aminoethanol reacts with the carbonyl of the thioester to yield an ester, *D*.

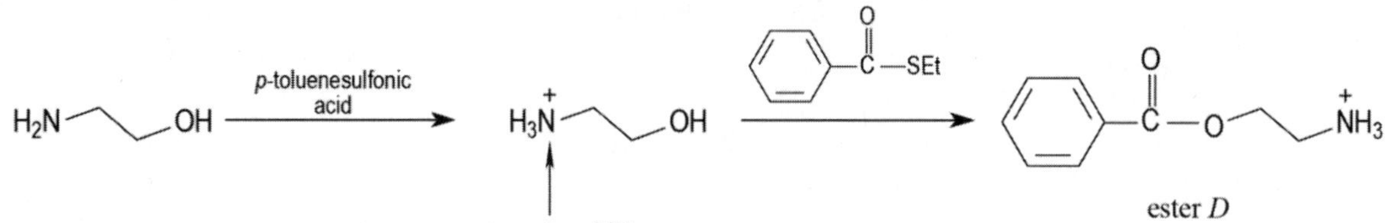

ester *D*

not nucleophilic

25.26 The first hydride equivalent reacts with the thioester to convert it to an aldehyde by displacing thiolate as a leaving group. A second equivalent of hydride reduces the aldehyde to an alcohol. The overall process is analogous to the reduction of esters (Sec. 21.9A of the text).

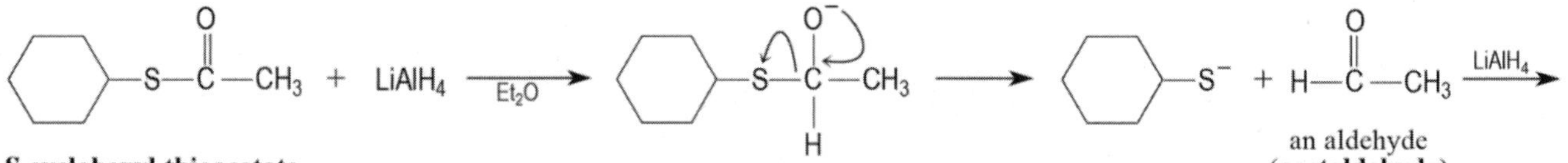

S-cyclohexyl thioacetate

an aldehyde
(acetaldehyde)

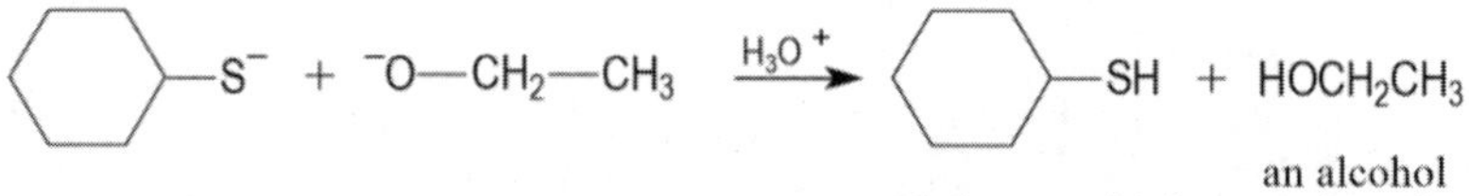

cyclohexanethiol   an alcohol
(ethanol)

When boron trifluoride is added, a boron-oxygen bond is formed that stabilizes the tetrahedral intermediate. Instead of the carbonyl reforming and displacing the leaving group, as above, the oxygen leaves instead. The electrons on sulfur help to stabilize the resulting carbocation, similar to the way that nitrogen operates in the reduction of amides, as shown in Eq. 21.55d on text p. 1081. The boron-containing leaving group likely breaks down into the lithium salt of BOF_2 and LiF.

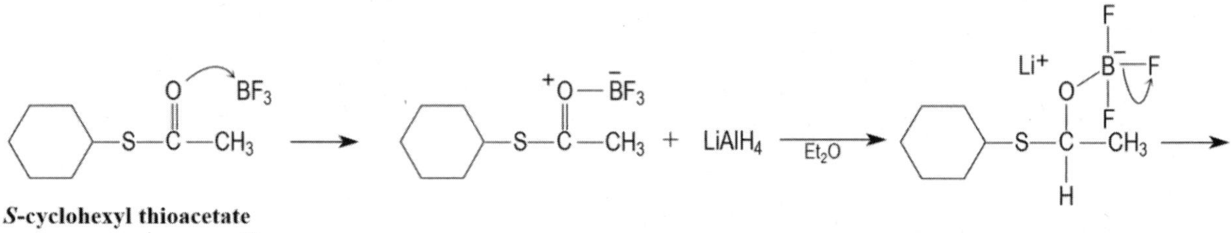

S-cyclohexyl thioacetate

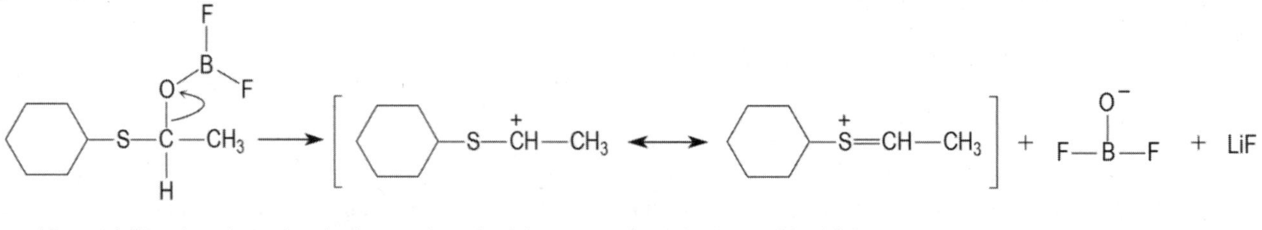

The sulfur-stabilized carbocation is then reduced with a second equivalent of hydride, again analogous to the reduction of the imine intermediate in the case with amides (see Eq. 21.55e, text p. 1082).

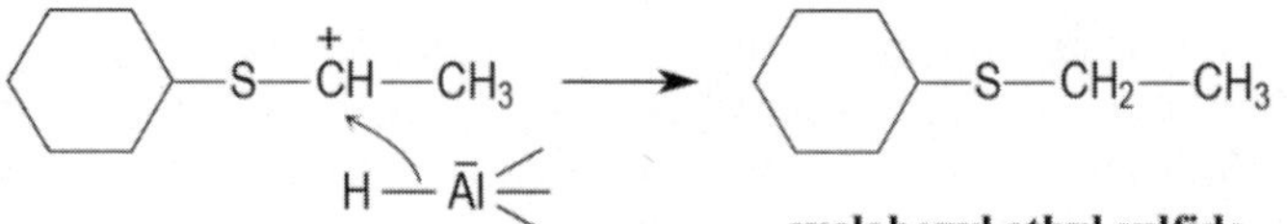

cyclohexyl ethyl sulfide

25.27 (a)

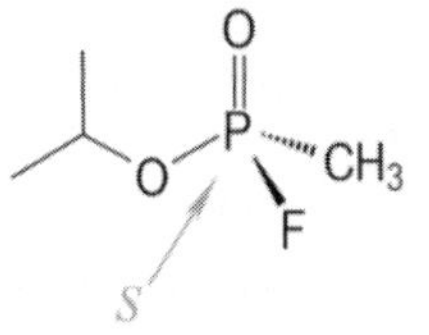

(b)

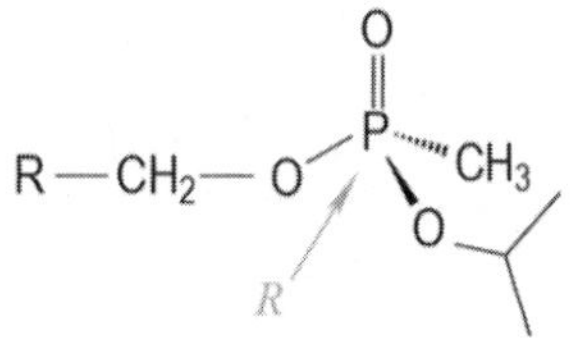

25.28 (a)

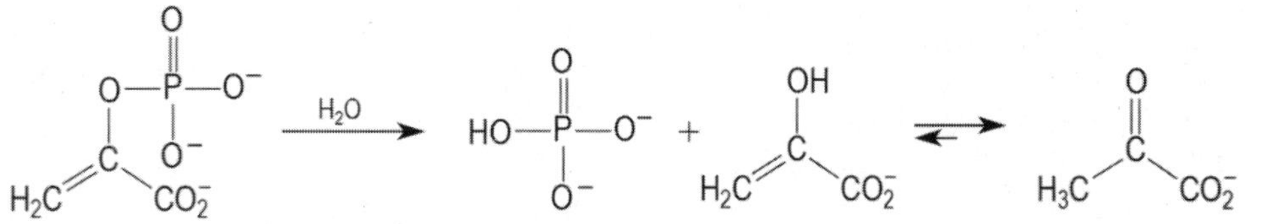

(b) This reaction is driven to completion by the fact that the "alcohol" leaving group is actually an enol, which equilibrates into a much more stable ketone. Since these products are so much lower in energy than the products of the hydrolysis of an ordinary phosphate ester, the reaction releases much more energy.

25.29 (a)

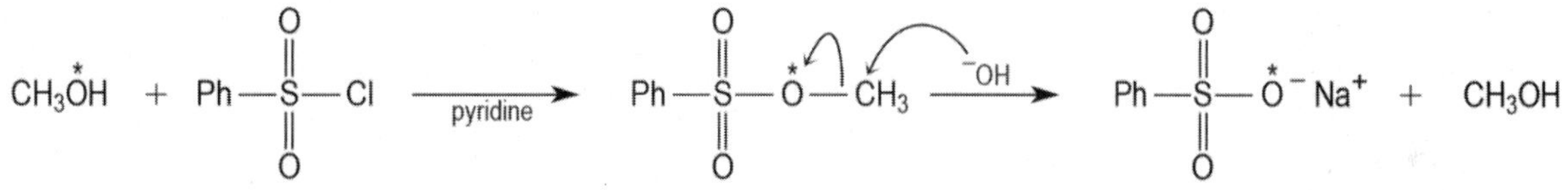

A *B*

(b)

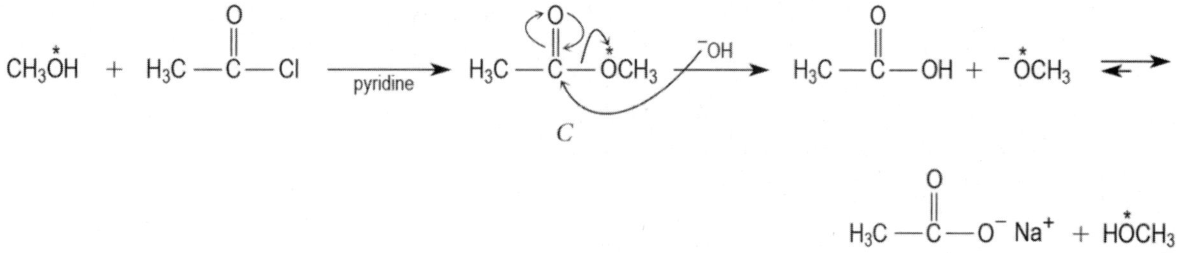

C

D

(c)

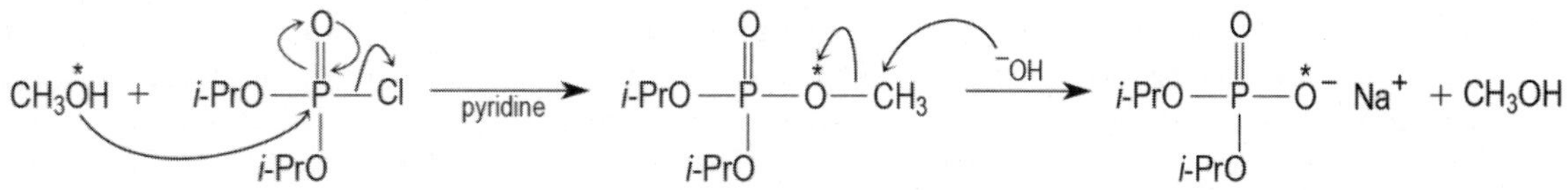

E *F*

Now consider the final part of the question. Sulfonate ions are much less basic than carboxylate ions, and are therefore better leaving groups than carboxylate ions, which is why the nucleophile reacts with the methyl group in reaction (a) and with the carbonyl in reaction (b). For reaction (c), the reaction could occur at either phosphorus (Sec. 25.7A) or the methyl group (Sec. 25.7C). Since we know that the isotope is incorporated into compound F, we could argue that steric interference of the isopropyl groups precludes reaction at the phosphorus, and reaction at the methyl carbon is preferred.

25.30

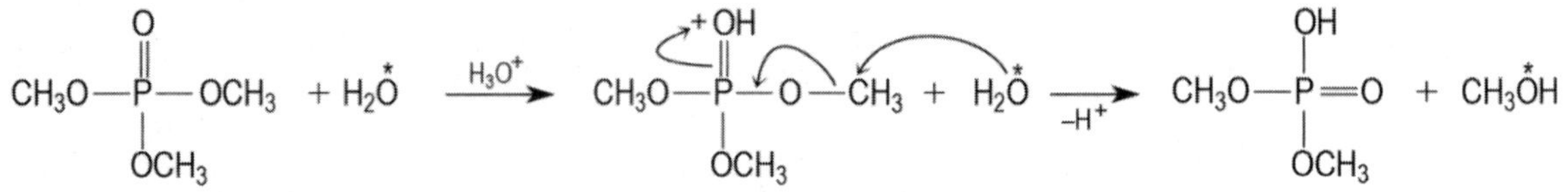

25.31   Racemization of the product suggests that a single inversion by an S_N2-type mechanism is not operating. Instead, the formation of a planar metaphosphate ion (Eq. 25.19a, text p. 1300) could be trapped by the acetonitrile solvent from either side, racemizing the phosphorous. With the stereochemical information of the starting material lost, the *tert*-butyl alcohol would eventually react with the metaphosphate, yielding a racemic mixture of the labeled *tert*-butyl phosphate.

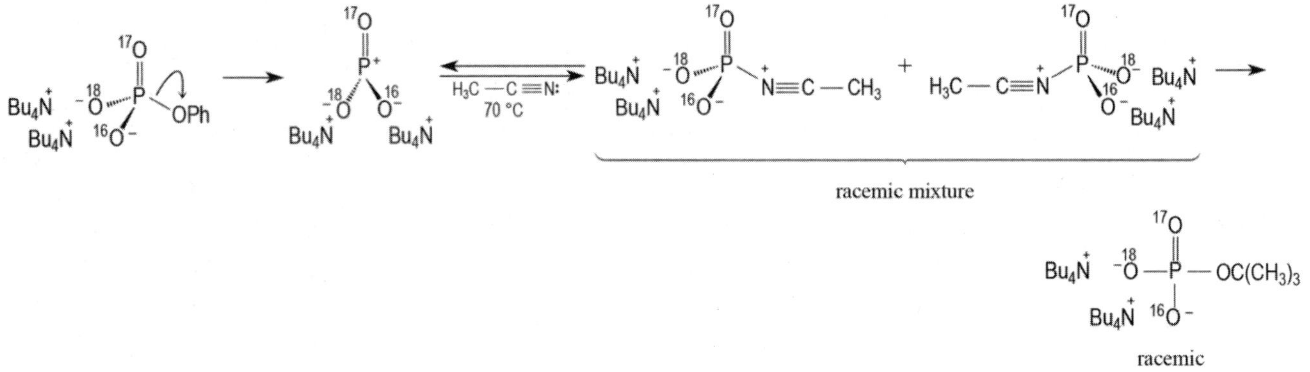

racemic mixture

racemic

25.32   Nitrate ion can react with hydroxide to form orthonitrate, but the reaction is unfavorable. Nitrate ion is stabilized by resonance, which distributes the negative charge around to all three oxygen atoms equally. After reaction with hydroxide at the nitrogen, the product is very unstable for a few reasons. First, the resonance of the nitrate ion is lost—it is tetrahedral and the three oxygens bear three formal negative charges because no delocalization is possible. Also, its formation is unfavorable, because approach of the hydroxide is repelled by the nearby negative charges on the oxygens. Additionally, the product has a 2– charge, with negative charges localized on three oxygens. Finally, the three negative charges are pushed closer together in a tetrahedral arrangement than in the trigonal planar nitrate ion, further destabilizing the product.

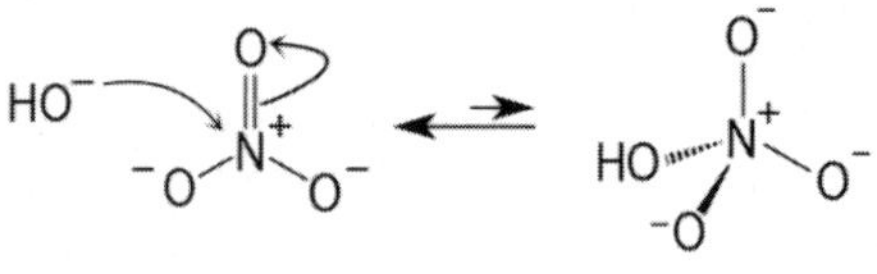

25.33   The active site of the enzyme would likely look very similar to the one shown in Eq. 25.32 on text p. 1312. There would need to be a glutamate or aspartate residue to remove a proton from the nucleophile; phenol in this case. The phosphate negative charges would need to be stabilized by metal ions such as manganese, calcium or magnesium, and residues such as arginine or lysine. Arginine or lysine could also provide a proton to the UDP leaving group. Draw the active site using the structures shown in Fig. P25.33, with Eq. 25.32 as your guide.

25.34   A double inversion mechanism must be operating if the product shows retention of stereochemistry. However, the problem states that no phosphoenzyme intermediate is formed. Therefore, we could postulate that another nucleophile does the initial displacement, inverting the stereochemistry at phosphorus, and the labeled water inverts the phosphorus back to its original configuration to yield the product.

25.35   (a)

(b)   Since inversion occurs twice at phosphorus, the product would contain phosphorous with retention of configuration.

25.36   (a)   Use a strategy similar to the one shown in the solution to problem 25.17(a) in this manual.

a fatty acyl-CoA + H_2O ⇌ RCO_2^- + HSCoA + H^+ $\Delta G^{\circ\prime} = -31.4$ kJ mol^{-1}

RCO_2^- + carnitine + H^+ ⇌   a fatty acyl carnitine + H_2O $\Delta G^{\circ\prime} = +17.4$ kJ mol^{-1}

a fatty acyl-CoA + carnitine   ⇌   a fatty acyl carnitine + HSCoA $\Delta G^{\circ\prime} = -14.0$ kJ mol^{-1}

(b)   Since $\Delta G^{\circ\prime} = -2.3RT \log K_{eq}$, and assuming a temperature of 37 °C, $K_{eq} = 229$.

(c)   This reaction is favorable because the overall conversion is from a thioester to a more stable ester. See the discussion in Sec. 25.8C, and Fig. 25.10 on text p. 1318.

25.37   Use a strategy similar to the ones shown in the solutions to problem 25.17(a) and 25.36(a) in this manual. The free-energy diagram is shown in Fig. SG25.2 on the next page.

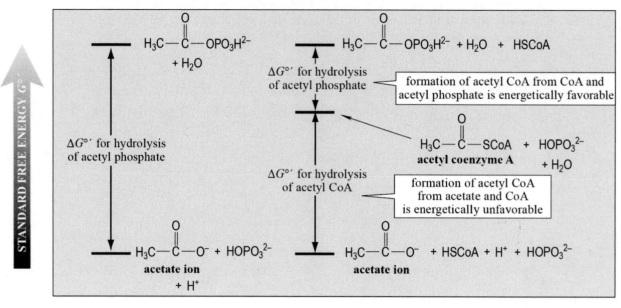

Figure SG25.2　A free-energy diagram to accompany the solution to problem 25.37. Note: CoA and CoASH are equivalent representations of coenzyme A.

$$\text{acetyl phosphate} + H_2O \rightleftharpoons CH_3CO_2^- + \text{phosphate} + H^+ \quad \Delta G^{\circ\prime} = -43.1 \text{ kJ mol}^{-1}$$

$$RCO_2^- + HSCoA + H^+ \rightleftharpoons \text{acetyl CoA} + H_2O \quad \Delta G^{\circ\prime} = +31.4 \text{ kJ mol}^{-1}$$

$$\text{acetyl phosphate} + HSCoA \rightleftharpoons \text{acetyl CoA} + \text{phosphate} \quad \Delta G^{\circ\prime} = -11.7 \text{ kJ mol}^{-1}$$

25.38　As noted in Sec. 25.5A, thioesters are about as reactive as the corresponding oxygen esters, that is, not very reactive. At pH = 7, hydrolysis can proceed only with water, which is not a very good nucleophile, and the rate of the reaction is very slow. At pH = 10, hydroxide is present, and it is a much more reactive nucleophile than water. Additionally, after reaction at the carbonyl and expulsion of the leaving group, CoAS⁻ reacts with the acetic acid which drives the equilibrium to the right (see Eq. 25.3, text p. 1291, and Sec. 21.7A, text p. 1061). Therefore, ΔG° would be more negative at pH = 11, and the reaction would be more favorable.

25.39　(a)　The nucleophilic hydroxyl group reacts at the β-phosphate group, and the by-product is adenosine monophosphate, or AMP.

　　　　(b)

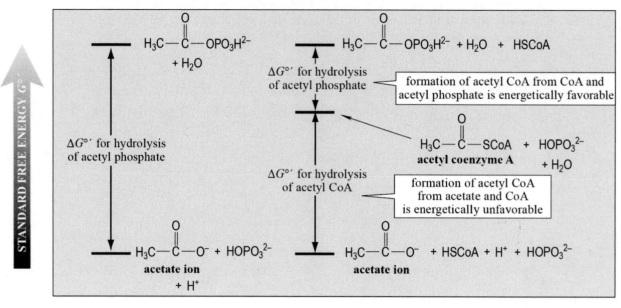

　　　　(c)　Pyrophosphate is the by-product in the answer to part (b). Pyrophosphatase catalyzes the irreversible conversion of pyrophosphate to phosphate (Eq. 25.24, and discussion on text pp. 1305–6), which drives the reaction to completion.

25.40　(a)　In the first step, aspartic acid is adenylated by ATP, as illustrated by Eq. 25.25c on text p. 1308.

The reaction schemes showing aspartic acid, ATP, and the product with pyrophosphate.

aspartic acid

ATP

+ **pyrophosphate**

Next, glutamine displaces AMP through a nucleophilic acyl substitution reaction.

glutamine

AMP

Finally, hydrolysis of the intermediate generated in the previous step yields glutamic acid and asparagine.

glutamic acid   **asparagine**

You may have written an alternative mechanism to the one shown above where the hydrolysis of glutamine to glutamate and ammonia happens first, and the ammonia reacts with the aspartyl phosphate carbonyl. That is also a valid way to write the mechanism and may be the way the reaction actually happens.

(b)   As seen in many previous examples, the high energetic favorability of ATP hydrolysis is used to help otherwise less favorable reactions occur in high yield. In this case, aspartic acid is adenylated, and the AMP group is a much better leaving group than the OH group would have been. This substitution favors the amide formation shown in the second step of part (a).

(c)   Use a strategy similar to the ones shown in the solutions to problem 25.17(a) and 25.36(a) in this manual.

$$ATP \;+\; H_2O \;\rightleftharpoons\; AMP \;+\; pyrophosphate \qquad \Delta G°' = -45.6 \text{ kJ mol}^{-1}$$

$$aspartic\ acid \;\rightleftharpoons\; asparagine \;+\; H_2O \qquad \Delta G°' = +15.1 \text{ kJ mol}^{-1}$$

$$glutamine \;+\; H_2O \;\rightleftharpoons\; glutamic\ acid \qquad \Delta G°' = -14.2 \text{ kJ mol}^{-1}$$

$$
\begin{array}{l}
ATP \;+\; aspartic\ acid \\
+\; H_2O \;+\; glutamine
\end{array}
\;\rightleftharpoons\;
\begin{array}{l}
AMP \;+\; pyrophosphate \\
+\; asparagine \;+\; glutamic\ acid
\end{array}
\qquad \Delta G°' = -44.7 \text{ kJ mol}^{-1}
$$

(d)   This reaction is coupled to ATP hydrolysis, which makes it much more favorable. If you inspect the calculation made in part (c), you will see that the reaction is unfavorable ($+0.9$ kJ mol^{-1}) if ATP is removed from the process. Although ATP itself isn't actually hydrolyzed directly in the mechanism [see part (a)] we can say that the reaction is coupled to ATP hydrolysis because that process indirectly provides the favorable energy release needed to drive the reaction towards completion.

(e)   Pyrophosphate is one of the by-products generated during the process. Pyrophosphatase catalyzes the irreversible conversion of pyrophosphate to phosphate (Eq. 25.24, and discussion on text pp. 1305–6), which drives the reaction to completion.

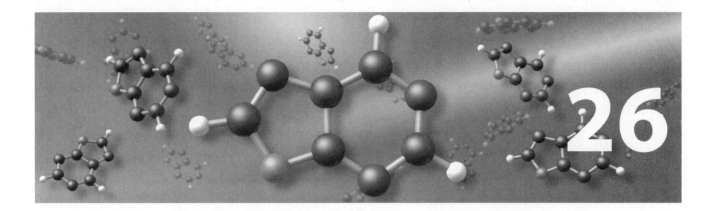

The Chemistry of the Aromatic Heterocycles and Nucleic Acids

FURTHER EXPLORATIONS

 26.1   Relative Acidities of 1,3-Cyclopentadiene and Pyrrole

It is interesting that pyrrole, a nitrogen acid, is somewhat *less* acidic than 1,3-cyclopentadiene, a carbon acid.

1,3-cyclopentadiene
$pK_a \approx 15$

pyrrole
$pK_a \approx 17$

This result stands in contrast to the general expectation from the *element effect* (Scc. 3.6A of the text) that hydrogens on more electronegative atoms are more acidic than those on less electronegative atoms. (For example, the pK_a of NH_3 is about 35; the pK_a of CH_4 is about 55; the N—H pK_a of amides is about 16; the α-hydrogen pK_a of a ketone is about 19.)

The greater acidity of 1,3-cyclopentadiene reflects the importance of aromaticity. 1,3-Cyclopentadiene itself is not aromatic, but its conjugate-base anion is. Consequently, 1,3-cyclopentadiene "gains aromaticity" when it ionizes. This effect lowers the pK_a of 1,3-cyclopentadiene significantly from what it otherwise would be. (The effect of aromaticity might be worth as much as 20 pK_a units.) In contrast, both pyrrole and its conjugate-base anion are aromatic; ionization of pyrrole does not gain any aromatic stability for the molecule. In summary, then, the increase in stability resulting from the formation of an aromatic species causes the pK_a of the carbon acid to be lower than that of the nitrogen acid.

REACTION REVIEW

I. ELECTROPHILIC AROMATIC SUBSTITUTION REACTIONS OF AROMATIC HETEROCYCLIC COMPOUNDS

A. FURAN, PYRROLE, AND THIOPHENE

1. Furan, thiophene, and pyrrole undergo electrophilic substitution predominantly at the 2-position of the ring.
   a. The carbocation resulting from substitution at carbon-2 has more important resonance structures, and is therefore more stable, than the carbocation resulting from reaction at carbon-3.

   b. Some carbon-3 substitution product accompanies the major carbon-2 substitution product in many cases.
2. Furan, pyrrole, and thiophene are all much more reactive than benzene in electrophilic aromatic substitution reactions.
   a. The reactivity order of the heterocycles is a consequence of the relative abilities of the heteroatoms to stabilize positive charge in the intermediate carbocations.

   pyrrole > furan > thiophene >> benzene

   - *i.* Both pyrrole and furan have heteroatoms from the first row of the periodic table.
   - *ii.* Because nitrogen is better than oxygen at delocalizing positive charge (it is less electronegative), pyrrole is more reactive than furan.
   - *iii.* The sulfur of thiophene is a second row element and, although it is less electronegative than oxygen, its $3p$ orbitals overlap less efficiently with the $2p$ orbitals of the aromatic π-electron system.
   b. The reactivity order of the heterocycles in aromatic substitution parallels the reactivity order of the corresponding substituted benzene derivatives:

   $(CH_3)_2N-Ph > CH_3O-Ph > CH_3S-Ph$

3. Milder reaction conditions must be used with more reactive compounds.

4. The usual activating and directing effects of substituents in aromatic substitution apply; superimposed on these effects is the normal effect of the heterocyclic atom in directing substitution to the 2-position.

a. Count around the carbon framework of the heterocycle, not through the heteroatom, when using the ortho, meta, para analogy.

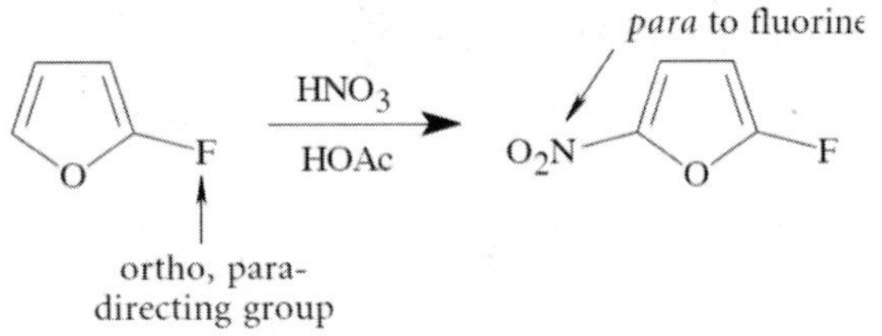

para to fluorine

ortho, para-
directing group

b. When the directing effects of substituents and the ring compete, it is not unusual to observe mixtures of products.

c. If both 2-positions are occupied, 3-substitution takes place.

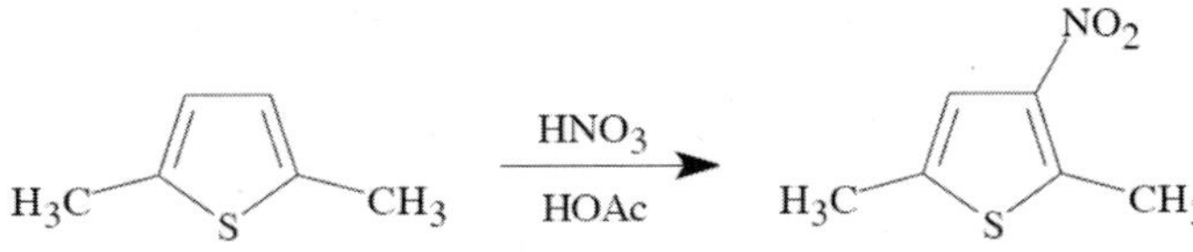

B. PYRIDINE

1. In general, pyridine has very low reactivity in electrophilic aromatic substitutions; it is much less reactive than benzene.
   a. An important reason for this low reactivity is that pyridine is protonated under the very acidic conditions of most electrophilic aromatic substitution reactions.
   b. The resulting positive charge on nitrogen makes it difficult to form a carbocation intermediate, which would place a second positive charge within the same ring.

2. Pyridine rings substituted with activating groups such as methyl groups do undergo electrophilic aromatic substitution reactions.
   a. When substitution in pyridine does occur, it generally takes place at the 3-position.
   b. The preference for 3-substitution in pyridines can be understood by considering the resonance structures for the possible carbocation intermediates.
      i. Substitution in the 3-position gives a carbocation with three resonance structures.

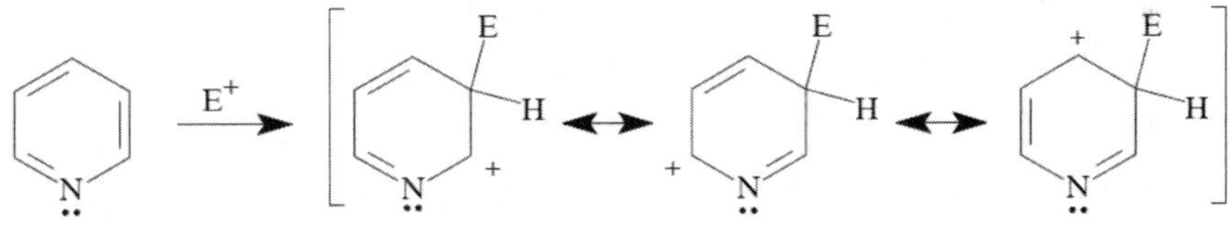

ii. Substitution at the 4-position also involves an intermediate with three resonance structures, but the one shown in gray is particularly unfavorable because the nitrogen, an electronegative atom, is electron deficient.

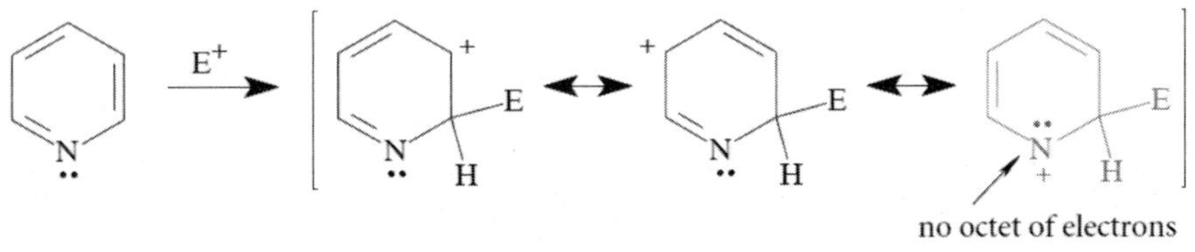

no octet of electrons

3. Pyridine can be oxidized to give pyridine-*N*-oxide.

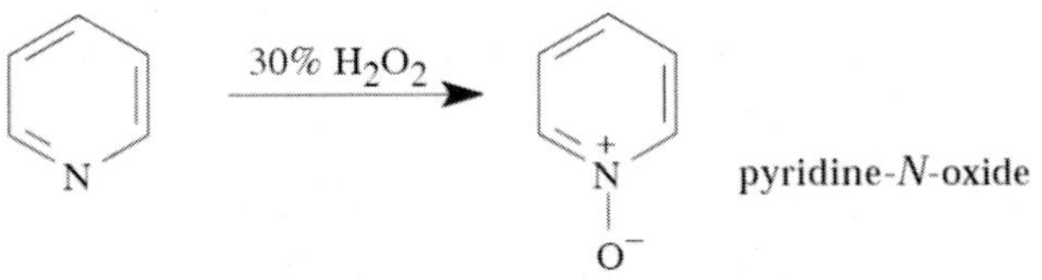

pyridine-*N*-oxide

4. Pyridine-*N*-oxide is more reactive than pyridine and undergoes nitration in the 4-position.

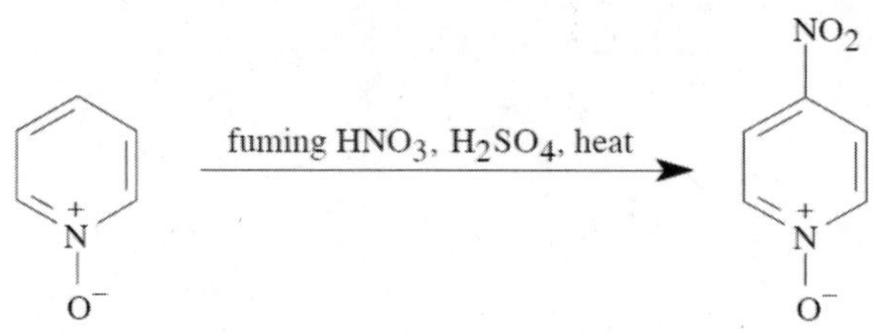
fuming HNO_3, H_2SO_4, heat
NO_2
N
O

a. The *N*-oxide function can be removed by catalytic hydrogenation, which will also reduce any nitro groups present.

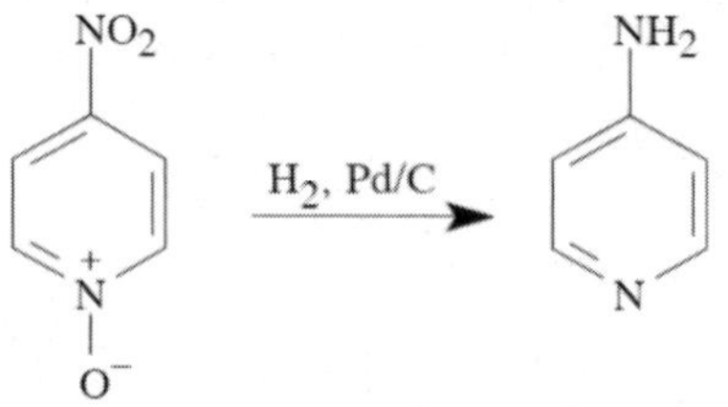
NO_2
NH_2
H_2, Pd/C
N
O

b. Reaction with trivalent phosphorous compounds, such as PCl_3, removes the *N*-oxide function without reducing nitro groups.

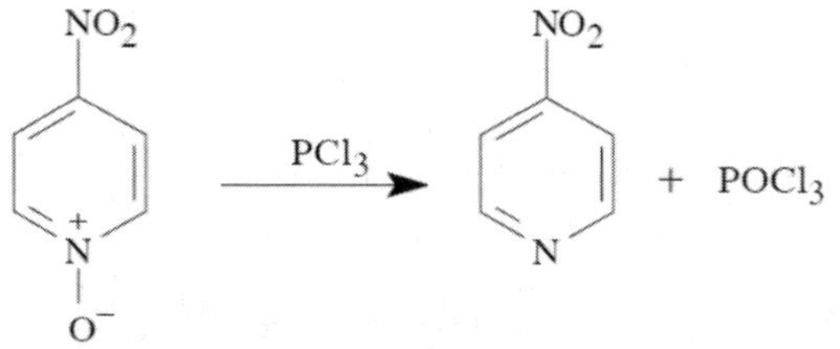
NO_2
PCl_3
+ $POCl_3$
N
O

II. NUCLEOPHILIC AROMATIC SUBSTITUTION REACTIONS OF AROMATIC HETEROCYCLIC COMPOUNDS

A. Pyridine

1. Treatment of a pyridine derivative with the strong base sodium amide ($Na^{+}\,{}^{-}NH_2$) brings about the direct substitution of an amino group for a ring hydrogen; this reaction is called the Chichibabin reaction.

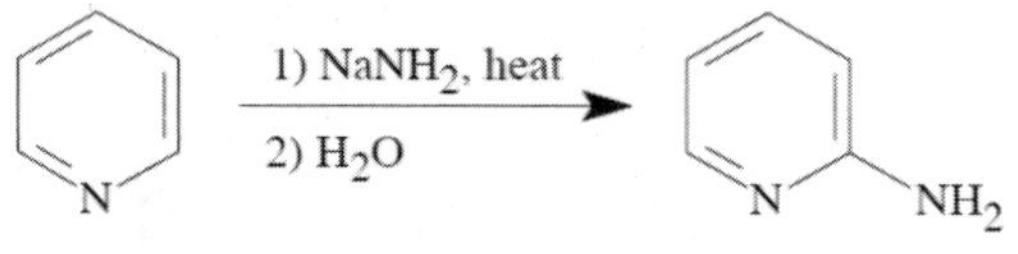
1) $NaNH_2$, heat
2) H_2O
N
NH_2

2. A reaction similar to the Chichibabin reaction occurs with organolithium reagents.

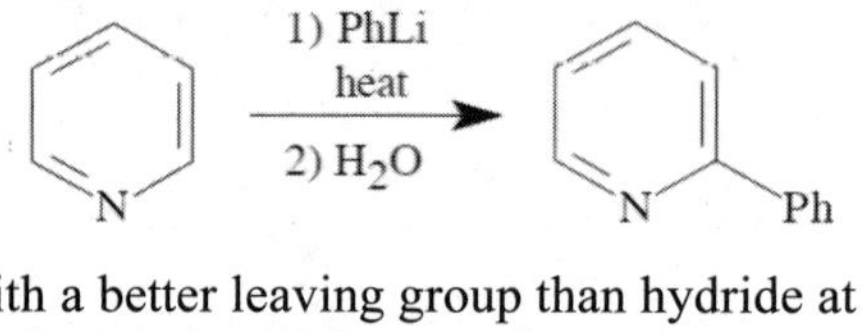
1) PhLi
heat
2) H_2O
N
Ph

3. When pyridine is substituted with a better leaving group than hydride at the 2-position, it reacts more rapidly with nucleophiles; thus, the 2-halopyridines readily undergo substitution of the halogen by other nucleophiles under conditions that are much milder than those used in the Chichibabin reaction.

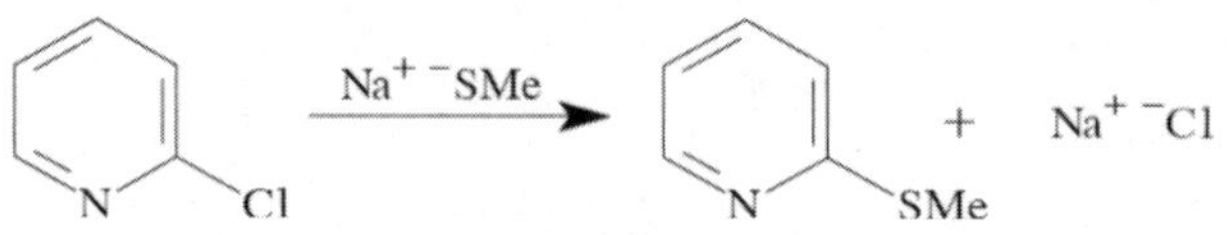
$Na^{+}\,{}^{-}SMe$
+ $Na^{+}\,{}^{-}Cl$
N
Cl
SMe

4. Nucleophilic substitution reactions on pyridine rings can be classified as nucleophilic aromatic substitution reactions.
   a. The "electron-withdrawing group" in the reaction of pyridines is the pyridine nitrogen itself; consider the ring nitrogen of pyridine as if it were a nitro group attached to a benzene ring.
   b. The tetrahedral addition intermediate is analogous to the Meisenheimer complex of nucleophilic aromatic substitution. (See Sec. 18.4, of the text.)

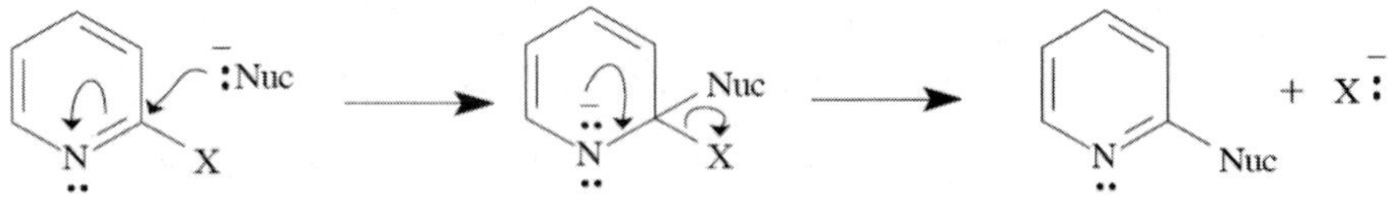

tetrahedral
addition intermediate

5. 3-Substituted pyridines are not reactive in nucleophilic substitution because negative charge in the addition intermediate cannot be delocalized onto the electronegative nitrogen.
6. 2-Pyridone contains a significant amount of both carbonyl and hydroxy isomers, and it often undergoes some reactions reminiscent of hydroxy compounds.

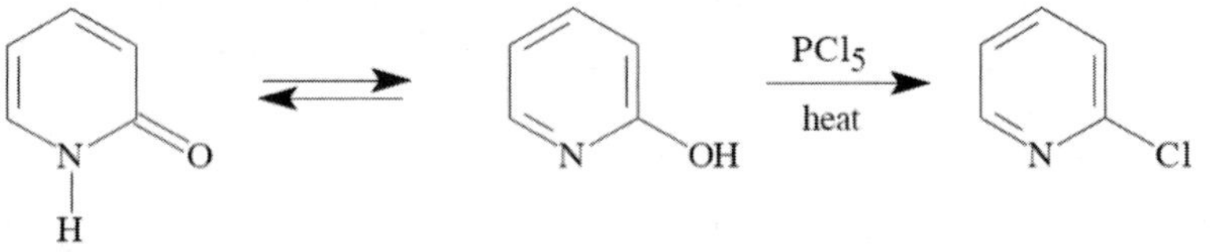

2-pyridone

B. PYRIDINIUM SALTS AND THEIR REACTIONS

1. Pyridine is a nucleophile and reacts in S_N2 reactions with alkyl halides or sulfonate esters to form quaternary ammonium salts, called pyridinium salts.

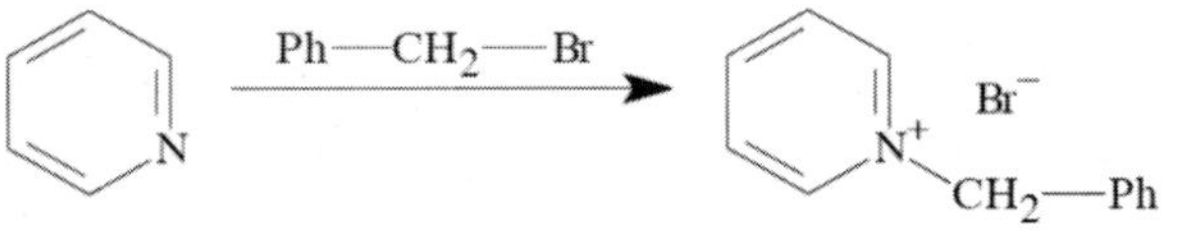

2. Pyridinium salts are activated toward nucleophilic displacement of groups at the 2- and 4-positions of the ring much more than are pyridines themselves, because the positively charged nitrogen is much more electronegative than the neutral nitrogen of a pyridine.

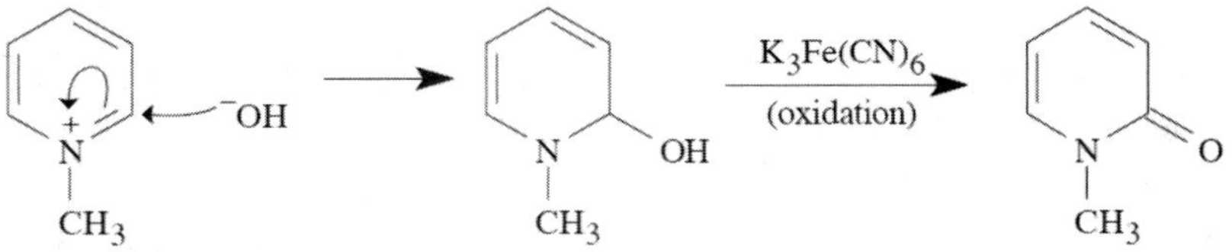

a. When the nucleophiles in such displacement reactions are anions, charge is neutralized.
b. Pyridine-N-oxides are in one sense pyridinium ions, and they react with nucleophiles in much the same way as quaternary pyridinium salts.

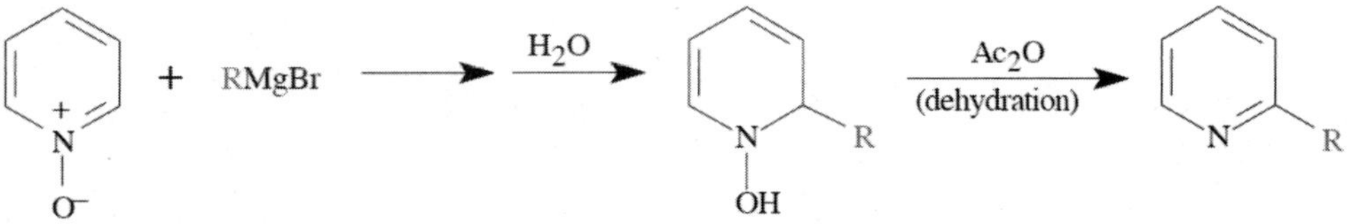

III. OTHER REACTIONS OF HETEROCYCLIC AROMATIC COMPOUNDS

A. ADDITION REACTIONS TO FURAN

1. Furan, pyrrole, or thiophene can be viewed as a 1,3-butadiene with its terminal carbons "tied down" by a heteroatom bridge.

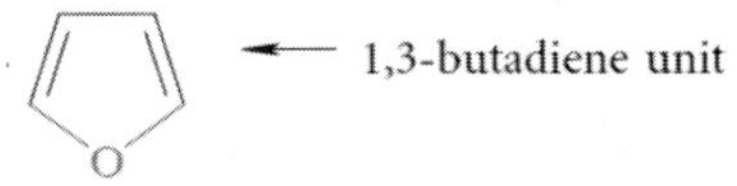

← 1,3-butadiene unit

2. Of the three heterocyclic compounds furan, pyrrole, and thiophene, furan has the least resonance energy.
   a. Furan has the greatest tendency to behave like a conjugated diene.

b. Furan undergoes some conjugate-addition reactions.
c. Furan undergoes Diels–Alder reactions with reactive dienophiles.

B. SIDE-CHAIN REACTIONS OF FURAN, PYRROLE, AND THIOPHENE

1. Many reactions occur at the side chains of heterocyclic compounds without affecting the rings.

C. SIDE-CHAIN REACTIONS OF PYRIDINE DERIVATIVES

1. The "benzylic" hydrogens of an alkyl group at the 2- or 4-position of a pyridine ring are more acidic than ordinary benzylic hydrogens because the electron pair (and charge) in the conjugate-base anion is delocalized onto the electronegative pyridine nitrogen.
   a. Strongly basic reagents such as organolithium reagents or $NaNH_2$ abstract the benzylic proton from 2- or 4-alkylpyridines.

b. The anion formed in this way has a reactivity much like that of other organolithium reagents.
   c. These anions are somewhat analogous to enolate ions, and thus they undergo some of the reactions of enolate anions, such as aldol condensations.
2. The "benzylic" hydrogens of 2- or 4-alkylpyridinium salts are much more acidic than those of the analogous pyridines.
   a. One resonance form of the conjugate-base "anion" is an uncharged species.

b. The conjugate base "anion" can be formed in useful concentrations by aqueous NaOH or amines.
3. Many side-chain reactions of pyridines, for example, side-chain oxidation, are analogous to those of the corresponding benzene derivatives.

D. PYRIDINIUM IONS IN BIOLOGY: PYRIDOXAL PHOSPHATE

1. Pyridoxal phosphate is one of the several forms of vitamin B_6.

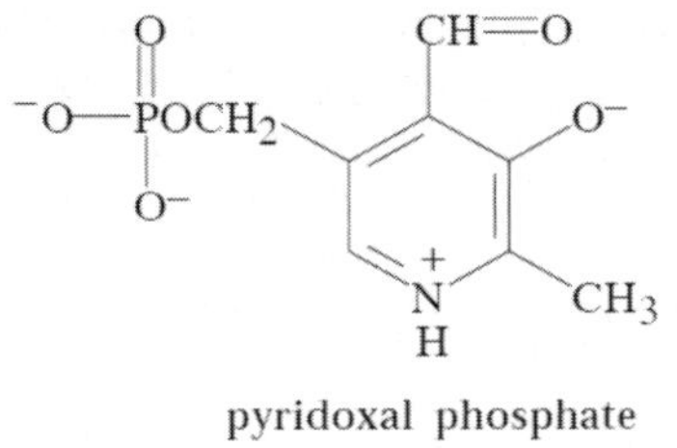

pyridoxal phosphate

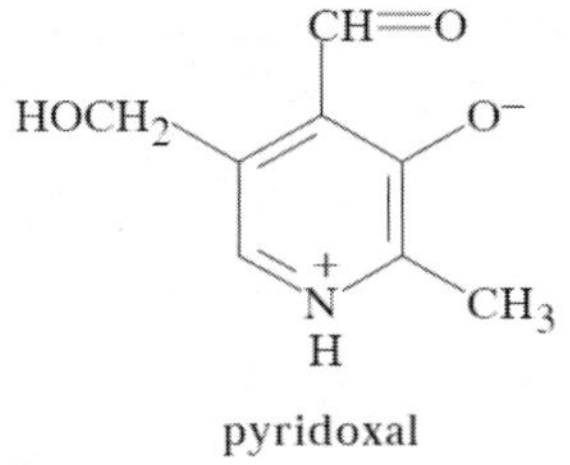

pyridoxal

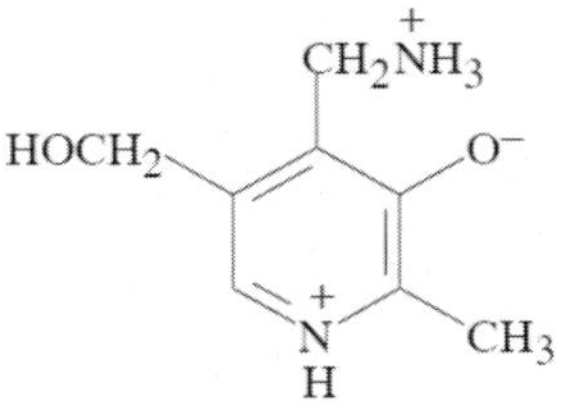

pyridoxamine

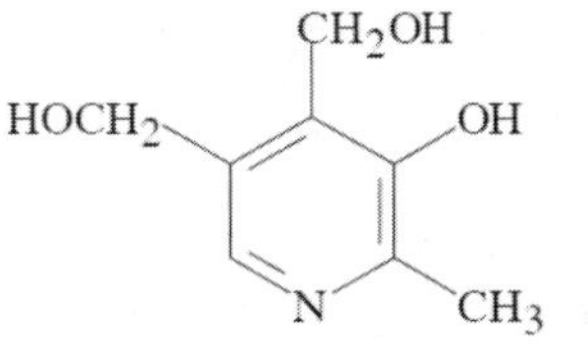

pyridoxol
(pyridoxine)

2. Pyridoxal phosphate is an essential reactant in several important biochemical transformations.
   a. interconversion of α-amino acids and α-keto acids

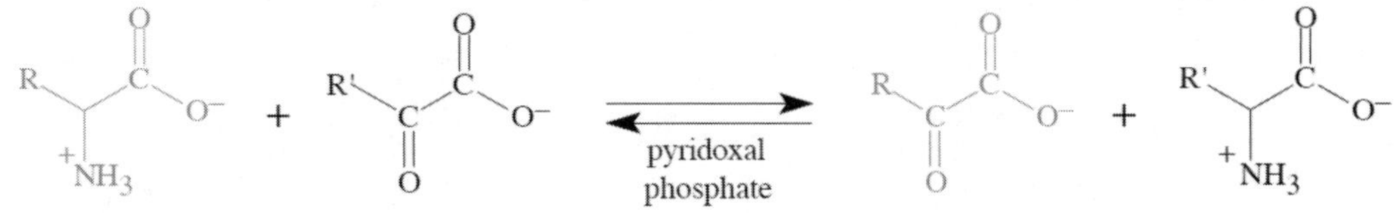

an α-amino acid   an α-keto acid   an α-keto acid   an α-amino acid

   b. decarboxylation of amino acids

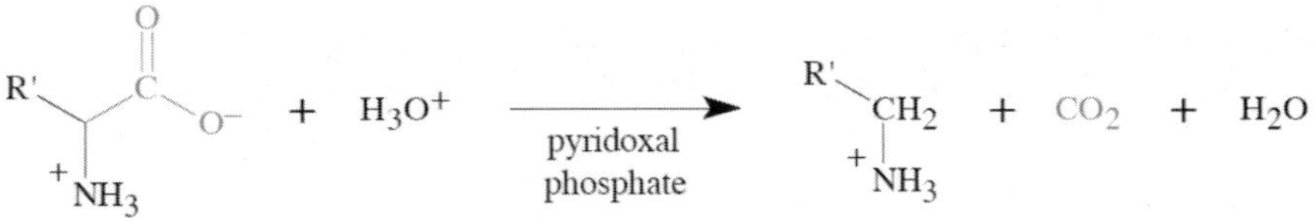

an α-amino acid

   c. loss of formaldehyde from serine

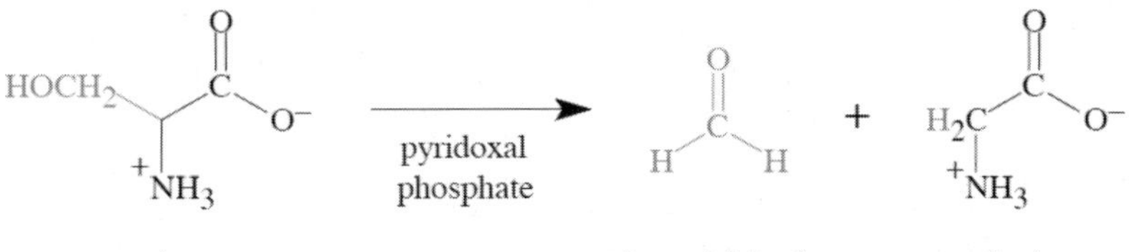

serine
(an α-amino acid)   formaldehyde   glycine
(an α-amino acid)

3. In the biological world, pyridoxal phosphate exists as imine derivatives of various amines, including enzymes.

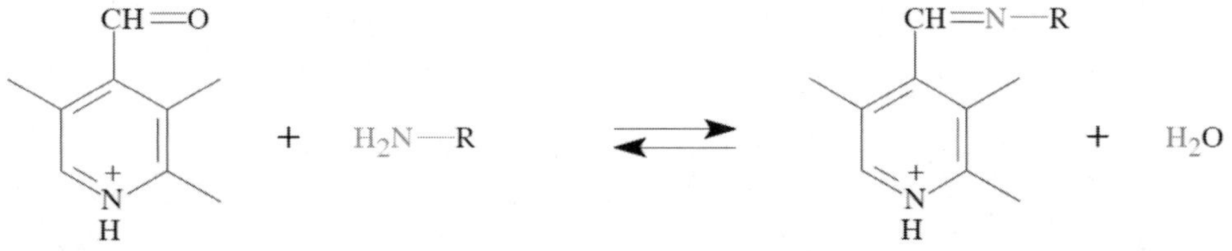

pyridoxal phosphate
(abbreviated structure)   an imine form of
pyridoxal phosphate

4. Biological reactions involving pyridoxal phosphate hinge on the ability of the pyridinium ion to stabilize carbon anions.

uncharged resonance structure

5. The *N*-protonated form of pyridoxal phosphate is thus very important to its mechanism of action.
   a. The predominant form of pyridoxal phosphate at physiological pH is the form in which the phenol is ionized and the pyridine is protonated.
   b. The —OH group in the 3-position and its ortho relationship to the aldehyde makes the phenolic —OH group of pyridoxal phosphate unusually acidic.
   c. Ionization of the phenolic —OH group raises the pK_a of the pyridinium ion.
   d. The pyridinium form is stabilized when pyridoxal phosphate is bound to the enzymes that catalyze its reactions.

SOLUTIONS TO PROBLEMS

Solutions to In-Text Problems

26.1   (a)   (b)

(CH$_3$)$_2$N—⟨pyridine⟩

4-(dimethylamino)pyridine

CH$_3$CH$_2$—⟨imidazole⟩—NO$_2$

4-ethyl-2-nitroimidazole

26.2   (a)   5-bromo-2-methylthiazole
        (b)   2-bromo-4-nitropyrrole
        (c)   pyridine-4-carboxaldehyde or 4-pyridinecarboxaldehyde
        (d)   8-methoxyquinoline

26.3   The resonance structures of pyrrole:

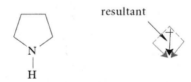

26.4   (a)   Because nitrogen is an electronegative atom, the C—N bond dipoles in pyrrolidine are directed towards the nitrogen, and their resultant is also directed towards the nitrogen:

resultant

It is given that the dipole moment of pyrrole is in the opposite direction, that is, directed away from the nitrogen. Although there is a dipole component directed towards the nitrogen as shown for pyrrolidine, the major component is due to the charge separation shown in the resonance structures given in the solution to Problem 26.3. In all resonance structures but the first, positive charge resides on the nitrogen and negative charge is on one of the carbons. Evidently, these contributions significantly outweigh those from the carbon–nitrogen bond dipoles.

        (b)   Oxygen is more electronegative than nitrogen. Consequently, the contribution of the carbon–oxygen bond dipoles in furan is greater than the contribution of the carbon–nitrogen bond dipoles in pyrrole, and the contribution of the charge-separated resonance structures for furan is smaller than the contribution of the analogous structures for pyrrole. Notice that the resonance energy of furan is less than that of pyrrole (Table 26.1 on text p. 1331); this is consistent with the idea that the resonance structures of furan are less important than those of pyrrole.

                In summary: In furan, the sum of the C—O bond dipoles outweighs the dipole–moment contribution of π-electron delocalization (see Eq. 26.2 on text p. 1329). In pyrrole, the dipole–moment contribution of π-electron delocalization outweighs the sum of its C—N bond dipoles.

(c)   Because the dipole vector of pyrrole is directed away from the nitrogen, and because the resultant of the two carbon-chlorine bond dipoles is also directed away from the nitrogen, the dipole moment of 3,4-dichloropyrrole should be greater than that of pyrrole:

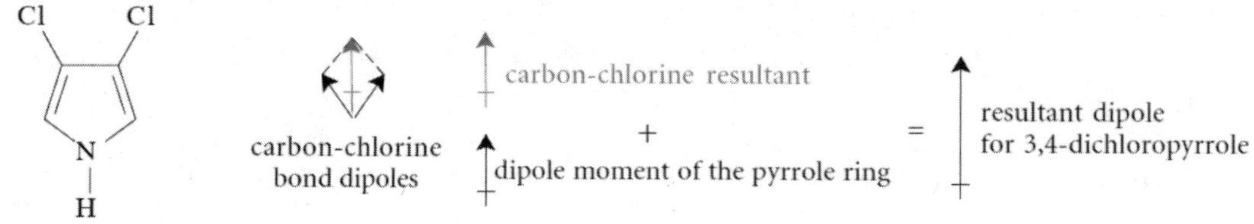

26.5   Both pyrrole and pyridine are aromatic, and therefore both should have ring current-induced chemical shifts. Consequently, the two larger shifts are associated with the two aromatic compounds, and the δ 2.82 resonance is that of pyrrolidine. This chemical shift is not far from the value predicted for the α-proton resonance of an alkylamine. (See Fig. 13.4 on text p. 621.) The resonance structures for pyrrole (see the solution to Problem 26.3, above) as well as its dipole moment (see the solution to Problem 26.4(a), above) show that pyrrole has partial negative charge on carbon-2; in contrast, the resonance structures of pyridine (Eq. 26.3, text p. 1329) show that carbon-2 has partial positive charge. Since electron deficiency is associated with large chemical shifts, it follows that the δ 8.51 resonance is that of pyridine and the δ 6.41 resonance is that of pyrrole.

26.6   (a)   The pyridine nitrogen is basic. Since bases can accept hydrogen bonds, pyridine can accept hydrogen bonds from water. This hydrogen-bonding capability increases the solubility of pyridine in water. Pyrrole, in contrast, is not basic, and the N—H hydrogen is not particularly acidic. Hence, pyrrole does not form strong hydrogen bonds with water, and it is therefore not very soluble.

(b)   Imidazole has one nitrogen that has the electronic character of the nitrogen in pyrrole, and one that is like the nitrogen in pyridine, except that it is somewhat more basic than the nitrogen of pyridine (Study Problem 26.1 on text p. 1332). Because of its basicity, imidazole can accept hydrogen bonds from water and, like pyridine, it is very soluble in water.

26.7   (a)   Protonation on the pyridine nitrogen occurs because the conjugate acid is resonance-stabilized.

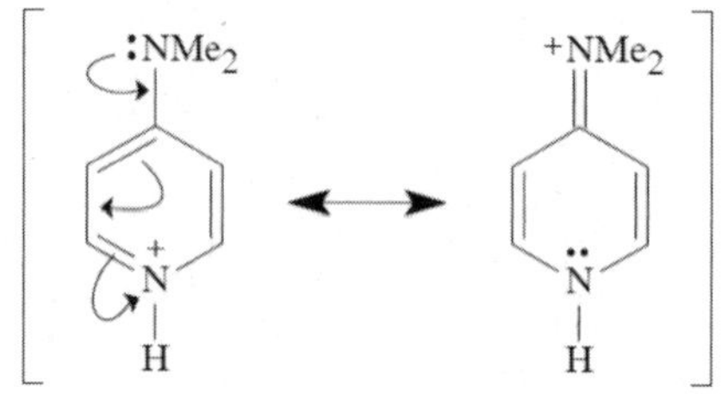

Protonation on the dimethylamino nitrogen, in contrast, gives a conjugate acid that is not resonance-stabilized. (Be sure to demonstrate this for yourself.) 4-(Dimethylamino)pyridine is more basic than pyridine because of the electron-donating resonance interaction of the dimethylamino substituent in the conjugate-base cation, shown above. The conjugate-base cation of pyridine itself, of course, lacks this stabilizing interaction.

(b)   Because the pyridine nitrogen is the more basic nitrogen, it is also the more nucleophilic nitrogen. Consequently, alkylation occurs on this nitrogen.

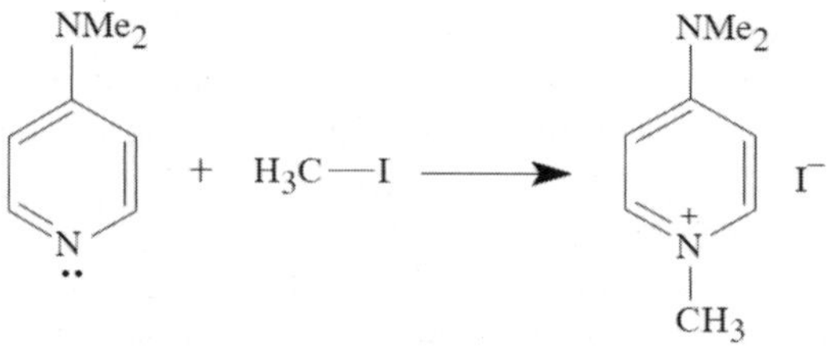

26.8   The unshared electron pair on the nitrogen of aniline is conjugated with the π-electron system of the ring. When aniline is protonated, this electron pair is no longer available for conjugation with the ring. Consequently, the reduction in conjugation causes a change in the UV spectrum. (See the solution to Problem 23.69 on p. 865 of this manual.) The pyridine unshared electron pair, in contrast, is not conjugated with the π-electron system of the ring because its orbital is in the plane of the ring. Consequently, protonation does not affect the conjugated system and therefore does not affect the UV spectrum.

26.9   (a)   The bromine directs the incoming nitro group to the "ortho" position, and the thiophene ring is more activated at the same position.

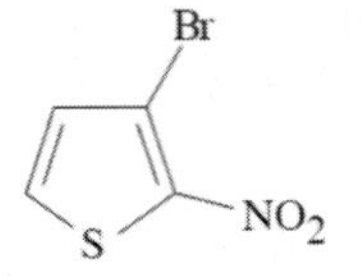

3-bromo-2-nitrothiophene

(b)   This is a Friedel–Crafts acylation; both the methyl group and the ring direct substitution to the open "para" position (5-position).

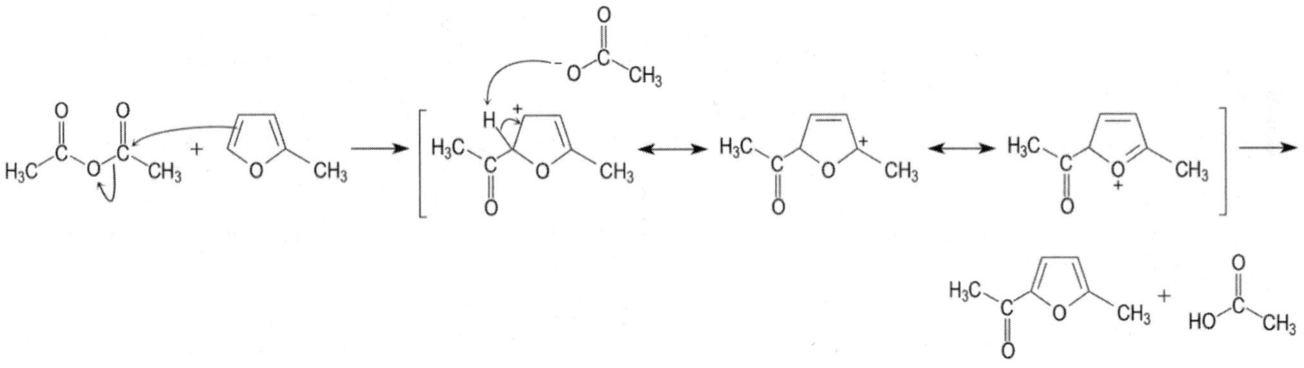

(c)   This is an aldol condensation.

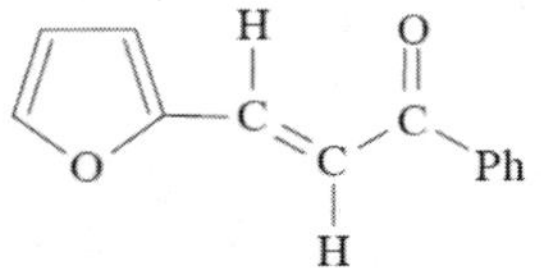

(d)   This is essentially a "benzylic" bromination: a free-radical bromination at a carbon adjacent to an aromatic ring (Sec. 17.2).

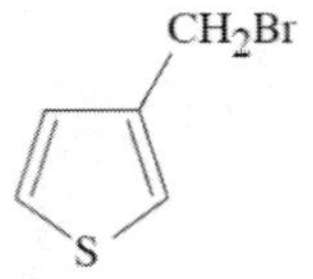

3-(bromomethyl)thiophene

26.10   This is an electrophilic aromatic substitution reaction in which the electrophile is the protonated aldehyde. (The carbonyl-protonated aldehyde shown in the mechanism below is undoubtedly in equilibrium with its nitrogen-protonated isomer, but the latter is not shown since it is not involved directly in the mechanism.)

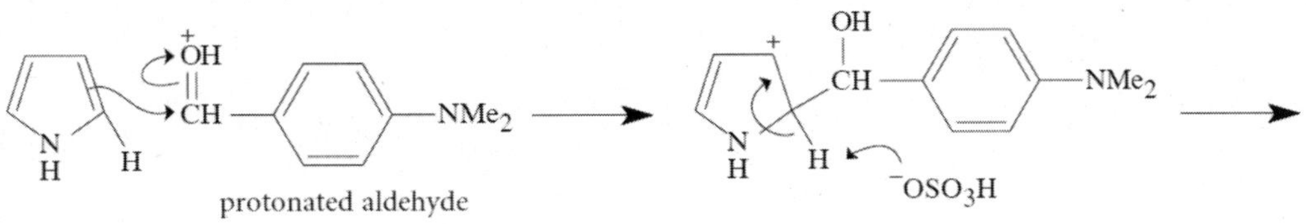

protonated aldehyde

26.11 The structures of the picolines are given in Sec. 26.4A in the text. α-Picoline is more reactive because both the methyl substituent group and the ring direct substitution to the 3-position. The carbocation intermediate for substitution at the 3-position has a resonance structure in which the positive charge is on the carbon bearing the methyl group:

tertiary carbocation resonance structure

product of nitration at carbon-3

In contrast, nitration of β-picoline gives a carbocation intermediate for which all the resonance structures are secondary carbocations.

(Nitration at the 2- or 4-position gives a carbocation that has a tertiary resonance structure; however, one of the resonance structures of this carbocation is unimportant because it has an electron-deficient nitrogen.)

26.12 Substitution at the 4-position gives a carbocation that is stabilized by resonance interaction with an unshared electron pair on the oxygen. Substitution at the 3-position gives a carbocation that is not stabilized by such an interaction.

two resonance structures of the carbocation intermediate that results from substitution at the 4-position

26.13   (a)

(b)   Follow the steps described in Sec. 10.6A of the text. Calculate the oxidation level of each atom involved in the reaction, in this case, nitrogen and phosphorus. Nitrogen is reduced (it gains two electrons) and phosphorus is oxidized (it loses two electrons). This is a two-electron redox reaction.

26.14   The curved-arrow mechanism for the nucleophilic aromatic substitution reaction of 4-chloropyridine with methoxide ion is as follows. Notice that negative charge in the anionic intermediate is delocalized onto the nitrogen, an electronegative atom.

26.15   4-Bromopyridine undergoes nucleophilic aromatic substitution by phenolate ion. The anionic intermediate is stabilized because negative charge is delocalized onto the nitrogen, as shown in Eq. 26.43b on text p. 1345 (with Y = Br, ⁻:Nuc = ⁻OPh). In the analogous substitution reaction of 3-bromopyridine, negative charge cannot be delocalized onto the nitrogen in the anionic intermediate; consequently, the intermediate is less stable and the reaction doesn't occur.

4-bromopyridine   4-phenoxypyridine

26.16   (a)   (b)   (c)

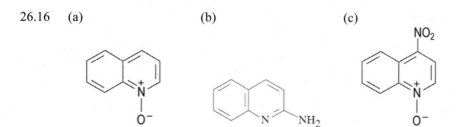

26.17   (a)   Because nitration of a pyridine ring generally occurs in a 3- or 5-position, the pyridine must first be oxidized to a pyridine-*N*-oxide so that nitration is directed to the 4-position. Once nitrated, the *N*-oxide is reduced back to a substituted pyridine. (See Eq. 26.31 on text p. 1341.)

3-picoline
(**β**-picoline)
$\xrightarrow{\text{30\% } H_2O_2}$
$\xrightarrow[H_2SO_4]{CH_3 \ HNO_3}$
$\xrightarrow{PCl_3}$
3-methyl-4-nitropyridine

(b)   In this case, the pyridine ring is nitrated on the 3-position by direct nitration.

4-picoline
(**γ**-picoline)
$\xrightarrow[H_2SO_4]{HNO_3}$
4-methyl-
3-nitropyridine

(c)   The "anion" formed by abstraction of a methyl proton by butyllithium can be carbonated with CO_2 much as any other organolithium reagent or Grignard reagent can be carbonated.

2-picoline
(**α**-picoline)
$\xrightarrow{LiCH_2CH_2CH_2CH_3}$
$\xrightarrow[2) \ H_3O^+]{1) \ CO_2}$

$+ \ CH_3CH_2CH_2CH_3$

(d)   A Hofmann or Curtius rearrangement can be used to introduce an amino group with loss of a carbon atom.

3-picoline
(**β**-picoline)
$\xrightarrow{HNO_3, \ heat}$
$\xrightarrow[2) \ NH_3]{1) \ SOCl_2}$
$\xrightarrow{Br_2, \ NaOH}$
3-aminopyridine

26.18   (a)   The formula of the product shows that one methyl has been introduced. The conjugate-base anion of 3,4-dimethylpyridine is formed by proton removal from the 4-methyl group, and this anion is alkylated by methyl iodide. A 4-methyl hydrogen is considerably more acidic than a 3-methyl hydrogen because the anion *A* that results from removal of a 4-methyl hydrogen is more stable. It is more stable because the negative charge can be delocalized to the pyridine nitrogen. (Draw the appropriate resonance structures.) The negative charge in the anion derived from ionization of the 3-methyl group cannot be delocalized to the nitrogen.

3,4-dimethylpyridine   *A*   4-ethyl-3-methylpyridine

(b)   The bromine at the 4-position is displaced by ammonia for the reasons discussed in the solution to Problem 26.14 in this supplement.

3,4-dibromopyridine   4-amino-3-bromopyridine

26.19   (a)   The pyridinium ion can stabilize a negative charge on the carbon α to the carboxylate (see Eq. 22.57b, text p. 1351) and thus facilitates the reaction.

pyridoxamine phosphate
(protonated form)

an α-keto acid

(b)   Begin with the imine product of Eq. 26.55, text p. 1350, formed from pyridoxal phosphate and the amino group of serine. (We abbreviate the structure of pyridoxal phosphate as in the text.)
        The first step is a β-elimination that produces formaldehyde and a resonance-stabilized "carbanion" in which the charge is essentially neutralized by delocalization of electrons onto the nitrogen of the pyridinium ring. (Be sure to show this!) This "carbanion" is then protonated to give an imine, which, in turn, hydrolyzes to glycine and pyridoxal phosphate.

resonance-stabilized "anion"
(Draw the structures!)

+ $H_2C{=}O$ formaldehyde

imine hydrolysis

pyridoxal phosphate

glycine

26.20   Isoniazid is a type of amide called a *hydrazide*. Semicarbazide (Table 19.3, text p. 985) is the same type of compound. As Table 19.3 shows, such compounds form imine derivatives called *hydrazones*. In this case the structure of the hydrazone is as follows:

It is reasonable to ask why isoniazid does not harm the host organism (*i.e.*, humans) by destroying pyridoxal phosphate in the host. The reason is that the bacterium has a very low level of the enzyme that converts pyridoxal into pyridoxal phosphate, whereas the host has a much higher level. Hence, replenishing the pool of pyridoxal phosphate takes a relatively long time in the Mycobacterium. Isoniazid is typically co-dosed with large amounts of Vitamin B_6 (that is, unphosphorylated pyridoxal and its congeners), which the human body recruits to form pyridoxal phosphate much faster than the Mycobacterium does. Hence, the Mycobacterium's reserve of available pyridoxal phosphate is destroyed by the drug and, unlike the human host, it cannot replenish pyridoxal phosphate fast enough to survive.

26.21   (a)

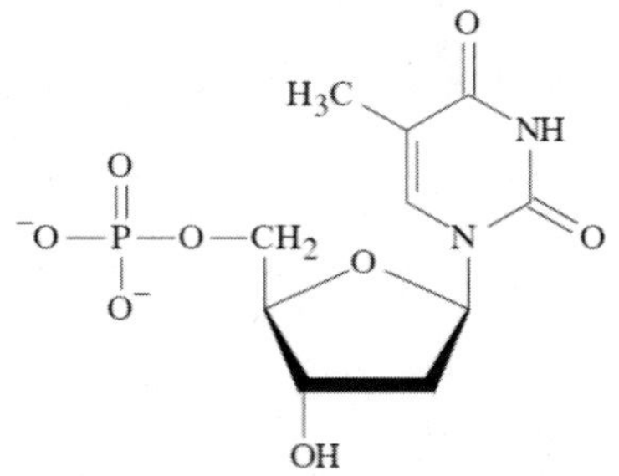

deoxythymidine monophosphate (*d*TMP)

(b)

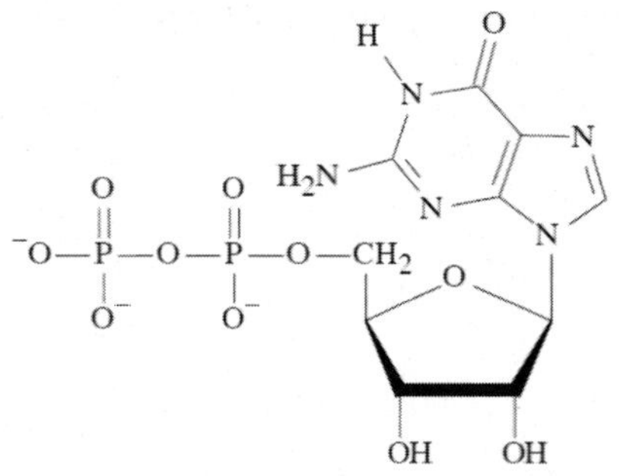

guanosine diphosphate (GDP)

26.22

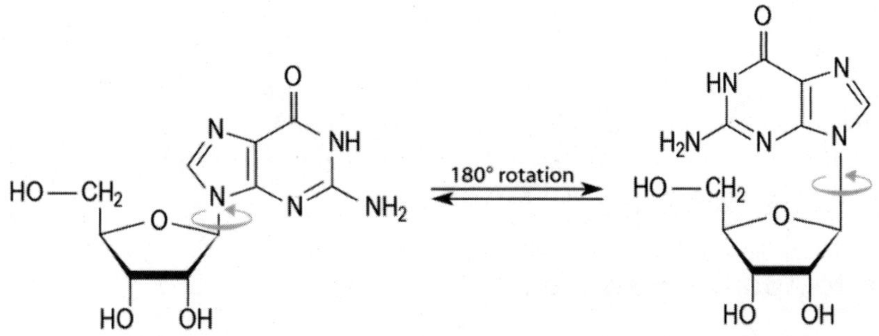

180° rotation

guanosine
(anti conformation
about C1′–N9)

guanosine
(syn conformation
about C1′–N9)

26.23   The structure of a tetranucleotide (a four-residue segment of RNA) with the sequence A-U-C-G is:

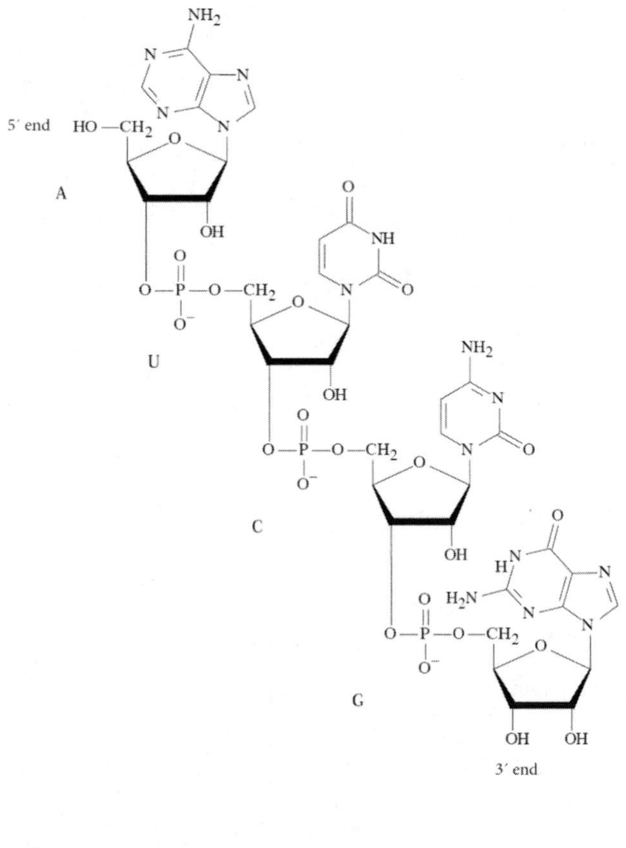

If you find an error in this manual, please visit http://people.pharmacy.purdue.edu/~loudonm/teaching/

26.24 Chargaff's first parity rule does not apply to individual DNA strands because the rule is a consequence of the base–pair complementarity that occurs *between* strands.

26.25 (a) The structure of an *O*-4-methylated thymine residue in DNA is as follows. (For the numbering of purine and pyrimidine rings see text p. 1354.)

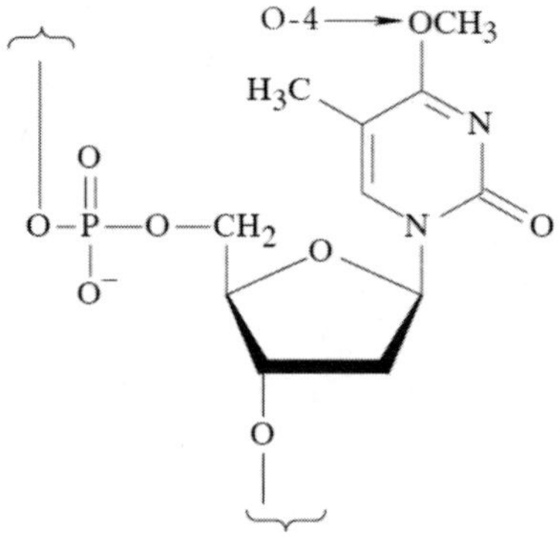

(b) Refer to Fig. 26.6(b), text p. 1359, and to Eq. 26.61 on text p. 1362. The answer to part (a) shows that *O*-methylation eliminates the hydrogen on nitrogen-3 of thymine that is involved in hydrogen bonding with nitrogen-1 of adenine. Because this hydrogen bonding is an essential part of base pairing, disruption of this hydrogen bond also disrupts base pairing.

Solutions to Additional Problems

26.26 (a) (b) 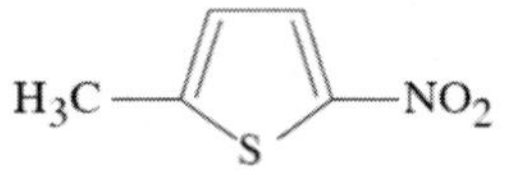 (c) 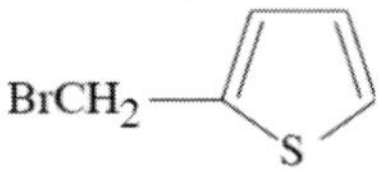 (d) no reaction

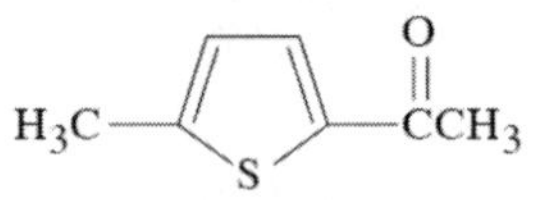

(e) no reaction (f) (g)

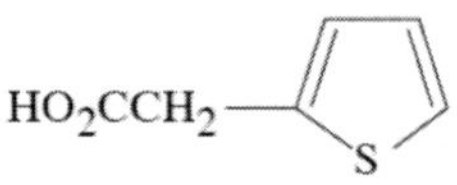

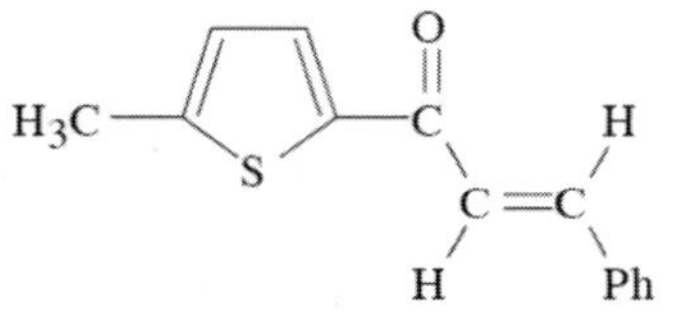

26.27 (a) 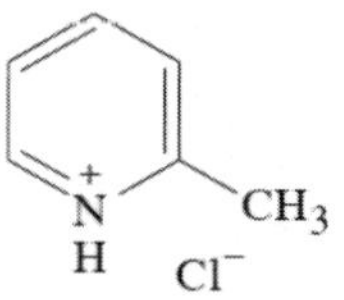 (b) no reaction (c) (d)

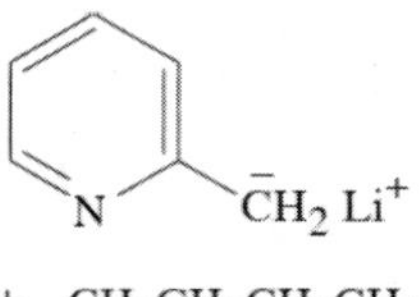

+ $CH_3CH_2CH_2CH_3$

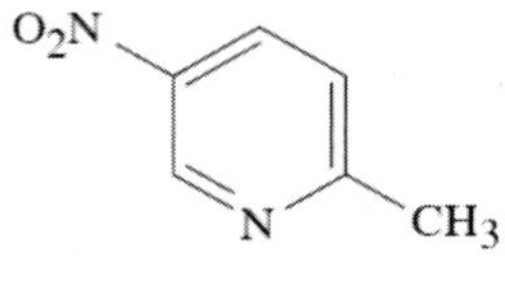

(e) 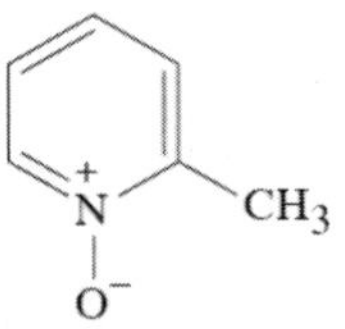 (f) 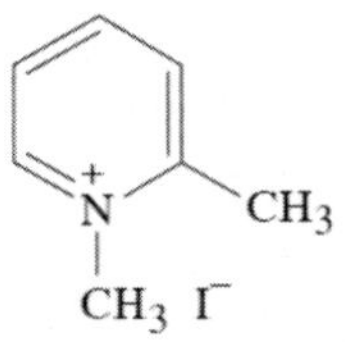 (g) 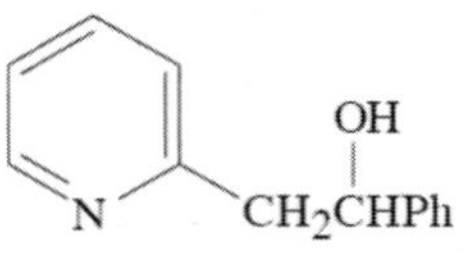 (h)

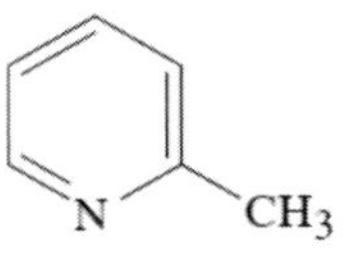

26.28 The order of increasing reactivity is

benzene $\ll$ thiophene $<$ 3-methylthiophene $<$ 2-methylfuran

The relative reactivity of thiophene and benzene is discussed on text pp. 1334–5. Another reason that thiophene is more reactive is that it has a smaller aromatic resonance energy than benzene, and therefore it sacrifices less aromatic stabilization when reacting with electrophile. 3-Methylthiophene is more reactive than thiophene because it contains an activating substituent, the methyl group. Furan is even more reactive for reasons also discussed on text p. 1334, and 2-methylfuran is even more reactive because the methyl group is an activating substituent.

26.29 (a) The order of increasing S_N1 reactivity is as follows:

$$B < A < D < C$$

These reactivities parallel the corresponding reactivities of the respective ring systems themselves in electrophilic aromatic substitution reactions. For example, furan is more reactive than benzene in electrophilic aromatic substitution, and thus compound C is more reactive than compound A in S_N1 reactions, because of the involvement of an oxygen unshared electron pair in the resonance stabilization of the carbocation intermediate:

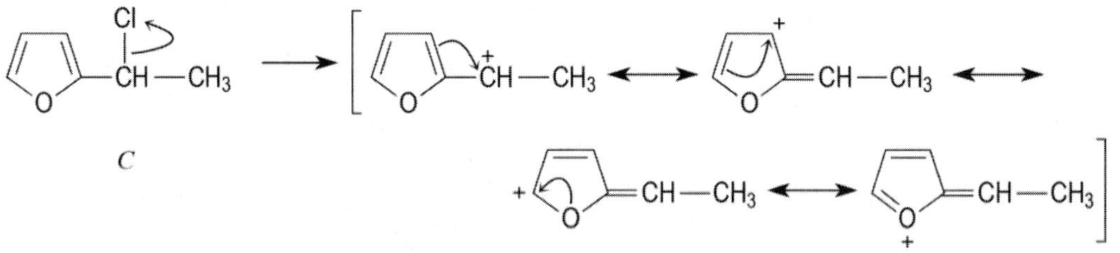

Just as a *para*-methoxy group accelerates solvolysis reactions (see Eq. 17.6 on text p. 839), atoms that donate electrons by resonance within the ring also accelerate solvolysis reactions. In contrast, a pyridine ring retards solvolysis reactions because the nitrogen unshared electron pair *cannot* be used to stabilize the carbocation by resonance; this electron pair is in the plane of the ring and cannot overlap with the rest of the π-electron system. Furthermore, one resonance structure of the carbocation intermediate places positive charge and electron deficiency on the nitrogen, an electronegative atom. Such a structure is not important. Thus, the nitrogen exerts only its rate-retarding polar effect. The reactivity of the pyridine derivative is further reduced by the formation of HCl as a by-product of the solvolysis reaction. The HCl protonates the pyridine nitrogen and further enhances its electronegativity, thus retarding the solvolysis rate even more.

The carbocation intermediate in the solvolysis of the thiophene derivative has resonance structures analogous to those shown above for the furan derivative. However, because sulfur uses orbitals from period 3, its overlap is less effective in stabilizing the carbocation.

(b) Compound E is the same as compound C, in part (a). As shown in the solution to part (a), compound E has four resonance structures to stabilize the cation after solvolysis. Compound F has only three resonance structures. Thus, compound E undergoes solvolysis faster than F, because the carbocation intermediate is more stable.

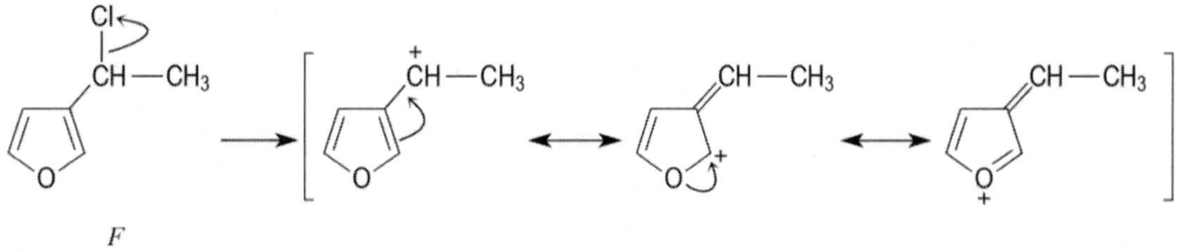

26.30 The hydroxy isomer is aromatic, and the carbonyl isomer is not. The principle to apply is that the compound with the greater resonance energy (Table 26.1 on text p. 1331) has the greater amount of hydroxy isomer.

(a)   Because pyrrole has more resonance energy (that is, more stabilization due to aromaticity) than furan, 2-hydroxypyrrole contains a greater percentage of hydroxy isomer than 2-hydroxyfuran.

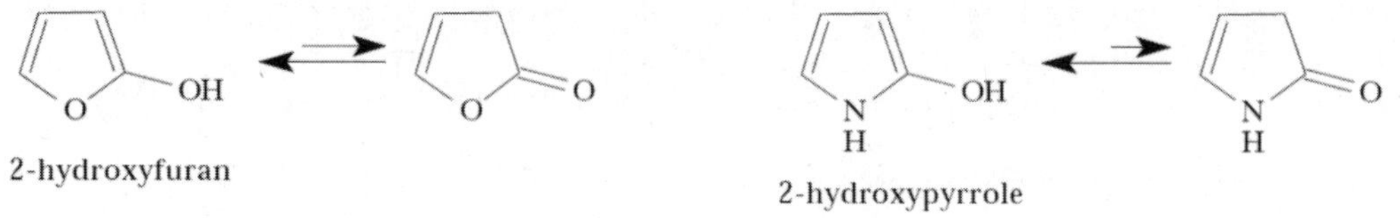

2-hydroxyfuran

2-hydroxypyrrole

(b)   Because benzene has greater aromatic stabilization (greater resonance energy) than pyridine, phenol contains the greater percentage of hydroxy isomer and 4-hydroxypyridine contains more carbonyl isomer.

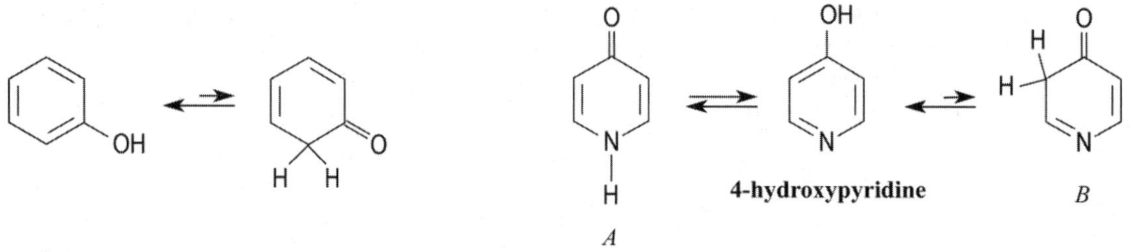

A

4-hydroxypyridine

B

(A second carbonyl isomer of phenol, shown in Eq. 22.14 on text p. 1111, is not shown above.) Pyridine also has two carbonyl isomers; isomer *A* is undoubtedly the major one of the two because the unshared pair on nitrogen can participate in amide-like resonance:

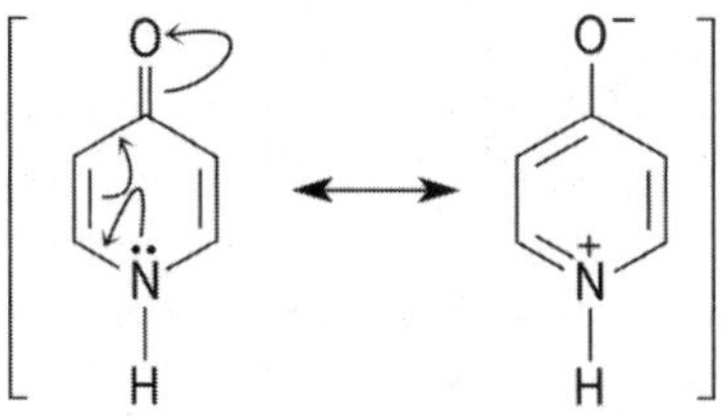

Because the structure on the right is aromatic, this isomer of 4-hydroxypyridine is stabilized. This type of resonance is possible neither for isomer *B* nor for phenol. This is a second reason that 4-hydroxypyridine has more of the carbonyl isomer than phenol. (See the similar discussion of amide resonance in 2-hydroxypyridine on text p. 1343.)

26.31   (a)   The order of increasing basicity (increasing conjugate-acid pK_a) is as follows:

5-methoxyindole $\ll$ 3-methoxypyridine $<$ pyridine $<$ 4-methoxypyridine

All pyridines are much more basic than the indole, which is not basic at nitrogen. 4-Methoxypyridine is more basic than either of the other pyridines, because an unshared electron pair of the methoxy substituent is available to stabilize the conjugate acid by resonance:

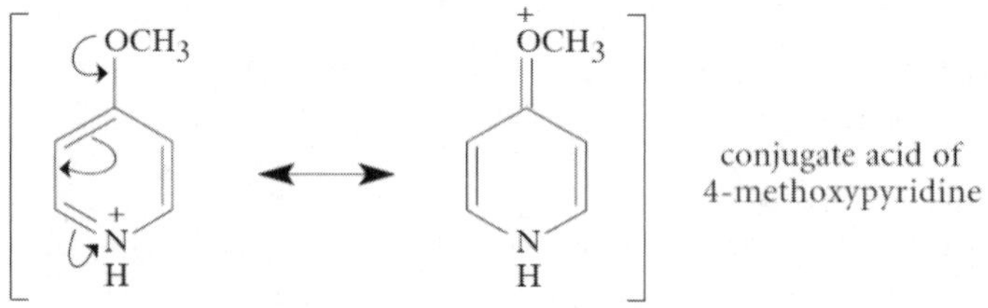

conjugate acid of
4-methoxypyridine

Such resonance is not possible when the methoxy group is in the 3-position; hence, the effect of a 3-methoxy substituent is limited to its acid-strengthening (base weakening) polar effect. Pyridine itself has neither the base-strengthening resonance effect nor the base-weakening polar effect of a methoxy substituent. Hence, its conjugate-acid pK_a is between those of the methoxypyridines.

(b)   The order of increasing basicity is

3-nitropyridine < 3-chloropyridine < pyridine

The order is a consequence of the relative polar effects of the nitro and chloro substituents, both of which are electron-withdrawing and base-weakening. As shown by the pK_a values in Table 23.1, text p. 1190, the electron-withdrawing polar effect of the nitro group on the pK_a of an ammonium ion is greater than that of the chloro group.

(c)   The first "anion" in the problem is actually a neutral compound.

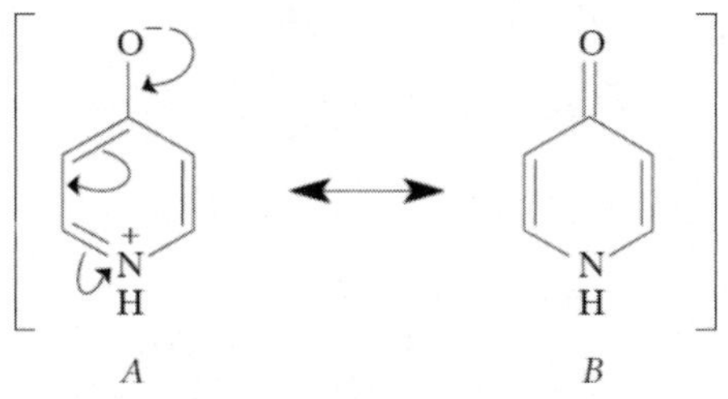

A B

That is, it is a carbonyl compound. It has much of the characteristics of an amide, because, as structures A and B above show, a nitrogen unshared electron pair is delocalized into the carbonyl group, as in an amide. This analogy suggests that the pK_a of its conjugate acid is near -1. (In fact, its conjugate-acid pK_a is 0.79.) The other anion in the problem is a phenolate ion; its conjugate acid is a phenol, with a pK_a of about 10. Consequently, the phenolate ion is considerably more basic.

(d)   The resonance structures of protonated imidazole are shown in Eq. 26.6 on text p. 1331. Compare the structure of protonated imidazole with the structure of protonated oxazole:

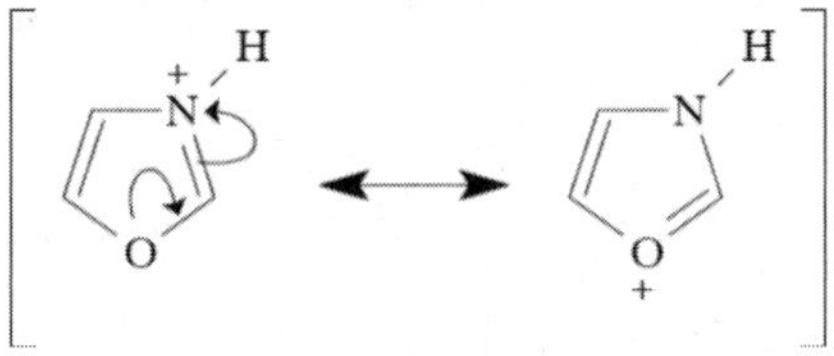

conjugate acid
of oxazole

In protonated oxazole, the positive charge is shared between an oxygen and a nitrogen; in protonated imidazole, the positive charge is shared between two nitrogens. Because oxygen is more electronegative than nitrogen, it supports positive charge less effectively; consequently, oxazole is less basic, and imidazole is more basic. (The conjugate-acid pK_a of imidazole is 6.95; that of oxazole is 0.38.)

(e)   See Study Problem 26.1 on text p. 1332 for the structure of protonated imidazole. In the case of thiazole, an electron pair of a sulfur is involved in the resonance stabilization of the conjugate acid:

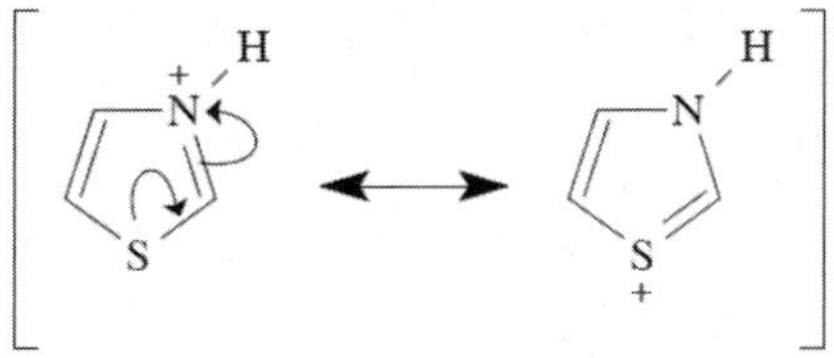

conjugate acid
of thiazole

The orbitals of sulfur overlap more poorly than those of nitrogen with carbon $2p$ orbitals because they are derived from a different quantum level; consequently, the resonance structure on the right is less important than the corresponding structure of protonated imidazole. The sulfur in thiazole is actually less electronegative than the nitrogen in imidazole, but, as in electrophilic aromatic substitution, the electron-donating resonance effect of nitrogen far outweighs its electron-withdrawing polar effect. Consequently, imidazole is considerably more basic than thiazole. (The actual conjugate-acid pK_a of imidazole is 6.95, and that of thiazole is 2.44.)

26.32   The conjugate acid is formed by protonation on the nitrogen of the double bond. Protonation on either of the other nitrogens does not yield a resonance-stabilized cation.

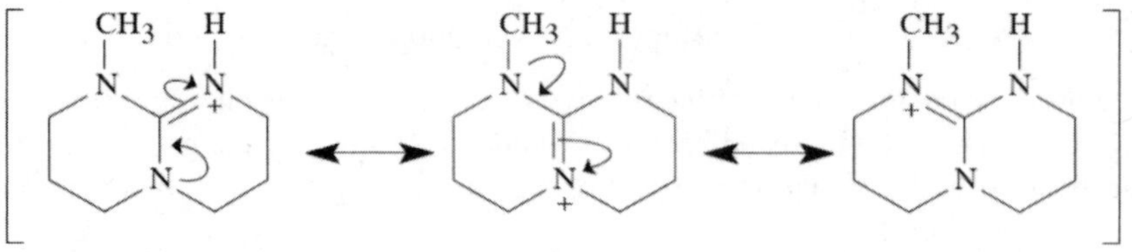

26.33 (a) Because alkylamines are more basic than pyridines (or quinolines), the conjugate acid of quinine is protonated on the nitrogen of the bridged bicyclic ring:

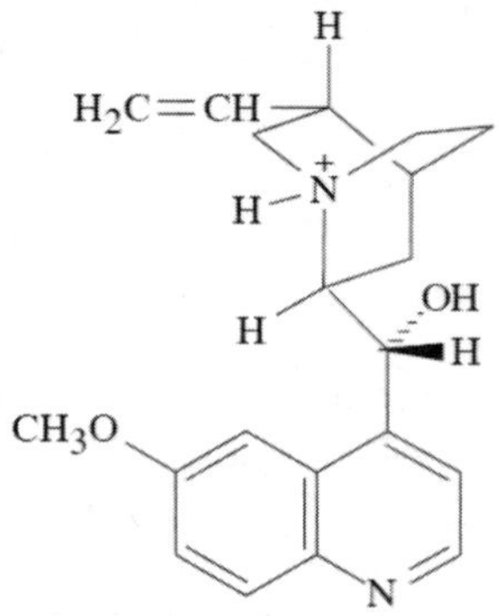

conjugate acid
of quinine

(b) The alkylamine group is more basic than the pyridine:

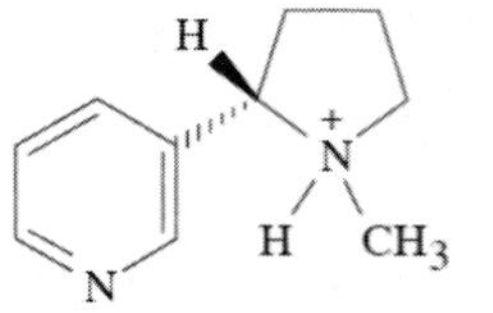

conjugate acid
of nicotine

(c) Indoles are not appreciably basic; tryptamine protonates on the nitrogen of the primary amino group.

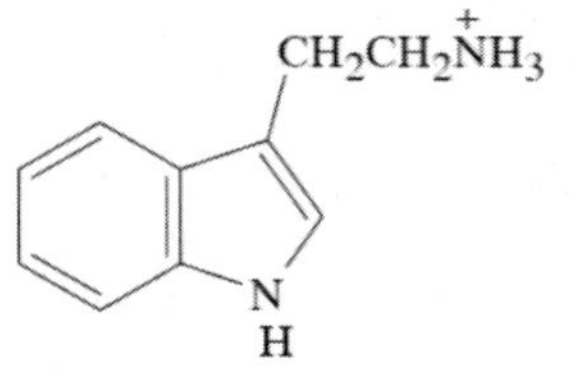

conjugate acid
of tryptamine

(d) Although it might seem that the nitrogen of an amino group might be more basic than the nitrogen of a pyridine ring, the pyridine nitrogen protonates because the conjugate acid is resonance-stabilized. (See the solution to Problem 26.7(a) on page 670 of this manual for a similar situation.)

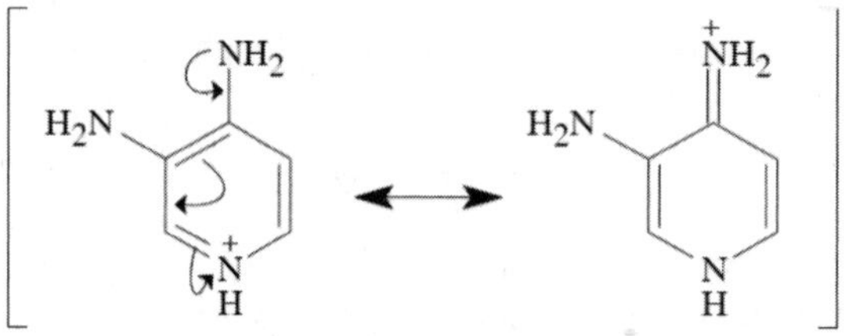

conjugate acid of
3,4-diaminopyridine

(e) One nitrogen resembles the nitrogen of indole, which is not very basic; the other resembles the nitrogen of pyridine, which is more basic. Furthermore, protonation of the pyridine nitrogen gives a conjugate-acid cation that is stabilized by resonance interaction with the unshared pair on the "indole" nitrogen. (Draw the appropriate resonance structures.) In fact, the conjugate-acid pK_a of 1,4-diazaindene is 6.92, whereas the conjugate-acid pK_a of pyridine is 5.2. The greater basicity of 1,4-diazaindene reflects the resonance stabilization of its conjugate acid.

conjugate acid of 1,4-diazaindene

(f)   Think of a 1,2,3-triazole as an imidazole with an extra nitrogen. Protonation occurs on the terminal nitrogen of the double bond as it does in imidazole because a resonance-stabilized cation is formed:

conjugate acid of
1-methyl-1,2,3-benzotriazole

This resonance stabilization notwithstanding, the conjugate-acid pK_a of benzotriazole is 1.6; the 1-methyl group will raise that somewhat, but not a great deal. Compare this with the conjugate-acid pK_a of benzimidazole, which is 5.52. Why is 1-methyl-1,2,3-benzotriazole so much less basic?

(g)   The tertiary alkyl nitrogens are more basic than both the aromatic nitrogens and amide nitrogen. The two alkyl nitrogens have very similar pK_a values, and the hydrogen is likely shared by both tertiary nitrogens to some degree.

26.34 (a)   The text has stressed that many reactions of pyridines can be rationalized by thinking of the N=C bond of the pyridine as if it were a carbonyl group. Thus, the hydrogens of the methyl group, like the α-hydrogens of a methyl ketone, are acidic enough to exchange for deuterium in basic D_2O.

(b)   The indole N—H is acidic enough to be removed by strong bases such as phenyllithium:

+   Ph—H (*i.e.,* benzene)

(c)   The fact that the amino group is part of a five-membered ring does not alter the fact that it is an activating, ortho, para-directing substituent. Under the conditions shown, the aromatic ring brominates once. (See Eq. 18.84, text p. 925, for the bromination of phenol under similar conditions.)

(d)   The product results from hydrogenation of the double bond that is not part of the aromatic system:

(e) The ethoxycarbonylamino substituent and the pyridine ring itself direct aromatic nitration to the "para" position (carbon-5). The stringent conditions are necessary because the pyridine ring is protonated, and thus is highly deactivated, under the acidic conditions of the reaction.

(f) Because furan is more activated than thiophene towards electrophilic aromatic substitution (Eq. 26.15 on text p. 1334), substitution occurs predominantly in the furan ring. Notice that the directing effect of the electron-releasing ring oxygen overrides the directing effect of the carbonyl group, which, by itself, would result in substitution at the 4-position of the furan ring.

(g) Pyridine-N-oxides are nitrated at carbon-4. (See Eq. 26.30, text p. 1341.) The directing effect of methyl is not strong enough to override the effect of an OH group. (Note that the N-oxide oxygen is protonated under the conditions of nitration; this point was not mentioned in the text.)

(h) Wolff–Kishner reductions occur on acylated pyridines just as they do on acylated benzenes.

26.35 (a) Doreen is attempting to apply a reaction of pyridines and quinolines to indole. Because attack on a double bond of indole by the amide ion cannot result in charge delocalization to the nitrogen, the reaction does not take place. The reaction that occurs instead is removal of the N—H proton of indole.

(b) Doreen is attempting the Chichibabin reaction on a pyridine derivative, but this derivative contains a chlorine in the 2-position, which, as chloride ion, is a vastly superior leaving group to hydride. The chlorine is lost rather than hydride to give 2-aminopyridine.

2-aminopyridine

26.36   The conjugate-base anion of 2-aminopyridine, formed as shown in Eq. 26.33d, text p. 1342, reacts with another pyridine molecule in a Chichibabin-like reaction:

26.37   (a)   Protonation of one pyrrole molecule at carbon-2 gives a carbocation that reacts with a second indole. The resulting product *X* undergoes an acid-catalyzed double-bond shift to give *A*. This rearrangement is favorable because, in compound *A*, the double bond is conjugated with a nitrogen unshared electron pair whereas, in compound *X*, such conjugation is absent.

(b)   Compound A from part (a) is protonated to give a resonance-stabilized carbocation *Y*, which serves as the electrophile in an electrophilic aromatic substitution reaction at carbon-2 of pyrrole.

26.38   The *p*-nitrobenzenediazonium ion, which is formed in the diazotization reaction, acts as the electrophile in an electrophilic aromatic substitution at carbon-3 of indole:

26.39 (a) Prepare pyridine-2-carbonitrile and reduce it to the amine.

(b) Use the *N*-oxide to direct nitration to the 4-position. Then convert the nitrated *N*-oxide into 4-chloropyridine, which is then displaced with ethanethiolate anion.

(c) Prepare 1-(2-furyl)-1-ethanone ("2-acetylfuran") and carry out an aldol condensation with furfural.

(d)   Oxidize furfural to the carboxylic acid and convert the acid into the ester. Notice that the acid chloride method of esterification is preferable to acid-catalyzed esterification with propanol because furans are somewhat sensitive to strongly acidic conditions. (See, for example, Problem 26.40, text p. 1368.)

(e)   Oxidize the methyl group to carboxylic acid, convert it into an acyl azide, and carry out a Curtius rearrangement to the isocyanate. Divide the isocyanate in half; convert one part into the amine, and then let the amine react with the remaining isocyanate to give the urea.

(f)   Use the enhanced "benzylic" acidity of *N*-methylpyridinium salts to form a nucleophilic anion at the benzylic carbon of the ethyl group, which is then cyanoethylated twice. (Compare to the reactions in Eqs. 26.50 and 26.51, text p. 1347. Cyanoethylation is discussed at the top of text p. 1158).

(g)   Alkylate the conjugate-base anion of 2-methylpyridine with propyl iodide; then convert the resulting 2-butylpyridine into its conjugate-base anion, which is carboxylated with CO_2 like any other organolithium or Grignard reagent.

26.40   A compound with two degrees of unsaturation that has an NMR spectrum consisting entirely of singlets which is completely obliterated in base, and that gives hexane as a Clemmensen reduction product, must be a six-carbon unbranched diketone in which all hydrogens are α-hydrogens, that is, 2,5-hexanedione. A mechanism for the formation of this compound from 2,5-dimethylfuran is as follows.

a resonance-stabilized carbocation

enol of A

2,4-hexanedione (compound A)

The final step, formation of the ketone from the enol, occurs by the reverse of the mechanism shown in Eq. 22.17b on text p. 1112.

26.41   The unsaturation number of compound A is 4; its oxidation to nicotinic acid shows that it contains a pyridine ring with a 3-substituent. The pyridine ring accounts for all four degrees of unsaturation. Because pyridine contains five carbons, the side-chain has three carbons. Compound A cannot be a primary alcohol, because the oxidation product of such an alcohol would be an aldehyde, which could not have five exchangeable α-hydrogens. (An aldehyde of the form $RCH_2CH{=}O$ has only two α-hydrogens.) Therefore, compound A is a secondary alcohol, and ketone B is its oxidation product.

A

B

26.42   (a)   By the principle of microscopic reversibility (see the bottom of text p. 175), the mechanism for the desulfonation of an arylsulfonic acid will be the reverse of the mechanism for sulfonation (see Eq. 16.13, text p. 804). For sulfonation to take place, a high concentration of the electrophilic agent (SO_3) is used in the presence of H_2SO_4, a powerful acid and dehydrating agent to provide a low concentration of water. For desulfonation to take place, a high concentration of water is required to reverse the equilibrium of the reaction. (SO_3 reacts with water to give H_2SO_4.)

The reaction scheme showing benzenesulfonic acid deprotonation and desulfonation.

(b) During the nitration, the sulfonyl chloride is converted into the sulfonic acid by the aqueous reaction conditions.

**2-thiophene-
sulfonyl chloride
*A***

**4-nitro-2-thiophene-
sulfonic acid
*B***

**3-nitrothiophene
*C***

(c) 3-Nitrothiophene cannot be made directly from thiophene because thiophene nitrates mostly in the 2-position to give 2-nitrothiophene (see Eq. 26.14, text p. 1334).

26.43 (a) Diethylamine and formaldehyde react to form an imminium ion by a mechanism completely analogous to the one shown in Eq. 23.28 on text p. 1201. This ion then serves as the electrophile in an electrophilic aromatic substitution reaction at the 3-position of indole.

imminium ion

 Indole, benzofuran, and benzothiophene tend to give a much greater percentage of 3-substitution than pyrrole, furan, and thiophene.

(b) Exchange is initiated with protonation by the deuterated acid at carbon-3 of the indole. In any exchange reaction of this sort, the experiment is set up so that there is a large excess of deuterium in the solvent; thus, the hydrogen that washes out into the solvent is swamped by the large excess of deuterium. In the last step, hydrogen is lost much more rapidly than deuterium because of the primary isotope effect (Sec. 9.5D, text p. 408). However, the relatively few molecules from which deuterium is lost are continually recycled through the mechanism until hydrogen is eventually washed out into the solvent.

(c) This reaction is a carbocation rearrangement known as the *indolenine rearrangement*. The driving force for this reaction is the aromaticity of the product.

(d) Alkylation of the nitrogen to give compound *A* (a quaternary ammonium salt) activates the carbon–nitrogen double bond toward a nucleophilic reaction with hydroxide ion, which, in turn, leads to ring opening.

(e) Sulfonylation of the *N*-oxide by the sulfonyl chloride activates the carbon–nitrogen double bond toward a nucleophilic reaction with chloride ion, which is produced in the sulfonation reaction. Hydroxide-promoted elimination forms a double bond and liberates chloride ion, which then reacts as a nucleophile at the terminal carbon of the double bond and displaces tosylate ion in an allylic variation of an S_N2 reaction (sometimes called an S_N2' reaction). This last step is undoubtedly driven by the aromatic stability of the product. The mechanism below begins with the sulfonate ester and chloride ion, produced in the formation of the sulfonate ester. (For the mechanism of sulfonate ester formation, see Further Exploration 10.2 on p. 269XR of this manual.) The toluenesulfonate group is abbreviated — OTs.

(f)   Because of the three chlorines on its α-carbon, trichloroacetyl chloride is unusually reactive in nucleophilic acyl substitution reactions. (Recall that reactivity toward nucleophiles of a carbonyl compound is enhanced by electron-withdrawing substituents on the α-carbon; see Sec. 19.7B.) The π electrons of pyrrole are nucleophilic enough to be acylated by this acid chloride to give compound *B*. The remaining mechanistic steps are the same as the final steps of a haloform reaction (see Eq. 22.27e on text p. 1116), except that the base is ethoxide rather than hydroxide. Ethoxide displaces the trichloromethyl anion, which reacts with ethanol to give chloroform and ethoxide.

(g)   This reaction is a nucleophilic aromatic substitution reaction by the benzenethiolate anion (PhS$^-$), which is formed by the reaction of benzenethiol with triethylamine. This reaction is like a nucleophilic aromatic substitution reaction on pyridine in the sense that a pyridine nitrogen stabilizes the anionic intermediate.

26.44   Hydrolysis proceeds through the standard acyl addition-elimination mechanism (Sec. 21.7). The rate of reaction can be determined by considering the stability of the leaving group. Compound *C* releases an amide base, which is a strong base. Compounds *A* and *B* release aromatic leaving groups, however *A* releases an aromatic ion, and *B* releases a neutral aromatic ion. The neutral aromatic leaving group is more stable, so the order of reactivity from least to most reactive is $C < A < B$.

26.45   (a)   For simplicity, represent the histidine side-chain as R; that is,

In this mechanism, the pyruvyl group is acting as a "surrogate" of pyridoxal phosphate. The first step is formation of the imine A of the pyruvyl group. This forms at the ketone carbonyl (why?). The mechanism of imine formation is outlined in Eq. 19.60a–b on text p. 984. This imine then loses CO_2 to give carbanion B, which is stabilized because its electron pair is delocalized into the other carbonyl group.

Protonation of this carbanion at carbon yields a new imine, which hydrolyses by a mechanism that is the reverse of imine formation to give histamine:

(b)   The carbonyl group of carbanion B shown in part (a) serves as the electron sink for decarboxylation.

26.46   (a)   First, serine forms an imine with pyridoxal phosphate as shown in Eq. 26.56 on text p. 1350. This imine undergoes a β-elimination of the elements of water to give the imine of dehydroalanine (compound A in the problem). The role of imine formation is to make the proton that is eliminated quite acidic; recall (Sec. 17.3B) that E2 reactions involving acidic β-protons are particularly rapid. (Why is this β proton acidic?) A base on the enzyme B: abstracts the proton, and an acidic group on the enzyme B—H protonates the —OH group as it leaves to form water.

imine derivative of serine and
pyridoxal phosphate

compound A

$+ \; H_2O + \overset{+}{B}{-}H \; + \; :B$

(b)   Delocalization of electrons from the double bond on the colored carbon to the pyridinium nitrogen reveals that this carbon has carbocation character, and should thus be electrophilic:

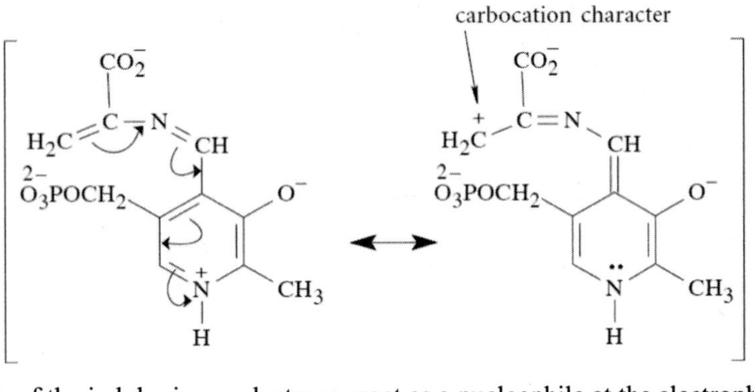

carbocation character

(c)   An electron pair of the indole ring π electrons react as a nucleophile at the electrophilic carbon of the imine of β-alanine:

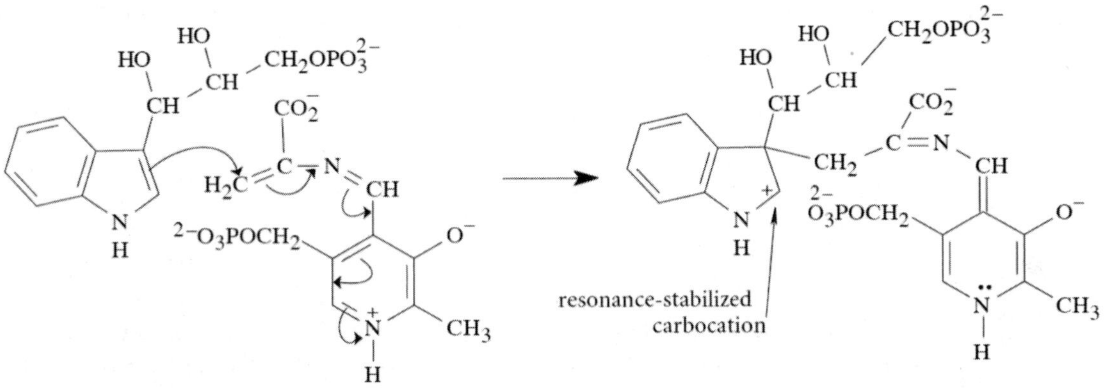

resonance-stabilized carbocation

The resulting carbocation intermediate is resonance-stabilized by the electrons of the indole nitrogen. A basic group removes an O—H proton to start an electronic cascade that generates glyceraldehyde-3-phosphate and an imine C:

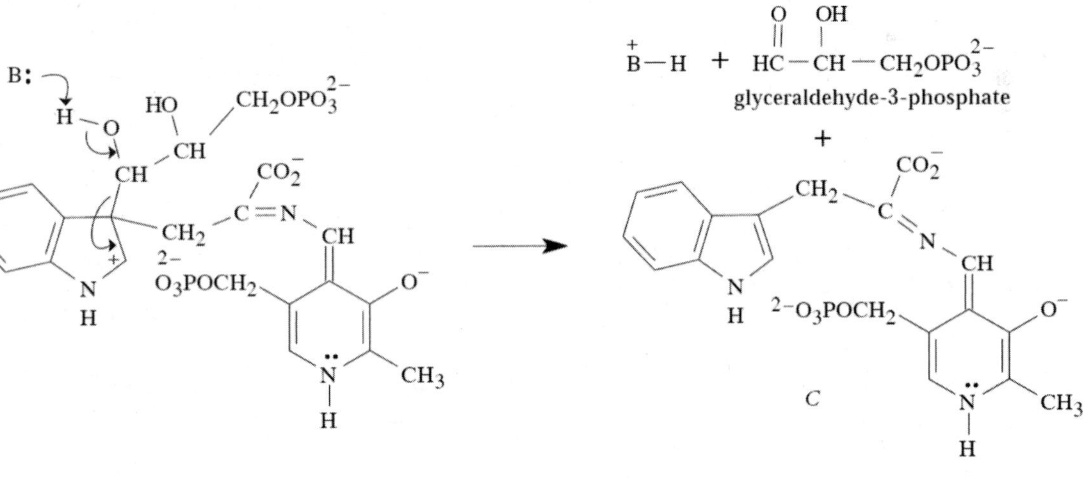

glyceraldehyde-3-phosphate

C

Finally, compound C is protonated on carbon to form an imine D between pyridoxal phosphate and tryptophan. Hydrolysis of this imine gives pyridoxal phosphate and the amino acid.

imine of pyridoxal phosphate and trypotophan

troptophan

pyridoxal phosphate

 The stereochemical issues have been ignored in this problem. Tryptophan produced in this way is solely the *S* enantiomer. At what step is the configuration of tryptophan established?

26.47 Form the carbanion intermediate as shown in Eqs. 26.56 and 26.57a, pp. 1350–1 of the text. This carbanion can be protonated from either face to give enantiomers. The enantiomeric amino acids are freed by imine hydrolysis.

imine hydrolysis

H₂O, H₃O⁺

enantiomeric amino acids

imine hydrolysis

26.48 (a) Exchange occurs when the conjugate-base anion is formed by abstraction of one of the colored protons, and this anion removes a deuteron (a deuterium nucleus) from the solvent. The anionic intermediate involved in this exchange is resonance-stabilized in such a way that negative charge is delocalized onto two electronegative atoms—one of the nitrogens and the carbonyl oxygen. In contrast, removal of a proton from the other methyl group gives an anion in which the negative charge is not delocalized onto electronegative atoms; consequently, this anion is less stable. Thus, exchange of a colored hydrogen occurs because this exchange involves the more stable anionic intermediate.

anionic intermediate involved in the exchange of the colored protons

The imide hydrogen is also exchanged by a base-catalyzed mechanism that involves the conjugate-base anion of the imide. (Imides, like other β-dicarbonyl compounds, are acidic; see Problem 22.2 on text p. 1108.)

(b)   Removal of a hydrogen from the black methyl group gives an anion that is stabilized by the polar effect of the attached positively charged nitrogen and by resonance interaction with the adjacent double bond. However, the resonance structure of this anion (draw it) delocalizes charge to a carbon atom. In contrast, removal of a hydrogen from the colored methyl group gives an "anion" which is actually a neutral compound; that is, the negative charge is delocalized to the positively charged nitrogen. Such charge neutralization is a significant stabilizing effect, and is the reason for the greater acidity of the colored hydrogen.

resonance structures of the "anion" formed by removal of a hydrogen from the colored methyl group

(c)   The positive charge on the nitrogen activates the carbon–nitrogen double bond towards a nucleophilic reaction with hydroxide ion. This forms a tetrahedral addition intermediate that breaks down to a substituted formamide A, which then undergoes amide hydrolysis to the products shown in the problem.

A

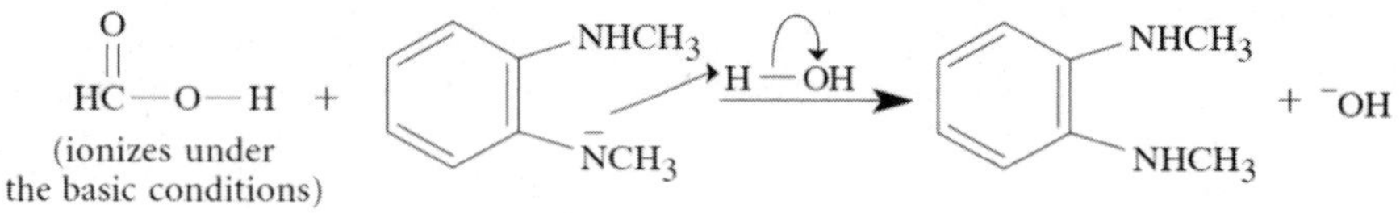

(d)   2-Pyridone does not hydrolyze because it undergoes a different reaction: ionization to its conjugate-base anion. This ionization occurs because the resulting anion is aromatic and is therefore very stable. The ion doesn't hydrolyze because the ionized amide is resonance-stabilized *and aromatic*. Remember that the more stable a carbonyl compound is, the slower it hydrolyzes (Sec. 21.7E).

Furthermore, hydrolysis would require that a second negative charge be introduced into the molecule when ⁻OH reacts at the carbonyl group. γ-Butyrolactam is much less acidic; consequently, ionization does not compete with the hydrolysis reaction.

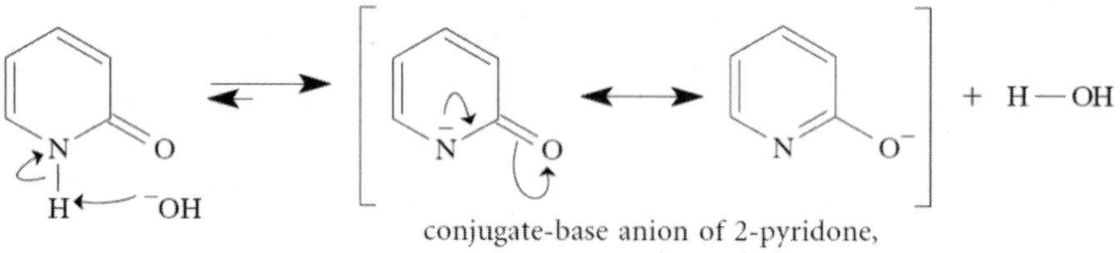

conjugate-base anion of 2-pyridone,
an aromatic species

(e)   The reaction of 4-chloropyridine with ammonia is a nucleophilic aromatic substitution reaction that occurs readily because the negative charge in the addition intermediate is delocalized to the pyridine nitrogen, an electronegative atom, as shown in Eq. 26.43b, text p. 1345 (with :Nuc = :NH₃ and Y = Cl). In contrast, the negative charge in the corresponding intermediate for nucleophilic aromatic substitution of 3-chloropyridine can only be delocalized to carbon atoms. (See Eq. 26.43c, text page 1345.) As a result, 3-chloropyridine is much less reactive.

26.49   Step (1) is a nitration reaction. Because 2-pyridone derivatives such as this starting material contain significant amounts of their aromatic hydroxypyridine isomers (Eq. 26.38 on text p. 1343), the nitro group can be introduced by aromatic nitration. The conditions of step (1) are HNO₃ and H₂SO₄.

Step (2) can be effected with PCl₅ and heat. (See Eq. 26.40 on text p. 1344.)

Step (3) is a reduction of both the nitro group and the nitrile to amine groups; catalytic hydrogenation is the method of choice. Evidently, the chlorine is also removed under the conditions of this hydrogenation. To see why this is reasonable, recall that 2-chloropyridines have some of the reactivity characteristics of acid chlorides, and that acid chlorides are reduced to aldehydes (that is, the chlorine is replaced by a hydrogen) in a type of catalytic hydrogenation (Rosenmund reduction; see Sec. 21.9D of the text).

Step (4) is an ether cleavage, which can be effected with aqueous H₂SO₄; water is the nucleophile in this reaction. This ether is activated toward cleavage by either an S_N1 or an S_N2 mechanism because it is benzylic. Neutralization with aqueous hydroxide gives the free amine shown.

Finally, step (5) can be realized by diazotization with NaNO₂ in aqueous H₂SO₄, which converts the benzylic amine into a benzylic diazonium ion, and the arylamine into an aryldiazonium ion; nitrogen is rapidly displaced from the benzylic diazonium ion by water. Elimination, usually a competing reaction in the decomposition of alkyldiazonium ions, is not observed here because there are no β-hydrogens. The aryldiazonium ion is also converted into the phenolic hydroxy group by heating the aqueous solution. (See the discussion above Eq. 23.45 on text p. 1208.)

26.50   (a)   Bromine adds to the double bond, and KOH saponifies the resulting dibromo lactone. An internal nucleophilic substitution reaction then closes the five-membered ring with loss of one bromine as bromide ion. (Recall that α-halo carbonyl compounds are particularly reactive in nucleophilic substitution reactions; see Sec. 22.3D.) An E2 reaction forms the double bond with loss of the second bromine as bromide ion. (This is a very favorable E2 reaction; why? See Sec. 17.3B.) The mechanism of bromine addition, not shown below, is exactly analogous to the mechanism shown in Eqs. 5.10–5.13 on text p. 185; and the mechanism of saponification, also not shown below, is exactly analogous to the mechanism shown in Eqs. 21.9a–b on text pp. 1061–2.)

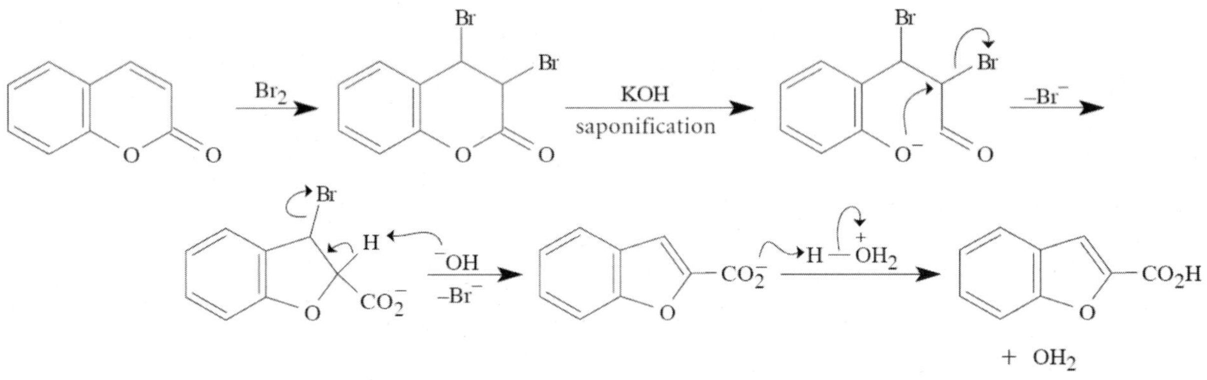

(b)   This reaction begins as an aldol condensation between the conjugate-base enolate ion *A* of the diester and the diketone. Normally esters undergo Claisen condensations, but, because 1,2-diketones are particularly reactive (why?), the aldol condensation is observed instead. Once the two molecules are joined by the first aldol condensation, a second aldol condensation by an analogous mechanism closes the ring.

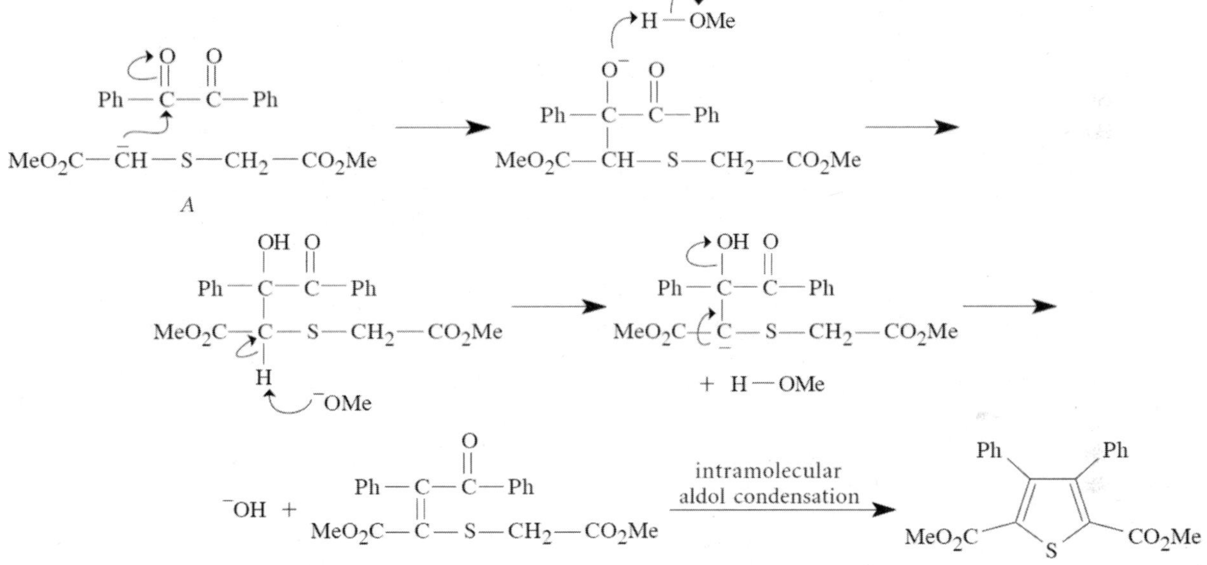

(c)   The mechanism shown below starts with the enamine formed between the amino group and the carbonyl group of 2-butanone. (The mechanism of enamine formation is discussed in Sec. 19.11B of the text.) A key step is the nucleophilic reaction of this enamine at the carbon of the protonated carbonyl group. As this reaction demonstrates, enamines, like enolate ions, are nucleophilic at the α-carbon. (See part (e) for a more extensive discussion of this point.) Note that, although primary amines form imines with ketones, imines are in equilibrium with enamines just as ketones are in equilibrium with enols. (See Eq. 19.67a–b on text p. 988.) Water is produced as a by-product of enamine formation. The conjugate acid of water, H_3O^+, is shown as the catalyzing acid, but any of the other acids present could also catalyze the reaction.

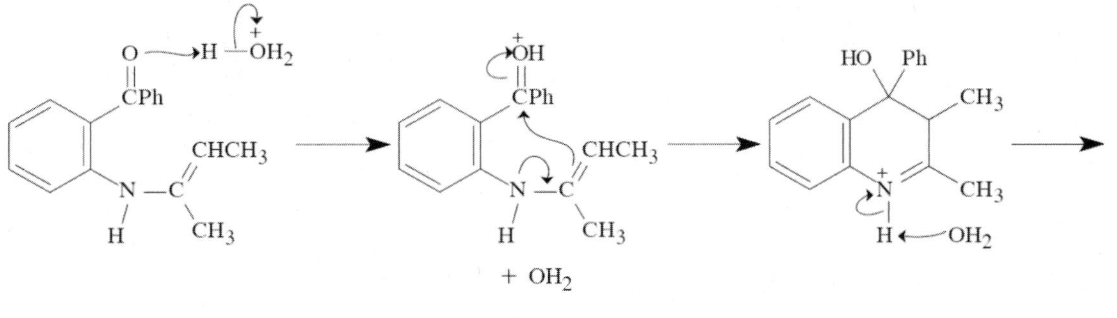

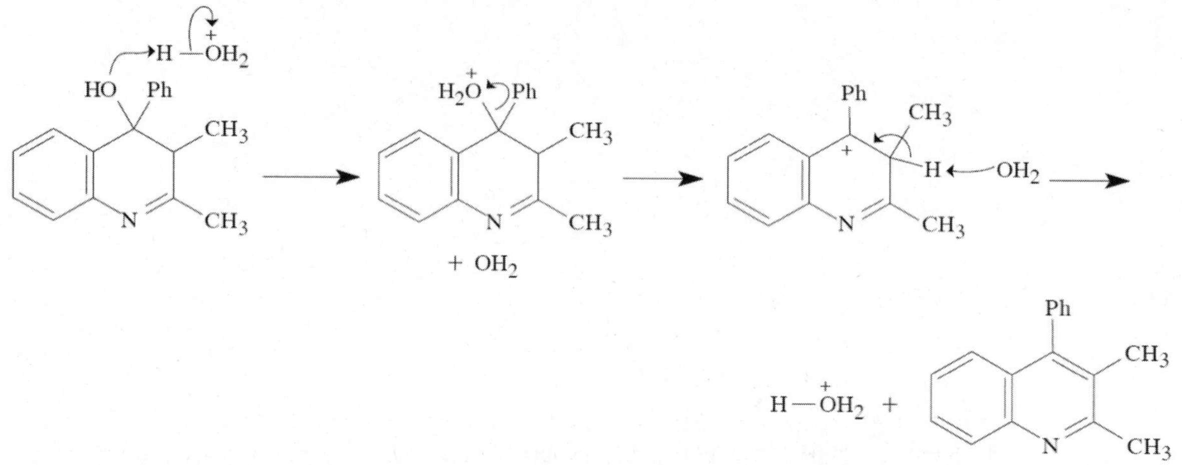

(d)   Imine formation between the amine and one of the ketone groups is followed by an intramolecular electrophilic aromatic substitution reaction in which the protonated ketone serves as the electrophile. The mechanism below begins with the imine; the mechanism of imine formation is discussed in Sec. 19.11S in the text.

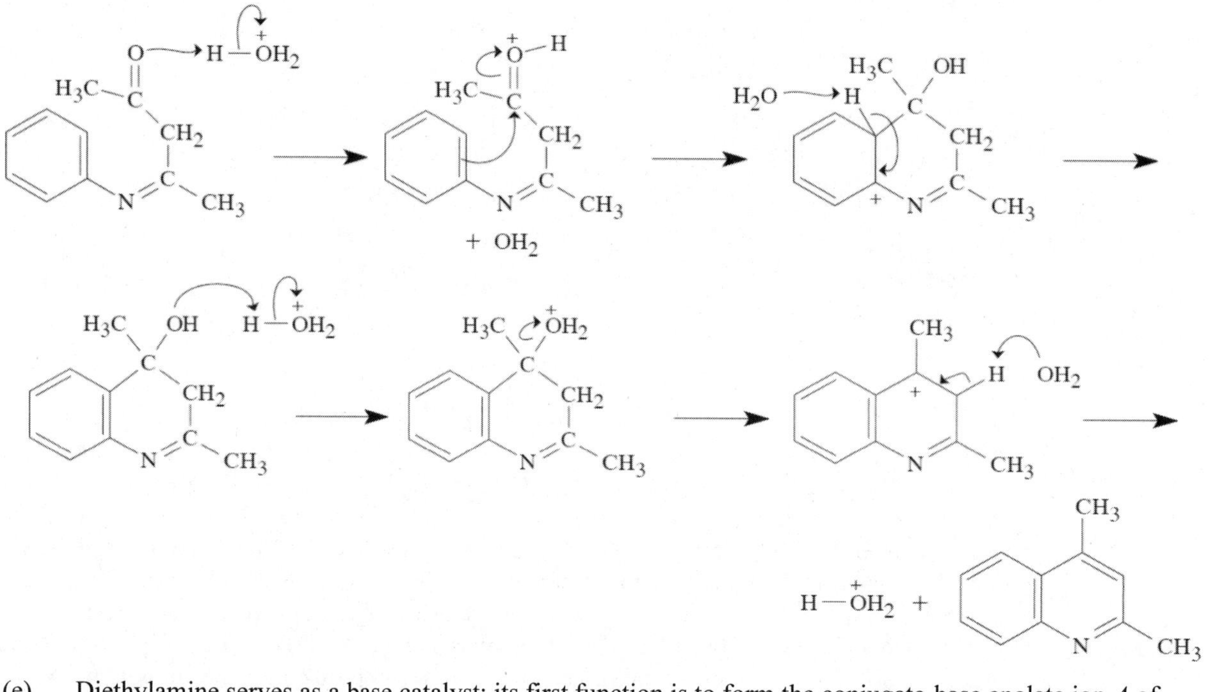

(e)   Diethylamine serves as a base catalyst; its first function is to form the conjugate-base enolate ion *A* of ethyl acetoacetate:

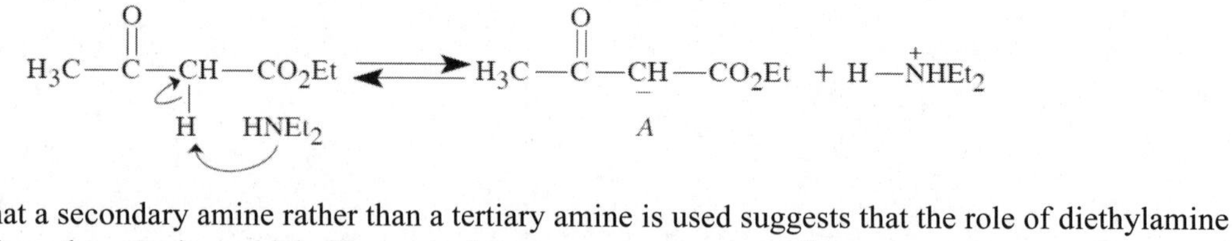

The fact that a secondary amine rather than a tertiary amine is used suggests that the role of diethylamine is also to form the enamine, which, like an enol, can serve as a nucleophile:

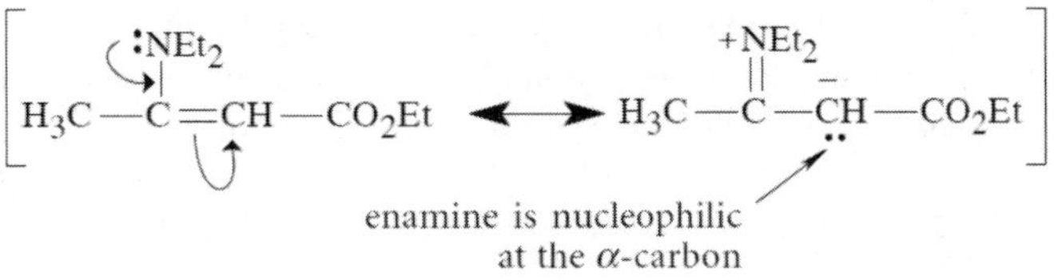

$$\left[\; H_3C-\underset{}{C}=CH-CO_2Et \quad \longleftrightarrow \quad H_3C-\overset{+NEt_2}{\underset{}{C}}-\overset{..}{C}H-CO_2Et \;\right]$$

enamine is nucleophilic
at the α-carbon

Diethylamine can also form an imminium ion by reaction with formaldehyde (Eq. 23.28, text p. 1134):

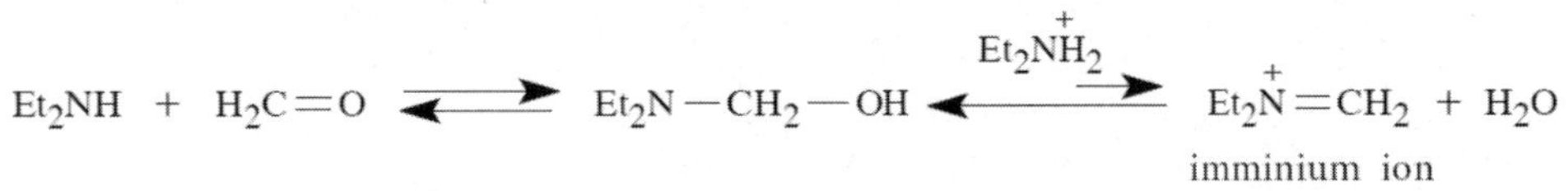

$$Et_2NH \;+\; H_2C=O \;\rightleftarrows\; Et_2N-CH_2-OH \;\overset{Et_2\overset{+}{N}H_2}{\underset{}{\longleftarrow}}\; Et_2\overset{+}{N}=CH_2 \;+\; H_2O$$

imminium ion

The first stage of the mechanism is a crossed-aldol condensation in which enolate *A* (or the enamine shown above) condenses with formaldehyde, (or with the imminium ion shown above). The mechanism of the aldol condensation is given in Eq. 22.40b, text p. 1120, and Eq. 22.44, text p. 1121. The product of this aldol condensation then undergoes a conjugate-addition reaction (that is, a Michael addition) with a second enolate ion *A* (or a second enamine). Ammonia reacts with one of the ketone functional groups to form an enamine, and the ring is closed when the amino group of that enamine forms an intramolecular enamine with the second ketone group. (The mechanism of enamine formation is discussed in Sec. 19.11B of the text; see the note about the equilibration of imines and enamines in the solution to part (c) of this problem.)

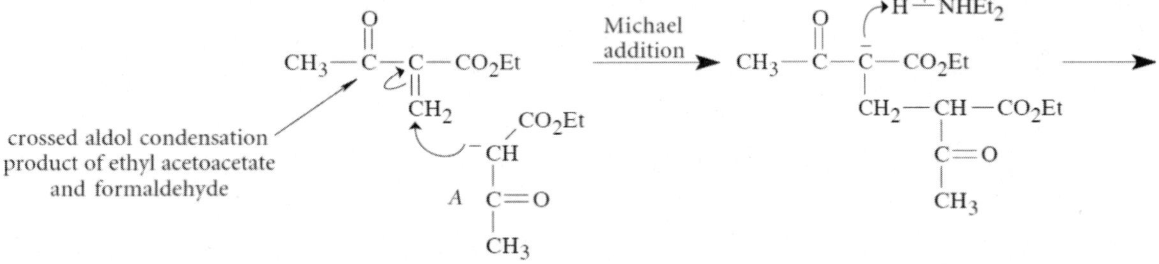

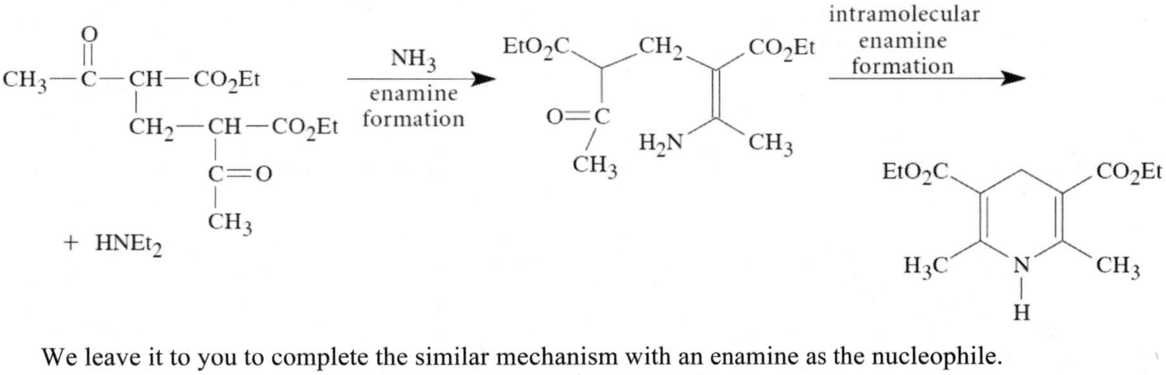

We leave it to you to complete the similar mechanism with an enamine as the nucleophile.

(f)   The *o*-nitro group is an essential element in the success of this reaction because its presence makes the benzylic methyl hydrogens acidic enough to be remove by ethoxide; the resulting anion is resonance-stabilized:

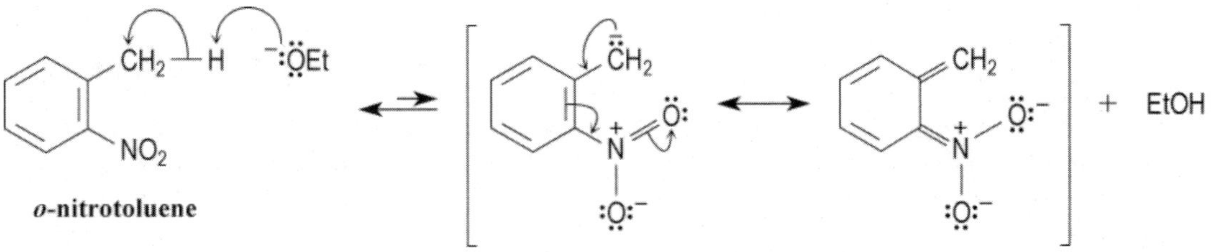

o-nitrotoluene

In a variation of the Claisen condensation (see Sec. 22.6A), this nucleophilic anion reacts with a carbonyl group of diethyl oxalate, displacing ethanol. Like the Claisen condensation, this reaction is driven to completion by ionization of the product. For this reason, at least one equivalent of the base must be used:

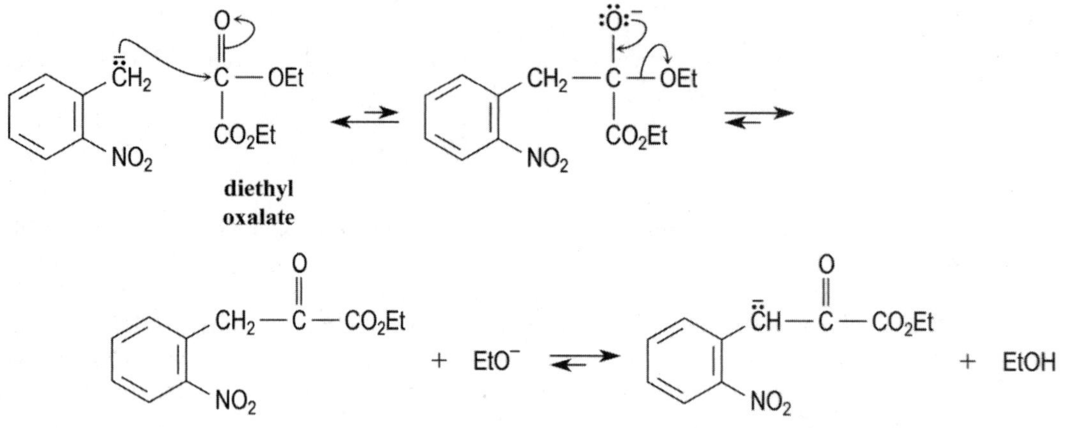

diethyl oxalate

The anion is neutralized by protonation in acetic acid, and the nitro group is converted into an amino group in a separate reduction step. [Catalytic hydrogenation (see Sec. 23.11B) is the reduction method used here.] The amino group thus formed reacts with the neighboring ketone to yield, after acid–base equilibria, an "enamine," that is, the aromatic indole.

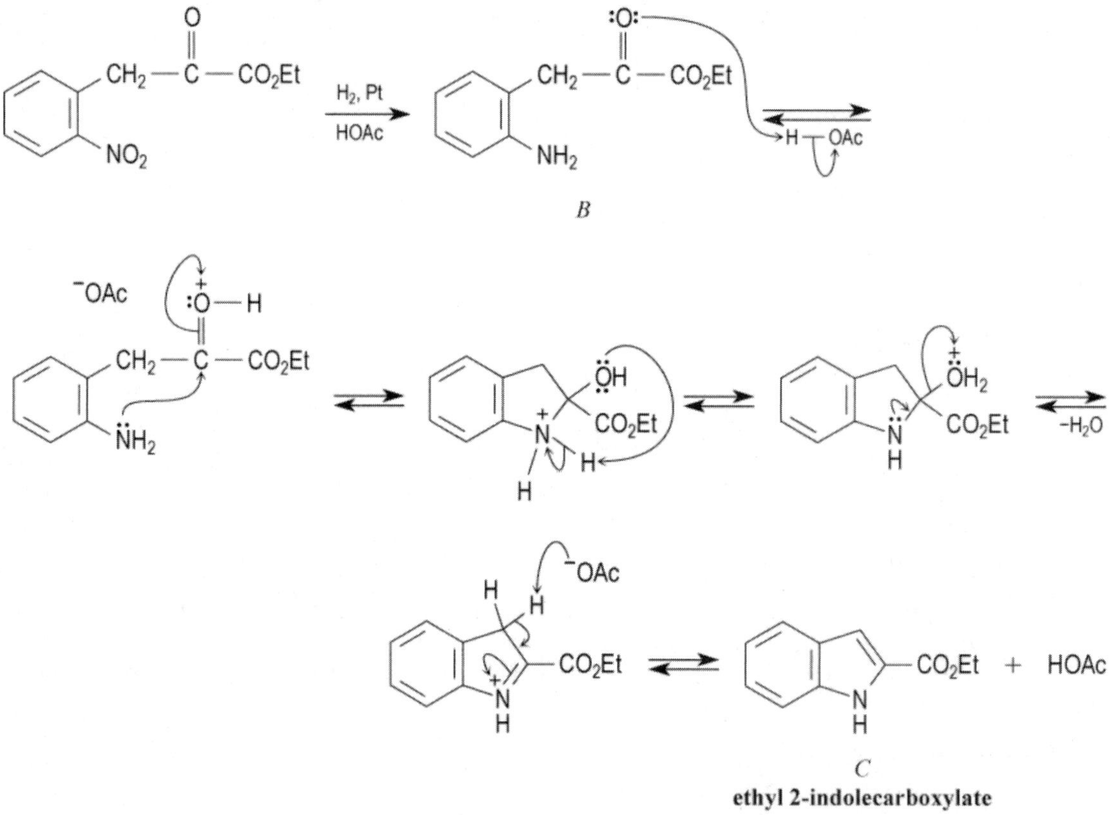

B

C
ethyl 2-indolecarboxylate

(g)   The reaction proceeds by enamine formation (see Sec. 19.11B) followed by an intramolecular Heck reaction (see Sec. 18.6A) in the presence of a bridged amine base (DABCO, 1,4-diazabicyclo[2,2,2]octane).

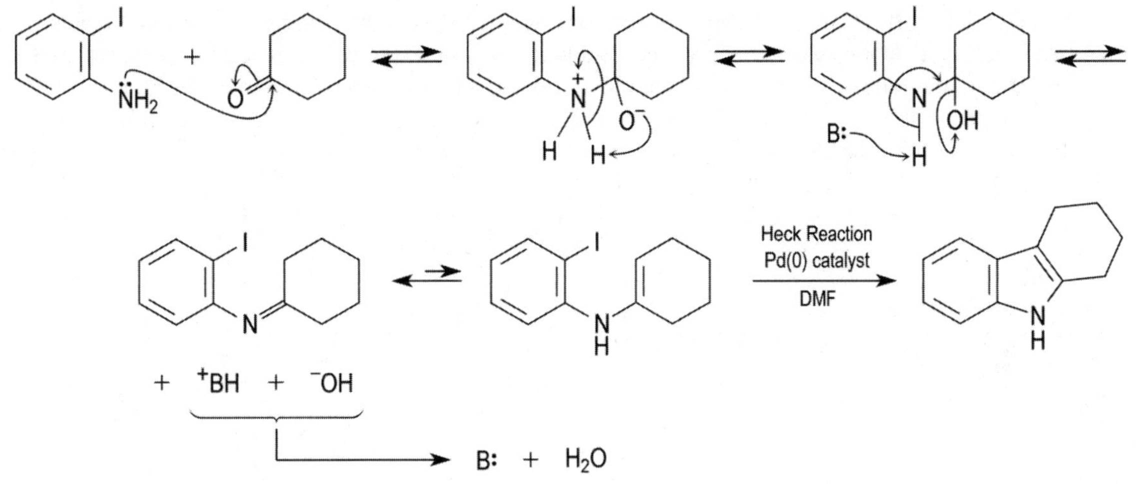

26.51 (a) The pairing of an imine isomer of C with A:

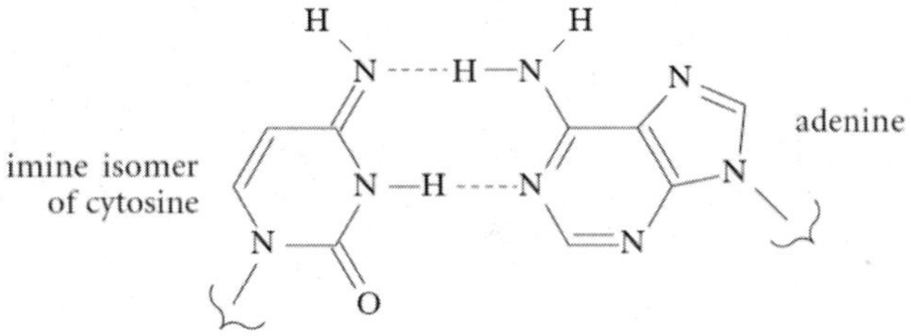

(b) The pairing of an enol isomer of T with G:

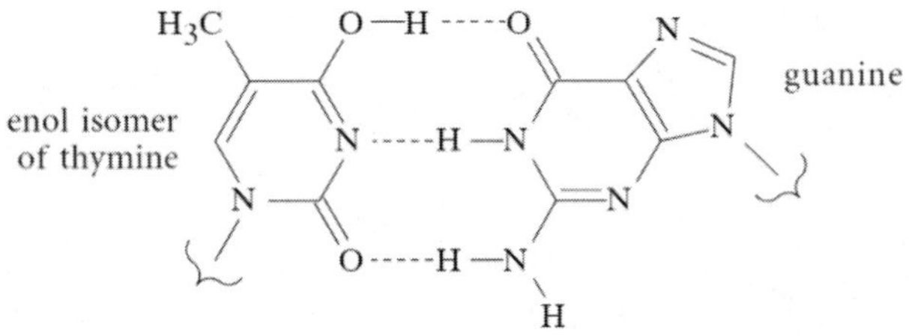

26.52 (a) Periodic acid cleaves the *cis*-2′,3′-diol group at the 3′-end of RNA to form a dialdehyde. (Note that only the 3′ residue of RNA contains a 2′,3′-diol.) The presence of an aldehyde carbonyl group makes the α-hydrogen at carbon-4′ more acidic. Removal of this hydrogen sparks an E2 reaction in which the internucleotide bond is cleaved.

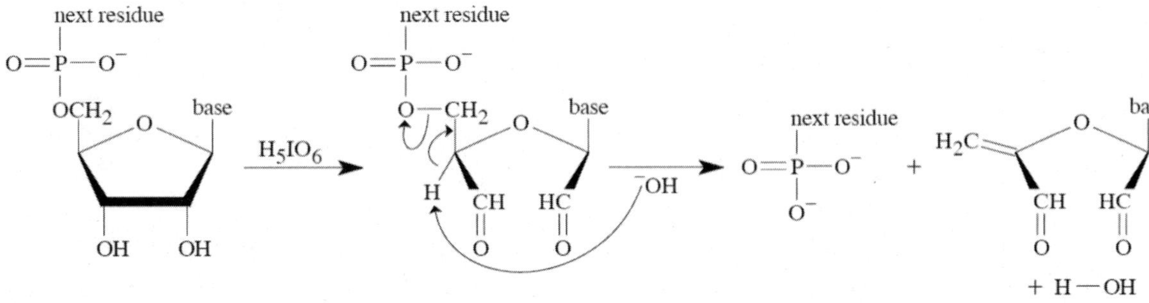

(b) DNA cannot undergo this reaction because a cis 2′,3′-diol group is needed for the periodate cleavage reaction. DNA, of course, has no 2′-hydroxy group at the 3′ end, and thus, no 2′,3′-diol.

26.53   The reaction of RNA involves ionization of the 2′-hydroxy group within each residue and reaction of the resulting alkoxide ion as a nucleophile at the neighboring phosphorus. This reaction is particularly rapid because it is intramolecular; it is an example of neighboring-group participation. (See Sec. 11.8.) Loss of the leaving group splits the internucleotide bond. DNA cannot undergo this reaction because the reaction depends on the presence of the 2′-hydroxy group to act as an intramolecular nucleophile; DNA lacks this hydroxy group. (That is why it is called deoxyribonucleic acid.)

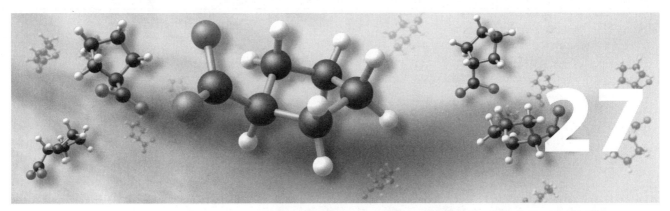

Amino Acids, Peptides, and Proteins

FURTHER EXPLORATIONS

 ## 27.1   Solid Phase Peptide Synthesis

The reactions of solid-phase peptide synthesis must be carried out in solvents that *solvate* the Merrifield resin—that is, solvents that interact well with the groups of the resin. If these groups are not solvated, reagents in solution cannot penetrate the resin and no reaction takes place. Because the polystyrene-based Merrifield resin is essentially a polymeric ethylbenzene, it is mostly a hydrocarbon. Yet many of the reactions of solid-phase peptide synthesis involve ionic compounds and intermediates. Water and alcohols do not work as solvents in solid-phase peptide synthesis because they do not solvate the resin. Dimethylformamide, *N*-methylpyrrolidone, and methylene chloride (dichloromethane) have been found empirically to solvate the Merrifield resin adequately and to dissolve the reagents used in solid-phase peptide synthesis. In fact, when such solvents are added to a Merrifield resin, a visually perceptible swelling of the resin bed occurs; this is caused by incorporation of the solvent into the resin.

The use of a cesium salt in Eq. 27.21 on text p. 1393 is related to the requirement for aprotic solvents. In DMF, ionic compounds tend to form ion pairs and other aggregates, and carboxylate salts have metal–oxygen bonds with a significant amount of covalent character. Yet the alkylation shown in this equation is fastest when the carboxylate oxygen has as much *ionic character* as possible, because a carboxylate oxygen with a full negative charge is more nucleophilic and therefore reacts much more rapidly in S_N2 reactions than an ion-paired (or covalently bonded) carboxylate oxygen. Because cesium is the most electropositive of the readily available alkali metals, its carboxylate salts have more ionic character than lithium, sodium, or potassium salts and hence are most reactive in S_N2 reactions.

 ## 27.2   Reactions of α-Amino Acids with Ninhydrin

To understand the reaction of ninhydrin with α-amino acids, first notice that *two* molecules of ninhydrin react per molecule of amino acid. Second, notice that ninhydrin is the hydrate of a tricarbonyl compound.

ninhydrin   (FE27.1)

The ninhydrin reaction with α-amino acids is a combination of simpler reactions that you have already studied. The first step is the formation of an imine (Sec. 19.11A). In this reaction, the amino group of the amino acid reacts with the central carbonyl group of the tricarbonyl form of ninhydrin. (Why is the central carbonyl group more reactive than the other carbonyl groups? See Sec. 19.7B in the text.)

an imine   (FE 27.2)

This imine is converted into a different imine by loss of CO_2 (decarboxylation), and the new imine hydrolyzes to an aldehyde and an amine. (Recall that imine formation is reversible.) Note that the aldehyde bears the side-chain R of the original amino acid.

(FE27.3)

The resulting amine then forms Ruhemann's purple, the final product, by reacting with a second molecule of ninhydrin in another imine-forming reaction.

Ruhemann's purple   (FE27.4)

Simple primary amines also react with ninhydrin to give Ruhemann's purple. The only alteration in the mechanism is the loss of a proton instead of decarboxylation in Eq. FE27.3 on the previous page.

REACTION REVIEW

I. SYNTHESIS AND OPTICAL RESOLUTION OF α-AMINO ACIDS

A. ALKYLATION OF AMMONIA

1. Some α-amino acids can be prepared by alkylation of ammonia with α-bromo carboxylic acids.

$$\underset{Br}{-}\overset{\overset{\displaystyle Br}{|}}{\underset{|}{C}}-CO_2H \xrightarrow{\text{NH}_3 \text{ (large excess)}} -\overset{\overset{\displaystyle \overset{+}{N}H_3}{|}}{\underset{|}{C}}-CO_2^- \ + \ \overset{+}{N}H_4 \ + \ Br^-$$

   a. This is an S_N2 reaction in which ammonia acts as the nucleophile.
   b. The use of a large excess of ammonia in the synthesis favors monoalkylation.
2. Multiple alkylation is not a problem for two reasons:
   a. Amino acids are less reactive toward alkylating agents than simple alkylamines because the amino groups of amino acids are less basic (nucleophilic) than ammonia and simple alkylamines.
   b. Branching in amino acids provides a steric effect that retards further alkylation.

B. ALKYLATION OF AMINOMALONATE DERIVATIVES

1. Another method for preparing α-amino acids is a variation of the malonic ester synthesis.

$$H_3C-\overset{\overset{\displaystyle O}{||}}{C}-NH-\overset{\overset{\displaystyle CO_2Et}{|}}{\underset{\underset{\displaystyle CO_2Et}{|}}{C}}-H \ + \ X-R \ \Rightarrow \ H_3\overset{+}{N}-\overset{\overset{\displaystyle |}{}}{\underset{\underset{\displaystyle CO_2Et}{|}}{CH}}-R \ + \ CO_2 \ + \ 2\,EtOH \ + \ CH_3CO_2H$$
$$X^-$$

2. The malonic ester derivative used is one in which a protected amino group is already in place: diethyl α-acetamidomalonate.
   a. Treatment of diethyl α-acetamidomalonate with sodium ethoxide in ethanol forms the conjugate base enolate ion, which is then alkylated with an alkyl halide.
   b. The resulting compound is then treated with hot aqueous HCl or HBr which accomplishes three things:
      i. The ester groups are hydrolyzed to give a substituted malonic acid.
      ii. The malonic acid derivative is decarboxylated under the reaction conditions.
      iii. The acetamido group, an amide, is also hydrolyzed.
   c. Neutralization affords the α-amino acid.

$$AcNH-\overset{\overset{\displaystyle CO_2Et}{|}}{\underset{\underset{\displaystyle H}{|}}{C}}-CO_2Et \xrightarrow[\text{2) R-X}]{\text{1) NaOEt/HOEt}} AcNH-\overset{\overset{\displaystyle CO_2Et}{|}}{\underset{\underset{\displaystyle R}{|}}{C}}-CO_2Et \xrightarrow[\text{heat}]{H_3O^+}$$

$$AcNH-\overset{}{\underset{\underset{\displaystyle R}{|}}{CH}}-CO_2^- \ + \ 2\,EtOH \ + \ CO_2\uparrow \ + \ HOAc$$

C. STRECKER SYNTHESIS

1. Hydrolysis of α-amino nitriles gives α-amino acidsis called the Strecker synthesis.

$$H_2N-\overset{}{\underset{\underset{\displaystyle R}{|}}{CH}}-C\equiv N \xrightarrow[\text{heat}]{H_3O^+} H_3\overset{+}{N}-\overset{}{\underset{\underset{\displaystyle R}{|}}{CH}}-CO_2^- \ + \ \overset{+}{N}H_4$$

2. α-Amino nitriles are prepared by treatment of aldehydes with ammonia in the presence of cyanide ion, a reaction that probably involves an imine intermediate.
   a. The conjugate acid of the imine reacts with cyanide under the conditions of the reaction to give the α-amino nitrile.

b. The addition of cyanide to an imine is analogous to the formation of a cyanohydrin from an aldehyde or ketone.

$$O{=}CH{-}R \xrightarrow[\text{NaCN}]{\text{NH}_4\text{Cl}} HN{=}CH{-}R \longrightarrow H_2N{-}CH(R){-}C{\equiv}N$$

D. ENANTIOMERIC RESOLUTION OF α-AMINO ACIDS

1. α-Amino acids synthesized by common laboratory methods are racemic; since many applications require the pure enantiomers, the racemic compounds must be resolved.
2. An alternative approach to the preparation of enantiomerically pure amino acids is the synthesis of amino acids by microbiological fermentation.
3. Certain enzymes can be used to resolve racemic amino acid derivatives by selectively catalyzing a reaction of one enantiomer.

II. REACTIONS OF AMINO ACIDS AND SYNTHESIS OF PEPTIDES

A. ACYLATION AND ESTERIFICATION

1. Amino acids undergo many of the characteristic reactions of both amines and carboxylic acids.
   a. They can be acylated by on the amino group by acid chlorides or anhydrides.
   b. Their carboxylic acid groups are easily esterified by heating with an alcohol and a strong acid catalyst.

B. SOLID-PHASE PEPTIDE SYNTHESIS—GENERAL

1. In a method called solid-phase peptide synthesis, the carboxy-terminal amino acid is covalently anchored to an insoluble polymer, and the peptide is "grown" by adding one residue at a time to this polymer.
   a. Solutions containing the appropriate reagents are allowed to contact the resin with shaking.
   b. At the conclusion of each step, the resin containing the peptide is simply filtered away from the solution, which contains soluble by-products and impurities.
   c. The completed peptide is removed from the resin by a reaction that cleaves the peptide–resin bond.
2. The advantage of the solid-phase method is the ease with which dissolved impurities and by-products are removed from the resin-bound peptide by simple filtration.
   a. The same reagents used in solid-phase peptide synthesis can also be used for peptide synthesis in solution, but removal of by-products from the product peptide is sometimes difficult.
   b. In order to avoid impurities, each step in the solid-phase synthesis must occur with virtually 100% yield, an ideal that is often approached in practice.

C. SOLID-PHASE PEPTIDE SYNTHESIS—FMOC PROTECTING GROUP

1. In a solid-phase peptide synthesis, the amino group of the amino acids is protected with a special acyl group, the (9-fluorenylmethyloxy)carbonyl (Fmoc) group. (For complete details, see Sec. 27.6A of the text.)
   a. Fmoc-Amino acids are prepared by allowing an *N*-hydroxysuccinimide ester derivative, Fmoc-NHS, to react with the amino group of the amino acid.

Fmoc-NHS   an α-amino acid

Fmoc-amino acid   *N*-hydroxysuccinimide

b. An S$_N$2 reaction between the cesium salt of an Fmoc-amino acid and the chloromethyl group of the resin results in the formation of an ester linkage to the resin.

2. Once the Fmoc-amino acid is anchored to the resin, the Fmoc-protecting group is removed by treatment with an amine base, piperidine, in an E2 reaction.

resin-bound amino acid

a. The product of the β-elimination is a carbamate anion.
b. The carbamate anion decarboxylates under the reaction conditions.
3. Coupling of the next Fmoc-amino acid to the free amino group of the resin-bound amino acid is effected by the reagent 1,3-diisopropylcarbodiimide (DIC) in the presence of 1-hydroxy-1,2,3-benzotriazole (HOBt).

Fmoc-amino acid **DIC** **amino acid-resin**

peptide-resin **N,N′-dicyclohexylurea (DCU)**

   a. Addition of the carboxylic acid group to a double bond of DIC gives a derivative called an
      O-acylisourea.
   b. The HOBt ester reacts with the amino group of the resin-bound amino acid to form the peptide bond (an
      amide linkage).
  4. The ester linkage used to bind the peptide to the resin is broken by a carbocation mechanism using 50–60%
     trifluoroacetic acid (TFA) in dichloromethane.

resin-bound peptide **peptide**

  D. SOLID-PHASE PEPTIDE SYNTHESIS—BOC PROTECTING GROUP

    1. In another method of solid-phase peptide synthesis, the amino group of the amino acids is protected with
       another special acyl group, the *tert*-butyloxycarbonyl (Boc) group.
      a. The Boc group is introduced by allowing an amino acid to react with the anhydride di-*tert*-butyl
         dicarbonate.

      b. The amino group of the amino acid rather than the carboxylate group reacts with the anhydride because
        the amino group is the more basic, and therefore the more nucleophilic, group.
    2. At the start of the solid-phase peptide synthesis, a Boc-protected amino acid is anchored to the insoluble
       solid support (called a Merrifield resin) using the reactivity of its free carboxylic acid group.
      a. An S_N2 reaction between the cesium salt of the Boc-amino acid and the resin results in the formation of
        an ester linkage to the resin.
      b. Once the Boc-amino acid is anchored to the resin, the Boc protecting group is removed with anhydrous
        trifluoroacetic acid. This deprotection step, followed by neutralization of the resulting ammonium salt,
        exposes the free amino group of the resin-bound amino acid, which is used as a nucleophile in the next
        reaction.

3. Coupling of another Boc-protected amino acid to the free amino group of the resin-bound amino acid is similar to that of an Fmoc-protected amino acid.

4. Completion of the peptide synthesis requires deprotection of the resin-bound peptide in the usual way, a final coupling step with the amino-terminal Boc-amino acid and DCC, and removal of the peptide from the resin.
   a. The ester linkage that connects the peptide to the resin, like most esters, is more easily cleaved than the peptide (amide) bonds, and is typically broken by liquid HF.
   b. This acidic reagent also removes the Boc group from the product peptide.

III. ANALYTICALLY IMPORTANT REACTIONS OF PEPTIDES

A. HYDROLYSIS OF PEPTIDES; AMINO ACID ANALYSIS

1. An important reaction used to determine the structures of unknown peptides is hydrolysis of the peptide (amide) bonds of a peptide to give its constituent amino acids.
2. When a peptide or protein is hydrolyzed, the product amino acids can be separated, identified, and quantitated by a technique called amino acid analysis.
   a. In the method most commonly used today, the amino acids in the hydrolyzed mixture are allowed to reacted with AQC-NHS to form AQC-amino acids.

1-[[(6-quinolylamino)carbonyl]-
oxy]-2,5-pyrrolidinedione
(AQC-NHS)

α-amino acid

AQC-amino acid   *N*-hydroxysuccinimide

   i. The mixture of AQC-amino acids are separated by passing them through a C18 HPLC column under very carefully defined conditions; the time at which each derivatized amino acid emerges from the column is accurately known.
   ii. The AQC-amino acids with the greatest hydrocarbon character are eluted last.
   iii. The AQC group is fluorescent; the absorption of light at 254 nm (UV detector) or emission of light at 395 nm (fluorescence detector) is used to determine the relative amounts of each amino acid present.

b. In a second and older method, the amino acids in the hydrolyzed mixture are separated by passing them through a cation-exchange column under very carefully defined conditions; the time at which each amino acid emerges from the column is accurately known.

   i. As each amino acid emerges, it is mixed with ninhydrin to give an intense blue-violet dye called Ruhemann's purple; the intensity of the resulting color is proportional to the amount of the amino acid present.

ninhydrin

Ruhemann's purple

   ii. The color intensity is recorded as a function of time; the area of the peak is proportional to the amount of the amino acid.

c. Mass spectrometry may be used to determine the sequence of short peptides since peptides undergo residue-by-residue fragmentation, in many cases from both ends simultaneously, in a mass spectrometer.

3. Amino acid analysis can determine the identity and relative amounts of amino acid residues, but not their relative orders within a peptide.

B. SEQUENTIAL DEGRADATION OF PEPTIDES

1. The actual arrangement, or sequential order, of amino acid residues in a peptide is called the amino acid sequence, or primary sequence, of the peptide.

2. It is possible to remove one residue at a time from the amino end of the peptide, identify it, and then repeat the process sequentially on the remaining peptide; the standard method for implementing this strategy is called the Edman degradation.

   a. The peptide is treated with phenyl isothiocyanate (Edman reagent), with which it reacts at its amino groups to give a thiourea derivative. (Only the reaction at the terminal amino group is relevant to the degradation.)

   b. Any remaining phenyl isothiocyanate is removed, and the modified peptide is then treated with anhydrous trifluoroacetic acid.

   c. The sulfur of the thiourea, which is nucleophilic, displaces the amino group of the adjacent residue to yield a five-membered heterocycle called a thiazolinone; the other product of the reaction is a peptide that is one residue shorter.

   d. When treated subsequently with aqueous acid, the thiazolinone derivative forms an isomer called a phenylthiohydantoin (PTH).

   e. Because the PTH derivative carries the characteristic side chain of the amino-terminal residue, the structure of the PTH derivative identifies the amino acid residue that was removed.

Ph—N=C=S   + H_2N—CH—C—NH—Pep^C

phenyl isothiocyanate
(Edman reagent)

R

1) pyridine/H_2O/Me_2NPh
   (gives the thiourea derivative)

2) CF_3CO_2H

a phenylthiohydantoin (PTH) derivative of the amino terminal residue

a thiazolinone

3. In practice, this type of analysis is limited to 20–60 consecutive residues because the yields at each step are not perfectly quantitative; hence, an increasingly complex mixture of peptides is formed with each successive step in the cleavage, and after a number of such steps the results become ambiguous.

4. The most common sequencing method used today involves proteolytic cleavage of larger peptides into smaller peptides (see C. and D. below) followed by sequencing of the smaller peptides with tandem (MS–MS) mass spectrometry.

C. SPECIFIC CLEAVAGE OF PEPTIDES: INTRODUCTION

1. The amino acid sequence of most large proteins is determined by breaking the protein into smaller peptides and sequencing these peptides individually. (The sequence of the protein is reconstructed from the sequences of the peptides.)

2. When breaking a larger protein into smaller peptides, it is desirable to use reactions that cleave the protein in high yield at well-defined points so that a relatively small number of peptides are obtained.

3. One method uses ordinary chemical reagents; another method involves the use of enzymes to catalyze peptide-bond hydrolysis.

D. PEPTIDE CLEAVAGE WITH PROTEOLYTIC ENZYMES

1. A number of enzymes (called proteases, peptidases, or proteolytic enzymes) catalyze the hydrolysis of peptide bonds at specific points in an amino acid sequence.

2. One of the most widely used proteases is the enzyme trypsin, which catalyzes the hydrolysis of peptides or proteins at the carbonyl group of arginine or lysine residues provided that:
   a. these residues are not at the amino end of the protein.
   b. these residues are not followed by a proline residue.

3. Because trypsin catalyzes the hydrolysis of peptides at internal rather than terminal residues, it is called an endopeptidase; enzymes that cleave peptides only at terminal residues are termed exopeptidases.

4. Chymotrypsin, a protein related to trypsin, is used to cleave peptides at amino acid residues with aromatic side chains and, to a lesser extent, residues with large hydrocarbon side chains; thus, chymotrypsin cleaves peptides at Phe, Trp, Tyr, and in some cases Leu and Ile residues.

SOLUTIONS TO PROBLEMS

Solutions to In-Text Problems

27.1   The peptide is drawn in the form that exists at neutral pH.

(a)   The structure of tryptophylglycylisoleucylaspartic acid:

tryptophylglycylisoleucylaspartic acid
(Trp-Gly-Ile-Asp, or W-G-I-D)

(b)   The structure of Glu-Gln-Phe-Arg:

Glu-Gln-Phe-Arg
(glutamylglutaminylphenylalanylarginine, or E-Q-F-R)

27.2   This peptide is His-Ile-Tyr-Met-Ser (histidylisoleucyltyrosylmethionylserine, or H-I-Y-M-S).

27.3   (a)   (b)

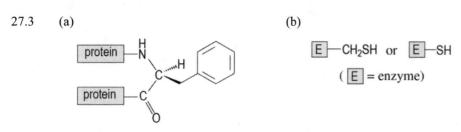

E—CH₂SH or E—SH

(E = enzyme)

27.4   Three more of the many possible configurations of α-amino acids in addition to the ones shown at the beginning of Sec. 27.2 on text p. 1378.

27.5   (a)   L-isoleucine:

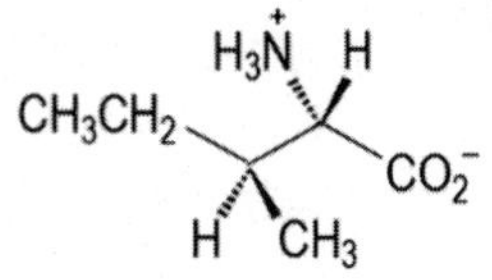

L-isoleucine
(2S, 3S)

(b)   As noted on text p. 1379, the "allo" version of amino acids retains configuration at the α-carbon. So, it must be the side chain of isoleucine that has the opposite configuration from the side chain of L-isoleucine. L-Alloisoleucine has the 2S configuration, that is, the same configuration at the α-carbon as L-isoleucine, but has a different configuration at the β-carbon.

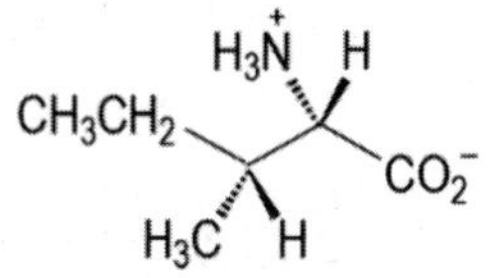

L-alloisoleucine
(2S, 3R)

27.6   (a)   The α-carbon of L-cysteine has the R configuration.

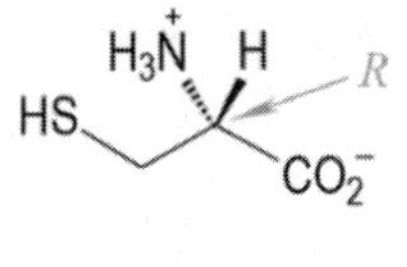

L-cysteine

(b)   In terms of the sequence rules for determining configuration, the two carbons attached to the α-carbons of cysteine or serine can be represented as C(O,O,O) and C(X,H,H), where X = S or O, respectively. When X = S (cysteine), the latter carbon receives priority. When X = O (serine), the former receives priority. Sulfur receives priority over oxygen because it has higher atomic number, and the priority is decided *at the first point of difference*. (See Sec. 4.2B of the text.)

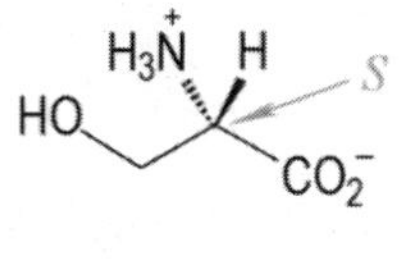

L-serine

27.7   (a)

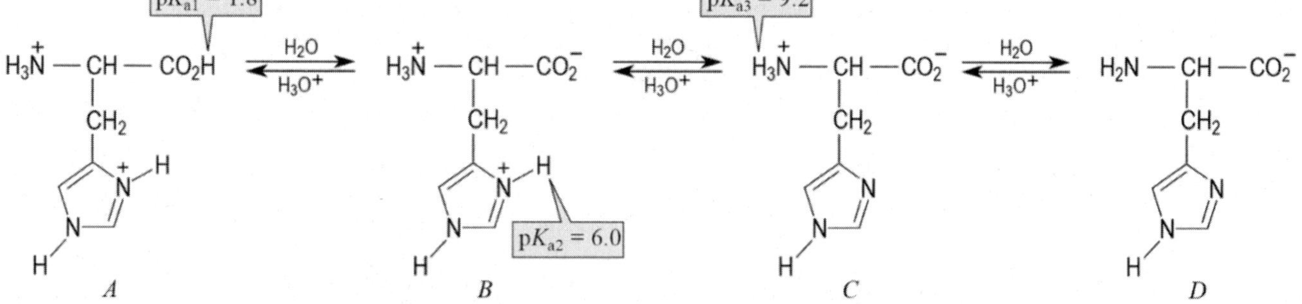

(b)   A plot of the fraction of each species shown in part (a) as a function of pH is shown on the next page. The pH at which any two species are present in equal concentrations is noted.

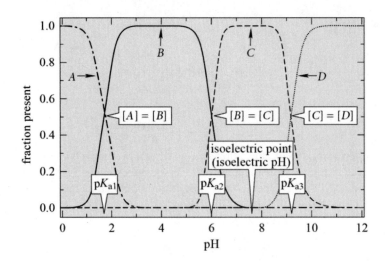

27.8 (a) The amino group of tyrosine can be protonated, and both the carboxy group and the phenolic O—H group can be ionized.

(b) At pH 6, the net charge on tyrosine is zero. A pH value of 6 is below the pK_a of the conjugate acid of the amino group, and the amino group is therefore protonated; a pH value of 6 is above the pK_a of the carboxy group, which is therefore ionized; and a pH value of 6 is below the pK_a of the phenolic O—H group, which is therefore un-ionized.

(c) The structure of tyrosine in aqueous solution at pH 6:

$$\text{H}_3\overset{+}{\text{N}}-\text{CH}-\overset{\displaystyle\text{O}}{\overset{\displaystyle\|}{\text{C}}}-\text{O}^-$$

with CH$_2$ and a para-hydroxyphenyl (OH) ring substituent.

27.9 (a) The isoelectric point of A-K-V-I-M (Ala-Lys-Val-Ile-Met) is approximately the same as lysine, since it is the only ionizable side chain in the peptide. Therefore, its isoelectric point is about 9.82. The isoelectric point of G-D-G-L-F (Gly-Asp-Gly-Leu-Phe) is about 2.76, because it contains an acidic side chain, aspartic acid.

(b) In the major neutral form of A-K-V-I-M, the amino terminal amino group is un-ionized and the side chain on the Lys is ionized because it has the lower pK_a of the two amino groups (Table 27.1, text p. 1376). This form would exist at a pH between the pK_a values of the two amino groups. That is, the pH at which this neutral form predominates is relatively basic. (See Sec. 27.3B.)

$$\text{H}_2\text{N}-\text{CH}-\overset{\text{O}}{\overset{\|}{\text{C}}}-\text{NH}-\text{CH}-\overset{\text{O}}{\overset{\|}{\text{C}}}-\text{NH}-\text{CH}-\overset{\text{O}}{\overset{\|}{\text{C}}}-\text{NH}-\text{CH}-\overset{\text{O}}{\overset{\|}{\text{C}}}-\text{NH}-\text{CH}-\overset{\text{O}}{\overset{\|}{\text{C}}}-\text{O}^-$$

with side chains: CH$_3$; (CH$_2$)$_4$ — $^+$NH$_3$; CH(CH$_3$)$_2$; CH$_2$ — CH(CH$_3$)$_2$; (CH$_2$)$_2$ — SCH$_3$

In the major neutral form of G-D-G-L-F, the side-chain carboxy group of Asp is un-ionized and the carboxy-terminal carboxy group is ionized because it has the lower pK_a of the two carboxy groups (Table 27.1, text p. 1376). This form would exist at a pH between the pK_a values of the two carboxy groups. That is, the pH at which this neutral form predominates is relatively acidic. (See Sec. 27.3B.)

$$H_3\overset{+}{N}-CH_2-\overset{O}{\overset{\|}{C}}-NH-CH-\overset{O}{\overset{\|}{C}}-NH-CH_2-\overset{O}{\overset{\|}{C}}-NH-CH-\overset{O}{\overset{\|}{C}}-NH-CH-\overset{O}{\overset{\|}{C}}-O^-$$

with side chains: CH_2 / CO_2H ; $CH_2 / CH(CH_3)_2$; CH_2 / Ph

Notice that the major neutral form of a peptide (such as these) may not be the form that predominates at *neutral pH*.

27.10 The general rule of thumb is that if the peptide contains more acidic than basic groups it is an acidic peptide; if it contains more basic than acidic groups it is a basic peptide; and if the number of acidic and basic groups are equal, the peptide is neutral. (An acidic group is a group that is in its conjugate-base form at neutral pH, such as a carboxy group; a basic group is a group that is in its conjugate-acid form at neutral pH, such as an amino group.)

(a) The peptide is neutral because it contains one basic group (the terminal amino group, which is protonated at pH 6) and one acidic group (the terminal carboxy group, which is ionized at pH 6). Its net charge at pH 6 is 0.

(b) This peptide contains three basic groups: the α-amino group of the amino-terminal residue Leu and the side-chain amino groups of the Lys residues. The peptide contains one acidic group: the carboxy group of the carboxy-terminal Lys residue. Because the peptide contains more basic residues than acidic residues; it is a basic peptide. Its net charge at pH 6 is +2.

(c) Acetylation eliminates the basicity of the terminal amino group. The peptide contains two basic groups (the Arg residues) and two acidic groups (the Asp residue and the terminal carboxy group). Thus, the peptide is neutral and its net charge is 0.

(d) This peptide contains three acidic groups: the side-chain carboxy groups of the Asp and Glu residues and the carboxy group of the carboxy-terminal Ile residue. The peptide contains two basic groups: the α-amino group of the amino-terminal residue Glu and the side-chain amino group of the Lys residue. Because the peptide contains more acidic residues than basic residues, it is an acidic peptide. Its net charge at pH 6 is –1.

27.11 (a) A resin containing a cationic group will serve as an anion exchanger. A quaternary ammonium salt is a suitable cationic group:

$$\cdots-CH_2-CH-CH_2-CH-CH_2-CH-\cdots$$

with phenyl groups bearing $+NMe_3$ substituents.

(b) The ion-exchange column is positively charged, as shown by the solution to part (a). At pH 6, the peptide A-V-G has a net charge of zero, is not retained by the column, and will emerge first; the peptide D-N-N-G has a net charge of –1, is retained by the column and will emerge next; and the peptide D-E-E-G has a charge of –3, is most strongly retained by the column, and will emerge last.

27.12 (a) Alkylation of ammonia by α-bromophenylacetic acid will work, particularly since the alkyl halide is benzylic.

$$\text{PhCHCO}_2\text{H} \xrightarrow{\text{NH}_3\ (\text{large excess})} \text{PhCHCO}_2^-$$

with Br below the reactant (α-bromophenyl-acetic acid) and $+NH_3$ below the product (α-phenylglycine).

The acetamidomalonate method would not work because it would require alkylation of a malonate anion by bromobenzene. Bromobenzene does not undergo S_N2 reactions. The Strecker synthesis would work.

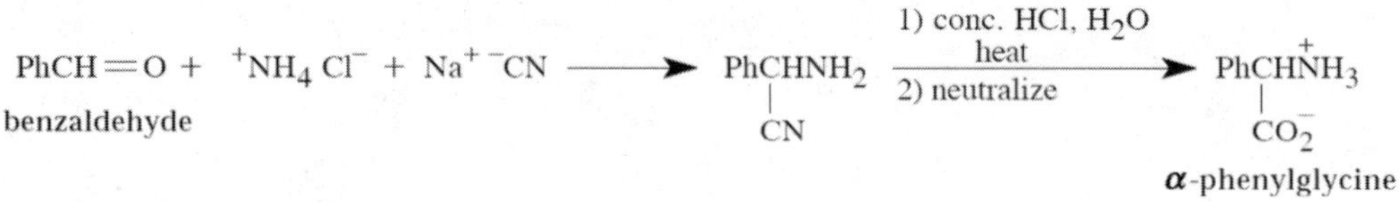

PhCH=O + $^+$NH$_4$ Cl$^-$ + Na$^+$ $^-$CN $\longrightarrow$ PhCHNH$_2$ $\xrightarrow[\text{2) neutralize}]{\substack{\text{1) conc. HCl, H}_2\text{O} \\ \text{heat}}}$ PhCH$\overset{+}{\text{N}}$H$_3$

benzaldehyde |CN |CO$_2^-$

α-phenylglycine

(b)   All three methods would work in principle for leucine.

Alkylation of ammonia:

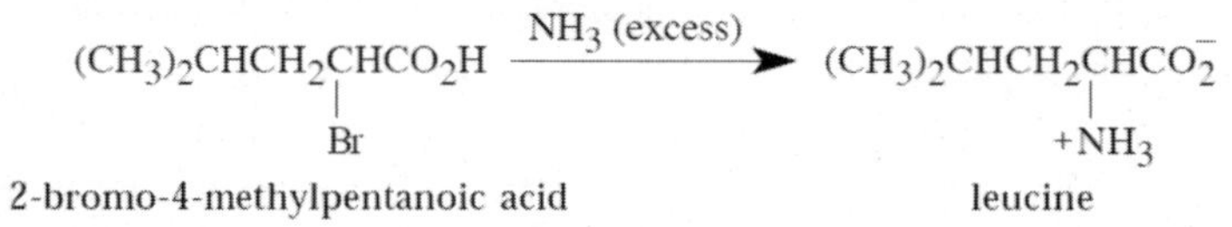

(CH$_3$)$_2$CHCH$_2$CHCO$_2$H $\xrightarrow{\text{NH}_3 \text{ (excess)}}$ (CH$_3$)$_2$CHCH$_2$CHCO$_2^-$

           |Br |$^+$NH$_3$

2-bromo-4-methylpentanoic acid   leucine

Acetamidomalonate method:

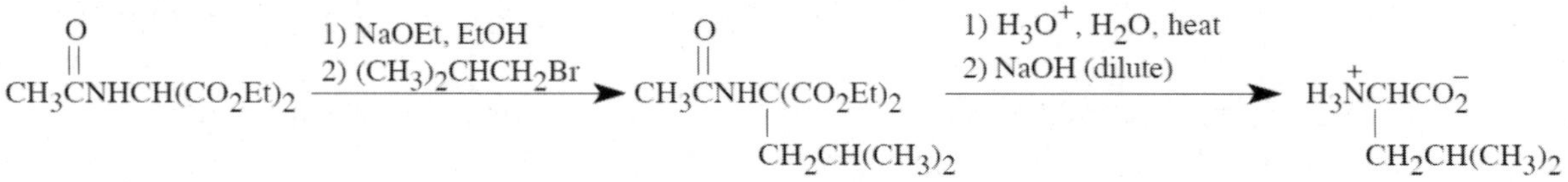

CH$_3$CNHCH(CO$_2$Et)$_2$ $\xrightarrow[\text{2) (CH}_3)_2\text{CHCH}_2\text{Br}]{\text{1) NaOEt, EtOH}}$ CH$_3$CNHC(CO$_2$Et)$_2$ $\xrightarrow[\text{2) NaOH (dilute)}]{\text{1) H}_3\text{O}^+, \text{H}_2\text{O, heat}}$ H$_3$$\overset{+}{\text{N}}$CHCO$_2^-$

                         |CH$_2$CH(CH$_3$)$_2$ |CH$_2$CH(CH$_3$)$_2$

Strecker synthesis:

(CH$_3$)$_2$CHCH$_2$CH=O + $^+$NH$_4$ Cl$^-$ + Na$^+$ $^-$CN $\longrightarrow$ (CH$_3$)$_2$CHCH$_2$CHNH$_2$ $\xrightarrow[\text{2) neutralize}]{\substack{\text{1) conc. HCl, H}_2\text{O} \\ \text{heat}}}$

3-methylbutanal |CN

(CH$_3$)$_2$CHCH$_2$CH$\overset{+}{\text{N}}$H$_3$

                         |CO$_2^-$

27.13   (a)   Tosyl chloride, the acid chloride of a typical sulfonic acid, reacts in the same way as other acid chlorides.

(CH$_3$)$_2$CHCH$_2$CH$\overset{+}{\text{N}}$H$_3$ + Cl—S(=O)$_2$—⟨C$_6$H$_4$⟩—CH$_3$ $\xrightarrow{\text{base}}$ (CH$_3$)$_2$CHCH$_2$CHNH—S(=O)$_2$—⟨C$_6$H$_4$⟩—CH$_3$

     |CO$_2^-$ |CO$_2$H

leucine   tosyl chloride   *N*-tosylleucine

(b)   Alanine is esterified to give its methyl ester:

CH$_3$CH$\overset{+}{\text{N}}$H$_3$

|CO$_2$CH$_3$

alanine methyl ester

27.14   Polyglycine results from aminolysis of the ester group on one molecule by the amine on another:

H$_2$NCH$_2$COCH$_3$ H$_2$NCH$_2$COCH$_3$ $\xrightarrow{-\text{CH}_3\text{OH}}$ H$_2$NCH$_2$C—NHCH$_2$COCH$_3$ $\xrightarrow[\text{H}_2\text{NCH}_2\text{CO}_2\text{CH}_3]{-\text{CH}_3\text{OH}}$

H$_2$NCH$_2$C—NHCH$_2$C—NHCH$_2$COCH$_3$ $\longrightarrow$ etc.

In this reaction the amino group of glycine acts as a nucleophile. In acidic solution the amino group of glycine is protonated and cannot act as a nucleophile; hence, the reaction does not occur.

27.15   If the average yield of each step is *Y*, then the yield of the first step (assuming it is average) is *Y*; the yield of the second is *Y·Y*; and that of the *n*th step is Y^n. Hence,

$$Y^{369} = 0.17$$

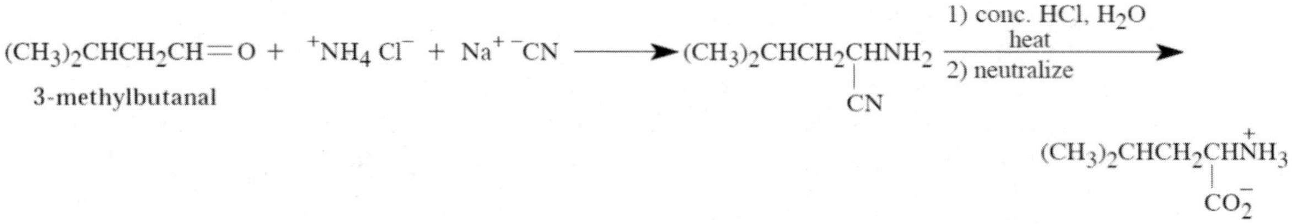

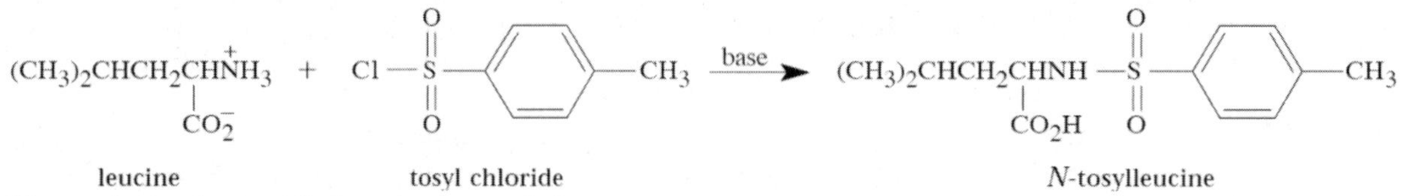

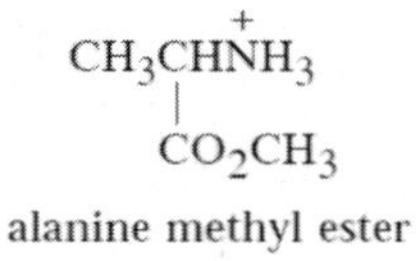

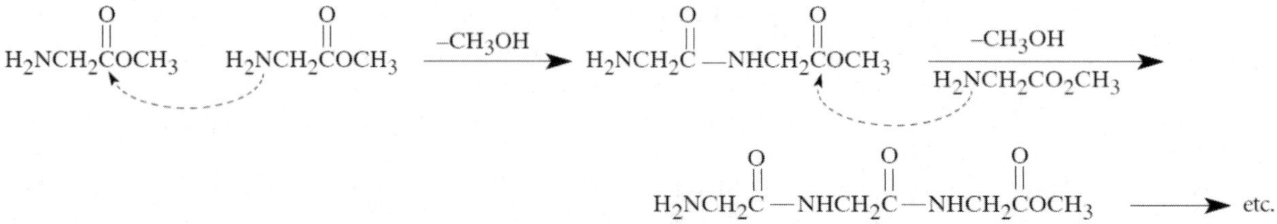

$$369 \log Y = \log 0.17 = -0.7696$$

$$\log Y = -0.0021, \text{ or } Y = 0.995$$

Therefore, the average yield of each step is 99.5%!

The high yield of each step demonstrates two points about peptide synthesis. First, it has been developed into a remarkably efficient process. The second point, however, is that even with yields in excess of 99%, the overall yields of large proteins prepared by this method will be extremely small. Furthermore, these will be contaminated by large numbers of impurities that will be difficult to separate from the desired material. Fortunately, genetic-engineering methods (which you will study if you take biochemistry) allow scientists to prepare pure naturally occurring proteins in large amounts. At the present time, the chemical synthesis of peptides is most useful for the preparation of peptides containing about 2–50 residues, although a few noteworthy successes have been achieved in the synthesis of larger peptides and some proteins. Furthermore, chemical synthesis can provide peptides that contain modified or unnatural amino acids that cannot be produced by the biological methods.

27.16   The box on text p. 1391 shows that solid-phase peptide synthesis requires about three synthetic steps per residue. Hence, about 300 steps are required for the synthesis of a protein containing 100 amino acid residues. If the average yield of each step is Y, then

$$Y^{300} = 0.50$$

$$300 \log Y = \log 0.50 = -0.3010$$

$$\log Y = -0.001003, \text{ or } Y = 0.998$$

Thus, an average yield of 99.8% per step would be required.

If you used 100 steps for your calculation, the average yield is 99.3%. Either way, the yield must be nearly perfect at each step to achieve even a 50% overall yield. Notice that, because there are so many steps, a drop of a small fraction of a percent in average yield per step has a large effect on the overall yield.

27.17   (a)   The reactivity of the esters is related to the stability of the leaving group. The more stable the leaving group, the more reactive the ester. The less basic the leaving group, the more stable it is. Weaker bases come from stronger acids. Therefore, the base (leaving group) with the conjugate acid that has the lowest pK_a will be the weakest base, and the ester of that base will be the most reactive. Therefore, the order of reactivity (least reactive to most reactive) is $C < D < A < B$.

(b)

$$CH_3(CH_2)_2 \overset{\overset{\displaystyle O}{\|}}{C}-OBt \;+\; H_2N-CH_2-\overset{\overset{\displaystyle O}{\|}}{C}-NH_2 \;\longrightarrow\; CH_3(CH_2)_2\overset{\overset{\displaystyle O}{\|}}{C}-NH-CH_2-\overset{\overset{\displaystyle O}{\|}}{C}-NH_2 \;+\; HOBt$$

glycinamide

27.18   (a)   Because lysine contains two amino groups, both must be protected in order to prevent nucleophilic side reactions, such as the reaction of the amino group of one lysine molecule with the DIC-activated carboxy group of another.

(b)   The 20% piperidine step removes both Fmoc protecting groups. Hence, both amino groups react in subsequent acylation reactions. (The Lys residue is drawn in more structural detail for clarity.)

$$\underset{\substack{| \\ (CH_2)_4 \\ | \\ NH_2}}{Gly-NHCHC\overset{\overset{\displaystyle O}{\|}}{}-Ala} \qquad \underset{\substack{| \\ (CH_2)_4 \\ | \\ NH-Gly}}{H_2NCHC\overset{\overset{\displaystyle O}{\|}}{}-Ala} \qquad \underset{\substack{| \\ (CH_2)_4 \\ | \\ NH-Gly}}{Gly-NHCHC\overset{\overset{\displaystyle O}{\|}}{}-Ala}$$

27.19   (a)   Compound *A* results from the S$_N$2 coupling reaction to the resin (analogous to Eq. 27.21 on text p. 1393), and compound *B* results from deprotection of *A*:

$$\underset{A}{\text{FmocNHCH}_2\overset{\displaystyle O}{\overset{\|}{\text{C}}}-\text{O}-\text{CH}_2-}} \qquad \underset{B}{\text{H}_2\text{NCH}_2\overset{\displaystyle O}{\overset{\|}{\text{C}}}-\text{O}-\text{CH}_2-}$$

Compound *C* is the coupling product that results from coupling of the lysine derivative reacting at its carboxylic acid group with the free amino group of *B*, and compound *D* results from removal of the Fmoc protecting group, *but not the Boc group*, from compound *C*.

$$\underset{C}{\text{FmocNH}-\underset{\underset{\text{NHBoc}}{\overset{|}{\underset{|}{(\text{CH}_2)_4}}}}{\overset{|}{\text{CH}}}-\overset{\overset{\displaystyle O}{\|}}{\text{C}}-\text{NHCH}_2\overset{\overset{\displaystyle O}{\|}}{\text{C}}-\text{O}-\text{CH}_2-} \qquad \underset{D}{\text{H}_2\text{N}-\underset{\underset{\text{NHBoc}}{\overset{|}{\underset{|}{(\text{CH}_2)_4}}}}{\overset{|}{\text{CH}}}-\overset{\overset{\displaystyle O}{\|}}{\text{C}}-\text{NHCH}_2\overset{\overset{\displaystyle O}{\|}}{\text{C}}-\text{O}-\text{CH}_2-}$$

Compound *E* results from coupling of the carboxy terminus of Boc-Val with the free amino group of *D*. Peptide *P* results from both removal of *E* from the resin and removal of the Boc group.

$$\underset{E}{\text{BocVal}-\text{NH}-\underset{\underset{\text{NHBoc}}{\overset{|}{\underset{|}{(\text{CH}_2)_4}}}}{\overset{|}{\text{CH}}}-\overset{\overset{\displaystyle O}{\|}}{\text{C}}-\text{NHCH}_2\overset{\overset{\displaystyle O}{\|}}{\text{C}}-\text{O}-\text{CH}_2-}$$

$$\underset{P}{\text{Val}-\text{Lys}-\text{Gly}} \quad + \quad \text{F}_3\text{C}-\overset{\overset{\displaystyle O}{\|}}{\text{C}}-\text{O}-\text{CH}_2-$$

$$+ \quad \text{CO}_2 \quad + \quad \text{F}_3\text{C}-\overset{\overset{\displaystyle O}{\|}}{\text{C}}-\text{O}-\text{C(CH}_3)_3$$

(from reaction of *tert*-butyl cation of the Boc group with trifluoroacetic acid)

Note that the tripeptide *P* is in the di-cationic, amino-protonated form as a result of the acidic deprotection conditions. The neutral (zwitterionic) form would be obtained by neutralization.

(b)   The Boc group prevents the side-chain amino group of lysine from reacting with the carboxylic acid group of another molecule of itself under the coupling conditions. Also, it remains intact when the amino-terminal Fmoc group is removed with piperidine. The lysine side-chain thus remains protected in the subsequent coupling step as well.

(c)   Boc-Val introduces a protecting group that can be removed at the same time as the peptide is released from the resin. Use of Fmoc-Val would require an additional deprotection step with 20% piperidine.

27.20   (a)   As shown by Table 27.2 on text p. 1401, there often exists more than one mRNA code for each individual amino acid. So, for any given peptide, there is likely to be multiple mRNA sequences. Thus, the code for this peptide would start with either of the two codes for Phe, followed by any of the six codes for Arg, followed by any of the four codes for Gly, followed by either of the two codes for His, and terminated by the single code for Trp. Therefore, there are 96 possible codes ($2 \times 6 \times 4 \times 2 \times 1$) for this peptide. An example is UUU-CGU-GGU-CAU-UGG.

(b)   The DNA codes are complimentary to every RNA code, as listed on text p. 1400. Thus, there are 96 possible DNA codes. The complimentary DNA strand to the mRNA strand listed in part (a) is AAA-GCA-CCA-GTA-ACC.

(c)   The mRNA code for Trp, from 5' → 3', is UGG. The tRNA code is then ACC.

27.21 (a)   (b)

hypoxanthine   uracil   hypoxanthine   adenine

C1′ of deoxyribose   C1′ of deoxyribose   C1′ of deoxyribose

C1′ of deoxyribose

27.22 (a) The amide group of Asn hydrolyzes to give ammonium ion and Asp. Consequently, Asn appears in amino acid analysis as Asp.

(b) Glu and Gln are not differentiated by amino acid analysis because the side-chain carboxamide group of Gln, like that of Asn in part (a), is hydrolyzed; that is, Gln is converted into Glu.

27.23 The indole side-chain of tryptophan is very hydrocarbonlike, more so than the side chain of any other amino acid. [Recall (Sec. 26.2 of the text) that pyrrole and indole are neither basic nor acidic.] Hence, we would expect to see AQC-Trp emerge on the far right of the amino acid analysis chromatogram, beyond AQC-Tyr.

27.24 The side chains of these amino acids as well as the α-amino groups are nucleophilic. Hence, they are acylated as well by AQC-NHS. The structures of these derivatives are as follows, with the carboxylic acid group in its ionized form:

AQC-lysine   AQC-cysteine

27.25 Trypsin catalyzes the hydrolysis of peptides or proteins at the carbonyl group of arginine (R) or lysine (K) residues. Therefore, the peptide E-R-G-A-N-I-K-K-H-E-M would produce the following peptides upon trypsin-catalyzed hydrolysis:

(1) E-R   (2) G-A-N-I-K   (3) K   (4) K-H-E-M   (5) H-E-M

Peptides (2) and (4) originate by hydrolysis at the peptide bond between the two K residues. Peptides (3) that is, K, and (5) originate from hydrolysis at the peptide bond between the K and H residues to give G-A-N-I-K-K followed by hydrolysis at the C-terminal peptide bond of this peptide. However, peptide (5) does not originate from (4), because, when a lysine residue is at the amino terminus of a peptide, the hydrolysis of its peptide bond is not catalyzed by trypsin. (Trypsin is an endopeptidase.)

27.26 Trypsin catalyzes the hydrolysis of peptides or proteins at the carbonyl group of arginine (Arg, R) or lysine (Lys, K) residues, provided that these residues are (a) not at the amino end of the protein, and (b) not followed by a proline residue. Since *T1* contains an R residue, the carbonyl terminal of *T1* must be an R and *T1* must be

(A,F,S)-R; since *T2* contains a K residue, the carbon terminal of *T2* must be a K and *T2* must be (G,I)-K; and thus *T3* must be at the carbonyl terminal of *Q* and the amino terminal cannot be a P.

$$Q \xrightarrow{\text{trypsin}} \underset{T1}{\text{(A,F,S)-R}} + \underset{T2}{\text{(G,I)-K}} + \underset{T3}{\text{(G,N,P,Y)}}$$

Chymotrypsin cleaves peptides at phenylalanine (Phe, F), tryptophan (Trp, W), tyrosine (Tyr, Y), and occasionally, at leucine (Leu, L) and isoleucine (Ile, I). Since *C1* contains a Y residue, the carbonyl terminal of *C1* must be a Y and *C1* must be (A,N,R,S)-Y; since the dipeptide *C2* contains an F residue, the carbonyl terminal of *C2* must be an F and *C2* must be K-F; and since *C3* contains an I residue, the carbonyl terminal of *C3* must be an I and *C3* must be G-I, and thus *T2* must be G-I-K.

$$Q \xrightarrow{\text{chymotrypsin}} \underset{C1}{\text{(A,N,R,S)-Y}} + \underset{C2}{\text{K-F}} + \underset{C3}{\text{G,I}} + \underset{C4}{\text{(G,P)}}$$

Since *C2* is K-F and *T1* contains an F residue while *T2* contains a K residue, that means that those two peptides are linked as *T2-T1* to give G-I-K-F-(A,S)-R. With that information, *C1* must be (A,S)-R-N-Y. Thus, *T3* must be N-Y-(G,P) and *Q* is G-I-K-F-(A,S)-R-N-Y-(G,P), where the order of (A,S) and (G,P) are uncertain. (Although it is not mentioned in the text, chymotrypsin does not catalyze the hydrolysis of peptides in which proline is at the C-terminal side of the cleavage point. Hence, it is likely that the sequence of peptide *C4* is G-P.)

27.27   (a)   b-Type fragmentation of the peptide N-F-E-S-G-K would produce the following *m/z* values. In this notation, each amino acid letter represents a residue (that is, —NH—CHR—CO—). Therefore, the *H*— must be added to indicate the additional hydrogen at the amino terminus H$_2$N—, and one mass unit must be added to the corresponding residue mass.

H—N ⇨ $1 + 114.0 = 115.0$

H—N-F ⇨ $1 + 114.0 + 147.1 = 262.1$

H—N-F-E ⇨ $1 + 114.0 + 147.1 + 129.0 = 391.1$

H—N-F-E-S ⇨ $1 + 114.0 + 147.1 + 129.0 + 87.0 = 478.1$

H—N-F-E-S-G ⇨ $1 + 114.0 + 147.1 + 129.0 + 87.0 + 57.0 = 535.1$

H—N-F-E-S-G-K—OH ⇨ $1 + 114.0 + 147.1 + 129.0 + 87.0 + 57.0 + 128.1 + 17.0 = 620.2$

(b)   The fragmentation occurs in the following way.

$$\overset{y_5 \quad y_4 \quad y_3 \quad y_2 \quad y_1}{\text{H—N} \mid \text{F} \mid \text{E} \mid \text{S} \mid \text{G} \mid \text{K—OH}}$$

The "*H$_2$*" at the amino terminus of each fragment that follows represents two protons added to the residue itself to give the protonated terminal amino group indicated in the problem. For example, the residue mass for K includes a terminal —NH—; a K residue with an amino group at the amino terminus is then *H*—K; and a protonated amino-terminal K residue is *H$_2$*K, with a positive charge on the nitrogen. The following fragments should be observed:

$[H_2\text{K—OH}]^+$ ⇨ $147.1 = y_1$

$[H_2\text{G—K—OH}]^+$ ⇨ $204.1 = y_2$

$[H_2\text{S—G—K—OH}]^+$ ⇨ $291.1 = y_3$

$[H_2\text{E—S—G—K—OH}]^+$ ⇨ $420.1 = y_4$

$[H_2\text{F—E—S—G—K—OH}]^+$ ⇨ $567.2 = y_5$

$[H_2\text{N—F—E—S—G—K—OH}]^+$ ⇨ $681.2 = \text{M} + 1$

27.28   (a)

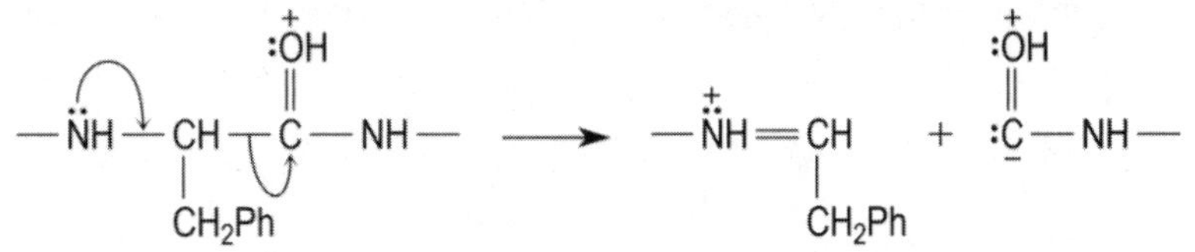

$$m/z = 551.94$$

(b)   The M + 1 ion (shown here protonated on the —OH group of the carboxylic acid) loses water to give an acylium ion, which is the ion of interest. Although carboxylic acids are normally protonated on the carbonyl group, the small amount of the form shown here, once formed, is very unstable.

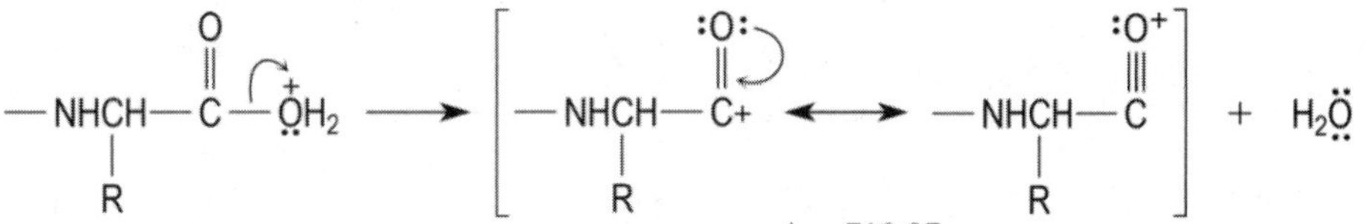

$$m/z = 710.97$$

(c)   Two conceivable mechanisms for forming H_2 are the following:

Mechanism (1):

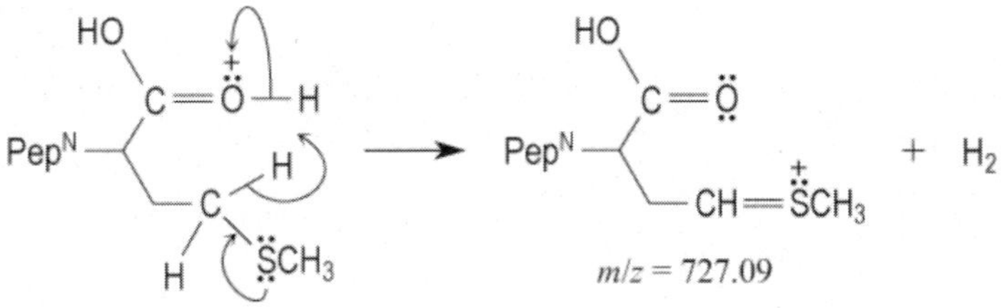

$$m/z = 727.09$$

The mechanism could be tested by labeling the CH_2 group adjacent to the sulfur with deuterium. Then, if this fragmentation occurs, a loss of DH ($m/z = 3$) would confirm the hypothesis.

Mechanism (2):

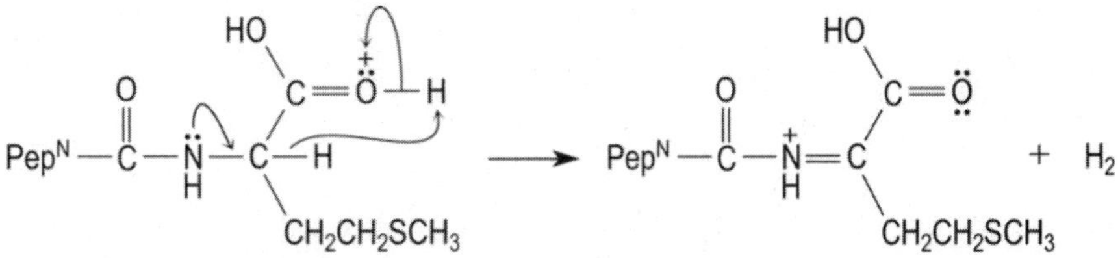

$$m/z = 727.09$$

This mechanism could be tested with a peptide in which the α-hydrogen is replaced by D. The indicated fragmentation would result in the loss of 3 mass units.

It is relevant that the loss of H_2 occurs in many peptides; a C-terminal methionine is not necessary. This observation makes mechanism (2) more likely but does not rule out mechanism (1) as a contributor for this particular peptide.

27.29   (a)   The amino group serves as a nucleophile in an addition to the isothiocyanate.

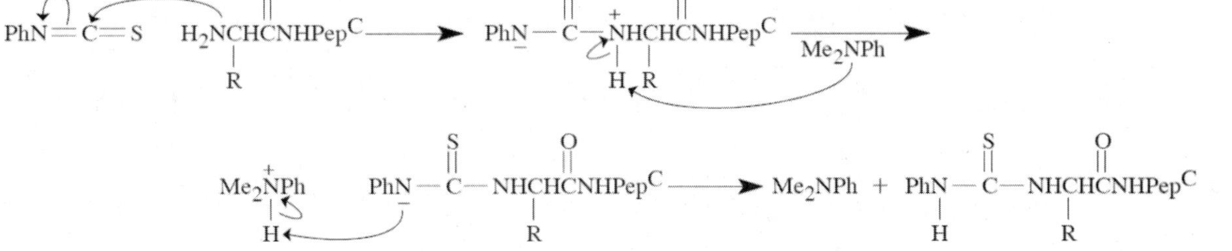

(b)   The sulfur serves as the nucleophile in an intramolecular nucleophilic acyl substitution reaction. The following mechanism begins with the carbonyl-protonated thiourea derivative.

(c)   Water serves as a nucleophile in opening the thiazolinone, and nitrogen of the resulting thiourea serves as the nucleophile in closing the ring to the PTH. The mechanism below begins with the carbonyl-protonated thiazolinone.

27.30 (a)   As Eq. 27.42a on text p. 1415 as well as its mechanism in the solution to Problem 27.29(a) shows, the Edman degradation depends on the presence of a free terminal amino group in the peptide. Because this group is blocked as an amide in acetylated peptides, such peptides cannot undergo the Edman degradation.

(b)   *N*-Acylation of peptides and proteins has no adverse effect on sequencing by MS–MS because it just another amide bond. Thus, typical b-type and y-type cleavage will still occur.

27.31   (a)

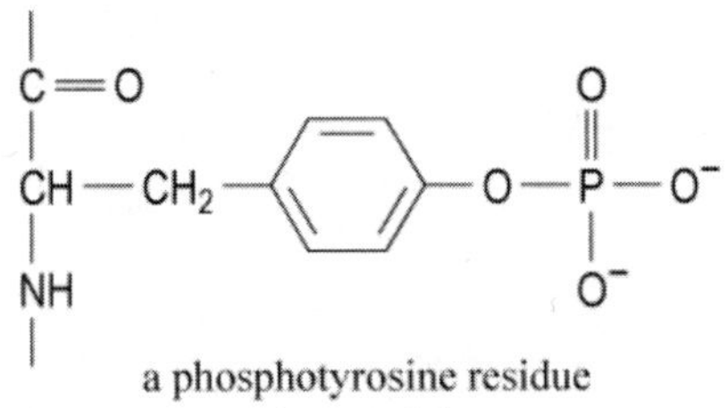

a phosphotyrosine residue
in a protein

(b)   Tyrosine contains an acidic phenolic group unlike serine and threonine. The equilibrium constants for the formation of ordinary aryl esters are considerably less favorable than the equilibrium constants for the formation of alkyl esters, and the same is true of phosphate esters. The reason is that the conjugate base of a phenol is less basic and is therefore a poorer nucleophile and a better leaving group. Therefore, the equilibrium constant for the formation of a phosphotyrosine residue from ATP would be less than the K_{eq} for the phosphorylation of a serine residue.

27.32   The compound is shown in its ionization state at pH = 7.4 (physiological pH). (Refer to the configuration of L-threonine on text p. 1379.) The configuration that is naturally occurring in proteins is assumed.

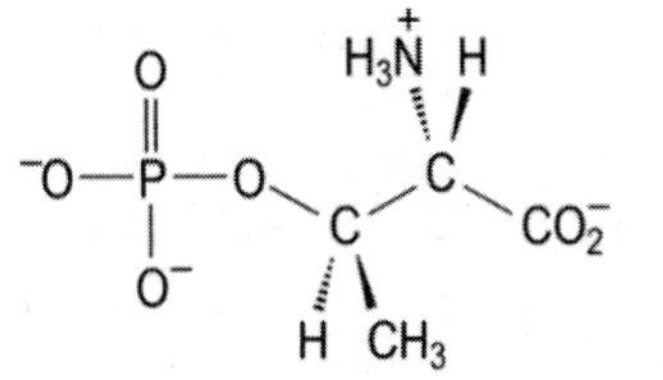

L-phosphothreonine

27.33

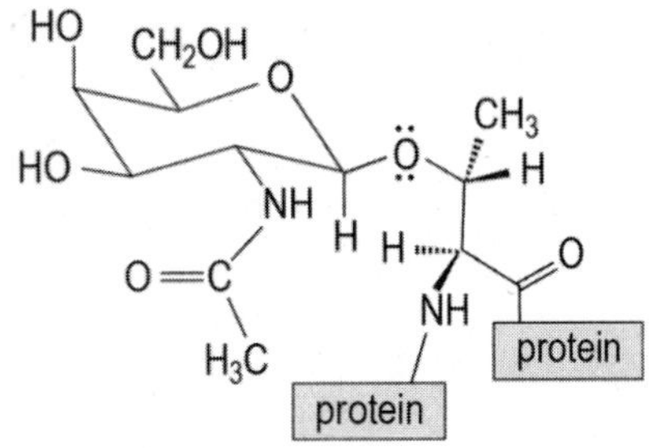

27.34   (a)

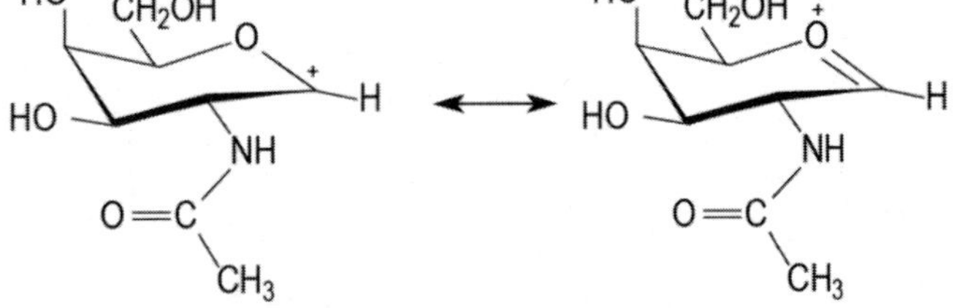

(b)   The Mn^{2+} ion likely stabilizes the phosphate anhydride group in the UDP-leaving group (see Sec. 25.7D of the text).

27.35   Reference the mechanism in Eqs. 24.24a–b on text p. 1252, and apply it to the glucose-Hb imine in Eq. 27.50. The net result is that the alcohol is converted into a carbonyl group.

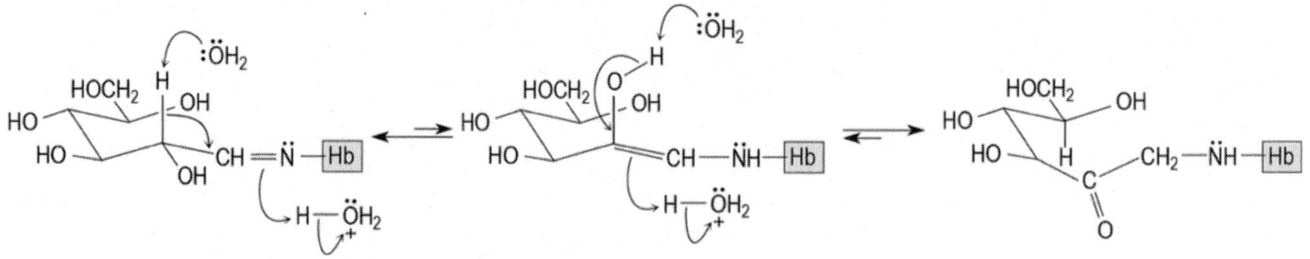

27.36 As described in Sec. 19.11A, primary amines react with aldehydes to form imines. An example is shown in Eq. 27.49 on text p. 1424, with the aldehyde of glucose reacting with the amino terminus of hemoglobin. A simplified version is represented below.

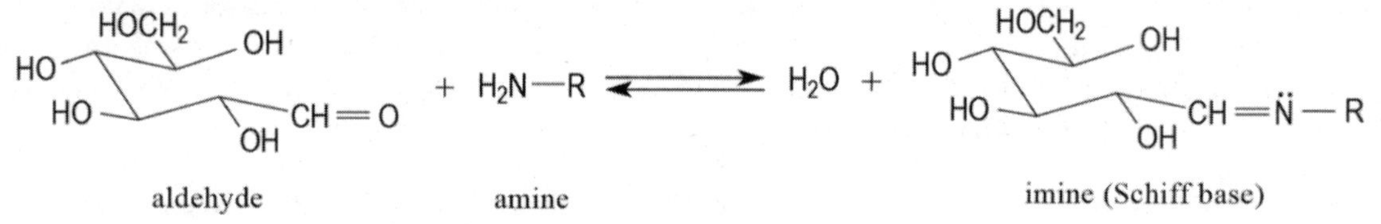

aldehyde   amine   imine (Schiff base)

The question asks why the nitrogen couldn't be supplied from an amide, such as from the side chain of an asparagine amino acid. If you consider the reaction mechanism of imide formation, you'll see that an intermediate forms after loss of water through an E1 mechanism. This intermediate is very unstable however, due to the nearby carbonyl carbon. It is unlikely that this unstable structure forms, and so imide formation between aldehydes and amides is much less favorable than imide formation from aldehydes and amines, where this unfavorable interaction does not exist.

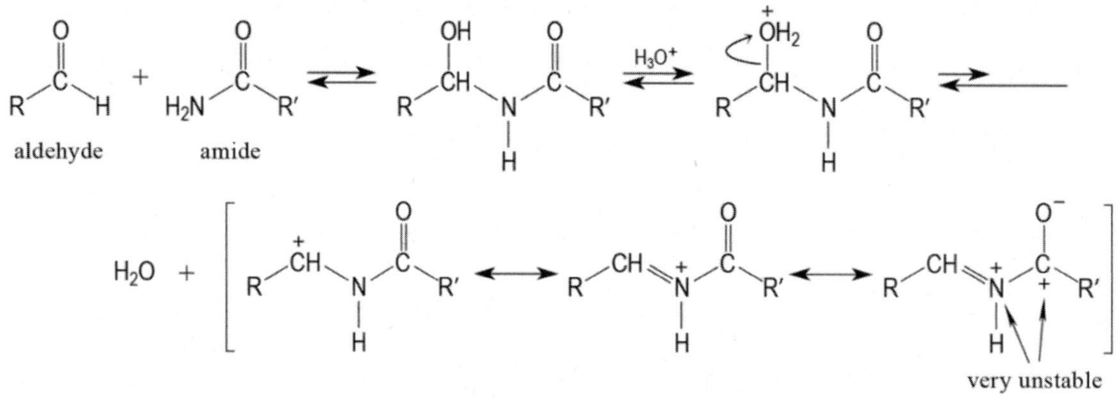

aldehyde   amide

very unstable

27.37 In the first step of the Amadori rearrangement (see solution to 27.35 in this manual), the C2 carbon is deprotonated. The reversible deprotonation can reprotonate from the opposite side, inverting the stereochemistry at C2. This process converts the glucose derivative into a mannose derivative.

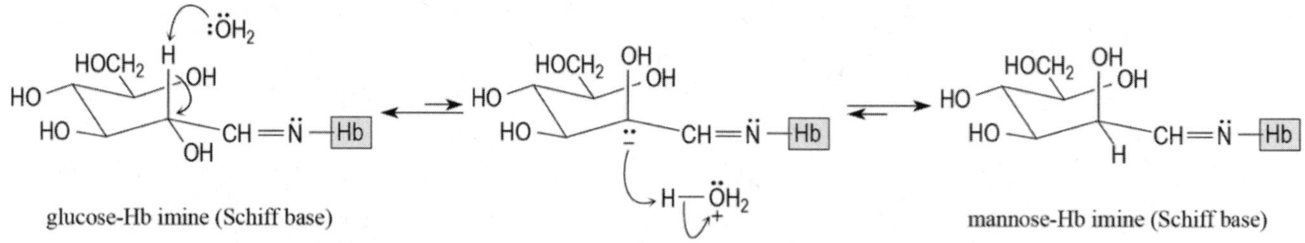

glucose-Hb imine (Schiff base)   mannose-Hb imine (Schiff base)

The reversible hydrolysis of the mannose imine yields the aldehyde form of mannose, which equilibrates into the cyclic form of mannose.

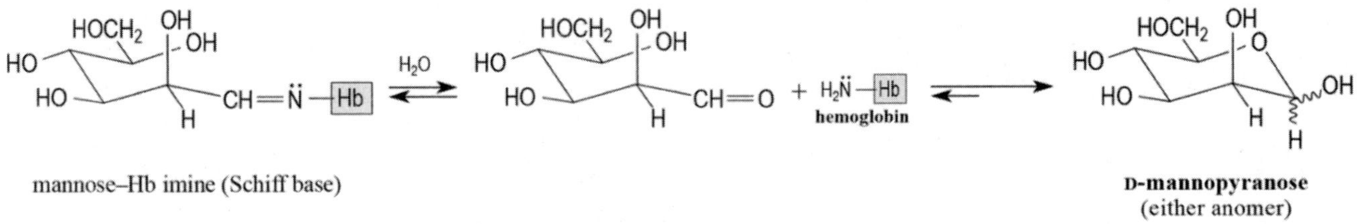

mannose–Hb imine (Schiff base)

D-mannopyranose
(either anomer)

27.38   Because the four chains of hemoglobin are held together by the same noncovalent forces that account for the tertiary structures of the individual chains, the subunits of hemoglobin would dissociate and would unfold into two random-coil α chains and two random-coil β chains.

27.39   One end of aeruginosin-B resembles an arginine side chain; one can hypothesize the following correspondences:

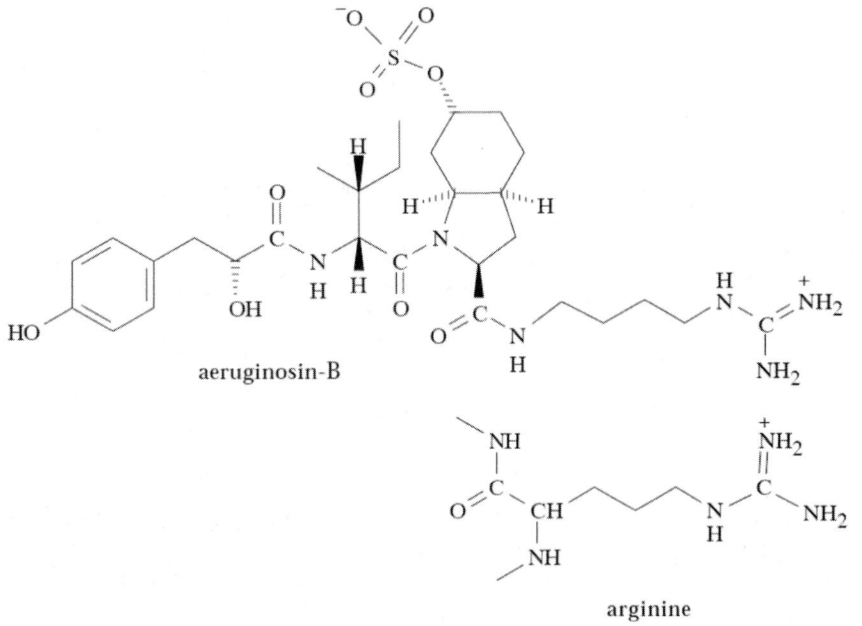

aeruginosin-B

arginine

Aeruginosin-B has a positively charged guanidino group, like arginine, that can interact with the ionized Asp-189 in the binding site; and this is attached to a hydrocarbon chain that can interact favorably with the "walls" of the hydrophobic pocket.

27.40   In the following mechanism, the enzyme groups are shown in shaded type.

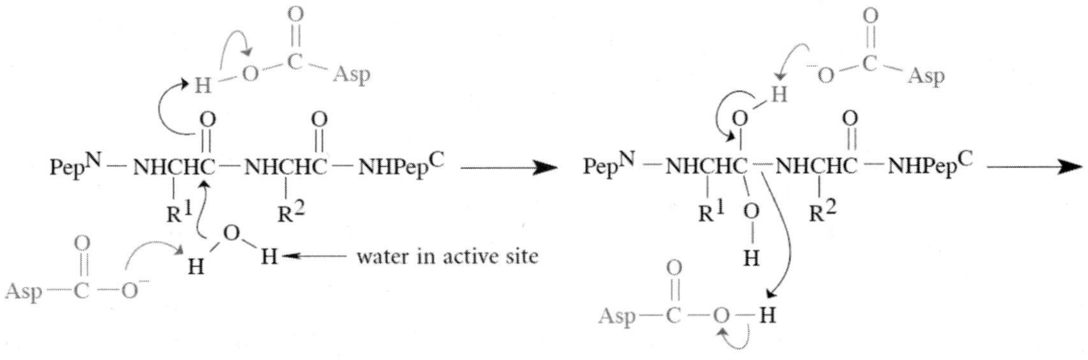

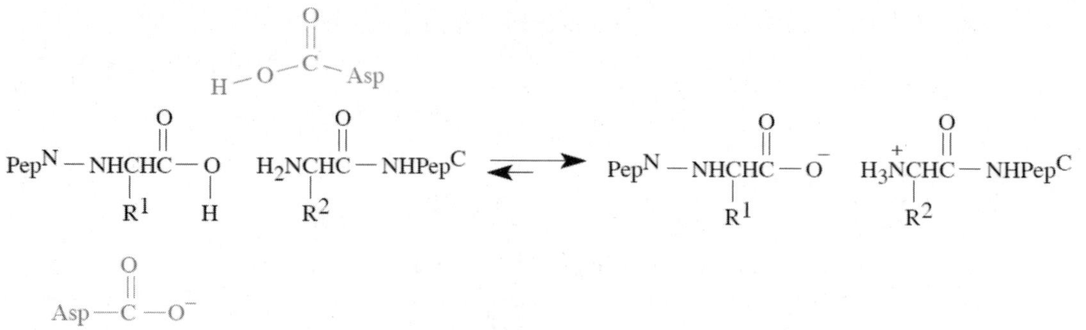

27.41   Although the sulfur has two unshared pairs, the nitrogen of the thiazole is like the nitrogen of pyridine or imidazole and is the most basic site on the ring; hence, it is likely to be the hydrogen-bond acceptor. An amide backbone N—H is the only hydrogen bond donor site on a peptide backbone.

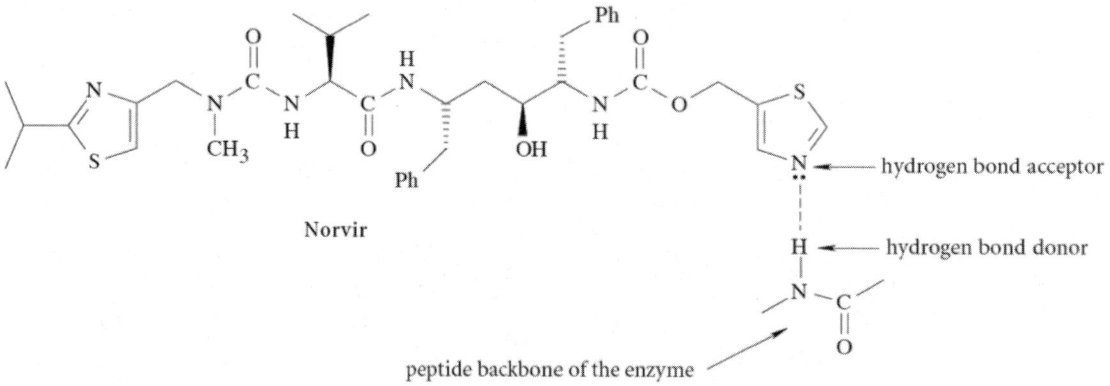

Norvir

27.42   We use the criteria for hydrogen-bond donors and acceptors discussed in Sec. 8.5C of the text.

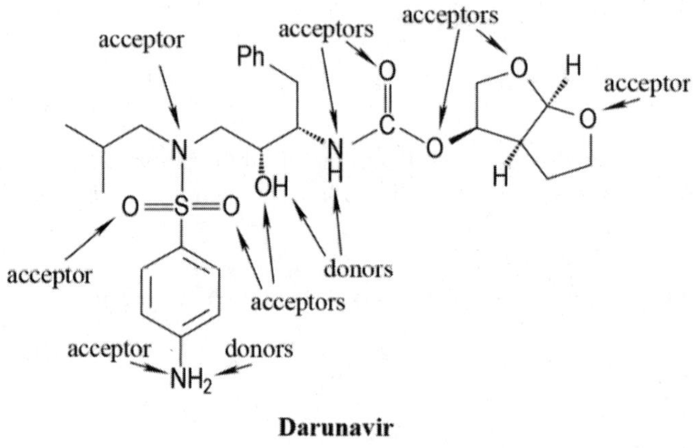

Darunavir

Solutions to Additional Problems

27.43   The results for valine:

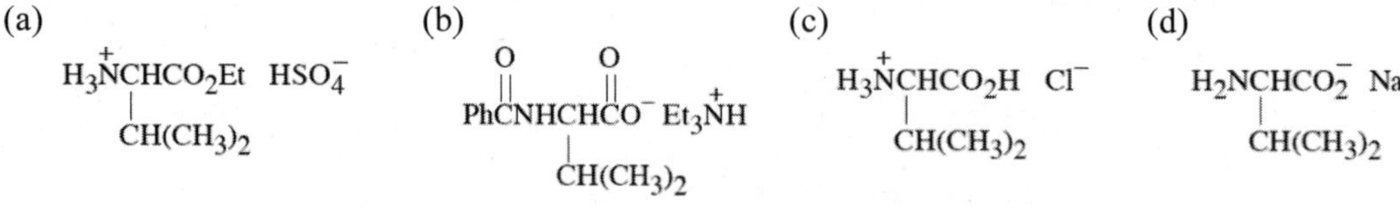

(a)

$$H_3\overset{+}{N}CHCO_2Et \ \ HSO_4^-$$
$$\overset{|}{CH(CH_3)_2}$$

(b)

$$PhCNHCHCO^- \ Et_3\overset{+}{N}H$$
O   O (above)
$$\overset{|}{CH(CH_3)_2}$$

(c)

$$H_3\overset{+}{N}CHCO_2H \ \ Cl^-$$
$$\overset{|}{CH(CH_3)_2}$$

(d)

$$H_2NCHCO_2^- \ \ Na^+$$
$$\overset{|}{CH(CH_3)_2}$$

The reaction in (e) is a variation of the first reaction of the Strecker synthesis in which an amine—in this case, the amino acid valine—rather than ammonia is the source of nitrogen. (See Eq. 27.12 on text p. 1388.)

(e)

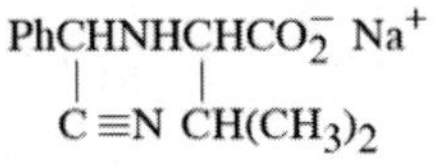

PhCHNHCHCO$_2^-$ Na$^+$
| |
C≡N CH(CH$_3$)$_2$

(f)

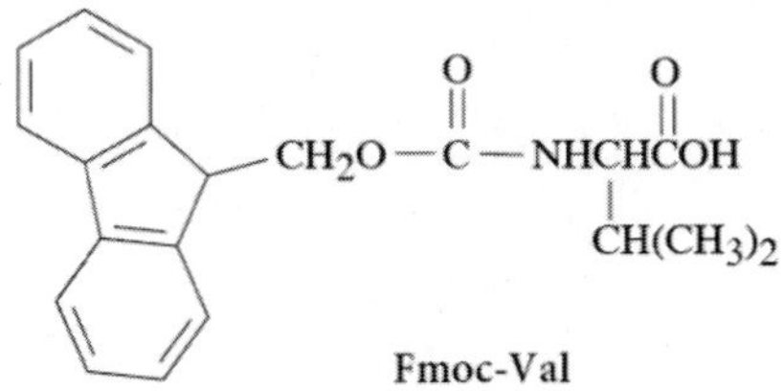

Fmoc-Val

(g)

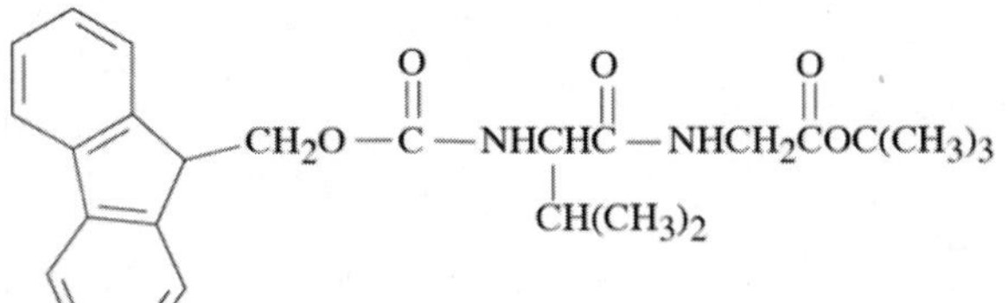

Fmoc-Val-Gly *tert*-butyl ester

(h)

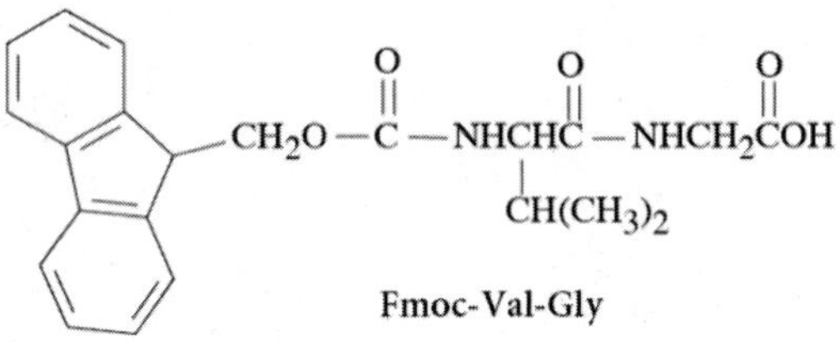

Fmoc-Val-Gly

(i)

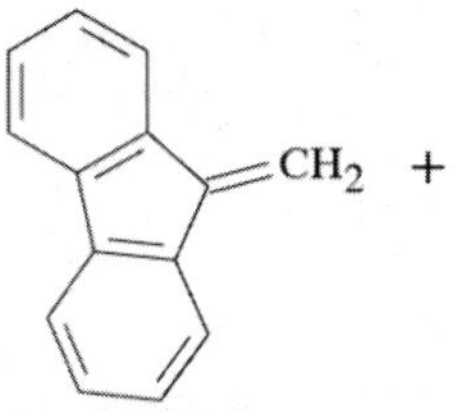

(reacts further with piperidine)

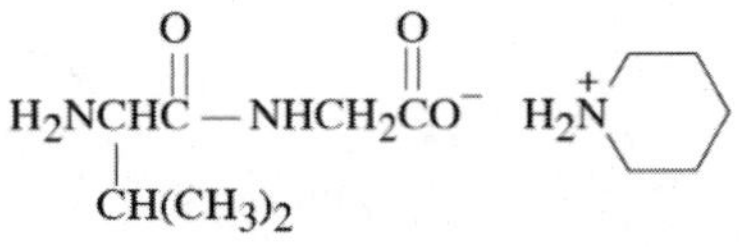

Val-Gly (piperidinium salt)

(j)

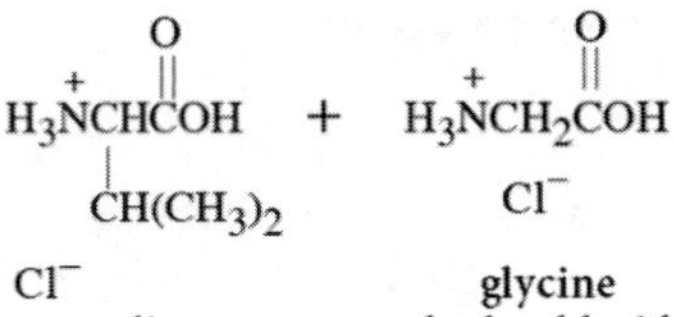

H$_3$NCHCOH + H$_3$NCH$_2$COH
  |
  CH(CH$_3$)$_2$ Cl$^-$

Cl$^-$
valine hydrochloride   **glycine hydrochloride**

The results for proline:

(a)

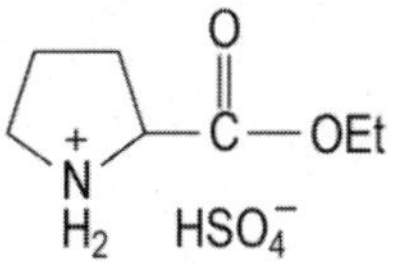

(b)

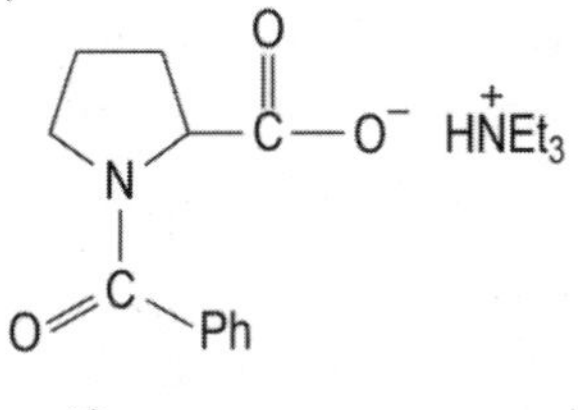

(c)

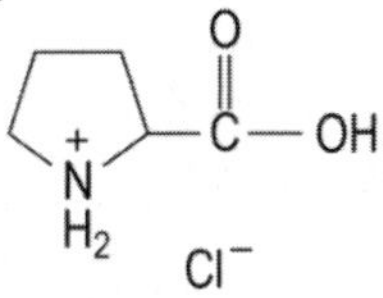

(d)

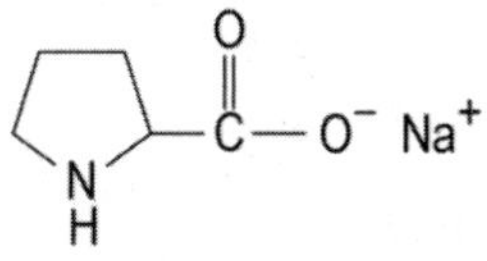

(e)

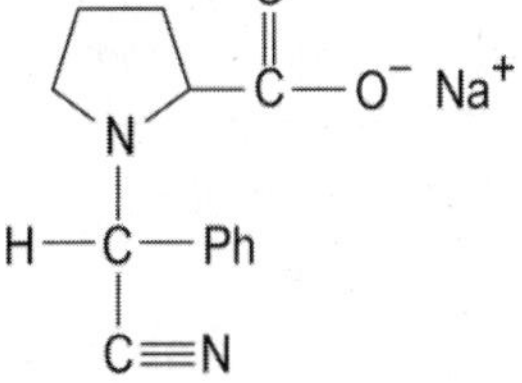

(f)

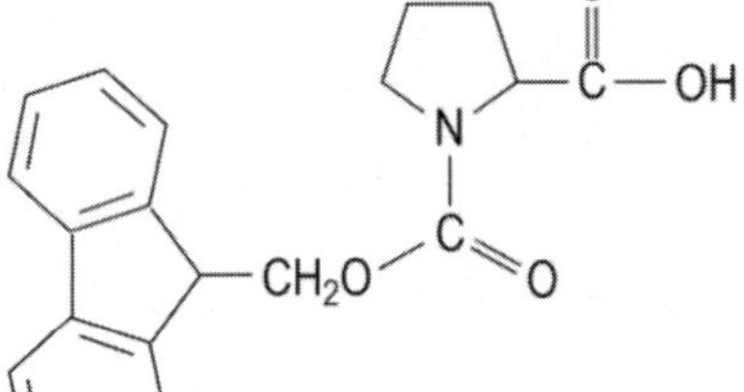

Fmoc-Pro

(g)

Fmoc-Pro-Gly *tert*-butyl ester

(h)

Fmoc-Pro-Gly

+ *tert*-butyl trifluoroacetate

(i)

+

(reacts further
with piperidine)

Pro-Gly
(piperidinium salt)

(j)

proline hydrochloride   glycine hydrochloride

27.44 (a) Aspartic acid. It has the lowest isoelectric point.
  (b) The most basic amino acid is arginine. It has the highest isoelectric point.
  (c) Isoleucine and threonine. Each has two asymmetric carbon stereocenters.
  (d) Glycine is not chiral and therefore it cannot be optically active.
  (e) Asparagine and glutamine. Their side-chain amides are hydrolyzed to give aspartic acid and glutamic acid, respectively.

27.45 (a) Because the amino group of leucine is not protected, a certain amount (maybe most) of attachment to the resin will occur by reaction with the α-amino group on the chloromethyl groups of the resin. Subsequent peptide synthesis would then not proceed as planned.
  (b) Fmoc-Leu, once attached to the resin, must be deprotected before the subsequent coupling reaction can be expected to succeed. Polly forgot the deprotection step with 20% piperidine, which is essential for removing the Fmoc group.

27.46 Since lysozyme has many more basic residues (Lys and Arg) than acidic residues (Asp and Glu), lysozyme is expected to be (and is) a basic protein. Its isoelectric point should be >>6. (Its isoelectric point is about 12.)

27.47 (a) As discussed on the top of text p. 1385, the general rule of thumb is that if the peptide contains more acidic than basic groups it is an acidic peptide; if it contains more basic than acidic groups it is a basic peptide; and if the number of acidic and basic groups are equal, the peptide is neutral. Peptide *B* is more basic. It contains a basic Lys and no other acidic side chains. Peptide *A* contains a basic side chain (Lys), but two acidic side chains (Glu and Asp).
  (b) As discussed in Sec 26.3D of the text, the more basic peptide should come off of the column first, so peptide *B* comes off before peptide *A*.
  (c) First, rule out sequences 1 and 4 since they are not DNA sequences; they are RNA sequences, because they contain U and not T. Next, translate sequences 2 and 3 one at a time. Take the DNA sequence and divide it in to 3-letter pieces, or codons. Translate each codon into its mRNA anticodon, changing G to C, C to G, T to A, and A to U (there is no T in RNA, so use U). Next, use the genetic code (Table 27.2 on text p. 1401) to translate each 3-letter mRNA anticodon into the amino acid sequence. Finally, convert each three-letter amino acid abbreviation into its one-letter abbreviation (Table 27.1, text p. 1376), and check to see which sequence corresponds to peptide *A*. Sequence 2 is the answer (note that there is an extra codon that codes for STOP at the end of the sequence).

(d)

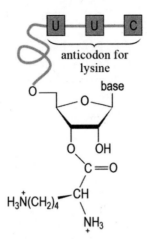

27.48   Cysteic acid is essentially an aspartic acid in which the side-chain carboxy group has been substituted with a sulfonic acid group. Because sulfonic acids are more acidic than carboxylic acids, cysteic acid is expected to be (and is) more acidic than aspartic acid. Thus, the correct answer is (1): cysteic acid has a lower isoelectric point than aspartic acid.

27.49   The amino-terminal residue is lysine. Both the side-chain amino group and the α-amino group react with phenyl isothiocyanate (the Edman reagent). Thus, the side-chain amino group in the resulting PTH derivative (shown in the problem) is present as a thiourea derivative.

27.50   The amino-terminal residue of the peptide P is valine. Because dansyl-valine is obtained, the α-amino group of valine must have been free in the peptide, and hence, valine must have been the amino-terminal residue. The sequence of the other residues cannot be determined from the data given.

   "Dansylation" is a method for determining the amino-terminal residue of a peptide.

27.51   Number the residues of Q from the amino terminus as 1,2,3, … First, the Edman degradation of Q shows that leucine is the amino-terminal residue. The formation of only dipeptides, including the dipeptide Leu-Val, shows that DPAP catalyzes the hydrolysis of peptide Q at every even-numbered residue from the amino terminus. Thus, this enzyme also catalyzes the hydrolysis of peptide R at every other residue — the odd residues in the numbering of Q:

Q: 1—2—3—4—5—6—7—8—9   = DPAP cleavage site

Edman

R: 2—3—4—5—6—7—8—9

This analysis shows that Gly is residue 9, the carboxy-terminal residue of Q. To get the order of the dipeptides, work back and forth between the dipeptides derived from Q and those derived from R. Since Gly has to be at the carboxy terminus of one of the dipeptides from R, and since the only dipeptide that meets this criterion is Ala-Gly, then Ala is the next residue in from the carboxy terminus. If Ala is in position 8, then position 7, from the Q dipeptides, must be Gln; position 6, from the R dipeptides, must be Asp; and so on. Alternatively, work from the amino terminus of Q: Leu-Val of Q and Val-Arg of R establish the sequence Leu-Val-Arg; Arg-Gly of Q shows that the next residue is Gly; and so on. The final sequence of peptide R, then, is:

Leu-Val-Arg-Gly-Val-Asp-Gln-Ala-Gly

(Remember that the amino terminus by convention is on the left when peptides are written this way.) The ammonia arises from the hydrolysis of Gln to glutamic acid.

27.52 (a) Trypsin should catalyze the cleavage of glucagon at Lys and Arg residues. The following five fragments are expected. (Remember that, by convention, peptides written in this way have their amino ends on the left.) We assume that all cleavages proceed to completion, although in practice intermediate cleavage fragments might be formed as well.

(1) His-Ser-Gln-Gly-Thr-Phe-Thr-Ser-Asp-Tyr-Ser-Lys
(2) Tyr-Leu-Asp-Ser-Arg
(3) Arg
(4) Ala-Gln-Asp-Phe-Val-Gln-Trp-Leu-Met-Asn-Thr
(5) Arg-Ala-Gln-Asp-Phe-Val-Gln-Trp-Leu-Met-Asn-Thr

If cleavage occurs after the first Arg in the Arg-Arg sequence of glucagon, then the peptide (5) is formed, and the second Arg becomes the amino terminus of this peptide. Cleavage does not occur at the peptide bond of this Arg because trypsin is an endopeptidase. (Eq. 27.37b, text p. 1408.) If cleavage of glucagon occurs after the second Arg, then Tyr-Leu-Asp-Ser-Arg-Arg is formed. Subsequent cleavage of this peptide at the Arg-Arg bond gives (2) and (3).

(b) The products of this Edman cleavage would be the PTH derivative of histidine along with a new peptide, which is glucagon without its amino-terminal residue.

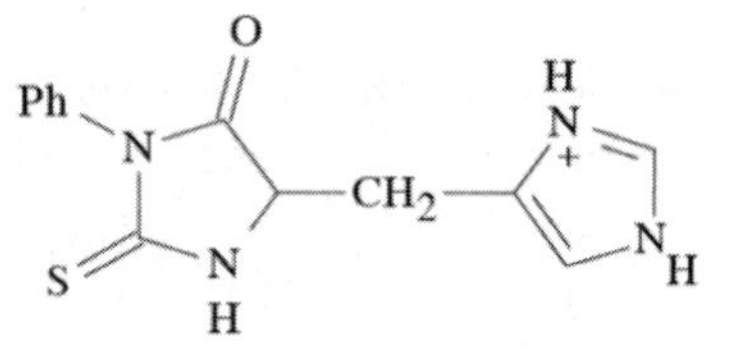

+ Ser-Gln-Gly- ··· (remainder of glucagon sequence)

PTH derivative of histidine

27.53 (a) A molecular mass of about 1000 is consistent with the actual composition (Ala_4,Arg_2,Gly_2,Ser_2); amino acid analysis gives only the *relative* amounts of each amino acid. The absence of a reaction with the Edman reagent suggests that there is no terminal amino group. A *cyclic* peptide would give these results. Cleavage of a cyclic peptide at two arginine residues would give two peptides with arginine at the C-terminus of each; but if the two peptides were identical, two equivalents of a single peptide would be obtained. The sequencing results are in accord with the following structure:

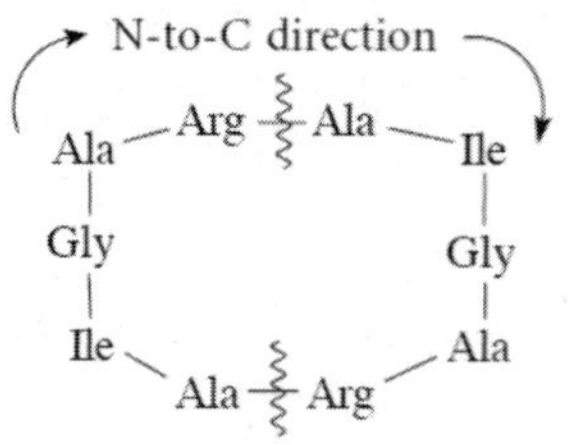

= trypsin cleavage site

(b) Peptide *D* is an ordinary (noncyclic) peptide that results from two cleavages at the R residues. The MS–MS b-type fragmentation of the M + 1 ion of peptide *D* would be typical of any linear peptide. (The positive charge on the fragments represents the charge on the acylium ion, and the *H*— is the extra proton that ends up as one of the H_2N— protons on the terminal amino group; see Eq. 27.39 on text p. 1412.) The residue masses are from Table 27.3 on text p. 1413.

H—A-I-G-A-R•H^+—OH ⇨ $m/z = 1.0 + 71.0 + 113.1 + 57.0 + 71.0 + 156.1 + 18.0 = 487.2 = M + 1$

$[H$—A-I-G-A-R$]^+$ ⇨ $m/z = 1.0 + 71.0 + 113.1 + 57.0 + 71.0 + 156.1 = 469.2$

$[H$—A-I-G-A$]^+$ ⇨ $m/z = 1.0 + 71.0 + 113.1 + 57.0 + 71.0 = 313.1$

$[H$—A-I-G$]^+$ ⇨ $m/z = 1.0 + 71.0 + 113.1 + 57.0 = 242.1$

$[H$—A-I$]^+$ ⇨ $m/z = 1.0 + 71.0 + 113.1 = 185.1$

$$[H\!-\!A]^+ \quad \Rightarrow \quad m/z = 1.0 + 71.0 = 72.0$$

The cyclic peptide C, on the other hand, would undergo fragmentation at every possible peptide bond to give five different fragments.

$$H\!-\!A\text{-}I\text{-}G\text{-}A\text{-}R\text{-}A\text{-}I\text{-}G\text{-}A\text{-}R^+$$

$$H\!-\!R\text{-}A\text{-}I\text{-}G\text{-}A\text{-}R\text{-}A\text{-}I\text{-}G\text{-}A^+$$

Peptide C $\xrightarrow[\text{of cyclic peptide}]{\text{fragmentation}}$ $H\!-\!A\text{-}R\text{-}A\text{-}I\text{-}G\text{-}A\text{-}R\text{-}A\text{-}I\text{-}G^+$

$$H\!-\!G\text{-}A\text{-}R\text{-}A\text{-}I\text{-}G\text{-}A\text{-}R\text{-}A\text{-}I^+$$

$$H\!-\!I\text{-}G\text{-}A\text{-}R\text{-}A\text{-}I\text{-}G\text{-}A\text{-}R\text{-}A^+$$

However, these all have identical masses because they differ only in the positions of the amino acids and not their identity. The m/z is M + 1, where M is the mass of the parent cyclic peptide; this peak occurs at $m/z = 937.4$.

27.54   The presence of two amino-terminal residues suggests that either *(a)* insulin is an approximately equimolar mixture of two proteins, or *(b)* it is a single protein containing two peptide chains. In fact, the latter is correct; insulin is two polypeptide chains connected by disulfide bonds. Glycine is the amino-terminal residue of one chain, and phenylalanine is the amino-terminal residue of the other.

27.55   Amino acid A is 2,4-diaminobutanoic acid. The key piece of structural data is the Hofmann rearrangement, which converts the side-chain amide of glutamine into an amino group. Compound A is expected to be basic (and thus have a high isoelectric point) because it has two amino groups and only one carboxy group.

27.56   This amino acid is really a substituted malonic acid. When heated in acid, it, like most malonic acid derivatives, decarboxylates. In this case, the product is glutamic acid. (See Eq. 20.41, text p. 1031.)

27.57   (a)   Reaction of chloromethylated polystyrene with trimethylamine would give the resin shown:

The alkyl halide is benzylic and is thus particularly reactive in S_N2 reactions.

(b)   This resin is an *anion-exchange resin*. (See the solution to Problem 27.11 on p. 983 of this manual.) At pH 6, Arg is positively charged and is therefore repelled by the column; it emerges earliest from the column. Leu has zero charge at pH 6; it emerges next. Glu is negatively charged at pH 6 and is therefore attracted to the positively-charged groups on the column; it emerges last from the column.

27.58   The principle is that a peptide with a given charge migrates to the electrode of opposite charge; a peptide with zero charge does not migrate. Because Gly-Lys is a basic peptide, it has a net positive charge at pH 6, and it therefore migrates to the cathode, the negatively charged electrode. Because Gly-Asp is an acidic peptide, it has a net negative charge at pH 6, and it migrates to the anode, the positively charged electrode. Gly-Ala is a neutral peptide; it has net zero charge at pH 6, and therefore it does not migrate.

27.59   The resin contains a large number of pendant phenyl groups. Phenylalanine also contains a phenyl group. The resin has dual characteristics. Not only is it an anion-exchange resin, but also it is like an HPLC resin in which the side-chains are phenyl instead of C-18 hydrocarbons. The "like-dissolves-like" principle suggests that Phe, the amino acid containing the phenyl side chain will be partitioned into the resin phase more extensively than Gly. The selective adsorption of phenylalanine is due to this effect.

27.60   Such a resin should be less effective in partitioning the amino acids with hydrocarbon side chains, because they will be "less soluble" in the resin phase. In effect, there are fewer $—CH_2—$ groups on the resin to provide favorable interactions with the amino acid side chains. Furthermore, the amino acid side-chains would be forced to be closer to the more polar and hydrophilic resin surface, with which they might not interact as favorably. The separation should be less effective. The separation should be less effective with he C8-silica.

27.61   (a)   Alanine (like the other amino acids) is a different compound in HCl, NaOH, and neutral H_2O because its ionization state is different, and different compounds have different optical rotations. In acidic solution, the optical rotation is that of the acidic form; in base, the optical rotation is that of the basic form; and in water, the optical rotation is that of the neutral (zwitterion) form.

(b)   Lysine has two amino groups, and acetylation of each is possible.

the two mono-*N*-acetylated lysine derivatives

(c)   The peptide is cleaved at the Arg-Ala bond at pH 8. However, in the presence of 8 *M* urea, trypsin, like most enzymes, is denatured. A denatured enzyme is devoid of catalytic activity because its tertiary structure (and therefore its active site) is disrupted. Thus, trypsin, when denatured, cannot catalyze cleavage of the peptide.

(d)   The 2-mercaptoethanol treatment ensures that disulfide bonds are reduced. Aziridine reacts much like an epoxide when the nitrogen is protonated. Aziridine is basic enough that a significant amount of it is protonated in aqueous solution. Reaction of protonated aziridine with the thiol group of a cysteine residue gives an amine:

The amino group of the modified residue is protonated at pH 8, the pH at which trypsin digestion is carried out. The side chain of the modified residue resembles the side chain of lysine in both length and charge, and trypsin hydrolyzes peptides at the modified residue much as it hydrolyzes peptides at lysine residues.

(e)   The two sulfoxides are diastereomers. They both have the same configuration at the α-carbon, but differ in configuration at the sulfur of the sulfoxide group. This sulfur is an asymmetric atom and stereocenter because it has four different "groups" attached: the CH_2, the CH_3, the O, and the electron pair. Evidently, inversion at sulfur in a sulfoxide, unlike amine inversion, is very slow, because the individual diastereomers can be isolated.

27.62 (a)   At a pH of 8–9, the thiol group of cysteine is partially deprotonated (pK_a of the thiol group is 10.28, see Table 27.1, text p. 1376); the resulting thiolate is a good nucleophile and the iodine of iodoacetic acid is an excellent leaving group. Masking the thiol group of any cysteine residues present in the protein is necessary to prevent the thiol groups from reoxidizing to disulfides.

$$\text{protein} - CH_2 - SH \overset{\text{pH 8–9}}{\rightleftharpoons} \text{protein} - CH_2 - S^- \xrightarrow{I - CH_2CO_2^-}$$

cysteine residue

$$\text{protein} - CH_2 - S - CH_2CO_2^- \ + \ I^-$$

(b)   Hydrogen peroxide oxidizes the disulfide bonds to sulfenic acids, then to sulfinic acids, and finally to sulfonic acids (see Fig. 10.3, text p. 496).

27.63 The coenzyme involved in the conversion of aspartic acid residues to Asp-methyl ester residues is *S*-adenosylmethionine (SAM). (See Figure 11.1, text p. 537, for the structure of SAM.)

$$-NH-CH-\overset{\displaystyle O}{\overset{\displaystyle \|}{C}}-$$
$$\underset{\displaystyle CH_2-CO_2CH_3}{|}$$

Asp-methyl ester residue

27.64 Trypsin catalyzes the hydrolysis of peptides and proteins at the carbonyl group of arginine and lysine, amino acids with side chains containing strongly basic groups that are positively charged at physiological pH. Chymotrypsin, in contrast, catalyzes the hydrolysis of peptides and proteins at the carbonyl group of phenylalanine, tryptophan, tyrosine, leucine, and isoleucine. As explained in the text (see Sec 27.10) and illustrated in Fig. 27.18, text p. 1435, the positively charged group in the peptide or protein substrate form a complex in the trypsin active site by electrostatic attractions and hydrogen bonding. The addition of a specific trypsin inhibitor such as benzamidinium chloride (text p. 1436) would block the trypsin active site without appreciably affecting chymotrypsin. (Considerably more potent trypsin inhibitors are known and are used for this purpose.)

27.65 The mechanism of cleavage in trifluoroacetic acid involves a benzylic carbocation intermediate. This carbocation is resonance-stabilized not only by the *para*-oxygen, but also by the *ortho*-oxygen as well. (Draw the resonance structures.) Because the carbocation intermediate is more stable, the cleavage reaction is faster. Hence, milder conditions (lower acid concentration) can be used for the cleavage.

27.66 The two Asp residues would be expect to interact with this hydroxy group by hydrogen bonding. Perhaps one Asp in protonated form could serve as a donor and the other, in ionized form, could serve as an acceptor:

(The interaction with Crixivan is similar.)

27.67 Figure 27.11 on text p. 1426 shows that the amino acid side chains extend outward from the periphery of the helix and are actually rather close to each other in space. The pK_a of a protonated lysine side-chain amino group is about 10.5. At pH values below 10, the side-chain amino groups in polylysine are protonated; thus, these side chains are positively charged. The peptide avoids the α-helical conformation at pH values below 10 to avoid the repulsive electrostatic interactions between the adjacent positively charged groups. At pH values above 11, the side-chain amino groups are unprotonated and hence uncharged. As a result, there are no charge–charge repulsions to destabilize the α-helix.

　　The same principles operate in polyglutamic acid, but at the other end of the pH scale. In polyglutamic acid, the side chains are negatively charged at pH values above the pK_a of the carboxy groups. Hence, the repulsions between negative charges cause the helix to be destabilized, and it unfolds at high pH. At low pH, the carboxy groups arc un-ionized and hence uncharged. Consequently, the helix can form at low pH because there are no charge–charge repulsions to destabilize it.

27.68 (a) A membrane is a very hydrophobic environment. If a protein is embedded in a membrane, then the surface of that protein that interacts with the membrane environment must also be largely hydrophobic. These favorable hydrophobic interactions hold the protein together in its three dimensional shape. If a membrane protein is extracted into an aqueous environment, it will denature because the polar water molecules cause the protein to adopt a different structure—one where the more hydrophilic residues make contact with the surrounding solvent molecules. The protein will assume the most stable conformation it can in this non-native, hydrophilic, environment, likely a very different one that it had inside the membrane, its native, hydrophobic environment.

　　(b) Detergents are very similar to the fatty acids that make up membranes: they have a long, hydrocarbon "tail" and a polar "head group". A membrane protein extracted into an aqueous environment containing a high concentration of detergent may find an environment similar to that of the membrane, which may prevent it from denaturing.

27.69 The interior of a membrane is a very hydrophobic environment. Therefore, amino acids containing side chains that are very non-polar typically populate the surface of the protein that comes in contact with the membrane. (Refer to Table 27.1, text p. 1376–7, for the structures of the amino acid side chains.) Group A and B contain acidic and basic side chains, respectively, and would be more likely to be on the surface of a protein that came into contact with water. Similarly, group F contains polar side chains, which are not hydrophobic. Group D, Pro, is rarely found in α-helix regions because of its cyclic, and thus rigid, structure. Group E contains very hydrophobic side chains, and is thus the most likely group engage in favorable hydrocarbon-hydrocarbon interactions with the interior of the membrane. Group C, Gly, contains only hydrogen as a side chain and which is not as hydrophobic as the side chains in group E.

27.70   (a)   This is a reductive amination of formaldehyde by the amino group of the lysine residue. Because excess formaldehyde is present, the amino group of lysine is methylated twice.

(b)   The side-chain amino group of lysine serves as a nucleophile to open the anhydride. The resulting carboxy group is ionized at the pH of the reaction.

(c)   The thiol of cysteine serves as a nucleophile in a conjugate-addition reaction to maleimide. The pH must be high enough to form a small amount of the conjugate-base thiolate anion of cysteine, which is the actual nucleophilic species.

(d)   The carbodiimide promotes a condensation between the side-chain carboxy group of aspartic acid and the amino group of the glycine ester.

(e)   The phenol ring of the tyrosine residue undergoes electrophilic substitution by the diazonium ion (see Sec. 23.10B, text p. 1209). Because the para position is occupied, substitution occurs at the position ortho to the hydroxy group.

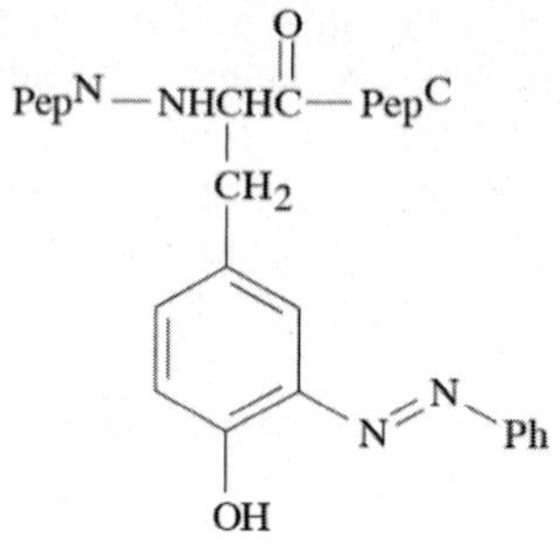

27.71 **(a)**

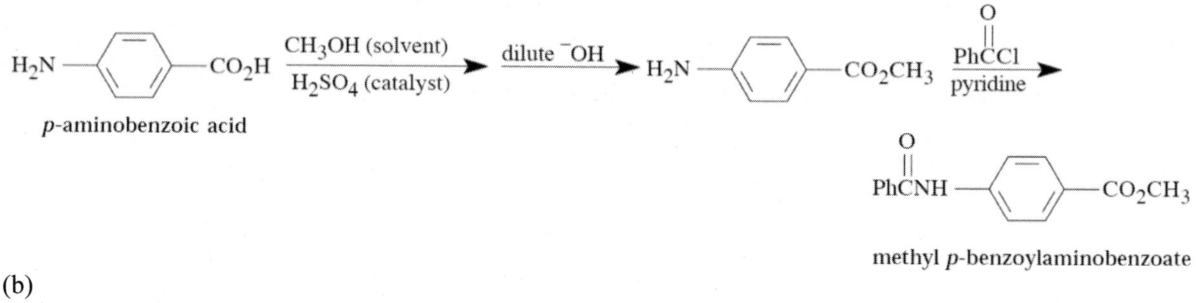

p-aminobenzoic acid

methyl *p*-benzoylaminobenzoate

(b)

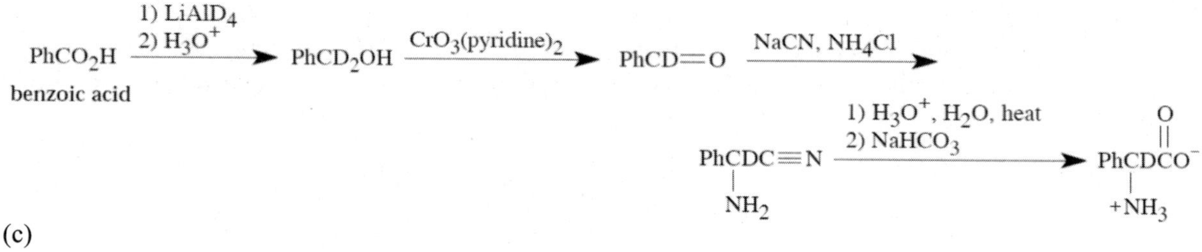

benzoic acid

(c)

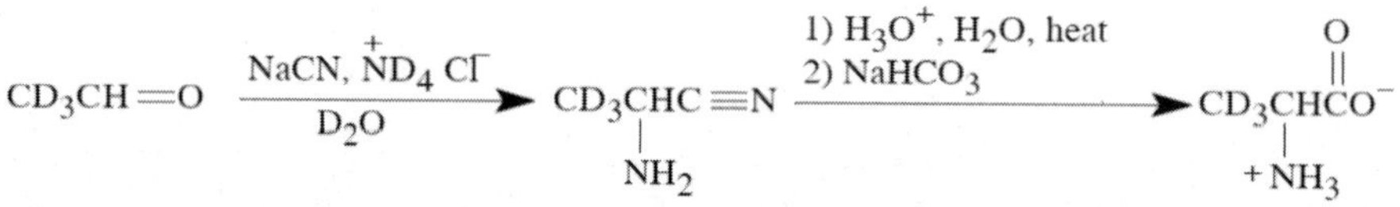

The isotopically substituted reagents ($^+ND_4\ Cl^-$, D_2O) are necessary in the first step because the pH of the reaction is sufficiently basic that the α-deuteriums of the starting trideuteroacetaldehyde would exchange for hydrogens if the analogous protic reagents $^+NH_4\ Cl^-$ and H_2O were used. (See Eq. 22.6 on text p. 1108.)

(d)

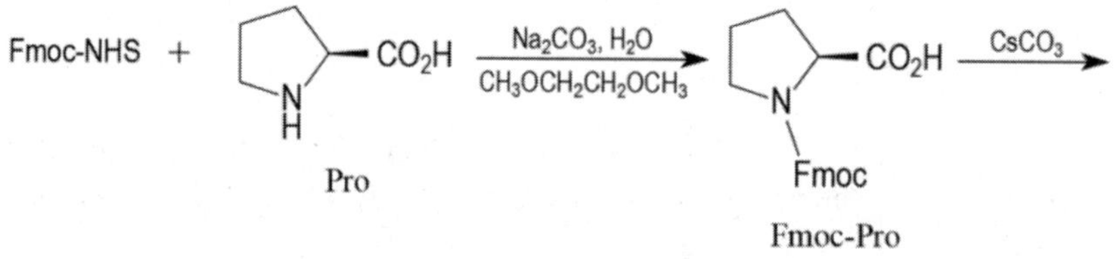

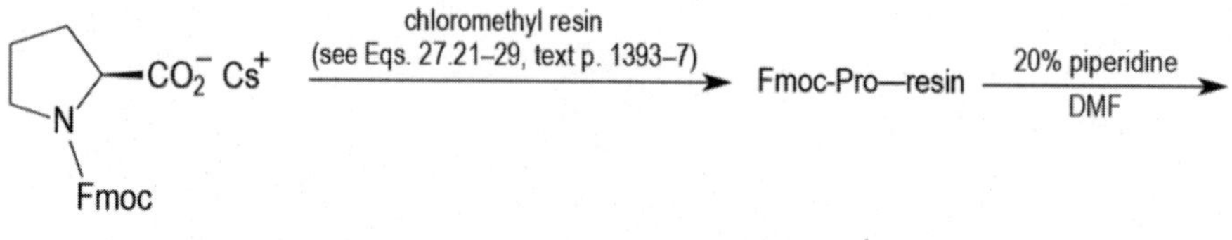

Pro

Fmoc-Pro

$$\text{Pro—resin} \xrightarrow{\text{Fmoc-Ala, DIC, HOBt}} \text{Fmoc-Ala—Pro—resin} \xrightarrow[\text{DMF}]{\text{20\% piperidine}}$$

$$\text{Ala—Pro—resin} \xrightarrow{\text{di-Boc-lysine, DIC, HOBt}} \text{di-Boc-Lys—Ala—Pro—resin} \xrightarrow[\text{CH}_2\text{Cl}_2]{\text{50\% TFA}} \text{Lys—Ala—Pro}$$

(e)

$$\text{EtO}_2\text{CCH}=\text{O} \xrightarrow{\text{NaCN, NH}_4\text{Cl}} \underset{\underset{\text{NH}_2}{|}}{\text{EtO}_2\text{CCHC}\equiv\text{N}} \xrightarrow{\text{Ac}_2\text{O}} \underset{\underset{\underset{\text{O}}{\overset{||}{\text{NHCCH}_3}}}{|}}{\text{EtO}_2\text{CCHC}\equiv\text{N}}$$

(f)   First prepare 3-bromocyclopentene from cyclopentene:

3-bromocyclopentene

Then use it in the following synthesis. Because the α-hydrogen of the starting material [the product of part (e)] is about as acidic as an α-hydrogen of diethyl malonate, the techniques of the acetamidomalonate method can be used. Notice that 3-bromocyclopentene is an allylic halide and is therefore very reactive in S_N2 reactions. (See Sec. 17.4 of the text.)

(g)

27.72   Every synthesis has the same final steps:

$$\underset{\text{CH}_3\overset{\overset{\text{O}}{||}}{\text{C}}\text{NHCH(CO}_2\text{Et)}_2}{} \xrightarrow[\text{2) RBr}]{\text{1) NaOEt, EtOH}} \underset{\underset{\text{R}}{|}}{\text{CH}_3\overset{\overset{\text{O}}{||}}{\text{C}}\text{NHC(CO}_2\text{Et)}_2} \xrightarrow[\text{2) NaOH (dilute)}]{\text{1) H}_3\text{O}^+\text{, H}_2\text{O, heat}} \underset{\underset{\text{R}}{|}}{\text{H}_3\overset{+}{\text{N}}\text{CHCO}_2^-}$$

The challenge in each case is to prepare the alkyl halide R—Br from the given starting material.

(a)

$$\underset{\substack{\text{2-methylpropene}\\\text{(isobutylene)}}}{(\text{CH}_3)_2\text{C}=\text{CH}_2} \xrightarrow{\text{DBr, peroxides}} (\text{CH}_3)_2\text{CD—CH}_2\text{Br}$$

Alternatively, the same alkene could be hydroborated with BD_3, the resulting organoborane oxidized to the primary alcohol by alkaline H_2O_2, and the alcohol converted into the alkyl bromide with concentrated HBr and H_2SO_4 catalyst.

(b)

$$PhCH{=}O \xrightarrow[\text{2) H}_3\text{O}^+]{\text{1) LiAlD}_4} PhCHD{-}OH \xrightarrow{HBr} PhCHD{-}Br$$

benzaldehyde

(c)

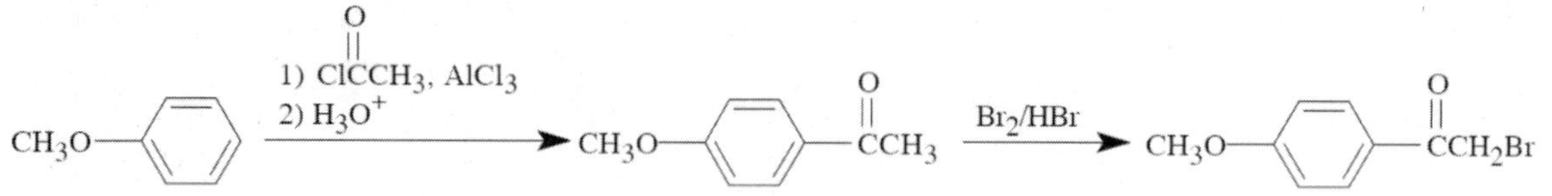

anisole

Alkylate the acetamidomalonate ion with this halide. Then, in the final hydrolysis step, use concentrated HBr, which will also cleave the methyl ether. Recall that α-bromo carbonyl compounds are very reactive in S_N2 reactions (Sec. 22.3 of the text).

27.73   In the first step, the amine adds to the Edman reagent; for the mechanism of this reaction, see the solution to Problem 27.29(a) on p. 989 of this chapter. The sulfur of the thiourea then serves as a nucleophile to close a six-membered ring and cleave the peptide. The mechanism below begins with the product of the reaction between the Edman reagent and the peptide, which is protonated on the carbonyl oxygen by the catalyzing acid.

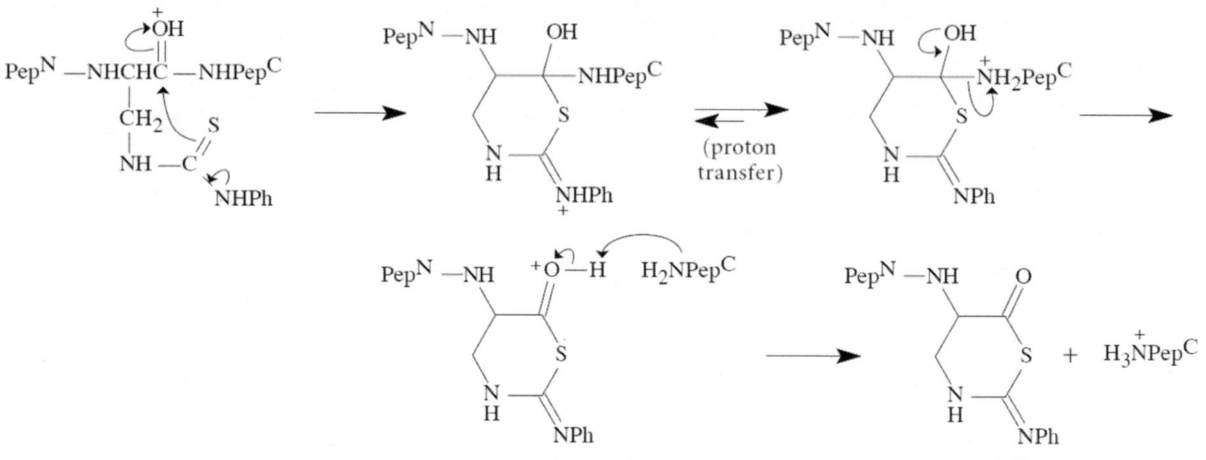

27.74   This reaction is an intramolecular ester aminolysis. The mechanism below begins with the conjugate base of the peptide formed by loss of a proton from the terminal amino group.

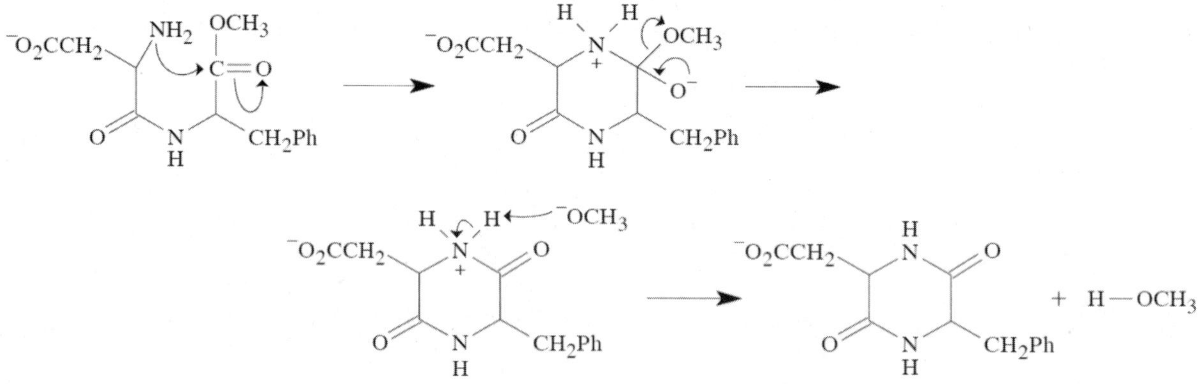

You may have noticed that diet soft drinks that go on sale in the supermarket are not quite as sweet as really fresh ones. This is because some of the sweetener has degraded by the mechanism just depicted.

27.75 (a) This is like the reaction of an amine with an isocyanate; see Eq. 23.73 on text p. 1218.

$$\text{EtNH} - \overset{\overset{\text{S}}{\|}}{\text{C}} - \text{NHPh}$$

(b) This is a modified Strecker synthesis in which methylamine is used in place of ammonia.

$$\text{Ph} - \underset{\underset{+}{\overset{|}{\text{H}_2\text{NCH}_3}}}{\overset{|}{\text{CH}}} - \text{CO}_2^-$$

(c) This is a reductive amination in which the amine is the α-amino group of the amino acid.

$$(\text{CH}_3)_2\text{CHCH}_2\text{NHCH} - \overset{}{\underset{\overset{|}{\text{CH}_3}}{}}\text{CO}_2^-$$

(d) The methyl ester is saponified much more rapidly that the *tert*-butyl ester for two reasons. First, the methyl branches of the *tert*-butyl group impede the approach of hydroxide to the carbonyl group. Second, the *tert*-butyl "ester" is also an amide, and amides hydrolyze much more slowly than esters.

$$(\text{CH}_3)_3\text{CO}\overset{\overset{\text{O}}{\|}}{\text{C}}\text{NHCH}\overset{\overset{\text{O}}{\|}}{\underset{\underset{\text{CH}_3}{|}}{\text{C}}}\text{CO}^- \text{ Na}^+ \;+\; \text{CH}_3\text{OH}$$

(e) This is a diazotization to form the *N*-diazo compound, that is, the acyl azide. (See Eq. 23.75 on text p. 1218.) [*Note:* The products of parts (e), (f), and (g) are summarized in the solution to part (g).]

(f) Heating the acyl azide gives a Curtius rearrangement to the isocyanate.

(g) The amino group of the amino acid ester adds to the isocyanate to give a urea. To summarize parts (e)–(g):

$$\underset{\text{benzoyl azide}\atop\text{product of part (e)}}{\text{Ph}\overset{\overset{\text{O}}{\|}}{\text{C}} - \overset{-}{\text{N}} - \overset{+}{\text{N}} \equiv \text{N}} \xrightarrow[-\text{N}_2]{\text{heat}} \underset{\text{phenyl isocyanate}\atop\text{product of part (f)}}{\text{PhN} = \text{C} = \text{O}} \xrightarrow{\underset{\text{valine methyl ester}}{\overset{\text{H}_2\text{NCHCO}_2\text{CH}_3}{\underset{|}{\text{CH(CH}_3)_2}}}} \underset{\substack{\text{N-phenylaminocarbonylvaline}\\\text{methyl ester}\\\text{product of part (g)}}}{\text{PhNH}\overset{\overset{\text{O}}{\|}}{\text{C}}\text{NHCHCO}_2\text{CH}_3\atop\underset{|}{\text{CH(CH}_3)_2}}$$

(h) This is a formylamidomalonate reaction, a variation of the acetamidomalonate synthesis.

$$\overset{+}{\text{H}_3\text{NCHCO}_2\text{H}} \;+\; \text{HCO}_2\text{H} \;+\; \text{EtOH}$$

(i) Hydrazine ($\text{H}_2\text{N}-\text{NH}_2$) displaces the ethoxy group of the ester to form a hydrazide *A*; diazotization forms the acyl azide *B*; heating this in ethanol gives a Curtius rearrangement in which the isocyanate is trapped by reaction with the solvent to give the ethyl carbamate *C*; and acidic hydrolysis generates the amino-protonated amino acid *D*.

$$\underset{A}{\text{N} \equiv \text{CCHCNHNH}_2 \atop \underset{|}{\text{CH(CH}_3)_2}} \xrightarrow{\text{NaNO}_2/\text{HCl}} \underset{B}{\text{N} \equiv \text{CCHCN}_3 \atop \underset{|}{\text{CH(CH}_3)_2}} \xrightarrow{\text{EtOH, heat}} \underset{C}{\text{N} \equiv \text{CCHNHCOEt} \atop \underset{|}{\text{CH(CH}_3)_2}} \xrightarrow[\text{heat}]{\text{HCl/H}_2\text{O}} \underset{D}{\text{HO}_2\text{CCHNH}_3 \text{ Cl}^- \atop \underset{|}{\text{CH(CH}_3)_2}}$$

27.76 The chemistry is very similar to that of Problem 27.75(i), text p. 1446. Compound *A* is the hydrazide; compound *B* is the acyl azide; compound *C* is the conjugate acid of an *N*-(1-aminoalkyl)amide, which is a nitrogen analog of a hemiacetal. This hydrolyzes in base to the aldehyde shown, ammonia, and acetamide *D*. (If the conditions were harsh enough, acetamide *D* would be converted into acetate ion and ammonia. If you postulated that compound *D* is acetate ion, your answer is equally satisfactory.)

27.77 **(a)** The first step of the mechanism is formation of an imine. (This is discussed in Sec. 19.11A of the text.) Reaction of this imine with the conjugate base of the thiol gives an addition product which is eventually transformed into the product. The following mechanism begins with the imine.

(b) In the first part of the mechanism, ⁻SH displaces chloride in an S_N2 reaction. Recall that α-halo carbonyl compounds are particularly reactive in S_N2 reactions (Sec. 22.3D of the text). Ammonia then reacts to forms an enamine. (Enamine formation is discussed in Sec. 19.11B of the text.) Although imine formation is favored, imines and enamines are in equilibrium just as aldehydes and enols are in equilibrium. Addition of both the —SH group and the —NH_2 group of this enamine to acetone (in a reaction much like acetal formation) and proton transfers give the first product *A*.

The isomerization leading to compound *A* is analogous to the conversion of an enol into a carbonyl compound. Compound *A* then reacts with cyanide ion in a variation of the Strecker synthesis. Hydrolysis in acid liberates acetone and the conjugate acid *B* of the α-cyano amine. The cyano group hydrolyzes in acid to give the conjugate acid of cysteine. (The mechanism of nitrile hydrolysis is given in Eqs. 21.20a–c, text p. 1066.)

(c)　Formaldehyde and the amino group of tryptophan react to give the conjugate acid of an imine, which acts as an electrophile in an intramolecular electrophilic aromatic substitution reaction to give the product. The mechanism below begins with the conjugate acid of the imine.

(d)   This reaction is a Curtius rearrangement (with concomitant loss of N_2) to give an isocyanate, which reacts intramolecularly with the side-chain hydroxy group. The mechanism of the Curtius rearrangement is given in Eq. 23.69 on text p. 1217. The mechanism below begins with the isocyanate.

27.78   Peptide *I* results from intramolecular nucleophilic reaction with the conjugate-base anion of the neighboring amide bond. This occurs mostly at Asn-Gly because glycine has no carbon side chain; in residues other than glycine, the carbon side chains can cause rate-retarding van der Waals repulsions in the ring-closure step.

Derivative *I* is an imide, which is the nitrogen analog of an anhydride. Hydroxide can react with either carbonyl carbon of the imide ring to open the ring, thus generating either peptide *J* or peptide *K*. The mechanism for the formation of peptide *K* is as follows:

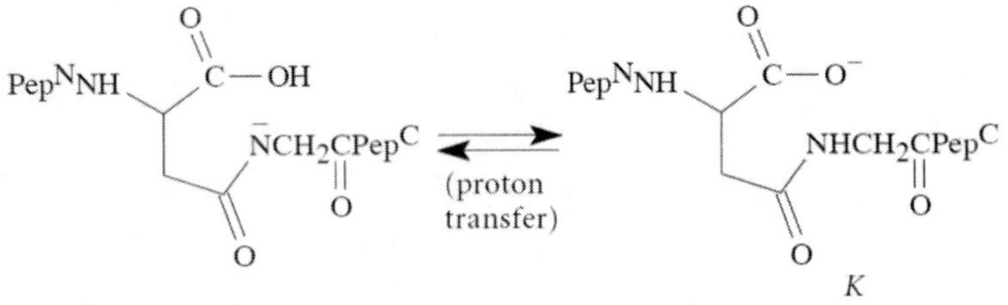

The structure of peptide J is as follows; you should show the mechanism for its formation.

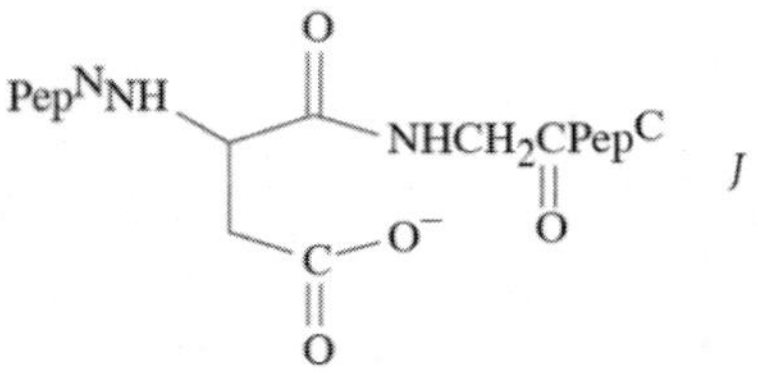

27.79   (a)   With the mass M of peptide P equaling the mass of $(A + B) - 17$ mass units, the mass of C equaling the mass of $(T\text{-}K\text{-}N\text{-}L + N\text{-}S\text{-}L) - 17$ mass units, and the mass of D equaling the mass of $(V\text{-}T\text{-}K\text{-}N\text{-}L + N\text{-}S\text{-}L) - 17$ mass units, the evidence suggests that the N-S-L fragment is somehow connected in each of the fragments but minus 17 mass units. The residues common to A, C, and D are K, N, and L (lysine, asparagine, and leucine, respectively) and the 17 mass units could correspond to a loss of NH_3, indicating that perhaps lysine and/or asparagine may be the common factor. If asparagine undergoes a transamination with lysine, the resulting product would link the K-N with the loss of NH_3. Therefore, the structure of peptide P is:

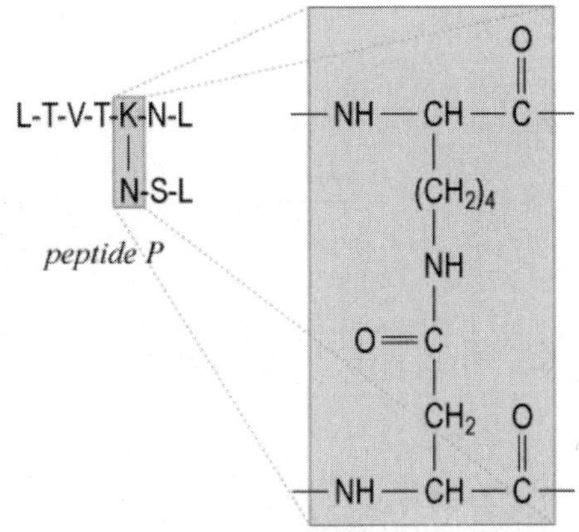

(b)   The asparagine residue is protonated by a neighboring acidic residue (aspartic acid or glutamic acid, RCO_2H) and a transamination ensues between the asparagine and lysine residues to produce the new amide linking the two peptides.

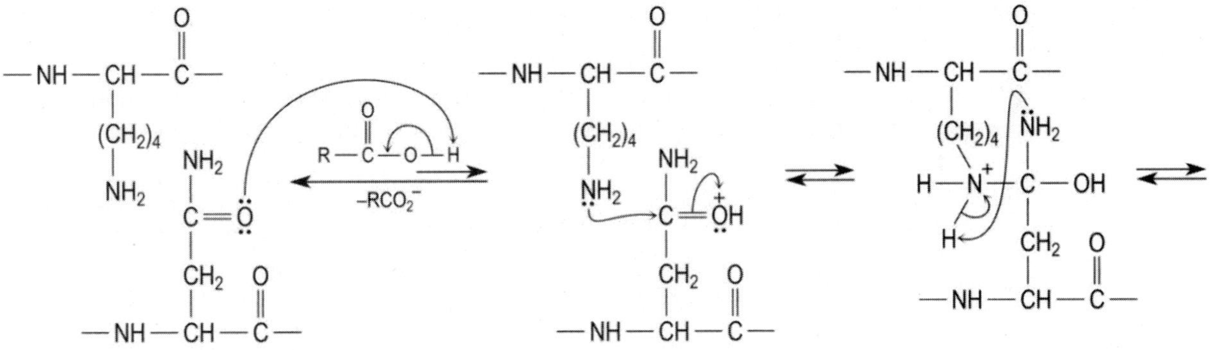

(The reaction scheme showing the lysine-asparagine substitution mechanism appears at the top of the page.)

(c)   The structure of the pilus evidently brings the nucleophilic amino group of the lysine and the carbonyl carbon of the asparagine into proximity. As discussed in Sec. 11.8, proximity can accelerate a substitution reaction by many orders of magnitude. This acceleration would turn an ordinarily sluggish reaction into a very rapid one.

27.80   (a)   In the Z conformation, the two large groups Pep^N and the N-alkyl group Pep^C are trans, and thus van der Waals repulsions between these groups cannot occur. In the E conformation, these two groups are close enough that van der Waals repulsions can result. These raise the energy of the peptide; hence, the Z conformation is energetically preferred.

(b)   The proline nitrogen bears two N-alkyl groups, whereas the peptide-bond nitrogens of other residues have only one. Thus, in either the E or the Z conformation, an alkyl group is cis to the Pep^N group. Although the group that is cis to Pep^N in the E configuration is larger, there is a much smaller difference in energy between E and Z conformations for proline residues than there is for other residues. Hence, E conformations in peptide bonds are sometimes found at proline residues in proteins and peptides.

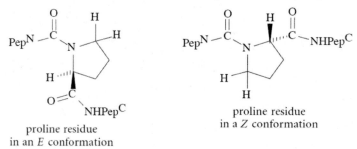

Notice that the E conformation of proline in a peptide forms a turn or bend in the peptide chain. Proline residues with E conformations are often found within turns in the three-dimensional structures of proteins.

27.81   (a)   The two monomethyl esters result from ester formation at each of the two carboxy groups.

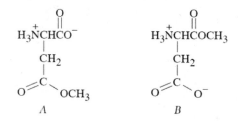

(b)   According to Table 27.1, text p. 1376, the pK_a values of the two carboxy groups in aspartic acid are about 1.9 (for the α-carboxy group) and 3.6 (for the β-carboxy group). (How do we know which is which?) Assume that the pK_a values for the respective esters are similar. At pH 3.0, the carboxy group of compound A is largely ionized; at the same pH, the carboxy group of B is mostly un-ionized. Hence, at pH 3.0, compound B carries a positive charge, whereas compound A is neutral. Therefore, compound B will be retained by a cation-exchange column (which contains negatively charged groups) and will be eluted more slowly than compound A. At pH 7, however, both compounds are negatively charged. Since they have the same charge, neither is retained by a cation-exchange column; consequently, these compounds are therefore not separated at pH 7.

27.82   The formula of compound A is the same as that of N-acetylaspartic acid minus the elements of water. Since treatment of a dicarboxylic acid with acetic anhydride is a good way to prepare a cyclic anhydride (see Eq. 20.31, text p. 1027), it is reasonable to suppose that compound A is the cyclic anhydride of aspartic acid. When this anhydride is treated with L-alanine, the amino group of alanine can react with either of the two carbonyl groups to give either of the two isomeric peptides B and C. Both would hydrolyze to give alanine and aspartic acid (as well as acetic acid).

27.83   (a)   The carboxylic acid groups of the Glu$_{35}$ and Asp$_{52}$ are positioned in the active site of lysozyme near the reducing end of the hexasaccharide oligomer. (In the structures that follow, R^2 represents two N-acetylglucosamine units and R^3 represents three N-acetylglucosamine units.) The side-chain carboxylic acid group of Glu$_{35}$, which is un-ionized at the pH optimum of the enzyme, acts as an acid catalyst and protonates the saccharide linkage. The side-chain carboxylate group of the neighboring Asp$_{52}$, which is ionized at the pH optimum of the enzyme, provides solvation for the carbocation intermediate at the face opposite to that of the leaving group. (Or, perhaps a weak covalent bond is formed.) Because the back side of the carbocation is obstructed, water approaches the carbocation at the same side from which the leaving saccharide departs. This mechanism accounts for the retention of stereochemistry that is observed.

(b)　The lysozyme reaction with triethyloxonium fluoroborate, Et$_3$O$^+$ $^-$BF$_4$, results in the formation of the ethyl ester of the carboxylate of Asp$_{52}$. As shown in Sec. 11.7A of the text, oxonium salts are powerful alkylating agents.

**triethyloxonium
fluoroborate**

The obliteration of enzyme activity could be due to either one or both of the following reasons. First, the ethyl group of the ethyl ester may block the active site between the Asp$_{52}$ and Glu$_{35}$ residues and thus prevent the substrate from binding. Second, the mechanism shows that the negatively charged Asp$_{52}$ residue plays a key role in stabilizing the cationic intermediates, possibly by nucleophilic involvement. Formation of the ethyl ester eliminates the nucleophilic capability of the carboxylate group.

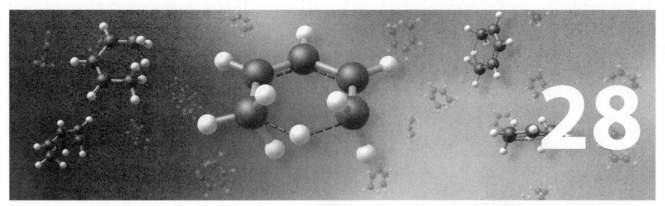

Pericyclic Reactions

STUDY GUIDE LINKS

28.1 Frontier Orbitals

You've seen that the analysis of pericyclic reactions in the text has focused on frontier orbitals. The discussion of electrocyclic reactions focused on one of the frontier orbitals, the HOMO, and the analysis of cycloadditions focused on the HOMO of one component and the LUMO of the other. Yet each of the frontier orbitals houses only two of the many reacting π electrons. Because *all* π electrons in the reacting molecules are involved in pericyclic reactions, it can perhaps be appreciated that all of the π molecular orbitals are involved as well. Why focus, then, on frontier orbitals? This focus is justified because it has been shown that frontier-orbital interactions play the *major* role in determining the energies of transition states for concerted pericyclic reactions. However, these are not the only orbital interactions that occur; they are simply the *most important* interactions.

Some students try to associate frontier molecular orbital interactions with formation of *a particular bond* in the reactant or product. This temptation is understandable because of the association of a π molecular orbital with the double bond in a simple alkene, or the association of a σ molecular orbital with a particular single bond in saturated molecules. These early encounters with molecular orbitals fail to take account of the fact that all molecular orbitals within a molecule must have certain symmetry properties, namely, they must be symmetric or antisymmetric with respect to the symmetry elements of the molecule itself. (For example, if a molecule contains an internal mirror plane, molecular orbitals of the molecule must be either antisymmetric or symmetric with respect to that plane.) For this reason, a one-to-one correspondence in general does not exist between molecular orbitals and the localized bonds used to write Lewis structures. Molecular orbitals are orbitals of *molecules,* and changes in a particular molecular orbital may involve more than one bond, and vice versa.

28.2 Orbital Analysis of Sigmatropic Reactions

Sigmatropic reactions are concerted and do not involve free ions, but the orbital relationships are nevertheless understood by thinking of the two components of the transition state as ions of opposite charge. However, for the reaction on text p. 1452 we could just as well think of the reaction as the migration of a hydrogen atom across a 2,4-pentadien-1-yl radical, or as the migration of a hydrogen anion (hydride ion) across a 2,4-pentadien-1-yl cation; these different approaches give the same predictions. Let's verify this point by viewing the transition state of the [1,5] hydrogen migration as the migration of a hydrogen atom across a 2,4-pentadien-1-yl radical:

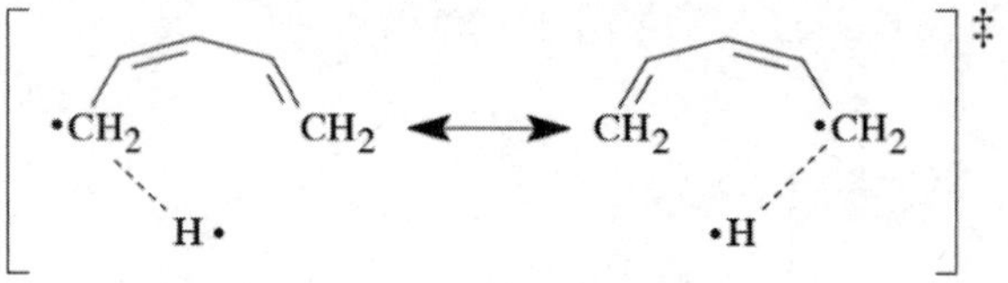

In this case, analyze the symmetries of the orbitals containing the unpaired electrons, that is, the SOMO of each radical. For a hydrogen atom, this orbital is just the $1s$ orbital. For the 2,4-pentadien-1-yl radical, this is π_3 (see Fig. 28.5 on text p. 1457). Because this molecular orbital is symmetric, the migration is predicted to occur suprafacially.

Now let's analyze the reaction as if it were the migration of a hydrogen anion (hydride ion, $H:^-$) across the termini of a 2,4-pentadien-1-yl cation and show that such an analysis makes the same prediction. In this case, the $1s$ orbital of the migrating hydride contains two electrons; thus, this orbital is classified as the HOMO of the hydride ion. It interacts with the LUMO of the 2,4-pentadien-1-yl cation, which, from Fig. 28.5 on text p. 1457, is π_3. Because this molecular orbital is symmetric, the migration hydride is predicted to occur suprafacially.

Notice that whether we deal with the migration of a proton across an anion, a radical across another radical, or an anion across a cation, the relevant orbitals of the species involved are the same: the $1s$ orbital of the hydrogen and π_3 of the pentadienyl system. Hence, the predictions of orbital symmetry must also be the same.

REACTION REVIEW

I. ELECTROCYCLIC REACTIONS

1. When an electrocyclic reaction takes place, the carbons at each end of the conjugated π system turn in a concerted fashion so that the $2p$ orbitals can overlap (and rehybridize) to form the σ bond that closes the ring.

2. This turning can occur in two stereochemically distinct ways.
    a. In a conrotatory closure the two carbon atoms turn in the same direction, either both clockwise or both counterclockwise.

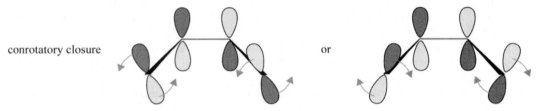

conrotatory closure   or

    b. In a disrotatory closure the two carbon atoms turn in opposite directions so that either the upper lobes of the $2p$ orbitals overlap or the lower lobes of the $2p$ orbitals overlap.

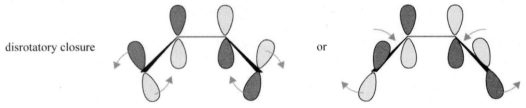

disrotatory closure   or

3. The HOMO of the conjugated alkene contains the π electrons of highest energy and governs the course of pericyclic reactions.
    a. When the ring closure takes place, the two $2p$ orbitals on the ends of the π system must overlap in phase.
    b. The wave peak on one carbon must overlap with the wave peak on the other, or a wave trough must overlap with a wave trough.

4. The relative orbital phase at the terminal carbon atoms of the HOMO (the orbital symmetry) determines whether the reaction is conrotatory or disrotatory.
    a. Conjugated alkenes with $4n$ π electrons (n = any integer) have antisymmetric HOMOs and undergo conrotatory ring closure. Conrotatory ring closure is allowed for systems with $4n$ π electrons; it is forbidden for systems with $4n + 2$ π electrons.
    b. Conjugated alkenes with $4n + 2$ π electrons have symmetric HOMOs and undergo disrotatory ring closure. Disrotatory ring closure is allowed for systems with $4n + 2$ π electrons; it is forbidden for systems with $4n$ π electrons.

5. For photochemical (excited-state) electrocyclic reactions, the selection rules are reversed; that is, conrotatory closure is allowed for a $4n + 2$ π-electron reaction and disrotatory closure is allowed for a $4n$ π-electron reaction.

II. CYCLOADDITION REACTIONS

1. Cycloadditions are classified, first, by the number of electrons involved in the reaction with respect to each component.
    a. The number of electrons involved is determined by writing the reaction mechanism in the curved-arrow notation.
    b. The number of electrons contributed by each given reactant is equal to twice the number of curved arrows originating from that component (two electrons per arrow).

2. Cycloaddition reactions are also classified by their stereochemistry with respect to the plane of each reacting molecule and may in principle occur either across the same face, or across opposite faces, of the planes in each reacting component.
   a. If the reaction occurs across the same face of a π system, the reaction is said to be suprafacial with respect to that π system; that is, a *syn*-addition that occurs in a single mechanistic step.
   b. If the reaction bridges opposite faces of a π system, it is said to be antarafacial; that is, it is an *anti*-addition that occurs in one mechanistic step.
3. The reaction classification is written as a combination of numbers and letters. Thus, a [4s + 4a] cycloaddition involves 4 π electrons on each component and is suprafacial (*s*) on one component and antarafacial (*a*) on the other.
4. In order for a cycloaddition to occur, bonding overlap must take place between the 2p orbitals at the terminal carbons of each π electron system.
   a. The bonding overlap begins when the HOMO of one component interacts with the LUMO of the other.
   b. The LUMO of the other component is the empty orbital of lowest energy into which the electrons from the HOMO must flow.
   c. These two frontier MOs involved in the interaction must have matching phases if bonding overlap is to be achieved.
   d. It does not matter which component provides the HOMO and which provides the LUMO.
5. All-suprafacial cycloadditions are allowed thermally for systems in which the total number of reacting electrons is $4n + 2$, and they are allowed photochemically for systems in which the number is $4n$.

$$\text{(H, CO_2CH_3, CH_3O_2C, H)} + \text{(O furan)} \xrightarrow[\text{(a Diels-Alder reaction)}]{[2s + 4s]} \text{product}$$

III. THERMAL SIGMATROPIC REACTIONS

A. STEREOCHEMISTRY OF SIGMATROPIC REACTIONS

1. Sigmatropic reactions can be classified by their stereochemistry according to whether the migrating bond moves over the same face, or between opposite faces, of the π-electron system.
   a. If the migrating bond moves across one face of the π system the reaction is said to be suprafacial.
   b. If the migrating bond moves from one face of the π system to the other the reaction is said to be antarafacial.
   c. When both ends of a σ bond migrate, the reaction can be suprafacial or antarafacial with respect to each π system.
2. The stereochemistry of a sigmatropic reaction is revealed experimentally only if the molecules involved have stereocenters at the appropriate carbons.
3. Molecular orbital theory provides the connection between the type of sigmatropic reaction and its stereochemistry.

B. [1,3] AND [1,5] SIGMATROPIC REARRANGEMENTS

1. The interaction of the LUMO of the migrating group with the HOMO of the π system, or vice-versa, controls the stereochemistry of the reaction.
   a. In the migration of a hydrogen, the orbital involved is a 1s orbital, which has no nodes.
   b. In the migration of a carbon, the orbital involved is a 2p orbital, which has one node.
   c. The shift of a carbon can occur in two stereochemically distinct ways:
      i. Migration with retention of configuration.
      ii. Migration with inversion of configuration.
   d. In the allyl anion, the HOMO is antisymmetric:

allyl anion

      i. The suprafacial [1,3] shift of a hydrogen is forbidden by orbital symmetry.

      ii. The antarafacial [1,3] migration of hydrogen is allowed by orbital symmetry but is virtually nonexistent in organic chemistry because unreasonably long bonds would be required in the transition state.

      iii. The suprafacial [1,3] shift of a carbon with inversion of configuration is allowed by orbital symmetry.

      iv. The antarafacial [1,3] shift of a carbon with retention of configuration is allowed by orbital symmetry but is rarely observed.

   e. In the 2,4-pentadienyl anion, the HOMO is symmetric:

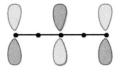

2,4-pentadienyl anion

      i. The suprafacial [1,5] migration of a hydrogen is allowed by orbital symmetry.

      ii. The antarafacial [1,5] migration of a hydrogen is forbidden by orbital symmetry.

      iii. The suprafacial [1,5] migration of a carbon with retention of configuration is allowed by orbital symmetry.

      iv. The antarafacial [1,5] migration of a carbon with inversion of configuration is allowed by orbital symmetry.

C. SIGMATROPIC REARRANGEMENTS; COPE AND CLAISEN REARRANGEMENTS

   1. A [3,3] sigmatropic rearrangement is a reaction in which both ends of a σ bond change positions (migrate).

     a. The transition state of a [3,3] sigmatropic rearrangement can be visualized as the interaction of two allylic systems, one a cation and one an anion.

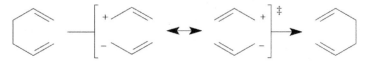

     b. The two MOs involved achieve bonding overlap when the [3,3] sigmatropic rearrangement occurs suprafacially on both components.

   2. The Cope rearrangement is a [3,3] sigmatropic rearrangement in which a 1,5-diene isomerizes.

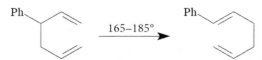

   3. The oxyCope reaction is a variation of the Cope rearrangement that involves the initial formation of an enol.

     a. Isomerization of the enol into the corresponding carbonyl compound is a very favorable equilibrium that drives the reaction to completion.

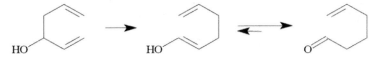

     b. The oxyCope reaction is strongly accelerated by conversion of the —OH group into its conjugate-base alkoxide (anionic oxyCope).

4. In the Claisen rearrangement, an ether that is both allylic and vinylic (or an allylic aryl ether) undergoes a [3,3] sigmatropic rearrangement. If both *ortho* positions of an aryl allylic ether are blocked by substituent groups, the *para*-substituted derivative is obtained by a sequence of two Claisen rearrangements followed isomerization of the product to the phenol.

SOLUTIONS TO PROBLEMS

Solutions to In-Text Problems

28.1   (a)   This is an electrocyclic reaction; six electrons are involved.

(b)   This is a sigmatropic reaction; two electrons are involved.

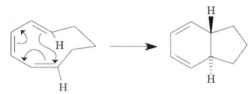

(c)   This is an intramolecular cycloaddition reaction; four electrons are involved.

(d)   This is a sigmatropic reaction; six electrons are involved.

(e)   This is an electrocyclic reaction; two electrons are involved.

28.2   (a)   1,3,5-Hexatriene has six π molecular orbitals.
        (b)   The symmetry of the 1,3,5-hexatriene MOs alternates; that is, π_1, π_3, and π_5^* are symmetric, and π_2, π_4^*, and π_6^* are antisymmetric.
        (c) π_1, π_2, and π_3 are bonding and π_4^*, π_5^*, and π_6^* are antibonding.
        (d)   Because 1,3,5-hexatriene has six π electrons, the highest occupied molecular orbital (HOMO) is π_3; the lowest unoccupied molecular orbital (LUMO) is π_4^*.
        (e)   As indicated in the solution to part (b), the HOMO (π_3) is symmetric; therefore, the phase of the HOMO at the terminal carbons is the same.
        (f)   The phase of the LUMO (π_4^*) at the terminal carbons is different.

28.3   Because there are ten π electrons, there are five occupied MOs in this alkene, each containing two electrons. The MOs alternate in symmetry; the odd-numbered ones are symmetric, and the even-numbered ones are antisymmetric. Hence, π_6^* is antisymmetric (A); it is antibonding; and it is the LUMO and is thus one of the frontier orbitals.

28.4 (a) The MO π_4 is nonbonding. (If there is an odd number n of molecular orbitals in an unbranched acyclic system, the nonbonding MO is always the one with the number $(n + 1)/2$.)

(b) The symmetry of the MOs alternates; that is, π_1, π_3, π_5^*, and π_7^* are symmetric, and π_2, π_4, and π_6^* are antisymmetric.

(c) According to its resonance structures, this cation has sites of positive charge at alternating carbons:

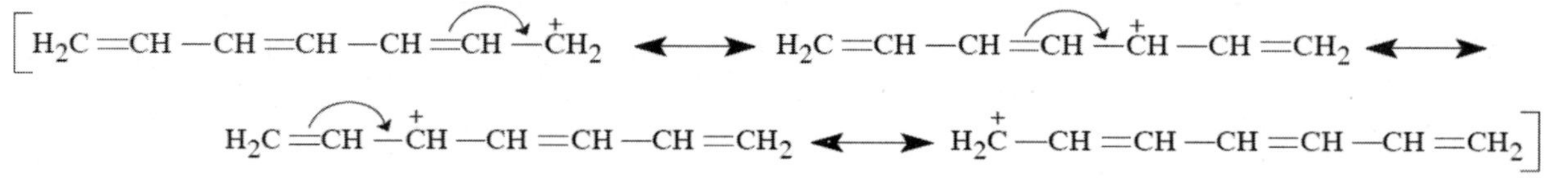

In molecular orbital terms, the positive charge in the cation results from the absence of an electron in the LUMO; the LUMO *is* the nonbonding MO. Positive charge is shared at alternating carbons because the LUMO has nodes at the other carbons. That is, positive charge can only exist on carbons at which there is no node in the LUMO. If this is so, the LUMO must have three nodes: one at each carbon that does *not* share the positive charge. As shown in the solution to part (a), π_4 is the nonbonding MO and is therefore the LUMO, and it has three nodes, one at each carbon that does not share positive charge in the above structures.

28.5 (a) The resonance structures of the allyl radical show that the unpaired electron is shared between the terminal carbons:

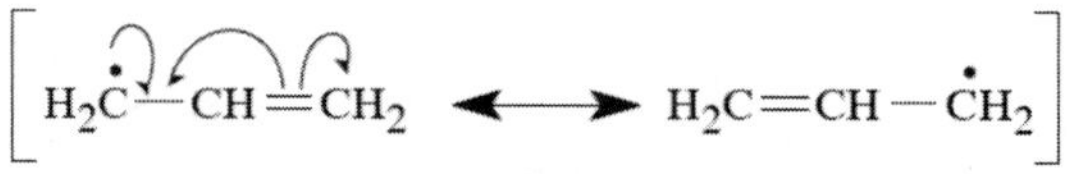

(b) Figure 28.4, text p. 1456, shows that the unpaired electron of the allyl radical resides in the nonbonding MO, which has a node at the central carbon. The unpaired electron density therefore exists only at the carbons on either side of a node, that is, at the terminal carbons, as the resonance structures show.

28.6 *Reaction 1* is a disrotatory reaction involving $4n + 2$ electrons, and is allowed. Therefore it can occur readily by a concerted mechanism. The reaction is disrotatory because the hydrogens shown in the product are both on the outside of the starting material. (Draw these hydrogens in the starting material, if necessary.) In order for these hydrogens to end up on the same face of the product, the ends of the π-electron system must turn in opposite directions.

Reaction 2 is a conrotatory reaction involving $4n + 2$ electrons, and is not allowed; it does not occur readily.

28.7 The two different conrotatory processes are as follows: (The curved arrows indicate atomic motion, not electron flow.)

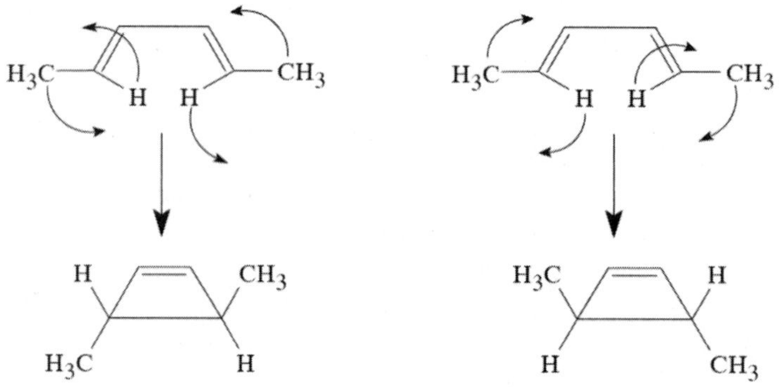

They are equally likely because they are enantiomeric; enantiomers have equal energies, and enantiomeric pathways have identical energies at all points and therefore identical rates (Sec. 7.7A).

28.8 The thermal ring opening of *trans*-3,4-dimethylcyclobutene must be a conrotatory process. The two possible conrotatory processes are as follows:

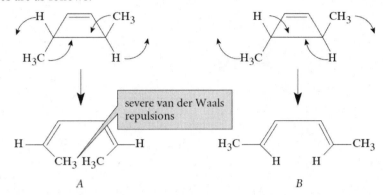

Product *A* has severe van der Waals repulsions between the methyl groups; product *B* does not. The repulsions raise the energy of product *A* and the transition state for its formation; consequently, product *B* is observed.

28.9 The opening of the cyclobutene is conrotatory and therefore gives (1*E*,3*Z*)-cyclodecadiene. The trans double bond does not introduce as much strain into the large ring as was present in the starting cyclobutene.

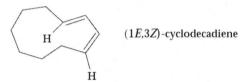

 Here is one way to use models to visualize this transformation. To avoid having to build two models, construct a cyclodecane (the cyclic *alkane* with a ten-membered ring); don't worry about the missing double bond. Now connect two carbons to make a cis-fused four-membered ring. Identify the "up" hydrogens on these two carbons. Using a pencil, draw a single line across the bond that is a double bond in the starting material (even though it is a single bond in your model). Now draw two lines across each of the bonds that will become double bonds as a result of the reaction. Now break the cyclobutane ring and turn the carbons bearing the "up" hydrogens in a conrotatory manner. Examine the stereochemistry at the bonds marked with two lines; one should have an *E* configuration and the other a *Z* configuration. Voila!

28.10 The HOMO of the 2π-electron component is symmetric, and the LUMO of the 4π-electron component is symmetric. As the diagram below shows, the ends of these two MOs have matching phases in a [4*s* + 2*s*] cycloaddition.

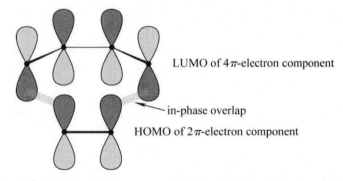

Interaction of the LUMO of the 4π-electron component and the HOMO of the 2π-electron component in a [4*s* + 2*s*] cycloaddition gives in-phase overlap at both ends of the two π-electron systems.

28.11 Use the HOMO of the 4π-electron component and the LUMO of the 2π-electron component. The [4*a* + 2*s*] process would involve the overlap shown in (a) below. The [4*s* + 2*a*] process would involve the overlap shown in (b). In both cases an out-of-phase overlap between the π-electron systems is required at one end. Consequently, these processes are not allowed.

HOMO of 4π-electron component

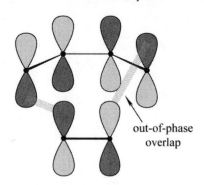

out-of-phase overlap

LUMO of 2π-electron component

(a)
[4a + 2s]

HOMO of 4π-electron component

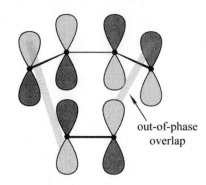

out-of-phase overlap

LUMO of 2π-electron component

(b)
[4s + 2a]

28.12 The [8s + 2s] cycloaddition and its product:

28.13 The four products are the diastereomeric 1,2,3,4-tetramethylcyclobutanes. Each of the pure alkene stereoisomers can undergo an allowed photochemical [2s + 2s] cycloaddition in two distinguishable ways. *Cis*-2-butene reacts to give compounds *A* and B; *trans*-2-butene reacts to give compounds *B* and *C*; and the mixture of *cis*- and *trans*-2-butene can give these three compounds plus a fourth, compound D, that results from the [2s + 2s] cycloaddition of *cis*-2-butene to *trans*-2-butene.

cis + cis → A

cis + cis → B

trans + trans → B

trans + trans → C

trans + cis → D

28.14   (a)   Since exchanging the positions of two groups at a stereocenter changes the configuration of the stereo-center, exchange the positions of the T and the D in the starting material; this will lead to product with the opposite, that is, the *R*, configuration at the indicated carbon.

(b)   Two other starting materials that would give the same product as in part (a):

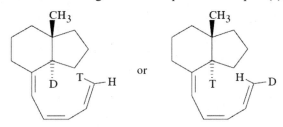

28.15   (a)   This is a [1,4] sigmatropic rearrangement.

migration of this end of the bond does not occur, that is, it "migrates" from carbon-1* to carbon-1*.

migration of this end of the bond occurs from carbon-1 to carbon-4.

(b)   This is a [2,3] sigmatropic rearrangement.

One end of this bond moves from carbon-1 to carbon-3; the other end moves from the sulfur (1*) to the oxygen (2*).

(c)   This is a [5,5] sigmatropic rearrangement.

This bond migrates from carbons 1 and 1* to carbons 5 and 5*, respectively.

28.16   Reaction (1) occurs readily because it is a [1,5] sigmatropic rearrangement, which is allowed when the hydrogen migrates suprafacially over the π-electron system. This reaction pathway does not introduce significant strain or twist into the π-electron system. Reaction (2) does not occur because it is a [1,3] sigmatropic rearrangement, which is allowed only if the hydrogen migrates antarafacially. Such an antarafacial migration requires that the hydrogen simultaneously bridge the upper and lower faces of the π-electron system. This requirement cannot be met while at the same time maintaining normal bond lengths and angles.

28.17   If the migration were antarafacial, the opposite stereochemical result would have been observed; that is, the 3*E* stereoisomer of the product would have the *S* configuration at the asymmetric carbon stereocenter, and the 3*Z* stereoisomer of the product would have the *R* configuration at the asymmetric carbon stereocenter. This is the result that is *not* observed.

28.18   (a)   Think of the five-carbon π-electron system as a 2,4-pentadienyl cation, whose MOs are shown in Fig. 28.5, text p. 1342, and think of the migrating group as a carbon anion in which the unshared electron pair resides in a *2p* orbital. The orbitals involved in the rearrangement are the LUMO of the pentadienyl system and the HOMO of the carbon anion. The LUMO of the pentadienyl system, as shown by Fig. 28.5, is symmetric; that is, at each end of the system, the LUMO has the same phase on a given face. The HOMO of the carbon anion is simply the filled *2p* orbital. The migration is allowed only if the carbon anion migrates suprafacially such that each end of the pentadienyl LUMO interacts with the same lobe of the carbon anion *2p* orbital.

HOMO of migrating carbon anion

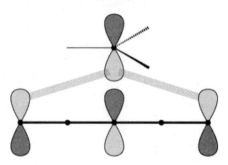

Migration occurs suprafacially on both systems (and thus with retention of configuration at the migrating carbon)

LUMO of 2,4-pentadienyl cation

 Alternatively, you can think of this problem in terms of the 2,4-pentadienyl *anion* and a methyl *cation*. The orbitals involved are now the HOMO of the anion and the LUMO of the cation. But these are the same orbitals used in the foregoing solution (only their occupancies are different); hence, the stereochemical result is the same.

(b)   The first step (conversion of the starting material into compound *A*) is a [1,5] sigmatropic rearrangement. Because the stereochemistry of the methyl group on the migrating carbon does not change, the rearrangement occurs with retention of configuration. This is the result predicted for an allowed suprafacial process. The remaining two steps are [1,5] sigmatropic hydrogen migrations, which presumably occur in a suprafacial manner, although the stereochemical course of the reaction cannot be determined from the structures of the products.

28.19 (a)   This transformation involves two successive [1,5] sigmatropic hydrogen migrations—sort of a "hydrogen walk," as shown by the following curved arrows. These are allowed suprafacial processes. (See the solution to Problem 28.18b, text p. 1473.)

Why does the equilibrium favor the two products over the starting material? See Sec. 4.5B, text p. 149.

(b)   The "hydrogen walk" on 2,3-dimethyl-1,3-cyclopentadiene should proceed as shown in the following equation. Although a hydrogen that starts out on carbon-1 could migrate to four other carbons, only two of the possible products are unique. It can be argued that compounds *A* and *B* should be the predominant alkenes at equilibrium, and that compound *C* should be present in least amount. (Why? See Sec. 4.5B, text p. 149.) Remember that the selection rules have nothing to say about which compound is favored at equilibrium, only whether the equilibrium can be attained at a reasonable rate.)

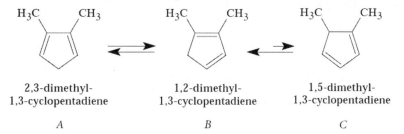

| 2,3-dimethyl-
1,3-cyclopentadiene | 1,2-dimethyl-
1,3-cyclopentadiene | 1,5-dimethyl-
1,3-cyclopentadiene |
| *A* | *B* | *C* |

28.20 (a)   Deduce the product by using the curved-arrow notation:

3,3-dimethyl-4-pentenal

(b)   Deduce the starting material by drawing the curved arrows for the reverse of a Claisen rearrangement:

28.22 (a)   Compound *A* undergoes the oxyCope reaction because the allylic group attached to the carbon of the bicycle[2,2,2]octene ring bearing the alcohol group has endo stereochemistry (see text p. 740) and positioned to interact with the allylic group in the bicycle[2,2,2]octene ring. The allylic group attached to the carbon bearing the alcohol group in compound *B*, on the other hand, has exo stereochemistry and cannot interact with the allylic group in the bicyclo[2,2,2]octene ring.

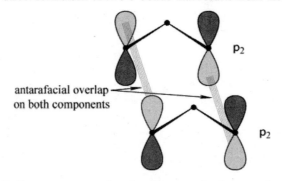

28.23 An orbital symmetry analysis involves the interaction of π_2 of one allylic system with π_2 of the other as shown below. Certainly the all-antarafacial rearrangement gives positive overlap at both ends of the allylic systems, but it requires either the initial formation of π bonds rather than σ bonds or introduction of significant strain in the two systems akin to having two trans double bonds in a six-membered ring. Neither alternative is as energetically favorable as the direct formation of two σ bonds that results from the all-suprafacial process.

antarafacial overlap on both components

p_2

p_2

Antarafacial overlap of both allylic components gives bonding overlap but requires side-to-side (π) rather than head-to-head (σ) overlap. The formation of σ bonds would require severe distortions of both systems.

28.24 **(b)** This is a carbocation rearrangement in which either methyl group migrates. (This is the "methyl walk" discussed in the solution to Problem 15.85(c) on p. 467 of this manual.)

This can also be thought of as an all-suprafacial [1,6] sigmatropic rearrangement, which is an allowed process.

28.25   As stated in the problem, previtamin D_2 is the same as previtamin D_3 except for the R-group. Comparing the structures of previtamin D_3 (or D_2) in Eq. 28.38 on text p. 1478, ergosterol at the top of text p. 1479, and lumisterol in the problem reveals that the latter two compounds result from conrotatory electrocyclic ring closures, ergosterol being formed by a clockwise conrotatory motion (as viewed edge-on by the "eye" below) and the lumisterol from a counterclockwise conrotatory motion. Photochemical conrotatory electrocyclic reactions involving $4n + 2$ electrons are allowed.

28.26   Heating should promote a disrotatory reaction; that is, the reaction should have a stereochemical course which is opposite to that of the photochemical reaction.

28.27   First determine why irradiation of *A* or *B* does not give back previtamin D_2. One reason is mechanistic. Since *A* and *B* originate from *heating* previtamin D_2, then previtamin D_2 can only result from the exact reverse of the same reaction, which must occur as a disrotatory process. However, ring opening by irradiation, if it were to occur, would have to take place with opposite stereochemistry, that is, as a conrotatory process. Either of the two possible conrotatory processes would result in a product containing a trans double bond within a six-membered ring. Even irradiation does not provide enough energy for this to occur.

the two possible products of conrotatory ring opening

The only alternative to this process is a four-electron disrotatory process that gives a cyclobutene:

A **light→** C

B **light→** D

28.28   A [1,7] thermal sigmatropic hydrogen shift should be antarafacial. That is, a hydrogen from the methyl group migrates from a conformation in which it is "up" to the lower face of the ring, or a hydrogen from the methyl group migrates from a conformation in which it is "down" to the upper face of the ring, or both.

 Problem 28.51 (text p. 1484) describes an effort to elucidate the stereochemistry of the previtamin D rearrangement.

Solutions to Additional Problems

28.29   (a)   Use the MOs of 1,3-butadiene as a "model" for the orbital phases. In all $4n$ π electrons systems the LUMO has opposite phases at each end of the π-electron system. For this reason, the allowed thermal electrocyclic reaction must be conrotatory.

(b)   Think of the Diels–Alder reaction as a prototype for an allowed thermal cycloaddition. If $4n + 2$ electrons are involved, a cycloaddition is allowed thermally. If $4n$ electrons are involved, then the thermal cycloaddition is forbidden and the photochemical cycloaddition is allowed. Therefore, a photochemical [8s + 4s] process (12 electrons) is allowed.

(c)   The highest occupied molecular orbital (HOMO) of a conjugated triene is π_3. Since orbitals of conjugated dienes alternate in symmetry, this MO is symmetric. (See the solution to Problem 28.2 on p. 1021XR of this manual.) The methyl groups, to a useful approximation, have no effect on the nodal properties of the molecular orbitals.

28.30   The pericyclic selection rules say *absolutely nothing* about the position of equilibrium in each case. These rules refer to *rates* of reactions, not to equilibrium constants. Other considerations must be used to decide on the position of equilibrium. The point of the problem is for you to use what you have learned about the relative energies of molecules; the molecule of lower energy is favored in each equilibrium.

(a)   The right side of the equation is favored at equilibrium because the double bonds are conjugated with the phenyl rings; conjugation is a stabilizing effect.

(b)   The left side of the equation is favored at equilibrium because the double bonds have more alkyl branches. (Alkyl substitution stabilizes a double bond.)

(c)   The right side of the equation is favored at equilibrium because a C=O double bond is formed at the expense of a C=C double bond. (C=O bonds are stronger than C=C bonds; see text page 1161.)

(d)   The right side of the equation is favored at equilibrium because one product is aromatic and therefore particularly stable, and the other product is volatile (ethylene is a gas).

(e) The left side of the equation is favored at equilibrium because the double bonds are conjugated not only with each other but also with the carbonyl group.

28.31 (a) Because it involves $4n$ π electrons and thermal conditions, the electrocyclic reaction that converts compound A into compound B is conrotatory. (A "thermal" process can occur at low temperature if the process is particularly rapid.) Hence, the methyl groups in compound B are trans. The thermal process that converts compound B into compound C involves six ($4n + 2$) electrons and is therefore disrotatory. (Both compounds B and C are, of course, formed as racemates.)

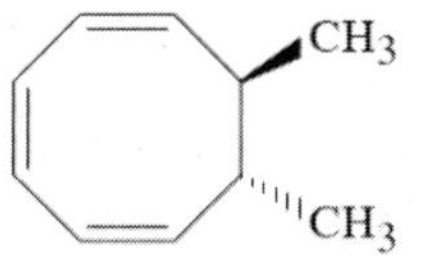

B 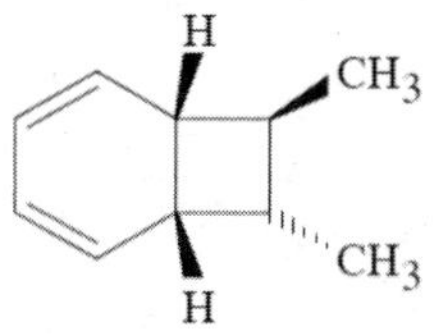 C

(b) The stereoisomer of compound A that also gives compound C on heating is the one in which both of the terminal double bonds have the Z configuration.

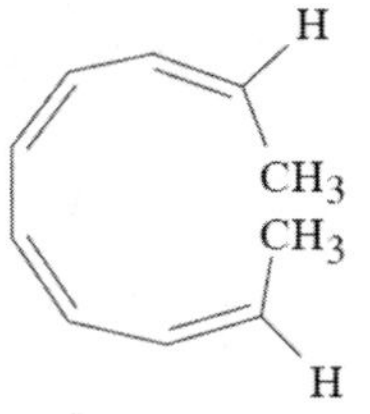

28.32 (a) This reaction can be viewed as either a [1,9] or a [1,13] sigmatropic rearrangement. In either of these classifications the reaction is an allowed process that is suprafacial on the π-electron system and occurs with retention at the migrating carbon. The [1,9] migration is as follows:

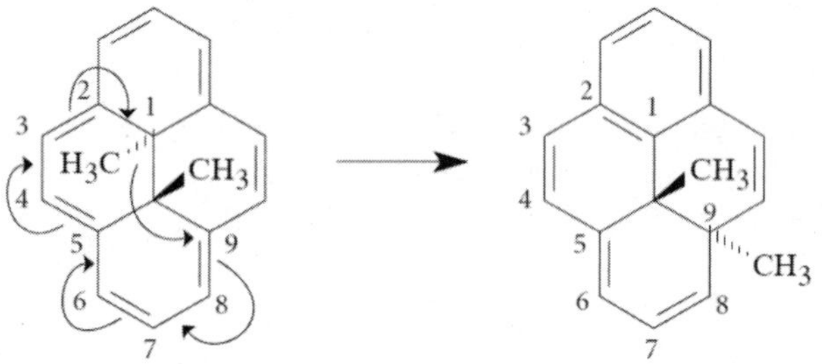

And the [1,13] migration is as follows:

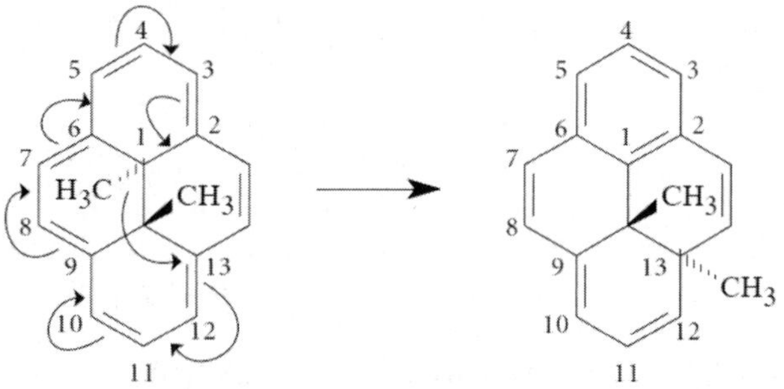

The two final structures shown above differ only in the positions of double bonds in the benzene ring; that is, they are resonance structures and hence are the same compound.

(b) Because a thermal suprafacial [1,9] or [1,13] methyl migration must occur with retention of configuration at the migrating carbon, the methyl group that migrates, when isotopically substituted, must have the S configuration in both starting material and product.

28.33 (a) The formula indicates that an addition has taken place. This is an allowed photochemical [2s + 2s] cycloaddition:

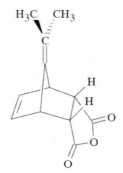

(b)   This allylic vinylic ether undergoes a [3,3] sigmatropic rearrangement (that is, a Claisen rearrangement) on heating to give ketone *A*, which isomerizes to its aromatic "enol" (that is, the phenol) form *B*.

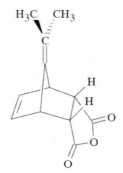

(c)   This is a [4s + 2s] cycloaddition, that is, a Diels–Alder reaction. The exocyclic double bond is not involved in the reaction, and *endo* stereochemistry is assumed:

 Two other reactions are allowed in principle that do involve the exocyclic double bond. One is a [2s + 6a] cycloaddition to give product *A*, and the other is a [2a + 6s] cycloaddition to give product *B*:

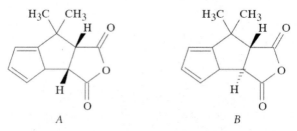

If you construct models of the starting materials, and if you bring them together so that the appropriate orbital interactions can occur, you will see that the transition states required to form these products (as well as the products themselves) are considerably strained. Hence, the reactions that form these products are much slower than the Diels–Alder process.

(d)   This reaction is very similar to the reaction in Eq. 28.33, text p. 1474, but with two allylic alcohol π systems instead of one. The anionic oxyCope reaction forms a product *A* that is both an enol and an enolate; on protonation, the diketone product *B* is formed.

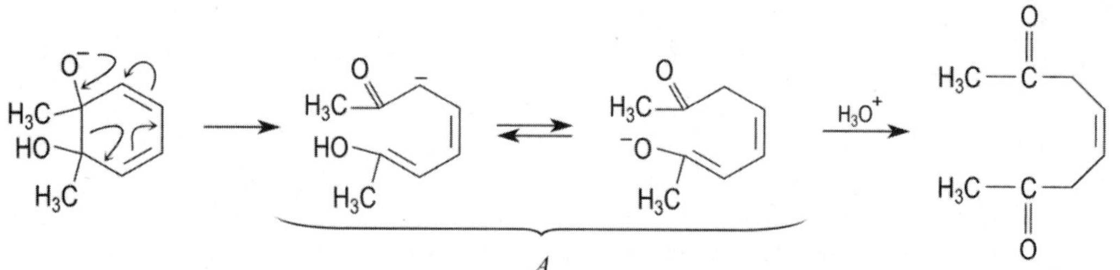

28.34 (a)   The structure of the ozonolysis product *C* shows that compound *B* is a cyclobutene, which must be formed in a disrotatory photochemical electrocyclic reaction. This defines the stereochemistry of *B*, which, in turn, defines the stereochemistry of *C*:

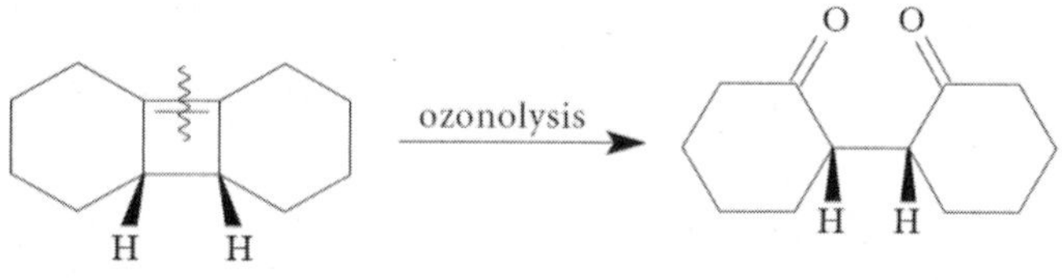

(b)   Compound *D* is the trans-fused stereoisomer of the cyclobutene derivative, which undergoes conrotatory opening upon heating to give compound *A*.

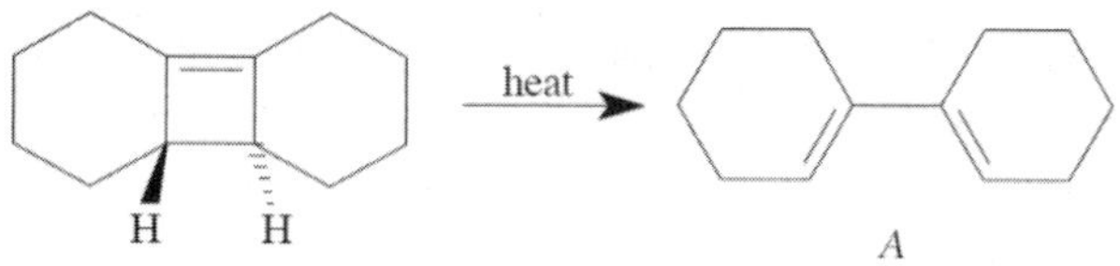

The reason that compound *B* is inert under the same conditions is that conrotatory ring opening of *B* would give a stereoisomer of *A* in which one of the cyclohexene rings would contain a trans double bond within a six-membered ring. This introduces so much strain that the reaction does not occur, even though it is allowed by the selection rules.

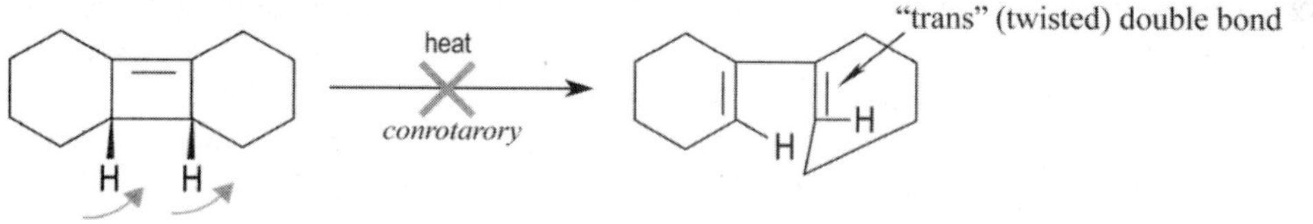

28.35   As usual, it helps to analyze first the origins of the carbons. The carbonyl groups are a good place to start, as they can have originated only from the quinone. The carbons marked with an asterisk are derived from the quinone. The remaining carbons, marked with a dagger (†) are derived from 1,3-cyclopentadiene. The new bonds are shown with heavy shading.

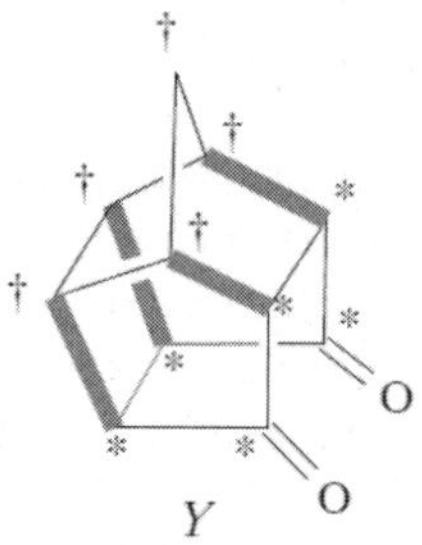

The first reaction is a Diels–Alder reaction because we have a very reactive Diels–Alder diene and dienophile:

1,3-cyclopentadiene   *p*-benzoquinone   X

Compound *X* is all set up for a [2*s* + 2*s*] allowed photochemical cycloaddition to complete the formation of compound *Y*:

light
[2*s* + 2*s*]

X   Y

28.36   This cycloaddition reaction involves sixteen electrons; hence, it must be suprafacial on one component and antarafacial on the other. Thus, it must be a [14*s* + 2*a*] or a [14*a* + 2*s*] cycloaddition. The heptafulvene molecule is large enough that its π-electron system can twist without introducing too much strain or without losing too much π-electron overlap; hence, the cycloaddition is a [14*a* + 2*s*] process. In either case, the product has the following stereochemistry:

28.37   (a)   Toluene has added stability associated with aromaticity; compound *A*, although conjugated, is not aromatic. Because any equilibrium favors the more stable compound, the added stability of toluene causes it to be strongly favored in the equilibrium.

(b)   If the reaction were concerted, it would have to be a [1,3] thermal sigmatropic rearrangement, which must occur in an antarafacial manner:

compound *A*

antarafacial
thermal [1,3]

In this diagram, the "up" hydrogen (asterisk) is shown migrating to the lower face of the $=CH_2$ group. (Migration of the "down" hydrogen to the upper face of the $=CH_2$ group is equally likely.) As discussed on text p. 1357, such an antarafacial migration requires that the migrating hydrogen bridge too great a distance for proper orbital overlap. Hence, this process is so slow that compound *A* has significant thermal stability.

(c)   We would expect the conversion of compound *C* to compound *D* under thermal conditions to be extremely rapid, because this is a [1,5] sigmatropic rearrangement that could (and must) occur suprafacially. (Likewise, we would expect compound *A* to be rapidly transformed into *B* under *photochemical* conditions; why?)

28.38 **(a)** This is a [3,3] sigmatropic rearrangement, that is, a Claisen rearrangement.

> The solution to the foregoing part and the solution to part (c) that follows illustrate a useful two-step process for drawing the products of complicated rearrangements. It usually makes sense to draw the bond connections first and thus obtain a highly distorted structure of the product. Then convert the distorted structure into a more conventional structure, using models if necessary.

(b) This reaction superficially appears to involve internal rotation about both double bonds, a process that is virtually impossible. However, a more reasonable pericyclic mechanism is available: a sequence of two conrotatory electrocyclic reactions, the first one a ring closing, and the second one a ring opening. (The curved arrows refer to the clockwise rotations of groups and not to electron flow.)

(c) This is a [3,3] sigmatropic rearrangement in which the product requires a little redrawing!

28.39 This is a [1,3] sigmatropic rearrangement with inversion in the migrating carbon, as required for the thermal [1,3] rearrangement.

28.40 Take a cue from the solution to Problem 28.31(a) on p. 1031XR of this manual, in which the last step involves a thermal reaction of 1,3,5-cyclooctatriene with two methyl groups. Replace the methyl groups with hydrogens and the result is the same. The resulting compound *B* undergoes a Diels–Alder reaction (that is, a [4s + 2s] cycloaddition) with the alkyne, and the product of that reaction, *C*, undergoes a reverse Diels–Alder to generate the final products.

1,3,5-cyclooctatriene (compound *A*) → disrotatory electrocyclic → *B* → CH$_3$O$_2$C—C≡C—CO$_2$CH$_3$ Diels-Alder →

C → dimethyl phthalate + cyclobutene

28.41 These transformations are very much like the last two in the solution to Problem 28.40, text p. 1482. A Diels–Alder reaction is followed by a reverse Diels–Alder reaction. The driving force for the reverse Diels–Alder is the formation of a product that is stabilized by aromaticity.

EtO$_2$C—C≡C—CO$_2$Et Diels-Alder

compound *A*

28.42 Heating an allylic vinylic ether generally results in a [3,3] sigmatropic (Claisen) rearrangement.

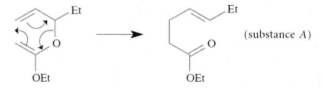

(substance *A*)

Why the product should have *E* stereochemistry is not obvious from the mechanism shown above, and an answer that gives either the *E* or the *Z* stereoisomer (or a mixture of both) is satisfactory. If you work Problem 28.48(a), text p. 1483, you will find that the Claisen rearrangement proceeds through a chairlike conformation. If the ethyl group is placed in the more favorable equatorial position of the chairlike conformation, the product is the *E* stereoisomer of substance *A* shown above.

28.43 (a) The approach to solving this type of problem is described in the solution to Problems 15.81 and 15.82 on p. 467XR of this manual. The intermediate *X* is formed from the starting material by an allowed thermal disrotatory ring closure. Because *X* is a conjugated diene, it reacts in a Diels–Alder reaction with maleic anhydride, a good dienophile, to give the product shown in the problem.

X

reacts with maleic anhydride in a Diels-Alder reaction

(b)   In this case maleic anhydride is used to trap the trienone intermediate *Y* formed in the first of what would, in the absence of maleic anhydride, be two successive [3,3] sigmatropic rearrangements. (See Eqs. 28.36a–b on text p. 1481.) Maleic anhydride reacts with the conjugated-diene unit of intermediate *Y*.

[3,3]

reacts with maleic anhydride in a Diels-Alder reaction

Y

28.44   Compound *A* is a secondary alcohol, and the aldehyde *B* is formed from an enol intermediate (shown in parentheses) that results from a [3,3] sigmatropic (oxyCope) rearrangement.

1) H_2C=CH—CH_2MgCl
2) H_3O^+

2-methyl-2-propenal

A

heat

an enol

B

H_2C=CHCH_2CH_2CHCH with CH_3

    Evidently in this case the Grignard reagent undergoes mostly 1,2-addition to the α,β-unsaturated aldehyde; none of the 1,4-addition product is formed. (See Sec. 22.11.)

28.45   The structure of the ozonolysis product *D* and the formula of elemicin show that elemicin has the structure *C*. Compound *B* and elemicin differ by one methyl group, and compound *B* ionizes in base. Consequently, a reasonable hypothesis is that one of the three oxygens of compound *B* is part of a phenol —OH group, and that this is methylated by dimethyl sulfate treatment to give elemicin. Compound *A* is *not* a phenol because it is insoluble in base, but forms a phenol *B* when heated. This series of observations suggests that compound *A* is an allyl ether that undergoes a Claisen rearrangement when heated. Which oxygen is part of the allyl ether? If one of the two outer oxygens of *A* is an allyl ether, its rearrangement would result in transfer of the allyl group to the adjacent ortho position. Since the allyl group in *C* is not ortho to one of the methoxy groups, the rearrangement must have occurred from the central oxygen. Since the ortho positions are blocked, rearrangement occurs to the para position. The structures of *A* and *B* are thus defined.

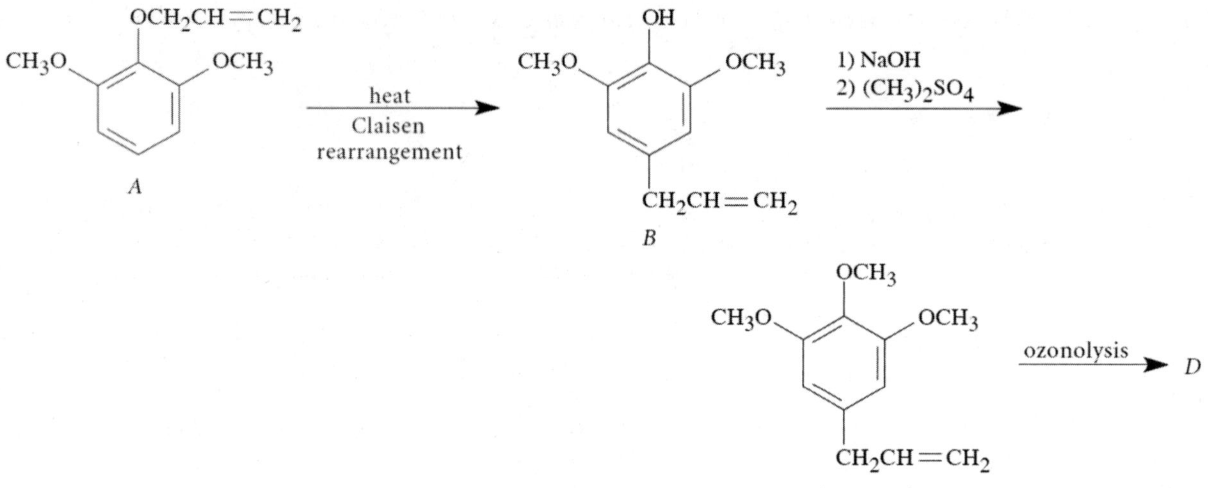

28.46 Claisen rearrangements are key steps in both parts (a) and (b).

(a)

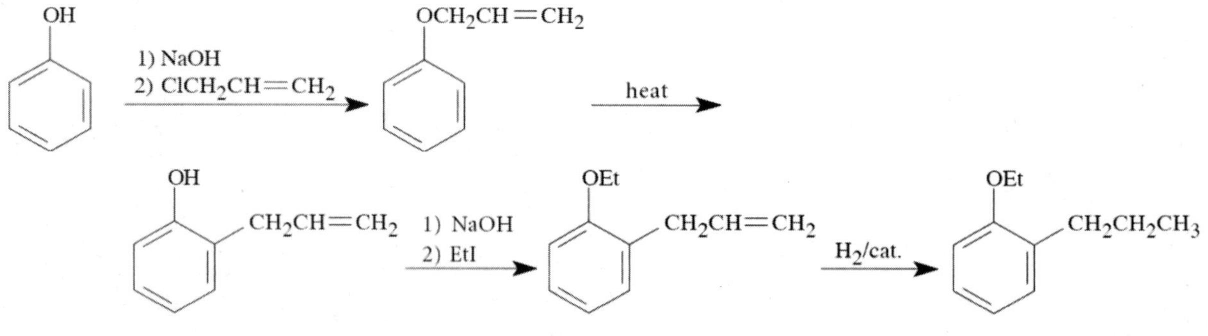

1-ethoxy-2-propylbenzene

(b) A key step in this synthesis is a Claisen rearrangement of the starting material, which is prepared in the solution to part (a)

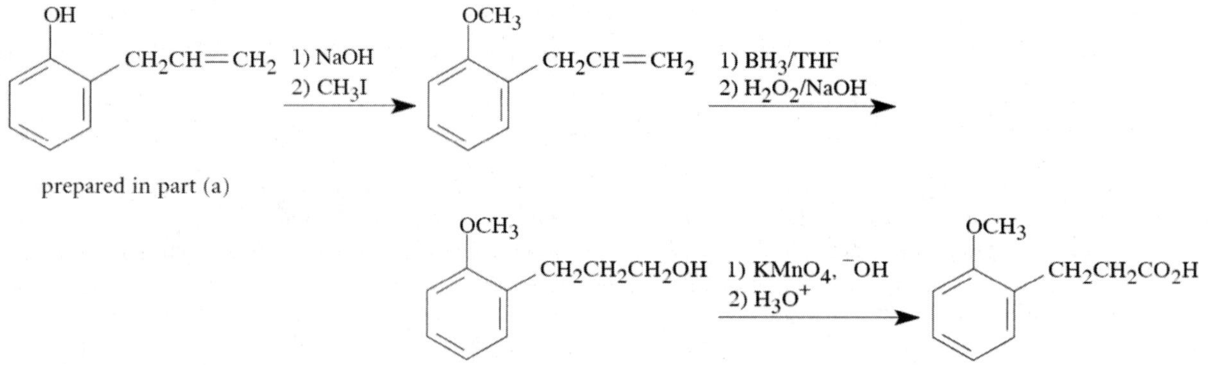

prepared in part (a)

28.47 (a) The first transformation is a thermal conrotatory electrocyclic ring opening, and the second transformation is a thermal disrotatory electrocyclic ring closure.

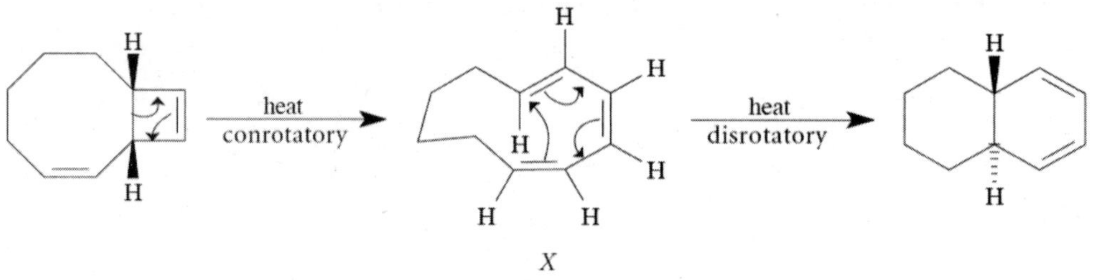

X

(b)   Two successive photochemical **disrotatory electrocyclic reactions account for the products.**

Y

The product, the very unstable **antiaromatic cyclic diene 1,3-cyclobutadiene (see text p. 771–2),** was generated in 1972 by **this sequence of reactions carried out at a temperature of 8 K in the** laboratory of Professor **Orville L. Chapman (1932–2004),** then at Iowa State University, and later at the University of **California, Los Angeles. This technique allowed the observation and** characterization of 1,3-cyclobutadiene **by infrared spectroscopy.**

28.48   (a)   Place the starting material in **a chairlike conformation, carry out the [3,3] sigmatropic rearrangement, and** compare the stereochemistry of **the product with that obtained experimentally. (Models are useful to** ensure that the asymmetric carbons **have the proper configuration.)**

One double bond is *E* and one is *Z*; this is identical to the product obtained experimentally.

This verifies that the product **could be formed from a chairlike conformation. However, the boatlike** conformation must also be **examined:**

Both double bonds are *Z*; this is different from the product obtained experimentally.

(A second boatlike conformation **is possible, and it would lead to formation of the all-*E* product, which is** also different from that observed **experimentally.) Thus, of the two possible conformations, only a** chairlike transition state fits the **observations. You should use similar reasoning to verify that a chairlike** conformation of the second reactant **in the problem gives the all-*E* product observed and that a boatlike** conformation does not.

Notice that experiments with *both* alkene stereoisomers are required to establish that a chairlike transition state is a general **phenomenon.**

(b)   Given that germacrone assumes a **chairlike conformation in its transition state when it reacts to give** β-elemenone, deduce the structure **of germacrone by mentally reversing the Cope rearrangement.**

β-elemenone   germacrone

Now re-examine (or work) Problem 28.42, text p. 1482, to see why the product in that problem has *E* stereochemistry.

28.49 A thermal electrocyclic reaction of the carbocation, which is formed by protonation of the ketone starting material, is followed by loss of a proton to give an enol, which spontaneously isomerizes to the corresponding ketone. Since four electrons are involved, the reaction is predicted to be conrotatory, and the methyl groups in the product are therefore trans:

conjugate acid of
the starting ketone

an enol

28.50 (a) A (presumably) conrotatory ring opening gives intermediate *A*. Although this compound is no longer aromatic, its formation is driven by the relief of strain in the four-membered ring. Compound *A* undergoes an *intramolecular* Diels–Alder reaction that restores aromaticity and gives the product with precisely the desired stereochemistry.

$[4s + 2s]$

(b) Oxidation of the secondary alcohol and cleavage of the ether gives estrone:

PCC

HI, heat

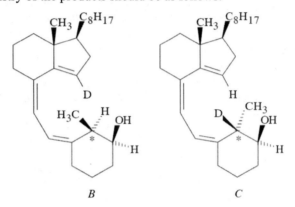

estrone

28.51   This is a [1,7] sigmatropic rearrangement, which, according to the selection rules, should be antarafacial. (See Problem 28.28, text p. 1479, and its solution on p. 1030XR of this chapter.) Migration of hydrogen gives *B* and migration of deuterium gives *C*. (Presumably, the predominance of *B* is due in part to a primary isotope effect, which should favor more rapid migration of hydrogen, all things being equal.) If the rearrangements are indeed antarafacial, the stereochemistry of the products should be as follows:

This was indeed the stereochemistry observed.

This work was carried out to adduce evidence for the stereochemistry of the previtamin D rearrangement to vitamin D (Eq. 28.39, text p. 1482). Compound *A* was constructed to be a "model" for previtamin D. As noted in Problem 28.28, the structure of the previtamin itself does not reveal the stereochemistry of the rearrangement.

28.52   First deduce the structure of compound *C* from the structure of the Diels–Alder adduct. Then deduce the structures of compound *A* and *B* from the reactions. Compound *A* is the product of bromine addition; compound *B* is a conjugated diene that results from two successive E2 eliminations of compound *A*. Compound *C* is formed from compound *B* by a disrotatory electrocyclic ring opening.

Compound *C* reacts as a diene with tetracyanoethylene in a Diels–Alder reaction to give the final product showm in the problem.

Compound *C* is interesting because if it were planar, it would have antiaromatic character. Like 1,3,5,7-cyclooctatetraene (COT; text p. 772), it avoids antiaromaticity by avoiding planarity. However, this means that it reacts like an alkene.

Compound *C* reacts as a diene with tetracyanoethylene rather than as a triene in a [6*a* + 2*s*] or [6*s* + 2*a*] cycloaddition. Why?

28.53   (a)   If compound *B* is aromatic, then it must be planar. If it is planar, the "inner" hydrogens (the ones shown explicitly in the problem) fall essentially on top of each other. (A model can be used to demonstrate this point.) The resulting van der Waals repulsions are so severe that compound *B* does not exist in spite of its aromaticity. (Compound *B* was also the subject of Problem 15.48 on text p. 780.)

(b)   The formation of compound *B* from compound *A* involves $4n + 2$ (= 6) electrons and is one of two possible allowed conrotatory photochemical electrocyclic ring-opening reactions. (The other is discussed in the solution to part (c).)

(c)   Compound *C* could be formed in two ways. First, compound *B* could be formed as an unstable intermediate, and then compound *C* could be formed from *B* by an allowed *thermal* disrotatory electrocyclic reaction. Such a reaction would be very rapid even at low temperature because of the instability of compound *B*. The second pathway is that the all-*cis* pentaene *D* could be formed from compound *A* in an allowed photochemical conrotatory electrocyclic reaction. (Note that two different allowed conrotatory reactions of compound *A* are possible: one gives compound *B*, and one gives compound *D*.) Compound *C* could then be formed from compound *D* by a thermal disrotatory electrocyclic reaction.

compound *D*

One could potentially distinguish between the two pathways for the formation of compound *C* by preparing compound *D* by another route and determining whether it spontaneously closes to compound *C*. If it does not, then the first pathway must be operating. In fact, compound *D* was prepared in 1971 by chemists at the University of Alberta. They found that, indeed, compound *D* is spontaneously transformed into compound *C* in about 3.5 hr at 6 °C. The rate of this process was found to be at least ten times grater than the rate of the photochemical conversion of *A* to *C* described in the problem. Hence, compound *D* is a competent intermediate. Furthermore, when they carried out the photochemical reaction of *A* described in the problem at temperatures so low that compound *D* would not react further, compound *D* accumulated in the reaction mixture. The simplest interpretation of these observations is that compound *B* is not an intermediate, and the pathway for the observations recorded in the problem is the one involving compound *D*.

28.54   (a)   If the formation of benzene were concerted, it would have to be a thermal disrotatory electrocyclic ring-opening reaction involving four electrons.

(b)   The reaction that gives benzene is not allowed by the selection rules; therefore, it does not occur. In spite of the high degree of strain of Dewar benzene, and in spite of the great stability of benzene, Dewar benzene is effectively constrained to exist by the selection rules!

Prismane, or Ladenburg benzene, which is discussed on text p. 759 and 1462, is another very unstable constitutional isomer of benzene that is effectively trapped into existence because its concerted conversion into benzene would violate the selection rules for pericyclic reactions.

28.55   (a)   Bromine addition to the starting material is followed by four successive base-promoted β-elimination reactions to give compound *A*, which spontaneously forms compound *B* by an allowed disrotatory thermal electrocyclic reaction. Notice that compound *B* is a "tied-down" variation on hydrocarbon *B* in Problem 28.53. Because the offending inner hydrogens are absent in *B*, its continuous π-electron system can nearly achieve planarity.

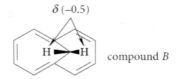

(b)   As discussed in part (a), compound *B* contains a planar (or nearly planar) $4n + 2$ π-electron system. Consequently, it shows the typical ring-current effects expected of aromatic compounds in its NMR spectra. (See Fig. 16.2 on text p. 794 and Problem 16.5 on text p. 795.) The protons on the double bonds, like those in benzene, show typical aromatic chemical shifts near $\delta\,7.1$. The two methylene protons, however, occupy the region that is strongly shielded by the ring current. Thus, it is these protons that have the negative chemical shift.